Advances in Modal Logic

Volume 10

Advances in Modal Logic

Volume 10

Edited by

Rajeev Goré

Barteld Kooi

and

Agi Kurucz

ISBN 978-1-84890-151-3

College Publications
Scientific Director: Dov Gabbay
Managing Director: Jane Spurr

http://www.collegepublications.co.uk

Printed by Lightning Source, Milton Keynes, UK

Contents

Preface

Advances in Modal Logic (AiML) is an initiative founded in 1995 and aimed at presenting an up-to-date picture of the state of the art in modal logic and its many applications. It consists of a conference series together with volumes based on the conferences. The conference is the main international forum at which research on all aspects of modal logic is presented. The first one was held in 1996 in Berlin, Germany, and since then it has been organised biennially, with meetings in 1998 in Uppsala, Sweden; in 2000 in Leipzig, Germany (jointly with ICTL-2000); in 2002 in Toulouse, France; in 2004 in Manchester, UK; in 2006 in Noosa, Australia; in 2008 in Nancy, France; in 2010 Moscow, Russia, and in 2012 in Copenhagen, Denmark. Information about AiML and related events, including conference proceedings, is available at the website `www.aiml.net`.

The tenth conference in the AiML series was organised by Barteld Kooi (Department of Theoretical Philosophy at the University of Groningen) with the assistance of Dave Gilbert, Bouke Kuijer, Paolo Maffezioli and Allard Tamminga. It was held on 5-8 August, 2014; the conference web page can be found at `http://www.philos.rug.nl/AiML2014/`.

This volume contains invited and contributed papers from the conference. The conference included invited lectures from the following people:

- Franz Baader (TU Dresden, Germany)
- Stéphane Demri (New York University, US & CNRS, France)
- Joseph Halpern (Cornell University, US)
- Sara Negri (University of Helsinki, Finland)
- Frank Wolter (University of Liverpool, UK).

The workshop on Reasoning About Other Minds: logical and cognitive perspectives, organised by Rineke Verbrugge, was co-located with AiML 2014 and was held on Monday 4th of August, with all participants invited to also attend the morning session of AiML 2014 on Tuesday 5th of August.

The Programme Committee received 55 regular paper submissions. Of these, 31 were selected for this volume by a rigorous reviewing process where every paper received 3 independent expert reviews. The volume includes papers on general problems in model theory, proof theory and algorithmic properties of modal logics, on systems for temporal and epistemic reasoning, on related kinds of logics - description, paraconsistent, substructural, intuitionistic, and on related topics in algebraic logic. In addition, there were many submissions for short presentations at the conference, and a large majority of these were accepted.

The Steering Committee of AiML for 2012–2014 consisted of:

- Carlos Areces (FaMAF, Universitad Nacional de Cordoba, Argentina)
- Silvio Ghilardi (Università degli Studi di Milano, Italy)

- Robert Goldblatt (Victoria University of Wellington, New Zealand)
- Valentin Goranko (Technical University of Denmark)
- Rajeev Goré (The Australian National University)
- Barteld Kooi (University of Groningen, The Netherlands)
- Agi Kurucz (King's College London, UK)
- Lawrence Moss (Indiana University, USA)
- Valentin Shehtman (Russian Academy of Sciences).

Here are the members of the Programme Committee for the conference:

- Carlos Areces (FaMAF, Universitad Nacional de Cordoba, Argentina)
- Alexandru Baltag (ILLC, University of Amsterdam, The Netherlands)
- Nick Bezhanishvili (Imperial College London, UK)
- Patrick Blackburn (Roskilde University, Denmark)
- Hans van Ditmarsch (LORIA, France)
- David Fernández-Duque (ITAM, Mexico)
- Melvin Fitting (Lehman College, CUNY, USA)
- Mai Gehrke (LIAFA, Université Paris Diderot, France)
- Silvio Ghilardi (Università degli Studi di Milano, Italy)
- Robert Goldblatt (Victoria University of Wellington, New Zealand)
- Valentin Goranko (Technical University of Denmark)
- Guido Governatori (NICTA Queensland, Australia)
- Andreas Herzig (IRIT, Toulouse, France)
- Rosalie Iemhoff (Utrecht University, The Netherlands)
- Roman Kontchakov (Birkbeck College London, UK)
- Barteld Kooi (University of Groningen, The Netherlands)
- Marcus Kracht (Universität Bielefeld, Germany)
- Alexander Kurz (University of Leicester, UK)
- Carsten Lutz (Universität Bremen, Germany)
- Jakub Michaliszyn (Imperial College London, UK)
- Larry Moss (Indiana University, USA)
- Hiroakira Ono (Japan Advanced Institute of Science and Technology)
- Revantha Ramanayake (Vienna University of Technology, Austria)
- Mark Reynolds (University of Western Australia)
- Vladimir Rybakov (Manchester Metropolitan University, UK)
- Renate Schmidt (University of Manchester, UK)
- Jeremy Seligman (University of Auckland, New Zealand)
- Ilya Shapirovsky (Institute for Information Transmission Problems, Moscow, Russia)
- Valentin Shehtman (Moscow State University, Russia)
- Dimiter Vakarelov (Sofia University, Bulgaria)

- Yde Venema (ILLC, University of Amsterdam, The Netherlands)
- Heinrich Wansing (Ruhr University Bochum, Germany)
- Michael Zakharyaschev (Birkbeck College, London, UK).

Programme committee co-chairs:

- Rajeev Goré (The Australian National University)
- Agi Kurucz (King's College London, UK).

Many other people assisted with the reviewing process, including: Jesse Alama, Philippe Balbiani, Guram Bezhanishvili, Marta Bilkova, Felix Bou, Balder ten Cate, Roberto Ciuni, Evgenij Dashkov, Raul Fervari, Tim French, Stefan Göller, Sam van Gool, Jeroen Goudsmit, Paula Henk, Guillaume Hoffmann, Ullrich Hustadt, Mohammad Khodadadi, Stanislav Kikot, Sophia Knight, Andrey Kudinov, Hidenori Kurokawa, Antti Kuusisto, Ori Lahav, Francois Laroussinie, Bjoern Lellmann, Arne Meier, Pierluigi Minari, Ezequiel Orbe, Jan Otop, Fedor Pakhomov, Francesco Paoli, Fabio Papacchini, Mattia Petrolo, Bryan Renne, Mikhail Rybakov, Joshua Sack, Mehrnoosh Sadrzadeh, Katsuhiko Sano, Thomas Schneider, François Schwarzentruber, Viorica Sofronie-Stokkermans, Sara Uckelman, Fernando Velázquez-Quesada, Dirk Walther, Chunlai Zhou, Evgeny Zolin. We apologise to anyone whose name was inadvertently left off this list.

We thank the organizers of the conference for their hard and dedicated work. We thank the members of the Programme Committee and all other reviewers for the time, professional effort and the expertise that they invested in ensuring the high scientific standards of the conference and its proceedings. We thank the authors for their excellent contributions. We also thank Thomas Bolander for assistance with formatting the proceedings, and Jane Spurr for bringing this volume to publication.

Thanks are also due to the Netherlands Organisation for Scientific Research (NWO) who have financially supported the conference through the VIDI project of Barteld Kooi and the VICI project of Rineke Verbrugge. We also thank the Faculty of Philosophy of the University of Groningen for providing organizational support.

June 2014

Rajeev Goré, Barteld Kooi and Agi Kurucz

Bisimulation Safe Fixed Point Logic

Faried Abu Zaid [1]

Mathematical Foundations of Computer Science, RWTH Aachen University
D-52056 Aachen

Erich Grädel [2]

Mathematical Foundations of Computer Science, RWTH Aachen University
D-52056 Aachen

Stephan Jaax [3]

Mathematical Foundations of Computer Science, RWTH Aachen University
D-52056 Aachen

Abstract

We define and investigate a new modal fixed-point logic, called bisimulation safe fixed-point logic BSFP, which is a calculus of binary relations that extends both PDL and the modal μ-calculus. The logic is motivated by concepts and results due to van Benthem and Hollenberg on bisimulation safety which plays a similar role for binary relations as the more familiar notion of bisimulation invariance plays for monadic ones. We prove that BSFP is indeed bisimulation invariant for state formulae and bisimulation safe for action formulae. We investigate the expressive power of BSFP and show that it is not limited to monadic second-order definability. Further, we reveal a close relationship of BSFP with context-free languages. We identify a fragment of BSFP that is equivalent to the extension of PDL by context-free grammars. Although BSFP is far more expressive than the modal μ-calculus, its model-checking problem has the same complexity. On the other side, the satisfiability problem for BSFP is highly undecidable.

Keywords: Modal Logic, Dynamic logic, Fixed Point Logic. Bisimulation Invariance, Safety for Bisimulation.

1 Introduction

Bisimulation is a fundamental notion for the analysis of modal logics and the behaviour of transition systems. Intuitively, two states v, v' in transition systems

[1] abuzaid@logic.rwth-aachen.de

[2] graedel@logic.rwth-aachen.de

[3] stephan.jaax@rwth-aachen.de

$\mathcal{T}, \mathcal{T}'$ are bisimilar if the set of possible traces from these states are equivalent in a strong sense. Bisimilar states must share the same local properties and any transition from v to w in $\mathcal{T}$ must have a transition of the same kind from v' to w' in $\mathcal{T}'$ (and vice versa) such that w and w' are again bisimilar.

Modal logics are invariant under bisimulation. This means that for any pair of bisimilar nodes $v \in \mathcal{T}$ and $v' \in \mathcal{T}'$, and for any modal formula ψ we have that $\mathcal{T}, v \models \psi$ if, and only if $\mathcal{T}', v' \models \psi$. This *bisimulation invariance* holds not only for the basic propositional modal logic ML, but also for the extensions to stronger logics used in program analysis and verification such as the computation tree logics CTL, CTL*, the propositional dynamic logic PDL, and fixed-point logics such as the modal μ-calculus L_μ and the modal iteration calculus MIC [3]. Since every pointed transition system $\mathcal{T}, v$ can be unraveled from v to a tree $\mathcal{T}^*$ with root v, such that $\mathcal{T}, v \sim \mathcal{T}^*, v$ it follows that every bisimulation-invariant logic has the *tree model property*: every satisfiable formula is true at the root of a tree model. The tree model property is also algorithmically very important since it paves the way to the use of automata-based methods for satisfiability testing.

The relationship between modal logic and bisimulation can in fact be taken an important step further to *model-theoretic characterization theorems.* It is a classical result by van Benthem [12] that modal logic is precisely the bisimulation-invariant fragment of first-order logic. This means that an arbitrary first-order formula $\varphi(x)$ (in a vocabulary of unary and binary relations) is invariant under bisimulation if, and only if, it is equivalent to a formula of ML. An important counterpart of van Benthem's characterization is the Theorem by Janin and Walukiewicz [10] saying that, in precisely the same sense, the modal μ-calculus L_μ is the bisimulation-invariant fragment of monadic second-order logic MSO. For more details, including characterizations theorems for several other variants of bisimulations we refer to the survey [5] and the references there.

In this paper we study a related notion, called *bisimulation safety*, that has been introduced by van Benthem (see [13,9]). To motivate this notion, we have a closer look at the propositional dynamic logic PDL [7]. Recall that PDL is a logic with a two-sorted syntax that distinguishes between state formulae and programs, defined by the mutual induction

$$\begin{aligned}\varphi &::= P \mid \varphi \vee \varphi \mid \neg\varphi \mid \langle\alpha\rangle\varphi \\ \alpha &::= E \mid \varphi? \mid \alpha \cup \alpha \mid \alpha;\alpha \mid \alpha^*\end{aligned}$$

In a given transition system $\mathcal{T}$ over a set of states V, a state formula φ defines a set of states, $[\![\varphi]\!]^{\mathcal{T}} = \{v : \mathcal{T}, v \models \varphi\} \subseteq V$, whereas a PDL-program α defines a set of transitions, i.e. a binary relation $[\![\alpha]\!]^{\mathcal{T}} \subseteq V \times V$. State formulae and programs are linked in one direction by using programs as modalities to form state formulae $\langle\alpha\rangle\varphi$, saying that there is a transition in α leading to a new state w at which φ holds, and in he other direction by the possibility to form test programs $\varphi?$ defining transitions (v, v) at states where φ holds.

When one says that PDL can be embedded into the modal μ-calculus L_μ and that PDL is bisimulation-invariant, one just considers the state formulae. The

PDL-programs have no direct counterpart in L_μ for the trivial reason that L_μ is a logic of state formulae only and the extension of any L_μ formula is a set of states rather than a set of transitions. Thus, the notion of bisimulation invariance applies to state formulae only, not to programs. However PDL-programs are *bisimulation-safe* in the sense that they do not destroy bisimulations.

Definition 1.1 A binary global relation φ that associates with every transition system $\mathcal{T}$ (of a fixed vocabulary τ) a set of transitions $[\![\varphi]\!]^{\mathcal{T}}$ is *safe for bisimulations* if every bisimulation Z between two transition systems $\mathcal{T}$ and $\mathcal{T}'$ is also a bisimulation between the expansions $(\mathcal{T}, [\![\varphi]\!]^{\mathcal{T}})$ and $(\mathcal{T}', [\![\varphi]\!]^{\mathcal{T}'})$.

Typical bisimulation safe operations are the union and composition of two binary relations whereas intersection and complementation are unsafe for bisimulation. In the same sense as ML is the bisimulation-invariant fragment of first-order logic, van Benthem [13] also proved a similar correspondence between the bisimulation-safe fragment of first-order logic and the class of PDL-programs that do not contain the Kleene star: A first-order formula $\varphi(x, y)$ is bisimulation-safe if, and only if it is equivalent to some $*$-free PDL-program.

This result, together with the Janin-Walukiewicz Theorem raises the following questions.

(1) Can one characterize in a similar way the bisimulation-safe fragment of monadic second-order logic?

(2) Is there an embedding of full PDL (state formulae and programs) into a natural fixed-point logic L that is not only bisimulation-invariant for state formulae but also bisimulation-safe for action formulae?

To the first question, an answer has been given by Marco Hollenberg [9] who considered so-called *μ-programs.* These can be defined by applying the program constructions of PDL not just to state formulae of PDL but to formulae of the modal μ-calculus. As for PDL, one can define μ-formulae and μ-programs by a mutual induction

$$\varphi ::= P \mid X \mid \varphi \vee \varphi \mid \neg\varphi \mid \langle\alpha\rangle\varphi \mid \mu X.\varphi$$
$$\alpha ::= E \mid \varphi? \mid \alpha \cup \alpha \mid \alpha ; \alpha \mid \alpha^*$$

It is not difficult to see that μ-formulae (defined in this slightly nonstandard way) are bisimuation-invariant, μ-programs are bisimulation-safe and that this definition does not take us outside of monadic second-order logic. In particular, this way of defining μ-formulae is equivalent to the standard definition of the μ-calculus which does not refer to μ-programs at all. The main result of Hollenberg says that μ-programs coincide with the bisimulation-safe fragment of monadic second-order logic [9, Corollary 3.5.5]: An MSO-formula $\varphi(x, y)$ is safe for bisimulations if, and only if, it is equivalent to a μ-program.

It should be noted that the only enrichment of μ-programs with respect to PDL-programs concerns the application of test-instructions $\varphi?$ which now refer to μ-formulae rather that just PDL-formulae. The iteration mechanism of μ-programs, however, remains limited to the Kleene star; in particular μ-programs do not have a full least (or greatest) fixed-point mechanism for sets of transitions.

In this paper, we shall address the second question and define a modal fixed-point logic for defining sets of transitions, which we call *bisimulation-safe fixed-point logic* BSFP. We shall analyse its expressive power and its model-theoretic and algorithmic properties. In particular we shall prove that BSFP is indeed safe for bisimulations whereas previously known extensions of the modal μ-calculus either remain limited to monadic fixed-points or are not bisimulation-safe. In particular, this is the case for the binary fragment of the least fixed point logic LFP and for the two-dimensional μ-calculus by Otto [11].

We shall provide several presentations of our logic. The first has a minimal syntax as a pure calculus of binary relations, with a projection operator to recover monadic relations. The second presentation is based on a two-sorted syntax, as for PDL and μ-programs, distinguishing between state formulae and action formulae. The equivalence of the two presentations will reveal that BSFP is the generalization of PDL- and μ-programs by admitting full binary fixed point definitions rather than just the Kleene star. We shall see that while this construction remains bisimulation-safe and does not increase the complexity of the model-checking problem, it nevertheless makes the logic much stronger. Contrary to the modal μ-calculus, BSFP admits infinity axioms, is not restricted to MSO-definability and is intimately connected to context-free languages. We shall see that all Boolean combinations of context-free languages are definable in BSFP, and we shall identify a fragment of BSFP that is equivalent to the extension of PDL by context-free grammars. As a consequence, the satisfiability problem for BSFP is highly undecidable.

2 Background from logic

We assume that the reader is familiar with modal logic, first-order logic (FO), monadic second-order logic (MSO), the extension of first-order logic by second-order quantification $\exists X$ and $\forall X$ over *sets* of elements of the structure on which the formula is evaluated. In contrast to second-order logic (SO), where quantification over arbitrary relations (or functions) is admitted, MSO is a much more manageable formalism; it is decidable on many interesting classes of structures (on words and on trees in particular) and amenable to automata-based methods.

We further assume that the reader is familiar with the modal μ-calculus L_μ, briefly described in the introduction of this paper, which extends propositional modal logic ML by least (and greatest) fixed points, and which plays a fundamental role in many areas of logic in computer science, in particular for the specification and verification of computing systems. In finite model theory, descriptive complexity and database theory, other fixed-point logics are of central importance (see [6]). Relevant for the purpose of this paper is the least fixed-point logic LFP which augments the power of first order logic by least and greatest fixed points of definable relational operators and thus extends FO in a similar way as the μ-calculus extends propositional modal logic. The bisimulation safe fixed point logic BSFP that we are studying in this paper lies between L_μ and LFP. We will briefly recall some basic definitions for LFP here. For a more detailed account, we refer to [6].

Every formula $\psi(R,\overline{x})$, where R is a relation symbol of arity k and $\overline{x}$ is a tuple of k variables, defines, for any structure $\mathfrak{A}$ of appropriate vocabulary, an update operator $F : \mathcal{P}(A^k) \to \mathcal{P}(A^k)$ on the class of k-ary relations over the universe A of $\mathfrak{A}$, namely $F : R \mapsto \{\overline{a} : (\mathfrak{A}, R) \models \psi(R,\overline{a})\}$. If ψ is positive in R, that is, if every occurrence of R falls under an even number of negations, this operator is monotone in the sense that $R \subseteq R'$ implies $F(R) \subseteq F(R')$. It is well known that every monotone operator has a least fixed point and a greatest fixed point, which can be defined as the intersection and union, respectively, of all fixed points, but which can also be constructed by transfinite induction.

LFP is defined by adding to the syntax of first order logic the following *fixed point formation rule:* If $\psi(R,\overline{x})$ is a formula with a relational variable R occurring only positively and a tuple of first-order variables $\overline{x}$, and if $\overline{t}$ is a tuple of terms (such that the lengths of $\overline{x}$ and $\overline{t}$ match the arity of R), then $[\mathbf{lfp}\, R\overline{x}.\psi](\overline{t})$ and $[\mathbf{gfp}\, R\overline{x}.\psi](\overline{t})$ are also formulae, binding the occurrences of the variables R and $\overline{x}$ in ψ.

The semantics of least fixed-point formulae in a structure $\mathfrak{A}$, providing interpretations for all free variables in the formula, is the following: $\mathfrak{A} \models [\mathbf{lfp}\, R\overline{x}.\psi](\overline{t})$ if $\overline{t}^{\mathfrak{A}}$ belongs to the least fixed point of the update operator defined by ψ on $\mathfrak{A}$. Similarly for greatest fixed points.

Note that in formulae $[\mathbf{lfp}\, R\overline{x}.\psi](\overline{t})$ one may allow ψ to have other free variables besides $\overline{x}$.

The duality between least and greatest fixed point implies that for any ψ,

$$[\mathbf{gfp}\, R\overline{x}.\psi](\overline{t}) \equiv \neg[\mathbf{lfp}\, R\overline{x}.\neg\psi[R/\neg R]](\overline{t}).$$

The *width* of an LFP-formula is the maximal number of free variables in its subformulae. Further, an LFP-formula is *parameter-free* if in all its fixed-point expressions $[\mathbf{lfp}\, R\overline{x}\,.\,\varphi(R,\overline{x})](\overline{x})$ and $[\mathbf{gfp}\, R\overline{x}\,.\,\varphi(R,\overline{x})](\overline{x})$ the only free variables occurring in φ are those in $\overline{x}$. It is well-known that every LFP-formula can be translated into an equivalent one that is parameter-free, but this does, in general, increase the arity of the fixed-point variables and the width of the formulae.

Notice that any property of finite structures that is expressible by a fixed LFP-formula can be decided in polynomial time. In fact, on *linearly ordered* finite structures, precisely the polynomial-time decidable properties are LFP-definable, but this is not true in the absence of a linear order (although certain P-complete problems, such as winning regions of reachability games, remain definable in LFP and even in the modal μ-calculus). Indeed, it is a major open problem in finite model theory and descriptive complexity theory whether there exists an extension of LFP that precisely captures the polynomial-time properties of arbitrary (ordered or unordered) finite structures (see [6]).

Evaluation problems in logic, where the formula is not fixed, but part of the input, are more difficult to analyze. The model checking problem for a logic L is the problem to decide, given a formula $\psi \in L$ and a finite structure $\mathcal{K}$ (with elements instantiating the free variables of ψ) whether the formula is true in $\mathcal{K}$. Concerning the complexity of the model-checking problem for LFP and its fragments the following is known (see [6, Chapter 3.3] for details and references).

- For LFP-formulae of unbounded width, model-checking is EXPTIME-complete.
- For LFP-formulae of bounded width that may contain parameters it is PSPACE-complete.
- For parameter-free LFP-formulae of bounded width, as well as for the modal μ-calculus, the model-checking problem is in UP $\cap$ Co-UP and PTIME-hard. It is open whether it is solvable in polynomial time, and this is equivalent to the question whether winning regions of parity games are computable in polynomial time.

3 Bisimulation Safe Fixed Point Logic

In this section, we introduce several presentations of bisimulation safe fixed-point logic BSFP. We shall see that BSFP does not have the finite model property and that it is bisimulation invariant for state formulae and bisimulation safe for action formulae. This will also imply that BSFP is not contained in monadic second-order logic. Finally we will discuss simultaneous fixed points and present a normal form for BSFP.

We start by giving a minimal syntax for BSFP as a pure calculus of binary relations.

Minimal syntax. Let τ be a vocabulary of monadic predicates P_i and binary action predicates E_a, and let $Z_1, Z_2 \dots$ be a collection of binary predicate variables. Formulae of BSFP in minimal syntax are build by the grammar

$$\alpha ::= \bot \mid P_i? \mid Z_j \mid E_a \mid \alpha \cup \alpha \mid\sim \alpha \mid \alpha \circ \alpha \mid \mu Z_j.\alpha$$

where, for formulae $\mu Z_j.\alpha$, we require that every free occurrence of Z_j in α is in the scope of an even number of $\sim$ symbols.

Semantics. Let $\mathcal{T} = (V, (P_i^{\mathcal{T}})_i, (E_a^{\mathcal{T}})_a)$ be a transition system (which interprets all monadic predicates P_i by $P_i^{\mathcal{T}} \subseteq V$, all transitions relations E_a and all variables Z that occur free in α as subsets of $V \times V$ denoted by $E_a^{\mathcal{T}}$ and $Z^{\mathcal{T}}$, respectively. When it is clear from the context, we will often omit the superscripts in the notation.) The extension $[\![\alpha]\!]^{\mathcal{T}}$ of a formula α in $\mathcal{T}$ is defined inductively by:

- $[\![\bot]\!]^{\mathcal{T}} := \varnothing$.
- $[\![P_i?]\!]^{\mathcal{T}} := \{(v, v) \in V^{\mathcal{T}} \times V^{\mathcal{T}} : v \in P_i{}^{\mathcal{T}}\}$.
- $[\![E_a]\!]^{\mathcal{T}} := E_a{}^{\mathcal{T}}$ for every $a \in$ ACT.

- $[\![\alpha_1 \cup \alpha_2]\!]^{\mathcal{T}} := [\![\alpha_1]\!]^{\mathcal{T}} \cup [\![\alpha_2]\!]^{\mathcal{T}}$.
- $[\![\alpha_1 \circ \alpha_2]\!]^{\mathcal{T}} := \{(u,w) \in V^{\mathcal{T}} \times V^{\mathcal{T}} : \exists v\,(u,v) \in [\![\alpha_1]\!]^{\mathcal{T}} \wedge (v,w) \in [\![\alpha_2]\!]^{\mathcal{T}}\}$.
- $[\![\sim \alpha]\!]^{\mathcal{T}} := \{(v,v) \in V^{\mathcal{T}} \times V^{\mathcal{T}} : \forall v'\,(v,v') \notin [\![\alpha]\!]^{\mathcal{T}}\}$.
- The μ-operator is a binary least-fixed-point operator:
  $[\![\mu Z.\alpha]\!]^{\mathcal{T}} := \bigcap \left\{ R \subseteq V^{\mathcal{T}} \times V^{\mathcal{T}} : [\![\alpha]\!]^{\mathcal{T}[Z:=R]} \subseteq R \right\}$

Some simple but important definable relations are the diagonal $D :=\sim \bot$ and the projection to the first component, denoted $\downarrow \alpha :=\sim\sim \alpha$. By definition $[\![\downarrow \alpha]\!]^{\mathcal{T}} = \{(v,v) \in V^{\mathcal{T}} \times V^{\mathcal{T}} : \exists v'(v,v') \in [\![\alpha]\!]^{\mathcal{T}}\}$.

We next present an extended syntax for BSFP which relates this logic to PDL and μ-programs in the sense that it defines state formulae and action formulae by mutual induction. In fact BSFP can be seen as the extension of PDL by the possibility to form unary and binary fixed points.

Two-sorted syntax. For a set $X_1, X_2, \ldots$ of monadic variables and a set $Z_1 Z_2, \ldots$ of binary variables, the state and action formulae are defined by

$$\varphi ::= P_i \mid X_i \mid \varphi \vee \varphi \mid \neg\varphi \mid \langle\alpha\rangle\varphi \mid \mu X_i.\varphi$$
$$\alpha ::= D \mid \varnothing \mid E_a \mid Z_k \mid \alpha \circ \alpha \mid \alpha \cup \alpha \mid \varphi? \mid \mu Z_j.\alpha$$

Again we require that for fixed-point formulae $\mu X_i.\varphi$ and $\mu Z_j.\alpha$, every free occurrence of X_i or Z_j is the scope of an even number of $\neg$ symbols.

Semantics. For state formulae the extension $[\![\varphi]\!]^{\mathcal{T}}$ is defined in the standard way, as for PDL and μ-programs. For action formulae, the extensions are defined as in the minimal syntax. We use the expression $[\alpha]\varphi$ as shorthand for $\neg\langle\alpha\rangle\neg\varphi$ and $\varphi \wedge \psi$ as a shorthand for $\neg(\neg\varphi \vee \neg\psi)$. As usual we write $\mathcal{T}, v \models \varphi$ to denote that $v \in [\![\varphi]\!]^{\mathcal{T}}$ and $\mathcal{T},(v,w) \models \alpha$ to denote that $(v,w) \in [\![\alpha]\!]^{\mathcal{T}}$.

It is not difficult to see that the two presentations of BSFP are equivalent.

Theorem 3.1 *For every* BSFP *state formula φ in two-sorted syntax there is a formula $\hat{\varphi}$ in minimal syntax such that $\mathcal{T}, v \models \varphi \Leftrightarrow \mathcal{T},(v,v) \models \downarrow \hat{\varphi}$ and for every action formula α there is an equivalent formula $\hat{\alpha}$ in minimal syntax.*

Proof. The translations from φ to $\hat{\varphi}$ and from α to $\hat{\alpha}$ leave the atomic predicates and variables invariant, but monadic variables X_i of φ and α are considered as binary variables in $\hat{\varphi}$ and $\hat{\alpha}$. We then inductively translate the formulae by the following rules:

- if $\varphi = \varphi_1 \vee \varphi_2$, we set $\hat{\varphi} := \hat{\varphi}_1 \cup \hat{\varphi}_2$,
- if $\varphi = \neg\varphi_1$, we set $\hat{\varphi} :=\sim \hat{\varphi}_1$,
- if $\varphi = \langle\alpha_1\rangle\varphi_1$, we set $\hat{\varphi} := \hat{\alpha}_1 \circ \hat{\varphi}_1$,
- if $\varphi = \mu X.\varphi_1$, we set $\hat{\varphi} := \mu X. \downarrow \hat{\varphi}_1$.
- if $\alpha = \alpha_1 \otimes \alpha_2$ with $\otimes \in \{\circ, \cup\}$ simply set $\hat{\alpha} := \hat{\alpha}_1 \otimes \hat{\alpha}_2$,
- if $\alpha = \varphi_1?$ set $\hat{\alpha} :=\downarrow \hat{\varphi}_1$, and

Fig. 1. Sketch of $\mathcal{T}$

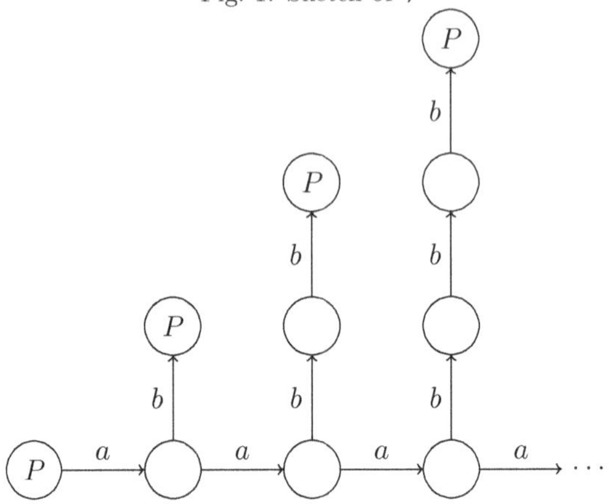

- if $\alpha = \mu Z.\alpha_1$ set $\hat{\alpha} := \mu Z.\hat{\alpha}_1$

It is easily verified that this translation gives us a formula with the desired properties in minimal syntax. □

Example 3.2 (i) The action formula $\mu Z.(D \cup (Z \circ E_a))$ defines the set of pairs of states connected by a path of the form a^*.

(ii) The action formula $\mu Z.(D \cup (E_a \circ Z \circ E_b))$ defines the set of pairs of states connected by a path of the form $a^n b^n$ for $n \geq 0$.

We generalize these examples to show that BSFP admits formulae that only have infinite models. We use a construction taken essentially from [8] for $\mathrm{PDL}_{\mathrm{CFG}}$, a logic that is in fact closely related to BSFP (see Sect. 5 below).

Theorem 3.3 BSFP *does not have the finite model property.*

Proof. For BSFP action formulae α, β let $\alpha^* := \mu Z.(D \cup Z \circ \alpha)$ and $\alpha^\Delta \beta^\Delta := \mu Z.(D \cup \alpha \circ Z \circ \beta)$. We claim that the formula

$$\varphi = (P \wedge [E_a^*]\langle E_a \circ E_b^* \rangle P) \wedge [(E_a \cup E_b)^* \circ E_b \circ E_a]\bot$$
$$\wedge [E_a^* \circ E_a \circ E_a^\Delta E_b^\Delta]\neg P \wedge [E_a^\Delta E_b^\Delta \circ E_b]\bot$$

is satisfiable but has no finite model. Consider the structure

$$\mathcal{T} = (\{w \in \{a,b\}^* \mid w = a^n b^m \text{ with } n \geq m\}, E_a, E_b, P) \text{ with}$$
$$E_a = \{(a^n, a^{n+1}) \mid n \geq 0\},$$
$$E_b = \{(a^n b^m, a^n b^{m+1}) \mid n > m\} \text{ and}$$
$$P = \{a^n b^n \mid n \geq 0\}.$$

Obviously $\mathcal{T}$ fulfils all conjuncts of φ from the node ε (c.f. Figure 1), hence $\mathcal{T}, \varepsilon \models \varphi$. Now suppose $\mathcal{T}', v \models \varphi$ for some finite transition system $\mathcal{T}'$ over the signature $\{P, E_a, E_b\}$. We can interpret $\mathcal{T}'$ as a finite automaton with initial state v and accepting states P. The regular language L accepted by this automaton is determined by the labels of the paths connecting v to a state in P. Therefore, the second conjunct enforces that $L \subseteq a^*b^*$ and the third and fourth conjunct enforce that $L \subseteq \{a^n b^n \mid n \geq 0\}$. Given the other parts of the formula, the first conjunct enforces that for every $n \geq 0$ $a^n b^n \in L$. Thus $L = \{a^n b^n \mid n \geq 0\}$ which is not regular. A contradiction. □

Corollary 3.4 BSFP *is strictly more expressive than the modal* μ*-calculus.*

We are now ready to show that BSFP has the desired properties with respect to bisimulation.

Theorem 3.5 *State formulae of* BSFP *are bisimulation invariant, and action formulae of* BSFP *are safe for bisimulation.*

Proof. We have to prove that, for every bisimulation S between two transition systems $\mathcal{T}$ and $\mathcal{T}'$ with $(v, v') \in S$, it holds that

(1) v and v' satisfy the same BSFP state formulae, and

(2) whenever $(v, w) \in [\![\alpha]\!]^{\mathcal{T}}$ for an action formula α, then there exists a w' such that $(w, w') \in S$ and $(v', w') \in [\![\alpha]\!]^{\mathcal{T}'}$.

By Theorem 3.1 it suffices to establish (2) for formulae in minimal syntax. Claim (1) then also follows. Indeed, suppose that there is BSFP state formula φ such that $\mathcal{T}, v \models \varphi$ but $\mathcal{T}', v' \not\models \varphi$. By Theorem 3.1 there is a formula $\hat{\varphi}$ such that $\mathcal{T}, (v, v) \models \downarrow \hat{\varphi}$ but $\mathcal{T}', (v', v') \not\models \downarrow \hat{\varphi}$. But then $\downarrow \hat{\varphi}$ would be unsafe for bisimulation.

Apart from the least fixed-point operator μ, every BSFP-operator has an analogous counterpart in PDL and PDL-operators are known to be safe for bisimulation. It thus suffices to show that if α is safe for bisimulation, then so is $\mu Z.\alpha$. But this follows by a straightforward induction over the stages α^η of the least fixed point induction defined by α. Indeed, for all ordinals η and all transition systems $\mathcal{T}, \mathcal{T}'$ it holds holds that if $(v, w) \in [\![\alpha^\eta]\!]^{\mathcal{T}}$, then there exists a state $w' \in V^{\mathcal{T}'}$ such that $(v', w') \in S$ and $(w, w') \in [\![\alpha^\eta]\!]^{\mathcal{T}'}$.

Zero case: For $\eta = 0$ the claim is trivial.

Successor case: Let $(v, w) \in [\![\varphi^{\eta+1}]\!]^{\mathcal{T}}$. Hence, by definition, we have that $(v, w) \in [\![\varphi]\!]^{\mathcal{T}[X:=[\![\varphi^\eta]\!]^{\mathcal{T}}]}$. Applying the induction hypothesis, we obtain that there exists a $w' \in V^{\mathcal{T}}$ with $(v', w') \in S$ and $(w, w') \in [\![\varphi]\!]^{\mathcal{T}'[X:=[\![\varphi^\eta]\!]^{\mathcal{T}'}]}$ which is, by definition, equivalent to $(w, w') \in [\![\varphi^{\eta+1}]\!]^{\mathcal{T}'}$.

Limit case: Let λ be a limit ordinal such that $(v, v') \in [\![\varphi^\lambda]\!]^{\mathcal{T}}$. Thus we have that $(v, v') \in [\![\varphi^\eta]\!]^{\mathcal{T}}$ for some $\eta < \lambda$. Applying the induction hypothesis, we obtain that there exists a $w' \in V^{\mathcal{T}'}$ with $(v', w') \in S$ and $(w, w') \in [\![\varphi^\eta]\!]^{\mathcal{T}}$ and thus, $(w, w') \in [\![\varphi^\lambda]\!]^{\mathcal{T}'}$. □

Corollary 3.6 BSFP *is not a fragment of* MSO.

Proof. Consider the infinity axiom in BSFP presented in the proof of Theorem 3.3. If it were equivalent to an MSO-formula it would, being bisimulation invariant, also be equivalent to a formula in L_μ. But this is impossible since L_μ has the finite model property. □

Simultaneous Fixed Points. As for the μ-calculus and other fixed-point logics one can generalize also BSFP to admit systems of simultaneous fixed points. These do not increase the expressive power but sometimes allow for more straightforward formalisations. Here one associates with any tuple $\overline{\psi} = (\psi_1, \ldots, \psi_k)$ of formulae $\psi_i(\overline{X}) = \psi_i(X_1, \ldots, X_k)$, in which all occurrences of all X_i are positive, a new formula $\varphi = \mu\overline{X} \,.\, \overline{\psi}$. The semantics of φ is induced by the least fixed point of the monotone operator $\psi^{\mathcal{T}}$ mapping $\overline{X}$ to $\overline{X}'$ where $X_i' = [\![\psi_i]\!]^{(\mathcal{T},\overline{X})}$. More precisely, $\mathcal{K}, (v, w) \models \varphi$ iff (v, w) is an element of the first component of the least fixed point of the above operator. It is known that simultaneous least fixed points can be eliminated in favour of nested individual fixed points by the so-called Békic principle (see e.g. [1, page 27]). Indeed, $\mu XY \,.\, [\psi(X, Y), \varphi(X, Y)]$ is equivalent to $\mu X.\psi(X, \mu Y.\varphi(X, Y))$, and this equivalence generalizes to larger systems in the obvious way.

On this basis, we now introduce a normal form for BSFP action formulae which will be helpful when we investigate the expressive power of BSFP.

Definition 3.7 A action formula α is in normal form if $\alpha = \mu\overline{Z}(\alpha_{Z_1}, \ldots, \alpha_{Z_k})$ with $\alpha_{Z_\ell} = \bigcup_i \beta_i$ and $\beta_i = \gamma_{i1} \circ \cdots \circ \gamma_{in_i}$ where γ_{ij} is either a binary predicate (or binary predicate variable) or $\varphi?$.

Lemma 3.8 *For every* BSFP *action formula* α *there is an equivalent formula* $\hat{\alpha}$ *in normal form.*

Proof. Let α be a action formula. We obtain $\hat{\alpha}$ by the following procedure: By applying the Békic principle we get that $\alpha \equiv \mu\overline{Z}(\alpha_{Z_1}, \ldots, \alpha_{Z_k})$ where the α_{Z_i} are $\cup, \circ$ combinations formed from binary predicate symbols (and variables) and tests. Such an α_{Z_ℓ} can be transformed into the form $\bigcup_i(\beta_{i1} \circ \cdots \circ \beta_{in_i})$ by the equivalences $\alpha \circ (\beta \cup \gamma) \equiv \alpha \circ \beta \cup \alpha \circ \gamma$ and $(\alpha \cup \beta) \circ \gamma \equiv \alpha \circ \gamma \cup \beta \circ \gamma$. □

4 Relationship with other fixed-point logics and model-checking

Clearly bisimulation safe fixed-point logic BSFP extends the modal μ-calculus L_μ and can be embedded into the least fixed-point logic LFP, in short $L_\mu \leq$ BSFP $\leq$ LFP. Hence every property expressible in BSFP can be checked in polynomial time, and there exist P-complete properties that are definable in BSFP. Modal fixed-point logics with a similar status are the k-dimensional μ-calculi L_μ^k by Martin Otto [11], for any $k \geq 1$. We investigate the relationship with these other fixed-point logics more closely.

It is known that formulae of the μ-calculus can be translated into parameter-free LFP-formulae of width two. We observe that there is similar embedding of BSFP into LFP which, however, produces formulae of width three.

Proposition 4.1 *There is a linear translation mapping every* BSFP*-formula* φ *to an equivalent LFP-formula* $\varphi^{\#}(x, y)$ *which is parameter-free and of width at most three.*

The translation is straightforward; it maps P_i to $x = y \wedge P_i y$ and $\sim \varphi$ to $x = y \wedge \neg\exists y : \varphi^{\#}(x, y)$, translates $\vee$ and least fixed-points literally, and only needs to introduce a third variable for expressing composition: $(\varphi \circ \psi)^{\#}(x, y) := \exists z(\varphi^{\#}(x, z) \wedge \psi^{\#}(z, y))$

Corollary 4.2 *The model-checking problem for* BSFP *is in UP*$\cap$ *Co-UP and* PTIME*-hard. It is polynomial time equivalent to the model-checking problem of the modal* μ*-calculus.*

We next consider the relationship of BSFP with the logic L^2_μ from [11] which is also a modal fixed-point calculus of binary relations. On transition systems $\mathcal{T}$ with universe V and a vocabulary of monadic relations P_i and, for simplicity, just one binary relation E, the k-dimensional μ-calculus L^k_μ is defined by taking the usual μ-calculus L_μ on an expanded system $\mathcal{T}^k$ with universe V^k monadic relations P_{ij} and binary relations E_j, for $j = 1, \ldots, k$ and additional binary relations E_σ, for every substitution $\sigma : \{1, \ldots, k\} \rightarrow \{1, \ldots, k\}$. The relations P_{ij} and E_j on V^k are given by P_i and E on the jth component and the relations E_σ contain the transitions from $(v_1, \ldots, v_k)$ to $(v_{\sigma(1)}, \ldots, v_{\sigma(k)})$. The meaning of an L^k_μ-formula ψ on $\mathcal{T}$ is given as the k-ary relation of all tuples $\overline{v}$ such that $\mathcal{T}^k, \overline{v} \models \psi$ (in the sense of L_μ).

A typical relation expressible in L^2_μ is bisimilarity. Two nodes v_1, v_2 are bisimilar in $\mathcal{T}$ if $\mathcal{T}, v_1, v_2 \models \nu Z.(\bigwedge_i (P_{i1} \leftrightarrow P_{i2}) \wedge [1]\langle 2\rangle Z \wedge [2]\langle 1\rangle Z)$.

Martin Otto proved that the multi-dimensional μ-calculus $L^\omega_\mu = \bigcup_{k\in\omega} L^k_\mu$ captures precisely the bisimulation-invariant fragment of polynomial time. Given that BSFP and L^2_μ both are fixed-point calculi that extend the modal μ-calculus to binary relations while respecting bisimulation in some sense, the question arises of how the expressive power of L^2_μ and BSFP compare. A closer look reveals that the two logics respect bisimulations in a rather different sense. First of all we observe that L^2_μ is closed under all Boolean operations and can therefore not be bisimulation safe. For instance, the formula that defines bisimilarity of two nodes in a given transition system is clearly not safe for bisimulation. On the other side L^2_μ is *component-wise invariant under bisimulations*: For any two pairs $v, w \in \mathcal{T}$ and $v', w' \in \mathcal{T}'$ such that v and v' but also w and w' are bisimilar, and any formula $\psi \in L^2_\mu$ it follows that $\mathcal{T}, v, w \models \psi$ if, and only if $\mathcal{T}', v', w' \models \psi$ (see [11]). However, there are quite simple BSFP-formulae, such as for instance the diagonal D, that violate this component-wise bisimulation invariance.

Proposition 4.3 *Concerning expressive power, the two logics* BSFP *and* L^2_μ *are incomparable.*

5 Flat BSFP

In this section we define the flat fragment of BSFP and show that it is equivalent to $\mathrm{PDL}_{\mathrm{CFG}}$, the extension of PDL by context-free grammars. We first recall the definition of $\mathrm{PDL}_{\mathrm{CFG}}$ from [8] which extends the definition of PDL by a more powerful construction for programs: The set of programs of $\mathrm{PDL}_{\mathrm{CFG}}$ consists of all context-free grammars α whose terminals are atomic actions and test formulae. Such a grammar defines a language $L(\alpha) \subseteq (A \cup \{\varphi_1?, \ldots, \varphi_n?\})^*$. The binary relation, defined by α on a transition system $\mathcal{T}$ is the set of pairs (u, v) such that there is path from u to v in $\mathcal{T}$ (expanded by $[\![\varphi_1?]\!]^{\mathcal{T}}, \ldots, [\![\varphi_n?]\!]^{\mathcal{T}}$) labelled by a word in $L(\alpha)$. $\mathrm{PDL}_{\mathrm{CFG}}$ is known to be much more powerful than PDL. For details, see [8]. We next define Flat BSFP.

Definition 5.1 Flat BSFP is the fragment of all BSFP formulae formed by the following rules:

$$\varphi ::= P_i \mid \neg\varphi \mid \varphi \vee \varphi \mid \langle\alpha\rangle\varphi$$
$$\alpha ::= E_a \mid Z_i \mid \alpha \circ \alpha \mid \alpha \cup \alpha \mid \varphi? \mid \mu Z_i.\alpha$$

For tests $\varphi?$ we additionally demand that φ is closed, which means φ does not contain any free variables.

Theorem 5.2 Flat BSFP $\equiv \mathrm{PDL}_{\mathrm{CFG}}$

Proof. Since the building rules for state formulae of flat BSFP and $\mathrm{PDL}_{\mathrm{CFG}}$ coincide, we only need to show how to translate action formulae. First we show how to translate $\mathrm{PDL}_{\mathrm{CFG}}$ programs into BSFP action formulae. We need to show that for every set $\{\varphi_1, \ldots, \varphi_n\} \subseteq \mathrm{PDL}_{\mathrm{CFG}}$ and every context-free language L over the alphabet $A \cup \{\varphi_1?, \ldots, \varphi_n?\}$ the global relation defined by L is definable by a flat BSFP action formula. Let G be a context-free grammar with non-terminals $\Gamma = \{Z_1, \ldots, Z_n\}$, terminals $\Sigma = A \cup \{\varphi_1?, \ldots, \varphi_n?\}$ and start-symbol Z_1. We may assume that we have already constructed BSFP formulae $\hat{\varphi}_1, \ldots, \hat{\varphi}_n$ such that $\hat{\varphi}_i$ is equivalent to φ_i. With every string $s = s_0 s_1 \cdots s_j \in (\Sigma \cup \Gamma)^*$ we associate the BSFP-formula $\alpha(s) := \alpha(s_0) \circ \alpha(s_1) \circ \ldots \circ \alpha(s_j)$ where $\alpha(s_k) = s_k$, if $s_k \in \Gamma$, $\alpha(s_k) = E_{s_k}$ for $s_k \in A$ and $\alpha(s_k) = \hat{\varphi}_i?$ for $s_k = \varphi_i?$. For instance $\alpha(aZ\varphi_1?Z) = E_a \circ Z \circ \hat{\varphi}_1? \circ Z$. Furthermore, with every production-rule

$$Z_i \longrightarrow A_1 | A_2 | \ldots | A_k \text{ with } A_j \in (\Sigma \cup \Gamma)^*$$

we associate a BSFP action formula $\alpha_{Z_i} := \bigcup_{1 \le j \le k} \alpha(A_j)$, and claim that $\alpha = \mu \overline{Z}.(\alpha_{Z_1}, \ldots, \alpha_{Z_n})$ defines the same relation as G. To see this one recalls that $L(G)$ is the simultaneous least fixed-point, projected on the start symbol Z_1 of the system of equations defined by the production rules of G [2]. For a given transition system $\mathcal{T}$ let $(Z_1^\eta, \ldots, Z_n^\eta)$ denote the η-fold approximation of the least fixed point of the formulae $(\alpha_{Z_1}, \ldots, \alpha_{Z_n})$ over $\mathcal{T}$ and $(X_1^\eta, \ldots, X_n^\eta)$ be the η-fold approximation of the fixed point of the operator associated with the grammar G. One can show via induction that for all ordinals η we have $(u, v) \in Z_\ell^\eta$ if, and only if there is a path from u to v labelled by a word in X_ℓ^η.

For the other direction, it suffices to find for every flat action formula α a context-free grammar G_α over atomic actions and $\mathrm{PDL}_{\mathrm{CFG}}$ tests that defines the same relation as α. By Lemma 3.8 we have that $\alpha \equiv \mu\overline{Z}(\alpha_{Z_1},\dots,\alpha_{Z_k})$ with $\alpha_{Z_\ell} = \bigcup_i \beta_i$. Every β_i is a composition of atomic formulae and tests $\beta_i = \beta_{i1} \circ \cdots \circ \beta_{in_i}$. We assign to each such composition a word $w(\beta_i) = w(\beta_{i1})w(\beta_{i2})\dots w(\beta_{in_i})$. Here we set $w(E_a) = a, w(Z_i) = Z_i$ and $w(\vartheta?) = \hat{\vartheta}?$ where $\hat{\vartheta}$ is an $\mathrm{PDL}_{\mathrm{CFG}}$ formula equivalent to ϑ (which exists by the induction hypothesis). For every $\alpha_{Z_\ell} = \bigcup_{1\le i\le n_\ell} \beta_i$ we add the production rule

$$Z_\ell \longrightarrow w(\beta_1)|\dots|w(\beta_{n_\ell}).$$

Again by an induction over the stages of the fixed point iteration one shows that for every transition system $\mathcal{T}$ and every pair of nodes (u, v) we have that $\mathcal{T},(u, v) \models \alpha$ if, and only if, there is a path from u to v labelled by a word in $L(G_\alpha)$. □

We therefore know that $\mathrm{PDL}_{\mathrm{CFG}} \le$ BSFP and even $\mathrm{PDL}_{\mathrm{CFG}} \lneq$ BSFP since it is known that $\mathrm{PDL}_{\mathrm{CFG}}$ is incomparable to the modal μ-calculus, which is a fragment of BSFP. The satisfiability problem for $\mathrm{PDL}_{\mathrm{CFG}}$ is known to be Σ^1_1-complete [8].

Corollary 5.3 *The satisfiability problem for* BSFP *is Σ^1_1-hard.*

The precise complexity level of Sat(BSFP) remains open. It is known that the satisfiability problem for LFP is in the stronger class Σ^2_1 [4], and we do not know whether satisfiablity for BSFP is as hard as for LFP, or Σ^1_1-complete.

We quickly turn our attention to another fragment of BSFP. A BSFP action formula is test-free if it does not contain any test φ?. Obviously every test-free action formula is flat. An inspection of the proof of Theorem 5.2 reveals that every test-free action formula can be translated into a context-free grammar over terminals in A (i.e. without tests) and vice versa. Hence we obtain the following result for test-free action formulae.

Corollary 5.4 *A global binary relation R is definable by a test-free* BSFP *action formula if, and only if there is a context-free language $L \subseteq A^*$ such that for all transition systems $\mathcal{T}$ it holds that $(u, v) \in R^{\mathcal{T}}$ iff there is a path from u to v that is labelled by some word in L.*

6 Definability of Languages

An important aspect of the expressive power of a logic is the question which classes of languages it can define. There are several possibilities to model the specification of a language by a BSFP formula. The standard way is to identify words with certain structures and associate with a BSFP-formula the language of all words such that the corresponding word structure is a model of the formula. With a finite word $w = w_0w_2\dots w_{n-1} \in \Sigma^n$ of length $n \ge 0$ we may associate the (unlabelled) transition system $\mathcal{T}(w) := (\{0,\dots,n\},(E_a)_{a\in\Sigma})$ where $(i, j) \in E_a$ iff $j = i + 1$ and $w_i = a$. A BSFP state formula φ then defines the language

$L(\varphi) := \{w \in \Sigma^* : \mathcal{T}(w), 0 \models \varphi\}$. It is also possible to specify a language by an action formula α, by defining $L(\alpha) := \{w \in \Sigma^* : \mathcal{T}(w), (0, |w|) \models \alpha\}$. It is not hard to see that these two definitions capture the same class of languages.

Lemma 6.1 *A language $L \subseteq \Sigma^*$ is definable by a* BSFP *state formula iff it is definable by a* BSFP *action formula.*

Proof. Let φ be a BSFP formula over the signature $\tau = \{E_a \mid a \in \Sigma\}$. Consider the action formula $\alpha = \mu Z.(\varphi? \cup Z \circ E)$ where E is shorthand for $\bigcup_{a\in\Sigma} E_a$. Obviously for a word w we have

$$[\![\alpha]\!]^{\mathcal{T}(w)} = \{(i,j) \mid 0 \leq i \leq j \leq |w| \text{ and } \mathcal{T}(w), i \models \varphi\}$$

and therefore $\mathcal{T}(w), (0, |w|) \models \alpha$ iff $\mathcal{T}(w), 0 \models \varphi$.

Now consider an action formula α over τ. For $\varphi = \langle\alpha\rangle[E]\bot$ we have by definition that $\mathcal{T}(w), 0 \models \varphi$ iff there is a j such that $(0, j) \in [\![\alpha]\!]^{\mathcal{T}(w)}$ and j has no successor in $\mathcal{T}(w)$ which means $j = |w|$. □

While the modal μ-calculus and PDL have on words the same expressive power as MSO and therefore capture exactly the regular languages, from Corollary 5.4 we know that even the test-free BSFP formulae capture a much richer class of languages.

Corollary 6.2 *A language $L \subseteq \Sigma^*$ is context-free if, and only if, it is definable by a test-free* BSFP *action formula. As a consequence, every Boolean combination of context-free languages is* BSFP*-definable.*

Example 6.3 For the context-free languages

$$L_1 = \{a^n b^n c^m \mid n, m \in \mathbb{N}\} \text{ and } L_2 = \{a^m b^n c^n \mid n, m \in \mathbb{N}\}$$

let $\varphi_{L_1}, \varphi_{L_2}$ be BSFP formulae that define the respective languages. Then $\varphi_{L_1} \wedge \varphi_{L_2}$ defines the context-sensitive language $\{a^n b^n c^n \mid n \in \mathbb{N}\}$.

We remark that the way we associated structures with words differs slightly from the way this is usually done when one proves that MSO or L_μ capture exactly the regular languages. There one associates with every word w the structure $\mathcal{T}'(w) = (\{0, \ldots, |w| - 1\}, (P_a)_{a\in\Sigma}, E)$ with $P_a = \{i \mid w_i = a\}$ and $E = \{(i, i+1) \mid 0 \leq i \leq |w| - 2\}$. However, it is not hard to see that on this class of structures BSFP is not less expressive.

Lemma 6.4 *For every* BSFP *action formula α there is a formula $\hat{\alpha}$ such that*

$$\mathcal{T}(w), (0, |w|) \models \alpha \Leftrightarrow \mathcal{T}'(w), (0, |w|) \models \hat{\alpha}.$$

Proof. From α construct $\hat{\alpha}$ in the following way: first replace every sub-formula E_a by $P_a?$ and then replace every formula of the form $\alpha_1 \circ \alpha_2$ by $\alpha_1 \circ E \circ \alpha_2$.□

A different approach for defining a language with BSFP is to consider the structure $\mathcal{T}_{\Sigma^*} = (\Sigma^*, (E_a)_{a\in\Sigma})$ with $E_a^{\mathcal{T}_{\Sigma^*}} = \{(w, wa) \mid w \in \Sigma^*\}$. We say a BSFP formula α defines a language L in $\mathcal{T}_{\Sigma^*}$ if $L = \{w \in \Sigma^* \mid \mathcal{T}_{\Sigma^*}, (\varepsilon, w) \models \alpha\}$

Theorem 6.5 *A language L is* BSFP *definable in $\mathcal{T}_{\Sigma^*}$ if, and only if, it is context-free.*

Proof. We claim that for every action formula α there is a test-free action formula $\hat{\alpha}$ such that $[\![\alpha]\!]^{\mathcal{T}_{\Sigma^*}} = [\![\hat{\alpha}]\!]^{\mathcal{T}_{\Sigma^*}}$. Since BSFP state formulae are invariant under bisimulation and every pair of nodes in $\mathcal{T}_{\Sigma^*}$ is bisimilar, every test $\varphi?$ holds either for every node or for no node at all. We can therefore replace every test in α either by D or by $\emptyset$ and arrive at a test-free BSFP formula with the same extension in $\mathcal{T}_{\Sigma^*}$. With Corollary 5.4 we get that $\hat{\alpha}$ corresponds to some context-free language L and therefore $\{w \in \Sigma^* \mid \mathcal{T}_{\Sigma^*}, (\varepsilon, w) \models \alpha\} = L$. □

References

[1] Arnold, A. and D. Niwiński, "Rudiments of μ-calculus," North Holland, 2001.

[2] Bertoni, A., C. Choffrut and R. Radicioni, *The inclusion problem of context-free languages: Some tractable cases*, in: *Developments in language theory*, Springer, 2009, pp. 103–112.

[3] Dawar, A., E. Grädel and S. Kreutzer, *Inflationary fixed points in modal logic*, ACM Transactions on Computational Logic **5** (2004), pp. 282 – 315.

[4] Dawar, A. and Y. Gurevich, *Fixed point logics*, Bulletin of Symbolic Logic **8** (2002), pp. 65–88.

[5] Grädel, E. and M.Otto, *The freedoms of (guarded) bisimulation*, in: *Johan F.A. K.van Benthem on Logical and Informational Dynamics*, Springer, 2014 p. To appear.

[6] Grädel, E. et al., "Finite Model Theory and Its Applications," Springer, 2007.

[7] Harel, D., D. Kozen and J. Tiuryn, "Dynamic Logic," MIT Press, 2000.

[8] Harel, D., A. Pnueli and J. Stavi, *Propositional dynamic logic of nonregular programs*, Journal of Computer and System Sciences **26** (1983), pp. 222–243.

[9] Hollenberg, M., "Logic and Bisimulation," Ph.D. thesis, Utrecht University (1998).

[10] Janin, D. and I. Walukiewicz, *On the expressive completeness of the propositional mu-calculus with respect to monadic second order logic*, in: *Proceedings of 7th International Conference on Concurrency Theory CONCUR '96*, number 1119 in Lecture Notes in Computer Science (1996), pp. 263–277.

[11] Otto, M., *Bisimulation-invariant Ptime and higher-dimensional mu-calculus*, Theoretical Computer Science **224** (1999), pp. 237–265.

[12] van Benthem, J., "Modal Correspondence Theory," Ph.D. thesis, University of Amsterdam (1976).

[13] van Benthem, J., *Program constructions that are safe for bisimulation*, Studia Logica **60** (1998), pp. 311–330.

Definability and Computability for $PRSPDL$

Philippe Balbiani [1]

Institut de recherche en informatique de Toulouse
CNRS — Université de Toulouse

Tinko Tinchev [2]

Department of Mathematical Logic and Applications
Sofia University

Abstract

$PRSPDL$ is a variant of PDL with parallel composition. In the Kripke models in which $PRSPDL$-formulas are evaluated, states have an internal structure. We devote this paper to the definability issue of several classes of frames by means of the language of $PRSPDL$ and to the computability issue of $PRSPDL$-validity for various fragments of the $PRSPDL$-language and for various classes of $PRSPDL$-frames.

Keywords: Propositional dynamic logic; parallel composition; definability; computability.

1 Introduction

Propositional dynamic logic (PDL) is a non-classical logic designed for reasoning about the behaviour of programs [11,16,19]. Its syntax is based on the idea of associating with each program α of some programming language the modal operator $[\alpha]$, formulas $[\alpha]\phi$ being read "every execution of α from the present state leads to a state bearing the formula ϕ". Syntactically, PDL is a modal logic with a structure in the set of modal operators: composition $(\alpha;\beta)$ of programs α and β corresponds to the composition of the accessibility relations $R(\alpha)$ and $R(\beta)$; test $\phi?$ on formula ϕ corresponds to the partial identity relation in the subsets of the Kripke models in which the formula ϕ is true; iteration $\alpha^\star$ corresponds to the reflexive and transitive closure of $R(\alpha)$. A number

[1] Address: Institut de recherche en informatique de Toulouse, CNRS — Université de Toulouse, 118 route de Narbonne, 31062 Toulouse Cedex 9, FRANCE; balbiani@irit.fr.
Partially supported by the "French National Research Agency" (contract ANR-11-BS02-011) and the "Bulgarian National Science Fund" (contract DID02/32/2009).

[2] Address: Department of Mathematical Logic and Applications, Sofia University, Blvd James Bouchier 5, 1126 Sofia, Bulgaria; tinko@fmi.uni-sofia.bg.
Partially supported by the "Centre international de mathématiques et d'informatique" (contract ANR-11-LABX-0040-CIMI within the program ANR-11-IDEX-0002-02).

of variants have been obtained by extending or restricting the syntax or the semantics of PDL [3,4,5,9,10,15,17,21,22].

The problem with most of these variants is that the states of the Kripke models in which formulas are evaluated have no internal structure. However, in the field of non-classical logics, it seems natural to propose formalisms with which one can cope with structured data such as heaps, pointers, etc. In addition to the standard Boolean constructs, separation logics are based on the formula construct $(\cdot \circ \cdot)$ of separating conjunction, formulas $(\phi \circ \psi)$ being read "the memory model can be split into two disjoint models respectively satisfying ϕ and ψ", and the formula construct $(\cdot\!-\!\!\circ\cdot)$ of adjoint implication, formulas $(\phi-\circ\psi)$ being read "if the memory model is extended with a model satisfying ϕ, then the resulting model satisfies ψ" [7,8,14,25]. In order to illustrate the significance of these constructs, one may consider the set of all words on an alphabet and its associated operation of concatenation, the set of all binary trees and its associated operation of join and the set of all heaps (partially defined functions mapping locations to values) and its associated operation of union (undefined when domains overlap).

$PRSPDL$, the propositional dynamic logic with storing, recovering and parallel composition introduced by Benevides *et al.* [5], is a separation-based non-classical logic too. Benevides *et al.* [5] extend the semantics of PDL by considering Kripke models structured by means of a function $*$: the state x is the result of applying the function $*$ to the states y, z iff x can be separated in a first part y and a second part z. They extend the syntax of PDL as well by adding the program construct $(\cdot \parallel \cdot)$ of parallel composition, the storing programs s_1 and s_2 and the recovering programs r_1 and r_2. In this variant, parallel composition $(\alpha \parallel \beta)$ corresponds to the fork $R(\alpha)\nabla R(\beta)$ of $R(\alpha)$ and $R(\beta)$ defined as follows:

- whenever x and y are related via $R(\alpha)$ and z and t are related via $R(\beta)$, $x * z$ and $y * t$ are related via $R(\alpha)\nabla R(\beta)$.

About s_1 and s_2, x is related, by s_1, to the states $x * z$ and, by s_2, to the states $z * x$. As for r_1 and r_2, the states $x * z$, by r_1, and the states $z * x$, by r_2, are related to x. Hence, s_1, s_2, r_1 and r_2 enable us to view states as ordered pairs of states. The function $*$ considered in [5] has its origin in the addition of an extra binary operation of fork denoted ∇ in relation algebras [12,13].

It appears that $(\cdot \parallel \cdot)$ can be eliminated from the language of $PRSPDL$ extended with $(\cdot \cap \cdot)$. To see this, it suffices to consider the equivalence between $(\alpha \parallel \beta)$ and $((r_1; \alpha; s_1) \cap (r_2; \beta; s_2))$ in all Kripke models structured by means of a function $*$ as above. On one hand, the decidability of PDL with intersection [9] seems to indicate that $PRSPDL$-validity is decidable as well. The problem is that the language of $PRSPDL$ contains two programs, namely r_1 and r_2, interpreted in [5] by deterministic binary relations. Hence, Danecki's result cannot be directly applied. On the other hand, the undecidability of PDL with intersection and at least two program variables interpreted by deterministic binary relations [18] seems to indicate that $PRSPDL$-validity is undecidable as well. The problem is that $(\cdot \cap \cdot)$ cannot be defined in the language of $PRSPDL$. Thus, Harel's result cannot be directly applied.

Nevertheless, following the line of reasoning suggested in [18], it is possible to reduce the Σ_1^1-hard $\mathbb{N} \times \mathbb{N}$ recurring tiling problem to satisfiability of $PRSPDL$-formulas when r_1 and r_2 are interpreted by deterministic binary relations as in [5]. Hence, $PRSPDL$-validity is Π_1^1-hard. The section-by-section breakdown of this article is as follows. In Section 2, we present the syntax and semantics of $PRSPDL$. The aim of Section 3 is to investigate the definability of several classes of frames. In Section 4, we demonstrate that neither the program construct $(\cdot \parallel \cdot)$, nor the storing programs s_1 and s_2, nor the recovering programs r_1 and r_2 can be eliminated from the language of $PRSPDL$. For various fragments of the $PRSPDL$-language and for various classes of $PRSPDL$-frames, we will devote Sections 5 and 6 to the computability of $PRSPDL$-validity.

2 Syntax and semantics

2.1 Syntax

Programs and formulas are inductively defined as follows:

- $\alpha ::= a \mid (\alpha; \beta) \mid \phi? \mid \alpha^\star \mid (\alpha \parallel \beta) \mid s_1 \mid s_2 \mid r_1 \mid r_2$;
- $\phi ::= p \mid \bot \mid \neg\phi \mid (\phi \vee \psi) \mid [\alpha]\phi$;

where a ranges over a countably infinite set of program variables and p ranges over a countably infinite set of propositional variables. The other Boolean constructs for formulas are defined as usual. The modal construct $\langle\cdot\rangle\cdot$ for formulas is defined as follows:

- $\langle\alpha\rangle\phi ::= \neg[\alpha]\neg\phi$.

We will follow the standard rules for omission of the parentheses.

Example 2.1 If α, β are programs and ϕ, ψ are formulas, then $\langle\alpha \parallel \beta\rangle\phi \rightarrow \langle r_1; \alpha; s_1\rangle(\phi \wedge \psi) \vee \langle r_2; \beta; s_2\rangle(\phi \wedge \neg\psi)$ is a formula as well.

Let the level of an expression exp (either a program, or a formula), in symbols $lev(exp)$, be the number of occurrences of the program construct $(\cdot \parallel \cdot)$ of parallel composition in exp.

2.2 Frames

A frame is a 3-tuple $\mathcal{F} = (W, R, *)$ where

- W is a nonempty set of states,
- R is a function from the set of all program variables into the set of all binary relations between states,
- $*$ is a function from the set of all pairs of states into the set of all sets of states.

We will use $x, y, \ldots$ for states. In $\mathcal{F}$, W is to be regarded as the set of all possible states in a computation process, R associates with each program variable a the binary relation $R(a)$ on W with $xR(a)y$ meaning "y can be reached from x by performing program variable a" and $*$ associates with each pair (y, z) of states the subset $y * z$ of W with $x \in y * z$ meaning "x can be obtained as a result of the combination of y and z". We shall say that a frame $\mathcal{F} = (W, R, *)$ is functional iff for all $x, y, z \in W$, if $xR(a)y$ and $xR(a)z$, then $y = z$ for every program variable a. We will also be interested in the following types of frames:

- $*$-distributive frames, i.e. frames $\mathcal{F} = (W, R, *)$ such that for all $x, y, z, t \in W$, $(x * y) \cap (z * t) = (x * t) \cap (z * y)$,
- $*$-separated frames, i.e. frames $\mathcal{F} = (W, R, *)$ such that for all $x, y, z, t \in W$, if $(x * y) \cap (z * t) \neq \emptyset$, then $x = z$ and $y = t$,
- $*$-deterministic frames, i.e. frames $\mathcal{F} = (W, R, *)$ such that for all $x, y, z, t \in W$, if $x \in z * t$ and $y \in z * t$, then $x = y$,
- $*$-serial frames, i.e. frames $\mathcal{F} = (W, R, *)$ such that for all $x, y \in W$, $x * y \neq \emptyset$.

Remark that every $*$-separated frame is $*$-distributive. Moreover, each frame considered in [5] is $*$-separated and $*$-deterministic. In order to illustrate the significance of these types of frames, we present the following:

Example 2.2 Let W_1 be the set of all words on an alphabet and $*_1$ be the operation of concatenation. The structure $\mathcal{F}_1 = (W_1, *_1)$ is not $*$-distributive. Nevertheless, it is $*$-deterministic and $*$-serial.

Let W_2 be the set of all binary trees and $*_2$ be the operation of join. The structure $\mathcal{F}_2 = (W_2, *_2)$ is $*$-separated, $*$-deterministic and $*$-serial.

Let W_3 be the set of all heaps (partially defined functions mapping locations to values) and $*_3$ be the operation of union (undefined when domains overlap). The structure $\mathcal{F}_3 = (W_3, *_3)$ is neither $*$-distributive, nor $*$-serial. Nevertheless, it is $*$-deterministic.

2.3 Models

A model on the frame $\mathcal{F} = (W, R, *)$ is a 4-tuple $\mathcal{M} = (W, R, *, V)$ where

- V is a valuation on $\mathcal{F}$, i.e. a function from the set of all propositional variables into the set of all sets of states.

In $\mathcal{M}$, V associates with each propositional variable p the subset $V(p)$ of W with $x \in V(p)$ meaning "propositional variable p is true at x". In a model $\mathcal{M} = (W, R, *, V)$, we inductively define the properties "y can be reached from x by performing program α" (in symbols $xR_{\mathcal{M}}(\alpha)y$) and "formula ϕ is true at x" (in symbols $x \in V_{\mathcal{M}}(\phi)$) as follows:

- $xR_{\mathcal{M}}(a)y$ iff $xR(a)y$;
- $xR_{\mathcal{M}}(\alpha; \beta)y$ iff there exists $z \in W$ such that $xR_{\mathcal{M}}(\alpha)z$ and $zR_{\mathcal{M}}(\beta)y$;
- $xR_{\mathcal{M}}(\phi?)y$ iff $x = y$ and $y \in V_{\mathcal{M}}(\phi)$;
- $xR_{\mathcal{M}}(\alpha^{\star})y$ iff there exists $n \in \mathbb{N}$ and there exists $z_0, \ldots, z_n \in W$ such that $z_0 = x$, $z_0 R_{\mathcal{M}}(\alpha) z_1$, …, $z_{n-1} R_{\mathcal{M}}(\alpha) z_n$ and $z_n = y$;
- $xR_{\mathcal{M}}(\alpha \parallel \beta)y$ iff there exists $z, t, u, v \in W$ such that $x \in z * t$, $zR_{\mathcal{M}}(\alpha)u$,

$tR_{\mathcal{M}}(\beta)v$ and $y \in u * v$;

- $xR_{\mathcal{M}}(s_1)y$ iff there exists $z \in W$ such that $y \in x * z$;
- $xR_{\mathcal{M}}(s_2)y$ iff there exists $z \in W$ such that $y \in z * x$;
- $xR_{\mathcal{M}}(r_1)y$ iff there exists $z \in W$ such that $x \in y * z$;
- $xR_{\mathcal{M}}(r_2)y$ iff there exists $z \in W$ such that $x \in z * y$;
- $x \in V_{\mathcal{M}}(p)$ iff $x \in V(p)$;
- $x \notin V_{\mathcal{M}}(\bot)$;
- $x \in V_{\mathcal{M}}(\neg\phi)$ iff $x \notin V_{\mathcal{M}}(\phi)$;
- $x \in V_{\mathcal{M}}(\phi \vee \psi)$ iff either $x \in V_{\mathcal{M}}(\phi)$, or $x \in V_{\mathcal{M}}(\psi)$;
- $x \in V_{\mathcal{M}}([\alpha]\phi)$ iff for all $y \in W$, if $xR_{\mathcal{M}}(\alpha)y$, then $y \in V_{\mathcal{M}}(\phi)$.

As a result, $x \in V_{\mathcal{M}}(\langle\alpha\rangle\phi)$ iff there exists $y \in W$ such that $xR_{\mathcal{M}}(\alpha)y$ and $y \in V_{\mathcal{M}}(\phi)$. A formula ϕ is said to be true in the model $\mathcal{M} = (W, R, *, V)$, in symbols $\mathcal{M} \models \phi$, iff $V_{\mathcal{M}}(\phi) = W$. We shall say that a formula ϕ is satisfied in $\mathcal{M}$ iff $V_{\mathcal{M}}(\phi) \neq \emptyset$. A formula ϕ is said to be valid in the frame $\mathcal{F}$, in symbols $\mathcal{F} \models \phi$, iff for all models $\mathcal{M}$ on $\mathcal{F}$, $\mathcal{M} \models \phi$. We shall say that a formula ϕ is satisfied in $\mathcal{F}$ iff there exists a model $\mathcal{M}$ on $\mathcal{F}$ such that ϕ is satisfied in $\mathcal{M}$. A formula ϕ is said to be satisfied in a class $\mathcal{C}$ of frames iff there exists a frame $\mathcal{F}$ in $\mathcal{C}$ such that ϕ is satisfied in $\mathcal{F}$.

Example 2.3 The formula $\langle\alpha \parallel \beta\rangle\phi \to \langle r_1; \alpha; s_1\rangle(\phi \wedge \psi) \vee \langle r_2; \beta; s_2\rangle(\phi \wedge \neg\psi)$ considered in Example 2.1 is valid in every $*$-separated frame.

2.4 A decision problem

Let $\mathcal{L}$ be a fragment of the $PRSPDL$-language and $\mathcal{C}$ be a class of $PRSPDL$-frames. The set of all $\mathcal{L}$-formulas that are valid in every $\mathcal{C}$-frame will be denoted $VAL(\mathcal{L}, \mathcal{C})$. For various fragments $\mathcal{L}$ of the $PRSPDL$-language and for various classes $\mathcal{C}$ of $PRSPDL$-frames, we will devote Sections 5 and 6 of this paper to the computability of the following decision problem:

- input: an $\mathcal{L}$-formula ϕ;
- output: determine whether ϕ is valid in every $\mathcal{C}$-frame.

3 Definability

A class $\mathcal{C}$ of frames is said to be modally defined by a set Σ of formulas iff for all frames $\mathcal{F}$, $\mathcal{F}$ is in $\mathcal{C}$ iff $\mathcal{F} \models \Sigma$. We shall say that a class of frames is modally definable iff it is modally defined by a set of formulas. Obviously, the class of all functional frames is modally defined by the formulas $\langle a\rangle p \to [a]p$ for every program variable a. About the class of all $*$-distributive frames, the class of all $*$-separated frames and the class of all $*$-deterministic frames, we have the following:

Proposition 3.1 1) *The class of all $*$-distributive frames is modally defined by the formula* $\langle p? \parallel \top?\rangle\top \wedge \langle\top? \parallel q?\rangle\top \to \langle p? \parallel q?\rangle\top$.

2) *The class of all $*$-separated frames is modally defined by the formulas* $\langle p? \parallel \top?\rangle\top \to [\neg p? \parallel \top?]\bot$, $\langle \top? \parallel q?\rangle\top \to [\top? \parallel \neg q?]\bot$.

3) *The class of all $*$-deterministic frames is modally defined by the formula* $p \to [\top? \parallel \top?]p$.

Proof. We only give the proof of 1), leaving the proof of 2) and 3) to the reader. Let $\mathcal{F} = (W, R, *)$ be a frame.
Suppose $\mathcal{F}$ is $*$-distributive. If $\mathcal{F} \not\models \langle p? \parallel \top?\rangle\top \wedge \langle \top? \parallel q?\rangle\top \to \langle p? \parallel q?\rangle\top$, then there exists a model $\mathcal{M} = (W, R, *, V)$ on $\mathcal{F}$ and there exists $x \in W$ such that $x \notin V_{\mathcal{M}}(\langle p? \parallel \top?\rangle\top \wedge \langle \top? \parallel q?\rangle\top \to \langle p? \parallel q?\rangle\top)$. Hence, $x \in V_{\mathcal{M}}(\langle p? \parallel \top?\rangle\top)$, $x \in V_{\mathcal{M}}(\langle \top? \parallel q?\rangle\top)$ and $x \notin V_{\mathcal{M}}(\langle p? \parallel q?\rangle\top)$. Thus, there exists $y, z, s, t \in W$ such that $x \in y * z$, $y \in V(p)$, $x \in s * t$ and $t \in V(q)$. Therefore, $x \in (y*z) \cap (s*t)$. Since $\mathcal{F}$ is $*$-distributive, then $(y*z) \cap (s*t) = (y*t) \cap (s*z)$. Since $x \in (y * z) \cap (s * t)$, then $x \in (y * t) \cap (s * z)$. Consequently, $x \in y * t$. Since $y \in V(p)$ and $t \in V(q)$, then $x \in V_{\mathcal{M}}(\langle p? \parallel q?\rangle\top)$: a contradiction.
Suppose $\mathcal{F} \models \langle p? \parallel \top?\rangle\top \wedge \langle \top? \parallel q?\rangle\top \to \langle p? \parallel q?\rangle\top$. If $\mathcal{F}$ is not $*$-distributive, then there exists $y, z, s, t \in W$ such that $(y*z) \cap (s*t) \neq (y*t) \cap (s*z)$. Hence, there exists $x \in W$ such that either $x \in (y * z) \cap (s * t)$ and $x \notin (y * t) \cap (s * z)$, or $x \in (y * t) \cap (s * z)$ and $x \notin (y * z) \cap (s * t)$. Without loss of generality, assume $x \in (y * z) \cap (s * t)$ and $x \notin (y * t) \cap (s * z)$. Thus, $x \in y * z$, $x \in s * t$ and either $x \notin y * t$, or $x \notin s * z$. Without loss of generality, assume $x \notin y * t$. Let V be a valuation on $\mathcal{F}$ such that $V(p) = \{y\}$ and $V(q) = \{t\}$. Let $\mathcal{M} = (W, R, *, V)$. Since $\mathcal{F} \models \langle p? \parallel \top?\rangle\top \wedge \langle \top? \parallel q?\rangle\top \to \langle p? \parallel q?\rangle\top$, then $x \in V_{\mathcal{M}}(\langle p? \parallel \top?\rangle\top \wedge \langle \top? \parallel q?\rangle\top \to \langle p? \parallel q?\rangle\top)$. Therefore, if $x \in V_{\mathcal{M}}(\langle p? \parallel \top?\rangle\top)$ and $x \in V_{\mathcal{M}}(\langle \top? \parallel q?\rangle\top)$, then $x \in V_{\mathcal{M}}(\langle p? \parallel q?\rangle\top)$ Since $x \in y * z$, $V(p) = \{y\}$, $x \in s * t$ and $V(q) = \{t\}$, then $x \in V_{\mathcal{M}}(\langle p? \parallel \top?\rangle\top)$ and $x \in V_{\mathcal{M}}(\langle \top? \parallel q?\rangle\top)$. Since if $x \in V_{\mathcal{M}}(\langle p? \parallel \top?\rangle\top)$ and $x \in V_{\mathcal{M}}(\langle \top? \parallel q?\rangle\top)$, then $x \in V_{\mathcal{M}}(\langle p? \parallel q?\rangle\top)$, then $x \in V_{\mathcal{M}}(\langle p? \parallel q?\rangle\top)$. Consequently, there exists $u, v \in W$ such that $x \in u * v$, $u \in V(p)$ and $v \in V(q)$. Since $V(p) = \{y\}$ and $V(q) = \{t\}$, then $u = y$ and $v = t$. Since $x \in u * v$, then $x \in y * t$: a contradiction. □

As for the class of all $*$-serial frames, we have the following:

Proposition 3.2 *The class of all $*$-serial frames is not modally definable.*

Proof. Suppose there exists a set Σ of formulas that modally defines the class of all $*$-serial frames. Let $\mathcal{F} = (W, R, *)$ and $\mathcal{F}' = (W', R', *')$ be the frames defined as follows:

- $W = \{x_1, x_2\}$,
- R is the empty function,
- $x_1 * x_1 = \{x_1\}$, $x_2 * x_2 = \{x_2\}$ and otherwise $*$ is the empty function,
- $W' = \{x'\}$,
- R' is the empty function,
- $x' *' x' = \{x'\}$.

Obviously, $\mathcal{F}$ is not $*$-serial and $\mathcal{F}'$ is $*$-serial. Since Σ modally defines the class of all $*$-serial frames, then $\mathcal{F} \not\models \Sigma$ and $\mathcal{F}' \models \Sigma$. Hence, there exists a formula $\phi \in \Sigma$ such that $\mathcal{F} \not\models \phi$. Since $\mathcal{F}' \models \Sigma$, then $\mathcal{F}' \models \phi$. Since $\mathcal{F} \not\models \phi$, then there exists a model $\mathcal{M} = (W, R, *, V)$ on $\mathcal{F}$ such that either $x_1 \notin V_{\mathcal{M}}(\phi)$, or $x_2 \notin V_{\mathcal{M}}(\phi)$. Without loss of generality, assume $x_1 \notin V_{\mathcal{M}}(\phi)$. Let $\mathcal{M}' = (W', R', *', V')$ be the model on $\mathcal{F}'$ defined as follows:

- $V'(p) =$ if $x_1 \in V(p)$, then $\{x'\}$, else $\emptyset$ for every propositional variable p.

Since $\mathcal{F}' \models \phi$, then $x' \in V_{\mathcal{M}'}(\phi)$.

Claim 3.3 *Let α be a program and ψ be a formula from the language of $PRSPDL$. Then,*

- *not $x_1 R_{\mathcal{M}}(\alpha) x_2$,*
- *$x_1 R_{\mathcal{M}}(\alpha) x_1$ iff $x' R_{\mathcal{M}'}(\alpha) x'$,*
- *$x_1 \in V_{\mathcal{M}}(\psi)$ iff $x' \in V_{\mathcal{M}'}(\psi)$.*

Proof. By induction on α and ψ. Left to the reader. □

Since $x_1 \notin V_{\mathcal{M}}(\phi)$, then $x' \notin V_{\mathcal{M}'}(\phi)$: a contradiction. □

4 Expressivity

In the class of all $*$-separated frames, remark that the formula construct of separating conjunction $(\cdot \circ \cdot)$ and the formula construct of adjoint implication $(\cdot{-\!\circ}\cdot)$ evoked in the introduction can be defined in the language of $PRSPDL$ as follows:

- $(\phi \circ \psi) ::= \langle r_1 \rangle \phi \wedge \langle r_2 \rangle \psi$,
- $(\phi {-\!\circ} \psi) ::= [s_2](\langle r_1 \rangle \phi \rightarrow \psi)$.

Here are results proving that the program construct $(\cdot \parallel \cdot)$ of parallel composition, the storing programs s_1 and s_2 and the recovering programs r_1 and r_2 cannot be eliminated from the language of $PRSPDL$.

Proposition 4.1 *For all $\parallel$-free formulas ϕ from the language of $PRSPDL$, $\langle a \parallel a \rangle \top \leftrightarrow \phi$ is not valid in the class of all $*$-separated $*$-deterministic frames for every program variable a.*

Proof. Suppose there exists a $\parallel$-free formula ϕ from the language of $PRSPDL$ such that $\langle a \parallel a \rangle \top \leftrightarrow \phi$ is valid in the class of all $*$-separated $*$-deterministic frames for some program variable a. Without loss of generality, assume a is the only program variable in ϕ and ϕ contains no propositional variable. Let $\mathcal{F} = (W, R, *)$ and $\mathcal{F}' = (W', R', *')$ be the $*$-separated $*$-deterministic frames defined as follows:

- $W = \{x, y, z, t, u\}$,
- $R(a) = \{(y, z), (y, t)\}$ and otherwise R is the empty function,
- $y * y = \{x\}$, $z * t = \{u\}$ and otherwise $*$ is the empty function,
- $W' = \{x', y', z'_1, z'_2, t'_1, t'_2, u'_1, u'_2\}$,

- $R'(a) = \{(y', z_1'), (y', t_2')\}$ and otherwise R' is the empty function,
- $y' *' y' = \{x'\}$, $z_1' *' t_1' = \{u_1'\}$, $z_2' *' t_2' = \{u_2'\}$ and otherwise $*'$ is the empty function.

Since $\langle a \parallel a \rangle \top \leftrightarrow \phi$ is valid in the class of all $*$-separated $*$-deterministic frames, then $\mathcal{F} \models \langle a \parallel a \rangle \top \leftrightarrow \phi$ and $\mathcal{F}' \models \langle a \parallel a \rangle \top \leftrightarrow \phi$. Let $Z = \{(x, x'), (y, y'), (z, z_1'), (z, z_2'), (t, t_1'), (t, t_2'), (u, u_1'), (u, u_2')\}$. Let $\mathcal{M} = (W, R, *, V)$ be a model on $\mathcal{F}$ and $\mathcal{M}' = (W', R', *', V')$ be the model on $\mathcal{F}'$ corresponding to it with respect to Z. Obviously, $x \in V_{\mathcal{M}}(\langle a \parallel a \rangle \top)$ and $x' \notin V_{\mathcal{M}'}(\langle a \parallel a \rangle \top)$. Since $\mathcal{F} \models \langle a \parallel a \rangle \top \leftrightarrow \phi$ and $\mathcal{F}' \models \langle a \parallel a \rangle \top \leftrightarrow \phi$, then $x \in V_{\mathcal{M}}(\langle a \parallel a \rangle \top \leftrightarrow \phi)$ and $x' \in V_{\mathcal{M}'}(\langle a \parallel a \rangle \top \leftrightarrow \phi)$. Since $x \in V_{\mathcal{M}}(\langle a \parallel a \rangle \top)$ and $x' \notin V_{\mathcal{M}'}(\langle a \parallel a \rangle \top)$, then $x \in V_{\mathcal{M}}(\phi)$ and $x' \notin V_{\mathcal{M}'}(\phi)$.

Claim 4.2 *Let α be a $\parallel$-free program and ψ be a $\parallel$-free formula from the language of PRSPDL. For all $v \in W$ and for all $v' \in W'$, if vZv', then*

- *for all $w \in W$, if $vR_{\mathcal{M}}(\alpha)w$, then there exists $w' \in W'$ such that wZw' and $v'R_{\mathcal{M}'}(\alpha)w'$,*
- *for all $w' \in W'$, if $v'R_{\mathcal{M}'}(\alpha)w'$, then there exists $w \in W$ such that wZw' and $vR_{\mathcal{M}}(\alpha)w$,*
- *$v \in V_{\mathcal{M}}(\psi)$ iff $v' \in V_{\mathcal{M}'}(\psi)$.*

Proof. By induction on α and ψ. Left to the reader. □

Since xZx' and $x \in V_{\mathcal{M}}(\phi)$, then $x' \in V_{\mathcal{M}'}(\phi)$: a contradiction. □

Proposition 4.3 *For all storing-free formulas ϕ from the language of PRSPDL, $\langle s_i \rangle \top \leftrightarrow \phi$ is not valid in the class of all functional $*$-separated $*$-deterministic frames for every $i \in \{1, 2\}$.*

Proof. Suppose there exists a storing-free formula ϕ from the language of *PRSPDL* such that $\langle s_i \rangle \top \leftrightarrow \phi$ is valid in the class of all functional $*$-separated $*$-deterministic frames for some $i \in \{1, 2\}$. Without loss of generality, assume ϕ contains neither program variable, nor propositional variable. Moreover, we can assume $i = 1$. Let $\mathcal{F} = (W, R, *)$ and $\mathcal{F}' = (W', R', *')$ be the functional $*$-separated $*$-deterministic frames defined as follows:

- $W = \{x, y\}$,
- R is the empty function,
- $x * x = \{y\}$ and otherwise $*$ is the empty function,
- $W' = \{x', y'\}$,
- R' is the empty function,
- $*'$ is the empty function.

Since $\langle s_1 \rangle \top \leftrightarrow \phi$ is valid in the class of all functional $*$-separated $*$-deterministic frames, then $\mathcal{F} \models \langle s_1 \rangle \top \leftrightarrow \phi$ and $\mathcal{F}' \models \langle s_1 \rangle \top \leftrightarrow \phi$. Let $\mathcal{M} = (W, R, *, V)$ be a model on $\mathcal{F}$. Let $\mathcal{M}' = (W', R', *', V')$ be the model on $\mathcal{F}'$ defined as follows:

- $V'(p) =$ if $x \in V(p)$, then $\{x'\}$, else $\emptyset$ for every propositional variable p.

Obviously, $x \in V_{\mathcal{M}}(\langle s_1 \rangle \top)$ and $x' \notin V_{\mathcal{M}'}(\langle s_1 \rangle \top)$. Since $\mathcal{F} \models \langle s_1 \rangle \top \leftrightarrow \phi$ and $\mathcal{F}' \models \langle s_1 \rangle \top \leftrightarrow \phi$, then $x \in V_{\mathcal{M}}(\langle s_1 \rangle \top \leftrightarrow \phi)$ and $x' \in V_{\mathcal{M}'}(\langle s_1 \rangle \top \leftrightarrow \phi)$. Since $x \in V_{\mathcal{M}}(\langle s_1 \rangle \top)$ and $x' \notin V_{\mathcal{M}'}(\langle s_1 \rangle \top)$, then $x \in V_{\mathcal{M}}(\phi)$ and $x' \notin V_{\mathcal{M}'}(\phi)$.

Claim 4.4 *Let α be a storing-free program and ψ be a storing-free formula from the language of $PRSPDL$. Then,*

- *not $xR_{\mathcal{M}}(\alpha)y$,*
- *$xR_{\mathcal{M}}(\alpha)x$ iff $x'R_{\mathcal{M}'}(\alpha)x'$,*
- *$x \in V_{\mathcal{M}}(\psi)$ iff $x' \in V_{\mathcal{M}'}(\psi)$.*

Proof. By induction on α and ψ. Left to the reader. □

Since $x \in V_{\mathcal{M}}(\phi)$, then $x' \in V_{\mathcal{M}'}(\phi)$: a contradiction. □

Proposition 4.5 *For all recovering-free formulas ϕ from the language of $PRSPDL$, $[r_i^\star]\langle \top? \parallel \top? \rangle \top \leftrightarrow \phi$ is not valid in the class of all functional $*$-separated $*$-deterministic frames for every $i \in \{1, 2\}$.*

Proof. Suppose there exists a recovering-free formula ϕ from the language of $PRSPDL$ such that $[r_i^\star]\langle \top? \parallel \top? \rangle \top \leftrightarrow \phi$ is valid in the class of all functional $*$-separated $*$-deterministic frames for some $i \in \{1, 2\}$. Let $n = lev(\phi)$. Without loss of generality, assume ϕ contains neither program variable, nor propositional variable. Moreover, we can assume $i = 1$. Let $\mathcal{F} = (W, R, *)$ and $\mathcal{F}' = (W', R', *')$ be the functional $*$-separated $*$-deterministic frames defined as follows:

- $W = \{x, y\} \times \mathbb{N}$,
- R is the empty function,
- $(x, k+1) * (y, k+1) = \{(x, k)\}$ and otherwise $*$ is the empty function,
- $W' = \{x', y'\} \times \{0, \ldots, n\}$,
- R' is the empty function,
- $(x', 1) *' (y', 1) = \{(x', 0)\}$, …, $(x', n) *' (y', n) = \{(x', n-1)\}$ and otherwise $*'$ is the empty function.

Since $[r_1^\star]\langle \top? \parallel \top? \rangle \top \leftrightarrow \phi$ is valid in the class of all functional $*$-separated $*$-deterministic frames, then $\mathcal{F} \models [r_1^\star]\langle \top? \parallel \top? \rangle \top \leftrightarrow \phi$ and $\mathcal{F}' \models [r_1^\star]\langle \top? \parallel \top? \rangle \top \leftrightarrow \phi$. Let $Z = \{((x, 0), (x', 0)), \ldots, ((x, n), (x', n)), ((y, 0), (y', 0)), \ldots, ((y, n), (y', n))\}$. Let $\mathcal{M} = (W, R, *, V)$ be a model on $\mathcal{F}$ and $\mathcal{M}' = (W', R', *', V')$ be the model on $\mathcal{F}'$ corresponding to it with respect to Z. Obviously, $(x, 0) \in V_{\mathcal{M}}([r_1^\star]\langle \top? \parallel \top? \rangle \top)$ and $(x', 0) \notin V_{\mathcal{M}'}([r_1^\star]\langle \top? \parallel \top? \rangle \top)$. Since $\mathcal{F} \models [r_1^\star]\langle \top? \parallel \top? \rangle \top \leftrightarrow \phi$ and $\mathcal{F}' \models [r_1^\star]\langle \top? \parallel \top? \rangle \top \leftrightarrow \phi$, then $(x, 0) \in V_{\mathcal{M}}([r_1^\star]\langle \top? \parallel \top? \rangle \top \leftrightarrow \phi)$ and $(x', 0) \in V_{\mathcal{M}'}([r_1^\star]\langle \top? \parallel \top? \rangle \top \leftrightarrow \phi)$. Since $(x, 0) \in V_{\mathcal{M}}([r_1^\star]\langle \top? \parallel \top? \rangle \top)$ and $(x', 0) \notin V_{\mathcal{M}'}([r_1^\star]\langle \top? \parallel \top? \rangle \top)$, then $(x, 0) \in V_{\mathcal{M}}(\phi)$ and $(x', 0) \notin V_{\mathcal{M}'}(\phi)$.

Claim 4.6 *Let α be a recovering-free program from the language of $PRSPDL$. For all $k \in \{0, \ldots, n\}$,*

- $R_{\mathcal{M}}(\alpha)((x,k)) \subseteq \{(x,0),\ldots,(x,k)\}$ *and* $R_{\mathcal{M}'}(\alpha)((x',k)) \subseteq \{(x',0),\ldots,(x',k)\}$,
- $R_{\mathcal{M}}(\alpha)((y,k)) \subseteq \{(x,0),\ldots,(x,k-1)\} \cup \{(y,k)\}$ *and* $R_{\mathcal{M}'}(\alpha)((y',k)) \subseteq \{(x',0),\ldots,(x',k-1)\} \cup \{(y',k)\}$.

Proof. By induction on α. Left to the reader. □

Claim 4.7 *Let α be a recovering-free program and ψ be a recovering-free formula from the language of PRSPDL. For all $k \in \{0,\ldots,n\}$, if $k + lev(\alpha) \leq n$ and $k + lev(\psi) \leq n$, then*

- $(x,k)R_{\mathcal{M}}(\alpha)(x,l)$ *iff* $(x',k)R_{\mathcal{M}'}(\alpha)(x',l)$,
- $(y,k)R_{\mathcal{M}}(\alpha)(x,l)$ *iff* $(y',k)R_{\mathcal{M}'}(\alpha)(x',l)$,
- $(y,k)R_{\mathcal{M}}(\alpha)(y,l)$ *iff* $(y',k)R_{\mathcal{M}'}(\alpha)(y',l)$,
- $(x,k) \in V_{\mathcal{M}}(\psi)$ *iff* $(x',k) \in V_{\mathcal{M}'}(\psi)$,
- $(y,k) \in V_{\mathcal{M}}(\psi)$ *iff* $(y',k) \in V_{\mathcal{M}'}(\psi)$.

Proof. By induction on α and ψ. Left to the reader. □

Since $(x,0) \in V_{\mathcal{M}}(\phi)$, then $(x',0) \in V_{\mathcal{M}'}(\phi)$: a contradiction. □

Now, let us extend *PRSPDL* with the program construct $(\cdot \cap \cdot)$ of intersection. In this variant, intersection $(\alpha \cap \beta)$ of programs α and β corresponds to the intersection of the accessibility relations $R(\alpha)$ and $R(\beta)$. We have the following:

Proposition 4.8 *Let α, β be programs from the language of PRSPDL extended with $(\cdot\cap\cdot)$. For all models $\mathcal{M}$, $R_{\mathcal{M}}(\alpha \parallel \beta) = R_{\mathcal{M}}((r_1;\alpha;s_1)\cap(r_2;\beta;s_2))$.*

Proof. Left to the reader. □

Hence, the program construct $(\cdot \parallel \cdot)$ of parallel composition can be eliminated from the language of *PRSPDL* extended with $(\cdot \cap \cdot)$. Nevertheless,

Proposition 4.9 *For all formulas ϕ from the language of PRSPDL, $\langle a \cap b\rangle\top \leftrightarrow \phi$ is not valid in the class of all functional $*$-separated $*$-deterministic frames for every distinct program variables a, b.*

Proof. Suppose there exists a formula ϕ from the language of *PRSPDL* such that $\langle a \cap b\rangle\top \leftrightarrow \phi$ is valid in the class of all functional $*$-separated $*$-deterministic frames for some distinct program variables a, b. Without loss of generality, assume a, b are the only program variables in ϕ and ϕ contains no propositional variable. Let $\mathcal{F} = (W, R, *)$ and $\mathcal{F}' = (W', R', *')$ be the functional $*$-separated $*$-deterministic frames s defined as follows:

- $W = \{x, y\}$,
- $R(a) = \{(x,y)\}$, $R(b) = \{(x,y)\}$ and otherwise R is the empty function,
- $*$ is the empty function,
- $W' = \{x', y'_1, y'_2\}$,

- $R'(a) = \{(x', y'_1)\}$, $R'(b) = \{(x', y'_2)\}$ and otherwise R' is the empty function,
- $*'$ is the empty function.

Since $\langle a \cap b \rangle \top \leftrightarrow \phi$ is valid in the class of all functional $*$-separated $*$-deterministic frames, then $\mathcal{F} \models \langle a \cap b \rangle \top \leftrightarrow \phi$ and $\mathcal{F}' \models \langle a \cap b \rangle \top \leftrightarrow \phi$. Let $Z = \{(x, x'), (y, y'_1), (y, y'_2)\}$. Let $\mathcal{M} = (W, R, *, V)$ be a model on $\mathcal{F}$ and $\mathcal{M}' = (W', R', *', V')$ be the model on $\mathcal{F}'$ corresponding to it with respect to Z. Obviously, $x \in V_{\mathcal{M}}(\langle a \cap b \rangle \top)$ and $x' \notin V_{\mathcal{M}'}(\langle a \cap b \rangle \top)$. Since $\mathcal{F} \models \langle a \cap b \rangle \top \leftrightarrow \phi$ and $\mathcal{F}' \models \langle a \cap b \rangle \top \leftrightarrow \phi$, then $x \in V_{\mathcal{M}}(\langle a \cap b \rangle \top \leftrightarrow \phi)$ and $x' \in V_{\mathcal{M}'}(\langle a \cap b \rangle \top \leftrightarrow \phi)$. Since $x \in V_{\mathcal{M}}(\langle a \cap b \rangle \top)$ and $x' \notin V_{\mathcal{M}'}(\langle a \cap b \rangle \top)$, then $x \in V_{\mathcal{M}}(\phi)$ and $x' \notin V_{\mathcal{M}'}(\phi)$.

Claim 4.10 *Let α be a program and ψ be a formula from the language of $PRSPDL$. For all $v \in W$ and for all $v' \in W'$, if vZv', then*

- *for all $w \in W$, if $vR_{\mathcal{M}}(\alpha)w$, then there exists $w' \in W'$ such that wZw' and $v'R_{\mathcal{M}'}(\alpha)w'$,*
- *for all $w' \in W'$, if $v'R_{\mathcal{M}'}(\alpha)w'$, then there exists $w \in W$ such that wZw' and $vR_{\mathcal{M}}(\alpha)w$,*
- *$v \in V_{\mathcal{M}}(\psi)$ iff $v' \in V_{\mathcal{M}'}(\psi)$.*

Proof. By induction on α and ψ. Left to the reader. □

Since xZx' and $x \in V_{\mathcal{M}}(\phi)$, then $x' \in V_{\mathcal{M}'}(\phi)$: a contradiction. □

Hence, the program construct $(\cdot \cap \cdot)$ of intersection cannot be defined in the language of $PRSPDL$.

5 Decidability

Let $\mathcal{L}_{PDL}^{s_1,s_2}$ be the set of all $\|$-free recovering-free formulas. Let $\mathcal{C}_{*sep}$ be the class of all $*$-separated frames and $\mathcal{C}_{*sep}^{*det}$ be the class of all $*$-separated $*$-deterministic frames. The tree model property of PDL enables us to prove the following:

Proposition 5.1 (i) *$VAL(\mathcal{L}_{PDL}^{s_1,s_2}, \mathcal{C}_{*sep})$ is $EXPTIME$-complete.*

(ii) *$VAL(\mathcal{L}_{PDL}^{s_1,s_2}, \mathcal{C}_{*sep}^{*det})$ is $EXPTIME$-complete.*

Proof. The key thing to note about $\|$-free recovering-free formulas is the following:

Claim 5.2 *Let $\phi \in \mathcal{L}_{PDL}^{s_1,s_2}$. The following conditions are equivalent:*

a) ϕ, where s_1 and s_2 are considered as ordinary program variables, is satisfied in a PDL-frame.

b) ϕ, where s_1 and s_2 are considered as ordinary program variables, is satisfied in a tree-like PDL-frame.

c) ϕ, where s_1 and s_2 are considered as storing programs, is satisfied in a $*$-separated $*$-deterministic $PRSPDL$-frame.

d) ϕ, where s_1 and s_2 are considered as storing programs, is satisfied in a $*$-separated $PRSPDL$-frame.

Proof. *a)* $\Rightarrow$ *b)* Suppose ϕ, where s_1 and s_2 are considered as ordinary program variables, is satisfied in a PDL-frame $\mathcal{F} = (W, R)$. Hence, there exists a model $\mathcal{M} = (W, R, V)$ on $\mathcal{F}$ and there exists $x \in W$ such that $x \in V_{\mathcal{M}}(\phi)$. Let $\mathcal{F}' = (W', R')$ be the *Unravelling* of $\mathcal{F}$ around x and $\mathcal{M}' = (W', R', V')$ be the model on $\mathcal{F}'$ corresponding to $\mathcal{M}$. See [6, Pages 63, 218 and 219] for precise definitions. Obviously, $\mathcal{F}'$ is a tree-like PDL-frame. Since $x \in V_{\mathcal{M}}(\phi)$, by [6, Proposition 2.14 and Lemma 4.52], then $(x) \in V_{\mathcal{M}'}(\phi)$. Thus, ϕ, where s_1 and s_2 are considered as ordinary program variables, is satisfied in a tree-like PDL-frame.

b) $\Rightarrow$ *c)* Suppose ϕ, where s_1 and s_2 are considered as ordinary program variables, is satisfied in a tree-like PDL-frame $\mathcal{F} = (W, R)$. Hence, there exists a model $\mathcal{M} = (W, R, V)$ on $\mathcal{F}$ and there exists $x \in W$ such that $x \in V_{\mathcal{M}}(\phi)$. Let $\mathcal{F}' = (W', R', *')$ be the $*$-separated $*$-deterministic $PRSPDL$-frame defined as follows:

- $W' = W \cup \{(y, z, i)\colon\ y, z \in W,\ i \in \{1, 2\}$ and $yR(s_i)z\}$,
- $R'(a) = R(a)$ for every program variable a,
- $y *' (y, z, 1) = \{z\}$ for every $(y, z, 1) \in W'$, $(y, z, 2) *' y = \{z\}$ for every $(y, z, 2) \in W'$ and otherwise $*'$ is the empty function.

Let $\mathcal{M}' = (W', R', *', V')$ be the model on $\mathcal{F}'$ defined as follows:

- $V'(p) = V(p)$ for every propositional variable p.

Claim 5.3 *Let* $\psi \in \mathcal{L}_{PDL}^{s_1,s_2}$. *For all* $y \in W$, $y \in V_{\mathcal{M}}(\psi)$ *iff* $y \in V_{\mathcal{M}'}(\psi)$.

Proof. By induction on ψ. Left to the reader. □

Since $x \in V_{\mathcal{M}}(\phi)$, then $x \in V_{\mathcal{M}'}(\phi)$. Thus, ϕ, where s_1 and s_2 are considered as storing programs, is satisfied in a $*$-separated $*$-deterministic $PRSPDL$-frame.

c) $\Rightarrow$ *d)* Obvious.

d) $\Rightarrow$ *a)* Suppose ϕ, where s_1 and s_2 are considered as storing programs, is satisfied in a $*$-separated $PRSPDL$-frame $\mathcal{F} = (W, R, *)$. Hence, there exists a model $\mathcal{M} = (W, R, *, V)$ on $\mathcal{F}$ and there exists $x \in W$ such that $x \in V_{\mathcal{M}}(\phi)$. Let $\mathcal{F}' = (W', R')$ be the PDL-frame defined as follows:

- $W' = W$,
- $R'(a) = R(a)$ for every program variable a,
- $R'(s_1) = \{(x, y)\colon\ x, y, z \in W$ and $y \in x * z\}$,
- $R'(s_2) = \{(x, y)\colon\ x, y, z \in W$ and $y \in z * x\}$.

Let $\mathcal{M}' = (W', R', V')$ be the model on $\mathcal{F}'$ defined as follows:

- $V'(p) = V(p)$ for every propositional variable p.

Claim 5.4 *Let* $\psi \in \mathcal{L}_{PDL}^{s_1,s_2}$. *For all* $y \in W$, $y \in V_{\mathcal{M}}(\psi)$ *iff* $y \in V_{\mathcal{M}'}(\psi)$.

Proof. By induction on ψ. Left to the reader. □

Since $x \in V_{\mathcal{M}}(\phi)$, then $x \in V_{\mathcal{M}'}(\phi)$. Thus, ϕ, where s_1 and s_2 are considered as ordinary program variables, is satisfied in a PDL-frame. □

Since satisfiability in a PDL-frame of $\mathcal{L}_{PDL}^{s_1,s_2}$-formulas where s_1 and s_2 are considered as ordinary program variables is $EXPTIME$-complete [11,24], then satisfiability in a $*$-separated $PRSPDL$-frame and satisfiability in a $*$-separated $*$-deterministic $PRSPDL$-frame of $\mathcal{L}_{PDL}^{s_1,s_2}$-formulas where s_1 and s_2 are considered as storing programs are $EXPTIME$-complete. Hence, $VAL(\mathcal{L}_{PDL}^{s_1,s_2}, \mathcal{C}_{*sep})$ and $VAL(\mathcal{L}_{PDL}^{s_1,s_2}, \mathcal{C}_{*sep}^{*det})$ are $EXPTIME$-complete. □

Let $\mathcal{L}_{;}^{s_1,s_2}$ be the set of all ?-free $\star$-free $\|$-free recovering-free formulas. Claim 5.2 enables us to prove the following:

Proposition 5.5 1) *$VAL(\mathcal{L}_{;}^{s_1,s_2}, \mathcal{C}_{*sep})$ is $PSPACE$-complete.*

2) *$VAL(\mathcal{L}_{;}^{s_1,s_2}, \mathcal{C}_{*sep}^{*det})$ is $PSPACE$-complete.*

Proof. By Claim 5.2, ϕ, where s_1 and s_2 are considered as ordinary program variables, is satisfied in a PDL-frame iff ϕ, where s_1 and s_2 are considered as storing programs, is satisfied in a $*$-separated $*$-deterministic $PRSPDL$-frame iff ϕ, where s_1 and s_2 are considered as storing programs, is satisfied in a $*$-separated $PRSPDL$-frame for every $\phi \in \mathcal{L}_{;}^{s_1,s_2}$. Since satisfiability in a PDL-frame of $\mathcal{L}_{;}^{s_1,s_2}$-formulas where s_1 and s_2 are considered as ordinary program variables is $PSPACE$-complete [20], then satisfiability in a $*$-separated $PRSPDL$-frame and satisfiability in a $*$-separated $*$-deterministic $PRSPDL$-frame of $\mathcal{L}_{;}^{s_1,s_2}$-formulas where s_1 and s_2 are considered as storing programs are $PSPACE$-complete. Hence, $VAL(\mathcal{L}_{;}^{s_1,s_2}, \mathcal{C}_{*sep})$ and $VAL(\mathcal{L}_{;}^{s_1,s_2}, \mathcal{C}_{*sep}^{*det})$ are $PSPACE$-complete. □

6 Undecidability

Together with the decidability of PDL with intersection obtained by Danecki [9], Propositions 4.8 and 4.9 seems to indicate that $PRSPDL$ is decidable. It is interesting to observe that this assertion is false. Let $\mathcal{L}_{PDL}^{\|,r_1,r_2}$ be the set of all storing-free formulas. Solving an open problem put forward in [5], with the aid of the $\mathbb{N} \times \mathbb{N}$ recurring tiling problem, let us prove the following:

Proposition 6.1 *$VAL(\mathcal{L}_{PDL}^{\|,r_1,r_2}, \mathcal{C})$ is Π_1^1-hard for the following classes $\mathcal{C}$ of frames:*

- *the class $\mathcal{C}_{fun,*sep}^{*det,*ser}$ of all functional $*$-separated $*$-deterministic $*$-serial frames,*
- *the class $\mathcal{C}_{fun,*sep}^{*det}$ of all functional $*$-separated $*$-deterministic frames,*
- *the class $\mathcal{C}_{fun,*sep}^{*ser}$ of all functional $*$-separated $*$-serial frames,*
- *the class $\mathcal{C}_{*sep}^{*det,*ser}$ of all $*$-separated $*$-deterministic $*$-serial frames,*
- *the class $\mathcal{C}_{fun,*sep}$ of all functional $*$-separated frames,*
- *the class $\mathcal{C}_{*sep}^{*det}$ of all $*$-separated $*$-deterministic frames,*
- *the class $\mathcal{C}_{*sep}^{*ser}$ of all $*$-separated $*$-serial frames,*
- *the class $\mathcal{C}_{*sep}$ of all $*$-separated frames.*

Proof. Let $\mathcal{C}$ be one of the classes of frames considered in Proposition 6.1. A tile type t is a square, fixed in orientation, each side of which has a color: $left(t)$, $right(t)$, $down(t)$ and $up(t)$. A finite set T of tile types is said to tile $\mathbb{N} \times \mathbb{N}$ iff there exists a function f from $\mathbb{N} \times \mathbb{N}$ into T such that for all $x, y \in \mathbb{N}$, $right(f(x,y)) = left(f(x+1,y))$ and $up(f(x,y)) = down(f(x,y+1))$. The $\mathbb{N} \times \mathbb{N}$ recurring tiling problem is the following decision problem:

- input: a finite set T of tile types which includes some distinguished tile type t_1;
- output: determine whether T can tile $\mathbb{N} \times \mathbb{N}$ in such a way that t_1 occurs infinitely often in the first row.

It is a well-known fact that the $\mathbb{N} \times \mathbb{N}$ recurring tiling problem is Σ_1^1-hard [18]. Given pairwise distinct tile types $t_1, \ldots, t_n$, let $p_1, \ldots, p_n$ be pairwise distinct propositional variables. We associate to $t_1, \ldots, t_n$, the conjunction $\phi(t_1, \ldots, t_n)$ of the following formulas:

$B1$ $[r_1^\star \parallel r_2^\star]\langle r_1 \parallel \top?\rangle\top$;

$B2$ $[r_1^\star \parallel r_2^\star]\langle \top? \parallel r_2\rangle\top$;

$B3$ $[r_1^\star \parallel r_2^\star]\neg(p_i \wedge p_j)$ for every $i, j \in \{1, \ldots, n\}$ such that $i \neq j$;

$B4$ $[r_1^\star \parallel r_2^\star](p_1 \vee \ldots \vee p_n)$;

$B5$ $[r_1^\star \parallel r_2^\star](p_i \rightarrow [\top? \parallel \top?]p_i)$ for every $i \in \{1, \ldots, n\}$;

$B6$ $[r_1^\star \parallel r_2^\star](p_i \rightarrow \langle r_1 \parallel \top?\rangle(p_{k_1} \vee \ldots \vee p_{k_l}))$ for every $i \in \{1, \ldots, n\}$, $t_{k_1}, \ldots, t_{k_l}$ being the tile types in $t_1, \ldots, t_n$ horizontally matching with t_i;

$B7$ $[r_1^\star \parallel r_2^\star](p_i \rightarrow \langle \top? \parallel r_2\rangle(p_{k_1} \vee \ldots \vee p_{k_l}))$ for every $i \in \{1, \ldots, n\}$, $t_{k_1}, \ldots, t_{k_l}$ being the tile types in $t_1, \ldots, t_n$ vertically matching with t_i.

Let $\psi(t_1, \ldots, t_n) ::= \langle \top? \parallel \top?\rangle\top \wedge \phi(t_1, \ldots, t_n) \wedge [r_1^\star \parallel \top?]\langle r_1^\star \parallel \top?\rangle p_1$.

Claim 6.2 *The following conditions are equivalent:*

a) $\{t_1, \ldots, t_n\}$ *can tile* $\mathbb{N} \times \mathbb{N}$ *in such a way that* t_1 *occurs infinitely often in the first row.*

b) $\psi(t_1, \ldots, t_n)$ *is satisfied in* $\mathcal{C}$.

Proof. *a)* $\Rightarrow$ *b)* Suppose $\{t_1, \ldots, t_n\}$ can tile $\mathbb{N} \times \mathbb{N}$ in such a way that t_1 occurs infinitely often in the first row. Hence, there exists a function f from $\mathbb{N} \times \mathbb{N}$ into T such that for all $x, y \in \mathbb{N}$, $right(f(x,y)) = left(f(x+1,y))$ and $up(f(x,y)) = down(f(x,y+1))$. Moreover, for all $x \in \mathbb{N}$, there exists $z \in \mathbb{N}$ such that $x \leq z$ and $f(z,0) = t_1$. Let $\mathcal{F} = (W, R, *)$ be the functional $*$-separated $*$-deterministic $*$-serial $PRSPDL$-frame defined as follows:

- $W = (\mathbb{N} \times \mathbb{N}) \cup (\mathbb{N} \times \{l_1, l_2, l_3, l_4\})$ where l_1, l_2, l_3, l_4 are new distinct elements,
- R is the empty function,
- $*$ is a one-to-one correspondence between the elements of $W \times W$ and the singletons over W such that $(x, l_1) * (y, l_2) = \{(x,y)\}$, $(x+1, l_1) * (x, l_3) = \{(x, l_1)\}$ and $(y, l_4) * (y+1, l_2) = \{(y, l_2)\}$.

Let $\mathcal{M} = (W, R, *, V)$ be the model on $\mathcal{F}$ defined as follows:

- $V(p_i) = \{(x, y)\colon f(x, y) = t_i\}$ and otherwise V is the empty function.

Obviously, for all $x, y \in \mathbb{N}$, $(x + 1, y)$ is the only state in W accessible from (x, y) by means of $R_{\mathcal{M}}(r_1 \parallel \top?)$ and $(x, y + 1)$ is the only state in W accessible from (x, y) by means of $R_{\mathcal{M}}(\top? \parallel r_2)$. Thus, $(0, 0) \in V_{\mathcal{M}}(\psi(t_1, \ldots, t_n))$. Therefore, $\psi(t_1, \ldots, t_n)$ is satisfied in a functional $*$-separated $*$-deterministic $*$-serial $PRSPDL$-frame. Consequently, $\psi(t_1, \ldots, t_n)$ is satisfied in $\mathcal{C}$.

b) $\Rightarrow$ *a)* Suppose $\psi(t_1, \ldots, t_n)$ is satisfied in $\mathcal{C}$. Hence, $\psi(t_1, \ldots, t_n)$ is satisfied in a $*$-separated $PRSPDL$-frame $\mathcal{F} = (W, R, *)$. Thus, there exists a model $\mathcal{M} = (W, R, *, V)$ on $\mathcal{F}$ and there exists $u \in W$ such that $u \in V_{\mathcal{M}}(\psi(t_1, \ldots, t_n))$. Obviously, thanks to the formulas $B1$ and $B2$, for all $x, y \in \mathbb{N}$, there exists $v \in W$ such that $uR_{\mathcal{M}}(r_1^x \parallel r_2^y)v$; the set of all such v will be denoted $P(x, y)$. Moreover, thanks to the formulas $B3$, $B4$ and $B5$, for all $x, y \in \mathbb{N}$, there exists $i \in \{1, \ldots, n\}$ such that $P(x, y) \subseteq V(p_i)$ and for all $j \in \{1, \ldots, n\}$, if $i \neq j$, then $P(x, y) \cap V(p_j) = \emptyset$. In other respect, thanks to the formulas $B6$ and $B7$, for all $x, y \in \mathbb{N}$ and for all $i, j \in \{1, \ldots, n\}$, if $P(x, y) \subseteq V(p_i)$ and $P(x + 1, y) \subseteq V(p_j)$, then $right(t_i) = left(t_j)$ and if $P(x, y) \subseteq V(p_i)$ and $P(x, y + 1) \subseteq V(p_j)$, then $up(t_i) = down(t_j)$. Finally, thanks to the formula $[r_1^\star \parallel \top?]\langle r_1^\star \parallel \top?\rangle p_1$, for all $x \in \mathbb{N}$, there exists $z \in \mathbb{N}$ such that $x \leq z$ and $P(z, 0) \subseteq V(p_1)$. Let f be the function from $\mathbb{N} \times \mathbb{N}$ into $\{t_1, \ldots, t_n\}$ defined as follows:

- $f(x, y) = t_i$ iff $P(x, y) \subseteq V(p_i)$.

Obviously, for all $x, y \in \mathbb{N}$, $right(f(x, y)) = left(f(x + 1, y))$ and $up(f(x, y)) = down(f(x, y + 1))$. Moreover, for all $x \in \mathbb{N}$, there exists $z \in \mathbb{N}$ such that $x \leq z$ and $f(z, 0) = t_1$. Therefore, $\{t_1, \ldots, t_n\}$ can tile $\mathbb{N} \times \mathbb{N}$ in such a way that t_1 occurs infinitely often in the first row. □

Hence, the $\mathbb{N} \times \mathbb{N}$ recurring tiling problem is reducible to satisfiability in the class $\mathcal{C}$ of $\mathcal{L}_{PDL}^{\parallel, r_1, r_2}$-formulas. Since the $\mathbb{N} \times \mathbb{N}$ recurring tiling problem is Σ_1^1-hard [18], satisfiability in the class $\mathcal{C}$ of $\mathcal{L}_{PDL}^{\parallel, r_1, r_2}$-formulas is Σ_1^1-hard. Thus, $VAL(\mathcal{L}_{PDL}^{\parallel, r_1, r_2}, \mathcal{C})$ is Π_1^1-hard. □

Corollary 6.3 *$VAL(\mathcal{L}_{PDL}^{\parallel, r_1, r_2}, \mathcal{C})$ is Π_1^1-complete for all classes $\mathcal{C}$ of frames considered in Proposition 6.1.*

Proof. It suffices to prove that if a $PRSPDL$-formula is satisfied in a frame in $\mathcal{C}$, then it is satisfied in a finite or countable frame in $\mathcal{C}$. By means of the so-called *Standard Translation,* one can prove that $PRSPDL$ is a fragment of $L_{\omega_1\omega}$, the infinitary logic in which one is allowed to consider countable conjunctions in addition to the usual first-order constructs. See [6, Pages 83–86 and 496] for precise definitions. By the Löwenheim-Skolem theorem for $L_{\omega_1\omega}$, if the standard translation of a $PRSPDL$-formula is satisfied in a frame in $\mathcal{C}$, then it is satisfied in a finite or countable frame in $\mathcal{C}$. Hence, if a $PRSPDL$-formula is satisfied in a frame in $\mathcal{C}$, then it is satisfied in a finite or countable frame in $\mathcal{C}$. □

7 Conclusion

We present our computability results in the following tables.

	$\mathcal{C}^{*det,*ser}_{fun,*sep}$	$\mathcal{C}^{*det}_{fun,*sep}$	$\mathcal{C}^{*ser}_{fun,*sep}$	$\mathcal{C}^{*det,*ser}_{*sep}$
$\mathcal{L}^{s_1,s_2}_{PDL}$				
$\mathcal{L}^{s_1,s_2}_{;}$				
$\mathcal{L}^{\|,r_1,r_2}_{PDL}$	Π^1_1-c	Π^1_1-c	Π^1_1-c	Π^1_1-c

	$\mathcal{C}_{fun,*sep}$	$\mathcal{C}^{*det}_{*sep}$	$\mathcal{C}^{*ser}_{*sep}$	$\mathcal{C}_{*sep}$
$\mathcal{L}^{s_1,s_2}_{PDL}$		$EXPTIME$-c		$EXPTIME$-c
$\mathcal{L}^{s_1,s_2}_{;}$		$PSPACE$-c		$PSPACE$-c
$\mathcal{L}^{\|,r_1,r_2}_{PDL}$	Π^1_1-c	Π^1_1-c	Π^1_1-c	Π^1_1-c

Let $\mathcal{C}$ be one of the classes of frames considered in the above tables.

As a consequence of Corollary 6.3, $VAL(\mathcal{L}^{\|,r_1,r_2}_{PDL}, \mathcal{C})$ is Π^1_1-complete. Nevertheless, one may try to axiomatize $VAL(\mathcal{L}^{\|,r_1,r_2}_{PDL}, \mathcal{C})$ by means of an infinitary derivation rule similar to the one used in [4]. The result stated in Proposition 4.8 suggests that an unorthodox derivation rule similar to the one used in [3] could be considered as well.

As for the set $\mathcal{L}^{\|,s_1,s_2}_{PDL}$ of all recovering-free formulas, the decidability/undecidability status of $VAL(\mathcal{L}^{\|,s_1,s_2}_{PDL}, \mathcal{C})$ is not known.

In other respect, seeing that in a frame $\mathcal{F} = (W, R, *)$, W is to be regarded as the set of all possible states in a computation process, it seems natural to consider the restriction $\mathcal{C}\,|\mathrm{wf}$ of $\mathcal{C}$ to those frames $\mathcal{F} = (W, R, *)$ in which the transitive closure of the binary relation $\longrightarrow_{\mathcal{F}}$ defined as follows is well-founded:

- $x \longrightarrow_{\mathcal{F}} y$ iff there exists $z \in W$ such that either $x \in y * z$, or $x \in z * y$.

Remark that the transitive closures of the binary relations $\longrightarrow_{\mathcal{F}_1}$, $\longrightarrow_{\mathcal{F}_2}$ and $\longrightarrow_{\mathcal{F}_3}$ associated to the frames $\mathcal{F}_1$, $\mathcal{F}_2$ and $\mathcal{F}_3$ considered in Example 2.2 are well-founded. The computability status of $VAL(\mathcal{L}^{\|,r_1,r_2}_{PDL}, \mathcal{C}\,|\mathrm{wf})$ is not known.

Finally, following the line of reasoning suggested in [12], the accessibility relation associated to $(\alpha \parallel \beta)$ can be defined as follows in the class of all $*$-deterministic frames:

- whenever x and y are related via $R(\alpha)$ and x and z are related via $R(\beta)$, x and $y * z$ are related via $R(\alpha \parallel \beta)$.

Seeing that this variant of $PRSPDL$ is appealing in computer science, especially in system specification and program construction [13], it seems natural to consider, with respect to it, the computability status of satisfiability in the class $\mathcal{C}$ of $\mathcal{L}^{\|,r_1,r_2}_{PDL}$-formulas. The equivalence, in the class of all $*$-separated frames, between $(\alpha \parallel \beta)$ — when interpreted by Benevides *et al.* [5] — and $((r_1; \alpha) \parallel (r_2; \beta))$ — when interpreted by Frias [12] and Frias *et al.* [13] — suggests that satisfiability in the class $\mathcal{C}$ of $\mathcal{L}^{\|,r_1,r_2}_{PDL}$-formulas is highly undecidable.

Acknowledgements

Special acknowledgement is heartly granted to Joseph Boudou who made several helpful comments for improving the correctness and the readability of this article. The work of Philippe Balbiani was partially supported by the "French National Research Agency" (contract ANR-11-BS02-011) and the "Bulgarian National Science Fund" (contract DID02/32/2009). The work of Tinko Tinchev was partially supported by the "Centre international de mathématiques et d'informatique" (contract ANR-11-LABX-0040-CIMI within the program ANR-11-IDEX-0002-02).

References

[1] Balbiani, P., Boudou, J.: Decidability of iteration-free *PDL* with parallel composition. Submitted for publication.

[2] Balbiani, P., Boudou, J.: Iteration-free *PDL* with storing, recovering and parallel composition: a complete axiomatization. Submitted for publication.

[3] Balbiani, P., Vakarelov, D.: Iteration-free *PDL* with intersection: a complete axiomatization. Fundamenta Informaticæ**45** (2001) 173–194.

[4] Balbiani, P., Vakarelov, D.: *PDL* with intersection of programs: a complete axiomatization. Journal of Applied Non-Classical Logics **13** (2003) 231–276.

[5] Benevides, M., de Freitas, R., Viana, P.: Propositional dynamic logic with storing, recovering and parallel composition. Electronic Notes in Theoretical Computer Science **269** (2011) 95–107.

[6] Blackburn, P., de Rijke, M., Venema, Y.: Modal Logic. Cambridge University Press (2001).

[7] Brotherston, J., Calcagno, C.: Classical *BI*: its semantics and proof theory. Logical Methods in Computer Science **6** (2010) 1–42.

[8] Courtault, J.-R., Galmiche, D.: A modal *BI* logic for dynamic resource properties. In Artemov, S., Nerode, A. (Editors): Logical Foundations of Computer Science. Springer (2013) 134–148.

[9] Danecki, R.: Nondeterministic propositional dynamic logic with intersection is decidable. In Skowron, A. (Editor): Computation Theory. Springer (1985) 34–53.

[10] Fariñas del Cerro, L., Orłowska, E.: *DAL* — a logic for data analysis. Theoretical Computer Science **36** (1985) 251–264.

[11] Fisher, M., Ladner, R.: Propositional dynamic logic of regular programs. Journal of Computer and System Sciences **18** (1979) 194–211.

[12] Frias, M.: Fork Algebras in Algebra, Logic and Computer Science. World Scientific (2002).

[13] Frias, M., Veloso, P., Baum, G.: Fork algebras: past, present and future. Journal of Relational Methods in Computer Science **1** (2004) 181–216.

[14] Galmiche, D., Larchey-Wendling, D.: Expressivity properties of Boolean *BI* through relational models. In Arun-Kumar, S., Garg, N. (Editors): FSTTCS 2006: Foundations of Software Technology and Theoretical Computer Science. Springer (2006) 357–368.

[15] Gargov, G., Passy, S.: A note on Boolean modal logic. In Petkov, P. (Editor): Mathematical Logic. Plenum Press (1990) 299–309.

[16] Goldblatt, R.: Logics of Time and Computation. Center for the Study of Language and Information (1987).

[17] Göller, S., Lohrey, M., Lutz, C.: *PDL* with intersection and converse: satisfiability and infinite-state model checking. The Journal of Symbolic Logic **74** (2009) 279–314.

[18] Harel, D.: Recurring dominoes: making the highly undecidable highly understandable. Annals of Discrete Mathematics **24** (1985) 51–72.

[19] Harel, D., Kozen, D., Tiuryn, J.: Dynamic Logic. MIT Press (2000).

[20] Ladner, R.: The computational complexity of provability in systems of modal propositional logic. SIAM Journal on Computing **6** (1977) 467–480.

[21] Massacci, F.: Decision procedures for expressive description logics with intersection, composition, converse of roles and role identity. In Nebel, B. (Editor): Proceedings of the Seventeenth International Joint Conference on Artificial Intelligence. Morgan Kaufmann (2001) 193–198.

[22] Mirkowska, G.: *PAL* — propositional algorithmic logic. Fundamenta Informaticæ**4** (1981) 675–760.

[23] O'Hearn, P., Pym, D.: The logic of bunched implications. The Bulletin of Symbolic Logic **5** (1999) 215–244.

[24] Pratt, V.: Models of program logics. In: 20th Annual Symposium on Foundations of Computer Science. IEEE (1979) 115–122.

[25] Reynolds, VJ.: Separation logic: a logic for shared mutable data structures. In: 17th Annual IEEE Symposium on Logic in Computer Science. IEEE (2002) 55–74.

A Parametrized Propositional Dynamic Logic with Application to Service Synthesis

Walid Belkhir

INRIA Nancy–Grand Est & LORIA, France

Gisela Rossi

National University of Córdoba, Argentina

Michael Rusinowitch

INRIA Nancy–Grand Est & LORIA, France

Abstract

We extend propositional dynamic logic (PDL) with variables ranging over an infinite domain. This extension, called *parametrized* PDL or PPDL for short, is interpreted over parametrized transitions systems whose edges are labeled with letters or variables and whose states are labeled with non-parametrized propositions. We show that the satisfiability problem for PPDL is decidable.

We apply these results to the composition problem of web services in presence of constraints on the global ordering of the message-exchange events between the agents. We express the client specification and the available services as parametrized transitions systems and we express the behavioral constraints as a PPDL formula that the generated orchestrator must fulfill. It turns out that the model of such a formula represents the desired orchestrator.

Keywords: parametrised propositional dynamic logic, infinite domain, satisfiability, service synthesis, games.

1 Introduction

The synthesis problem has initially been introduced by Church [8] in the context of digital circuits and amounts to construct from a given specification an automaton satisfying this specification whenever it exists and return a negative answer otherwise. Wolper [23] has considered the synthesis problems for communicating processes. Controller synthesis in Ramadge and Wonham theory of discrete event processes [22] aims to generate supervisors that restrict the behavior of a plant so that a given specification is fulfilled. Synthesis in the context of Service Oriented Computing can be defined as the automated derivation of a specification of how to coordinate some available component services to fulfill the client requests [7]. In several interesting cases this composition specification can be derived automatically and can be turned into a program that monitors the flow among the component services and the client.

PDL is a logic that was introduced in [12] to reason about programs and was successfully applied in several areas in computer science: program verification, agent-based system specification (e.g. [14]), planning, knowledge representation. It also admits strong connections with Description Logics [3], making it even more interesting. PDL combines two entities: formulas to be interpreted in the nodes of a Kripke structure, and programs to be interpreted by binary relations over the set of nodes of a Kripke structure. PDL is well-adapted to describe transition systems, and the model checking problem for PDL remains PTIME-complete even when the logic is extended by looping and repeat operators [15].

The satisfiability problem for PDL asks whether a given formula has a model and to construct it whenever it exists. One of the applications of the satisfiability problem is the automatic program synthesis that, roughly speaking, consists in the automatic generation of programs out of their specification. When service behaviors can be represented with finite-state transition systems then the composition synthesis problem for Web services can be reduced to PDL satisfiability [10]. In this approach the existence of an orchestrator, that is a transition system that delegates any requested action from the client to one of the available community of services, is expressed with a PDL formula. If this one is satisfiable then from any of its finite models (known to exist) we can extract a transition system that solves the synthesis problem.

However computational models based on finite-state transition systems over finite alphabets (i.e. over finite set of actions) are inefficient and even insufficient to accurately describe systems that need to deal with an arbitrary large amount of data. This observation has motivated many works to introduce and study models over infinite alphabets e.g. [20,19,11,6] since large datasets can be abstracted as infinite domains. On the other hand, these models have not been applied to service composition except in [6] where the composition problem for Web services is proved to be decidable and is reduced to compute a simulation preorder. Such simulation preorder can be turned into an orchestrator that suitably schedules the actions of the available community of services to fulfill the client requests.

Besides, we would like also to specify different interaction modes between the client and the available services, as well as additional constraints on the global ordering of the message-exchange events between the composite services and the client. Among the possible interaction modes between the agents we can mention orchestration, choreography [1,21], and distributed orchestration [2]. An advantage of PDL-based synthesis over the simulation-based synthesis is that, on the one hand, PDL provides a systematic approach for the description of interactions modes between the agents. And, on the other hand, it is possible in this framework to express behavioral constraints (as PDL formulas) that the synthesized composition must fulfill. Such constraints can not be expressed and taken into account, at least in a straightforward way, in the simulation-based framework. Here are examples of such constraints: preventing data exchange between competing agents (Chinese wall security policies), preventing data update conflicts from different services (critical section paradigm), saving agents' data at the end of a session, closing every open file after usage.

An interesting potential application of PPDL is parametrized system verification by model-checking. Moreover, in some cases PPDL might be more advantageous than PDL even for specifying finite (non-parametrized) systems since PPDL formulas can be exponentially more succinct than their equivalent PDL formulas, leading to better readable formulas. The complexity of PPDL model-checking on parametrized systems is worth studying and requires a separate work.

Contributions. We introduce parametrized transition systems (PTS) and parametrized PDL (PPDL). The transitions of PTS are labeled by variables that can be assigned the read letter. A variable binding can be released at some states: in that case we say that the variable is *refreshed.* This mechanism is natural to express iteration processes, for instance when a service has to scan a list of item identifiers, or sessions. It is useful for real-world applications where service actions are parameterized by terms built with data taken from infinite alphabets (identifiers, codes, addresses ...). Besides, the states are labeled with (non-parametrized) propositional constants, i.e. the propositional constants being true at these states. On the other hand, standard PDL incorporates two entities: formulas and programs, where the programs are regular expressions built over a finite set of atomic actions using concatenation, union and Kleene star. For PPDL we consider *parametrized* regular expressions. In this setting we allow variables in the regular expressions. Such variables range over the infinite set of actions Σ. Besides, in order to free a variable after being bound to an action we shall introduce the *reset* operator $\mathsf{res}(.)$. For instance, the expression $(x;\mathsf{res}(x))^\star$, where $\mathsf{res}(x)$ denotes the resetting of the variable x, stands for all possible finite traces in $\Sigma^\star$ i.e. traces of the form $a_1a_2\dots a_n$, where $a_i \in \Sigma$ and $n \in \mathbb{N}$. The PPDL formula $\phi_1 = \forall x.[(x;\mathsf{res}(x))^\star]\mathsf{p}$, where $[-]$ stands for the *necessity* modal operator, p is a propositional constant and x is a variable, states that p holds globally, i.e. p holds in every state of the model. Then, we introduce satisfiability games for PPDL and prove its completeness, then we show the decidability of the satisfiability problem of PPDL. As an application, we show how to use these results to the synthesis of parametrized services.

Related works. Many extensions of PDL were developed e.g. with propositional assignments [4] with intersection operator [5], with converse operators [18], with context-free programs [16]. Model checking algorithms as well as their complexity for PDL with looping, repeat, test intersection, converse operators and context-free programs were developed in [15]. Model checking problems of PDL, and its extension with intersection, over various classes of infinite state systems (basic parallel processes, basic process algebra, pushdown systems, prefix-recognizable systems and Petri nets) were studied in [13]. The nominal automata that are used for resource usage control in [11] subsumes our parametrized transition systems with refreshing. However model checking technique, rather than satisfiability and synthesis in our case, were considered in this work. Web Services Choreography Description Language (WS-CDL) [1] provides another approach in describing the global ordering of the message-exchange events between the communicating agents. Our proofs are inspired by [17]: they rely on a game-theoretic formulation of satisfiability together with the *focus* mechanism, rather than automata-theoretic techniques. It is shown in [17] how the focus technique solved satisability for the temporal logics LTL and CTL, and at the same time led to simple completeness proofs.

When service behaviors are represented with finite-state transition systems i.e. they are not data-aware, the composition synthesis problem was reduced to PDL satisfiability in [10]. The composition problem can also be reduced to computing a simulation preorder as in [7]. This approach cannot be extended to the data-centric Colombo model for services since simulation is undecidable in this case. Known decidable cases of the composition synthesis problem in Colombo framework need restrictions such as determinism or finite domain for values or empty database. A theory of contracts that formalizes the compatibility of a client to a service, and the safe replacement of a service with another service has been developed in [9]. Contracts ensures that every possible interaction between compatible clients and services can be completed successfully. Access control policies can be expressed by graphs. We believe that it is possible to translate such graphs into PDL and PPDL and to integrate them in synthesis problems.

Paper organization. The paper is organized as follows. Section 2 introduces parametrized transition systems (PTS) and parametrized PDL (PPDL). Section 3 introduces satisfiability games for PPDL. Section 4 proves the decidability of the satisfiability problem of PPDL. Section 5 applies these results to the synthesis of parametrized services.

Preliminaries. Let $\mathcal{X}$ be a finite set of variables, Σ an infinite set of atomic actions. A substitution is an idempotent mapping $\{x_1 \mapsto \alpha_1, \ldots, x_n \mapsto \alpha_n\} \cup \bigcup_{a \in \Sigma}\{a \mapsto a\}$ with variables $x_1, \ldots, x_n$ in $\mathcal{X}$ and $\alpha_1, \ldots, \alpha_n$ in $\mathcal{X} \cup \Sigma$. We call $\{x_1, \ldots, x_n\}$ its *proper domain*, and denote it by $dom(\sigma)$. We denote by $Dom(\sigma)$ the set $dom(\sigma) \cup \Sigma$. We denote by $codom(\sigma)$ the set $\{a \in \Sigma \mid \exists x \in dom(\sigma) \text{ s.t. } \sigma(x) = a\}$. The empty substitution (*i.e.*, with an empty proper domain) is denoted by $\emptyset$. The set of substitutions from $\mathcal{X} \cup \Sigma$ to a set A is denoted by $\zeta_{\mathcal{X},A}$, or by $\zeta_{\mathcal{X}}$, or simply by ζ if there is no ambiguity. If σ_1 and σ_2 are substitutions that coincide on the domain $dom(\sigma_1) \cap dom(\sigma_2)$, then $\sigma_1 \cup \sigma_2$ denotes their union in the usual sense. If $dom(\sigma_1) \cap dom(\sigma_2) = \emptyset$ then we denote by $\sigma_1 \uplus \sigma_2$ their *disjoint* union. We define the function $\mathcal{V} : \Sigma \cup \mathcal{X} \longrightarrow \mathcal{P}(\mathcal{X})$ by $\mathcal{V}(\alpha) = \{\alpha\}$ if $\alpha \in \mathcal{X}$, and $\mathcal{V}(\alpha) = \emptyset$, otherwise. For a function $F : A \to B$, and $A' \subseteq A$, the restriction of F on A' is denoted by $F_{|A'}$.

2 Parametrized PDL

In this section we define parametrized propositional dynamic logic (PPDL). Formulas of PPDL are interpreted over *parametrized transitions systems* whose edges are labeled with variables or atomic actions and whose states are labeled with atomic (non-parametrized) propositions. Firstly we introduce the syntax of PPDL and the main ideas behind it, then we introduce parametrized transitions systems and define their traces. Finally, the semantics of PPDL over parametrized transition systems is defined. But first let us illustrate our ideas through a practical example.

A motivating example. Figure 1 represents an e-commerce Web site allowing clients to search and to buy plane tickets with prior authentication. For each action the client performs, the services save the data in a file. The agents in this example are: CLIENT, AUTHENTICATION, FLIGHT, PAYMENT and FILE, they communicate with messages ranging over a possibly infinite set of terms. The problem is to check whether the services AUTHENTICATION, FLIGHT, PAYMENT and FILE can *collaborate* to satisfy the CLIENT requests in presence of certain global constraints expressed by PPDL formulas to be introduced in the following. Services collaborate by exchanging messages before answering a client request. This notion will be formalized by a so-called $\star$-*simulation* relation, which is a variant of the classical simulation preorder. For saving space, a transition labeled by a term, say `write(m,n)`, abbreviates successive transitions labeled by the root symbol and its arguments, here `write`, `m` and `n`, respectively. Besides, while composing these agents, there are some requirements that must be fulfilled. We impose that every open file has to be closed and that the flight data of the client have to be stored in an appropriate file. These requirements can be turned into a PPDL formula, Appendix B.2.

Syntax of PPDL. In standard PDL the programs are regular expressions built over a finite set of atomic actions using concatenation, union and Kleene star. For PPDL we consider *parametrized* regular expressions. In this setting we allow variables in the regular expressions. These variables range over the infinite set of actions. Besides, in order to free a variable after being bound to an action we shall introduce the *reset* operator res(.). For example the expression $x^\star$, where x is a variable, stands for all traces a^* in $\Sigma^\star$, where a is an atomic action in Σ. While the expression $(x; \mathsf{res}(x))^\star$ stands for all possible finite traces in $\Sigma^\star$, where $\mathsf{res}(x)$ denotes the resetting of the variable x, i.e. traces of the form $a_1 a_2 \ldots a_n$, where $a_i \in \Sigma$ and $n \in \mathbb{N}$.

Let $\mathbb{P}$ be a finite set of propositional constants containing tt and ff, and Σ an infinite set of atomic actions (or atomic programs), and $\mathcal{X}$ a finite set of variables ranging over Σ. The syntax of PPDL formula is given by the following grammar:

$$\phi ::= [\alpha]\phi \mid \phi \wedge \phi \mid \phi \vee \phi \mid \mathtt{p} \mid \forall x.\phi \mid \neg\phi$$
$$\alpha ::= a \mid x \mid \alpha;\alpha \mid \alpha \cup \alpha \mid \alpha^* \mid \mathsf{res}(x) \mid \phi?$$

where $a \in \Sigma$, $x \in \mathcal{X}$ and $\mathtt{p} \in \mathbb{P}$.

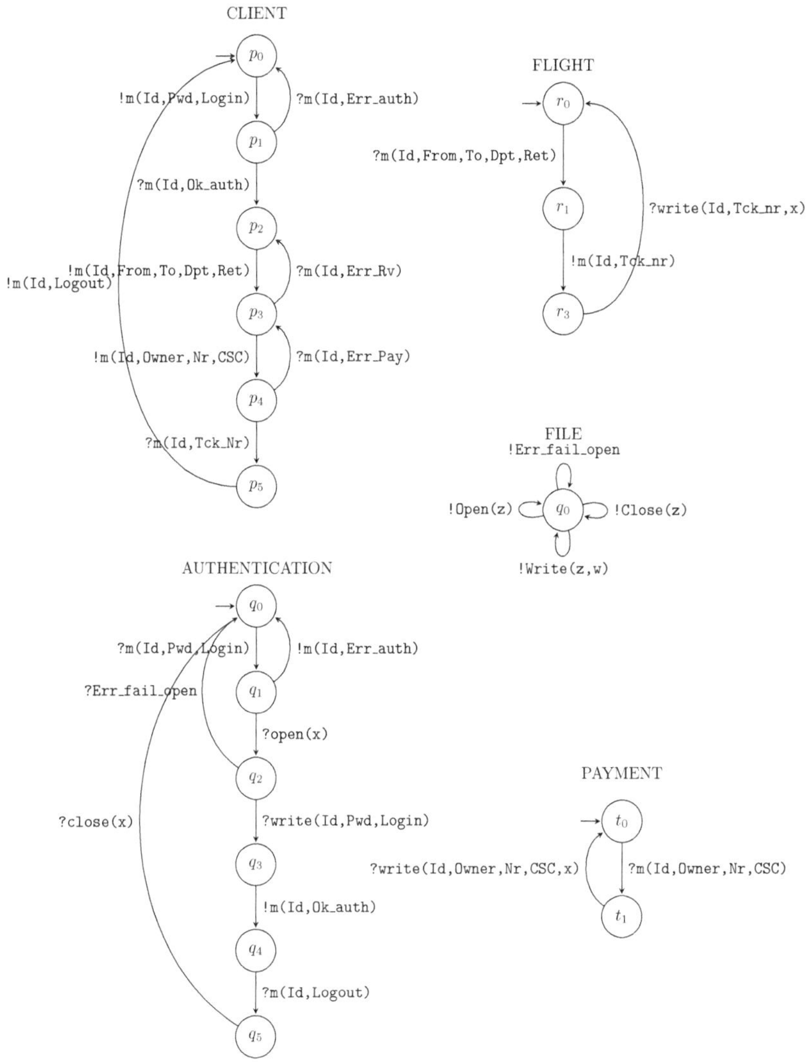

Fig. 1. Flight reservation example where "!" (resp. "?") stands for sending (resp. receiving) a message.

The diamond operator $\langle\cdot\rangle$ (resp. existential quantifier $\exists$) can be defined in terms of the box operator $[\cdot]$ (resp. universal quantifier $\forall$) in the standard way as follows: $\langle\alpha\rangle\phi \stackrel{def}{=} \neg(\neg[\alpha]\neg\phi)$ and $\exists x.\phi \stackrel{def}{=} \neg(\forall x.\neg\phi)$.

For a formula ϕ, we define the finite set of atomic actions appearing in ϕ, denoted by $\Sigma(\phi)$, inductively as follows: $\Sigma([\alpha]\psi) = \Sigma(\alpha) \cup \Sigma(\psi)$, $\Sigma(\psi_1 \vee \psi_2) = \Sigma(\psi_1 \wedge \psi_2) = \Sigma(\psi_1) \cup \Sigma(\psi_2)$, $\Sigma(\exists x.\psi) = \Sigma(\psi?) = \Sigma(\neg\psi) = \Sigma(\psi)$, $\Sigma(\alpha^\star) = \Sigma(\alpha)$, $\Sigma(\alpha_1;\alpha_2) = \Sigma(\alpha_1 \cup \alpha_2) = \Sigma(\alpha_1) \cup \Sigma(\alpha_2)$, $\Sigma(a) = \{a\}$ where

$a \in \Sigma$, and $\Sigma(x) = \Sigma(\mathsf{res}(x)) = \Sigma(\mathsf{p}) = \emptyset$ where $x \in \mathcal{X}$ and $\mathsf{p} \in \mathbb{P}$. For an occurrence x of a variable in a formula, we let $\tau(x) = \exists$ (resp. $\tau(x) = \forall$) if x is existentially (resp. universally) quantified. If λ is a formula or a program, we shall denote by $\mathtt{Res}(\lambda)$ the set of variables being reset in λ. Throughout this paper, formulas are presented in positive form (i.e. the negation only operates on propositional constants in $\mathbb{P}$). This is possible since De Morgan laws can be proved.

Parametrized transition systems. PPDL formulas are interpreted over *parametrized transition systems* (PTS). Before introducing them formally, let us first explain the main ideas behind them. The transitions of a parametrized transition system are labeled with actions or variables. We have a finite number of variables ranging over an infinite set of actions. On a transition labeled with a variable x and an input action l, if x is not bound then taking the transition amounts to binding x to l. On the other hand if x is already bound then the transition can be taken only if x is already bound to l. Since we would like to reuse variables, we add an additional mechanism which will free the variables depending on the states of the automaton. That is, some variables are refreshed in some states, i.e. variables can be freed in these states so that new actions can be assigned to them. The formal definition follows.

Definition 2.1 A parametrized transition system (PTS for short) is a tuple $\mathcal{M} = \langle \Sigma, \mathcal{X}, Q, q_0, \delta, \pi, \kappa \rangle$ where:

- Σ is a infinite set of actions, $\mathcal{X}$ is a finite set of variables,
- Q is a finite set of states, $q_0 \in Q$ is a the initial state,
- $\delta : Q \times (\Sigma_{\mathcal{A}} \cup \mathcal{X}) \to 2^Q$ is a transition function where $\Sigma_{\mathcal{A}}$ is a finite subset of Σ,
- $\pi : Q \to 2^{\mathbb{P}}$ assigns truth values to each propositional constant in $\mathbb{P}$ for each state, and
- $\kappa : \mathcal{X} \to 2^Q$ is the refreshing function that associates to every variable the (possibly empty) set of states where it is refreshed.

We shall denote by $\Sigma_{\mathcal{A}}$ the finite subset of actions from Σ appearing in the PTS $\mathcal{A}$. For a refreshing function $\kappa : \mathcal{X} \to 2^Q$, we define the function $\kappa^{-1} : Q \to 2^{\mathcal{X}}$ by $\kappa^{-1}(q) = \{x \mid q \in \kappa(x)\}$. A LTS is a tuple $(\Sigma, S, s_0, \Delta, \Pi)$ where S is a (possibly infinite) set of states, $s_0 \in S$, $\Delta : S \times \Sigma \to 2^S$ and $\Pi : S \to 2^{\mathbb{P}}$.
The formal definition of configurations and trace for PTSs.

Definition 2.2 Let $\mathcal{A} = \langle \Sigma, \mathcal{X}, Q, q_0, \delta, \pi, \kappa \rangle$ be a PTS. A *configuration* is a pair (q, γ) where $q \in Q$ and γ is a substitution. We define a transition relation over the configurations as follows: $(q_1, \gamma_1) \xrightarrow{a} (q_2, \gamma_2)$, where $a \in \Sigma$, iff there exists a substitution σ such that $dom(\sigma) \cap dom(\gamma_1) = \emptyset$ and there exists a label $\alpha \in \Sigma \cup \mathcal{X}$ such that $q_2 \in \delta(q_1, \alpha, g)$, $(\gamma_1 \uplus \sigma)(\alpha) = a$ and $\gamma_2 = (\gamma_1 \uplus \sigma)_{|D}$, with $D = Dom(\gamma_1 \uplus \sigma) \setminus \kappa^{-1}(q_2)$. A trace of a PTS is a sequence $a_1 a_2 \ldots a_n$ such that there exist states q_i and substitutions σ_i, $i = 1, \ldots, n$ such that $(q_0, \emptyset) \xrightarrow{a_1} (q_1, \sigma_1) \ldots \xrightarrow{a_n} (q_n, \sigma_n)$.

Example 2.3 Let $\mathcal{A}$ and $\mathcal{A}'$ be the PTS depicted in Figure 2 where the variable x is refreshed in q_0 and the variable z is refreshed in q'_0, q'_1 and q'_2.

The behavior of $\mathcal{A}$ is as follows. Being in the initial state q_0:

- Makes the transition $q_0 \to q_1$ by making an action and bounding the variable x to it, then enters the state q_1,
- Makes the transition $q_1 \to q_0$ by making an action that equals to the value of x, then enters the state q_0,
- From the state q_0, refresh the variable x, that is, it is no longer bound to the input symbol. Then, start again.

We illustrate the run of $\mathcal{A}$ on the trace $w = aabb$, starting from the initial configuration $(\emptyset, q_0)$ as follows:

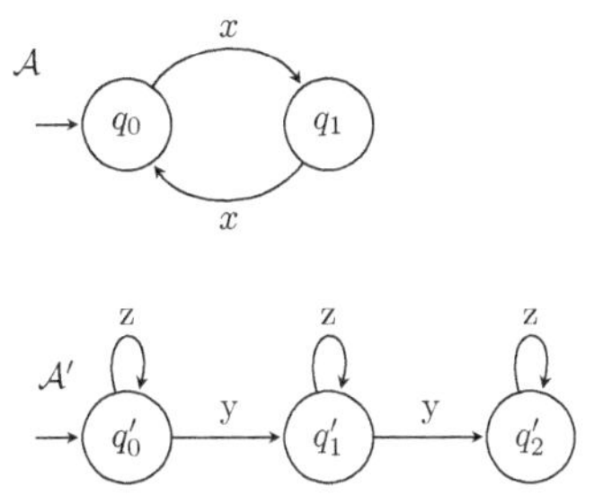

Fig. 2: Two PTS $\mathcal{A}$ and $\mathcal{A}'$ where the variable x is refreshed in q_0 and the variable z is refreshed in q'_0, q'_1 and q'_2.

$$(\emptyset, q_0) \xrightarrow{a} (\{x \mapsto a\}, q_1) \xrightarrow{a} (\emptyset, q_0) \xrightarrow{b} (\{x \mapsto b\}, q_1) \xrightarrow{b} (\emptyset, q_0) \xrightarrow{c} (\{x_1 \mapsto c\}, p').$$

We next define the instantiation of a PTS, it consists in instantiating its variables with all possible actions in Σ, yielding a system with possibly infinite number of states and transitions.

Definition 2.4 [Instantiation of a PTS] Let $\mathcal{M} = \langle \Sigma, \mathcal{X}, Q, q_0, \delta, \pi, \kappa \rangle$ be a PTS. The *instantiation* of $\mathcal{M}$, denoted by $\mathcal{C}(\mathcal{M})$, is the LTS $(\Sigma, S, s_0, \Delta, \Pi)$, where:

$$\begin{aligned} S &= Q \times \xi_{\mathcal{X},\Sigma}, \\ s_0 &= (q_0, \emptyset), \\ (q', \sigma') \in \Delta(a, (q, \sigma)) &\text{ iff } (q, \sigma) \xrightarrow{a} (q', \sigma'), \text{ and} \\ \Pi((q, \sigma)) &= \pi(q), \quad \text{for all } \sigma \in \xi_{\mathcal{X},\Sigma} \text{ and } q \in Q. \end{aligned}$$

Semantics of PPDL over parametrized transition systems. PPDL formulas are interpreted over *configurations* of parametrized transition systems, or equivalently over the states of the instantiation of PTSs. That is, given a structure $\mathcal{M} = \langle \Sigma, \mathcal{X}, Q, q_0, \delta, \pi, \kappa \rangle$. The interpretation of a formula ϕ over the LTS which is the instantiation of $\mathcal{M}$: $\mathcal{C}(\mathcal{M}) = (\Sigma, S, s_0, \Delta, \Pi)$, will be denoted by $[\![\phi]\!]_{\mathcal{M}}$, or simply $[\![\phi]\!]$, such that $[\![\phi]\!] \subseteq S$, if ϕ is a formula; and $[\![\alpha]\!] \subseteq S \times S$, if α is a program, is defined as follows:

$$
\begin{aligned}
[\![\mathsf{ff}]\!] &= \emptyset \\
[\![\mathtt{p}]\!] &= \Pi^{-1}(\mathtt{p}), \quad \text{where } \mathtt{p} \in \mathbb{P} \\
[\![\psi_1 \wedge \psi_2]\!] &= [\![\psi_1]\!] \cap [\![\psi_2]\!] \\
[\![\psi_1 \vee \psi_2]\!] &= [\![\psi_1]\!] \cup [\![\psi_2]\!] \\
[\![[\alpha]\psi]\!] &= \{s \mid \forall s' \text{ if } (s, s') \in [\![\alpha]\!] \text{ then} \\
&\qquad s' \in [\![\psi]\!]\}, \text{ if } \alpha \notin \mathcal{X} \\
[\![[x]\psi]\!] &= [\![\forall x.[x]\psi]\!], \text{ if } x \in \mathcal{X} \\
[\![\forall x.\psi]\!] &= \bigcup_{a \in \Sigma} [\![\psi[x := a]]\!],
\end{aligned}
\qquad
\begin{aligned}
[\![\neg\psi]\!] &= S \setminus [\![\psi]\!] \\
[\![\alpha;\beta]\!] &= [\![\alpha]\!] \circ [\![\beta]\!] \\
[\![\alpha \cup \beta]\!] &= [\![\alpha]\!] \cup [\![\beta]\!] \\
[\![\alpha^\star]\!] &= [\![\alpha]\!]^\star = \bigcup_{n \geq 0} [\![\alpha]\!]^n \\
[\![\mathsf{res}(x)]\!] &= \mathtt{Id} \\
[\![a]\!] &= \{(s, s') | s' \in \Delta(a, s)\}, \\
&\qquad \text{if } a \in \Sigma,
\end{aligned}
$$

where $\phi[x := a]$ stands for the application of the substitution $\{x \mapsto a\}$ to ϕ. It is inductively defined as follows:

$$
\begin{aligned}
\mathtt{p}[x := a] &= \mathtt{p}, \quad \text{if } \mathtt{p} \in \mathbb{P} \\
(\psi_1 \wedge \psi_2)[x := a] &= \psi_1[x := a] \wedge \psi_2[x := a] \\
([\alpha]\psi)[x := a] &= ([\alpha[x := a]]\psi[x := a]) \\
(\forall y.\psi)[x := a] &= (\forall y.\psi[x := a]), \text{ if } x \neq y \\
(\neg\psi)[x := a] &= \neg(\psi[x := a]) \\
(\alpha \cup \beta)[x := a] &= (\alpha[x := a] \cup \beta[x := a]), \\
\alpha^\star[x := a] &= (\alpha[x := a])^\star \\
(\mathtt{res}(y))[x := a] &= \mathtt{res}(y),
\end{aligned}
\qquad
\begin{aligned}
&(\alpha;\beta)[x := a] = \\
&\begin{cases} (\alpha[x := a]; \beta), & \text{if } x \in \mathtt{Res}(\alpha) \\ (\alpha[x := a]; \beta[x := a]), & \text{other.} \end{cases} \\
&\beta[x := a] = \\
&\begin{cases} a \text{ if } \beta = x \\ \beta \text{ if } \beta \in \Sigma \cup \mathcal{X} \text{ and } \beta \neq x \end{cases}
\end{aligned}
$$

A formula ϕ is *satisfiable* if there is a PTS $\mathcal{M}$ together with its instantiation $\mathcal{C}(\mathcal{M}) = (\Sigma, S, s_0, \Delta, \Pi)$ and a state $s \in S$, s.t. $s \in [\![\phi]\!]_{\mathcal{M}}$. The formula ϕ is said to be valid, denoted $\models \phi$, if it is true in every state of every parametrized transition system. Notice that $\not\models \phi$ iff $\neg\phi$ is satisfiable.

Example 2.5 Firstly, we give some PPDL formula to illustrate the combination of the quantifiers with the modalities. Let $\phi_1 = \forall x.[x]\phi'_1$, $\phi_2 = \exists x.[x]\phi'_2$, $\phi_3 = \forall x.\langle x\rangle\phi'_3$, and $\phi_4 = \exists x.\langle x\rangle\phi'_4$. where ϕ'_i are PPDL formula. The formula ϕ_1 holds in a state q iff for every instantiation of the variable x, say with an action $a \in \Sigma$, each transition outgoing from q and labeled with a yields a state where $\phi'_1[x := a]$ holds. The formula ϕ_2 holds in a state q iff there exists an instantiation of the variable x, say with an action $a \in \Sigma$, such that each transition outgoing from q and labeled with a yields a state where $\phi'_2[x := a]$ holds. The formula ϕ_3 holds in an state q iff for every instantiation of the variable x, say with an action $a \in \Sigma$, there exists a transition outgoing from q and labeled with a that yields a state where $\phi'_3[x := a]$ holds. The formula ϕ_4 holds in a state q iff there exists an instantiation of the variable x, say with an action $a \in \Sigma$, such that there exists a transition outgoing from q and labeled with a that yields a state where $\phi'_4[x := a]$ holds.

Secondly, we give some examples of useful properties. The property "there is a transition that is labeled with the action a and that can be reached from the current state" can be expressed by the PPDL formula $\exists x.\langle(x;\mathsf{res}(x))^\star\rangle\langle a\rangle\mathsf{tt}$. The formula $\langle((\exists x.\langle x\rangle\mathsf{tt})?)^\star\rangle\mathsf{p}$ states that either p holds in the current state, or there exist transitions for which p holds in the outgoing states.

□

It is worth mentioning that if the alphabet is finite, the three properties above can be expressed by PDL formulas whose size depends on the size of the (finite) set of actions. However this is not the case if they are expressed by PPDL formulas (in which the variables range over a finite set of actions).

3 Satisfiability games

A satisfiability game $\mathcal{G}_S(\phi)$ on a PPDL formula ϕ where the variables are instantiated from the set of actions $S \subseteq \Sigma$, is an infinite duration game played between two players: player I (or Abelard) and player II (or Eloise). Player II is looking to prove that ϕ is satisfiable and player I is trying to prove that it is not satisfiable. Besides, the game positions are configurations (i.e. a pair composed of a state and a substitution). The name of the rules is indicated on the left of the inference rule; and the name of the player is indicated on the right.

$$\mathsf{R}_\wedge\colon \frac{\Big[(\phi_0 \wedge \phi_1, \sigma)\Big], \Gamma}{\Big[(\phi_i, \sigma)\Big], (\phi_{1-i}, \sigma), \Gamma}\ \mathrm{I} \qquad \mathsf{R}_\vee\colon \frac{\Big[(\phi_0 \vee \phi_1, \sigma)\Big], \Gamma}{\Big[(\phi_i, \sigma)\Big], \Gamma}\ \mathrm{II}$$

The rules of the formulas starting with modalities follow:

$$\mathsf{R}_1\colon \frac{\Big[(\langle\alpha_0 \cup \alpha_1\rangle\phi, \sigma)\Big], \Gamma}{\Big[(\langle\alpha_i\rangle\phi, \sigma)\Big], \Gamma}\ \mathrm{II} \qquad \mathsf{R}_2\colon \frac{\Big[([\alpha_0 \cup \alpha_1]\phi, \sigma)\Big], \Gamma}{\Big[([\alpha_i]\phi, \sigma)\Big], ([\alpha_{1-i}]\phi, \sigma), \Gamma}\ \mathrm{I}$$

$$\mathsf{R}_3\colon \frac{\Big[(\langle\alpha_0;\alpha_1\rangle\phi, \sigma)\Big], \Gamma}{\Big[(\langle\alpha_0\rangle\langle\alpha_1\rangle\phi, \sigma)\Big], \Gamma} \qquad \mathsf{R}_4\colon \frac{\Big[([\alpha_0;\alpha_1]\phi, \sigma)\Big], \Gamma}{\Big[([\alpha_0][\alpha_1]\phi, \sigma)\Big], \Gamma}$$

$$\mathsf{R}_5\colon \frac{\Big[(\langle\alpha^\star\rangle\phi, \sigma)\Big], \Gamma}{\Big[(\phi, \sigma) \vee (\langle\alpha\rangle\langle\alpha^\star\rangle\phi, \sigma)\Big], \Gamma} \qquad \mathsf{R}_6\colon \frac{\Big[([\alpha^\star]\phi, \sigma)\Big], \Gamma}{\Big[(\phi, \sigma) \wedge ([\alpha][\alpha^\star]\phi, \sigma)\Big], \Gamma}$$

The rules for the reset operator and the test operator follow:

$$\mathsf{R}^1_{\mathtt{r}}\colon \frac{\Big[([\mathsf{res}(x);\alpha]\phi, \sigma)\Big], \Gamma}{\Big[([\alpha]\phi, \sigma_{|Dom(\sigma)\setminus\{x\}}\Big], \Gamma} \qquad \mathsf{R}^2_{\mathtt{r}}\colon \frac{\Big[(\langle\mathsf{res}(x);\alpha\rangle\phi, \sigma)\Big], \Gamma}{\Big[(\langle\alpha\rangle\phi, \sigma_{|Dom(\sigma)\setminus\{x\}}\Big], \Gamma}$$

The rules for free variables appearing in a modality:

$$\mathrm{R}_x^1\colon \frac{\Big[([x]\phi,\sigma)\Big],\Gamma \quad \text{if } \tau(x)=\forall}{\Big[([x]\phi,\sigma\uplus\{x\mapsto a\}\Big],\Gamma}\ \mathrm{I} \qquad \mathrm{R}_x^2\colon \frac{\Big[([x]\phi,\sigma)\Big],\Gamma \quad \text{if } \tau(x)=\exists}{\Big[([x]\phi,\sigma\uplus\{x\mapsto a\}\Big],\Gamma}\ \mathrm{II}$$

$$\mathrm{R}_?^1\colon \frac{\Big[(\langle\psi?\rangle\phi,\sigma)\Big],\Gamma}{(\psi,\sigma),\Big[(\phi,\sigma)\Big],\Gamma} \qquad \mathrm{R}_?^2\colon \frac{\Big[([\psi?]\phi,\sigma)\Big],\Gamma}{\Big[(\neg\psi,\sigma)\vee(\phi,\sigma)\Big],\Gamma}$$

The rules for the universally and existentially quantified formula follow:

$$\mathrm{R}_\forall\colon \frac{\Big[(\forall x\,\phi,\sigma)\Big],\Gamma}{\Big[(\phi,\sigma_{|Dom(\sigma)\setminus\{x\}}\uplus[x\mapsto a])\Big],\Gamma}\ \mathrm{I} \qquad \mathrm{R}_\exists\colon \frac{\Big[(\exists x\,\phi,\sigma)\Big],\Gamma}{\Big[(\phi,\sigma_{|Dom(\sigma)\setminus\{x\}}\uplus[x\mapsto a])\Big],\Gamma}\ \mathrm{II}$$

The successive applications of the above rules might yield a configuration in which all formulas are either propositional constants or of the form $(\langle\alpha\rangle\phi,\sigma)$ or $([\alpha]\phi,\sigma)$ where $\alpha\in\Sigma\cup\mathcal{X}$ and $\sigma(\alpha)\in\Sigma$.

$$\mathrm{X}_1\colon \frac{\Big[(\langle\alpha_1\rangle\phi_1,\sigma_1)\Big],\ldots,(\langle\alpha_n\rangle\phi_n,\sigma_n),([\beta_1]\psi_1,\gamma_1),\ldots,([\beta_m]\psi_m,\gamma_m),\ldots,p_1,\ldots,p_l}{\Big[(\phi_1,\sigma_1)\Big],(\psi_{j_1},\gamma_{j_1}),\ldots,(\psi_{j_q},\gamma_{j_q})}$$

where $\alpha_i,\beta_i\in\Sigma\cup\mathcal{X}$ and $\forall i=1,\ldots,q:\sigma_1(\alpha_1)=\gamma_{j_i}(\beta_{j_i}), j_i\in\{1,\ldots,m\}$ and $\sigma_l(\alpha_l),\gamma_{l'}(\beta_{l'})\in\Sigma$ for all l,l'.

$$\mathrm{X}_2\colon \frac{(\langle\alpha_1\rangle\phi_1,\sigma_1),\ldots,(\langle\alpha_n\rangle\phi_n,\sigma_n),\Big[([\beta_1]\psi_1,\gamma_1)\Big],\ldots,([\beta_m]\psi_m,\gamma_m),\ldots,p_1,\ldots,p_l}{(\phi_k,\sigma_k),\Big[(\psi_{j_1},\gamma_{j_1})\Big],\ldots,(\psi_{j_q},\gamma_{j_q})}\ \mathrm{I}$$

where $\alpha_i,\beta_i\in\Sigma\cup\mathcal{X}$ and $\forall i=1,\ldots,q:\sigma_k(\alpha_k)=\gamma_{j_i}(\beta_{j_i}), j_i\in\{1,\ldots,m\}$, and $\sigma_l(\alpha_l),\gamma_{l'}(\beta_{l'})\in\Sigma$ for all l,l'.

Moreover, there is a rule allowing player I to change his mind w.r.t. to the focus:

$$\mathrm{FC}\colon \frac{\Big[(\phi,\sigma)\Big],(\psi,\gamma),\Gamma}{(\phi,\sigma),\Big[(\psi,\gamma)\Big],\Gamma}\ \mathrm{I}$$

We notice that the main difference with the satisfiability games for standard PDL is that our games have (possibly) an infinite number of positions and infinite branching. This is due to the fact that the size of our configurations is unbounded. Thus we modify accordingly the winning conditions of to deal with this new setting as follows.

Firstly, the winning conditions have to deal with the fact that a least fixed-point construct is fulfilled and there is no contradiction in the propositional constants. Player I wins the (possibly infinite) play $\pi = C_0, \ldots, C_n, \ldots$ iff

(i) $C_m = \left[(q, \sigma)\right], \Gamma$ and ($q = \mathsf{ff}$ or $\bar{q} \in \Gamma$), for some m, or

(ii) The formula $\langle \alpha^\star \rangle \phi$ appears infinitely often under the focus in π and player I has applied the focus rule (FC) only a finite number of times. That is, there exists an infinite sequence $i_1, i_2, \ldots$ of integers such that $C_{i_j} = \left[(\langle \alpha^\star \rangle \phi, \sigma_{i_j})\right], \Gamma_{i_j}$ for all $j = 1, 2, \ldots$ and there exists some i_k such that no focus rule is applied from the configurations C_m for all $m \geq i_k$.

Player II wins the (possibly infinite) play $\pi = C_0, \ldots, C_n, \ldots$ if

(iii) $C_n = \left[(q_1, \sigma_1)\right], \ldots, (q_k, \sigma_k)$ and $\{q_1, \ldots, q_k\}$ is satisfiable, where q_i are propositional constants, or

(iv) The formula $[\alpha^\star]\phi$ appears infinitely often in π under the focus, that is, there exists an infinite sequence $i_1, i_2, \ldots$ of integers such that $C_{i_j} = \left[([\alpha^\star]\phi, \sigma_{i_j})\right], \Gamma_{i_j}$ appears in π, or

(v) The formula ϕ appears infinitely often in π under the focus and Player I has applied the focus rule (FC) infinitely often.

One can argue that these winning conditions are mutually exclusive, hence:

Lemma 3.1 *The game $\mathcal{G}_S(\phi)$ has a unique winner, where $S \subseteq \Sigma$.*

Now we will prove that the game theoretic characterization of PPDL is sound and complete.

Theorem 3.2 *The following hold:*

(Soundness). If player II wins the game $\mathcal{G}(\Phi_0)$ then Φ_0 is satisfiable.

(Completeness). If Φ_0 is satisfiable then player II wins the game $\mathcal{G}(\Phi_0)$.

The proof of the completeness relies on the following Lemma:

Lemma 3.3 *We have that $\phi \wedge \langle \alpha \rangle \psi$ is satisfiable iff $\phi \wedge (\psi \vee \langle \alpha \rangle (\langle \alpha^\star \rangle \psi \vee \neg \phi))$ is satisfiable.*

4 Decidability of PPDL

The idea of the proof of the decidability of PPDL relies on reducing the satisfiability problem of a PPDL formula into the satisfiability problem of the same formula in which the actions range over a *finite* set of actions. It turns out that solving the resulting game is decidable since it is a finite 2-players game with Büchi winning conditions. The construction of this finite set of actions follows.

Definition 4.1 Let ϕ be a PPDL formula and let $\Sigma(\phi) = \{a_1, \dots, a_n\}$ and $\mathcal{X} = \{x_1, \dots, x_k\}$. Define $\mathcal{G}_C(\phi)$ to be the satisfiability game in which the variables are instantiated from the finite set of constants C:

$$C = \{a_1, \dots, a_n, c_1, \dots, c_k\} \tag{1}$$

The idea is that the game $\mathcal{G}_C(\phi)$ is used to simulate the game $\mathcal{G}_\Sigma(\phi)$. Before showing that, we need to introduce a variant of satisfiability games in which Player II can duplicate a formula under the focus a finite number of times.

Definition 4.2 Define $\mathcal{G}_S^{\mathcal{D}}$ to be satisfiability game in which we add the duplication rule for player II:

$$\text{DP:} \frac{\left[(\phi, \sigma)\right], \Gamma}{\left[(\phi, \sigma)\right], (\phi, \sigma), \Gamma} \text{ II}$$

that has to be applied a finite number of times.

The main result of this section follows, as well as the structure of the proof:

Theorem 4.3 *Satisfiability for PPDL is decidable.*

On the one hand, Corollary 4.8 shows that $\mathcal{G}_\Sigma$ and $\mathcal{G}_\Sigma^{\mathcal{D}}$ are equivalent. On the other hand, Lemma 4.4 shows that $\mathcal{G}_\Sigma$ and $\mathcal{G}_C^{\mathcal{D}}$ are equivalent, and hence $\mathcal{G}_C$ and $\mathcal{G}_C^{\mathcal{D}}$ are equivalent as well. It follows that $\mathcal{G}_C$ and $\mathcal{G}_\Sigma$ are equivalent. The equivalence must be understood as player II wins in one game iff she wins in the other game.

Lemma 4.4 *The games $\mathcal{G}_\Sigma$ and $\mathcal{G}_\Sigma^{\mathcal{D}}$ are equivalent. That is, Player II wins in $\mathcal{G}_\Sigma(\phi)$ iff she wins in $\mathcal{G}_\Sigma^{\mathcal{D}}(\phi)$.*

In order to relate the configurations of $\mathcal{G}_C(\phi)$ to the ones of $\mathcal{G}_\Sigma(\phi)$, we define a relation $\lhd$ between the configurations of $\mathcal{G}_C(\phi)$ and the configurations of $\mathcal{G}_\Sigma(\phi)$. We shall denote by Ψ the (finite) set of formulas appearing in a configuration, i.e. $\Psi((\psi, \sigma)) = \{\psi\}$ and $\Psi([C_1], \dots, C_n) = \bigcup_i \Psi(C_i)$. Firstly, we define the *coherence* relation between substitutions.

Definition 4.5 Let C be a finite subset of Σ. The coherence relation $\bowtie_C \subseteq \zeta \times \zeta$ between substitutions is defined by $\bar{\sigma} \bowtie_C \sigma$ iff the three following conditions hold:

(i) $dom(\bar{\sigma}) = dom(\sigma)$,

(ii) If $\bar{\sigma}(x) \in C$ then $\bar{\sigma}(x) = \sigma(x)$, and if $\sigma(x) \in C$, then $\bar{\sigma}(x) = \sigma(x)$, for any variable $x \in dom(\sigma)$, and

(iii) for any variables $x, y \in dom(\sigma)$, $\bar{\sigma}(x) = \bar{\sigma}(y)$ iff $\sigma(x) = \sigma(y)$.

In order to relate the configurations of $\mathcal{G}_\Sigma$ and $\mathcal{G}_C$, where C is the set of letters defined in Eq (1), we define next a binary relation, denoted by $\lhd$, between configurations.

Definition 4.6 Let ϕ be a PPDL formula with $\Sigma(\phi) = \{a_1, \ldots, a_n\}$. Let Γ (resp. $\widehat{\Gamma}$) be a list of configurations in $\mathcal{G}_\Sigma(\phi)$ (resp. $\mathcal{G}_C(\phi)$) of the form: $\Gamma = (\psi_1, \sigma_1), \ldots, (\psi_m, \sigma_m), \cdots$, and $\widehat{\Gamma} = (\widehat{\psi}_1, \widehat{\sigma}_1), \ldots, (\widehat{\psi}_m, \widehat{\sigma}_m)$, where ψ_i and $\widehat{\psi}_i$ are PPDL formulas, and σ_i and $\widehat{\sigma}_i$ are substitutions. Let f be a total surjective function from the set of configurations of $\mathcal{G}_\Sigma(\phi)$ to the set of configurations of $\mathcal{G}_C^{\mathcal{D}}(\phi)$. We define $\Gamma \lhd_f^{\Sigma,C} \widehat{\Gamma}$ iff the following hold:

(i) $\Psi(\Gamma) = \Psi(\widehat{\Gamma})$.

(ii) If $f((\psi_i, \sigma_i)) = (\widehat{\psi}_j, \widehat{\sigma}_j)$ then $\psi_i = \widehat{\psi}_j$ and $\sigma_i \bowtie_C \widehat{\sigma}_j$.

(iii) If $(\widehat{\psi}, \widehat{\sigma})$ is under the focus in $\widehat{\Gamma}$ and (ψ, σ) is under the focus in Γ, then $f((\psi, \sigma)) = (\widehat{\psi}, \widehat{\sigma})$.

Besides, we write $\mathcal{G}_\Sigma(\phi) \lhd \mathcal{G}_C(\phi)$ iff there exists a surjective function f such that $\left[\phi\right], \emptyset \lhd_f^{\Sigma,C} \left[\phi\right], \emptyset$.

In what follows we shall write $\lhd_f$ instead of $\lhd_f^{\Sigma,C}$ if there is no ambiguity. The following Lemma shows that the inference rules of section 3 preserve the relation $\lhd$:

Lemma 4.7 *Let ϕ be a PPDL formula with the finite set of actions $C = \{a_1, \ldots, a_n, c_1, \ldots, c_k\}$ as defined in Eq. (1). Let Γ (resp. $\widehat{\Gamma}$) be a list of configurations in $\mathcal{G}_\Sigma(\phi)$ (resp. $\mathcal{G}_C^{\mathcal{D}}(\phi)$). If $\Gamma \lhd_f \widehat{\Gamma}$ then*

(i) *for every Γ' such that $\Gamma \xrightarrow{I} \Gamma'$ is a move in $\mathcal{G}_\Sigma(\phi)$, there exists a total surjective function f' and a list of configurations $\widehat{\Gamma}'$ such that $\widehat{\Gamma} \xrightarrow{I} \widehat{\Gamma}'$ is a possible move in $\mathcal{G}_C^{\mathcal{D}}(\phi)$ and $\Gamma' \lhd_{f'} \widehat{\Gamma}'$, and*

(ii) *for every $\widehat{\Gamma}'$ such that $\widehat{\Gamma} \xrightarrow{II} \widehat{\Gamma}'$ is a move in $\mathcal{G}_C^{\mathcal{D}}(\phi)$, there exists a total surjective function f' and a list of configurations Γ' such that $\Gamma \xrightarrow{II} \Gamma'$ is a possible move in $\mathcal{G}_\Sigma(\phi)$ and $\Gamma' \lhd_{f'} \widehat{\Gamma}'$,*

Corollary 4.8 *Let ϕ be a PPDL formula. If $\mathcal{G}_\Sigma(\phi) \lhd \mathcal{G}_C(\phi)$ then Player II wins in $\mathcal{G}_\Sigma(\phi)$ iff she wins in $\mathcal{G}_C(\phi)$.*

5 Application to service composition

We can formulate service composition problem under policy constraints as a PPDL satisfiability problem as in [10]. The client and services are firstly specified by parametrized systems which can be translated into PPDL formulas.

Modeling services collaboration with $\star$-simulation. In previous works [7,6] a simulation relation is used to express that a combination of services jointly satisfies a client. In this setting a client request is met by an elementary action of a single service. A $\star$-simulation relation is a variant of the previous simulation relation that allows the services to communicate with each others before answering the client request. Its formal definition is in Annex B.

Composition synthesis as a PPDL satisfiability problem. Web-service composition via $\star$-simulation in presence of policy constraints can be expressed as a satisfiability problem of a PPDL formula by following the same construction as [10] for PDL. Let $S = (S_1, ..., S_n)$ be a community of available services over the shared actions Σ. Each available service S_i is represented by a parametrized transition system TS_i. Let TS_0 be the client specification. The detailed construction of the formula is given in Annex B.

Extraction of a mediator. It is possible to extract a model, as a PTS, for a satisfiable PPDL formula out of a winning strategy ρ for player II in the game $\mathcal{G}_C(\phi)$. We firstly take the sub-game $(\mathcal{G}_C(\phi))_{|\rho}$ induced by the strategy ρ. Then, we turn the game $(\mathcal{G}_C(\phi))_{|\rho}$ into a symbolic game in which the moves are labeled with variables, i.e. the transitions labeled with variables correspond to the moves in which the rules X1 and X2 are applied; the remaining transitions are ε-transitions and correspond to the other rules (i.e. $R_\vee$, $R_\wedge$, R_1, R_2, etc).

6 Conclusion

We have introduced an extension of PDL, called PPDL, in which the actions can be letters or variables ranging over an infinite domain. We have proved that the satisfiability problem of PPDL is decidable when it is interpreted over the subclass of parametrized transition systems in which the variables can be refreshed. As an application, we have shown how to formulate services composition problem of parametrized services as a synthesis problem of PPDL.

References

[1] *Web service choreography description language (WS-CDL)*, www.w3.org/TR/ws-cdl-10.

[2] Avanesov, T., Y. Chevalier, M. A. Mekki, M. Rusinowitch and M. Turuani, *Distributed orchestration of Web services under security constraints*, in: *DPM'11*, pp. 235–252.

[3] Baader, F., D. Calvanese, D. McGuinness, D. Nardi and P. Patel-Schneider, "The Description Logic Handbook: Theory, Implementation and Applications," CUP, 2003.

[4] Balbiani, P., A. Herzig and N. Troquard, *Dynamic logic of propositional assignments: A well-behaved variant of PDL*, in: *LICS* (2013), pp. 143–152.

[5] Balbiani, P. and D. Vakarelov, *PDL with intersection of programs: A complete axiomatization*, Journal of Applied Non-Classical Logics **13** (2003), pp. 231–276.

[6] Belkhir, W., Y. Chevalier and M. Rusinowitch, *Fresh-variable automata: Application to service composition*, in: *15th International Symposium on Symbolic and Numeric Algorithms for Scientific Computing, SYNASC* (2013), pp. 153–160.

[7] Berardi, D., F. Cheikh, G. D. Giacomo and F. Patrizi, *Automatic service composition via simulation*, Int. J. Found. Comput. Sci. **19** (2008), pp. 429–451.

[8] Büchi, J. R. and L. H. Landweber, *Solving sequential conditions by finite-state strategies*, Transactions of the American Mathematical Society **138** (1969), pp. 295–311.

[9] Castagna, G., N. Gesbert and L. Padovani, *A theory of contracts for web services*, ACM Trans. Program. Lang. Syst. **31** (2009), pp. 19:1–19:61.

[10] Cheikh, F., G. D. Giacomo and M. Mecella, *Automatic web services composition in trustaware communities*, in: *SWS*, 2006, pp. 43–52.

[11] Degano, P., G. L. Ferrari and G. Mezzetti, *Nominal automata for resource usage control*, in: *CIAA*, 2012, pp. 125–137.

[12] Fischer, M. J. and R. E. Ladner, *Propositional dynamic logic of regular programs*, J. Comput. Syst. Sci. **18** (1979), pp. 194–211.

[13] Göller, S. and M. Lohrey, *Infinite state model-checking of propositional dynamic logics*, in: Z. Ésik, editor, *CSL*, Lecture Notes in Computer Science **4207** (2006), pp. 349–364.

[14] Herzig, A., E. Lorini, F. Moisan and N. Troquard, *A dynamic logic of normative systems*, in: *IJCAI 2011*, 2011, pp. 228–233.
[15] Lange, M., *Model checking propositional dynamic logic with all extras*, Journal of Applied Logic **4** (2006), pp. 39 – 49.
[16] Lange, M. and R. Somla, *Propositional dynamic logic of context-free programs and fixpoint logic with chop*, Inf. Process. Lett. **100** (2006), pp. 72–75.
[17] Lange, M. and C. Stirling, *Focus games for satisfiability and completeness of temporal logic*, in: *LICS*, 2001, pp. 357–365.
[18] Lutz, C., *PDL with intersection and converse is decidable.*, in: C.-H. L. Ong, editor, *CSL*, Lecture Notes in Computer Science **3634** (2005), pp. 413–427.
[19] Manuel, A. and R. Ramanujam, *Automata over infinite alphabets*, World Scientific Review **9** (2011), pp. 329–363.
[20] Neven, F., T. Schwentick and V. Vianu, *Finite state machines for strings over infinite alphabets*, ACM Trans. Comput. Log. **5** (2004), pp. 403–435.
[21] Peltz, C., *Web services orchestration and choreography*, Computer **36** (2003), pp. 46–52.
[22] Ramadge, P. J. and W. M. Wonham, *Supervisory control of a class of discrete event processes*, SIAM J. Control Optim. **25** (1987), pp. 206–230.
[23] Wolper, P., *Specification and synthesis of communicating processes using an Extended Temporal Logic*, in: *POPL* (1982), pp. 20–33.

Appendix

A Proofs for Section 4

The claims in the following remark are not hard to prove.

Remark A.1 Let $C \subseteq \Sigma$ be a finite set of letters, $\bar{\sigma}$ and σ two substitutions, x a variable, and a a letter in C. The following hold.

(i) If $\bar{\sigma} \bowtie_C \sigma$ then $|codom(\bar{\sigma})| = |codom(\sigma)|$ and $\bar{\sigma}_{|D} \bowtie_C \sigma_{|D}$, where $D \subseteq Dom(\sigma)$.

(ii) Consequently, if $(\bar{\sigma}_1 \uplus \bar{\sigma}_2) \bowtie (\sigma_1 \uplus \sigma_2)$ with $dom(\bar{\sigma}_i) = dom(\sigma_i)$, then $\bar{\sigma}_i \bowtie \sigma_i$, for $i = 1, 2$.

Lemma A.2 *[6] Let $C_1 \subseteq \Sigma$ and $C_2 \subseteq \Sigma$ be two sets of actions. Let $C = C_1 \cap C_2$ be s.t. $|C| > |\mathcal{X}|$. Let a_1 be an action in C_1 and $x \in \mathcal{X}$ and let $\sigma_i : \mathcal{X} \to C_i$, $i = 1, 2$ be two substitutions where $\sigma_1 \bowtie_C \sigma_2$. Then, there exists a function $\mathbf{\Theta}^{C_1,C_2}$ satisfying $\sigma_1 \uplus \{x \mapsto a_1\} \bowtie_C \mathbf{\Theta}(\sigma_1, x, a_1, \sigma)$.*

Lemma A.3 (i.e. Lemma 4.7) *Let ϕ be a PPDL formula with the finite set of actions $C = \{a_1, \ldots, a_n, c_1, \ldots, c_k\}$ as defined in Eq. (1). Let Γ (resp. $\widehat{\Gamma}$) be a list of configurations in $\mathcal{G}_\Sigma(\phi)$ (resp. $\mathcal{G}_C^{\mathcal{D}}(\phi)$). If $\Gamma \lhd_f \widehat{\Gamma}$ then*

(i) *for every Γ' such that $\Gamma \xrightarrow{I} \Gamma'$ is a move in $\mathcal{G}_\Sigma(\phi)$, there exists a total surjective function f' and a list of configurations $\widehat{\Gamma}'$ such that $\widehat{\Gamma} \xrightarrow{I} \widehat{\Gamma}'$ is a possible move in $\mathcal{G}_C^{\mathcal{D}}(\phi)$ and $\Gamma' \lhd_{f'} \widehat{\Gamma}'$, and*

(ii) *for every $\widehat{\Gamma}'$ such that $\widehat{\Gamma} \xrightarrow{II} \widehat{\Gamma}'$ is a move in $\mathcal{G}_C^{\mathcal{D}}(\phi)$, there exists a total surjective function f' and a list of configurations Γ' such that $\Gamma \xrightarrow{II} \Gamma'$ is a possible move in $\mathcal{G}_\Sigma(\phi)$ and $\Gamma' \lhd_{f'} \widehat{\Gamma}'$,*

Proof. Assume $\mathcal{X} = \{x_1, \ldots, x_k\}$. We discuss many cases depending on the applied rule.

(i) The applied rules here for player I can be $R_\wedge$, R_2, R_3, R_4, R_5, R_6, R_Υ^1, R_Υ^2, R_x^1, $R_?^1$, $R_?^2$, $R_\forall$, X_1, X_2 and FC. For the rule $R_\wedge$, Let:

$$\begin{cases} \Gamma = \Big[(\phi_0 \wedge \phi_1, \sigma)\Big], \Upsilon \quad \text{and} \\ \Gamma' = \Big[(\phi_i, \sigma)\Big], (\phi_{1-i}, \sigma), \Upsilon \quad \text{and} \\ \widehat{\Gamma} = \Big[(\phi_0 \wedge \phi_1, \widehat{\sigma})\Big], \widehat{\Upsilon} \end{cases}$$

where $\sigma \bowtie_{\Sigma(\phi)} \widehat{\sigma}$. In this case we let

$$\begin{cases} \widehat{\Gamma}' \stackrel{def}{=} \Big[(\phi_i, \widehat{\sigma})\Big], (\phi_{1-i}, \widehat{\sigma}), \widehat{\Upsilon} \\ \sigma \stackrel{def}{=} \widehat{\sigma}, \text{ and} \\ f' \stackrel{def}{=} f_{|\Gamma} \cup \{((\phi_i, \sigma), (\phi_i, \widehat{\sigma}))\} \cup \{((\phi_{1-i}, \sigma), (\phi_{1-i}, \widehat{\sigma}))\} \end{cases}$$

Thus, $\Gamma \lhd_{f'} \widehat{\Gamma}'$.

The rules R_2, R_3, R_4, R_5, R_6, $R_?^1$ and $R_?^2$ can be handled similarly.

For the rules R_Γ^i, $i = 1, 2$ the claim follows from the fact that if $\sigma \bowtie_{\Sigma(\phi)} \widehat{\sigma}$ then $\sigma_{|Dom(\sigma)\setminus\{x\}} \bowtie_{\Sigma(\phi)} \widehat{\sigma}_{|Dom(\widehat{\sigma})\setminus\{x\}}$, see Item 1 of remark A.1.

For the rule $R_\forall$, assume that

$$\begin{cases} \Gamma &= \Big[(\forall x\phi, \sigma)\Big], \Lambda \quad \text{and} \\ \Gamma' &= \quad \Big[(\phi, \sigma_{x\mapsto a})\Big], \Lambda \quad \text{and} \\ \widehat{\Gamma} &= \Big[(\forall x\phi, \widehat{\sigma})\Big], \widehat{\Lambda} \end{cases}$$

We distinguish two cases

Case 1. If $\nexists C \in \Lambda$ s.t. $f(C) = (\forall x\phi, \widehat{\sigma})$, then in this case the related move in $\widehat{\mathcal{G}}_{\mathcal{D}}$ is

$$\underbrace{\Big[(\forall x\phi, \widehat{\sigma})\Big], \widehat{\Lambda}}_{\widehat{\Gamma}} \overset{(R_\forall)}{\longrightarrow} \underbrace{\Big[(\phi, \widehat{\sigma}')\Big], \widehat{\Lambda}}_{\widehat{\Gamma}'}$$

where the substitution $\widehat{\sigma}'$ is defined by $\widehat{\sigma}' \overset{def}{=} \boldsymbol{\Theta}^{\Sigma, \Sigma_f}(\sigma, x, a, \widehat{\sigma})$.

Case 2. If $\exists C \in \Lambda$ s.t. $f(C) = (\forall x\phi, \widehat{\sigma})$, then in this case the related moves in $\widehat{\mathcal{G}}_{\mathcal{D}}$ are

$$\underbrace{\Big[(\forall x\phi, \widehat{\sigma})\Big], \widehat{\Lambda}}_{\widehat{\Gamma}} \overset{(DP)}{\longrightarrow} \Big[(\forall x\phi, \widehat{\sigma})\Big], (\forall x\phi, \widehat{\sigma}), \widehat{\Lambda} \overset{(R_\forall)}{\longrightarrow} \underbrace{\Big[(\phi, \widehat{\sigma}')\Big], (\forall x\phi, \widehat{\sigma}), \widehat{\Lambda}}_{\widehat{\Gamma}'}$$

where $\widehat{\sigma}'$ is defined by $\widehat{\sigma}' \overset{def}{=} \boldsymbol{\Theta}^{\Sigma, \Sigma_f}(\sigma, x, a, \widehat{\sigma})$.

Notice that in both cases we have that $\widehat{\sigma}' \bowtie_{\Sigma(\phi)} \sigma$, Lemma A.2. Besides, in both cases the function f' is defined by $f' = f_{|Dom(f)\setminus\{(\forall x\phi, \sigma)\}}$, and $f'((\phi, \sigma_{x\mapsto a})) = (\phi, \widehat{\sigma}')$. Thus $\Gamma \lhd_{f'} \widehat{\Gamma}'$.

For the rule X1, assume that

$$\begin{cases} \Gamma & = \Big[(\langle\alpha_1\rangle\phi_1, \sigma_1)\Big], \ldots, (\langle\alpha_n\rangle\phi_n, \sigma_n), ([\beta_1]\psi_1, \gamma_1), \ldots, \\ & \quad ([\beta_m]\psi_m, \gamma_m), \ldots, p_1, \ldots, p_l \\ \text{and} & \\ \widehat{\Gamma} & = \Big[(\langle\widehat{\alpha}_1\rangle\widehat{\phi}_1, \widehat{\sigma}_1)\Big], \ldots, (\langle\widehat{\alpha}_k\rangle\widehat{\phi}_k, \widehat{\sigma}_k), ([\widehat{\beta}_1]\widehat{\psi}_1, \widehat{\gamma}_1), \ldots, \\ & \quad ([\widehat{\beta}_r]\widehat{\psi}_r, \widehat{\gamma}_r), \ldots, p_1, \ldots, p_{l'} \end{cases}$$

Hence,

$$\begin{cases} \Gamma' &= \Big[(\phi_1, \sigma_1)\Big], (\psi_{j_1}, \gamma_{j_1}), \ldots, (\psi_{j_q}, \gamma_{j_q}) \quad \text{and} \\ \widehat{\Gamma}' &= \Big[(\widehat{\phi}_1, \widehat{\sigma}_1)\Big], (\widehat{\psi}_{j'_1}, \widehat{\gamma}_{j'_1}), \ldots, (\widehat{\psi}_{j'_p}, \widehat{\gamma}_{j'_p}) \end{cases}$$

where, on the one hand, $\alpha_i, \beta_i \in \Sigma \cup \mathcal{X}$ and $\forall i = 1, \ldots, q : \sigma_1(a_1) = \gamma_{j_i}(b_{j_i}), j_i \in \{1, \ldots, m\}$ and, on the other hand, $\widehat{\alpha}_i, \widehat{\beta}_i \in \Sigma \cup \mathcal{X}$ and $\forall i = 1, \ldots, p : \widehat{\sigma}_1(\widehat{\alpha}_1) = \widehat{\gamma}_{j'_i}(\widehat{\beta}_{j'_i}), j'_i \in \{1, \ldots, r\}$. Finally, we let $f' \stackrel{def}{=} f_{|\Gamma'}$. Therefore, $\Gamma' \lhd_{f'} \widehat{\Gamma}'$.

For the rule FC, assume that

$$\begin{cases} \Gamma &= \Big[(\phi, \sigma)\Big], (\psi, \gamma), \Upsilon \quad \text{and} \\ \Gamma' &= \Big[(\psi, \gamma)\Big], (\phi, \sigma), \Upsilon \quad \text{and} \\ \widehat{\Gamma} &= \Big[(\widehat{\phi}, \widehat{\sigma})\Big], \widehat{\Upsilon} \end{cases}$$

The idea is to choose a formula $(\widehat{\psi}, \widehat{\gamma})$ from Υ such that $\widehat{\Gamma}' = \Big[(\widehat{\psi}, \widehat{\gamma})\Big], \widehat{\Upsilon}$. We define $(\widehat{\psi}, \widehat{\gamma}) := f((\psi, \gamma))$. Thus we have $\Gamma' \lhd_f \widehat{\Gamma}'$ since $\widehat{\psi} = \psi$ and $\widehat{\gamma} \bowtie \gamma$.

(ii) The possible rules for player II are $\mathtt{R}_\exists$, X_1 and $\mathtt{R}^2_x$. The rule $\mathtt{R}_\exists$ (resp. $\mathtt{R}^2_x$) is exactly like the rule $\mathtt{R}_\forall$ (resp. $\mathtt{R}^1_x$) apart that player II who moves instead of Player I.

□

B Proofs and definitions for Section 5

B.1 Modeling services collaboration with ⋆-simulation

Definition B.1 Let $\mathcal{A}_i = \mathcal{M} = \langle \Sigma, \mathcal{X}_i, Q^i, q^i_0, \delta_i, \pi_i, \kappa_i \rangle$, $i = 1, 2$ be two parametrized labeled transition systems. A ⋆-simulation is a relation $\trianglelefteq \subseteq (Q_1 \times \zeta_{\mathcal{X}_1,\Sigma}) \times (Q_2 \times \zeta_{\mathcal{X}_2,\Sigma})$ such that:

- $(q^1_0, \emptyset) \trianglelefteq (q^2_0, \emptyset)$.
- If $(q_1, \sigma_1) \trianglelefteq (q_2, \sigma_2)$ and if $(q_1, \sigma_1) \stackrel{a}{\to} (q'_1, \sigma'_1)$ for some action $a \in \Sigma$, then there exist states $q^0_2, \ldots, q^n, p^0_2, \ldots, p^m_2 \in Q_2$ and substitutions $\sigma^0_2, \ldots, \sigma^n, \gamma^0_2, \ldots, \gamma^m_2$ and actions $a_0, \ldots, a_{n-1}, b_0, \ldots, b_{m-1} \in \Sigma$ such that

$$\begin{aligned} &(q_2, \sigma_2) \stackrel{a_0}{\to} (q^0_2, \sigma^0_2) \stackrel{a_0}{\to} (q^1_2, \sigma^1_2) \stackrel{\star}{\to} \ldots \stackrel{a_{n-1}}{\to} (q^n_2, \sigma^n_2) \stackrel{a_{n-1}}{\to} (q^n_2, \sigma^n_2) \\ &\stackrel{a}{\to} (q'_2, \sigma'_2) \\ &\stackrel{b_0}{\to} (p^0_2, \gamma^0_2) \stackrel{b_0}{\to} (p^1_2, \gamma^1_2) \stackrel{\star}{\to} \ldots \stackrel{b_{m-1}}{\to} (p^m_2, \gamma^m_2) \stackrel{b_{m-1}}{\to} (p^m_2, \gamma^m_2) \end{aligned}$$

and $(\sigma'_1, q'_1) \trianglelefteq (p^m_2, \gamma^m_2)$.

B.2 Composition synthesis as a PPDL satisfiability problem

Let $S = (S_1, ..., S_n)$ be a community of available services over the shared actions Σ. Each variable services S_i is represented by a parametrized transition system $TS_i = \langle \Sigma, \mathcal{X}_i, S_i, s_{i0}, \delta_i, \pi_i, \kappa_i \rangle$ defined as above. Let $TS_0 = \langle \Sigma, \mathcal{X}_0, S_0, s_{00}, \delta_0, \pi_0, \kappa_0 \rangle$ be the client specification.

Then we build a PPDL formula Φ to be checked for satisfiability as follows. As propositional constants, we have:

- One propositional constant s for each $i \in \{0, 1, ..., n\}$ and each state s of TS_i, which intuitively denotes that TS_i is in a final state.
- Propositional constants $exec_{ix}$, for $i \in \{0, 1, ..., n\}$ and $x \in \mathcal{A}$, denoting that x will be executed next by the available service S_i.
- One propositional constant *undef* denoting that we are in an "illegal" situation, where the orchestrator program can be left undefined.

For representing the transitions of each available service S_i, we construct a formula Φ_i as the conjunction of:

- $\forall x(s \rightarrow \bigwedge_{(s',x)\in\varepsilon}(\langle x\rangle s') \wedge [x](\bigvee_{(s',x)\in\varepsilon} s'))$. Where $\varepsilon = \{(s', x)|(s, x, s') \in \delta_i\}$, for each s of S_i and $x \in \mathcal{A}$.
- $\forall x(s \wedge exec_{ix} \rightarrow [x]\mathbf{false})$, for each s of S_i such that for no g and s' we have that $(s, g, x, s') \in \delta_i$.
- $\forall x(s \wedge exec_{ix} \rightarrow [x]s)$ for each s of S_i and $x \in \mathcal{A}$.

In addition, we have the formula Φ_{add} obtained as the conjunction of:

- $s \rightarrow \neg s'$ for all pairs of states s, s' of S_i, and for $i \in \{0, 1, ..., n\}$.
- $F_i \leftrightarrow \bigvee_{s\in F_i} s$, for $i \in \{0, 1, ..., n\}$.
- $\forall x(undef \rightarrow [x]undef)$, for $x \in \mathcal{A}$.
- $\forall x(\neg undef \rightarrow \langle x\rangle\mathbf{true} \rightarrow \bigvee_{i\in\{1,\ldots,n\}} exec_{ix})$, for $x \in \mathcal{A}$.
- $\forall x(exec_{ix} \rightarrow \neg exec_{jx})$, for each $i, j \in \{1, ..., n\}, i \neq j$ and each $a \in \mathcal{A}$
- $F_0 \rightarrow \bigvee_{i\in\{1,\ldots,n\}} F_i$.

The requirements that every open file has to be closed and that the flight data of the client have to be stored in an appropriate file can be respectively expressed by the two PPDL formula ψ_1 and ψ_1 as follows:

$$\begin{cases} \psi_1 &= \forall x\forall y[(\mathsf{res}(x); x)^\star; \mathtt{Open(x)}]\big(\exists z\langle(\mathsf{res}(z); z)^\star\rangle\langle\mathtt{Close(x)}\rangle\mathtt{tt}\big), \text{ and} \\ \psi_2 &= \forall x\forall y[(\mathsf{res}(x); x)^\star; \mathtt{!m(Id,Owner,Nbr,CSC}] \\ & \qquad \big(\exists f\ \exists z\langle(\mathsf{res}(z); z)^\star\rangle\langle\mathtt{Write(m(Id,Owner,Nbr,CSC,\ f)}\rangle\mathtt{tt}\big) \end{cases}$$

Finally, we describe Φ as

$$Init \wedge \forall z.\,[\mathbf{u}](\Phi_0 \wedge \bigwedge_{i\in\{1,\ldots,n\}} \Phi_i \wedge \Phi_{add}) \wedge \psi_1 \wedge \psi_2,$$

where *Init* stands for $s_{00} \wedge s_{10} \wedge ... \wedge s_{n0}$ and represents the initial state of all services S_i (including the target) and $(\mathbf{u} = (z; \mathsf{res}(z))^*)$, which is used to force $(\Phi_0 \wedge \bigwedge_{i\in\{1,\ldots,n\}} \Phi_i \wedge \Phi_{add})$ to be true in every point of the model.

Multiple-conclusion Rules, Hypersequents Syntax and Step Frames

Nick Bezhanishvili [1]

University of Amsterdam,
The Netherlands

Silvio Ghilardi [2]

Università degli Studi
Milano, Italy

Abstract

We investigate proof theoretic properties of logical systems via algebraic methods. We introduce a calculus for deriving multiple-conclusion rules and show that it is a Hilbert style counterpart of hypersequent calculi. Using step-algebras we develop a criterion establishing the bounded proof property and finite model property for these systems. Finally, we show how this criterion can be applied to universal classes axiomatized by certain canonical rules, thus recovering and extending known results from both semantically and proof-theoretically inspired modal literature.

Keywords: Multiple-conclusion rules, hypersequents, step algebras, step frames.

1 Introduction

In this paper we continue the proof theoretic investigations of modal logic via algebraic methods which started in [4,5]. In [4,5] the *bounded proof property* (the *bpp*), which is a kind of analytic subformula property, was introduced as a measurement of robustness of proof systems. An algebraic criterion was developed in [4, 5] establishing whether a modal system axiomatized by standard rules possesses the bpp. Here we extend this research in two directions. First, we investigate more expressive proof systems axiomatized by multiple-conclusion rules for which we develop equivalent systems via hypersequent calculi and prove for them an algebraic criterion for the bpp. Second, for a large class of logics (stable logics) we systematically design proof systems that have the bpp (see Section 5). Thus, we are at a position to not only check whether a system is robust, but also to *design robust proof systems*, by finding appropriate rules.

[1] Partially supported by the Rustaveli Foundation of Georgia grant FR/489/5-105/11.

[2] Supported by the PRIN 2010-2011 project "Logical Methods for Information Management" funded by the Italian Ministry of Education, University and Research (MIUR).

Multiple-conclusion rules recently gained attention in the modal logic literature (see e.g., [3, 16, 18]), because they constitute an essential tool for investigating classes of algebras beyond varieties and because canonical formulae axiomatizations can be nicely developed within this framework. On the other hand and from a completely different research perspective, the proof-theoretic oriented community realized that standard sequent formalisms are insufficient to handle complex logics and moved to more expressive hypersequent calculi (compare for instance the simplicity of communication rules used for the logics of linear frames developed in [1] with the more complex systems needed for cut elimination in the traditional context [10, 14]).

In this paper we connect multiple-conclusion rules and hypersequent calculi. To our best knowledge, no explicit calculus for multiple-conclusion rules has been proposed so far. Note that for semantic investigations such as [3, 16], it is in fact sufficient to specify *abstractly* the properties that a rule system (seen just as a set of rules) should satisfy. On the other hand, a specific calculus for multiple-conclusion rules is needed if we want to make a close comparison with the hypersequent approach. This calculus will play the role of a Hilbert calculus for hyperformulae, i.e., for the syntactic components of a hypersequent. We will introduce such a calculus and investigate it using the techniques developed in [4, 5]. These techniques, based on semantic analysis of 'step' structures, have been shown to be rather effective in establishing the bpp. Our long-term proposal is to apply these techniques to obtain the bpp and the *finite model property* (the fmp), thus also decidability, for logics axiomatized by canonical formulae. In this paper, we report a first success in this direction, already covering the bpp and fmp for a continuum of logics, including some of those recently analyzed in [19] via the hypersequent approach.

Proofs of the results from Section 2 will be deferred to the appendix. Proofs of the results from Sections 3 and 4 (requiring routine adjustments from the corresponding proofs in [4, 5]) are included only in the Technical Report [6].

2 A calculus for derived multiple-conclusion rules

Modal formulae are built from propositional variables $x, y, \ldots$ by using the Booleans $(\neg, \wedge, \vee, \rightarrow, 0, 1)$ and modal operators $(\Diamond, \Box)$. For simplicity, we take $\neg$, $\wedge$, $\Diamond$ as primitive connectives, the remaining ones being defined in the customary way (in particular, $\Box$ is defined as $\neg\Diamond\neg$). We shall also use parameters $a, b, \ldots$ instead of variables whenever we want to stress that uniform substitution does not apply to them. Underlined letters stand for tuples of unspecified length and formed by distinct elements, thus for instance, we may use $\underline{x}$ for a tuple such as $x_1, \ldots, x_n$. When we write $\phi(\underline{x})$ we want to stress that ϕ contains at most the variables $\underline{x}$ (and no parameters) and similarly when we write $\phi(\underline{a})$ we want to stress that ϕ contains at most the parameters $\underline{a}$ (and no variables). The same convention applies to sets of formulae: if Γ is a set of formulae and we write $\Gamma(\underline{a})$, we mean that all formulae in Γ are of the kind $\phi(\underline{a})$, etc. We may occasionally replace variables with parameters in a formula: for this, we use the following self-explanatory notation. For a formula $\phi(\underline{x})$ we write $\phi(\underline{a})$

to mean that $\phi(\underline{a})$ is obtained from $\phi(\underline{x})$ by replacing $\underline{x} = x_1, \ldots, x_n$ (simultaneously and respectively) by $\underline{a} = a_1, \ldots, a_n$. The modal complexity (or the modal degree) of a formula ϕ counts the maximum number of nested modal operators in ϕ (the precise definition is by an obvious induction).

We recall some background on modal algebras, see e.g., [8, Sec. 5.2] or [9, Sec. 7.6] for more details. A *modal algebra* $\mathfrak{A} = (A, \Diamond)$ is a Boolean algebra A endowed with a unary operator $\Diamond$ satisfying $\Diamond(x \vee y) = \Diamond x \vee \Diamond y, \Diamond 0 = 0$. Notice that, here and elsewhere, we use the same name for a connective and the corresponding operator in modal algebras (thus, for instance, 0 is zero, $\vee$ is join, etc.). In this way, propositional formulae can be identified with *terms* in the first order language of modal algebras.

From the semantic side, we have the notion of a frame. A *frame* $\mathfrak{F} = (W, R)$ is a set W endowed with a binary relation R. The *dual* of a frame $\mathfrak{F} = (W, R)$ is the modal algebra $\mathfrak{F}^* = (\wp(W), \Diamond_R)$, where $\wp(W)$ is the powerset Boolean algebra and $\Diamond_R$ is the semilattice morphism associated with R. The latter is defined as follows: for $S \subseteq W$, we have $\Diamond_R(S) = \{w \in W \mid R(w) \cap S \neq \emptyset\}$ (here $R(w)$ denotes $\{v \in W \mid (w, v) \in R\}$). It should be noticed that there is a real duality (in the categorical sense) between modal algebras and frames only if we restrict to finite modal algebras and finite frames. If we want a full duality working for arbitrary modal algebras, we must introduce some topological structures on our frames (see, e.g., [8, Sec. 5.5], [9, Sec. 7.5], [17, Ch. 4] or [20]). For the purposes of this paper, however, the duality between finite frames and finite modal algebras will suffice.

2.1 Multiple-conclusion rules

Normal modal logics are an adequate formalism to describe equational classes of modal algebras. However, in this paper we are interested in more general classes. A class of modal algebras is said to be:

(i) *a variety* iff it is the class of models of a set of equations, i.e., of sentences of the kind $\forall \underline{x} \bigwedge_{i=1}^n \phi_i(\underline{x}) = 1$, where the ϕ_i are modal formulae (aka terms in the first order language of modal algebras);

(ii) *a quasi-variety* iff it is the the class of models of a set of implications of equations, i.e., of sentences of the kind $\forall \underline{x}\ (\bigwedge_{i=1}^n \phi_i(\underline{x}) = 1 \to \psi(\underline{x}) = 1)$, where $\phi_1, \ldots, \phi_n, \psi$ are modal formulae;

(iii) *a universal class* iff it is the class of models of a set of clauses, i.e., of sentences of the kind $\forall \underline{x}\ (\bigwedge_{i=1}^n \phi_i(\underline{x}) = 1 \to \bigvee_{j=1}^m \psi_j(\underline{x}) = 1)$, where $\phi_1, \ldots, \phi_n$, $\psi_1, \ldots, \psi_m$ are modal formulae.

In order to describe universal classes within a propositional modal language, we shall use multiple-conclusion rules. A *multiple-conclusion rule* (or just a *rule*) is a pair of finite sets of formulae $\langle \Gamma, S \rangle$. If $\Gamma = \{\gamma_1, \ldots, \gamma_n\}, S = \{\delta_1, \ldots, \delta_m\}$, we write the rule $\langle \Gamma, S \rangle$ as Γ / S or as

$$\frac{\gamma_1, \ \ldots, \ \gamma_n}{\delta_1 \mid \cdots \mid \delta_m} \ (R)$$

The formulae $\Gamma = \{\gamma_1, \dots . \gamma_n\}$ are said to be the *premises* of the rule (R) and the formulae $S = \{\delta_1, \dots, \delta_m\}$ are said to be the *conclusions* of the rule (R). The multiple-conclusion rule (R) is said to be an *inference rule* or a *single-conclusion rule* iff $m = 1$, i.e., iff it has a single conclusion. A modal algebra $\mathfrak{A} = (A, \Diamond)$ validates the multiple-conclusion rule (R) iff it is a model of the clause $\forall \underline{x}\ (\bigwedge_{i=1}^{n} \phi_i(\underline{x}) = 1 \rightarrow \bigvee_{j=1}^{m} \psi_j(\underline{x}) = 1)$. A frame $\mathfrak{F} = (W, R)$ validates (R) iff its dual algebra $\mathfrak{F}^*$ does.

We recall the notion of a rule system from [16]:

Definition 2.1 A set of multiple-conclusion rules K is said to be a *rule system* iff it satisfies the following conditions for every formula ϕ and for every finite sets of formulae Γ, Γ', S, S':

(i) $\phi/\phi \in K$;

(ii) if $\Gamma/S, \phi \in K$ and $\Gamma, \phi/S \in K$, then $\Gamma/S \in K$;

(iii) if $\Gamma/S \in K$ then $\Gamma, \Gamma'/S, S' \in K$;

(iv) if $\Gamma/S \in K$ then for every substitution σ, we have that $\Gamma\sigma/S\sigma \in K$.

Above we used obvious conventions about set-theoretic union of finite sets of formulae (e.g., Γ, ϕ stands for $\Gamma \cup \{\phi\}$, moreover Γ, Γ' stands for $\Gamma \cup \Gamma'$, etc.). In addition, we used $\Gamma\sigma$ to denote the set resulting from the application of σ to all members of Γ.

Definition 2.2 A (normal) *modal rule system* is a rule system containing classical tautologies and the distribution schema $\Box(\alpha_1 \rightarrow \alpha_2) \rightarrow (\Box\alpha_1 \rightarrow \Box\alpha_2)$ (as single-conclusion 0-premises rules) as well as necessitation $(\alpha/\Box\alpha)$ and modus ponens $(\alpha, \alpha \rightarrow \beta/\beta)$ rules.

We say that a set of rules K *entails* or *derives* a rule Γ/S (written $K \vdash \Gamma/S$) iff Γ/S belongs to the smallest modal rule system $[K]$ containing K. The following algebraic completeness theorem is proved in [16] (but follows also from our results below):

Theorem 2.3 *Let K be a set of multiple-conclusion rules. Then $K \vdash \Gamma/S$ iff every modal algebra validating all rules in K validates also Γ/S.*

2.2 Hyperformulae and hyperproofs

We now design a calculus for recognizing syntactically the relation $K \vdash \Gamma/S$. We shall actually give two equivalent versions of such a calculus, the latter to be seen just as a Hilbert-style analogue of the well-known hypersequent calculi [1].

A *hyperformula* is a finite set of propositional formulae written in the form

$$\alpha_1 \mid \cdots \mid \alpha_n. \tag{1}$$

We use letters $S, S_1, S', \dots$ for hyperformulae; the notation $S \mid S'$ means set union and $S \mid \alpha$ and $\alpha \mid S$ stand for $S \mid \{\alpha\}$ and $\{\alpha\} \mid S$, respectively.

Definition 2.4 Let Γ be a set of propositional modal formulae and let K be a set of multiple-conclusion rules. A *K-hyperproof* (or a *K-derivation* or just a *derivation*) under assumptions Γ is a finite list of hyperformulae $S_1, \ldots, S_n$ such that each S_i in it matches one of the following requirements:

(i) S_i is of the kind $\alpha \mid S$, where $\alpha \in \Gamma$ or α is a tautology or α is an instance of the distribution schema;

(ii) S_i is obtained from hyperformulae preceding it by applying a rule from K or the necessitation rule or the modus ponens rule.

We write $\Gamma \vdash_K S$ to mean that there is a K-derivation ending with S.

An important remark is in order for (ii): when we say that S_i is obtained by applying an inference rule, *we include uniform substitution and weakening in the application of the rule.* Thus, if the rule is

$$\frac{\gamma_1, \ \ldots, \ \gamma_n}{\delta_1 \mid \cdots \mid \delta_m} \ (R)$$

when we say that S_i is obtained from (R), we mean that there are a hyperformula S and a substitution σ such that S_i is of the kind $S \mid \delta_1\sigma \mid \cdots \mid \delta_m\sigma$ and that there are $j_1, \ldots, j_n < i$ such that S_{j_1} is of the kind $S \mid \gamma_1\sigma$, and ... and S_{j_n} is of the kind $S \mid \gamma_n\sigma$ (of course, this applies also to the case $n = 0$, i.e., to zero-premisses rules).

Theorem 2.5 *Let K be a set of multiple-conclusion rules. Then $\Gamma \vdash_K S$ iff the multiple-conclusion rule Γ/S is valid in every modal algebra validating K.*

Corollary 2.6 *Let K be a set of multiple-conclusion rules. For each multiple-conclusion rule Γ/S, we have $K \vdash \Gamma/S$ iff $\Gamma \vdash_K S$.*

Notice that Theorem 2.3 follows from Corollary 2.6 and Theorem 2.5.

2.3 Hypersequent syntax

A *sequent* is a pair of finite sets of formulae written as $\Gamma \Rightarrow \Delta$ and a hypersequent is a finite set of sequents written as

$$\Gamma_1 \Rightarrow \Delta_1 \mid \cdots \mid \Gamma_n \Rightarrow \Delta_n. \tag{2}$$

In this paper, we are investigating proof theoretic facts that only depends on the modal degree of formulae and on the modal degree of formulae occurring within proofs, thus *we view a sequent $\Gamma \Rightarrow \Delta$ as the formula $\bigwedge \Gamma \to \bigvee \Delta$ and a hypersequent* (2) *as the hyperformula*

$$\bigwedge \Gamma_1 \to \bigvee \Delta_1 \mid \cdots \mid \bigwedge \Gamma_n \to \bigvee \Delta_n. \tag{3}$$

Still, there is an important difference between hyperproofs according to Definition 2.4 and hypersequent calculi e.g., in [1]: once translated into our formalism, the difference is in the possibility of using rules having hyperformulae (and not just formulae) as premises. We show here that this difference is immaterial because we can translate these more general rules and proofs into our formalism. The translation is effective, does not increase the modal degree of formulae involved in the proofs, but might be harmful for complexity.

We first introduce the definitions needed to make the comparison. A *hyperrule* is a $n*1$-tuple of hyperformulae, written as $S_1,\ldots,S_k/S$. If H is a set of hyperrules, Γ is a set of hyperformulae and S is a hyperformula, we say that S is *provable from* Γ *in* H, written $\Gamma \Vdash_H S$ iff there exists a finite list of hyperformulae $S_1,\ldots,S_n$ (called a *derivation*) such that each S_i in it matches one of the following requirements:

(i) S_i is a hyperformula containing a member of Γ, or a tautology, or a formula of the form $\Box(\alpha_1 \to \alpha_2) \to (\Box\alpha_1 \to \Box\alpha_2)$;

(ii) S_i is obtained from hyperformulae preceding it by applying modus ponens rule $\alpha,\ \alpha \to \beta/\beta$, necessitation rule $\alpha/\Box\alpha$, or a hyperrule from H.

Again, 'to apply a rule $S_1,\ldots,S_k/S$ to get S_i' means that there is a substitution σ such that S_i is of the kind $\tilde{S} \mid S\sigma$ and that there are $j_1,\ldots,j_n < i$ such that S_{j_1} is of the kind $\tilde{S} \mid S_1\sigma$ and ... and S_{j_k} is of the kind $\tilde{S} \mid S_k\sigma$.[3]

Proposition 2.7 *Let H be a finite set of hyperrules. Then it is possible to produce a set of rules K such that for all $\Gamma, \tilde{S}$ we have $\Gamma \Vdash_H \tilde{S}$ iff $\Gamma \vdash_K \tilde{S}$.*

Proof. (Sketch, see the appendix for full details) Consider a hyperrule $S_1,\ldots,S_k/S$ from H: to obtain K, we simply replace it with the set of rules $\gamma(S_1),\ldots,\gamma(S_n)/S$, varying γ among the functions that pick one formula from each S_i, for each $i = 1,\ldots,n$. □

Next we give a few examples. In order to make a more direct link with the current literature, we will use the hypersequent syntax (Gentzen standard sequent rules for classical logic, as well as external structural rules will be always implicitly assumed below). Since in this paper we are interested only in investigating modal degrees of formulae and proofs, in most cases the meta-variables $\Gamma, \Delta, \ldots$ occurring in the sequent notation below can be replaced by single formulae (hence the rules in Examples 2.8-2.9 can be seen as single rules,[4] not as schemata standing for infinitely many rules).

Example 2.8 An adequate calculus for **S4** comprises the following two rules (taken from [15])

$$\frac{\Box\Gamma \Rightarrow A_1 \mid \cdots \mid \Box\Gamma \Rightarrow A_n}{\Gamma', \Box\Gamma \Rightarrow \Delta, \Box A_1, \cdots, \Box A_n}\ (\Rightarrow\Box)$$

$$\frac{\Box A, A, \Gamma \Rightarrow \Delta}{\Box A, \Gamma \Rightarrow \Delta}\ (T)$$

where, if $\Gamma = \{\phi_1,\ldots,\phi_n\}$, then $\Box\Gamma$ stands for $\{\Box\phi_1,\ldots,\Box\phi_n\}$.

[3] This notion of a derivation avoids the introduction of side components (in the sense of [1]) when specifying rules: in fact, the side component S is introduced directly when applying the rule.

[4] This is not the case for Example 2.10 because $\Box\Delta$ on the right of $\Rightarrow$ cannot be replaced by a single formula.

Example 2.9 Let us now consider the universal class of *prime* **S4.3** algebras: these are the modal algebras validating the above rules and satisfying in addition the clause

$$\forall x\, \forall y\, (\Box x \leq \Box y \text{ or } \Box y \leq \Box x).$$

To axiomatize this class, we can add to the above rules the further rule

$$\frac{\tilde{\Gamma}, \Box\Gamma, \Box\Gamma' \Rightarrow \Delta \qquad \tilde{\Gamma}', \Box\Gamma', \Box\Gamma \Rightarrow \Delta'}{\tilde{\Gamma}, \Box\Gamma \Rightarrow \Delta \mid \tilde{\Gamma}', \Box\Gamma' \Rightarrow \Delta'}\ (Dich)$$

taken from [15]. Rule (*Dich*) is nothing but a variant of the communication rule introduced in [1].

Example 2.10 For prime **S5** algebras, we can add to **S4**-rules the following rule taken from [1]

$$\frac{\Box\Gamma, \Gamma' \Rightarrow \Box\Delta, \Delta'}{\Box\Gamma \Rightarrow \Box\Delta \mid \Gamma' \Rightarrow \Delta'}\ (S5)$$

3 Bounded proofs and step frames

From now on, we shall make exclusive reference to the calculus explained in Definition 2.4. We call a *modal calculus* (or simply a *calculus*) a set of multiple-conclusion rules *where only formulae of modal degree at most one occur.* [5]

When we write $\Gamma \vdash^n_K S$ we mean that there is a K-hyperproof under assumptions Γ (see Definition 2.4) in which only formulae of modal complexity at most n occur. We are mostly interested in the semantic characterization of the following property:

Definition 3.1 We say that a calculus K has the *bounded proof property* (the bpp, for short) iff for every hyperformula S of modal complexity at most n and for every Γ containing only formulae of modal complexity at most n, we have

$$\Gamma \vdash_K S \quad \Rightarrow \quad \Gamma \vdash^n_K S.$$

A remarkable consequence of the bpp is explained in the following:

Proposition 3.2 *If a modal calculus K consisting of finitely many rules enjoys the bpp, then the relation $\Gamma \vdash_K S$ (as well as the derivability of rules in K, see Corollary 2.6) is decidable.*

Proof. Since K has the bpp, it is sufficient to prove that $\Gamma \vdash^n_K S$ is decidable, where n is as in Definition 3.1. We show the decidability of the relation $\Gamma \vdash^n_K S$, by bounding the set of formulae that may occur as components $\psi_1, \dots, \psi_m$ of a hyperformula $\psi_1 \mid \cdots \mid \psi_m$ included in a derivation witnessing $\Gamma \vdash^n_K S$. Notice

[5] This property can be assumed without loss of generality, by applying the transformation suggested in [4] (that transformation does not increase the modal degree of proofs). In [4] another property is assumed on rules (namely that variables occurring in them have occurrences inside a modal operator). This property was assumed there to simplify the definition of evaluation in step algebras, but in the present more general context it can have unclear side effects, so we prefer not to assume it anymore.

first that we can freely suppose that only the variables X occurring in Γ, S appear in such a derivation. This is because extra variables can be uniformly replaced by, say, 0.

Let us say that ϕ is equivalent to ψ (written $\phi \approx \psi$) iff $\phi \leftrightarrow \psi$ is provable in the minimum normal modal system (i.e., by using tautologies, modus ponens and necessitation). Notice that the relation $\approx$ is decidable and that, whenever $\phi \approx \psi$ holds, the replacement rule

$$\frac{\phi}{\psi} \ (\mathit{Repl})$$

is derivable in K. In fact, (*Repl*) can be simulated by a derivation having modal degree at most n in case the modal degrees of ϕ, ψ are at most n. In addition, it is well-known (e.g., from the theory of normal forms [12]) that there are finitely many $\approx$-equivalence classes of the formulae ϕ having at most degree n and built up from the finite set of propositional variables X. We can effectively fix a representative for each of these classes. Let C_n be the set of such representatives.

A canonical substitution σ for a rule $R \in K$ is a substitution σ associating with a variable x a formula $\sigma(x)$ in $C_n \cup C_{n-1}$ in such a way that the formulae occurring as components of hyperformulae from $R\sigma$ are of complexity at most n. Thus, recalling that rules have complexity one, $\sigma(x)$ must be in C_{n-1} if x has an occurrence in R inside a modal operator and $\sigma(x)$ must be in C_n, otherwise. A canonical instance of $R \in K$ is an instance $R\sigma$ of R via a canonical substitution σ. Notice that there are only finitely many canonical instances.

We let Θ be the set of formulae which either occur in $\Gamma \cup S$ or are in C_n or occur in a canonical instance of a rule in K. Again, Θ is finite and has modal complexity at most n.

By induction, we transform a derivation π from K in which formulae of degree at most n occur into a derivation π' in which only members of Θ occur. When building π', we make use also of the replacement rule (*Repl*) introduced above.

The construction of π' is easy for the base case of derivations of length 1. Let us consider the inductive case of a derivation π ending with the application of a rule R from K (we include also the case in which R is modus ponens or necessitation). For simplicity, let R have a single premise, i.e., R is ϕ/T. Suppose that, in π, the rule is used to infer $T\sigma \mid \tilde{S}$ from $\phi\sigma \mid \tilde{S}$. In π' we derive by induction $\psi \mid \tilde{S}'$, where the formulae in $\psi \mid \tilde{S}'$ belongs to Θ and are equivalent to the formulae in $\phi\sigma \mid \tilde{S}$. Let σ' be a canonical substitution such that $\sigma(x)$ is equivalent to $\sigma'(x)$ for every variable x occurring in R. Then $\phi\sigma$ is equivalent to $\phi\sigma'$ and the formulae in $T\sigma$ are equivalent to the formulae in $T\sigma'$, respectively. By the replacement rule, we can infer $\phi\sigma' \mid \tilde{S}'$ in π' (because $\phi'\sigma \approx \phi\sigma \approx \psi$). Then we infer $T\sigma' \mid \tilde{S}'$ via the rule R. The latter is a hyperformula whose components are all in Θ and are equivalent to the components of the hyperformula $T\sigma \mid \tilde{S}$ inferred by π.

Thus, checking whether $\Gamma \vdash^n_K S$ holds is reduced to checking whether there is a derivation using formulas from a finite set Θ. Note also that we can assume that a given hyperformula occurs at most once in a derivation (because occurrences following the first one can be removed). The result follows. □

The following proposition shows that we can limit our consideration to formulae of complexity 1 when checking the bpp.

Proposition 3.3 *A calculus K has the bounded proof property iff for every hyperformula S of modal complexity at most 1 and for every Γ containing only formulae of modal complexity at most 1, we have $\Gamma \vdash_K S \Rightarrow \Gamma \vdash^1_K S$.*

In the following, we shall adopt the equivalent formulation of the bpp suggested by the above proposition. We shall call finite sets Γ of formulae of modal complexity at most 1, *finite presentations.* It is useful to use parameters (see Section 2) to name the variables occurring in a finite presentation Γ: this is because in a K-hyperproof under assumptions Γ, the formulae in Γ are introduced in the derivation as they are (no substitution applies to them), whereas substitutions are applied to rules in K. Thus, we write $\Gamma(\underline{a})$ to emphasize that at most the parameters $\underline{a}$ occur in Γ and $\Gamma(\underline{a}) \vdash_K S(\underline{a})$ to emphasize that the tuple $\underline{a}$ includes all parameters occurring in both Γ, S.

3.1 Conservative one-step algebras and one-step frames

We first recall the definition of one-step modal algebras and one-step frames from [11] and [7], and define conservative one-step modal algebras and one-step frames.

Definition 3.4 A *one-step modal algebra* is a quadruple $\mathcal{A} = (A_0, A_1, i_0, \Diamond_0)$, where A_0, A_1 are Boolean algebras, $i_0 : A_0 \to A_1$ is a Boolean morphism, and $\Diamond_0 : A_0 \to A_1$ is a semilattice morphism (i.e., it preserves only $0, \vee$). The algebras A_0, A_1 are called the *source* and the *target* Boolean algebras of the one-step modal algebra $\mathcal{A}$. We say that $\mathcal{A}$ is *conservative* iff i_0 is injective and the union of the images $i_0(A_0) \cup \Diamond(A_0)$ generates A_1 as a Boolean algebra.

From the dual semantic point of view we have the following:

Definition 3.5 A *one-step frame* is a quadruple $\mathcal{S} = (W_1, W_0, f, R)$, where W_0, W_1 are sets, $f : W_1 \to W_0$ is a map and $R \subseteq W_1 \times W_0$ is a relation between W_1 and W_0. We say that $\mathcal{S}$ is *conservative* iff f is surjective and the following condition is satisfied for all $w_1, w_2 \in W_1$:

$$f(w_1) = f(w_2) \ \&\ R(w_1) = R(w_2) \quad \Rightarrow \quad w_1 = w_2. \tag{4}$$

Similarly to the case of Kripke frames, above we used the notation $R(w_1)$ to mean the set $\{v \in W_0 \mid (w_1, v) \in R\}$ (and similarly for $R(w_2)$). The dual of a finite one-step frame $\mathcal{S} = (W_1, W_0, f, R)$ is the one-step modal algebra $\mathcal{S}^* = (\wp(W_0), \wp(W_1), f^*, \Diamond_R)$, where f^* is the inverse image operation and $\Diamond_R$ is the semilattice morphism associated with R. The latter is defined as follows: for $S \subseteq W_0$, we have $\Diamond_R(S) = \{w \in W_1 \mid R(w) \cap S \neq \emptyset\}$. Conservativity also carries over from one-step frames to one-step modal algebras (see [4] for a proof of the following proposition):

Proposition 3.6 *A finite one-step frame $\mathcal{S}$ is conservative iff its dual one-step modal algebra $\mathcal{S}^*$ is conservative.*

To complete our list of definitions, let us observe that a one-step modal algebra $\mathcal{A} = (A_0, A_1, i_0, \Diamond_0)$ in which we have $A_0 = A_1$ and $i_0 = id$ is nothing but a modal algebra. Similarly, a one-step frame $\mathcal{S} = (W_1, W_0, f, R)$ where we have $W_0 = W_1$ and $f = id$ is just a frame. For clarity, we shall sometimes call modal algebras and frames *standard* or *plain* modal algebras and frames, respectively.

3.2 Inference validation in step algebras

We spell out what it means for a one-step modal algebra and a one-step frame to validate a modal calculus K and a finite presentation Γ (the definition requires little modifications with respect to [4, 5] because we do not restrict to reduced rules).

Let us fix two finite sets of variables $\underline{x} = x_1, \dots, x_n$, $\underline{y} = y_1, \dots, y_m$ and a finite set of parameters $\underline{a} = a_1, \dots, a_m$ (either $\underline{x}, \underline{y}$ or $\underline{a}$ can be empty). An $\underline{a}$-*augmented* one-step modal algebra $\mathcal{A} = (A_0, A_1, i_0, \Diamond_0, \underline{\mathtt{a}})$ is a one-step modal algebra together with displayed elements $\underline{\mathtt{a}} = \mathtt{a}_1, \dots, \mathtt{a}_m \in A_0$ (these elements will interpret parameters).

Given an $\underline{a}$-augmented one-step modal algebra as above, an $\mathcal{A}$-*valuation* is a map associating with each variable $x_i \in \underline{x}$ an element $\mathtt{v}(x_i) \in A_0$ and with each variable $y_j \in \underline{y}$ an element $\mathtt{v}(y_j) \in A_1$. For every formula $\phi(\underline{x})$ of complexity 0, we define $\phi^{\mathtt{v}0} \in A_0$ as follows:

$$x_i^{\mathtt{v}0} = \mathtt{v}(x_i) \quad (\text{for every variable } \ x_i \in \underline{x}); \qquad a_i^{\mathtt{v}0} = \mathtt{a_i} \quad (a_i \in \underline{a});$$
$$(\phi_1 * \phi_2)^{\mathtt{v}0} = \phi_1^{\mathtt{v}0} * \phi_2^{\mathtt{v}0} \quad (* = \wedge, \vee); \qquad (\neg\phi)^{\mathtt{v}0} = \neg(\phi^{\mathtt{v}0}).$$

For every $\psi(\underline{x}, \underline{y})$ of complexity at most 1 in which the $\underline{y}$ do not have occurrences within the scope of a modal operator, $\psi^{\mathtt{v}1} \in A_1$ is defined as follows:

$$x_i^{\mathtt{v}1} = i_0(\mathtt{v}(x_i)) \quad (\text{for every variable } \ x_i \in \underline{x}); \qquad a_i^{\mathtt{v}1} = i_0(\mathtt{a_i}) \quad (a_i \in \underline{a});$$
$$y_j^{\mathtt{v}1} = \mathtt{v}(y_j) \quad (\text{for every variable } \ y_j \in \underline{y}); \qquad (\Diamond\phi(\underline{x}))^{\mathtt{v}1} = \Diamond(\phi^{\mathtt{v}0});$$
$$(\psi_1 * \psi_2)^{\mathtt{v}1} = \psi_1^{\mathtt{v}1} * \psi_2^{\mathtt{v}1} \quad (* = \wedge, \vee); \qquad (\neg\psi)^{\mathtt{v}1} = \neg(\psi^{\mathtt{v}1}).$$

It is immediate to see by induction that for every formula $\phi(\underline{x})$ of complexity 0 (in which the $\underline{y}$ do not occur), we have $\phi^{\mathtt{v}1} = i_0(\phi^{\mathtt{v}0})$.

Definition 3.7 Suppose that the formulae $\delta_1(\underline{x}, \underline{y}), \dots, \delta_k(\underline{x}, \underline{y})$, $\gamma_1(\underline{x}, \underline{y}), \dots, \gamma_l(\underline{x}, \underline{y})$ have modal degree at most one and that the variables $\underline{y}$ are the variables not occurring in them inside the scope of a modal operator. We say that a one-step modal algebra $\mathcal{A}$ *validates the multiple-conclusion rule*

$$\frac{\gamma_1, \ \ \dots, \ \gamma_n}{\delta_1 \mid \cdots \mid \delta_m} \ (R)$$

iff for every $\mathcal{A}$-valuation v, we have that if ($\phi_1^{v1} = 1$ and $\cdots$ and $\phi_m^{v1} = 1$), then ($\gamma_1^{v1} = 1$ or $\cdots$ or $\gamma_n^{v1} = 1$). We say that $\mathcal{A}$ *validates a modal calculus* K (written $\mathcal{A} \models K$) iff $\mathcal{A}$ validates all inferences from K.

Notice that it might well be that K_1 and K_2 are equivalent (in the sense that rules from K_1 are derivable in K_2 and vice versa), but that only one of them is validated by a given $\mathcal{A}$. This phenomenon, however, cannot happen in case $\mathcal{A}$ is standard (i.e., it is a modal algebra).

For formulae $\phi(\underline{a})$ where the variables $\underline{x}, \underline{y}$ do not occur, the valuation v is not relevant. Thus, in such cases, we may write $\phi^{\underline{\mathsf{a}}0}, \phi^{\underline{\mathsf{a}}1}$ instead of ϕ^{v0}, ϕ^{v1}, respectively, to stress the fact that the augmentation $\underline{\mathsf{a}}$ is the essential part of the definition. We write $\mathcal{A} \models \phi(\underline{a})$ for $\phi^{\underline{\mathsf{a}}1} = 1$ and $\mathcal{A} \models S(\underline{a})$ iff there is a $\phi \in S$ such that $\mathcal{A} \models \phi$. We say that $\mathcal{A}$ *validates the presentation* Γ (in symbols, $\mathcal{A} \models \Gamma(\underline{a})$) iff we have that $\mathcal{A} \models \phi(\underline{a})$ for all $\phi(\underline{a}) \in \Gamma$.

The notion of an $\mathcal{S}$-*valuation* for a one-step frame $\mathcal{S}$ is the expected one, namely v is an $\mathcal{S}$-*valuation* iff it is an $\mathcal{S}^*$-*valuation*. In the same way the other notions introduced above (augmentation, ϕ^{v0}, ϕ^{v1}, validation of a presentation, of an inference, of an axiomatic system) can be extended by duality to one-step frames.

Example 3.8 For the systems **S4, S4.3, S5**, it can be shown (by applying the 'step' variant of modal correspondence theory [4, 5]) that a conservative one-step frame $\mathcal{S} = (W_1, W_0, f, R)$

- validates the rules of Example 2.8 iff it is step-transitive and step-reflexive, where the latter means $f \subseteq R$ and the former means $R \subseteq f \circ \geq_R$ (here $\circ$ is relation composition and $w_1 \geq_R w_2$ is defined to be $R(w_1) \supseteq R(w_2)$);
- validates the rules of Example 2.9 iff it is step-transitive, step-reflexive and step-linear, where the latter means $\forall w_1, w_2 \in W_1$ ($R(w_1) \subseteq R(w_2)$ or $R(w_2) \subseteq R(w_1)$);
- validates the rules of Example 2.10 iff we have $R(w) = W_0$ for all $w \in W_1$.

We can specialize our notions to standard modal algebras and frames. An $\underline{a}$-augmentation in a modal algebra $\mathfrak{A} = (A, \Diamond)$ is a tuple $\underline{\mathsf{a}}$ of elements from the support of A, matching the length of $\underline{a}$. For frames $\mathfrak{F} = (W, R)$, we dually take a tuple from $\wp(W)$, i.e., a tuple of subsets. Given such $\underline{a}$-augmentation, we can define $\mathfrak{A} \models \Gamma(\underline{a})$ and $\mathfrak{F} \models \Gamma(\underline{a})$ for a presentation $\Gamma(\underline{a})$, just specializing the above definitions (standard modal algebras and frames are special one-step modal algebras and frames). Notice that $\mathfrak{F} \models \Gamma(\underline{a})$ is *global* validity in terms of the Kripke forcing from the modal logic literature, see e.g., [17, Sec. 3.1].

Proposition 3.9 *Let* $\mathcal{A} = (A, B, i, \Diamond, \underline{\mathsf{a}})$ *be an augmented conservative one-step modal algebra that validates the modal calculus* K *and the presentation* $\Gamma(\underline{a})$. *Then, for every hyperformula* $S(\underline{a})$, *we have that* $\Gamma \vdash_K^1 S$ *implies* $\mathcal{A} \models S$.

4 Semantic characterizations of the bpp and the fmp

In this section we first introduce the morphisms of one-step modal algebras and one-step frames.

Definition 4.1 An *embedding* between one-step modal algebras $\mathcal{A} = (A_0, A_1, i_0, \Diamond_0)$ and $\mathcal{A}' = (A_0', A_1', i_0', \Diamond_0')$ is a pair of injective Boolean morphisms $h : A_0 \to A_0'$, $k : A_1 \to A_1'$ such that

$$k \circ i_0 = i_0' \circ h \qquad \text{and} \qquad k \circ \Diamond_0 = \Diamond_0' \circ h \ . \tag{5}$$

$$\begin{array}{ccc} A_0 & \xrightarrow{\ h\ } & A_0' \\ {\scriptstyle i_0}\big\downarrow & & \big\downarrow{\scriptstyle i_0'} \\ A_1 & \xrightarrow{\ k\ } & A_1' \end{array}$$

$$\begin{array}{ccc} A_0 & \xrightarrow{\ h\ } & A_0' \\ {\scriptstyle \Diamond_0}\big\downarrow & & \big\downarrow{\scriptstyle \Diamond_0'} \\ A_1 & \xrightarrow{\ k\ } & A_1' \end{array}$$

Notice that, when $\mathcal{A}'$ is standard (i.e. $A_1' = A_0' =$ and $i_0' = id$), h must be $k \circ i_0$ and (5) reduces to

$$k \circ \Diamond_0 = \Diamond_0' \circ k \circ i_0. \tag{6}$$

For frames we have the dual definition. In the definition below, we use $\circ$ to denote relational composition: for $R_1 \subseteq X \times Y$ and $R_2 \subseteq Y \times Z$, we have $R_2 \circ R_1 := \{(x, z) \in X \times Z \mid \exists y \in Y \ ((x, y) \in R_1 \ \& \ (y, z) \in R_2)\}$. Notice that the relational composition applies also when one or both of R_1, R_2 is a function.

Definition 4.2 A *p-morphism* between step frames $\mathcal{F}' = (W_1', W_0', f', R')$ and $\mathcal{F} = (W_1, W_0, f, R)$ is a pair of surjective maps $\mu : W_1' \to W_1$, $\nu : W_0' \to W_0$ such that

$$f \circ \mu = \nu \circ f' \qquad \text{and} \qquad R \circ \mu = \nu \circ R'. \tag{7}$$

$$\begin{array}{ccc} W_1' & \xrightarrow{\ \mu\ } & W_1 \\ {\scriptstyle f'}\big\downarrow & & \big\downarrow{\scriptstyle f} \\ W_0' & \xrightarrow{\ \nu\ } & W_0 \end{array}$$

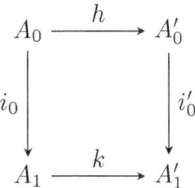

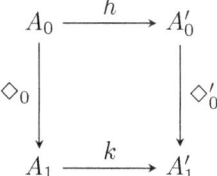

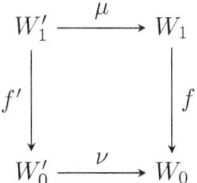

$$\begin{array}{ccc} W_1' & \xrightarrow{\mu} & W_1 \\ {\scriptstyle R'}\downarrow & & \downarrow{\scriptstyle R} \\ W_0' & \xrightarrow{\nu} & W_0 \end{array}$$

Notice that, when $\mathcal{F}'$ is standard (i.e., $W_1' = W_0'$ and $f' = id$), ν must be $f \circ \mu$ and (7) reduces to

$$R \circ \mu = f \circ \mu \circ R'. \tag{8}$$

The following definitions introduce the semantic notions needed for our characterization of the bpp.

Definition 4.3 Let $\mathcal{A}_0 = (A_0, A_1, i_0, \Diamond_0)$ be a one-step modal algebra. A *one-step extension* of $\mathcal{A}_0$ is a one-step modal algebra $\mathcal{A}_1 = (A_1, A_2, i_1, \Diamond_1)$ satisfying $i_1 \circ \Diamond_0 = \Diamond_1 \circ i_0$. Dually, a *one-step extension* of the one-step frame $\mathcal{S}_0 = (W_1, W_0, f_0, R_0)$ is a one-step frame $\mathcal{S}_1 = (W_2, W_1, f_1, R_1)$ satisfying $R_0 \circ f_1 = f_0 \circ R_1$.

Definition 4.4 A class of one-step modal algebras has the *extension property* iff every conservative one-step modal algebra $\mathcal{A}_0 = (A_0, A_1, i_0, \Diamond_0)$ in the class has an extension $\mathcal{A}_1 = (A_1, A_2, i_1, \Diamond_1)$ such that i_1 is injective and $\mathcal{A}_1$ is also in the class. A class of one-step modal frames has the *extension property* iff every conservative one-step frame $\mathcal{S}_0 = (W_1, W_0, f_0, R_0)$ in the class has an extension $\mathcal{S}_1 = (W_2, W_1, f_1, R_1)$ such that f_1 is surjective and $\mathcal{S}_1$ is also in the class.

Theorem 4.5 *A modal calculus K has the bpp iff the class of finite one-step modal algebras (equivalently, the class of finite one-step frames) validating K has the extension property.*

The characterization of the bpp from Theorem 4.5 may not be easy to handle, because in practical cases one would like to avoid managing one-step extensions and would prefer to work with standard frames instead. This is possible, if we combine the bpp with the finite model property.

Definition 4.6 A modal calculus K has the (global) *finite model property*, the *fmp* for short, if for every tuple $\underline{a}$ of parameters, for every finite set of formulae $\Gamma(\underline{a})$ and for every hyperformula $S(\underline{a})$ we have $\Gamma \not\vdash_K S$ iff there exists a finite $\underline{a}$-augmented modal algebra $\mathfrak{A}$ such that $\mathfrak{A} \models K$, $\mathfrak{A} \models \Gamma(\underline{a})$ and $\mathfrak{A} \not\models S(\underline{a})$ (equivalently, iff there exists a finite $\underline{a}$-augmented Kripke frame $\mathfrak{F}$ such that $\mathfrak{F} \models K$, $\mathfrak{F} \models \Gamma(\underline{a})$ and $\mathfrak{F} \not\models S(\underline{a})$).

We are ready for a characterization result:

Theorem 4.7 *A modal calculus K has both the bpp and the fmp iff every finite conservative one-step frame validating K is a p-morphic image of a finite frame validating K (equivalently, iff every finite conservative one-step modal algebra validating K has an embedding into a finite modal algebra validating K).*

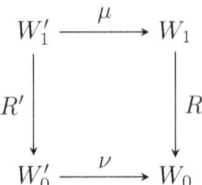

Example 4.8 Theorem 4.7 applies to all Examples 2.8-2.10. The construction is the same in all cases and it is rather straightforward: given a finite conservative step frame $\mathcal{S} = (W_1, W_0, f, R)$ validating the rules of the calculus, we can define $\mathfrak{F}' = (W', R')$ and μ so that condition (8) is satisfied as follows:

$$W' := W, \qquad \mu := id, \qquad w_1 R' w_2 :\Leftrightarrow R(w_1) \supseteq R(w_2).$$

5 Modal stable rules

Canonical formulae for transitive modal logics and intuitionistic logic were introduced by Zakharyaschev (see [9] for an overview) who proved that all transitive modal logics and all intermediate logics are axiomatizable by canonical formulae. Jeřábek [16] defined *canonical rules*, which are multiple-conclusion rules generalizing canonical formulas. Jeřábek used these rules for an alternative proof of decidability of admissible rules for intuitionistic logic and transitive modal logics **K4**, **S4**, **S4.3**, etc. However, there are non-transitive modal logics not axiomatizable by canonical formulae and rules. [3] defines *stable canonical rules*, which differ from Zakharyaschev's canonical formulae and Jeřábek's canonical rules and proves that every modal logic (including non-transitive ones) is axiomatizable by these rules. In this section we will concentrate on logics axiomatizable by a special subclass of stable canonical rules.

Subframe logics are the logics whose frames are closed under taking subframes. Transitive subframe logics are axiomatizable by a special subclass of canonical formulae called *subframe formulae*, see, e.g., [9]. A similar restriction to stable canonical rules gives a class of *stable logics*. But stable logics are not necessarily transitive. Logics in this class are exactly the logics that are closed under relation-preserving (following [3] we will call such maps *stable* [6]) onto maps. Transitive subframe logics and stable logics enjoy the fmp. Transitive subframe logics enjoy the fmp because they admit selective filtration, and stable logics enjoy the fmp because they admit the standard filtration (see [3] for the details).

In this section we show that all stable logics admit an axiomatization that has the bounded proof property. As we will see below, stable canonical rules will not produce an axiomatization that has the bpp. However, we will modify these rules so that the obtained rules do possess the bpp. This provides a systematic method of producing infinitely many proof calculi that are good (enjoying the bpp) from the proof-theoretic point of view. We remark that Lahav [19] also considers a class of modal logics whose Kripke frames satisfy special first-order conditions. He introduces hypersequent calculi for these logics and proves that these calculi admit cut elimination. It is easy to see that the non-transitive logics studied in [19] are stable logics – their frame classes are closed under stable onto maps. Thus, the class of logics we investigate in this section extends the class of logics studied in [19] in the non-transitive

[6] In [13] these maps are called *continuos*.

case. [7] Note, however, that [19] studies cut elimination, whereas we work with the bpp only. Now, if cut elimination gives the subformula property as a by-product, the bpp follows trivially. The converse is not true: we might have the bpp without the subformula property. [8] However, it should be noticed that the bpp is a strong evidence about the proof-theoretic robustness of a system and supplies a loose notion of analiticity which is sufficient for decidability and which can hold for a wide class of calculi, including cases where the design of cut-eliminating systems looks very problematic.

We start by recalling the definition of modal stable rules. Let $\mathfrak{F} = (F, R_F)$ be a finite frame. For every $a \in F$ we introduce a new propositional variable x_a. The *modal stable rule* of $\mathfrak{F}$ is

$$\frac{\bigvee_{i=1}^{n} x_{a_i}, \quad \bigwedge_{i \neq j} \neg(x_{a_i} \wedge x_{a_j}), \quad \bigwedge_{i=1}^{n}(x_{a_i} \to \Box \bigvee_{b \in R_F(a_i)} x_b)}{\neg x_{a_1} \mid \cdots \mid \neg x_{a_n}} \quad (r_{\mathfrak{F}})$$

where we suppose that $F = \{a_1, \ldots, a_n\}$.

A *stable embedding* of a modal algebra $\mathfrak{A} = (A, \Diamond)$ into a modal algebra $\mathfrak{B} = (B, \Diamond)$ is an injective Boolean morphism $\mu : A \to B$ such that we have $\Diamond\mu(x) \leq \mu(\Diamond x)$ for all $x \in A$. For a frame $\mathfrak{F}$ we denote by $\mathfrak{F}^*$ its dual modal algebra and for an algebra $\mathfrak{A}$ we denote by $\mathfrak{A}_*$ the descriptive frame dual to $\mathfrak{A}$. Recall that a map $f : W \to W'$ between standard frames (W, R) and (W', R') is called *stable* if for each $x, y \in W$ we have xRy implies $f(x)R'f(y)$.

The following proposition is proved in [3].

Proposition 5.1 *Let $\mathfrak{A} = (A, \Diamond)$ be a modal algebra. Then*

(i) *$\mathfrak{A}$ does not validate $(r_{\mathfrak{F}})$ iff there is a stable embedding of $\mathfrak{F}^*$ into $\mathfrak{A}$.*

(ii) *$\mathfrak{A}$ does not validate $(r_{\mathfrak{F}})$ iff there is a surjective stable map from $\mathfrak{A}_*$ onto $\mathfrak{F}$.*

Our aim is to show that all modal calculi axiomatized by rules of the kind $(r_{\mathfrak{F}})$ have the bounded proof property. Rules $(r_{\mathfrak{F}})$, however, are not good for the bpp, see the counterexample below. We replace rules $(r_{\mathfrak{F}})$ by modified versions.

For each $a \in F$ we just add an extra propositional variable r_a and define the new rule $(r_{\mathfrak{F}}^+)$ by [9]

$$\frac{\bigvee_{i=1}^{n} x_{a_i}, \quad \bigwedge_{i \neq j} \neg(x_{a_i} \wedge x_{a_j}), \quad \bigwedge_{i=1}^{n}(x_{a_i} \to \Box r_{a_i}), \quad \bigwedge_{i=1}^{n}(r_{a_i} \to \bigvee_{b \in R_F(a_i)} x_b)}{\neg x_{a_1} \mid \cdots \mid \neg x_{a_n}}$$

[7] The transitive logic **K4** is not stable. The investigation of the bounded proof property of stable logics over **K4** is a topic for future research.

[8] For example, we show that all stable logics have the bpp. This class contains a continuum of logics [2,3]. Whether all these logics admit natural calculi with cut elimination is an open question.

[9] For uniformity, we prefer all the r_a to have at least one occurrence located inside a modal operator in the rule $(r_{\mathfrak{F}}^+)$. In order to obtain this, one might add premisses such as $\Box(r_a \vee \neg r_a)$.

Lemma 5.2 *Rules* $(r_{\mathfrak{F}}^{+})$ *and* $(r_{\mathfrak{F}})$ *are inter-derivable.*

Proof. On the one hand, $(r_{\mathfrak{F}})$ can be obtained from $(r_{\mathfrak{F}}^{+})$ by applying the substitution $r_{a_i} \mapsto \bigvee_{b \in R_F(a_i)} x_b$. On the other hand, we apply necessitation and distribution to the premise $\bigwedge_{i=1}^{n}(r_{a_i} \to \bigvee_{b \in R(a_i)} x_b)$ and then transitivity of implication to obtain $\bigwedge_{i=1}^{n}(x_{a_i} \to \Box \bigvee_{b \in R_F(a_i)} x_b)$. □

Notice that the above fragment of a derivation, when plugged into a hyperproof, may increase the modal degree (if the substitution used to apply the rule $(r_{\mathfrak{F}})$ replaces the x_b with formulae of, say, modal degree 1, we obtain formulae of modal degree 2 when we use $(r_{\mathfrak{F}})$ to simulate $(r_{\mathfrak{F}}^{+})$). This is why $(r_{\mathfrak{F}}^{+})$ is preferable to $(r_{\mathfrak{F}})$ from the point of view of the modal complexity analysis of proofs.

Theorem 5.3 *A modal calculus comprising only rules of the kind* $(r_{\mathfrak{F}}^{+})$ *enjoys the bpp and fmp.*

Proof. We use Theorem 4.7. Let $\mathcal{S} = (W_1, W_0, f, R)$ be a finite conservative one-step frame validating $(r_{\mathfrak{F}}^{+})$. Consider the standard frame $(W_1, \tilde{R})$ where $\tilde{R}$ is defined by

$$w\tilde{R}w' \text{ iff } wRf(w') \tag{9}$$

(i.e. we have $\tilde{R} = f^o \circ R$, where f^o is the converse of f, seen as a relation). This is a finite Kripke frame having $\mathcal{S}$ as a p-morphic image. In fact, (8) is satisfied by taking $\mu := id$ because $f \circ \tilde{R} = f \circ f^o \circ R = R$, (we used that $f \circ f^o = id$, which holds by the surjectivity of f).

We now show that $(W_1, \tilde{R})$ validates $(r_{\mathfrak{F}})$ (recall that $(r_{\mathfrak{F}}^{+})$ is equivalent to it in standard frames because the two rules are inter-derivable): to this aim, we prove that if there is a surjective R-preserving map μ from $(W_1, \tilde{R})$ onto $\mathfrak{F} = (F, R_F)$, then $\mathcal{S}$ does not validate $(r_{\mathfrak{F}}^{+})$, contrary to the hypothesis. Suppose there is such a μ. Define now a valuation $\mathtt{v}$ by taking $\mathtt{v}(x_a) = \{w \mid \mu(w) = a\} \subseteq W_1$ and

$$\mathtt{v}(r_a) = \{v \in W_0 \mid \forall w\ (f(w) = v \Rightarrow aR_F\mu(w))\}.$$

The definition is well defined because the variables having at least an occurrence inside a modal operator are precisely the r_a's, so these variables are evaluated as subsets of W_0 and the other ones as subsets of W_1. Thus $\mathtt{v}$ evaluates to 1 the formulae $\bigvee_{i=1}^{n} x_{a_i}$ and $\bigwedge_{i \neq j} \neg(x_{a_i} \wedge x_{a_j})$, whereas $\neg x_{a_1}, \dots, \neg x_{a_n}$ are not evaluated to 1 (because μ is surjective). It remains to check that for every $a \in F$, we have (i) $x_a^{\mathtt{v}1} \subseteq \Box r_a^{\mathtt{v}1}$ and (ii) $r_a^{\mathtt{v}1} \subseteq (\bigvee_{b \in R_F(a)} x_b)^{\mathtt{v}1}$. Now (i) holds by (9) and because μ is stable: if $w \in x_a^{\mathtt{v}1}$ and wRv then $v \in r_a^{\mathtt{v}1}$ because if $f(w') = v$ then $w\tilde{R}w'$ and consequently $a = \mu(w)R_F\mu(w')$. To prove (ii), pick $w \in f^*(r_a)$; we have in particular $aR_F\mu(w)$, thus $w \in (\bigvee_{b \in R_F(a)} x_b)^{\mathtt{v}1}$. □

From Lemma 5.2, we immediately obtain the following result from [3]:

Corollary 5.4 *A modal calculus comprising only rules of the kind* $(r_{\mathfrak{F}})$ *enjoys the finite model property.*

The following counter-example shows that we really need to replace $(r_{\mathfrak{F}})$ by $(r^+_{\mathfrak{F}})$ to obtain the bpp.

Example 5.5 Consider the two element reflexive chain

$$\mathfrak{F} := \boxed{b \circlearrowleft \longrightarrow \circlearrowleft a}$$

The rule $(r_{\mathfrak{F}})$ simplifies to

$$\frac{x_a \to \Box x_a}{x_a \mid \neg x_a}$$

This rule is validated in a step frame $\mathcal{S} = (W_1, W_0, f, R)$ iff for every proper subset $a \subseteq W_0$ (i.e., for every subset different from $\emptyset, W_0$) there is $w \in W_1$ such that $f(w) \in a$ and $R(w) \not\subseteq a$. In a standard frame (W, S) this means that every pair of elements of W are connected via an S-path (to see this, consider as a the set of points which are reachable in $n \geq 0$ steps by any given point and show that such an a must be total). It is not difficult to check that putting $W_1 := \{w_1, w_2\}, W_0 := \{v\}, f(w_1) := f(w_2) := v, R(w_1) := \{v\}, R(w_2) := \emptyset$, we obtain a finite conservative one-step frame that validates $(r_{\mathfrak{F}})$ but cannot be a p-morphic image of a standard frame validating it (because in the latter there cannot be terminal points and any pre-image of w_2 along a p-morphism must be such by (8)). Since the fmp holds for the modal calculus axiomatized by the rule $(r_{\mathfrak{F}})$ according to Corollary 5.4, it is clear that it is the bpp that fails for it (failure of the bpp can also be directly checked by using Theorem 4.5 instead of Theorem 4.7 and Corollary 5.4).

References

[1] Avron, A., *The method of hypersequents in the proof theory of propositional non-classical logics*, in: Hodges and et al., editors, *Logic: from foundations to applications*, The Clarendon Press, 1996 pp. 1–32.

[2] Bezhanishvili, G. and N. Bezhanishvili, *Locally finite reducts of Heyting algebras and canonical formulas*, Notre Dame Journal of Formal Logic (2014), to appear.

[3] Bezhanishvili, G., N. Bezhanishvili and R. Iemhoff, *Stable canonical rules* (2014), ILLC Prepublication Series Report PP-2014-08.

[4] Bezhanishvili, N. and S. Ghilardi, *The bounded proof property via step algebras and step frames*, Technical Report 306, Department of Philosophy, Utrecht University (2013), to appear in Annals of Pure and Applied Logic.

[5] Bezhanishvili, N. and S. Ghilardi, *Bounded proofs and step frames*, in: *Proc. Tableaux 2013*, number 8123 in Lecture Notes in Artificial Intelligence, 2013, pp. 44–58.

[6] Bezhanishvili, N. and S. Ghilardi, *Multiple-conclusion rules, hypersequents syntax and step frames* (2014), ILLC Prepublication Series Report PP-2014-05.

[7] Bezhanishvili, N., S. Ghilardi and M. Jibladze, *Free modal algebras revisited: the step-by-step method*, in: *Leo Esakia on Duality in Modal and Intuitionistic Logics*, Trends in Logic, Springer, 2014 .

[8] Blackburn, P., M. de Rijke and Y. Venema, "Modal Logic," Cambridge Uni. Press, 2001.

[9] Chagrov, A. and M. Zakharyaschev, "Modal Logic," The Clarendon Press, 1997.

[10] Corsi, G., *A cut-free calculus for Dummett's* LC *quantified*, Z. Math. Logik Grundlag. Math. **35** (1989), pp. 289–301.

[11] Coumans, D. and S. van Gool, *On generalizing free algebras for a functor*, Journal of Logic and Computation **23** (2013), pp. 645–672.
[12] Ghilardi, S., *An algebraic theory of normal forms*, Annals of Pure and Applied Logic **71** (1995), pp. 189–245.
[13] Ghilardi, S., *Continuity, freeness, and filtrations*, J. Appl. Non-Classical Logics **20** (2010), pp. 193–217.
[14] Goré, R., *Cut-free sequent and tableaux systems for propositional diodorean modal logics*, Technical report, Dept. of Comp. Sci., Univ. of Manchester (1993).
[15] Indrzejczak, A., *Cut-Free hypersequent calculus for S4.3*, Bulletin of the Section of Logic (2012), pp. 89–104.
[16] Jeřábek, E., *Canonical rules*, J. Symbolic Logic **74** (2009), pp. 1171–1205.
[17] Kracht, M., "Tools and techniques in modal logic," Studies in Logic and the Foundations of Mathematics **142**, North-Holland Publishing Co., 1999, xiv+559 pp.
[18] Kracht, M., *Modal consequence relations*, in: P. Blackburn, J. van Benthem and F. Wolter, editors, *Handbook of Modal Logic*, Elsevier, Amsterdam, 2007 pp. 92–130.
[19] Lahav, O., *From frame properties to hypersequent rules in modal logics*, in: *Proc. of (LICS 2013)*, IEEE Publ., 2013.
[20] Sambin, G. and V. Vaccaro, *Topology and duality in modal logic*, Annals of Pure and Applied Logic **37** (1988), pp. 249–296.

Appendix

For the proof of the algebraic completeness Theorem 2.5, we need a couple of lemmas:

Lemma .6 *Weakening is admissible: we have* $\Gamma \vdash_K S \Rightarrow \Gamma \vdash_K S \mid S'$, *for every* S'.

Proof. Trivial by induction on the length of derivation. □

Lemma .7 *Let* Γ *be a set of formulae,* α *a formula,* S *a hyperformula and* K *a set of multiple-conclusion rules. If* $\Gamma \cup \{\alpha\} \vdash_K S$ *and* $\Gamma \vdash_K \alpha \mid S$, *then* $\Gamma \vdash_K S$.

Proof. Assume $\Gamma \vdash_K \alpha \mid S$. Using weakening, by induction on proof length, it is easy to see that $\Gamma \cup \{\alpha\} \vdash_K \tilde{S}$ implies $\Gamma \vdash_K S \mid \tilde{S}$ for every $\tilde{S}$. The claim now follows because $S \mid S$ is equal to S (hyperformulae are defined as sets of formulae). □

Theorem 2.5 *Let* K *be a set of multiple-conclusion rules. Then* $\Gamma \vdash_K S$ *iff the multiple-conclusion rule* Γ/S *is valid in every modal algebra validating* K.

Proof. One direction is trivial. For the other direction, let us suppose that $\Gamma \vdash_K S$ does not hold. By Zorn's lemma, pick $\tilde{\Gamma}$ to be a maximal set of formulae containing Γ such that $\tilde{\Gamma} \not\vdash_K S$. We claim that for every hyperformula $\alpha_1 \mid \cdots \mid \alpha_n$

$$\tilde{\Gamma} \vdash_K \alpha_1 \mid \cdots \mid \alpha_n \mid S \qquad \Rightarrow \qquad \exists i \;\; \tilde{\Gamma} \vdash_K \alpha_i. \tag{.1}$$

In fact, if this does not hold, by the maximality of $\tilde{\Gamma}$, we have both that $\tilde{\Gamma} \vdash_K \alpha_1 \mid \cdots \mid \alpha_n \mid S$ and that $\tilde{\Gamma} \cup \{\alpha_1\} \vdash_K S$. By the above lemma, this implies $\tilde{\Gamma} \vdash_K \alpha_2 \mid \cdots \mid \alpha_n \mid S$. Repeating the argument n-times, we obtain $\tilde{\Gamma} \vdash_K S$, contradiction.

Now notice that Lemma .7 and the maximality of $\tilde{\Gamma}$ imply that if $\tilde{\Gamma} \vdash_K \alpha$, then $\alpha \in \tilde{\Gamma}$ and $\Box\alpha \in \tilde{\Gamma}$ (the latter is because necessitation rule is mentioned in condition (ii) of Definition (2.4)). In addition, $\tilde{\Gamma}$ contains Γ and is disjoint from S, by condition (i) of Definition (2.4). Thus, if we put

$$\alpha_1 \approx \alpha_2 \quad \Leftrightarrow \quad \alpha_1 \leftrightarrow \alpha_2 \in \tilde{\Gamma}$$

we can introduce on the set of equivalence classes a modal algebra structure $\mathfrak{A} = (A, \Diamond)$. Since Γ is included in $\tilde{\Gamma}$ and is disjoint from S, $\mathfrak{A}$ does not validate Γ/S. By the claim (.1) and condition (ii) of Definition (2.4), it is evident that $\mathfrak{A}$ validates all rules from K. □

Corollary 2.6 *Let* K *be a set of multiple-conclusion rules. For each multiple-conclusion rule* Γ/Δ, *we have* $K \vdash \Gamma/\Delta$ *iff* $\Gamma \vdash_K \Delta$.

Proof. If is sufficient to observe that (I) if $\Gamma \vdash_K \Delta$, then Γ/Δ belongs to every modal rule system $\mathcal{K}$ containing K and that (II) $\{\Gamma/\Delta \mid \Gamma \vdash_K \Delta\}$ is a modal rule system extending K.

Claim (II) is immediate from Lemmas .6, .7.

Claim (I) is by induction on the length of the K-hyperproof witnessing $\Gamma \vdash_K \Delta$: for instance, if the K-hyperproof ends with an application of the necessitation rule according to Definition 2.4(ii), then from $\Gamma/\alpha, S \in \mathcal{K}$ (this holds by induction hypothesis) and from the fact that the necessitation rule belongs to every modal rule system, from conditions (iii) and (ii) of Definition 2.1, we obtain $\Gamma/\Box\alpha, S \in \mathcal{K}$. [10] □

We now fill the missing details for the proof of Proposition 2.7:

Proposition 2.7 *Let H be a finite set of hyperrules. Then it is possible to produce a set of rules K such that for all $\Gamma, \tilde{S}$ we have $\Gamma \Vdash_H \tilde{S}$ iff $\Gamma \vdash_K \tilde{S}$.*

Proof. Consider a hyperrule $S_1, \dots, S_k/S$ from H: to obtain K, we simply replace it with the set of rules $\gamma(S_1), \dots, \gamma(S_n)/S$, varying γ among the functions that pick one formula from each S_i, for each $i = 1, \dots, n$.

The right-to-left claim of the proposition is immediate by weakening. To show the left-to-right direction, we use the argument below. Suppose H' is obtained from H by replacing the hyperrule $S_1, \dots, S_n/S$ with the pair of rules

$$S_1', S_2, \dots, S_n/S, \qquad S_1'', S_2, \dots, S_n/S \tag{.2}$$

where we suppose that S_1', S_1'' are both not empty and such that $S_1 = S_1' \cup S_1''$. We claim that we have $\Gamma \Vdash_H S$ iff $\Gamma \Vdash_{H'} S$ (clearly, the statement of the proposition follows from an iterated application of this claim). Again that $\Gamma \Vdash_H \tilde{S} \Leftarrow \Gamma \Vdash_{H'} \tilde{S}$ holds is trivial by weakening. Now suppose that we have $\Gamma \Vdash_H \tilde{S}$. In the derivation witnessing this, there will possibly be lines labelled by $S_1\sigma \mid T, \dots, S_n\sigma \mid T$ justifying a line labelled $S\sigma \mid T$ via the use of the hyperrule $S_1, \dots, S_n/S$. The derivation can be corrected so to use the rules (.2) instead (iterated corrections will eliminate any use of the rule $S_1, \dots, S_n/S$). We first produce (by weakening) derivations of $S_2\sigma \mid S_1''\sigma \mid T$ and $\cdots$ and $S_n\sigma \mid S_1''\sigma \mid T$. These hyperformulae, combined with $S_1'\sigma \mid S_1''\sigma \mid T$ yield a derivation of $S_1''\sigma \mid S\sigma \mid T$ by applying the first hyperrule from (.2). By weakening again, we produce now derivations of $S_2\sigma \mid S\sigma \mid T$ and $\cdots$ and $S_n\sigma \mid S\sigma \mid T$. These hyperformulae, combined with $S_1''\sigma \mid S\sigma \mid T$ yields a derivation of $S\sigma \mid S\sigma \mid T$ by applying the second hyperrule from (.2) and we are done because $S\sigma \mid S\sigma \mid T$ is equal to $S\sigma \mid T$ (hyperformulae are sets, not multisets). □

[10] Notice that we added S to α because, according to the remark following Definition 2.4, when we apply the necessitation rule $\alpha/\Box\alpha$, then we deduce $\Box\alpha \mid S$ from a proof line containing the hyperformula $\alpha \mid S$.

PDL Inside the μ-calculus: A Syntactic and an Automata-theoretic Characterization

Facundo Carreiro [1] Yde Venema [2]

Institute for Logic, Language and Computation
Universiteit van Amsterdam, The Netherlands

Abstract

It is well known that Propositional Dynamic Logic (PDL) can be seen as a fragment of the modal μ-calculus. In this paper we provide an exact syntactic characterization of the fragments of the μ-calculus that correspond to PDL and to test-free PDL. In addition we give automata-theoretic characterizations for PDL, with and without tests, which shed light on the relation between these logics and the modal μ-calculus and provide a new framework for the development of the theory of PDL.

Keywords: propositional dynamic logic, automata theory, modal μ-calculus.

1 Introduction

The language now called Propositional Dynamic Logic was first investigated by Fisher and Ladner [3] as a logic to reason about computer program execution. PDL extends the basic modal logic with an infinite collection of diamonds $\langle\pi\rangle$ where the intended intuitive interpretation of $\langle\pi\rangle\varphi$ is that "some terminating execution of the program π from the current state leads to a state satisfying φ".

The inductive structure of programs is made explicit in PDL's syntax, as complex programs are built out of atomic programs using four program constructors. Formally, the formulas of full PDL are given by a mutual induction:

$$\varphi ::= p \mid \neg\varphi \mid \varphi \vee \varphi \mid \langle\pi\rangle\varphi$$
$$\pi ::= d \mid \pi;\pi \mid \pi \oplus \pi \mid \pi^* \mid \varphi?$$

where p is a proposition letter and d is an atomic action (or atomic program). Test-free PDL is a variant of full PDL excluding use of the test operator (?).

One of the most important and characteristic features of PDL is that the program construction π^* (corresponding to iteration) endows PDL with second-order capabilities while still keeping it computationally well-behaved. For an extensive treatment of PDL we refer the reader to [7].

[1] E-mail: `fcarreiro@dc.uba.ar`

[2] E-mail: `y.venema@uva.nl`

Modal μ-calculus. The modal μ-calculus (μML) was introduced in its present form by Dexter Kozen [10]. It is highly expressive, corresponding to the bisimulation-invariant fragment of monadic second-order logic [9]. The modal μ-calculus is a very expressive language, subsuming a vast amount of dynamic and temporal logics such as PDL, CTL* and Game Logic. Yet μML is computationally well-behaved, and enjoys some excellent meta-logical properties, such as *uniform interpolation* [2].

The language of the modal μ-calculus on a set of propositions P and atomic actions D is given by the following grammar:

$$\varphi \ ::= q \mid \neg\varphi \mid \varphi \vee \varphi \mid \langle d \rangle \varphi \mid \mu p.\varphi$$

where $p, q \in \mathsf{P}$, $d \in \mathsf{D}$ and p is positive in φ (i.e., all occurrences of p are under an even number of negations).

The semantics of this language is completely standard. Observe that, given a Kripke model $\mathbb{S}$ and a formula φ with p free, the extension $[\![\varphi]\!]^{\mathbb{S}}$ of φ in $\mathbb{S}$ depends on the set of points where p holds. This dependence can be formalized as a map $\varphi_p^{\mathbb{S}} : \wp(S) \to \wp(S)$. The semantics of the least fixpoint operator is then given by interpreting $[\![\mu p.\varphi]\!]^{\mathbb{S}}$ as the least fixpoint of $\varphi_p^{\mathbb{S}}$.

Relative expressive power. It is well known that PDL can be translated to μML. However, to the best of our knowledge the exact fragment of μML that corresponds to PDL has not been characterized. As we will see in this article, the key notion leading to such a characterization is that of *complete additivity*. A formula φ is said to be completely p-additive if for any family of subsets $\{P_i\}_i \in I$ with $P_i \subseteq S$ it satisfies

$$\varphi_p^{\mathbb{S}}(\bigcup_i P_i) = \bigcup_i \varphi_p^{\mathbb{S}}(P_i). \tag{1}$$

An equivalent characterization of complete additivity is the requirement that $\varphi_p^{\mathbb{S}}(P) = \bigcup_{t \in P} \varphi_p^{\mathbb{S}}(\{t\})$, which implies that if φ is completely p-additive and true at a point s in a Kripke model, it will remain so if we restrict the valuation of p to a singleton. Complete additivity has been studied (under the name 'continuity') by van Benthem [15], in the context of operations on relations that are *safe for* (that is, preserve) bisimulations. Hollenberg [8] linked the notion to the syntax of PDL, showing that any completely p-additive formula in μML can be equivalently rewritten as $\langle \pi \rangle p$, where π belongs to the set of so-called μ-programs, which extend PDL-programs by admitting tests of arbitrary formulas in μML.

Fontaine and Venema [4,5] gave a different syntactic characterization of complete additivity. One of our two main contributions builds on this work, by showing that PDL exactly corresponds to the fragment μ_{ca}ML of μML where the fixpoint operator $\mu x.\varphi$ is restricted to formulas φ which belong to this syntactic fragment characterizing completely p-additivity. Similarly, test-free PDL corresponds to a smaller fragment μ_{rca}ML of μML.

Automata characterizations for μ-calculus and PDL. It is difficult to overstress the importance of automata-theoretic techniques in the theory of the μ-calculus: many of the fundamental results on μML are proved by means of the so-called μ-automata introduced by Janin and Walukiewicz [9].

In the case of PDL, the first automata-based result was by Streett [13,14] who translated PDL (with additional looping and converse operators) to deterministic two-way automata on infinite trees, and obtained decidability for the satisfiability problem. Vardi and Wolper [17] proved that PDL can be translated to Büchi (tree) automata, thus obtaining sharper complexity results. Muller et al. [12] showed that many dynamic and temporal logics can be uniformly represented using so-called weak alternating automata.

While the mentioned papers use translations of PDL-formulas to some kind of automata, none of them provides a precise automata-theoretic characterization of PDL. That is, the classes of automata under consideration contain automata that do not correspond to an equivalent PDL formula. The second main contribution of our article is to define two classes of alternating parity automata which exactly correspond, respectively, to PDL and its test-free variant PDL^{tf}. These two types of automata are easily seen to be subclasses of the *alternating* automata corresponding to μML [6].

Due to space constraints, most of our proofs have been moved to the Appendix.

2 Preliminaries

We assume that the reader is familiar with the syntax and semantics of PDL and the modal μ-calculus, and with parity games. We fix some notation and terminology, and discuss parity automata.

2.1 Structures and Languages

Throughout the article we fix a set of proposition symbols P and a set of atomic actions D. The structures that we are considering are multi-modal Kripke models, i.e., tuples $\mathbb{S} = \langle S, R_{d \in \mathsf{D}}, V\rangle$ where $R_d \subseteq S \times S$ and $V : \mathsf{P} \to \wp S$.

Definition 2.1 Propositional dynamic logic (in negation normal form) is given by mutual induction on formulas and programs:

$$\varphi ::= p \mid \neg p \mid \varphi \wedge \varphi \mid \varphi \vee \varphi \mid \langle\pi\rangle\varphi \mid [\pi]\varphi$$
$$\pi ::= d \mid \pi;\pi \mid \pi \oplus \pi \mid \pi^* \mid \varphi?$$

where $p \in \mathsf{P}$, $d \in \mathsf{D}$. We denote this language by PDL(P,D) and drop P,D when clear from context. As an abuse of notation we write $\pi \in$ PDL(P,D) to mean that π is a program of PDL(P,D). Test-free propositional dynamic logic (denoted by PDL^{tf}) is PDL without the test operator (?). The set of proposition letters occurring in a formula φ (program π) is denoted by Var(φ) (respectively, Var(π)).

The accessibility relation induced by a program π in a model $\mathbb{S}$ is denoted as $R^{\mathbb{S}}_{\pi}$, and the truth relation is denoted by $\Vdash$.

2.2 Parity Games

A *parity game* $\mathcal{G}$ consists of a partitioned board $G = G_\exists \uplus G_\forall$, a relation $E \subseteq G \times G$ indicating the available moves $E[u]$ from a position $u \in G$, and a parity map, i.e., a map $\Omega : G \to \mathbb{N}$ of finite range. A match of such a parity game consists of the two players, $\exists$ and $\forall$, moving a token from one position to another over the board; matches can be represented as paths through the graph (G, E). A player who gets stuck during the match by having arrived at a position with no admissible moves, immediately looses. If the match goes on forever then the parity map Ω is used to call a winner. The winner is $\exists$ if the maximum parity which occurs infinitely often in the match is even, otherwise $\forall$ wins. An *initialized parity game* $\mathcal{G}@u$ is a pair $(\mathcal{G}, u)$ where $\mathcal{G}$ is a parity game and $u \in G$ is the *initial position* of the game.

A *strategy* for player $\Pi \in \{\exists, \forall\}$ is, intuitively, a specification of choices to be made in the positions belonging to Π. Strategies for parity games can be taken to be *positional* or memory-free and therefore can be represented as a function $\sigma : G_\Pi \to G$. A match is σ*-guided* if for each position $u \in G_\Pi$ player Π chooses $\sigma(u)$ as the next position. We say that σ is *surviving* for Π if for each σ-guided match, the moves suggested by σ are always available to Π, and *winning* if in addition, Π wins each σ-guided match of the game. A winning position is one from which Π has a winning strategy. Finally, a player Π has a surviving strategy in an initialized game $\mathcal{G}@u$ *taking her/leading to position* p if Π has a strategy σ such that for every σ-guided match of $\mathcal{G}@u$ she either wins or gets to position p in finitely many steps.

2.3 Parity automata

Parity automata are finite devices operating on possibly infinite structures. In our paper, the automata classify pointed Kripke structures, and the question whether an automaton accepts or rejects such a structure is determined by a certain parity game. This acceptance game proceeds in a (possibly infinite) number of rounds, during each of which a certain *one-step formula* is the focus of attention.

Definition 2.2 Let A be a set of names. The set $\mathrm{ML}_1(A)$ of one-step modal formulas is given by the following grammar.

$$\varphi ::= p \mid \neg p \mid \varphi \wedge \varphi \mid \varphi \vee \varphi \mid \langle d \rangle a \mid [d]a,$$

where $p \in \mathsf{P}$, $d \in \mathsf{D}$, $a \in A$.

Clearly, in order to interpret one-step formulas in Kripke models, we need, besides the valuation, which takes care of the proposition letters in P, an interpretation for the variables in A. It will be convenient to use *markings* for this purpose, that is, maps $m : S \to \wp A$. Such a marking can also be presented as a valuation, or as a relation $Z_m \subseteq A \times S$, defined by $Z_m := \{(a, s) \mid s \in S, a \in m(s)\}$. We use these perspectives interchangeably.

Definition 2.3 A *(modal) parity automaton* is a tuple $\mathbb{A} = \langle A, \Delta, \Omega \rangle$ such that A is a finite set of states of the automaton; $\Delta : A \to \mathrm{ML}_1(A)$ is the transition map; and $\Omega : A \to \mathbb{N}$ is the parity map. An *initialized automaton* is a pair $(\mathbb{A}, a_I)$ where $a_I \in A$ is the initial state.

The acceptance game associated with a parity automaton $\mathbb{A}$ and a Kripke model $\mathbb{S}$ is given as follows. A match of this game consists of two players, $\exists$ and $\forall$, moving a token from one position to another. When such a match arrives at a so-called *basic* position, i.e., a position of the form $(a, s) \in A \times S$, the players consider the sentence $\Delta(a) \in \mathrm{ML}_1(A)$. At this position $\exists$ has to come up with a marking $m : S \to \wp A$, such that the formula $\Delta(a)$ is true at $\mathbb{S}, s$ under m. After that, $\forall$ chooses an element of Z_m to continue the match.

Definition 2.4 Given a model $\mathbb{S}$ and an automaton $\mathbb{A}$ we define the *acceptance game* $\mathcal{A}(\mathbb{A}, \mathbb{S})$ as the parity game given by the following table:

Position	Player	Admissible moves	Parity
$(a, s) \in A \times S$	$\exists$	$\{m : S \to \wp A \mid \mathbb{S}, m, s \Vdash \Delta(a)\}$	$\Omega(a)$
$m : S \to \wp A$	$\forall$	$\{(a, s) \mid s \in S, a \in m(s)\}$	0

Positions of the form $(a, s) \in A \times S$ will be called *basic*.

If (a_I, s) is a winning position for $\exists$ in the game $\mathcal{A}(\mathbb{A}, \mathbb{S})$ we say that $(\mathbb{A}, a_I)$ *accepts* the pointed model $(\mathbb{S}, s)$, notation: $\mathbb{S}, s \Vdash (\mathbb{A}, a_I)$.

We say that an initialized automaton $(\mathbb{A}, a_I)$ is equivalent to a formula φ if $\mathbb{S}, s \Vdash (\mathbb{A}, a_I) \iff \mathbb{S}, s \Vdash \varphi$, for all $\mathbb{S}, s$. More generally, we use the symbol $\equiv$ to denote equivalence between automata or formulas. The following fact lies behind the automata-theoretic approach towards the modal μ-calculus.

Fact 2.5 ([18,6]) *There are effective procedures transforming a formula of* μML *into an equivalent parity automaton, and vice versa.*

We now turn to the definition of *weak* parity automata.

Definition 2.6 Given $\mathbb{A} = \langle A, \Delta, \Omega \rangle$, we define the relation $\rightsquigarrow \subseteq A \times A$ by putting $a \rightsquigarrow b$ if b occurs in the formula $\Delta(a)$; we let $\prec$ and $\preceq$ denote, respectively, the transitive and reflexive-transitive closure of $\rightsquigarrow$. A *strongly connected component* or *SCC* of $\mathbb{A}$ is a subset $C \subseteq A$ such that for every $b, c \in C$ we have $b \preceq c$ and $c \preceq b$. An SCC is called *maximal*, or an *MSCC*, if none of its proper supersets is an SCC.

Definition 2.7 A parity automaton $\mathbb{A} = \langle A, \Delta, \Omega \rangle$ is *weak* if Ω satisfies

(weakness) if $a \preceq a'$ and $a' \preceq a$ then $\Omega(a) = \Omega(a')$.

Since in this case all states of a strongly connected component C have the same parity we may speak of the *parity of* C and denote it by $\Omega(C)$.

Remark 2.8 Any weak parity automaton $\mathbb{A}$ is equivalent to a weak parity automaton $\mathbb{A}'$ with $\Omega : A' \to \{0, 1\}$. From now on we assume such a parity map for weak parity automata.

3 Syntactic characterization of PDL and PDL^{tf}

In this section we will provide a precise characterization of the fragments of μML that correspond to full and test-free PDL. It will be convenient for us to work with a version of PDL that includes the empty program ϵ (or `skip`), which is interpreted as the identity relation in any Kripke model. Observe that in full PDL, the role of ϵ can be taken by the test program $\top?$.

Definition 3.1 The formulas and programs of the language $\mathrm{PDL}^{tf\epsilon}$ is given by the following grammar:

$$\begin{aligned}\varphi &::= p \mid \neg\varphi \mid \varphi \vee \varphi \mid \langle\pi\rangle\varphi \\ \pi &::= d \mid \epsilon \mid \pi \oplus \pi' \mid \pi;\pi' \mid \pi^*\end{aligned}$$

Remark 3.2 It is not difficult to show that adding the skip program does not add expressive power to PDL^{tf}. To see this, think of the programs of PDL^{tf} and $\mathrm{PDL}^{tf\epsilon}$ as the sets of regular expressions over the set D that may and may not use the empty string symbol ϵ, respectively. Let $\equiv_\ell$ denote the relation of language equivalence between regular expressions, that is, write $\pi \equiv_\ell \pi'$ if π and π' denote the same regular language over D. One may show, by induction on programs, that for any $\pi \in \mathrm{PDL}^{tf\epsilon}$ either (a) $\pi \equiv_\ell \epsilon$, or there is a program $\overline{\pi} \in \mathrm{PDL}^{tf\epsilon}$ such that either (b) $\pi \equiv_\ell \overline{\pi}$ or (c) $\pi \equiv_\ell \epsilon \oplus \overline{\pi}$. Based on this observation we may inductively define a truth-preserving translation from $\mathrm{PDL}^{tf\epsilon}$-formulas to PDL^{tf}-formulas; the key clause of this translation uses that $\langle\pi\rangle\varphi$ is equivalent to either (a) φ, (b) $\langle\overline{\pi}\rangle\varphi$ or (c) $\varphi \vee \langle\overline{\pi}\rangle\varphi$.

Definition 3.3 Given a set X of propositional variables, we define the *completely additive fragment* with respect to X, notation: $\mathrm{CAF}(X)$, as follows:

$$\varphi ::= x \in X \mid \psi \mid \psi \wedge \varphi \mid \varphi \vee \varphi \mid \langle d\rangle\varphi \mid \mu y.\varphi'$$

Here we require that ψ belongs to the X-free fragment of the modal μ-calculus (i.e., none of the variables in X occurs freely in ψ), and $\varphi' \in \mathrm{CAF}(X \cup \{y\})$.

The *restricted completely additive fragment* with respect to X, notation: $\mathrm{RAF}(X)$, is defined, similarly, by:

$$\varphi ::= x \in X \mid \psi \mid \varphi \vee \varphi \mid \langle d\rangle\varphi \mid \mu y.\varphi'$$

We define $\mu_{ca}\mathrm{ML}$ and $\mu_{rca}\mathrm{ML}$ to be the fragments of the modal μ-calculus where the use of the least fixpoint operator is restricted to these fragments.

Remark 3.4 The fragment $\mathrm{CAF}(X)$ provides a syntactic characterization of a (minor) variant of complete additivity, where we require (1) to hold only for *non-empty* index sets I. It is proved in [5] that a formula $\varphi \in \mu\mathrm{ML}$ satisfies this property for each $x \in X$ iff φ is equivalent to a formula in the $\mathrm{CAF}(X)$.

Definition 3.5 Formulas of the fragment $\mu_{ca}\mathrm{ML}$ ($\mu_{rca}\mathrm{ML}$, respectively), are given by the following induction:

$$\alpha ::= p \mid \neg\alpha \mid \alpha \vee \alpha \mid \langle d\rangle\alpha \mid \mu x.\varphi,$$

where $\varphi \in \mu_{ca}\mathrm{ML} \cap \mathrm{CAF}(x)$ ($\varphi \in \mu_{rca}\mathrm{ML} \cap \mathrm{RAF}(x)$, respectively).

Theorem 3.6 PDL *and test-free* PDL *are effectively equivalent to the fragments* $\mu_{ca}\mathrm{ML}$ *and* $\mu_{rca}\mathrm{ML}$*, respectively.*

Proof of Theorem 3.6. The theorem follows directly from Propositions 3.8 and 3.9 below. □

We start with the direction from PDL and PDL^{tf} to $\mu_{ca}\text{ML}$ and $\mu_{rca}\text{ML}$.

Definition 3.7 By a simultaneous induction on PDL-formulas and -programs, we define, for each PDL-program π, a function $f_\pi : \mu\text{ML} \to \mu\text{ML}$ on the set of modal fixpoint formulas, and a map $(\cdot)^t$ from PDL to μML:

$$\begin{array}{lll} f_d(\alpha) & := & \langle d\rangle\alpha \\ f_{\varphi?}(\alpha) & := & \varphi\wedge\alpha \\ f_{\pi\oplus\pi'}(\alpha) & := & f_\pi(\alpha)\vee f_{\pi'}(\alpha) \\ f_{\pi;\pi'}(\alpha) & := & f_\pi(f_{\pi'}(\alpha)) \\ f_{\pi^*}(\alpha) & := & \mu x.\alpha\vee f_\pi(x) \end{array} \qquad \begin{array}{lll} p^t & := & p \\ (\neg\varphi)^t & := & \neg\varphi^t \\ (\varphi_0\vee\varphi_1)^t & := & \varphi_0^t\vee\varphi_1^t \\ (\langle\pi\rangle\varphi)^t & := & f_\pi(\varphi^t) \end{array}$$

where, in the clause for f_{π^*}, x is some canonically chosen fresh variable.

Proposition 3.8 *For any* PDL*-formula* φ, φ^t *belongs to the fragment* $\mu_{ca}\text{ML}$, *and is equivalent to* φ. *If* φ *belongs to* PDL^{tf}, *then* $\varphi^t\in\mu_{rca}\text{ML}$.

The translation in the other direction is provided by the following proposition.

Proposition 3.9

(i) There is an effective procedure rewriting any modal fixpoint formula $\alpha\in\mu_{ca}\text{ML}$ *into an equivalent* PDL*-formula* α^s. *Moreover, if* $\alpha\in\mu_{rca}\text{ML}$ *then* $\alpha^s\in\text{PDL}^{tf\epsilon}$.

(ii) There is an effective procedure that, given a formula $\varphi\in\mu_{ca}\text{ML}$ *and a set* X *of variables such that* $\varphi\in\text{CAF}(X)$, *returns an* X*-free* PDL*-formula* ψ, *a subset* $Y\subseteq X$, *and a collection* $\{\pi_x \mid x\in Y\}$ *of* X*-free* PDL*-programs, such that*

$$\varphi\equiv\psi\vee\bigvee_{x\in Y}\langle\pi_x\rangle x. \tag{2}$$

If $\varphi\in\mu_{rca}\text{ML}\cap\text{RAF}(X)$, *then* ψ *and each* π_x *belong to* $\text{PDL}^{tf\epsilon}$.

4 A characterization of PDL^{tf} by automata

In this section we give the automata-theoretic characterization of test-free PDL. In μ-automata, $\rightsquigarrow$-cycles naturally correspond to fixpoint operators being unfolded. When considering (test-free) PDL, in the light of the previous section it is obvious that cycles have to be restricted so that they can only occur when induced by programs involving the iteration operator. The slogan that drives the definition of PDL- and PDL^{tf}-automata is that *maximal strongly connected components correspond to programs.*

The automata corresponding to PDL^{tf} will be weak parity automata where the transition map is subject to an additional constraint for all elements belonging to the same SCC. We begin by defining additional one-step languages needed for these automata. Recall that $\text{ML}_1(A)$ is the set of one-step modal formulas in A.

Definition 4.1 Let $C, O \subseteq A$ be such that $C \cap O = \varnothing$. The sets $\mathrm{ADD}_1^{tf}(O, C)$ and $\mathrm{MUL}_1^{tf}(O, C)$ of one-step formulas that are *test-free additive*, resp. *test-free multiplicative* in C, are defined, respectively, by the following grammars:

$$\varphi ::= \varphi \vee \varphi \mid \langle d \rangle c \mid \psi \qquad \text{and} \qquad \varphi ::= \varphi \wedge \varphi \mid [d]c \mid \psi$$

where $c \in C$, $\psi \in \mathrm{ML}_1(O)$.

Definition 4.2 A PDL^{tf}*-automaton* is a weak parity automaton $\mathbb{A}$ satisfying, for every strongly connected component C, the following constraint:

(tf-additivity) If $\Omega(C) = 1$ then $\Delta(c) \in \mathrm{ADD}_1^{tf}(A \setminus C, C)$ for each $c \in C$. In case $\Omega(C) = 0$ then $\Delta(c) \in \mathrm{MUL}_1^{tf}(A \setminus C, C)$ for each $c \in C$.

The main theorem of this section states that these PDL^{tf}-automata characterize PDL^{tf}. It will be proved in the two following subsections.

Theorem 4.3

(i) For every PDL^{tf}*-formula* φ *one can compute an equivalent initialized* PDL^{tf}*-automaton* $(\mathbb{A}_\varphi, a_\varphi)$.

(ii) For every initialized PDL^{tf}*-automaton* $(\mathbb{A}, a_I)$ *one can compute an equivalent* PDL^{tf}*-formula* $\varphi_{\mathbb{A},a_I}$.

4.1 From formulas to automata

In this subsection we will prove Theorem 4.3(i), that is, given a PDL^{tf}-formula φ we construct an equivalent PDL^{tf}-automaton $(\mathbb{A}_\varphi, a_\varphi)$. To begin, we focus on formulas of the form $\varphi = \langle \pi \rangle \alpha$ and, for the moment, assume that we already have an automaton $(\mathbb{A}, a_A) \equiv \alpha$. It is of our interest to understand how the operation $\langle \pi \rangle$ changes $(\mathbb{A}, a_A)$ to get an automaton for $\varphi \langle \pi \rangle \alpha$. Our idea is to represent π itself as a PDL^{tf}-automaton, which we will then combine with $\mathbb{A}$.

In this subsection we will briefly use non-deterministic finite-state automata (NFA). Recall that an NFA is a tuple $\underline{A} = \langle A, \delta, F, a_I \rangle$ where $\delta : A \times \mathsf{D} \to \wp A$ is the transition map, $F \subseteq A$ are the final states and $a_I \in A$ is the initial state. Given a model $\mathbb{S}$, an NFA denotes a set of *paths* through $\mathbb{S}$. We formalize the acceptance of a path with the following game.

Definition 4.4 Given a model $\mathbb{S}$ and an NFA $\underline{A} = \langle A, \delta, F, a_I \rangle$ we define the *acceptance game* $\underline{\mathcal{A}}(\underline{A}, \mathbb{S})$ having as basic positions pairs $(a, s) \in A \times S$.

Position	Pl'r	Admissible moves
$(a, s) \in (A \setminus F) \times S$	$\exists$	$\{(b, t) \mid \exists d \in \mathsf{D}.\, b \in \delta(a, d)\ \&\ R_d(s, t)$
$(f, s) \in F \times S$	$\exists$	$\{\mathsf{end}\} \cup \{(b, t) \mid \exists d \in \mathsf{D}.\, b \in \delta(f, d)\ \&\ R_d(f, t)\}$
end	$\forall$	$\varnothing$

Finite matches are lost by the player who gets stuck, infinite matches are all won by $\forall$. A path $\vec{s}$ in $\mathbb{S}$ is accepted by $\underline{A}$ iff $\exists$ has a winning strategy σ for the initialized game $\underline{\mathcal{A}}(\underline{A}, \mathbb{S})@(a_I, s)$ such that every σ-guided match visits precisely (and in order) the states of $\vec{s}$.

Using this game-theoretic approach towards NFAs we can easily prove the following lemma.

Lemma 4.5 *For every π there exists an initialized* PDL^{tf}*-automaton* $(\mathbb{P}_\pi, a_\pi)$ *and a set $F \subseteq |\mathbb{P}_\pi|$ such that for all $\mathbb{S}$ and $s,t \in |\mathbb{S}|$ we have $R_\pi(s,t)$ iff $\exists$ has a surviving strategy in $\mathcal{A}(\mathbb{P}_\pi, \mathbb{S})@(a_\pi, s)$ taking her to (f,t) for some $f \in F$.*

The above lemma gives us an automaton $(\mathbb{P}_\pi, a_\pi, F) = \langle P, \Delta_P, \Omega_P, a_\pi, F\rangle$ which works as a representation of π. We now combine this automaton with the representation of α given by $(\mathbb{A}, a_A) = \langle A, \Delta_A, \Omega_P, a_A\rangle$ yielding an automaton $(\mathbb{A}_\varphi, a_\varphi)$ for $\varphi = \langle\pi\rangle\alpha$. Define $(\mathbb{A}_\varphi, a_\varphi) := \langle A \uplus P, \Delta, \Omega, a_\varphi\rangle$ where $a_\varphi := a_\pi$, $\Omega := \Omega_A \cup \Omega_P$, and the transition map is defined as

$$\Delta(e) := \begin{cases} \Delta_P(e) & \text{if } e \in P \setminus F \\ \Delta_P(e) \vee \Delta(a_A) & \text{if } e \in F \\ \Delta_A(e) & \text{if } e \in A. \end{cases}$$

Remark 4.6 Observe that the construction has the following properties

(i) $\mathbb{A}_\varphi$ is a well-defined PDL^{tf}-automaton,
(ii) You can only go from the $\mathbb{P}_\pi$ part to the $\mathbb{A}$ part from a state in $F \subseteq P$,
(iii) Once you leave the $\mathbb{P}_\pi$ part you cannot come back.

Now we prove that $(\mathbb{A}_\varphi, a_\varphi)$ is an automaton representation of $\varphi = \langle\pi\rangle\alpha$.

Proposition 4.7 $\mathbb{S}, s \Vdash \langle\pi\rangle\alpha$ *iff $\exists$ has a winning strategy in $\mathcal{A}(\mathbb{A}_\varphi, \mathbb{S})@(a_\varphi, s)$.*

This finishes the proof of Theorem 4.3(i) for the particular case of $\varphi = \langle\pi\rangle\alpha$ when we already have an automaton for α. The general case can be proved by induction on $\varphi \in$ PDL^{tf}. The propositional and Boolean cases are easy and Proposition 4.7 gives us the required automaton for $\varphi = \langle\pi\rangle\alpha$. If $\varphi = [\pi]\alpha$ the construction and proofs are dual to the diamond case.

4.2 From automata to formulas

In this subsection we prove Theorem 4.3(ii), that is, for every initialized PDL^{tf}-automaton $(\mathbb{A}, a_I)$ we give an equivalent PDL^{tf}-formula $\varphi_{\mathbb{A},a_I}$. The key idea underlying our construction is to turn MSCCs of $\mathbb{A}$ into sets of PDL^{tf}-equations, and use the properties of PDL^{tf}-automata to solve these equations *inside* PDL^{tf}. In more detail, for every MSCC C and entry point $b \in C$ we will show how to get an equivalent formula $\varphi_{C,b} \in$ PDL$^{tf}(\mathsf{P} \uplus O, \mathsf{D})$ where the propositional variables in $O := A \setminus C$ correspond to the states outside C.

Definition 4.8 A *set of B-incomplete τ-equations* is a tuple $\mathbf{E} = (E, \xi, \tau)$ where E is a non-empty, finite set of equations specified by the map $\xi : E \to$ PDL$^{tf}(\mathsf{P} \uplus B \uplus E, \mathsf{D})$ and $\tau \in \{\mu, \nu\}$ is the type of $\mathbf{E}$. We sometimes specify a set of equations using the notation $\mathbf{E} := \{e_1 \approx \psi_1, \ldots, e_n \approx \psi_n\}_\tau$.

Definition 4.9 Given a model $\mathbb{S}$ and a set of B-incomplete τ-equations $\mathbf{E} = (E, \xi, \tau)$ we define the *solution game* $\mathcal{S}(\mathbf{E}, \mathbb{S})$ having as basic positions pairs $(x, s) \in (E \cup B) \times S$.

Position	Player	Admissible moves
$(e, s) \in E \times S$	$\exists$	$\{m : S \to \wp(E \cup B) \mid \mathbb{S}, m, s \Vdash \xi(e)\}$
$m : S \to \wp(E \cup B)$	$\forall$	$\{(x, s) \mid s \in S, x \in m(s)\}$

Whenever a position of the form $(b, s) \in B \times S$ is reached, the match is declared a tie; finite matches not ending in a tie are lost by the player that got stuck and infinite matches are won by $\exists$ if $\tau = \nu$, and by $\forall$ if $\tau = \mu$.

Let C be an MSCC of $\mathbb{A}$. First we consider the case where the parity of C is 1. We turn the information of C into a set of O-incomplete μ-equations $\mathbf{C} = (C, \xi, \mu)$ given by $\xi(c) := \Delta(c)$ for all $c \in C$. Observe that, by construction, the set of equations satisfies

$$\xi(c) = \alpha \vee \bigvee_{u \in U} \langle \pi_u \rangle u \quad \text{for } U \subseteq C \text{ and } \alpha, \pi_u \in \mathrm{PDL}^{tf}(\mathsf{P} \uplus O, \mathsf{D}). \qquad (*)$$

This set of equations is equivalent to C in the following sense:

Proposition 4.10 *Let $b \in C$, $o \in O$, and $s, t \in |\mathbb{S}|$, the following are equivalent*

(i) $\exists$ has a surviving strategy in $\mathcal{A}(\mathbb{A}, \mathbb{S})@(b, s)$ taking her to (o, t),
(ii) $\exists$ has a surviving strategy in $\mathcal{S}(\mathbf{C}, \mathbb{S})@(b, s)$ taking her to (o, t).

Proof. Straightforward from the definition of the games. □

For a moment, we forget about MSCCs and focus on sets of equations. We show that if a set of equations satisfies $(*)$ we can solve it inside PDL^{tf}. The proof is basically a game-theoretic version of the one found in [16], which is also reminiscent of the transformation of linear grammars into regular expressions.

Lemma 4.11 *Let $\mathbf{E} = (E, \xi, \mu)$ be a set of B-incomplete μ-equations satisfying $(*)$. For all $e \in E$ there exists $\varphi_{E,e} \in \mathrm{PDL}^{tf}(\mathsf{P} \uplus B, \mathsf{D})$ such that for all $b \in B$ and $s, t \in |\mathbb{S}|$ the following are equivalent:*

(i) $\exists$ has a surviving strategy in $\mathcal{S}(\mathbf{E}, \mathbb{S})@(e, s)$ taking her to (b, t),
(ii) $\mathbb{S}, m, s \Vdash \varphi_{E,e}$ where $m : S \to \wp B$ is such that $Z_m = \{(b, t)\}$.

It is only left to apply the above results to $\mathbf{C}$ to get the required formula.

Corollary 4.12 *For every MSCC C and $b \in C$ there is $\varphi_{C,b} \in \mathrm{PDL}^{tf}(\mathsf{P} \uplus O, \mathsf{D})$ such that for all $o \in O$ and $s, t \in |\mathbb{S}|$ the following are equivalent:*

(i) $\exists$ has a surviving strategy in $\mathcal{A}(\mathbb{A}, \mathbb{S})@(b, s)$ taking her to (o, t),
(ii) $\mathbb{S}, m, s \Vdash \varphi_{C,b}$ where $m : S \to \wp O$ is such that $Z_m = \{(o, t)\}$.

Proof. Combination of Proposition 4.10 and Lemma 4.11 applied to $\mathbf{C}$. □

The above corollary provides a formula $\varphi_{C,b}$ when the parity of the connected component C is 1. The case where the parity of C is 0 is solved in a dual but completely similar way.

Now that we can get a formula for every point of an MSCC we turn to the general case. In order to create a formula from an initialized automaton we introduce the following concept.

Definition 4.13 Given a PDL^{tf}-automaton $\mathbb{A}$, the *DAG of connected components of* $\mathbb{A}$ is the pair $\mathrm{DCC}(\mathbb{A}) = (G, E)$ where G is the set of $\preceq$-MSCCs of $\mathbb{A}$ and $(C_1, C_2) \in E$ if $C_1 \neq C_2$ and $a \rightsquigarrow b$ for some $a \in C_1$ and $b \in C_2$.

Remark 4.14 Observe that a node of DCC($\mathbb{A}$) is either a $\prec$-connected component or a single element $a \in A$ which does not belong to any $\prec$-cycle. Another observation is that, even though DCC($\mathbb{A}$) may not be a tree, it certainly contains no cycles. Therefore E is well-founded and, given $C \in G$, we can associate a notion of height to the subgraph generated by C.

We are now ready to prove the main theorem of this section.

Proof of Theorem 4.3(ii). For every initialized PDL^{tf}-automaton $(\mathbb{A}, a_I)$ we give an equivalent PDL^{tf}-formula $\varphi_{\mathbb{A},a_I}$. The proof will be done by induction on the height of the subgraph of DCC($\mathbb{A}$) generated by a_I.

If the height of the subgraph is 1, then it is composed of a single MSCC C and $a_I \in C$. By Corollary 4.12, we get a formula $\varphi_{C,a_I} \in \mathrm{PDL}^{tf}(\mathsf{P} \uplus O, \mathsf{D})$. We only have to observe that, because C is not connected to any other MSCC then $O = \varnothing$. Therefore $\varphi_{C,a_I} \in \mathrm{PDL}^{tf}(\mathsf{P}, \mathsf{D})$ and is equivalent to $(\mathbb{A}, a_I)$.

Suppose the height of the subgraph is n and $a_I \in C$ for some MSCC C. Again by Corollary 4.12 we get a formula $\varphi_{C,a_I} \in \mathrm{PDL}^{tf}(\mathsf{P} \uplus O, \mathsf{D})$ where $O = \{o_1, \ldots, o_k\}$ and $o_i \in C_i$ for some MSCCs C_i. By inductive hypothesis, we get a formula $\varphi_{\mathbb{A},o_i} \in \mathrm{PDL}^{tf}(\mathsf{P}, \mathsf{D})$ for each o_i. It is straightforward to check that $\varphi_{\mathbb{A},a_I} := \varphi_{C,a_I}[o_i \mapsto \varphi_{\mathbb{A},o_i} \mid i \leq k]$ is equivalent to $(\mathbb{A}, a_I)$. □

5 A characterization of PDL by automata

This section concerns the automata-theoretic characterization of full PDL. Since our approach here is an adaptation of the one taken in the previous section we will be a bit more sketchy.

Definition 5.1 Let $C, O \subseteq A$ be such that $C \cap O = \varnothing$. The sets $\mathrm{ADD}_1(O, C)$ and $\mathrm{MUL}_1(O, C)$ of one-step formulas that are additive, resp. multiplicative in C, are defined, respectively, by the grammars

$$\varphi ::= \varphi \vee \varphi \mid \langle d \rangle c \mid \psi \mid \varphi \wedge \psi \qquad \text{and} \qquad \varphi ::= \varphi \wedge \varphi \mid [d]c \mid \psi \mid \varphi \vee \psi$$

where $c \in C$, $\psi \in \mathrm{ML}_1(O)$.

Note that the difference with the sets one-step formulas for PDL^{tf}-automata given in Definition 4.1 lies in the fact that here, in order to take care of tests, we allow conjunctions with C-free formulas.

Definition 5.2 A PDL-*automaton* is a weak parity automaton $\mathbb{A}$ satisfying, for every strongly connected component C, the following constraint:

(additivity) If $\Omega(C) = 1$ then $\Delta(c) \in \mathrm{ADD}_1(A \setminus C, C)$ for each $c \in C$. In case $\Omega(C) = 0$ then $\Delta(c) \in \mathrm{MUL}_1(A \setminus C, C)$ for each $c \in C$.

The main theorem of this section states that PDL-automata characterize PDL.

Theorem 5.3

(i) For every PDL-*formula* φ *one can compute an equivalent initialized automaton* $(\mathbb{A}_\varphi, a_\varphi)$.

(ii) For every initialized automaton $(\mathbb{A}, a_I)$ *one can compute an equivalent* PDL-*formula* $\varphi_{\mathbb{A},a_I}$.

5.1 From formulas to automata

In this section we will prove Theorem 5.3(i), that is, given a PDL-formula φ we create an equivalent PDL-automaton $(\mathbb{A}_\varphi, a_\varphi)$. We give a proof by induction on φ. If φ is a test-free formula (that is, $\varphi \in \mathrm{PDL}^{tf}$) we can get the corresponding PDL-automaton from Theorem 4.3 by observing that every PDL^{tf}-automaton is also a PDL-automaton. It is also easy to check that the class of PDL-automata is closed under the Boolean operators.

The interesting case, therefore, is where $\varphi = \langle\pi\rangle\alpha$, with $\alpha \equiv (\mathbb{A}_\alpha, a_\alpha)$ and π involving tests. To treat this case we use the following strategy: first we will consider tests as additional atomic actions and get an NFA for π, similar to what we did in Section 4; after that we show how to convert this NFA to a PDL-automaton and merge it with the automata for the tested formulas to get a PDL-automaton $\mathbb{P}_\pi$ for π. To finish, we combine $\mathbb{P}_\pi$ and $\mathbb{A}_\alpha$ to get an automaton for φ.

In the process of creating a PDL-automaton for π we encounter new complexities because of the presence of tests. To be able to properly define a merging operation we need to introduce the following concepts.

Definition 5.4 Let B be a set of names such that $A \cap B = \varnothing$ and $\mathsf{P} \cap B = \varnothing$. A *$B$-incomplete* PDL-*automaton* $\mathbb{A}$ is a PDL-automaton based on the set of propositions $\mathsf{P} \cup B$ such that the elements of B occur only positively in the transition map of $\mathbb{A}$. The acceptance games of Definition 2.4 are extended to B-incomplete automata with the intention to interpret the elements of B as names (as opposed to propositions). Basic positions are then taken from $(A \cup B) \times S$ and markings are of the type $m : S \to \wp(A \cup B)$. Whenever a position from $B \times S$ is reached, the match is declared a tie.

Definition 5.5 The *completion of a B-incomplete automaton $\mathbb{A}$ with a* PDL-*automaton* $\mathbb{A}' = \langle A', \Delta', \Omega'\rangle$ is defined as $(\mathbb{A} \rtimes \mathbb{A}') = \langle C, \Delta_C, \Omega_C\rangle$ where $C := A \uplus A'$, $\Omega_C := \Omega \cup \Omega'$ and the transition map is given by

$$\Delta_C(c) := \begin{cases} \Delta'(c) & \text{if } c \in A', \\ \Delta(c)[\mathsf{b} \mapsto \Delta'(\mathsf{b}) \mid \mathsf{b} \in B \cap A'] & \text{if } c \in A. \end{cases}$$

Note that the completion can be partial if $B \not\subseteq A'$, in this case the outcome will be $(B \setminus A')$-incomplete. If $B \subseteq A'$, the outcome will be a (complete) PDL-automaton. Also observe that a completion cannot generate new cycles.

Definition 5.6 Given $\pi \in \mathrm{PDL}(\mathsf{P}, \mathsf{D})$ we use $\pi^\flat \in \mathrm{PDL}^{tf}(\mathsf{P}, \mathsf{D} \cup \mathsf{T})$ to denote the version of π where its top-level tests T are considered as atomic actions. The T-extension of a model $\mathbb{S} = \langle S, R_{d\in\mathsf{D}}, V\rangle$ is defined as $\mathbb{S}^\mathsf{T} = \langle S, R_{d\in\mathsf{D}}, R_{\chi\in\mathsf{T}}, V\rangle$ where $R_\chi := \{(s,s) \in S \times S \mid \mathbb{S}, s \Vdash \chi\}$.

Lemma 5.7 *For every $\pi \in$ PDL there exists an $\mathtt{x}$-incomplete initialized* PDL-*automaton $(\mathbb{P}_\pi, a_\pi)$ such that for all models $\mathbb{S}$ and $s,t \in |\mathbb{S}|$ we have that $R^{\mathbb{S}}_\pi(s,t)$ iff $\exists$ has a surviving strategy in $\mathcal{A}(\mathbb{P}_\pi, \mathbb{S})@(a_\pi, s)$ taking her to $(\mathtt{x}, t)$.*

Proof. Let T be the top-level tests appearing in π. We claim the following:

CLAIM 1 For every model $\mathbb{S}$ and $s, t \in |\mathbb{S}|$ we have that $R^{\mathbb{S}}_{\pi}(s,t)$ iff $R^{\mathbb{S}^{\mathsf{T}}}_{\pi^{\flat}}(s,t)$.

As in Section 4, we can construct an NFA $\underline{\mathrm{A}}_{\pi} = \langle A, \delta, F, a_I \rangle$ which recognizes $\pi^{\flat}$. By definition of $\underline{\mathrm{A}}_{\pi}$ recognizing $\pi^{\flat}$, we have the following claim.

CLAIM 2 For every model $\mathbb{S}$, $\underline{\mathrm{A}}_{\pi}$ accepts the path $s, \ldots, t$ in $\mathbb{S}^{\mathsf{T}}$ iff $R^{\mathbb{S}^{\mathsf{T}}}_{\pi^{\flat}}(s,t)$.

CLAIM 3 Without loss of generality we can assume these properties on $\underline{\mathrm{A}}_{\pi}$:

(i) Each state has either exiting action transitions or test transitions (but not both), and will accordingly be called *action state* or *test state.*

(ii) Every cycle contains at least one action state.

(iii) The initial state has no incoming transitions.

(iv) Test transitions always arrive into an action state.

These properties are reminiscent of the work by Kozen [11]. For reasons of space limitations we have to omit the proofs.

Let $T = \{\mathsf{a}_{\chi} \mid \chi \in \mathsf{T}\} \cup \{\mathtt{x}\}$ be a set of names. From $\underline{\mathrm{A}}_{\pi}$ we define a T-incomplete initialized PDL-automaton $\mathbb{A}_{\pi} := \langle A_{\pi}, \Delta_{\pi}, a_{\pi} \rangle$ by setting $A_{\pi} := A$, $\Omega(a) := 1$ for all $a \in A$,

$$\Delta_{\pi}(a) := \begin{cases} \bigvee\{\langle d \rangle b \mid d \in \mathsf{D}, b \in \delta(a,d)\} & \text{if } a \notin F \text{ is an action state,} \\ \mathtt{x} \vee \bigvee\{\langle d \rangle b \mid d \in \mathsf{D}, b \in \delta(a,d)\} & \text{if } a \in F \text{ is an action state,} \\ \bigvee\{\mathsf{a}_{\chi} \wedge \Delta_{\pi}(b) \mid \chi \in \mathsf{T}, b \in \delta(a,\chi)\} & \text{if } a \notin F \text{ is a test state,} \\ \mathtt{x} \vee \bigvee\{\mathsf{a}_{\chi} \wedge \Delta_{\pi}(b) \mid \chi \in \mathsf{T}, b \in \delta(a,\chi)\} & \text{if } a \in F \text{ is a test state.} \end{cases}$$

Note that this is well defined since we first define $\Delta_{\pi}(a)$ for action states and then for test states.

CLAIM 4 $\mathbb{A}_{\pi}$ is a well-defined T-incomplete PDL-automaton.

Let $(\mathbb{A}_{\chi}, a_{\chi})_{\chi \in \mathsf{T}}$ be the family of PDL-automata for $\mathsf{T} = \{\chi_1, \ldots, \chi_k\}$, provided by the inductive hypothesis. To finish the construction define $(\mathbb{P}_{\pi}, a_{\pi}) := (\mathbb{A}_{\pi} \rtimes \mathbb{A}_{\chi_1} \rtimes \cdots \rtimes \mathbb{A}_{\chi_k}, a_{\pi})$.

CLAIM 5 For every model $\mathbb{S}$ and $s, t \in |\mathbb{S}|$, the following are equivalent.

(i) $\exists$ has a surviving strategy in $\mathcal{A}(\underline{\mathrm{A}}_{\pi}, \mathbb{S}^{\mathsf{T}})@(a_{\pi}, s)$ leading to $(f \in F, t)$.

(ii) $\exists$ has a surviving strategy in $\mathcal{A}(\mathbb{P}_{\pi}, \mathbb{S})@(a_{\pi}, s)$ taking her to $(\mathtt{x}, t)$.

Due to lack of space we have to omit the proof of this claim.

Finally, combining the above claims we find the following equivalences:

$$\begin{aligned} R^{\mathbb{S}}_{\pi}(s,t) &\iff R^{\mathbb{S}^{\mathsf{T}}}_{\pi^{\flat}}(s,t) && \text{(Claim 1)} \\ &\iff \underline{\mathrm{A}}_{\pi} \text{ accepts the path } s, \ldots, t \text{ in } \mathbb{S}^{\mathsf{T}} && \text{(Claim 2)} \\ &\iff \exists \text{ has a surviving strategy in} \\ &\qquad \mathcal{A}(\mathbb{P}_{\pi}, \mathbb{S})@(a_{\pi}, s) \text{ leading to } (\mathtt{x}, t) && \text{(Claim 5)} \end{aligned}$$

This finishes the proof of the lemma. □

To conclude we have to give an automaton for $\varphi = \langle \pi \rangle \alpha$. Let $(\mathbb{A}_{\alpha}, \mathtt{x})$ be the automaton for α, given by the inductive hypothesis. Define $(\mathbb{A}_\varphi, a_\varphi) := (\mathbb{P}_\pi \rtimes \mathbb{A}_\alpha, a_\pi)$. We prove that $(\mathbb{A}_\varphi, a_\varphi)$ is an automaton representation of φ.

Proposition 5.8 $\mathbb{S}, s \Vdash \langle\pi\rangle\alpha$ *iff* $\exists$ *has a winning strategy in* $\mathcal{A}(\mathbb{A}_\varphi, \mathbb{S})@(a_\varphi, s)$.

Proof. The proof is similar to Proposition 4.7 but using Lemma 5.7. □

5.2 From automata to formulas

The proof is basically the same as for PDL^{tf}. The new challenge lies in showing that when solving the system of equations (i.e., an analogue of Lemma 4.11) we can provide a normal form which functions as (∗). The key observation is that one-step formulas in $\mathrm{ADD}_1(O, C)$ can be assumed to be in the form

$$\alpha \vee \bigvee_{u \in U} \langle\pi_u\rangle u \quad \text{with} \quad U \subseteq C;\ \alpha, \pi_u \in \mathrm{PDL}(\mathsf{P} \cup O, \mathsf{D})$$

This can be proved by a straightforward induction on the complexity of $\mathrm{ADD}_1(O, C)$-formulas, where in the inductive step we use that $\psi \wedge \varphi$ is equivalent to $\langle\psi?\rangle\varphi$. Further details are left to the reader.

6 Conclusions

In this paper we have clarified the relation between PDL and the modal μ-calculus, by (1) providing explicit syntactic translations between PDL and the fragment of μML to which it corresponds, and (2) giving an automata-theoretic characterization of PDL. Both results were obtained in versions for full and for test-free PDL, respectively.

Although we have treated the syntactic and the automata-theoretic characterizations separately, the two results are in fact closely related. This is witnessed, in the case of full PDL [3], by the close syntactic similarities between the completely additive fragment of μML (Definition 3.3) and the additive one-step formulas (Definition 5.1), in that both use the same set of operators. Using these similarities it would not be very hard to also give direct transformations between formulas in the fragment μ_{ca}ML and PDL-automata. Another similarity between the two characterizations surfaces in the proof: in both cases, the heart of the argument showing one direction of the equivalence lies in the fact that certain sets of fixpoint equations can be solved *inside* PDL. In fact, we could have proved our results for PDL in the form 'PDL-formulas ⟶ PDL-automata ⟶ μ_{ca}ML-formulas'. We chose to discuss the syntactic and the automata-theoretic result separately because we believe that the two proofs have some merits in their own right: the syntactic transformations are simple and straightforward, and the automata-theoretic proof sheds some light on the automata-theoretic nature of PDL-programs.

We hope that our characterizations of PDL provide new tools for proving results on PDL. In particular, it is an interesting question to find a natural logic of which PDL is the bisimulation-invariant fragment. In this light, note

[3] The case of PDL^{tf} is analogous.

that in a related paper [1] with Facchini and Zanasi we used similar automata to the ones in this paper to prove that a variant of μ_{rca}ML corresponds to the bisimulation-invariant fragment of weak monadic second-order logic. Other interesting questions would be to find semantic properties that set PDL apart as a fragment of μML, to prove the (un-)decidability of the question whether a given μ-calculus formula has an equivalent in PDL, to give a characterization of PDL-programs (cf. [8, p. 91]), and to give a constructive proof of the Craig interpolation property for PDL.

References

[1] Carreiro, F., A. Facchini, Y. Venema and F. Zanasi, *Weak MSO: Automata and expressiveness modulo bisimilarity*, in: *CSL/LICS 2014*, 2014, to appear.

[2] D'Agostino, G. and M. Hollenberg, *Logical Questions Concerning the μ-Calculus: Interpolation, Lyndon and Loś-Tarski*, The Journal of Symbolic Logic **65** (2000), pp. 310–332.

[3] Fischer, M. J. and R. E. Ladner, *Propositional dynamic logic of regular programs*, J. Comput. Syst. Sci. **18** (1979), pp. 194–211.

[4] Fontaine, G., "Modal fixpoint logic: some model-theoretic questions," Ph.D. thesis, ILLC (University of Amsterdam) (2010).

[5] Fontaine, G. and Y. Venema, *Some model theory for the modal μ-calculus: syntactic characterizations of semantic properties* (2012), submitted.

[6] Grädel, E., W. Thomas and T. Wilke, editors, "Automata, Logics, and Infinite Games: A Guide to Current Research," Lecture Notes in Computer Science **2500**, Springer, 2002.

[7] Harel, D., J. Tiuryn and D. Kozen, "Dynamic Logic," MIT Press, Cambridge, MA, USA, 2000.

[8] Hollenberg, M., "Logic and Bisimulation," Ph.D. thesis, University of Utrecht (1998).

[9] Janin, D. and I. Walukiewicz, *On the expressive completeness of the propositional mu-calculus with respect to monadic second order logic*, in: U. Montanari and V. Sassone, editors, *CONCUR*, Lecture Notes in Computer Science **1119** (1996), pp. 263–277.

[10] Kozen, D., *Results on the propositional mu-calculus*, Theor. Comput. Sci. **27** (1983), pp. 333–354.

[11] Kozen, D., *Automata on guarded strings and applications*, Technical report, Ithaca, NY, USA (2001).

[12] Muller, D. E., A. Saoudi and P. E. Schupp, *Weak alternating automata give a simple explanation of why most temporal and dynamic logics are decidable in exponential time*, in: *Proceedings of the Third Annual IEEE Symposium on Logic in Computer Science (LICS 1988)* (1988), pp. 422–427.

[13] Streett, R. S., *Propositional dynamic logic of looping and converse*, in: *STOC* (1981), pp. 375–383.

[14] Streett, R. S., *Propositional dynamic logic of looping and converse is elementarily decidable*, Information and Control **54** (1982), pp. 121 – 141.

[15] van Benthem, J., "Exploring Logical Dynamics," CSLI Publications, Stanford, California, 1996.

[16] van Benthem, J. and D. Ikegami, *Modal fixed-point logic and changing models*, in: A. Avron, N. Dershowitz and A. Rabinovich, editors, *Pillars of Computer Science*, Lecture Notes in Computer Science **4800** (2008), pp. 146–165.

[17] Vardi, M. Y. and P. Wolper, *Automata-theoretic techniques for modal logics of programs*, Journal of Computer and System Sciences **32** (1986), pp. 183 – 221.

[18] Venema, Y., *Lecture notes on the modal μ-calculus* (2011).

Appendix: Proofs

Proof of Proposition 3.8. By a simultaneous induction on formulas and programs, we prove that
(1) for any PDL-program π, and any formula $\alpha \in \mu$ML:
(1a) $f_\pi(\alpha)$ belongs to μ_{ca}ML if $\alpha \in \mu_{ca}$ML;
(1b) $f_\pi(\alpha) \in \text{CAF}(X)$ if $\alpha \in \text{CAF}(X)$ and $\text{Var}(\pi) \cap X = \varnothing$;
(1c) $\langle \pi \rangle \alpha \equiv f_\pi(\alpha)$.
(2) for any PDL-formula α:
(2a) $\alpha^t \in \mu_{ca}$ML;
(2b) $\alpha \equiv \alpha^t$.
Analogous statements can be proved for PDL$^{tf\epsilon}$, RAF(X) and μ_{rca}ML.

For the proof of (1), we confine our attention to the case where $\pi = \rho^*$. Take an arbitrary formula $\alpha \in \mu$ML. Recall that $f_\pi(\alpha)$ is of the form $\mu x.\alpha \vee f_\rho(x)$, where x does not occur in either α or ρ. For (1a), suppose that $\alpha \in \mu_{ca}$ML. By the inductive hypothesis (1b), applied to ρ and the formula $x \in \text{CAF}(x)$, we have that $f_\rho(x) \in \text{CAF}(x)$. Thus we see that $\alpha \vee f_\rho(x) \in \text{CAF}(x)$ as well, and so $\mu x.\alpha \vee f_\rho(x)$ belongs to the set μ_{ca}ML indeed. For (1b), let X be a set of variables that do not occur in ρ, and are such that $\alpha \in \text{CAF}(X)$. Since x does not occur in α this means that $\alpha \in \text{CAF}(X \cup \{x\})$. Since x, as a formula, also belongs to the set $\text{CAF}(X \cup \{x\})$, and $\text{Var}(\pi) \cap (X \cup \{x\}) = \varnothing$, an application of the inductive hypothesis shows that $f_\rho(x) \in \text{CAF}(X \cup \{x\})$ as well. Hence the disjunction $\alpha \vee f_\rho(x) \in \text{CAF}(X \cup \{x\})$, and from this we may conclude that, indeed, $f_\pi(\alpha)$ belongs to the set $\text{CAF}(X)$. For (1c), it is obvious that $\langle \rho^* \rangle \alpha \equiv \mu x.\alpha \vee \langle \rho \rangle x \equiv \mu x.\alpha \vee f_\rho(x) \equiv f_\pi(\alpha)$.

For the proof of (2), we only consider the inductive case where α is of the form $\langle \pi \rangle \beta$. Inductively, we may assume that $\beta^t \in \mu_{ca}$ML, and that $\beta \equiv \beta^t$. But then it follows from the inductive hypothesis, applied to the program π, that $\alpha^t = f_\pi(\beta^t)$ belongs to μ_{ca}ML as well (1a), and that $\alpha^t = f_\pi(\beta^t)$ is equivalent to the formula $\alpha = \langle \pi \rangle \beta$. This suffices to prove the proposition. □

Proof of Proposition 3.9. We prove the proposition via a mutual induction on the fragments μ_{ca}ML and CAF. The stronger statements concerning formulas in the restricted fragments follow from an easy inspection.

We first consider item (i). Leaving the other cases as exercises for the reader, we focus on the interesting case of the inductive step, where we are dealing with a formula $\mu x.\varphi$, with $\varphi \in \mu_{ca}\text{ML} \cap \text{CAF}(x)$. In order to find the right translation for this formula, we use the induction hypothesis of item (ii). That is, we assume that φ has been rewritten as an equivalent disjunction $\psi \vee \langle \pi_x \rangle x$, where x does not occur in either ψ or π_x. Hence, if we put

$$(\mu x.\varphi)^s := \langle \pi_x^* \rangle \psi,$$

it is easy to verify that this definition satisfies the required properties.

This leaves the proof of item (ii). In case $\varphi = x$, simply take $\psi := \bot$,

$Y := \{x\}$, and $\pi_x := \top?$. (In the case $\varphi \in \mu_{rca}\mathrm{ML} \cap \mathrm{RAF}(X)$ and we need to land in test-free PDL, we put $\pi_x := \epsilon$. This is the reason for adding the skip program ϵ to PDL^{tf}.) Clearly $\varphi \equiv \bot \vee \langle d \rangle x$. In case φ is X-free, inductively we may assume that we have applied item 1 to φ, obtaining the equivalent PDL-formula φ^s. Take $\psi := \varphi^s$, and put $Y := \varnothing$. Clearly then $\bigvee_{x \in Y} \langle \pi_x \rangle x \equiv \bot$, so that indeed we find $\varphi \equiv \varphi^s \vee \bigvee_{x \in Y} \langle \pi_x \rangle x$.

We leave the inductive cases where $\varphi = \psi \wedge \varphi'$, $\varphi = \varphi' \vee \varphi''$ and $\varphi = \langle d \rangle \varphi'$ as exercises, and turn to the case where $\varphi = \mu z.\varphi'$. We may apply the inductive hypothesis with respect to φ' and $X \cup \{z\}$, which gives an $X \cup \{z\}$-free (and hence, X-free) formula ψ', a subset $Y \subseteq X \cup \{z\}$, and a program π'_y for each $y \in Y$, such that

$$\varphi' \equiv \psi' \vee \bigvee_{y \in Y} \langle \pi'_y \rangle y.$$

Now distinguish cases. If $z \notin Y$ then $Y \subseteq X$ and the formula $\psi' \vee \bigvee_{y \in Y} \langle \pi'_y \rangle y$ is z-free, so that $\varphi = \mu z.\varphi' \equiv \psi' \vee \bigvee_{y \in Y} \langle \pi'_y \rangle y$ and we are done. If, on the other hand, $z \in Y$, then define $Y' := Y \setminus \{z\}$, so that we have $Y' \subseteq X$ and

$$\varphi' \equiv \psi' \vee \bigvee_{y \in Y'} \langle \pi'_y \rangle y \vee \langle \pi'_z \rangle z.$$

From this it is immediate that

$$\mu z.\varphi' \equiv \langle (\pi'_z)^* \rangle (\psi' \vee \bigvee_{y \in Y'} \langle \pi'_y \rangle y),$$

and so we find that

$$\mu z.\varphi' \equiv \langle (\pi'_z)^* \rangle \psi' \vee \bigvee_{y \in Y'} \langle (\pi'_z)^* ; \pi'_y \rangle y,$$

from which we can read off the formula $\psi := \langle (\pi'_z)^* \rangle \psi'$ and the programs $\pi_y := (\pi'_z)^* ; \pi'_y$, for each $y \in Y'$. □

Proof of Lemma 4.5. A PDL^{tf}-program π is nothing but a regular expression over D. Regular expressions over D can be given semantics over Kripke models such that they denote a set of paths. Using this approach, we know that there is a non-deterministic finite-state automaton (NFA) which recognizes the same language as π. Let $\underline{\mathrm{A}}_\pi = \langle A, \delta, F, a_I \rangle$ be such an automaton. Since $\underline{\mathrm{A}}_\pi$ accepts the language denoted by the regular expression π, it is straightforward to verify that $\underline{\mathrm{A}}_\pi$ accepts the path $s, \ldots, t$ iff $R_\pi(s,t)$.

Next we define $\mathbb{P}_\pi = \langle A, \Delta, \Omega \rangle$ where for all $a \in A$ the transition map is

$$\Delta(a) := \bigvee_{d \in \mathsf{D}, b \in \delta(a,d)} \langle d \rangle b \ ,$$

and the parity map is $\Omega(a) := 1$ for every element. It is not difficult to see that $\mathbb{P}_\pi$ is a well-defined PDL^{tf}-automaton: it satisfies the PDL^{tf} restrictions for cycles because every state appears under a diamond and there are only disjunctions in the transition map, and the other conditions (e.g., weakness) are trivially satisfied.

Furthermore, it is clear from the definition of $\mathbb{P}_\pi$ that the following are equivalent, for any $(a, s) \in A \times S$:

(i) (b, t) is an admissible move for $\exists$ in $\underline{\mathcal{A}}(\underline{\mathbb{A}}_\pi, \mathbb{S})@(a, s)$,
(ii) $\{(b, t)\}$ is an admissible move for $\exists$ in $\mathcal{A}(\mathbb{P}_\pi, \mathbb{S})@(a, s)$.

Now consider the triple $(\mathbb{P}_\pi, a_\pi, F)$ consisting of the initialized automaton $(\mathbb{P}_\pi, a_\pi)$ where $a_\pi := a_I$ and F is the set of final states. Combining the above claims we get that $R_\pi(s, t)$ iff $\exists$ has a surviving strategy in $\mathcal{A}(\mathbb{P}_\pi, \mathbb{S})@(a_\pi, s)$ taking her to (f, t) for some $f \in F$. This finishes the proof of the lemma. □

Proof of Proposition 4.7. ($\Rightarrow$) Suppose $\mathbb{S}, s \Vdash \langle\pi\rangle\alpha$. By definition there is $t \in S$ such that $R_\pi(s, t)$ and $\mathbb{S}, t \Vdash \alpha$. Using Lemma 4.5 we know that therefore $\exists$ has a surviving strategy in $\mathcal{A}(\mathbb{P}_\pi, \mathbb{S})@(a_\pi, s)$ taking her to (f, t) for some $f \in F$. Now $\exists$ can use that strategy to play a match in $\mathcal{A}(\mathbb{A}_\varphi, \mathbb{S})@(a_\varphi, s)$ and get to the same position (f, t). By inductive hypothesis (as $\mathbb{S}, t \Vdash \alpha$) we know that $\exists$ has a winning strategy in $\mathcal{A}(\mathbb{A}, \mathbb{S})@(a_A, t)$. Because of the way the transition map Δ is defined, she can use that same strategy to win $\mathcal{A}(\mathbb{A}_\varphi, \mathbb{S})@(f, t)$.

($\Leftarrow$) Suppose that $\exists$ has a winning strategy in $\mathcal{A}(\mathbb{A}_\varphi, \mathbb{S})@(a_\varphi, s)$. As the parity of $\mathbb{P}_\pi$ is 1 for every element this means that $\exists$ plays finitely many moves in $\mathbb{P}_\pi$ which get her to some position (f, t) and then makes a move which takes her to the $\mathbb{A}$ part of the automaton. Observe that this can only happen if $f \in F$. Using Lemma 4.5 we get that $R_\pi(s, t)$. As $\exists$ has a winning strategy in $\mathcal{A}(\mathbb{A}_\varphi, \mathbb{S})@(f, t)$ and because of how Δ is defined, she can use that same strategy to win the game $\mathcal{A}(\mathbb{A}, \mathbb{S})@(f, t)$ and thus by inductive hypothesis we get that $\mathbb{S}, t \Vdash \alpha$. By definition, this means that $\mathbb{S}, s \Vdash \langle\pi\rangle\alpha$. □

Proof of Lemma 4.11. By induction on $|E|$, we solve this set of equations while preserving $(*)$ and finally get a formula in $\mathrm{PDL}^{tf}(\mathsf{P} \uplus B, \mathsf{D})$.

For the base case let $E = \{e\}$, we have to consider two cases: if $e \notin \xi(e)$ then $\xi(e) = \alpha$ with $\alpha \in \mathrm{PDL}^{tf}(\mathsf{P} \uplus B, \mathsf{D})$ and we are done. Otherwise, the equation should be of the form $\xi(e) = \alpha \vee \langle\pi\rangle e$. Let $\varphi_{E,e} := \langle\pi^*\rangle\alpha$, it is easy to see that the formula belongs to the right fragment. The following claim states that $\varphi_{E,e}$ is equivalent to E.

Claim 1 The following are equivalent:

- $\exists$ has a surviving strategy in $\mathcal{S}(\{e \approx \alpha \vee \langle\pi\rangle e\}_\mu, \mathbb{S})@(e, s)$ taking her to (b, t).
- $\mathbb{S}, m, s \Vdash \langle\pi^*\rangle\alpha$ where $m : S \to \wp B$ is such that $Z_m = \{(b, t)\}$.

Proof of Claim. ($\Rightarrow$) As the set of equations is of type μ this means that $\exists$ plays only a finite number of moves, otherwise she would lose. Because of the shape of the set of equations she has to play markings $m_1, \ldots, m_k$ such that

$e \in m_i(s_i)$ for some s_i and in each turn $\forall$ chooses (e, s_i). After that $\exists$ plays a marking m such that $e \in m(t)$ and $\forall$ must choose (b, t). It is clear to observe that the first k rounds induce a π^*-path $s, s_1, \ldots, s_k$ and the last round implies that $\mathbb{S}, m, s_k \Vdash \alpha$. It is only left to observe that as $\exists$ can *force* $\forall$ to choose (b, t) then it must be the case that $Z_m = \{(b, t)\}$.

($\Leftarrow$) Assume $\mathbb{S}, m, s \Vdash \langle\pi^*\rangle\alpha$, then by definition there is an s_k such that $R^*_\pi(s, s_k)$ and $\mathbb{S}, m, s_k \Vdash \alpha$. Moreover this means that there are $s_1, \ldots, s_k$ such that $R_\pi(s_i, s_{i+1})$. We can give a surviving strategy for $\exists$ as follows: first she plays, in order, markings $m_1, \ldots, m_k$ such that $Z_{m_i} = \{(e, s_i)\}$. These markings constitute legitimate moves for $\exists$ and constrain $\forall$ to follow the path $s, s_1, \ldots, s_k$. Finally, she plays the marking m which by hypothesis makes α true at s_k and leaves $\forall$ only one choice, namely (b, t). ◀

For the inductive case let $E = \{e, e_1, \ldots, e_n\}$ with $n > 0$. If $e \notin \xi(e)$ we skip to the next step, otherwise we need to treat this equation first. Let $\xi(e) = \alpha \vee \langle\pi\rangle e \vee \bigvee_{u \in U}\langle\pi_u\rangle u$ be such that $e \notin U$. In order to eliminate e from $\xi(e)$ we create a slightly modified version of $\mathbf{E}$.

Claim 2 Let $\mathbf{E}' := (E, \xi', \mu)$ with $\xi'(e) := \langle\pi^*\rangle\alpha \vee \bigvee_{u \in U}\langle\pi^*;\pi_u\rangle u$ and let $\xi'(e_i) := \xi(e_i)$ for all i. For all $s, t \in |\mathbb{S}|$ and $b \in B$, the following are equivalent,

(i) $\exists$ has a surviving strategy in $\mathcal{S}(\mathbf{E}, \mathbb{S})@(e, s)$ taking her to (b, t),
(ii) $\exists$ has a surviving strategy in $\mathcal{S}(\mathbf{E}', \mathbb{S})@(e, s)$ taking her to (b, t).

Proof of Claim. As the two sets only differ on e, it will be enough to show that, given a strategy for $\exists$, we can simulate the moves made by $\exists$ (when standing at e) in one set of equations using the other set of equations.

($\Rightarrow$) The type of $\mathbf{E}$ is μ, therefore $\exists$ will only play a finite amount of moves. Assume $\exists$ plays, in order, markings $m_1, \ldots, m_k$ such that $\forall$ chooses (e, s_i) on each round and finally plays a marking m such that $\forall$ chooses (x, s') with $x \neq e$. It is easy to check that in $\mathbf{E}'$ she can play m and will also get to (x, s').

($\Leftarrow$) Suppose $\exists$ plays a marking m such that it actually makes $\langle\pi^*\rangle\alpha$ true (the case for $\langle\pi^*;\pi_u\rangle u$ is analogous) and $\forall$ chooses (x, s'). This means that there is an R_{π^*} path $s, s_1, \ldots, s_k$ and a marking m_α with $\mathbb{S}, m_\alpha, s_k \Vdash \alpha$. She can simulate this play in $\mathbf{E}$ by playing as follows: first she plays, in order, markings m_i such that $Z_{m_i} = \{(e, s_i)\}$; after that she plays m_α. ◀

Having removed e from $\xi(e)$, we still have a formula where other elements of E may occur. We first substitute $\xi'(e)$ into the other equations, setting $\xi'(e_i) := \xi(e_i)[e \mapsto \xi'(e)]$ for all i. It is easy to see that this substitution preserves the behaviour of $\mathbf{E}'$.

Using the distribution laws of the diamond and PDL^{tf} identities the new formulas can be taken to the normal form in $(*)$. To illustrate the process suppose $\xi(e_i) = \alpha \vee \langle\pi_e\rangle e \vee \bigvee_{u \in U}\langle\pi_u\rangle u$ with $e \notin U$ and $\xi'(e) = \alpha' \vee \bigvee_{u \in U}\langle\pi'_u\rangle u$.[4]

[4] To simplify the presentation we assume that U is the same in $\xi(e_i)$ and $\xi(e)$. This need not be this way but the process can be easily adjusted to work for the general case.

The formula $\xi'(e_i)$ is then obtained as follows:

$$\alpha \vee \langle \pi_e \rangle e \vee \bigvee_{u \in U} \langle \pi_u \rangle u \qquad \text{(before replacement)}$$

$$\alpha \vee \langle \pi_e \rangle \big(\alpha' \vee \bigvee_{u \in U} \langle \pi'_u \rangle u\big) \vee \bigvee_{u \in U} \langle \pi_u \rangle u \qquad \text{(after replacement)}$$

$$\big(\alpha \vee \langle \pi_e \rangle \alpha'\big) \vee \bigvee_{u \in U} \langle \pi_e \rangle \langle \pi'_u \rangle u \vee \langle \pi_u \rangle u \quad \text{(distribution of diamonds, regrouping)}$$

$$\big(\alpha \vee \langle \pi_e \rangle \alpha'\big) \vee \bigvee_{u \in U} \langle \pi_e ; \pi'_u \oplus \pi_u \rangle u \qquad \text{(program identities)}$$

We inductively solve the smaller set of equations $\mathbf{E}'' := (E \setminus \{e\}, \xi', \mu)$ and get formulas ψ_u for every $u \in E \setminus \{e\}$. Finally we give a solution for e setting $\varphi_{E,e} := \xi'(e)[u \mapsto \psi_u \mid u \in E \setminus \{e\}]$. Observe that $\varphi_{E,e} \in \mathrm{PDL}^{tf}(\mathsf{P} \uplus B, \mathsf{D})$ because it is of the form $\alpha \vee \bigvee_{u \in U} \langle \pi_u \rangle \psi_u$ where (by induction and hypothesis) we have $\alpha, \psi_u \in \mathrm{PDL}^{tf}(\mathsf{P} \uplus B, \mathsf{D})$. □

Modalities in the Realm of Questions: Axiomatizing Inquisitive Epistemic Logic

Ivano Ciardelli [1]

ILLC, University of Amsterdam
Science Park 107
1098 XG Amsterdam

Abstract

Building on ideas from inquisitive semantics, the recently proposed framework of *inquisitive epistemic logic* (IEL) provides the tools to model and reason about scenarios in which agents do not only have information, but also entertain issues. This framework has been shown to allow for a generalization to issues of important notions, such as common knowledge and public announcements, and it has been argued to form a suitable basis for the analysis of information exchange as an interactive process of raising and resolving issues. From an abstract point of view, the system is interesting, in that it implies extending the logical operations, including the modalities, beyond the truth-conditional realm, in such a way that they can embed not only standard declarative formulas, but also interrogatives. The present paper investigates the logic of IEL, building up to a completeness result. It is shown that the standard logical features of the logical constants extend smoothly beyond the truth-conditional realm, except for double negation, which is the hallmark of truth-conditionality. In particular, while the modalities of IEL operate in a crucially richer semantic space than Kripke modalities do, they retain entirely standard logical features.

Keywords: Epistemic logic, inquisitive semantics, logic of questions.

1 Introduction

Standard epistemic logic provides a framework to reason about scenarios comprising facts and information. In a large body of work (see [7], [1] for recent surveys), this framework has been taken to provide a basis for dynamic logics that aim at describing information exchange. However, an exchange of information is not a mere sequence of informative utterances; rather, it is best regarded as an orderly process in which participants try to achieve certain epistemic goals by raising and addressing issues. In order to formalize this idea, we need a framework that describes not only the information that agents have, but also the issues that they entertain, and that allows us to reason about them.

[1] I am indebted to Jeroen Groenendijk and Floris Roelofsen for useful discussions of the ideas presented here. Financial support from the Netherlands Organization for Scientific Research (NWO) is gratefully acknowledged.

Providing such a framework is the aim of the *inquisitive epistemic logic* proposed in [6], which extends epistemic logic with issues and interrogative formulas, building on ideas and techniques recently developed in *inquisitive semantics* ([5], [3], [4], among others). From an abstract perspective, this framework is interesting in that it shows that the standard account of the logical constants, including the modalities, can be extended smoothly and conservatively beyond the truth-conditional realm, to a richer setting where both declaratives and interrogatives receive a natural interpretation. Moreover, such a generalization has been shown in [6] to extend to other key notions of epistemic and dynamic logics, such as common knowledge and public announcements.

In the present paper we investigate the logic to which this framework gives rise, illustrating the significance of entailment in this richer semantic context, and building up to a completeness result. The paper is organized as follows: section 2 introduces inquisitive epistemic logic; section 3 discusses the combined notion of entailment that arises from this framework; finally, in section 4 a proof system for this logic is provided, and a completeness result is established.

2 Inquisitive epistemic logic

This section provides a concise overview of inquisitive epistemic logic. For a more detailed introduction to the system, and for proofs of the results stated in this section, the reader is referred to [6]. Our presentation will make use of two fundamental ingredients, the notions of *information states* and *issues*. The former notion is standard, while the latter comes from work on inquisitive semantics (for motivation, see e.g. [3], [6]). An information state represents a piece or body of information, identified with the set of worlds compatible with it. Similarly, an issue represents a certain desire, or request, for information, identified with those states in which it is *resolved*.

Definition 2.1 [States and issues] If $\mathcal{W}$ is a set of possible worlds, then:

- an *information state* is a subset $s \subseteq \mathcal{W}$;
- an *issue* is a non-empty set I of information states which is *downward closed*: if $s \in I$ and $t \subseteq s$, then $t \in I$. We denote by $\mathcal{I}$ the set of all issues.

Intuitively, downward closure corresponds to the following persistence property of resolution conditions: if I is resolved in s, and t is more informed than s, then I is resolved in t as well. Now, an issue I can only be truthfully resolved if the actual world w is located in some resolving state $s \in I$, that is, if $w \in \bigcup I$. Hence, an issue I assumes the information corresponding to $\bigcup I$: we will say that it is an issue *over* the state $s = \bigcup I$.

With these notions in place, we are ready to define the models for our logic. Standard epistemic logic allows us to model and reason about certain facts, together with what certain agents know about these facts. Accordingly, a possible world w is fully specified by two aspects: (i) a propositional valuation $V(w)$, which specifies which atomic sentences are true at w; and (ii) for every agent a, an information state $\sigma_a(w)$, representing the information available to a in w. Thus, a model for epistemic logic is a triple $\langle \mathcal{W}, V, \{\sigma_a \,|\, a \in \mathcal{A}\}\rangle$, where the functions $\sigma_a : \mathcal{W} \to \wp(\mathcal{W})$, called *epistemic maps*, are constrained by certain requirements, the most standard being factivity and introspection.

In *inquisitive* epistemic logic, what matters is not only the information that agents have, but also the issues that they entertain. Thus, the description of a possible world will comprise a third aspect: for every agent a, our models will have to specify an issue $\Sigma_a(w)$ over $\sigma_a(w)$, which represents the inquisitive state of a, the agent's desire to locate the actual world more precisely inside her information state. Intuitively, if the inquisitive state of a is $\Sigma_a(w)$, this means that a's epistemic goals are to reach some information state $t \in \Sigma_a(w)$.

Now, since $\Sigma_a(w)$ is required to be an issue over $\sigma_a(w)$, we must have that $\sigma_a(w) = \bigcup \Sigma_a(w)$. Hence, the map Σ_a by itself describes both information and issues of a, and we do not need σ_a as a separate component of our models. Like in standard epistemic logic, the maps Σ_a may be constrained by specific requirements. Since the choice of these requirements is rather orthogonal to the main novelties introduced, IEL builds on the most standard version of epistemic logic, requiring the maps to satisfy factivity and introspection, where the latter now concerns both information and issues.

Definition 2.2 [Inquisitive epistemic models] An inquisitive epistemic model for a set $\mathcal{P}$ of atoms and a set $\mathcal{A}$ of agents is a triple $M = \langle \mathcal{W}, V, \Sigma_\mathcal{A} \rangle$ where:

- $\mathcal{W}$ is a set, whose elements we refer to as *possible worlds*.
- $V : \mathcal{W} \to \wp(\mathcal{P})$ is a *valuation map* that specifies for every world w which atomic sentences are true at w.
- $\Sigma_\mathcal{A} = \{\Sigma_a \mid a \in \mathcal{A}\}$ is a set of *state maps* $\Sigma_a : \mathcal{W} \to \mathcal{I}$, each of which assigns to any world w an issue $\Sigma_a(w)$, in accordance with:

  Factivity : for any $w \in \mathcal{W}$, $w \in \sigma_a(w)$
  Introspection : for any $w, v \in \mathcal{W}$, if $v \in \sigma_a(w)$, then $\Sigma_a(v) = \Sigma_a(w)$

  where $\sigma_a(w) := \bigcup \Sigma_a(w)$ represents the *information state* of agent a in w.

Now that issues have entered the stage, it seems natural to equip the logical language with the means to talk about them. Following *dichotomous inquisitive semantics* ([4]), this is done in IEL by augmenting a standard logical language of *declaratives* with a new syntactic category, the category of *interrogatives*. The set $\mathcal{L}_!$ of declarative formulas and the set $\mathcal{L}_?$ of interrogative formulas of IEL are defined by simultaneous recursion as follows.

Definition 2.3 [Syntax]
Let $\mathcal{P}$ be a set of atomic sentences and let $\mathcal{A}$ be a set of agents.

(i) For any $p \in \mathcal{P}$, $p \in \mathcal{L}_!$

(ii) $\bot \in \mathcal{L}_!$

(iii) If $\alpha_1, \dots, \alpha_n \in \mathcal{L}_!$, then $?\{\alpha_1, \dots, \alpha_n\} \in \mathcal{L}_?$

(iv) If $\varphi \in \mathcal{L}_\circ$ and $\psi \in \mathcal{L}_\circ$, then $\varphi \wedge \psi \in \mathcal{L}_\circ$, where $\circ \in \{!, ?\}$

(v) If $\varphi \in \mathcal{L}_! \cup \mathcal{L}_?$ and $\psi \in \mathcal{L}_\circ$, then $\varphi \to \psi \in \mathcal{L}_\circ$, where $\circ \in \{!, ?\}$

(vi) If $\varphi \in \mathcal{L}_! \cup \mathcal{L}_?$ and $a \in \mathcal{A}$, then $K_a\varphi \in \mathcal{L}_!$

(vii) If $\varphi \in \mathcal{L}_! \cup \mathcal{L}_?$ and $a \in \mathcal{A}$, then $E_a\varphi \in \mathcal{L}_!$

Importantly, conjunction, implication, and the modalities are allowed to apply to interrogatives as well as declaratives. Also, notice how the two syntactic categories are intertwined: from a sequence of declaratives, clause (iii) allows us to form a *basic interrogative*, from which more complex interrogatives may be formed by means of clauses (iv) and (v). On the other hand, clauses (vi) and (vii) allow us to embed an interrogative under a modality, resulting in a new declarative. In this way, we can form sentences such as $E_a?K_b?p$ (which will express the fact that a wants to get to know whether b knows whether p).

Besides our primitive connectives, we also make use of some defined ones. We write $\varphi \leftrightarrow \psi$ for $(\varphi \to \psi) \wedge (\psi \to \varphi)$ and $\neg\varphi$ for $\varphi \to \bot$. Moreover, for α and β declaratives, we write $\alpha \vee \beta$ for $\neg(\neg\alpha \wedge \neg\beta)$, and $?\alpha$ for $?\{\alpha, \neg\alpha\}$.

Throughout the paper, we adopt the following notational convention: α, β, γ range over declaratives, μ, ν, λ over interrogatives, and φ, ψ, χ over the whole language. Moreover, Γ ranges over sets of declaratives, Λ over sets of interrogatives, and Φ over sets of formulas in the whole language.

We now have to specify a semantics for this language. Standardly, this means giving truth-conditions with respect to worlds in a model. However, our language now contains interrogatives as well as declaratives. We do not lay out the meaning of an interrogative by specifying at which worlds it is true, but rather by specifying what information is needed to resolve it. Thus, the natural evaluation points for interrogatives are not worlds, but rather information states. One option would then be to define by simultaneous recursion truth for declaratives and resolution for interrogatives. However, IEL adopts a solution which is both more practical and conceptually more insightful: it lifts the interpretation of *all* formulas from worlds to information states. This brings out the interesting fact that the logical operations—conjunction, implication, and the modalities—make a uniform semantic contribution, and display uniform logical properties, whether they apply to declaratives or to interrogatives. The semantics of IEL is thus defined by a relation of *support* between information states and formulas. Intuitively, for a declarative being supported amounts to being *established*, while for an interrogative it amounts to being *resolved*.

Definition 2.4 [Support] Let M be a model and s an information state in M.

(i) $M, s \models p \iff p \in V(w)$ for all worlds $w \in s$

(ii) $M, s \models \bot \iff s = \emptyset$

(iii) $M, s \models ?\{\alpha_1, \ldots, \alpha_n\} \iff M, s \models \alpha_1$ or ... or $M, s \models \alpha_n$

(iv) $M, s \models \varphi \wedge \psi \iff M, s \models \varphi$ and $M, s \models \psi$

(v) $M, s \models \varphi \to \psi \iff$ for every $t \subseteq s$, if $M, t \models \varphi$ then $M, t \models \psi$

(vi) $M, s \models K_a\varphi \iff$ for every $w \in s$, $M, \sigma_a(w) \models \varphi$

(vii) $M, s \models E_a\varphi \iff$ for every $w \in s$ and every $t \in \Sigma_a(w)$, $M, t \models \varphi$

The set of states in M that support φ is called the *proposition expressed* by φ

and denoted $[\varphi]_M$. Reference to the model M will be dropped when possible.

Before turning to an explanation of the clauses, let us review some fundamental facts and notions. A first, crucial feature of IEL is that support is *persistent.*

Fact 2.5 (Persistence) *If $M, s \models \varphi$ and $t \subseteq s$, then $M, t \models \varphi$.*

Thus, more formulas are supported as information grows. In the limit, the empty state supports all formulas. Thus, we refer to $\emptyset$ as the *absurd* state.

Fact 2.6 *$M, \emptyset \models \varphi$ for any M and φ.*

Together, these two properties guarantee that $[\varphi]_M$ is always an issue, in the sense of Definition 2.1 (or, an *inquisitive proposition*, as non-empty downward closed sets of states are also called in inquisitive semantics).

Although our semantics is defined in terms of support, *truth* at a world can be recovered by defining it as support at the corresponding singleton state.

Definition 2.7 [Truth]
We say that φ is *true* at a world w, notation $M, w \models \varphi$, in case $M, \{w\} \models \varphi$. The set of worlds at which φ is true is called the *truth-set* of φ, notation $|\varphi|_M$.

Writing out the support clauses for singletons, it is easy to see that the connectives get their standard truth-conditional clauses, even when their constituents are interrogative. Moreover, persistence implies that a world makes a formula true iff it is contained in some supporting state.

Fact 2.8 *$M, w \models \varphi \iff w \in s$ for some state s s.t. $M, s \models \varphi$.*

This fact also tells us how truth should be viewed for interrogatives: an interrogative μ is true in w iff $w \in s$ for some s which resolves μ, that is, iff there is some body of information that resolves μ and is true at w. In other words, an interrogative is true at those worlds where it can be truthfully resolved.

In general, truth conditions do not determine support conditions. For instance, the polar interrogatives $?p$ and $?q$ are both true everywhere, but clearly, in general they have different support conditions. However, the semantics of declaratives *is*, as usual, fully determined by truth-conditions: for, a declarative is supported in a state iff it is true at all the worlds in the state.

Fact 2.9 *For any M, s, and α: $M, s \models \alpha \iff (M, w \models \alpha$ for all $w \in s)$.*

Importantly, this does *not* mean that truth for declaratives can be defined *independently* of support. For, the truth of a modal formula depends crucially on the *support* conditions of its argument, as the following truth-clauses show.

- $M, w \models K_a\varphi \iff M, \sigma_a(w) \models \varphi$
- $M, w \models E_a\varphi \iff$ for all $t \in \Sigma_a(w)$, $M, t \models \varphi$

Another notion that will play a crucial role below is that of *resolutions* of a formula; intuitively, the resolutions of a formula are declaratives that correspond to the different *ways* in which the formula may be settled.

Definition 2.10 [Resolutions]
The set $\mathcal{R}(\varphi)$ of resolutions of a formula φ is defined recursively as follows:

- $\mathcal{R}(\alpha) = \{\alpha\}$ if α is a declarative
- $\mathcal{R}(?\{\alpha_1, \ldots, \alpha_n\}) = \{\alpha_1, \ldots, \alpha_n\}$
- $\mathcal{R}(\mu \wedge \nu) = \{\alpha \wedge \beta \,|\, \alpha \in \mathcal{R}(\mu) \text{ and } \beta \in \mathcal{R}(\nu)\}$
- $\mathcal{R}(\varphi \to \mu) = \{\bigwedge_{\alpha \in \mathcal{R}(\varphi)} \alpha \to f(\alpha) \,|\, f : \mathcal{R}(\varphi) \to \mathcal{R}(\mu)\}$

We may think of the resolutions of an interrogative as syntactic answers to it. [2] The next fact, provable by induction, says that to resolve an interrogative is to establish some resolution of it; so, each resolution provides sufficient information to resolve the interrogative, and, taken together, the resolutions exhaust the ways in which an interrogative may be resolved.

Fact 2.11 *For any M, s and φ, $M, s \models \varphi \iff M, s \models \alpha$ for some $\alpha \in \mathcal{R}(\varphi)$*

As a corollary, we get the following normal form result: every formula φ is equivalent to a basic interrogative having the resolutions of φ as constituents. [3]

Corollary 2.12 (Normal form) *For any φ, $\varphi \equiv ?\mathcal{R}(\varphi)$.*

In terms of resolutions we define the notion of *presupposition* of an interrogative.

Definition 2.13 [Presupposition of an interrogative]
The presupposition of an interrogative μ is the declarative $\pi_\mu = \bigvee \mathcal{R}(\mu)$.

Since the interrogative operator has the same truth conditions as a disjunction, it follows from Corollary 2.12 that μ and π_μ have the same truth conditions. The notion of resolution can be generalized to sets of formulas as follows.

Definition 2.14 [Resolutions of a set]
The set $\mathcal{R}(\Phi)$ of resolutions of a set Φ contains those sets Γ of declaratives s.t.:

- for all $\varphi \in \Phi$ there is an $\alpha \in \Gamma$ such that $\alpha \in \mathcal{R}(\varphi)$
- for all $\alpha \in \Gamma$ there is a $\varphi \in \Phi$ such that $\alpha \in \mathcal{R}(\varphi)$

That is, a resolution of a set Φ is a set of declaratives obtained by replacing every formula in Φ by one or more resolutions. Notice that, since a declarative has itself as unique resolution, the declarative component of Φ is inherited by any resolution. In particular, if Γ is a set of declaratives, $\mathcal{R}(\Gamma) = \Gamma$. Fact 2.11 generalizes to sets: writing $M, s \models \Phi$ for '$M, s \models \varphi$ for all $\varphi \in \Phi$', we have:

Fact 2.15 *For any M, s and Φ, $M, s \models \Phi \iff M, s \models \Gamma$ for some $\Gamma \in \mathcal{R}(\Phi)$*

Equipped with these basic facts and notions, we are ready to briefly explain the support clauses. Clause (i) simply says that an atom is established in a state iff it is true everywhere in the state, a fact that we have seen to hold for

[2] Indeed, our notion of *resolutions* is a general version of the notion of *basic answers* in the interrogative frameworks of Hintikka [8,9] and Wiśniewski [11]. We use the term *resolutions* as a reminder that this is only a specific technical notion, sufficient for the present purposes. More can be said, also in a logical framework, on the complex phenomenon of *answerhood*. Our notion of *presupposition* of a question is also shared by the mentioned theories.

[3] It may seem strange that a declarative α can be equivalent to an interrogative. However, the corresponding interrogative is the trivial interrogative $?\{\alpha\}$, which has a unique resolution α. Also, while α and $?\{\alpha\}$ are semantically equivalent, they differ in pragmatics: see [6].

declaratives in general. Similarly for (ii), which says that the falsum is only established in the absurd state $\emptyset$. Clause (iii) lays out the resolution conditions for basic interrogatives: $?\{\alpha_1, \ldots, \alpha_n\}$ is resolved in a state s iff some α_i is established in s. Clause (iv) says that a conjunction is established (resolved) in a state s iff both conjuncts are established (resolved) in s.

The clause for implication requires some more explanation. First, if the antecedent is a declarative α, clause (v) amounts to the simpler clause (v′):

(v′) $M, s \models \alpha \to \varphi \iff M, s \cap |\alpha| \models \varphi$

The conditional $\alpha \to \varphi$ is established (resolved) in s iff φ is established (resolved) in the state $s \cap |\alpha|$ which results from augmenting s with the assumption that α is true. For a conditional *declarative*, this delivers a standard material implication. At the same time, this also yields conditional interrogatives like $p \to ?q$, which is resolved in a state iff either $p \to q$ or $p \to \neg q$ is established.

Now consider the case in which the antecedent is an interrogative μ. If the consequent is a declarative α, then $\mu \to \alpha$ is a declarative, and may be seen to be equivalent with $\pi_\mu \to \alpha$: so, interrogative antecedents may be substituted by their presuppositions when the consequent is declarative, and add no expressive power to the language. If the consequent is itself an interrogative ν, on the other hand, the clause says that $\mu \to \nu$ is resolved in s in case, if we extend s so as to resolve μ, the resulting state will resolve ν. So, we can resolve $\mu \to \nu$ if we can resolve ν *conditionally* on having a resolution of μ, i.e., if we have enough information to turn any resolution of μ into a resolution of ν. E.g., the conditional interrogative $?p \to ?q$ is resolved precisely in case at least one of the following declaratives is established:

1. $(p \to q) \land (\neg p \to q) \equiv q$
2. $(p \to q) \land (\neg p \to \neg q) \equiv q \leftrightarrow p$
3. $(p \to \neg q) \land (\neg p \to q) \equiv q \leftrightarrow \neg p$
4. $(p \to \neg q) \land (\neg p \to \neg q) \equiv \neg q$

which are precisely the four ways to link the answer to $?q$ to the answer to $?p$.

To conclude the tour, let us consider the modalities, starting with K_a. Since $K_a\varphi$ is a declarative, Fact 2.9 guarantees that we need only look at truth-conditions. Now, $K_a\varphi$ is true at w in case the information state $\sigma_a(w)$ of a at w supports φ. If φ is a declarative α, this simply means that α is true everywhere in $\sigma_a(w)$, and we recover the standard truth conditions familiar from epistemic logic: $K_a\alpha$ is true iff α is true everywhere in a's information state. At the same time, K_a is more general in IEL, since it also embeds interrogatives. For an interrogative μ, we read $\sigma_a(w) \models \mu$ as "μ is resolved in $\sigma_a(w)$". Thus, $K_a\mu$ is true at w in case the information available to a at w resolves μ. For instance, we have: $w \models K_a?\alpha \iff \sigma_a(w) \subseteq |\alpha|$ or $\sigma_a(w) \subseteq |\neg\alpha| \iff w \models K_a\alpha \lor K_a\neg\alpha$.

Now consider the *entertain* modality E_a. The clause says that $E_a\varphi$ is true at w iff any $t \in \Sigma_a(w)$ supports φ, that is, if φ is supported in all the states where a's epistemic goals are achieved. If φ is a declarative α, it is not difficult to see using Fact 2.9 that this holds iff $\sigma_a(w)$ supports α, which means that $E_a\alpha$ boils down to $K_a\alpha$. On the other hand, if φ is an interrogative μ, then the clause says that $E_a\mu$ holds in case every state that a wants to reach is a state where μ is resolved. Thus, $E_a\mu$ expresses that a wants resolve μ.

Notice that, if $K_a\mu$ holds, i.e., if $\sigma_a(w)$ *already* resolves μ, then all enhancements of $\sigma_a(w)$ will resolve μ as well, so $E_a\mu$ will hold too. However, combining K_a and E_a, we can define a modality W_a which rules out this case, expressing *not knowing and wanting to know*. We read $W_a\varphi$ as "*a wonders about* φ".

- $W_a\varphi := \neg K_a\varphi \wedge E_a\varphi$

Finally, a remark about the mathematical workings of the modalities. In standard EL, the modality K_a expresses a relation (inclusion) between two semantic objects of the same kind, namely, two sets of worlds: a state $\sigma_a(w)$ associated with the evaluation world, and a proposition $|\varphi|$ expressed by its argument. In general, Kripke modalities may be seen as expressing such a relation. The modalities of our system are not Kripke modalities—they are not quantifiers over accessible worlds—yet in a sense they behave in just the same way. Now both the state $\Sigma_a(w)$ associated with the evaluation world and the proposition $[\varphi]$ expressed by a sentence are more structured objects than simple sets of worlds, embodying both information and issues. Accordingly, more types of relations between them are possible. Our modalities express two such relations, as shown by the next reformulation of their truth-conditions:

- $M, w \models K_a\varphi \iff \bigcup \Sigma_a(w) \in [\varphi]$
- $M, w \models E_a\varphi \iff \Sigma_a(w) \subseteq [\varphi]$

To sum up, IEL is a conservative extension of standard EL, in that all EL-formulas in the language receive their standard truth conditions. At the same time, IEL allows us to talk not only about the facts that agents know, but also about the issues that they can resolve and that they entertain, including higher-order ones. Moreover, containing both declarative and interrogative sentences, IEL provides a suitable ground for a dynamics in which announcements may both provide information and raise issues, as spelled out in detail in [6].

3 Entailment

Entailment in IEL is defined in the natural way, as preservation of support.

Definition 3.1 [Entailment]
$\Phi \models \psi \iff$ for any model M and state s, if $M, s \models \Phi$ then $M, s \models \psi$.

To see what this notion captures, consider first entailment towards a declarative. Fact 2.9 implies that, in this case, only truth-conditional content matters.

Fact 3.2 $\Phi \models \alpha \iff$ *for any* M *and world* w : *if* $M, w \models \Phi$ *then* $M, w \models \alpha$.

Thus, entailment among declaratives amounts as usual to preservation of truth. In particular, since formulas in the language of epistemic logic get their usual truth conditions, for them entailment amounts to entailment in epistemic logic.

Fact 3.3 (Conservativity over epistemic logic)

Let Γ, α *consist of formulas in* $\mathcal{L}_{\mathsf{EL}}$. *Then* $\Gamma \models \alpha \iff \Gamma \models_{\mathsf{EL}} \alpha$

However, the declarative fragment of IEL is strictly richer than epistemic logic, encompassing in particular a logic of entertaining issues. As an example, consider:

$$E_a?\{p, q, r\} \models K_a\neg p \rightarrow E_a?\{q, r\}$$

This reads: suppose a wants to establish at least one of p, q, and r; it follows that if a knows that $\neg p$, then a wants to establish one of q and r.

What about the case in which we also have some interrogative assumptions? Well, since only truth-conditions matter when the conclusion is a declarative, each interrogative assumption μ may be replaced by its presupposition π_μ, which shares the same truth conditions. Thus, entailment towards declaratives is essentially a declarative business.

Let us now consider the case in which the conclusion is an interrogative. We first establish an important characterization of entailment in IEL. Recall that to support a formula, or a set, is to support some resolution of it (facts 2.11 and 2.15). From this, we get the following characterization, which shows how cross-categorial entailment is grounded in declarative entailment: Φ entails ψ iff every resolution of Φ entails some resolution of ψ.

Fact 3.4 $\Phi \models \psi \iff$ *for all* $\Gamma \in \mathcal{R}(\varphi)$ *there is an* $\alpha \in \mathcal{R}(\psi)$ *s.t.* $\Gamma \models \alpha$.

Now, decomposing Φ into a set Γ of declaratives and a set Λ of interrogatives, and assuming ψ is an interrogative μ, this tells us that $\Gamma, \Lambda \models \mu$ holds iff any resolution of all interrogatives in Λ, together with Γ, entails some resolution of μ; that is, if *given* Γ, *any resolution of the interrogatives in* Λ *determines some resolution of* μ. For instance, the following entailment is valid

$$p \leftrightarrow q \wedge r,\ ?q \wedge ?r \models ?p$$

since, given $p \leftrightarrow q \wedge r$, any resolution of the conjunctive question $?q \wedge ?r$ determines a resolution of $?p$: the resolution $q \wedge r$ determines the resolution p, the resolution $q \wedge \neg r$ determines the resolution $\neg p$, and so on. Thus, entailment involving an interrogative conclusion and interrogative assumptions captures the notion of *interrogative dependency.*[4]

Finally, how about the case in which we have an interrogative conclusion and *no* interrogative assumption? Well, since a set of declaratives Γ is the only resolution of itself, it follows form Fact 3.4 that Γ entails an μ iff it establishes some resolution of μ, i.e., in case it settles μ in a particular way.

Fact 3.5 *If* Γ *is a set of declaratives,* $\Gamma \models \psi \iff \Gamma \models \alpha$ *for some* $\alpha \in \mathcal{R}(\psi)$.

Summing up, the combined notion of entailment of IEL unifies three crucial and seemingly independent notions of a logic of information and issues: *standard declarative entailment*, *answerhood*, and *interrogative dependency.*

[4] For a discussion of this interesting aspect of inquisitive logic, and the way in which it relates to the framework of dependence logic ([10,12]), see [2].

Having clarified the significance of entailment, let us turn to the formal properties of the modalities. First, the modalities of IEL are distributive. This is easy to check, but not obvious, since our modalities are not Kripke modalities.

Fact 3.6 $\Box(\varphi \to \psi) \to (\Box\varphi \to \Box\psi)$ *is valid for* $\Box \in \{K_a, E_a \mid a \in \mathcal{A}\}$.

Moreover, it has been already mentioned above that Fact 2.9 implies that the two modalities are equivalent when applied to declaratives.

Fact 3.7 *For any declarative* α, $K_a\alpha \equiv E_a\alpha$

The next fact says that the knowledge modality distributes over the resolutions of a formula: that is, to know a formula is to know some resolution of it.

Fact 3.8 *For any formula* φ, $K_a\varphi \equiv \bigvee_{\alpha\in\mathcal{R}(\varphi)} K_a\alpha$

Proof. Since we are dealing with declaratives, we just need to check identity of truth conditions. Now, $K_a\varphi$ is true at a world w just in case φ is supported by $\sigma_a(w)$. By Fact 2.11, this is the case iff $\sigma_a(w)$ supports some $\alpha \in \mathcal{R}(\varphi)$, which in turn is precisely what is needed for $\bigvee_{\alpha\in\mathcal{R}(\varphi)} K_a\alpha$ to be true at w. □

Taken together, the last two facts imply that the knowledge modality can be completely paraphrased away from our language: for, Fact 3.8 tell us that a K_a-formula is always equivalent to one in which K_a applies only to declaratives; then, Fact 3.7 states that, on declaratives, K_a may be replaced by E_a. Thus, any formula is equivalent to a K_a-free one. Notice, however, that K_a is not *uniformly* definable in terms of E_a and the connectives: for, the paraphrase of $K_a\varphi$ depends on the specific formula φ; moreover, it is possible to show that the size of the paraphrase may grow exponentially relative to the size of φ.

As for the modality E_a, it is worth remarking that $E_a\varphi$ is *not* in general equivalent to a formula where the modalities are applied only to declaratives. For, on declaratives, E_a coincides with K_a: thus, the truth-conditions of any formula in which the modalities occur only on declaratives depend exclusively on the epistemic states of the agents, as well as the propositional valuation; but of course, in general, the truth of a formula $E_a\varphi$ depends crucially on a's issues, not just on a's information; hence, $E_a\varphi$ cannot in general be equivalent to any formula in which the modalities occur only applied to declaratives.

Intuitively, this witnesses that *entertaining* is a relation between an agent and an issue, which is not reducible to a more basic relation between the agent and a proposition, as in the case of *knowing*. Formally, it shows that the enrichment that comes about by letting modalities embed interrogatives is substantial, as we can express things that we could not express in epistemic logic.

Finally, it is easy to verify that the factivity and introspection conditions required from state maps render valid the usual schemes for both modalities.

Fact 3.9 *The following are valid for* $\Box \in \{K_a, E_a \mid a \in \mathcal{A}\}$, *and* α *declarative:*

- $\Box\alpha \to \alpha$
- $\Box\varphi \to \Box\Box\varphi$
- $\neg\Box\varphi \to \Box\neg\Box\varphi$

Conjunction

$$\frac{\varphi \quad \psi}{\varphi \wedge \psi} \qquad \frac{\varphi \wedge \psi}{\varphi} \qquad \frac{\varphi \wedge \psi}{\psi}$$

Implication

$$\frac{\begin{array}{c}[\varphi]\\ \vdots \\ \psi\end{array}}{\varphi \to \psi} \qquad \frac{\varphi \quad \varphi \to \psi}{\psi}$$

Interrogative

$$\frac{\alpha_i}{?\{\alpha_1, \dots, \alpha_n\}} \qquad \frac{\begin{array}{c}[\alpha_1]\\ \vdots \\ \varphi\end{array} \quad \dots \quad \begin{array}{c}[\alpha_n]\\ \vdots \\ \varphi\end{array} \quad ?\{\alpha_1, \dots, \alpha_n\}}{\varphi}$$

Falsum

$$\frac{\bot}{\varphi}$$

Kreisel-Putnam axiom

$$(\alpha \to ?\{\beta_1, \dots, \beta_n\}) \to ?\{\alpha \to \beta_1, \dots, \alpha \to \beta_n\}$$

Double negation

$$\neg\neg\alpha \to \alpha$$

This concludes our short discussion of the significance and of the features of IEL-entailment. In the next section, the insights gained in the present section will be put to use to provide a sound and complete proof system.

4 Axiomatization

Since the propositional fragment of IEL coincides with dichotomous inquisitive semantics, let us start out with a proof system for this fragment. The table above describes a natural deduction system, proved in [4] to be sound and complete. The standard connectives—conjunction, implication, and falsum—are all assigned their standard inference rules. These inference rules are generalized to apply not only when the constituents are declarative, but also when they are interrogative. Thus, the core proof-theoretic features of the connectives are preserved when these operations are generalized to interrogatives.

There *is*, however, one element of the system which is restricted to declaratives, namely, the double negation axiom. Indeed, the following fact says that the double negation axiom is valid for φ iff φ enjoys the fundamental property of declaratives (Fact 2.9), for which support amounts to truth at each world.

Fact 4.1 (Double negation characterizes truth-conditionality)
$\neg\neg\varphi \to \varphi$ *is valid iff for all* M, s*:* $M, s \models \varphi \iff (M, w \models \varphi$ *for all* $w \in s)$

The rules for the interrogative operator are simply the usual ones for a disjunction. This is hardly surprising, since the semantics of ? is disjunctive. Intuitively, the introduction rule says that if we have established α_i for some i, then we have resolved $?\{\alpha_1, \dots, \alpha_n\}$. The elimination rule says that if we can

E-distributivity $E_a(\varphi \to \psi) \to (E_a\varphi \to E_a\psi)$	K-distributivity $K_a(\varphi \to \psi) \to (K_a\varphi \to K_a\psi)$
E-factivity $E_a\alpha \to \alpha$	K-E equivalence on declaratives $E_a\alpha \leftrightarrow K_a\alpha$
E-positive introspection $E_a\varphi \to E_aE_a\varphi$	K distributes over interrogatives $K_a?\{\alpha_1, \ldots, \alpha_n\} \to K_a\alpha_1 \vee \cdots \vee K_a\alpha_n$
E-negative introspection $\neg E_a\varphi \to E_a\neg E_a\varphi$	Necessitation, for $\Box \in \{K_a, E_a \mid a \in \mathcal{A}\}$ $\dfrac{\begin{matrix}\emptyset \\ \vdots \\ \varphi\end{matrix}}{\Box\varphi}$

infer φ from the assumption that α_i is established for each i, then we can infer φ from the assumption that $?\{\alpha_1, \ldots, \alpha_n\}$ is resolved.[5] The last component of the system is the Kreisel-Putnam axiom, which distributes an implication over an interrogative consequent, provided the antecedent is a declarative.

These ingredients provide a complete axiomatization of the propositional fragment of IEL. We then need to extend this system with axioms and rules for the modalities, which are described in the table above. Each of these corresponds to some property discussed in the previous section: for the entertain modalities we have the distributivity axiom (valid by Fact 3.6) and the axioms rendered valid by the constraints of factivity and introspection (Fact 3.9); for the knowledge modalities we have again distributivity (Fact 3.6), coincidence of K_a and E_a on declaratives (Fact 3.7) and distributivity over the interrogative operator (a special case of Fact 3.8). Finally, we have a standard necessitation rule for both modalities: if φ has been derived without undischarged assumptions, infer $\Box\varphi$. This completes the description of our deduction system.

Definition 4.2 We write $P : \Phi \vdash \psi$ if P is a proof in our deduction system, whose conclusion is ψ and whose set of assumptions is included in Φ. As usual, we then write $\Phi \vdash \psi$ if some proof $P : \Phi \vdash \psi$ exists. We say that two formulas φ and ψ are *provably equivalent*, notation $\varphi \dashv\vdash \psi$, in case $\varphi \vdash \psi$ and $\psi \vdash \varphi$.

As customary, it is a tedious but straightforward matter to check that the proof system is *sound* for entailment in IEL.

[5] The standard rules for negation and disjunction, which are derived connectives in IEL, are admissible, with one caveat: a disjunction may only be eliminated towards a declarative. This restriction marks the difference between $\vee$ and $?$ and prevents unsound derivations such as $p \vee \neg p \vdash ?p$ (remember that $?p$ is defined as $?\{p, \neg p\}$).

Theorem 4.3 (Soundness) *If* $\Phi \vdash \psi$ *then* $\Phi \models \psi$.

We will now prove some facts about the proof system which, besides providing important insights into the logic, play a crucial role in the completeness proof. First, the normal form result of Corollary 2.12 is provable in the system.

Lemma 4.4 *For any* φ, $\varphi \dashv\vdash ?\mathcal{R}(\varphi)$.

Proof. The lengthy but straightforward proof is essentially the same as that given for the propositional fragment in [4]. We omit it in the interest of space.□

As a corollary, a formula is always derivable from any of its resolutions.

Corollary 4.5 *If* $\alpha \in \mathcal{R}(\varphi)$, *then* $\alpha \vdash \varphi$.

Proof. If $\alpha \in \mathcal{R}(\varphi)$ then by a simple application of ?-introduction we have $\alpha \vdash ?\mathcal{R}(\varphi)$, whence by the previous lemma, $\alpha \vdash \varphi$. □

Lemma 4.4 also implies that interrogatives always derive their presupposition.

Corollary 4.6 *For any interrogative* μ, $\mu \vdash \pi_\mu$.

Let us now mention a few basic facts about the modalities. First, we can prove as usual that the distributivity axioms and the necessitation rules together ensure that the modalities are monotonic.

Lemma 4.7
If $\varphi_1, \ldots, \varphi_n \vdash \psi$ *then* $\Box\varphi_1, \ldots, \Box\varphi_n \vdash \Box\psi$ *for* $\Box \in \{E_a, K_a \,|\, a \in \mathcal{A}\}$.

Moreover, the equivalences we have seen in facts 3.7 and 3.8 are provable in our system: that is, K_a distributes over resolutions and coincides with E_a on declaratives, which means that it can be paraphrased away from the language.

Lemma 4.8 *For any* φ, $K_a\varphi \dashv\vdash \bigvee_{\alpha\in\mathcal{R}(\varphi)} K_a\alpha \dashv\vdash \bigvee_{\alpha\in\mathcal{R}(\varphi)} E_a\alpha$

Proof. Immediate from Lemma 4.4, using the axiom of K-distributivity over the interrogative operator and the axiom of K-E equivalence on declaratives.□

The next thing to show is that derivability shares the fundamental property of entailment expressed by Fact 3.4: from Φ we can derive ψ iff from any specific resolution Γ of Φ we can derive some resolution α of ψ.

Theorem 4.9 (Resolution theorem)
$\Phi \vdash \psi \iff$ *for all* $\Gamma \in \mathcal{R}(\Phi)$ *there exists some* $\alpha \in \mathcal{R}(\psi)$ *s.t.* $\Gamma \vdash \alpha$.

The substantial proof is given in the appendix. In particular, the proof of the left-to-right direction has an interesting computational interpretation. For, suppose we have a proof $P : \Phi \vdash \psi$. By soundness, $\Phi \models \psi$, which by Fact 3.4 means that any resolution Γ of Φ entails some resolution α of ψ: the proof of the theorem tells us to use P to find such an α and to produce a proof $Q : \Gamma \vdash \alpha$. Thus, a proof $P : \Phi \vdash \psi$ essentially encodes how a resolution of ψ may be obtained from a resolution of Φ.

Notice that, since a set of declaratives has itself as unique resolution, the resolution theorem has the following corollary.

Corollary 4.10 (Split)
Let Γ be a set of declaratives. If $\Gamma \vdash \psi$ then $\Gamma \vdash \alpha$ for some $\alpha \in \mathcal{R}(\psi)$.

On our way to the proof of the resolution theorem, in the appendix we will also establish the following fact.

Lemma 4.11 *If $\Phi \not\vdash \psi$ then there exists some $\Gamma \in \mathcal{R}(\Phi)$ such that $\Gamma \not\vdash \psi$.*

Since consistency with a declarative α amounts to not deriving $\neg\alpha$, an immediate consequence of this lemma is that if Φ is consistent with α, then some resolution Γ of Φ is consistent with α.

We will prove completeness by constructing a canonical model. The construction is similar to the familiar one from standard modal logic, but slightly more sophisticated, since the model we need to build has a richer structure than a standard Kripke model.

The possible worlds in our canonical model will be *complete theories of declaratives* (CTD), defined as sets Γ of declaratives which are (i) closed under deduction of declaratives; (ii) consistent; and (iii) complete, in the sense that for any declarative α, either α or $\neg\alpha$ is in Γ. The following features of CTDs are familiar from classical logic and modal logic.

Fact 4.12 (Disjunction property)
If Γ is a CTD and $\alpha_1 \vee \cdots \vee \alpha_n \in \Gamma$, then $\alpha_i \in \Gamma$ for some i.

Fact 4.13 (Lindenbaum's lemma)
If Θ is a consistent set of declaratives, then $\Theta \subseteq \Gamma$ for some CTD Γ.

We are now ready to define our canonical model for inquisitive epistemic logic.

Definition 4.14 [Canonical model for IEL]
The canonical model for IEL is the model $M^c = \langle \mathcal{W}^c, V^c, \Sigma^c_{\mathcal{A}} \rangle$, where:

- the elements of $\mathcal{W}^c$ are the complete theories of declaratives
- $V^c(\Gamma) = \{p \in \mathcal{P} \mid p \in \Gamma\}$
- $\Sigma^c_a(\Gamma)$ is the set of states $S \subseteq \mathcal{W}^c$ defined as follows:
  $S \in \Sigma^c_a(\Gamma) \iff \bigcap S \vdash \varphi$ whenever $E_a\varphi \in \Gamma$ [6]

Recall that the information state $\sigma^c_a(\Gamma)$ of an agent a at a world Γ is defined as the union of the inquisitive state $\Sigma^c_a(\Gamma)$ of the agent. The following lemma gives a direct characterization of $\sigma^c_a(\Gamma)$ in terms of the theories that it contains. The proof is given in the appendix.

Lemma 4.15 $\sigma^c_a(\Gamma) = \{\Delta \mid \alpha \in \Delta$ *whenever* $E_a\alpha \in \Gamma\}$

We then have to show that the structure we defined is a proper inquisitive epistemic model, in the sense that the state maps Σ^c_a satisfy the factivity and introspection requirements. This is precisely what the axioms of factivity and positive and negative introspection for E are intended to enforce.

Lemma 4.16 *M^c is an inquisitive epistemic model.*

---

[6] The intersection of the empty state is defined as the set of all formulas, $\bigcap \emptyset = \mathcal{L}_{\mathcal{P}}$.

Proof. It is an easy exercise to check that the axiom $E_a\alpha \to \alpha$ ensures that the state maps are satisfy factivity, while the axioms $E_a\varphi \to E_aE_a\varphi$ and $\neg E_a\varphi \to E_a\neg E_a\varphi$ ensure that the state maps satisfy introspection. □

The bridge between derivability and semantics in the canonical model is usually provided by a *truth lemma* equating truth at a world in the canonical model with derivability from that world. In IEL, the fundamental semantic relation is not truth at a world, but support at a state. Accordingly, the bridge between derivability and semantics is given by the following *support lemma*, stating that support at a state S in the canonical model amounts to derivability from the intersection $\bigcap S$ of all the theories in S. The proof is given in the appendix.

Lemma 4.17 (Support lemma)
For any $S \subseteq \mathcal{W}^c$ and any φ, $M^c, S \models \varphi \iff \bigcap S \vdash \varphi$.

We can then rely on the support lemma to prove the completeness theorem.

Theorem 4.18 (Completeness theorem) *If $\Phi \models \psi$, then $\Phi \vdash \psi$.*

Proof. Suppose $\Phi \nvdash \psi$. By Theorem 4.9, there is a resolution Θ of Φ which does not derive any resolution of ψ. Let $\mathcal{R}(\psi) = \{\alpha_1, \dots, \alpha_n\}$: for each i, since $\Theta \nvdash \alpha_i$, the set $\Theta \cup \{\neg\alpha_i\}$ is consistent, and thus extendible to a CTD $\Gamma_i \in W^c$. Now let $S = \{\Gamma_1, \dots, \Gamma_n\}$: we claim that $S \models \Phi$ but $S \not\models \psi$.

To see that $S \models \Phi$, notice that by construction, $\Theta \in \bigcap S$, whence by the support lemma $M^c, S \models \Theta$. But since $\Theta \in \mathcal{R}(\Phi)$, by Fact 2.15 we also have $M^c, S \models \Phi$. However, suppose S supported ψ: then by Fact 2.11 it should also support α_i for some i. By the support lemma, that would mean that $\bigcap S \vdash \alpha_i$, and so also $\alpha_i \in \Gamma_i$, since $\bigcap S \subseteq \Gamma_i$ and Γ_i is closed under declarative deduction. But that is impossible, since Γ_i is consistent and contains $\neg\alpha_i$ by construction. Hence, $M^c, S \models \Phi$ but $M^c, S \not\models \psi$, which witnesses that $\Phi \not\models \psi$. □

Conclusion In this paper we have investigated and axiomatized the notion of entailment arising from *inquisitive epistemic logic.* Concretely, this gives us a conservative extension of standard epistemic logic in which we can reason not only about the agents' information, but also about the agents' issues, that is, their epistemic goals, thus providing the ground for a logical account of information exchange as a directed process of raising and resolving issues.

From an abstract standpoint, the main finding is that propositional and modal logics generalize smoothly beyond the truth-conditional realm, to a setting where both declaratives and interrogatives receive a uniform semantics. In the proof system, the connectives are handled by their standard rules. What does *not* generalize is the double negation axiom, which characterizes truth-conditionality, and must be restricted to declaratives. Finally, while the modalities of our system are not Kripke modalities, in that they operate on crucially richer semantic objects, they enjoy completely standard logical properties: distributivity holds, and the frame conditions of factivity and introspection are characterized by the familiar schemes.

Appendix

Proof of theorem 4.9 Let us first show the left-to-right direction of the theorem: if Φ derives ψ, any resolution Γ of Φ derives some resolution α of ψ. The proof goes by induction on the complexity of the proof $P : \Phi \vdash \psi$. We distinguish a number of cases depending on the last rule applied in P. In the interest of space, the most straightforward inductive cases are omitted.

- ψ is an undischarged assumption, $\psi \in \Phi$. In this case, any resolution Γ of Φ contains a resolution α of ψ by definition, so $\Gamma \vdash \alpha$.
- ψ is an axiom. If ψ is declarative, the claim is trivially true. If ψ is interrogative, it must be an instance of the Kreisel-Putnam axiom, $(\beta \to ?\{\gamma_1, \ldots, \gamma_n\}) \to ?\{\beta \to \gamma_1 \ldots, \beta \to \gamma_n\}$, since all other axioms are declaratives. In this case, take $\alpha = \bigwedge_{1 \leq i \leq n}((\beta \to \gamma_i) \to (\beta \to \gamma_i))$: α is a resolution of ψ and, being a classical tautology, we have $\Gamma \vdash \alpha$ for any set Γ whatsoever.
- $\psi = \chi \to \mu$ was obtained by an implication introduction rule. Then the immediate subproof of P is a proof of μ from the set of assumptions $\Phi \cup \{\chi\}$. Take any resolution Γ of Φ. Suppose $\alpha_1, \ldots, \alpha_n$ are the resolutions of χ. For any $1 \leq i \leq n$, then, $\Gamma \cup \{\alpha_i\}$ is a resolution of $\Phi \cup \{\chi\}$, whence by induction hypothesis we have a proof $Q_i : \Gamma \cup \{\alpha_i\} \vdash \beta_i$ for some resolution β_i of μ. But then, extending Q_i by an application of implication introduction, we derive $\alpha_i \to \beta_i$ from Γ. And since this is the case for $1 \leq i \leq n$, from Γ we can derive $(\alpha_1 \to \beta_1) \wedge \cdots \wedge (\alpha_n \to \beta_n)$, which is a resolution of $\chi \to \mu = \psi$.
- ψ was obtained by an implication elimination rule from χ and $\chi \to \psi$. Then the immediate subproofs of P are a proof of χ from Φ, and a proof of $\chi \to \psi$ from Φ. Consider any $\Gamma \in \mathcal{R}(\Phi)$. By induction hypothesis we have a proof $Q_1 : \Gamma \vdash \beta$ where $\beta \in \mathcal{R}(\chi)$, and a proof $Q_2 : \Gamma \vdash \gamma$, where $\gamma \in \mathcal{R}(\chi \to \psi)$. Now, if $\mathcal{R}(\chi) = \{\beta_1, \ldots, \beta_n\}$, then $\beta = \beta_i$ for some i, and, by definition of the resolutions of an implication, $\gamma = (\beta_1 \to \gamma_1) \wedge \cdots \wedge (\beta_n \to \gamma_n)$ for some $\{\gamma_1, \ldots, \gamma_n\} \subseteq \mathcal{R}(\psi)$. Extending Q_2 with a conjunction elimination rule we get a proof of $\beta_i \to \gamma_i$ from Γ. So, from Γ we can derive both β_i and $\beta_i \to \gamma_i$ whence, eliminating the implication, we can derive γ_i, a resolution of ψ.
- ψ was obtained by a ?-elimination rule from $?\{\beta_1, \ldots, \beta_m\}$. Then the immediate subproofs of P are a proof $P_0 : \Phi \vdash ?\{\beta_1, \ldots, \beta_m\}$ and, for $1 \leq i \leq n$ a proof $P_i : \Phi \cup \{\beta_i\} \vdash \psi$. Now consider a resolution Γ of Φ. By induction hypothesis we have a proof $Q_0 : \Gamma \vdash \beta$ for some $\beta \in \mathcal{R}(?\{\beta_1, \ldots, \beta_m\})$. Moreover, for any $1 \leq i \leq n$, since $\Gamma \cup \{\beta_i\}$ is a resolution of $\Phi \cup \{\beta_i\}$, by induction hypothesis we have a proof $Q_i : \Gamma \cup \{\beta_i\} \vdash \alpha_i$ where $\alpha_i \in \mathcal{R}(\psi)$. Now since β is a resolution of $?\{\beta_1, \ldots, \beta_m\}$, by definition $\beta = \beta_i$ for some i. But then, combining the proof $Q_0 : \Gamma \vdash \beta_i$ with the proof $Q_i : \Gamma \cup \{\beta_i\} \vdash \alpha_i$ (more precisely, substituting any undischarged assumption of β_i in Q_i with an occurrence of the proof Q_0 with conclusion β_i) we obtain a proof of α_i from Γ, which is what we needed, since α_i is a resolution of ψ.

This case-by-case examination proves the left-to-right direction of the theorem. In order to establish the converse, let us make a detour to prove Fact 4.11.

Proof of Lemma 4.11 First let us prove this for the case in which Φ is finite. We will prove by induction on the number of formulas in Φ the claim that *for any* ψ, if $\Phi \nvdash \psi$ there is some $\Gamma \in \mathcal{R}(\Phi)$ such that $\Gamma \nvdash \psi$.

If $\Phi = \emptyset$, the claim is trivially true. Now make the inductive hypothesis that the claim is true for sets of n formulas, and let us consider a set Φ of $n+1$ formulas. Then Φ is of the form $\Psi \cup \{\chi\}$ for some set Ψ of n formulas and some formula χ. Now consider a formula ψ such that $\Psi, \chi \nvdash \psi$. By Lemma 4.4, we must also have $\Psi, ?\mathcal{R}(\chi) \nvdash \psi$ whence, by the ?-introduction rule, we must have $\Psi, \alpha \nvdash \psi$ for some $\alpha \in \mathcal{R}(\chi)$. By the rules for implication, we must then have $\Psi \nvdash \alpha_i \to \psi$, and so by induction hypothesis there is a $\Gamma \in \mathcal{R}(\Psi)$ such that $\Gamma \nvdash \alpha_i \to \psi$. Finally, again by the rules for implication we have $\Gamma, \alpha_i \nvdash \psi$, which proves the claim since $\Gamma \cup \{\alpha\}$ is a resolution of $\Psi \cup \{\chi\}$.

Our inductive proof is thus complete, and the claim is proved for the case in which Φ is finite. Now let us suppose that Φ is infinite and choose an enumeration of Φ, so that $\Phi = \{\varphi_n \,|\, n \geq 1\}$. Now, for $n \in \mathbb{N}$, put:

$$T_n = \{\ \langle \alpha_1, \ldots, \alpha_n \rangle \mid \alpha_i \in \mathcal{R}(\varphi_i) \text{ for } 1 \leq i \leq n\}$$

Now let $T = \bigcup_{n \in \mathbb{N}} T_n$ and, for $a, b \in T$, let $a \leq b$ in case a is an initial segment of b. Clearly, $\langle T, \leq \rangle$ is a tree. Moreover, T is finitely branching: this is because the immediate successors of $a = \langle \alpha_1, \ldots, \alpha_n \rangle$ are $a' = \langle \alpha_1, \ldots, \alpha_n, \alpha_{n+1} \rangle$ where $\alpha_{n+1} \in \mathcal{R}(\varphi_{n+1})$, and the set of resolutions of a sentence is always finite.

Now consider a formula ψ such that $\Phi \nvdash \psi$. To find a resolution $\Gamma \in \mathcal{R}(\Phi)$ such that $\Gamma \nvdash \psi$, we first divide T into two parts:

- $T_{\vdash\psi} = \{\ \langle \alpha_1, \ldots, \alpha_n \rangle \in T \mid \{\alpha_1, \ldots, \alpha_n\} \vdash \psi\}$
- $T_{\nvdash\psi} = \{\ \langle \alpha_1, \ldots, \alpha_n \rangle \in T \mid \{\alpha_1, \ldots, \alpha_n\} \nvdash \psi\}$

Cearly, $T_{\nvdash\psi}$ and $T_{\vdash\psi}$ form a partition of T. Notice that $T_{\vdash\psi}$ is upward closed, that is, if $a \leq b$ and $a \in T_{\vdash\psi}$, then $b \in T_{\vdash\psi}$ as well: for, if ψ is provable from a certain set, it is also provable from any superset. Conversely, $T_{\nvdash\psi}$ is downward closed, that is, if $a \leq b$ and $b \in T_{\nvdash\psi}$ then $a \in T_{\nvdash\psi}$.

We claim that $T_{\nvdash\psi}$ is infinite. For, if it were finite, it would only intersect finitely many of the T_n's. For an index k such that $T_{\nvdash\psi} \cap T_k = \emptyset$, this would mean that $T_k \subseteq T_{\vdash\psi}$. But, recalling the definition of T_k, this means that every resolution Γ of the set $\{\varphi_1, \ldots, \varphi_k\}$ derives ψ. Since we have already proved our claim for finite sets, we can conclude that $\{\varphi_1, \ldots, \varphi_k\}$ must derive ψ as well. But this is a contradiction, since $\{\varphi_1, \ldots, \varphi_k\} \subseteq \Phi$ and $\Phi \nvdash \psi$ by assumption.

So, $T_{\nvdash\psi}$ must be infinite, and $\langle T_{\nvdash\psi}, \leq \rangle$ is an infinite tree. Since $\langle T_{\nvdash\psi}, \leq \rangle$ is finitely branching, by König's lemma it must have an infinite branch. But this means precisely that there exists an infinite sequence $\alpha_n, n \geq 1$ of declaratives (the limit of the finite sequences on the infinite branch of $T_{\nvdash\psi}$) such that (i) $\alpha_n \in \mathcal{R}(\varphi_n)$ for any $n \geq 1$ and (ii) for any $n \geq 1$, $\{\alpha_1, \ldots, \alpha_n\} \nvdash \psi$.

Consider $\Gamma = \{\alpha_n \,|\, n \geq 1\}$. Since $\alpha_n \in \mathcal{R}(\varphi_n)$ for every $n \geq 1$, $\Gamma \in \mathcal{R}(\Phi)$. Moreover, since for every $n \geq 1$, $\{\alpha_1, \ldots, \alpha_n\} \nvdash \psi$, Γ cannot derive ψ. So we have found a resolution $\Gamma \in \mathcal{R}(\Phi)$ such that $\Gamma \nvdash \psi$. □

Proof of theorem 4.9, right-to-left direction. Suppose $\Phi \nvdash \psi$. By Lemma

4.11 we have a $\Gamma \in \mathcal{R}(\Phi)$ such that $\Gamma \not\vdash \psi$. Since for any $\alpha \in \mathcal{R}(\psi)$ we have $\alpha \vdash \psi$ (Corollary 4.5), Γ cannot derive any $\alpha \in \mathcal{R}(\psi)$, otherwise it would derive ψ. So, Γ is a resolution of Φ which does not derive any resolution of ψ. □

Proof of Lemma 4.15. First assume $\Delta \in \sigma_a^c(\Gamma)$. Since $\sigma_a^c(\Gamma) = \bigcup \Sigma_a^c(\Gamma)$, this means that $\Delta \in S$ for some state S such that $\bigcap S \vdash \varphi$ whenever $E_a\varphi \in \Gamma$. In particular, then, if $E_a\alpha \in \Gamma$ we have $\bigcap S \vdash \alpha$, whence also $\Delta \vdash \alpha$. Since Δ is closed under deduction of declaratives, this implies $\alpha \in \Delta$.

Conversely, suppose $\alpha \in \Delta$ whenever $E_a\alpha \in \Gamma$. We claim that the singleton state $\{\Delta\}$ belongs to $\Sigma_a^c(\Gamma)$, so that $\Delta \in \bigcup \Sigma_a^c(\Gamma) = \sigma_a^c(\Gamma)$. Since $\bigcap\{\Delta\} = \Delta$, to show that $\{\Delta\} \in \Sigma_a^c(\Gamma)$ we must show that $\Delta \vdash \varphi$ whenever $E_a\varphi \in \Gamma$.

So, suppose $E_a\varphi \in \Gamma$ and let us show $\Delta \vdash \varphi$. If φ is a declarative, this is true by assumption. Now consider an interrogative μ such that $E_a\mu \in \Gamma$. Corollary 4.6 gives $\mu \vdash \pi_\mu$, whence by Lemma 4.7 we have $E_a\mu \vdash E_a\pi_\mu$. Since $E_a\mu \in \Gamma$ and Γ is closed under deduction of declaratives, $E_a\pi_\mu \in \Gamma$. But since, unlike μ, π_μ is a declarative, by assumption we have $\pi_\mu \in \Delta$. By definition, $\pi_\mu = \bigvee \mathcal{R}(\mu)$. Now, since Δ is a CTD, it has the disjunction property (Fact 4.12), which means that we must have $\alpha \in \Delta$ for some $\alpha \in \mathcal{R}(\mu)$. But then, since $\alpha \vdash \mu$ by Corollary 4.5, it follows $\Delta \vdash \mu$, as required. □

In preparation to the support lemma, we will prove three intermediate lemmata.

Lemma A.19 *For any state $S \subseteq \mathcal{W}^c$ and any $\alpha \in \mathcal{L}_!$, $\bigcap S \vdash \alpha \iff \alpha \in \bigcap S$*

Proof. If $\alpha \in \bigcap S$ then obviously $\bigcap S \vdash \alpha$. For the converse, suppose $\bigcap S \vdash \alpha$. For any $\Gamma \in S$ we have $\bigcap S \subseteq \Gamma$, so also $\Gamma \vdash \alpha$. But then, because Γ is closed under deduction of declaratives, we must have $\alpha \in \Gamma$. So, $\alpha \in \bigcap S$. □

Lemma A.20
Let $\Gamma \in \mathcal{W}^c$. If $E_a\varphi \notin \Gamma$ there exists a state $T \in \Sigma_a^c(\Gamma)$ such that $\bigcap T \not\vdash \varphi$.

Proof. Put $\Gamma^{E_a} = \{\psi \mid E_a\psi \in \Gamma\}$ (notice that Γ^{E_a} does not only contain declaratives but also interrogatives). We claim that Γ^{E_a} does not entail φ. Towards a contradiction, suppose $\Gamma^{E_a} \vdash \varphi$. Let $\psi_1, \ldots, \psi_n \in \Gamma^{E_a}$ be assumptions such that $\psi_1, \ldots, \psi_n \vdash \varphi$. By Lemma 4.7 we have $E_a\psi_1, \ldots, E_a\psi_n \vdash E_a\varphi$. But the fact that $\psi_1, \ldots, \psi_n$ are in Γ^{E_a} means that $E_a\psi_1, \ldots, E_a\psi_n$ are in Γ. Hence, we would also have $\Gamma \vdash E_a\varphi$, and so also $E_a\varphi \in \Gamma$, contrary to assumption.

We have thus proved $\Gamma^{E_a} \not\vdash \varphi$. But then, by Lemma 4.9 we know that there must be a resolution Θ pf Γ^{E_a} which entails no resolution α of φ. But then, for any $\alpha \in \mathcal{R}(\varphi)$, the set $\Theta \cup \{\neg\alpha\}$ is a consistent set of declaratives, and so by Lindenbaum's lemma it can be extended to some CTD $\Delta_\alpha \in \mathcal{W}^c$.

Now consider the state $T = \{\Delta_\alpha \mid \alpha \in \mathcal{R}(\varphi)\}$. We claim that T has the properties we need. First, since $\Theta \subseteq \Delta_\alpha$ for each α, we have $\Theta \subseteq \bigcap T$. Now suppose $E_a\psi \in \Gamma$: then $\psi \in \Gamma^{E_a}$, and since Θ is a resolution of Γ^{E_a}, it contains some resolution β of ψ. But then, since $\beta \in \Theta \subseteq \bigcap T$ and $\beta \vdash \psi$, we must also have $\bigcap T \vdash \psi$. So, $\bigcap T \vdash \psi$ whenever $E_a\psi \in \Gamma$, which means that $T \in \Sigma_a^c(\Gamma)$.

On the other hand, $\bigcap T \not\vdash \varphi$. For, if we had $\bigcap T \vdash \varphi$, by Corollary 4.10 we should have $\bigcap T \vdash \alpha$ for some resolution α of φ, which would entail $\Delta_\alpha \vdash \alpha$, since $\bigcap T \subseteq \Delta_\alpha$. But this is impossible, since by construction Δ_α contains $\neg\alpha$ and is a consistent theory. Hence $\bigcap T \nvdash \varphi$ and our lemma is proven. □

Lemma A.21 *Let* $\Gamma \in \mathcal{W}^c$. *If* $K_a\varphi \notin \Gamma$, *then* $\bigcap \sigma_a^c(\Gamma) \nvdash \varphi$.

Proof. Suppose $K_a\varphi \notin \Gamma$. Since $K_a\varphi \dashv\vdash \bigvee_{\alpha\in\mathcal{R}(\varphi)} E_a\alpha$, the latter formula is not in Γ either. Since Γ is closed under declarative deduction, this implies $E_a\alpha \notin \Gamma$ for every $\alpha \in \mathcal{R}(\varphi)$. Now consider any $\alpha \in \mathcal{R}(\varphi)$: since $E_a\alpha \notin \Gamma$, by Lemma A.20 there is a state $T_\alpha \in \Sigma_a^c(\Gamma)$ such that $\bigcap T_\alpha \nvdash \alpha$. Now since $T_\alpha \in \Sigma_a^c(\Gamma)$ we have $T_\alpha \subseteq \bigcup \Sigma_a^c(\Gamma) = \sigma_a^c(\Gamma)$, whence $\bigcap \sigma_a^c(\Gamma) \subseteq \bigcap T_\alpha$. And since $\bigcap T_\alpha \nvdash \alpha$, a fortiori $\bigcap \sigma_a^c(\Gamma) \nvdash \alpha$. But as $\bigcap \sigma_a^c(\Gamma)$ does not derive any resolution of φ, by Corollary 4.10 it cannot derive φ either: $\bigcap \sigma_a^c(\Gamma) \nvdash \varphi$. □

Proof of Lemma 4.17. The proof goes by induction on the complexity of φ. The straightforward cases for atoms, falsum, and conjunction are omitted.

Implication Suppose $\bigcap S \vdash \varphi \to \psi$. Take any $T \subseteq S$: if $T \models \varphi$ then by induction hypothesis $\bigcap T \vdash \varphi$. Since $T \subseteq S$, we have $\bigcap T \supseteq \bigcap S$, and since $\bigcap S \vdash \varphi \to \psi$, also $\bigcap T \vdash \varphi \to \psi$. But from $\bigcap T \vdash \varphi \to \psi$ and $\bigcap T \vdash \varphi$ it follows $\bigcap T \vdash \psi$, which by induction hypothesis implies $T \models \psi$. So, every substate of S that supports φ also supports ψ, which proves that $S \models \varphi \to \psi$.

Viceversa, suppose $\bigcap S \nvdash \varphi \to \psi$. By the introduction rule for implication, this means that $\bigcap S, \varphi \nvdash \psi$. Now by Lemma 4.11 there is a a resolution of $(\bigcap S) \cup \{\varphi\}$ which does not derive ψ. Since $\bigcap S$ is a set of declaratives, this resolution must include a set of the form $(\bigcap S) \cup \{\alpha\}$ where α is a resolution of φ. Hence, there must exist a resolution α of φ such that $\bigcap S, \alpha \nvdash \psi$.

Now let $T = \{\Gamma \in S \mid \alpha \in \Gamma\}$. First, by definition we have $\alpha \in \bigcap T$, whence $\bigcap T \vdash \varphi$ by Corollary 4.5. By induction hypothesis we then have $T \models \varphi$. Now, if we can show that $\bigcap T \nvdash \psi$ we are done. For then, the induction hypothesis gives $T \not\models \psi$, which means that T is a substate of S that supports φ but not ψ, which shows that $S \not\models \varphi \to \psi$.

So, we are left to show that $\bigcap T \nvdash \psi$. Towards a contradiction, suppose that $\bigcap T \vdash \psi$. Since $\bigcap T$ is a set of declaratives, Corollary 4.10 tells us that $\bigcap T \vdash \beta$ for some resolution β of ψ, which by Lemma A.19 amounts to $\beta \in \bigcap T$. So, for any $\Gamma \in T$ we have $\beta \in \Gamma$ and thus also $\alpha \to \beta \in \Gamma$, since Γ is closed under deduction of declaratives and $\beta \vdash \alpha \to \beta$. Now consider any $\Gamma \in S - T$: this means that $\alpha \notin \Gamma$; then since Γ is complete we have $\neg\alpha \in \Gamma$, whence $\alpha \to \beta \in \Gamma$, because Γ is closed under deduction of declaratives and $\neg\alpha \vdash \alpha \to \beta$. We have thus shown that $\alpha \to \beta \in \Gamma$ for any $\Gamma \in S$, whether $\Gamma \in T$ or $\Gamma \in S - T$. We can then conclude $\alpha \to \beta \in \bigcap S$, whence $\bigcap S, \alpha \vdash \beta$. And since β is a resolution of ψ we also have $\bigcap S, \alpha \vdash \psi$. But this is a contradiction since by assumption α is such that $\bigcap S, \alpha \nvdash \psi$.

Question mark If $S \models ?\{\alpha_1, \ldots, \alpha_n\}$, then $S \models \alpha_i$ for some i, so by induction hypothesis we have $\bigcap S \vdash \alpha_i$ and by ?-introduction also $\bigcap S \vdash ?\{\alpha_1, \ldots, \alpha_n\}$. Conversely, suppose $\bigcap S \vdash ?\{\alpha_1, \ldots, \alpha_n\}$. Since $\bigcap S$ is a set of declaratives, it follows from Corollary 4.10 that $\bigcap S \vdash \alpha_i$ for some $1 \le i \le n$. By induction hypothesis we then have $S \models \alpha_i$, and thus also $S \models ?\{\alpha_1, \ldots, \alpha_n\}$.

E_a **modality** Suppose $\bigcap S \vdash E_a\varphi$. Now consider any $\Gamma \in S$. Since $\bigcap S \subseteq \Gamma$, we have $\Gamma \vdash E_a\varphi$, and since Γ is closed under deduction of declaratives, $E_a\varphi \in \Gamma$. By definition of Σ_a^c, then, for any $T \in \Sigma_a^c(\Gamma)$ we must have $\bigcap T \vdash \varphi$, which by induction hypothesis entails $T \models \varphi$. Since this is true for any $\Gamma \in S$ and any $T \in \Sigma_a^c(\Gamma)$, it follows that $S \models E_a\varphi$.

For the converse, suppose $\bigcap S \nvdash E_a\varphi$. Then $E_a\varphi \notin \bigcap S$, which means that $E_a\varphi \notin \Gamma$ for some $\Gamma \in S$. Then, Lemma A.20 ensures that there exists a state $T \in \Sigma_a^c(\Gamma)$ such that $\bigcap T \nvdash \psi$, that is, by induction hypothesis, such that $T \not\models \psi$. Therefore, we do not have $T \models \varphi$ for every $\Gamma \in S$ and $T \in \Sigma_a^c(\Gamma)$, which means that $S \not\models E_a\varphi$.

K_a **modality** Suppose $\bigcap S \vdash K_a\varphi$, which by Lemma A.19 implies $K_a\varphi \in \bigcap S$. Since $K_a\varphi \dashv\vdash \bigvee_{\alpha\in\mathcal{R}(\varphi)} E_a\alpha$ (Lemma 4.8), we have $\bigvee_{\alpha\in\mathcal{R}(\varphi)} E_a\alpha \in \bigcap S$. Now consider any $\Gamma \in S$. Since $\bigvee_{\alpha\in\mathcal{R}(\varphi)} E_a\alpha \in \bigcap S$, also $\bigvee_{\alpha\in\mathcal{R}(\varphi)} E_a\alpha \in \Gamma$. Since complete theories have the disjunction property, $E_a\alpha \in \Gamma$ for some $\alpha \in \mathcal{R}(\varphi)$. Since $E_a\alpha \in \Gamma$, Lemma 4.15 tells us that $\alpha \in \Delta$ for any $\Delta \in \sigma_a^c(\Gamma)$, so $\alpha \in \bigcap \sigma_a^c(\Gamma)$. Since $\alpha \vdash \varphi$ (Corollary 4.5) we then have $\sigma_a^c(\Gamma) \vdash \varphi$, which by induction hypothesis means that $\sigma_a^c(\Gamma) \models \varphi$. Summing up, for any $\Gamma \in S$ we have $\sigma_a^c(\Gamma) \models \varphi$, and so $S \models K_a\varphi$.

Conversely, suppose $\bigcap S \nvdash K_a\varphi$. Then obviously $K_a\varphi \notin \bigcap S$, so there is a $\Gamma \in S$ such that $K_a\varphi \notin \Gamma$. But then, Lemma A.21 establishes that $\bigcap \sigma_a^c(\Gamma) \nvdash \varphi$, which by induction hypothesis amounts to $\sigma_a^c(\Gamma) \not\models \varphi$. So, it is not the case that $\sigma_a^c(\Gamma) \models \varphi$ for every $\Gamma \in S$, which means that $S \not\models K_a\varphi$.□

References

[1] van Benthem, J., "Logical dynamics of information and interaction," Cambridge University Press, 2011.

[2] Ciardelli, I., *Interrogative dependencies and the constructive content of inquisitive proofs* (2014), in the proceedings of *WoLLIC 21*.

[3] Ciardelli, I., J. Groenendijk and F. Roelofsen, *Inquisitive semantics: a new notion of meaning*, Language and Linguistics Compass **7** (2013), pp. 459–476.

[4] Ciardelli, I., J. Groenendijk and F. Roelofsen, *On the semantics and logic of declaratives and interrogatives*, Synthese (2013).

[5] Ciardelli, I. and F. Roelofsen, *Inquisitive logic*, Journal of Philosophical Logic **40** (2011), pp. 55–94.

[6] Ciardelli, I. and F. Roelofsen, *Inquisitive dynamic epistemic logic*, Synthese (2014).

[7] van Ditmarsch, H., W. van der Hoek and B. Kooi, "Dynamic Epistemic Logic," Springer, 2007.

[8] Hintikka, J., "Inquiry as inquiry: A logic of scientific discovery," Kluwer Academic Publishers, 1999.

[9] Hintikka, J., "Socratic epistemology: explorations of knowledge-seeking by questioning," Cambridge University Press, 2007.

[10] Väänänen, J., "Dependence Logic: A New Approach to Independence Friendly Logic," Cambridge University Press, Cambridge, 2007.

[11] Wiśniewski, A., *The logic of questions as a theory of erotetic arguments*, Synthese **109** (1996), pp. 1–25.

[12] Yang, F., "On extensions and variants of dependence logic: A study of intuitionistic connectives in the team semantics setting," Ph.D. thesis, University of Helsinki (2014).

More on Empirical Negation

Michael De [1]

Department of Philosophy, Universität Konstanz
Universitätsstraße 10, 78464 Konstanz Germany

Hitoshi Omori [2]

The Graduate Center, City University of New York
365 Fifth Avenue, New York, NY 10016 USA

Abstract

Intuitionism can be seen as a verificationism restricted to mathematical discourse. An attempt to generalize intuitionism to empirical discourse presents various challenges. One of those concerns the logical and semantical behavior of what has been called 'empirical negation'. An extension of intuitionistic logic with empirical negation was given by Michael De and a labelled tableaux system was there shown sound and complete. However, a Hilbert-style axiom system that is sound and complete was missing. In this paper we provide the missing axiom system which is shown sound and complete with respect to its intended semantics. Along the way we consider some further applications of empirical negation.

Keywords: Intuitionistic logic, empirical negation, completeness.

1 Introduction

A verificationism along the lines of Michael Dummett and Neil Tennant seeks to generalize a constructive interpretation of mathematical discourse to empirical discourse. A simplified strategy along these lines can be thought to be carried out as follows. First, take the famous Brouwer-Heyting-Kolmogorov (BHK) interpretation of the base (propositional) logical connectives, where provability is taken informally:

(i) a conjunction $A \wedge B$ is provable iff A and B are;

(ii) a disjunction $A \vee B$ is provable iff either A or B is;

[1] This research was funded in part by the European Research Council under the European Community's Seventh Framework Programme (FP7/2007-2013)/ERC Grant agreement nr 263227. Email: `mikejde@gmail.com`.

[2] Postdoctoral fellow for research abroad of Japan Society for the Promotion of Science (JSPS). Email: `hitoshiomori@gmail.com`

(iii) a conditional $A \to B$ is provable iff there is a method transforming any proof of A into a proof of B;

(iv) $\bot$ isn't provable.

As usual the negation $\neg A$ of A is defined by $A \to \bot$. Second, replace in those clauses 'provable' with 'verifiable', where a verification is intended to cover not only mathematical statements but empirical ones as well. Third and finally, provide a workable account of what verification amounts to. E.g. one such account might hold that a statement is verifiable iff it could be warranted, where warrants are proofs in the mathematical case and some other sort of evidence in the empirical case, depending on the domain of discourse.

Of course filling in the details of such an account is far from trivial. Besides meeting the onerous task of giving a plausible account of warrant or evidence, there are already significant worries concerning the logical language. Will $\wedge$, $\vee$, and $\to$ and $\bot$ suffice as propositional connectives for an empirical language? There are good reasons to think not, especially concerning negation. Dummett says:

> Negation ... is highly problematic. In mathematics, given the meaning of "if ... then", it is trivial to explain "Not A" as meaning "If A, then 0 = 1"; by contrast, a satisfactory explanation of "not", as applied to empirical statements for which bivalence is not, in general, taken as holding, is very difficult to arrive at. Given that the sentential operators cannot be thought of as explained by means of the two-valued truth-tables, the possibility that the laws of classical logic will fail is evidently open: but it is far from evident that the correct logical laws will always be the intuitionistic ones. More generally, it is by no means easy to determine what should serve as the analogue, for empirical statements, of the notion of proof as it figures in intuitionist semantics for mathematical statements. [7, p.473]

One problem is that the "arrow-falsum" definition of negation is often too strong to serve as the negation for empirical statements. Suppose we attempt to express in our generalized intuitionistic language the fact that Goldbach's conjecture is not decided. We obtain the statement that says that any warrant for 'Goldbach's conjecture is decided' can be transformed into warrant for an absurdity (say '0=1'). Since there could be no proof of an absurdity, this statement says that Goldbach's conjecture is undecidable! But it might turn out in the future that someone prove or refute Goldbach's conjecture. So the fact that Goldbach's conjecture is not decided does not imply that it is undecid*able*, as our translation gives us. What we rather wished to say was merely that there is no sufficient evidence at present for the truth of the conjecture.

Couldn't we translate the original statement in a way that avoids this problem? For example, couldn't we translate 'Goldbach's conjecture is not decided' as 'If Goldbach's conjecture is decided *at present*, then 0=1'? Since the conjecture is not decided at present, the idea would be that we can (vacuously) turn any warrant or evidence for the antecedent of the conditional into evidence for an absurdity, for there is no evidence for the antecedent! The problem is that if there were *any* evidence at all to a sufficient degree that Goldbach's conjecture is decided and if that evidence can be fallible (which is likely given that the statement is empirical), then it is not at all clear that that evidence can be transformed into a proof of '0=1'. Suppose in the future some reliable quantum computer has returned a very long but flawed proof of the conjecture. How exactly could that evidence be turned into evidence for '0=1'? That would depend on the nature of the mistake. If the conjecture is true and decidable and if from the mistake alone (with Peano arithmetic) one cannot infer that '0=1', then it's hard to see how that evidence could be turned into evidence for '0=1'. If the conjecture is undecidable it's true, so again having evidence for or against it won't likely convert into evidence for an absurdity since the conjecture and its negation are both consistent with Peano arithmetic. If the conjecture is false (hence decidable), then there is still no reason to think that a flawed proof of a false conjecture will convert into evidence for an absurdity. [3]

There are cases that are more problematic than the one just mentioned, as they involve no mathematical content. Consider the purely empirical statement 'There are ten thousand leaves in my garden'. If that statement is false and yet after spending the entire day counting the leaves I (mistakenly) arrive at ten thousand, how precisely will my evidence be converted into evidence that 0=1? One surely cannot use that evidence in a *derivation* in Peano arithmetic to '0=1'. We could choose a different absurdity since we are in a different domain of discourse, but exactly what would that absurdity be—one concerning features of my garden? Could any evidence concerning the wrong number of leaves in my garden be converted into evidence for this absurdity? This all seems unlikely. We are best to conclude that the fact that an empirical statement is false does not imply that *any* (fallible) evidence for it, whether that evidence be got now or in the future, can be converted into evidence for some given absurdity. [4]

We need, therefore, a *sui generis* empirical negation that is not definable from the standard (generalized) intuitionistic connectives. There have been some proposals as to how such a negation should behave logically and semantically in a verificationist setting. Two such accounts may be found in [4] and [6], with some general philosophical difficulties for such a project raised in [21]. In [4], intuitionistic logic is enriched by an empirical negation given the intuitive reading 'It is not the case that there is sufficient evidence at present that'. A tableaux system for the logic was there shown sound and complete, as no Hilbert-style axiom formulation could be found. We build on the work of [4] by

[3] If the conjecture is false, then its negation is provable, so if we had a "proof" of the conjecture, could we not turn that into proof of an absurdity? Only if we also had the proof of the negation of the conjecture! Since we do not, it is not at all obvious that an alleged proof of the conjecture could *at any time* be converted into a proof of an absurdity.

[4] Interestingly, empirical negation was discussed as far back as [10, p. 18] under the guise of 'factual' negation. Timothy Williamson [21] argues that sentences of the form 'A will never be decided' are in fact inconsistent in a generalized intuitionistic language, even one with what he considers a plausible empirical negation. We will have more to say about this argument in section §4.

providing a Hilbert-style proof system for the logic which is shown sound and complete with respect to its intended semantics. We conclude by addressing some of the worries raised in [21].

2 Semantics and proof theory

After setting up the language, we first present the semantics, and then turn to the proof theory.

Definition 2.1 The language $\mathcal{L}$ consists of a finite set $\{\sim, \wedge, \vee, \rightarrow\}$ of propositional connectives and a countable set Prop of propositional variables which we denote by p, q, etc. Furthermore, we denote by Form the set of formulas defined as usual in $\mathcal{L}$. We denote a formula of $\mathcal{L}$ by A, B, C, etc. and a set of formulas of $\mathcal{L}$ by Γ, Δ, Σ, etc.

2.1 Semantics

Definition 2.2 A model for the language is a quadruple $\langle W, g, \leq, V\rangle$, where W is a non-empty set (of states); $g \in W$ (the base state); $\leq$ is a partial order on W with g being the least element; and $V : W \times \mathsf{Prop} \rightarrow \{0,1\}$ an assignment of truth values to state-variable pairs with the condition that $V(w_1, p) = 1$ and $w_1 \leq w_2$ only if $V(w_2, p) = 1$ for all $p \in \mathsf{Prop}$ and all $w_1, w_2 \in W$. [5] Valuations V are then extended to interpretations I to state-formula pairs by the following conditions:

- $I(w,p) = V(w,p)$
- $I(w, \sim A) = 1$ iff $I(g, A) = 0$
- $I(w, A \wedge B) = 1$ iff $I(w, A) = 1$ and $I(w, B) = 1$
- $I(w, A \vee B) = 1$ iff $I(w, A) = 1$ or $I(w, B) = 1$
- $I(w, A \rightarrow B) = 1$ iff for all $x \in W$: if $w \leq x$ and $I(x, A) = 1$ then $I(x, B) = 1$.

For a philosophical interpretation of the semantics, the reader is referred to [4].

Semantic consequence is now defined in terms of truth preservation at g: $\Sigma \models A$ iff for all models $\langle W, g, \leq, I\rangle$, $I(g, A) = 1$ if $I(g, B) = 1$ for all $B \in \Sigma$.

2.2 Proof Theory

Definition 2.3 The system $\mathbf{IPC}^{\sim}$ consists of the following axiom schemata and rules of inference:

$$A \rightarrow (B \rightarrow A) \tag{Ax1}$$

$$(A \rightarrow (B \rightarrow C)) \rightarrow ((A \rightarrow B) \rightarrow (A \rightarrow C)) \tag{Ax2}$$

$$(A \wedge B) \rightarrow A \tag{Ax3}$$

$$(A \wedge B) \rightarrow B \tag{Ax4}$$

$$(C \rightarrow A) \rightarrow ((C \rightarrow B) \rightarrow (C \rightarrow (A \wedge B))) \tag{Ax5}$$

[5] Note here that in the semantics presented in [4], it is not assumed that the distinguished element g is a least element with respect to the partial ordering. This constraint doesn't affect the consequence relation but is needed for the completeness proof below.

$$A \to (A \vee B) \tag{Ax6}$$
$$B \to (A \vee B) \tag{Ax7}$$
$$(A \to C) \to ((B \to C) \to ((A \vee B) \to C)) \tag{Ax8}$$
$$A \vee {\sim}A \tag{Ax9}$$
$${\sim}A \to ({\sim}{\sim}A \to B) \tag{Ax10}$$
$$\frac{A \quad A \to B}{B} \tag{MP}$$
$$\frac{A \vee B}{{\sim}A \to B} \tag{RP}$$

Following the usual convention, we define $A \leftrightarrow B$ as $(A \to B) \wedge (B \to A)$. Finally, we write $\Gamma \vdash A$ if there is a sequence of formulas $B_1, \ldots, B_n, A$, $n \geq 0$, such that every formula in the sequence $B_1, \ldots, B_n, A$ either (i) belongs to Γ; (ii) is an axiom of $\mathbf{IPC}^{\sim}$; (iii) is obtained by (MP) or (RP) from formulas preceding it in sequence.

Remark 2.4 We will refer to the subsystem of $\mathbf{IPC}^{\sim}$ which consists of axiom schemata from (Ax1) to (Ax8) and a rule of inference (MP) as $\mathbf{IPC}^{+}$. Note that the deduction theorem does not hold with respect to $\to$ in this system, as observed by De. However, we do have a deduction theorem in a slightly different form, given below as Theorem 2.11.

Our goal now is to prove a variant of the deduction theorem. For this purpose, we begin with some preparations.

Fact 2.5 *The following formulas are provable in* $\mathbf{IPC}^{+}$ *and thus in* $\mathbf{IPC}^{\sim}$.

$$A \to A \tag{1}$$
$$(A \vee B) \to (B \vee A) \tag{2}$$
$$(A \to (B \to C)) \to (B \to (A \to C)) \tag{3}$$
$$(A \vee B) \to ((B \to C) \to (A \vee C)) \tag{4}$$
$$(A \to (B \to C)) \to ((A \wedge B) \to C) \tag{5}$$
$$(A \wedge (B \vee C)) \to ((A \wedge B) \vee C) \tag{6}$$

Lemma 2.6 *The following formulas are provable in* $\mathbf{IPC}^{\sim}$.

$$(A \to {\sim}A) \to {\sim}A \tag{7}$$
$$({\sim}A \to A) \to A \tag{8}$$
$${\sim}{\sim}A \to A \tag{9}$$
$$({\sim}A \to B) \to (A \vee B) \tag{10}$$
$$({\sim}{\sim}A \to B) \leftrightarrow ({\sim}A \vee B) \tag{11}$$

Proof. (7) can be derived by making use of (Ax8), (1), (3) and (Ax9); and (8) is similarly derived. (9) is proved by (Ax9), (2) and (RP). (10) is also easy to derive by combining (4) and (Ax9). For (11), the left-to-right direction is (10). For the other way around, we only need to apply (Ax8) to (Ax1) and (Ax10).□

Now, we can prove one direction of the deduction theorem.

Proposition 2.7 *For* $\Gamma \cup \{A, B\} \subseteq \mathsf{Form}$, *if* $\Gamma, A \vdash B$ *then* $\Gamma \vdash {\sim}{\sim}A \to B$.

Proof. By the induction on the length n of the proof of $\Gamma, A \vdash B$. If $n = 1$, then we have the following three cases.

- If B is one of the axioms of $\mathbf{IPC}^{\sim}$, then we have $\vdash B$. Therefore, by (Ax1), we obtain $\vdash {\sim}{\sim}A \to B$ which implies the desired result.
- If $B \in \Gamma$, then we have $\Gamma \vdash B$, and thus we obtain the desired result by (Ax1).
- If $B = A$, then by (9), we have ${\sim}{\sim}A \to B$ which implies the desired result.

For $n > 1$, then there are two additional cases to be considered.

- If B is obtained by applying (MP), then we will have $\Gamma, A \vdash C$ and $\Gamma, A \vdash C \to B$ lengths of the proof of which are less than n. Thus, by induction hypothesis, we have $\Gamma \vdash {\sim}{\sim}A \to C$ and $\Gamma \vdash {\sim}{\sim}A \to (C \to B)$, and by (Ax2) and (MP), we obtain $\Gamma \vdash {\sim}{\sim}A \to B$ as desired.
- If B is obtained by applying (RP), then $B = {\sim}C \to D$ and we will have $\Gamma, A \vdash C \vee D$ length of the proof of which is less than n. Thus, by induction hypothesis, we have $\Gamma \vdash {\sim}{\sim}A \to (C \vee D)$. By (11) and (RP), we have $\Gamma \vdash {\sim}C \to ({\sim}A \vee D)$. Another application of (11) gives us $\Gamma \vdash {\sim}C \to ({\sim}{\sim}A \to D)$ and thus by exchange, we obtain $\Gamma \vdash {\sim}{\sim}A \to ({\sim}C \to D)$, i.e. $\Gamma \vdash {\sim}{\sim}A \to B$ as desired.

This completes the proof. □

For the purpose of proving the other direction of the deduction theorem, we need another lemma.

Lemma 2.8 *The following formulas are provable and rules are derivable in* $\mathbf{IPC}^{\sim}$.

$$\frac{A \to B}{{\sim}B \to {\sim}A} \tag{RC}$$

$${\sim}(A \to A) \to B \tag{12}$$

$${\sim}{\sim}(A \to A) \tag{13}$$

$$\frac{A}{{\sim}{\sim}A} \tag{RD}$$

Proof. For (RC), assume $A \to B$. Then by making use of (4) and (Ax9), we have ${\sim}A \vee B$ which is equivalent to $B \vee {\sim}A$ by (2). Thus by applying (RP), we obtain ${\sim}B \to {\sim}A$. For (12), note first that ${\sim}B \to (A \to A)$ is derivable by (1) and (Ax1). Then by (RC), we obtain ${\sim}(A \to A) \to {\sim}{\sim}B$. This together with (9) implies the desired result. Then, (13) follows by (12), taking ${\sim}{\sim}(A \to A)$ in place of B, and (7). For (RD), assume A. Then by (Ax1), we obtain $(A \to A) \to A$. By applying (RC) twice, we get ${\sim}{\sim}(A \to A) \to {\sim}{\sim}A$. The desired result follows by this and (13). □

Remark 2.9 Note that $A \to {\sim}{\sim}A$, a stronger form of (RD), is *not* derivable, although we have the other way around (cf. (9)). In this sense the behavior of double empirical negation is dual to the behavior of double intuitionistic negation. Note also that based on (12), we may define the bottom element $\perp$ by $\sim(A \to A)$, instead of taking it as primitive. In view of (Ax10) and (5), $\sim A \wedge \sim\sim A$ also serves as a suitable definition of $\perp$. We can thus define intuitionistic negation $\neg$ by the usual "arrow-falsum" definition: $\neg A := A \to \perp$.

Proposition 2.10 *For* $\Gamma \cup \{A, B\} \subseteq \mathsf{Form}$, *if* $\Gamma \vdash \sim\sim A \to B$ *then* $\Gamma, A \vdash B$.

Proof. By the assumption $\Gamma \vdash \sim\sim A \to B$, we have $\Gamma, \sim\sim A \vdash B$ by (MP). Moreover, we have $\Gamma, A \vdash \sim\sim A$ by (RD). Thus, we obtain the desired result.□

By combining Propositions 2.7 and 2.10, we obtain the following theorem.

Theorem 2.11 *For* $\Gamma \cup \{A, B\} \subseteq \mathsf{Form}$, $\Gamma, A \vdash B$ *iff* $\Gamma \vdash \sim\sim A \to B$.

Corollary 2.12 *For* $\Gamma \cup \{A, B\} \subseteq \mathsf{Form}$, *we have* $\Gamma, A \vdash B$ *iff* $\Gamma \vdash \sim A \vee B$.

Proof. Immediate in view of the above result and (11). □

Remark 2.13 The deduction theorem formulated in terms of $\sim$ and $\vee$ is already discussed by De in [4, p.63] in which he takes the formula $\sim A \vee \sim\sim B$ instead of $\sim A \vee B$. However, these formulas are equivalent in the sense that both $\sim A \vee B \vdash \sim A \vee \sim\sim B$ and $\sim A \vee \sim\sim B \vdash \sim A \vee B$ hold.[6] Thus, the deduction theorem discussed in [4] is equivalent to the one presented here although the version here is slightly simpler.

The following proposition shows that the de Morgan laws with respect to empirical negation are fully provable in $\mathbf{IPC}^{\sim}$.

Proposition 2.14 *The following formulas are provable in* $\mathbf{IPC}^{\sim}$.

$$\sim(A \vee B) \to (\sim A \wedge \sim B) \tag{14}$$
$$(\sim A \vee \sim B) \to \sim(A \wedge B) \tag{15}$$
$$\sim(A \wedge B) \to (\sim A \vee \sim B) \tag{16}$$
$$(\sim A \wedge \sim B) \to \sim(A \vee B) \tag{17}$$

Proof. (14) and (15) are essentially by (Ax3), (Ax4) and (Ax6), (Ax7) respectively together with (RC). For (16), it runs as follows.

1	$\sim(\sim A \vee \sim B) \to (\sim\sim A \wedge \sim\sim B)$	[(14)]
2	$\sim(\sim A \vee \sim B) \to (A \wedge B)$	[1, (9)]
3	$\sim(A \wedge B) \to \sim\sim(\sim A \vee \sim B)$	[2, (RC)]
4	$\sim(A \wedge B) \to (\sim A \vee \sim B)$	[3, (9)]

Finally, for (17), it suffices to prove $\sim\sim(A \vee B) \to \sim(\sim A \wedge \sim B)$, and in view of (11), it suffices to prove $\sim(A \vee B) \vee \sim(\sim A \wedge \sim B)$. But then by (15) and (16), the formulas is equivalent to $\sim((A \vee B) \wedge \sim A) \vee \sim\sim B$. And in view of the deduction theorem, it suffices to show $(A \vee B) \wedge \sim A \vdash \sim\sim B$. And this can

[6] For the latter, something stronger holds, i.e. $(\sim A \vee \sim\sim B) \to (\sim A \vee B)$. The former is derivable in view of (RD) and Proposition 3.4 which is proved later. Note that, semantically, only $\sim A \vee \sim\sim B$ defines material implication at the base state in the sense that $\sim A \vee \sim\sim B$ is true at an arbitrary state iff A is false or B is true at the base state.

be proved as follows. Assume $(A \vee B) \wedge {\sim}A$. Then by applying (RP) to the first conjunct, we obtain ${\sim}A \to B$. Since we have ${\sim}A$ as our second conjunct of the assumption, we obtain B by (MP). Finally by applying (RD), we obtain the desired result. This completes the proof. □

Aside from having proved some important validities, it is helpful to consider some invalidities. The following proposition provides a list of such notables.

Proposition 2.15 *The following formulas are invalid in* **IPC**$^{\sim}$.

$$\begin{array}{ll} (A \wedge {\sim}B) \to {\sim}(A \to B) & ({\sim}A \to {\sim}B) \to (B \to A) \\ (A \wedge {\sim}A) \to B & {\sim}(A \to B) \to (A \wedge {\sim}B) \\ \neg(A \wedge {\sim}A) & {\sim}A \to \neg A \\ (A \to B) \to ({\sim}B \to {\sim}A) & \end{array}$$

Note, however, that the rule-forms of all formulas in the left-hand-column are valid.

Proof. To see the invalidity of the above formulas, note that the invalid $(A \wedge {\sim}A) \to B$ follows from the others. The details are left to the interested reader.

As for the validity of the rule-forms, let us briefly sketch their proofs. Since the last one is (RC) of Lemma 2.8, we deal with the other three rules. For the second one, assume A and ${\sim}A$. Then, by applying (RD) to A, we get ${\sim}{\sim}A$. Therefore, this together with the other assumption ${\sim}A$ and (Ax10) gives us B as desired. Proof of third rule is exactly parallel. For the first, assume $A \wedge {\sim}B$ and $A \to B$. Then we obtain B and ${\sim}B$ by (Ax3) and (MP). So, by the rule we just proved, we can in particular derive ${\sim}(A \to B)$. That is, we have $A \wedge {\sim}B, A \to B \vdash {\sim}(A \to B)$. Then, by the variant of deduction theorem (Theorem 2.11), we get $A \wedge {\sim}B \vdash {\sim}{\sim}(A \to B) \to {\sim}(A \to B)$. Thus by (8), we obtain $A \wedge {\sim}B \vdash {\sim}(A \to B)$ as desired. □

We have seen that the rule (RP) plays an important role. One may wonder whether this rule is derivable or not.[7] The following proposition shows that the rule is independent of other axioms and rule (MP).

Proposition 2.16 *The rule* (RP) *is independent of the other axioms and rule* (MP).

Proof. Consider the following truth tables which characterize the logic $\mathbf{P}^1$ introduced by Antonio Sette in [18]:

$\wedge$	1	i	0
1	1	1	0
i	1	1	0
0	0	0	0

$\vee$	1	i	0
1	1	1	1
i	1	1	1
0	1	1	0

$\to$	1	i	0
1	1	1	0
i	1	1	0
0	1	1	1

	$\sim$
1	0
i	1
0	1

Note here that designated values are 1 and i. It is straightforward to verify that the above truth tables validate all the axiom schemata of **IPC**$^{\sim}$, and that (MP) preserves designationhood. However, if we assign the values i and 0 to

[7] We thank a referee for asking us to clarify this point.

A and B of (RP), then we see that designated values are not preserved. Thus, we obtain the desired result. □

Remark 2.17 Before turning to further results on $\mathbf{IPC}^{\sim}$, let us briefly mention some related systems. There are at least two closely related systems in the literature. One of them is the $\mathbf{TCC}_{\omega}$ of A. B. Gordienko which is introduced and studied in [9]. This extends Richard Sylvan's $\mathbf{CC}_{\omega}$, studied in detail in [19] which is motivated by the fact that $\mathbf{C}_{\omega}$ of Newton da Costa (cf. [3]) lacks intersubstitutivity of provable equivalents:

$$\frac{A \leftrightarrow B}{D(A) \leftrightarrow D(B)}$$

where $D(A)$ is some wff containing A and $D(B)$ results from $D(A)$ by replacing one (derivatively, zero or more) occurrence of A by B. Now, $\mathbf{C}_{\omega}$ is obtained by adding (Ax9) and (9) to $\mathbf{IPC}^{+}$. Then, $\mathbf{CC}_{\omega}$ is obtained by adding (RC) to $\mathbf{C}_{\omega}$, and $\mathbf{TCC}_{\omega}$ extends $\mathbf{CC}_{\omega}$ by adding (Ax10). This shows that $\mathbf{TCC}_{\omega}$ is a subsystem of $\mathbf{IPC}^{\sim}$, and since (RP) is not valid in $\mathbf{TCC}_{\omega}$, it is a strict subsystem of $\mathbf{IPC}^{\sim}$.

The other closely related system is WECQ of Thomas Ferguson considered in [8]. This extends Graham Priest's da Costa Logic daC introduced and studied in [15]. Although the proof theory of daC is given in terms of natural deduction by Priest, we may easily observe that the subsystem of $\mathbf{IPC}^{\sim}$ obtained by eliminating (Ax10) is *weakly* complete with respect to the semantics developed in [15].[8] Then one of the observations of Ferguson shows that the addition of (Ax10) in the proof theory corresponds to the addition of the following condition on R dubbed "backwards convergence": if tRv and uRv, then there exists an w such that wRt and wRu. Note here that besides backward convergence, the relation R satisfies the usual conditions deployed in Kripke semantics for intuitionistic logic.[9]

3 Soundness and completeness

We now proceed to the proof of soundness and completeness. Our proof follows an idea employed in [16]. Since the system we deal with is not a relevant logic, the proof is rather simplified compared to that for $\mathbf{B}^{+}$ and its related systems.

3.1 Soundness

Theorem 3.1 *For* $\Gamma \cup \{A\} \subseteq \mathsf{Form}$, *if* $\Gamma \vdash A$ *then* $\Gamma \models A$.

Proof. By induction on the length of the proof, as usual. □

3.2 Preliminaries and key notions for completeness

As a preliminary for the completeness proof, we prove some rules that will be used in the following, and also some rules that emphasize some differences from

[8] There is another axiomatization in terms of Hilbert-style calculus in [2], although the consequence relation is defined in an idiosyncratic way.

[9] More on the relation between these systems can be found in [14].

$\mathbf{B}^+$.

Lemma 3.2 *The following rules are derivable in* $\mathbf{IPC}^{\sim}$.

$$\frac{C \to D}{(A \to C) \to (A \to D)} \quad \text{(Prefixing)}$$

$$\frac{A \to B}{(B \to C) \to (A \to C)} \quad \text{(Suffixing)}$$

$$\frac{A \to B \quad B \to C}{A \to C} \quad \text{(Transitivity)}$$

$$\frac{C \vee (A \to B) \quad C \vee A}{C \vee B} \quad \text{(D-MP)}$$

$$\frac{C \vee (A \vee B)}{C \vee ({\sim}A \to B)} \quad \text{(D-RP)}$$

Proof. We only deal with the last two rules as the others are obvious by the fact that $\mathbf{IPC}^{\sim}$ extends $\mathbf{IPC}^+$.
For (D-MP): Assume $C \vee (A \to B)$ and $C \vee A$. Then, by applying (RP), we obtain ${\sim}C \to (A \to B)$ and ${\sim}C \to A$ respectively. Combining these with (Ax2) will give us ${\sim}C \to B$. Finally, we apply (10) to obtain the desired result.
For (D-RP): Assume $C \vee (A \vee B)$. Then, by applying (RP), we obtain ${\sim}(C \vee A) \to B$, Therefore, we obtain $({\sim}C \wedge {\sim}A) \to B$ by using (17) which is equivalent to ${\sim}C \to ({\sim}A \to B)$. Finally, we apply (10) to obtain the desired result. □

Remark 3.3 (D-MP) and (D-RP) show that disjunctive forms of (MP) and (RP) can be proved in $\mathbf{IPC}^{\sim}$, and thus we do not have to assume them as rules of inference as in $\mathbf{B}^+$.

Since we have the deduction theorem, we can prove the following metatheorem in a rather simple manner.

Proposition 3.4 *If* $A \vdash C$ *and* $B \vdash C$*, then* $A \vee B \vdash C$.

Proof. Assume $A \vdash C$ and $B \vdash C$. Then, by Theorem 2.11, we obtain $\vdash {\sim}{\sim}A \to C$ and $\vdash {\sim}{\sim}B \to C$ respectively. By (Ax7), we get $\vdash ({\sim}{\sim}A \vee {\sim}{\sim}B) \to C$, and therefore $\vdash {\sim}{\sim}(A \vee B) \to C$ by (14) and (17). Finally, we obtain the desired result by another application of Theorem 2.11. □

We now state some definitions that will play important roles in the proof.

(i) $\Sigma \vdash_\pi A$ iff $\Sigma \cup \Pi \vdash A$.

(ii) Σ is a Π-*theory* iff:
  (a) if $A, B \in \Sigma$ then $A \wedge B \in \Sigma$
  (b) if $\vdash_\pi A \to B$ then (if $A \in \Sigma$ then $B \in \Sigma$).

(iii) Σ is *prime* iff (if $A \vee B \in \Sigma$ then $A \in \Sigma$ or $B \in \Sigma$).

(iv) $\Sigma \vdash_\pi \Delta$ iff for some $D_1, \ldots, D_n \in \Delta, \Sigma \vdash_\pi D_1 \vee \cdots \vee D_n$.

(v) $\vdash_\pi \Sigma \to \Delta$ iff for some $C_1, \ldots, C_n \in \Sigma$ and $D_1, \ldots, D_m \in \Delta$:

$$\vdash_\pi C_1 \wedge \cdots \wedge C_n \to D_1 \vee \cdots \vee D_n.$$

(vi) Σ is Π-*deductively closed* iff (if $\Sigma \vdash_{\pi} A$ then $A \in \Sigma$).

(vii) Let Form be the set of formulas. Then, $\langle \Sigma, \Delta \rangle$ is a Π-*partition* iff:
   (a) $\Sigma \cup \Delta = \mathsf{Form}$
   (b) $\nvdash_{\pi} \Sigma \to \Delta$

(viii) Σ is *non-trivial* iff $A \notin \Sigma$ for some formula A.

In all of the above, if Π is $\emptyset$, then the prefix 'Π-' will simply be omitted.

Remark 3.5 One point of departure from [16] is that we do not need the definition of the set $\Pi_{\to}$ which is defined as the set of all members of Π of the form $A \to B$ where Π is a set of sentences. Moreover, we added a definition of non-triviality that is not required in the proof of [16] as there is no bottom element in $\mathbf{B}^{+}$. The definition is necessary for our proof since a bottom element is available in $\mathbf{IPC}^{\sim}$.

We here note the following useful fact which is not the case in $\mathbf{B}^{+}$ as it relies on (Ax1).

Lemma 3.6 *If* Γ *is* Π-*theory, then* $\Pi \subseteq \Gamma$.

Proof. Take $A \in \Pi$. Then, we have $\Pi \vdash A$. Now, take any $C \in \Gamma$. Then, by (Ax1), we obtain $\Pi \vdash C \to A$, i.e. $\vdash_{\pi} C \to A$. Thus, combining this together with $C \in \Gamma$ and the assumption that Γ is Π-theory, we conclude that $A \in \Gamma$.□

3.3 Extension lemmas

We now prove a number of lemmas. The first group concerns extensions of sets with various properties.

Lemma 3.7 *If* $\langle \Sigma, \Delta \rangle$ *is a* Π-*partition then* Σ *is a prime* Π-*theory.*

Proof. We need to prove the following three facts:

(i) if $A, B \in \Sigma$ then $A \wedge B \in \Sigma$.

(ii) if $\vdash_{\pi} A \to B$ then (if $A \in \Sigma$ then $B \in \Sigma$).

(iii) if $A \vee B \in \Sigma$ then $A \in \Sigma$ or $B \in \Sigma$.

For (1): Assume $A, B \in \Sigma$ and $A \wedge B \notin \Sigma$. Then since $\Sigma \cup \Delta = \mathsf{Form}$, we have $A \wedge B \in \Delta$. This immediately implies $\vdash_{\pi} \Sigma \to \Delta$ which is a contradiction in view of $\nvdash_{\pi} \Sigma \to \Delta$.
For (2): Assume $\vdash_{\pi} A \to B$ and $A \in \Sigma$ and $B \notin \Sigma$. Then since $\Sigma \cup \Delta = \mathsf{Form}$, we have $B \in \Delta$. This means $\vdash_{\pi} \Sigma \to \Delta$ which is a contradiction in view of $\nvdash_{\pi} \Sigma \to \Delta$.
For (3): Assume $A \vee B \in \Sigma$ and $A \notin \Sigma$ and $B \notin \Sigma$. Then, since $\Sigma \cup \Delta = \mathsf{Form}$, we have $A, B \in \Delta$. This immediately implies $\vdash_{\pi} \Sigma \to \Delta$ which is a contradiction in view of $\nvdash_{\pi} \Sigma \to \Delta$. □

Lemma 3.8 *If* $\nvdash_{\pi} \Sigma \to \Delta$ *then there are* $\Sigma' \supseteq \Sigma$ *and* $\Delta' \supseteq \Delta$ *such that* $\langle \Sigma', \Delta' \rangle$ *is a* Π-partition.

Proof. The details are spelled out in the appendix. □

Corollary 3.9 *Let Σ be a Π-theory, Δ be closed under disjunction, and $\Sigma \cap \Delta = \emptyset$. Then there is $\Sigma' \supseteq \Sigma$ such that $\Sigma' \cap \Delta = \emptyset$ and Σ' is a prime Π-theory.*

Proof. First, it follows that $\not\vdash_\pi \Sigma \to \Delta$. For otherwise there would be some $C_1, \ldots, C_n \in \Sigma$ and $D_1, \ldots, D_m \in \Delta$:

$$\vdash_\pi C_1 \wedge \cdots \wedge C_n \to D_1 \vee \cdots \vee D_n.$$

Then, since Σ be a Π-theory, Σ is closed under conjunction, so $C_1 \wedge \cdots \wedge C_n \in \Sigma$. Moreover, if Σ be a Π-theory and $C_1 \wedge \cdots \wedge C_n \in \Sigma$ and $\vdash_\pi C_1 \wedge \cdots \wedge C_n \to D_1 \vee \cdots \vee D_n$, then it follows that $D_1 \vee \cdots \vee D_n \in \Sigma$. On the other hand, Δ be closed under disjunction so $D_1 \vee \cdots \vee D_n \in \Delta$. By combining these, we obtain $D_1 \vee \cdots \vee D_n \in \Sigma \cap \Delta$ which is a contradiction in view of $\Sigma \cap \Delta = \emptyset$.

Now since we have $\not\vdash_\pi \Sigma \to \Delta$, we obtain, by the previous lemma, that there are $\Sigma' \supseteq \Sigma$ and $\Delta' \supseteq \Delta$ such that $\langle \Sigma', \Delta' \rangle$ is a Π-partition. And by Lemma 3.7, it follows that Σ' is a prime Π-theory. It also follows that $\Sigma' \cap \Delta = \emptyset$, for otherwise we will have a formula $A_0 \in \Sigma' \cap \Delta \subseteq \Sigma' \cap \Delta'$. This will immediately imply that $\vdash_\pi \Sigma' \to \Delta'$ which cannot be the case in view of the fact that $\langle \Sigma', \Delta' \rangle$ is a Π-partition. □

Lemma 3.10 *If $\Sigma \not\vdash \Delta$ then there are $\Sigma' \supseteq \Sigma$ and $\Delta' \supseteq \Delta$ such that $\langle \Sigma', \Delta' \rangle$ is a partition, and Σ' is deductively closed.*

Proof. Similar to the proof of Lemma 3.8. The details are spelled out in the appendix. □

Corollary 3.11 *If $\Sigma \not\vdash A$ then there are $\Pi \supseteq \Sigma$ such that $A \notin \Pi$, Π is a prime Π-theory and is Π-deductively closed.*

Proof. Let Δ be $\{A\}$ and Σ' be Π in Lemma 3.10. Then by the lemma, we obtain a Π such that $\Pi \supseteq \Sigma$, $\langle \Pi, \Delta' \rangle$ is a partition, and Π is deductively closed. By Lemma 3.7, it follows that Π is a prime theory. Furthermore, $A \notin \Pi$ is obvious by the construction of Π, and Π is Π-deductively closed since if $\Pi \vdash_\pi A$ then $\Pi \vdash A$. Since Π is deductively closed, if follows that $A \in \Pi$, as desired. It remains to be shown that Π is a Π-theory. Suppose that $\vdash_\pi C \to D$ and $C \in \Pi$. Then clearly $\Pi \vdash D$, and hence $D \in \Pi$ by deductive closure. □

3.4 Counter-example lemma

The second lemma establishes that there are certain theories with properties that are crucial in the recursion case for $\to$ in the proof of the main theorem.

Lemma 3.12 *If Π is a prime Π-theory that is Π-deductively closed and $A \to B \notin \Pi$, then there is a prime Π-theory Γ, such that $A \in \Gamma$ and $B \notin \Gamma$.*

Proof. Let $\Sigma = \{C : A \to C \in \Pi\}$. Then Σ is a Π-theory. For suppose that $C_1, C_2 \in \Sigma$. Then $A \to C_1, A \to C_2 \in \Pi$. Thus $\vdash_\pi A \to (C_1 \wedge C_2)$ by (Ax5), so $A \to (C_1 \wedge C_2) \in \Pi$ since Π is Π-deductively closed, and thus $C_1 \wedge C_2 \in \Sigma$ by the definition of Σ. Now suppose that $\vdash_\pi C \to D$ and $C \in \Sigma$. Then $A \to C \in \Pi$ and so $\vdash_\pi A \to C$. By (Transitivity) we have $\vdash_\pi A \to D$. Since Π is Π-deductively closed, we have $A \to D \in \Pi$, and by the definition of Σ, we obtain $D \in \Sigma$ as desired.

Clearly $A \in \Sigma$ and $B \vee \cdots \vee B \notin \Sigma$. Based on this, let Δ be the closure of $\{B\}$ under disjunction. Then, $\Sigma \cap \Delta = \emptyset$. The result then follows from Corollary 3.9. □

3.5 Completeness

We are finally ready to prove the completeness.

Theorem 3.13 *For* $\Gamma \cup \{A\} \subseteq \mathsf{Form}$, *if* $\Gamma \models A$ *then* $\Gamma \vdash A$.

Proof. We prove the contrapositive. Suppose that $\Gamma \nvdash A$. Then, by the above lemma, there is a $\Pi \supseteq \Gamma$ such that Π is a prime theory and $A \notin \Pi$. Define the interpretation $\mathfrak{A} = \langle \Pi, X, \leq, I \rangle$, where $X = \{\Delta : \Delta \text{ is a non-trivial prime } \Pi\text{-theory}\}$, $\Delta \leq \Sigma$ iff $\Delta \subseteq \Sigma$ and I is defined thus. For every state Σ and propositional parameter p:

$$I(\Sigma, p) = 1 \text{ iff } p \in \Sigma$$

We show that this condition holds for any arbitrary formula B:

$$I(\Sigma, B) = 1 \text{ iff } B \in \Sigma \qquad (*)$$

It then follows that $\mathfrak{A}$ is a counter-model for the inference, and hence that $\Gamma \not\models A$. The proof of $(*)$ is by induction on the complexity of B.

Disjunction:

$$\begin{aligned} I(\Sigma, C \vee D) = 1 &\text{ iff } I(\Sigma, C) = 1 \text{ or } (I(\Sigma, D) = 1 \\ &\text{ iff } C \in \Sigma \text{ or } D \in \Sigma && \text{IH} \\ &\text{ iff } C \vee D \in \Sigma && \Sigma \text{ is a prime theory} \end{aligned}$$

Conjunction:

$$\begin{aligned} I(\Sigma, C \wedge D) = 1 &\text{ iff } I(\Sigma, C) = 1 \text{ and } (I(\Sigma, D) = 1 \\ &\text{ iff } C \in \Sigma \text{ and } D \in \Sigma && \text{IH} \\ &\text{ iff } C \wedge D \in \Sigma && \Sigma \text{ is a theory} \end{aligned}$$

Negation:

$$\begin{aligned} I(\Sigma, {\sim}C) = 1 &\text{ iff } I(\Pi, C) \neq 1 \\ &\text{ iff } C \notin \Pi && \text{IH} \\ &\text{ iff } {\sim}C \in \Sigma \end{aligned}$$

For the last equivalence, assume $C \notin \Pi$ and ${\sim}C \notin \Sigma$. Then, by the latter and $\Pi \subseteq \Sigma$, we obtain ${\sim}C \notin \Pi$. This together with $C \notin \Pi$ and the primeness of Π implies $C \vee {\sim}C \notin \Pi$, and thus $\Pi \nvdash C \vee {\sim}C$, which is a contradiction. For the other way around, assume ${\sim}C \in \Sigma$ and $C \in \Pi$. Then, by the latter and $\Pi \subseteq \Sigma$, we obtain $C \in \Sigma$. Therefore, together with ${\sim}C \in \Sigma$, it follows that $\Sigma \vdash C \wedge {\sim}C$ and thus $\Sigma \vdash B$ for any B. This contradicts that Σ is non-trivial.

Conditional:

$$
\begin{aligned}
I(\Sigma, C \to D) = 1 &\text{ iff for all } \Delta \text{ s.t. } \Sigma \subseteq \Delta, \text{ if } I(\Delta, C) = 1 \text{ then } I(\Delta, D) = 1 \\
&\text{ iff for all } \Delta \text{ s.t. } \Sigma \subseteq \Delta, \text{ if } C \in \Delta \text{ then } D \in \Delta \qquad \text{IH} \\
&\text{ iff } C \to D \in \Sigma
\end{aligned}
$$

For the last equivalence, assume $C \to D \in \Sigma$ and $C \in \Delta$ for any Δ s.t. $\Sigma \subseteq \Delta$. Then by $\Sigma \subseteq \Delta$ and $C \to D \in \Sigma$, we obtain $C \to D \in \Delta$. Therefore, we have $\Delta \vdash C \to D$ and $\Delta \vdash C \to D$, so by (MP), we obtain $\Delta \vdash D$, i.e. $D \in \Delta$, as desired. On the other hand, suppose $C \to D \notin \Sigma$. Then by the lemma above, there is a $\Sigma' \supseteq \Sigma$ such that $C \in \Sigma'$, $D \notin \Sigma'$ and Σ' is a prime Π-theory. And if Σ' is a prime Π-theory, then it follows that $\Pi \subseteq \Sigma'$. Furthermore, non-triviality of Σ' is obvious by $D \notin \Sigma'$. Thus, we obtain the desired result. □

4 Some reflections

We now consider some related issues. First, we deal with a variant of the semantics we presented in §2. Second, we respond to some remarks of Williamson on empirical negation.

4.1 A "many distinguished states" semantics

Instead of taking a single base state as representing our current state of evidence, we may also think of having a set of base states representing ways our current state of evidence might be.[10] In practice we can't always be certain about what our evidential state is because e.g. whether something counts as evidence or exactly what the evidence is might in principle be indeterminate. All we can do is rule out certain states as ours if they make false statements we know are supported by our current evidence. Since ruling out certain states as our own won't always secure a unique state, we might be more cautious to consider several states that are, for all we know or even could know, our current evidential state. This leads to the following modification of our semantics as follows.

Definition 4.1 A model for the language is a quadruple $\langle W, D, \leq, V\rangle$, where W is a non-empty set (of states); $D \subseteq W$ (the distinguished states); $\leq$ is a partial order on W such that for all $\delta \in D$ and all $w \in W \setminus D$, $\delta \leq w$. Moreover, $V(w, p) \mapsto \{0, 1\}$ an assignment of truth values to state-variable pairs with the condition that $V(w_1, p) = 1$ and $w_1 \leq w_2$ only if $V(w_2, p) = 1$ for all $p \in \mathsf{Prop}$ and all $w_1, w_2 \in W$. Valuations V are then extended to interpretations I to state-formula pairs as in Definition 2.2 except for the empirical negation, we extend it by the following conditions:

$$I(w, {\sim}A) = 1 \text{ iff for some } \delta \in D, I(\delta, A) = 0$$

Finally, semantic consequence is now defined in terms of truth preservation at all the elements of D.

---

[10] We would like to thank Thomas Müller for suggesting that models have a set of base states.

Definition 4.2 We say that A is true in a model $\mathcal{M} = \langle W, D, \leq, V\rangle$ iff for all $\delta \in D$, $V(\delta, A) = 1$. Consequence is then defined as truth preservation in a model: $\Gamma \models A$ iff for all models $\mathcal{M}$ and all $B \in \Gamma$, if B is true in $\mathcal{M}$ then A is true in $\mathcal{M}$.

It turns out, interestingly, that the "many base states" interpretation of empirical negation is equivalent to the original interpretation in the sense that they generate the same consequence relation.

Theorem 4.3 *For* $\Gamma \cup \{A\} \subseteq \mathsf{Form}$, $\Gamma \models' A$ *iff* $\Gamma \vdash A$.

Proof. The only non-trivial direction concerns soundness, as the completeness proof is exactly the same. (We only ever need a counter model in which D is a singleton.) Showing soundness is not hard, which we leave to the interested reader. □

Remark 4.4 If we change the truth conditions of $\sim$ in Definition 4.1 by replacing the existential quantifier by a universal one, we obtain a strictly weaker logic. For every original model is a "universal model", but universal models give us additional countermodels. For instance, $A \vee {\sim}A$ is no longer valid over the class of universal models, though it was valid on the original semantics.

4.2 Williamson on empirical negation

Williamson argues in [21] that using empirical negation $\sim$ to block the argument to the **IPL**-inconsistency of 'A will never be decided' won't work because ${\sim}A \to \neg A$ ought to be valid. But then the statement that A will never be decided, ${\sim}(KA \vee K\neg A)$, implies $\neg(KA \vee K\neg A)$ and we're back in inconsistency. [11]

Why think empirical negation is stronger than intuitionistic negation, i.e. that ${\sim}A \to \neg A$? Williamson says:

> For if $\sim$ is to count intuitionistically as any sort of negation at all, ${\sim}A$ should at least be inconsistent with A in the ordinary intuitionistic sense. A warrant for $A \wedge {\sim}A$ should be impossible. That is, we should have $\neg(A \wedge {\sim}A)$. [21, p. 139]

There is good reason to believe everything he says *except for the last line*: $A \wedge {\sim}A$ may be impossible while $\neg(A \wedge {\sim}A)$ not be valid. But how?

Suppose that ${\sim}A$ is read 'A is not warranted by our current state of evidence', as we have been interpreting it. Then $A \wedge {\sim}A$ can only be warranted by our current evidential state if A is currently warranted and is currently not, which is a contradiction. So $A \wedge {\sim}A$ could never be warranted, *except in some merely possible, non-current evidential state.* We will never be in such a state, since we ever only have our current evidence to work with, but there's no reason to dismiss such states as playing a role in the truth conditions of empirical discourse; for they represent ways in which the current evidential state might evolve. There is a good sense, then, in which $A \wedge {\sim}A$ is impossible—it could

[11] Williamson gives a standard Fitch argument for the validity of $\neg KA \to \neg A$. With this $\neg KA \wedge \neg K\neg A$, which is equivalent to $\neg(KA \vee K\neg A)$, implies $\neg A \wedge \neg\neg A$.

never be warranted by our current state of evidence—though this does not imply that $\neg(A \wedge {\sim}A)$ which says that $A \wedge {\sim}A$ could never be warranted even by *future* evidential states. But surely it could! For in some future state, we may have that A is warranted together with it being warranted that A is not warranted at our current state. In sum, we may have both $A, {\sim}A \vdash \bot$ and $\nvdash \neg(A \wedge {\sim}A)$.[12]

Empirical negation therefore provides us with a well-motivated way of blocking a Fitch-like argument against the sort of verificationism we have been considering.

5 Conclusion

Here is a brief summary of the paper. We first discussed the motivation for empirical negation (§1) in a broadly verificationist setting. We then introduced the semantics and the proof theory (§2) which was followed by the proof of strong completeness using techniques from relevant logics (§3). The completeness proof was followed by some reflections on the semantics we presented, and on some arguments of Williamson against the kind of empirical negation we have been considering (§4). Before closing, we wish to briefly sketch two future directions of this research.

Adding empirical negation to Nelson's logic. We have expanded intuitionistic logic by empirical negation, but there is another well-known expansion of intuitionistic logic by strong negation which results in Nelson's logic **N3**. (cf. [20,13]) The motivation here is to consider not only the verification but also the falsification of a sentence. While empirical negation allows us to extend our constructive interpretation from the mathematical to the empirical domain, the focus is on verification. It is of interest, however, to see how we might further extend that interpretation by treating verification and falsification separately.

In formulating the semantics of Nelson's logic, we need cover not only the truth but also the falsity conditions of sentences. Since we already have the truth condition for empirical negation, what we need in carrying out our extension are adequate falsity conditions for empirical negation. We shall, however, have to leave these details for another occasion.

Empirical negation as classical negation. One way of looking at empirical negation is as classical negation; in other words, $\mathbf{IPC}^{\sim}$ can be seen as an expansion of intuitionistic logic by classical negation. Some expansions in a similar vein may be found in [5,11,12].[13] To see the difference between our account and others, consider Kripke semantics for intuitionistic logic along with the most straightforward truth condition for classical negation:

$$I(w, {\sim}A) = 1 \text{ iff } I(w, A) = 0. \qquad (\star)$$

Validity is defined as truth preservation at *all* states, not just some distinguished state(s). A natural question then is to ask whether extending intu-

[12] This is part of the result proved in Proposition 2.15.

[13] See also [1] in which intuitionistic and classical implications are combined.

itionistic logic this way, in terms of $(\star)$, is better or worse compared with extending it by empirical negation. While this is a question we wish to address in a more general setting on another occasion, let us quickly point out a fact that favors the empirical negation approach in an intuitionistic context.

If we define classical negation by $(\star)$, the resulting system conservatively extends classical logic but it is not sound for intuitionistic logic (since truth is no longer preserved up the order). On the other hand, if we define classical negation as empirical negation, then the resulting system conservatively extends the intuitionistic fragment while preserving soundness. Empirical negation may then be seen as providing a simple way of adding classical negation to intuitionistic logic or any logic whose semantics makes use of a similar kind of requirement that truth be preserved up an order, such as a simplified semantics for relevant logic. Indeed, we may regard the semantics for empirical negation as a special case of so-called star semantics for relevant logics given by Richard and Val Routley in [17].[14] In this case, the star function maps every world to a single world, namely the base world g. We leave the task of investigating this connection to relevant logic for another occasion.

Appendix

Proof of Lemma 3.8:

Let $A_0, A_1, \ldots$ be an enumeration of the set of formulas Form. Then we define Σ_i and Δ_i $(i \in \omega)$ by induction as follows:

- $\Sigma_0 := \Sigma$; $\Delta_0 := \Delta$.
- For Σ_i, the definition is as follows:

$$\Sigma_{i+1} := \begin{cases} \Sigma_i \cup \{A_i\} & \text{if } \nvdash_\pi \Sigma_i \cup \{A_i\} \to \Delta_i \\ \Sigma_i & \text{otherwise} \end{cases}$$

And for Δ_i, the definition is as follows:

$$\Delta_{i+1} := \begin{cases} \Delta_i & \text{if } \nvdash_\pi \Sigma_i \cup \{A_i\} \to \Delta_i \\ \Delta_i \cup \{A_i\} & \text{otherwise} \end{cases}$$

Finally, we define Σ', Δ' as follows:

$$\Sigma' := \bigcup_{i<\omega} \Sigma_i \text{ and } \Delta' := \bigcup_{i<\omega} \Delta_i.$$

We now prove that $\langle \Sigma', \Delta' \rangle$ is a Π-*partition*. For this purpose, we need to prove the following two facts:

(i) $\Sigma' \cup \Delta' = \mathsf{Form}$

(ii) $\nvdash_\pi \Sigma' \to \Delta'$

[14] We thank a referee for this interesting remark.

Since the former is satisfied by the construction, it need only show the latter. By the compactness of $\vdash_\pi$, if $\vdash_\pi \Sigma' \to \Delta'$ then $\vdash_\pi \Sigma_k \to \Delta_k$ for some $k \in \omega$. Therefore, it suffices to show that for no $i \in \omega$, $\nvdash_\pi \Sigma_i \to \Delta_i$. This can be proved by induction on i. For the case when $i = 0$, it is true by the definition. Suppose now that it is true for $i = j$ but not $i = j+1$. Then, we have $\nvdash_\pi \Sigma_j \to \Delta_j$ and $\vdash_\pi \Sigma_{j+1} \to \Delta_{j+1}$.

Now, if $\vdash_\pi \Sigma_{j+1} \to \Delta_{j+1}$ and $\nvdash_\pi \Sigma_j \cup \{A_j\} \to \Delta_j$, then $\Sigma_{j+1} = \Sigma_j \cup \{A_j\}$ and $\Delta_{j+1} = \Delta_j$ by the latter, and thus by applying this to the former we obtain $\vdash_\pi \Sigma_j \cup \{A_j\} \to \Delta_j$. Therefore, $\vdash_\pi \Sigma_{j+1} \to \Delta_{j+1}$ implies $\vdash_\pi \Sigma_j \cup \{A_j\} \to \Delta_j$ by reductio. So, we obtain the following from $\vdash_\pi \Sigma_{j+1} \to \Delta_{j+1}$:

$$\vdash_\pi \Sigma_j \cup \{A_j\} \to \Delta_j$$

On the other hand, if $\vdash_\pi \Sigma_{j+1} \to \Delta_{j+1}$ and $\vdash_\pi \Sigma_j \cup \{A_j\} \to \Delta_j$, then $\Sigma_{j+1} = \Sigma_j$ and $\Delta_{j+1} = \Delta_j \cup \{A_j\}$ by the latter, and thus by applying this to the former we obtain $\vdash_\pi \Sigma_j \to \Delta_j \cup \{A_j\}$. Therefore, by the above result and $\vdash_\pi \Sigma_{j+1} \to \Delta_{j+1}$, we get the following:

$$\vdash_\pi \Sigma_j \to \Delta_j \cup \{A_j\}$$

So, for some conjunctions C_1, C_2 of members of Σ_j and some disjunctions D_1, D_2 of members of Δ_j, we obtain the following:

$$C_1 \wedge A_j \to D_1 \quad C_2 \to D_2 \vee A_j.$$

But then, this leads to contradiction. Indeed,

1	$C_1 \wedge A_j \to D_1$	[sup.]
2	$C_2 \to D_2 \vee A_j$	[sup.]
3	$C_1 \wedge C_2 \to C_2$	[(Ax4)]
4	$C_1 \wedge C_2 \to D_2 \vee A_j$	[3, 2, (Transitivity)]
5	$C_1 \wedge C_2 \to C_1$	[(Ax3)]
6	$C_1 \wedge C_2 \to (D_2 \vee A_j) \wedge C_1$	[4, 5, (Ax5), (MP)]
7	$C_1 \wedge C_2 \to D_2 \vee (A_j \wedge C_1)$	[6, (6), (MP)]
8	$C_1 \wedge A_j \to D_1 \vee D_2$	[1, (Ax6), (MP)]
9	$D_2 \to D_1 \vee D_2$	[(Ax7)]
10	$D_2 \vee (C_1 \wedge A_j) \to D_1 \vee D_2$	[8, 9, (Ax8), (MP)]
11	$C_1 \wedge C_2 \to D_1 \vee D_2$	[7, 10, (Transitivity)]

And this last formula shows that $\vdash_\pi \Sigma_j \to \Delta_j$ which is a contradiction in view of $\nvdash_\pi \Sigma_j \to \Delta_j$.

Proof of Lemma 3.10:

Let $A_0, A_1, \ldots$ be an enumeration of the set of formulas Form. Then we define Σ_i and Δ_i $(i \in \omega)$ by induction as follows:

- $\Sigma_0 := \Sigma$; $\Delta_0 := \Delta$.
- For Σ_i, the definition is as follows:

$$\Sigma_{i+1} := \begin{cases} \Sigma_i \cup \{A_i\} & \text{if } \Sigma_i \cup \{A_i\} \nvdash \Delta_i \\ \Sigma_i & \text{otherwise} \end{cases}$$

And for Δ_i, the definition is as follows:

$$\Delta_{i+1} := \begin{cases} \Delta_i & \text{if } \Sigma_i \cup \{A_i\} \nvdash \Delta_i \\ \Delta_i \cup \{A_i\} & \text{otherwise} \end{cases}$$

Finally, we define Σ', Δ' as follows:

$$\Sigma' := \bigcup_{i<\omega} \Sigma_i \text{ and } \Delta' := \bigcup_{i<\omega} \Delta_i.$$

Then, we now prove that $\langle \Sigma', \Delta' \rangle$ is a *partition*. For this purpose, we need to prove the following two facts:

(i) $\Sigma' \cup \Delta' = \mathsf{Form}$

(ii) $\nvdash \Sigma' \to \Delta'$

Since the former is satisfied by the construction, only the latter remains to be shown. By the compactness of $\vdash$, if $\vdash \Sigma' \to \Delta'$ then $\vdash \Sigma_k \to \Delta_k$ for some $k \in \omega$. Furthermore, it holds that $\vdash \Sigma \to \Delta$ implies $\Sigma \vdash \Delta$ for any Σ and Δ. Therefore, it suffices to show that for no $i \in \omega$, $\Sigma_k \nvdash \Delta_k$. This can be proved by induction on i. For the case when $i = 0$, it is true by the definition. Suppose now that it is true for $i = j$ but not $i = j + 1$. Then, we have $\Sigma_j \nvdash \Delta_j$ and $\Sigma_{j+1} \vdash \Delta_{j+1}$.
Case 1. if $\Sigma_j \cup \{A_j\} \nvdash \Delta_j$, then $\Sigma_{j+1} = \Sigma_j \cup \{A_j\}$ and $\Delta_{j+1} = \Delta_j$, and thus by applying this to $\Sigma_{j+1} \vdash \Delta_{j+1}$ we obtain $\Sigma_j \cup \{A_j\} \vdash \Delta_j$. Therefore, $\Sigma_j \cup \{A_j\} \vdash \Delta_j$ by reductio.
Case 2. if $\Sigma_j \cup \{A_j\} \vdash \Delta_j$, then $\Sigma_{j+1} = \Sigma_j$ and $\Delta_{j+1} = \Delta_j \cup \{A_j\}$ by the latter, and thus by applying this to $\Sigma_{j+1} \vdash \Delta_{j+1}$ we obtain $\Sigma_j \vdash \Delta_j \cup \{A_j\}$. Therefore, by the above result, we get $\Sigma_j \vdash \Delta_j \cup \{A_j\}$.

So, for some conjunctions C_1, C_2 of members of Σ_j and some disjunctions D_1, D_2 of members of Δ_j, we obtain the following:

$$C_1 \wedge A_j \vdash D_1 \quad C_2 \vdash D_2 \vee A_j.$$

But then, this leads to contradiction. Indeed,

1	$C_1 \wedge A_j \vdash D_1$	[sup.]
2	$C_2 \vdash D_2 \vee A_j$	[sup.]
3	$C_1 \wedge C_2 \vdash C_2$	[(Ax4)]
4	$C_1 \wedge C_2 \vdash D_2 \vee A_j$	[3, 2, (Transitivity)]
5	$C_1 \wedge C_2 \vdash C_1$	[(Ax3)]
6	$C_1 \wedge C_2 \vdash (D_2 \vee A_j) \wedge C_1$	[4, 5, (Ax5), (MP)]
7	$C_1 \wedge C_2 \vdash D_2 \vee (A_j \wedge C_1)$	[6, (6), (MP)]
8	$C_1 \wedge A_j \vdash D_1 \vee D_2$	[1, (Ax6), (MP)]
9	$D_2 \vdash D_1 \vee D_2$	[(Ax7)]
10	$D_2 \vee (C_1 \wedge A_j) \vdash D_1 \vee D_2$	[8, 9, Proposition 3.4, (MP)]
11	$C_1 \wedge C_2 \vdash D_1 \vee D_2$	[7, 10, (Transitivity)]

And this last formula shows that $\Sigma_j \vdash \Delta_j$ which is a contradiction in view of $\Sigma_j \nvdash \Delta_j$.

All that remains is to show that Σ' is deductively closed. Suppose that $\Sigma' \vdash A$ but $A \notin \Sigma'$. Since $\Sigma' \cup \Delta' = \mathsf{Form}$, we have $A \in \Delta'$, and thus $\Sigma' \vdash \Delta'$. But this is a contradiction. □

References

[1] Caleiro, C. and J. Ramos, *Combining classical and intuitionistic implications*, in: B. Konev and F. Wolter, editors, *Frontiers of Combining Systems*, LNCS 4720 (2007), pp. 118–132.

[2] Castiglioni, J. L. and R. C. E. Biraben, *Strict paraconsistency of truth-degree preserving intuitionistic logic with dual negation*, Logic Journal of the IGPL **22** (2014), pp. 268–273.

[3] da Costa, N. C. A., *On the theory of inconsistent formal systems*, Notre Dame Journal of Formal Logic **15** (1974), pp. 497–510.

[4] De, M., *Empirical Negation*, Acta Analytica **28** (2013), pp. 49–69.

[5] del Cerro, L. F. and A. Herzig, *Combining classical and intuitionistic logic*, in: F. Baader and K. Schulz, editors, *Frontiers of Combining Systems* (1996), pp. 93–102.

[6] DeVidi, D. and G. Solomon, *Empirical negation in intuitionisitic logic*, in: D. DeVidi and T. Kenyon, editors, *A logical approach to philosophy*, Springer, Dordrecht, The Netherlands, 2006 pp. 151–168.

[7] Dummett, M., "The seas of language," Oxford University Press, 1996.

[8] Ferguson, T. M., *Extensions of Priest-da Costa Logic*, Studia Logica **102** (2014), pp. 145–174.

[9] Gordienko, A. B., *A Paraconsistent Extension of Sylvan's Logic*, Algebra and Logic **46** (2007), pp. 289–296.

[10] Heyting, A., "Intuitionism: An Introduction," North-Holland Publishing, 1971.

[11] Humberstone, L., *Interval semantics for tense logic: some remarks*, Journal of Philosophical Logic **8** (1979), pp. 171–196.

[12] Humberstone, L., *The pleasures of anticipation: enriching intuitionistic logic*, Journal of Philosophical Logic **30** (2001), pp. 395–438.

[13] Kamide, N. and H. Wansing, *Proof theory of Nelson's paraconsistent logic: A uniform perspective*, Theoretical Computer Science **415** (2012), pp. 1–38.

[14] Omori, H., *A note on da Costa Logic and* CC_ω, in preparation.

[15] Priest, G., *Dualising Intuitionistic Negation*, Principia **13** (2009), pp. 165–184.

[16] Priest, G. and R. Sylvan, *Simplified Semantics for Basic Relevant Logic*, Journal of Philosophical Logic **21** (1992), pp. 217–232.

[17] Routley, R. and V. Routley, *Semantics for first degree entailment*, Noûs **6** (1972), pp. 335–359.

[18] Sette, A., *On the propositional calculus* P^1, Mathematica Japonicae **16** (1973), pp. 173–180.

[19] Sylvan, R., *Variations on da Costa C Systems and Dual-Intuitionistic Logics I. Analyses of* C_ω *and* CC_ω, Studia Logica **49** (1990), pp. 47–65.

[20] Wansing, H., *Negation*, in: L. Goble, editor, *The Blackwell Guide to Philosophical Logic*, Basil Blackwell Publishers, Cambridge/MA, 2001 pp. 415–436.

[21] Williamson, T., *Never say never*, Topoi **13** (1994), pp. 135–145.

The Effects of Modalities in Separation Logics (Extended Abstract)

Stéphane Demri [1]

New York University, USA & CNRS, France

Morgan Deters [2]

New York University, USA

Abstract

Like modal logic, temporal logic, or description logic, separation logic has become a popular class of logical formalisms in computer science, conceived as assertion languages for Hoare-style proof systems with the goal to perform automatic program analysis. We present similarities with modal and temporal logics, and we present landmark results about decidability, complexity and expressive power.

Keywords: Separation logic, decidability, computational complexity, expressive power, temporal logic, modal logic, first-order logic, second-order logic

When separation logic joins the club Introducing new logics is always an uncertain enterprise since there must be sufficient interest to use new formalisms. In spite of this hurdle, we know several recent success stories. For instance, even though a pioneering work on symbolic modal logic by Lewis appeared in 1918 [20], the first monographs on symbolic modal logic appear some fifty years later, see e.g. [16]. Nowadays, modal logic is divided into many distinct branches and remains one of the most active research fields in logic and computer science. Additionally, the introduction of temporal logic to computer science, due to Pnueli [25], has been a major step in the development of model-checking techniques, see e.g. [10,3]. This is now a well-established approach for the formal verification of computer systems: one models the system to be verified by a mathematical structure (typically a directed graph) and expresses behavioral properties in a logical formalism (typically a temporal logic). Verification by model-checking [10] consists of developing algorithms whose goal is

[1] demri@cs.nyu.edu. Work partially supported by the EU Seventh Framework Programme under grant agreement No. PIOF-GA-2011-301166 (DATAVERIF).

[2] mdeters@cs.nyu.edu. Work partially supported by the Air Force Office of Scientific Research under award FA9550-09-1-0596, and the National Science Foundation under grant 0644299.

to verify whether the logical properties are satisfied by the abstract model. The development of tools is done in parallel with the design of techniques to optimize the verification process. Apart from the development of model-checkers such as Cadence SMV, SPIN, or Uppaal, the transfer towards industrial applications is also present in research and development units. The development of description logics for knowledge representation has also followed a successful path, thanks to a permanent interaction between theoretical works, pushing ever further the high complexity and undecidability borders, and more applied works dedicated to the design of new tools and the production of more and more applications, especially in the realm of ontology languages. The wealth of research on description logic is best illustrated by [2], in which can be found many chapters on theory, implementations, and applications. By contrast, Chapter 1 of [2] provides a gentle introduction to description logics and recalls that its roots can be traced back a few decades. It is well-known that modal logic, temporal logic, and description logic have many similarities even though each family has its own research agenda. For instance, models can be (finite or infinite) graphs, the classes of models range from concrete ones to more abstract ones, and any above-mentioned class includes a wide range of logics and fragments. In this work, we deal with another class of logics, separation logic, that has been introduced quite recently (see e.g. [17,26]). Separation logic is the subject of tremendous interest, leading to many works on theory, tools and applications (mainly for the automatic program analysis). Any resemblance to modal, temporal, or description logic is certainly not purely coincidental—but separation logic also has its own assets.

In the possible-world semantics for modal logic, the modal operator $\Box$ [resp. $\Diamond$] corresponds to universal [resp. existential] quantification on successor worlds, and these are essential properties to be stated, partly explaining the impact of Kripke's discovery [18,12]. Similarly, the ability to divide a model into two disjoint parts happens to be a very natural property. This might explain the success of separation logic: disjoint memory states can be considered, providing an elegant means to perform local reasoning. *Separation* is a key concept that has been already introduced in interval temporal logic (ITL) [23] with the *chop* operator (and probably in many other logical formalisms such as in graph logics [13]) and therefore, the development of separation logic can be partly explained by the relevance of the separation concept. Its impressive development can be also justified by the fact that separation logic extends Hoare logic for reasoning about programs with dynamic data structures, meeting also industrial needs as witnessed by the recent acquisition of Monoidics Ltd. by Facebook.

Separation and composition Separation logic has been introduced as an extension of Hoare logic [15] to verify programs with mutable data structures [17,26]. A major feature is the ability to reason locally in a modular way, which can be performed thanks to the separating conjunction $*$ that allows one to state properties in disjoint parts of the memory. Moreover, the adjunct implication $\mathrel{-\!\!*}$ asserts that whenever a fresh heap satisfies a property, its composition with the current heap satisfies another property. This is particularly useful when a piece of code mutates memory locally, and we want to state some property of the entire memory (such as the preservation of data structure invariants). In a sense, if modal logic is made for reasoning about necessity and possibility, separation logic is made for reasoning about separation and composition. Of course this type of statement is an oversimplification—apart from the fact that it may appear a bit old-fashioned to most modal logicians—but this may help to get a first picture. As a taste of separation logic, it is worth observing that models can be finite graphs and the classes of models range from concrete ones (with heaps for instance) to very abstract ones (e.g., cancellative partial commutative monoids). While evaluating a formula, models can be updated as in public announcement logics.

Smallfoot was the first implementation to use separation logic, its goal to verify the extent to which proofs and specifications made by hand could be treated automatically [4]. The automatic part is related to assertion checking, but the user has to provide preconditions, postconditions, and loop invariants. A major step has been then to show that the method is indeed scalable [28]. In a sense, the legitimate question about the practical utility of separation logic was quickly answered, leading to a new generation of tools including Slayer developed by Microsoft Research, Space Invader [14,28], and Infer [7]. Actually, nowadays, many tools support separation logic as an assertion language and, more importantly, in order to produce interactive proofs with separation logic, several proof assistants encode the logic, see e.g. [27].

From the very beginning, the theory of separation logic has been an important research thread even if not always related to automatic verification. This is not very surprising since separation logic can be understood as a concretization of bunched logic BI which is a general logic of resources with a nice proof theory [24]. Besides, as for modal and temporal logics, the relationships between separation logic, and first-order or second-order logics have been the source of many characterizations and works. This is particularly true since the separation connectives are second-order in nature, see e.g. [21,19,8,5]. For instance, separation logic is equivalent to a Boolean propositional logic [22,21] if first-order quantifiers are disabled. Similarly, the complexity of satisfiability and model-checking problems for fragments of separation logic has been examined [9,26,11,1,6]. In [9], the model-checking and satisfiability problems for propositional separation logic are shown PSPACE-complete; this is done by proving a small model property.

Content The goal of this work is twofold. First, we would like to emphasize the similarities between separation logic and modal and temporal logics. Our intention is to pinpoint the common features in terms of models, proof techniques, motivations, and so forth. Second, we wish to present landmark results about decidability, complexity, and expressive power, providing a survey on the theoretical side of separation logic. These are standard themes for studying logics in computer science, and we deliberately focus on the logical aspects of separation logic.

References

[1] Antonopoulos, T., N. Gorogiannis, C. Haase, M. Kanovich and J. Ouaknine, *Foundations for decision problems in separation logic with general inductive predicates*, in: *FOSSACS'14*, Lecture Notes in Computer Science **8412** (2014), pp. 411–425.

[2] Baader, F., D. Calvanese, D. McGuinness, D. Nardi and P. Patel-Schneider, editors, "The Description Logic Handbook: Theory, Implementation and Applications," Cambridge University Press, 2003.

[3] Bérard, B., M. Bidoit, A. Finkel, F. Laroussinie, A. Petit, L. Petrucci and P. Schnoebelen, "Systems and Software Verification, Model-Checking Techniques and Tools," Springer, 2001.

[4] Berdine, J., C. Calcagno and P. O'Hearn, *Smallfoot: Modular automatic assertion checking with separation logic*, in: *FMCO'05*, Lecture Notes in Computer Science **4111** (2005), pp. 115–137.

[5] Brochenin, R., S. Demri and E. Lozes, *On the almighty wand*, Information and Computation **211** (2012), pp. 106–137.

[6] Brotherston, J., C. Fuhs, N. Gorogiannis and J. Navarro Perez, *A decision procedure for satisfiability in separation logic with inductive predicates*, in: *LICS'14*, 2014, to appear.

[7] Calcagno, C. and D. Distefano, *Infer: An automatic program verifier for memory safety of C programs*, in: *NASA Formal Methods*, Lecture Notes in Computer Science **6617** (2011), pp. 459–465.

[8] Calcagno, C., P. Gardner and M. Hague, *From separation logic to first-order logic*, in: *FOSSACS'05*, Lecture Notes in Computer Science **3441** (2005), pp. 395–409.

[9] Calcagno, C., P. O'Hearn and H. Yang, *Computability and complexity results for a spatial assertion language for data structures*, in: *FSTTCS'01*, Lecture Notes in Computer Science **2245** (2001), pp. 108–119.

[10] Clarke, E., O. Grumberg and D. Peled, "Model checking," The MIT Press Books, 2000.

[11] Cook, B., C. Haase, J. Ouaknine, M. Parkinson and J. Worrell, *Tractable reasoning in a fragment of separation logic*, in: *CONCUR'11*, Lecture Notes in Computer Science **6901** (2011), pp. 235–249.

[12] Copeland, J., *The genesis of possible worlds semantics*, Journal of Philosophical Logic **31** (2002), pp. 99–137.

[13] Dawar, A., P. Gardner and G. Ghelli, *Expressiveness and complexity of graph logic*, Information and Computation **205** (2007), pp. 263–310.

[14] Distefano, D., P. O'Hearn and H. Yang, *A local shape analysis based on separation logic*, in: *TACAS'06*, Lecture Notes in Computer Science **3920** (2006), pp. 287–302.

[15] Hoare, C. A. R., *An axiomatic basis for computer programming*, Communications of the ACM **12** (1969), pp. 576–580.

[16] Hughes, G. and M. Cresswell, "An introduction to modal logic," Methuen and Co., 1968.

[17] Ishtiaq, S. and P. O'Hearn, *BI as an assertion language for mutable data structures*, in: *POPL'01* (2001), pp. 14–26.

[18] Kripke, S., *A completeness theorem in modal logic*, The Journal of Symbolic Logic **24** (1959), pp. 1–14.

[19] Kuncak, V. and M. Rinard, *On spatial conjunction as second-order logic*, Technical Report MIT–CSAIL–TR–2004–067, MIT CSAIL (2004).

[20] Lewis, C. I., "A survey of symbolic logic," University of California Press, Berkeley, 1918.

[21] Lozes, E., "Expressivité des Logiques Spatiales," Phd thesis, ENS Lyon (2004).

[22] Lozes, E., *Separation logic preserves the expressive power of classical logic*, in: *SPACE'04*, 2004.

[23] Moszkowski, B., *Reasoning about digital circuits*, Technical Report STAN-CS-83-970, Dept. of Computer Science, Stanford University, Stanford, CA (1983).

[24] O'Hearn, P. and D. Pym, *The logic of bunched implications*, Bulletin of Symbolic Logic **5** (1999), pp. 215–244.

[25] Pnueli, A., *The temporal logic of programs*, in: *FOCS'77* (1977), pp. 46–57.

[26] Reynolds, J., *Separation logic: a logic for shared mutable data structures*, in: *LICS'02* (2002), pp. 55–74.

[27] Tuerk, T., "A separation logic fragment for HOL," Ph.D. thesis, University of Cambridge (2011).

[28] Yang, H., O. Lee, J. Berdine, C. Calcagno, B. Cook, D. Distefano and P. O'Hearn, *Scalable shape analysis for systems code*, in: *CAV'08*, Lecture Notes in Computer Science **5123** (2008), pp. 385–398.

Some Exponential Lower Bounds on Formula-size in Modal Logic

Hans van Ditmarsch [1]

CNRS, LORIA Université de Lorraine

Jie Fan [2]

Department of Philosophy
Peking University

Wiebe van der Hoek [3]

Department of Computer Science
University of Liverpool

Petar Iliev [4]

CNRS, LORIA, Université de Lorraine

Abstract

We present two families of exponential lower bounds on the size of modal formulae and use them to establish the following succinctness results. We show that the logic of contingency (ConML) is exponentially more succinct than basic modal logic (ML). We strengthen the known proofs that the so-called public announcement logic (PAL) in a signature containing at least two different diamonds and one propositional symbol is exponentially more succinct than ML by showing that this is already true for signatures that contain only one diamond and one propositional symbol. As a corollary of these results, we obtain an alternative proof of the fact that modal circuits are exponentially more succinct than ML-formulae.

Keywords: lower bounds on formula size, succinctness of modal logics, public announcement logic, modal logic of contingency.

---

[1] hans.van-ditmarsch@loria.fr

[2] fanjie@pku.edu.cn

[3] wiebe@csc.liv.ac.uk

[4] petar.iliev@loria.fr

1 Introduction

Unlike computational complexity theory where the question of proving lower bounds on the size of Boolean formulae and Boolean circuits computing a given Boolean function is a central challenge of the whole field, it seems that the general problem of proving lower bounds on formula and circuit-size in modal logic (ML) has not attracted much attention outside the field of temporal logics.

As far as we know, the first lower bound on the size of ML-formulae comes from [17] where it is shown that any ML-formula locally corresponding to the first-order condition $\forall y \forall z((xRy \wedge xRz) \rightarrow (yRz \vee zRy \vee y = z))$ contains at least two different propositional variables. After that, the interest in proving such bounds was focussed mainly on temporal logics with one of the first results, derivable from [15], being that there is a sequence of first-order formulae with three variables $\varphi_1, \varphi_2, \ldots$ for which there is a polynomial p with the property that, for $n \geq 1$, the length of φ_n is less or equal to $p(n)$, but there is no sequence of temporal formulae $\psi_1, \psi_2, \ldots$ such that φ_i is equivalent on ω-words to ψ_i and the lengths of the formulae $\psi_1, \psi_2, \ldots$ can be bounded from above by an elementary function of their indices, i.e., first-order logic is non-elementarily more succinct than temporal logic on ω-words. Two lower bounds on the size of temporal formulae deserve special mention. The first one [18] is that every computation tree logic (or μ-calculus) formula expressing the property *there is a path along which there are n positions* v_1, v_2, ..., v_n *(not necessary in this order) satisfying the propositions* p_1, p_2, ..., p_n *respectively* must have size at least $\frac{2^n}{\sqrt{n}}$. This estimate was improved in [1] to $n!$. The second is from [4], in which it is shown that every formula of the linear-time temporal logic (LTL) expressing the property *for any two positions* v *and* w *on a path* π, *if* $\pi, v \models p_i$ *iff* $\pi, w \models p_i$ *for any* $1 \leq i \leq n$, *then* $\pi, v \models p_{n+1}$ *if and only if* $\pi, w \models p_{n+1}$, *i.e., if* v *and* w *agree on the first* n *propositions, then they agree on* p_{n+1} *too* has size at least 2^n. Using the property *any position* v *on a path* π *that agrees with the initial position* v_0 *on* p_1, p_2, ..., p_n *must also agree on* p_{n+1} it was proven in [9] that any LTL formula that expresses this property, contains only future temporal operators, and is evaluated at the initial position of the path has size at least 2^n.

In contrast to temporal logic, results on lower bounds on formula size in the general setting of modal logic seem to be scarce. Besides [17], we would like to mention the following articles. In [11], a modal language with 5 boxes: $[R]$, $[id]$, $[\neg S]$, $[S_1 \cap S_2]$, and $[S^-]$ is studied. As usual, R is an atomic binary relation on the underlying Kripke structure. The more complex modalities are obtained from the identity relation id, the complement $\neg S$ of S, the intersection $S_1 \cap S_2$ of S_1 and S_2, and the converse S^- of the relation S. The authors proved that, for any n, any formula in this language that defines the property *the carrier set of the Kripke model has cardinality at least* 2^n has size at least 2^{n-1}.

Ehrenfeucht-Fraïssé games were used in [13] to show that every ML-formula that "says" *there is a point that satisfies the proposition* p *reachable from the current point in at most* 2^n *steps* must have modal depth of at least 2^n.

In [16], it is shown, among other things, that, for any $n \geq 1$ and any k such that $2^n < k \leq 2^{n+1}$, any ML-formula that modally defines the property *the current point has less than k successors* contains at least $n+1$ different propositional symbols.

Two other papers that establish lower bounds on the size of ML-formulae are [5] and [10]. In the former, using a technique developed in [1], the authors proved an exponential lower bound on the size of certain ML-formulae. In the latter, it is established that, on the class of all Kripke models (**K**), there is no equivalence-preserving translation from public announcement logic (a conservative extension of ML that has a popular epistemic interpretation and is usually denoted by PAL) to ML that produces an equivalent ML-formula of sub-exponential length for every PAL-formula, i.e., PAL is exponentially more succinct than ML on **K** provided that there are at least two different boxes $[a]$, $[b]$ and one propositional symbol in both logics. If there are at least four boxes and four propositional symbols in both logics, it was shown in [6] that PAL is exponentially more succinct than ML on $\mathbf{S}_5$-models, too.

In the present paper,

- we strengthen the result from [10] in yet another way by proving that the presence of two different boxes is not necessary to show that PAL is exponentially more succinct than ML on **K** (one box is enough);
- we prove that the logic of contingency [12] which is strictly less expressive on **K** than ML is nevertheless exponentially more succinct than ML on **K**;
- we use the above results to give an alternative proof of the fact that modal circuits are exponentially more succinct than modal formulae [8].

2 Technical Preliminaries

Let P be a countable set of propositional symbols. We fix a modal signature $S = \{P, \neg, \wedge, \vee, \Box, \Diamond\}$. The definition of modal formulae over S, Kripke models and the truth of a formula φ in a point m of a Kripke model $\mathcal{M}$, written $(\mathcal{M}, m) \models \varphi$ is standard [2]. *Boolean* formulae are modal formulae that do not contain the $\Box$ and $\Diamond$ operators. We call the pair $(\mathcal{M}, m)$ a *pointed model.* Sets of pointed models are denoted $\mathbb{A}, \mathbb{B} \dots$. We write $\mathbb{M} \models \varphi$ to mean that $(\mathcal{M}, w) \models \varphi$ for all $(\mathcal{M}, w) \in \mathbb{M}$. We would like to stress that $\mathbb{M}$ may be empty. In this case, it is trivially true that, for all $(\mathcal{M}, w) \in \emptyset$, we have $(\mathcal{M}, w) \models \varphi$ and therefore, $\emptyset \models \varphi$. Abusing notation, we write $m \in \mathcal{M}$ to mean that m is a node or a point of the carrier set of $\mathcal{M}$. The binary relation in the Kripke model $\mathcal{M}$ that is used to interpret the operators $\Box$ and $\Diamond$ is denoted $R^{\mathcal{M}}$ or simply R when no confusion arises. We assume that the reader is familiar with the notion of bisimulation between pointed models and that two bisimilar pointed models satisfy the same formulae over S [2]. We proceed now to the definition of our main technical tool that is motivated as follows.

Our goal is to prove exponential lower bounds on the size of formulae φ that express a certain property P of pointed models. By "size", we mean the length of φ as a string over the alphabet S. Intuitively, it will be helpful if we have a tool that is tailored simultaneously to some useful approximation of our notion of size of the formulae φ and to the fact that they can differentiate between models that have the property P and those that do not. One such tool can be found by defining the length of any φ to be the number of nodes of its syntax tree T_φ. In addition, we must add some new features to syntax trees in order to be able to reason about formulae that differentiate between Kripke models that have a given property from the ones that do not. Extended syntax trees were introduced in [7] in the setting of first order logic and can be used as a formalisation of the above intuition [5].

As its name suggests, an extended syntax tree of a modal formula φ is just the usual syntax tree of φ where, apart from a syntax label that is a symbol from S, each node has a semantic label that is a pair of sets of pointed models $\langle \mathbb{M}, \mathbb{N} \rangle$. A node η with a semantic label $\langle \mathbb{M}, \mathbb{N} \rangle$ will be denoted $\mathbb{M} \circ \mathbb{N}$ when no confusion arises. The pointed models in $\mathbb{M}$ are called *the models on the left of* η. Similarly, the pointed models in $\mathbb{N}$ are called *the models on the right of* η. To simplify our exposition, we write $\mathtt{lft}(\eta)$ to mean the set of models on the left of η (in this case, $\mathtt{lft}(\eta) = \mathbb{M}$) and, similarly, $\mathtt{rght}(\eta)$ to mean the set of models on the right of η (in the present case, $\mathtt{rght}(\eta) = \mathbb{N}$).

We begin by defining a number of useful operations on pointed models.

Definition 2.1 Let $(\mathcal{M}, w)$ be a pointed model and $\mathbb{M}$ be a (not necessarily non-empty) set of pointed models. Then

- $$\Box(\mathcal{M}, w) = \{(\mathcal{M}, v) \mid v \in \mathcal{M} \text{ and } wR^{\mathcal{M}}v\}.$$

  Intuitively, $\Box(\mathcal{M}, w)$ is the set of all pointed models that can be reached from w by making one $R^{\mathcal{M}}$-step. Note that if there is no point $v \in \mathcal{M}$ such that $wR^{\mathcal{M}}v$, then $\Box(\mathcal{M}, w) = \emptyset$.

- If $(\mathcal{M}, w) \models \Diamond\psi$, then there is at least one $v \in \mathcal{M}$ such that $wR^{\mathcal{M}}v$ and $(\mathcal{M}, v) \models \psi$. We construct the non-empty set of all such pointed models

  $$\Diamond\psi(\mathcal{M}, w) = \{(\mathcal{M}, v) \mid v \in \mathcal{M} \text{ such that } wR^{\mathcal{M}}v \text{ and } (\mathcal{M}, v) \models \psi\}.$$

- $\Box(\mathbb{M})$ is defined as

  $$\Box(\mathbb{M}) = \bigcup_{(\mathcal{M},w)\in\mathbb{M}} \Box(\mathcal{M}, w).$$

  It is obvious that $\Box(\mathbb{M})$ is empty when $\Box(\mathcal{M}, w) = \emptyset$ for each $(\mathcal{M}, w) \in \mathbb{M}$ or when $\mathbb{M} = \emptyset$.

- If $\mathbb{M} \models \Diamond\psi$, then, we form the set of pointed models

  $$\Diamond\psi(\mathbb{M}) = \bigcup_{(\mathcal{M},w)\in\mathbb{M}} \Diamond\psi(\mathcal{M}, w).$$

[5] Readers familiar with [1] can easily see that these trees can be thought of as closed game trees for suitably defined versions of the Adler-Immerman games introduced in [1]. Essentially the same tool was used in [5] under the name *uniform strategy trees*.

It is easy to see that $\Diamond\psi(\mathbb{M}) = \emptyset$ iff $\mathbb{M} = \emptyset$.

We are ready to define extended syntax trees of ML-formulae. For convenience, we are working with *formulae in negation normal form*, i.e., formulae in which $\neg$ can appear only in front of propositional symbols. From now on, and unless otherwise stated, a *formula* means an ML-formula in negation normal form. As usual, a *literal* l is a propositional symbol p or its negation $\neg p$. We denote the set of all literals by LIT. For a symbol $s \in LIT \cup \{\wedge, \vee, \Box, \Diamond\}$, we write $synl(\eta) = s$ to mean that the node η has the syntax label s.

Definition 2.2 [Extended Syntax Trees] For any formula φ and any sets of pointed models $\mathbb{M}$ and $\mathbb{N}$ such that $\mathbb{M} \models \varphi$ and $\mathbb{N} \models \neg\varphi$, the extended syntax tree $T_{\varphi}^{\mathbb{M}\circ\mathbb{N}}$ is defined recursively on the structure of φ as follows:

(φ is a literal l): $T_{l}^{\mathbb{M}\circ\mathbb{N}}$ has a single node $r = \mathbb{M} \circ \mathbb{N}$ such that $synl(r) = l$.

(φ is $\psi_1 \wedge \psi_2$): $T_{\psi_1\wedge\psi_2}^{\mathbb{M}\circ\mathbb{N}}$ has a root $r = \mathbb{M}\circ\mathbb{N}$ and $synl(r) = \wedge$. The left successor of r is the root $\mathbb{M} \circ \mathbb{N}_1$ of $T_{\psi_1}^{\mathbb{M}\circ\mathbb{N}_1}$. The right successor of r is the root $\mathbb{M} \circ \mathbb{N}_2$ of $T_{\psi_2}^{\mathbb{M}\circ\mathbb{N}_2}$ where the sets $\mathbb{N}_1$ and $\mathbb{N}_2$ are defined as follows. $\mathbb{N}_1 = \{(\mathcal{N}, v) \in \mathbb{N} \mid (\mathcal{N}, v) \models \neg\psi_1\}$ and $\mathbb{N}_2 = \{(\mathcal{N}, v) \in \mathbb{N} \mid (\mathcal{N}, v) \models \neg\psi_2\}$. Hence, while $\mathbb{M} \models (\psi_1 \wedge \psi_2)$, we have $\mathbb{N}_1 \models \neg\psi_1$ and $\mathbb{N}_2 \models \neg\psi_2$ and thus $\mathbb{N} \models \neg(\psi_1 \wedge \psi_2)$. We would like to stress that $\mathbb{N} = \mathbb{N}_1 \cup \mathbb{N}_2$ does not imply $\mathbb{N}_1 \cap \mathbb{N}_2 = \emptyset$.

(φ is $\psi_1 \vee \psi_2$): $T_{\psi_1\vee\psi_2}^{\mathbb{M}\circ\mathbb{N}}$ has a root $r = \mathbb{M}\circ\mathbb{N}$ and $synl(r) = \vee$. The left successor of r is the root $\mathbb{M}_1 \circ \mathbb{N}$ of $T_{\psi_1}^{\mathbb{M}_1\circ\mathbb{N}}$. The right successor of r is the root $\mathbb{M}_2 \circ \mathbb{N}$ of $T_{\psi_2}^{\mathbb{M}_2\circ\mathbb{N}}$ where $\mathbb{M}_1 = \{(\mathcal{M}, v) \in \mathbb{M} \mid (\mathcal{M}, v) \models \psi_1\}$ and $\mathbb{M}_2 = \{(\mathcal{M}, v) \in \mathbb{M} \mid (\mathcal{M}, v) \models \psi_2\}$.

Therefore, $\mathbb{M}_1 \models \psi_1$ and $\mathbb{M}_2 \models \psi_2$ while $\mathbb{N} \models \neg(\psi_1 \vee \psi_2)$. Again, $\mathbb{M}_1$ and $\mathbb{M}_2$ may have a non-empty intersection.

(φ is $\Box\psi$): $T_{\Box\psi}^{\mathbb{M}\circ\mathbb{N}}$ has a root $r = \mathbb{M} \circ \mathbb{N}$ and $synl(r) = \Box$. The unique successor of r is the root $\Box(\mathbb{M}) \circ \Diamond\neg\psi(\mathbb{N})$ of $T_{\psi}^{\Box(\mathbb{M})\circ\Diamond\neg\psi(\mathbb{N})}$. It is obvious that $\Box(\mathbb{M}) \models \psi$ and $\Diamond\neg\psi(\mathbb{N}) \models \neg\psi$.

(φ is $\Diamond\psi$): $T_{\Diamond\psi}^{\mathbb{M}\circ\mathbb{N}}$ has a root $r = \mathbb{M} \circ \mathbb{N}$ and $synl(r) = \Diamond$. The unique successor of r is the root $\Diamond\psi(\mathbb{M}) \circ \Box(\mathbb{N})$ of $T_{\psi}^{\Diamond\psi(\mathbb{M})\circ\Box(\mathbb{N})}$. It is obvious that $\Diamond\psi(\mathbb{M}) \models \psi$ and $\Box(\mathbb{N}) \models \neg\psi$.

As we said above, the size of a formula φ is its length as a word over S or, equivalently, the number of nodes in it syntax tree. However, for our purposes, defining the size of φ as the number of leaves of its syntax tree will suffice.

Definition 2.3 The size of a formula φ, denoted $|\varphi|$, is the number of leaves of an extended syntax tree $T_{\varphi}^{\mathbb{M}\circ\mathbb{N}}$ for some (any) sets of pointed models $\mathbb{M}$ and $\mathbb{N}$ such that $\mathbb{M} \models \varphi$ and $\mathbb{N} \models \neg\varphi$.

It is obvious that $|\varphi|$ is the number of not necessarily different literals occurring in φ. Note that, for every φ (even for contradictions like $p \wedge \neg p$), we can find a pair $\langle \mathbb{M}, \mathbb{N} \rangle$ such that $\mathbb{M} \models \varphi$ and $\mathbb{N} \models \neg\varphi$. This, of course, is done by taking $\mathbb{M} = \mathbb{N} = \emptyset$. In fact, we can identify the usual syntax tree T_{φ} of φ with the extended syntax tree $T_\varphi^{\emptyset\circ\emptyset}$. It is clear that extended syntax trees contain finitely many nodes; in particular, they do not have infinitely long branches.

Example 2.4 The extended syntax tree $T_{\Diamond b\wedge\Diamond\neg b}^{\langle\mathbb{A}\circ\mathbb{D}\rangle}$ is shown in Figure 1. Pointed models occurring in the semantic labels of the tree-nodes are the pairs consisting of the relevant Kripke model $\mathcal{A}$, $\mathcal{B}$ or $\mathcal{C}$ and the nodes marked by $\triangleright$ and $\triangleleft$. Hence, $\mathbb{A}$ consists of the pointed model on the left of the root of the tree which has syntax label $\wedge$ while $\mathbb{D}$ is on the right and contains the pointed models based on the Kripke models $\mathcal{B}$ and $\mathcal{C}$. Black circles denote the points where the atom **b** is true; white circles denote points that do not satisfy any proposition.

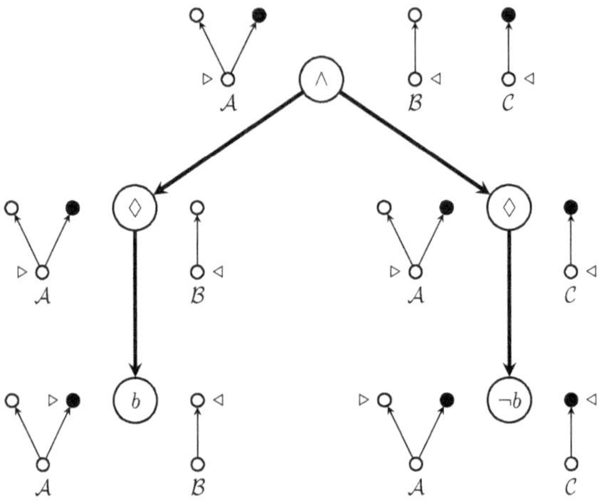

Fig. 1. An extended syntax tree of $\Diamond b \wedge \Diamond\neg b$.

The meaning of the next obvious proposition is that even if the formula $\neg\varphi$ is not in negation normal form, we can always find an equivalent to it formula in negation normal form that has the same number of literals.

Proposition 2.5 *For any* ML*-formula in negation normal form* φ*, there is a formula in negation normal form* $\overline{\varphi}$ *such that* $\overline{\varphi}$ *is equivalent to* $\neg\varphi$ *on* **K** *and* $|\varphi| = |\overline{\varphi}|$.

Proof. The desired formula $\overline{\varphi}$ is obtained from $\neg\varphi$ by pushing $\neg$ inside φ using DeMorgan's laws and the equivalences $\neg\Diamond\psi \equiv \Box\neg\psi$ and $\neg\Box\psi \equiv \Diamond\neg\psi$. $\square$

Proposition 2.6 *For any pair of sets of pointed models* $\langle\mathbb{M},\mathbb{N}\rangle$*, any formula* φ *such that* $\mathbb{M} \models \varphi$ *and* $\mathbb{N} \models \neg\varphi$ *has size at least* n *iff any formula* ψ *such that* $\mathbb{N} \models \psi$ *and* $\mathbb{M} \models \neg\psi$ *has size at least* n.

Proof. Let us assume that $|\varphi| \geq n$ for any formula φ for which $\mathbb{M} \models \varphi$ and $\mathbb{N} \models \neg\varphi$ but that there is a ψ such that $\mathbb{N} \models \psi$ and $\mathbb{M} \models \neg\psi$ and $|\psi| < n$. Then $\mathbb{M} \models \overline{\psi}$ and $\mathbb{N} \models \neg\overline{\psi}$. Using Proposition 2.5, we see that $|\overline{\psi}| = |\psi| < n$ and, thus, arrive at a contradiction. The other direction is similar. $\square$

Proposition 2.7 *Let $\langle \mathbb{M}, \mathbb{N} \rangle$ and $\langle \mathbb{M}_1, \mathbb{N}_1 \rangle$ be two pairs of sets of pointed models such that, for every $(\mathcal{M}, m) \in \mathbb{M}$, there is a bisimilar model $(\mathcal{M}_1, m_1) \in \mathbb{M}_1$ and, for every $(\mathcal{N}, n) \in \mathbb{N}$, there is a bisimilar model $(\mathcal{N}_1, n_1) \in \mathbb{N}_1$. If every formula φ such that $\mathbb{M} \models \varphi$ and $\mathbb{N} \models \neg\varphi$ has size at least n, then every formula ψ for which $\mathbb{M}_1 \models \psi$ and $\mathbb{N}_1 \models \neg\psi$ has size at least n.*

Proof. The proof follows immediately from the fact that for every formula ψ for which $\mathbb{M}_1 \models \psi$ and $\mathbb{N}_1 \models \neg\psi$, we have that $\mathbb{M} \models \psi$ and $\mathbb{N} \models \neg\psi$. □

Proposition 2.8 *If the pointed models $(\mathcal{A}, a)$ and $(\mathcal{B}, b)$ are bisimilar, then there is no formula φ such that its extended syntax tree T contains a node η for which $(\mathcal{A}, a) \in \mathtt{lft}(\eta)$ and $(\mathcal{B}, b) \in \mathtt{rght}(\eta)$.*

Proof. According to Definition 2.2, every node η of T is a root of a sub-tree T_1 that is an extended syntax tree of a sub-formula ψ of φ such that $\mathtt{lft}(\eta) \models \psi$ and $\mathtt{rght}(\eta) \models \neg\psi$. Thus, assuming that there are two bisimilar pointed models $(\mathcal{A}, a) \in \mathtt{lft}(\eta)$ and $(\mathcal{B}, b) \in \mathtt{rght}(\eta)$ leads to a contradiction because bisimilar pointed models satisfy the same formulae. □

Proposition 2.9 *If every formula φ such that $\mathbb{M} \models \varphi$ and $\mathbb{N}_1 \cup \ldots \cup \mathbb{N}_k \models \neg\varphi$ has size at least n, then $(|\varphi_1| + |\varphi_2| + \ldots + |\varphi_k|) \geq n$ for any k formulae $\varphi_1, \ldots, \varphi_k$ such that $\mathbb{M} \models \varphi_1 \wedge \ldots \wedge \varphi_k$ and $\mathbb{N}_i \models \neg\varphi_i$.*

Proof. Let $\varphi_1, \ldots \varphi_k$ be such that $\mathbb{M} \models \varphi_1 \wedge \ldots \wedge \varphi_k$ and $\mathbb{N}_i \models \neg\varphi_i$. Therefore, $\mathbb{M} \models \varphi_1 \wedge \varphi_2 \wedge \ldots \wedge \varphi_k$ and $\mathbb{N}_1 \cup \ldots \cup \mathbb{N}_k \models \neg(\varphi_1 \wedge \varphi_2 \wedge \ldots \wedge \varphi_k)$. The result follows immediately from our assumption. □

3 Main Results

In this section, we are going to prove some lower bounds on the size of formulae which we will later use to show that public announcement logic (PAL) and contingency logic (ConML) are exponentially more succinct than ML. To this end, we begin with a very brief introduction to these two logics.

PAL [14] is a conservative extension of ML in which formulae of the form $[\varphi]\psi$ are allowed. In a natural way, we can introduce a dual $\langle\varphi\rangle$ of the operator $[\varphi]$ by stipulating that $\langle\varphi\rangle$ is an abbreviation of $\neg[\varphi]\neg$. Intuitively, a formula $[\varphi]\psi$ is true in a pointed model $(\mathcal{M}, m)$ if after removing all points that do not satisfy the formula φ, the formula ψ is true at the point m in the resulting new model. For every PAL-formula, there is an equivalent ML-formula that can be obtained by following the rewriting rules below [14]. For any pointed model $(\mathcal{M}, m)$,

$$\begin{array}{lll}
(\mathcal{M}, m) \models \langle\varphi\rangle p & \text{iff} & (\mathcal{M}, m) \models \varphi \wedge p; \\
(\mathcal{M}, m) \models \langle\varphi\rangle(\psi_1 \wedge \psi_2) & \text{iff} & (\mathcal{M}, m) \models \langle\varphi\rangle\psi_1 \wedge \langle\varphi\rangle\psi_2; \\
(\mathcal{M}, m) \models \langle\varphi\rangle\neg\psi & \text{iff} & (\mathcal{M}, m) \models \varphi \wedge \neg\langle\varphi\rangle\psi; \\
(\mathcal{M}, m) \models \langle\varphi\rangle\Diamond\psi & \text{iff} & (\mathcal{M}, m) \models \varphi \wedge \Diamond\langle\varphi\rangle\psi; \\
(\mathcal{M}, m) \models \langle\varphi_1\rangle\langle\varphi_2\rangle\psi & \text{iff} & (\mathcal{M}, m) \models \langle\langle\varphi_1\rangle\varphi_2\rangle\psi.
\end{array}$$

ConML was introduced in [12]. It extends Boolean logic with formulae $\triangle\varphi$. For our purposes, it is enough to say that for every pointed model $(\mathcal{M}, m)$, we have $(\mathcal{M}, m) \models \triangle\varphi$ iff $(\mathcal{M}, m) \models \Box\varphi \vee \Box\neg\varphi$. It is known [3] that CONML is strictly less expressive on **K** than ML.

Here, we are going to show that on **K**, there is no equivalence-preserving translation from either PAL or CONML to ML that produces an equivalent ML-formula of sub-exponential length. To this end, we are going to exhibit two infinite sequences $\delta_1, \delta_2, \ldots$ and $\theta_1, \theta_2, \ldots$ of PAL and CONML-formulae respectively and show that every ML-formula ψ_n that is equivalent to δ_n or θ_n on **K** has size at least 2^n whereas the lengths of δ_n and θ_n are linear in n.

Definition 3.1 Let the sequences $\delta_1, \delta_2, \ldots$ and $\theta_1, \theta_2, \ldots$ be defined as follows.

$$\begin{array}{llll} \delta_1 \stackrel{\text{def}}{=} & \Diamond\mathbf{b} \wedge \Diamond\neg\mathbf{b} & \theta_1 \stackrel{\text{def}}{=} & \triangle\mathbf{b} \\ \vdots & & \vdots & \\ \delta_{n+1} \stackrel{\text{def}}{=} & \langle\delta_n\rangle\delta_1 & \theta_n \stackrel{\text{def}}{=} & \underbrace{\triangle\ldots\triangle}_{n \text{ times}}\mathbf{b} \\ \vdots & & \vdots & \end{array}$$

In order to apply extended syntax trees to show that every ML-formula ψ that is equivalent to δ_n or θ_n on **K** has size at least 2^n, we must exhibit two sets of pointed models $\mathbb{M}$ and $\mathbb{N}$ such that $\mathbb{M} \models \psi$ whereas $\mathbb{N} \models \neg\psi$ and show that the extended syntax tree of ψ with root $\mathbb{M} \circ \mathbb{N}$ has at least 2^n leaves. We begin by defining the required sets of pointed models for the formulae δ_n.

Definition 3.2 The set $\mathbb{A}^1$ consists of the pointed model $(\mathcal{A}^1, a^1)$ shown on the left of the leftmost dotted line in Figure 2. The set $\mathbb{B}^1$ contains the two pointed models $(\mathcal{B}^1_1, b^1_1)$ and $(\mathcal{B}^1_2, b^1_2)$ between the leftmost dotted line and the thick vertical line. **B**lack nodes satisfy the proposition **b** whereas white nodes

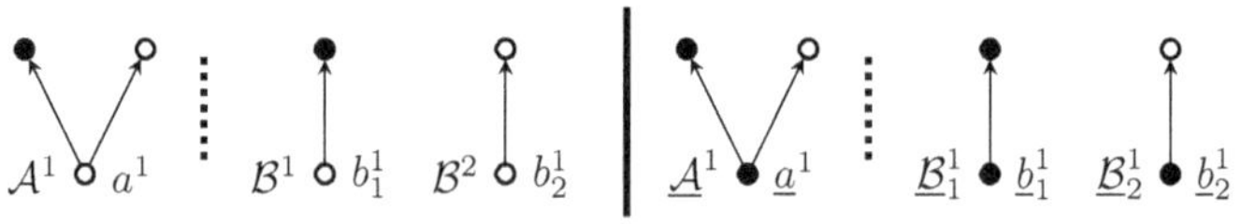

Fig. 2. The sets of models $\mathbb{A}^1$, $\mathbb{B}^1$, $\underline{\mathbb{A}}^1$, and $\underline{\mathbb{B}}^1$.

do not satisfy any proposition. An arrow in a model $\mathcal{M}$ coming from a point m_1 and pointing to a point m_2 means that $m_1 R^{\mathcal{M}} m_2$. The sets $\underline{\mathbb{A}}^1$, and $\underline{\mathbb{B}}^1$ are shown on the right of the thick vertical line. The only difference between the pointed model $(\mathcal{A}^1, a^1)$ and $(\underline{\mathcal{A}}^1, \underline{a}^1)$ is that in the latter the point $\underline{a}^1$ satisfies **b**; similarly for the pointed models $(\mathcal{B}^1_i, b^1_i)$ and $(\underline{\mathcal{B}}^1_i, \underline{b}^1_i)$ where $1 \leq i \leq 2$.

Let us suppose that the sets $\mathbb{A}^n$, $\mathbb{B}^n$, $\underline{\mathbb{A}}^n$, and $\underline{\mathbb{B}}^n$ have been constructed. For any pointed model $(\mathcal{A}^n, a^n)$, $(\mathcal{B}^n_k, b^n_k)$, $(\underline{\mathcal{A}}^n, \underline{a}^n)$, and $(\underline{\mathcal{B}}^n_k, \underline{b}^n_k)$, we call the points a^n, b^n_k, $\underline{a}^n$, and $\underline{b}^n_k$ the *root* of the respective model. The set $\mathbb{A}^{n+1}$ consists of the pointed model $(\mathcal{A}^{n+1}, a^{n+1})$ (shown in the Figure 3 on the left of the dotted vertical line) and built from the models in $\mathbb{A}^n \cup \mathbb{B}^n \cup \underline{\mathbb{A}}^n \cup \underline{\mathbb{B}}^n$ as follows. We take the pointed models $(\mathcal{A}^n, a^n), (\underline{\mathcal{A}}^n, \underline{a}^n), (\mathcal{B}_i^n, b_i^n), (\underline{\mathcal{B}}_i^n, \underline{b}_i^n)$, where $1 \leq i \leq 2$, and connect each of the roots of these models to the point a^{n+1} as shown.

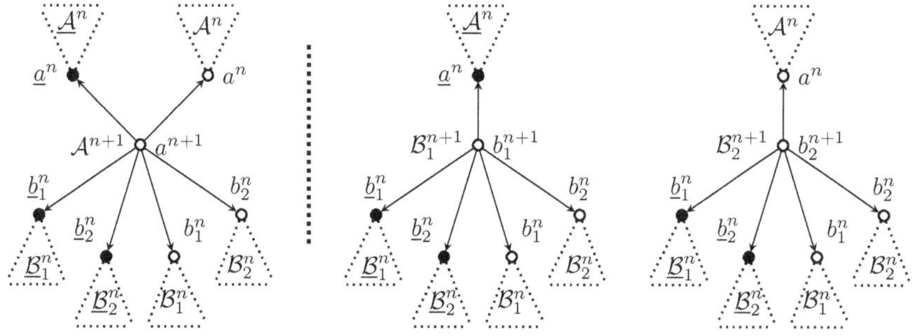

Fig. 3. The sets of models $\mathbb{A}^{n+1}$ and $\mathbb{B}^{n+1}$.

The set $\mathbb{B}^{n+1}$ contains the pointed models $(\mathcal{B}_1^{n+1}, b_1^{n+1})$ and $(\mathcal{B}_2^{n+1}, b_2^{n+1})$ shown on the right of the dotted line.

The sets of pointed models $\underline{\mathbb{A}}^{n+1} = \{(\underline{\mathcal{A}}^{n+1}, \underline{a}^{n+1})\}$ and $\underline{\mathbb{B}}^{n+1} = \{(\underline{\mathcal{B}}_1^{n+1}, \underline{b}_1^{n+1}), (\underline{\mathcal{B}}_2^{n+1}, \underline{b}_2^{n+1})\}$ are obtained from the models in the sets $\mathbb{A}^{n+1}$ and $\mathbb{B}^{n+1}$ by making the roots of the relevant models satisfy the proposition $\mathbf{b}$.

Example 3.3 The sets $\mathbb{A}^2$, $\mathbb{B}^2$ (on the left of the thick vertical line) and $\underline{\mathbb{A}}^2$, $\underline{\mathbb{B}}^2$ (on the right) are given below.

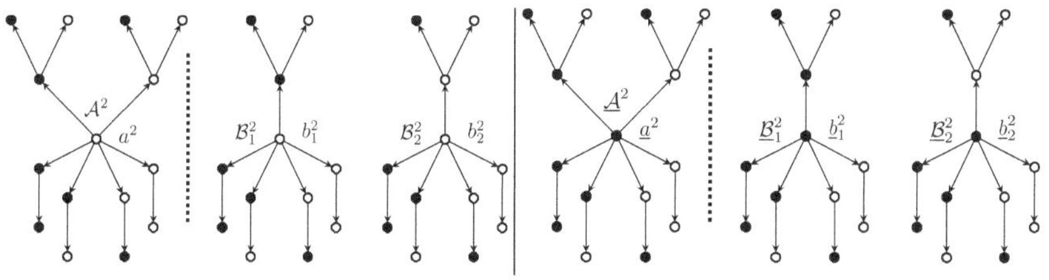

Note how the models from $\mathbb{A}^1$, $\mathbb{B}^1$ and $\underline{\mathbb{A}}^1$, $\underline{\mathbb{B}}^1$ shown in Figure 2 were used in the construction of the models $(\mathcal{A}^2, a^2)$, $(\mathcal{B}_1^2, b_1^2)$ and $(\mathcal{B}_2^2, b_2^2)$. Intuitively, $\underline{\mathbb{A}}^2$ consists of the pointed model obtained from $(\mathcal{A}^2, a^2)$ by making its root a^2 black. Similarly, $\underline{\mathbb{B}}^2$ consists of the models obtained from $(\mathcal{B}_1^2, b_1^{n+1})$ and $(\mathcal{B}_2^2, b_2^2)$ by making b_1^2 and b_2^2 black.

Proposition 3.4 *Consider the formulae $\delta_1, \delta_2, \ldots$ from Definition 3.1. Then $\mathbb{A}^n \models \delta_n$ and $\mathbb{B}^n \models \neg\delta_n$.*

Proof. Given the geometric shape of the models, it is easy to establish by induction the following facts.

(i) For every δ_n, there is an equivalent ML-formula δ'_n that can be obtained from δ_n by applying the rewriting rules for PAL-formulae. Namely, we have the recursively defined sequence of formulae

$$\delta'_1 = \delta_1 \text{ and } \delta'_{n+1} = \delta'_n \wedge \Diamond(\mathbf{b} \wedge \delta'_n) \wedge \Diamond(\neg\mathbf{b} \wedge \delta'_n).$$

(ii) $\mathbb{A}^1 \models \delta'_1$ and $\underline{\mathbb{A}}^1 \models \delta'_1$ whereas $\mathbb{B}^1 \models \neg\delta'_1$ and $\underline{\mathbb{B}}^1 \models \neg\delta'_1$.

(iii) If $n > 1$, then

- $\mathbb{A}^n \models \delta'_j$ and $\underline{\mathbb{A}}^n \models \delta'_j$ for every j such that $1 \leq j \leq n$;
- $(\mathcal{B}_1^n, b_1^n) \models \neg\Diamond(\neg\mathbf{b} \wedge \delta'_n)$ and $(\underline{\mathcal{B}}_1^n, \underline{b}_1^n) \models \neg\Diamond(\neg\mathbf{b} \wedge \delta'_n)$;
- $(\mathcal{B}_2^n, b_2^n) \models \neg\Diamond(\mathbf{b} \wedge \delta'_n)$ and $(\underline{\mathcal{B}}_2^n, \underline{b}_2^n) \models \neg\Diamond(\mathbf{b} \wedge \delta'_n)$.

The result follows immediately from the items above.

□

Next, we define suitable sets of pointed models for the formulae $\theta_1, \theta_2, \ldots$ from Definition 3.1.

Definition 3.5 Using the conventions established in Definition 3.2, namely, that black nodes satisfy the proposition **b**, white nodes do not satisfy any proposition, and arrows represent relations, the sets of pointed models $\mathbb{C}^n$ and $\mathbb{D}^n$ are defined recursively as shown in Figures 4, 5, and 6 where $(\mathcal{D}_1^n, d_1^n), \ldots, (\mathcal{D}_k^n, d_k^n)$ are all the pointed models in $\mathbb{D}^n$.

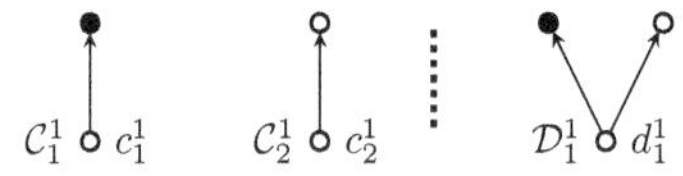

Fig. 4. The sets of models $\mathbb{C}^1$ (on the left of the dotted line) and $\mathbb{D}^1$ (on the right).

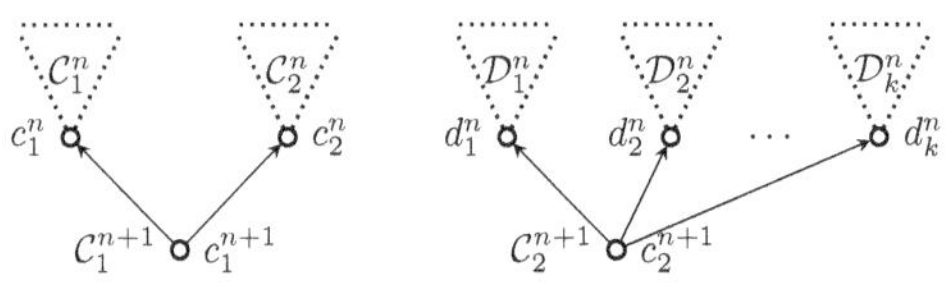

Fig. 5. The set of models $\mathbb{C}^{n+1}$.

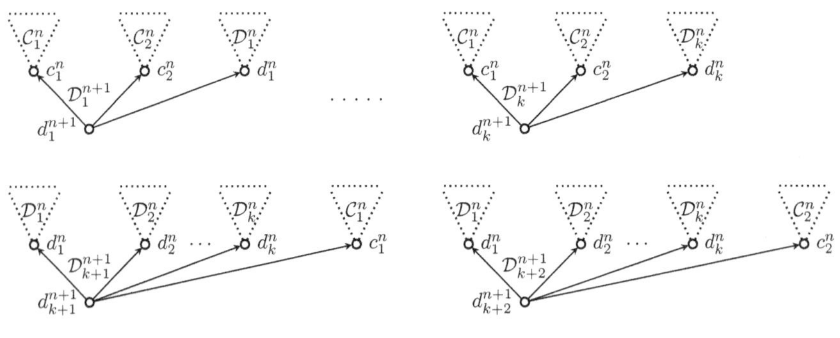

Fig. 6. The set of models $\mathbb{D}^{n+1}$.

Example 3.6 Figure 7 shows the sets $\mathbb{C}^2$, consisting of the pointed models $(\mathcal{C}_1^2, c_1^2)$ and $(\mathcal{C}_2^2, c_2^2)$ and $\mathbb{D}^2$, consisting of $(\mathcal{D}_1^2, d_1^2)$, $(\mathcal{D}_2^2, d_2^2)$, and $(\mathcal{D}_3^2, d_3^2)$. Note how the models in $\mathbb{C}^1$ and $\mathbb{D}^1$ from Figure 4 are used in the construction of the sets $\mathbb{C}^2$ and $\mathbb{D}^2$.

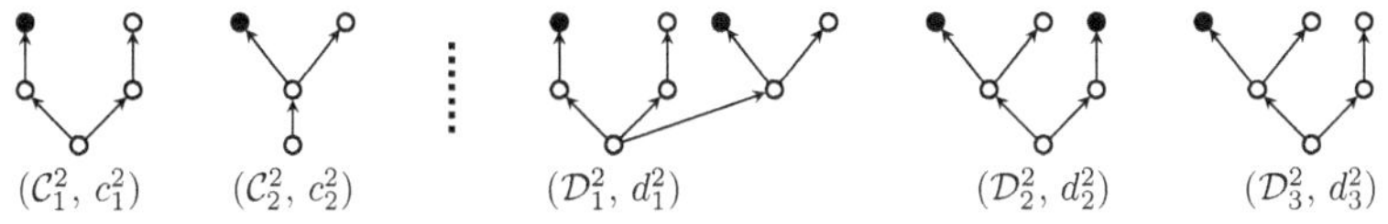

Fig. 7. The set of models $\mathbb{C}^2$ and $\mathbb{D}^2$.

Proposition 3.7 *For any formula θ_n as defined in Definition 3.1, we have $\mathbb{C}^n \models \theta_n$ and $\mathbb{D}^n \models \neg\theta_n$.*

Proof. The truth of the statement follows easily from the geometric shape of the models in $\mathbb{C}^n$ and $\mathbb{D}^n$ and from the fact that, for every θ_n, there is an equivalent ML-formula θ'_n defined recursively as follows $\theta'_1 = \Box\mathbf{b} \vee \Box\neg\mathbf{b}$ and $\theta'_{n+1} = \Box\theta'_n \vee \Box\neg\theta'_n$. □

We are ready now to prove our main lower-bound results formulated in Theorem 3.8 and Theorem 3.9 below.

Theorem 3.8 (First Lower Bound on ML-formulae) *Let the sets $\mathbb{A}^n$ and $\mathbb{B}^n$ be as defined in Definition 3.2. Any formula ψ such that $\mathbb{A}^n \models \psi$ and $\mathbb{B}^n \models \neg\psi$ has size at least 2^n.*

Theorem 3.9 (Second Lower Bound on ML-formulae) *Let the sets $\mathbb{C}^n$ and $\mathbb{D}^n$ be as defined in Definition 3.5. Any formula φ such that $\mathbb{C}^n \models \varphi$ and $\mathbb{D}^n \models \neg\varphi$ has size at least 2^n.*

The proofs of both theorems rely on a number of preliminary statements that revolve around similar ideas in both cases. We begin by establishing a convention that will simplify our arguments. Consider, for example, the Kripke model $\mathcal{C}_1^{n+1}$ from Figure 5. It is a tree with root c_1^{n+1}. The left successor of c_1^{n+1} is the root c_1^n of the model $\mathcal{C}_1^n$. It is obvious that the point c_1^n in $\mathcal{C}_1^{n+1}$ and the point c_1^n in $\mathcal{C}_1^n$ are bisimilar and satisfy the same ML-formulae. Since, in what follows, we are mainly interested in formulae-satisfiability, to increase readability, we are going to identify bisimilar pointed models. This allows us to substitute, e.g., the clearer $(\mathcal{C}_2^n, c_2^n)$ for the hard to read $(\mathcal{D}_{k+2}^{n+1}, c_2^n)$.

Let us first prove Theorem 3.8

Proposition 3.10 *For any $i \in \{1, 2\}$, no extended syntax tree contains a node η such that $synl(\eta) \in LIT \cup \{\Box\}$, $\mathbb{A}^n \subseteq \mathtt{lft}(\eta)$, and $(\mathcal{B}_i^n, b_i^n) \in \mathtt{rght}(\eta)$. The statement remains true if $\mathbb{A}^n$ and $(\mathcal{B}_i^n, b_i^n)$ are replaced with $\underline{\mathbb{A}}^n$ and $(\underline{\mathcal{B}}_i^n, \underline{b}_i^n)$, respectively.*

Proof. It is obvious that $(\mathcal{A}^n, a^n)$ and $(\mathcal{B}_i^n, b_i^n)$ satisfy the same Boolean formulae. Hence, there is no extended syntax tree that contains a node η such that $synl(\eta) \in LIT$, $\mathbb{A}^n \subseteq \mathtt{lft}(\eta)$, and $(\mathcal{B}_i^n, b_i^n) \in \mathtt{rght}(\eta)$.

Let us suppose that there is an extended syntax tree containing a node η such that $synl(\eta) = \Box$, $\mathbb{A}^n \subseteq \mathtt{lft}(\eta)$, and $(\mathcal{B}_i^n, b_i^n) \in \mathtt{rght}(\eta)$. We consider only the case $n > 1$ (the case $n = 1$ is similar). Then, the set $\Box(\mathbb{A}^n) = \{(\underline{\mathcal{A}}^{n-1}, \underline{a}^{n-1}), (\mathcal{A}^{n-1}, a^{n-1}), (\underline{\mathcal{B}}_1^{n-1}, \underline{b}_1^{n-1}), (\underline{\mathcal{B}}_2^{n-1}, \underline{b}_2^{n-1}), (\mathcal{B}_1^{n-1}, b_1^{n-1}), (\mathcal{B}_2^{n-1}, b_2^{n-1})\} \subseteq \mathtt{lft}(\eta_1)$, where η_1 is the successor of η. The geometry of $(\mathcal{B}_i^n, b_i^n)$ is such that at least one of the models in $\Box(\mathbb{A}^n)$ must appear on the right of η_1. Thus, we arrive at a contradiction with the help of Proposition 2.8. The proof for $\underline{\mathbb{A}}^n$ and $(\underline{\mathcal{B}}_i^n, \underline{b}_i^n)$ is the same. □

Proposition 3.11 *There is no extended syntax tree that contains a node η such that $synl(\eta) \in LIT \cup \{\Diamond, \Box\}$, $\mathbb{A}^n \subseteq \mathtt{lft}(\eta)$ and $\mathbb{B}^n \subseteq \mathtt{rght}(\eta)$. The statement remains true if $\mathbb{A}^n$ and $\mathbb{B}^n$ are replaced with $\underline{\mathbb{A}}^n$ and $\underline{\mathbb{B}}^n$, respectively.*

Proof. The fact that no node η for which $\mathbb{A}^n \subseteq \mathtt{lft}(\eta)$ and $\mathbb{B}^n \subseteq \mathtt{rght}(\eta)$ (or ($\underline{\mathbb{A}}^n \subseteq \mathtt{lft}(\eta)$ and $\underline{\mathbb{B}}^n \subseteq \mathtt{rght}(\eta)$) can have a syntax label that is either a literal or $\Box$ follows from Proposition 3.10. If $synl(\eta) = \Diamond$, then it is easily seen by consulting the relevant items from Definition 2.2 that the successor node of η would contain two bisimilar models one on the left and the other on the right which is impossible according to Proposition 2.8. □

Proposition 3.12 *Any extended syntax tree T with root r such that $\mathbb{A}^n \subseteq \mathtt{lft}(r)$ and $\mathbb{B}^n \subseteq \mathtt{rght}(r)$ contains a node η for which $synl(\eta) = \wedge$, $\mathbb{A}^n \subseteq \mathtt{lft}(\eta)$ and $\mathbb{B}^n \subseteq \mathtt{rght}(\eta)$; moreover, if η_1 and η_2 are the two successor of η, then $\mathbb{A}^n \subseteq \mathtt{lft}(\eta_1)$, $(\mathcal{B}_1^n, b_1^n) \in \mathtt{rght}(\eta_1)$, $(\mathcal{B}_2^n, b_2^n) \notin \mathtt{rght}(\eta_1)$ while $\mathbb{A}^n \subseteq \mathtt{lft}(\eta_2)$, $(\mathcal{B}_2^n, b_2^n) \in \mathtt{rght}(\eta_2)$, and $(\mathcal{B}_1^n, b_1^n) \notin \mathtt{rght}(\eta_2)$. The statement remains true if $\mathbb{A}^n$, $\mathbb{B}^n$, and $(\mathcal{B}_i^n, b_i^n)$ for $i \in \{1, 2\}$ are replaced with $\underline{\mathbb{A}}^n$, $\underline{\mathbb{B}}^n$, and $(\underline{\mathcal{B}}_i^n, \underline{b}_i^n)$, respectively.*

Proof. Let us assume that T does not have such a node η. We will show that T contains an infinite branch which is absurd. We saw already in Proposition 3.11 that $synl(r) \notin LIT \cup \{\Diamond, \Box\}$. Therefore, $synl(r) \in \{\vee, \wedge\}$. If $synl(r) = \vee$, then, since $\mathbb{A}^n$ contains just one model, we see that at least one of the successors r_1 and r_2, say r_1, of r is such that $\mathbb{A}^n \subseteq \mathtt{lft}(r_1)$ and $\mathbb{B}^n \subseteq \mathtt{rght}(r_1)$. If $r = \wedge$, since $\mathbb{B}^n$ contains two models, our assumption and the second item from Definition 2.2 imply that, again, at least one of the successors r_1 and r_2, say r_1, of r is such that $\mathbb{A}^n \subseteq \mathtt{lft}(r_1)$ and $\mathbb{B}^n \subseteq \mathtt{rght}(r_1)$. In either case, we can find a successor r_1 of the root r of T such that $\mathbb{A}^n \subseteq \mathtt{lft}(r_1)$ and $\mathbb{B}^n \subseteq \mathtt{rght}(r_1)$. It is obvious that this reasoning can be applied to the node r_1. Hence, we can find the desired infinite branch by starting at the root and "following" the nodes that contain the models $\mathbb{A}^n$ on the left and the models $\mathcal{B}^n$ on the right. □

Lemma 3.13 *For any extended syntax tree T with root r, the following hold.*

(i) *If $\mathbb{A}^{n+1} \subseteq \mathtt{lft}(r)$ and $(\mathcal{B}_1^{n+1}, b_1^{n+1}) \in \mathtt{rght}(r)$, then T contains a node η such that $synl(\eta) = \Diamond$, $\mathbb{A}^{n+1} \subseteq \mathtt{lft}(\eta)$ and $(\mathcal{B}_1^{n+1}, b_1^{n+1}) \in \mathtt{rght}(\eta)$; moreover, if η_1 is its successor, then $(\mathcal{A}^n, a^n) \in \mathtt{lft}(\eta_1)$ and $\{(\mathcal{B}_1^n, b_1^n), (\mathcal{B}_2^n, b_2^n)\} \subseteq \mathtt{rght}(\eta_1)$. The statement remains true if $\mathbb{A}^{n+1}$ and $(\mathcal{B}_1^{n+1}, b_1^{n+1})$ are replaced with $\underline{\mathbb{A}}^{n+1}$ and $(\underline{\mathcal{B}}_1^{n+1}, \underline{b}_1^{n+1})$, respectively.*

(ii) *If* $\mathbb{A}^{n+1} \subseteq \texttt{lft}(r)$ *and* $(\mathcal{B}_2^{n+1}, b_2^{n+1}) \in \texttt{rght}(r)$*, then* T *contains a node* η *such that* $synl(\eta) = \Diamond$*,* $\mathbb{A}^{n+1} \subseteq \texttt{lft}(\eta)$ *and* $(\mathcal{B}_2^{n+1}, b_2^{n+1}) \in \texttt{rght}(\eta)$*; moreover, if* η_1 *is its successor, then* $(\underline{\mathcal{A}}^n, \underline{a}^n) \in \texttt{lft}(\eta_1)$ *and* $\{(\underline{\mathcal{B}}_1^n, \underline{b}_1^n), (\underline{\mathcal{B}}_2^n, \underline{b}_2^n)\} \subseteq \texttt{rght}(\eta_1)$*. The statement remains true if* $\mathbb{A}^{n+1}$ *and* $(\mathcal{B}_2^{n+1}, b_2^{n+1})$ *are replaced with* $\underline{\mathbb{A}}^{n+1}$ *and* $(\underline{\mathcal{B}}_2^{n+1}, \underline{b}_2^{n+1})$*, respectively.*

Proof.

(i) Let us suppose that T does not have a node η such that $synl(\eta) = \Diamond$, $\mathbb{A}^{n+1} \subseteq \texttt{lft}(\eta)$ and $(\mathcal{B}_1^{n+1}, b_1^{n+1}) \in \texttt{rght}(\eta)$. We are going to show that, in this case, T contains an infinite branch which is absurd. Using Proposition 3.10 and our assumption, we see that $synl(r) \notin LIT \cup \{\Box, \Diamond\}$. Thus, $synl(r) \in \{\vee, \wedge\}$. Using reasoning identical to the one in the proof of Proposition 3.12, we can find at least one successor r_1 of r such that $\mathbb{A}^{n+1} \subseteq lft(r_1)$ and $(\mathcal{B}_1^{n+1}, b_1^{n+1}) \in rght(r_1)$. It is obvious that the same considerations can be applied to r_1, too. Thus the desired infinite branch is constructed by starting at r and following the nodes that contain $\mathbb{A}^{n+1}$ on the left and $(\mathcal{B}_1^{n+1}, b_1^{n+1})$ on the right. Hence, T must contain a node η, such that $synl(\eta) = \Diamond$, $\mathbb{A}^n \subseteq \texttt{lft}(\eta)$, and $(\mathcal{B}_1^{n+1}, b_1^{n+1}) \in \texttt{rght}(\eta)$. Let η_1 be its successor. According to the fifth item from Definition 2.2, we have $\Box(\mathcal{B}_1^{n+1}, b_1^{n+1}) \subseteq \texttt{rght}(\eta_1)$. Since $\Box(\mathcal{B}_1^{n+1}, b_1^{n+1}) = \{(\underline{\mathcal{A}}^n, \underline{a}^n), (\underline{\mathcal{B}}_1^n, \underline{b}_1^n), (\underline{\mathcal{B}}_2^n, \underline{b}_2^n), (\mathcal{B}_1^n, b_1^n), (\mathcal{B}_2^n, b_2^n)\}$, we see that none of these pointed models can appear on the right of η_1 according to Proposition 2.8. Given the geometry of $(\mathcal{A}^{n+1}, a^{n+1})$, we see that $(\mathcal{A}^n, a^n) \in \texttt{lft}(\eta_1)$. The proof for $\underline{\mathbb{A}}^{n+1}$ and $(\underline{\mathcal{B}}_1^{n+1}, \underline{b}_1^{n+1})$ is the same.

(ii) The proof of this item is completely analogous.

□

We are now ready to prove Theorem 3.8. We proceed by proving, with induction on n, the stronger statement below.

Any formula φ *such that* $\mathbb{A}^n \models \varphi$ *and* $\mathbb{B}^n \models \neg\varphi$ *has size at least* 2^n *and any formula* ψ *such that* $\underline{\mathbb{A}}^n \models \psi$ *and* $\underline{\mathbb{B}}^n \models \neg\psi$ *has size at least* 2^n*.*

Proof.

Base step: The fact that the extended syntax tree T of any formula φ such that $\mathbb{A}^1 \models \varphi$ and $\mathbb{B}^1 \models \neg\varphi$ has at least two leaves follows immediately from Proposition 3.12. The same is true about $\underline{\mathbb{A}}^1$ and $\underline{\mathbb{B}}^1$.

Induction step: The induction step depends on the following claim.

Claim For any $i \in \{1, 2\}$ and any $n \geq 1$, any formula φ such that $\mathbb{A}^{n+1} \models \varphi$ and $(\mathcal{B}_i^{n+1}, b_i^{n+1}) \models \neg\varphi$ has size at least 2^n. The statement remains true if $\mathbb{A}^{n+1}$ and $(\mathcal{B}_i^{n+1}, b_i^{n+1})$ are replaced with $\underline{\mathbb{A}}^{n+1}$ and $(\underline{\mathcal{B}}_i^{n+1}, \underline{b}_i^{n+1})$, respectively.

Proof. Let us consider the case $i = 2$. Let T be an extended syntax tree with root r for which $\mathbb{A}^{n+1} \subseteq \texttt{lft}(r)$ and $(\mathcal{B}_2^{n+1}, b_2^{n+1}) \in \texttt{rght}(r)$. Using the second item from Lemma 3.13, we see that T contains a node η_1 such that $\underline{\mathbb{A}}^n \subseteq \texttt{lft}(\eta_1)$ and $\{(\underline{\mathcal{B}}_1^n, \underline{b}_1^n), (\underline{\mathcal{B}}_2^n, \underline{b}_2^n)\} = \underline{\mathbb{B}}^n \subseteq \texttt{rght}(\eta_1)$. Applying the induction hypothesis and Proposition 2.7 to the subtree T_1 of T with root η_1, we see that T_1 and thus T has size at least 2^n. The proof of the case $i = 1$ is analogous modulo the fact that we use the first item of Lemma 3.13. This completes the proof of the claim. □

Let us complete now the proof of the induction step. Using Proposition 3.12, we see that the extended syntax tree T with root r, where $\mathbb{A}^{n+1} = \texttt{lft}(r)$ and $\mathbb{B}^{n+1} = \texttt{rght}(r)$, of any formula φ such that $\mathbb{A}^{n+1} \models \varphi$ and $\mathbb{B}^{n+1} \models \neg\varphi$ contains two different sub-trees T_1 and T_2 with roots r_1 and r_2 such that $\mathbb{A}^{n+1} = \texttt{lft}(r_1)$, $\{(\mathcal{B}_1^{n+1}, b_1^{n+1})\} = \texttt{rght}(r_1)$ and $\mathbb{A}^{n+1} = \texttt{lft}(r_2)$, $\{(\mathcal{B}_2^{n+1}, b_2^{n+1})\} = \texttt{rght}(r_2)$. It follows from the above Claim that both T_1 and T_2 have size at least 2^n. Thus, the size of T must be at least 2^{n+1}. Again, the proof about $\underline{\mathbb{A}}^{n+1}$ and $\underline{\mathbb{B}}^{n+1}$ is the same.

□

Next, we prove Theorem 3.9.

Consider the sets of pointed models from Definition 3.5.

Proposition 3.14 *For any $n \geq 0$,*

(i) *if $1 \leq i \leq k$, then no extended syntax tree contains a node η such that $synl(\eta) \in LIT \cup \{\Diamond\}$, $(\mathcal{C}_1^{n+1}, c_1^{n+1}) \in \texttt{lft}(\eta)$, and $(\mathcal{D}_i^{n+1}, d_i^{n+1}) \in \texttt{rght}(\eta)$;*

(ii) *if $k+1 \leq i \leq k+2$, then no extended syntax tree contains a node η such that $synl(\eta) \in LIT \cup \{\Diamond\}$, $(\mathcal{C}_2^{n+1}, c_2^{n+1}) \in \texttt{lft}(\eta)$ and $(\mathcal{D}_i^{n+1}, d_i^{n+1}) \in \texttt{rght}(\eta)$.*

Proof. It is obvious that all the pointed models in $\mathbb{C}^n$ and $\mathbb{D}^n$ satisfy the same Boolean formulae and, therefore, η cannot have a syntax label that is a literal. We consider only the case $synl(\eta) = \Diamond$ for the first item and we assume $n \geq 1$. The proofs for $(\mathcal{C}_1^1, c_1^1) \in \texttt{lft}(\eta)$ and $(\mathcal{D}_i^1, d_i^1) \in \texttt{rght}(\eta)$ and (ii) are analogous. Suppose that there is an extended syntax tree containing such a node η. Let η_1 be its successor. According to the fourth item from Definition 2.2, we have that $\Box(\mathcal{D}_i^{n+1}, d_i^{n+1}) \subseteq \texttt{rght}(\eta_1)$. Since $1 \leq i \leq k$, it is obvious that $\Box(\mathcal{D}_i^{n+1}, d_i^{n+1}) = \{(\mathcal{C}_1^n, c_1^n), (\mathcal{C}_2^n, c_2^n), (\mathcal{D}_i^n, d_i^n)\}$. Given the geometry of the pointed model $(\mathcal{C}_1^{n+1}, c_1^{n+1})$, either $(\mathcal{C}_1^n, c_1^n)$ or $(\mathcal{C}_2^n, c_2^n)$ must appear on the left of η_1. Using Proposition 2.8, we arrive at a contradiction. □

Proposition 3.15 *For any $(\mathcal{D}_i^n, d_i^n) \in \mathbb{D}^n$, there is no extended syntax tree that contains a node η such that $synl(\eta) \in LIT \cup \{\Diamond, \Box\}$, $\mathbb{C}^n \subseteq \texttt{lft}(\eta)$, and $(\mathcal{D}_i^n, d_i^n) \in \texttt{rght}(\eta)$.*

Proof. In the case of $synl(\eta) = \Diamond$ or $synl(\eta) \in LIT$, the statement follows from Proposition 3.14. If $synl(\eta) = \Box$, then it is easily seen that its successor η_1 has two bisimilar models one on the left and the other on the right. Hence, we arrive at a contradiction with the help of Proposition 2.8. □

Proposition 3.16 *For any $(\mathcal{D}_i^n, d_i^n) \in \mathbb{D}^n$, if T is an extended syntax tree with root r for which $\mathbb{C}^n \subseteq \texttt{lft}(r)$ and $(\mathcal{D}_i^n, d_i^n) \in \texttt{rght}(r)$, then T has a node η such that $synl(\eta) = \vee$, $\mathbb{C}^n \subseteq \texttt{lft}(\eta)$, and $(\mathcal{D}_i^n, d_i^n) \in \texttt{rght}(\eta)$; moreover, if η_1 and η_2 are the two successor of η, then $(\mathcal{C}_1^n, c_1^n) \in \texttt{lft}(\eta_1)$, $(\mathcal{C}_2^n, c_2^n) \notin \texttt{lft}(\eta_1)$,* *and* $(\mathcal{D}_i^n, d_i^n) \in \mathtt{rght}(\eta_1)$ *while* $(\mathcal{C}_2^n, c_2^n) \in \mathtt{lft}(\eta_2)$, $(\mathcal{C}_1^n, c_1^n) \notin \mathtt{lft}(\eta_2)$, *and* $(\mathcal{D}_i^n, d_i^n) \in \mathtt{rght}(\eta_2)$.

Proof. Let us assume that T does not have such a node η. We are going to show that T contains an infinite branch which is absurd. Indeed, using Proposition 3.15, we see that $synl(r) \notin LIT \cup \{\Box, \Diamond\}$. Therefore, either $synl(r) = \wedge$ or $synl(r) = \vee$. In the first case, at least one of the successors r_1 and r_2 of r, say r_1, will be such that $\mathbb{C}^n \subseteq \mathtt{lft}(r_1)$ and $(\mathcal{D}_i^n, d_i^n) \in \mathtt{rght}(r_1)$. In the second case, since the two successors r_1 and r_2 of r do not have the properties described in the statement, at least one of them, say r_1, must be such that $\mathbb{C}^n \subseteq \mathtt{lft}(r_1)$ and $(\mathcal{D}_i^n, d_i^n) \in \mathtt{rght}(r_1)$. In either case, we can find a successor r_1 of the root r of T such that $\mathbb{C}^n \subseteq \mathtt{lft}(r_1)$ and $(\mathcal{D}_i^n, d_i^n) \in \mathtt{rght}(r_1)$. It is obvious that this reasoning can be applied to the node r_1. Hence, we can find the desired infinite branch by starting at the root and "following" the nodes that contain the models $\mathbb{C}^n$ on the left and the model $(\mathcal{D}_i^n, d_i^n)$ on the right. □

To simplify the exposition of the proofs below, we write $\Box(\eta)$ to mean the successor of a node η in an extended syntax tree such that $synl(\eta) = \Box$.

Lemma 3.17 *For any extended syntax tree T with root r,*

(i) *if* $(\mathcal{C}_1^{n+1}, c_1^{n+1}) \in \mathtt{lft}(r)$ *and* $(\mathcal{D}_i^{n+1}, d_i^{n+1}) \in \mathtt{rght}(r)$, *where* $1 \leq i \leq k$, *then T contains a node η such that* $synl(\eta) = \Box$, $(\mathcal{C}_1^{n+1}, c_1^{n+1}) \in \mathtt{lft}(\eta)$, *and* $(\mathcal{D}_i^{n+1}, d_i^{n+1}) \in \mathtt{rght}(\eta)$;

(ii) *if* $k+1 \leq i \leq k+2$, $(\mathcal{C}_2^{n+1}, c_2^{n+1}) \in \mathtt{lft}(r)$ *and* $(\mathcal{D}_i^{n+1}, d_i^{n+1}) \in \mathtt{rght}(r)$, *then T contains a node η such that* $synl(\eta) = \Box$, $(\mathcal{C}_2^{n+1}, c_2^{n+1}) \in \mathtt{lft}(\eta)$, *and* $(\mathcal{D}_i^{n+1}, d_i^{n+1}) \in \mathtt{rght}(\eta)$;

(iii) *if T contains a node η such that* $synl(\eta) = \Box$, $(C_1^{n+1}, c_1^{n+1}) \in \mathtt{lft}(\eta)$, *and* $\{(\mathcal{D}_i^{n+1}, d_i^{n+1}), \ldots, (\mathcal{D}_j^{n+1}, d_j^{n+1})\} \subseteq \mathtt{rght}(\eta)$, *where* $1 \leq i \leq j \leq k$, *then* $\{(C_1^n, c_1^n), (C_2^n, c_2^n)\} \subseteq \mathtt{lft}(\Box(\eta))$ *and* $\{(\mathcal{D}_i^n, d_i^n), \ldots, (\mathcal{D}_j^n, d_j^n)\} \subseteq \mathtt{rght}(\Box(\eta))$;

(iv) *if T contains a node η such that* $synl(\eta) = \Box$, $(C_2^{n+1}, c_2^{n+1}) \in \mathtt{lft}(\eta)$, *and* $\{(\mathcal{D}_i^{n+1}, d_i^{n+1}), (\mathcal{D}_j^{n+1}, d_j^{n+1})\} \subseteq \mathtt{rght}(\eta)$, *where* $k+1 \leq i \leq j \leq k+2$, *then* $\{(\mathcal{D}_1^n, d_1^n), \ldots, (\mathcal{D}_k^n, d_k^n)\} \subseteq \mathtt{lft}(\Box(\eta))$ *and* $\{(C_l^n, c_l^n), (C_m^n, c_m^n)\} \subseteq \mathtt{rght}(\Box(\eta))$.

Proof.

(i) Let us assume that there is a syntax tree T that does not have a node η with the desired properties. We show that T contains an infinite branch which is absurd. Indeed, using the first item from Proposition 3.14, we see that $synl(r) \notin LIT \cup \{\Diamond\}$. According to our assumption $synl(r) \neq \Box$. Hence, either $synl(r) = \vee$ or $synl(r) = \wedge$. In either case, r has at least one successor r_1 such that $(\mathcal{C}_1^{n+1}, c_1^{n+1}) \in \mathtt{lft}(r_1)$ and $(\mathcal{D}_i^{n+1}, d_i^{n+1}) \in \mathtt{rght}(r_1)$. Thus, we can find the desired infinite branch by starting at r and "following" the nodes that contain the models $(\mathcal{C}_1^{n+1}, c_1^{n+1})$ on the left and $(\mathcal{D}_i^{n+1}, d_i^{n+1})$ on the right.

(ii) The proof is the same as the one above modulo using (ii) from Proposition 3.14.

(iii) It is obvious that $\Box(\mathcal{C}_1^{n+1}, c_1^{n+1}) = \{(C_1^n, c_1^n), (C_2^n, c_2^n)\}$. Using Definition 2.2, we see that $\{(C_1^n, c_1^n), (C_2^n, c_2^n)\} \subseteq \mathtt{lft}(\Box(\eta))$. It follows immediately from Proposition 2.8 that $(C_1^n, c_1^n) \notin \mathtt{rght}(\Box(\eta))$ and $(C_2^n, c_2^n) \notin \mathtt{rght}(\Box(\eta))$. Given the geometry of the models $(\mathcal{D}_i^{n+1}, d_i^{n+1}), \ldots, (\mathcal{D}_j^{n+1}, d_j^{n+1})$, we obtain $\{(\mathcal{D}_i^n, d_i^n), \ldots, (\mathcal{D}_j^n, d_j^n)\} \subseteq \mathtt{rght}(\Box(\eta))$.

(iv) Obviously, $\Box(\mathcal{C}_2^{n+1}, c_2^{n+1}) = \{(\mathcal{D}_1^n, d_1^n), \ldots, (\mathcal{D}_k^n, d_k^n)\}$. The fourth item from Definition 2.2 implies that $\Box((\mathcal{C}_2^{n+1}, c_2^{n+1})) \subseteq \mathtt{lft}(\Box(\eta))$. Using Proposition 2.8, we see that none of the pointed models in $\Box(\mathcal{C}_2^{n+1}, c_2^{n+1})$ can appear on the right of $\Box(\eta)$. Given the geometry of the models $(\mathcal{D}_{k+1}^{n+1}, d_{k+1}^{n+1})$ and $(\mathcal{D}_{k+2}^{n+1}, d_{k+2}^{n+1})$, we see that $\{(C_l^n, c_l^n), (C_m^n, c_m^n)\} \subseteq \mathtt{rght}(\Box(\eta))$.

□

We are ready now to prove Theorem 3.9. It follows immediately from the stronger statement below.

Theorem 3.18 *For any $n \geq 1$, any formula φ such that $\mathbb{C}^n \models \varphi$ and $\mathbb{D}^n \models \neg\varphi$ has size at least 2^n and any formula ψ such that $\mathbb{D}^n \models \psi$ and $\mathbb{C}^n \models \neg\psi$ has size at least 2^n.*

Proof. The proof proceeds by induction on n.

Base step: The fact that the extended syntax tree T of any formula φ such that $\mathbb{C}^1 \models \varphi$ and $\mathbb{D}^1 \models \neg\varphi$ has at least two leaves follows immediately from Propositions 3.16. Using this and Proposition 2.6, we see that the size of any formula ψ such that $\mathbb{D}^1 \models \psi$ and $\mathbb{C}^1 \models \neg\psi$ is at least 2.

Induction step: Let us consider the extended syntax tree T with a root $r = \mathbb{C}^{n+1} \circ \mathbb{D}^{n+1}$ of a formula φ such that $\mathbb{C}^{n+1} \models \varphi$ and $\mathbb{D}^{n+1} \models \neg\varphi$. It follows from Proposition 3.16, that we have two types of nodes.

(i) For the pointed model $(\mathcal{C}_1^{n+1}, c_1^{n+1})$ and any pointed model $(\mathcal{D}_i^{n+1}, d_i^{n+1})$, where $1 \leq i \leq k$, there is a node η in T such that $\{(\mathcal{C}_1^{n+1}, c_1^{n+1})\} = \mathtt{lft}(\eta)$ and $(\mathcal{D}_i^{n+1}, d_i^{n+1}) \in \mathtt{rght}(\eta)$.

(ii) For the pointed model $(\mathcal{C}_2^{n+1}, c_2^{n+1})$ and any pointed model $(\mathcal{D}_j^{n+1}, d_j^{n+1})$, where $k+1 \leq j \leq k+2$, there is a node ζ in T such that $\{(\mathcal{C}_2^{n+1}, c_2^{n+1})\} = \mathtt{lft}(\zeta)$ and $(\mathcal{D}_j^{n+1}, d_j^{n+1}) \in \mathtt{rght}(\zeta)$.

It is obvious that a node η and a node ζ cannot coincide. Hence η and ζ are the roots of two sub-trees T_η and T_ζ of T with no common nodes. Using item (i) from Lemma 3.17, we see that T_η contains a node η_1 such that $synl(\eta_1) = \Box$, $\{(\mathcal{C}_1^{n+1}, c_1^{n+1})\} = \mathtt{lft}(\eta_1)$, and $(\mathcal{D}_i^{n+1}, d_i^{n+1}) \in \mathtt{rght}(\eta_1)$; similarly, it follows from Lemma 3.17 (ii), that T_ζ contains a node ζ_2 such that $synl(\zeta_2) = \Box$, $\{(\mathcal{C}_2^{n+1}, c_2^{n+1})\} = \mathtt{lft}(\zeta_2)$, and $(\mathcal{D}_j^{n+1}, d_j^{n+1}) \in \mathtt{rght}(\zeta_2)$; Therefore, T has the shape shown in Figure 8. Namely,

- there are nodes $\eta_1^1 = (\mathcal{C}_1^{n+1}, c_1^{n+1}) \circ \mathbb{G}_1, \ldots, \eta_1^l = (\mathcal{C}_1^{n+1}, c_1^{n+1}) \circ \mathbb{G}_l$ such that each one of them has a syntax label □; what is more,

$\{(\mathcal{D}_1^{n+1}, d_1^{n+1}), \ldots, (\mathcal{D}_k^{n+1}, d_k^{n+1})\} \subseteq \mathbb{G}_1 \cup \ldots \cup \mathbb{G}_l$;

- there are nodes $\zeta_2^1 = (\mathcal{C}_2^{n+1}, c_2^{n+1}) \circ \mathbb{H}_1, \ldots, \zeta_2^m = (\mathcal{C}_2^{n+1}, c_2^{n+1}) \circ \mathbb{H}_m$ such that each one of them has a syntax label $\Box$; moreover, $\{(\mathcal{D}_{k+1}^{n+1}, d_{k+1}^{n+1}), (\mathcal{D}_{k+2}^{n+1}, d_{k+2}^{n+1})\} \subseteq \mathbb{H}_1 \cup \ldots \cup \mathbb{H}_m$.

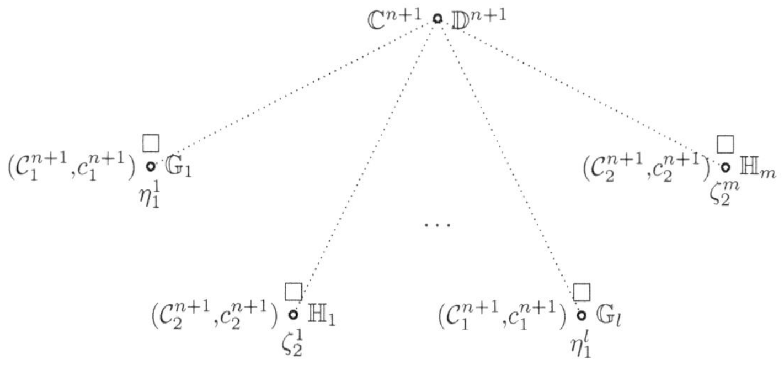

Fig. 8. An extended syntax tree T with root $\mathbb{C}^{n+1} \circ \mathbb{D}^{n+1}$.

Using item (iii) of Lemma 3.17, we see that, for any η_1^i, we have $\mathbb{C}^n = \{(C_1^n, c_1^n), (C_2^n, c_2^n)\} = \Box((\mathcal{C}_1^{n+1}, c_1^{n+1})) = \mathtt{lft}(\Box(\eta_1^i))$ while $\mathbb{D}^n = \{(\mathcal{D}_1^n, d_1^n), \ldots, (\mathcal{D}_k^n, d_k^n)\} \subseteq \mathtt{rght}(\Box(\eta_1^1)) \cup \ldots \cup \mathtt{rght}(\Box(\eta_1^l))$. According to the induction hypothesis, any formula φ such that $\mathbb{C}^n \models \varphi$ and $\mathbb{D}^n \models \neg\varphi$ has size at least 2^n. Applying Proposition 2.7 and Proposition 2.9, we see that, for any formulae $\varphi_1, \ldots, \varphi_l$ for which $\mathtt{lft}(\Box(\eta_1^1)) \models \varphi_1^1, \ldots, \mathtt{lft}(\Box(\eta_1^l)) \models \varphi_l$ whereas $\mathtt{rght}(\Box(\eta_1^1)) \models \neg\varphi_1, \ldots, \mathtt{rght}(\Box(\eta_1^l)) \models \neg\varphi_l$, we have $(|\varphi_1| + \ldots + |\varphi_l|) \geq 2^n$.

Similarly, using item (iii) of Lemma 3.17, for any ζ_2^i, we see that $\mathbb{D}^n = \{(\mathcal{D}_1^n, d_1^n), \ldots, (\mathcal{D}_k^n, d_k^n)\} = \Box(\mathcal{C}_2^{n+1}, c_2^{n+1}) = \mathtt{lft}(\Box(\zeta_2^i))$ while $\mathbb{C}^n \subseteq \mathtt{rght}(\Box(\zeta_2^1)) \cup \ldots \cup \mathtt{rght}(\Box(\zeta_2^m))$. Again, according to the induction hypothesis, any formula ψ such that $\mathbb{D}^n \models \psi$ and $\mathbb{C}^n \models \neg\psi$ has size at least 2^n. Applying Proposition 2.7 and Proposition 2.9, we see that, for any formulae $\psi_1, \ldots, \psi_m$ for which $\mathtt{lft}(\Box(\zeta_2^1)) \models \psi_1, \ldots, \mathtt{lft}(\Box(\zeta_m)) \models \psi_m$ whereas $\mathtt{rght}(\Box(\zeta_2^1)) \models \neg\psi_1, \ldots, \mathtt{rght}(\Box(\zeta_2^m)) \models \neg\psi_m$, we have $(|\psi_1| + \ldots + |\psi_m|) \geq 2^n$.

Thus, the number of leaves of any extended syntax tree with root $\mathbb{C}^{n+1} \circ \mathbb{D}^{n+1}$ is at least 2^{n+1}. Hence, any formula φ such that $\mathbb{C}^{n+1} \models \varphi$ and $\mathbb{D}^{n+1} \models \neg\varphi$ has size at least 2^{n+1}. It follows from Proposition 2.6, that any formula ψ such that $\mathbb{D}^{n+1} \models \psi$ and $\mathbb{C}^{n+1} \models \neg\psi$ has size at least 2^{n+1}.

$\Box$

As usual, we can represent ML-formulae compactly as *directed acyclic graphs* (*DAGs*) or *modal circuits*. The size of such a graph is the number of its edges. To the best of our knowledge, the fact that ML-formulae represented as *DAGs* are exponentially more succinct than ML-formulae in their tree representation seems to be taken for granted but we were unable to find a published proof although it can be easily obtained from the results in, e.g., [5], [6] or [10]. Nevertheless, we consider the corollary below to be yet another confirmation of a widely known folklore fact rather than an original new result.

Corollary 3.19 *Modal circuits are exponentially more succinct than* ML-*formulae on the class of all Kripke models* **K**.

Proof. According to Theorem 3.8, every ML-formula that is equivalent to the formula δ_n from Definition 3.1 has size 2^n. In the proof of Proposition 3.4, we defined, for every δ_n, an equivalent ML-formula δ'_n as follows. $\delta'_1 = \delta_1$ and $\delta'_{n+1} = \delta'_n \wedge \Diamond(\mathbf{b} \wedge \delta'_n) \wedge \Diamond(\neg\mathbf{b} \wedge \delta'_n)$. Obviously, the formulae δ'_n can be represented as linearly growing *DAGs* as shown below. □

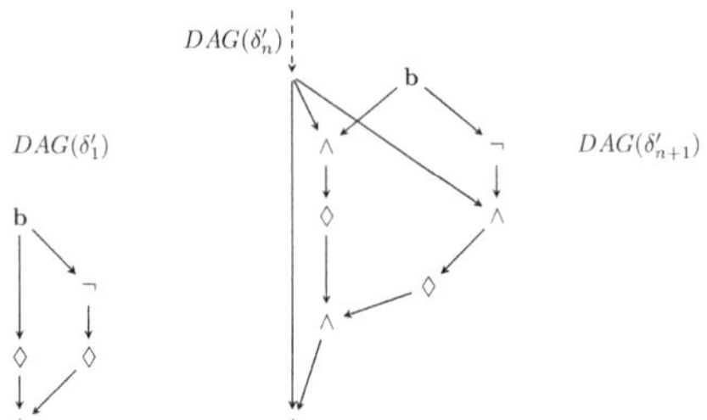

4 Conclusion

With the benefit of hindsight, we can say that our Theorem 3.9 is related to Theorem 4.1 from [5]. The latter can be interpreted as showing that there is no sub-exponential equivalence preserving translation from the recursively defined formulae $\varphi_1 = \Diamond(\mathbf{b} \vee \neg\mathbf{b})$ and $\varphi_{n+1} = \Diamond\neg \triangle \varphi_n$ to ML, i.e., extending ML with formulae $\triangle\varphi$ leads to an exponential increase of succinctness with respect to ML. However, since $\Diamond$ is not definable in ConML on many classes of frames [3], this result cannot in general be used to show that ConML is exponentially more succinct than ML.

More results on the comparison between modal formulae and modal circuits for extensions of ML and a list of open problems about lower bounds on formula and circuit-size in modal logics can be found in [8].

A very general problem was pointed out in [16]. It consists of finding the shortest possible modal equivalents of modally definable first-order conditions. The potential importance of this question is witnessed by the fact that its initial study has led to an extension of the class of Sahlqvist formulae [16].

As far as our present work is concerned, it would be nice to compare in terms of succinctness PAL and ConML. We conjecture that PAL is exponentially more succinct than ConML on **K** even for modal languages with one diamond and one propositional symbol. We think that this remains true on the class of $\mathbf{S}_5$-models commonly used in epistemic logic but we do not know how many different diamonds and propositional symbols are needed. Additionally, we conjecture that ConML is not exponentially more succinct than PAL but that ML-circuits are exponentially more succinct than PAL-formulae.

Acknowledgements

Hans van Ditmarsch, Jie Fan, and Petar Iliev acknowledge support from European Research Council grant EPS 313360.

Hans van Ditmarsch is also affiliated to the Institute of Mathematical Sciences, Chennai, India as a research associate.

Jie Fan gratefully acknowledges support from the China Scholarship Council.

References

[1] Adler, M. and N. Immerman, *An n! lower bound on formula size*, ACM Trans. Comput. Logic **4** (2003), pp. 296–314.

[2] Blackburn, P., M. de Rijke and Y. Venema, "Modal Logic," CUP, 2001.

[3] Cresswell, M. J., *Necessity and contingency*, Studia Logica **47** (1988), pp. 145–149.

[4] Etessami, K., M. Vardi and T. Wilke, *First-order logic with two variables and unary temporal logic*, in: *Proceedings of LICS*, 1997, pp. 228–236.

[5] Figueira, S. and D. Gorín, *On the size of shortest modal descriptions*, Advances in Modal Logic **8** (2010), pp. 114–132.

[6] French, T., W. van der Hoek, P. Iliev and B. Kooi, *On the succinctness of some modal logics*, Artificial Intelligence **197** (2013), pp. 56–85.

[7] Grohe, M. and N. Schweikardt, *The succinctness of first-order logic on linear orders*, Logical Methods in Computer Science **1** (2005), pp. 1–25.

[8] Iliev, P., "On The Relative Succinctness of Some Modal Logics," PhD thesis, Department of Computer Science, University of Liverpool, 2013.

[9] Laroussinie, F., N. Markey and P. Schnoebelen, *Temporal logic with forgettable past*, in: *Proceedings of LICS*, 2002, pp. 383–392.

[10] Lutz, C., *Complexity and succinctness of public announcement logic*, in: P. Stone and G. Weiss, editors, *Proceedings of AAMAS*, 2006, pp. 137–144.

[11] Lutz, C., U. Sattler and F. Wolter, *Modal logic and the two-variable fragment*, in: L. Fribourg, editor, *Proceedings of CSL*, 2001, pp. 247–261.

[12] Montgomery, H. and R. Routley, *Contingency and non-contingency bases for normal modal logics*, Logique et Analyse **9** (1966), pp. 318–328.

[13] Otto, M., *Bisimulation invariance and finite models*, in: W. P. Z. Chatzidakis, P. Koepke, editor, *Lecture Notes in Logic, Logic Colloquium 2002*, 2006 pp. 276–298.

[14] Plaza, J., *Logics of public communications*, in: M. Emrich, M. Pfeifer, M. Hadzikadic and Z. Ras, editors, *Proceedings of the 4th International Symposium on Methodologies for Intelligent Systems: Poster Session Program* (1989), pp. 201–216.

[15] Stockmeyer, L. J., "The Complexity of Decision Problems in Automata Theory and Logic," PhD thesis MIT, 1974.

[16] Vakarelov, D., *Modal definability in languages with a finite number of propositional variables and a new extension of the Sahlqvist's class*, Advances in Modal Logic **4** (2003), pp. 499–518.

[17] van Benthem, J., "Modal logic and Classical Logic," Bibliopolis, 1983.

[18] Wilke, T., *CTL^+ is exponentially more succinct than CTL*, in: *Proceedings of FSTTCS*, 1999, pp. 110–121.

Epistemic Probability Logic Simplified

Jan van Eijck [1]

CWI and ILLC
Science Park 123
1098 XG Amsterdam

François Schwarzentruber [2]

ENS Rennes
Campus de Ker Lann
35170 BRUZ

Abstract

We propose a simplified logic for reasoning about (multi-agent) epistemic probability models, and for epistemic probabilistic model checking. Epistemic probability models are multi-agent Kripke models that assign to each agent an equivalence relation on worlds, together with a function from worlds to positive rationals (a lottery). The difference with the usual approach is that probability is linked to knowledge rather than belief, and that knowledge is equated with certainty.

A first contribution of the paper is a comparison of a semantics for epistemic probability in terms of models with multiple lotteries and models with a single lottery. We give a proof that multiple lottery models can always be replaced by single lottery models. As multiple lotteries represent multiple subjective probabilities, our result connects subjective and intersubjective probability.

Next, we define an appropriate notion of bisimulation, and use it to prove an adaptation of the Hennessy-Milner Theorem and to prove that some finite multiple lottery models only have infinite single lottery counterparts. We then prove completeness, and state results about model checking complexity. In particular, we show the PSPACE-completeness of the model checking in the dynamic version with action models.

The logic is designed with model checking for epistemic probability logic in mind; a prototype model checker for it exists. This program can be used to keep track of information flow about aleatory acts among multiple agents.

Keywords: Probability. Epistemic modal logic. Lottery. Hennessy Milner Theorem. Dynamic epistemic logic. Complexity theory.

---

[1] jve@cwi.nl

[2] francois.schwarzentruber@ens-rennes.fr

1 Probability as a function of degree of information

A classical view of probability theory is that probability measures degree of information. Here is a characteristic quote from [17]:

> *Dans les choses qui ne sont que vraisemblables, la différence des données que chaque homme a sur elles, est une des causes principales de la diversité des opinions que l'on voit régner sur les mêmes objects.*
> (Laplace)

We present a multi-agent logic of probability and knowledge, with a very natural product update, yielding a simplification of the logic proposed in [7], which is in turn based on [15] and [6]. We show how probability measures on Kripke models can be defined in a straightforward way from lotteries. We propose a complete logic for lottery models, define an appropriate notion of bisimulation (different from the notion in [14,15]), and prove a Hennessey-Milner result for this notion. We prove that every model with lotteries is equivalent to a single-lottery model, where all agents share the same lottery. Finally, we investigate the model checking complexity of the logic.

This paper presents a logic of probability and knowledge where the two are related as follows:

Agent a knows ϕ if and only if the probability a assigns to ϕ equals 1.

Our proposal has obvious relations to earlier proposals on combining knowledge and probability [10,15,14,7,5,13] and many more. A key difference is that these proposals do not equate knowledge with certainty. An exception to this is [1].

A possible reason for not equating knowledge with certainty is the well-known difference between impossible in practice and impossible in theory which arises when measuring probabilities in uncountable spaces, where one equates "the probability of ϕ equals 1" with "ϕ is almost certain". An infinite process of fair coin throwing that results in an infinite sequence of 1s is practically imposssible (its probability is 0), but the sequence *is* in the sample space. Since we are careful to work with countable models and with lotteries that are bounded (Definition 2.2), this difficulty does not arise for us.

In real applications, knowledge and certainty are strongly related. We present our simplified framework of epistemic probability logic in Section 2. In particular, we will present models with a single lottery and in Section 3 we prove that the semantics with a single lottery and the semantics with several lotteries are equivalent, by constructing single lottery models from multiple lottery models. This throws light on the relation between subjective probability (modeled by multiple lotteries) and intersubjective probability (modeled by single lotteries). In Section 4 we define the appropriate notion of bisimulation, and use it to prove a Hennessy-Milner Theorem for epistemic probability logic. Section 5 gives an axiomatization for epistemic probability logic based on [10] and proves that the $S5$ axioms for certainty can be derived. In Section 6, we deal with the model checking procedure, and show that it runs in polynomial time. Section 7 explains how to add action model update in DEL style (but simplified), and gives a PSPACE-completeness proof for the model checking problem that results from adding a dynamic operator to the language.

2 Epistemic probabilistic logic

We present our epistemic lottery models (with the variant with a single lottery even for the multi-agent case). We then present the language of our version of epistemic probabilistic logic and its semantics and finally we show how to embed standard epistemic logic in our framework.

2.1 Epistemic lottery models

We start out from the definition of standard epistemic models.

Definition 2.1 A standard epistemic model $\mathcal{M}$ for a set $\mathbf{P}$ of propositions and a set A of agents is a tuple (W, V, R) where

- W is a non-empty, finite or countable set of worlds,
- V is a valuation function that assigns to every $w \in W$ a subset of $\mathbf{P}$.
- R is a function that assigns to every agent $a \in A$ an equivalence relation R_a on W.

To turn a standard epistemic model into an epistemic probability model, we assign to each agent a lottery, representing the subjective probabilities the agent assigns.

Definition 2.2 A W**-lottery** L is a function from a (finite or countable) set W to the set of positive (non-zero) rationals, i.e., $L : W \rightarrow \mathbb{Q}^+$. A W-lottery L is **bounded** on $V \subseteq W$ if $\sum_{v \in V} L(v) < \infty$.

Definition 2.3 An **epistemic multiple lottery model** $\mathcal{M}$ is a tuple $(W, V, R, \mathbb{L})$ where W, V, R are as in Definition 2.1 and $\mathbb{L}$ is a function that assigns to every agent $a \in A$ a W-lottery that is bounded on every R_a equivalence class.

We say that an epistemic lottery model is *normalized* if $\mathbb{L}_a$ restricted to E is a probability measure for all agents a and for all R_a-equivalence classes E. By the boundedness condition, all epistemic lottery models can be normalized.

Now, we define an epistemic lottery model where the lotteries are the same for each agent, that is $\mathbb{L}_a = \mathbb{L}_b$ for all agents a, b. We will write L instead of $\mathbb{L}_a$ for a given agent a. Models where there is a single lottery seem easier to manipulate. Formally:

Definition 2.4 An **epistemic single lottery model** $\mathcal{M}$ is a tuple (W, V, R, L) where W, V, R are as in Definition 2.1 and L is a W-lottery that is bounded on every R_a equivalence class, for every agent a.

2.2 Epistemic probability logic language

The language $\mathcal{L}$ of multi-agent epistemic probability logic is defined as follows.

Definition 2.5 Let p range over $\mathbf{P}$, a over A, q over $\mathbb{Q}$. Then $\mathcal{L}$ is given by:

$$\begin{array}{lcl} \phi & ::= & \top \mid p \mid \neg\phi \mid \phi \wedge \phi \mid t_a \geq 0 \mid t_a = 0 \\ t_a & ::= & q \mid q \cdot P_a\phi \mid t_a + t_a \end{array}$$

The intention in $t_a + t_a$ is that both indices are the same.

Some useful abbreviations:

- $\bot, \phi_1 \vee \phi_2, \phi_1 \rightarrow \phi_2, \phi_1 \leftrightarrow \phi_2$.

- $t \geq t'$ for $t + (-1)t' \geq 0$.
- $t < t'$ for $\neg t \geq t'$.
- $t > t'$ for $\neg t' \geq t$.
- $t \leq t'$ for $t' \geq t$.
- $t \neq t'$ for $\neg t = t'$.
- $P_a(\phi_1|\phi_2) = q$ for $P_a(\phi_2) > 0 \wedge q \cdot P_a(\phi_2) = P_a(\phi_1 \wedge \phi_2)$.

t_a generates linear expressions dealing with subjective probabilities of agent a. A formula of the form $t_a = 0$ or $t_a > 0$ is called an *a-probability formula.*

Given these, we have:

- $P_a\phi = q$ expresses that the probability of ϕ according to a equals q.
- $P_a(\phi_1|\phi_2) = q$ expresses that according to a, the probability of ϕ_1, conditional on ϕ_2, equals q.

The truth definition for $\mathcal{L}$ is given below.

Definition 2.6 Let $\mathcal{M} = (W, V, R, \mathbb{L})$ be an epistemic lottery model and let $w \in W$.

$$\begin{aligned}
\mathcal{M}, w &\models \top \quad \text{always} \\
\mathcal{M}, w &\models p \text{ iff } p \in V(w) \\
\mathcal{M}, w &\models \neg\phi \text{ iff it is not the case that } \mathcal{M}, w \models \phi \\
\mathcal{M}, w &\models \phi_1 \wedge \phi_2 \text{ iff } \mathcal{M}, w \models \phi_1 \text{ and } \mathcal{M}, w \models \phi_2 \\
\mathcal{M}, w &\models t_a \geq 0 \text{ iff } [\![t_a]\!]_w^{\mathcal{M}} \geq 0 \\
\mathcal{M}, w &\models t_a = 0 \text{ iff } [\![t_a]\!]_w^{\mathcal{M}} = 0.
\end{aligned}$$

$$\begin{aligned}
[\![q]\!]_w^{\mathcal{M}} &:= q \\
[\![q \cdot P_a\phi]\!]_w^{\mathcal{M}} &:= q \times P_{a,w}^{\mathcal{M}}(\phi) \\
[\![t_a + t'_a]\!]_w^{\mathcal{M}} &:= [\![t_a]\!]_w^{\mathcal{M}} + [\![t'_a]\!]_w^{\mathcal{M}}
\end{aligned}$$

$$P_{a,w}^{\mathcal{M}}(\phi) = \frac{\sum\{\mathbb{L}_a(u) \mid wR_a u \text{ and } \mathcal{M}, u \models \phi\}}{\sum\{\mathbb{L}_a(u) \mid wR_a u\}}.$$

Notice that $\mathbb{L}_a(u) > 0$ for all $u \in W$ so that there is no division by zero. Also, $\sum\{\mathbb{L}_a(u) \mid wR_a u\} < \infty$, by the boundedness condition on $\mathbb{L}_a$. So $P_{a,w}$ is well-defined. The interpretation of formulas in epistemic single lottery models is similar except that we directly use L instead of $\mathbb{L}_a$ for a given agent a.

2.3 Relating Knowledge to Certainty

We use $K_a(\phi)$ as an abbreviation for $P_a(\phi) = 1$. This interprets knowledge as certainty and makes K_a behave as an $S5$-operator.[3]

Example 2.7 [Agents with different priors]

---

[3] If you have still qualms about this, then please read: "This interprets knowledge as almost-certainty."

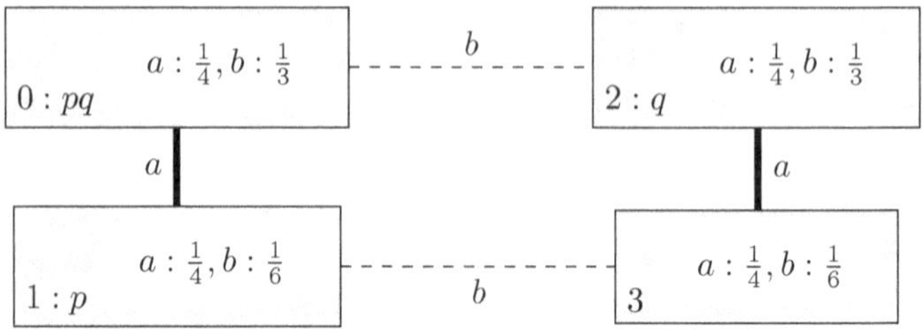

In the model of this example, loterries are $\mathbb{L}_a = \{0 : \frac{1}{4}, 1 : \frac{1}{4}, 2 : \frac{1}{4}, 3 : \frac{1}{4}\}$ and $\mathbb{L}_b = \{0 : \frac{1}{3}, 1 : \frac{1}{6}, 2 : \frac{1}{3}, 3 : \frac{1}{6}\}$. At world 0, the probability that a (represented by solid lines) assigns to p is 1, so $K_a p$ is true at 0. $K_a q$ is false at 0, for the probability that a assigns to q is less than 1. In fact, we have:

Proposition 2.8 *Let ϕ be a formula of standard epistemic logic. The following statements are equivalent:*

(i) *ϕ is satisfiable in a standard epistemic model;*

(ii) *$tr(\phi)$ is satisfiable in an epistemic single lottery model*

(iii) *$tr(\phi)$ is satisfiable in an epistemic lottery model*

where tr is defined by $tr(K_a\phi) = P_a tr(\phi) = 1$.

Proof. $(iii) \Rightarrow (i)$. If $tr(\phi)$ is satisfiable in an epistemic lottery model, we extract a standard epistemic model by dropping the lotteries and we prove that ϕ is true.

$(i) \Rightarrow (ii)$. Suppose that ϕ is satisfiable in a standard epistemic model. As $S5_n$ has the finite model property [8], there is a *finite* standard epistemic model for ϕ. We transform this standard epistemic model into an epistemic single lottery model $\mathcal{M} = (W, V, R, L)$ by adding a fake single lottery L that assigns 1 to all worlds. As the model is finite, it is guaranteed that the W-lottery L is bounded on every R_a equivalence class. We prove that $tr(\phi)$ is true in $\mathcal{M}$.

$(ii) \Rightarrow (iii)$. Because an epistemic single lottery model is an epistemic lottery model. □

Proposition 2.8 can be generalized to set of formulas.

Proposition 2.9 *Let Σ be a formula of standard epistemic logic. The following statements are equivalent:*

(i) *Σ is satisfiable in a standard epistemic model;*

(ii) *$\{tr(\phi) \mid \phi \in \Sigma\}$ is satisfiable in an epistemic single lottery model*

(iii) *$\{tr(\phi) \mid \phi \in \Sigma\}$ is satisfiable in an epistemic lottery model*

where tr is defined by $tr(K_a\phi) = P_a tr(\phi) = 1$.

Proof. The proof is essentially the same one than for proposition 2.8 except for $(i) \Rightarrow (ii)$. Suppose that Σ is satisfiable in a standard epistemic model. The finite model property argument does not work anymore. Nevertheless, we suppose that the standard epistemic model has at most a countable number of worlds. We transform this standard epistemic model into an epistemic lottery model $\mathcal{M} = (W, V, R, \mathbb{L})$

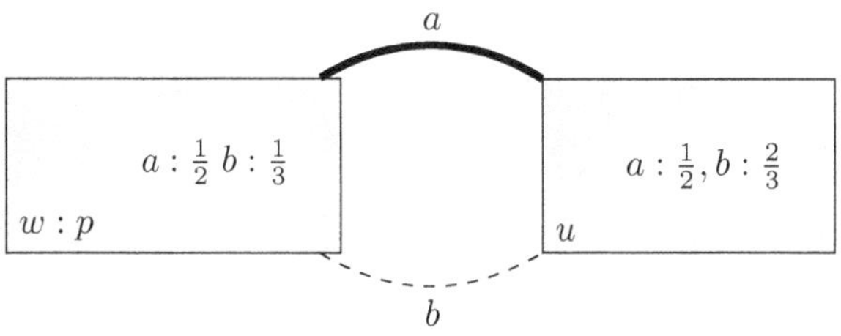

Fig. 1. Example of an epistemic probability model $\mathcal{M}$

by adding a fake lottery $\mathbb{L}$ as follows: we consider $\{w_0, w_1, \dots\}$ a (possibly finite) enumeration of worlds in W. We define $\mathbb{L}_a(w_k) = \frac{1}{k^2}$. As $\sum_{k=0}^{+\infty} \frac{1}{k^2} < +\infty$, it is guaranteed that the W-lottery $\mathbb{L}_a$ is bounded on every R_a equivalence class. We prove that $\{tr(\phi) \mid \phi \in \Sigma\}$ is true in $\mathcal{M}$. □

Remark 2.10 Notice that if we define a belief operator $B_a\phi$ by $P_a(\phi) > \alpha$ for some $\alpha \in (\frac{1}{2}, 1]$, the formula $B_a p \wedge B_a q \wedge \neg B_a(p \wedge q)$ is satisfiable. That is, B_a behaves as a non-normal operator and not as a $KD45$ operator. This provides a way out of the so-called *lottery paradox* [16].

3 Single lottery versus multiple lottery models

In this section, we prove that the semantics given in terms of epistemic multiple lottery models (definition 2.3) and the semantics given in terms of epistemic *single* lottery models (definition 2.4) are equivalent, in the sense that for each multiple lottery model there is an equivalent single lottery model.

Philosophically, this suggests that objective probability, or at least intersubjective probability, can be defined from subjective probabilities. In any case, epistemic single lottery models are easier to handle because we attribute the same value to a world for each agent.

Proposition 3.1 *Given an epistemic lottery model $\mathcal{M} = (W, V, R, \mathbb{L})$, given a world w, there exists an epistemic single lottery model $\mathcal{M}' = (W', V', R', L)$ and a world $w' \in W'$ such that for all formulas ϕ, $\mathcal{M}, w \models \phi$ iff $\mathcal{M}', w' \models \phi$.*

Before starting the proof, let us consider an example. We start with the model $\mathcal{M}$ depicted in Figure 1 consisting in two worlds w and u. p is true in w and false in u. The lottery for agent a assigns probability $\frac{1}{2}$ to w and $\frac{1}{2}$ to u. The lottery for agent b assigns probability $\frac{1}{3}$ to w and $\frac{2}{3}$ to u.

In order to get a model with a single lottery, we unravel the model $\mathcal{M}$ and we obtain the infinite epistemic lottery model of Figure 2. The proof formalizes this transformation.

Proof. The construction goes as follows. The set of worlds W' is the set of all sequences of the form:

$$w_0 a_0 w_1 a_1 \dots w_{n-1} a_{n-1} w_n$$

such that, $n \geq 0$, $w_0 = w$, w_i are worlds, a_i are agents, $(w_i, w_{i+1}) \in R_{a_i}$ and $a_i \neq a_{i+1}$. For any sequence s, we write $end(s)$ for the last world in the sequence

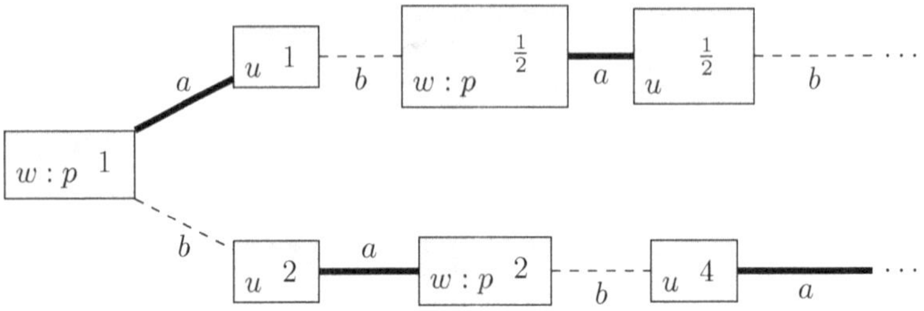

Fig. 2. Epistemic lottery model $\mathcal{M}'$ obtained by unraveling the model $\mathcal{M}$

that is, $end(w_0a_0w_1a_1 \dots w_{n-1}a_{n-1}w_n) = w_n$. The valuation V' is defined by $V'(\boldsymbol{s}) = V(end(\boldsymbol{s}))$. The relation R'_a is defined as follows:

$$
\begin{aligned}
R'_a = \; & \{(\boldsymbol{s}, \boldsymbol{s}) \mid \boldsymbol{s} \in W'\} \cup \\
& \{(\boldsymbol{s}, \boldsymbol{s}au), (\boldsymbol{s}au, \boldsymbol{s}) \mid \boldsymbol{s}, \boldsymbol{s}au \in W'\} \cup \\
& \{(\boldsymbol{s}au_1, \boldsymbol{s}au_2) \mid s, \boldsymbol{s}au_1, \boldsymbol{s}au_2 \in W'\}
\end{aligned}
$$

The lottery is defined by induction on sequences as follows:

- $L(w) = 1$;
- $L(sau) = \frac{\mathbb{L}_a(u)L(s)}{\mathbb{L}_a(end(s))}$.

It can now be proved by induction on ϕ that for all sequences s, $\mathcal{M}, end(s) \models \phi$ iff $\mathcal{M}', s \models \phi$. □

Notice that our construction produces models with infinite sets of worlds. We will prove in the next Section (Proposition 4.3) that this is unavoidable. What this means is that the logic, when interpreted over the class of single lottery models, does not have the finite model property. Also, it suggests that the logic is not expressive enough to characterize models that are built from a single lottery. What is needed to make such characterization possible? We leave this question for future work.

4 Bisimulation

In this section, we pick up a yarn in the story about bisimulation from [14], we modify (simplify) the definition so that it suits our logic, and we prove a Hennessy-Milner result for our new version. Next, we use our notion of bisimulation to prove that some finite multiple lottery models cannot have finite single lottery counterparts.

If X is a set of worlds with $X \subseteq R_a(w)$ then we use $\mathbb{L}_{a,w}(X)$ for $\frac{\sum_{x \in X} \mathbb{L}_a(x)}{\sum_{v \in R_a(w)} \mathbb{L}_a(v)}$.

Given two epistemic lottery models $\mathcal{M} = (W, V, R, \mathbb{L}), \mathcal{M}' = (W', V', R', \mathbb{L}')$, we say that a relation B is a bisimulation over $W \times W'$ if wBw' implies:

(i) w and w' satisfy the same atomic propositions;

(ii) for every set $E \subseteq R_a(w)$ there exists a set $E' \subseteq R'_a(w')$ such that:
  - for all $u' \in E'$, there exists $u \in E$ such that uBu';
  - and $\mathbb{L}_{a,w}(E) \leq \mathbb{L}_{a,w'}(E')$.

(iii) for every set $E' \subseteq R'_a(w')$ there exists a set $E \subseteq R_a(w)$ such that:

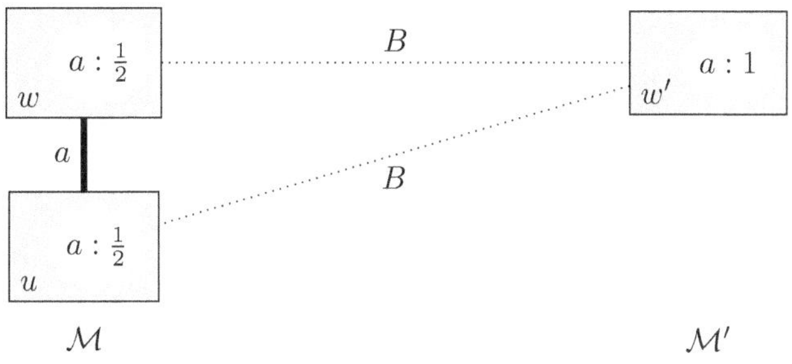

Fig. 3. Two models $\mathcal{M}$, $\mathcal{M}'$ and a bisimulation relation B

- for all $u \in E$, there exists $u' \in E'$ such that uBu';
- and $\mathbb{L}_{a,w'}(E') \leq \mathbb{L}_{a,w}(E)$.

If there exists a bisimulation B such that wBw' we say that w and w' are bisimilar, notation $w \leftrightarroweq w'$ or $\mathcal{M}, w \leftrightarroweq \mathcal{M}', w'$ if there is danger of ambiguity.

Figure 3 shows two models $\mathcal{M}$, $\mathcal{M}'$ and a bisimulation B. We see that condition (ii) and condition (iii) require inequalities and not equalities.

Proposition 4.1 *Let $\mathcal{M}, \mathcal{M}'$ be two models and w and w' be worlds of $\mathcal{M}$ and $\mathcal{M}'$, respectively. If $\mathcal{M}, w \leftrightarroweq \mathcal{M}', w'$ then $\mathcal{M}, w$ and $\mathcal{M}', w'$ satisfy the same formulas.*

Proof. We will prove by induction on ϕ and t_a:

(a) for all w, w' with $\mathcal{M}, w \leftrightarroweq \mathcal{M}', w'$: $\mathcal{M}, w \models \phi$ iff $\mathcal{M}', w' \models \phi$,

(b) for all w, w' with $\mathcal{M}, w \leftrightarroweq \mathcal{M}', w'$: $[\![t_a]\!]_w^{\mathcal{M}} = [\![t_a]\!]_{w'}^{\mathcal{M}'}$.

Let C be a bisimulation that witnesses $\mathcal{M}, w \leftrightarroweq \mathcal{M}', w'$.

For $\top$ (a) holds trivially. For atoms p, use property (i) of C. The cases of $\neg\phi$ and $\phi_1 \wedge \phi_2$ are straightforward, using the induction hypothesis for (a).

For the cases of $t_a \geq 0$ and $t_a = 0$, we assume that $[\![t_a]\!]_w^{\mathcal{M}} = [\![t_a]\!]_{w'}^{\mathcal{M}'}$, from which the statements $\mathcal{M}, w \models t_a \geq 0$ iff $\mathcal{M}', w' \models t_a \geq 0$ and $\mathcal{M}, w \models t_a = 0$ iff $\mathcal{M}', w' \models t_a = 0$ follow.

Next, we show (b). The key issue here is to show that $P_{a,w}^{\mathcal{M}}(\phi) = P_{a,w'}^{\mathcal{M}'}(\phi)$. From this (b) easily follows.

Let

$$E = \{v \in R_a^{\mathcal{M}}(w) \mid \mathcal{M}, v \models \phi\}.$$

By property (ii) of C there is a set $E' \subseteq R_a^{\mathcal{M}'}(w')$ such that:

(i) for all $u' \in E'$ there exists $u \in E$ with uCu';

(ii) $\mathbb{L}_{a,w}(E) \leq \mathbb{L}'_{a,w'}(E')$.

From (i) we get that for all $u' \in E'$: $\mathcal{M}', u' \models \phi$. Use this, plus (ii) and the fact that $[\![P_a\phi]\!]_w^{\mathcal{M}} = \mathbb{L}_{a,w}(E)$ to get:

$$[\![P_a\phi]\!]_w^{\mathcal{M}} \leq \mathbb{L}'_{a,w'}(E') \leq [\![P_a\phi]\!]_{w'}^{\mathcal{M}'}.$$

Let

$$E' = \{v' \in R_a^{\mathcal{M}'}(v) \mid \mathcal{M}', v' \models \phi\}.$$

By property (iii) of C there is a set $E \subseteq R_a^{\mathcal{M}}(w)$ such that:

(i) for all $u \in E$ there exists $u' \in E'$ with uCu';

(ii) $\mathbb{L}'_{a,w'}(E') \leq \mathbb{L}_{w,a}(E)$.

From (i) we get that for all $u \in E$: $\mathcal{M}, u \models \phi$. Use this, plus (ii) and the fact that $[\![P_i\phi]\!]_{w'}^{\mathcal{M}'} = \mathbb{L}'_{a,w'}(E')$ to get:

$$[\![P_i\phi]\!]_{w'}^{\mathcal{M}'} \leq \mathbb{L}_{a,w}(E) \leq [\![P_i\phi]\!]_w^{\mathcal{M}}.$$

Together this gives $[\![P_i\phi]\!]_w^{\mathcal{M}} = [\![P_i\phi]\!]_{w'}^{\mathcal{M}'}$. □

Now we adapt the proof of the Hennessy-Milner Theorem [8, p. 69] to our epistemic lottery logic. We say that a model $\mathcal{M}$ is image-finite iff for all worlds w in $\mathcal{M}$, and for all agents a, $R_a(w)$ is finite.

Proposition 4.2 *Let $\mathcal{M}, \mathcal{M}'$ be two image-finite models and w and w' be respectively two worlds of $\mathcal{M}$ and $\mathcal{M}'$. $\mathcal{M}, w$ and $\mathcal{M}', w'$ are bisimilar if, and only if $\mathcal{M}, w$ and $\mathcal{M}', w'$ satisfy the same formulas.*

Proof. We show the right to left direction and for that, we prove that the relation $\leftrightsquigarrow$ of modal equivalence on the two models is itself a bisimulation.

Condition (i) is immediate: if w and w' satisfy the same formulas, they satisfy the same atomic propositions. Assume that $w \leftrightsquigarrow w'$ and let E be a subset of $R_a(w)$. We will prove condition (ii) by arriving at a contradiction by assuming that there is no $E' \subseteq R'_a(w')$ such that

- for all $u' \in E'$, there exists $u \in E$ such that $u \leftrightsquigarrow u'$;
- and $\mathbb{L}_a(E) \leq \mathbb{L}_a(E')$.

That is, we assume that for every set $E' \subseteq R'_a(w')$ such that $\mathbb{L}_a(E) \leq \mathbb{L}_a(E')$ there exists $u' \in E'$ such that for all $u \in E$ we have $u \not\leftrightsquigarrow u'$.

Let $\mathcal{S}' = \{E'_1, \ldots, E'_n\}$ be an enumeration of sets $E' \subseteq R'_a(w')$ such that $\mathbb{L}_a(E) \leq \mathbb{L}_a(E')$. For all $i \in \{1, \ldots, n\}$, there exists $u' \in E'_i$ and a (finite) collection of formulas $(\psi_{i,u})_{u \in E}$ such that for all $u \in E$, $\mathcal{M}, u \models \psi_{i,u}$ and $\mathcal{M}', u' \not\models \psi_{i,u}$.

Let $\phi = \bigwedge_{i=1..n} \bigvee_{u \in E} \psi_{i,u}$. On the one hand, we have that for all $u \in E$, $\mathcal{M}, u \models \phi$. Thus, if we pose $\alpha = \mathbb{L}_a(E)$, we have $\mathcal{M}, w \models P_a(\phi) \geq \alpha$.

On the other hand, for all $i \in \{1, \ldots, n\}$, there exists a world $u' \in E'_i$ such that $\mathcal{M}', u' \models \bigwedge_{u \in E} \neg\psi_{i,u}$. That is $\mathcal{M}', u' \models \bigvee_{i=1..n} \bigwedge_{u \in E} \neg\psi_{i,u}$, that is to say $\mathcal{M}', u' \not\models \phi$. In particular, the set $\{u' \in R_a(w') \mid \mathcal{M}', u' \models \phi\}$ is not in $\mathcal{S}'$ and is therefore of probability strictly lower that α. So, $\mathcal{M}, w' \not\models P_a(\phi) \geq \alpha$.

So w and w' do not satisfy the same formulas and there is a contradiction hence condition (ii) holds. Condition (iii) is symmetrical and may be checked in a similar way. □

Proposition 4.3 *There is no finite epistemic single lottery model $\mathcal{M}'$ that is bisimilar to the model $\mathcal{M}$ from Figure 1.*

Proof. Suppose there is a finite epistemic single lottery model $\mathcal{M}' = (W', V', R', L)$ that is bisimilar to the model $\mathcal{M}$ from Figure 1.

Let $R'_a(w_1), \ldots R'_a(w_\ell)$ be an enumeration of a-equivalence classes in $\mathcal{M}'$. Let $R'_b(w_1), \ldots R'_b(w_n)$ be an enumeration of b-equivalence classes in $\mathcal{M}'$.

As $\mathcal{M}$ and $\mathcal{M}'$ are bisimilar, we have:

- $\sum_{u \in R'_a(w_i) | p \in V(u)} L(u) = \sum_{u \in R'_a(w_i) | p \notin V(u)} L(u)$;
- $2 \sum_{u \in R'_b(w_i) | p \in V(u)} L(u) = \sum_{u \in R'_b(w_i) | p \notin V(u)} L(u)$.

Now:

- On the one hand,
$$\begin{aligned} \textstyle\sum_{u \in W' | p \notin V(u)} L(u) &= \textstyle\sum_{i=1}^{\ell} \sum_{u \in R'_a(w_i) | p \notin V(u)} L(u) \\ &= \textstyle\sum_{i=1}^{\ell} \sum_{u \in R'_a(w_i) | p \in V(u)} L(u) \\ &= \textstyle\sum_{u \in W' | p \in V(u)} . \end{aligned}$$
- On the other hand,
$$\begin{aligned} \textstyle\sum_{u \in W' | p \notin V(u)} &= \textstyle\sum_{i=1}^{n} \sum_{u \in R'_b(w_i) | p \notin V(u)} L(u) \\ &= 2 \textstyle\sum_{i=1}^{n} \sum_{u \in R'_b(w_i) | p \in V(u)} L(u) \\ &= 2 \textstyle\sum_{u \in W' | p \in V(u)} . \end{aligned}$$

Thus, $\sum_{u \in W' | p \notin V(u)} L(u) = 0$ which contradicts the definition of model $\mathcal{M}'$. We have proved by contradiction that there is no finite epistemic single lottery model $\mathcal{M}'$ that is bisimilar to $\mathcal{M}$.

□

5 Axiomatization

Figure 4 shows a complete axiomatization of epistemic probabilistic logic. We show in subsection 5.1 that the principles of standard epistemic logic $S5$ (where $K_a\phi$ is replaced by $P_a\phi = 1$) are derivable from the axiomatization. In subsection 5.2, we adapt the proof of completeness of [10] to our simplified logic.

5.1 Principles of $S5$ are derivable

Principles of $S5$ are the following:

- the necessitation rule: if $\vdash_{S5} \phi$ then $\vdash_{S5} K_a\phi$;
- the K-principle: $K_a\phi_1 \wedge K_a(\phi_1 \rightarrow \phi_2) \rightarrow K_a\phi_2$;
- the T-axiom or truth axiom : $K_a\phi \rightarrow \phi$;
- the 4-axiom (positive introspection): $K_a\phi \rightarrow K_aK_a\phi$;
- the 5-axiom (negative introspection): $\neg K_a\phi \rightarrow K_a\neg K_a\phi$.

First, remark that **ProbaT** corresponds to the T-axiom. Now, we prove that all other $S5$-principles are derivable.

Proposition 5.1 *The necessitation rule for certainty is derivable:*

If $\vdash \phi$ *then* $\vdash P_a\phi = 1$.

Proof. From $\vdash \phi$ derive $\vdash \phi \leftrightarrow \top$. From this and **ProbaRule**, $\vdash P_a(\phi) = P_a(\top)$ and with **ProbaTrue**, gives $\vdash P_a(\phi) = 1$. □

Propositional Logic Axioms

All instances of tautologies of propositional logic are axioms **(CPL)**

From $\vdash \phi$ and $\vdash \phi_1 \rightarrow \phi_2$ conclude $\vdash \phi_2$. **(ModusPonens)**

Probability Rule

If $\vdash \phi_1 \leftrightarrow \phi_2$ then $\vdash P_a\phi_1 = P_a\phi_2$. **(ProbaRule)**

Probability Axioms

$\vdash \quad P_a\phi \geq 0$ **(ProbaNonNeg)**

$\vdash \quad P_a\top = 1$ **(ProbaTrue)**

$\vdash \quad P_a(\phi_1 \wedge \phi_2) + P_a(\phi_1 \wedge \neg\phi_2) = P_a\phi_1$ **(ProbaAdditivity)**

$\vdash \quad \psi \rightarrow P_a\psi = 1$ for all a-probability formulas ψ **(ProbaProba)**

Certainty Axioms

$\vdash \quad P_a\phi = 1 \rightarrow \phi$ **(ProbaT)**

Linear (in)equality axioms

All instances of valid formulas about linear inequalities **(Linear)**

Fig. 4. Axiomatization

Proposition 5.2 *The K-principle for certainty is derivable:*

$$\vdash P_a\phi_1 = 1 \wedge P_a(\phi_1 \rightarrow \phi_2) = 1 \rightarrow P_a\phi_2 = 1 \tag{*}$$

Proof. From **ProbaRule**, **ProbaAdditivity** and **Linear**, we have: $\vdash P_a(\top \wedge \phi) + P_a(\top \wedge \neg\phi) = 1$. With **ProbaRule** this gives: $\vdash P_a\phi + P_a(\neg\phi) = 1$. And therefore: $\vdash P_a(\neg\phi) = 1 - P_a\phi$. Using this, we derive $P_a(\phi_1 \rightarrow \phi_2) = P_a(\neg(\phi_1 \wedge \neg\phi_2)) = 1 - P_a(\phi_1 \wedge \neg\phi_2)$. From this: $P_a(\phi_1 \rightarrow \phi_2) = 1 \leftrightarrow P_a(\phi_1 \wedge \neg\phi_2) = 0$. On the other hand, using **ProbaAdditivity**: $P_a(\neg\phi_2) = P_a(\phi_1 \wedge \neg\phi_2) + P_a(\neg\phi_1 \wedge \neg\phi_2)$. Therefore: $P_a\phi_1 = 1 \wedge P_a(\phi_1 \rightarrow \phi_2) = 1 \rightarrow P_a(\neg\phi_2) = P_a(\neg\phi_1 \wedge \neg\phi_2) = 0$ By Propositional logic axioms, this proves (*). □

Formula (*) is a theorem of epistemic probability logic; it can be viewed as a probabilistic version of the K-axiom in epistemic logic.

Note that the following is derivable from **ProbaT** and **ProbaAdditivity**:

$$\vdash P_a\phi = 0 \rightarrow \neg\phi \qquad \textbf{(ProbaTfalse)}$$
$$\vdash \phi \rightarrow P_a\phi > 0 \qquad \textbf{(ProbaGeq0)}$$

Note that the following formulas are theorem because they are instantiation of **ProbaProba**:

$$\vdash P_a\phi = 1 \rightarrow P_a(P_a\phi = 1) = 1 \qquad \textbf{(4)}$$
$$\vdash P_a\phi > 0 \rightarrow P_a(P_a\phi > 0) = 1 \qquad \textbf{(5)}$$

They corresponds respectively to axiom 4 (positive introspection) and axiom 5 (negative introspection) in standard epistemic logic $S5$.

5.2 Soundness and Completeness

Theorem 5.3 *The EPL calculus is sound.*

Proof. All axioms are valid in all EPL models. All rules preserve validity. □

Theorem 5.4 *The EPL calculus is complete.*

The proof is given in the appendix.

6 Model checking

Here is the algorithm for model checking, where again it is assumed that the input model $\mathcal{M}$ is a normalized epistemic lottery model.

```
function mc(M = (W, V, R, 𝕃), φ)
    if T[φ] is defined then
        return T[φ];
    endIf
    match φ do
        case ⊤:
            T[φ] := W;
            return T[φ];
        case p:
            T[φ] := {w ∈ W | p ∈ V(w)};
            return T[φ];
        case ¬φ:
            T[φ] := W \ mc(φ);
            return T[φ];
        case φ1 ∧ φ2:
            T[φ] := mc(M, φ1) ∩ mc(M, φ2);
            return T[φ];
        case t_a ≥ q:
            T[φ] := {w ∈ W | get(M, t_a, w, i) ≥ q};
            return T[φ];
    endMatch
endFunction
```

```
function get(M, t, w, i)
    match t do
        case q:
            return q;
        case q · P_a(φ):
            Σ := mc(M, φ);
            v := Σ_{u∈Σ|wR_a u} 𝕃_a(u)
            return q × v;
        case t1 + t2:
            return get(M, t1, w) + get(M, t2, w);
    endMatch
endFunction
```

Theorem 6.1 *A call to $mc(\mathcal{M}, \phi)$ returns the set $\{w \in W \mid \mathcal{M}, w \models \phi\}$.*

Proof. By induction on ϕ. □

Theorem 6.2 *A call to $mc(\mathcal{M}, \phi)$ requires $O(|\phi|^2 \times |W|^3)$ elementary operations.*

Proof. A call to $mc(\mathcal{M}, \phi)$ calls $mc(\mathcal{M}, \psi)$ where ψ is a subformula of ϕ. As the algorithm $mc(\mathcal{M}, \phi)$ is based on memoization: for a given ψ, the call $mc(\mathcal{M}, \psi)$ is called at most once. So it is sufficient to compute an upper bound of the number of elementary operations performed in one call $mc(\mathcal{M}, \psi)$. Then we multiply this upper bound by an upper bound of the number of calls, that is the number of subformulas of ϕ which is $O(|\phi|)$.

We may represent subsets of W by an array of Booleans $W \to \{0, 1\}$. The case $\phi_1 \times \phi_2$ uses the intersection operation that requires $O(|W|)$ operations. Let us study the case $t_a \geq q$. We first browse all the $O(|W|)$ worlds w and we check whether $\sum_k get(\mathcal{M}, t_k, w) \geq q$ holds or not. We make n calls to get and $n = O(\phi)$. Each call to get requires at most $O(|W|)$ for browsing successors of w by R_a. We count the call to $mc(\mathcal{M}, \phi)$ as $O(1)$ since it is done once with the memoïzation and its effective computation is counted apart.

Conclusion: each call to $mc(\mathcal{M}, \psi)$ costs at most $O(|\phi| \times |W|^2)$. There are at most $|\phi|$ such calls so the global complexity is bounded by $O(|\phi|^2 \times |W|^3)$. □

7 Updates

7.1 Example

Consider the following story. An urn contains a single marble, either white or black. Mr A and Mrs B know this, and they also know that both possibilities are equally likely. Next, Mr A looks in the urn, while Mrs B is watching. Mr A puts another marble in the urn, a white one, and Mrs B sees this. The urn now contains two marbles. Next, Mrs B draws one of the two marbles from the urn. It turns out to be white. What is the probability, according to Mr A, that the other marble is also white? What is the probability, according to Mrs B, that the other marble is also white? (This is a multi-agent variation on a puzzle by Lewis Carroll, see [12].)

Call the first white marble p and the second one q. We start with a situation where there is nothing in the urn, and both agents know this. Update this with the action of tossing a fair coin and making p true in case the coin shows heads. It is assumed that the two agents a and b see that the action happens, but do not see what the outcome is. The action model for this (solid arrows for a, dashed arrows for b):

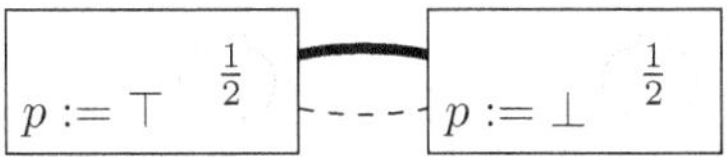

The update of the initial model looks like this:

The action where a takes a look, while b sees this but does not observe what a sees:

The situation after a has taken a look:

The action of putting another white marble (represented by q) in the urn:

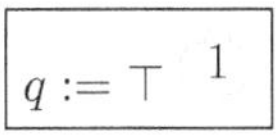

The result of updating with this:

Extracting a white marble from the box is represented as either the act of removing p or the act of removing q, with neither a nor b seeing the difference. The act of removing p (making p false) has as precondition that p is true, the act of removing q has as precondition that q is true.

The result of updating with this is the following model:

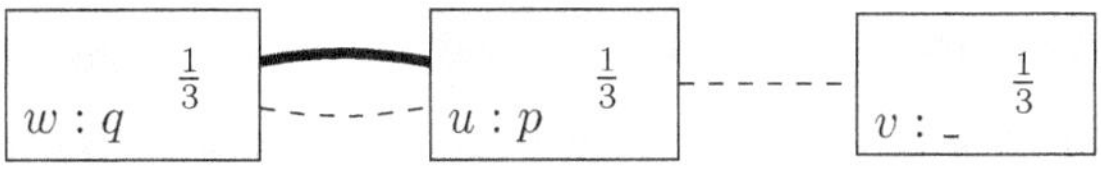

We see that in w it holds that $P_a(p \vee q) = 1$, $P_b(p \vee q) = \frac{2}{3}$. Same values in u, while in v it holds that $P_a(p \vee q) = 0$, $P_b(p \vee q) = \frac{2}{3}$.

7.2 Definitions

Formally, an action model $\mathcal{E}$ for epistemic probability logic is the result of replacing the valuation function in an epistemic lottery model by a pair of functions PRE and

POST that assign to every world (or: event) a precondition and a postcondition, where the precondition ϕ is a formula of the epistemic probability language, and the postcondition is a finite set of bindings $p := \phi$, with p in the set of basic proposition letters of the epistemic probability language, and ϕ a formula of the language.

Update is defined as the product construction of [4], with the extra proviso that $L_a(w, e) = L_a(w) \times L'_a(e)$, where L_a is the a-lottery of the input model and L'_a is the a-lottery of the update model. Let $\mathcal{M} \times \mathcal{E}$ denote the product of $\mathcal{M}$ and $\mathcal{E}$. If the initial epistemic lottery model and the update model are both normalized, then the product defines an epistemic lottery model (not necessarily normalized, for some (w, e) pairs may drop out by the update restriction). Our update definition is a considerable simplification of the update defined in [7].

We consider a probabilistic version of the language extended with the dynamic operator of [4] $[\pi]\psi$ where π defined by $\pi ::= \mathcal{E}, e \mid \pi \cup \pi$. This allows updates with pointed action models and choice between such updates, by means of the union operator $\cup$. Call the new language DEPL. The truth conditions are defined as follows:

- $\mathcal{M}, w \models [\mathcal{E}, e]\psi$ iff $\mathcal{M}, w \models PRE(e)$ implies $\mathcal{M} \otimes \mathcal{E}, (w, e) \models \psi$;
- $\mathcal{M}, w \models [\pi \cup \pi']\psi$ iff $\mathcal{M}, w \models [\pi]\psi$ and $\mathcal{M}, w \models [\pi']\psi$.

7.3 Model checking with updates

To study model checking in DEPL, we adapt the model checking procedure written in Section 6. Now, array T is replaced by $T_{\mathcal{M}}$ where $\mathcal{M}$ is the current epistemic lottery model. The implemented version works as follows:

case $[\mathcal{E}, e]\psi$:

$$T_{\mathcal{M}}(PRE(e)) = mc(\mathcal{M}, PRE(e));$$

$$T_{\mathcal{M}}([\mathcal{E}, e]\psi) := \left\{ w \in W \mid \begin{array}{l} w \notin T_{\mathcal{M}}(PRE(e)) \\ (w, e) \in mc(\mathcal{M} \times \mathcal{E}, \psi) \end{array} \right\};$$

return $T_{\mathcal{M}}([\mathcal{E}, e]\psi)$;

This leads to an algorithm which is running in exponential time and that uses an exponential amount of memory. We may write an algorithm that only use a polynomial amount of memory in the size of the initial model and the size of the formula, that is inspired by the algorithm provided in [2]: we browse the product models *on the fly*. Thus our model checking in the dynamic case is in PSPACE. Nevertheless, the PSPACE-hardness bound for DEL without probability, with $\cup$ operator (and where preconditions are all $\top$) shown in [2] does not provide a lower bound because we can not reduce DEL without probability on models without constraints to the model checking problem in DEPL. Nevertheless the idea of the proof of [2] can be adapted and it provides the following lemma.

Lemma 7.1 *The model checking problem when the initial models and action models are $S5$-models, when we have the $\cup$-operator in the language and when there are at least two agents is PSPACE-hard. The result holds even if all preconditions of the action models are propositional formulas*

Proof. Without loss of generality, we only consider in this proof quantified Boolean formulas of the form $\forall p_1 \exists p_2 \forall p_3 \ldots \forall p_{2k-1} \exists p_{2k} \psi(p_1, \ldots p_{2k})$, where $\psi(p_1, \ldots, p_{2k})$ is a Boolean formula over the atomic propositions $p_1, \ldots, p_{2k}$.

The *quantified Boolean formula satisfiability problem* takes as an input a natural number k and a quantified Boolean formula $\phi \triangleq \forall p_1 \exists p_2 \forall p_3 \ldots \forall p_{2k-1} \exists p_{2k} \psi(p_1, \ldots, p_{2k})$. It returns yes iff ϕ is true in quantified Boolean logic.

Let ϕ be such a quantified Boolean formula. We define a pointed epistemic model $\mathcal{M}, w^0$, $2k$ pointed event models $\mathcal{E}_1, e_1, \ldots, \mathcal{E}_{2k}, e_{2k}$, a pointed event model $\mathcal{E}_{\circlearrowleft}, e_{\circlearrowleft}$ and an epistemic formula ψ' that are computable in polynomial time in the size of ϕ such that:

$$\begin{gathered}\phi \text{ is satisfiable in quantified Boolean logic} \\ \text{iff} \\ \mathcal{M}, w^0 \models [\mathcal{E}_1, e_1 \cup \mathcal{E}_{\circlearrowleft}, e_{\circlearrowleft}]\langle \mathcal{E}_2, e_2 \cup \mathcal{E}_{\circlearrowleft}, e_{\circlearrowleft}\rangle \ldots \\ [\mathcal{E}_{2k-1}, e_{2k-1} \cup \mathcal{E}_{\circlearrowleft}, e_{\circlearrowleft}]\,\langle \mathcal{E}_{2k}, e_{2k} \cup \mathcal{E}_{\circlearrowleft}, e_{\circlearrowleft}\rangle \psi'.\end{gathered}$$

where

- $\mathcal{M}$ is depicted below:

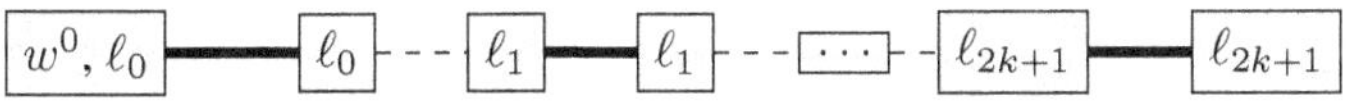

  where $\ell_0, \ell_1, \ldots, \ell_{2k+1}$ are distinct propositional letters.
- For all $i \in \{1, \ldots, 2k\}$, $\mathcal{E}_i, e_i$ is the action model depicted below:

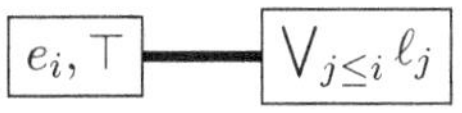

- $\mathcal{E}_{\circlearrowleft}, e_{\circlearrowleft}$ is the action model made up of a single event $e_{\circlearrowleft}$ with precondition $\top$;
- ψ' is the formula ψ where all p_i occurrences are substituted by $(\hat{K}_a \hat{K}_b)^i K_a K_b \ell_i$.

p_i is true is interpreted as the existence of a branch that stop at ℓ_i world. Making the product with $\mathcal{E}_i, e_i$ will add such a branch in the model whereas making the product with $\mathcal{E}_{\circlearrowleft}, e_{\circlearrowleft}$ will leave the epistemic model as it is. The universal and existential choices of values for the p_i's are simulated by the dynamic epistemic operators. □

Proposition 7.2 *Model checking for DEPL with at least two agents and with the* $\cup$ *operator is PSPACE-complete.*

Proof. Membership in PSPACE comes from the remark above. We polynomially reduce the model checking of $S5$-models that is PSPACE-hard (Lemma 7.1) in the model checking of DEPL. To do so, we add to the models 'artificial' lotteries and we recursively replace all subformulas $K_a\phi$ by $P_a(\phi) = 1$. Thus, we obtain the lower-bound. □

8 Connections, Further Work

The assumption that agents have a common prior, widely used in epistemic game theory, is not built into our concept of an epistemic probability model. It follows from Proposition 3.1 that "having a common prior" does *not* coincide with "having the same lottery."

If we want to impose common prior conditions, say for proposition p, then a natural way to express this would be by means of:

$$\bigwedge_{a,b \in A} P_a p = P_b p.$$

Currently, this is not in our language, but if we allow such expressions, then this formula rules out models like the following:

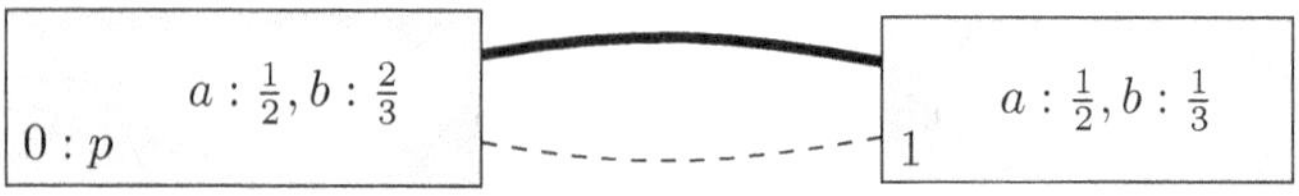

This model describe a situation where a and b 'agree to disagree' on the probability of p. If they are both willing to take bets on the truth of p, they are not rational, for then they make themselves vulnerable to a pair of bets that forms a Dutch book [3]. In finite models with a single lottery Dutch books cannot occur.

Question 1 *Can we strengthen the language to allow for an axiom that forces lotteries to be single?*

If we want to allow lotteries with unknowns in our models, then the language should be extended with expressions B_p with meaning: the (unknown) probability of p, and lotteries should allow for factors B_p. To handle cases where it is *given* that no probability distribution for an event exists, we can allow lotteries with *unknown factors*. A W-lottery with unknowns $X \subseteq \mathbf{P}$ (or: a W-lottery functional over X) is a function from $(0..1)^X$ to W-lotteries, where $(0..1)$ is the open unit interval $\subseteq \mathbb{Q}$. Thus, the type of a W-lottery with unknowns X is:

$$(X \to (0..1)) \to W \to \mathbb{Q}^+$$

Let B be a function that assigns probabilities to the members of X, i.e., $B : X \to (0..1)$. Let l be a W-lottery with values summing up to 1 over W, and let V be a valuation for W. Then $L_{l,V,B}$ is the W-lottery given by:

$$\begin{aligned} L_{l,V,B}(w) = & \, l(w) \\ & \times \prod\{B(p) \mid p \in Q, p \in V(w)\} \\ & \times \prod\{1 - B(p) \mid p \in \mathbf{P}, p \notin V(w)\}. \end{aligned}$$

Then for all $w \in W$, $L_{l,V,B}(w) \in (0..1) \subseteq \mathbb{Q}$, so $L_{l,V,B}$ is a W-lottery. The function $B \mapsto L_{l,V,B}$ is a lottery functional.

Example 8.1 [Von Neumann's Trick] How to obtain fair results from a coin with unknown bias [18]:

> Toss the coin twice. If the results match, forget both results and start over. If the results differ, use the first result.

Here is the explanation. Represent the coin as a lottery functional for the set $\{h\}$. Let B assign a probability to h. That is, $B_h = b$ is the coin bias. Then the probabilities of the four possible outcomes of Von Neumann's procedure are represented by the following lottery:

$$\{hh : b^2, ht : b - b^2, th : b - b^2, tt : (1 - b)^2\}.$$

This shows that the cases ht and th are equally likely, so interpreting the first as h and the second as t gives indeed a model of a fair coin.

Example 8.2 [Model representing a coin with unknown bias]

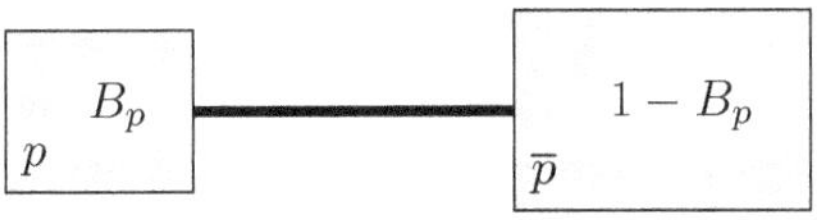

Model checking and model update for the epistemic probability logic of this paper is implemented in [9]. This allows to solve urn problems in a multi-agent setting by means of epistemic model checking. This extension generates lots of further logical questions. Also, it can serve as a solid basis for the design and analysis of probabilistic protocol languages for epistemic probability updating. Hooking up to more sophisticated model checkers like NuSMV (`nusmv.fbk.eu`) is future work. Finally, we would like to further explore the obvious connections with Bayesian learning.

Acknowledgement

Thanks to Alexandru Baltag, Bryan Renne and Joshua Sack for illuminating conversations on the topic of this paper. We also thank the reviewers for their support and comments.

References

[1] Achimescu, A. C., "Games and Logics for Informational Cascades," Master's thesis, ILLC, Amsterdam (2014).

[2] Aucher, G. and F. Schwarzentruber, *On the complexity of dynamic epistemic logic*, in: *Proceedings of TARK 2013*, 2013.

[3] Aumann, R., *Agreeing to disagree*, Annals of Statistics **4(6)** (1976), pp. 1236–1239.

[4] Baltag, A., L. Moss and S. Solecki, *The logic of public announcements, common knowledge, and private suspicions*, in: I. Bilboa, editor, *Proceedings of TARK'98*, 1998, pp. 43–56.

[5] Baltag, A. and S. Smets, *Probabilistic dynamic belief revision*, Synthese **165** (2008), pp. 179–202.

[6] Benthem, J. v., *Conditional probability meets update logic*, Journal of Logic, Language and Information **12** (2003), pp. 409–421.

[7] Benthem, J. v., J. Gerbrandy and B. Kooi, *Dynamic update with probabilities*, Studia Logica **93** (2009), pp. 67–96.

[8] Blackburn, P., M. de Rijke and Y. Venema, "Modal Logic," Cambridge Tracts in Theoretical Computer Science, Cambridge University Press, 2001.

[9] Eijck, J. v., *Learning about probability* (2013), available from `homepages.cwi.nl:~/jve/software/prodemo`.

[10] Fagin, R. and J. Halpern, *Reasoning about knowledge and probability*, Journal of the ACM (1994), pp. 340–367.

[11] Fagin, R., J. Y. Halpern and N. Megiddo, *A logic for reasoning about probabilities*, Information and computation **87** (1990), pp. 78–128.

[12] Gartner, M., "Mathematical Circus," Vintage, 1981.

[13] Gierasimszuk, *Bridging learning theory and dynamic epistemic logic*, Synthese **169** (2009), pp. 371–374.

[14] Kooi, B. P., "Knowledge, Chance, and Change," Ph.D. thesis, Groningen University (2003).
[15] Kooi, B. P., *Probabilistic dynamic epistemic logic*, Journal of Logic, Language and Information **12** (2003), pp. 381–408.
[16] Kyburg, H., "Probability and the Logic of Rational Belief," Wesleyan University Press, Middletown, CT, 1961.
[17] Laplace, M. l. C., "Essai Philosophique sur les Probabilitės," Courcier, Paris, 1814.
[18] von Neumann, J., *Various techniques used in connection with random digits*, Technical Report 12: 36, National Bureau of Standards Applied Math Series (1951).

Appendix: Completeness

The completeness works as follows. We prove that:

Proposition 8.3 *If ϕ is consistent, then ϕ is satisfiable.*

We adapt the proof from [10]. First we construct a canonical epistemic probability model. Contrary to the proof in [10], the epistemic relations are inferred from probabilities.

Let $SF(\phi)$ be the set of all subformulas of ϕ augmented with the negations of subformulas. Let us define the canonical model $\mathcal{M} = (W, V, R, \mathbb{L})$. W is the set of all maximal consistent subsets of $SF(\phi)$. W is not empty because ϕ is supposed to be consistent. Valuations are defined as follows: $V(w) = \mathbf{P} \cap w$.

Let $sat(w) = \{\psi \mid w \vdash \psi\}$, that is, $sat(w)$ is the set of formulas that are provable from w. Relations are defined as follows: $wR_a u$ iff $sat(w)$ and $sat(u)$ contain the same a-probability formulas.

Now it remains to define $\mathbb{L}$. Let us consider an agent a and an equivalence class $R_a(w)$ in the canonical model $\mathcal{M}$. All worlds u of $R_a(w)$ contain the same a-probability formulas. In the sequel, we are transforming all the a-probability formulas in a system of linear inequations that is consistent.

For all $u \in W$, we write ϕ_u the conjunction of all formulas in u. We have:

- $\vdash \phi_u \to \neg\phi_v$ if $u \neq v$ by **CPL**.

Given any formulas ψ of $SF(\phi)$, we have

- $\vdash \psi \leftrightarrow \bigvee_{u \in W \mid \psi \in u} \phi_u$ by **CPL**.

Let ψ be any formula of $SF(\phi)$. By axioms **ProbaRule** and **ProbaAdditivity**, we have:

- $\vdash P_a(\psi) = \sum_{u \in W \mid \psi \in u} P_a(\phi_u)$.

Thus, if we take any a-probability formula ψ, and we replace any term $P_a(\chi)$ by $\sum_{u \in W \mid \psi \in u} P_a(\phi_u)$, we obtain

$$\sum_{u \in W} c_u P_a(\phi_u) \geq b$$

where $c_u, b \in \mathbb{Q}$. Now, when we evaluate the value of $P_a(\phi_u)$ in w, we should obtain non-zero if, and only if, $u \in R_a(w)$. Let us prove it.

- If $u \in R_a(w)$, we have:

(i) $\vdash \phi_u \to P_a(\phi_u) > 0$ by **ProbaGeq0**;
(ii) $P_a(\phi_u) > 0 \in sat(u)$;

(iii) $P_a(\phi_u) > 0 \in sat(w)$ because $u \in R_a(w)$.
There, $P_a(\phi_u) > 0$ should be also true in w.

- If $u \notin R_a(w)$, u and w differ by at least one a-probability formula $\psi \in SF(\phi)$ such that $\psi \in w$ and $\psi \notin u$ without loss of generality. We have:

(i) $\vdash \phi_w \to \psi$ by **CPL**;
(ii) $\vdash \psi \to \neg\phi_u$ by **CPL**;
(iii) $\vdash \psi \to P_a(\psi) = 1$ axiom **ProbaProba**;
(iv) $\vdash \phi_w \to P_a(\psi) = 1$ by **CPL** and **ModusPonens**;
(v) $\vdash \phi_w \to P_a(\neg\phi_u) = 1$ by 2. and 4.
(vi) $\vdash \phi_w \to P_a(\phi_u) = 0$.
Therefore $P_a(\phi_u) = 0$ should be true in w.

Thus, ψ should be equivalent to

$$\sum_{u \in R_a(w)} c_u P_a(\phi_u) \geq b$$

where $c_u, b \in \mathbb{Q}$. This yields a system of linear inequations made up of inequations $\sum_{u \in R_a(w)} c_u x_u \geq b$ when $\psi \in w$ or $\sum_{u \in R_a(w)} c_u x_u < b$ when $\psi \notin w$, plus $\sum_{u \in R_a(w)} x_u = 1$ and $x_u > 0$ for all $u \in R_a(w)$. The set $sat(w)$ is consistent so the above system, which is a rephrasing of some inequations that are in $sat(w)$, is also consistent and therefore satisfiable [11, Theorem 2.2]. Let $(x_u*)_{u \in R_w(a)}$ be a solution. We define $\mathbb{L}_a(u) = x_u*$.

Lemma 8.4 (truth lemma) *For all formulas $\psi \in SF(\phi)$, we have $\mathcal{M}, w \models \psi$ iff $\psi \in w$.*

Proof. By induction on ψ. □

Almost Necessary

Jie Fan [1] Yanjing Wang [2]

Department of Philosophy, Peking University, China

Hans van Ditmarsch [3]

LORIA-CNRS/University of Lorraine

Abstract

A formula is contingent if it is possibly true and possibly false. A formula is non-contingent if it is not contingent, i.e., if it is necessarily true or necessarily false. In an epistemic setting, 'a formula is contingent' means that you are ignorant about its value, whereas 'a formula is non-contingent' means that you know whether it is true. Although non-contingency is definable in terms of necessity as above, necessity is not always definable in terms of non-contingency, as studied in the literature. We propose an 'almost-definability' schema AD for non-contingency logic, the logic with the non-contingency operator as the only modality, making precise when necessity is definable with non-contingency. Based on AD we propose a notion of bisimulation for non-contingency logic, and characterize non-contingency logic as the (non-contingency) bisimulation invariant fragment of modal logic and of first-order logic. A known pain for non-contingency logic is the absence of axioms characterizing frame properties. This makes it harder to find axiomatizations of non-contingency logic over given frame classes. In particular, no axiomatization over symmetric frames is known, despite the rich results about non-contingency logic obtained in the literature since the 1960s. We demonstrate that the 'almost-definability' schema AD can guide our search for proper axioms for certain frame properties, and help us in defining the canonical models. Following this idea, as the main result, we give a complete axiomatization of non-contingency logic over symmetric frames.

Keywords: Non-contingency, modal logic, completeness, definability, bisimulation.

1 Introduction

Contingency is an important concept in philosophical logic; the notion goes back to Aristotle [2]. In [10], Montgomery and Routley define contingency in modal logic. A proposition φ is *non-contingent*, if it is necessary that φ or it

[1] fanjiechina@gmail.com

[2] y.wang@pku.edu.cn

[3] hans.van-ditmarsch@loria.fr

is necessary that $\neg\varphi$. Otherwise, it is *contingent.* For 'φ is non-contingent' we write $\Delta\varphi$ and for 'φ is contingent' we write $\nabla\varphi$.

One theme is how to define necessity from non-contingency. Non-contingency is definable in terms of necessity as above, i.e., as $\Delta\varphi =_{df} \Box\varphi \vee \Box\neg\varphi$. But necessity cannot always be defined with non-contingency. In [10] it is proposed to define necessity as $\Box\varphi =_{df} \Delta\varphi \wedge \varphi$. Intuitively, necessity is non-contingent truth. However, this definition is only available in the systems containing $\Box\varphi \to \varphi$ [13, page 128]. When else is necessity definable in terms of non-contingency? In [3], it is shown that $\Box$ can only be defined in terms of Δ in the Verum system (i.e. the minimal modal logic extended with $\Box\varphi$), or the systems containing $\Box\varphi \to \Diamond\varphi$.

To provide the definability of $\Box$ in the general case, researchers extend the language with the introduction of extra operators. In [11], based on the postulate that some proposition is contingent, the author uses propositional quantifiers when defining necessity: $\Box\varphi =_{df} \forall p(\Delta(p \wedge \varphi) \to \Delta p)$. This says that a proposition is necessary if adding it cannot change the contingency of any contingent proposition. In the subsequent papers such as [12] the author introduces a *propositional constant* τ instead of propositional quantifiers to define necessity based on the axiom $\nabla\tau$: φ is necessary, if it is non-contingent, and it is non-contingently implied by τ, formally, $\Box\varphi =_{df} \Delta\varphi \wedge \Delta(\tau \to \varphi)$. In [17], inspired by the similarity between the definition of canonical relation in the completeness proof for the minimal non-contingency logic and that for the minimal modal logic, the author defines an infinitary operator in terms of Δ, and shows that this new operator behaves like, but differs from $\Box$. Such methods are compared in detail in [8]. In this paper we propose the 'almost-definability' schema $\nabla\psi \to (\Box\varphi \leftrightarrow \Delta\varphi \wedge \Delta(\psi \to \varphi))$: necessity is definable by non-contingency on a world, when some contingent proposition holds on that world. Note that we do not require $\nabla\psi$ to be valid. This schema also guides our proposals for bisimulation for non-contingency logic, and to characterize it within modal logic and within first-order logic using this new notion of bisimulation.

Another theme is axiomatizing the logic with the non-contingency operator as the only modality. A well-known difficulty is the absence of axioms characterizing frame properties in this logic, which makes it highly non-trivial to find axiomatizations of non-contingency logics over given frame classes. An unpublished axiomatization for non-contingency-based **S5** was proposed by Lemmon and Gjertsen in 1959 [7, note 10]. The non-contingency logics over reflexive frames and its extensions are axiomatized in [10]. In [6], Humberstone presents an infinite axiomatization for non-contingency logic over arbitrary frames and over serial frames. A finite axiomatization is given in [9]. This also provides a finite axiomatization for transitive non-contingency logic. In [16], an axiomatization for Euclidean non-contingency logic is proposed. However, to our knowledge, the axiomatization for non-contingency logic over symmetric frames is still open, due to technical difficulties. In this paper we solve this open problem by using the 'almost-definability' schema as a guiding clue.

Non-contingency logic also arose in the area of epistemic logic but with different terminology: 'φ is non-contingent' there means 'the agent knows whether φ', so that 'φ is contingent' means 'the agent is ignorant about φ'. Apparently unaware of the non-contingency logic literature, in [4] the author provides an axiomatization on **S5** frames. In [15] *a logic of ignorance* is presented and this logic is axiomatized over arbitrary frames. In [14] a topological completeness on the class of **S4** models is shown for the logic of ignorance. In [5], knowing whether logic is axiomatized over transitive frames and other frame classes (except symmetric frames), employing other than the traditional methods in the non-contingency literature. A novel result in [5] is the extension of knowing whether logic with public announcements, and its axiomatization.

As the main technical contributions of this work, we characterize the non-contingency logic within modal logic and within first-order logic using a novel notion of bisimulation, and give a complete axiomatization of non-contingency logic over symmetric frames. Both results are inspired by the almost-definability schema.

In Section 2 we define non-contingency logic and the almost-definability schema. In Section 3 we propose a notion of bisimulation on Kripke models that is suitable for non-contingency logic (called Δ-bisimulation), and also a suitable notion of bisimulation contraction. These are non-trivially different from standard bisimulation and contraction. In Section 4 we then characterize non-contingency logic as the Δ-bisimulation invariant fragment of modal logic and of first-order logic. Section 5 axiomatizes non-contingency logic over the class of symmetric frames. We conclude with some discussions in Section 6.

2 Non-contingency logic and almost-definability

Let us first recall the language and semantics of non-contingency logic as a fragment of the following logical language with both the necessity operator and the non-contingency operator:

Definition 2.1 (Logical languages NCL□, NCL and ML) *Given a set* $\boldsymbol{P}$ *of propositional variables, the logical language* $\boldsymbol{NCL}\Box$ *is defined as:*

$$\varphi ::= \top \mid p \mid \neg\varphi \mid (\varphi \wedge \varphi) \mid \Delta\varphi \mid \Box\varphi$$

where $p \in \boldsymbol{P}$*. Without the* $\Box\varphi$ *construct, we have the* language **NCL** of non-contingency logic. *Without the* $\Delta\varphi$ *construct, we have the* language **ML** of modal logic. *If* $\varphi \in \boldsymbol{NCL}$*, then we say* φ *is an* ***NCL****-formula, Similarly we say* φ *is an* ***ML****-formula for* $\varphi \in \boldsymbol{ML}$*.*

In the rest of the paper, we will be mostly focusing on **NCL** which has Δ as the only primitive modality.

The formula $\Box\varphi$ says 'it is necessary that φ' and $\Delta\varphi$ expresses 'it is non-contingent that φ'.[4] As usual, we define $\bot$, $(\varphi \vee \psi)$, $(\varphi \to \psi)$, $(\varphi \leftrightarrow \psi)$,

[4] In [6], Humberstone suggested to rephrase it as 'it is non-contingent *whether* φ' to avoid

$\nabla\varphi$ and $\Diamond\varphi$ as the abbreviations of, respectively, $\neg\top$, $\neg(\neg\varphi\wedge\neg\psi)$, $(\neg\varphi\vee\psi)$, $((\varphi\to\psi)\wedge(\psi\to\varphi))$, $\neg\Delta\varphi$ and $\neg\Box\neg\varphi$. We omit parentheses from formulas unless confusion. Note that $\nabla\varphi$ is *not* the dual but the negation of $\Delta\varphi$, which expresses 'it is contingent that φ'.

Definition 2.2 (Model) *A* model *is a triple $\mathcal{M}=\langle S,R,V\rangle$ where S is a non-empty set of possible worlds, R is a binary relation over S, and V is a valuation function assigning a set of worlds $V(p)\subseteq S$ to each $p\in \boldsymbol{P}$. Given a world $s\in S$, the pair $(\mathcal{M},s)$ is a* pointed model. *We will omit parentheses around pointed models $(\mathcal{M},s)$ whenever convenient. A* frame *is a pair $\mathcal{F}=\langle S,R\rangle$, i.e. a model without a valuation.*

Definition 2.3 (Semantics) *Given a model $\mathcal{M}=\langle S,R,V\rangle$, the semantics of $\boldsymbol{NCL}\Box$ is defined as follows:*

$\mathcal{M},s\vDash\top$	$\Leftrightarrow$	*true*
$\mathcal{M},s\vDash p$	$\Leftrightarrow$	$s\in V(p)$
$\mathcal{M},s\vDash\neg\varphi$	$\Leftrightarrow$	$\mathcal{M},s\nvDash\varphi$
$\mathcal{M},s\vDash\varphi\wedge\psi$	$\Leftrightarrow$	$\mathcal{M},s\vDash\varphi$ *and* $\mathcal{M},s\vDash\psi$
$\mathcal{M},s\vDash\Delta\varphi$	$\Leftrightarrow$	*for any* t_1,t_2 *such that* sRt_1,sRt_2 : $(\mathcal{M},t_1\vDash\varphi\Leftrightarrow\mathcal{M},t_2\vDash\varphi)$
$\mathcal{M},s\vDash\Box\varphi$	$\Leftrightarrow$	*for all* t *such that* $sRt:\mathcal{M},t\vDash\varphi$

If $\mathcal{M},s\vDash\varphi$ we say that φ is true in $(\mathcal{M},s)$, and sometimes write $s\vDash\varphi$ if $\mathcal{M}$ is clear; if for all s in $\mathcal{M}$ we have $\mathcal{M},s\vDash\varphi$ we say that φ is valid on *$\mathcal{M}$ and write $\mathcal{M}\vDash\varphi$; if for all $\mathcal{M}$ based on $\mathcal{F}$ with $\mathcal{M}\vDash\varphi$ we say that φ is* valid on *$\mathcal{F}$ and write $\mathcal{F}\vDash\varphi$; if for all $\mathcal{F}$ with $\mathcal{F}\vDash\varphi$, φ is* valid *and we write $\vDash\varphi$. If there exists an $(\mathcal{M},s)$ such that $\mathcal{M},s\vDash\varphi$, then φ is* satisfiable. *Given any two pointed models $(\mathcal{M},s)$ and $(\mathcal{N},t)$, we say they are Δ-equivalent, notation: $(\mathcal{M},s)\equiv_\Delta(\mathcal{N},t)$, if they satisfy the same $\boldsymbol{NCL}$-formulas; we say they are $\Box$-equivalent, notation: $(\mathcal{M},s)\equiv_\Box(\mathcal{N},t)$, if they satisfy the same $\boldsymbol{ML}$-formulas.*

We are now ready to propose the *almost-definability* of the necessity operator.

Definition 2.4 *Let $\varphi,\psi\in\boldsymbol{NCL}$. Almost-definability is the schema $\nabla\psi\to(\Box\varphi\leftrightarrow\Delta\varphi\wedge\Delta(\psi\to\varphi))$ for which we write AD.*

Proposition 2.5 *Almost-definability AD is a validity of $\boldsymbol{NCL}\Box$.* [5]

Proof Given any pointed model $(\mathcal{M},s)$, suppose that $\mathcal{M},s\vDash\nabla\psi$. We need to show $\mathcal{M},s\vDash\Box\varphi\leftrightarrow\Delta\varphi\wedge\Delta(\psi\to\varphi)$.

First, assume that $\mathcal{M},s\vDash\Box\varphi$. It follows that for all t such that sRt, we have $t\vDash\varphi$ (thus $t\vDash\psi\to\varphi$). Then $\mathcal{M},s\vDash\Delta\varphi$ and $\mathcal{M},s\vDash\Delta(\psi\to\varphi)$, and thus $\mathcal{M},s\vDash\Delta\varphi\wedge\Delta(\psi\to\varphi)$.

---

ambiguity. Here we follow the traditional reading of $\Delta\varphi$ in the literature.

[5] From Proposition 2.5 it follows "$\vDash\nabla\psi$ implies $\vDash\Box\varphi\leftrightarrow(\Delta\varphi\wedge\Delta(\psi\to\varphi))$". This validates Pizzi's definition of $\Box\varphi$ using the new proposition constant τ (let ψ be that τ, see the Introduction). But note that there are no ψ for which $\nabla\psi$ is valid.

Next, assume that $\mathcal{M}, s \vDash \Delta\varphi \wedge \Delta(\psi \to \varphi)$. From the supposition that $\mathcal{M}, s \vDash \nabla\psi$, it follows that there exist t_1 and t_2 such that sRt_1 and sRt_2 and $t_1 \vDash \psi$ and $t_2 \vDash \neg\psi$. By $t_2 \vDash \neg\psi$, it is clear that $t_2 \vDash \psi \to \varphi$. Then using the fact that $s \vDash \Delta(\psi \to \varphi)$, sRt_1 and sRt_2, we obtain that $t_1 \vDash \psi \to \varphi$, thus $t_1 \vDash \varphi$. Since $s \vDash \Delta\varphi$, we have $u \vDash \varphi$ for each u in $\mathcal{M}$ such that sRu. Therefore $s \vDash \Box\varphi$. □

With the almost-definability schema, we are able to find the proper notions of bisimulation and of bisimulation contraction for non-contingency logic, as shown in the next section. Also, almost-definability can guide us to search for proper axioms for certain frame properties, and help us in defining the canonical models. This will be seen more clearly in Section 5.

3 Bisimulation

The standard notion of bisimulation (□-bisimulation) is too refined for non-contingency logic. In this section we propose a suitable weaker notion of Δ-bisimulation.

Definition 3.1 (□-Bisimulation) *Let $\mathcal{M} = \langle S, R, V\rangle$ and $\mathcal{M}' = \langle S', R', V'\rangle$ be two models. A binary relation Z is a* □-bisimulation *between $\mathcal{M}$ and $\mathcal{M}'$, if Z is non-empty and whenever sZs':*

(Invariance) s and s' satisfy the same propositional variables;

(□-Zig) if sRt, then there is a t' in $\mathcal{M}'$ such that $s'R't'$ and tZt';

(□-Zag) if $s'R't'$, then there is a t in $\mathcal{M}$ such that sRt and tZt'.

We say that $(\mathcal{M}, s)$ and $(\mathcal{M}', s')$ are □-bisimilar, *if there is a □-bisimulation linking two states s in $\mathcal{M}$ and s' in $\mathcal{M}'$, and we write $(\mathcal{M}, s) \leftrightarrows_\Box (\mathcal{M}', s')$.*

Example 3.2 The models $(\mathcal{M}, s)$ and $(\mathcal{M}', s')$ below satisfy the same **NCL**-formulas but they are not □-bisimilar.

$$\mathcal{M}: \quad s : p \longrightarrow t : p \qquad\qquad \mathcal{M}': \quad s' : p \longrightarrow t' : \neg p$$

Inspired by the almost-definability schema, we can obtain the notion of Δ-bisimulation by revising the □-Zig and □-Zag conditions in the definition of □-bisimulation. Recall that almost-definability says □ is definable in terms of Δ, given a condition $\nabla\psi$ for some ψ. This condition corresponds to a precondition that the current world can see two non-**NCL**-equivalent successors. Note that for technical convenience, we define Δ-bisimulation within a single model, since the new Zig and Zag conditions require a precondition about 'sibling' worlds, i.e. the structural counterpart of non-**NCL**-equivalency. Based on Δ-bisimulation we can define Δ-bisimilarity between different models.

Definition 3.3 (Δ-Bisimulation) *Let $\mathcal{M} = \langle S, R, V\rangle$ be a model. A binary relation Z over S is a* Δ-bisimulation *on $\mathcal{M}$, if Z is non-empty and whenever sZs':*

(Invariance) s and s' satisfy the same propositional variables;

(Δ-Zig) if there are two successors t_1, t_2 of s such that $(t_1, t_2) \notin Z$ and sRt for some t, then there is a t' such that $s'Rt'$ and tZt';

(Δ-Zag) if there are two successors t_1', t_2' of s' such that $(t_1', t_2') \notin Z$ and $s'Rt'$ for some t', then there is a t such that sRt and tZt'.

We say $(\mathcal{M}, s)$ and $(\mathcal{M}', s')$ are Δ-bisimilar, *notation: $(\mathcal{M}, s) \leftrightarrows_\Delta (\mathcal{M}', s')$, if there is a Δ-bisimulation linking s and s' in the disjoint union of $\mathcal{M}$ and $\mathcal{M}'$.*

We observe (without proof) that the notion of Δ-bisimilarity is an equivalence relation.

The following result indicates the relationship between the notion of Δ-bisimilarity and that of $\Box$-bisimilarity: Δ-bisimilarity is strictly weaker than that of $\Box$-bisimilarity. This corresponds to the fact that non-contingency logic is strictly weaker than modal logic.

Proposition 3.4 *$(\mathcal{M}, s) \leftrightarrows_\Box (\mathcal{N}, t)$ implies $(\mathcal{M}, s) \leftrightarrows_\Delta (\mathcal{N}, t)$ for any pointed models $(\mathcal{M}, s)$ and $(\mathcal{N}, t)$, but the converse is not true.*

Proof (Sketch) Collect all the $\Box$-bisimilar pairs (s, t) with s in $\mathcal{M}$ and t in $\mathcal{N}$ to construct a relation Z. We can check that Z is a Δ-bisimulation on the disjoint union of $\mathcal{M}$ and $\mathcal{N}$. For the converse, Example 3.2 yields two Δ-bisimilar but not $\Box$-bisimilar pointed models $(\mathcal{M}, s)$ and $(\mathcal{M}', s')$. $\Box$

The following result says that **NCL**-formulas are invariant under Δ-bisimilarity.

Proposition 3.5 *Let $\mathcal{M} = \langle S, R, V\rangle$ and $\mathcal{M}' = \langle S', R', V'\rangle$. Then, for every $s \in S$ and $s' \in S'$, if $(\mathcal{M}, s) \leftrightarrows_\Delta (\mathcal{M}', s')$, then $(\mathcal{M}, s) \equiv_\Delta (\mathcal{M}', s')$. In other words, Δ-bisimilarity implies Δ-equivalence.*

Proof Assume that $(\mathcal{M}, s) \leftrightarrows_\Delta (\mathcal{M}', s')$. We need to show that for any $\varphi \in$ **NCL**, we have $\mathcal{M}, s \vDash \varphi$ iff $\mathcal{M}', s' \vDash \varphi$.

By induction on φ. The non-trivial case is $\Delta\varphi$.

Suppose $\mathcal{M}, s \nvDash \Delta\varphi$. Then there exist t_1, t_2 such that sRt_1, sRt_2 and $t_1 \vDash \varphi$ and $t_2 \nvDash \varphi$. As $(\mathcal{M}, s) \leftrightarrows_\Delta (\mathcal{M}', s')$, there exists a Δ-bisimulation Z linking s and s'. By the fact that $t_1 \vDash \varphi$ and $t_2 \nvDash \varphi$ and the induction hypothesis, $(t_1, t_2) \notin Z$. From sRt_1 we obtain by (Δ-Zig) that there exists t_1' such that $s'R't_1'$ and t_1Zt_1', thus $(\mathcal{M}, t_1) \leftrightarrows_\Delta (\mathcal{M}', t_1')$. Similarly, from sRt_2 we have that there exists t_2' such that $s'R't_2'$ and $(\mathcal{M}, t_2) \leftrightarrows_\Delta (\mathcal{M}', t_2')$. From $t_1 \leftrightarrows_\Delta t_1'$ and $t_1 \vDash \varphi$, by the induction hypothesis, $t_1' \vDash \varphi$. Analogously, we can get $t_2' \nvDash \varphi$. Therefore $\mathcal{M}', s' \nvDash \Delta\varphi$. For the other direction use (Δ-Zag).

$\Box$

The notion of Δ-bisimulation has many applications. First, it can be used to show that some properties of models definable in **ML** cannot be defined in **NCL**; second, it can show undefinability for usual frame properties; moreover, it can help to show that **NCL** is less expressive than **ML** on symmetric (and many other) models. For the definitions of expressivity and definability, we refer the reader to, e.g. [1].

Proposition 3.6 *The property "is an endpoint" is undefinable in* ***NCL****, while it can be defined in* ***ML****.*

Proof The property "is an endpoint" is defined by $\Box\bot$.

For the other part, consider pointed models $(\mathcal{M}, s)$ and $(\mathcal{N}, t)$, where s is an endpoint, t has only one successor, and s, t agree on proposition variables. If the property in question were defined by a set of **NCL**-formulas, say Φ. Since s is an endpoint, we have $\mathcal{M}, s \vDash \Phi$. Moreover, $(\mathcal{M}, s) \underline{\leftrightarrow}_\Delta (\mathcal{N}, t)$. By Proposition 3.5 we obtain $\mathcal{N}, t \vDash \Phi$, thus t is an endpoint, contradiction. □

The undefinability results below were presented in the literature ([6,16,5]). With Δ-bisimulation, we can give them simpler proofs.

Proposition 3.7 *The frame properties of seriality, reflexivity, transitivity, symmetry, and Euclidicity are not definable in* ***NCL***.

Proof Consider the following frames:

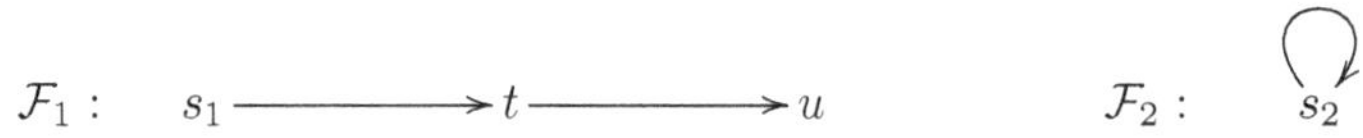

We first show that, for any $\varphi \in$ **NCL**, $\mathcal{F}_1 \models \varphi$ iff $\mathcal{F}_2 \models \varphi$. Fix a φ. If $\mathcal{F}_1 \nvDash \varphi$, then there exists $\mathcal{M}_1 = \langle \mathcal{F}_1, V_1 \rangle$ and s in $\mathcal{M}_1$ such that $\mathcal{M}_1, s \nvDash \varphi$. Let V_2 be a valuation based on $\mathcal{F}_2$ such that $p \in V_2(s_2)$ iff $p \in V_1(s)$ for all $p \in \mathbf{P}$. By definition of Δ-bisimilarity, $(\mathcal{M}_1, s) \underline{\leftrightarrow}_\Delta (\mathcal{M}_2, s_2)$ where $\mathcal{M}_2 = \langle \mathcal{F}_2, V_2 \rangle$. From Proposition 3.5 follows that $\mathcal{M}_2, s_2 \nvDash \varphi$, thus $\mathcal{F}_2 \nvDash \varphi$. The converse is similar.

If seriality were to be defined by a set of **NCL**-formulas, say Γ, then since $\mathcal{F}_2$ is serial, we have $\mathcal{F}_2 \vDash \Gamma$. Then we should also have $\mathcal{F}_1 \vDash \Gamma$, i.e., $\mathcal{F}_1$ should also be serial, contradiction. The proof for other properties are similar. □

Proposition 3.8 was also shown in [5]. With Δ-bisimulation, we get a simpler proof.

Proposition 3.8 ***NCL*** *is less expressive than* ***ML*** *on the class of symmetric models.*

Proof Since $\Delta\varphi =_{df} \Box\varphi \vee \Box\neg\varphi$, **ML** is at least as expressive as **NCL**. Consider the following symmetric models which can be distinguished by $\Box p$:

However, by definition of Δ-bisimilarity, $(\mathcal{M}, s) \underline{\leftrightarrow}_\Delta (\mathcal{N}, t)$. Due to Proposition 3.5, $(\mathcal{M}, s) \equiv_\Delta (\mathcal{N}, t)$, thus no **NCL**-formulas can distinguish the two. □

A model $\mathcal{M}$ is said to be ***NCL**-saturated*, if given any s in $\mathcal{M}$, and any set $\Sigma \subseteq$ **NCL**, if every finite subset of Σ is satisfiable in the set of successors of s, then Σ is satisfiable in the set of successors of s. In what follows, we show that $\equiv_\Delta$ and $\underline{\leftrightarrow}_\Delta$ coincide on **NCL**-saturated models. The proof is similar to its modal counterpart but it makes a crucial use of the **NCL**-formulas.

Proposition 3.9 *For any* **NCL***-saturated pointed models* $(\mathcal{M}, s)$ *and* $(\mathcal{N}, t)$, $(\mathcal{M}, s) \equiv_\Delta (\mathcal{N}, t)$ *iff* $(\mathcal{M}, s) \leftrightarroweq_\Delta (\mathcal{N}, t)$.

Proof Based on Proposition 3.5, we only need to show the direction from left to right. Let $\mathcal{M} = \langle S, R, V \rangle$ and $\mathcal{N} = \langle S', R', V' \rangle$ be two **NCL**-saturated models. Suppose $(\mathcal{M}, s) \equiv_\Delta (\mathcal{N}, t)$, we need to show $\equiv_\Delta$ is a Δ-bisimulation on the disjoint union of $\mathcal{M}$ and $\mathcal{N}$, which entails $(\mathcal{M}, s) \leftrightarroweq_\Delta (\mathcal{N}, t)$. It suffices to show the condition (Δ-Zig) holds, as the proof for (Δ-Zag) is similar.

For this, assume that there exist s_1, s_2 such that sRs_1, sRs_2 and $s_1 \not\equiv_\Delta s_2$, and assume sRs', to show there exists t' such that $tR't'$ and $s' \equiv_\Delta t'$. Let $\Sigma = \{\psi \in \mathbf{NCL} \mid s' \vDash \psi\}$. Clearly, $s' \vDash \Sigma$. Then for any finite $\Gamma \subseteq \Sigma$, $s' \vDash \bigwedge \Gamma$. If for all t' such that $tR't'$, $t' \nvDash \bigwedge \Gamma$, then $t \vDash \Delta \bigwedge \Gamma$, and by supposition we derive $s \vDash \Delta \bigwedge \Gamma$. In the meantime, as $s_1 \not\equiv_\Delta s_2$, there exists $\varphi \in \mathbf{NCL}$ such that $s_1 \vDash \varphi$ and $s_2 \nvDash \varphi$. Then by the fact that $s' \vDash \bigwedge \Gamma$ and $s \vDash \Delta \bigwedge \Gamma$, it is not hard to get $s_1 \vDash \bigwedge \Gamma \to \varphi$ and $s_2 \nvDash \bigwedge \Gamma \to \varphi$, and thus $s \nvDash \Delta(\bigwedge \Gamma \to \varphi)$. On the other hand, since for every t' such that $tR't'$ it holds that $t' \nvDash \bigwedge \Gamma$, we have that $t' \vDash \bigwedge \Gamma \to \varphi$, and thus $t \vDash \Delta(\bigwedge \Gamma \to \varphi)$, contradicting to $s \equiv_\Delta t$ and $s \nvDash \Delta(\bigwedge \Gamma \to \varphi)$. Hence there exists t_Γ such that $tR't_\Gamma$ and $t_\Gamma \vDash \bigwedge \Gamma$. By **NCL**-saturation, there exists t' with $tR't'$ and $t' \vDash \Sigma$. Moreover, $s' \equiv_\Delta t'$: given any $\psi \in \mathbf{NCL}$, if $s' \vDash \psi$, then $\psi \in \Sigma$, and thus $t' \vDash \psi$; if $s' \nvDash \psi$, then $s' \vDash \neg\psi$, and thus $\neg\psi \in \Sigma$, and hence $t' \vDash \neg\psi$, i.e., $t' \nvDash \psi$. □

If we remove the condition of **NCL**-saturation, then $\leftrightarroweq_\Delta$ does not coincide with $\equiv_\Delta$, as illustrated below.

Example 3.10 Consider two models $\mathcal{M} = \langle S, R, V \rangle$ and $\mathcal{M}' = \langle S', R', V' \rangle$, where $S = \mathbb{N} \cup \{s\}$, $R = \{(s, n) \mid n \in \mathbb{N}\}, V(p_n) = \{n\}$ and $S' = \mathbb{N} \cup \{s', \omega\}$, $R' = \{(s', n) \mid n \in \mathbb{N}\} \cup \{(s', \omega)\}$, and $V'(p_n) = \{n\}$. In pictures:

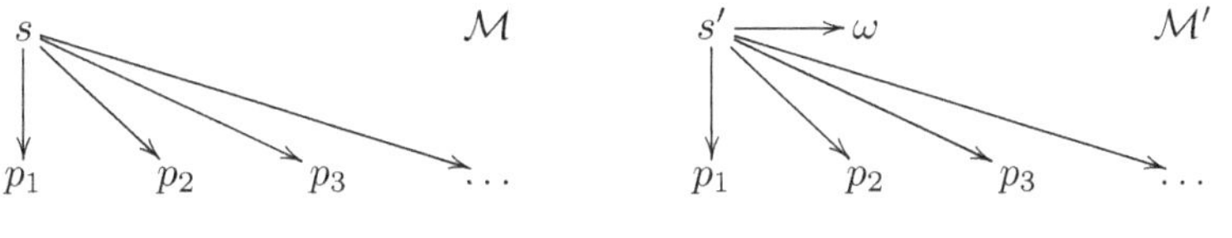

Now $\mathcal{M}$ is not **NCL**-saturated. We can also check that $(\mathcal{M}, s) \equiv_\Delta (\mathcal{M}', s')$ but $(\mathcal{M}, s)$ is not Δ-bisimilar to $(\mathcal{M}', s')$.

Bisimulation contraction The □-bisimulation contraction is defined as a quotient model modulo □-bisimilarity (as the equivalence relation) such that one equivalence class is accessible from another equivalence class if a world in the first is accessible from a world in the second. However, if we just replace □-bisimilarity with Δ-bisimilarity in this definition, the contracted model may not be Δ-bisimilar to the original one, as the following example shows. Therefore we propose a novel notion of Δ-bisimulation contraction, which features an extra condition in the definition of the quotient relation.

Example 3.11 Model $\mathcal{M}'$ is the 'Δ-bisimulation contraction' of $\mathcal{M}$ if we just replace □-bisimilarity with Δ-bisimilarity as above, but $(\mathcal{M}', [s_1])$ is not Δ-

bisimilar to $(\mathcal{M}, s_1)$: for example, Δp is true in $(\mathcal{M}, s_1)$ but false in $(\mathcal{M}', [s_1])$.

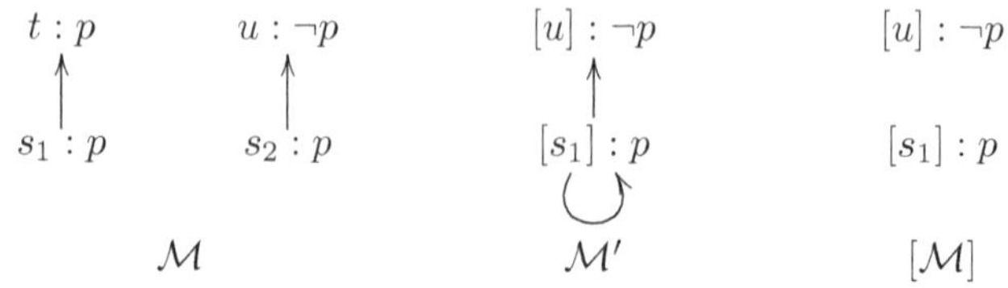

To overcome the above problem, we propose a novel notion of Δ-bisimulation contraction. In particular, we would like to have $[\mathcal{M}]$ as the contracted model of $\mathcal{M}$ in the above example.

Definition 3.12 (Δ-Bisimulation Contraction) *Given a model $\mathcal{M} = \langle S, R, V\rangle$, recall that Δ-bisimilarity ($\leftrightarroweq_\Delta$) within $\mathcal{M}$ is an equivalence relation. Let $[s]$ be the equivalence class of s w.r.t. $\leftrightarroweq_\Delta$ within $\mathcal{M}$. The* Δ-bisimulation contraction *of $\mathcal{M}$ is the quotient structure $[\mathcal{M}] = \langle [S], [R], [V]\rangle$, where*

- *$[S] = \{[s] \mid s \in S\}$;*
- *$[s][R][t]$ iff there exist $s' \in [s]$ and $t' \in [t]$ such that $s'Rt'$ and there exist t_1, t_2 such that $s'Rt_1$ and $s'Rt_2$ and $t_1 \not\leftrightarroweq_\Delta t_2$;*
- *For all propositional variables p, $[V](p) = \{[s] \mid s \in V(p)\}$.*

According to the above definition, we do get $[\mathcal{M}]$ in the earlier example. We show that the contracted model is Δ-bisimilar to the original model.

Proposition 3.13 *Let $\mathcal{M} = \langle S, R, V\rangle$ and $[\mathcal{M}]$ be the Δ-bisimulation contraction of $\mathcal{M}$. Then for all $s \in S$, we have $([\mathcal{M}], [s]) \leftrightarroweq_\Delta (\mathcal{M}, s)$.*

Proof Define

$$Z = \{([s], s) \mid s \in S\} \cup \{(t, t') \mid t \in S,\ t' \in S \text{ and } t \leftrightarroweq_\Delta t'\}.$$

We show Z is a Δ-bisimulation on the disjoint union of $[\mathcal{M}]$ and $\mathcal{M}$, which implies that $([\mathcal{M}], [s]) \leftrightarroweq_\Delta (\mathcal{M}, s)$. First of all, Z is clearly non-empty due to the fact that S is non-empty.

- Invariance: by the definition of $[V]$.
- Δ-Zig: We prove a stronger version ($\Box$-Zig in fact) that for any $u, v \in [S]$, if $u[R]v$ and uZs (i.e. $s \in u$) then there exists $t \in S$ such that vZt (i.e. $t \in v$) and sRt. Now suppose $u[R]v$ and uZs, then according to the definition of $[R]$ there is an $s' \in u$ and a $t' \in v$ such that $s'Rt'$ and there are t_1 and t_2 such that $s'Rt_1$ and $s'Rt_2$ and $t_1 \not\leftrightarroweq_\Delta t_2$. Since $s \in u$ and $s' \in u$, $s \leftrightarroweq_\Delta s'$. Note that $\leftrightarroweq_\Delta$ is also a Δ-bisimulation,[6] thus since $s'Rt'$ there is a t such that sRt and $t \leftrightarroweq_\Delta t'$ thus $t \in v$.
- Δ-Zag: Suppose that sRt and there exist two R-successors t_1, t_2 of s such that $(t_1, t_2) \notin Z$. By the definition of Z, we have $t_1 \not\leftrightarroweq_\Delta t_2$, thus $[s][R][t]$ by

[6] In fact, $\leftrightarroweq_\Delta$ is the largest Δ-bisimulation.

the definition of $[R]$. Now since $t \in [t]$, we have $[t]Zt$. We have thus proved that there exists $[t] \in [S]$ such that $[s][R][t]$ and $[t]Zt$, as desired.

□

4 Characterization results via Δ-bisimulation

The non-contingency logic **NCL** can be seen as a fragment of modal logic **ML**, as $\Delta\varphi =_{df} \Box\varphi \vee \Box\neg\varphi$. In this section we characterize the fragment of **NCL** within **ML** and within first-order logic. Given a model $\mathcal{M}$ and s in $\mathcal{M}$, we use $ue(\mathcal{M})$ and π_s to denote the ultrafilter extension of $\mathcal{M}$ and the principle ultrafilter generated by s, respectively. We refer the reader to [1] for the definitions about these notions, as well as m-saturation. We first state (but not prove) a standard result in modal logic (cf. e.g., [1]).

Proposition 4.1 *For any $\mathcal{M}$ and s in $\mathcal{M}$, $ue(\mathcal{M})$ is an m-saturated model and $(\mathcal{M}, s) \equiv_\Box (ue(\mathcal{M}), \pi_s)$.*

Since **NCL** is a fragment of **ML**, any set of **NCL**-formulas can be viewed as a set of **ML**-formulas, thus we have:

Lemma 4.2 *Let $(\mathcal{M}, s)$ be a pointed model. Then $ue(\mathcal{M})$ is an **NCL**-saturated model and $(\mathcal{M}, s) \equiv_\Delta (ue(\mathcal{M}), \pi_s)$.*

Lemma 4.3 *Let $(\mathcal{M}, s)$ and $(\mathcal{N}, t)$ be pointed models. Then*

$$(\mathcal{M}, s) \equiv_\Delta (\mathcal{N}, t) \textit{ implies } (ue(\mathcal{M}), \pi_s) \leftrightarroweq_\Delta (ue(\mathcal{N}), \pi_t).$$

Proof By Lemma 4.2 and Proposition 3.9. □

We are now ready to prove the two characterization results: non-contingency logic is the Δ-bisimulation-invariant fragment of modal logic and of first-order logic.

Let φ be an **ML**-formula and α be a first-order formula. Call φ (resp. α) *invariant under Δ-bisimulation*, if for any models $(\mathcal{M}, s), (\mathcal{N}, t)$ such that $(\mathcal{M}, s) \leftrightarroweq_\Delta (\mathcal{N}, t)$, we have $\mathcal{M}, s \vDash \varphi$ iff $\mathcal{N}, t \vDash \varphi$ (resp. $\mathcal{M}, s \vDash \alpha$ iff $\mathcal{N}, t \vDash \alpha$).

Theorem 4.4 *An **ML**-formula is equivalent to an **NCL**-formula iff it is invariant under Δ-bisimulation.*

Proof It is clear for the direction 'only if' from Proposition 3.5.

For the converse direction, suppose that an **ML**-formula φ is invariant under Δ-bisimulation. Define

$$MOC(\varphi) = \{t(\psi) \mid \psi \text{ is an } \mathbf{NCL}\text{-formula and } \varphi \vDash t(\psi)\}$$ [7]

If we can show $MOC(\varphi) \vDash \varphi$, then by the compactness theorem for modal logic, for some finite subset T of $MOC(\varphi)$ such that $T \vDash \varphi$, i.e., $\vDash \bigwedge T \to \varphi$. In the meantime, the definition of $MOC(\varphi)$ implies that $\varphi \vDash \bigwedge T$, i.e., $\vDash \varphi \to \bigwedge T$.

---

[7] Here t is a translation function which recursively translates every **NCL** formula into the corresponding equivalent **ML** formula. In particular, for every **NCL** formula φ of the form $\Delta\psi$, $t(\varphi) = \Box t(\psi) \vee \Box t(\neg\psi)$.

Thus $\models \varphi \leftrightarrow \bigwedge T$. As each $\chi \in T$ is the translation of an **NCL**-formula, so is $\bigwedge T$. Then we are done.

So it remains to show $MOC(\varphi) \models \varphi$. Assume that $\mathcal{M}, s \models MOC(\varphi)$, we need to show $\mathcal{M}, s \models \varphi$. Let $X = \{t(\psi) \mid \mathcal{M}, s \models t(\psi), \psi \in \mathbf{NCL}\}$. We claim that $X \cup \{\varphi\}$ is satisfiable: if not, then by compactness theorem for modal logic again, for some finite $X' \subseteq X$ such that $\models \varphi \to \neg \bigwedge X'$, viz. $\varphi \models \neg \bigwedge X'$, then $\neg \bigwedge X' \in MOC(\varphi)$, and thus $\mathcal{M}, s \models \neg \bigwedge X'$, contradicting $\mathcal{M}, s \models X$ and $X' \subseteq X$.

Since $X \cup \{\varphi\}$ is satisfiable, it may as well assume that there is a pointed model $(\mathcal{N}, t)$ with $\mathcal{N}, t \models X \cup \{\varphi\}$. We can show that $(\mathcal{M}, s) \equiv_\Delta (\mathcal{N}, t)$: given any **NCL**-formula ψ, if $\mathcal{M}, s \models \psi$, then $\mathcal{M}, s \models t(\psi)$, and then $t(\psi) \in X$, thus $\mathcal{N}, t \models t(\psi)$, therefore $\mathcal{N}, t \models \psi$; if $\mathcal{M}, s \not\models \psi$, then $\mathcal{M}, s \models \neg\psi$, and then $\mathcal{M}, s \models t(\neg\psi)$, thus $t(\neg\psi) \in X$, hence $\mathcal{N}, t \models t(\neg\psi)$, and therefore $\mathcal{N}, t \models \neg\psi$, i.e., $\mathcal{N}, t \not\models \psi$.

Now construct the ultrafilter extensions of $\mathcal{M}$ and $\mathcal{N}$, denoted by $ue(\mathcal{M})$ and $ue(\mathcal{N})$, respectively. According to the fact that $(\mathcal{M}, s) \equiv_\Delta (\mathcal{N}, t)$ and Lemma 4.3, we have $(ue(\mathcal{M}), \pi_s) \rightleftharpoons_\Delta (ue(\mathcal{N}), \pi_t)$. From $\mathcal{N}, t \models \varphi$ and Proposition 4.1, it follows that $ue(\mathcal{N}), \pi_t \models \varphi$. By supposition and $(ue(\mathcal{M}), \pi_s) \rightleftharpoons_\Delta (ue(\mathcal{N}), \pi_t)$, we get $ue(\mathcal{M}), \pi_s \models \varphi$. By Proposition 4.1 again, one may conclude that $\mathcal{M}, s \models \varphi$, as desired. □

By proposition 3.4, if a formulas is invariant under Δ-bisimulation then it is invariant under $\Box$-bisimulation. The theorem below follows from Theorem 4.4 and Van Benthem Characterization Theorem (cf. e.g., [1]).

Theorem 4.5 *A first-order formula is equivalent to an* ***NCL****-formula iff it is invariant under* Δ*-bisimulation.*

5 Axiomatization of NCL over symmetric frames

In this section, we propose an axiomatization for non-contingency logic over the symmetric frames, a result so far not obtained in the extensive literature on the topic of non-contingency.

As mentioned at the end of Section 2, the schema AD can guide us to search for proper axioms in **NCL** for certain frame properties. For example, AD can help us in finding T-like axiom in **NCL** in a precise way.

$$\begin{aligned} & \nabla\neg\psi \to (\Box\neg\varphi \to \neg\varphi) && (1)\\ \Leftrightarrow\ & \nabla\neg\psi \wedge \Box\neg\varphi \to \neg\varphi && (2)\\ \Leftrightarrow\ & \nabla\neg\psi \wedge \Delta\neg\varphi \wedge \Delta(\neg\psi \to \neg\varphi) \to \neg\varphi && (3)\\ \Leftrightarrow\ & \Delta\varphi \wedge \Delta(\varphi \to \psi) \wedge \varphi \to \Delta\psi && (4) \end{aligned}$$

We write $\nabla\neg\psi \to (\Box\neg\varphi \to \neg\varphi)$ rather than $\Box\neg\varphi \to \neg\varphi$, since $\Box$ is definable in terms of Δ under the condition $\nabla\neg\psi$ for some $\neg\psi$. The above transition from (2) to (3) follows from Proposition 2.5. By using TAUT, ΔEqu and REΔ below and Def∇, we then get the desired axiom (4), which was used in [10] to axiomatize **NCL** over the reflexive frames.

Similar idea guides us to the right direction of finding the B-like axiom (axiom ΔB below) in **NCL**, but further fine-tunings are needed.

Definition 5.1 (Proof system $\mathbb{NCLB}$) *The proof system $\mathbb{NCLB}$ consists of the following axiom schemas and inference rules.*

TAUT	*all instances of tautologies*
Δ*Con*	$\Delta(\chi \to \varphi) \wedge \Delta(\neg\chi \to \varphi) \to \Delta\varphi$
Δ*Dis*	$\Delta\varphi \to \Delta(\varphi \to \psi) \vee \Delta(\neg\varphi \to \chi)$
Δ*Equ*	$\Delta\varphi \leftrightarrow \Delta\neg\varphi$
Δ*B*	$\varphi \to \Delta((\Delta\varphi \wedge \Delta(\varphi \to \psi) \wedge \neg\Delta\psi) \to \chi)$
MP	*From* φ *and* $\varphi \to \psi$ *infer* ψ
REΔ	*From* $\varphi \leftrightarrow \psi$ *infer* $\Delta\varphi \leftrightarrow \Delta\psi$

In the following, we write $\vdash \varphi$ *if there is a proof of* φ *in* $\mathbb{NCLB}$.

Proposition 5.2 *The rule (NECΔ):* $\dfrac{\varphi}{\Delta\varphi}$ *is derivable in* $\mathbb{NCLB}$:

Proof First, we show that $\vdash \Delta\top$:

(i)	$\top \to \Delta((\Delta\top \wedge \Delta(\top \to \psi) \wedge \neg\Delta\psi) \to \chi)$	ΔB
(ii)	$\Delta((\Delta\top \wedge \Delta(\top \to \psi) \wedge \neg\Delta\psi) \to \chi)$	TAUT(i)
(iii)	$\Delta((\Delta\top \wedge \Delta\psi \wedge \neg\Delta\psi) \to \chi)$	RE$\Delta(ii)$
(iv)	$\Delta(\bot \to \chi)$	RE$\Delta(iii)$
(v)	$\Delta\top$	RE$\Delta(iv)$

Next, suppose that $\vdash \varphi$, to show that $\vdash \Delta\varphi$. From the supposition and TAUT it follows that $\vdash \varphi \leftrightarrow \top$. Then using RE$\Delta$ we get $\vdash \Delta\varphi \leftrightarrow \Delta\top$. Since we have shown $\vdash \Delta\top$, we conclude that $\vdash \Delta\varphi$. □

When we drop the axiom ΔB and add the rule NECΔ, we get the logical system $\mathbb{PLKW}$, which is shown to be sound and complete with respect to the class of arbitrary frames in [5, Section 4] (by taking *Kw* there as Δ). But note that NECΔ is indispensable in $\mathbb{PLKW}$, since NECΔ is *not* admissible in $\mathbb{PLKW}$ − NECΔ (equivalently, $\mathbb{NCLB}$ − ΔB). To see this, we can show that $\Delta\top$ is not provable in $\mathbb{PLKW}$ − NECΔ: define an auxiliary semantics $\Vdash$, which is the same as $\vDash$ except that wherein each $\Delta\varphi$ is interpreted always *false*, then we can show that $\mathbb{PLKW}$ − NECΔ is sound with respect to $\Vdash$, but $\nVdash \Delta\top$, thus $\Delta\top$ is not provable in $\mathbb{PLKW}$ − NECΔ, therefore NECΔ cannot be admissible in $\mathbb{PLKW}$ − NECΔ.

Proposition 5.3 $\mathbb{NCLB}$ *is sound with respect to the class of symmetric frames.*

Proof Since $\mathbb{PLKW}$ is sound, we only need to show the validity of Axiom ΔB.

Given any symmetric model $\mathcal{M} = \langle S, R, V\rangle$ and $s \in S$, suppose that $\mathcal{M}, s \vDash \varphi$. Let t be an arbitrary world with sRt. By the symmetry of R, we have tRs. We show that $\mathcal{M}, t \vDash (\Delta\varphi \wedge \Delta(\varphi \to \psi) \wedge \neg\Delta\psi) \to \chi$. If $\mathcal{M}, t \vDash \Delta\varphi \wedge \Delta(\varphi \to \psi) \wedge \neg\Delta\psi$, then there exist t_1, t_2 such that tRt_1 and tRt_2 and $t_1 \vDash \psi$ and

$t_2 \models \neg\psi$. From $t \models \Delta\varphi$, tRs and the supposition, it follows that $t_1 \models \varphi$ and $t_2 \models \varphi$. Thus $t_1 \models \varphi \to \psi$ and $t_2 \models \neg(\varphi \to \psi)$, contrary to the fact that $t \models \Delta(\varphi \to \psi)$ and tRt_1 and tRt_2. Therefore $\mathcal{M}, t \not\models \Delta\varphi \wedge \Delta(\varphi \to \psi) \wedge \neg\Delta\psi$, which implies that $\mathcal{M}, t \models (\Delta\varphi \wedge \Delta(\varphi \to \psi) \wedge \neg\Delta\psi) \to \chi$, as desired.

□

We now proceed with the completeness of $\mathbb{NCLB}$. First, some preparations.

Lemma 5.4

$$\vdash (\Delta(\varphi \to (\chi \to \psi)) \wedge \Delta(\neg\varphi \to \psi) \wedge \neg\Delta\psi \wedge \Delta\varphi) \to \Delta(\chi \to \psi)$$

Proof

(i)	$\Delta(\neg\varphi \to \psi) \wedge \neg\Delta\psi \to \neg\Delta(\varphi \to \psi)$	ΔCon
(ii)	$\Delta\varphi \to \Delta(\varphi \to \psi) \vee \Delta(\neg\varphi \to (\chi \to \psi))$	ΔDis
(iii)	$\Delta(\varphi \to (\chi \to \psi)) \wedge \Delta(\neg\varphi \to (\chi \to \psi)) \to \Delta(\chi \to \psi)$	ΔCon
(iv)	$\Delta(\neg\varphi \to \psi) \wedge \neg\Delta\psi \wedge \Delta\varphi \to \Delta(\neg\varphi \to (\chi \to \psi))$	TAUT$(i)(ii)$
(v)	$(\Delta(\varphi \to (\chi \to \psi)) \wedge \Delta(\neg\varphi \to \psi) \wedge \neg\Delta\psi \wedge \Delta\varphi) \to \Delta(\chi \to \psi)$	TAUT$(iii)(iv)$

□

Proposition 5.5 *For all $k \geq 1$:*

$$\vdash \Delta(\bigwedge_{j=1}^{k} \varphi_j \to \neg\psi) \wedge \bigwedge_{j=1}^{k} \Delta\varphi_j \wedge \bigwedge_{j=1}^{k} \Delta(\psi \to \varphi_j) \to \Delta\psi$$

Proof By induction on k.

- $k = 1$. We need to show that $\vdash \Delta(\varphi_1 \to \neg\psi) \wedge \Delta\varphi_1 \wedge \Delta(\psi \to \varphi_1) \to \Delta\psi$. This is clear from TAUT, REΔ, ΔCon and ΔEqu.
- Inductive step. Assume by induction hypothesis (IH) that the proposition holds for $k = n$. We now need to show that:

$$\vdash \Delta(\bigwedge_{j=1}^{n+1} \varphi_j \to \neg\psi) \wedge \bigwedge_{j=1}^{n+1} \Delta\varphi_j \wedge \bigwedge_{j=1}^{n+1} \Delta(\psi \to \varphi_j) \to \Delta\psi$$

The proof is as follows.

(i)	$\Delta(\bigwedge_{j=1}^{n} \varphi_j \to \neg\psi) \wedge \bigwedge_{j=1}^{n} \Delta\varphi_j \wedge \bigwedge_{j=1}^{n} \Delta(\psi \to \varphi_j) \to \Delta\psi$	IH
(ii)	$(\Delta(\varphi_{n+1} \to (\bigwedge_{j=1}^{n} \varphi_j \to \neg\psi)) \wedge \Delta(\neg\varphi_{n+1} \to \neg\psi) \wedge \neg\Delta\neg\psi \wedge \Delta\varphi_{n+1}) \to \Delta(\bigwedge_{j=1}^{n} \varphi_j \to \neg\psi)$	Lemma 5.4
(iii)	$(\Delta(\bigwedge_{j=1}^{n+1} \varphi_j \to \neg\psi)) \wedge \Delta(\psi \to \varphi_{n+1}) \wedge \neg\Delta\psi \wedge \Delta\varphi_{n+1}) \to \Delta(\bigwedge_{j=1}^{n} \varphi_j \to \neg\psi)$	TAUT, REΔ, ΔEqu, (ii)
(iv)	$(\Delta(\bigwedge_{j=1}^{n+1} \varphi_j \to \neg\psi)) \wedge \bigwedge_{j=1}^{n+1} \Delta\varphi_j \wedge \neg\Delta\psi \wedge \bigwedge_{j=1}^{n+1} \Delta(\psi \to \varphi_j) \to \Delta\psi$	TAUT$(i)(iii)$
(v)	$\Delta(\bigwedge_{j=1}^{n+1} \varphi_j \to \neg\psi) \wedge \bigwedge_{j=1}^{n+1} \Delta\varphi_j \wedge \bigwedge_{j=1}^{n+1} \Delta(\psi \to \varphi_j) \to \Delta\psi$	TAUT(iv)

□

Next, we turn to the canonical model. The model defined below will be used to construct the desired canonical model for $\mathbb{NCLB}$, though it is not suitable for the system in question.

Definition 5.6 (Pseudo-canonical model) *Define* $\mathcal{M}^c = \langle S^c, R^c, V^c \rangle$ *as follows:*

- $S^c = \{s \mid s$ *is a maximal consistent set of* $\mathbb{NCLB}\}$
- *For all* $s, t \in S^c$, sR^ct *iff there exists* χ *such that:*
  - · $\neg\Delta\chi \in s$, *and*
  - · *for all* φ, $\Delta\varphi \wedge \Delta(\chi \to \varphi) \in s$ *implies* $\varphi \in t$.
- $V^c(p) = \{s \in S^c \mid p \in s\}$.

This definition of R^c can be viewed as a simplification of the canonical relation in [5, Definition 20] based on the almost-definability (Proposition 2.5). In the construction of canonical model for standard modal logic, the canonical relation R^c is usually defined by sR^ct holds iff for all φ, $\Box\varphi \in s$ implies $\varphi \in t$. According to the almost-definability, $\Box\varphi \in s$ can be replaced by $\Delta\varphi \wedge \Delta(\chi \to \varphi) \in s$ provided that $\neg\Delta\chi \in s$.

Analogous to the proof of Truth Lemma of $\mathbb{PLKW}$ ([5, Lemma 21]), we have

Lemma 5.7 *For all* $\varphi \in$ ***NCL*** *and* $s \in S^c$, $\mathcal{M}^c, s \vDash \varphi$ *iff* $\varphi \in s$.

Proof By induction on φ. The only non-trivial case is when $\varphi = \Delta\psi$.

'If': Assume that $\Delta\psi \in s$, we need to show $\mathcal{M}^c, s \vDash \Delta\psi$. Suppose not, then there exist $t_1, t_2 \in S^c$ such that sR^ct_1, sR^ct_2 and $t_1 \vDash \psi$ and $t_2 \nvDash \psi$. From $t_1 \vDash \psi$ and $t_2 \nvDash \psi$, and induction hypothesis, we have that $\psi \in t_1$ and $\psi \notin t_2$, respectively. From sR^ct_1 we infer that there is a χ_1 such that $\neg\Delta\chi_1 \in s$ and $(*)$: for all φ, $\Delta\varphi \wedge \Delta(\chi_1 \to \varphi) \in s$ implies $\varphi \in t_1$. Since $\Delta\psi \in s$ and $\psi \in t_1$, $\Delta\neg\psi \in s$ and $\neg\psi \notin t_1$. Now from $(*)$, it follows that $\neg\Delta(\chi_1 \to \neg\psi) \in s$, thus $\neg\Delta(\psi \to \neg\chi_1) \in s$ by $\mathtt{RE}\Delta$. Similarly, from sR^ct_2 we derive that there exists χ_2 such that $\neg\Delta(\chi_2 \to \psi) \in s$, i.e., $\neg\Delta(\neg\psi \to \neg\chi_2) \in s$. By the axiom $\Delta\mathtt{Dis}$, we obtain that $\neg\Delta\psi \in s$, contradiction.

'Only if': Suppose that $\Delta\psi \notin s$. Then $\neg\Delta\psi \in s$ and $\neg\Delta\neg\psi \in s$. We need to construct two points $t_1, t_2 \in S^c$ such that sR^ct_1 and sR^ct_2 and $\psi \in t_1$ and $\neg\psi \in t_2$. First, we have to show

(i) $\{\varphi \mid \Delta\varphi \wedge \Delta(\psi \to \varphi) \in s\} \cup \{\psi\}$ is consistent.

(ii) $\{\varphi \mid \Delta\varphi \wedge \Delta(\neg\psi \to \varphi) \in s\} \cup \{\neg\psi\}$ is consistent.

We prove item (i). Suppose the set is inconsistent. Then there exist $\varphi_1, \cdots, \varphi_n$ such that $\vdash \varphi_1 \wedge \cdots \wedge \varphi_n \to \neg\psi$ and $\Delta\varphi_k \wedge \Delta(\psi \to \varphi_k) \in s$ for all $k \in [1, n]$. From $\mathtt{NEC}\Delta$ follows that $\Delta(\varphi_1 \wedge \cdots \wedge \varphi_n \to \neg\psi) \in s$. Now from Proposition 5.5, we infer that $\Delta\psi \in s$, contradiction.

From item (i), the definition of R^c, and the observation that every consistent set can be extended to a maximal consistent set (Lindenbaum Lemma), we conclude that there is a t_1 such that $sR^c t_1$ and $\psi \in t_1$.

The proof of item (ii) is similar to item (i), and similarly, from item (ii), we conclude that there is a t_2 such that $sR^c t_2$ and $\neg\psi \in t_2$. □

We show that the relation R^c is *almost* symmetric.

Proposition 5.8 *For any $s, t \in S^c$, if $sR^c t$ and there exists a χ such that $\neg\Delta\chi \in t$ then $tR^c s$.*

Proof Assume that $sR^c t$ and $\neg\Delta\chi \in t$, we need to show $tR^c s$. Suppose not, then there exists φ such that $\Delta\varphi \wedge \Delta(\chi \to \varphi) \in t$ but $\varphi \notin s$ (thus $\neg\varphi \in s$). Since $sR^c t$, by definition, there is a ψ such that $\neg\Delta\psi \in s$ and ($\star$): for all θ : $\Delta\theta \wedge \Delta(\psi \to \theta) \in s$ implies $\theta \in t$. Thanks to ΔB, since $\neg\varphi \in s$, we have $\Delta((\Delta\neg\varphi \wedge \Delta(\neg\varphi \to \neg\chi) \wedge \neg\Delta\neg\chi) \to \neg\psi) \in s$ and $\Delta(\Delta\varphi \wedge \Delta(\neg\varphi \to \neg\chi) \wedge \neg\Delta\neg\chi) \in s$. By ΔEqu and REΔ, $\Delta(\Delta\varphi \wedge \Delta(\chi \to \varphi) \wedge \neg\Delta\chi) \wedge \Delta((\Delta\varphi \wedge \Delta(\chi \to \varphi) \wedge \neg\Delta\chi) \to \neg\psi) \in s$, finally we have $\Delta\neg(\Delta\varphi \wedge \Delta(\chi \to \varphi) \wedge \neg\Delta\chi) \wedge \Delta(\psi \to \neg(\Delta\varphi \wedge \Delta(\chi \to \varphi) \wedge \neg\Delta\chi)) \in s$. By ($\star$), we have $\neg(\Delta\varphi \wedge \Delta(\chi \to \varphi) \wedge \neg\Delta\chi) \in t$, contradiction. □

The above proposition tells us that R^c is *almost* symmetric: $sR^c t$ implies $tR^c s$, given that there exists a χ such that $\neg\Delta\chi \in t$. However, there are some states in $\mathcal{M}^c$ which include $\Delta\chi$ for any χ. In the sequel, we call them 'dead ends' due to the fact that those states cannot have outgoing transitions according to the definition of R^c. To turn $\mathcal{M}^c$ into a symmetric model, we handle these special points based on the following two crucial observations.

(i) if $sR^c t$ and t includes no $\neg\Delta\chi$ formula, then t has no outgoing R^c in $\mathcal{M}^c$ by definition. Now if we add a *unique* transition from t back to s to make it symmetric, then it does not change the truth values of formulas on both s and t.

(ii) However, there might be two or more transitions to a dead end t, e.g., $sR^c t$ and $uR^c t$, then adding both back arrows will change the truth values of formulas on t, since now t has two different successors and some $\Delta\varphi$ may not hold any more. We can fix this by replacing those dead ends with some new copies of themselves such that each copy has only one incoming transition. Essentially, we just split those dead ends.

Based on the above observations, we build the canonical model of $\mathbb{NCLB}$ formally. First let $D = \{t \mid t \in S^c, \Delta\chi \in t \text{ for all } \chi, \text{ and there exists an } s \in S^c \text{ such that } sR^c t\}$, where S^c and R^c are defined as in Definition 5.6. Let $\overline{D} = S^c \backslash D$.

Definition 5.9 (Canonical Model of $\mathbb{NCLB}$) *The canonical model $\mathcal{M}^+$ of $\mathbb{NCLB}$ is a tuple $\langle S^+, R^+, f, V^+ \rangle$ where:*

- $S^+ = \overline{D} \cup \{(s,t) \mid t \in D, sR^c t\}$
- *$sR^+ t$ iff one of the following cases holds:*

(i) *$s, t \in \overline{D}$ and $sR^c t$,*
(ii) *$s \in \overline{D}$ and $t = (s, s') \in S^+$,*
(iii) *$t \in \overline{D}$ and $s = (t, t') \in S^+$.*

- *f is a function assigning each state in S^+ to a maximal consistent set in S^c such that $f(s) = s$ for $s \in \overline{D}$, and $f((s,t)) = t$ for $(s,t) \in S^+$.*
- $V^+(p) = \{s \in S^+ \mid p \in f(s)\}$

The function f is introduced to label the maximal consistent sets in $\mathcal{M}^+$, since there can be multiple states in $\mathcal{M}^+$ sharing the same maximal consistent set.

Proposition 5.10

(i) *f is surjective.*

(ii) *if $s \in \overline{D}$ then sR^+t implies $f(s)R^cf(t)$.*

(iii) *if $f(s)R^ct$ then there exists $u \in S^+$ such that $f(u) = t$ and sR^+u.*

Proof For item (i), we need to show that for every $t \in S^c$, there exists a $u \in S^+$ such that $f(u) = t$. Given any $t \in S^c$, there are two cases to consider:

- $t \notin D$, i.e., $t \in \overline{D}$. By the definition of S^+, we have $t \in S^+$; by the definition of f, we have $f(t) = t$.
- $t \in D$. By the definition of D, there exists an $s \in S^c$ such that sR^ct. By the definition of S^+, we have $(s,t) \in S^+$. Then by the definition of f, we have $f((s,t)) = t$.

Either case implies that there exists a $u \in S^+$ such that $f(u) = t$.

For item (ii), suppose $s \in \overline{D}$ (thus $f(s) = s$) and sR^+t. If $t \in \overline{D}$ then sR^ct and $f(t) = t$ by definitions, thus $f(s)R^cf(t)$; if $t \notin \overline{D}$ then $t = (s,s') \in S^+$ and sR^cs', thus $f(t) = f((s,s')) = s'$, and hence $f(s)R^cf(t)$.

For item (iii), suppose $f(s)R^ct$. By the definition of D, we have $f(s) \notin D$. Since $f(s) \in S^c$, it follows that $f(s) \in \overline{D}$, thus $f(s) = s$: otherwise, $f(s) = t'$ and $s = (t,t') \in S^+$, then $t' \in \overline{D}$ and $t' \in D$, which is impossible. If $t \in \overline{D}$ then $f(t) = t$ and sR^+t. If $t \in D$ then $(s,t) \in S^+$ and $f((s,t)) = t$, and $sR^+(s,t)$.□

Notice that the condition $s \in \overline{D}$ in (ii) above is indispensable. For instance, suppose that $s = (t,t') \in S^+$ and $t \in \overline{D}$. By definition, sR^+t. By the definition of f, we have $f(s) = f((t,t')) = t'$ and $f(t) = t$. Since $t' \in D$, we do not have $t'R^ct$, i.e., it is not the case that $f(s)R^cf(t)$.

Lemma 5.11 *$\mathcal{M}^+$ is symmetric.*

Proof Suppose for any $s,t \in S^+$ that sR^+t. We need to show that tR^+s. According to the definition of R^+, we have three cases:

(i) $s,t \in \overline{D}$ and sR^ct. Then $t \notin D$. From the definition of D, we can see that there exists a χ such that $\neg\Delta\chi \in t$. According to Proposition 5.8 we have tR^cs thus tR^+s.

(ii) $s \in \overline{D}$ and $t = (s,s')$. By the third condition of the definition of R^+ we have tR^+s.

(iii) $t \in \overline{D}$ and $s = (t,t')$. By the second condition of the definition of R^+ we have tR^+s.

□

We show that $\mathcal{M}^+$ preserves the truth values of formulas w.r.t. f:

Proposition 5.12 *For any $s \in S^+$ and any $\varphi \in$ **NCL**, we have*

$$\mathcal{M}^+, s \vDash \varphi \iff \mathcal{M}^c, f(s) \vDash \varphi.^8$$

Proof By induction on φ.

- $\varphi = p \in \mathbf{P}$. For any $s \in S^+$, $s \in V^+(p)$ iff $p \in f(s)$ iff $f(s) \in V^c(p)$.
- Boolean cases are immediate.
- $\varphi = \Delta\psi$. We show that $\mathcal{M}^+, s \nvDash \Delta\psi \iff \mathcal{M}^c, f(s) \nvDash \Delta\psi$.
  $\Rightarrow$: Suppose $\mathcal{M}^+, s \nvDash \Delta\psi$ then there are two points t and t' such that sR^+t, sR^+t', $\mathcal{M}^+, t \vDash \psi$ and $\mathcal{M}^+, t' \nvDash \psi$. First note that $s \in \overline{D}$, for otherwise s cannot have two different successors according to the definition of R^+. Now due to item (ii) of Proposition 5.10, $f(s)R^cf(t)$ and $f(s)R^cf(t')$. By IH, $\mathcal{M}^c, f(t) \vDash \psi$ and $\mathcal{M}^c, f(t') \nvDash \psi$. Thus $\mathcal{M}^c, f(s) \nvDash \Delta\psi$.
  $\Leftarrow$: Suppose $\mathcal{M}^c, f(s) \nvDash \Delta\psi$ then there are two points t and t' such that $f(s)R^ct$, $f(s)R^ct'$, $\mathcal{M}^c, t \vDash \psi$ and $\mathcal{M}^c, t' \nvDash \psi$. Now according to item (iii) of Proposition 5.10, there are $u, u' \in S^+$ such that $t = f(u)$, $t' = f(u')$, sR^+u and sR^+u'. By IH, we have $\mathcal{M}^+, u \vDash \psi$ and $\mathcal{M}^+, u' \nvDash \psi$, therefore $\mathcal{M}^+, s \nvDash \Delta\psi$.

□

From Lemma 5.7 and Proposition 5.12 we have:

Lemma 5.13 *For any $\varphi \in$ **NCL** and any $s \in S^+$, $\mathcal{M}^+, s \vDash \varphi$ iff $\varphi \in f(s)$.*

Now due to item (i) of Proposition 5.10, every $s \in S^c$ is an image of some u in $\mathcal{M}^+$ under f, thus each maximal consistent set is satisfiable in $\mathcal{M}^+$, which gives us the completeness theorem based on Lemma 5.11.

Theorem 5.14 (Soundness and Completeness of $\mathbb{NCLB}$) $\mathbb{NCLB}$ *is sound and strongly complete with respect to the class of symmetric frames.*

6 Conclusions and future work

We showed that necessity is almost definable in terms of non-contingency, which is demonstrated by the valid principle $\nabla\psi \to (\Box\varphi \leftrightarrow \Delta\varphi \wedge \Delta(\psi \to \varphi))$ (AD). We proposed notions of Δ-bisimulation and of Δ-bisimulation contraction. We also characterized non-contingency logic as the Δ-bisimulation invariant fragment of modal logic and of first-order logic. Inspired again by almost-definability, we axiomatized non-contingency logic over symmetric frames. This completes the spectrum of complete systems in the literature for non-contingency logic over the usual frame classes.

[8] We can also prove this proposition by showing that the function f is the graph of a Δ-bisimulation.

As mentioned in the introduction, in an epistemic setting, $\Delta\varphi$ is read as "the agent knows whether φ is true". There, it is natural to consider multi-agent scenarios which require multiple Δ-operators. For future work, we conjecture that the multi-agent version of the proof system $\mathbb{NCLB}$ also axiomatizes multi-agent non-contingency logic, where instead of unlabelled Δ-operators we employ labeled Δ_a-operators.

Other future work involves the dynamics of non-contingency logic. In [5] non-contingency logic with public announcements is axiomatized. This can be straightforwardly generalized to action models with non-contingency operators. A reduction axiom for knowing whether (non-contingency) consequences after update is $[\mathbf{M},\mathbf{s}]\Delta\psi \leftrightarrow (\mathbf{pre}(\mathbf{s}) \rightarrow \bigwedge_{\mathbf{sRt}}(\Delta[\mathbf{M},\mathbf{t}]\psi \vee \Delta[\mathbf{M},\mathbf{t}]\neg\psi))$. Such dynamics can also be added to the axiomatizations for non-contingency logic over various other frames, such as the underlying one on symmetric frames.

Acknowledgements

We acknowledge support from European Research Council grant EPS 313360. Jie Fan gratefully acknowledges the support from China Scholarship Council. Yanjing Wang is supported by SSFC grant 11CZX054 and SSFC major project 12&ZD119. Hans van Ditmarsch is also affiliated to IMSc Chennai, as a research associate. We are also grateful to the anonymous referees of AiML2014 for their insightful comments.

References

[1] Blackburn, P., M. de Rijke and Y. Venema, "Modal Logic," Cambridge University Press, 2002.

[2] Brogan, A., *Aristotle's logic of statements about contingency*, Mind **76** (1967), pp. 49–61.

[3] Cresswell, M., *Necessity and contingency*, Studia Logica **47** (1988), pp. 145–149.

[4] Demri, S., *A completeness proof for a logic with an alternative necessity operator*, Studia Logica **58** (1997), pp. 99–112.

[5] Fan, J., Y. Wang and H. van Ditmarsch, *Knowing whether*, CoRR **abs/1312.0144** (2013), under submission.

[6] Humberstone, L., *The logic of non-contingency*, Notre Dame Journal of Formal Logic **36** (1995), pp. 214–229.

[7] Humberstone, L., *The modal logic of agreement and noncontingency*, Notre Dame Journal of Formal Logic **43** (2002), pp. 95–127.

[8] Humberstone, L., *Zolin and Pizzi: Defining necessity from noncontingency*, Erkenntnis **78** (2013), pp. 1275–1302.

[9] Kuhn, S., *Minimal non-contingency logic*, Notre Dame Journal of Formal Logic **36** (1995), pp. 230–234.

[10] Montgomery, H. and R. Routley, *Contingency and non-contingency bases for normal modal logics*, Logique et Analyse **9** (1966), pp. 318–328.

[11] Pizzi, C., *Contingency logics and propositional quantifications*, Manuscrito **22** (1999), pp. 283–303.

[12] Pizzi, C., *Necessity and relative contingency*, Studia Logica **85** (2007), pp. 395–410.

[13] Segerberg, K., "Classical Propositional Operators," Oxford, Clarendon Press, 1982.

[14] Steinsvold, C., *A note on logics of ignorance and borders*, Notre Dame Journal of Formal Logic **49(4)** (2008), pp. 385–392.

[15] van der Hoek, W. and A. Lomuscio, *A logic for ignorance*, Electronic Notes in Theoretical Computer Science **85(2)** (2004), pp. 117–133.

[16] Zolin, E., *Completeness and definability in the logic of noncontingency*, Notre Dame Journal of Formal Logic **40** (1999), pp. 533–547.

[17] Zolin, E., *Infinitary expressibility of necessity in terms of contingency*, Proceedings of the sixth ESSLLI student session (2001), pp. 325–334.

A Composable Language for Action Models

Tim French James Hales [1] Edwin Tay

Computer Science and Software Engineering
The University of Western Australia
Perth, Australia

Abstract

Action models are semantic structures similar to Kripke models that represent a change in knowledge in an epistemic setting. Whereas the language of action model logic [8,7] embeds the semantic structure of an action model directly within the language, this paper introduces a language that represents action models using syntactic operators inspired by relational actions [11,12,13]. This language admits an intuitive description of the action models it represents, and we show in several settings that it is sufficient to represent any action model up to a given modal depth and to represent the results of action model synthesis [18], and give a sound and complete axiomatisation in some of these settings.

Keywords: Modal logic, Epistemic logic, Doxastic logic, Temporal epistemic logic, Multi-agent system, Action model logic.

1 Introduction

Dynamic epistemic logic describes the way knowledge can change in multi-agent systems subject to informative actions taking place. For example, if Tim were to announce "I like cats", then everyone in the room would know the proposition *Tim likes cats* is true, and furthermore, everybody would know that this fact is common knowledge among the people in the room. This simple informative action is what is referred to as a public announcement [21], and such actions of these have been extensively studied in epistemic logics. More complex actions can include private announcements (where some agents are oblivious to the informative action occurring), or a group announcement (where members of a group simultaneously make a truthful announcement to every other member of the group [1]). These complex actions may be modelled and reasoned about using action models [8] which are effectively a semantic model of the change caused by an informative action. Consequently they are very useful for reasoning about the consequences of an informative action, but less well suited to reasoning about the action itself.

[1] Acknowledges the support of the Prescott Postgraduate Scholarship.

We present a language for describing epistemic actions syntactically. Complex actions may be built as an expression upon simpler primitive actions. This approach is a generalisation of the relational actions introduced by van Ditmarsch [12]. We show in several settings that this language is sufficient to represent any informative action represented by an action model (up to a given model depth), we present a synthesis result, and give a a sound and complete axiomatisation for some of these settings. The synthesis result is an important application of this work: given a desired state of knowledge among a group of agents, we are able to compute a complex informative action that will achieve that particular knowledge state (given it is consistent with the current knowledge of agents). We provided these results in a variety of modal logics suited to epistemic reasoning: $\mathcal{K}$, $\mathcal{K}45$ and $\mathcal{S}5$.

Example 1.1 James, Ed and Tim submit a research grant proposal, and eagerly await the outcome. Is there a series of actions that will result in:

(i) Ed knowing the grant application was successful;

(ii) James not knowing whether the grant application was successful, but knowing that either Ed or Tim does know;

(iii) Tim does not know whether the grant application was successful, but knows that if the grant application was unsuccessful, then James knows that it was unsuccessful.

Such an epistemic state may be achieved by a series of messages: Ed is sent a message congratulating him on a successful application, James is sent a message informing him that at least one applicant on each grant has been informed of the outcome, and Tim is sent a message informing him that the first investigator of all unsuccessful grants has been notified.

In the following sections we present a syntactic approach for describing informative actions (Sections 3 and 4), provide a sound and complete axiomatisation of the language for $\mathcal{K}$ and $\mathcal{S}5$ (Section 5) provide a correspondence result between this language and action models (Section 6), and give a computational method for synthesising actions to achieve an epistemic goal (Section 7).

2 Technical Preliminaries

We recall definitions from modal logic, the action model logic of Baltag, Moss and Solecki [8,7] the refinement modal logic of van Ditmarsch, French and Pinchinat [14] and the arbitrary action model logic of Hales [18]. We direct the reader to the extended version of this paper [16] for lemmas and propositions related to these definitions.

Let P be a non-empty, countable set of propositional atoms, and let A be a non-empty, finite set of agents.

Definition 2.1 [Kripke model] A *Kripke model* $M = (S, R, V)$ consists of a *domain* S, which is a non-empty set of states (or possible worlds), an *accessibility* function $R : A \to \mathcal{P}(S \times S)$, which is a function from agents to accessibility relations on S, and a *valuation* function $V : P \to \mathcal{P}(S)$, which is a function from states to sets of propositional atoms.

The *class of all Kripke models* is called $\mathcal{K}$. A *multi-pointed Kripke model* $M_T = (M, T)$ consists of a Kripke model M along with a designated set of states $T \subseteq S$.

We write R_a to denote $R(a)$. Given two states $s, t \in S$, we write sR_at to denote that $(s,t) \in R_a$. We write TR_a to denote the set of states $\{s \in S \mid t \in T, tR_as\}$ and write R_aT to denote the set of states $\{s \in S \mid t \in T, sR_at\}$. We write M_s as an abbreviation for $M_{\{s\}}$, and write tR_a and R_at as abbreviations for $\{t\}R_a$ and $R_a\{t\}$ respectively. As we will often be required to discuss several models at once, we will use the convention that $M_T = ((S, R, V), T)$, $M'_{T'} = ((S', R', V'), T')$, $M^\gamma_{T^\gamma} = ((S^\gamma, R^\gamma, V^\gamma), T^\gamma)$, etc.

Definition 2.2 [Action model] Let $\mathcal{L}$ be a logical language. An *action model* $\mathsf{M} = (\mathsf{S}, \mathsf{R}, \mathsf{pre})$ with preconditions defined on $\mathcal{L}$ consists of a *domain* S, which is a non-empty, finite set of action points, an *accessibility* function $\mathsf{R} : A \to \mathcal{P}(\mathsf{S} \times \mathsf{S})$, which is a function from agents to accessibility relations on S, and a *precondition* function $\mathsf{pre} : \mathsf{S} \to \mathcal{L}$, which is a function from action points to formulae from $\mathcal{L}$.

The *class of all action models* is called $\mathcal{AM}$. A *multi-pointed action model* $\mathsf{M_T} = (\mathsf{M}, \mathsf{T})$ consists of an action model M along with a designated set of action points $\mathsf{T} \subseteq \mathsf{S}$.

We use the same abbreviations and conventions for action models as are used for Kripke models. We use the convention of using sans-serif fonts for action models, as in $\mathsf{M_T}$ and italic fonts for Kripke models, as in M_T.

In addition to the class $\mathcal{K}$ of all Kripke models, and the class $\mathcal{AM}$ of all action models we will be referring to several other classes of Kripke models and action models.

Definition 2.3 [Classes of Kripke models and action models] The class of all Kripke models / action models with transitive and Euclidean accessibility relations is called $\mathcal{K}45$ / $\mathcal{AM}_{\mathcal{K}45}$.

The class of all Kripke models / action models with serial, transitive and Euclidean accessibility relations is called $\mathcal{KD}45$ / $\mathcal{AM}_{\mathcal{KD}45}$.

The class of all Kripke models / action models with reflexive, transitive and Euclidean accessibility relations is called $\mathcal{S}5$ / $\mathcal{AM}_{\mathcal{S}5}$.

Definition 2.4 [Language of arbitrary action model logic] The language $\mathcal{L}_{\otimes\forall}$ of arbitrary action model logic is inductively defined as:

$$\varphi ::= p \mid \neg\varphi \mid (\varphi \land \varphi) \mid \Box_a\varphi \mid [\mathsf{M_T}]\varphi \mid \forall\varphi$$

where $p \in P$, $a \in A$, and $\mathsf{M_T} \in \mathcal{AM}$ is a multi-pointed action model with preconditions defined on the language $\mathcal{L}_{\otimes\forall}$.

We use all of the standard abbreviations for propositional logic, in addition to the abbreviations $\Diamond_a\varphi ::= \neg\Box_a\neg\varphi$, $\langle\mathsf{M_T}\rangle\varphi ::= \neg[\mathsf{M_T}]\neg\varphi$, and $\exists\varphi ::= \neg\forall\neg\varphi$.

We also use the cover operator of Janin and Walukiewicz [20], following the definitions given by Bílková, Palmigiano and Venema [9]. The cover operator, $\nabla_a \Gamma$ is an abbreviation defined by $\nabla_a \Gamma ::= \Box_a \bigvee_{\gamma \in \Gamma} \gamma \wedge \bigwedge_{\gamma \in \Gamma} \Diamond_a \gamma$, where $\Gamma \subseteq \mathcal{L}_{\otimes\forall}$ is a finite set of formulae. We note that the modal operators $\Box_a$, $\Diamond_a$ and ∇_a are interdefineable as $\Box_a \varphi \leftrightarrow \nabla_a\{\varphi\} \vee \nabla_a \emptyset$ and $\Diamond_a \varphi \leftrightarrow \nabla_a\{\varphi, \top\}$. This is the basis for the axiomatisations of refinement modal logic and arbitrary action model logic, and plays an important part in our correspondence and synthesis results. This was previously used as the basis of several axiomatisations of refinement modal logics [14,19,10,18].

We refer to the language $\mathcal{L}_\otimes$ of action model logic, which is $\mathcal{L}_{\otimes\forall}$ without the $\forall$ operator, the language $\mathcal{L}_\forall$ of refinement modal logic, which is $\mathcal{L}_{\otimes\forall}$ without the $[\mathsf{M}_\mathsf{T}]$ operator, the language $\mathcal{L}$ of modal logic, which is $\mathcal{L}_\otimes$ without the $[\mathsf{M}_\mathsf{T}]$ operator, and the language $\mathcal{L}_0$ of propositional logic, which is $\mathcal{L}$ without the $\Box_a$ operator.

Definition 2.5 [Semantics of modal logic] Let $\mathcal{C}$ be a class of Kripke models and let $M = (S, R, V) \in \mathcal{C}$ be a Kripke model. The interpretation of $\varphi \in \mathcal{L}$ in the logic C is defined inductively as:

$$\begin{aligned}
M_s \models p &\text{ iff } s \subseteq V(p)\\
M_s \models \neg\varphi &\text{ iff } M_s \not\models \varphi\\
M_s \models \varphi \wedge \psi &\text{ iff } M_s \models \varphi \text{ and } M_s \models \psi\\
M_s \models \Box_a \varphi &\text{ iff for every } t \in sR_a : M_t \models \varphi\\
M_T \models \varphi &\text{ iff for every } t \in T : M_t \models \varphi
\end{aligned}$$

Definition 2.6 [Bisimilarity and n-bisimilarity of Kripke models] Let $n \in \mathbb{N}$, and let $M_s = ((S, R, V), s) \in \mathcal{K}$ and $M'_{s'} = ((S', R', V'), s') \in \mathcal{K}$ be Kripke models. We say that M_s is *n-bisimilar* to $M'_{s'}$, and write $M_s \leftrightarroweq_n M'_{s'}$, if and only if for every $a \in A$ the following conditions hold:

atoms For every $p \in P$: $s \in V(p)$ if and only if $s' \in V'(p)$.

forth-n-a If $n > 0$ then for every $t \in sR_a$ there exists $t' \in s'R'_a$ such that $M_t \leftrightarroweq_{(n-1)} M'_{t'}$

back-n-a If $n > 0$ then for every $t' \in s'R'_a$ there exists $t \in sR_a$ such that $M_t \leftrightarroweq_{(n-1)} M'_{t'}$

We say that M_s is *bisimilar* to $M'_{s'}$, and write $M_s \leftrightarroweq M'_{s'}$, if and only if for every $n \in \mathbb{N}$: $M_s \leftrightarroweq_n M'_{s'}$.

Definition 2.7 [Modal depth] Let $\varphi \in \mathcal{L}$. The *modal depth of* φ, written as $d(\varphi)$, is defined recursively as follows:

$$\begin{aligned}
d(p) &= 0 \text{ for } p \in P\\
d(\neg\psi) &= d(\psi)\\
d(\psi \wedge \chi) &= max(d(\psi), d(\chi))\\
d(\Box_a \psi) &= 1 + d(\psi)
\end{aligned}$$

Definition 2.8 [B-bisimilarity of Kripke models] Let $M_s = ((S, R, V), s) \in \mathcal{K}$

and $M'_{s'} = ((S', R', V'), s') \in \mathcal{K}$ be Kripke models. We say that M_s is B-*bisimilar* to $M'_{s'}$, and write $M_s \leftrightarroweq_B M'_{s'}$, if and only if for every $b \in B$ the following conditions hold:

atoms For every $p \in P$: $s \in V(p)$ if and only if $s' \in V'(p)$.

forth-b For every $t \in sR_b$ there exists $t' \in s'R'_b$ such that $M_t \leftrightarroweq M'_{t'}$.

back-b For every $t' \in s'R'_b$ there exists $t \in sR_b$ such that $M_t \leftrightarroweq M'_{t'}$.

Definition 2.9 [B-restricted formulae] Let $B \subseteq A$. A *B-restricted formula* is defined by the following abstract syntax:

$$\varphi ::= p \mid \neg\varphi \mid (\varphi \wedge \varphi) \mid \Box_b \psi$$

where $p \in P$, $b \in B$, $\psi \in \mathcal{L}$.

We recall the semantics of action model logic of Baltag, Moss and Solecki [8,7].

Definition 2.10 [Semantics of action model logic] Let $\mathcal{C}$ be a class of Kripke models, let $M = (S, R, V) \in \mathcal{C}$ be a Kripke model and let $\mathsf{M} \in \mathcal{AM}$ be an action model.

We first define *action model execution.* We denote the result of executing the action model M on the Kripke model M as $M \otimes \mathsf{M}$, and we define the result as $M \otimes \mathsf{M} = M' = (S', R', V')$ where:

$$\begin{aligned} S' &= \{(s, \mathsf{s}) \mid s \in S, \mathsf{s} \in \mathsf{S}, M_s \models \mathsf{pre}(\mathsf{s})\} \\ (s, \mathsf{s})R'_a(t, \mathsf{t}) &\text{ iff } sR_a t \text{ and } \mathsf{s}\mathsf{R}_a\mathsf{t} \\ (s, \mathsf{s}) \in V'(p) &\text{ iff } s \in V(p) \end{aligned}$$

We also define *multi-pointed action model execution* as $M_T \otimes \mathsf{M}_\mathsf{T} = M'_{T'} = ((S', R', V'), T') = ((M \otimes \mathsf{M}), (T \times \mathsf{T}) \cap S')$.

Then the interpretation of $\varphi \in \mathcal{L}_\otimes$ in the logic $C_\otimes$ is the same as its interpretation in the modal logic C given in Definition 2.5, with the additional inductive case:

$$M_s \models [\mathsf{M}_\mathsf{T}]\varphi \text{ iff } M_s \otimes \mathsf{M}_\mathsf{T} \in \mathcal{C} \text{ implies } M_s \otimes \mathsf{M}_\mathsf{T} \models \varphi$$

Definition 2.11 [Sequential execution of action models] Let $\mathsf{M}, \mathsf{M}' \in \mathcal{AM}$. We define the *sequential execution of* M *and* M' as $\mathsf{M} \otimes \mathsf{M}' = \mathsf{M}'' = (\mathsf{S}'', \mathsf{R}'', \mathsf{pre}'')$ where:

$$\begin{aligned} \mathsf{S}'' &= \mathsf{S} \times \mathsf{S}' \\ (\mathsf{s}, \mathsf{s}')R''_a(\mathsf{t}, \mathsf{t}') &\text{ iff } \mathsf{s}\mathsf{R}_a\mathsf{t} \text{ and } \mathsf{s}'\mathsf{R}'_a\mathsf{t}' \\ \mathsf{pre}''((\mathsf{s}, \mathsf{s}')) &= \langle \mathsf{M}_\mathsf{s} \rangle \mathsf{pre}'(\mathsf{s}') \end{aligned}$$

We also define *sequential action of* M_T *and* $\mathsf{M}'_{\mathsf{T}'}$ as $\mathsf{M}_\mathsf{T} \otimes \mathsf{M}'_{\mathsf{T}'} = \mathsf{M}''_{\mathsf{T}''} = ((\mathsf{S}'', \mathsf{R}'', \mathsf{pre}''), \mathsf{T} \times \mathsf{T}')$.

Definition 2.12 [n-bisimilarity of action models] Let $n \in \mathbb{N}$, and let $\mathsf{M}_\mathsf{s} = ((\mathsf{S}, \mathsf{R}, \mathsf{pre}), \mathsf{s}) \in \mathcal{AM}$ and $\mathsf{M}'_{\mathsf{s}'} = ((\mathsf{S}', \mathsf{R}', \mathsf{pre}'), \mathsf{s}') \in \mathcal{AM}$ be action models. We say that M_s is *n-bisimilar* to $\mathsf{M}'_{\mathsf{s}'}$, and write $\mathsf{M}_\mathsf{s} \leftrightarroweq_n \mathsf{M}'_{\mathsf{s}'}$, if and only if for every $a \in A$ the following conditions hold:

atoms $\vdash \mathsf{pre}(\mathsf{s}) \leftrightarrow \mathsf{pre}'(\mathsf{s}')$

forth-n-a If $n > 0$ then for every $\mathsf{t} \in \mathsf{sR}_a$ there exists $\mathsf{t}' \in \mathsf{s}'\mathsf{R}'_a$ such that $\mathsf{M}_\mathsf{t} \underline{\leftrightarrow}_{(n-1)} \mathsf{M}'_{\mathsf{t}'}$

back-n-a If $n > 0$ then for every $\mathsf{t}' \in \mathsf{s}'\mathsf{R}'_a$ there exists $\mathsf{t} \in \mathsf{sR}_a$ such that $\mathsf{M}_\mathsf{t} \underline{\leftrightarrow}_{(n-1)} \mathsf{M}'_{\mathsf{t}'}$

We say that M_s is *bisimilar* to $\mathsf{M}'_{\mathsf{s}'}$, and write $\mathsf{M}_\mathsf{s} \underline{\leftrightarrow} \mathsf{M}'_{\mathsf{s}'}$, if and only if for every $n \in \mathbb{N}$: $\mathsf{M}_\mathsf{s} \underline{\leftrightarrow}_n \mathsf{M}'_{\mathsf{s}'}$,

Definition 2.13 [B-bisimilarity of action models] Let $\mathsf{M}_\mathsf{s} = ((\mathsf{S}, \mathsf{R}, \mathsf{pre}), \mathsf{s}) \in \mathcal{K}$ and $\mathsf{M}'_{\mathsf{s}'} = ((\mathsf{S}', \mathsf{R}', \mathsf{pre}'), \mathsf{s}') \in \mathcal{K}$ be Kripke models. We say that M_s is B-bisimilar to $\mathsf{M}'_{\mathsf{s}'}$ and write $\mathsf{M}_\mathsf{s} \underline{\leftrightarrow}_B \mathsf{M}'_{\mathsf{s}'}$, if and only if for every $b \in B$ the following conditions hold:

atoms For every $p \in P$: $\mathsf{s} \in V(p)$ if and only if $\mathsf{s}' \in V'(p)$.

forth-b For every $\mathsf{t} \in \mathsf{sR}_b$ there exists $\mathsf{t}' \in \mathsf{s}'\mathsf{R}'_b$ such that $\mathsf{M}_\mathsf{t} \underline{\leftrightarrow} \mathsf{M}'_{\mathsf{t}'}$.

back-b For every $\mathsf{t}' \in \mathsf{s}'\mathsf{R}'_b$ there exists $\mathsf{t} \in \mathsf{sR}_b$ such that $\mathsf{M}_\mathsf{t} \underline{\leftrightarrow} \mathsf{M}'_{\mathsf{t}'}$.

Definition 2.14 [Simulation and refinement] Let $M, M' \in \mathcal{K}$ be Kripke models. A non-empty relation $\mathfrak{R} \subseteq S \times S'$ is a *simulation* if and only if it satisfies **atoms**, **forth-a** for every $a \in A$. If $(s, s') \in \mathfrak{R}$ then we call $M'_{s'}$ a *simulation* of M_s and call M_s a *refinement* of $M'_{s'}$. We write $M'_{s'} \underline{\rightarrow} M_s$ or equivalently $M_s \underline{\leftarrow} M'_{s'}$.

Definition 2.15 [Semantics of arbitrary action model logic] Let $\mathcal{C}$ be a class of Kripke models and let $M \in \mathcal{C}$ be a Kripke model. The interpretation of $\varphi \in \mathcal{L}_{\otimes\forall}$ in the logic $C_{\otimes\forall}$ is the same as its interpretation in the action model logic $C_\otimes$ given in Definition 2.10 with the additional inductive case:

$$M_s \models \forall\varphi \text{ iff for every } M'_{s'} \in \mathcal{C} \text{ such that } M'_{s'} \underline{\leftarrow} M_s : M'_{s'} \models \varphi$$

3 Syntax

Definition 3.1 [Language of arbitrary action formula logic] The language $\mathcal{L}_{?\forall}$ of arbitrary action formula logic is inductively defined as:

$$\varphi ::= p \mid \neg\varphi \mid (\varphi \wedge \varphi) \mid \Box_a\varphi \mid [\alpha]\varphi \mid \forall\varphi$$

where $p \in P$, $a \in A$ and $\alpha \in \mathcal{L}_{?\forall}{}^{\mathrm{act}}$, and where the language $\mathcal{L}_{?\forall}{}^{\mathrm{act}}$ of arbitrary action formulae is inductively as:

$$\alpha ::= ?\varphi \mid \alpha \sqcup \alpha \mid \alpha \otimes \alpha \mid L_B(\alpha, \alpha)$$

where $\varphi \in \mathcal{L}_{?\forall}$ and $\emptyset \subset B \subseteq A$.

We use all of the standard abbreviations for arbitrary action model logic, in addition to the abbreviations $L_B\alpha ::= L_B(\alpha, \alpha)$ and $L_a(\alpha, \beta) ::= L_{\{a\}}(\alpha, \beta)$.

We denote non-deterministic choice ($\sqcup$) over a finite set of action formula $\Delta \subseteq \mathcal{L}_{?\forall}{}^{\mathrm{act}}$ by $\bigsqcup\Delta$ and we denote sequential execution ($\otimes$) of a finite, non-empty sequence of action formulae $(\alpha_i)_{i=0}^n \in \mathbb{N}^{\mathcal{L}_{?\forall}{}^{\mathrm{act}}}$ by $\bigotimes(\alpha_i)_{i=0}^n$ and define them in the obvious way.

We refer to the languages $\mathcal{L}_?$ of action formula logic and $\mathcal{L}_?^{\text{act}}$ of action formulae, which are $\mathcal{L}_{?\forall}$ and $\mathcal{L}_{?\forall}^{\text{act}}$ respectively, both without the $\forall$ operator,

As in the action model logic [7], the intended meaning of the operator $[\alpha]\varphi$ is that "φ is true in the result of any successful execution of the action α". In the following section we define the semantics of the action formula logic in terms of action model execution. For each setting of $\mathcal{K}$, $\mathcal{K}45$ and $\mathcal{S}5$ we provide a function $\tau_C : \mathcal{L}_?^{\text{act}} \to \mathcal{AM}$ of translating action formulae from $\mathcal{L}_?^{\text{act}}$ into action models. The result of executing an action $\alpha \in \mathcal{L}_?^{\text{act}}$ is determined by translating α into an action model $\tau_C(\alpha) \in \mathcal{AM}_C$, and then executing the action model in the usual way.

In each setting we have attempted to define the translation from action formulae into action models in such a way that the action formulae carry an intuitive description of the action that is performed by the corresponding action model. We call the ? operator the test operator, and describe the action $?\varphi$ as a test for φ. A test is intended to restrict the states in which an action can successfully execute to states where the condition φ is true initially, but otherwise leaves the state unchanged. We call the $\sqcup$ operator the non-deterministic choice operator, and describe the action $\alpha \sqcup \beta$ as a non-deterministic choice between α and β. We call the $\otimes$ operator the sequential execution operator, and describe the action $\alpha \otimes \beta$ as an execution of α followed by β. Finally we call L_B the learning operator, and describe the action $L_B(\alpha, \beta)$ as the agents in B learning that the actions α or β occurred.

Example 3.2 If p stands for the proposition "the grant application was successful" then the action described in Example 1.1 might be written in the form of an action formula as:

$$\begin{aligned}\alpha = &L_{Ed}(?p)\otimes\\ &L_{James}(L_{Ed}?p \sqcup L_{Ed}?\neg p \sqcup L_{Tim}?p \sqcup L_{Tim}?\neg p)\otimes\\ &L_{Tim}((?\neg p \otimes L_{James}?\neg p) \sqcup ?\top)\end{aligned}$$

4 Semantics

We now define the semantics of arbitrary action formula logic. As mentioned earlier, the semantics are defined by translating action formulae into action models. The translation used varies in each class of $\mathcal{K}$, $\mathcal{K}45$ and $\mathcal{S}5$ that we work in, according to the frame conditions in each class. Therefore our semantics are parameterised by a function $\tau_C : \mathcal{L}_?^{\text{act}} \to \mathcal{AM}$ that will vary according to the class of Kripke models.

Definition 4.1 [Semantics of arbitrary action formula logic] Let $\mathcal{C}$ be a class of Kripke models, let $\tau_C : \mathcal{L}_?^{\text{act}} \to \mathcal{AM}$ be a function from action formulae to multi-pointed action models, and let $M = (S, R, V) \in \mathcal{C}$ be a Kripke model.

Then the interpretation of $\varphi \in \mathcal{L}_{?\forall}$ in the logic $\mathcal{C}_{?\forall}$ is the same as its interpretation in modal logic given in Definition 2.5, with the additional inductive cases:

$$M_s \models [\alpha]\varphi \text{ iff } M_s \otimes \tau_C(\alpha) \in C \text{ implies } M_s \otimes \tau_C(\alpha) \models \varphi$$
$$M_s \models \forall\varphi \text{ iff for every } M'_{s'} \in C \text{ such that } M'_{s'} \preceq M_s : M'_{s'} \models \varphi$$

where action model execution $\otimes$ is as defined in Definition 2.10 and the refinement relation is defined in Definition 2.14.

We note that the semantics of arbitrary action formula logic $C_{?\forall}$ are very similar to the semantics of arbitrary action model logic $C_{\otimes\forall}$ [18]. We generalise the semantics to the classes of $\mathcal{K}$, $\mathcal{K}45$ and $\mathcal{S}5$ by introducing the parameterised class C and restricting successful updates to those that result in C models as in the approach of Balbiani, et al [5]. The difference is that as actions are specified in $\mathcal{L}_{?\forall}$ formulae as action formulae, then the semantics must first translate the action formulae into action models before performing action model execution. As such there is a semantically correct translation from $\mathcal{L}_{?\forall}$ formulae to $\mathcal{L}_{\otimes\forall}$ formulae (by replacing occurrences of α with $\tau_C(\alpha)$), and any validities, axioms or results from arbitrary action model logic also apply in this setting if the language is restricted to action models that are defineable by action formulae. Therefore for the current section and the following sections concering the axiomatisations (Section 5) and correspondence results (Section 6), we will deal only with the action formula logic, rather than the full arbitrary action formula logic, focussing on the differences and correspondences between action formulae and action models, rather than getting distracted by the refinement quantifiers which behave identically between each logic. We return to the full arbitrary action formula logic in Section 7 for the synthesis results.

We give the following general result.

Proposition 4.2 *Let C be a class of Kripke models. For every $\varphi \in \mathcal{L}_{?\forall}$ there exists $\varphi' \in \mathcal{L}_{\otimes\forall}$ such that for every $M_T \in C$: $M_T \models_{C_{?\forall}} \varphi$ if and only if $M_T \models_{C_{\otimes\forall}} \varphi'$.*

In the following subsections we will give definitions for $\tau_{\mathcal{K}}$, $\tau_{\mathcal{K}45}$ and $\tau_{\mathcal{S}5}$. These functions vary according to the class of Kripke models being used. When the class is clear from context, then we will simply write τ instead of τ_C.

We begin by giving a definition of τ for translating actions involving non-deterministic choice and sequential execution. These definitions are common to all of the settings we are working in.

Definition 4.3 [Non-deterministic choice] Let $C \in \{\mathcal{K}, \mathcal{K}45, \mathcal{S}5\}$ and let $\alpha, \beta \in \mathcal{L}_?^{\text{act}}$ where $\tau_C(\alpha) = \mathsf{M}^\alpha_{\mathsf{T}^\alpha} = ((\mathsf{S}^\alpha, \mathsf{R}^\alpha, \mathsf{pre}^\alpha), \mathsf{T}^\alpha)$ and $\tau_C(\beta) = \mathsf{M}^\beta_{\mathsf{T}^\beta} = ((\mathsf{S}^\beta, \mathsf{R}^\beta, \mathsf{pre}^\beta), \mathsf{T}^\beta)$ such that S^α and S^β are disjoint. We define $\tau_C(\alpha \sqcup \beta) = \mathsf{M}_\mathsf{T} = ((\mathsf{S}, \mathsf{R}, \mathsf{pre}), \mathsf{T})$ where:

$$\mathsf{S} = \mathsf{S}^\alpha \cup \mathsf{S}^\beta$$
$$\mathsf{R}_a = \mathsf{R}^\alpha_a \cup \mathsf{R}^\beta_a \text{ for } a \in A$$
$$\mathsf{pre} = \mathsf{pre}^\alpha \cup \mathsf{pre}^\beta$$
$$\mathsf{T} = \mathsf{T}^\alpha \cup \mathsf{T}^\beta$$

Definition 4.4 [Sequential execution] Let $C \in \{\mathcal{K}, \mathcal{K}45, \mathcal{S}5\}$, and let $\alpha, \beta \in$

$\mathcal{L}_?{}^{\mathrm{act}}$ where $\tau_C(\alpha) = \mathsf{M}^{\alpha}_{\mathsf{T}^{\alpha}} = ((\mathsf{S}^{\alpha}, \mathsf{R}^{\alpha}, \mathsf{pre}^{\alpha}), \mathsf{T}^{\alpha})$ and $\tau_C(\beta) = \mathsf{M}^{\beta}_{\mathsf{T}^{\beta}} = ((\mathsf{S}^{\beta}, \mathsf{R}^{\beta}, \mathsf{pre}^{\beta}), \mathsf{T}^{\beta})$. We define $\tau_C(\alpha \otimes \beta) = \mathsf{M}^{\alpha}_{\mathsf{T}^{\alpha}} \otimes \mathsf{M}^{\beta}_{\mathsf{T}^{\beta}}$.

We give some properties of non-deterministic choice and sequential execution of action formulae.

Proposition 4.5 *Let $\alpha, \beta, \gamma \in \mathcal{L}_?{}^{act}$ and $\varphi \in \mathcal{L}_?$. Then the following are valid in $K_?$, $K45_?$ and $S5_?$:*

$$\models [\alpha \sqcup \beta]\varphi \leftrightarrow ([\alpha]\varphi \wedge [\beta]\varphi) \qquad \models [\alpha \otimes \beta]\varphi \leftrightarrow [\alpha][\beta]\varphi$$

These validities follow trivially from the semantics of $C_{?\forall}$ and Definitions 4.3 and 4.4.

In the following subsections we give definitions of τ_C for translating action formulae involving tests and learning in the settings of $\mathcal{K}$, $\mathcal{K}45$ and $\mathcal{S}5$. We note that in each subsection the constructions of action models used to define tests and learning closely resemble the constructions of refinements used to show the soundness of axioms in refinement modal logic [10,19].

4.1 $\mathcal{K}$

Definition 4.6 [Test] Let $\varphi \in \mathcal{L}_?$. We define $\tau(?\varphi) = \mathsf{M}_{\mathsf{T}} = ((\mathsf{S}, \mathsf{R}, \mathsf{pre}), \mathsf{T})$ where:

$$\begin{aligned} \mathsf{S} &= \{\mathsf{test}, \mathsf{skip}\} & \mathsf{R}_a &= \{(\mathsf{test}, \mathsf{skip}), (\mathsf{skip}, \mathsf{skip})\} \text{ for } a \in A \\ \mathsf{pre} &= \{(\mathsf{test}, \varphi), (\mathsf{skip}, \top)\} & \mathsf{T} &= \{\mathsf{test}\} \end{aligned}$$

Definition 4.7 [Learning] Let $\alpha \in \mathcal{L}_?{}^{\mathrm{act}}$ where $\tau(\alpha) = \mathsf{M}^{\alpha}_{\mathsf{T}^{\alpha}} = ((\mathsf{S}^{\alpha}, \mathsf{R}^{\alpha}, \mathsf{pre}^{\alpha}), \mathsf{T}^{\alpha})$. Let test and skip be new states not appearing in S^{α}. We define $\tau(L_B(\alpha, \alpha)) = \mathsf{M}_{\mathsf{T}} = ((\mathsf{S}, \mathsf{R}, \mathsf{pre}), \mathsf{T})$ where:

$$\begin{aligned} \mathsf{S} &= \mathsf{S}^{\alpha} \cup \{\mathsf{test}, \mathsf{skip}\} \\ \mathsf{R}_a &= \mathsf{R}^{\alpha}_a \cup \{(\mathsf{skip}, \mathsf{skip})\} \cup \{(\mathsf{test}, \mathsf{t}^{\alpha}) \mid \mathsf{t}^{\alpha} \in \mathsf{T}^{\alpha}\} \text{ for } a \in B \\ \mathsf{R}_a &= \mathsf{R}^{\alpha}_a \cup \{(\mathsf{test}, \mathsf{skip}), (\mathsf{skip}, \mathsf{skip})\} \text{ for } a \notin B \\ \mathsf{pre} &= \mathsf{pre}^{\alpha} \cup \{(\mathsf{test}, \top), (\mathsf{skip}, \top)\} \\ \mathsf{T} &= \{\mathsf{test}\} \end{aligned}$$

We define $\tau(L_B(\alpha, \beta)) = \tau(L_B(\alpha \sqcup \beta, \alpha \sqcup \beta))$.

We note that the syntax of action formula logic defines the learning operator as a binary operator that can be applied to two different action formulae, however in the setting of $\mathcal{K}$ and $\mathcal{K}45$ we only give a direct definition of τ for actions of the form $L_B(\alpha, \alpha)$ and define the more general case in terms of this. Intuitively $L_B(\alpha, \beta)$ is intended to represent an action where the agents in B learn that α or β have occurred (i.e. that $\alpha \sqcup \beta$ has occurred). The setting of $\mathcal{S}5$ corresponds to a notion of *knowledge*, where anything that an agent *knows* must be true, and therefore anything that an agent *learns* must also be true. So in an action where agents learn that α or β have occurred, one of those actions must have actually occurred. Therefore in $\mathcal{S}5$ we describe the action $L_B(\alpha, \beta)$

as the agents in B learning that α or β have occurred, when in reality α has actually occurred. On the other hand, the settings of $\mathcal{K}$ and $\mathcal{K}45$ correspond more closely to a notion of *belief*, where there is no requirement that what an agent *believes* is true. So in an action where agents learn that α or β have occurred, neither of these actions must actually have occurred. Therefore in the settings of $\mathcal{K}$ and $\mathcal{K}45$ we make no distinction between α and β in a description of the action $L_B(\alpha, \beta)$, hence the definition of τ given in these settings.

4.2 $\mathcal{K}45$

Definition 4.8 [Test] Let $\varphi \in \mathcal{L}_?$. We define $\tau(?\varphi)$ as in Definition 4.6 for $\mathcal{K}$.

Definition 4.9 [Learning] Let $\alpha \in \mathcal{L}_?{}^{\mathrm{act}}$ where $\tau(\alpha) = \mathsf{M}^{\alpha}_{\mathsf{T}^{\alpha}} = ((\mathsf{S}^{\alpha}, \mathsf{R}^{\alpha}, \mathsf{pre}^{\alpha}), \mathsf{T}^{\alpha})$. Let test and skip be new states not appearing in S^{α}. For every $\mathsf{t}^{\alpha} \in \mathsf{T}^{\alpha}$ let $\bar{\mathsf{t}}^{\alpha}$ be a new state not appearing in S^{α}. We call each $\bar{\mathsf{t}}^{\alpha}$ a *proxy state* for t^{α}. We define $\tau(L_B(\alpha, \alpha)) = \mathsf{M}_{\mathsf{T}} = ((\mathsf{S}, \mathsf{R}, \mathsf{pre}), \mathsf{T})$ where:

$$
\begin{aligned}
\mathsf{S} &= \mathsf{S}^{\alpha} \cup \{\mathsf{test}, \mathsf{skip}\} \cup \{\bar{\mathsf{t}}^{\alpha} \mid \mathsf{t}^{\alpha} \in \mathsf{T}^{\alpha}\} \\
\mathsf{R}_a &= \mathsf{R}^{\alpha}_a \cup \{(\mathsf{skip}, \mathsf{skip})\} \cup \{(\mathsf{test}, \bar{\mathsf{t}}^{\alpha}) \mid \mathsf{t}^{\alpha} \in \mathsf{T}^{\alpha}\} \cup \\
&\qquad \{(\bar{\mathsf{t}}^{\alpha}, \bar{\mathsf{u}}^{\alpha}) \mid \mathsf{t}^{\alpha}, \mathsf{u}^{\alpha} \in \mathsf{T}^{\alpha}\} \text{ for } a \in B \\
\mathsf{R}_a &= \mathsf{R}^{\alpha}_a \cup \{(\mathsf{test}, \mathsf{skip}), (\mathsf{skip}, \mathsf{skip})\} \cup \\
&\qquad \{(\bar{\mathsf{t}}^{\alpha}, \mathsf{u}^{\alpha}) \mid \mathsf{t}^{\alpha} \in \mathsf{T}^{\alpha}, \mathsf{u}^{\alpha} \in \mathsf{t}^{\alpha}\mathsf{R}^{\alpha}_a\} \text{ for } a \notin B \\
\mathsf{pre} &= \mathsf{pre}^{\alpha} \cup \{(\mathsf{test}, \top), (\mathsf{skip}, \top)\} \cup \{(\bar{\mathsf{t}}^{\alpha}, \mathsf{pre}^{\alpha}(\mathsf{t}^{\alpha})) \mid \mathsf{t}^{\alpha} \in \mathsf{T}^{\alpha}\} \\
\mathsf{T} &= \{\mathsf{test}\}
\end{aligned}
$$

As in Definition 4.7, we define $\tau(L_B(\alpha, \beta)) = \tau(L_B(\alpha \sqcup \beta, \alpha \sqcup \beta))$.

Lemma 4.10 *Let $\alpha \in \mathcal{L}_?{}^{act}$. Then $\tau(\alpha) \in \mathcal{AM}_{\mathcal{K}45}$.*

Lemma 4.11 *Let $\alpha \in \mathcal{L}_?{}^{act}$ and let $M_T \in \mathcal{K}45$. Then $M_T \otimes \tau(\alpha) \in \mathcal{K}45$.*

We note that the definition for τ given here varies considerably from the definition given in the setting of $\mathcal{K}$ due to the presence of the proxy states. The proxy states are introduced due to the additional frame constraints in $\mathcal{K}45$ and the desire that the action models constructed by τ be $\mathcal{AM}_{\mathcal{K}45}$ action models. In constructing $\tau(L_B\alpha)$ we wish to construct an action model with a root state whose B-successors are the root states of $\tau(\alpha)$, so that the result of executing the action $L_B\alpha$ is that the agents B believe that the action α has occurred. However in order for this construction to result in a $\mathcal{AM}_{\mathcal{K}45}$ action model, we must take the transitive, Euclidean closure of the B-successors of the root state. If we were to perform a construction similar to that used in the setting of $\mathcal{K}$ where proxy states are not used, then this would mean that the for every $b \in B$, the b-successors of the root state would include all of the b-successors of the root states, and not just the root states themselves. To show why this is not desireable, consider the simple example of the action $L_a?\varphi$. The intention is that this action represents a private announcement to a that φ is true, as it is in the setting of $\mathcal{K}$. Without using proxy states, if we wanted to include the state test in the a-successors of the root state of $\tau(\alpha)$ then in order to construct a $\mathcal{AM}_{\mathcal{K}45}$ action model we would need to take the transitive, Euclidean closure of the a-successors of test. As skip is an a-successor of test in the action $?\varphi$, then this would mean that a would not be able to distinguish between the actions states test and skip and so the result of executing $\tau(\alpha)$ would be that a learns nothing. With the construction provided, the action $L_a?\varphi$ gives the desired result that a learns that φ is true.

We also note that the results presented in this paper for $\mathcal{K}45$ can be extended to $\mathcal{KD}45$ by modifying Definition 4.9 so that $\mathsf{pre}(\mathsf{test}) = \bigwedge_{a\in B}\bigvee_{\mathsf{t}^\alpha\in\mathsf{T}^\alpha}\Diamond_a\mathsf{pre}^\alpha(\mathsf{t}^\alpha)$, which guarantees that the result of successfully executing an action formula has the seriality property of $\mathcal{KD}45$.

4.3 $\mathcal{S}5$

Definition 4.12 [Test] Let $\varphi \in \mathcal{L}_?$. We define $\tau(?\varphi) = \mathsf{M}_\mathsf{T} = ((\mathsf{S},\mathsf{R},\mathsf{pre}),\mathsf{T})$ where:

$$\mathsf{S} = \{\mathsf{test},\mathsf{skip}\} \qquad \mathsf{R}_a = \mathsf{S}^2 \text{ for } a \in A$$
$$\mathsf{pre} = \{(\mathsf{test},\varphi),(\mathsf{skip},\top)\} \quad \mathsf{T} = \{\mathsf{test}\}$$

Definition 4.13 [Learning] Let $\alpha,\beta \in \mathcal{L}_?{}^{act}$ where $\tau(\alpha) = \mathsf{M}^\alpha_{\mathsf{T}^\alpha} = ((\mathsf{S}^\alpha,\mathsf{R}^\alpha,\mathsf{pre}^\alpha),\mathsf{T}^\alpha)$ and $\tau(\beta) = \mathsf{M}^\beta_{\mathsf{T}^\beta} = ((\mathsf{S}^\beta,\mathsf{R}^\beta,\mathsf{pre}^\beta),\mathsf{T}^\beta)$. For every $\mathsf{t} \in \mathsf{T}^\alpha\cup\mathsf{T}^\beta$ let $\bar{\mathsf{t}}$ be a new state not appearing in $\mathsf{S}^\alpha\cup\mathsf{S}^\beta$. We define $\tau(L_B(\alpha,\beta)) = \mathsf{M}_\mathsf{T} = ((\mathsf{S},\mathsf{R},\mathsf{pre}),\mathsf{T})$ where:

$$\mathsf{S} = \mathsf{S}^\alpha\cup\mathsf{S}^\beta\cup\{\bar{\mathsf{t}} \mid \mathsf{t}\in\mathsf{T}^\alpha\cup\mathsf{T}^\beta\}$$
$$\mathsf{R}_a = \mathsf{R}^\alpha_a\cup\mathsf{R}^\beta_a\cup\{(\bar{\mathsf{t}},\bar{\mathsf{u}}) \mid \mathsf{t},\mathsf{u}\in\mathsf{T}^\alpha\cup\mathsf{T}^\beta\} \text{ for } a\in B$$
$$\mathsf{R}_a = \mathsf{R}^\alpha_a\cup\mathsf{R}^\beta_a\cup\bigcup_{\mathsf{t}\in\mathsf{T}^\alpha\cup\mathsf{T}^\beta}(\{\bar{\mathsf{t}}\}\cup\mathsf{t}(\mathsf{R}^\alpha_a\cup\mathsf{R}^\beta_a))^2 \text{ for } a\notin B$$
$$\mathsf{pre} = \mathsf{pre}^\alpha\cup\mathsf{pre}^\beta\cup\{(\bar{\mathsf{t}},(\mathsf{pre}^\alpha\cup\mathsf{pre}^\beta)(\mathsf{t})) \mid \mathsf{t}\in\mathsf{T}^\alpha\cup\mathsf{T}^\beta\}$$
$$\mathsf{T} = \{\bar{\mathsf{t}} \mid \mathsf{t}\in\mathsf{T}^\alpha\}$$

Lemma 4.14 *Let $\alpha \in \mathcal{L}_?{}^{act}$. Then $\tau(\alpha)\in\mathcal{AM}_{\mathcal{S}5}$.*

Lemma 4.15 *Let $\alpha \in \mathcal{L}_?{}^{act}$ and let $M_T \in \mathcal{S}5$. Then $M_T\otimes\tau(\alpha)\in\mathcal{S}5$.*

We note that as in the setting of $\mathcal{K}45$ the definition of τ uses proxy states to construct action models from learning operators. However unlike in the settings of $\mathcal{K}$ and $\mathcal{K}45$ this construction does not introduce the new states test and skip. As discussed earlier this is because in the setting of $\mathcal{S}5$, in an action where agents learn that α or β have occurred, one of those actions must have actually occurred. Unlike in the settings of $\mathcal{K}$ and $\mathcal{K}45$ we have distinguished between the actions α and β, designating that α is the action that has actually occurred. We also note that the definition of τ for test operators is different from that used in $\mathcal{K}$ and $\mathcal{K}45$, simply to account for the additional frame constraints of $\mathcal{S}5$.

5 Axiomatisation

In the following subsections we give sound and complete axiomatisations for the action formulae logic in the settings of $\mathcal{K}$ and $\mathcal{K}45$. We note that axiomatisations for arbitrary action formula logic in these settings can be derived trivially from these axiomatisations by adding the additional axioms and rules from refinement modal logic.

5.1 $\mathcal{K}$

Definition 5.1 [Axiomatisation $\mathbf{AFL_K}$] The axiomatisation $\mathbf{AFL_K}$ is a substitution schema consisting of the rules and axioms of $\mathbf{K}$ along with the axioms:

$$\begin{array}{llll}
\mathbf{LT} & \vdash [?\varphi]\psi \leftrightarrow (\varphi \rightarrow \psi) \text{ for } \psi \in \mathcal{L} & \mathbf{LN} & \vdash [L_B(\alpha,\beta)]\neg\varphi \leftrightarrow \neg[L_B(\alpha,\beta)]\varphi \\
\mathbf{LU} & \vdash [\alpha \sqcup \beta]\varphi \leftrightarrow ([\alpha]\varphi \wedge [\beta]\varphi) & \mathbf{LC} & \vdash [L_B(\alpha,\beta)](\varphi \wedge \psi) \leftrightarrow \\
 & & & \quad ([L_B(\alpha,\beta)]\varphi \wedge [L_B(\alpha,\beta)]\psi) \\
\mathbf{LS} & \vdash [\alpha \otimes \beta]\varphi \leftrightarrow [\alpha][\beta]\varphi & \mathbf{LK1} & \vdash [L_B(\alpha,\beta)]\Box_a\varphi \leftrightarrow \Box_a[\alpha \sqcup \beta]\varphi \\
 & & & \quad \text{for } a \in B \\
\mathbf{LP} & \vdash [L_B(\alpha,\beta)]p \leftrightarrow p & \mathbf{LK2} & \vdash [L_B(\alpha,\beta)]\Box_a\varphi \leftrightarrow \Box_a\varphi \text{ for } a \notin B
\end{array}$$

and the rule:

$$\mathbf{NecL} \text{ From } \vdash \varphi \text{ infer } \vdash [\alpha]\varphi$$

Proposition 5.2 *The axiomatisation* $\mathbf{AFL_K}$ *is sound in the logic* $K_\otimes$.

Proof. **LT** follows from applying the reduction axioms of $\mathbf{AML_K}$ inductively to $[?\varphi]\psi$.

LU and **LS** follow from Proposition 4.5.

Let $\tau(L_b(\alpha,\beta)) = \mathsf{M_s} = ((\mathsf{S}, \mathsf{R}, \mathsf{pre}), \mathsf{s})$. **LP**, **LN** and **LC** follow trivially from the $\mathbf{AML_K}$ axioms **AP**, **AN** and **AC** respectively, noting from Definition 4.7 that $\mathsf{pre}(\mathsf{s}) = \top$. **LK1** follows trivially from the $\mathbf{AML_K}$ axiom **AK**, noting from Definition 4.7 that as $a \in A$ then $\mathsf{M}_{\mathsf{sR}_a} \underline{\leftrightarrow} \tau(\alpha \sqcup \beta)$. **NecL** follows trivially from the $\mathbf{AML_K}$ rule **NecA**. **LK2** follows trivially from the $\mathbf{AML_K}$ axiom **AK**, noting from Definition 4.7 that as $a \notin A$ then $\mathsf{M}_{\mathsf{sR}_a} \underline{\leftrightarrow} \tau(?\top)$. □

Proposition 5.3 *The axiomatisation* $\mathbf{AFL_K}$ *is complete for the logic* $K_\otimes$.

We note that the axiomatisation $\mathbf{AFL_K}$ forms a set of reduction axioms that gives a provably correct translation from $\mathcal{L}_?$ to $\mathcal{L}$.

Example 5.4 We give an example derivation that the action formula α given in Example 3.2 does indeed satisfy (part of) the epistemic goal stated in Example 1.1. We get $\vdash [L_{Ed}?p]\Box_{Ed}p$ from **LT**, **NecK** and **LK1**. Similarly we have $\vdash [L_{Ed}?\neg p]\Box_{Ed}\neg p$, $\vdash [L_{Tim}?p]\Box_{Tim}p$ and $\vdash [L_{Tim}?\neg p]\Box_{Tim}\neg p$

Let $\varphi = \Box_{Ed}p \vee \Box_{Ed}\neg p \vee \Box_{Tim}p \vee \Box_{Tim}\neg p$. Then:

$$\vdash [L_{Ed}?p \sqcup L_{Ed}?\neg p \sqcup L_{Tim}?p \sqcup L_{Tim}?\neg p]\varphi \tag{1}$$

$$\vdash \Box_{James}[L_{Ed}?p \sqcup L_{Ed}?\neg p \sqcup L_{Tim}?p \sqcup L_{Tim}?\neg p]\varphi \tag{2}$$

$$\vdash [L_{James}(L_{Ed}?p \sqcup L_{Ed}?\neg p \sqcup L_{Tim}?p \sqcup L_{Tim}?\neg p)]\Box_{James}\varphi \tag{3}$$

$$\vdash [\alpha]\Box_{James}\varphi \tag{4}$$

(1) follows from **LU**, (2) follows from **NecK** and (3) follows from **LK1**. (4) follows from **LS** and **LK2**.

5.2 $\mathcal{K}45$

Definition 5.5 [Axiomatisation $\mathbf{AFL_{K45}}$] The axiomatisation $\mathbf{AFL_{K45}}$ is a substitution schema consisting of the rules and axioms of **K45** along with the rules and axioms of $\mathbf{AFL_K}$, but substituting the $\mathbf{AFL_K}$ axiom **LK1** for the axiom:

$$\mathbf{LK1} \vdash [L_B(\alpha, \beta)]\Box_a\chi \leftrightarrow \Box_a[\alpha \sqcup \beta]\chi \text{ for } a \in B$$

and the rule:

$$\mathbf{NecL} \text{ From } \vdash \varphi \text{ infer } \vdash [\alpha]\varphi$$

where χ is a $(A \setminus \{a\})$-restricted formula.

Proposition 5.6 *The axiomatisation* $\mathbf{AFL_{K45}}$ *is sound in the logic* $K45_\otimes$.

Proof. Soundness of **LT**, **LU**, **LS**, **LP**, **LN**, **LC**, **LK2** and **NecL** follow from the same reasoning as in the proof of Proposition 5.2.

LK1 follows from the $\mathbf{AML_{K45}}$ axiom **AK**. We note that as $a \in B$, from Definition 4.9 we have $\mathsf{M}_{\mathsf{sR}_a} \leftrightarrows_{(A\setminus\{a\})}\tau(\alpha \sqcup \beta)$, and as χ is $(A \setminus \{a\})$-restricted formula then $\models [\mathsf{M}_{\mathsf{sR}_a}]\chi \leftrightarrow [\tau(\alpha \sqcup \beta)]\chi$. □

Proposition 5.7 *The axiomatisation* $\mathbf{AFL_{K45}}$ *is complete for the logic* $K45_\otimes$.

We note that the axiomatisation $\mathbf{AFL_{K45}}$ forms a set of reduction axioms that gives a provably correct translation from $\mathcal{L}_?$ to $\mathcal{L}$. To translate a subformula $[\alpha]\varphi$, where $\varphi \in \mathcal{L}$, we must first translate φ to the alternating disjunctive normal form of [19], which gives the property that for every subformula $\Box_a\psi$, the formula ψ is $(A \setminus \{a\})$-restricted, and therefore **LK1** is applicable.

6 Correspondence

In the following subsections we show the correspondence between action formulae and action models in the settings of $\mathcal{K}$, $\mathcal{K}45$ and $\mathcal{S}5$. In each setting we show that action formulae are capable of representing any action model up to n-bisimilarity.

6.1 $\mathcal{K}$

To begin we give two lemmas to simplify the construction that we will use for our correspondence result in $\mathcal{K}$.

Lemma 6.1 *Let* $\varphi \in \mathcal{L}_?$ *and* $\mathsf{M_s} = ((\mathsf{S}, \mathsf{R}, \mathsf{pre}), \mathsf{s}) \in \mathcal{AM}$. *Then let* $\mathsf{M}'_{\mathsf{s}'} = ((\mathsf{S}', \mathsf{R}', \mathsf{pre}'), \mathsf{s}') \in \mathcal{AM}$ *where:*

$$\mathsf{S}' = \mathsf{S} \cup \{\mathsf{s}'\}$$
$$\mathsf{R}'_a = \mathsf{R}_a \cup \{(\mathsf{s}', \mathsf{t}) \mid \mathsf{t} \in \mathsf{sR}_a\} \text{ for } a \in A$$
$$\mathsf{pre}' = \mathsf{pre} \cup \{(\mathsf{s}', \varphi \wedge \mathsf{pre}(\mathsf{s}))\}$$

Then $\tau(?\varphi) \otimes \mathsf{M_s} \leftrightarrows \mathsf{M}'_{\mathsf{s}'}$.

Lemma 6.2 *Let* $\alpha \in \mathcal{L}_?{}^{act}$ *where* $\tau(\alpha) = \mathsf{M}^\alpha_{\mathsf{T}^\alpha} = ((\mathsf{S}^\alpha, \mathsf{R}^\alpha, \mathsf{pre}^\alpha), \mathsf{T}^\alpha)$, $a \in A$ *and* $\mathsf{M_s} = ((\mathsf{S}, \mathsf{R}, \mathsf{pre}), \mathsf{s}) \in \mathcal{AM}$ *such that* $\mathsf{sR}_a = \{\mathsf{t}\}$ *for some* $\mathsf{t} \in \mathsf{S}$ *and* $\mathsf{tR}_a = \{\mathsf{t}\}$ *Then let* $\mathsf{M}'_{\mathsf{s}'} = ((\mathsf{S}', \mathsf{R}', \mathsf{pre}'), \mathsf{s}') \in \mathcal{AM}$ *where:*

$$\begin{aligned}
\mathsf{S}' &= \mathsf{S} \cup \mathsf{S}^\alpha \cup \{\mathsf{s}'\}\\
\mathsf{R}'_a &= \mathsf{R}_a \cup \mathsf{R}^\alpha_a \cup \{(\mathsf{s}', \mathsf{t}^\alpha) \mid \mathsf{t}^\alpha \in \mathsf{T}^\alpha\}\\
\mathsf{R}'_b &= \mathsf{R}_b \cup \mathsf{R}^\alpha_b \cup \{(\mathsf{s}', \mathsf{t}) \mid \mathsf{t} \in \mathsf{s}\mathsf{R}_b\} \textit{ for } b \in A \setminus \{a\}\\
\mathsf{pre}' &= \mathsf{pre} \cup \{(\mathsf{s}', \mathsf{pre}(\mathsf{s}))\}
\end{aligned}$$

Then $\tau(L_a\alpha) \otimes \mathsf{M}_\mathsf{s} \leftrightarrows \mathsf{M}'_{\mathsf{s}'}$.

Proposition 6.3 *Let* $\mathsf{M}_\mathsf{s} \in \mathcal{AM}$ *and let* $n \in \mathbb{N}$. *Then there exists* $\alpha \in \mathcal{L}_?{}^{act}$ *such that* $\mathsf{M}_\mathsf{s} \leftrightarrows_n \tau(\alpha)$.

Proof. By induction on n.

Suppose that $n = 0$. Let $\alpha = ?\mathsf{pre}(\mathsf{s})$ and $\tau(\alpha) = \mathsf{M}'_{\mathsf{s}'} = ((\mathsf{S}', \mathsf{R}', \mathsf{pre}'), \mathsf{s}')$. From Definition 4.6 we have that $\mathsf{pre}(\mathsf{s}) = \mathsf{pre}'(\mathsf{s}')$, so $(\mathsf{M}_\mathsf{s}, \mathsf{M}'_{\mathsf{s}'})$ satisfies **atoms** and therefore $\mathsf{M}_\mathsf{s} \leftrightarrows_0 \mathsf{M}'_{\mathsf{s}'}$.

Suppose that $n > 0$. By the induction hypothesis, for every $a \in A$, $\mathsf{t} \in \mathsf{s}\mathsf{R}_a$ there exists $\alpha^{a,\mathsf{t}} \in \mathcal{L}_?{}^{\mathsf{act}}$ such that $\mathsf{M}_\mathsf{t} \leftrightarrows_{(n-1)} \tau(\alpha^{a,\mathsf{t}})$, where $\tau(\alpha^{a,\mathsf{t}}) \leftrightarrows \mathsf{M}^{a,\mathsf{t}}_{\mathsf{s}^{a,\mathsf{t}}} = ((\mathsf{S}^{a,\mathsf{t}}, \mathsf{R}^{a,\mathsf{t}}, \mathsf{pre}^{a,\mathsf{t}}), \mathsf{s}^{a,\mathsf{t}})$.

Let $\alpha = ?\mathsf{pre}(\mathsf{s}) \otimes \bigotimes_{a \in A} L_a(\bigsqcup_{\mathsf{t} \in \mathsf{s}\mathsf{R}_a} \alpha^\mathsf{t})$. Then from Lemmas 6.1 and 6.2: $\tau(\alpha) \leftrightarrows \mathsf{M}'_{\mathsf{s}'} = ((\mathsf{S}', \mathsf{R}', \mathsf{pre}'), \mathsf{s}')$ where:

$$\begin{aligned}
\mathsf{S}' &= \bigcup_{a \in A, \mathsf{t} \in \mathsf{s}\mathsf{R}_a} (\mathsf{S}^{a,\mathsf{t}}) \cup \{\mathsf{s}'\}\\
\mathsf{R}'_a &= \bigcup_{b \in A, \mathsf{t} \in \mathsf{s}\mathsf{R}_b} (\mathsf{R}^{b,\mathsf{t}}_a) \cup \{(\mathsf{s}', \mathsf{s}^{a,\mathsf{t}}) \mid \mathsf{t} \in \mathsf{s}\mathsf{R}_a\} \text{ for } a \in A\\
\mathsf{pre}' &= \bigcup_{a \in A, \mathsf{t} \in \mathsf{s}\mathsf{R}_a} (\mathsf{pre}^{a,\mathsf{t}}) \cup \{(\mathsf{s}', \mathsf{pre}(\mathsf{s}))\}
\end{aligned}$$

We note for every $a \in A$, $\mathsf{t} \in \mathsf{s}\mathsf{R}_a$ that $\mathsf{M}'_{\mathsf{s}^{a,\mathsf{t}}} \leftrightarrows \mathsf{M}^{a,\mathsf{t}}_{\mathsf{s}^{a,\mathsf{t}}}$ as for every $a \in A$, $\mathsf{u} \in \mathsf{S}^{a,\mathsf{t}}$ we have $\mathsf{u}\mathsf{R}'_a = \mathsf{u}\mathsf{R}^{a,\mathsf{t}}_a$.

We show that $(\mathsf{M}_\mathsf{s}, \mathsf{M}'_{\mathsf{s}'})$ satisfies **atoms**, **forth-n-a** and **back-n-a** for every $a \in A$.

atoms By construction $\mathsf{pre}'(\mathsf{s}') = \mathsf{pre}(\mathsf{s})$.

forth-n-a Let $\mathsf{t} \in \mathsf{s}\mathsf{R}_a$. By construction $\mathsf{s}^{a,\mathsf{t}} \in \mathsf{s}'\mathsf{R}'_a$, by the induction hypothesis $\mathsf{M}_\mathsf{t} \leftrightarrows_{(n-1)} \mathsf{M}^{a,\mathsf{t}}_{\mathsf{s}^{a,\mathsf{t}}}$ and from above $\mathsf{M}^{a,\mathsf{t}}_{\mathsf{s}^{a,\mathsf{t}}} \leftrightarrows \mathsf{M}'_{\mathsf{s}^{a,\mathsf{t}}}$. Therefore by transitivity $\mathsf{M}_\mathsf{t} \leftrightarrows_{(n-1)} \mathsf{M}'_{\mathsf{s}^{a,\mathsf{t}}}$.

back-n-a Follows from similar reasoning to **forth-n-a**.

Therefore $\mathsf{M}_\mathsf{s} \leftrightarrows_n \tau(\alpha)$. □

Corollary 6.4 *Let* $\mathsf{M}_\mathsf{s} \in \mathcal{AM}$. *Then for every* $\varphi \in \mathcal{L}_\otimes$ *there exists* $\alpha \in \mathcal{L}_?{}^{act}$ *such that* $\models_{K_\otimes} [\mathsf{M}_\mathsf{s}]\varphi \leftrightarrow [\tau(\alpha)]\varphi$.

Corollary 6.5 *Let* $\varphi \in \mathcal{L}_\otimes$. *Then there exists* $\varphi' \in \mathcal{L}_?$ *such that for every* $M_s \in \mathcal{K}$: $M_s \models_{K_\otimes} \varphi$ *if and only if* $M_s \models_{K_?} \varphi'$.

6.2 $\mathcal{K}45$

We note that we can introduce similar results to Lemma 6.1 and Lemma 6.2 to simplify the construction that we will use for $\mathcal{K}45$. We omit the details. For full details refer to the extended version of this paper [16]

Proposition 6.6 *Let* $\mathsf{M}_\mathsf{s} \in \mathcal{AM}_{\mathcal{K}45}$ *and let* $n \in \mathbb{N}$. *Then there exists* $\alpha \in \mathcal{L}_?{}^{act}$ *such that* $\mathsf{M}_\mathsf{s} \leftrightarroweq_n \tau(\alpha)$.

Proof. By induction on n.

Suppose that $n = 0$. Let $\alpha = ?\mathsf{pre}(\mathsf{s})$ and $\tau(\alpha) = \mathsf{M}'_{\mathsf{s}'} = ((\mathsf{S}', \mathsf{R}', \mathsf{pre}'), \mathsf{s}')$. From Definition 4.8 we have that $\mathsf{pre}(\mathsf{s}) = \mathsf{pre}'(\mathsf{s}')$, so $(\mathsf{M}_\mathsf{s}, \mathsf{M}'_{\mathsf{s}'})$ satisfies **atoms** and therefore $\mathsf{M}_\mathsf{s} \leftrightarroweq_0 \mathsf{M}'_{\mathsf{s}'}$.

Suppose that $n > 0$. By the induction hypothesis, for every $a \in A$, $\mathsf{t} \in \mathsf{s}\mathsf{R}_a$ there exists $\alpha^{a,\mathsf{t}} \in \mathcal{L}_?{}^{act}$ such that $\mathsf{M}_\mathsf{t} \leftrightarroweq_{(n-1)} \tau(\alpha^{a,\mathsf{t}})$. For every $a \in A$, $\mathsf{t} \in \mathsf{s}\mathsf{R}_a$ let $\tau(\alpha^{a,\mathsf{t}}) = \mathsf{M}^{a,\mathsf{t}}_{\mathsf{s}^{a,\mathsf{t}}} = ((\mathsf{S}^{a,\mathsf{t}}, \mathsf{R}^{a,\mathsf{t}}, \mathsf{pre}^{a,\mathsf{t}}), \mathsf{s}^{a,\mathsf{t}})$.

Let $\alpha = ?\mathsf{pre}(\mathsf{s}) \otimes \bigotimes_{a \in A} L_a(\bigsqcup_{\mathsf{t} \in \mathsf{s}\mathsf{R}_a} \alpha^{a,\mathsf{t}})$. Then $\tau(\alpha) \leftrightarroweq \mathsf{M}'_{\mathsf{s}'} = ((\mathsf{S}', \mathsf{R}', \mathsf{pre}'), \mathsf{s}')$ where:

$$\mathsf{S}' = \bigcup_{a \in A, \mathsf{t} \in \mathsf{s}\mathsf{R}_a} (\mathsf{S}^{a,\mathsf{t}}) \cup \{\bar{\mathsf{s}}^{a,\mathsf{t}} \mid a \in A, \mathsf{t} \in \mathsf{s}\mathsf{R}_a\} \cup \{\mathsf{s}'\}$$

$$\mathsf{R}'_a = \bigcup_{b \in A, \mathsf{t} \in \mathsf{s}\mathsf{R}_b} (\mathsf{R}^{b,\mathsf{t}}_a) \cup \{(\mathsf{s}', \bar{\mathsf{s}}^{a,\mathsf{t}}) \mid \mathsf{t} \in \mathsf{s}\mathsf{R}_a\} \cup \{(\bar{\mathsf{s}}^{a,\mathsf{t}}, \bar{\mathsf{s}}^{a,\mathsf{u}}) \mid \mathsf{t}, \mathsf{u} \in \mathsf{s}\mathsf{R}_a\} \cup$$
$$\{(\bar{\mathsf{s}}^{b,\mathsf{t}}, \mathsf{u}) \mid b \in A \setminus \{a\}, \mathsf{t} \in \mathsf{s}\mathsf{R}_b, \mathsf{u} \in \mathsf{s}^{b,\mathsf{t}}\mathsf{R}^{b,\mathsf{t}}_a\} \text{ for } a \in A$$

$$\mathsf{pre}' = \bigcup_{a \in A, \mathsf{t} \in \mathsf{s}\mathsf{R}_a} (\mathsf{pre}^{a,\mathsf{t}}) \cup \{(\bar{\mathsf{s}}^{a,\mathsf{t}}, \mathsf{pre}^{a,\mathsf{t}}(\mathsf{s}^{a,\mathsf{t}})) \mid a \in A, \mathsf{t} \in \mathsf{s}\mathsf{R}_a\} \cup \{(\mathsf{s}', \mathsf{pre}(\mathsf{s}))\}$$

As in the proof of Proposition 6.3, we note for every $a \in A$, $\mathsf{t} \in \mathsf{s}\mathsf{R}_a$ that $\mathsf{M}'_{\mathsf{s}^{a,\mathsf{t}}} \leftrightarroweq \mathsf{M}^{a,\mathsf{t}}_{\mathsf{s}^{a,\mathsf{t}}}$.

We need to show that $(\mathsf{M}_\mathsf{s}, \mathsf{M}'_{\mathsf{s}'})$ satisfies **atoms**, **forth-n-a** and **back-n-a** for every $a \in A$. We use reasoning similar to the proof of Proposition 6.3, however noting that the successors of s' in M' are not the same as in the construction used previously. We claim that each $\bar{\mathsf{s}}^{a,\mathsf{t}}$ state is $(n-1)$-bisimilar to the corresponding $\mathsf{s}^{a,\mathsf{t}}$ state. We show this by showing for every $0 \leq i \leq n-1$, $a \in A$, $\mathsf{t} \in \mathsf{s}\mathsf{R}_a$ that $\mathsf{M}'_{\bar{\mathsf{s}}^{a,\mathsf{t}}} \leftrightarroweq_i \mathsf{M}'_{\mathsf{s}^{a,\mathsf{t}}}$. We proceed by induction on i and omit the details, as they are straight-forward. For full details refer to the extended version of this paper [16]

Therefore for every $a \in A$, $\mathsf{t} \in \mathsf{s}\mathsf{R}_a$ we have that $\mathsf{M}'_{\bar{\mathsf{s}}^{a,\mathsf{t}}} \leftrightarroweq_{(n-1)} \mathsf{M}'_{\mathsf{s}^{a,\mathsf{t}}}$.

We can now show that $\mathsf{M}_\mathsf{s} \leftrightarroweq_n \mathsf{M}'_{\mathsf{s}'}$ by using the same reasoning as the proof for Proposition 6.3, using the $(n-1)$-bisimilar $\mathsf{M}'_{\bar{\mathsf{s}}^{a,\mathsf{t}}}$ states in place of corresponding $\mathsf{M}'_{\mathsf{s}^{a,\mathsf{t}}}$ states.

Therefore $\mathsf{M}_\mathsf{s} \leftrightarroweq_n \tau(\alpha)$. □

Corollary 6.7 *Let* $\mathsf{M}_\mathsf{s} \in \mathcal{AM}_{\mathcal{K}45}$. *Then for every* $\varphi \in \mathcal{L}_\otimes$ *there exists* $\alpha \in \mathcal{L}_?{}^{act}$ *such that* $\models_{K45_\otimes} [\mathsf{M}_\mathsf{s}]\varphi \leftrightarrow [\tau(\alpha)]\varphi$.

Corollary 6.8 *Let* $\varphi \in \mathcal{L}_\otimes$. *Then there exists* $\varphi' \in \mathcal{L}_?$ *such that for every* $M_s \in \mathcal{K}45$: $M_s \models_{K45_\otimes} \varphi$ *if and only if* $M_s \models_{K45_?} \varphi'$.

6.3 $\mathcal{S}5$

As in $\mathcal{K}45$, we note that we can introduce similar results to Lemma 6.1 and Lemma 6.2 to simplify the construction that we will use for $\mathcal{S}5$. We omit the details. For full details refer to the extended version of this paper [16]

Proposition 6.9 *Let* $\mathsf{M_s} \in \mathcal{AM}_{S5}$ *and let* $n \in \mathbb{N}$. *Then there exists* $\alpha \in \mathcal{L}_?{}^{act}$ *such that* $\mathsf{M_s} \leftrightarrows_n \tau(\alpha)$.

Proof. By induction on n.

Suppose that $n = 0$. Let $\alpha = ?\mathsf{pre(s)}$ and $\tau(\alpha) = \mathsf{M}'_{\mathsf{s}'} = ((\mathsf{S}', \mathsf{R}', \mathsf{pre}'), \mathsf{s}')$. From Definition 4.12 we have that $\mathsf{pre(s)} = \mathsf{pre}'(\mathsf{s}')$, so $(\mathsf{M_s}, \mathsf{M}'_{\mathsf{s}'})$ satisfies **atoms** and therefore $\mathsf{M_s} \leftrightarrows_0 \mathsf{M}'_{\mathsf{s}'}$.

Suppose that $n > 0$. By the induction hypothesis, for every $a \in A$, $\mathsf{t} \in \mathsf{sR}_a$ there exists $\alpha^{a,\mathsf{t}} \in \mathcal{L}_?{}^{\mathrm{act}}$ such that $\mathsf{M_t} \leftrightarrows_{(n-1)} \tau(\alpha^{a,\mathsf{t}})$. For every $a \in A$, $\mathsf{t} \in \mathsf{sR}_a$ let $\tau(\alpha^{a,\mathsf{t}}) = \mathsf{M}^{a,\mathsf{t}}_{\mathsf{s}^{a,\mathsf{t}}} = ((\mathsf{S}^{a,\mathsf{t}}, \mathsf{R}^{a,\mathsf{t}}, \mathsf{pre}^{a,\mathsf{t}}), \mathsf{s}^{a,\mathsf{t}})$.

Let $\alpha = ?\mathsf{pre(s)} \otimes \bigotimes_{a\in A} L_a(?\top, \bigsqcup_{\mathsf{t}\in\mathsf{sR}_a} \alpha^{a,\mathsf{t}})$. Then $\tau(\alpha) = \mathsf{M}'_{\mathsf{s}'} = ((\mathsf{S}', \mathsf{R}', \mathsf{pre}'), \mathsf{s}')$ where:

$$
\begin{aligned}
\mathsf{S}' &= \bigcup_{a\in A,\mathsf{t}\in\mathsf{sR}_a} (\mathsf{S}^{a,\mathsf{t}}) \cup \{\bar{\mathsf{s}}^{a,\mathsf{t}} \mid a \in A, \mathsf{t} \in \mathsf{sR}_a\} \cup \{\mathsf{s}'\} \\
\mathsf{R}'_a &= \bigcup_{b\in A,\mathsf{t}\in\mathsf{sR}_b} (\mathsf{R}^{b,\mathsf{t}}_a) \cup (\{\mathsf{s}'\} \cup \{\bar{\mathsf{s}}^{a,\mathsf{t}} \mid \mathsf{t} \in \mathsf{sR}_a\})^2 \cup \\
&\qquad \bigcup_{b\in A\setminus\{a\},\mathsf{t}\in\mathsf{R}_b} (\{\bar{\mathsf{s}}^{b,\mathsf{t}}\} \cup \mathsf{s}^{b,\mathsf{t}}\mathsf{R}^{b,\mathsf{t}}_a)^2 \text{ for } a \in A \\
\mathsf{pre}' &= \bigcup_{a\in A,\mathsf{t}\in\mathsf{sR}_a} (\mathsf{pre}^{a,\mathsf{t}}) \cup \{(\bar{\mathsf{s}}^{a,\mathsf{t}}, \mathsf{pre}^{a,\mathsf{t}}(\mathsf{s}^{a,\mathsf{t}})) \mid a \in A, \mathsf{t} \in \mathsf{sR}_a\} \cup \{(\mathsf{s}', \mathsf{pre(s)})\}
\end{aligned}
$$

We note that unlike the constructions used for Proposition 6.3 and Proposition 6.6, this construction does not have $\mathsf{M}'_{\mathsf{u}} \leftrightarrows \mathsf{M}^{a,\mathsf{t}}_{\mathsf{u}}$, as we do not have that $\mathsf{s}^{a,\mathsf{t}}\mathsf{R}'_a = \mathsf{s}^{a,\mathsf{t}}\mathsf{R}^{a,\mathsf{t}}_a$. Similar to the proof of Proposition 6.6 we claim that each $\bar{\mathsf{s}}^{a,\mathsf{t}}$ state is $(n-1)$-bisimilar to the corresponding s^{t} state. However in lieu of bisimilarity of $\mathsf{S}^{a,\mathsf{t}}$ states we need another result for these states. We also need to consider the additional state s', which due to reflexivity is also a successor of itself.

We need to show for every $0 \leq i \leq n-1$:

(i) For every $a \in A$: $\mathsf{M}'_{\mathsf{s}'} \leftrightarrows_i \mathsf{M}'_{\bar{\mathsf{s}}^{a,\mathsf{s}}}$.

(ii) For every $a \in A$, $\mathsf{t} \in \mathsf{sR}_a$: $\mathsf{M}'_{\bar{\mathsf{s}}^{a,\mathsf{t}}} \leftrightarrows_i \mathsf{M}'_{\mathsf{s}^{a,\mathsf{t}}}$.

(iii) For every $a \in A$, $\mathsf{t} \in \mathsf{sR}_a$, $\mathsf{u} \in \mathsf{S}^{a,\mathsf{t}}$, $\mathsf{v} \in \mathsf{S}$: if $\mathsf{M}^{a,\mathsf{t}}_{\mathsf{u}} \leftrightarrows_i \mathsf{M_v}$ then $\mathsf{M}'_{\mathsf{u}} \leftrightarrows_i \mathsf{M_v}$.

We proceed by induction on i and omit the details, as they are straightforward. For full details refer to the extended version of this paper [16]

Therefore for every $a \in A$, $\mathsf{t} \in \mathsf{sR}_a$ we have that $\mathsf{M}'_{\mathsf{s}'} \leftrightarrows_{(n-1)} \mathsf{M_s}$ and $\mathsf{M}'_{\bar{\mathsf{s}}^{a,\mathsf{t}}} \leftrightarrows_{(n-1)} \mathsf{M_t}$. We can now show that $\mathsf{M}_{\mathsf{s}'} \leftrightarrows_n \mathsf{M_s}$ by using the same reasoning as the proof for Proposition 6.3, using the $(n-1)$-bisimilar $\mathsf{M}'_{\bar{\mathsf{s}}^{a,\mathsf{t}}}$ in place of corresponding $\mathsf{M}'_{\mathsf{s}^{\mathsf{t}}}$ states. □

Corollary 6.10 *Let* $\mathsf{M_s} \in \mathcal{AM}_{S5}$. *Then for every* $\varphi \in \mathcal{L}_\otimes$ *there exists* $\alpha \in \mathcal{L}_?{}^{act}$ *such that* $\models_{S5_\otimes} [\mathsf{M_s}]\varphi \leftrightarrow [\tau(\alpha)]\varphi$.

Corollary 6.11 *Let $\varphi \in \mathcal{L}_{\otimes}$. Then there exists $\varphi' \in \mathcal{L}_{?}$ such that for every $M_s \in \mathcal{S}5$: $M_s \models_{S5_{\otimes}} \varphi$ if and only if $M_s \models_{S5_?} \varphi'$.*

7 Synthesis

In the following subsections we give a computational method for synthesising action formulae to achieve epistemic goals, whenever those goals are achievable. We note that the notion of when an epistemic goal is achievable is captured by the refinement quantifiers of refinement modal logic [10], which are also included in the arbitrary action formula logic, and so in this section we will refer to the full arbitrary action formula logic, keeping in mind the correspondence with arbitrary action model logic mentioned in Section 4.

7.1 $\mathcal{K}$

Proposition 7.1 *For every $\varphi \in \mathcal{L}_?$ there exists $\alpha \in \mathcal{L}_?{}^{act}$ such that $\vdash [\alpha]\varphi$ and $\vdash \exists\varphi \to \langle\alpha\rangle\varphi$.*

Proof. We note that we use the axioms of $\mathbf{RML_K}$ for refinement modal logic [10] in this proof, as validities in the arbitrary action formula logic. This is similar to the approach used in the arbitrary action model logic of Hales [18].

Without loss of generality we assume that φ is in the disjunctive normal form of [10]. We proceed by induction on the structure of φ.

Suppose that $\varphi = \psi \vee \chi$. By the induction hypothesis there exists $\alpha^\psi, \alpha^\chi \in \mathcal{L}_?{}^{\mathrm{act}}$ such that $\vdash [\alpha^\psi]\psi$, $\vdash \exists\psi \to \langle\alpha^\psi\rangle\psi$, $\vdash [\alpha^\chi]\chi$ and $\vdash \exists\chi \to \langle\alpha^\chi\rangle\chi$. Let $\alpha = \alpha^\psi \sqcup \alpha^\chi$. Then $\vdash [\alpha^\psi \sqcup \alpha^\chi](\psi \vee \chi)$ and $\vdash \exists(\psi \vee \chi) \to \langle\alpha^\psi \sqcup \alpha^\chi\rangle(\psi \vee \chi)$ follows trivially from **LU** and the $\mathbf{RML_K}$ axiom **R**.

Suppose that $\varphi = \pi \wedge \bigwedge_{b \in B \subseteq A} \nabla_b \Gamma_b$. By the induction hypothesis for every $b \in B$, $\gamma \in \Gamma_b$ there exists $\alpha^\gamma \in \mathcal{L}_?{}^{\mathrm{act}}$ such that $\vdash [\alpha^\gamma]\gamma$ and $\vdash \exists\gamma \to \langle\alpha^\gamma\rangle\gamma$. Let $\alpha = ?\exists\varphi \otimes \bigotimes_{b \in B} L_b(\bigsqcup_{\gamma \in \Gamma_b} \alpha^\gamma)$.

Then for every $b \in B$:

$$\vdash \Box_b[\bigsqcup_{\gamma \in \Gamma_b} \alpha^\gamma] \bigvee_{\gamma \in \Gamma} \gamma \tag{5}$$

$$\vdash [L_b(\bigsqcup_{\gamma \in \Gamma_b} \alpha^\gamma)]\Box_b \bigvee_{\gamma \in \Gamma} \gamma \tag{6}$$

$$\vdash [?\exists\varphi \otimes \bigotimes_{c \in B} L_c(\bigsqcup_{\gamma \in \Gamma_c} \alpha^\gamma)]\Box_b \bigvee_{\gamma \in \Gamma} \gamma \tag{7}$$

$$\vdash \exists\varphi \to \bigwedge_{b \in B, \gamma \in \Gamma_b} \Diamond_b \exists\gamma \tag{8}$$

$$\vdash \exists\varphi \to \bigwedge_{b \in B, \gamma \in \Gamma_b} \Diamond_b \langle \bigsqcup_{\gamma' \in \Gamma_b} \alpha^{\gamma'}\rangle\gamma \tag{9}$$

$$\vdash [?\exists\varphi \otimes \bigotimes_{c \in B} L_c(\bigsqcup_{\gamma \in \Gamma_c} \alpha^\gamma)] \bigwedge_{b \in B, \gamma \in \Gamma_b} \Diamond_b \gamma \tag{10}$$

$$\vdash [?\exists\varphi \otimes \bigotimes_{c \in B} L_c(\bigsqcup_{\gamma \in \Gamma_c} \alpha^\gamma)](\pi \wedge \bigwedge_{b \in B} \nabla_b \Gamma_b) \tag{11}$$

(5) follows from the induction hypothesis, **LU** and **NecK**, (6) follows from **LK1**, (7) follows from **LK2**, **LS** and **NecL**. (8) follows from **RK**, (9) follows from the induction hypothesis and **LU**, (10) follows from **LK1**, **LK2** and **LS** and **LT**, (11) follow from (7), **LT**, **RP LC** and the definition of the cover operator.

Therefore $\vdash [\alpha]\varphi$.

Finally:

$$\vdash \langle ?\exists\varphi \otimes \bigotimes_{c\in B} L_c(\bigsqcup_{\gamma\in\Gamma_c} \alpha^{\gamma})\rangle\top \leftrightarrow \exists\varphi \quad (12)$$

$$\vdash \exists\varphi \to \langle\alpha\rangle\varphi \quad (13)$$

(12) follows from **LS**, **LP** and **LT**, (13) follows from (11) and (12).

Therefore $\vdash \exists\varphi \to \langle\alpha\rangle\varphi$. □

Corollary 7.2 *For every $M_s \in \mathcal{K}$ and $\varphi \in \mathcal{L}_{\otimes\forall}$: $M_s \models \exists\varphi$ if and only if there exists $\mathsf{M}_\mathsf{s} \in \mathcal{AM}$ such that $M_s \models \langle\mathsf{M}_\mathsf{s}\rangle\varphi$.*

7.2 $\mathcal{K}45$

Proposition 7.3 *For every $\varphi \in \mathcal{L}_?$ there exists $\alpha \in \mathcal{L}_?{}^{act}$ such that $\vdash [\alpha]\varphi$ and $\vdash \exists\varphi \to \langle\alpha\rangle\varphi$.*

Proof. Hales, French and Davies previously provided an axiomatisation $\mathbf{RML_{KD45}}$ of refinement modal logic in the setting of $\mathcal{KD}45$. We note that this axiomatisation can be adapted to the setting of $\mathcal{K}45$ with minor modifications and is sound and complete following essentially the same reasoning.

Without loss of generality we assume that φ is in the alternating disjunctive normal form of [19]. We use the same reasoning as in the proof of Proposition 7.1, substituting $\mathbf{AFL_{K45}}$ axioms for the corresponding $\mathbf{AFL_K}$ axioms, noting that the alternating disjunctive normal form gives the $(A\backslash\{a\})$-restricted properties required for **LK1** and the $\mathbf{RML_{K45}}$ axioms **RK45**, **RComm** and **RDist** to be applicable. □

Corollary 7.4 *For every $M_s \in \mathcal{K}45$ and $\varphi \in \mathcal{L}_{\otimes\forall}$: $M_s \models \exists\varphi$ if and only if there exists $\mathsf{M}_\mathsf{s} \in \mathcal{AM}_{\mathcal{K}45}$ such that $M_s \models \langle\mathsf{M}_\mathsf{s}\rangle\varphi$.*

7.3 $\mathcal{S}5$

Proposition 7.5 *For every $\varphi \in \mathcal{L}_?$ there exists $\alpha \in \mathcal{L}_?{}^{act}$ such that $\vdash [\alpha]\varphi$ and $\vdash \exists\varphi \to \langle\alpha\rangle\varphi$.*

Proof. Similar to Proposition 7.1 we use the axioms of $\mathbf{RML_{S5}}$ for refinement modal logic [19] in this proof.

Without loss of generality, assume that φ is a disjunction of explicit formulae of [19]. We proceed by induction on the structure of φ.

Suppose that $\varphi = \psi \vee \chi$. As in Proposition 7.1 this is trivial.

Suppose that $\varphi = \pi \wedge \gamma^0 \wedge \bigwedge_{a\in A} \nabla_a\Gamma_a$ is an explicit formula. By the induction hypothesis for every $a \in A$, $\gamma \in \Gamma_a$ there exists $\alpha^{a,\gamma} \in \mathcal{L}_?{}^{\mathsf{act}}$ such that $\vdash [\alpha^{a,\gamma}]\gamma$ and $\vdash \exists\gamma \to \langle\alpha^{a,\gamma}\rangle\gamma$, where $\tau(\alpha^{a,\gamma}) = \mathsf{M}^{a,\gamma}_{\mathsf{s}^{a,\gamma}} = ((\mathsf{S}^{a,\gamma}, \mathsf{R}^{a,\gamma}, \mathsf{pre}^{a,\gamma}), \mathsf{s}^{a,\gamma})$.

Let $\alpha = ?\exists\gamma^0 \otimes \bigotimes_{a\in A} L_a(?\top, \bigsqcup_{\gamma\in\Gamma_a} \alpha^{a,\gamma})$. We note that the construction of $\tau(\alpha)$ has essentially the same structure as the construction used in Theorem 6.9, and the only differences are in state naming and preconditions. We omit the details. For full details refer to the extended version of this paper [16]

Let $\Psi = \{\psi \leq \gamma \mid a \in A, \gamma \in \Gamma_a\}$. We need to show for every $\psi \in \Psi$:

(i) For every $a \in A$: $\vdash [\mathsf{M}_{\mathsf{s}}]\psi \leftrightarrow [\mathsf{M}_{\mathsf{s}^{a,\gamma^0}}]\psi$.

(ii) For every $a \in A$, $\gamma \in \Gamma_a$: $\vdash [\mathsf{M}_{\bar{\mathsf{s}}^{a,\gamma}}]\psi \leftrightarrow [\mathsf{M}_{\mathsf{s}^{a,\gamma}}]\psi$.

(iii) For every $a \in A$, $\gamma \in \Gamma_a$, $\mathsf{u} \in \mathsf{S}^{a,\gamma}$: $\vdash [\mathsf{M}_{\mathsf{u}}]\psi \leftrightarrow [\mathsf{M}^{a,\gamma}_{\mathsf{u}}]\psi$.

We proceed by induction on ψ, and omit the details, as they are straightforward. For full details refer to the extended version of this paper [16]

The remainder of the proof follows similar reasoning to that of Proposition 7.1, using $\mathbf{AML_{S5}}$ axioms in place of $\mathbf{AML_K}$ axioms, and noting that the $\mathbf{AFL_K}$ axioms **LT**, **LU** and **LS** are also sound for $\mathcal{S}5_?$. □

Corollary 7.6 *For every $M_s \in \mathcal{S}5$ and $\varphi \in \mathcal{L}_{\otimes\forall}$: $M_s \models \exists\varphi$ if and only if there exists $\mathsf{M}_{\mathsf{s}} \in \mathcal{AM}_{\mathcal{S}5}$ such that $M_s \models \langle\mathsf{M}_{\mathsf{s}}\rangle\varphi$.*

8 Related work

Several other papers have addressed the problem of describing and reasoning about epistemic actions. One of the most important works in this area is the work of Baltag, Moss and Solecki [8] which introduced the notion of action model logic, building on the earlier work of Gerbrandy and Groeneveld [17]. In later work Baltag and Moss extended action model logic to consider epistemic programs [7] which are expressions built from action models using such operators as sequential composition, non-deterministic choice and iteration. The atoms of these programs are action models, so the approach is still inherently semantic in nature. The logic is unable to decompose the program beyond the level of the atoms, which themselves may be complex semantic objects.

The relational actions of van Ditmarsch [12] provides a syntactic mechanism for describing an epistemic action, and provides the foundation for a lot of the work presented in this paper. The relational actions are constructed using essentially the same operators as in the language of action formulae. While the language is very similar, the semantics given are quite different [15]. In the logic of epistemic actions the semantics are given in such a way that worlds in a model are specified with respect to subsets of agents, so that the model is restricted to agents for whom the epistemic action was applied. The semantics were also specific to $\mathcal{S}5$, and non-trivial to generalise to other epistemic logics. A version of relational actions with concurrency is able to describe any $\mathcal{S}5$ action model, although it is unknown whether the expressivity of concurrent relational actions is greater than that of action models [6].

Related synthesis results have been given by Aucher, et al. [2,3,4] which presents an event model language and uses it to give a thorough exploration of the relationship between epistemic models, action models and epistemic goals. Aucher defines a logic for action models and provides calculi to describe epistemic progression (what is true after executing a given action in a given model) epistemic regression (what is the most general precondition for an epistemic action given an epistemic goal) and epistemic planning (what action is sufficient to achieve an epistemic goal given some precondition). In future work we hope to extend the correspondence between action formula logic and action models to include Aucher's event model language.

References

[1] Ågotnes, T., P. Balbiani, H. van Ditmarsch and P. Seban, *Group announcement logic*, Journal of Applied Logic **8** (2010), pp. 62–81.
[2] Aucher, G., *Del-squents for progression*, Journal of Applied Non-classical Logics **21** (2011), pp. 289–321.
[3] Aucher, G., *Del-squents for regression and epistemic planning*, Journal of Applied Non-classical Logics **22** (2012), pp. 337–367.
[4] Aucher, G. and T. Bolander, *Undecidability in epistemic planning*, in: *Proceedings of the Twenty-Third International Joint Conference on Artificial Intelligence*, 2013, pp. 27–33.
[5] Balbiani, P., H. van Ditmarsch, A. Herzig and T. de Lima, *Some truths are best left unsaid.*, Advances in Modal Logic **9** (2012), pp. 36–54.
[6] Baltag, A. and H. van Ditmarsch, *Relation between two dynamic epistemic logics* (2006), annual Conference of the Australasian Association for Logic.
[7] Baltag, A. and L. Moss, *Logics for epistemic programs*, Information, Interaction and Agency (2005), pp. 1–60.
[8] Baltag, A., L. S. Moss and S. Solecki, *The logic of common knowledge, public announcements and private suspicions*, in: *Proceedings of the 7th conference on theoretical aspects of rationality and knowledge*, 1998, pp. 43–56.
[9] Bílková, M., A. Palmigiano and Y. Venema, *Proof systems for the coalgebraic cover modality*, Advances in Modal Logic **7** (2008), pp. 1–21.
[10] Bozzelli, L., H. van Ditmarsch, T. French, J. Hales and S. Pinchinat, *Refinement modal logic*, Arxiv preprint arXiv:1202.3538 (2012).
[11] van Ditmarsch, H., *The logic of knowledge games: showing a card*, in: *Proceedings of the BNAIC*, 1999, pp. 35–42.
[12] van Ditmarsch, H., *Knowledge games*, Bulletin of Economic Research **53** (2001), pp. 249–273.
[13] van Ditmarsch, H., *Descriptions of game actions*, Journal of Logic, Language and Information **11** (2002), pp. 349–365.
[14] van Ditmarsch, H., T. French and S. Pinchinat, *Future event logic: axioms and complexity*, in: *Advances in Modal Logic*, 2010, pp. 24–27.
[15] van Ditmarsch, H., W. van der Hoek and B. Kooi, "Dynamic epistemic logic," Springer Verlag, 2007.
[16] French, T., J. Hales and E. Tay, *A composable language for action models*, Arxiv preprint arXiv:1406.2103 (2014), extended version.
[17] Gerbrandy, J. and W. Groeneveld, *Reasoning about information change*, Journal of logic, language and information **6** (1997), pp. 147–169.
[18] Hales, J., *Arbitrary action model logic and the synthesis of action models*, in: *Proceedings of the 2013 28th Annual IEEE/ACM Symposium on Logic in Computer Science*, IEEE Computer Society, 2013, pp. 253–262.
[19] Hales, J., T. French and R. Davies, *Refinement quantified logics of knowledge and belief for multiple agents*, Advances in Modal Logic **9** (2012), pp. 317–338.
[20] Janin, D. and I. Walukiewicz, *Automata for the modal μ-calculus and related results*, Mathematical Foundations of Computer Science 1995 (1995), pp. 552–562.
[21] Plaza, J., *Logics of public communications*, in: *Proceeding of the 4th International Symposium on Methodologies for Intelligent Systems*, 1989, pp. 102–216.

Free Algebras for Gödel-Löb Provability Logic

Sam J. van Gool [1]

Mathematical Institute, University of Bern, Switzerland

Abstract

We give a construction of finitely generated free algebras for Gödel-Löb provability logic, GL. On the semantic side, this construction yields a notion of canonical graded model for GL and a syntactic definition of those normal forms which are consistent with GL. Our two main techniques are incremental constructions of free algebras and finite duality for partial modal algebras. In order to apply these techniques to GL, we use a rule-based formulation of the logic GL by Avron (which we simplify slightly), and the corresponding semantic characterization that was recently obtained by Bezhanishvili and Ghilardi.

Keywords: Stone duality, provability logic, diagonalizable algebra, Magari algebra, partial modal algebra, one-step constructions.

1 Introduction

The provability logic GL is the axiomatic extension of the basic modal logic K by the Gödel-Löb axiom $\Box(\Box p \to p) \to \Box p$. The intended interpretation of the modal operator $\Box$ in GL is "it is provable in T that ...", where T is a sufficiently strong formal theory, such as, for example, Peano arithmetic. A classical theorem of Solovay [19] shows that, indeed, GL is exactly the logic of provability of Peano arithmetic. From a modal logic perspective, the logic GL is interesting because, on the one hand, it has reasonably nice model-theoretic properties (notably, GL is complete with respect to the class of finite irreflexive transitive frames), but on the other hand it fails to be canonical, that is, the canonical model of the logic GL does *not* validate the Gödel-Löb axiom.

The situation with normal forms for GL is similarly subtle: Boolos [7] gave normal forms for the fragment of GL-formulas containing no propositional variables, but in [8] Boolos showed that the same method does not apply to the

[1] The author would like to thank Nick Bezhanishvili, Mai Gehrke, and Silvio Ghilardi for many interesting and helpful discussions on the topic of this paper. The research reported in this paper was performed during a research stay at Université Paris 7 and INRIA, hosted by Paul-André Melliès, to whom the author is also grateful for providing the inspiration for the "graded" view in Section 6 of this paper. Finally, the author would like to thank the anonymous referees for their many useful suggestions.

fragment of GL-formulas in n variables for $n > 0$. This result has sometimes been cited in the literature as saying that *no* normal forms exist for GL on n variables, but this is not what Boolos proved, nor does it seem to be what he intended to claim[2]. Indeed, one of the contributions of this paper is to give an explicit construction of normal forms for GL on an arbitrary finite number of variables.

It has been known since the work of Fine [10] (which was put into algebraic perspective in [1] and [12]) that any modal formula of modal degree[3] n on k variables $\{p_1, \dots, p_k\}$ is equivalent in K to a finite (possibly empty) disjunction of *normal forms of degree n*. Here, the normal forms of degree 0 are the formulas of the form $\bigwedge_{i \in T} p_i \wedge \bigwedge_{j \notin T} \neg p_j$, for $T \subseteq \{p_1, \dots, p_k\}$, and normal forms of degree $n+1$ are formulas of the form

$$\phi \wedge \nabla \Psi, \tag{1}$$

where ϕ is a normal form of degree 0, Ψ is a finite set of normal forms of degree n, and $\nabla\Psi$ abbreviates $\Box \left(\bigvee_{\psi \in \Psi} \psi\right) \wedge \left(\bigwedge_{\psi \in \Psi} \Diamond \psi\right)$, a notation from coalgebraic logic [16] (we will make some remarks on the connection with coalgebraic logic in the conclusion of this paper).

In the modal logic K, each of the normal forms (1) is satisfiable. However, this is clearly no longer the case in GL. In this paper we give a bottom-up construction of the free GL-algebra (Section 5), and from the construction we extract a bottom-up definition of the normal forms that are consistent with GL (Definition 6.3). In Section 6, we also develop a notion of *graded model* for GL, and construct a canonical graded model for GL.

Let us now discuss the methods of this paper in some more detail. Recall that a *modal algebra* is a tuple $(A, \vee, \wedge, \neg, 0, 1, \Box)$ where $(A, \vee, \wedge, \neg, 0, 1)$ is a Boolean algebra and $\Box : A \to A$ is a unary operation on A which preserves $\wedge$ and 1. As is well-known, the variety of modal algebras is the algebraization of the basic modal logic K, in the sense that a formula ϕ is a tautology of K if, and only if, the equation $\phi = 1$ is valid in every modal algebra. By definition, a GL-*algebra*[4] is a modal algebra in which the equation $\Box(\Box a \to a) \leq \Box a$ is valid. The variety of GL-algebras algebraizes the logic GL. The *free* GL-*algebra* over variables $p_1, \dots, p_k$ is the GL-algebra consisting of GL-equivalence classes of modal formulas using variables from $p_1, \dots, p_k$. We shall denote this algebra by $\mathbb{F}_{\mathsf{GL}}(k)$.

The notion of modal degree allows one to approximate a finitely generated modal algebra by an infinite chain of finite algebras, in each of which the

---

[2] Note in particular that, in the title of [8], the word "certain" is under the scope of a negated existential quantifier.

[3] Recall that the *modal degree* of a modal formula is the maximum length of a string of nested occurrences of the modal operator $\Box$.

[4] These algebras are the same (although axiomatized slightly differently) as the algebras introduced by Magari [15] and are also called *Magari algebras* or *diagonalizable algebras* in the literature.

modal operator is partially defined. Concretely, if A is a modal algebra with generators $a_1, \ldots, a_k$, let

$$A_n := \{\phi^A(a_1, \ldots, a_k) \mid \phi(p_1, \ldots, p_k) \text{ an } k\text{-variable formula of degree} \leq n\}. \tag{2}$$

Then $A_0 \subseteq A_1 \subseteq \cdots$ is an increasing chain of finite Boolean subalgebras of A, and $A = \bigcup_{n \geq 0} A_n$, since A is generated by $a_1, \ldots, a_k$. Moreover, the operation $\Box$ on the modal algebra A restricts, for each $n \geq 0$, to an operation $\Box_n : A_n \to A_{n+1}$. The tuple $(A_{n+1}, A_n, \Box_n)$ is an example of a *partial modal algebra* (cf. Definition 2.1 below). Ghilardi's pioneering idea in [12], the germ of which appeared earlier in unpublished lectures by Abramsky [1], was to describe the increasing chain of Boolean algebras $A_0, A_1, \ldots$ and the operations $\Box_n$ between them *from the bottom up*. That is, for many classes V of modal algebras, it is possible to describe a functor F on partial modal algebras with the property that, if A is a finitely generated modal algebra in V and $(A_1, A_0, \Box_0)$ is the first algebra in the chain described above, then the nth such algebra is (isomorphic to) $(F_V)^n(A_1, A_0, \Box_0)$. In this case, one obtains an incremental, bottom-up construction for algebras in the class V, by starting from a partial modal algebra and taking an appropriate colimit of the chain of algebras $(F_V)^n(A_1, A_0, \Box_0)$. In Section 4, we give the definition of such a functor F in the case where V is the class of GL-algebras, and apply it to the particular case of the finitely generated free algebras $\mathbb{F}_{\mathsf{GL}}(k)$. Definition 4.1 is an instance of the general definition in [9, Section 2] of a free image-total functor associated to a set of quasi-equations of degree ≤ 1. It is important to note that the functor F depends on the particular axiomatization of the class V, also see the conclusion.
The free image-total functor F is often most conveniently described using duality. Dual to any partial modal algebra is a *partial frame* (called "q-frame" in [9]), which consists of a set equipped with an equivalence relation $\sim$ and a Kripke accessibility relation R which respects $\sim$, cf. Definition 3.1 below for the precise definition. Under this duality, the construction F on partial modal algebras corresponds to a dual construction G on partial frames. Since the construction that underlies F is a pushout in the category of Boolean algebras ([5, Prop. 5]), i.e., a quotient of a coproduct, the construction that underlies G is a pullback in the category of sets, i.e., a subset of a product. As a result, the action on objects of the functor G is usually easier to identify than that of the functor F. Indeed, in Definition 4.4 we will give a direct combinatorial description of the functor G.

Outline of the paper. Section 2 contains the necessary algebraic definitions, notably, the definition of partial GL-algebra. In Section 3 we recall duality for partial modal algebras, and specialize it to the case of partial GL-algebras. In Section 4 we define the free image-total functor for GL and characterize its dual. In Section 5 we apply the functor F to obtain a construction of the free finitely generated GL-algebra, and in Section 6 we use G to obtain a canonical graded model for GL.

2 Partial GL-algebras

In this paper, we make extensive use of a generalization of modal algebras, namely *partial* modal algebras.

Definition 2.1 A *partial modal algebra* is a tuple $(A, B, \Box)$, where A is a Boolean algebra, B is a Boolean subalgebra of A, and $\Box : B \to A$ is a function which preserves $\wedge$ and 1. The algebra A is called the *underlying Boolean algebra* of $(A, B, \Box)$. A *homomorphism* from a partial modal algebra $(A, B, \Box)$ to a partial modal algebra $(A', B', \Box')$ is a map $h : A \to A'$ such that $h(B) \subseteq B'$ and $h(\Box b) = \Box' h(b)$ for all $b \in B$. A homomorphism h is an *isomorphism* if h is bijective and $h(B) = B'$. A *congruence* on a partial modal algebra $(A, B, \Box)$ is a Boolean algebra congruence θ on A such that moreover, for all $b, b' \in B$, if $b\theta b'$, then $\Box b \theta \Box b'$. We denote by $(A/\theta, B/\theta, \Box/\theta)$ the quotient of $(A, B, \Box)$ by θ. Note that, for any congruence θ, $(A/\theta, B/\theta, \Box/\theta)$ is again a partial modal algebra. A partial modal algebra $(A, B, \Box)$ is called *total* if $B = A$.

Remark 2.2 Note that the category of partial modal algebras with homomorphisms is equivalent to the category whose objects are diagrams of shape $B \underset{i}{\overset{\Box}{\rightrightarrows}} A$ where A, B are Boolean algebras, i is an injective Boolean homomorphism, and $\Box$ is a meet-semilattice morphism, and whose morphisms are given by the obvious commuting diagrams. Therefore, the category of partial modal algebras is equivalent to a full subcategory of the category of one-step modal algebras introduced in [5].

Example 2.3 Let $(A, \Box)$ be a finitely generated modal algebra, with generators $a_1, \dots, a_k$. Let A_0 be the Boolean algebra generated by $a_1, \dots, a_k$, and let A_1 be the Boolean algebra generated by $a_1, \dots, a_k, \Box a_1, \dots, \Box a_k$. The operation $\Box$ on C restricts to an operation $\Box : A_0 \to A_1$, and it follows that $(A_0, A_1, \Box)$ is a partial modal algebra. Repeating this process, by taking all elements of A_1 as set of generators in the next step, one obtains, after n repetitions, the partial modal algebra $(A_{n+1}, A_n, \Box_n)$ defined in equation (2) in the introduction.

In particular, if $\mathbf{L}$ is a normal modal logic, we may take for A the k-variable Lindenbaum algebra for $\mathbf{L}$, i.e., the modal algebra of $\mathbf{L}$-equivalence classes of modal formulas in variables $c_1, \dots, c_k$. In this case, the partial modal algebras $(A_{n+1}, A_n, \Box)$ defined in the previous paragraph are the partial modal algebras of $\mathbf{L}$-equivalence classes of k-variable modal formulas of degree at most $n+1$.

We now identify a subclass $\mathbf{V}$ of the class of partial modal algebras such that the total algebras in $\mathbf{V}$ are exactly the GL-algebras. There is a choice to be made here: several such varieties $\mathbf{V}$ exist, and not all of them are suitable for our purposes (also cf. the conclusion).

Definition 2.4 A partial modal algebra $(A, B, \Box)$ is a *partial* GL-*algebra* if, for all $a, b \in B$:

$$\text{if } b \leq \Box a \to a \text{ then } \Box b \leq \Box a. \tag{3}$$

The proof of the following fact is a direct application of a procedure which is known in the literature as *Ackermann's lemma*, or *flattening*.

Lemma 2.5 *Let* $(A, \Box)$ *be a modal algebra. The partial modal algebra* $(A, A, \Box)$ *is a partial* GL*-algebra if, and only if,* $(A, \Box)$ *is a* GL*-algebra.*

Proof. If (3) holds in $(A, A, \Box)$, then, for any $a \in A$, we can instantiate (3) with $b := \Box a \to a$ to obtain $\Box(\Box a \to a) \leq \Box a$. Conversely, if $(A, \Box)$ is a GL-algebra and $a, b \in A$ are such that $b \leq \Box a \to a$, then we get

$$\Box b \leq \Box(\Box a \to a) \leq \Box a,$$

where we have used that $\Box : A \to A$ is order-preserving and that $(A, \Box)$ is a GL-algebra. □

Remark 2.6 Definition 2.4 was inspired by [3, Section 5], where a rule-based formulation for GL due to Avron [2] was used. Avron's rule, when translated to a quasi-equation, says that, for all $a, b \in B$,

$$\text{if } b \wedge \Box b \leq \Box a \to a \text{ then } \Box b \leq \Box a. \tag{4}$$

Avron [2] used the rule (4) to obtain a sequent calculus for GL which has cut-elimination, and used it to prove Kripke completeness of GL. It is possible to prove syntactically that the quasi-equations (3) and (4) are equivalent for any partial modal algebra. The reason why we used (3) rather than (4) in Definition 2.4 is that (3) is simply a flattening of the usual GL-axiom, while showing the equivalence of (4) with the usual GL-axiom requires some ingenuity.

3 Duality for partial GL-algebras

We now recall the facts about duality for finite partial modal algebras that we need in this paper, and we recall how this duality specializes to finite partial GL-algebras. For more details about duality for finite partial modal algebras, cf., e.g., [9, Sec. 4] or [5, Sec. 3.2].

Definition 3.1 A *partial frame*[5] is a tuple $(X, \sim, R)$, where X is a set, $\sim$ is an equivalence relation on X, and $R \subseteq X \times X$ is a relation such that $xRy \sim y'$ implies xRy'. A *bounded morphism* from a partial frame $(X, \sim_X, R)$ to a partial frame $(Y, \sim_Y, S)$ is a function f from X to Y such that $x \sim_X x'$ implies $f(x) \sim_Y f(x')$, and $f(x)Sy$ if, and only if, there exists $x' \in X$ such that xRx' and $f(x') \sim_Y y$. A *partial generated subframe* of a partial frame $(X, \sim, R)$ is a partial frame $(Y, \approx, Q)$ such that $Y \subseteq X$, the relations $\approx$ and Q are the restrictions to Y of the relations $\sim$ and R, respectively, and, for all $y \in Y$ and $x \in X$, if yRx, then there exists $x' \in Y$ such that $x' \sim x$. A partial frame is called *total* if the equivalence relation $\sim$ is the diagonal $\Delta = \{(x, x) \mid x \in X\}$.

[5] These were called q-frames in [9], and a generalization of these were called one-step frames in [5].

Notation. In a partial frame $(X, \sim, R)$, for $x \in X$, we use the notation $R(x)$ for the set $\{y \in X \mid xRy\}$ and $[x]_\sim$ for the $\sim$-equivalence class of y.

For any partial frame $(X, \sim, R)$, we define its *dual partial modal algebra* $(A, B, \Box)$ by

$$\begin{aligned} A &:= \mathcal{P}(X), \\ B &:= \mathcal{P}_\sim(X) = \{b \in \mathcal{P}(X) \mid \text{ if } x \in b \text{ and } x \sim x' \text{ then } x' \in b\}, \\ \text{for } b \in B,\ \Box b &:= \{x \in X \mid \text{ if } xRy \text{ then } y \in b\}. \end{aligned}$$

Note that indeed $(A, B, \Box)$ is a partial modal algebra. For a bounded morphism $f : (X, \sim_X, R) \to (Y, \sim_Y, S)$, if $(A, B, \Box)$, $(A', B', \Box')$ are the partial modal algebras dual to X and Y respectively, we define a homomorphism $h : A' \to A$ by $h(c) := f^{-1}(c)$. Note that h is indeed a homomorphism of partial modal algebras. Moreover, these assignments define a contravariant functor from the category of partial frames to the category of partial modal algebras. When restricted to *finite* partial frames, this functor becomes part of a dual equivalence, as is obvious by combining Remark 2.2 with the well-known fact that the category of finite Kripke frames is dually equivalent to the category of finite modal algebras.

Theorem 3.2 (Duality for finite partial modal algebras) *The category of finite partial frames is dually equivalent to the category of finite partial modal algebras.*

Proof. (Sketch) The functor from finite partial frames to finite partial modal algebras takes a finite partial frame to its (finite) dual partial modal algebra. The functor in the other direction takes a finite partial modal algebra to its set of atoms, equipped with the appropriate structure of a partial frame. See Theorem A.1 in the appendix for more details. □

Theorem 3.2 in particular implies that epimorphisms in the category of finite partial modal algebras dually correspond to monomorphisms of finite partial frames, and vice versa. We formulate these facts in some detail in the following two corollaries.

Corollary 3.3 *Let $(A, B, \Box)$ be a finite partial modal algebra with dual finite partial frame $(X, \sim, R)$. There is a Galois connection between $\mathcal{P}(A \times A)$ and $\mathcal{P}(X)$, given by the assignments:*

$$\begin{aligned} E \in \mathcal{P}(A \times A) &\mapsto Y_E := \{x \in X \mid \forall (a, b) \in E : x \le a \iff x \le b\} \\ Y \in \mathcal{P}(X) &\mapsto \theta_Y := \{(a, b) \in A \times A \mid \forall y \in Y \ : \ y \le a \iff y \le b\}. \end{aligned}$$

This Galois connection restricts to an order-reversing isomorphism between the poset of congruences on A and the poset of partial generated subframes of X, both ordered by inclusion. Moreover, for any $E \in \mathcal{P}(A \times A)$, the congruence θ_{Y_E} is the smallest congruence containing E.

Corollary 3.4 *Let $f : (X, \sim_X, R) \to (Y, \sim_Y, S)$ be a bounded morphism between partial frames and let $h : (A', B', \Box') \to (A, B, \Box)$ be the homomorphism* *of partial modal algebras dual to it. Then f is surjective if, and only if, h is injective.*

We now recall some constructions and facts about partial modal algebras that will be used later in this paper.[6] We first define, given a finite Boolean algebra A, a partial modal algebra $(A+V(A), A, \Box)$, and then describe its dual. Recall that the *coproduct* of any two Boolean algebras A and B exists, and can be characterized as the (up to isomorphism unique) Boolean algebra $A+B$ which contains A and B as subalgebras and such that any pair of homomorphisms $(A \to C,\ B \to C)$ factors uniquely through $A+B$.
In the rest of this section, let A and B be arbitrary finite Boolean algebras.

Fact 3.5 *The dual of $A+B$ is the Cartesian product* $\mathrm{At}(A) \times \mathrm{At}(B)$ *and the Boolean algebra $A+B$ is isomorphic to* $\mathcal{P}(\mathrm{At}(A) \times \mathrm{At}(B))$.

The *Vietoris algebra* $V(A)$ over a finite Boolean algebra A is defined to be the algebra of Boolean combinations of formal elements $\Box a$, for $a \in A$, quotiented by the equalities $\Box 1 = 1$ and $\Box(a \wedge b) = \Box a \wedge \Box b$.

Fact 3.6 *The dual of $V(A)$ is* $\mathcal{P}(\mathrm{At}(A))$ *and the Boolean algebra $V(A)$ is isomorphic to* $\mathcal{P}(\mathcal{P}(\mathrm{At}(A)))$.

Combining the above two constructions, for any Boolean algebra A we can form a partial modal algebra $(A+V(A), A, \Box)$, where $\Box$ sends an element $a \in A$ to $\Box a \in V(A)$.

Fact 3.7 *The dual of $A+V(A)$ is $X \times \mathcal{P}(X)$, where $X :=$* $\mathrm{At}(A)$ *and $A+V(A)$ is isomorphic to $\mathcal{P}(X \times \mathcal{P}(X))$. The subalgebra A of $A+V(A)$ is dual to the equivalence relation $\approx$ on $X \times \mathcal{P}(X)$ defined by $(x,T) \approx (y,S)$ if, and only if, $x = y$. The dual of the operation $\Box : A \to A+V(A)$ is the relation Q on $X \times \mathcal{P}(X)$ defined by $(x,T)Q(y,S)$ if, and only if, $y \in T$.*

We end this section by specializing the duality in Theorem 3.2 to finite partial GL-algebras. The following definition identifies the objects in the category of finite partial frames that are dual to finite partial GL-algebras.

Definition 3.8 A *partial* GL*-frame* is a partial frame $(X, \sim, R)$ such that, for any $x, y \in X$, if xRy, then there exists $y' \sim y$ such that $R(y') \subsetneq R(x)$.

It is easy (but not entirely trivial) to show that the finite partial GL-frames of the form $(X, =, R)$ are exactly the frames for which R is irreflexive and transitive. This also follows from Lemma 2.5 and Proposition 3.9 below.
The condition in Definition 3.8 was first derived in [4, Section 9.2] using general correspondence theory for one-step algebras. We give a direct proof of the correspondence between finite partial GL-algebras and finite partial GL-frames.

Proposition 3.9 *Let $(X, \sim, R)$ be a finite partial frame with dual partial modal algebra $(A, B, \Box)$. The following are equivalent:*

[6] The proofs of the well-known Facts 3.5–3.7, which amount to an algebraic explanation of Fine's normal forms (1), are given in the appendix (Facts A.2–A.4).

(i) $(A, B, \Box)$ *is a partial* GL*-algebra;*

(ii) $(X, \sim, R)$ *is a partial* GL*-frame.*

Proof. First suppose that $(A, B, \Box)$ is a partial GL-algebra, and let $x, y \in X$ with xRy. Define $b := R(x)$ and $a := R(x) \setminus [y]_\sim$. Note that $x \in \Box b$, but $x \notin \Box a$, since xRy and $y \notin a$. Thus, $\Box b \nleq \Box a$, and since $(A, B, \Box)$ is a partial GL-algebra we get that $b \nleq \Box a \to a$. Pick $y' \in b$ such that $y' \in \Box a$ and $y' \notin a$. We then have $y' \in b \setminus a = [y]_\sim$, and, since $y' \in \Box a$, we get $R(y') \subseteq a \subsetneq R(x)$.
Conversely, assume that $(X, \sim, R)$ is a partial GL-frame, and let $a, b \in B$ be such that $b \leq \Box a \to a$. Reasoning towards a contradiction, suppose that $\Box b \nleq \Box a$.
Pick $x_0 \in \Box b \setminus \Box a$. Since $x_0 \notin \Box a$, pick $x_1' \in X$ with $x_0 R x_1'$ and $x_1' \notin a$. Since $(X, \sim, R)$ is a partial GL-frame, pick $x_1 \sim x_1'$ such that $R(x_1) \subsetneq R(x_0)$. Since $x_0 R x_1' \sim x_1$, we have $x_0 R x_1$. Therefore, since $x_0 \in \Box b$, we have $x_1 \in b$. Using the assumption $b \leq \Box a \to a$, we get that $x_1 \in \Box a \to a$. Since $a \in B = \mathcal{P}_\sim(X)$, from $x_1' \notin a$ and $x_1' \sim x_1$ we get $x_1 \notin a$. Thus, since $x_1 \in \Box a \to a$, we must have $x_1 \notin \Box a$. Also, since $R(x_1) \subseteq R(x_0)$ and $x_0 \in \Box b$, we have $x_1 \in \Box b$.
In the preceding paragraph, starting from a point $x_0 \in \Box b \setminus \Box a$, we have constructed a point $x_1 \in \Box b \setminus \Box a$ such that $R(x_1) \subsetneq R(x_0)$. Repeating this argument at most $n = |R(x_0)|$ times, we will obtain a point x_n with $x_n \notin \Box a$, but $R(x_n) = \emptyset$, which is the desired contradiction. □

Combining Proposition 3.9 with Theorem 3.2, we see that the category of finite partial GL-algebras is dually equivalent to the category of finite partial GL-frames.

4 The free image-total GL-algebra and its dual

In this section, we will apply the construction of the *free image-total algebra* for a variety of partial modal algebras [9, Section 3] to the particular case of partial GL-algebras. The idea is to construct, given a partial GL-algebra $(A, B, \Box)$, in which $\Box a$ is only defined for $a \in B$, a larger partial GL-algebra in which the value of $\Box a$ is defined *for all* $a \in A$. To this end, we first build a partial modal algebra which consists of all Boolean combinations of elements from A with formal elements '$\Box a$' for all $a \in A$. Next, we take the largest possible quotient of this partial modal algebra which is a partial GL-algebra. A crucial fact, to be proved in Theorem 4.7 below, is that this quotient does not identify any elements from the original partial GL-algebra $(A, B, \Box)$.

Definition 4.1 Let $(A, B, \Box)$ be a partial modal algebra. Let θ be the smallest congruence on the partial modal algebra $(A+V(A), A, \boxast)$ [7] such that θ contains $E := \{(\Box b, \boxast b) \mid b \in B\}$ and the quotient by θ is a partial GL-algebra. Let $F(A, B, \Box)$ be the partial GL-algebra $((A + V(A))/\theta, A/\theta, \boxast/\theta)$, and let i be the natural homomorphism $(A, B, \Box) \to F(A, B, \Box)$ which sends $a \in A$ to $[a]_\theta \in (A + V(A))/\theta$.

---

[7] We use the notation $\boxast$ to distinguish the formal box operation $A \to A + V(A)$ from the already existing box operation $B \to A$.

The following proposition says that $F(A, B, \Box)$ defined in the above definition is the *free image-total* GL-*algebra* over $(A, B, \Box)$, cf. [9, Definition 2.10].

Proposition 4.2 *Let $(A, B, \Box)$ be a partial modal algebra. For any homomorphism $h : (A, B, \Box) \to (C, D, \Box)$ such that $(C, D, \Box)$ is a partial* GL*-algebra and $h(A) \subseteq D$, there exists a unique homomorphism $\bar{h} : F(A, B, \Box) \to (C, D, \Box)$ such that $\bar{h} \circ i = h$.*

Proof. This is a special case of [9, Lemma 3.12]. □

Our next aim is to directly describe the dual of the construction F in Definition 4.1, by giving a construction which associates to any finite partial frame $(X, \sim, R)$ a partial GL-frame $G(X, \sim, R)$ and a bounded morphism $p : G(X, \sim, R) \to (X, \sim, R)$. Recall from Fact 3.7 that the dual of the construction $(A, B, \Box) \mapsto (A + V(A), A, \boxast)$ is $(X, \sim, R) \mapsto (X \times \mathcal{P}(X), \approx, Q)$. Since $F(A, B, \Box)$ is a certain quotient of $(A + V(A), A, \boxast)$, we have that $G(X, \sim, R)$ is a certain partial generated subframe of $(X \times \mathcal{P}(X), \approx, Q)$, by Corollary 3.3. We now give a definition which will be seen to characterize exactly which points are in $G(X, \sim, R)$.

Definition 4.3 Let $(X, \sim, R)$ be a finite partial frame. An element $(x, T) \in X \times \mathcal{P}(X)$ will be called GL-*suitable*[8] if $R(x) = [T]_\sim$ and, for any $y \in T$, there exists $S \subsetneq T$ such that (y, S) is GL-suitable.

Note that Definition 4.3 is recursive: in order to determine whether (x, T) is GL-suitable, one first needs to know whether (y, S) is GL-suitable for $y \in T$ and $S \subsetneq T$. The recursion terminates: for any $x \in X$, the element $(x, \emptyset)$ is GL-suitable if, and only if, $R(x) = \emptyset$.

Definition 4.4 Let $(X, \sim, R)$ be a finite partial frame. Let $G(X, \sim, R)$ be the partial frame $(Y, \approx, Q)$ defined by:

$$\begin{aligned} Y &:= \{(x, T) \in X \times \mathcal{P}(X) \mid (x, T) \text{ is GL-suitable}\}, \\ (x, T) \approx (y, S) &\iff x = y, \\ (x, T) Q (y, S) &\iff y \in T. \end{aligned}$$

Proposition 4.5 *Let $(X, \sim, R)$ be a finite partial frame with dual partial modal algebra $(A, B, \Box)$. Then the partial frame $G(X, \sim, R)$ is dual to the partial* GL*-algebra $F(A, B, \Box)$.*

Proof. By Fact 3.7, $(A + V(A), A, \boxast)$ is dual to $(X \times \mathcal{P}(X), \approx, Q)$. Let θ be the congruence from Definition 4.1; by definition, $F(A, B, \Box)$ is the quotient of $(A + V(A), A, \boxast)$ by θ. Combining Corollary 3.3 and Proposition 3.9, this quotient is dual to the largest partial generated sub-GL-frame of $(X \times \mathcal{P}(X), \approx, Q)$ that is contained in the subset $Y_E \subseteq X$ corresponding to $E = \{(\Box b, \boxast b) \mid b \in B\}$. Therefore, we need to prove the following two properties for $G(X, \sim, R) = (Y, \approx, Q)$ defined in Definition 4.4:

[8] Our terminology is inspired by terms such as "K4-suitable", as used in [10].

(i) The partial frame $(Y, \approx, Q)$ is a partial generated subframe of the partial frame $(X \times \mathcal{P}(X), \approx, Q)$, is a partial GL-frame, and is contained in Y_E.

(ii) If $Z \subseteq X \times \mathcal{P}(X)$ is such that $(Z, \approx, Q)$ is a partial generated subframe of the partial frame $(X \times \mathcal{P}(X), \approx, Q)$, is a partial GL-frame, and is contained in Y_E, then $Z \subseteq Y$.

For (i), first note that $(Y, \approx, Q)$ is a partial generated subframe of $X \times \mathcal{P}(X)$: if $(x, T) \in Y$ and $y \in T$, then there exists S with $(y, S) \in Y$, by GL-suitability of (x, T). To see that $(Y, \approx, Q)$ is a partial GL-frame (Definition 3.8), let $(x, T), (y, S) \in Y$ with $(x, T)Q(y, S)$, i.e., $y \in T$. Since (x, T) is GL-suitable, there exists $S' \subsetneq T$ such that (y, S') is GL-suitable. Now $(y, S) \approx (y, S')$, and $Q((y, S')) \subsetneq Q((x, T))$.
We now prove that $(Y, \approx, Q)$ satisfies all equalities in E. Note first that, for $b \in B$ and $(x, T) \in X \times \mathcal{P}(X)$, we have $(x, T) \in \boxast b$ if, and only if, $T \subseteq b$, and $(x, T) \in \Box b$ if, and only if, $R(x) \subseteq b$. Hence, a pair $(x, T) \in X \times \mathcal{P}(X)$ satisfies an equality $(\Box b, \boxast b) \in E$ iff $T \subseteq b \iff R(x) \subseteq b$. Now, since the elements of B are the $\sim$-saturated subsets, it easily follows that a pair (x, T) satisfies *all* equalities in E iff $[T]_\sim = R(x)$ (cf. [9, Lemma 5.11]).
To prove (ii), let $Z \subseteq X \times \mathcal{P}(X)$ be as in the assumptions of (ii). We need to show that all elements in Z are GL-suitable. First of all, by the argument in the previous paragraph, since Z satisfies all equalities in E, any pair $(x, T) \in Z$ satisfies $[T]_\sim = R(x)$. Now, since X is finite, it suffices to prove the following statement for all n:

$$\text{For any } (x, T) \in X \times \mathcal{P}(X) \text{ such that } |T| \leq n, \text{ if } (x, T) \in Z \text{ then } (x, T) \in Y. \qquad (H_n)$$

To prove (H_0), note that if $(x, \emptyset) \in Z$ then $R(x) = \emptyset$, so $(x, \emptyset)$ is GL-suitable. Now assume (H_n) holds. Let $(x, T) \in Z$ with $|T| = n + 1$. To prove that (x, T) is GL-suitable, we have already noted that $[T]_\sim = R(x)$. Let $y \in T$ be arbitrary. Since Z is a partial generated subframe of $X \times \mathcal{P}(X)$, there exists S such that $(y, S) \in Z$. Since Z is a partial GL-frame and $(x, T)Q(y, S)$, there exists $(y, S') \in Z$ such that $S' \subsetneq T$. By the induction hypothesis (H_n) applied to (y, S'), we see that (y, S') is GL-suitable. Thus, (x, T) is GL-suitable, since y was arbitrary. This concludes the proof of (H_{n+1}). □

Note that, by Proposition 4.5, the homomorphism $i : (A, B, \Box) \to F(A, B, \Box)$ is dual to the function $p : G(X, \sim, R) \to (X, \sim, R)$ which sends (x, T) to x.

Proposition 4.6 *Let $(X, \sim, R)$ be a finite partial* GL*-frame. Then the bounded morphism $p : G(X, \sim, R) \to (X, \sim, R)$ is surjective.*

Proof. We will prove that, for any $x \in X$, the pair (x, T_x) is in $G(X, \sim, R)$, where

$$T_x := \{y \mid xRy \text{ and } R(y) \subsetneq R(x)\}.$$

To this end, we use induction on the number of elements in $R(x)$. Specifically, we will show that, for each n,

$$\text{For any } x \in X \text{ such that } |R(x)| \leq n, (x, T_x) \text{ is } \mathsf{GL}\text{-suitable.} \qquad (I_n)$$

To prove (I_0), note that if $R(x) = \emptyset$ then $T_x = \emptyset$ and $(x, \emptyset)$ is GL-suitable. Now assume (I_n) holds. Let $x \in X$ such that $|R(x)| = n+1$. Towards proving that (x, T_x) is GL-suitable, we first show that $R(x) = [T_x]_\sim$. If $y \in R(x)$, then since $(X, \sim, R)$ is a partial GL-frame, there exists $y' \sim y$ such that $R(y') \subsetneq R(x)$. We then also have xRy', since xRy, and therefore $y' \in T_x$, so $y \in [T_x]_\sim$. Conversely, since $R(x)$ is $\sim$-saturated and contains T_x, it also contains $[T_x]_\sim$. Now let $y \in T_x$. We need to show that there exists $S \subsetneq T_x$ such that (y, S) is GL-suitable; we will show that T_y is an instance of such an S. Since $y \in T_x$, we have $R(y) \subsetneq R(x)$, so $|R(y)| \leq n$, and by the induction hypothesis (I_n) applied to y, we see that (y, T_y) is GL-suitable. We now prove that $T_y \subsetneq T_x$. If $z \in T_y$, then yRz, and since $y \in T_x$ we have $R(y) \subsetneq R(x)$, so xRz. Also, $R(z) \subsetneq R(y) \subsetneq R(x)$, proving that $z \in T_x$. Finally, we have that $y \in T_x$ but clearly $y \notin T_y$, so T_y is strictly contained in T_x. This concludes the proof of (I_{n+1}). We now easily deduce that p is surjective: for any $x \in X$, we have $p((x, T_x)) = x$, and (x, T_x) is GL-suitable by $I_{|R(x)|}$. □

The following is now an easy but important consequence of the results in this section.

Theorem 4.7 *Let $(A, B, \Box)$ be a partial GL-algebra. Then the homomorphism $i : (A, B, \Box) \to F(A, B, \Box)$ is injective.*

Proof. Since $p : G(X, \sim, R) \to (X, \sim, R)$ is surjective by Proposition 4.6, the result follows by Corollary 3.3. □

In the terminology of [3], Theorem 4.7 shows that $F(A, B, \Box)$ is an *injective one-step extension* of $(A, B, \Box)$, and (hence) that the class of partial GL-algebras has the *extension property* [3, Def. 9]. A proof of a closely related fact is given by different methods in [3, Sec. 9.2, Thm. 4]. The latter theorem, however, shows that the dual map p is surjective without identifying exactly which points are in the frame $G(X, \sim, R)$, but just showing that there are enough points. As such, the arguments given in the proofs of Propositions 4.5 and 4.6 are the main technical contributions of this paper. In the following sections, we will apply these results to construct free algebras and graded models for GL.

5 Application: free algebras for GL

Throughout the rest of the paper, fix a finite set $P = \{p_1, \ldots, p_k\}$ of propositional variables.

Definition 5.1 Let $\mathbb{F}_{\mathsf{K}}(P)$ denote the free modal algebra over P and let $(A_{n+1}, A_n, \Box_n)$ be the increasing chain of sub-partial modal algebras of $\mathbb{F}_{\mathsf{K}}(P)$ where A_n is the Boolean algebra of K-equivalence classes of modal formulas of degree $\leq n$, as in Example 2.3.

Let $(B_1, B_0, \Box_0) \overset{i_0}{\hookrightarrow} (B_2, B_1, \Box_1) \overset{i_1}{\hookrightarrow} \cdots$ be a countable chain of embeddings of partial modal algebras with $i_n(B_{n+1}) \subseteq B_{n+1}$. For any $v : P \to B_0$, there exists, for each $n \geq 0$, a natural *interpretation function* v_n from A_n to B_n, defined as follows:

- v_0 is the unique Boolean algebra homomorphism extending v;

- v_{n+1} is the unique Boolean algebra homomorphism such that, for all ϕ of modal degree $\leq n$, $v_{n+1}(\Box\phi) = \Box_n v_n(\phi)$ and $v_{n+1}(\phi) = i_n(v_n(\phi))$.

Note that each interpretation function v_n is a well-defined homomorphism of partial modal algebras $(B_{n+1}, B_n, \Box_n) \to (A_{n+1}, A_n, \Box_n)$.

We now apply the results in the previous section to give an incremental construction of the free k-generated GL-algebra, for any $k \geq 0$. Let B_0 be the free Boolean algebra over P. Let $B_1 := B_0 + V(B_0)$. Then B_0 is a subalgebra of B_1, and we define $\Box_0 : B_0 \to B_1$ to be the map which sends $a \in B_0$ to the formal element $\Box a \in V(B_0)$. We thus obtain a finite partial modal algebra $(B_1, B_0, \Box_0)$. The finite partial frame $(X_1, \sim_1, R_1)$ that is dual to $(B_1, B_0, \Box_0)$ for the 1-generated case is depicted in the figure below. Note that, for any k, the partial frame $(X_1, \sim_1, R_1)$ is a partial GL-frame, simply because each of the 2^k equivalence classes contains a blind point. Therefore, $(B_1, B_0, \Box_0)$ is a partial GL-algebra by Proposition 3.9.

Now, for $n \geq 1$, we inductively define $(B_{n+1}, B_n, \Box_n) := F(B_n, B_{n-1}, \Box_{n-1})$, and we let i_n be the natural map $B_n \to B_{n+1}$. Then, for each n, $(B_{n+1}, B_n, \Box_n)$ is a partial GL-algebra, and, hence, by Theorem 4.7, each i_n is injective.

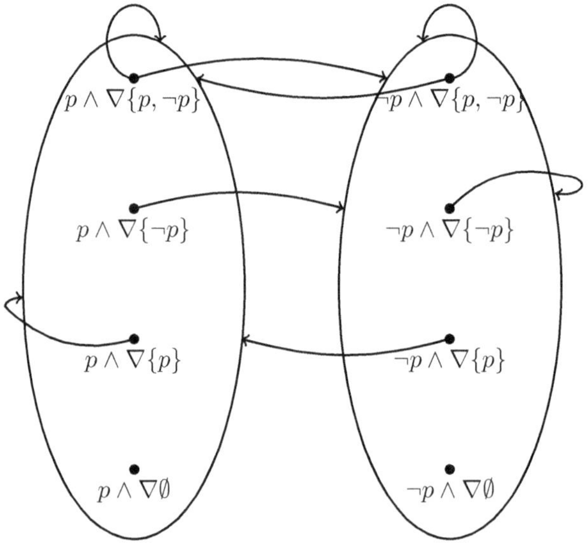

Fig. 1. The partial frame $(X_1, \sim_1, R_1)$ in the case $P = \{p\}$. In this diagram, the points are labelled by the atoms of the Boolean algebra A_1 that they represent. The classes of the equivalence relation $\sim_1$ are depicted by the two ellipses. The relation R_1 is depicted by arrows from points to classes; for instance, the arrow from the point $x := p \wedge \nabla\{p\}$ to the class $[x]_{\sim_1}$ indicates that xR_1y for any point y in $[x]_{\sim_1}$.

Theorem 5.2 *For each $n \geq 0$, the partial* GL*-algebra $(B_{n+1}, B_n, \Box_n)$ is isomorphic to the partial modal subalgebra $(A'_{n+1}, A'_n, \Box_n)$ of the free* GL*-algebra which consists of the* GL*-equivalence classes of modal formulas of degree $\leq n$.*

Theorem 5.2 is a straightforward consequence of the results in Section 3 of [9], in particular cf. Thm. 3.15. The isomorphism $A'_n \to B_n$ is given by factoring the natural interpretation function $A_n \to B_n$ (Definition 5.1) through the quotient $A_n \twoheadrightarrow A'_n$ of A_n. The reader is referred to [9, Section 2–3] for a more detailed proof.

Corollary 5.3 *The free k-generated* GL*-algebra is isomorphic to* $(B_\omega, \Box)$*, where* B_ω *is the Boolean algebra colimit of the chain* $(i_n : B_n \hookrightarrow B_{n+1})_{n\geq 0}$ *of embeddings of Boolean algebras, and* $\Box$ *is the unique operation* $B_\omega \to B_\omega$ *such that, for all* $n \geq 0$, $\Box \circ j_n = j_{n+1} \circ \Box_n$ *(here,* $j_n : B_n \hookrightarrow B_\omega$ *denotes the colimit embedding).*

6 Application: Graded models for GL

In this section, we give a semantic interpretation of the above results on finitely generated free GL-algebras.

We first translate the algebraic definition of the interpretation functions (Definition 5.1) into a dual definition of a notion of *graded model* and a satisfaction relation for it.

Definition 6.1 A *graded frame* X is a chain $\{(X_n, \sim_n, R_n)\}_{n\geq 1}$ of partial frames such that $X_{n+1}/\sim_{n+1} \cong X_n$ for all $n \geq 1$. A *graded model* is a pair (X, f) where X is a graded model and $f : X_1 \to \mathcal{P}(P)$ is a function such that, for each $p \in P$, $f^{-1}(p)$ is $\sim_1$-saturated.

We inductively define, for any formula ϕ of modal degree $\leq n$, and any $x \in X_n$, the relation "$X, x \models \phi$" as follows:

- For the case $n = 1$: let $x \in X_1$.
  - For $\phi = p_i$, we define $X, x \models p_i$ if, and only if, $p_i \in f(x)$, and we extend this definition to all $\Box$-free formulas in the usual way;
  - For $\phi = \Box\psi$, where ψ is $\Box$-free, we define $X, x \models \Box\phi$ if, and only if, for all $y \in X_1$ such that xR_1y, we have $X, y \models \phi$.
- For the induction step, assume that the relation "$X, x \models \phi$" has been defined for all formulas ϕ of modal degree $\leq n$ and all $x \in X_n$. Let $x \in X_{n+1}$.
  - For ϕ of modal degree $\leq n$, we define $X, x \models \phi$ if, and only if, $X, [x]_{\sim_{n+1}} \models \phi$;
  - For ϕ of modal degree $\leq n$, we define $X, x \models \Box\phi$ if, and only if, for all $y \in X_{n+1}$ such that $xR_{n+1}y$, we have $X, [y]_{\sim_{n+1}} \models \phi$;
  - Now, for an arbitrary ϕ of modal degree $\leq n+1$, we write ϕ as a Boolean combination of formulas $\phi_1 \ldots, \phi_r$, and $\Box\phi_{r+1}, \ldots, \Box\phi_{r+s}$, where each ϕ_i is of modal degree $\leq n$. We now define $X, x \models \phi$ in the usual way.

Proposition 6.2 *Let* (X, f) *be a graded model. For* $n \geq 1$, *let* $(A_n, A_{n-1}, \Box_n)$ *be the partial modal algebra dual to the partial frame* $(X_n, \sim_n, R_n)$*. For any* ϕ *of modal degree* $\leq n$ *and* $x \in X_n$*, we have*

$$x \in v_n(\phi) \iff X, x \models \phi.$$

Proof. Clear from the definitions of the functions v_n, the satisfaction relation, and the dual partial modal algebra of a partial frame. □

Definition 6.3 Fix $P = \{p_1, \ldots, p_k\}$. Let $X_0 := \mathcal{P}(P)$, $X_1 := X_0 \times \mathcal{P}(X_0)$. For $n \geq 1$, $(X_{n+1}, \sim_{n+1}, R_{n+1}) := G(X_n, \sim_n, R_n)$, where G is defined as in Definition 4.4. Let (X, f) be the graded model given by these partial frames and the valuation $f : X_1 \to \mathcal{P}(P)$ given by projection onto the first coordinate. We call (X, f) the *canonical graded* GL*-model* (on k generators).
For each $n \geq 0$, we now associate to each $x \in X_n$ a formula $\phi(x)$ of degree $\leq n$. For $x \in X_0$, let

$$\phi(x) := \bigwedge_{p_i \in x} p_i \wedge \bigwedge_{p_i \notin x} \neg p_i.$$

For $n \geq 0$, assume that the formulas $\phi(x)$ have been defined for all $x \in X_n$, and let $y \in X_{n+1}$. Since $y \in X_n \times \mathcal{P}(X_n)$, we may write $y = (x, T)$, and we define

$$\phi(y) := \phi(x) \wedge \nabla\{\phi(t) \mid t \in T\}.$$

The following corollary is a consequence of combining Proposition 4.5, Theorem 5.2 and Proposition 6.2.

Corollary 6.4 *Any modal formula ψ of degree $\leq n$ is* GL*-equivalent to the finite (possibly empty) disjunction of the set*

$$N_\psi := \{\phi(x) \mid x \in X_n \text{ such that } X, x \models \psi\}.$$

7 Conclusion

In this paper we applied the general theory of incremental constructions of free modal algebras using partial modal algebras ([12], [9], [5]) to the particular case of GL. We obtained a bottom-up construction of the free GL-algebra and normal forms for GL. Normal forms play a role in the proofs of interpolation and definability theorems for GL, cf. e.g. [13] and [18]. It is now perhaps also a natural and feasible question whether constructions like the one given in this paper could be applied to provability and interpretability logics [20] other than GL. We leave it as an open question for further research whether the results in this paper could yield new insights about such proofs. We have also not been able to discuss more proof-theoretical aspects of the incremental approach; more about this can be found [4] of Bezhanishvili and Ghilardi.
The reader will have noted that the normal forms that we defined for GL in Definition 6.3 are of a slightly different form than Fine's normal forms as in (1) above. In particular, while Fine's normal forms are the conjunction of a ∇-formula with a modal formula of degree 0, here we only obtain a conjunction of a ∇-formula with a modal formula of lesser degree. We expect that an equivalence with Fine-type normal forms for GL can be obtained by either a syntactic or semantic argument, but we leave the precise formulation of such a result to further research.
A phenomenon that seems to emerge from the approach taken in this paper is that either a *two-sorted* or a *bi-modal* variant of a logic can be better behaved

and easier to study than the one-sorted or uni-modal logic that one intended to study in the first place. We regard it as an important question to obtain a more structural understanding from this phenomenon, as it could also be interesting in the study of other modal logics than GL.

On a related topic, it remains rather mysterious why, of the possible quasi-equational flattenings of the Gödel-Löb axiom, one flattening (namely (3)) is "better" than another, namely

$$\text{if } b \leq \Box a \text{ then } \Box(b \rightarrow a) \leq \Box a, \tag{5}$$

which does not admit a construction as the one in this paper, cf. [4, Sec. 9.2, Ex. 3]. In order to understand this phenomenon, it would probably help to take the idea seriously that multi-sortedness/polymodality plays an important role here. To make the question a bit more concrete: is there is an apparent structural, maybe Sahlqvist-like, explanation for the fact that the quasi-equation (3) is better behaved than the quasi-equation (5)?

The results in this paper are close in spirit to [14], where a construction of a "canonical exact model" for GL is given, in an analogous way to the universal model for intuitionistic propositional logic. However, note that the approach used in this paper uses a different "slicing into finite parts" than the approach of exact/universal models. Whereas the latter organizes normal/canonical formulas according to the height of the corresponding point in the canonical frame, our method organizes formulas according to their modal degree. It is another interesting direction for further research to investigate how these two different methods are precisely related, both in the context of GL and in the context of intuitionistic propositional logic. We refer to [11, Section 8] for a more detailed study of the relationship between the universal model and the approximating chain in the context of intuitionistic propositional logic.

Moss [17] gives a one-sorted account of canonical formulas and compares his approach to filtrations and Fine's original approach. It would be equally interesting to compare the recent results of the "incremental construction" kind to the more common modal logic method of filtrations.

The nabla connective, ∇, is important in coalgebraic modal logic [16] and we think it is an important question how the methods discussed in this paper, or more generally in the algebraic theory of normal forms, relate to coalgebraic modal logic. As a first step in this direction, a coalgebraic account of Ghilardi's method for the incremental construction of free algebras, but only in the case of equations of rank exactly 1, was given in [6]. It would be an interesting research project to give a similar coalgebraic account of the incremental construction of free algebras for varieties axiomatized by quasi-equations of rank at most 1. A first pointer in this direction may be [9, Section 2], where one can find a category-theoretic formulation of the results about incremental constructions that were used in this paper.

Appendix

A Proofs of basic facts in Section 2

Theorem A.1 (Duality for finite partial modal algebras) *The category of finite partial frames is dually equivalent to the category of finite partial modal algebras.*

Proof. We have already defined a contravariant functor from partial frames to partial algebras, and this functor clearly sends *finite* partial frames to *finite* partial modal algebras. Conversely, given a finite partial modal algebra $(A, B, \Box)$, its dual finite partial frame $(X, \sim, R)$ is defined by:

$$X := \mathrm{At}(A) = \{x \in A \mid x \text{ is an atom of } A\},$$
$$\text{for } x, x' \in X,\ x \sim x' \overset{\text{def}}{\iff} \text{for all } b \in B, x \leq b \text{ if and only if } x' \leq b,$$
$$xRx' \overset{\text{def}}{\iff} \text{for all } b \in B, x \leq \Box b \text{ implies } x' \leq b.$$

Given a homomorphism of partial modal algebras $h : (A, B, \Box) \to (A', B', \Box')$, its dual bounded morphism $f : (Y, \sim_Y, S) \to (X, \sim_X, R)$ is defined by $f(y) := \bigwedge\{a \in A \mid y \leq h(a)\}$. It is standard to show that these assignments define a functor from finite partial modal algebras to finite partial frames which is an inverse, up to natural isomorphisms, of the functor from finite partial frames to finite partial modal algebras. (Cf., e.g., [9, Thm. 4.3].) $\Box$

Fact A.2 *The dual of $A + B$ is the Cartesian product $\mathrm{At}(A) \times \mathrm{At}(B)$ and the Boolean algebra $A + B$ is isomorphic to $\mathcal{P}(\mathrm{At}(A) \times \mathrm{At}(B))$.*

Proof. Immediate from finite duality. The isomorphism in the second part of the statement sends $u \in \mathcal{P}(\mathrm{At}(A) \times \mathrm{At}(B))$ to the element $\bigvee_{(x,y)\in u}(x \wedge y)$ in $A + B$. $\Box$

Fact A.3 *The dual of $V(A)$ is $\mathcal{P}(\mathrm{At}(A))$ and the Boolean algebra $V(A)$ is isomorphic to $\mathcal{P}(\mathcal{P}(\mathrm{At}(A)))$.*

Proof. Note that atoms of $V(A)$ correspond to homomorphisms $V(A) \to \mathbf{2}$, which correspond to meet-preserving functions $A \to \mathbf{2}$, which correspond to subsets of $\mathrm{At}(A)$ by finite duality. The isomorphism in the second part of the statement sends an element $u \in \mathcal{P}(\mathcal{P}(\mathrm{At}(A)))$ to the element $\bigvee_{T\in u} \nabla T$ of $V(A)$, where we use the 'nabla'-notation $\nabla T := \Box\left(\bigvee_{t\in T} t\right) \wedge \left(\bigwedge_{t\in T} \Diamond t\right)$. $\Box$

Fact A.4 *The dual of $A+V(A)$ is $X \times \mathcal{P}(X)$, where $X := \mathrm{At}(A)$ and $A+V(A)$ is isomorphic to $\mathcal{P}(X \times \mathcal{P}(X))$. The subalgebra A of $A + V(A)$ is dual to the equivalence relation $\approx$ on $X \times \mathcal{P}(X)$ defined by $(x, T) \approx (y, S)$ if, and only if, $x = y$. The dual of the operation $\Box : A \to A + V(A)$ is the relation Q on $X \times \mathcal{P}(X)$ defined by $(x, T)Q(y, S)$ if, and only if, $y \in T$.*

Proof. Combining the previous two facts, the isomorphism between $A+V(A)$ and $\mathcal{P}(X \times \mathcal{P}(X))$ is given by sending $u \in \mathcal{P}(X \times \mathcal{P}(X))$ to the element $\bigvee_{(x,T)\in u}(x \wedge \nabla T)$. For the other statements, cf., e.g., [9, Lemma 5.10]). $\Box$

References

[1] Abramsky, S., *A Cook's Tour of the Finitary Non-Well-Founded Sets*, in: S. Artemov, H. Barringer, A. d'Avila Garcez, L. C. Lamb and J. Woods, editors, *We Will Show Them*, Essays in honour of Dov Gabbay **1** (2005), pp. 1–18, based on unpublished lecture notes (1988).

[2] Avron, A., *On modal systems having arithmetical interpretations*, J. Symbolic Logic **49** (1984), pp. 935–942.

[3] Bezhanishvili, N. and S. Ghilardi, *Bounded proofs and step frames*, in: D. Galmiche and D. Larchey-Wendling, editors, *Automated Reasoning with Analytic Tableaux and Related Methods*, Lecture Notes in Computer Science **8123** (2013), pp. 44–58.

[4] Bezhanishvili, N. and S. Ghilardi, *Bounded proofs and step frames* (2013), preprint available at `http://users.mat.unimi.it/users/ghilardi/allegati/TR_NickSilvio.pdf`.

[5] Bezhanishvili, N., S. Ghilardi and M. Jibladze, *Free modal algebras revisited: the step-by-step method*, in: *Leo Esakia on duality in modal and intuitionistic logics*, Trends in Logic: Outstanding Contributions, Springer, 2014, to appear. Preprint available at `http://dspace.library.uu.nl/handle/1874/257814`.

[6] Bezhanishvili, N. and A. Kurz, *Free Modal Algebras: A Coalgebraic Perspective*, in: T. Mossakowski, U. Montanari and M. Haveraaen, editors, *Algebra and Coalgebra in Computer Science*, Lecture Notes in Computer Science **4624**, Springer, 2007 .

[7] Boolos, G., *On deciding the truth of certain statements involving the notion of consistency*, J. Symbolic Logic **41** (1976), pp. 779–781.

[8] Boolos, G., *On the nonexistence of certain normal forms in the logic of provability*, J. Symbolic Logic **47** (1982), pp. 638–640.

[9] Coumans, D. C. S. and S. J. van Gool, *On generalizing free algebras for a functor*, J. Logic Computation **23** (2013), pp. 645–672.

[10] Fine, K., *Normal forms in modal logic*, Notre Dame J. of Formal Logic **XVI** (1975).

[11] Gehrke, M., *Canonical extensions, Esakia spaces, and universal models*, in: *Leo Esakia on duality in modal and intuitionistic logics*, Trends in Logic: Outstanding Contributions, Springer, 2014, to appear. Preprint available at `http://www.liafa.univ-paris-diderot.fr/~mgehrke/Ge12.pdf`.

[12] Ghilardi, S., *An algebraic theory of normal forms*, Ann. Pure Appl. Logic. **71** (1995), pp. 189–245.

[13] Gleit, Z. and W. Goldfarb, *Characters and Fixed Points in Provability Logic*, Notre Dame J. of Formal Logic **31** (1990), pp. 26–36.

[14] Hendriks, L. and D. De Jongh, *Finitely generated magari algebras and arithmetic*, in: A. Ursini and P. Aglianò, editors, *Logic and Algebra*, Lecture Notes in Pure and Applied Mathematics **180**, Marcel Dekker, Inc., 1996 .

[15] Magari, R., *The diagonalizable algebras (the algebraization of the theories which express theor. ii)*, Boll. Un. Mat. Ital. **12** (1975), pp. 117–125.

[16] Moss, L. S., *Coalgebraic Logic*, Ann. Pure Appl. Logic. **96** (1999), pp. 277–317.

[17] Moss, L. S., *Finite models constructed from canonical formulas*, J. Philosophical Logic **36** (2007), pp. 605–640.

[18] Shavrukov, V. Y., *Subalgebras of diagonalizable algebras of theories containing arithmetic*, Dissertationes Mathematicae **323** (1993).

[19] Solovay, R. M., *Provability interpretations of modal logic*, Israel J. of Math. **25** (1976), pp. 287–304.

[20] Visser, A., *An overview of interpretability logic*, in: M. Kracht, M. de Rijke, H. Wansing and M. Zakharyaschev, editors, *Advances in modal logic, Volume 1*, CSLI Lecture Notes **87** (1998), pp. 307–359.

Optimal Decision Procedures for Satisfiability in Fragments of Alternating-time Temporal Logics

Valentin Goranko[a,b] Steen Vester[a 1]

[a] *Department of Applied Mathematics and Computer Science*
Technical University of Denmark
[b] *Department of Mathematics, University of Johannesburg*

Abstract

We consider several natural fragments of the alternating-time temporal logics ATL* and ATL with restrictions on the nesting between temporal operators and strategic quantifiers. We develop optimal decision procedures for satisfiability in these fragments, showing that they have much lower complexities than the full languages. In particular, we prove that the satisfiability problem for state formulae in the full 'strategically flat' fragment of ATL* is PSPACE-complete, whereas the satisfiability problems in the flat fragments of ATL and ATL$^+$ are Σ_3^P-complete. We note that the nesting hierarchies for fragments of ATL* collapse in terms of expressiveness above nesting depth 1, hence our results cover all such fragments with lower complexities.

Keywords: satisfiability, decision procedures, alternating-time temporal logics, flat fragments, complexity.

1 Introduction

The Alternating-time temporal logic ATL* was introduced and studied in [2] as a multi-agent extension of the branching time temporal logic CTL*, applied for specification and verification of properties of open systems. The most natural semantics for ATL* is defined in multi-agent transition systems, also known as concurrent game models, in which all agents take simultaneous actions at the current state and the resulting collective action effects the state transition. The language of ATL* involves expressions of the type $\langle\langle C\rangle\rangle\Phi$ meaning that the coalition of agents C has a collective strategy to guarantee – no matter how the other agents choose to act – achieving the goal Φ on all plays (computations) enabled by that collective strategy. The logic ATL* and its fragment ATL (analogous to CTL) have gradually become one of the most popular logical formalisms for reasoning about multi-agent systems, studied extensively during the past 10 years both from a logical and computational perspective.

While found to be quite useful and natural, however, the logic ATL*, and even its fragment ATL, turned out to have some problematic semantic features

[1] Emails: vfgo@dtu.dk, stve@dtu.dk

related to the nesting of strategic quantifiers $\langle\langle\cdot\rangle\rangle$, including:

- Conceptual difficulty in understanding the very meaning of nested expressions of the type $\langle\langle A\rangle\rangle \dots \langle\langle B\rangle\rangle\Phi$, especially, when the coalitions A and B share common agents. For instance, what exactly should $\langle\langle A\rangle\rangle\neg\langle\langle A\rangle\rangle\Phi$ mean?
- This problem is related to a technical problem built in the semantics of ATL^*, where e.g., in the truth evaluation of a formula of the type $\langle\langle A\rangle\rangle \dots \langle\langle B\rangle\rangle\Phi$ the strategy for A adopted to guarantee the success of the goal $\dots \langle\langle B\rangle\rangle\Phi$ does not have any effect when evaluating the truth of the subgoal $\langle\langle B\rangle\rangle\Phi$, which, arguably, goes against the intuitive understanding of what a strategy and its execution mean. Such problems have lead to several proposals of alternative semantics for ATL^*, with irrevocable commitment to strategies [1] or with strategy contexts, explicitly controllable within the formulae [3]. The latter comes at a high price, resulting in an undecidable satisfiability problem [15].
- The meaning of ATL^* formulae with nested strategic quantifiers is sensitive to the ability and capacity of the agents to use memory in their strategies, leading to essential variations of the semantics [4].
- The complexity of the full ATL^* is very high: 2EXPTIME-complete for both the model checking [2] and the satisfiability testing [12] problems.
- These problems are amplified when incomplete information is assumed. Then the basic temporal operators can no longer be naturally (if at all) characterized as fixed points of suitable operators, the semantics becomes truly non-computational and even model checking of ATL becomes undecidable [2].

So, there are several good reasons to consider *flat* fragments of ATL^*, where nesting of strategic quantifiers and temporal operators is restricted or completely disallowed, thus avoiding the problems listed above at the cost of reduced expressiveness. There are two natural kinds of 'flatness' in the language of ATL^*: with respect to the temporal operators and with respect to strategic quantifiers. The former comes naturally from purely temporal logics and has been investigated before, see e.g., [9], [5], and [13] from a more general, coalgebraic perspective. Here we will mainly consider the latter type of flatness.

The objective of the present paper is to develop optimal algorithmic methods for solving the satisfiability problem for the variety of naturally definable flat fragments of ATL^* and to analyze their computational complexity. Our main results and the contributions of this paper are as follows:

(i) The algorithmic problem of satisfiability testing in the full fragment of ATL^* where nesting between strategic quantifiers is not allowed (but temporal operators can be nested in strategic quantifiers and between each other) is PSPACE-complete, in contrast to the 2EXPTIME-completeness of satisfiability in the full ATL^*.

(ii) The algorithmic problem of satisfiability testing in the flat fragments of ATL and ATL^+, where only nesting of temporal operators in the scope of strategic quantifiers is allowed, are Σ_3^P-complete, in contrast to the

2EXPTIME-completeness of that problem in the full ATL^+ (as subsuming CTL^+, see [10]) and its EXPTIME-completeness in the full ATL [8].

The structure of the paper is as follows: In Section 2 we summarize basics of the logics LTL, CTL and CTL^* as well as concurrent game models and the alternating-time temporal logics ATL, ATL^+ and ATL^*. In Section 3 we introduce various flat fragments of ATL^* and discuss their expressiveness. Section 4 contains the technical preparation for our algorithms, where we introduce some kinds of normal forms for ATL^* formulae and obtain some key technical results. In Section 5 we provide sound and complete decision procedures as well as matching lower bounds for the flat fragments of ATL^* considered in the paper. We end with brief concluding remarks in Section 6.

2 Preliminaries

2.1 Summary of LTL, CTL, CTL* and their flat fragments

We assume that the reader is familiar with the temporal logics LTL, CTL and CTL^*. A standard reference is e.g., [6].

Given a set of atomic propositions Prop, the set of *literals* over Prop is $\mathsf{Prop} \cup \{\neg p \mid p \in \mathsf{Prop}\}$. We assume that the primitive temporal operators in LTL and CTL^* are X ("at the next state") and U ("Until"), whereas F ("sometime in the future"), R ("Release"), and G ("always in the future") are definable as follows: $\mathsf{F}\,\varphi := \top \,\mathsf{U}\, \varphi$, $\psi \,\mathsf{R}\, \varphi := \neg((\neg\psi)\,\mathsf{U}\,(\neg\varphi))$, $\mathsf{G}\,\varphi := \bot \,\mathsf{R}\, \varphi$. Respectively, the primitive temporal operators in CTL are AX, AU and AR, whereas the rest are definable as follows: $\mathsf{EX}\,\varphi := \neg\mathsf{AX}\,\neg\varphi$, $\mathsf{E}(\psi\,\mathsf{U}\,\varphi) := \neg\mathsf{A}((\neg\psi)\,\mathsf{R}\,(\neg\varphi))$, $\mathsf{E}(\psi\,\mathsf{R}\,\varphi) := \neg\mathsf{A}((\neg\psi)\,\mathsf{U}\,(\neg\varphi))$, $\mathsf{AF}\,\varphi := \mathsf{A}(\top\,\mathsf{U}\,\varphi)$, $\mathsf{AG}\,\varphi := \mathsf{A}(\bot\,\mathsf{R}\,\varphi)$, $\mathsf{EF}\,\varphi := \mathsf{E}(\top\,\mathsf{U}\,\varphi)$, $\mathsf{EG}\,\varphi := \mathsf{E}(\bot\,\mathsf{R}\,\varphi)$.

The following LTL-equivalences characterize U and R as fixed points, where the formulae on the right hand side are called *the fixed point unfoldings* respectively of $\theta\,\mathsf{U}\,\eta$ and $\theta\,\mathsf{R}\,\eta$ (see e.g., [14], [6]):

$$\theta\,\mathsf{U}\,\eta \equiv \eta \vee (\theta \wedge \mathsf{X}\,(\theta\,\mathsf{U}\,\eta)), \quad \theta\,\mathsf{R}\,\eta \equiv \eta \wedge (\theta \vee \mathsf{X}\,(\theta\,\mathsf{R}\,\eta)).$$

We define the flat fragments LTL_1, CTL_1 and CTL^*_1 resp. as subsets of LTL, CTL and CTL^*. In LTL_1 no nesting of temporal operators is allowed, in CTL^*_1 no nesting of path quantifiers is allowed and in CTL_1 neither is allowed. They are generating as follows, where β is a Boolean formula and θ is an LTL formula:

LTL_1: $\theta ::= p \mid \neg\theta \mid \theta \wedge \theta \mid \mathsf{X}\,\beta \mid \beta\,\mathsf{U}\,\beta$;
CTL^*_1: $\varphi ::= p \mid \neg\varphi \mid \varphi \wedge \varphi \mid \mathsf{A}\theta$.
CTL_1: $\varphi ::= p \mid \neg\varphi \mid \varphi \wedge \varphi \mid \mathsf{AX}\,\beta \mid \mathsf{A}(\beta\,\mathsf{U}\,\beta) \mid \mathsf{A}(\beta\,\mathsf{R}\,\beta)$.

For instance:

- $p\,\mathsf{U}\,q \wedge \mathsf{X}\,(r \wedge (q \wedge \neg p))$ is in LTL_1 but $p\,\mathsf{U}\,(\mathsf{X}\,q)$ is not.
- $\mathsf{A}(\neg p\,\mathsf{U}\,(p \wedge \neg q)) \wedge \neg(\mathsf{EF}\,(q \wedge \neg p) \wedge \neg\mathsf{AF}\,\neg(p \wedge q))$ is in CTL_1 (and in CTL^*_1).
- $\mathsf{AG}\,\mathsf{F}\,p$ is in CTL^*_1 but not in CTL_1; $\mathsf{AG}\,\mathsf{EF}\,p$ is neither in CTL_1 nor in CTL^*_1.

2.2 Concurrent game models. The logic ATL* and fragments

A **concurrent game model** [2] (CGM) is a tuple
$\mathcal{M} = (\mathbb{A}, \mathsf{St}, \{\mathsf{Act_a}\}_{\mathsf{a}\in\mathbb{A}}, \{\mathsf{act_a}\}_{\mathsf{a}\in\mathbb{A}}, \mathsf{out}, \mathsf{Prop}, \mathsf{L})$ comprising:

- a finite, non-empty set of *players (agents)* $\mathbb{A} = \{1, \ldots, k\}$
- a set of actions $\mathsf{Act_a} \neq \emptyset$ for each $\mathsf{a} \in \mathbb{A}$. For any $A \subseteq \mathbb{A}$ we denote $\mathsf{Act}_A := \prod_{\mathsf{a}\in A} \mathsf{Act_a}$ and use $\boldsymbol{\alpha}_A$ to denote a tuple from Act_A. In particular, $\mathsf{Act}_{\mathbb{A}}$ is the set of all possible *action profiles* in $\mathcal{M}$.
- a non-empty set of *states* St,
- for each $\mathsf{a} \in \mathbb{A}$, a map $\mathsf{act_a} : \mathsf{St} \to \mathcal{P}(\mathsf{Act_a}) \setminus \{\emptyset\}$ setting for each state s the actions available to a at s,
- a *transition function* $\mathsf{out} : \mathsf{St} \times \mathsf{Act}_{\mathbb{A}} \to \mathsf{St}$ that assigns deterministically a *successor (outcome) state* $\mathsf{out}(s, \boldsymbol{\alpha}_{\mathbb{A}})$ to every state s and action profile $\boldsymbol{\alpha}_{\mathbb{A}} = \langle \alpha_1, \ldots, \alpha_{\mathsf{k}} \rangle$, provided that $\alpha_{\mathsf{a}} \in \mathsf{act_a}(s)$ for every $\mathsf{a} \in \mathbb{A}$ (i.e., every α_{a} that can be executed by player a in state s),
- a finite set of *atomic propositions* Prop and a *labelling* $\mathsf{L} : \mathsf{St} \to \mathcal{P}(\mathsf{Prop})$.

Concurrent game models represent multi-agent transition systems that function as follows: at any moment the system is in a given state, where each player select an action from those available to him at that state. All players execute their actions synchronously and the combination of these actions together with the current state determine a transition to a unique successor state in the model. A *play* in a CGM is an infinite sequence of such subsequent successor states. More formally, a play is an infinite sequence $s_0 s_1 \ldots \in \mathsf{St}^\omega$ of states such that for each $i \geq 0$ there exists an action profile $\boldsymbol{\alpha}_{\mathbb{A}} = \langle \alpha_1, \ldots, \alpha_{\mathsf{k}} \rangle$ such that $\mathsf{out}(s_i, \boldsymbol{\alpha}_{\mathbb{A}}) = s_{i+1}$. A *history* is a finite initial segment $s_0 s_1 \ldots s_\ell$ of a play. We denote by $\mathrm{Play}_{\mathcal{M}}$ and $\mathrm{Hist}_{\mathcal{M}}$ respectively the set of plays and set of histories in $\mathcal{M}$. For a state $s \in \mathsf{St}$ we define $\mathrm{Play}_{\mathcal{M}}(s)$ and $\mathrm{Hist}_{\mathcal{M}}(s)$ as the set of plays and set of histories with initial state s. For a sequence ρ of states ρ_0 is the initial state, ρ_i is the $(i+1)$th state, $\rho_{\leq i}$ is the prefix $\rho_0 \ldots \rho_i$ of ρ and $\rho_{\geq i}$ is the suffix $\rho_i \rho_{i+1} \ldots$ of ρ. When $\rho = \rho_0 \ldots \rho_\ell$ is finite, we say that it has length ℓ and write $|\rho| = \ell$. Further, we let $\mathrm{last}(\rho) = \rho_\ell$.

A *strategy* for a player a in $\mathcal{M}$ is a mapping $\sigma_{\mathsf{a}} : \mathrm{Hist}_{\mathcal{M}} \to \mathsf{Act_a}$ such that for all $h \in \mathrm{Hist}_{\mathcal{M}}$ we have $\sigma_{\mathsf{a}}(h) \in \mathsf{act_a}(\mathrm{last}(h))$. Intuitively, it assigns a legal action for player a after any history h of the game. If that action depends only only on the current state, the strategy is called *memoryless*. We denote by $\mathrm{Strat}_{\mathcal{M}}(\mathsf{a})$ the set of strategies of player a. A (collective) strategy of a coalition $C \subseteq \mathbb{A}$ is a tuple $(\alpha_{\mathsf{a}})_{\mathsf{a}\in C}$ of strategies, one for each player in C. When $C = \mathbb{A}$ this is called a *strategy profile*. We denote by $\mathrm{Strat}_{\mathcal{M}}(C)$ the set of collective strategies of coalition C. A play $\rho \in \mathrm{Play}_{\mathcal{M}}$ is consistent with a strategy $\sigma_C \in \mathrm{Strat}_{\mathcal{M}}(C)$ if for every $i \geq 0$ there exists an action profile $\boldsymbol{\alpha}_{\mathbb{A}} = \langle \alpha_1, \ldots, \alpha_{\mathsf{k}} \rangle$ such that $\mathsf{out}(\rho_i, \boldsymbol{\alpha}_{\mathbb{A}}) = \rho_{i+1}$ and $\alpha_{\mathsf{a}} = \sigma_{\mathsf{a}}(\rho_{\leq i})$ for all $\mathsf{a} \in C$. The set of plays with initial state s that are consistent with σ_C is denoted $\mathrm{Play}_{\mathcal{M}}(s, \sigma_C)$. In particular, we define $\mathrm{Play}_{\mathcal{M}}(s, \sigma_{\mathsf{a}}) = \mathrm{Play}_{\mathcal{M}}(s, \sigma_{\{\mathsf{a}\}})$ for any player a.

The **Alternating-time temporal logic** ATL^*, introduced in [2], is a logic, suitable for specifying and verifying qualitative objectives of players and coalitions in concurrent game models. The main syntactic construct of ATL^* is a formula of type $\langle\langle C\rangle\rangle\Phi$, intuitively meaning: *"The coalition C has a collective strategy to guarantee the satisfaction of the objective Φ on every play enabled by that strategy."* Formally, ATL^* is a multi-agent extension of the branching time logic CTL^* with *strategic quantifiers* $\langle\langle C\rangle\rangle$ indexed with sets (*coalitions*) C of players. There are two types of formulae in ATL^*, *state formulae*, that are evaluated at states, and *path formulae*, that are evaluated on plays. These are defined by mutual recursion as follows, where $C \subseteq \mathbb{A}$, $p \in \mathsf{Prop}$:

State formulae of ATL^*: $\varphi ::= p \mid \neg\varphi \mid \varphi \wedge \varphi \mid \langle\langle C\rangle\rangle\Phi$,
Path formulae of ATL^*: $\Phi ::= \varphi \mid \neg\Phi \mid \Phi \wedge \Phi \mid \mathsf{X}\,\Phi \mid \Phi\,\mathsf{U}\,\Phi \mid \Phi\,\mathsf{R}\,\Phi$.

All other Boolean connectives are defined as usual, and the temporal operators F and G are defined as in CTL^*, which can be regarded as the fragment of ATL^* only involving strategic quantifiers for the empty coalition $\langle\langle\emptyset\rangle\rangle$, identified with universal path quantifier A, and for the "grand coalition" of all players $\langle\langle\mathbb{A}\rangle\rangle$, identified with existential path quantifier E. Equivalently, by identifying all agents, CTL^* can be regarded as the 1-agent fragment of ATL^*. To keep the notation lighter, we will list the members of C in $\langle\langle C\rangle\rangle$ without using $\{\}$.

The fragment ATL^+ of ATL^* is obtained when the temporal operators may only be applied to state formulae, i.e. when path formulae are re-defined as

Path formulae of ATL^+: $\Phi ::= \varphi \mid \neg\Phi \mid \Phi \wedge \Phi \mid \mathsf{X}\,\varphi \mid \varphi\,\mathsf{U}\,\varphi \mid \varphi\,\mathsf{R}\,\varphi$

Another, technically simpler and computationally better behaved fragment of ATL^*, is the logic ATL, which is the multi-agent analogue of CTL, only involving state formulae defined as follows, for any $C \subseteq \mathbb{A}$, $p \in \mathsf{Prop}$:

Formulae of ATL: $\varphi ::= p \mid \neg\varphi \mid \varphi\wedge\varphi \mid \langle\langle C\rangle\rangle\mathsf{X}\,\varphi \mid \langle\langle C\rangle\rangle(\varphi\,\mathsf{U}\,\varphi) \mid \langle\langle C\rangle\rangle(\varphi\,\mathsf{R}\,\varphi)$

The combined operators $\langle\langle C\rangle\rangle\mathsf{F}\,\varphi$ and $\langle\langle C\rangle\rangle\mathsf{G}\,\varphi$ are defined respectively as $\langle\langle C\rangle\rangle\top\,\mathsf{U}\,\varphi$ and $\langle\langle C\rangle\rangle\bot\,\mathsf{R}\,\varphi$.

The semantics of ATL^* is given with respect to a concurrent game model $\mathcal{M} = (\mathbb{A}, \mathsf{St}, \{\mathsf{Act_a}\}_{\mathsf{a}\in\mathbb{A}}, \{\mathsf{act_a}\}_{\mathsf{a}\in\mathbb{A}}, \mathsf{out}, \mathsf{Prop}, \mathsf{L})$. The semantics of state formulae is given in terms of truth at a state s in $\mathcal{M}$, as follows, where $p \in \mathsf{Prop}$, φ_1 and φ_2 are state formulae, Φ is a path formula and $C \subseteq \mathbb{A}$:

$\mathcal{M}, s \models p$	if $p \in \mathsf{L}(s)$
$\mathcal{M}, s \models \neg\varphi_1$	if $\mathcal{M}, s \not\models \varphi_1$
$\mathcal{M}, s \models \varphi_1 \wedge \varphi_2$	if $\mathcal{M}, s \models \varphi_1$ and $\mathcal{M}, s \models \varphi_2$
$\mathcal{M}, s \models \langle\langle C\rangle\rangle\Phi$	if there exist a collective strategy $\sigma_C \in \mathrm{Strat}_{\mathcal{M}}(C)$, such that $\mathcal{M}, \rho \models \Phi$ for all $\rho \in \mathrm{Play}_{\mathcal{M}}(s, \sigma_C)$

The semantics of path formulae is given just like in LTL, in terms of truth on a path ρ in a CGM $\mathcal{M}$, as follows, where φ is a state formula, Φ_1 and Φ_2 are path formulae and $C \subseteq \mathbb{A}$:

$$
\begin{array}{ll}
\mathcal{M},\rho \models \varphi & \text{if } \mathcal{M},\rho_0 \models \varphi \\
\mathcal{M},\rho \models \neg\Phi_1 & \text{if } \mathcal{M},\rho \not\models \Phi_1 \\
\mathcal{M},\rho \models \Phi_1 \wedge \Phi_2 & \text{if } \mathcal{M},\rho \models \Phi_1 \text{ and } \mathcal{M},\rho \models \Phi_2 \\
\mathcal{M},\rho \models \mathsf{X}\,\Phi_1 & \text{if } \mathcal{M},\rho_{\geq 1} \models \Phi_1 \\
\mathcal{M},\rho \models \Phi_1 \,\mathsf{U}\, \Phi_2 & \text{if } \exists k.\mathcal{M},\rho_{\geq k} \models \Phi_2 \text{ and } \forall j < k.\mathcal{M},\rho_{\geq j} \models \Phi_1 \\
\mathcal{M},\rho \models \Phi_1 \,\mathsf{R}\, \Phi_2 & \text{if } \forall k.\mathcal{M},\rho_{\geq k} \models \Phi_2 \text{ or} \\
 & \exists k.\mathcal{M},\rho_{\geq k} \models \Phi_1 \text{ and } \forall j \leq k.\mathcal{M},\rho_{\geq j} \models \Phi_1
\end{array}
$$

We focus on the satisfiability problem for various fragments of ATL^* in this paper. We will distinguish between the state satisfiability and path satisfiability problems which are defined on a given fragment $\mathcal{L}$ of ATL^* as follows:

- Given a state formula φ in $\mathcal{L}$, does there exist a CGM $\mathcal{M}$ and a state s in $\mathcal{M}$ such that $\mathcal{M}, s \models \varphi$?
- Given a path formula Φ in $\mathcal{L}$, does there exist a CGM $\mathcal{M}$ and a play ρ in $\mathcal{M}$ such that $\mathcal{M}, \rho \models \Phi$?

Note that there are two variants of the satisfiability problem for formulae of ATL^*: *tight*, where it is assumed that all agents in the model are mentioned in the formula, and *loose*, where additional agents, not mentioned in the formula, are allowed in the model. It is easy to see that these variants are really different, but the latter one is immediately reducible to the former, by adding just one extra agent a to the language. Furthermore, this extra agent can be easily added superfluously to the formula, e.g., by adding a conjunct $\langle\langle \mathsf{a} \rangle\rangle \mathsf{X}\,\top$, so we hereafter only consider the tight satisfiability version. For further details and discussion on this issue, see e.g., [7,17].

We recall some important complexity results for the satisfiability problem: satisfiability in ATL is EXPTIME-complete [16,8], while satisfiability in ATL^* is 2EXPTIME-complete [12]. Since ATL^+ subsumes CTL^+, the satisfiability in which is also 2EXPTIME-complete [10], this is the optimal complexity for the satisfiability in ATL^+, too. All these results equally hold for satisfiability in concurrent game models and in alternating transition systems [2], as both semantics are equivalent (see e.g., [8]).

3 Flat fragments of ATL and ATL*

Here we define some *flat* fragments of ATL^* and ATL. Flatness generally means no nesting of non-Boolean operators. There are two natural notions of flatness in the languages of ATL and ATL^*: with respect to temporal operators and with respect to strategic quantifiers. We will be mostly concerned with the latter, but the former applies in the case of ATL, too.

We adopt the following notational conventions: we will typically denote Boolean formulae by β, γ; LTL formulae by θ, η, ζ; ATL formulae by φ, ψ; and ATL^* formulae – both state and path – by Θ, Φ, Ψ; all possibly with indices.

3.1 A hierarchy of flat fragments of ATL*

We will consider the following fragments of ATL^*, where p is any atomic proposition, $C \subseteq \mathbb{A}$, β is any Boolean formula and θ is any LTL formula:

(i) **Separated ATL^*,** denoted $\mathsf{ATL}^*_{\mathsf{Sep}}$, consists of those formulae of ATL^* in which there is no nesting of strategic quantifiers in the scope of temporal operators (but, any nesting of temporal operators within strategic quantifiers or temporal operators is allowed), so the (external) strategic and the (internal) temporal layers are separated. More precisely, the formulae of $\mathsf{ATL}^*_{\mathsf{Sep}}$ are generated as follows:

$$\Phi ::= \theta \mid \neg\Phi \mid (\Phi \land \Phi) \mid \langle\langle C\rangle\rangle\Phi$$

(ii) **Full (strategically) flat ATL^*,** denoted ATL^*_1, consists of those formulae of ATL^* in which there is no nesting of strategic quantifiers within strategic quantifiers (but, nesting of strategic quantifiers and temporal operators in temporal operators is allowed), formally generated as follows:

$$\Phi ::= p \mid \neg\Phi \mid (\Phi \land \Phi) \mid \langle\langle C\rangle\rangle\theta \mid \mathsf{X}\,\Phi \mid \Phi\,\mathsf{U}\,\Phi \mid \Phi\,\mathsf{R}\,\Phi$$

If the restriction $C \neq \emptyset$ is imposed, we denote the resulting fragment $\widehat{\mathsf{ATL}^*_1}$.

(iii) **State fragment of ATL^*_1,** denoted $\mathsf{St}(\mathsf{ATL}^*_1)$, consists of the state formulae of ATL^*_1, i.e. those formulae of ATL^* in which there is no nesting of strategic quantifiers in either temporal operators or strategic quantifiers (but, nesting between temporal operators is allowed). The formulae of $\mathsf{St}(\mathsf{ATL}^*_1)$ are explicitly generated as follows:

$$\Phi ::= p \mid \neg\Phi \mid (\Phi \land \Phi) \mid \langle\langle C\rangle\rangle\theta.$$

(iv) **Flat ATL^+** (or, **double-flat** ATL^*), denoted ATL^+_1, consists of those formulae of ATL^+ which are also in $\mathsf{St}(\mathsf{ATL}^*_1)$, e.g., with no nesting of either strategic quantifiers or temporal operators within temporal operators. The formulae of ATL^+_1 are generated as follows, where $\theta \in \mathsf{LTL}_1$:

$$\Phi ::= p \mid \neg\Phi \mid (\Phi \land \Phi) \mid \langle\langle C\rangle\rangle\theta$$

(v) **Flat ATL,** denoted ATL_1, consists of those formulae of ATL^+_1 which are in ATL, i.e., in which strategic quantifiers are followed immediately by temporal operators. The formulae of ATL_1 are generated as follows:

$$\varphi ::= p \mid \neg\varphi \mid (\varphi \land \varphi) \mid \langle\langle C\rangle\rangle\mathsf{X}\,\beta \mid \langle\langle C\rangle\rangle(\beta\,\mathsf{U}\,\beta) \mid \langle\langle C\rangle\rangle(\beta\,\mathsf{R}\,\beta)$$

Inclusions between the different flat fragments are illustrated in Figure 1. All inclusions shown in the figure are strict and there are no inclusions except the ones shown (where transitive closure is implicit). For example:

- $(\langle\langle 1\rangle\rangle\mathsf{G}\,\neg p \land \neg\langle\langle 2\rangle\rangle\mathsf{X}\,(p \lor \neg q)) \lor \langle\langle 1,2\rangle\rangle(p\,\mathsf{U}\,\neg q)$ is in ATL_1;
- $\langle\langle 1\rangle\rangle(\mathsf{G}\,\neg p \land \mathsf{F}\,q)$, $\neg\langle\langle 1,2\rangle\rangle((p\,\mathsf{R}\,\neg q) \lor (\neg p\,\mathsf{U}\,q))$ are in ATL^+_1 but not in ATL_1;
- $\langle\langle 1\rangle\rangle(\mathsf{G}\,\mathsf{F}\,\neg p \lor \mathsf{X}\,\neg(p\,\mathsf{U}\,\neg q))$ is in $\mathsf{St}(\mathsf{ATL}^*_1)$ but not ATL^+_1;
- $\mathsf{G}\,\langle\langle 1,2\rangle\rangle\mathsf{F}\,(\neg p \lor q)$ is in $\widehat{\mathsf{ATL}^*_1}$ but not in $\mathsf{St}(\mathsf{ATL}^*_1)$;

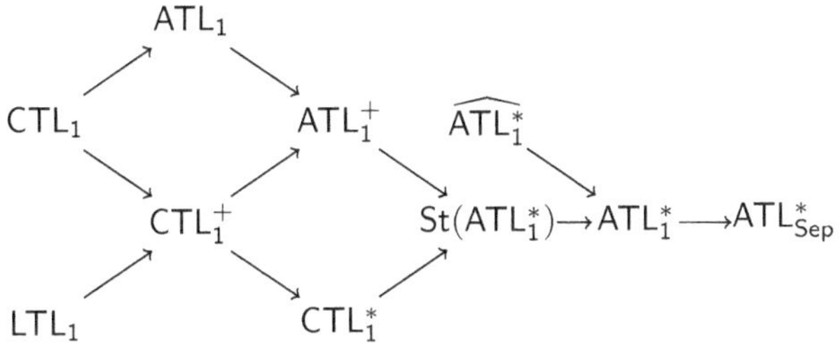

Fig. 1. Inclusions between flat fragments. An arrow from $\mathcal{L}_1$ to $\mathcal{L}_2$ means that every $\mathcal{L}_1$ formula is an $\mathcal{L}_2$ formula.

- $\langle\langle\emptyset\rangle\rangle \mathsf{G}\, p \wedge \mathsf{G}\, \langle\langle 1,2\rangle\rangle \mathsf{G}\,\mathsf{F}\, \neg p$ is in ATL_1^* but not in $\widehat{\mathsf{ATL}_1^*}$;
- $\langle\langle 2\rangle\rangle\langle\langle 1\rangle\rangle \neg(p\, \mathsf{U} \neg q)$ is in $\mathsf{ATL}^*_{\mathsf{Sep}}$ but not in ATL_1^*.

Even though ATL_1^* is included in $\mathsf{ATL}^*_{\mathsf{Sep}}$ they have the same expressive power and there is an efficient translation from $\mathsf{ATL}^*_{\mathsf{Sep}}$ to ATL_1^*.

Proposition 3.1 *Every formula of* $\mathsf{ATL}^*_{\mathsf{Sep}}$ *is logically equivalent to a formula of* $\mathsf{ATL}^*_{\mathsf{Sep}}$ *which is at most as long and has no nesting of strategic quantifiers. Such a formula is effectively computable in linear time.*

Proof. Because $\langle\langle C\rangle\rangle\Phi \equiv \Phi$ for every state formula Φ and coalition C. □

Thus, deciding satisfiablity in $\mathsf{ATL}^*_{\mathsf{Sep}}$ is reducible with no cost to satisfiablity in $\mathsf{St}(\mathsf{ATL}_1^*)$, so we will not discuss $\mathsf{ATL}^*_{\mathsf{Sep}}$ hereafter. On the other hand, due to the equivalence above, the fragment ATL_1^* can be extended even further by allowing nesting of strategic quantifiers, as long as there are no occurrences of temporal operators in between them. The equivalence of $\langle\langle C\rangle\rangle\Phi$ and Φ for state formulae Φ is an example of why nesting of strategic quantifiers in ATL^* can be considered unnatural. We note that this phenomenon is avoided in ATL^* with strategy context [3].

Between the full logics and their flat fragments, it is natural to consider the hierarchies of fragments with a bounded nesting depth of strategic quantifiers. However, the next result shows that the fragments with nesting depth 2 are essentially as expressive and computationally hard as the full logics.

Proposition 3.2 *For any logic* $\mathcal{L} \in \{\mathsf{LTL}, \mathsf{CTL}, \mathsf{CTL}^+, \mathsf{CTL}^*, \mathsf{ATL}, \mathsf{ATL}^+, \mathsf{ATL}^*\}$ *and formula* Φ *of* $\mathcal{L}$ *there is an equi-satisfiable formula* Φ' *in* $\mathcal{L}$ *with nesting depth 2 of strategic quantifiers (resp., temporal operators for* LTL*) and length* $|\Phi'| = O(|\Phi|)$ *that can be computed in linear time.*

Proof. In each of the cases, the flattening is done by repeated renaming of state subformulae with fresh atomic propositions. We illustrate the technique on ATL^*. Let Φ be an ATL^* formula. For any innermost subformula Ψ of Φ beginning with a strategic quantifier we introduce a fresh atomic proposition p_Ψ. Then Φ and $\Phi' = \Phi[p_\Psi/\Psi] \wedge \mathsf{AG}\,(p_\Psi \leftrightarrow \Psi)$ are equi-satisfiable. By repeated application of such renaming of strategically quantified subformulae we obtain an equi-satisfiable formula of nesting depth 2 that is linear in the size of Φ. Since $\mathsf{AG}\,(p_\Psi \leftrightarrow \Psi)$ is a CTL formula, this works for each logic $\mathcal{L} \in \{\mathsf{CTL}, \mathsf{CTL}^+, \mathsf{CTL}^*, \mathsf{ATL}, \mathsf{ATL}^+, \mathsf{ATL}^*\}$, while for LTL we use $\mathsf{G}\,(p_\Psi \leftrightarrow \Psi)$.□

Thus, the only complexity gain in restricting syntactic fragments of these logics with respect to nesting can occur when flat fragments are considered.

3.2 Some remarks on the expressiveness of the flat ATL*-fragments

The lower complexity of the satisfiability in the flat fragments of ATL* comes with a price, namely that various properties that require nesting of strategic quantifiers cannot be expressed anymore. However, many interesting and important properties of systems are still expressibe. For instance:

- $\langle\!\langle \mathtt{ctrl} \rangle\!\rangle \mathsf{G}\,\neg\mathtt{break}$ in ATL_1 specifies that a controller can make sure the system does not break no matter how the environment behaves,
- $\bigwedge_{i=1}^{n} \langle\!\langle \mathtt{proc}_i \rangle\!\rangle \mathsf{G}\,\mathsf{F}\,\mathtt{db\_access}_i$ expresses that each process can ensure database access infinitely often,
- $\langle\!\langle A \rangle\!\rangle(\theta_{\text{fair}} \to \theta)$ means that coalition A can make sure that the LTL property θ is satisfied on all fair paths (where fairness is defined by LTL formula θ_{fair}).

The semantics of ATL* is based on *unbounded memory strategies*, but it can be restricted and parameterized with the amount of memory that the proponent agents' strategies can use. The extreme case is the *memoryless semantics*, where the proponents may only use memoryless strategies. It turns out that satisfiability in ATL, is unaffected by such restrictions, but differences occur in ATL* and even in ATL$^+$. For discussion on these see e.g., [4]. In contrast, using our satisfiability decision procedures developed in Section 5, we will show that all semantics based on different memory restrictions yield the same satisfiable (resp., the same valid) formulae in the flat fragment ATL^*_1.

4 Normal forms and satisfiability of special sets in ATL*

4.1 Negation normal form of ATL* formulae

Definition 4.1 An ATL* formula Φ is in a *negation normal form* (NNF) if negations in Φ may only occur immediately in front of atomic propositions.

We now define the dual $[[\cdot]]$ to the strategic quantifier $\langle\!\langle\cdot\rangle\!\rangle$ as usual: $[[C]] ::= \neg\langle\!\langle C\rangle\!\rangle\neg$. If we consider $[[\cdot]]$ as a primitive operator in ATL*, then every ATL* formula can be transformed to an equivalent formula in NNF by driving all negations inwards, using the self-duality of X and the duality between U and R. However, using $[[\cdot]]$ formally breaks the syntax of the fragments ATL and ATL_1 because of inserting a $\neg$ between $\langle\!\langle\cdot\rangle\!\rangle$ and the temporal operator. Yet, this can be easily fixed by equivalently re-defining the applications of $[[\cdot]]$, using the following equivalences: $[[C]]\mathsf{X}\,\varphi \equiv \neg\langle\!\langle C\rangle\!\rangle\mathsf{X}\,\neg\varphi$, $[[C]](\varphi\,\mathsf{U}\,\psi) \equiv \neg\langle\!\langle C\rangle\!\rangle((\neg\varphi)\,\mathsf{R}\,(\neg\psi))$, $[[C]](\varphi\,\mathsf{R}\,\psi) \equiv \neg\langle\!\langle C\rangle\!\rangle((\neg\varphi)\,\mathsf{U}\,(\neg\psi))$.

Hereafter we assume that the language ATL* and each of its fragments introduced above are formally extended with the operator $[[\cdot]]$ applied just like $\langle\!\langle\cdot\rangle\!\rangle$ in the respective fragments. Due to the equivalences above, the resulting extensions preserve the expressiveness of these fragments. Formally:

Lemma 4.2 *Every formula of* ATL* *extended with the operator* $[[\cdot]]$ *can be transformed to an equivalent formula in NNF. Furthermore, each of the fragments* ATL, ATL_1, ATL_1^+, $\mathsf{St}(\mathsf{ATL}_1^*)$ *and* ATL_1^*, *extended with* $[[\cdot]]$, *is closed under this transformation, i.e. if a formula is in any of these fragments then its NNF-equivalent formula is in that fragment, too.*

4.2 Successor normal forms

Definition 4.3 [Successor formulae] An ATL* formula is a *successor formula (SF)* if it is of the type $\langle\langle C\rangle\rangle \mathsf{X}\,\Phi$ or $[[C]]\mathsf{X}\,\Phi$.

Definition 4.4 [Components] With every set of ATL* successor formulae

$\Gamma = \{\langle\langle A_0\rangle\rangle \mathsf{X}\,\Phi_0, \ldots, \langle\langle A_{m-1}\rangle\rangle \mathsf{X}\,\Phi_{m-1}, [[B_0]]\mathsf{X}\,\Psi_0, \ldots, [[B_{n-1}]]\mathsf{X}\,\Psi_{n-1}\}$
we associate the set of its

- $\langle\langle\cdot\rangle\rangle\mathsf{X}$ *-components*: $\langle\langle\cdot\rangle\rangle\mathsf{X}\,(\Gamma) = \{\Phi_0, \ldots, \Phi_{m-1}\}$,
- $[[\cdot]]\mathsf{X}$ *-components*: $[[\cdot]]\mathsf{X}\,(\Gamma) = \{\Psi_0, \ldots, \Psi_{n-1}\}$,
- *successor components*: $SC(\Gamma) = \langle\langle\cdot\rangle\rangle\mathsf{X}\,(\Gamma) \cup [[\cdot]]\mathsf{X}\,(\Gamma)$.

Definition 4.5 [Successor normal form]

(i) An LTL formula is in a LTL *successor normal form* (LSNF) if it is in NNF and is a Boolean combination of literals and successor formulae, i.e., LTL formulae beginning with X.

(ii) An ATL* formula is in a *successor normal form* (SNF) if it is in NNF and is a Boolean combination of literals and ATL* successor formulae.

Lemma 4.6 *Every* LTL*-formula* ζ *can be effectively transformed to an equivalent formula in* LTL *successor normal form* $LSNF(\zeta)$, *of length at most* $6|\zeta|$.

Proof. We can assume that ζ is already transformed to NNF (of length less than twice the original length). Consider all maximal subformulae of ζ of the types $(\theta\,\mathsf{U}\,\eta)$ and $(\theta\,\mathsf{R}\,\eta)$. Replace each of them with its LTL-equivalent fixpoint unfolding, respectively $\eta \vee (\theta \wedge \mathsf{X}\,(\theta\,\mathsf{U}\,\eta))$ and $\eta \wedge (\theta \vee \mathsf{X}\,(\theta\,\mathsf{R}\,\eta))$. Then, the same procedure is applied recursively to all respective subformulae θ, η occurring above and not in the scope of X, until all occurrences of U and R get in the scope of X. This procedure at most triples the length of the starting formula and the result is clearly a formula in LSNF. □

Definition 4.7 [Conjunctive formulae in SNF] An ATL* formula in SNF is *conjunctive* if it is of the form

$$\Theta = \Phi \wedge \langle\langle A_0\rangle\rangle \mathsf{X}\,\Phi_0 \wedge \ldots \wedge \langle\langle A_{m-1}\rangle\rangle \mathsf{X}\,\Phi_{m-1} \wedge [[B_0]]\mathsf{X}\,\Psi_0 \wedge \ldots \wedge [[B_{n-1}]]\mathsf{X}\,\Psi_{n-1}$$

With every such formula Θ we associate the set of its *successor conjuncts*:

$$C(\Theta) = \{\langle\langle A_0\rangle\rangle \mathsf{X}\,\Phi_0, \ldots, \langle\langle A_{m-1}\rangle\rangle \mathsf{X}\,\Phi_{m-1}, [[B_0]]\mathsf{X}\,\Psi_0, \ldots, [[B_{n-1}]]\mathsf{X}\,\Psi_{n-1}\}$$

4.3 Sets of distributed control of ATL* formulae

Definition 4.8 [Set of distributed control] A set of ATL* formulae Δ is a *set of distributed control* if $\Delta = \{\langle\langle A_0\rangle\rangle\Phi_0, \ldots, \langle\langle A_{m-1}\rangle\rangle\Phi_{l-1}, [[B]]\Psi\}$ where the coalitions $A_0, \ldots, A_{l-1}$ are pairwise disjoint, and $A_0 \cup \ldots \cup A_{l-1} \subseteq B$.

Lemma 4.9 *A set of* ATL* *successor formulae*

$$\Gamma = \{\langle\!\langle A_0 \rangle\!\rangle \mathsf{X}\, \Phi_0, \ldots, \langle\!\langle A_{m-1} \rangle\!\rangle \mathsf{X}\, \Phi_{m-1}, [[B_0]]\mathsf{X}\, \Psi_0, \ldots, [[B_{n-1}]]\mathsf{X}\, \Psi_{n-1}, [[\mathbb{A}]]\mathsf{X}\top\}$$

is satisfiable if and only if every subset of distributed control Δ *of* Γ *has a satisfiable set of successor components.*

Proof. First, note that the formula $[[\mathbb{A}]]\mathsf{X}\top$ is valid, so it plays no role in the satisfiability of Γ; it is only added there in order to enable sufficiently many subsets of distributed control.

Now, suppose Γ is true at a state s of a CGM $\mathcal{M}$. Then for every subset of distributed control $\Delta = \{\langle\!\langle A_0 \rangle\!\rangle \mathsf{X}\, \Phi_0, \ldots, \langle\!\langle A_{l-1} \rangle\!\rangle \mathsf{X}\, \Phi_{l-1}, [[B]]\mathsf{X}\, \Psi\}$ consider collective actions for the coalitions $A_0, \ldots, A_{l-1}$ at s that guarantee satisfaction of their respective nexttime objectives in Δ in any of the resulting successor states. Add arbitrarily fixed actions of the remaining agents in B and a respective collective action for $\mathbb{A} \setminus B$ dependent on the so fixed actions of the agents in B, that brings about satisfaction of Ψ in the resulting successor state s'. Then all successor components of Δ are true at s'.

Conversely, suppose that $\Delta_1, \ldots, \Delta_d$ are all subsets of Γ of distributed control and they are all satisfiable. For each Δ_i we fix a CGM $\mathcal{M}_i$ and a state s_i in it that satisfies $SC(\Delta_i)$. We can assume, w.l.o.g., that $\mathcal{M}_i$ is generated from s_i, i.e. consists only of states reachable by plays starting at s_i.

We will construct a CGM satisfying Γ by using a construction from [8]. The idea is to first create a root state s and supply all agents with sufficiently many actions at s in order to ensure the existence of all collective actions and respective successor states necessary for satisfying the successor components of Γ. We will show that it suffices to take care of the sets of successor components of each subset of distributed control Γ and then will use the CGMs satisfying these to complete the construction of the model satisfying Γ.

Now, the construction. Recall that $|\mathbb{A}| = k$ and let $r = m+n$ (the numbers of $\langle\!\langle \cdot \rangle\!\rangle$- and $[[\cdot]]$-components in Γ). Each agent will have r available actions $\{0, \ldots, r-1\}$ at the root state s, hence $\{0, \ldots, r-1\}^k$ is the set of all possible action profiles at s. The intuition is that every agent's action at s is a choice of that agent of a formula from Γ for the satisfaction of which the agent chooses to act. For every such action profile σ we denote by $N(\sigma)$ the set of agents $\{i \mid \sigma_i \geq m\}$ and then we define the number $\mathbf{neg}(\sigma)$ to be the remainder of $[\sum_{i \in N(\sigma)} (\sigma_i - m)]$ modulo n. (The idea of this definition is that, once all agents in any given proper subset of $N(\sigma)$ choose their actions, the remaining agents in $N(\sigma)$ can act accordingly to yield any value of $\mathbf{neg}(\sigma)$ between 0 and $n-1$ they wish, i.e., to set the "collective action" of all agents in $\mathbf{neg}(\sigma)$ on any $[[\cdot]]\mathsf{X}$-formula in Γ they choose.) Now, we consider the set

$$\Delta_\sigma = \{\langle\!\langle A_j \rangle\!\rangle \mathsf{X}\, \Phi_j \mid j < m \text{ and } \sigma_i = j \text{ for all } i \in A_j\} \cup \{[[B_l]]\mathsf{X}\, \Psi_l \mid \mathbf{neg}(\sigma) = l \text{ and } \mathbb{A} \setminus B_l \subseteq N(\sigma)\}$$

Note that Δ_σ is a subset of Γ of distributed control if it contains a formula

$[[B_j]]\mathsf{X}\,\Psi$, or else can be made a set of distributed control by adding $[[\mathbb{A}]]\mathsf{X}\top$ to it. Indeed, all agents in a A_j choose j, so all coalitions A_j must be pairwise disjoint. Besides, if $[[B_l]]\mathsf{X}\,\Psi_l \in \Delta_\sigma$ then it is clearly a unique $[[\cdot]]$-formula in Δ_σ and no agents from any $A_j \in \Delta_\sigma$ are in $N(\sigma)$, hence $A_j \subseteq B_l$ for each $\langle\langle A_j\rangle\rangle\mathsf{X}\,\Phi_j \in \Delta_\sigma$. Thus, Δ_σ is one of $\Delta_1, \ldots, \Delta_d$, say Δ_i. Then, we determine the successor state $\mathsf{out}(s, \sigma)$ to be s_i. To complete the definition of the CGM, at each successor state s_i of s we graft a copy of $\mathcal{M}_i$.

We will show that the resulting CGM $\mathcal{M}$ satisfies Γ at s. Indeed, for every $\langle\langle A_j\rangle\rangle\mathsf{X}\,\Phi_j \in \Gamma$ a collective strategy for A_j that guarantees the satisfaction of that formula at s consists in all agents from A_j acting j at s, following their strategy that guarantees in $\mathcal{M}_i$ the satisfaction of the objective Φ_j if the play enters the copy of $\mathcal{M}_i$, and acting in an arbitrarily fixed manner at all other states of $\mathcal{M}$. (Note that, if the strategy for A_j in $\mathcal{M}_i$ is positional, then the above described strategy is positional, too.) Lastly, every $[[B_l]]\mathsf{X}\,\Psi_l \in \Gamma$ is true at s, too, because if $B_l \neq \mathbb{A}$ then for every collective action of all agents from $[[B_l]]\mathsf{X}$ there is a suitable complementary action of $\mathbb{A} \setminus B_l$, where all agents choose actions greater than m and such that $\mathbf{neg}(\sigma)$ adds up to l modulo n. (In fact, this can be guaranteed by any agent in $\mathbb{A} \setminus B_l$ after all others have chosen their actions.) In the case when $B_l = \mathbb{A}$, every subset $\{\langle\langle A_j\rangle\rangle\mathsf{X}\,\Phi_j, [[B_l]]\mathsf{X}\,\Psi_l\}$ for $j < m$ is of distributed control, hence Ψ_l is true at the root s_i of $\mathcal{M}_i$ for each $i = 1, ..., d$. Thus, $\mathcal{M}, s \models \Gamma$, which completes the proof. □

A consequence of the proof above is that memoryless and memory-based semantics yield the same satisfiable state formulae in the flat fragments.

Corollary 4.10 *A* $\mathsf{St}(\mathsf{ATL}_1^*)$ *formula* Φ *is satisfiable in the memoryless semantics if and only if it is satisfiable in the memory-based semantics.*

Proof. Lemma 4.9 can be proved for memoryless semantics in the same way, but only for $\mathsf{St}(\mathsf{ATL}_1^*)$ formulae. This is because the successor components are LTL formulae which have the same semantics with and without memory. Further, for both semantics each subformula $\langle\langle A\rangle\rangle\theta$ or $[[A]]\theta$ of Φ with a strategic quantifier as main connective can be converted to SNF by converting θ to LSNF using Lemma 4.6. Then, we can use the memoryless and memory-based version of Lemma 4.9 and obtain that Φ is satisfiable in the memory-based semantics if and only if it is satisfiable in the memoryless semantics since the satisfiable sets of successor components are the same for the two types of semantics. □

5 Optimal decision procedures for satisfiability in fragments of ATL_1^*

5.1 Centipede models. Satisfiability in LTL_1, CTL_1 and CTL_1^+

Satisfiability of LTL_1 is analyzed in [5]. In particular, it is shown that if an LTL_1 formula θ is satisfiable then it is satisfiable in a model of the form $s_0 s_1 ... s_\ell^\omega$ where $\ell = |\theta|$. Consequently, it is shown that satisfiability of LTL_1 is NP-complete. We provide similar results for CTL_1^+ and CTL_1 here.

Proposition 5.1 *If a* CTL_1^+ *formula* φ *has a model, then it has a model with at most* $O(|\varphi|^2)$ *states.*

Proof. Suppose $\mathcal{M}, s_0 \models \varphi$ for a CTL_1^+ formula φ, a model $\mathcal{M}$ and a state s_0. Assume w.l.o.g. that φ is in NNF. We generate another model $\mathcal{M}'$ with $O(|\varphi|^2)$ states and a state s_0' such that $\mathcal{M}', s_0' \models \varphi$. Let Δ_Q be the set of subformulae of φ that has Q as main connective for $Q \in \{\mathsf{E}, \mathsf{A}\}$ and let Δ_B be the set of maximal Boolean subformulae of φ that do not occur in the scope of a path quantifier. For each $Z \in \{\mathsf{E}, \mathsf{A}, \mathsf{B}\}$ let $\Delta_Z^\top \subseteq \Delta_Z$ be the subsets satisfied in $\mathcal{M}, s_0$. Now, for each $\mathsf{E}\psi \in \Delta_\mathsf{E}^\top$ let $\rho^\psi = \rho_0^\psi \rho_1^\psi ...$ be a path in $\mathcal{M}$ starting in s_0 such that $\rho^\psi \models \psi$. Since $\mathcal{M}, s_0 \models \varphi$ we have for every $\mathsf{A}\psi' \in \Delta_\mathsf{A}^\top$ that $\rho^\psi \models \psi'$ because ψ' is satisfied along all paths from s_0. Further, $\rho_0^\psi = s_0$ implies that $\rho^\psi \models \psi \wedge \bigwedge_{\psi' \in \Delta_\mathsf{A}^\top} \psi' \wedge \bigwedge_{\beta \in \Delta_\mathsf{B}^\top} \beta$. Since this is an LTL_1 formula of size at most $|\varphi|$ it has a model π^ψ of the form $\pi_0^\psi ... (\pi_{|\varphi|}^\psi)^\omega$ where π_0^ψ is labelled as s_0. Now, by gluing together each path π^ψ (which is made finite by adding a self-loop to the state $\pi_{|\varphi|}^\psi$) in the initial state s_0' we obtain a transition system $\mathcal{M}'$ such that $\mathcal{M}', s_0' \models \varphi$. Since $|\Delta_\mathsf{E}^\top| \leq |\varphi|$ there are at most $O(|\varphi|^2)$ states in $\mathcal{M}'$. □

Further, we will see that for satisfiable formulae of CTL_1^*, ATL_1 and $\mathsf{St}(\mathsf{ATL}_1^*)$ there are models that can be obtained by gluing together ultimately periodic paths as in the proof of Proposition 5.1. We call such models *centipede models*, illustrated in Figure 2. Note that these models only branch in the initial state.

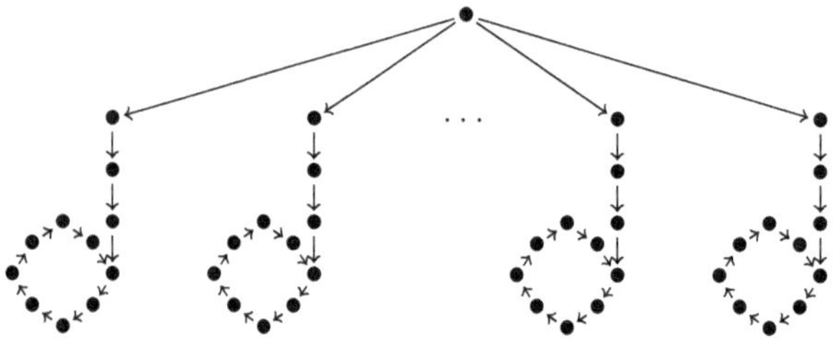

Fig. 2. A centipede model

However, for the flat fragments CTL_1^*, ATL_1, $\mathsf{St}(\mathsf{ATL}_1^*)$ models of polynomial size are not guaranteed to exist as for CTL_1^+. First, the length of the period and the prefix of the ultimately periodic paths can be exponential due to LTL subformulae in the case of CTL_1^* and $\mathsf{St}(\mathsf{ATL}_1^*)$. Second, in the cases of ATL_1 and $\mathsf{St}(\mathsf{ATL}_1^*)$ (but not for CTL_1^*) an exponential branching factor in the initial state may be forced by a formula. Indeed, consider the following ATL_1 formula

$$\varphi = \bigwedge_{i=1}^{n} \langle\langle i \rangle\rangle \mathsf{X}\, p_i \wedge \langle\langle i \rangle\rangle \mathsf{X} \neg p_i$$

over the propositions $\{p_1, ..., p_n\}$ and players $\{1, ..., n\}$. For a state s_0 to satisfy this formula there has to be a successor state for each possible truth assignment to the propositions $\{p_1, ..., p_n\}$, of which there are 2^n. This phenomenon does not occur in the branching-time logic CTL_1^*.

Proposition 5.2 *Satisfiability in* CTL_1^+ *is* NP*-complete.*

Proof. NP-hardness follows directly from Boolean satisfiability. An NP-algorithm for CTL_1^+ works as follows. It takes as input a CTL_1^+ formula φ in NNF, hence a positive Boolean combination of flat CTL_1^+ state formulae, and guesses non-deterministically a centipede model $\mathcal{M}, s_0$ of size $O(|\varphi|^2)$ for φ, as well as the disjuncts in φ that evaluate to true at s_0. (According to Proposition 5.1, if φ has a model then it has a model of this form and size.) After guessing, it checks whether the resulting formula of the form $\varphi' = \beta \wedge \bigwedge_{i=0}^{\ell} \varphi'_i$ is true in the guessed model where β is a Boolean formula and each φ'_i is of the form $\mathsf{A}\theta_i$ or $\mathsf{E}\theta_i$ for an LTL_1 formula θ_i. First, the model-checking of β can be done in linear time. Next, for each of the $O(|\varphi|)$ formulae θ_i it can checked whether it is true in each of the $O(|\varphi|)$ paths of the centipede model in polynomial time since LTL model-checking of an ultimately periodic path of length $O(|\varphi|)$ can be done in polynomial time in $|\varphi|$ [11]. Thus, the guess can be verified in polynomial time due to the small model property of Proposition 5.1 and the centipede shape of the model. □

Corollary 5.3 *Satisfiability in* CTL_1 *is* NP*-complete.*

5.2 Lower bound for satisfiability in ATL_1

Proposition 5.4 ATL_1*-SAT is* Σ_3^P*-hard.*

Proof. The proof is by reduction from the Σ_3^P-SAT problem, which is Σ_3^P-complete. This problem takes as input a quantified Boolean sentence

$$\gamma = \exists x_1, ..., x_m \forall x_{m+1}, ..., x_k \exists x_{k+1}, ..., x_n . \gamma'$$

where γ' is a Boolean formula over the Boolean variables $x_1, ..., x_n$. The output is true if and only if γ is true. Given γ, we construct an ATL_1 formula $\psi(\gamma)$ over the set $\mathsf{Prop} = \{x_1, ..., x_n\}$ of proposition symbols as follows

$$\psi = \bigwedge_{j=1}^{m} (\mathsf{AX}\, x_j \vee \mathsf{AX} \neg x_j) \wedge \bigwedge_{i=m+1}^{n} (\langle\!\langle \{i\} \rangle\!\rangle \mathsf{X}\, x_i \wedge \langle\!\langle \{i\} \rangle\!\rangle \mathsf{X} \neg x_i) \wedge \neg \langle\!\langle \{m+1, \dots, k\} \rangle\!\rangle \mathsf{X} \neg \gamma'$$

We now claim that γ is true if and only if $\psi(\gamma)$ is satisfiable.

First, suppose that γ is true. Then we construct a CGM $\mathcal{M} = (\mathbb{A}, \mathsf{St}, \{\mathsf{Act_a}\}_{\mathsf{a} \in \mathbb{A}}, \{\mathsf{act_a}\}_{\mathsf{a} \in \mathbb{A}}, \mathsf{out}, \mathsf{Prop}, \mathsf{L})$ and a state $s_0 \in \mathsf{St}$, such that $\mathcal{M}, s_0 \models \psi(\gamma)$, as follows. Let $\mathbb{A} = \{m+1, ..., n\}$, $\mathsf{St} = \{s_0\} \cup \{s_{v_{m+1}, ..., v_n} \mid v_i \in \{0,1\}$ for $m+1 \leq i \leq n\}$, $\mathsf{Act}_a = \{0,1\}$ for all $a \in \mathbb{A}$. Then, for every agent $a \in \mathbb{A}$ define $\mathsf{act}_a(s_0) = \{0,1\}$ and $\mathsf{act}_a(s) = \{0\}$ for all $s \neq s_0$. The transitions are defined by $\mathsf{out}(s_0, \langle v_{m+1}, \dots v_n \rangle) = s_{v_{m+1}, \dots v_n}$ and $\mathsf{out}(s, \alpha_{\mathbb{A}}) = s$ for all $s \neq s_0$ and all action profiles $\alpha_{\mathbb{A}}$. The set of proposition symbols is $\mathsf{Prop} = \{x_1, \dots, x_n\}$. The labelling is given by $L(s_0) = \emptyset$ and for every

$i \in \{m+1,\dots,n\}$ we have $x_i \in L(s_{v_{m+1},\dots,v_n})$ if and only if $v_i = 1$ when $s_{v_{m+1},\dots,v_n} \in \mathsf{St} \setminus \{s_0\}$. Finally, let $x'_1,\dots,x'_m$ be particular values such that $\forall x_{m+1},\dots,x_k \exists x_{k+1},\dots,x_n.\gamma'[x_1 \mapsto x'_1,\dots,x_m \mapsto x'_m]$ is true. Such values exist since γ is true. For $1 \le i \le m$ and every $s_{v_{m+1},\dots,v_n} \in \mathsf{St} \setminus \{s_0\}$, let $x_i \in L(s_{v_{m+1},\dots,v_n})$ if and only if $x'_i = 1$.

Intuitively, for all i such that $m+1 \le i \le n$ player i chooses the value of x_i in the successor state and then the play stays in that state forever. The value of x_i for $1 \le i \le m$ in the successor state is defined by the values $x'_1,\dots,x'_m$. The subformula $\bigwedge_{i=m+1}^{n} (\langle\!\langle \{i\} \rangle\!\rangle \mathsf{X}\, x_i \wedge \langle\!\langle \{i\} \rangle\!\rangle \mathsf{X} \neg x_i)$ is clearly true at s_0. The same is the case for $\bigwedge_{j=1}^{m} (\mathsf{AX}\, x_j \vee \mathsf{AX} \neg x_j)$. Next, since γ is true when x_i takes the values x'_i for $1 \le i \le m$, then no matter which values of x_i are chosen by players in $\{m+1,\dots,k\}$ there exists values of x_i for players in $\{k+1,\dots,n\}$ such that γ' is true in the successor state. Thus, coalition $\{m+1,\dots,k\}$ does not have a strategy to ensure that γ' is false in the successor state. Thus, $\mathcal{M}, s_0 \models \psi(\gamma)$.

For the converse direction, suppose that $\psi(\gamma)$ is satisfied by some model $\mathcal{M}, s_0$. For contradiction, suppose that γ is false. Then for all $x_1,\dots,x_m$ there exists $x_{m+1},\dots,x_k$ such that γ' is false for all $x_{k+1},\dots,x_n$. In particular, this must be the case when x_i take the unique values x'_i for $1 \le i \le m$ that are true in all successors of s_0. These are unique since s_0 satisfies $\bigwedge_{j=1}^{m} (\mathsf{AX}\, x_j \vee \mathsf{AX} \neg x_j)$. In this case there exists particular values x'_i for $m+1 \le i \le k$ such that γ' is false for all $x_{k+1},\dots,x_n$ when x_i take the values x'_i for $m+1 \le i \le k$. Consider the strategy for coalition $\{m+1,\dots,k\}$ that chooses these values for x_i in the successor state for $m+1 \le i \le k$. This strategy ensures that γ' is false in the successor state. However, this contradicts the fact that $\mathcal{M}, s_0 \models \neg \langle\!\langle \{m+1,\dots,k\} \rangle\!\rangle \mathsf{X} \neg \gamma'$. Thus, γ must be true. This completes the proof. □

Note that the hardness result only requires the use of the temporal operator X and neither U nor R. This is interesting since this lower bound will be shown to be an upper bound for the full ATL_1^+ in the following section. Thus, the $\langle\!\langle \cdot \rangle\!\rangle \mathsf{X}$ fragment of ATL_1 is as hard as the full ATL_1^+.

5.3 Deciding satisfiability in $\mathsf{St}(\mathsf{ATL}_1^*)$ and ATL^+

Lemma 5.5 *Let $\Phi = \langle\!\langle C \rangle\!\rangle \Psi$ be an ATL^* formula and let $\mathsf{Prop}(\Phi) = \{p_1, \dots p_r\}$ be the set of atomic propositions occurring in Φ. Consider any mapping $v : \mathsf{Prop}(\Phi) \to \{\top, \bot\}$ and let $v[\Phi]$ be the result of substitution of all occurrences of p_i in Φ which are not in the scope of a temporal operator by $v(p_i)$, for each $p_1, \dots p_r$. Further, let*

$$\delta(v) := \bigwedge_{v(p_i)=\top} p_i \wedge \bigwedge_{v(p_i)=\bot} \neg p_i$$

Then, $\delta(v) \wedge \Phi \equiv \delta(v) \wedge v[\Phi]$.

Proof. Consider any CGM $\mathcal{M}$ and a state s in it. If $\delta(v)$ is false at s then both sides are false. Suppose $\mathcal{M}, s \models \delta(v)$. Then $\mathcal{M}, s \models v(p_i) \leftrightarrow p_i$ for each $p_1, \dots p_r$. Then, Φ and $v[\Phi]$ are equally true or false at s, as they only differ in the occurrences of atomic propositions that are evaluated at s. □

Proposition 5.6 (i) *The satisfiability testing for* $\mathsf{St}(\mathsf{ATL}_1^*)$ *is in* PSPACE.
(ii) *The satisfiability testing for* ATL_1^+ *(and* ATL_1*) is in* Σ_3^P.

Proof. The decision procedures for both $\mathsf{St}(\mathsf{ATL}_1^*)$ and ATL_1^+ will be essentially the same, but in their last phases they work in different computational complexities. First, consider an $\mathsf{St}(\mathsf{ATL}_1^*)$ formula Φ and let $\mathsf{Prop}(\Phi) = \{p_1, \dots p_r\}$. The formula Φ is a Boolean combination of atomic propositions and subformulae of the type $\langle\langle C\rangle\rangle\theta$ where $\theta \in \mathsf{LTL}$. By Lemma 4.6, we can assume that each such θ is in a LSNF of linearly increased length, i.e., is a Boolean combination of atomic propositions and X-formulae (formulae beginning with X) of LTL. The algorithm now works as follows:

1. Guess a truth assignment τ for the atomic propositions in $\mathsf{Prop}(\Phi)$ at a state s of a CGM satisfying Φ, if any. Consider the unique map $v : \mathsf{Prop}(\Phi) \to \{\top, \bot\}$ for which $\delta(v)$ is true under τ. By Lemma 5.5, each maximal subformula $\langle\langle C\rangle\rangle\theta$ in Φ can be equivalently replaced by $v[\langle\langle C\rangle\rangle\theta]$, which is $\langle\langle C\rangle\rangle v[\theta]$.

2. After elementary Boolean simplifications (of the type $\top \wedge A \equiv A, \bot \wedge A \equiv \bot$, etc.) each $v[\theta]$ is transformed to a Boolean combination of X-formulae only. Using the LTL validities $\mathsf{X}\,\eta \wedge \mathsf{X}\,\zeta \equiv \mathsf{X}\,(\eta \wedge \zeta)$ and $\mathsf{X}\,\eta \vee \mathsf{X}\,\zeta \equiv \mathsf{X}\,(\eta \vee \zeta)$, it is further equivalently transformed into an X-formula which is at most as long.

The original formula is now (non-deterministically) transformed to an equisatisfiable Boolean combination of ATL^* formulae of type $\langle\langle C\rangle\rangle\mathsf{X}\,\theta$ and $[[C]]\mathsf{X}\,\theta$.

3. Now, assuming that the resulting formula is satisfiable, we further guess the true disjuncts in every $\vee$-subformula in a satisfying CGM and reduce the problem to checking satisfiability of a conjunctive formula of the type

$$\Theta = \langle\langle A_0\rangle\rangle\mathsf{X}\,\theta_0 \wedge \ldots \wedge \langle\langle A_{m-1}\rangle\rangle\mathsf{X}\,\theta_{m-1} \wedge [[B_0]]\mathsf{X}\,\eta_0 \wedge \ldots \wedge [[B_{n-1}]]\mathsf{X}\,\eta_{n-1}$$

Let $D(\Theta)$ be the union of the set $C(\Theta)$ of conjuncts of Θ and $\{[[\mathbb{A}]]\mathsf{X}\,\top\}$, i.e.

$$D(\Theta) = \{\langle\langle A_0\rangle\rangle\mathsf{X}\,\theta_0, \ldots, \langle\langle A_{m-1}\rangle\rangle\mathsf{X}\,\theta_{m-1}, [[B_0]]\mathsf{X}\,\eta_0, \ldots, [[B_{n-1}]]\mathsf{X}\,\eta_{n-1}, [[\mathbb{A}]]\mathsf{X}\,\top\}$$

4. By Lemma 4.9, the set $D(\Theta)$ is satisfiable iff every subset of distributed control of it has a satisfiable set of successor components. Since each of them is a set of LTL formulae, these checks can be done using standard techniques.

Each check in step 4. of the algorithm can be done in PSPACE when Φ is a $\mathsf{St}(\mathsf{ATL}_1^*)$ formula, since each successor component is an LTL formula. In the case of ATL_1^+ the checks can be done in NP according to [5], as in this case each successor component is an LTL_1 formula. Hence, checking that each of the (possibly exponentially many) subsets of distributed control is satisfiable can be done in $\mathrm{coNP}^{\mathrm{PSPACE}} = \mathrm{PSPACE}$ for $\mathsf{St}(\mathsf{ATL}_1^*)$ and in $\mathrm{coNP}^{\mathrm{NP}}$ for ATL_1^+. Thus, the whole procedure can be done respectively in $\mathrm{NP}^{\mathrm{PSPACE}} = \mathrm{PSPACE}$ for $\mathsf{St}(\mathsf{ATL}_1^*)$ and in $\mathrm{NP}^{\mathrm{coNP}^{\mathrm{NP}}}$ for ATL_1^+, by guessing the true propositions in the initial state and the true disjuncts in Φ, and then applying resp. a PSPACE-oracle and $\mathrm{coNP}^{\mathrm{NP}}$-oracle. Since $\mathrm{NP}^{\mathrm{coNP}^{\mathrm{NP}}} = \Sigma_3^P$ the proof is completed. □

This result, combined with Proposition 5.4 and the PSPACE-hardness of LTL satisfiability, yields the following.

Theorem 5.7 *The satisfiability problem of*

(i) $\mathsf{St}(\mathsf{ATL}_1^*)$ *is* PSPACE-*complete*

(ii) CTL_1^* *is* PSPACE-*complete*

(iii) ATL_1^+ *is* Σ_3^P-*complete*

(iv) ATL_1 *is* Σ_3^P-*complete*

Here is another consequence of the proof of Proposition 5.6:

Corollary 5.8 *Every satisfiable* $\mathsf{St}(\mathsf{ATL}_1^*)$ *formula* Φ *has a centipede model* $\mathcal{M}$ *with branching factor* $O(2^{|\Phi|})$ *in the root. Further, every ultimately periodic path in* $\mathcal{M}$ *has a prefix of length* $O(2^{|\Phi|})$ *and a period of length* $O(|\Phi| \cdot 2^{|\Phi|})$.

5.4 PSPACE decision procedure for the satisfiability in ATL_1^*

The decision procedure for $\mathsf{St}(\mathsf{ATL}_1^*)$ can be extended to a PSPACE-complete decision procedure for the whole ATL_1^*, by combining it with a PSPACE decision procedure for LTL and showing that every path-satisfiable ATL_1^* formula can be satisfied in a special type of CGMs described below. The proof of the latter is rather lengthy (see brief discussion further), so we only state and prove here the easier case of the slightly smaller fragment $\widehat{\mathsf{ATL}_1^*}$,where no strategic quantifiers $\langle\langle\emptyset\rangle\rangle$ (i.e, fully universal path quantifiers) are allowed. We only note that the procedure for the full ATL_1^* is essentially the same.

First, recall that every satisfiable LTL formula has an ultimately period linear model with prefix and period that both have length exponential in the size of the formula [14]. Further, according to Corollary 5.8, every satisfiable $\mathsf{St}(\mathsf{ATL}_1^*)$ formula can be satisfied at the root state of a centipede model of exponentially bounded number and length of legs. Combining these results leads to a new type of CGMs which we call *Lasso of Centipedes (LoC) models.* Such models consist of an ultimately periodic path (the lasso) where each state is the root of a centipede model. An illustration of a model like this is shown in Figure 3.

Proposition 5.9 *Every satisfiable* $\widehat{\mathsf{ATL}_1^*}$ *formula* Φ *is satisfied in a LoC model with size bounded exponentially in* $|\Phi|$.

Proof. Given an $\widehat{\mathsf{ATL}_1^*}$ formula Φ we define its LTL *skeleton* $\mathrm{Sk}_{\mathsf{LTL}}(\Phi)$ as follows: Let the state subformulae of Φ of type $\langle\langle C\rangle\rangle\theta$ or $[[C]]\theta$ be $\Psi_1, \ldots, \Psi_n$. For each of them Ψ we introduce a new (not in Prop) atomic proposition p_Ψ. Then we produce the LTL formula $\widehat{\Phi}$ by replacing every occurrence of such a subformula Ψ in Φ by p_Ψ. Now, define

$$\mathrm{Sk}_{\mathsf{LTL}}(\Phi) ::= \widehat{\Phi} \wedge \bigwedge_{i}^{n} \mathsf{G}\,(p_{\Psi_i} \rightarrow \Psi_i)$$

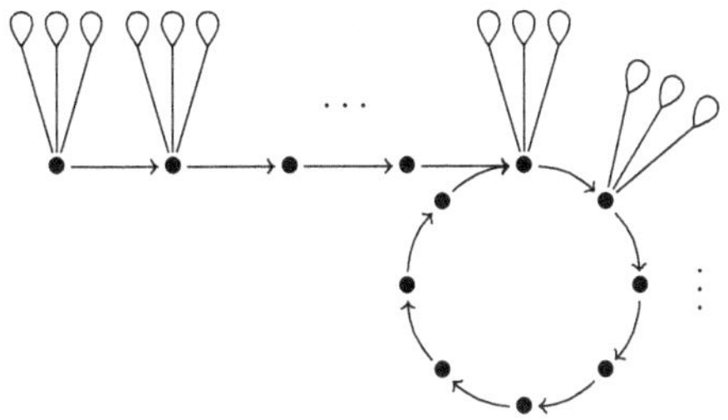

Fig. 3. A Lasso of Centipedes (LoC) model

We claim that any CGM $\mathcal{M}$ and a path π in it on which Φ is true can be expanded to a CGM $\widehat{\mathcal{M}}$ and a path $\widehat{\pi}$ in it satisfying $\mathrm{Sk}_{\mathsf{LTL}}(\Phi)$, by evaluating each new atomic proposition p_Ψ to be true at exactly those states of π at which Ψ is true in $\mathcal{M}$. Conversely, for any CGM $\mathcal{M}$ and a path π in it on which $\mathrm{Sk}_{\mathsf{LTL}}(\Phi)$ is true, the formula Φ is true on π, too, because all atomic propositions p_{Ψ_i} occur only positively in $\widehat{\Phi}$, so replacing them with the respective Ψ_i's will preserve the truth.

Thus, it suffices to show that if $\mathrm{Sk}_{\mathsf{LTL}}(\Phi)$ is path-satisfiable then it can be satisfied on the lasso path in some LoC model of size bounded exponentially in $|\Phi|$. Indeed, take any CGM $\mathcal{M}$ and a path π in it on which $\mathrm{Sk}_{\mathsf{LTL}}(\Phi)$ is true. Then, in particular, the path π alone is a linear model for $\widehat{\Phi}$. Now, take an ultimately periodic linear model $\widehat{\pi}$ of length bounded exponentially in $|\widehat{\Phi}|$, hence in $|\Phi|$. Such a model can be obtained from π by cutting its tail off at a suitable position and looping back to a suitable previous state. Thus, every state in $\widehat{\pi}$ has the label of a prototype state in π. Now, for every state $\widehat{s}$ on $\widehat{\pi}$, let s be its prototype in π. We do the following.

- Consider the set $\Gamma(s)$ of state subformulae Ψ of Φ such that p_Ψ is in the label of s in π. Since $\mathrm{Sk}_{\mathsf{LTL}}(\Phi)$ is true on π, every formula in $\Gamma(s)$ is true at s in $\mathcal{M}$. Thus, $\Gamma(s)$ is satisfiable, hence by Corollary 5.8, it can be satisfied at the root state of a centipede model $\mathcal{M}(\Gamma(s))$ of exponentially bounded in $|\Phi|$ number and length of legs.
- Now, we graft a copy of $\mathcal{M}(\Gamma(s))$ at the state $\widehat{s}$ in $\widehat{\pi}$ by identifying its root with $\widehat{s}$ and keeping all other states disjoint from $\widehat{\pi}$.
- Next, we add a special new action for every agent at the state $\widehat{s}$ and define the successor of the resulting action profile to be the successor of $\widehat{s}$ on the path $\widehat{\pi}$, while every other action profile involving some (but not all) of these special new actions leads to a successor of $\widehat{s}$ in the grafted copy of $\mathcal{M}(\Gamma(s))$, chosen so as not to affect the truth of any of the formulae from $\Gamma(s)$ at $\widehat{s}$. We omit the easy but tedious details of this construction.

After completing this procedure for each state of $\widehat{\pi}$, the result is a LoC model $\widetilde{\mathcal{M}}$ which, by construction, satisfies the formula $\widehat{\Phi}$ on $\widehat{\pi}$ and satisfies at each state $\widehat{s}$ on $\widehat{\pi}$ the set $\Gamma(s)$. Therefore, $\widetilde{\mathcal{M}}, \widetilde{\pi} \models \mathrm{Sk}_{\mathsf{LTL}}(\Phi)$, hence $\widetilde{\mathcal{M}}, \widetilde{\pi} \models \Phi$.□

For lack of space we only briefly indicate the additional complication in extending this result to ATL_1^*: if a subformula $\Psi = \langle\langle\emptyset\rangle\rangle\theta$ is true at some state of the path π in the CGM satisfying Φ, its effect cannot be constrained only on the centipede model grafted at the respective state of $\widetilde{\mathcal{M}}$, as done above, but it propagates through the path $\widetilde{\pi}$ to all centipede models grafted at all further states on $\widetilde{\pi}$. So, additional description in LTL is needed to describe and preserve this effect when converting π into the lasso $\widetilde{\pi}$. That is why we state the next result relativized to only what we have proved here.

Proposition 5.10 *The path-satisfiability problem in LoC models of size bounded exponentially in the length of input* ATL_1^* *formulae is in* PSPACE.

Proof. The algorithm begins like the PSPACE decision procedure for LTL satisfiability that guesses the lasso on the fly for an LTL input formula θ [14]. First, the length of the prefix and the length of the period are guessed. At each step around the lasso, the subformulae that are true from the current state are guessed non-deterministically and a local consistency check as well as a one-step consistency check are performed. Further, a set Δ (of at most polynomial size) of eventuality formulae is kept to make sure that all eventualities that are needed for θ to be true are actually true further on the lasso.

The algorithm for ATL_1^* works in the same way on an ATL_1^* formula Φ, but treats strategically quantified subformulae of Φ as atomic propositions and, at each step of the procedure, the local consistency check includes verifying these subformulae that have to be true at the current state. This amounts to checking satisfiability of an $\mathsf{St}(\mathsf{ATL}_1^*)$ formula and can be done in PSPACE, by Theorem 5.7. For the formulae of the form $\langle\langle A\rangle\rangle\theta$ where $A \neq \emptyset$ this can be done independently of the rest of the lasso, by ensuring that when agents in A commit to satisfying θ then the play goes into the centipede (and stays there). But, when $A = \emptyset$ then θ has to be true on all paths from the current state. This includes both the path around the lasso and those that enter one of the centipedes at some point. Note that the original set Δ of formulae we are keeping only needs to be satisfied around the lasso. To keep track of this we keep, in addition to Δ, an extra set of formulae Γ which must be satisfied both around the lasso and on paths that exits to a centipede. Thus, the formulae in Γ must be included in the $\mathsf{St}(\mathsf{ATL}_1^*)$ satisfiability check at each step. But since Γ is polynomial in size at each step, this check can still be performed in PSPACE. This means that the entire procedure can be performed in PSPACE. □

Corollary 5.11 *Satisfiability of* $\widehat{\mathsf{ATL}_1^*}$ *is* PSPACE*-complete.*

6 Concluding remarks and summary of results

We have developed optimal decision procedures for the satisfiability problems in flat fragments of ATL^*, and in particular CTL^* and have obtained exact complexity results for them. A summary of the main complexity results obtained in this paper is provided in the table in Fig. 4. It shows that these complexities are much lower than those for the full languages while, in view of Proposition 3.2, they are very tight with respect to syntactic extensions in terms of nesting depth of formulae.

$\mathcal{L}$	SAT($\mathcal{L}$)	SAT($\mathcal{L}_1$)
LTL	PSPACE [14]	NP [5]
CTL	EXPTIME [6]	NP (Cor. 5.3)
CTL^+	2EXPTIME [10]	NP (Prop. 5.2)
CTL^*	2EXPTIME [6]	PSPACE (Theo. 5.7)
ATL	EXPTIME [16]	Σ_3^P (Theo. 5.7)
ATL^+	2EXPTIME [12][10]	Σ_3^P (Theo. 5.7)
ATL^*	2EXPTIME [12]	PSPACE (Theo. 5.7, Cor. 5.11)

Fig. 4. Complexity of satisfiability. All results are completeness results. In the case of ATL^* the results refer to $\mathsf{St}(\mathsf{ATL}_1^*)$ and $\widehat{\mathsf{ATL}_1^*}$.

References

[1] Ågotnes, T., V. Goranko and W. Jamroga, *Alternating-time temporal logics with irrevocable strategies*, in: D. Samet, editor, *Proc. of TARK XI* (2007), pp. 15–24.

[2] Alur, R., T. A. Henzinger and O. Kuperman, *Alternating-time temporal logic*, Journal of the ACM **49** (2002), pp. 672–713.

[3] Brihaye, T., A. Da Costa, F. Laroussinie and N. Markey, *ATL with strategy contexts and bounded memory*, in: *Proc. of LFCS'2009, Springer LNCS 5407*, 2009, pp. 92–106.

[4] Bulling, N. and W. Jamroga, *Comparing variants of strategic ability: how uncertainty and memory influence general properties of games*, J. of AAMAS (2013), pp. 1–45.

[5] Demri, S. and P. Schnoebelen, *The complexity of propositional linear temporal logics in simple cases*, Inf. Comput. **174** (2002), pp. 84–103.

[6] Emerson, E. A., *Temporal and modal logic*, in: *Handbook of Theoretical Computer Science, Volume B: Formal Models and Sematics (B)*, 1990 pp. 995–1072.

[7] Goranko, V. and D. Shkatov, *Tableau-based decision procedures for logics of strategic ability in multiagent systems*, ACM Trans. Comput. Log. **11** (2009).

[8] Goranko, V. and G. van Drimmelen, *Complete axiomatization and decidablity of Alternating-time temporal logic*, Theor. Comp. Sci. **353** (2006), pp. 93–117.

[9] Halpern, J. Y., *The effect of bounding the number of primitive propositions and the depth of nesting on the complexity of modal logic*, Artif. Intell. **75** (1995), pp. 361–372.

[10] Johannsen, J. and M. Lange, *CTL$^+$ is Complete for Double Exponential Time*, in: *Proc. ICALP'03* (2003), pp. 767–775.

[11] Markey, N. and P. Schnoebelen, *Model checking a path*, in: *CONCUR*, 2003, pp. 248–262.

[12] Schewe, S., *ATL* satisfiability is 2EXPTIME-complete*, in: *Proc. of ICALP (2)*, 2008, pp. 373–385.

[13] Schröder, L. and Y. Venema, *Flat coalgebraic fixed point logics*, in: *Proc. of CONCUR'2010* (2010), pp. 524–538.

[14] Sistla, A. P. and E. M. Clarke, *The complexity of propositional linear temporal logics*, J. ACM **32** (1985), pp. 733–749.

[15] Troquard, N. and D. Walther, *On satisfiability of ATL with strategy contexts*, in: *Proc. of JELIA'12*, LNAI **7519** (2012), pp. 398–410.

[16] van Drimmelen, G., *Satisfiability in alternating-time temporal logic*, in: *Proc. of LICS'03*, 2003, pp. 208–217.

[17] Walther, D., C. Lutz, F. Wolter and M. Wooldridge, *ATL satisfiability is indeed ExpTime-complete*, Journal of Logic and Computation **16** (2006), pp. 765–787.

Subsumption Checking in Conjunctive Coalgebraic Fixpoint Logics

Daniel Gorín Lutz Schröder

Friedrich-Alexander-Universität Erlangen-Nürnberg

Abstract

While reasoning in a logic extending a complete Boolean basis is coNP-hard, restricting to conjunctive fragments of modal languages sometimes allows for tractable reasoning even in the presence of greatest fixpoints. One such example is the $\mathcal{EL}$ family of description logics; here, efficient reasoning is based on satisfaction checking in suitable small models that characterize formulas in terms of simulations. It is well-known, though, that not every conjunctive modal language has a tractable reasoning problem. Natural questions are then how common such tractable fragments are and how to identify them. In this work we provide sufficient conditions for tractability in a general way by considering unlabeled tableau rules for a given modal logic. We work in the framework of coalgebraic modal logics as unifying semantic setting. Apart from recovering known results for description logics such as $\mathcal{EL}$ and $\mathcal{FL}_0$, we obtain new ones for conjunctive fragments of relational and non-relational modal logics with greatest fixpoints. Most notably we find tractable fragments of game logic and the alternating-time μ-calculus.

Keywords: Materializers, convexity, tractable reasoning, fixpoints.

1 Introduction

The complexity of reasoning in logics extending a complete Boolean basis is at least coNP. For modal logics, it is typically even harder: already the basic (multi-)modal logic K_m is PSPACE-complete [21] and if fixed points are added to the mix, the complexity typically goes up to at least EXPTIME which, e.g., is the complexity of PDL and the μ-calculus [15]. Practical reasoning in these logics requires highly optimized heuristic strategies and will ultimately have only a limited degree of scalability.

This motivated the study of fragments in which core reasoning problems become tractable, i.e., decidable in polynomial time. Such fragments typically exclude negation and disjunction. Perhaps the best-known example is the $\mathcal{EL}$ family of *lightweight description logics*, where also universal restrictions (i.e. $\Box$-modalities) are excluded. In the absence of negation, satisfiability is no longer the central reasoning problem, being in fact often trivial (e.g., in $\mathcal{EL}$ every formula is satisfiable). Instead, one focuses on the entailment problem, alternatively called *subsumption checking* in the DL community. Indeed $\mathcal{EL}$

turns out to have a polynomial-time subsumption problem [4,6], even when extended with greatest fixed points [22]. Despite the limited syntax, $\mathcal{EL}$ can in practice accommodate large ontologies such as SNOMED CT.

Rather surprisingly, the subsumption problem of $\mathcal{FL}_0$, the counterpart of $\mathcal{EL}$ with universal instead of existential restriction, becomes intractable when greatest fixed points (or even just non-recursive global definitions, i.e. acyclic TBoxes) are added to the language [5,8,26]. This shows that there is more to lightweight logics than just dropping disjunctions. Here, we aim to develop conceptual tools to identify lightweight modal formalisms beyond the purely relational realm. For uniformity, we work in the setting of *coalgebraic modal logic* [28], where the notions of model and modal operators are suitably abstracted. We then state and prove a general version of each result and just obtain the featured instances as corollaries.

Tractability of relational lightweight logics exploits the existence of what are called *materializations* of formulas (which moreover need to be computable and small). (Alternatively, tractability can be shown by proof-theoretic methods [19].) A materialization for ϕ is a model that satisfies *only* the formulas that ϕ entails; thus, subsumption can be reduced to model-checking in a materialization [22]. Moreover, while there seems to be a strong connection between tractability of subsumption for a given fragment and *convexity* of its formulas, meaning that they imply at least one of the disjuncts of every disjunction they entail, the precise nature of such a connection is still partly unclear (see, e.g., [20,23]).

For coalgebraic logics, we show that a stronger (infinitary) version of convexity of their conjunctive fragments is actually equivalent to the existence of materializations (for the relational description logic $\mathcal{ALCFI}$, a more fine-grained connection at the level of TBoxes has been established by Lutz and Wolter [23]). As in the relational case, our materializations moreover have an even stronger property — they can be taken as complete replacements of the materialized formulas, in the sense that the satisfaction relation for the former corresponds to a similarity relation for the latter in a sense we developed recently [16]. However, the mere existence of materializations is not enough for tractability; one requires additional conditions that ensure that materializations for a given conjunctive fragment of a modal logic $\mathcal{L}$ can be obtained in polynomial time. For this, we develop a simple syntactic criterion based on the set of (unlabeled) tableau rules for $\mathcal{L}$, and show how to compute small materializations from them even in the presence of greatest fixpoints. With this result we can show tractability of several conjunctive modal (fixpoint) logics, including fragments of game logic [27] and the alternating-time μ-calculus [2].

Proofs are sketched or omitted; more technical details can be found in the arXiv preprint 1401.6359.

2 Preliminaries

We first present the various concrete logics that will serve as case studies throughout the paper and then briefly introduce the basic concepts of coalgebraic logic that are used in the generic development. For each concrete logic we also consider its reasoning principles, in the form of unlabeled tableau-rules $\Gamma_0/\Gamma_1 \mid \cdots \mid \Gamma_n$, where the Γ_i are sets of formulas. A set of rules $\mathcal{R}$ is meant to be used in the usual way: to show satisfiability of a set Γ (interpreted conjunctively), one needs to show the satisfiability of at least one conclusion of every rule in $\mathcal{R}$ applicable to Γ. All tableau systems are understood to extend a set of propositional rules.

Basic modal logic. We assume the reader to be familiar with the syntax and semantics of the basic modal logic K interpreted over Kripke models. We shall also consider its restriction KD to *serial* models, where every node has at least one successor, making $\Diamond\top$ valid. The set of rules $\mathcal{R}_K = \{K_n : n \geq 0\}$ (Fig. 1) induces a complete tableau system for K. For KD one needs to add to $\mathcal{R}_K$ the rules D_n for $n \geq 0$.

Monotone neighbourhood logic. The minimal monotone logic M uses the same language as K but is interpreted over monotone neighbourhoods, i.e., neighbourhood models where the set of neighbourhoods of each point is upwards closed w.r.t. set inclusion [13]. We read $\Box\phi$ as 'there is a neighbourhood where ϕ holds'. It is well known that this logic can be encoded in K, replacing $\Box$ with $\Diamond\Box$ and $\Diamond$ with $\Box\Diamond$ (e.g. [27]). A complete tableau system for M is obtained simply by taking rule K_1.

Here too, we will be interested in the *serial* case which corresponds to the case where $\Box\top$ and $\Diamond\top$ are taken as axioms; we shall denote the resulting logic by M_s. Serial monotone neighbourhood frames underlie the semantics of game logic [27], discussed in more detail in Section 6. Seriality means that each state has some neighbourhood and the empty set is never a neighbourhood. Notice that in the mentioned encoding of monotone modal logic into normal modal logic, serial monotone neighbourhood frames correspond exactly to serial Kripke frames. It is easy to see that the set of rules $\mathcal{R}_{M_s} = \{K_1, K_0, D_1\}$ is a complete tableau system for M_s (notice that K_0 and D_1 are just the instances of K_1 for $\Box\top$ and $\Diamond\top$, respectively).

Coalition logic and alternating-time logics. Coalition logic [29] is essentially the next-step fragment of the alternating-time μ-calculus AMC [2], discussed in Section 6. A *coalition* is a subset of a fixed set $N = \{1, \ldots, n\}$ of agents and one has a modal operator $[C]$ for each coalition C. Intuitively, we read formula $[C]\phi$ as 'coalition C has a joint strategy to enforce that ϕ shall hold in the next state'. Formally, the semantics is over *game frames*, where for each state x we have a function f_x with domain $S_1 \times \cdots \times S_n$, each S_q being a finite set of actions available to agent $q \in N$ in state x. Intuitively, the choice of an action by each agent determines a successor state as specified by the *outcome function* f_x. One then defines the semantics of $[C]$ by putting $x \models [C]\phi$ iff there exists a joint choice $(s_q)_{q \in C}$ of actions for the agents in C such that for each joint choice $(s_q)_{q \in N-C}$ for the agents outside C, $f((s_q)_{q\in N}) \models \phi$. Note that each choice of N defines a different logic CL_N (in the sense that extending CL_{N_0} to CL_{N_1} for $N_0 \subsetneq N_1$ does *not* preserve subsumption), since the seman-

$$K_n \; \frac{\Box a_1, \ldots, \Box a_n, \Diamond b}{a_1, \ldots, a_n, b} \qquad\qquad D_n \; \frac{\Box a_1, \ldots, \Box a_n}{a_1, \ldots, a_n}$$

$$C_{nm} \; \frac{[C_1]a_1, \ldots, [C_n]a_n, \langle D \rangle b, \langle N \rangle c_1, \ldots, \langle N \rangle c_m}{a_1, \ldots, a_n, b, c_1, \ldots, c_m} \; \dagger\ddagger \qquad C'_n \; \frac{[C_1]a_1, \ldots, [C_n]a_n}{a_1, \ldots, a_n} \; \dagger$$

Fig. 1. Tableau rules, with side conditions: (†) $i \neq j \Rightarrow C_i \cap C_j = \emptyset$, and (‡) $C_i \subseteq D$.

tics of $[C]$ depends on how many agents there are outside C. For a fixed N, the set of rules $\mathcal{R}_{\mathrm{CL}_N} = \{C_{ij}, C'_k : i, j \geq 0, k > 0\}$ (Fig. 1) yields a complete tableau system for CL [14,32].

We include only the basic definitions of coalgebraic logic, which is more comprehensively presented elsewhere [28,31,34]. The generality of the framework stems from the parametricity of its syntax and semantics. The language depends on a *similarity type* Λ, which may include atomic propositions, seen as modalities of arity 0. To simplify notation, we pretend that all modal operators are unary. The grammar for the set $L(\Lambda)$ of *positive* Λ*-formulas* is

$$\phi, \psi ::= \top \mid \bot \mid \phi \wedge \psi \mid \phi \vee \psi \mid \heartsuit\phi \qquad (\heartsuit \in \Lambda).$$

The set of *conjunctive* Λ-formulas is obtained by dropping the clauses for $\bot$ and $\vee$ from the grammar above. When $\mathcal{L}$ is a logic, we refer to the restriction of $\mathcal{L}$ to conjunctive formulas as *conjunctive* $\mathcal{L}$.

Given a modality $\heartsuit \in \Lambda$ we use $\bar{\heartsuit}$ to denote the *dual* of $\heartsuit$, with $\bar{\heartsuit}\phi$ interpreted as $\neg\heartsuit\neg\phi$ (under the usual meaning of $\neg$); we also use $\bar{\Lambda} := \{\bar{\heartsuit} : \heartsuit \in \Lambda\}$. We do *not* assume that Λ is closed under duals, as inclusion or non-inclusion of dual operators in Λ usually makes a big difference for the existence and size of materializations (Section 4).

The semantics is parametrized, first, in terms of an endofunctor T on the category Set of sets and maps, which determines the class of models. For a fixed T, a *model* is then just a T*-coalgebra* $C = (X, \xi)$, consisting of a set X (of *states*) and a *transition function* $\xi : X \to TX$. A *pointed model* is a pair (C, r), where r is a state of C, called the *point* or *root*. The intuition here is that $\xi(x)$ is the *local view* of the model standing on a state x; e.g., in a Kripke model, $\xi(w)$ would consist of the set of immediate successors of world w, plus the set of propositions that hold at w; thus, the class of all Kripke models arises as the class of all T-coalgebras for the functor $TX = \mathcal{P}(X) \times \mathcal{P}(\mathsf{Prop})$. As usual, we assume w.l.o.g. that T is non-trivial, i.e. $TX = \emptyset \implies X = \emptyset$ (otherwise, $TX = \emptyset$ for all X) and preserves subsets, i.e. $TX \subseteq TY$ whenever $X \subseteq Y$. (This is w.l.o.g. as we can assume that T preserves injective maps, possibly after changing $T\emptyset$ in a way that does not affect the class of coalgebras [9].)

The second parameter of the semantics is the interpretation of the modal operators, which relies on associating to each $\heartsuit \in \Lambda$ a *predicate lifting* $[\![\heartsuit]\!]$, i.e. a natural transformation $[\![\heartsuit]\!] : \mathcal{Q} \dot{\to} \mathcal{Q} \circ T^{op}$, where $\mathcal{Q} : \mathsf{Set}^{op} \to \mathsf{Set}$ is the contravariant powerset functor. That is, $\mathcal{Q}X = 2^X$ for every set X, and for a map f, Qf takes preimages under f. In particular, naturality of $[\![\heartsuit]\!]$ means that $[\![\heartsuit]\!]_X(f^{-1}[A]) = (Tf)^{-1}[[\![\heartsuit]\!]_Y(A)]$ for any map $f : X \to Y$.

Intuitively, a predicate lifting $[\![\heartsuit]\!]$ tells us what the local view of a state in X should be for it to satisfy a formula $\heartsuit\phi$ where ϕ has extension $A \subseteq X$; explicitly, the local view $\xi(x)$ of x should be an element of the set $[\![\heartsuit]\!]_X(A)$. E.g., one interprets $\Box$ on the Kripke functor T above using the predicate lifting

$$[\![\Diamond]\!]_X(A) := \{(S, V) : S \subseteq A, V \in \mathcal{P}(\mathsf{Prop})\}\,.$$

Formally, the notion of *satisfaction* of Λ-formulas ϕ at states x of C (denoted $x \models_C \phi$) is then defined by the expected clauses for Boolean operators, plus:

$$x \models_C \heartsuit\phi \iff \xi(x) \models \heartsuit[\![\phi]\!]_C$$

where $[\![\phi]\!]_C = \{x \in X : x \models_C \phi\}$ is the *extension* of ϕ in C, and, for $t \in TX$ and $A \subseteq X$,

$$t \models \heartsuit A$$

is a more suggestive notation for $t \in [\![\heartsuit]\!]_X(A)$. From $[\![\heartsuit]\!]$ we obtain the predicate lifting interpreting $\bar{\heartsuit}$ by $[\![\bar{\heartsuit}]\!]_X(A) = TX - [\![\heartsuit]\!]_X(X - A)$.

On positive formulas, the core reasoning task is *subsumption checking*: for formulas ϕ and ψ, we say that ψ *subsumes* ϕ, and write $\phi \sqsubseteq \psi$, if $[\![\phi]\!]_C \subseteq [\![\psi]\!]_C$ in all T-coalgebras C.

Abusing notation, we identify a similarity type Λ with this semantic structure $\langle T, [\![\heartsuit]\!]_{\heartsuit\in\Lambda}\rangle$ used to interpret it, and refer to both as Λ. We shall use T for the underlying functor throughout.

Example 2.1 All logics discussed above are coalgebraic; see, e.g., [34,16]. As an additional example, *graded (modal) logic*, which we call G, has the similarity type $\Lambda = \{\Diamond_k : k \in \mathbb{N}\}$, with $\Diamond_k\phi$ read 'ϕ holds in more than k successors', and is interpreted over the multiset functor $\mathcal{B}_\infty$, i.e., $\mathcal{B}_\infty X = X \to \mathbb{N} \cup \{\infty\}$. We regard $b \in \mathcal{B}_\infty X$ as an $\mathbb{N} \cup \{\infty\}$-valued measure on X, and correspondingly write $b(A) = \sum_{x\in A} b(x)$ for any subset $A \subseteq X$ (then, for a map f, $\mathcal{B}_\infty f$ acts by taking image measures, i.e. $\mathcal{B}_\infty f(\mu)(y) = \mu(f^{-1}[\{y\}]$.) Coalgebras for $\mathcal{B}_\infty$ are *multigraphs*, i.e. directed graphs whose edges are annotated with multiplicities from $\mathbb{N} \cup \{\infty\}$. Each $\Diamond_k$ is interpreted by the predicate lifting

$$[\![\Diamond_k]\!]_X(A) := \{b \in \mathcal{B}_\infty X : b(A) > k\}.$$

A multigraph (X, ξ) is essentially a more concise representation of a Kripke frame, with $\xi(x)(y) = n$ standing for n distinct successors of x, all of them isomorphic copies of y. Thus, $[\![\Diamond_k]\!]$ clearly captures the informal reading of $\Diamond_k$.

This framework is modular [34], and in particular supports fusion of modal logics by taking products of functors. For instance, the functor inducing Kripke models with m relations, supporting the interpretation of m relational modalities, can be seen as arising from the product $TX = \prod_{i=1}^m \mathcal{P}(X) \times 2^{\mathsf{Prop}}$ of m copies of the covariant powerset functor $\mathcal{P}$, and a copy of the constant functor

2 given by $2X = 2 = \{0, 1\}$ for each proposition symbol in Prop (the associated predicate liftings are derived in the obvious way).

Although coalgebraic logic supports non-monotone modalities, we assume operators to be *monotone* ($A \subseteq B \subseteq X \Rightarrow [\![\heartsuit]\!]_X A \subseteq [\![\heartsuit]\!]_X B$): to characterize formulas by simulations, we need monotonicity in inductive proofs, since simulations preserve but do not reflect satisfaction of formulas. Crucially, all monotone coalgebraic logics admit complete sets of tableau rules consisting (besides the standard propositional rules) of rules of the form $\Gamma_0/\Gamma_1 \mid \cdots \mid \Gamma_n$ where Γ_0 contains only formulas $\heartsuit a$, with $\heartsuit \in \Lambda \cup \bar{\Lambda}$, and $\Gamma_1, \ldots, \Gamma_n$ contain only variables, as in Fig. 1 [14]; we fix such a rule set $\mathcal{R}$ throughout.

In coalgebraic logic one exploits locality and reduces logical phenomena such as derivability or satisfiability from the full logic to the simpler setting of *one-step models*, which are, roughly, the result of forgetting the structure of a pointed model everywhere except at the root; see, e.g., [31]. With one-step models come *one-step formulas*, i.e. shallow modal formulas where propositional variables are introduced as placeholders for complex argument formulas under modal operators.

Definition 2.2 (One-step logic) Let V be a set of propositional variables (not fixed, and typically finite); a *one-step model over* V is just a tuple (X, τ, t) where X is a set (possibly empty), $\tau : V \to \mathcal{P}X$ interprets propositional variables, and $t \in TX$. The dual representation of τ is $\breve{\tau} : X \to \mathcal{P}V$, i.e. $\breve{\tau}(x) = \{p : x \in \tau(p)\}$. A *conjunctive one-step Λ-formula* is a finite conjunction of atoms $\heartsuit p$, where $\heartsuit \in \Lambda$, $p \in V$. The satisfaction relation is given by $(X, \tau, t) \models_\tau \bigwedge_{i \in I} \heartsuit_i p_i$ iff $t \models \heartsuit_i \tau(p_i)$ for all i. Similarly, a *positive one-step Λ-formula* is an element of $\mathsf{Pos}(\Lambda(\mathsf{Pos}(V)))$, where $\Lambda(W) = \{\heartsuit w : \heartsuit \in \Lambda, w \in W\}$ and Pos denotes positive propositional combinations (using $\top$, $\bot$, $\vee$, $\wedge$), with the expected semantics. We write $\sqsubseteq_1$ for the subsumption relation in the one-step logic: $\phi \sqsubseteq_1 \psi$ if $(X, \tau, t) \models \psi$ whenever $(X, \tau, t) \models \phi$.

The transfer of results between the one-step and the full logic is done by way *collages*, i.e., pasting pointed coalgebras into a one-step model to form a new coalgebra, and *décollages*, tearing away most of the structure of a pointed coalgebra to obtain a one-step model (see e.g. the construction of shallow models in [31,25]). Explicitly:

Definition 2.3 Given $t \in TX$, a family of pairwise disjoint pointed coalgebras $(C_x, x) = ((Y_x, \xi_x), x)$ for all $x \in X$, and a fresh root state r, the *collage* of these *collage data* is the pointed coalgebra (C, r), with $C = (Y, \xi)$, where Y is the (disjoint) union of $\{r\}$ and the Y_x, and

$$\xi(y) := \begin{cases} t & \text{if } y = r \\ \xi_x(y)) & \text{otherwise, for the } x \text{ such that } y \in Y_x \end{cases}$$

As indicated earlier, we assume that T preserves subsets, so, e.g., $TX \subseteq TY$.

In a nutshell, the collage is obtained from a root state r with successor structure $t \in TX$ by replacing every $x \in X$ with a pointed coalgebra (C_x, x). The following is immediate by construction:

Lemma 2.4 (Collage lemma) *For a collage (C, r) with collage data as in Definition 2.3, and all $x \in X$, $A \subseteq Y$ and $\heartsuit \in \Lambda$,*

(i) $x \models_C \phi \iff x \models_{C_x} \phi$, *and*

(ii) $t \in \heartsuit_X(A \cap X) \iff \xi(r) \in \heartsuit_Y A$.

Proof. The second equivalence follows directly from naturality of $\heartsuit$. For the first one, one proceeds by induction on ϕ; the relevant case is the modal one:

$$\begin{aligned} x \models_\xi \heartsuit\psi &\iff T(\hookrightarrow_{Y_x})(\xi_x(x)) \in \heartsuit_Y [\![\psi]\!]_\xi \\ &\iff \xi_x(x) \in \heartsuit_{Y_x}([\![\psi]\!]_\xi \cap Y_x) && \text{(naturality)} \\ &\iff \xi_x(x) \in \heartsuit_{Y_x}[\![\psi]\!]_{\xi_x} && \text{(IH)} \\ &\iff x \models_{\xi_x} \heartsuit\psi && \square \end{aligned}$$

One typically needs collages based on interpretations of propositional variables as modal formulas. Here, we will be interested in *preserving* the interpretation of the satisfied atoms; more precisely:

Definition 2.5 Given collage data as in Definition 2.3, a valuation $\tau : V \to \mathcal{P}(X)$ *(positively) matches* a substitution $\rho : V \to L(\Lambda)$ if for all $x \in X$, $x \models_{C_x} \rho(p)$ iff (if) $x \in \tau(p)$.

Lemma 2.6 *Let $\tau : V \to \mathcal{P}(X)$* (positively) match *$\rho : V \to L(\Lambda)$. Then*

(i) $x \in \tau(p)$ *iff (implies)* $x \models_C \rho(p)$, *and*

(ii) $t \models_\tau \heartsuit p$ *iff (implies)* $r \models_C \heartsuit\rho(p)$.

The converse process is as follows.

Definition 2.7 Given a pointed coalgebra (C, r) with $C = (X, \xi)$ and a substitution $\rho : V \to L(\Lambda)$, we say that (X, τ, t) is the *décollage of* (C, r) *by* ρ if $t = \xi(r)$ and $\tau(p) = [\![\rho(p)]\!]_C$.

Lemma 2.8 (Décollage lemma) *If (X, τ, t) is a décollage of (C, r) by $\rho : V \to L(\Lambda)$ then for all one-step formulas ϕ over V we have $(X, \tau, t) \models \phi \iff r \models_C \phi\rho$.*

3 Coalgebraic Simulations

We recall the notion of coalgebraic modal simulation from [16]. Given a binary relation $S \subseteq X \times Y$, we denote by S^- its relational inverse. Moreover, for $A \subseteq X$, the relational image of S over A is given by $S[A] := \{y : \exists x \in A.\, xSy\}$.

Definition 3.1 (Λ-Simulation) Let $C = (X, \xi)$ and $D = (Y, \zeta)$ be two given T-coalgebras. A *Λ-simulation* $S : C \to D$ *(of C by D)* is a relation $S \subseteq X \times Y$ such that xSy and $\xi(x) \models \heartsuit A$ imply $\zeta(y) \models \heartsuit S[A]$, for all $\heartsuit \in \Lambda$ and $A \subseteq X$. When xSy for a Λ-simulation S, we say that (D, y) *Λ-simulates* (C, x).

The properties of Λ-simulations that we need here are the following (cf. [16]):

Lemma 3.2 *Λ-simulations are stable under relational composition; moreover, (graphs of) identities are Λ-simulations.*

Lemma 3.3 *Let $S : C \to D$ be a Λ-simulation and ϕ be a positive Λ-formula. Then xSy and $x \models_C \phi$ imply $y \models_D \phi$.*

The effect of dualizing modal operators is to turn around the notion of simulation:

Proposition 3.4 *Let $\bar{\Lambda} := \{\bar{\heartsuit} : \heartsuit \in \Lambda\}$. A relation S between T-coalgebras is a $\bar{\Lambda}$-simulation iff S^- is a Λ-simulation.*

Example 3.5 (See [16] for details.)

(i) Over Kripke frames and for $\Lambda = \{\Diamond\}$, a Λ-simulation $S : C \to D$ is just a simulation $C \to D$ in the usual sense. By Proposition 3.4, for $\Lambda = \{\Box\}$, a Λ-simulation $S : C \to D$ is then a simulation $D \to C$ in the usual sense. Consequently, a $\{\Box, \Diamond\}$-simulation is just a standard bisimulation.

(ii) A $\{p\}$-simulation for a proposition p is just a relation that preserves p.

(iii) Over monotone neighbourhoods with $\Lambda = \{\Box\}$, $S \subseteq X \times Y$ is a Λ-simulation between $\mathcal{M}$-coalgebras (X, ξ) and (Y, ζ) iff xSy and $A \in \xi(x)$ imply $S[A] \in \zeta(y)$.

4 Weakly Simulation-Initial Models

In general, modal formulas need not have smallest models under the simulation preorder. In some cases, however, such smallest models do exist. Formally, we define this property as follows.

Definition 4.1 (wsi models) Let ϕ be a positive Λ-formula. A pointed model (C_ϕ, x_ϕ) is called *weakly simulation-initial (wsi) for* ϕ if for any other (D, y), $y \models_D \phi$ iff (D, y) Λ-simulates (C_ϕ, x_ϕ).

In the relational setting, the term *sim-initial* has been used for an analogous notion [23]. Initiality in this sense is rather weak, though, since the witnessing simulations are not necessarily unique.

Remark 4.2 Since identities are Λ-simulations, a wsi model for ϕ satisfies ϕ. Thus, by Lemma 3.3, (C_ϕ, x_ϕ) is wsi for ϕ iff (i) $x_\phi \models \phi$, and (ii) whenever (D, y) is such that $y \models_D \phi$, then (D, y) Λ-simulates (C_ϕ, x_ϕ).

Definition 4.3 ([22]) A pointed coalgebra (C, x) is a *materialization* of ϕ if for all positive Λ-formulas ψ, $x \models_C \psi$ iff $\phi \sqsubseteq \psi$. In this case, ϕ is *materializable.*

Of course, this definition implies that a materialization of ϕ is a model of ϕ. By Lemma 3.3, the following is immediate:

Lemma 4.4 *Every wsi model is a materialization.* □

Thus, subsumption reduces to model checking in wsi models when they exist.

Definition 4.5 (Convexity) [6] A satisfiable Λ-formula ϕ is *(strongly) convex* if whenever $\phi \sqsubseteq \bigvee_{i \in I} \psi_i$ for some (possibly infinite) index set I and positive

Λ-formulas ψ_i (with the expected semantics of $\bigvee$), then already $\phi \sqsubseteq \psi_i$ for some $i \in I$.

Lemma 4.6 *If ϕ is materializable then ϕ is strongly convex.* $\square$

Remark 4.7 Convexity is generally felt to be necessary for tractability; see, e.g., [20,23] (where it is considered w.r.t. *finite* disjunctions). It is not only an important structural property but also provides a good handle for showing that certain formulas are *not* materializable. E.g. a formula that is itself a disjunction can have a materialization only when it is equivalent to one of its disjuncts. It is thus no surprise that tractable logics such as $\mathcal{EL}$ and TBox-free $\mathcal{FL}_0$ exclude disjunction; also here, we will henceforth *restrict attention to conjunctive formulas.*

But even conjunctive formulas may fail to be materializable. E.g., in G with $\Lambda = \{\Diamond_k : k \in \mathbb{N}\}$ we have $\Diamond_1 a \wedge \Diamond_1 b \sqsubseteq \Diamond_2(a \vee b) \vee \Diamond_1(a \wedge b)$ but the left hand side is not subsumed by any of the disjuncts of the right hand side, so convexity fails (cf. [6]). Similarly, conjunctive $\{\Box_1\}$-formulas may fail to be convex, as witnessed by

$$\Box_1(a \wedge b) \wedge \Box_1(b \wedge c) \wedge \Box_1(c \wedge d) \wedge \Box_1(d \wedge a) \sqsubseteq \\ \Box_1(a \wedge b \wedge c) \vee \Box_1(b \wedge c \wedge d) \vee \Box_1(c \wedge d \wedge a) \vee \Box_1(d \wedge a \wedge b)\,.$$

Worse, with the wrong choice of Λ, even $\top$ may fail to be materializable: in K with $\Lambda = \{\Box, \Diamond\}$ we have $\top \sqsubseteq \Box\Diamond\top \vee \Diamond\top$ but $\top \not\sqsubseteq \Box\Diamond\top$ and $\top \not\sqsubseteq \Box\Diamond\top$. Similarly, in M with $\Lambda = \{\Box, \Diamond\}$, one has that $\top \sqsubseteq \Box\top \vee \Diamond\top$ and yet $\top \not\sqsubseteq \Box\top$ and $\top \not\sqsubseteq \Diamond\top$.

The existence of wsi models thus depends strongly on the chosen T-structure Λ, as well as on slight variations in the semantics (e.g. w.r.t. seriality). We now proceed to show that one can limit the study of the phenomenon to the level of the much simpler one-step logic (Section 2). As suggested by Remark 4.2, we define in this case:

Definition 4.8 (one-step wsi models) A one-step model (X, τ, t) is *weakly simulation-initial (wsi)* for a conjunctive one-step formula ϕ over V if (i) $t \models_\tau \phi$, and (ii) for every (Y, ϑ, s), $A \subseteq X$ and $\heartsuit \in \Lambda$, $t \in \heartsuit_X A$ implies $s \in \heartsuit_Y S[A]$, where $xSy \iff \check{\tau}(x) \subseteq \check{\vartheta}(y)$.

Remark 4.9 One-step wsi models are never unique. However, one can assume w.l.o.g. that if (X, τ, t) is wsi, then every $x \in X$ is uniquely determined by $\check{\tau}(x)$ (quotient (X, τ, t) by the equivalence relation induced by $\check{\tau}$), and hence that X is of at most exponential size on the number of variables.

Definition 4.10 We say that Λ *admits (one-step) wsi models* if every conjunctive (one-step) formula has a (one-step) wsi model.

The main technical result of this section is then the following.

Theorem 4.11 *Λ admits wsi models whenever it admits one-step wsi models.*

Proof (Sketch) Induction on ϕ. We have $\phi = \bigwedge_{i\in I} \heartsuit_i \chi_i$ for a finite (possibly empty) set I. Take $V_\phi := \{a_{\chi_i} : i \in I\}$ and decompose ϕ as $\phi = \phi^*\rho$ with $\phi^* := \bigwedge_{i\in I} \heartsuit_i a_{\chi_i}$ a one-step formula and $\rho(a_{\chi_i}) := \chi_i$ a substitution. Let (X, τ, t) be a wsi for ϕ^*. By IH, there is, for each $x \in X$, a wsi model (C_x, x) with $C_x = (Y_x, \xi_x)$ for $\bigwedge_{p\in\check{\tau}(x)} \rho(p)$ with root x; the Y_x can be assumed pairwise disjoint. Pick a fresh x_ϕ, and obtain (C_ϕ, x_ϕ) by taking $\xi(x_\phi) = t$ and attaching C_x at each $x \in X$ (cf. [31]). One easily shows that (C_ϕ, x_ϕ) is wsi for ϕ. □

We now analyze under which conditions one-step wsi models exist. To begin, note that at the one-step level, wsi models coincide with materializations (recall that $\sqsubseteq_1$ is the one-step subsumption relation of Def. 2.2):

Definition 4.12 A one-step model (X, τ, t) is a *one-step materialization* of a conjunctive one-step Λ-formula ϕ over V if for every literal $\heartsuit\rho$ with $\heartsuit \in \Lambda$ and $\rho \in \mathsf{Pos}(V)$, $t \models_\tau \heartsuit\rho$ iff $\phi \sqsubseteq_1 \heartsuit\rho$. In this case, ϕ is *materializable.*

Again, this implies that a one-step materialization of ϕ is a model of ϕ.

Lemma 4.13 *A one-step model is wsi for a conjunctive one-step Λ-formula ϕ iff it is a materialization of ϕ.*

Moreover, existence of materializations is equivalent to convexity:

Definition 4.14 A satisfiable one-step Λ-formula ϕ over V is *strongly convex* if whenever $\phi \sqsubseteq_1 \bigvee_{i\in I} \psi_i$ for positive one-step Λ-formulas ψ_i over V and a (possibly infinite) index set I, then already $\phi \sqsubseteq_1 \psi_i$ for some $i \in I$.

Remark 4.15 In case Λ is finite, strong convexity of one-step formulas is the same as convexity (the notion obtained by restricting I to be finite in Definition 4.14), as then there are, up to equivalence, only finitely many positive one-step Λ-formulas over V.

Lemma 4.16 *A one-step Λ-formula is materializable iff it is strongly convex.*

Remark 4.17 Summing up, at the one-step level the notions of *being materializable*, *having a wsi model* and *being strongly convex* coincide. For the full logic, we have already noted that wsi models are materializations and materializable formulas are strongly convex. We leave the equivalence of these notions for individual formulas, i.e. to show that every strongly convex formula has a wsi model, to future research (for some relational logics, this equivalence is known [1,11]). Under mild additional assumptions, it does follow at the current stage that the equivalence holds between the respective properties of the logic as a whole: assume for simplicity that Λ contains infinitely many proposition symbols (actually, it suffices that the logic is *non-trivial*, i.e. contains infinitely many propositionally independent formulas). If all conjunctive Λ-formulas are strongly convex, then this holds (emulating propositional variables by proposition symbols from Λ) also for conjunctive one-step Λ-formulas. By the above, it follows that Λ admits one-step wsi models, and hence admits wsi models.

Next, we show how to read off convexity from the structure of the tableau rules for Λ. At the same time, we obtain a description of the structure of one-step materializations.

Definition 4.18 We call a tableau rule *definite* if it has exactly one conclusion, i.e. is of the form Γ/Δ with $\Gamma \subseteq (\Lambda \cup \bar{\Lambda})(V)$ and $\Delta \subseteq V$. A set $\mathcal{R}$ of definite one-step rules *preserves* Λ*-convexity* if whenever a rule R over V in $\mathcal{R}$ can be written in the form $\Gamma_1, \Gamma_2/\Delta_1, \Delta_2 \in \mathcal{R}$ with $\Gamma_1 \subseteq \Lambda(V_1)$, $\Gamma_2 \subseteq \bar{\Lambda}(V_2)$, $\Delta_1 \subseteq V_1$, $\Delta_2 \subseteq V_2$, with V_1, V_2 a disjoint decomposition of V (we call this a Λ*-splitting* of R), then for each $\bar{\heartsuit} a \in \Gamma_2$, the rule $\Gamma_1, \bar{\heartsuit} a/\Delta_1, a$ is also in $\mathcal{R}$.

The next theorem will show that preservation of Λ-convexity is sufficient for convexity of conjunctive Λ-formulas. It is fairly clear that, in cases where all rules are definite, necessity also holds for a sufficiently carefully formulated weakening of preservation of Λ-convexity (e.g. in the above notation, it clearly suffices to have $\Gamma_1, \bar{\heartsuit} a/\Delta_1, a$ derivable from $\mathcal{R}$ in the obvious sense); we refrain from exploring details.

Remark 4.19 In case Λ is closed under duals, the rule set $\mathcal{R}$ preserves Λ-convexity iff whenever Γ/Δ is a rule over V in $\mathcal{R}$ and $\emptyset \neq V_0 \subseteq V$, then $(\Gamma \cap \Lambda(V_0))/(\Delta \cap V_0)$ is in $\mathcal{R}$ – that is, iff $\mathcal{R}$ is stable under deleting variables.

Theorem 4.20 *Let Λ be finite (for brevity; in fact it suffices to assume a more sophisticated form of completeness [33]). If $\mathcal{R}$ preserves Λ-convexity, then Λ admits wsi models. Moreover, a one-step materialization for a conjunctive one-step Λ-formula $\phi = \bigwedge_{i \in I} \heartsuit_i a_i$ (read also as the set $\{\heartsuit_i a_i : i \in I\}$) is then obtained as follows. First put $W = \{a_i : i \in I\}$, and define (X, τ) to consist of*

- *a state x with $\check{\tau}(x) = \Delta\sigma$, for each rule Γ/Δ over V in $\mathcal{R}$ and each renaming $\sigma : V \to W$ with $\Gamma\sigma \subseteq \phi$;*
- *a state x with $\check{\tau}(x) = \Delta_1\sigma$, for each rule $\Gamma, \bar{\heartsuit} b/\Delta_1, \Delta_2$ over $V \uplus \{b\}$ in $\mathcal{R}$ with $\Delta_2 \subseteq \{b\}$ and each renaming $\sigma : V \to W$ with $\Gamma\sigma \subseteq \phi$.*

(In both cases, we can restrict to rules and renamings for which $\Gamma\sigma$ becomes maximal.) Then there exists $t \in TX$ such that (X, τ, t) is a materialization of ϕ.

Remark 4.21 The rule sets in all examples are built in such a way that σ can be restricted to be injective in the construction of Theorem 4.20 [32]; however, it is easy to see that in such cases, this restriction does not actually affect the result of the construction.

Example 4.22 Over the proposition functor 2, $\Lambda = \{p\}$ and $\Lambda = \{\bar{p}\}$ (but not, of course, $\Lambda = \{p, \overline{p}\}$) are easily seen to admit wsi models; e.g. $(\emptyset, \emptyset, 1)$ is wsi for p. This is our only positive example not matching Theorem 4.20: the one-step rule $p, \bar{p}/\bot$ fails to be definite, having no conclusion.

Example 4.23 Over Kripke frames, we have the following.

(i) $\Lambda = \{\Diamond\}$ admits wsi models: a $\{\Diamond\}$-splitting $\Gamma_1, \Gamma_2/\Delta_1, \Delta_2$ of (K_n) in Fig. 1 is of the form $\Gamma_1 = \Diamond b$, $\Gamma_2 = \Box a_1, \ldots, \Box a_n$, and for each j we have a rule $\Diamond b, \Box a_j/b, a_j$ in $\mathcal{R}_K$, as required. The one-step wsi model for $\bigwedge_{i \in I} \Diamond a_i$ according to Theorem 4.20 is (I, τ, I) with $\tau(a_i) = \{i\}$. An example is depicted in Fig. 2(a). This extends to the multimodal case (see Remark 4.26), essentially, to $\mathcal{EL}$.

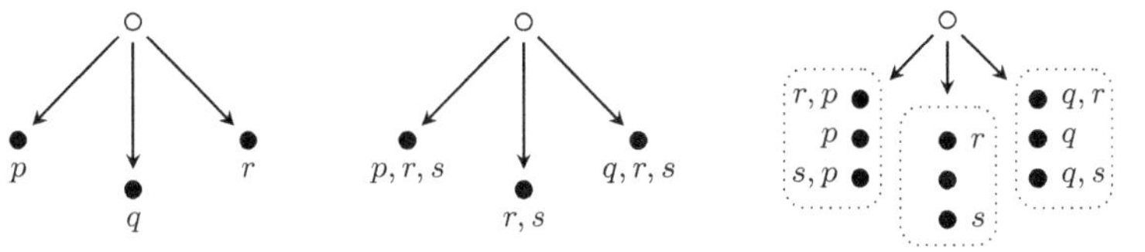

(a) K: $\Diamond p \wedge \Diamond q \wedge \Diamond r$ (b) KD: $\Diamond p \wedge \Diamond q \wedge \Box r \wedge \Box s$ (c) M_s: $\Box p \wedge \Box q \wedge \Diamond r \wedge \Diamond s$

Fig. 2. One-step wsi models for the indicated formulas. The white node is the implicit root, the black ones its domain. For M_s, minimal neighbourhoods are depicted (dotted boxes), not their supersets.

(ii) $\Lambda = \{\Box\}$ admits wsi models: any $\{\Box\}$-splitting $\Gamma_1, \Gamma_2/\Delta_1, \Delta_2$ of K_n already has Γ_2 of the form $\Diamond b$. Restricting to maximal rule matches, the one-step wsi model for $\bigwedge_{i\in I} \Box a_i$ according to (the second clause of) Theorem 4.20 is $(\{*\}, \tau, \{*\})$ with $\tau(a_i) = \{*\}$ for all i. This extends straightforwardly to the multi-modal case (Remark 4.26), of which $\mathcal{FL}_0$ is a syntactic variant.

(iii) In K, $\Lambda = \{\Box, \Diamond\}$ fails to be convex (Remark 4.7). Note that $\mathcal{R}_K$ fails to preserve convexity: deleting b from K_n yields $D_n \notin \mathcal{R}_K$ (Fig. 1, Remark 4.19). In KD, however, $\{\Box, \Diamond\}$ does admit wsi models, as the rules K_n and D_n together are stable under deleting occurrences of variables. Restricting to maximal matches, the one-step wsi model for $\bigwedge_{i\in I} \Box a_i \wedge \bigwedge_{j\in J} \Diamond b_j$ is $(J \cup \{*\}, \tau, J \cup \{*\})$ (with $* \notin J$) given by $\tau(a_i) = J \cup \{*\}$ and $\tau(b_j) = \{j\}$ (see Fig. 2(b)).

Example 4.24 Over monotone neighbourhoods, the situation is analogous as over Kripke frames, due to the similarity of the rule sets: both $\Lambda = \{\Box\}$ and $\Lambda = \{\Diamond\}$ admit wsi models in M, but not $\Lambda = \{\Box, \Diamond\}$ (M validates $\Box\top \vee \Diamond\top$ but none of the disjuncts, so no $\{\Box, \Diamond\}$-formula is convex). In M_s, $\{\Box, \Diamond\}$ does admit wsi models, though, for essentially the same reasons as in KD. The one-step wsi model from Theorem 4.20 for $\bigwedge_{i\in I} \Box a_i \wedge \bigwedge_{j\in J} \Diamond b_j$ (I, J disjoint) is $(X, \tau, \mathfrak{N})$:

$$X := \{K \subseteq I \cup J : |K \cap I| \leq 1, |K \cap J| \leq 1\} \qquad \tau(a_i) := \{K \in X : i \in K\}$$
$$\mathfrak{N} := \uparrow(\{\tau(a_i) : i \in I\} \cup \{\{K \in X : K \subseteq J\}\}) \quad \tau(b_j) := \{K \in X : j \in K\}$$

where $\uparrow$ is closure under taking supersets. Fig. 2(c) depicts the construction.

Example 4.25 In coalition logic, $\Lambda = \{[C], \langle C\rangle : C \subseteq N\}$ admits wsi models: its rules are stable under deleting occurrences of variables by Remark 4.19.

Remark 4.26 When one models fusion of modal logics by taking products of functors as noted in Sec. 2 (see [34]) this is reflected in the construction of one-step wsi models by just taking disjoint unions of the domains and pairing the transition structures (prolonged into the disjoint union). For instance, in the multimodal logic $\Lambda = \{\Box_1, \dots, \Box_n\}$ over Kripke frames, one-step wsi models for one-step formulas $\bigwedge_{i=1}^{n} \bigwedge_{j=1}^{m_i} \Box_i a_{ij}$ are formed by taking the disjoint union of the one-step wsi models for the formulas $\bigwedge_{j=1}^{m_i} \Box_i a_{ij}$ as described in Example 4.23, and thus have n states, with the i-th state satisfying the propositional variables $a_{i1}, \dots, a_{im_i}$. By Example 4.22, adding atomic propositions does not enlarge the carriers of wsi models at all.

For tractability, studied in the next section, we need wsi models to be small. However, existence of wsi models is of independent interest, even in those cases in which they may be exponentially large. For instance, from Example 4.23, we can already conclude that *conjunctive KD is convex.*

5 Tractability

Assume from now on that Λ admits one-step wsi models. Lemma 4.4 then allows us to reduce subsumption to satisfaction in such models. (This is also the principle underlying state-of-the-art consequence-based reasoning procedures, which for $\mathcal{EL}$ go back to [6].) In the previous sections, we have refrained from giving explicit descriptions of t when (X, τ, t) is wsi, and in fact it is not necessary to actually know t. Instead, we opt for a different representation of wsi models: in the recursive construction of a wsi model (C_ϕ, x_ϕ) for a conjunctive Λ-formula ϕ (see proof sketch of Theorem 4.11), we have calculated a one-step formula ϕ^* and used a one-step wsi model (X, τ, t) for it. For algorithmic purposes, we now drop t but store ϕ^*, X, and τ; we call the arising object an *abstract wsi model for* ϕ. We face then the following problem:

Definition 5.1 The *conjunctive one-step consequence problem* of Λ is to decide, given a conjunctive one-step Λ-formula ψ over V, $\heartsuit \in \Lambda$, and $\rho \in \mathsf{Pos}(V)$, if $\psi \sqsubseteq_1 \heartsuit\rho$.

If the conjunctive one-step consequence problem for Λ is in P, then we can check in time polynomial in the size of an abstract wsi model (C_ϕ, x_ϕ) for ϕ whether $x_\phi \models_{C_\phi} \psi$ for a positive Λ-formula ψ, e.g. by calculating extensions of subformulas of ψ bottom up. Now in the positive examples of the previous section, deciding whether, in the notation of the above definition, $\psi \sqsubseteq_1 \heartsuit\rho$ can be done using the respective rule sets to check whether $\psi \wedge \bar{\heartsuit}\neg\rho$ is satisfiable, which in turn will lead to checking satisfiability of a propositional formula of the form $\chi \wedge \neg\rho$ where χ is a conjunction over V, a trivial task given that ρ is positive. Thus, the conjunctive one-step consequence problem of Λ is in P provided that we can polynomially bound the number of rule matches to a given conjunction over $\Lambda(W)$, which is easily seen for all relevant examples.

Polynomial-time computability (entailing polynomially bounded size) of abstract wsi models will then imply tractability of subsumption. In some cases, tractability will hold only if we bound certain parameters. To avoid overformalization, we will call any set of conjunctive Λ-formulas a *conjunctive Λ-fragment* and apply notions defined so far w.r.t. the set of all conjunctive formulas, such as *admitting wsi models*, also to fragments. Note that sometimes restricting to a fragment will also restrict the relevant set of one-step formulas.

Definition 5.2 A conjunctive Λ-fragment $\mathcal{L}$ *admits polynomial wsi models* if every $\mathcal{L}$-formula has a polynomial-time computable abstract wsi model.

Lemma 5.3 *If $\mathcal{L}$ admits polynomial wsi models and the conjunctive one-step consequence problem of Λ is in P, then subsumption $\phi \sqsubseteq \psi$ between $\mathcal{L}$-formulas ϕ and positive Λ-formulas ψ is in P.*

We identify tractability criteria at the one-step level:

Definition 5.4 We say that one-step wsi models (X, τ, t) of one-step formulas $\phi = \bigwedge \heartsuit_i a_i$ are *linear* if $|\tau(a_i)| \leq 1$ for all i, *k-bounded* if $|\breve{\tau}(x)| \leq k$ for all $x \in X$, and *polynomial* if $|X|$ is polynomially bounded in the size of ϕ.

In words, linearity means that every propositional variable is satisfied in at most one state, while k-boundedness means that each state satisfies at most k propositional variables.

Proposition 5.5 a) *If a conjunctive Λ-fragment $\mathcal{L}$ admits linear or k-bounded one-step wsi models, then $\mathcal{L}$ admits polynomial wsi models.* b) *If Λ admits polynomial one-step wsi models, then conjunctive Λ-fragments defined by bounding the modal depth admit polynomial wsi models.*

(The complexity of bounded-depth fragments of modal logics over a complete Boolean basis has been studied, e.g., in [17].)

Proof (Sketch) Linearity implies that a wsi model for ϕ has at most as many states as ϕ has subformulas. On the other hand, k-boundedness ensures that wsi models, constructed as trees in the proof of Theorem 4.11, can be collapsed into polynomial-sized dags by identifying states realizing the same target formula; by k-boundedness, at most $|\phi|^k$ target formulas will arise in the construction.□

Example 5.6 (i) One-step wsi models for $\{\Diamond\}$ and for $\{\Box\}$ over Kripke frames (Example 4.23) are linear; those for $\{\Diamond\}$ are in addition 1-bounded. By Remark 4.26, this extends straightforwardly to the case with multiple modalities and atomic propositions. We thus recover the known results that subsumption checking in conjunctive multimodal K with only diamonds ($\mathcal{EL}$) or only boxes ($\mathcal{FL}_0$) is in P. As an aside, the conjunctive fragment of the co-contravariant modal logic of [1], which is essentially positive Hennessy-Milner logic with only diamonds for some actions and only boxes for the others, can be seen as a fusion of a logic of boxes with a logic of diamonds, and thus also has linear one-step wsi models, i.e. has a polynomial-time subsumption problem.

(ii) One-step wsi models for $\{\Box, \Diamond\}$ over serial Kripke frames, i.e. for conjunctive KD, are polynomial, so that *subsumption in bounded-depth fragments of conjunctive KD is in P* (with unboundedly many atomic propositions). This may be seen as a companion result to the (easily proved) coNP upper bound for bounded-depth fragments of full K [17].

Alternatively, if one restricts conjunctive KD formulas to use at most k boxes at each modal depth, then one-step wsi models for them become $k+1$-bounded, so that this restriction also ensures tractable reasoning. Again, this extends easily to the multimodal case with unboundedly many atomic propositions. Since one has a straightforward embedding of $\mathcal{EL}$ into multimodal KD (using a fresh propositional atom e marking 'existing' states to simulate arbitrary Kripke frames with serial ones), this result can be seen as generalizing the tractability of $\mathcal{EL}$ (which is just the case $k = 0$).

It is worth observing that the more specific problem of *satisfiability* but over unrestricted conjunctive KD extended with atomic negation (called *poor man's logic*) is known to be in P [18].

(iii) In M_s (Example 4.24), wsi models for $\Lambda = \{\Box, \Diamond\}$ are 2-bounded, so that *conjunctive M_s is tractable.* Similarly, wsi models for the structure $\Lambda = \{[C], \langle C \rangle : C \subseteq N\} - \{[\emptyset], \langle N \rangle\}$ in coalition logic / alternating-time logic are n-bounded, where n is the (fixed!) total number of agents (since n is also the maximal number of disjoint non-empty coalitions). Thus, for each finite set N of agents, *conjunctive coalition logic over N without $[\emptyset]$ and $\langle N \rangle$ is tractable.*

6 Greatest Fixpoints

We now proceed to extend the base logic with a fixpoint operator. This will allow us to cover global definitions (e.g.. classical terminological boxes, in DL parlance) and fragments of game logic and the alternating-time μ-calculus. We can only expect to get wsi models for formulas with *greatest* fixpoints, which are similar in flavour to infinite conjunctions, while least fixed points are disjunctive (e.g., $\nu x.(p \wedge \Diamond x)$ can be seen as the infinitary formula $p \wedge \Diamond(p \wedge \Diamond(p \wedge \dots)))$ which characterizes an infinite path of nodes satisfying p).

Following [22], we will actually allow for mutually recursive auxiliary definitions, as in the vectorial μ-calculus [3]. The resulting logic can be shown to be no more expressive than the one with only single-variable ν, but to admit exponentially more succinct definitions [22]. Syntactically, the grammar of *positive Λ-ν-formulas* extends that of positive Λ-formulas with fixpoint variables from a set Δ and, for $\alpha \in \{\nu, \mu\}$, formulas $\alpha(y; y_1, \dots y_n).(\phi, \phi_1 \dots \phi_n)$, where $y, y_1, \dots y_n \in \Delta$ must be distinct and $\phi, \phi_1, \dots \phi_n$ are positive Λ-ν-formulas. A formula $\nu(y; y_1, \dots y_n).(\phi; \phi_1, \dots, \phi_n)$ defines $y, y_1, \dots, y_n$ as a simultaneous greatest fixpoint, and then returns y; similarly for μ with least fixpoints. A *sentence* is a formula where every fixpoint variable is bound by a ν or μ. *Conjunctive* fixpoint Λ-formulas extend conjunctive Λ-formulas with ν only.

We define the semantics of this language over a T-coalgebra $C = (X, \zeta)$ and a valuation $\mathcal{V} : \Delta \to \mathcal{P}(X)$; by $[\![\phi]\!]_{C,\mathcal{V}}$ we denote the extension of ϕ in C assuming that the fixpoints variables are interpreted using $\mathcal{V}$. The propositional and modal cases are defined like before (with $[\![x]\!]_{C,\mathcal{V}} = \mathcal{V}(x)$); moreover, $[\![\nu(y_0; y_1, \dots y_n).(\phi_0; \phi_1, \dots \phi_n)]\!]_{C,\mathcal{V}}$ is the first projection of the greatest fixed point of the map taking $(A_0, \dots, A_n)$ to $([\![\phi_i]\!]_{C,\mathcal{V}[y_0 \mapsto A_0 \dots y_n \mapsto A_n]})_{i=1,\dots,n}$. The semantics of μ is dual. For a sentence ϕ, the initial $\mathcal{V}$ is irrelevant, so we may write just $[\![\phi]\!]_C$. Preservation of positive formulas by simulations extends to fixpoint formulas:

Lemma 6.1 *Let S be a Λ-simulation of a coalgebra $C = (X, \xi)$ by a coalgebra*

D, and let $\mathcal{V} : \Delta \to \mathcal{P}(X)$ be a valuation. Then for every positive Λ-ν-formula ϕ, $S[[\![\phi]\!]_{C,\mathcal{V}}] \subseteq [\![\phi]\!]_{D,S[\mathcal{V}]}$, where $S[\mathcal{V}]$ denotes the valuation taking x to $S[\mathcal{V}(x)]$.

Extending the definition of *wsi models* literally to positive Λ-ν-formulas, we thus obtain a generalization of Lemma 4.4, i.e. a wsi model for a fixpoint formula ϕ is a *materialization*, so that subsumption of ϕ by positive Λ-ν-formulas reduces to satisfaction in the wsi model.

Example 6.2 DLs are logics for knowledge representation, where terminologies are defined via axioms in *TBoxes* which effectively constrain the classes of models over which one reasons. In particular, one is sometimes interested in so-called *classical TBoxes with greatest fixpoint semantics* [7, Chapter 2]. Here, axioms of a TBox $\mathcal{T}$ are definitions of the form $a \equiv \phi$ with a a proposition symbol that is allowed to occur as a left-hand side of only one definition. Such an a is said to be a *derived* concept of $\mathcal{T}$. Each model C interpreting the non-derived propositions is extended to a unique model $C^{\mathcal{T}}$ which arises as the greatest fixpoint of the function mapping an extension C' of C interpreting also the derived propositions to the extension C'' where for each $a \equiv \phi \in \mathcal{T}$, $[\![a]\!]_{C''} = [\![\phi]\!]_{C'}$. One writes $\mathcal{T} \models \psi \sqsubseteq \chi$ if for each model C, $\psi \sqsubseteq \chi$ holds in $C^{\mathcal{T}}$. It is then clear that *subsumption over* $\mathcal{T}$, i.e. to decide whether $\mathcal{T} \models \psi \sqsubseteq \chi$, reduces to subsumption of fixpoint formulas: assume $\mathcal{T} = \{a_1 \equiv \phi_1, \dots a_n \equiv \phi_n\}$; we have $\mathcal{T} \models \psi \sqsubseteq \chi$ iff $\nu(z; a_1, \dots, a_n).(\psi; \phi_1, \dots, \phi_n) \sqsubseteq \nu(z; a_1, \dots, a_n).(\chi; \phi_1, \dots, \phi_n)$ where z is a fresh variable. Additional details are given by Lutz et al. [22].

Example 6.3 (Game logic) Model-checking a PDL formula $\langle\alpha\rangle\top$ can be seen as finding a winning strategy in a one-player game, where α describes the rules of the game and the model encodes the possible moves of the player on a fixed game board. In Game Logic (GL) [27], this notion is extended to two-player games (of perfect information). Composite games α are built from atomic games using the program constructors of PDL plus a *dualization* operator $(\cdot^d)$, which corresponds to players swapping roles, so that $\langle\alpha^d\rangle\phi \equiv [\alpha]\phi$ (and hence $[\alpha]$ can be omitted from the language). The two-player view disables normality (i.e. one no longer has $\langle\alpha\rangle(\phi \vee \psi) \to \langle\alpha\rangle\psi \vee \langle\alpha\rangle\psi$); hence, models of GL are products of monotone neighbourhood frames S_a, one per atomic game a. Intuitively, a set $A \in S_a(x)$ corresponds to (an upper bound on) positions that could be reached from x when following a fixed strategy for a; allowing for different responses of player II, we see that A need not be a singleton. As a notational infelicity, the predicate lifting interpreting $\langle a\rangle$ in GL (for a atomic) is that of $\Box$ in standard notation for monotone modal logic. Serial models are those where atomic games never get stuck, no matter which player begins. We note that GL has a well-known sublogic, concurrent propositional dynamic logic CPDL [30], which omits dualization $\cdot^d$ but retains $\cap$, the dual of $\cup$.

GL can be embedded into the fixpoint extension ${M_s}_m^{\nu}$ of multi-modal M_s (with duals of atomic propositions), much like PDL can be embedded into the relational μ-calculus. Two fixpoint variables suffice for this [10]. It is not hard to see that using fixpoint variables as a form of let-expressions, one can avoid the exponential blowup present in the original encoding. The *conjunctive* fragment of GL is swiftly defined as the preimage of the conjunctive fragment of ${M_{sm}}^{\nu}$ under this embedding.

Example 6.4 (Alternating time) The *alternating-time μ-calculus (AMC)* is essentially the extension of coalition logic with fixpoint operators (its actual notation is slightly different) [2]. The *conjunctive fragment* of the AMC can by defined in the obvious way excluding $\vee$, $\neg$, and μ. In this fragment, we can still express 'always' formulas from alternating-time temporal logic (ATL) such as $\langle\!\langle C\rangle\!\rangle\Box\phi$, which is read 'coalition C can maintain ϕ forever', and is equivalent to the fixpoint formula $\nu x.\,(\phi \wedge [C]x)$.

We proceed to show that if Λ admits one-step wsi models, we also obtain wsi models for conjunctive Λ-ν-formulas. We exploit the fact that any such sentence can be put, in polynomial time, in a *shallow* normal form, i.e. without nested occurrences of ν (using Bekič's law [3]) and without nesting of modal operators (using abbreviations for subformulas in analogy to standard TBox normalizations [4]).

Thus, let $\phi = \nu(x_0; x_1, \ldots, x_n)(\phi_0; \phi_1, \ldots, \phi_n)$ be a shallow sentence. We shall assume, for each conjunctive one-step Λ-formula ψ over $V = \Delta$, a fixed one-step wsi model $(X_\psi, \tau_\psi, t_\psi)$ which we then call *the* one-step wsi model for ψ. We assume w.l.o.g. that $X_\psi \subseteq \mathcal{P}(V(\psi))$, where $V(\psi)$ is the set of variables mentioned in ψ, and $\tau_\psi(x) = \{A \in X_\psi : x \in A\}$ (Remark 4.9). We then construct the carrier X_ϕ of C_ϕ as a subset of $\mathcal{P}(V)$. For $A \subseteq V$, we let ϕ_A denote the conjunctive one-step formula given by $\bigwedge_{x_i \in A} \phi_i$. Then, X_ϕ is the smallest subset of $\mathcal{P}(V)$ containing $r_\phi = \{x_0\}$ such that $X_{\phi_A} \subseteq X_\phi$, for each $A \in X_\phi$. We define a T-coalgebra structure ξ_ϕ on X_ϕ by $\xi_\phi(A) = T(i_A)t_{\phi_A}$, where i_A is the inclusion $X_{\phi_A} \hookrightarrow X_\phi$.

Theorem 6.5 *If Λ admits one-step wsi models, then for every shallow $L^\nu(\Lambda)$-sentence ϕ, (C_ϕ, r_ϕ) as constructed above is a wsi model.*

Proof (Sketch) Let ϕ have the form $\nu(x_0; x_1, \ldots x_n).(\phi_0; \phi_1, \ldots, \phi_n)$, so $V = \{x_0, \ldots x_n\}$. We have to show that (i) $r_\phi \models_{C_\phi} \phi$ and (ii) that if $d \models_D \phi$, then $r_\phi S d$ for some simulation $S : C_\phi \to D$ (Remark 4.2).

(i): By coinduction – taking $\mathcal{V}(x_i) = \{A \in X_\phi : x_i \in A\}$, one shows that $\mathcal{V}(x_i) \subseteq [\![\phi_i]\!]_{C_\phi,\mathcal{V}}$ for all $x_i \in V$. The gfp property of ϕ then implies $\mathcal{V}(x_0) \subseteq [\![\phi]\!]$, and clearly $r_\phi \in \mathcal{V}(x_0)$.

(ii): For $i = 0, \ldots n$, let $\phi^{(i)}$ denote the formula obtained by projecting the i-th component of ϕ:

$$\phi^{(i)} = \nu(x_i; x_0 \ldots x_{i-1}, x_{i+1} \ldots x_n).(\phi_i; \phi_0 \ldots \phi_{i-1}, \phi_{i+1} \ldots),$$

so in particular $\phi^{(0)} = \phi$. Assume $d \models_D \phi$ for some coalgebra $D = (Y, \zeta)$. Define a relation $S \subseteq X_\phi \times Y$ by

$$ASy \iff y \models_D \bigwedge_{x_i \in A} \phi^{(i)}.$$

Then clearly $r_\phi S d$, for by definition $r_\phi = \{x_0\}$ and $\phi^{(0)} = \phi$. One can show that S is a Λ-simulation. □

Clearly, all conjunctive logics listed as having one-step wsi models in the examples of Sec. 4 have wsi models when extended with greatest fixpoints, in particular remain convex. Of course, the wsi models constructed above may be exponentially large, even when Λ admits linear one-step wsi models. However, under k-boundedness, elements of $X_\phi \subseteq \mathcal{P}(\Delta)$ have at most k elements, leading to our main criterion for smallness of wsi models under greatest fixpoints:

Theorem 6.6 *If Λ admits k-bounded one-step wsi models for some k, then conjunctive Λ-ν-formulas have polymomial-size wsi models.*

By Theorem 6.6 and the description of one-step wsi models in Section 4, and using abstract wsi models as in Section 5, we regain the known result that subsumption checking over classical TBoxes with gfp semantics in $\mathcal{EL}$ is in P [4], and in fact can extend it to allow a bounded number of universal restrictions, always in conjunction with $\Diamond\top$. As new results, we obtain:

Corollary 6.7 *Subsumption checking for conjunctive Game Logic is in P.*

Corollary 6.8 *Subsumption checking for the conjunctive alternating-time μ-calculus (AMC) without $[\emptyset]$ and $\langle N\rangle$ is in P.*

Remark 6.9 There is one case where we do obtain polynomial-size wsi models without k-boundedness, namely $\Lambda = \{\Box\}$ over Kripke frames – here, one-step wsi models have only one state, so that wsi models for fixpoint formulas are lassos, i.e. chains of states ending in a loop. For smallness, one still needs to impose additional restrictions on shallow fixpoints $\nu(y; y_1, \dots, y_n).(\phi; \phi_1, \dots, \phi_n)$, e.g. that y_i always appears in ϕ_i, or that the fixpoint is acyclic, i.e. not actually recursive. This example does not extend to the multi-modal case since the property of one-step wsi models being singletons is not stable under taking disjoint sums (Remark 4.26). Indeed, the multimodal version is FL, and reasoning over even the most restrictive (i.e. acyclic) TBoxes in $\mathcal{FL}_0$ is known to be coNP-hard [26].

7 Conclusions

Representability of formulas by models in the sense that simulation of the model is equivalent to satisfaction of the formula is a highly useful phenomenon in conjunctive fragments of modal fixpoint logics. It implies, for instance, convexity of the formula (and is equivalent to it in the one-step case) and under a polynomial size bound on the model, tractability of reasoning. We have studied the question of existence of such *weakly simulation-initial (wsi) models*, in the framework of coalgebraic logic; in particular, we have proved a reduction of the problem to a local (*one-step*) version. We were able to derive a criterion for tractability from the shape of the tableau rules that enabled us to establish tractability in a number of key examples:

- we have recovered known tractability results for the description logics $\mathcal{EL}$ (over classical TBoxes with gfp semantics) and $\mathcal{FL}_0$ (without a TBox), and shown that reasoning over classical TBoxes with gfp semantics in $\mathcal{EL}$ (equivalently in the fragment of the multi-modal μ-calculus defined by restricting to conjunction, diamonds, and greatest fixed points) remains tractable when we allow a bounded number of universal restrictions (i.e. boxes);

- we established tractability of conjunctive monotone logic with greatest fixed points over serial models, which subsumes corresponding fragments of game logic [27];
- we have shown tractability of the conjunctive fragment (which has greatest but not least fixed points) of the alternating-time μ-calculus AMC [2]; this fragment still includes the game-based versions of *EG* and *AG* found in ATL.

Outside the large body of work on $\mathcal{EL}$, there has been only a limited amount of research on wsi models for conjunctive logics. Notable examples are the work on the relationship between relational modal logics and modal transition systems [11,1] where formulas in certain variants of positive Hennessy-Milner logic are shown to have wsi models iff they are convex. (*prime* in the cited works). We exhibited a similar equivalence at the level of conjunctive coalgebraic *logics*; we leave a generalization of the equivalence for individual *formulas* as future work. There is some work on sub-Boolean fragments of temporal logics, which however focuses on satisfiability rather than subsumption (e.g. [24]).

Further points for future investigation include the use of wsi models to calculate so-called *least common subsumers* [8], as well as covering *general* TBoxes (i.e. finite sets of arbitrary inclusion axioms), which is known to remain tractable in the case of $\mathcal{EL}$ [12].

Acknowledgments The authors wish to thank Carsten Lutz for useful discussions, and Erwin R. Catesbeiana for unsolicited remarks regarding the absence of unsatisfiable formulas in $\mathcal{EL}$.

References

[1] Aceto, L., I. Fábregas, D. de Frutos-Escrig, A. Ingólfsdóttir and M. Palomino, *Graphical representation of covariant-contravariant modal formulae*, in: *Expressiveness in Concurrency, EXPRESS 2011*, EPTCS **64**, 2011, pp. 1–15.

[2] Alur, R., T. A. Henzinger and O. Kupferman, *Alternating-time temporal logic*, J. ACM **49** (2002), pp. 672–713.

[3] Arnold, A. and D. Niwiński, "Rudiments of μ-calculus," Elsevier, 2001.

[4] Baader, F., *Terminological cycles in a description logic with existential restrictions*, in: *International Joint Conference on Artificial Intelligence, IJCAI 2003*, pp. 325–330.

[5] Baader, F., *Using automata theory for characterizing the semantics of terminological cycles*, Ann. Math. Artif. Intell. **18** (1996), pp. 175–219.

[6] Baader, F., S. Brandt and C. Lutz, *Pushing the $\mathcal{EL}$ envelope*, in: *International Joint Conference on Artificial Intelligence, IJCAI 2005.*

[7] Baader, F., D. Calvanese, D. L. McGuinness, D. Nardi and P. F. Patel-Schneider, editors, "The Description Logic Handbook," Cambridge University Press, 2007, second edition.

[8] Baader, F., R. Küsters and R. Molitor, *Computing least common subsumers in description logics with existential restrictions*, in: *International Joint Conference on Artificial Intelligence, IJCAI 1999*, pp. 96–101.

[9] Barr, M., *Terminal coalgebras in well-founded set theory*, Theoret. Comput. Sci. **114** (1993), pp. 299–315.

[10] Berwanger, D., E. Grädel and G. Lenzi, *The variable hierarchy of the μ-calculus is strict*, Theory Comput. Syst. **40** (2007), pp. 437–466.

[11] Boudol, G. and K. Larsen, *Graphical versus logical specifications*, Theoret. Comput. Sci. **106** (1992).
[12] Brandt, S., *Polynomial time reasoning in a description logic with existential restrictions, GCI axioms, and – what else?*, in: *Eureopean Conference on Artificial Intelligence, ECAI 2004* (2004), pp. 298–302.
[13] Chellas, B., "Modal Logic," Cambridge University Press, 1980.
[14] Cîrstea, C., C. Kupke and D. Pattinson, *EXPTIME tableaux for the coalgebraic μ-calculus*, Log. Methods Comput. Sci. **7** (2011).
[15] Emerson, E. and C. Jutla, *The complexity of tree automata and logics of programs*, SIAM J. Comput. **29** (1999), pp. 132–158.
[16] Gorín, D. and L. Schröder, *Simulations and bisimulations for coalgebraic modal logics*, in: *Algebra and Coalgebra in Computer Science, CALCO 2013*, LNCS **8089** (2013), pp. 253–266.
[17] Halpern, J., *The effect of bounding the number of primitive propositions and the depth of nesting on the complexity of modal logic*, Artif. Intell. **75** (1995), pp. 361–372.
[18] Hemaspaandra, E., *The complexity of poor man's logic*, Journal of Logic and Computation **11** (2001), pp. 609–622.
[19] Hofmann, M., *Proof-theoretic approach to description-logic*, in: *Logic in Computer Science, LICS 2005*, pp. 229–237.
[20] Krisnadhi, A. and C. Lutz, *Data complexity in the $\mathcal{EL}$ family of description logics*, in: *Logic for Programming, Artificial Intelligence, and Reasoning, LPAR 2007*, LNCS **4790** (2007), pp. 333–347.
[21] Ladner, R., *The computational complexity of provability in systems of modal propositional logic*, SIAM J. Comput. **6** (1977).
[22] Lutz, C., R. Piro and F. Wolter, *Enriching $\mathcal{EL}$-concepts with greatest fixpoints*, in: *European Conference on Artificial Intelligence, ECAI 2010*, pp. 41–46.
[23] Lutz, C. and F. Wolter, *Non-uniform data complexity of query answering in description logics*, in: *Principles of Knowledge Representation and Reasoning, KR 2012.*
[24] Meier, A., M. Thomas, H. Vollmer and M. Mundhenk, *The complexity of satisfiability for fragments of CTL and CTL**, Int. J. Found. Comput. Sci. **20** (2009), pp. 901–918.
[25] Myers, R., D. Pattinson and L. Schröder, *Coalgebraic hybrid logic*, in: *Foundations of Software Science and Computational Structures, FOSSACS 2009*, LNCS **5504** (2009), pp. 137–151.
[26] Nebel, B., *Terminological reasoning is inherently intractable*, Artif. Intell. **43** (1990), pp. 235–249.
[27] Parikh, R., *Propositional game logic*, in: *Foundations of Computer Science, FOCS 1983.*
[28] Pattinson, D., *Coalgebraic modal logic: Soundness, completeness and decidability of local consequence*, Theoret. Comput. Sci. **309** (2003), pp. 177–193.
[29] Pauly, M., *A modal logic for coalitional power in games*, J. Log. Comput. **12** (2002), pp. 149–166.
[30] Peleg, D., *Concurrent dynamic logic*, J. ACM **34** (1987), pp. 450–479.
[31] Schröder, L. and D. Pattinson, *Shallow models for non-iterative modal logics*, in: *Advances in Artificial Intelligence, KI 2008*, LNCS **5243** (2008), pp. 324–331.
[32] Schröder, L. and D. Pattinson, *PSPACE bounds for rank-1 modal logics*, ACM Trans. Comput. Log. **10** (2009), pp. 13:1–13:33.
[33] Schröder, L. and D. Pattinson, *Strong completeness of coalgebraic modal logics*, in: *Theoretical Aspects of Computer Science, STACS 2009*, LIPIcs **3** (2009), pp. 673–684.
[34] Schröder, L. and D. Pattinson, *Modular algorithms for heterogeneous modal logics via multi-sorted coalgebra*, Math. Structures Comput. Sci. **21** (2011), pp. 235–266.

One-dimensional Fragment of First-order Logic

Lauri Hella [1]

School of Information Sciences
University of Tampere
Finland

Antti Kuusisto [2]

Institute of Computer Science
University of Wrocław
Poland

Abstract

We introduce a novel decidable fragment of first-order logic. The fragment is *one-dimensional* in the sense that quantification is limited to applications of blocks of existential (universal) quantifiers such that at most one variable remains free in the quantified formula. The fragment is closed under Boolean operations, but additional restrictions (called *uniformity conditions*) apply to combinations of atomic formulae with two or more variables. We argue that the notions of *one-dimensionality* and *uniformity* together offer a novel perspective on the *robust decidability* of modal logics. We also show that the one-dimensional fragment is expressively equivalent to a polyadic modal logic with the capacity of permuting and forming Boolean combinations of accessibility relations. Furthermore, we establish that minor modifications to the restrictions of the syntax of the one-dimensional fragment lead to undecidable formalisms. Namely, the *two-dimensional* and *non-uniform one-dimensional* fragments are shown undecidable. Finally, we prove that with regard to expressivity, the one-dimensional fragment is incomparable with both the guarded negation fragment and two-variable logic with counting. Our proof of the decidability of the one-dimensional fragment is based on a technique involving a direct reduction to the monadic class of first-order logic. The novel technique is itself of an independent mathematical interest, and one of the principal contributions of the paper.

Keywords: Extensions of modal logic, fragments of first-order logic, Boolean modal logic, decidability.

[1] The research of Lauri Hella was partially funded by a Professor Pool grant awarded by the Finnish Cultural Foundation.

[2] Antti Kuusisto acknowledges that this work was carried out during a tenure of the ERCIM "Alain Bensoussan" Fellowship Programme, and that the research leading to these results has received funding from the European Union Seventh Framework Programme (FP7/2007-2013) under grant agreement number 246016.

1 Introduction

Decidability questions constitute one of the core themes in computer science logic. Decidability properties of several fragments of first-order logic have been investigated after the completion of the program concerning the classical decision problem. Currently perhaps the most important two frameworks studied in this context are the *guarded fragment* [1] and *two-variable logics*.

Two-variable logic FO^2 was introduced by Henkin in [10] and showed decidable in [14] by Mortimer. The satisfiability and finite satisfiability problems of two-variable logic were proved to be NEXPTIME-complete in [8]. The extension of two-variable logic with counting quantifiers, FOC^2, was shown decidable in [9], [15]. It was subsequently proved to be NEXPTIME-complete in [16].

Research concerning decidability of variants of two-variable logic has been very active in recent years. Recent articles in the field include for example [3] [5], [11], [17], and several others. The recent research efforts have mainly concerned decidability and complexity issues in restriction to particular classes of structures, and also questions related to different built-in features and operators that increase the expressivity of the base language.

Guarded fragment GF was originally conceived in [1]. It is a restriction of first-order logic that only allows quantification of "guarded" new variables—a restriction that makes the logic rather similar to modal logic.

The guarded fragment has generated a vast literature, and several related decidability questions have been studied. The fragment has recently been significantly generalized in [2]. The article introduces the *guarded negation first-order logic* GNFO. This logic only allows negations of formulae that are guarded in the sense of the guarded fragment. The guarded negation fragment has been shown complete for 2NEXPTIME in [2].

Two-variable logic and guarded-fragment are examples of decidable fragments of first-order logic that are not based on restricting the quantifier alternation patterns of formulae, unlike the prefix classes studied in the context of the classical decision problem. Surprisingly, not many such frameworks have been investigated in the literature.

In this paper we introduce a novel decidable fragment that essentially allows arbitrary quantifier alternation patterns. The *uniform one-dimensional fragment* UF_1 of first-order logic is obtained by restricting quantification to blocks of existential (universal) quantifiers that *leave at most one free variable* in the resulting formula. Additionally, a *uniformity condition* applies to the use of atomic formulae: if $n, k \geq 2$, then a Boolean combination of atoms $R(x_1, ..., x_k)$ and $S(y_1, ..., y_n)$ is allowed only if $\{x_1, ..., x_k\} = \{y_1, ..., y_n\}$. Boolean combinations of formulae with at most one free variable can be formed freely.

We establish decidability of the satisfiability and finite satisfiability problems of UF_1. We also show that if the uniformity condition is lifted, we obtain an undecidable logic. Furthermore, if we keep uniformity but go two-dimensional by allowing existential (universal) quantifier blocks that leave two variables free, we again obtain an undecidable formalism. Therefore, *if we lift either of the two restrictions that our fragment is based on, we obtain an undecidable logic.*

In addition to studying decidability, we also show that UF_1 is incomparable in expressive power with both FOC^2 and GNFO.

In [18], Vardi initiated an intriguing research effort that aims to understand phenomena behind the *robust decidability* of different variants of modal logic. In addition to [18], see also for example [7] and the introduction of [2]. Modal logic indeed has several features related to what is known about decidability. In particular, modal logic embeds into both FO^2 and GF.

However, there exist several important and widely applied decidable extensions of modal logic that do not embed into *both* FO^2 and GF. Such extensions include *Boolean modal logic* (see [6], [13]) and basic *polyadic modal logic*, i.e, modal logic containing accessibility relations of arities higher than two (see [4]). Boolean modal logic allows Boolean combinations of accessibility relations and therefore can express for example the formula $\exists y\big(\neg R(x,y) \wedge P(y)\big)$. Polyadic modal logic can express the formula $\exists x_2 ... \exists x_k\big(R(x_1, ..., x_k) \wedge P(x_2) \wedge ... \wedge P(x_k)\big)$. Boolean modal logic and polyadic modal logic are both inherently one-dimensional, and furthermore, satisfy the uniformity condition of UF_1. Both logics embed into UF_1. The notions of *one-dimensionality* and *uniformity* can be seen as novel features that can help, in part, *explain* decidability phenomena concerning modal logics.

Importantly, also the equality-free fragment of FO^2 embeds into UF_1. In fact, when attention is restricted to vocabularies with relations of arities at most two, the expressivities of UF_1 and the equality-free fragment of FO^2 coincide. Instead of seeing this as a weakness of UF_1, we in fact regard UF_1 as *a canonical generalisation* of (equality-free) FO^2 into contexts with arbitrary relational vocabularies. The fragment UF_1 can be regarded as a *vectorisation* of FO^2 that offers new possibilities for extending research efforts concerning two-variable logics. *It is worth noting that for example in database theory contexts, two-variable logics as such are not always directly applicable due to the arity-related limitations.* Thus we believe that the one-dimensional fragment is indeed a worthy discovery that *extends the scope of research on two-variable logics to the realm involving relations of arbitrary arities.*

Instead of extending basic techniques from the field of two-variable logic, our decidability proof is based on a direct satisfiability preserving translation of UF_1 into monadic first-order logic. The novel proof technique is mathematically interesting in its own right, and is in fact a central contribution of this article; the proof technique is clearly robust and can be modified and extended to give other decidability results. Furthermore, as a by-product of our proof, we identify a natural polyadic modal logic MUF_1, which is expressively equivalent to the one-dimensional fragment. This *modal normal form* for the one-dimensional fragments is also—we believe—a nice contribution.

2 Preliminaries

Let $\mathbb{Z}_+$ denote the set of positive integers. Let $\mathcal{T}$ denote a *complete relational vocabulary*, i.e., $\mathcal{T} := \bigcup_{k \in \mathbb{Z}_+} \tau_k$, where τ_k denotes a countably infinite set of

k-ary relation symbols. Each vocabulary τ we consider below is assumed to be a subset of $\mathcal{T}$. A τ-formula of first-order logic is a formula whose set of non-logical symbols is a subset of τ. A τ-model is a model whose set of interpreted non-logical symbols is τ.

Let VAR denote the countably infinite set $\{ x_i \mid i \in \mathbb{Z}_+ \}$ of *variable symbols.* We define the set of $\mathcal{T}$-formulae of first-order logic in the usual way, assuming that all variable symbols are from VAR. Below we use *meta-variables* x, y, z in order to denote variables in VAR. Also symbols of the type y_i and z_i, where $i \in \mathbb{Z}_+$, will be used as meta-variables. In addition to meta-variables, we also need to directly use the variables $x_i \in$ VAR below. Note that for example the meta-variables y_1 and y_2 may denote the same variable in VAR, while the variables $x_1, x_2 \in$ VAR of course simply *are* different variables.

Let R be a k-ary relation symbol, $k \in \mathbb{Z}_+$. An atomic formula $R(y_1, ..., y_k)$ is called *m-ary* if there are exactly m distinct variables in the set $\{y_1, ..., y_k\}$. For example, if x, y are distinct variables, then $S(x, y)$ and $T(y, x, y, y)$ are binary, and $U(x_1, x_6, x_3, x_2, x_1, x_6)$ is 4-ary. An *m-ary τ-atom* is an atomic formula that is m-ary, and the relation symbol of the formula is in τ.

Let $\tau \subseteq \mathcal{T}$. Let $\mathfrak{M}$ a τ-model with the domain M. A function f that maps some subset of VAR into M is an *assignment.* Let φ be a τ-formula with the free variables $y_1, ..., y_k$. Let f be an assignment that interprets the free variables of φ in M. We write $\mathfrak{M}, f \models \varphi$ if $\mathfrak{M}$ satisfies φ when the free variables of φ are interpreted according to f. Let $u_1, ..., u_k \in M$. Let φ be a τ-formula whose free variables are among $y_1, ..., y_k$. We write $\mathfrak{M}, \frac{(u_1, ..., u_k)}{(y_1, ..., y_k)} \models \varphi$ if $\mathfrak{M}, f \models \varphi$ for some assignment f such that $f(y_i) = u_i$ for each $i \in \{1, .., k\}$.

By a *non-empty conjunction* we mean a finite conjunction with at least one conjunct; for example $R(x, y) \wedge \exists y P(y)$ and $\top$ are non-empty conjunctions.

By *monadic first-order logic*, or MFO, we mean the fragment of first-order logic *without equality*, where formulae contain only unary relation symbols.

Let $k \in \mathbb{Z}_+$. A *k-permutation* is a bijection $\sigma : \{1, ..., k\} \to \{1, ..., k\}$. When k is irrelevant or clear from the context, we simply talk about permutations.

Let $k \in \mathbb{Z}_+$. We let $(u, ..., u)_k$ and u^k denote the k-tuple containing k copies of the object u. When $k = 1$, this tuple is identified with the object u.

Let l and $k \leq l$ be positive integers. Let K be a set, and let $(s_1, ..., s_l) \in K^l$ be a tuple. We let $(s_1, ..., s_l) \upharpoonright k$ denote the tuple $(s_1, ..., s_k)$. Let $R \subseteq K^l$ be an l-ary relation. We let $R \upharpoonright k$ denote the k-ary relation $R' \subseteq K^k$ defined such that for each $(s_1, ..., s_k) \in K^k$, we have $(s_1, ..., s_k) \in R'$ iff $(s_1, ..., s_k) = (u_1, ..., u_l) \upharpoonright k$ for some tuple $(u_1, ..., u_l) \in R$.

Recall that $\bigwedge \emptyset$ is assumed to be always true, while $\bigvee \emptyset$ is always false.

3 The one-dimensional fragment

We shall next define the *uniform one-dimensional* fragment UF_1 of first-order logic. Let $Y = \{y_1, ..., y_n\}$ be a set of variable symbols, and let R be a k-ary relation symbol. An atomic formula $R(y_{i_1}, ..., y_{i_k})$ is called a *Y-atom* if $\{y_{i_1}, ..., y_{i_k}\} = Y$. A finite set of Y-atoms is called a *Y-uniform set.* When Y is irrelevant or known from the context, we may simply talk about a *uniform set.*

For example, assuming that x, y, z are distinct variables, $\{T(x,y), S(y,x)\}$ and $\{R(x,x,y), R(y,y,x), S(y,x)\}$ are uniform sets, while $\{R(x,y,z), R(x,y,y)\}$ is not. The empty set is a $\emptyset$-uniform set.

Let $\tau \subseteq \mathcal{T}$. The set $\mathrm{UF}_1(\tau)$, or the set of τ-formulae of the one-dimensional fragment, is the smallest set $\mathcal{F}$ satisfying the following conditions.

(i) Every unary τ-atom is in $\mathcal{F}$, and $\bot, \top \in \mathcal{F}$.

(ii) If $\varphi \in \mathcal{F}$, then $\neg\varphi \in \mathcal{F}$. If $\varphi_1, \varphi_2 \in \mathcal{F}$, then $(\varphi_1 \wedge \varphi_2) \in \mathcal{F}$.

(iii) Let $Y = \{y_1, ..., y_k\}$ be a set of variable symbols. Let U be a finite set of formulae $\psi \in \mathcal{F}$ whose free variables are in Y. Let $V \subseteq Y$. Let F be a V-uniform set of τ-atoms. Let φ be any Boolean combination of formulae in $U \cup F$. Then $\exists y_2 ... \exists y_k\, \varphi \in \mathcal{F}$.

(iv) If $\varphi \in \mathcal{F}$, then $\exists y\, \varphi \in \mathcal{F}$.

Notice that there is no equality symbol in the language. Notice also that the formation rule (iv) is strictly speaking not needed since the rule (iii) covers it. Concerning the rule (i), notice that also atoms of the type $S(x, ..., x)_k$, where $k \neq 1$, are legitimate formulae. Let UF_1 denote the set $\mathrm{UF}_1(\mathcal{T})$.

3.1 Intuitions underlying the decidability proof

We show decidability of the satisfiability and finite satisfiability problems of UF_1 by translating UF_1-formulae into equisatisfiable MFO-formulae. We first translate UF_1 into a logic DUF_1. This logic is a normal form for UF_1 such that all literals of arities higher than one appear in simple conjunctions, as for example in the formula $\exists y \exists z \big(R(x,z,y,z) \wedge \neg S(y,x,z) \wedge \varphi(y)\big)$. The logic DUF_1 is then translated into a modal logic MUF_1, which is an essentially variable-free formalism for DUF_1. In Section 4 we show how formulae of the logic MUF_1 are translated into equisatisfiable formulae of MFO, which is well-known to have the finite model property.

The semantics of MUF_1 is defined (see Section 3.4) with respect to pointed models $(\mathfrak{M}, u)$, where $u \in M = Dom(\mathfrak{M})$. If φ is a formula of MUF_1, we let $\|\varphi\|^{\mathfrak{M}}$ denote the set $\{\, v \in M \mid (\mathfrak{M}, v) \models \varphi \,\}$. In Section 4 we fix a MUF_1-formula ψ and translate it to an MFO-formula $\psi^*(x)$. We prove that if $(\mathfrak{M}, v) \models \psi$, then $\psi^*(x)$ is satisfied in a model $\mathfrak{T}$, whose domain is $M \times T$, where T is the domain of an *m-dimensional hypertorus of arity l*. Such a hypertorus is a structure $(T, R_1, ..., R_m)$, where the m different relations R_i are all l-ary. Intuitively, the domain of $\mathfrak{T}$ consists of several copies of M, one copy for each point of the hypertorus. Let SUB_ψ denote the set of subformulae of ψ. The vocabulary of $\mathfrak{T}$ consists of monadic predicates P_α and P_t, where $\alpha \in \mathrm{SUB}_\psi$ and $t \in T$. The predicates are interpreted such that $P_\alpha^{\mathfrak{T}} := \|\alpha\|^{\mathfrak{M}} \times T$ and $P_t^{\mathfrak{T}} := M \times \{t\}$.

We will give a *rigorous and self-contained* proof of the decidability of UF_1, but to get an (admittedly very rough) initial idea of some of the related background intuitions, consider the following construction. (*It may also help to refer back to this section while internalizing the proof.*)

Consider a formula of ordinary unimodal logic φ and a Kripke model $\mathfrak{N}$.

We can *maximize* the accessibility relation R of $\mathfrak{N}$ by defining a new relation $S \subseteq N \times N$ such that $(u, v) \in S$ iff for all formulae $\Diamond\beta \in \mathrm{SUB}_\varphi$, we have

$$(\mathfrak{N}, v) \models \beta \Rightarrow (\mathfrak{N}, u) \models \Diamond\beta. \tag{1}$$

If we replace R by S in $\mathfrak{N}$, then each point w in the new model will satisfy exactly the same subformulae of φ as w satisfied in the old model. Thus we can *encode information concerning* R by using the (so-called filtration) condition given by Equation 1. The equation talks about the *sets* $\|\beta\|^{\mathfrak{N}}$ and $\|\Diamond\beta\|^{\mathfrak{N}}$, and thus it turns out that we can encode the information given by the equation by *monadic predicates* P_β and $P_{\Diamond\beta}$ corresponding to the sets $\|\beta\|^{\mathfrak{N}}$ and $\|\Diamond\beta\|^{\mathfrak{N}}$ (cf. the formulae $\mathit{PreCons}_\delta$ and Cons_δ in Section 4.1). *This way we can encode information concerning accessibility relations by using formulae of* MFO.

This construction does not work if one tries to maximize *both* a binary relation R *and its complement* $\overline{R}$ at the same time: the problem is that the maximized relations S and $\overline{S}$ will not necessarily be complements of each other. For this reason we need to *make enough room* for maximizing accessibility relations. Below we will simultaneously maximize several types of accessibility relations that cannot be allowed to intersect. Thus we need to use an *n-dimensional* hypertorus (rather than a usual 2D torus). Each k-ary accessibility relation type δ of the translated MUF_1-formula will be reserved a sequence $\overline{r} := (M \times \{t_1\}, ..., M \times \{t_k\})$ of copies of M from the domain of $\mathfrak{T}$. Information concerning δ will be encoded into this sequence $\overline{r}$ of models.

3.2 Diagrams

Let $\tau \subseteq \mathcal{T}$ be a *finite* vocabulary. Let $k \geq 2$ be an integer, and let $Y = \{y_1, ..., y_k\}$ be a set of *distinct* variable symbols. A *uniform k-ary τ-diagram* is a maximal satisfiable set of Y-atoms and negated Y-atoms of the vocabulary τ. (The empty set is *not* considered to be a uniform k-ary τ-diagram; this case is relevant when τ contains no relation symbols of the arity k or higher.)

For example, let $\tau = \{P, R, S\}$, where the arities of P, R, S are 1, 2, 3, respectively. Now $\{R(x,y), \neg R(y,x), S(y,x,x), S(x,y,x), \neg S(x,x,y), S(x,y,y), \neg S(y,x,y), S(y,y,x)\}$ is a uniform binary τ-diagram. Here we assume that x and y are distinct variables.

Let $\tau \subseteq \mathcal{T}$ be a *finite* vocabulary. The set $\mathrm{DUF}_1(\tau)$ is the smallest set $\mathcal{F}$ satisfying the following conditions.

(i) Every unary τ-atom is in $\mathcal{F}$. Also $\bot, \top \in \mathcal{F}$.

(ii) If $\varphi \in \mathcal{F}$, then $\neg\varphi \in \mathcal{F}$. If $\varphi_1, \varphi_2 \in \mathcal{F}$, then $(\varphi_1 \wedge \varphi_2) \in \mathcal{F}$.

(iii) Let δ be a uniform k-ary τ-diagram in the variables $y_1,...,y_k$, where $k \geq 2$. Let φ be a non-empty conjunction of a finite set U of formulae in $\mathcal{F}$ whose free variables are among $y_1,...,y_k$. Then $\exists y_2 ... \exists y_k \left(\bigwedge \delta \wedge \varphi \right) \in \mathcal{F}$.

(iv) If $\varphi \in \mathcal{F}$ has at most one free variable, y, then $\exists y\, \varphi \in \mathcal{F}$.

Let DUF_1 denote the set of formulae φ such that for some finite $\tau \subseteq \mathcal{T}$, we have $\varphi \in \mathrm{DUF}_1(\tau)$. UF_1 translates effectively into DUF_1; see the appendix for the proof. Here we briefly *sketch* the principal idea behind the translation.

Consider a UF_1-formula $\exists \overline{y}\, \psi$, where $\overline{y}$ is a tuple of variables. Put ψ into disjunctive normal form $\psi_1 \vee ... \vee \psi_k$. Thus $\exists \overline{y}\, \psi$ translates into the formula $\exists \overline{y}\, \psi_1 \vee ... \vee \exists \overline{y}\, \psi_k$, where each ψ_i is a conjunction. Each ψ_i is equivalent to a disjunction $\psi_{i,1} \vee ... \vee \psi_{i,m}$, where $\psi_{i,j}$ is of the desired type $\left(\bigwedge \delta \wedge \varphi\right)$.

3.3 Hypertori

We next define a class of hypertori. It may help to have a look at Lemma 3.1 before internalizing the definition. Let $l \geq 2$ and $n \geq 2$ be integers. Define $T := \{1, ..., n\} \times \{1, ..., l\} \times \{0, 1, 2\}$. Let $(t_1, ..., t_l) \in T^l$ be a tuple. Let $t_1 = (m, m', m'')$. Let $j \in \{1, ..., n\}$. A tuple $(t_1, ..., t_l) \in T^l$, where each $t_i = (p, q_i, r)$, is the *j-th good l-ary sequence originating from t_1*, if for each $i \in \{2, ..., l\}$, the following conditions hold.

(i) $p - m \equiv j - 1 \mod n$.

(ii) $q_i - m' \equiv i - 1 \mod l$.

(iii) $r - m'' \equiv 1 \mod 3$.

Define the relation $R_j \subseteq T^l$ such that $(s_1, ..., s_l) \in R_j$ iff $(s_1, ..., s_l)$ is the j-th good l-ary sequence originating from s_1. The structure $\left(T, R_1, ..., R_n\right)$ is the *n-dimensional hypertorus of the arity l*. The following lemma is easy to prove.

Lemma 3.1 *Let $\left(T, R_1, ..., R_n\right)$ be an n-dimensional hypertorus of the arity l. Let $j \in \{1, ..., n\}$ and $k \in \{2, ..., l\}$. Then the following conditions hold.*

(i) *For each $t \in T$, there exists exactly one tuple $(s_1, ..., s_k) \in R_j \upharpoonright k$ such that $t = s_1$. We have $s_i \neq s_j$ for all $i, j \in \{1, ..., k\}$ such that $i \neq j$.*

(ii) *Let $(s_1, ..., s_k) \in R_j \upharpoonright k$. Let σ be a k-permutation, and let $i \in \{1, ..., n\} \setminus \{j\}$. Then $(s_{\sigma(1)}, ..., s_{\sigma(k)}) \notin R_i \upharpoonright k$.*

(iii) *Let $(s_1, ..., s_k) \in R_j \upharpoonright k$. Let μ be any k-permutation other than the identity permutation. Then $(s_{\mu(1)}, ..., s_{\mu(k)}) \notin R_j \upharpoonright k$.*

In the rest of the article, we let $\mathfrak{T}(n, l)$ denote the n-dimensional hypertorus of the arity l. We let $T(n, l)$ and $R_j(n, l)$ denote, respectively, the domain and the relation R_j of $\mathfrak{T}(n, l)$.

3.4 Translation into a modal logic

Let $\tau \subseteq \mathcal{T}$ be a *finite* vocabulary, and let $k \geq 2$ be an integer. Let $\mathfrak{M}$ be a τ-model with the domain M. Let δ be a uniform k-ary τ-diagram in the variables $x_1, ..., x_k$. Notice that here we use the standard variables $x_1, ..., x_k$ from VAR. The diagram δ is a *standard uniform k-ary τ-diagram.* We define $\|\delta\|^{\mathfrak{M}}$ to be the relation $\{\, (u_1, ..., u_k) \in M^k \mid \mathfrak{M}, \frac{(u_1, ..., u_k)}{(x_1, ..., x_k)} \models \bigwedge \delta \,\}$. Standard variables are needed in order to uniquely specify the *order of elements* in tuples of $\|\delta\|^{\mathfrak{M}}$.

Let δ be a standard uniform k-ary τ-diagram. Let $q \leq k$ be a positive integer. Let $t : \{1, ..., k\} \rightarrow \{1, ..., q\}$ be a surjection. We let δ/t denote the set obtained from δ by replacing each variable x_i by $x_{t(i)}$.

Let k and q be positive integers such that $2 \leq q \leq k$. Let η and δ be standard uniform q-ary and k-ary τ-diagrams, respectively. Let $f : \{1, ..., k\} \rightarrow \{1, ..., q\}$

be a surjection. Assume that $\bigwedge \eta \models \bigwedge \delta/f$, i.e., the implication $\mathfrak{M}, h \models \eta \Rightarrow \mathfrak{M}, h \models \delta/f$ holds for each τ-model $\mathfrak{M}$ and each assignment h interpreting the variables $x_1, ..., x_q$ in the domain of $\mathfrak{M}$. Then we write $\eta \leq_f \delta$.

We then define a modal logic that provides an essentially variable-free representation of UF_1. Define the set $\text{MUF}_1(\tau)$ to be the smallest set $\mathcal{F}$ such that the following conditions are satisfied.

(i) If $S \in \tau$ is a relation symbol of any arity, then $S \in \mathcal{F}$. Also $\bot, \top \in \mathcal{F}$.

(ii) If $\varphi \in \mathcal{F}$, then $\neg\varphi \in \mathcal{F}$. If $\varphi_1, \varphi_2 \in \mathcal{F}$, then $(\varphi_1 \wedge \varphi_2) \in \mathcal{F}$.

(iii) If $\varphi_1, ..., \varphi_k \in \mathcal{F}$ and δ is a standard uniform k-ary τ-diagram, then $\langle\delta\rangle(\varphi_1, ..., \varphi_k) \in \mathcal{F}$.

(iv) If $\varphi \in \mathcal{F}$, then $\langle E\rangle\varphi \in \mathcal{F}$. (Here $\langle E\rangle$ denotes the *universal modality*; see below for the semantics.)

The semantics of $\text{MUF}_1(\tau)$ is defined with respect to *pointed σ-models* $(\mathfrak{M}, w)$, where $\mathfrak{M}$ is an ordinary σ-model of predicate logic for some vocabulary $\sigma \supseteq \tau$, and w is an element of the domain M of $\mathfrak{M}$. Obviously we define that $(\mathfrak{M}, w) \models \top$ always holds, and that $(\mathfrak{M}, w) \models \bot$ never holds. Let $S \in \tau$ be an n-ary relation symbol. We define $(\mathfrak{M}, w) \models S \Leftrightarrow w^n \in S^{\mathfrak{M}}$, where $S^{\mathfrak{M}}$ is the interpretation of the relation symbol S in the model $\mathfrak{M}$. The Boolean connectives $\neg$ and $\wedge$ have their usual meaning. For formulae of the type $\langle\delta\rangle(\varphi_1, ..., \varphi_k)$, we define that $(\mathfrak{M}, w) \models \langle\delta\rangle(\varphi_1, ..., \varphi_k)$ if and only if there exists a tuple $(u_1, ..., u_k) \in \|\delta\|^{\mathfrak{M}}$ such that $u_1 = w$ and $(\mathfrak{M}, u_i) \models \varphi_i$ for each $i \in \{1, ..., k\}$. For formulae $\langle E\rangle\varphi$, we define $(\mathfrak{M}, w) \models \langle E\rangle\varphi$ if and only if there exists some $u \in M$ such that $(\mathfrak{M}, u) \models \varphi$.

When φ is a $\text{MUF}_1(\tau)$-formula and $\mathfrak{M}$ a σ-model with the domain M, we let $\|\varphi\|^{\mathfrak{M}}$ denote the set $\{\, u \in M \mid (\mathfrak{M}, u) \models \varphi \,\}$. We let MUF_1 denote the union of all sets $\text{MUF}_1(\tau)$, where τ is a finite subset of $\mathcal{T}$.

It is very easy to show that there is an effective translation that turns any formula $\gamma(x) \in \text{DUF}_1$ into a formula $\chi \in \text{MUF}_1$ such that $(\mathfrak{M}, w) \models \chi \Leftrightarrow \mathfrak{M}, \frac{w}{x} \models \gamma(x)$ for all τ-models $\mathfrak{M}$, where τ is the set of non-logical symbols in $\gamma(x)$. (The set of non-logical symbols in χ is contained in τ, and the formula $\gamma(x)$ can either be a sentence or have the free variable x.)

4 UF_1 is decidable

Let us *fix* a formula ψ of MUF_1. We will first define a translation of ψ to an MFO-formula $\psi^*(x)$ in Section 4.1. We will then show in Sections 4.2 and 4.3 that the translation indeed preserves equivalence of satisfiability over finite models as well as over all models. Due to the above effective translations from UF_1 to DUF_1 and from DUF_1 to MUF_1, this implies that the satisfiability and finite satisfiability problems of UF_1 are decidable.

4.1 Translating MUF_1 into monadic first-order logic

We assume, w.l.o.g., that ψ contains at least one subformula of the type $\langle\delta\rangle(\chi_1, \chi_2)$. If not, we redefine ψ. The vocabulary of ψ may of course grow. We also assume, w.l.o.g., that ψ does not contain occurrences of the symbols $\top$,

$\perp$. Furthermore, we assume, w.l.o.g., that if R is a relation symbol occurring in some diagram of ψ, then $\neg R$ also occurs in ψ as a subformula: we can of course always add the conjunct $R \vee \neg R$ to ψ.

Let V_ψ be the set of all relation symbols in ψ, whether they occur in diagrams or as atomic subformulae; in fact, due to our assumptions above, the set of atomic formulae in ψ is equal to V_ψ. Let D_ψ be the set of relation symbols occurring in the diagrams of ψ. Let $V_\psi(k)$ denote the set of k-ary relation symbols in V_ψ. Define $D_\psi(k)$ analogously. Due to the assumption that ψ contains a subformula $\langle\delta\rangle(\chi_1, \chi_2)$, each relation symbol of some arity $m \geq 2$ that occurs as an atom in ψ, also occurs in the diagram δ. (This is due to the definition of MUF_1.) Thus $V_\psi(n) = D_\psi(n)$ for all $n > 1$.

Let $\mathcal{M}$ denote the maximum arity of all diagrams in ψ. For each $k \in \{2, ..., \mathcal{M}\}$, let Δ_k denote the set of exactly all standard uniform k-ary V_ψ-diagrams. Let Δ denote the union of the sets Δ_k, where $k \in \{2, ..., \mathcal{M}\}$. Let $\mathcal{N} := max\{\ |\Delta_k| \mid k \in \{2, ..., \mathcal{M}\}\ \}$. Recall that $T(\mathcal{N}, \mathcal{M})$ denotes the domain of the $\mathcal{N}$-dimensional hypertorus of the arity $\mathcal{M}$. For each $k \in \{2, ..., \mathcal{M}\}$, define an injection $b_k : \Delta_k \longrightarrow \{\, R_1(\mathcal{N}, \mathcal{M}), ..., R_{\mathcal{N}}(\mathcal{N}, \mathcal{M})\,\}$. For a k-ary diagram $\delta \in \Delta_k$, let T_δ denote the k-ary relation $\big(b_k(\delta)\big) \upharpoonright k$.

Let SUB_ψ denote the set of subformulae of the formula ψ. Fix fresh unary relation symbols P_α and P_t for each formula $\alpha \in \mathrm{SUB}_\psi$ and torus point $t \in T(\mathcal{N}, \mathcal{M})$. The vocabulary of the translation $\psi^*(x)$ of ψ will be the set $\{\, P_\alpha \mid \alpha \in \mathrm{SUB}_\psi\,\} \cup \{\, P_t \mid t \in T(\mathcal{N}, \mathcal{M})\,\}$. We let V^* denote this set.

We shall next define a collection of auxiliary formulae needed in order to define $\psi^*(x)$. If a pointed model $(\mathfrak{M}, u)$ satisfies ψ, then $\psi^*(x)$ will be satisfied in a larger model; the related model construction is defined in the beginning of Section 4.2. The predicates of the type P_α will be used to encode information about sets $\|\alpha\|^{\mathfrak{M}}$, while the predicates P_t encode information about the *diagrams* of ψ. The predicates P_t are crucial when defining a V_ψ-model $\mathfrak{B}$ that satisfies ψ based on a V^*-model $\mathfrak{A}$ of $\psi^*(x)$ in Section 4.3.

Let $\delta \in \Delta_k$. Define $PreCons_\delta(x_1, ..., x_k)$ to be the formula

$$\bigwedge_{\langle\delta\rangle(\chi_1,...,\chi_k)\ \in\ \mathrm{SUB}_\psi} \Big(P_{\chi_1}(x_1) \wedge ... \wedge P_{\chi_k}(x_k) \rightarrow P_{\langle\delta\rangle(\chi_1\,,\,...\,,\,\chi_k)}(x_1)\Big).$$

Let $\Delta(\delta)$ be the set of pairs (η, f), where $\eta \in \Delta$ is a p-ary diagram for some $p \geq k$, and $f : \{1, ..., p\} \rightarrow \{1, ..., k\}$ is a surjection such that we have $\delta \leq_f \eta$. The set $\Delta(\delta)$ is the set of *inverse projections of* δ *in* Δ. Define

$$Cons_\delta(x_1, ..., x_k) \ :=\ \bigwedge_{(\eta,f)\,\in\,\Delta(\delta)} PreCons_\eta(x_{f(1)}, ..., x_{f(p)}).$$

The following formula is the principal formula that encodes information about diagrams of δ (cf. Lemma 4.1).

$$Diag_\delta(x_1, ..., x_k) := \bigvee_{(t_1\,,\,...\,,\,t_k)\ \in\ T_\delta} P_{t_1}(x_1) \wedge ... \wedge P_{t_k}(x_k)\ \wedge\ Cons_\delta(x_1, ..., x_k).$$

Let $+(\delta)$ denote the set of relation symbols R that occur positively in δ, i.e., there exists some atom $R(y_1, ..., y_n) \in \delta$, where n is the arity of R. Let $-(\delta)$ be the relation symbols R that occur negatively in δ, i.e., $\neg R(y_1, ..., y_n) \in \delta$ for some atom $R(y_1, ..., y_n)$. The following three formulae encode information about atomic formulae in ψ. Define

$$\begin{aligned} \mathit{Local}_\delta(x) &:= \bigwedge_{R \in +(\delta)} P_R(x) \wedge \bigwedge_{R \in -(\delta)} P_{\neg R}(x), \\ \mathit{LocalDiag}_\delta(x) &:= \mathit{Local}_\delta(x) \rightarrow \mathit{PreCons}_\delta(x, ..., x)_k, \\ \psi_{local} &:= \bigwedge_{\delta \in \Delta} \forall x\, \mathit{LocalDiag}_\delta(x). \end{aligned}$$

The next formula is essential in the construction of a V_ψ-model of ψ from a V^*-model of $\psi^*(x)$ in Section 4.3. The two models have the same domain. The formula states that each tuple can be interpreted to satisfy *some* diagram δ such that information concerning the unary predicates in V^* is consistent with δ. See the way $\mathfrak{B}$ is defined based on $\mathfrak{A}$ in Section 4.3 for further details. Define

$$\psi_{total} := \bigwedge_{k \in \{2, ..., \mathcal{M}\}} \forall x_1 ... \forall x_k \bigvee_{\delta \in \Delta_k} \mathit{Cons}_\delta(x_1, ..., x_k).$$

Also the following formula is crucial for the definition of $\mathfrak{B}$.

$$\psi_{uniq} := \bigwedge_{t,\, s\, \in\, T(\mathcal{N}, \mathcal{M}),\ t \neq s} \neg \exists x \big(P_t(x) \wedge P_s(x) \big).$$

Let $\neg\alpha$, $(\beta \wedge \gamma)$, $\langle E \rangle \chi$, and $\langle \delta \rangle (\chi_1, ..., \chi_k)$ be formulae in SUB_ψ. The following formulae recursively encode information concerning subformulae of ψ. Define

$$\begin{aligned} \psi_{\neg\alpha} &:= \forall x \Big(P_{\neg\alpha}(x) \leftrightarrow \neg P_\alpha(x) \Big), \\ \psi_{(\beta \wedge \gamma)} &:= \forall x \Big(P_{(\beta \wedge \gamma)}(x) \leftrightarrow \big(P_\beta(x) \wedge P_\gamma(x) \big) \Big), \\ \psi_{\langle E \rangle \chi} &:= \forall x \Big(P_{\langle E \rangle \chi}(x) \leftrightarrow \exists y P_\chi(y) \Big), \\ \psi_{\langle \delta \rangle (\chi_1, ..., \chi_k)} &:= \forall x_1 \Big(P_{\langle \delta \rangle (\chi_1, ..., \chi_k)}(x_1) \\ &\qquad \leftrightarrow \exists x_2 ... x_k \big(\mathit{Diag}_\delta(x_1, ..., x_k) \\ &\qquad\qquad \wedge P_{\chi_1}(x_1) \wedge ... \wedge P_{\chi_k}(x_k) \big) \Big). \end{aligned}$$

Let $\psi_{sub} := \bigwedge_{\alpha \in \mathrm{SUB}_\psi} \psi_\alpha$. Finally, we define

$$\psi^*(x) := \psi_{total} \wedge \psi_{uniq} \wedge \psi_{local} \wedge \psi_{sub} \wedge P_\psi(x).$$

4.2 Satisfiability of ψ implies satisfiability of $\psi^*(x)$

Fix an arbitrary model V_ψ-model $\mathfrak{M}$ with the domain M. Fix a point $w \in M$. Assume $(\mathfrak{M}, w) \models \psi$. We shall next construct a model $\mathfrak{T}$ with the domain $M \times T(\mathcal{N}, \mathcal{M})$. We then show that $\mathfrak{T}, \frac{(w,t)}{x} \models \psi^*(x)$, where t is a torus point. If $\mathfrak{M}$ is a finite model, then so is $\mathfrak{T}$.

The domain $M \times T(\mathcal{N},\mathcal{M})$ of the V^*-model $\mathfrak{T}$ consists of copies of M, one copy for each torus point $t \in T(\mathcal{N},\mathcal{M})$. Let us define interpretations of the symbols in V^*. Consider a symbol P_α, where $\alpha \in \mathrm{SUB}_\psi$. If $(u,t) \in Dom(\mathfrak{T})$, then $(u,t) \in P_\alpha^{\mathfrak{T}} \Leftrightarrow u \in \|\alpha\|^{\mathfrak{M}}$. Consider then a symbol P_t, where $t \in T(\mathcal{N},\mathcal{M})$. If $(u,t') \in Dom(\mathfrak{T})$, then $(u,t') \in P_t^{\mathfrak{T}} \Leftrightarrow t' = t$.

Lemma 4.1 *Let $\langle\delta\rangle(\chi_1,...,\chi_k) \in \mathrm{SUB}_\psi$ and $(u,t) \in Dom(\mathfrak{T})$. Then $(\mathfrak{M},u) \models \langle\delta\rangle(\chi_1,...,\chi_k)$ iff $\mathfrak{T}, \frac{(u,t)}{x_1} \models \exists x_2...\exists x_k\big(\, Diag_\delta(x_1,...,x_k) \wedge P_{\chi_1}(x_1) \wedge ... \wedge P_{\chi_k}(x_k)\big)$.*

Proof. Define $u_1 := u$ and $t_1 := t$. Assume $(\mathfrak{M},u_1) \models \langle\delta\rangle(\chi_1,...,\chi_k)$. Thus $(u_1,...,u_k) \in \|\delta\|^{\mathfrak{M}}$ for some tuple $(u_1,...,u_k)$ such that $u_i \in \|\chi_i\|^{\mathfrak{M}}$ for each i. Hence $(u_i,s) \in P_{\chi_i}^{\mathfrak{T}}$ for each i and each torus point s. To conclude the first direction of the proof, it suffices to prove that $\mathfrak{T}, \frac{((u_1,t_1),...,(u_k,t_k))}{(x_1,...,x_k)} \models Diag_\delta(x_1,...,x_k)$ for some torus points $t_2,...,t_k$.

Let $t_2,...,t_k$ be the torus points such that $(t_1,...,t_k) \in T_\delta$. In order to establish that $\mathfrak{T}, \frac{((u_1,t_1),...,(u_k,t_k))}{(x_1,...,x_k)} \models Cons_\delta(x_1,...,x_k)$, assume that $\delta \leq_f \eta$, where $\eta \in \Delta_p$ and $p \geq k$. Assume that $\langle\eta\rangle(\gamma_1,...,\gamma_p) \in \mathrm{SUB}_\psi$, and that $\mathfrak{T}, \frac{((u_1,t_1),...,(u_k,t_k))}{(x_1,...,x_k)} \models P_{\gamma_1}(x_{f(1)}) \wedge ... \wedge P_{\gamma_p}(x_{f(p)})$. We must show that $(u_{f(1)},t_{f(1)}) \in P_{\langle\eta\rangle(\gamma_1,...,\gamma_p)}^{\mathfrak{T}}$.

For each $i \in \{1,...,p\}$, as $(u_{f(i)},t_{f(i)}) \in P_{\gamma_i}^{\mathfrak{T}}$, we have $u_{f(i)} \in \|\gamma_i\|^{\mathfrak{M}}$ by the definition of $P_{\gamma_i}^{\mathfrak{T}}$. As $(u_1,...,u_k) \in \|\delta\|^{\mathfrak{M}}$ and $\delta \leq_f \eta$, we have $(u_{f(1)},...,u_{f(p)}) \in \|\eta\|^{\mathfrak{M}}$. Therefore we have $u_{f(1)} \in \|\langle\eta\rangle(\gamma_1,...,\gamma_p)\|^{\mathfrak{M}}$. Thus $(u_{f(1)},t_{f(1)}) \in P_{\langle\eta\rangle(\gamma_1,...,\gamma_p)}^{\mathfrak{T}}$ by the definition of $P_{\langle\eta\rangle(\gamma_1,...,\gamma_p)}^{\mathfrak{T}}$.

We then deal with the converse implication of the lemma. Define $s_1 := t$ and $v_1 := u$. Assume $\mathfrak{T}, \frac{(v_1,s_1)}{x_1} \models \exists x_2...\exists x_k\big(\, Diag_\delta(x_1,...,x_k) \wedge P_{\chi_1}(x_1) \wedge ... \wedge P_{\chi_k}(x_k)\big)$. Hence $\mathfrak{T}, \frac{((v_1,s_1),...,(v_k,s_k))}{(x_1,...,x_k)} \models Diag_\delta(x_1,...,x_k)$ for some tuple $((v_1,s_1),...,(v_k,s_k))$ such that $(v_i,s_i) \in P_{\chi_i}^{\mathfrak{T}}$ for each i. As now $\mathfrak{T}, \frac{((v_1,s_1),...,(v_k,s_k))}{(x_1,...,x_k)} \models PreCons_\delta(x_1,...,x_k)$, we infer that $(v_1,s_1) \in P_{\langle\delta\rangle(\chi_1,...,\chi_k)}^{\mathfrak{T}}$. By the definition of $P_{\langle\delta\rangle(\chi_i,...,\chi_k)}^{\mathfrak{T}}$, we have $(\mathfrak{M},v_1) \models \langle\delta\rangle(\chi_1,...,\chi_k)$. □

Lemma 4.2 *Let t be any torus point. Under the assumption $(\mathfrak{M},w) \models \psi$, we have $\mathfrak{T}, \frac{(w,t)}{x} \models \psi^*(x)$.*

Proof. See the appendix. □

4.3 Satisfiability of $\psi^*(x)$ implies satisfiability of ψ

Let $\mathfrak{A}$ be a V^*-model with the domain A. Assume that $\mathfrak{A}, \frac{w}{x} \models \psi^*(x)$. We next define a V_ψ-model $\mathfrak{B}$ with the same domain A, and then show that $(\mathfrak{B},w) \models \psi$.

Let U be a non-empty set, and let $p \in \mathbb{Z}_+$. Let $(u_1,...,u_p) \in U^p$ be a tuple. We say that the tuple $(u_1,...,u_p)$ *spans the set* $\{u_1,...,u_p\}$.

Let $k \in \mathbb{Z}_+$, and let $S \in V_\psi$ be a k-ary symbol. We define $(u, ..., u)_k \in S^{\mathfrak{B}}$ iff $u \in P_s^{\mathfrak{A}}$. This settles the interpretation of the symbols $S \in V_\psi$ on tuples that span sets of size one. Interpretation of the symbols on tuples that span larger sets is more complicated. We begin with the following lemma, whose proof is straightforward by Lemma 3.1.

Lemma 4.3 *Let* $u_1, ..., u_k \in A$. *Assume* $\mathfrak{A}, \frac{(u_1,...,u_k)}{(x_1,...,x_k)} \models Diag_\delta(x_1, ..., x_k)$. *Then* $\mathfrak{A}, \frac{(u_{\sigma(1)},...,u_{\sigma(k)})}{(x_1,...,x_k)} \not\models Diag_\eta(x_1, ..., x_k)$ *holds for all all* k*-permutations* σ *and all* $\eta \in \Delta_k \setminus \{\delta\}$. *Also* $\mathfrak{A}, \frac{(u_{\mu(1)},...,u_{\mu(k)})}{(x_1,...,x_k)} \not\models Diag_\delta(x_1, ..., x_k)$ *holds for all* k*-permutations* μ *other than the identity permutation.*

Let $q \in \{2, ..., \mathcal{M}\}$. Consider subsets of A that have exactly $q \geq 2$ elements. Let us divide such sets into two classes. Let $U = \{u_1, ..., u_q\}$ be a set with q distinct elements. Assume first that there exists some q-permutation σ and some $\eta \in \Delta_q$ such that $\mathfrak{A}, \frac{(u_{\sigma(1)},...,u_{\sigma(q)})}{(x_1,...,x_q)} \models Diag_\eta(x_1, ..., x_q)$. Define $tuple(U) := (u_{\sigma(1)}, ..., u_{\sigma(q)})$ and $diagram(U) := \eta$. Define also $type(U) = 1$. Assume then that $\mathfrak{A}, \frac{(u_{\sigma(1)},...,u_{\sigma(q)})}{(x_1,...,x_q)} \not\models Diag_\eta(x_1, ..., x_q)$ holds for all $\eta \in \Delta_q$ and all q-permutations σ. As $\mathfrak{A} \models \psi_{total}$, there exists some diagram $\delta \in \Delta_q$ such that $\mathfrak{A}, \frac{(u_1,...,u_q)}{(x_1,...x_q)} \models Cons_\delta(x_1, ..., x_q)$. Define $tuple(U) = (u_1, ..., u_q)$ and $diagram(U) := \delta$. Define also $type(U) = 2$.

Notice that by our assumptions in Section 4.1, there are no relation symbols $S \in V_\psi \setminus D_\psi$ of any arity higher than one. Recall that $\mathcal{M}$ is the maximum arity of diagrams in Δ. We next define the relations $S^{\mathfrak{B}}$, where $S \in D_\psi$, on tuples of elements of A that span sets with $q \in \{2, ..., \mathcal{M}\}$ elements. The definition has the property—as Lemma 4.5 below establishes—that if $(u_1, ..., u_k) \in \|\delta\|^{\mathfrak{B}}$, where $\delta \in \Delta_k$, then $\mathfrak{A}, \frac{(u_1,...,u_k)}{(x_1,...,x_k)} \models PreCons_\delta(x_1, ..., x_k)$. In fact this holds also for tuples that span a singleton set, see Lemma 4.5.

Let $q \in \{2, ..., \mathcal{M}\}$, and let $U \subseteq A$ be a set of the size q. Assume first that $type(U) = 1$. Let $diagram(U) = \eta \in \Delta_q$ and $tuple(U) = (u_1, ..., u_q)$. We have $\mathfrak{A}, \frac{(u_1,...,u_q)}{(x_1,...,x_q)} \models Diag_\eta(x_1, ..., x_q)$. Let $k \geq q$ be an integer. Interpret each k-ary symbol $S \in D_\psi$ such that $\mathfrak{B}, \frac{(u_1,...,u_q)}{(x_1,...,x_q)} \models \eta$. This definition uniquely specifies the interpretation of S on each k-ary tuple that spans the set $\{u_1, ..., u_q\}$. To see this, let $f : \{1, ..., k\} \to \{1, ..., q\}$ be a surjection. Now we have $(u_{f(1)}, ..., u_{f(k)}) \in S^{\mathfrak{B}}$ iff $S(x_{f(1)}, ..., x_{f(k)}) \in \eta$.

Assume then that $type(U) = 2$. Let $diagram(U) = \delta \in \Delta_q$ and $tuple(U) = (v_1, ..., v_q)$. We have $\mathfrak{A}, \frac{(v_1,...,v_q)}{(x_1,...,x_q)} \models Cons_\delta(x_1, ..., x_p)$. Let $k \geq q$ be an integer. Interpret each k-ary symbol $S \in D_\psi$ such that $\mathfrak{B}, \frac{(v_1,...,v_q)}{(x_1,...,x_q)} \models \delta$.

We investigate each $q \in \{2, ..., \mathcal{M}\}$, and thereby obtain a definition of $\mathfrak{B}$; if there are symbols of arity $r > \mathcal{M}$ in D_ψ, we arbitrarily define the interpretations of such symbols on tuples that span sets with more than $\mathcal{M}$ elements.

Lemma 4.4 *If* $(u_1, ..., u_k) \in A^k$ *and* $\mathfrak{A}, \frac{(u_1,...,u_k)}{(x_1,...,x_k)} \models Diag_\delta(x_1, ..., x_k)$ *for some* $\delta \in \Delta_k$, *then* $(u_1, ..., u_k) \in \|\delta\|^{\mathfrak{B}}$.

Proof. Assume $\mathfrak{A}, \frac{(u_1,...,u_k)}{(x_1,...,x_k)} \models Diag_\delta(x_1, ..., x_k)$. Notice that $k \geq 2$, since diagrams have by definition an arity at least two. As $\mathfrak{A} \models \psi_{uniq}$, the set $U = \{u_1, ..., u_k\}$ has exactly k elements. We have $type(U) = 1$, and by Lemma 4.3, $tuple(U) = (u_1, ..., u_k)$. Thus $\mathfrak{B}, \frac{(u_1,...,u_k)}{(x_1,...,x_k)} \models \delta$. □

Lemma 4.5 *Let* $k \in \{1, ..., \mathcal{M}\}$. *If* $(u_1, ..., u_k) \in \|\delta\|^{\mathfrak{B}}$, *where* $\delta \in \Delta_k$, *then* $\mathfrak{A}, \frac{(u_1,...,u_k)}{(x_1,...,x_k)} \models PreCons_\delta(x_1, ..., x_k)$.

Proof. The case where $(u_1, .., u_k)$ spans a singleton set follows since $\mathfrak{A} \models \psi_{local}$. Let us consider the cases where $(u_1, .., u_k)$ spans a set of the size two or larger.

Assume that $(u_1, ..., u_k) \in \|\delta\|^{\mathfrak{B}}$ is a tuple such that $U = \{u_1, ..., u_k\}$ contains exactly $q \geq 2$ elements. Let $m : \{1, ..., q\} \to \{1, ..., k\}$ be an injection such that the tuple $(u_{m(1)}, ..., u_{m(q)})$ spans the set $\{u_1, ..., u_k\}$.

Assume first that we have $\mathfrak{A}, \frac{(u_{m(\sigma(1))},...,u_{m(\sigma(q))})}{(x_1,...,x_q)} \models Diag_\eta(x_1, ..., x_q)$ for some $\eta \in \Delta_q$ and some q-permutation σ. Thus $type(U) = 1$. By Lemma 4.3, we have $tuple(U) = (u_{m(\sigma(1))}, ..., u_{m(\sigma(q))})$ and $diagram(U) = \eta$. Let $s : \{1, ..., q\} \to \{1, ..., k\}$ be the injection such that $s(i) = m(\sigma(i))$ for each $i \in \{1, ..., q\}$. As $tuple(U) = (u_{s(1)}, ..., u_{s(q)})$, we have $\mathfrak{B}, \frac{(u_{s(1)},...,u_{s(q)})}{(x_1,...,x_q)} \models \eta$. As $\mathfrak{A}, \frac{(u_{s(1)},...,u_{s(q)})}{(x_1,...,x_q)} \models Diag_\eta(x_1, ..., x_q)$, we have $\mathfrak{A}, \frac{(u_{s(1)},...,u_{s(q)})}{(x_1,...,x_q)} \models Cons_\eta(x_1, ..., x_q)$.

The rest or the argument for the case where $type(U) = 1$, will be dealt with below. Let us next elaborate some details related to the case where $type(U) = 2$. So, assume $type(U) = 2$. Let $t : \{1, ..., q\} \to \{1, ..., k\}$ be an injection such that $tuple(U) = (u_{t(1)}, ..., u_{t(q)})$. Let $diagram(U) = \rho \in \Delta_q$. Thus $\mathfrak{A}, \frac{(u_{t(1)},...,u_{t(q)})}{(x_1,...,x_q)} \models Cons_\rho(x_1, ..., x_q)$ and $\mathfrak{B}, \frac{(u_{t(1)},...,u_{t(q)})}{(x_1,...,x_q)} \models \rho$.

We then complete the arguments for both cases $type(U) = 1$ and $type(U) = 2$. Let $(h, \nu) \in \{(s, \eta), (t, \rho)\}$, where s and t are the injections defined above, and of course η and ρ are the related diagrams.

Let $g : \{1, ..., k\} \to \{1, ..., q\}$ be the surjection such that $g(i) = j$ iff $u_i = u_{h(j)}$. Notice that $(u_{h(1)}, ..., u_{h(q)}) \in \|\nu\|^{\mathfrak{B}}$ and $(u_1, ..., u_k) \in \|\delta\|^{\mathfrak{B}}$, and these two tuples span the same set with q elements. Thus we have $\nu \leq_g \delta$.

We have $\mathfrak{A}, \frac{(u_{h(1)},...,u_{h(q)})}{(x_1,...,x_q)} \models Cons_\nu(x_1, ..., x_q)$. As $\nu \leq_g \delta$, we have $\mathfrak{A}, \frac{(u_{h(1)},...,u_{h(q)})}{(x_1,...,x_q)} \models PreCons_\delta(x_{g(1)}, ..., x_{g(k)})$. Recalling that $g(i) = j$ iff $u_i = u_{h(j)}$, we conclude that $\mathfrak{A}, \frac{(u_1,...,u_k)}{(x_1,...,x_k)} \models PreCons_\delta(x_1, ..., x_k)$, as required. □

Lemma 4.6 *Let* $\alpha \in \mathrm{SUB}_\psi$ *and* $u \in A$. *We have* $(\mathfrak{B}, u) \models \alpha$ *iff* $\mathfrak{A}, \frac{u}{x} \models P_\alpha(x)$.

Proof. See the appendix. □

Due to Lemma 4.6, we observe that since $\mathfrak{A}, \frac{w}{x} \models P_\psi(x)$, we must have $(\mathfrak{B}, w) \models \psi$. Together with Lemma 4.2, this establishes the following theorems.

Theorem 4.7 *The one dimensional fragment has the finite model property.*

Corollary 4.8 *The satisfiability and finite satisfiability problems of the one dimensional fragment are decidable.*

5 Undecidable extensions

The *general one-dimensional fragment* GF_1 of first-order logic is defined in the same way as UF_1, except that the uniformity condition is relaxed. The set of τ-formulae of GF_1 is the smallest set $\mathcal{F}$ satisfying the following conditions.

(i) If φ is a unary τ-atom, then $\varphi \in \mathcal{F}$. Also $\top, \bot \in \mathcal{F}$.

(ii) If $\varphi \in \mathcal{F}$, then $\neg\varphi \in \mathcal{F}$. If $\varphi_1, \varphi_2 \in \mathcal{F}$, then $(\varphi_1 \wedge \varphi_2) \in \mathcal{F}$.

(iii) Let $Y = \{y_1, ..., y_k\}$ be a set of variable symbols. Let U be a finite set of formulae $\psi \in \mathcal{F}$ with free variables in Y. Let F be a set of τ-atoms with free variables in Y. Let φ be any Boolean combination of formulae in $F \cup U$. Then $\exists y_2 ... \exists y_k\, \varphi \in \mathcal{F}$ and $\exists y_1 ... \exists y_k\, \varphi \in \mathcal{F}$.

There are different natural ways of generalizing UF_1 so that a two-dimensional logic is obtained. Here we consider a formalism which we call the *strongly uniform two-dimensional fragment* SUF_2 of first-order logic. The set of τ-formulae of SUF_2 is the smallest set $\mathcal{F}$ satisfying the following conditions.

(i) If φ is a unary or a binary τ-atom, then $\varphi \in \mathcal{F}$. Also $\top, \bot \in \mathcal{F}$.

(ii) If $\varphi \in \mathcal{F}$, then $\neg\varphi \in \mathcal{F}$. If $\varphi_1, \varphi_2 \in \mathcal{F}$, then $(\varphi_1 \wedge \varphi_2) \in \mathcal{F}$.

(iii) Let y_1 and y_2 be variable symbols. Let U be a finite set of formulae $\psi \in \mathcal{F}$ whose free variables are in $\{y_1, y_2\}$. Let φ be any Boolean combination of formulae in U. Then $\exists y_2\, \varphi \in \mathcal{F}$ and $\exists y_1 \exists y_2\, \varphi \in \mathcal{F}$.

(iv) Let $Y = \{y_1, ..., y_k\}$, $k \geq 3$, be a set of variable symbols. Let U be a finite set of formulae $\psi \in \mathcal{F}$ such that each ψ has at most one free variable, and the variable is in Y. Let F be a V-uniform set, $V \subseteq Y$, of τ-atoms. Let φ be any Boolean combination of formulae in $F \cup U$. Then $\exists y_3 ... \exists y_k\, \varphi \in \mathcal{F}$, $\exists y_2 ... \exists y_k\, \varphi \in \mathcal{F}$ and $\exists y_1 ... \exists y_k\, \varphi \in \mathcal{F}$.

Both of these extensions of UF_1 are Π_1^0-complete; see the appendix for the proofs. This shows that if we lift either of the two principal syntactic restrictions of UF_1, we obtain an undecidable formalism.

6 Expressivity

Guarded negation first-order logic GNFO is a novel fragment of first-order logic introduced in [2]. GNFO subsumes the guarded fragment GF. It turns out that UF_1 is incomparable in expressivity with both GNFO and the two-variable fragment with counting quantifiers FOC^2. This is proved in the appendix.

7 Conclusion

The main contribution of this paper is the discovery of the fragment UF_1 via the introduction of the notions of *uniformity* and *one-dimensionality*. The notions offer a new perspective on why modal logics are robustly decidable. Also, UF_1 extends equality-free FO^2 in a natural way, and thus provides a possible novel direction in the currently very active research on two-variable logics. Also, we believe that our satisfiability preserving translation of UF_1 into the monadic class is of independent mathematical interest. The translation is robust and can be altered and extended to give other decidability proofs.

In the future we intend to study variants of UF_1 with identity. It was observed in [2] that adding the formula $\forall x \forall y \big(Rxy \leftrightarrow x \neq y\big)$ to GNFO leads to an undecidable formalism. It is not immediately clear whether the extension of UF_1 with the free use of equality and inequality results in undecidability. We are currently working on related decidability and complexity questions.

We conjecture that our decidability result can be carried out by an alternative method combining a generalization of Scott normal form and the *dual Maslov class*. The alternative method does not involve a new proof technique, unlike the work above.

Appendix

A Translation $\mathrm{UF}_1 \to \mathrm{DUF}_1$

Proposition A.1 *There is an effective translation that transforms each formula in* UF_1 *to an equivalent formula in* DUF_1*.*

Proof. Let $\chi := \exists y_2 ... \exists y_k\, \varphi$ be a formula of UF_1 formed using the formation rule (iii) in the definition of UF_1. We may assume, w.l.o.g., that that the variables $y_1, ..., y_k$ are distinct, and that $k \geq 2$. Define $Y := \{y_1, ..., y_k\}$. Let τ_χ be the set of relation symbols in χ of the arity two and higher.

Put φ into disjunctive normal form. We obtain a formula $\exists y_2 ... \exists y_k \big(\varphi_1 \vee ... \vee \varphi_n\big)$. Now distribute the existential quantifier prefix $\exists y_2 ... \exists y_k$ over the disjunctions, obtaining the formula $\exists y_2 ... \exists y_k\, \varphi_1 \vee ... \vee \exists y_2 ... \exists y_k\, \varphi_n$.

Now consider the formula φ_j. Assume first that φ_j is of the type $\alpha \wedge \psi$, where α is a non-empty conjunction of atoms and negated atoms of the arity $m \geq 2$, and ψ is a non-empty conjunction of formulae that have at most one free variable. Let $z_2, ..., z_p \in Y$ denote the variables in $Y \setminus \{y_1\}$ that occur in α. Notice that $p = m$ if and only if y_1 occurs in α. Let z_1 denote y_1.

Let $z_{p+1}, ..., z_k \in Y$ be the variables in $Y \setminus \{z_1, ..., z_p\}$. Notice that the formula ψ is equivalent to the conjunction $\psi_1(z_1) \wedge ... \wedge \psi_k(z_k) \wedge \beta$, where each formula $\psi_i(z_i)$ is the conjunction of exactly all conjuncts of ψ with the free variable z_i, in the case such conjuncts exist, and $\psi_i(z_i)$ is the formula $\top$ otherwise; the formula β is the conjunction of the conjuncts of ψ without free variables. The formula φ_j is equivalent to the formula $\exists z_2 ... \exists z_p \big(\alpha \wedge \psi_1(z_1) \wedge ... \wedge \psi_p(z_p)\big) \wedge \exists z_{p+1} \psi_{p+1}(z_{p+1}) \wedge ... \wedge \exists z_k \psi_k(z_k) \wedge \beta$. Notice that for each i, the formula $\exists z_i \psi_i(z_i)$ is a DUF_1-formula if $\psi_i(z_i)$ is.

Consider the formula $\gamma := \exists z_2 ... \exists z_p \big(\alpha \wedge \psi_1(z_1) \wedge ... \wedge \psi_p(z_p)\big)$. The formula α is either equivalent to $\bot$, or equivalent to a non-empty disjunction $\delta_1 \vee ... \vee \delta_l$, where each δ_i denotes a conjunction over some uniform m-ary τ_χ-diagram. (Notice that since α is quantifier-free, the equivalence checking can be done effectively.) Assume first that α is equivalent to $\delta_1 \vee ... \vee \delta_l$. Therefore the formula γ is equivalent to the disjunction $\exists z_2 ... \exists z_p \big(\delta_1 \wedge \psi_1(z_1) \wedge ... \wedge \psi_p(z_p)\big) \vee ... \vee \exists z_2 ... \exists z_p \big(\delta_l \wedge \psi_1(z_1) \wedge ... \wedge \psi_p(z_p)\big)$. Notice that the disjunct $\exists z_2 ... \exists z_p \big(\delta_i \wedge \psi_1(z_1) \wedge ... \wedge \psi_p(z_p)\big)$ is a DUF_1-formula if the formulae $\psi_1(z_1), ..., \psi_p(z_p)$ are;

we may need to use the formation rule (iv) of DUF_1 in addition to rule (iii) if $\exists z_2 ... \exists z_p \big(\delta_i \wedge \psi_1(z_1) \wedge ... \wedge \psi_p(z_p) \big)$ does not contain the free variable z_1. In the case α is equivalent to $\bot$, then γ is equivalent to $\bot$.

We have now discussed the case where φ_j is of the type $\exists y_2 ... \exists y_k \big(\alpha \wedge \psi \big)$, where α is a non-empty conjunction of atoms and negated atoms of some arity higher than one, and ψ is a non-empty conjunction of formulae with at most one free variable. The case where φ_j is $\exists y_2 ... \exists y_k \, \alpha$, can be reduced to the case already discussed by considering the formula $\exists y_2 ... \exists y_k \big(\alpha \wedge \top \big)$. Assume thus that φ_j is the formula $\exists y_2 ... \exists y_k \, \psi$, where ψ is some conjunction $\psi_1(y_1) \wedge ... \wedge \psi_k(y_k) \wedge \beta$, where the formulae $\psi_i(y_i)$ have at most one free variable, and β has no free variables. Now φ_j is equivalent to the formula $\psi_1(y_1) \wedge \exists y_2 \psi_2(y_2) \wedge ... \wedge \exists y_k \psi_k(y_k) \wedge \beta$. Each conjunct $\exists y_i \psi_1(y_i)$ is a DUF_1-formula if $\psi_i(y_i)$ is.

All other cases the translation from UF_1 to DUF_1 are straightforward. □

B Proofs for Section 4

Proof of Lemma 4.2. We establish the claim of the lemma by showing that

$$\mathfrak{T}, \frac{(w,t)}{x} \models \psi_{total} \wedge \psi_{uniq} \wedge \psi_{local} \wedge \psi_{sub} \wedge P_{\psi}(x).$$

To show that $\mathfrak{T} \models \psi_{total}$, let $\big((u_1,t_1),...,(u_k,t_k)\big) \in (Dom(\mathfrak{T}))^k$, where $k \in \{2,...,\mathcal{M}\}$. We need to show that $\big((u_1,t_1),...,(u_k,t_k)\big)$ satisfies $Cons_{\delta}(x_1,...,x_k)$ for some $\delta \in \Delta_k$. Consider the tuple $(u_1,...,u_k) \in M^k$. Let η be the unique standard uniform k-ary V_{ψ}-diagram η such that $(u_1,...,u_k) \in \|\eta\|^{\mathfrak{M}}$. Let $p \in \{2,...,\mathcal{M}\}$, $p \geq k$. Let $\rho \in \Delta_p$. Let $f : \{1,...,p\} \to \{1,...,k\}$ be a surjection, and assume that $\eta \leq_f \rho$. Thus $(u_{f(1)},...,u_{f(p)}) \in \|\rho\|^{\mathfrak{M}}$. In order to conclude that $\mathfrak{T} \models \psi_{total}$, we need to show that $\mathfrak{T}, \frac{((u_1,t_1),...,(u_k,t_k))}{(x_1,...,x_k)} \models PreCons_{\rho}(x_{f(1)},...,x_{f(p)})$. Therefore we assume that $\mathfrak{T}, \frac{((u_1,t_1),...,(u_k,t_k))}{(x_1,...,x_k)} \models P_{\chi_1}(x_{f(1)}) \wedge ... \wedge P_{\chi_p}(x_{f(p)})$. Thus we have $(\mathfrak{M}, u_{f(i)}) \models \chi_i$ for each $i \in \{1,...,p\}$. As $(u_{f(1)},...,u_{f(p)}) \in \|\rho\|^{\mathfrak{M}}$, we therefore have $u_{f(1)} \in \|\langle\rho\rangle(\chi_1,...,\chi_p)\|^{\mathfrak{M}}$. Thus $(u_{f(1)},t_{f(1)}) \in P^{\mathfrak{T}}_{\langle\rho\rangle(\chi_1,...,\chi_p)}$, whence $\mathfrak{T}, \frac{((u_1,t_1),...,(u_k,t_k))}{(x_1,...,x_k)} \models P_{\langle\rho\rangle(\chi_1,...,\chi_p)}(x_{f(1)})$. Therefore $\mathfrak{T} \models \psi_{total}$.

It is immediate by the definition of the domain of $\mathfrak{T}$ and the predicates $P^{\mathfrak{T}}_t$, where t is a torus point, that $\mathfrak{T} \models \psi_{uniq}$.

To show that $\mathfrak{T} \models \psi_{local}$, assume $\mathfrak{T}, \frac{(u,t)}{x} \models Local_{\delta}(x)$ for some k-ary diagram $\delta \in \Delta$. Thus $(u,...,u)_k \in \|\delta\|^{\mathfrak{M}}$. To show that $\mathfrak{T}, \frac{(u,t)}{x} \models PreCons_{\delta}(x,...,x)_k$, let $\langle\delta\rangle(\chi_1,...,\chi_k) \in \mathrm{SUB}_{\psi}$ and assume that $\mathfrak{T}, \frac{(u,t)}{x} \models P_{\chi_1}(x) \wedge ... \wedge P_{\chi_k}(x)$. Therefore $u \in \|\chi_i\|^{\mathfrak{M}}$ for each $i \in \{1,...,k\}$, whence $u \in \|\langle\delta\rangle(\chi_1,...,\chi_k)\|^{\mathfrak{M}}$. Thus $(u,t) \in P^{\mathfrak{T}}_{\langle\delta\rangle(\chi_1,...,\chi_k)}$, as required.

The non-trivial part in proving that $\mathfrak{T} \models \psi_{sub}$ involves showing that $\mathfrak{T} \models \psi_{\langle\delta\rangle(\chi_1,...,\chi_k)}$ for formulae of the type $\langle\delta\rangle(\chi_1,...,\chi_k)$. This follows directly by Lemma 4.1, since $P^{\mathfrak{T}}_{\langle\delta\rangle(\chi_1,...,\chi_k)} = \|\langle\delta\rangle(\chi_1,...,\chi_k)\|^{\mathfrak{M}} \times Dom(\mathfrak{T})$.

Since $(\mathfrak{M}, w) \models \psi$ and $P_\psi^{\mathfrak{T}} = \|\psi\|^{\mathfrak{M}} \times Dom(\mathfrak{T})$, we have $\mathfrak{T}, \frac{(w,t)}{x} \models P_\psi(x)$.□

Proof of Lemma 4.6. We establish the claim by induction on the structure of α. For all atomic formulae $S \in \mathrm{SUB}_\psi$, the claim follows directly from the definition of the relations $S^{\mathfrak{B}}$ on tuples that span a singleton set. The cases where α is of form $\neg\beta$ or $(\beta \wedge \gamma)$ are straightforward since $\mathfrak{A} \models \psi_{sub}$.

Define $u_1 := u$ and $x_1 := x$. Assume that $\mathfrak{B}, \frac{u_1}{x_1} \models \langle\delta\rangle(\chi_1, ..., \chi_k)$, where $\langle\delta\rangle(\chi_1, ..., \chi_k) \in \mathrm{SUB}_\psi$. Thus $(u_1, ..., u_k) \in \|\delta\|^{\mathfrak{B}}$ for some tuple $(u_1, ..., u_k)$ such that $u_i \in \|\chi_i\|^{\mathfrak{B}}$ for each $i \in \{1, ..., k\}$. Now, for each $i \in \{1, ..., k\}$, we have $P_{\chi_i}^{\mathfrak{A}} = \|\chi_i\|^{\mathfrak{B}}$ by the induction hypothesis, and therefore $u_i \in P_{\chi_i}^{\mathfrak{A}}$. By Lemma 4.5, we have $\mathfrak{A}, \frac{(u_1,...,u_k)}{(x_1,...,x_k)} \models PreCons_\delta(x_1, ..., x_k)$. By the definition of the formula $PreCons_\delta(x_1, ..., x_k)$, we conclude that $\mathfrak{A}, \frac{u_1}{x_1} \models P_{\langle\delta\rangle(\chi_1,...,\chi_k)}(x_1)$.

For the converse, assume $\mathfrak{A}, \frac{u_1}{x_1} \models P_{\langle\delta\rangle(\chi_1,...,\chi_k)}(x_1)$. As $\mathfrak{A} \models \psi_{\langle\delta\rangle(\chi_1,...,\chi_k)}$, we have $\mathfrak{A}, \frac{u_1}{x_1} \models \exists x_2 ... \exists x_k \big(Diag_\delta(x_1, ..., x_k) \wedge P_{\chi_1}(x_1) \wedge ... \wedge P_{\chi_k}(x_k)\big)$. Hence there exists some tuple $(u_1, ..., u_k)$ such that $u_i \in P_{\chi_i}^{\mathfrak{A}}$ for each i and $\mathfrak{A}, \frac{(u_1,...,u_k)}{(x_1,...,x_k)} \models Diag_\delta(x_1, ..., x_k)$. By Lemma 4.4, we have $(u_i, ..., u_k) \in \|\delta\|^{\mathfrak{B}}$. As $\|\chi_i\|^{\mathfrak{B}} = P_{\chi_i}^{\mathfrak{A}}$ for each i by the induction hypothesis, we conclude that $(\mathfrak{B}, u_1) \models \langle\delta\rangle(\chi_1, ..., \chi_k)$.

Assume first that $(\mathfrak{B}, u) \models \langle E\rangle\chi$, where $\langle E\rangle\chi \in \mathrm{SUB}_\psi$. Thus $(\mathfrak{B}, v) \models \chi$ for some v, whence $\mathfrak{A}, \frac{v}{y} \models P_\chi(y)$ by the induction hypothesis. Thus $\mathfrak{A} \models \exists y P_\chi(y)$. As $\mathfrak{A} \models \psi_{sub}$, we have $\mathfrak{A}, \frac{u}{x} \models P_{\langle E\rangle\chi}(x)$. Assume then that $\mathfrak{A}, \frac{u}{x} \models P_{\langle E\rangle\chi}(x)$. As $\mathfrak{A} \models \psi_{sub}$, we have $\mathfrak{A} \models \exists y P_\chi(y)$, whence $\mathfrak{A}, \frac{v}{y} \models P_\chi(y)$ for some v. By the induction hypothesis, we have $(\mathfrak{B}, v) \models \chi$, whence $(\mathfrak{B}, u) \models \langle E\rangle\chi$. □

C Arguments concerning undecidable extensions

We recall the tiling problem of the infinite grid $\mathbb{N} \times \mathbb{N}$. A tile is a map $t : \{R, L, T, B\} \to C$, where C is a countably infinite set of colours. We use the notation $t_X := t(X)$ for $X \in \{R, L, T, B\}$. Intuitively, t_R, t_L, t_T and t_B are the colours of the right edge, left edge, top edge and bottom edge of the tile t.

Let $\mathbb{T}$ be a finite set of tiles. A $\mathbb{T}$-tiling of $\mathbb{N} \times \mathbb{N}$ is a function $f : \mathbb{N} \times \mathbb{N} \to \mathbb{T}$ that satisfies the following horizontal and vertical tiling conditions:

(T_H) For all $i, j \in \mathbb{N}$, if $f(i, j) = t$ and $f(i+1, j) = t'$, then $t_R = t'_L$.

(T_V) For all $i, j \in \mathbb{N}$, if $f(i, j) = t$ and $f(i, j+1) = t'$, then $t_T = t'_B$.

Thus, f is a proper tiling iff the colors on the matching edges of any two adjacent tiles coincide. The tiling problem for $\mathbb{N} \times \mathbb{N}$ asks whether for a finite set $\mathbb{T}$ of tiles, there is a $\mathbb{T}$-tiling of $\mathbb{N} \times \mathbb{N}$. It is well known that this problem is undecidable (Π_1^0-complete). Using the problem, it is easy to prove the following.

Proposition C.1 *The satisfiability problem of* GF_1 *is* Π_1^0*-complete.*

Proof. Let $\tau = \{H, V\}$ be a vocabulary, where H and V are binary relation symbols. The infinite grid $\mathbb{N} \times \mathbb{N}$ can be represented by a τ-structure $\mathfrak{G} := (\mathbb{N} \times \mathbb{N}, H^{\mathfrak{G}}, V^{\mathfrak{G}})$, where $H^{\mathfrak{G}} := \{((i,j),(i+1,j)) \mid i, j \in \mathbb{N}\}$ and $V^{\mathfrak{G}} := \{((i,j),(i,j+1)) \mid i, j \in \mathbb{N}\}$. Let Γ be the conjunction of the three τ-sentences $\eta_H := \forall x \exists y\, H(x,y)$, $\eta_V := \forall x \exists y\, V(x,y)$, and $\eta_{Com} := \forall x \forall y \forall z \forall w\, \big((H(x,y) \wedge V(x,z) \wedge H(z,w)) \to V(y,w)\big)$. It is easy to see that η_H, η_V and η_{Com} are in GF_1.

It is straightforward to show that if $\mathfrak{M}$ is a τ-model such that $\mathfrak{M} \models \Gamma$, then there exists a homomorphism $h : \mathfrak{G} \to \mathfrak{M}$.

Let $\mathbb{T}$ be a set of tiles. We simulate tiles by unary relation symbols P_t for each $t \in \mathbb{T}$. We denote the corresponding vocabulary $\tau \cup \{P_t \,|\, t \in \mathbb{T}\}$ by $\sigma_{\mathbb{T}}$. The tiling conditions (T_H) and (T_V) can be expressed by the $\sigma_{\mathbb{T}}$-sentences $\psi_H := \forall x \forall y \bigwedge_{t,t' \in \mathbb{T},\, t_R \neq t'_L} (P_t(x) \wedge P_{t'}(y)) \to \neg H(x,y)$ and $\psi_V := \forall x \forall y \bigwedge_{t,t' \in \mathbb{T},\, t_T \neq t'_B} (P_t(x) \wedge P_{t'}(y)) \to \neg V(x,y)$. Let $\Psi_{\mathbb{T}} := \psi_H \wedge \psi_V \wedge \psi_{part}$, where ψ_{part} is a sentence saying that every element is in exactly one of the relations P_t, $t \in \mathbb{T}$. Clearly ψ_{part} can be expressed in GF_1.

It is straightforward to show that the sentence $\Gamma \wedge \Psi_{\mathbb{T}}$ is satisfiable if and only if $\mathbb{N} \times \mathbb{N}$ is $\mathbb{T}$-tilable. Since the sentence $\Gamma \wedge \Psi_{\mathbb{T}}$ is in GF_1 for each finite set $\mathbb{T}$ of tiles, the tiling problem is effectively reducible to the satisfiability problem of GF_1. Hence the satisfiability problem is Π^0_1-hard. On the other hand, GF_1 is a fragment of first-order logic, whence its satisfiability problem is in Π^0_1. □

Let $\tau_+ = \{H_+, V_+, S\}$ be a vocabulary, where H_+ and V_+ are ternary relation symbols and S is a binary relation symbol. We will represent the infinite grid $\mathbb{N} \times \mathbb{N}$ as a τ_+-structure $\mathfrak{G}_+ := (\mathbb{N}, H_+^{\mathfrak{G}_+}, V_+^{\mathfrak{G}_+}, S^{\mathfrak{G}_+})$, where $H_+^{\mathfrak{G}_+} := \{(i, i+1, j) \,|\, i,j \in \mathbb{N}\}$, $V_+^{\mathfrak{G}_+} := \{(i,j,j+1) \,|\, i,j \in \mathbb{N}\}$, and $S_+^{\mathfrak{G}_+} := \{(i, i+1) \,|\, i \in \mathbb{N}\}$. Notice that $(u,v,w) \in V_+^{\mathfrak{G}_+}$ iff (u,v) connects to (u,w) via the vertical successor $V^{\mathfrak{G}}$ of the standard Cartesian grid $\mathfrak{G}$ defined in the proof of Proposition C.1. On the other hand, $(u,v,w) \in H_+^{\mathfrak{G}_+}$ iff $\big((u,w),(v,w)\big) \in H^{\mathfrak{G}}$. We shall next form a τ_+-sentence Γ_+ of SUF_2 such that $\mathfrak{G}_+ \models \Gamma_+$, and there is a homomorphism from $\mathfrak{G}_+$ to any model of Γ_+. Define Γ_+ to be the conjunction of the formulae $\theta_S := \forall x \exists y\, S(x,y)$, $\theta_H := \forall x_1 \forall x_2\, (S(x_1,x_2) \to \forall y\, H_+(x_1,x_2,y))$, and $\theta_V := \forall y_1 \forall y_2\, (S(y_1,y_2) \to \forall x\, V_+(x,y_1,y_2))$.

Lemma C.2 *If $\mathfrak{M}$ is a τ_+-model such that $\mathfrak{M} \models \Gamma_+$, then there exists a homomorphism $h : \mathfrak{G}_+ \to \mathfrak{M}$.*

Proof. We define a function $h : \mathbb{N} \to M$ by recursion as follows. Choose an arbitrary point $a_0 \in M$, and set $h(0) := a_0$. Assume that $h(i) = a$ has been defined. Since $\mathfrak{M} \models \theta_S$, there is $b \in M$ such that $(a,b) \in S^{\mathfrak{M}}$. Define $h(i+1) := b$. Observe first that $(h(i), h(i+1)) \in S^{\mathfrak{M}}$ for each $i \in \mathbb{N}$. Furthermore, since $\mathfrak{M} \models \theta_H \wedge \theta_V$, we have $(h(i), h(i+1), h(j)) \in H_+^{\mathfrak{M}}$ and $(h(i), h(j), h(j+1)) \in V_+^{\mathfrak{M}}$ for all $i,j \in \mathbb{N}$. Thus h is a homomorphism $\mathfrak{G}_+ \to \mathfrak{A}$. □

Theorem C.3 *The satisfiability problem of* SUF_2 *is Π^0_1-complete.*

Proof. By Lemma C.2, we know that if $\mathfrak{M}$ is a τ_+-model such that $\mathfrak{M} \models \Gamma_+$, then there exists a homomorphism $h : \mathfrak{G}_+ \to \mathfrak{M}$. (We also have $\mathfrak{G}_+ \models \Gamma_+$.)

Let $\mathbb{T}$ be a set of tiles. This time we simulate tiles by fresh ternary relation symbols $P_{X,t}$, where $X \in \{R, L, T, B\}$ and $t \in \mathbb{T}$. Let $\rho_{\mathbb{T}} := \tau_+ \cup \{P_{X,t} \,|\, X \in \{R,L,T,B\}, t \in \mathbb{T}\}$ be the corresponding vocabulary.

The idea here is that if $(a,b,c) \in P_{R,t}$ and $(a,b,c) \in P_{L,t'}$, then the right edge of (a,c) is coloured with t_R and the left edge of (b,c) is coloured with t'_L; recall that $(a,b,c) \in H_+^{\mathfrak{G}+}$ means that $\big((a,c),(b,c)\big) \in H^{\mathfrak{G}}$. Similarly, if $(a,b,c) \in P_{T,t}$ and $(a,b,c) \in P_{B,t'}$, then the top edge of (a,b) is coloured with t_T and the bottom edge of (a,c) is coloured with t'_B. Thus, we can express the tiling conditions (T_H) and (T_V) by the following SUF_2-sentences:

$$\varphi_H := \forall x_1 \forall x_2 \forall y \bigwedge_{t,t' \in \mathbb{T},\, t_R \neq t'_L} \Big(\big(P_{R,t}(x_1,x_2,y) \wedge P_{L,t'}(x_1,x_2,y)\big) \to \neg H_+(x_1,x_2,y)\Big),$$

$$\varphi_V := \forall x \forall y_1 \forall y_2 \bigwedge_{t,t' \in \mathbb{T},\, t_T \neq t'_B} \Big(\big(P_{T,t}(x,y_1,y_2) \wedge P_{B,t'}(x,y_1,y_2)\big) \to \neg V_+(x,y_1,y_2)\Big).$$

We also need a sentence φ_{prop} stating that each pair (a,b) is tiled by exactly one $t \in \mathbb{T}$. This amounts to stating, firstly, that the interpretation of each symbol $P_{R,t}$ depends only on the first and the last variable: $\bigwedge_{t \in \mathbb{T}} \forall x_1 \forall y\, (\exists x_2\, P_{R,t}(x_1,x_2,y) \to \forall x_2\, P_{R,t}(x_1,x_2,y))$, and analogously for $P_{L,t}$, $P_{T,t}$ and $P_{B,t}$. Secondly, the four colors of each pair correspond to the same tile, meaning that $\bigwedge_{t \in \mathbb{T}} \forall x_1 \forall y\, (\exists x_2\, P_{R,t}(x_1,x_2,y) \leftrightarrow \exists x_2\, P_{L,t}(x_2,x_1,y))$ holds, and similar conditions for the other pairs $(P_{X,t}, P_{Y,t})$ hold. Thirdly, for each $X \in \{L,R,B,T\}$, every triple is in exactly one of the relations $P_{X,t}$, $t \in \mathbb{T}$.

Clearly there is such a sentence φ_{prop} in SUF_2. Let $\Phi_{\mathbb{T}}$ be the conjunction of the sentences φ_H, φ_V and φ_{prop}. Thus the sentence $\Gamma_+ \wedge \Phi_{\mathbb{T}}$ is satisfiable if and only if $\mathbb{N} \times \mathbb{N}$ is $\mathbb{T}$-tilable. Hence we conclude that SUF_2 is Π^0_1-complete.□

D Expressivity

Theorem D.1 UF_1 *is incomparable in expressivity with both two-variable logic with counting* (FOC^2) *and guarded negation fragment* (GNFO).

Proof. The expressivity of FOC^2 is seriously limited when it comes to properties of relations of arities greater than two. It is easy to show that for example the UF_1-sentence $\exists x \exists y \exists z\, R(x,y,z)$ is not expressible in FOC^2. Thus UF_1 is not contained in FOC^2.

It is straightforward to show by using the bisimulation for GNFO, provided in [2], that the UF_1-sentence $\exists x \exists y \neg R(x,y)$ is not expressible in GNFO. This follows from the fact that structures $\big(\{a\},\{(a,a)\}\big)$ and $\big(\{a,b\},\{(a,a),(b,b)\}\big)$ are bisimilar in the sense of GNFO. Thus UF_1 is not contained in GNFO.

The FO^2-sentence $\forall x \forall y (x = y)$ cannot be expressed in UF_1. This can be seen (for example) by observing that the two directions of our decidability proof together entail that satisfiable sentences of the equality-free logic UF_1 can always be satisfied in a larger model. Thus UF_1 does not contain FO^2.

It follows immediately from the definition of UF_1 that the equality-free fragment of FO^2 is contained in UF_1. In fact, it is easy to prove that in restriction to models with relation symbols of arities at most two, the expressivities of UF_1 and the identity-free fragment of FO^2 coincide. (Consider for example the translation from UF_1 to MUF_1 in the case of such vocabularies.)

To see that UF_1 does not contain GNFO, consider the GNFO-sentence $\exists x \exists y \exists z(Rxy \wedge Ryz \wedge Rzx)$. It is easy to show (by a pebble game argument, see [12]), that this property is not expressible in FO^2. As UF_1 is contained in FO^2 when attention is restricted to models with only binary relations, UF_1 does not contain GNFO. □

References

[1] Andréka, H., J. van Benthem and I. Németi, *Modal languages and bounded fragments of predicate logic.*, Journal of Philosophical Logic **27** (1998), pp. 217–274.

[2] Bárány, V., B. ten Cate and L. Segoufin, *Guarded negation*, in: *Proceedings of Automata, Languages and Programming - 38th International Colloquium, (ICALP), Part II* (2011), pp. 356–367.

[3] Benaim, S., M. Benedikt, W. Charatonik, E. Kieroński, R. Lenhardt, F. Mazowiecki and J. Worrell, *Complexity of two-variable logic on finite trees*, in: *Proceedings of Automata, Languages and Programming - 40th International Colloquium, (ICALP), Part II* (2013), pp. 74–88.

[4] Blackburn, P., M. de Rijke and Y. Venema, "Modal Logic," Cambridge Tracts in Theoretical Computer Science, Cambridge University Press, 2001.

[5] Charatonik, W. and P. Witkowski, *Two-variable logic with counting and trees*, in: *Proceedings of the 28th Annual ACM/IEEE Symposium on Logic in Computer Science (LICS)*, 2013, pp. 73–82.

[6] Gargov, G. and S. Passy, *Modal environment for boolean speculations*, in: *Mathematical logic and its applications. Proc. of the Summer School and Conference dedicated to the 80th Anniversary of Kurt Gödel* (1987), pp. 253–263.

[7] Grädel, E., *Why are modal logics so robustly decidable?*, in: *Current Trends in Theoretical Computer Science*, World Scientific, 2001 pp. 393–408.

[8] Grädel, E., P. G. Kolaitis and M. Y. Vardi, *On the decision problem for two-variable first-order logic*, Bulletin of Symbolic Logic **3** (1997), pp. 53–69.

[9] Grädel, E., M. Otto and E. Rosen, *Two-variable logic with counting is decidable*, in: *12th Annual IEEE Symposium on Logic in Computer Science (LICS)*, 1997, pp. 306–317.

[10] Henkin, L., *Logical systems containing only a finite number of symbols.*, Presses De l'Université De Montréal (1967).

[11] Kieroński, E., J. Michaliszyn, I. Pratt-Hartmann and L. Tendera, *Two-variable first-order logic with equivalence closure*, in: *Proceedings of the 27th Annual IEEE Symposium on Logic in Computer Science (LICS)*, 2012, pp. 431–440.

[12] Libkin, L., "Elements of Finite Model Theory," Texts in Theoretical Computer Science. An EATCS Series, Springer, 2004.

[13] Lutz, C. and U. Sattler, *The complexity of reasoning with boolean modal logics*, in: *Proceedings of Advances in Modal Logic (AiML)*, 2000, pp. 329–348.

[14] Mortimer, M., *On languages with two variables.*, Mathematical Logic Quarterly **21** (1975), pp. 135–140.

[15] Pacholski, L., W. Szwast and L. Tendera, *Complexity of two-variable logic with counting*, in: *Proceedings of the 12th Annual IEEE Symposium on Logic in Computer Science (LICS)*, 1997, pp. 318–327.

[16] Pratt-Hartmann, I., *Complexity of the two-variable fragment with counting quantifiers*, Journal of Logic, Language and Information **14** (2005), pp. 369–395.

[17] Szwast, W. and L. Tendera, FO^2 *with one transitive relation is decidable*, in: *Proceedings of 30th International Symposium on Theoretical Aspects of Computer Science (STACS)*, 2013, pp. 317–328.

[18] Vardi, M. Y., *Why is modal logic so robustly decidable?*, in: *Descriptive Complexity and Finite Models, Proceedings of a DIMACS Workshop* (1996), pp. 149–184.

The Expressive Power of Modal Dependence Logic

Lauri Hella [1] Kerkko Luosto [1] Katsuhiko Sano [2]
Jonni Virtema [1]

[1] *School of Information Sciences*
University of Tampere

[2] *School of Information Science*
Japan Advanced Institute of Science and Technology

Abstract

We study the expressive power of various modal logics with team semantics. We show that exactly the properties of teams that are downward closed and closed under team k-bisimulation, for some finite k, are definable in modal logic extended with intuitionistic disjunction. Furthermore, we show that the expressive power of modal logic with intuitionistic disjunction and extended modal dependence logic coincide. Finally we establish that any translation from extended modal dependence logic into modal logic with intuitionistic disjunction increases the size of some formulas exponentially.

Keywords: Modal dependence logic, team semantics, bisimulation, expressive power.

1 Introduction

Dependence is a central notion in many scientific disciplines. For example in physics there are dependences in experimental data. Decision theory is concerned with identifying the variables on which the result depends. Furthermore, dependences between attributes is a key notion in database theory. In order to express such dependences in a formal framework, Väänänen [16] introduced first-order dependence logic. Dependence logic is based on team semantics, in which the truth of formulas is evaluated in sets of assignments instead of single assignments. Team semantics was originally defined by Hodges [10] as a means to obtain compositional semantics for the independence-friendly logic of Hintikka and Sandu [9].

[1] The research of Lauri Hella was partially funded by a Professor Pool grant awarded by the Finnish Cultural Foundation. The research of Jonni Virtema was supported by grant 266260 of the Academy of Finland, and grants by the Finnish Academy of Science and Letters and the University of Tampere.

[2] The research of Katsuhiko Sano was partially supported by JSPS KAKENHI, Grant-in-Aid for Young Scientists (B) 24700146.

With the aim to import dependences and team semantics to modal logic Väänänen [17] introduced *modal dependence logic* $\mathcal{MDL}$. In the context of modal logic a team is just a set of states in a Kripke model. Modal dependence logic extends standard modal logic with team semantics by modal dependence atoms, $=(p_1, \ldots, p_n, q)$. The intuitive meaning of the formula $=(p_1, \ldots, p_n, q)$ is that within a team the truth value of the proposition q is functionally determined by the truth values of the propositions $p_1, \ldots, p_n$.

Modal dependence logic is a first step toward combining functional dependences and modal logic. The logic however lacks the ability to express temporal dependences, only propositional dependences can be expressed. This is due to the restriction that only proposition symbols are allowed in the dependence atoms of $\mathcal{MDL}$. To overcome this defect Ebbing et al. [3] introduced the *extended modal dependence logic*, $\mathcal{EMDL}$, which is obtained from $\mathcal{MDL}$ by extending the scope of dependence atoms to arbitrary modal formulas, i.e., dependence atoms in $\mathcal{EMDL}$ are of the form $=(\varphi_1, \ldots \varphi_n, \psi)$, where $\varphi_1, \ldots, \varphi_n, \psi$ are $\mathcal{ML}$ formulas.

In recent years the research around modal dependence logic and other modal logics with team semantics has been active, see e.g. [3,4,5,6,12,13,15,18]. An important logic, closely related to modal dependence logic, is modal logic with intuitionistic disjunction, $\mathcal{ML}(\otimes)$. It was already observed by Väänänen [17] that dependence atoms can be defined by using the intuitionistic disjunction $\otimes$. Using this observation Ebbing et al. [3] showed that in terms of expressiveness, $\mathcal{EMDL}$ is contained in $\mathcal{ML}(\otimes)$. However, it was left open, whether the containment is strict, or whether $\mathcal{EMDL}$ and $\mathcal{ML}(\otimes)$ are actually equivalent with respect to expressive power.

Team semantics is also meaningful in the context of purely propositional logics. Propositional dependence logic was extensively studied in the recent Ph.D. thesis of Fan Yang [18]. As pointed out in [18], propositional dependence logic is closely related to the inquisitive logic of Groenendijk [8] (see also [2,14]). Like in the team semantics of propositional dependence logic, in inquisitive logic the meaning of formulas is defined on sets of assignments for proposition symbols. Ciardelli [1] proved that inquisitive logic is expressively complete in the sense that every downward closed property of teams (over a finite set of proposition symbols) is definable by a formula of inquisitive logic. Thus, we can say that the set of connectives used in inquisitive logic is complete in the same spirit as, e.g., $\{\neg, \wedge\}$ is a complete set of connectives for propositional logic. Fan Yang [18] proved that the same expressive completeness result holds for propositional dependence logic, and consequently, inquisitive logic and propositional dependence logic are equivalent with respect to expressive power.

It is well known that the expressive power of modal logic can be characterized via bisimulation: by the famous result of Gabbay and van Benthem, a class $\mathcal{K}$ of pointed Kripke models (K, w) is definable by a formula of modal logic if and only if $\mathcal{K}$ is closed under k-bisimulation, for some $k \in \mathbb{N}$. In this paper we prove a joint extension to this characterization and the characterization of the expressive power of inquisitive logic and propositional dependence logic mentioned above. We first define a canonical extension of bisimulation suitable for team semantics, called team bisimulation. Then we show that a class $\mathcal{K}$ of of pairs (K, T), where K is a Kripke model and T is a team, is definable by a sentence of $\mathcal{ML}(\otimes)$ if and only if $\mathcal{K}$ is downward closed and closed under team k-bisimulation, for some $k \in \mathbb{N}$.

Furthermore, we show that the expressive power of $\mathcal{EMDL}$ coincides with that of $\mathcal{ML}(\otimes)$, thus answering the open problem from [3] mentioned above. In particular, we obtain as a corollary that the expressive power of $\mathcal{EMDL}$ is also characterized by downward closure and closure under team k-bisimulation. Since team k-bisimulation is a natural adaptation of k-bisimulation to the context of team semantics, this result shows that $\mathcal{EMDL}$ can be regarded as a canonical extension of modal logic for expressing dependences between formulas.

In addition, we introduce two semantical invariants for formulas of $\mathcal{ML}(\otimes)$ and $\mathcal{EMDL}$, which we call lower dimension and upper dimension, respectively. We show that the truth of a formula in a team of a Kripke model can be determined by checking its truth on subteams of a fixed size n. The lower dimension of the formula in question is the least $n \in \mathbb{N}$ such that this holds. Thus, lower dimension gives rise to a natural classification of formulas with respect to their semantical complexity, and we believe that it can also be used for analyzing the computational complexity of the model checking problem of modal formulas.

The upper dimension of a formula is defined as the largest number of maximal teams satisfying the formula in any fixed Kripke model. We prove that the lower dimension of any formula is less than or equal to its upper dimension. Moreover, we show that the upper dimension admits well-behaved compositionally defined estimates. These estimates are very useful in establishing upper bounds for lower dimension as well, since finding good estimates for the lower dimension directly seems to be difficult.

Finally, we use the upper dimension for proving that any translation from $\mathcal{EMDL}$ into $\mathcal{ML}(\otimes)$ increases the size of some formulas exponentially. To prove this, we show that the upper dimension of a dependence atom $=(p_1, \ldots, p_n, q)$ is 2^{2^n}, while the upper dimension of any $\mathcal{ML}(\otimes)$-formula φ is at most 2^d, where d is the number of occurrences of $\otimes$ in φ.

2 Background

In this section we first give the syntax and team semantics for the modal logics studied in the paper. We then formulate the notions of definability and expressive power in team semantics. Finally we recall the basic results concerning bisimulation and definability in the context of standard Kripke semantics.

2.1 Modal logics with team semantics

The syntax of modal logic $\mathcal{ML}$ could be defined in any standard way. However, when we consider the extension of $\mathcal{ML}$ by dependence atoms, it is useful to assume that all formulas are in *negation normal form*, i.e., negations occur only in front of atomic propositions. Thus, we define the syntax of $\mathcal{ML}$ as follows:

Definition 2.1 Let Φ be a set of proposition symbols. The set of formulas of $\mathcal{ML}(\Phi)$ is generated by the following grammar

$$\varphi ::= p \mid \neg p \mid (\varphi \wedge \varphi) \mid (\varphi \vee \varphi) \mid \Diamond\varphi \mid \Box\varphi,$$

where $p \in \Phi$.

In this article we consider three extensions of $\mathcal{ML}$: *modal logic with intuitionistic disjunction* $\mathcal{ML}(\otimes)$, *modal dependence logic* $\mathcal{MDL}$, and *extended modal dependence logic* $\mathcal{EMDL}$.

Definition 2.2 (i) The syntax of modal logic with intuitionistic disjunction $\mathcal{ML}(\otimes)(\Phi)$ is obtained by extending the syntax of $\mathcal{ML}$ by the grammar rule

$$\varphi ::= (\varphi \otimes \varphi).$$

(ii) The syntax for modal dependence logic $\mathcal{MDL}(\Phi)$ is obtained by extending the syntax of $\mathcal{ML}$ by dependence atoms

$$\varphi ::= =(p_1, \ldots, p_n, q),$$

where $p_1, \ldots, p_n, q \in \Phi$.

(iii) The syntax for extended modal dependence logic $\mathcal{EMDL}(\Phi)$ is obtained by extending the syntax of $\mathcal{ML}$ by dependence atoms

$$\varphi ::= =(\psi_1, \ldots, \psi_n, \theta),$$

where $\psi_1, \ldots, \psi_n, \theta$ are $\mathcal{ML}$-formulas.

The notion of Kripke model is defined as usual. Thus, if Φ is a set of proposition symbols, a *Kripke model K over Φ* is a triple $K = (W, R, V)$, where W is a set of *states* or *(possible) worlds*, $R \subseteq W \times W$ is an *accessibility relation*, and V is a *valuation* $V: \Phi \to \mathcal{P}(W)$.

The semantics of $\mathcal{ML}$ is usually defined on pointed Kripke models. We write $K, w \models \varphi$ if $\varphi \in \mathcal{ML}(\Phi)$ is true in $w \in W$ according to the standard Kripke semantics. However, to give a meaningful semantics for dependence atoms and intuitionistic disjunction, we need to consider arbitrary sets of states instead of single states as points of evaluation.

Definition 2.3 Let $K = (W, R, V)$ be a Kripke model.

(i) Any subset T of W is called a *team* of K.

(ii) For any team $T \subseteq W$ we write $R[T] = \{v \in W \mid \exists w \in T : wRv\}$ and $R^{-1}[T] = \{w \in W \mid \exists v \in T : wRv\}$.

(iii) For teams $T, S \subseteq W$ we write $T[R]S$ if $S \subseteq R[T]$ and $T \subseteq R^{-1}[S]$.

Thus, $T[R]S$ holds if and only if for every $v \in S$ there is $w \in T$ such that wRv, and for every $w \in T$ there is $v \in S$ such that wRv. We are now ready to define *team semantics* for the modal logics studied in this paper.

Definition 2.4 The semantics for $\mathcal{ML}$, $\mathcal{ML}(\otimes)$, $\mathcal{MDL}$, and $\mathcal{EMDL}$ is defined as follows.

$$
\begin{array}{rcl}
K,T \models p & \Leftrightarrow & T \subseteq V(p). \\
K,T \models \neg p & \Leftrightarrow & T \cap V(p) = \emptyset. \\
K,T \models \varphi \wedge \psi & \Leftrightarrow & K,T \models \varphi \text{ and } K,T \models \psi. \\
K,T \models \varphi \vee \psi & \Leftrightarrow & K,T_1 \models \varphi \text{ and } K,T_2 \models \psi \\
& & \text{for some } T_1,T_2 \text{ such that } T_1 \cup T_2 = T. \\
K,T \models \Diamond\varphi & \Leftrightarrow & K,T' \models \varphi \text{ for some } T' \text{ such that } T[R]T'. \\
K,T \models \Box\varphi & \Leftrightarrow & K,T' \models \varphi, \text{ where } T' = R[T].
\end{array}
$$

For $\mathcal{ML}(\otimes)$ we have the following additional clause:

$$K,T \models \varphi \otimes \psi \quad \Leftrightarrow \quad K,T \models \varphi \text{ or } K,T \models \psi.$$

For $\mathcal{MDL}$ and $\mathcal{EMDL}$ we have the following additional clause:

$$
\begin{array}{rcl}
K,T \models {=}(\psi_1,\ldots,\psi_n,\theta) & \Leftrightarrow & \forall w,v \in T : \bigwedge_{i=1}^{n} (K,\{w\} \models \psi_i \Leftrightarrow K,\{v\} \models \psi_i) \\
& & \text{implies } (K,\{w\} \models \theta \Leftrightarrow K,\{v\} \models \theta).
\end{array}
$$

Note in particular that ${=}(\theta)$ is a formula saying that the truth value of θ is constant in the given team: $K,T \models {=}(\theta)$ if and only if either $K,\{w\} \models \theta$ for all $w \in T$, or $K,\{w\} \not\models \theta$ for all $w \in T$.

The team semantics for basic modal logic $\mathcal{ML}$ can be reduced to the usual Kripke semantics in the sense that a team T satisfies a formula φ if and only if every state in T satisfies φ:

Proposition 2.5 ([15, Theorem 1]) *Let K be a Kripke model, T a team of K, and φ an $\mathcal{ML}(\Phi)$-formula. Then*

$$K,T \models \varphi \quad \Leftrightarrow \quad K,w \models \varphi \textit{ for every } w \in T.$$

In particular, $K,\{w\} \models \varphi \Leftrightarrow K,w \models \varphi$.

2.2 Definability and expressive power

A Φ-*model with a team* is a pair (K,T), where K is a Kripke model over Φ and T is a team of K. We denote by $\mathcal{KT}(\Phi)$ the class of Φ-models with teams. If $\mathcal{L}$ is one of the logics $\mathcal{ML}, \mathcal{ML}(\otimes), \mathcal{MDL}, \mathcal{EMDL}$, then each formula $\varphi \in \mathcal{L}(\Phi)$ *defines* a class of Φ-models with teams:

$$\|\varphi\| := \{(K,T) \in \mathcal{KT}(\Phi) \mid K,T \models \varphi\}.$$

A class $\mathcal{K} \subseteq \mathcal{KT}(\Phi)$ is *definable* in $\mathcal{L}$, if there is a formula $\varphi \in \mathcal{L}(\Phi)$ such that $\mathcal{K} = \|\varphi\|$.

If $\mathcal{L}$ s a logic whose semantics is defined on Kripke models with teams, then the *expressive power* of $\mathcal{L}$ is just the collection of classes $\|\varphi\|$, $\varphi \in \mathcal{L}$, that are definable in $\mathcal{L}$. Accordingly, the expressive power of two such logics $\mathcal{L}$ and $\mathcal{L}'$ can be compared as follows:

- $\mathcal{L}'$ is *at least as expressive as* $\mathcal{L}$, $\mathcal{L} \leq \mathcal{L}'$, if for every $\varphi \in \mathcal{L}(\Phi)$ there is $\psi \in \mathcal{L}'(\Phi)$ such that $\|\varphi\| = \|\psi\|$.
- $\mathcal{L}$ is *less expressive than* $\mathcal{L}'$, $\mathcal{L} < \mathcal{L}'$, if $\mathcal{L} \leq \mathcal{L}'$, but $\mathcal{L}' \not\leq \mathcal{L}$.
- $\mathcal{L}$ and $\mathcal{L}'$ are *equally expressive*, $\mathcal{L} \equiv \mathcal{L}'$, if $\mathcal{L} \leq \mathcal{L}'$ and $\mathcal{L}' \leq \mathcal{L}$.

Clearly $\mathcal{ML} \leq \mathcal{MDL} \leq \mathcal{EMDL}$. Väänänen [17] gave a translation from $\mathcal{MDL}$ to $\mathcal{ML}(\varovee)$, and extending this translation to $\mathcal{EMDL}$, it was proved in [3] that $\mathcal{EMDL} \leq \mathcal{ML}(\varovee)$. Furthermore, it is easy to see that dependence atoms are not definable in $\mathcal{ML}$, and in [3] it was proved that the non-propositional dependence atom $=(\Diamond p)$ is not definable in $\mathcal{MDL}$. Summing up, the following relationships between the logics $\mathcal{ML}$, $\mathcal{MDL}$, $\mathcal{EMDL}$ and $\mathcal{ML}(\varovee)$ are known:

Proposition 2.6 ([3]) $\mathcal{ML} < \mathcal{MDL} < \mathcal{EMDL} \leq \mathcal{ML}(\varovee)$.

Moreover, it was proved in [3] that $\mathcal{EMDL} \equiv \mathcal{ML}(\varovee_{\mathcal{ML}})$, where $\mathcal{ML}(\varovee_{\mathcal{ML}})$ is the fragment of $\mathcal{ML}(\varovee)$ that does not allow nesting of the intuitionistic disjunction $\varovee$. However, it was left as an open problem in [3] whether the expressive power of $\mathcal{EMDL}$ is strictly weaker than that of $\mathcal{ML}(\varovee)$.

For any formula $\varphi \in \mathcal{L}(\Phi)$, the class $\|\varphi\|$ can be seen as its *global meaning.* But it is also useful to consider the meaning of formulas *locally*, i.e., with respect to a fixed Kripke model. For any Kripke model $K = (W, R, V)$ over Φ, each formula $\varphi \in \mathcal{L}(\Phi)$ *defines* a set of teams of K:

$$\|\varphi\|^K := \{T \subseteq W \mid K, T \models \varphi\}.$$

Note that it follows from Proposition 2.5 that the set $\|\varphi\|^K$ is *downward closed* for all $\varphi \in \mathcal{ML}$:

$$(*) \qquad \text{if } T \in \|\varphi\|^K \text{ and } S \subseteq T, \text{ then } S \in \|\varphi\|^K.$$

Although Proposition 2.5 fails for the extensions $\mathcal{ML}(\varovee)$, $\mathcal{MDL}$ and $\mathcal{EMDL}$ of $\mathcal{ML}$, downward closure still holds for all of these logics. We say that a logic $\mathcal{L}$ is *downward closed* if $(*)$ holds for every formula $\varphi \in \mathcal{L}$.

Proposition 2.7 ([17],[5]) *The logics $\mathcal{MDL}$, $\mathcal{EMDL}$ and $\mathcal{ML}(\varovee)$ are downward closed.*

Proof For $\mathcal{MDL}$ and $\mathcal{ML}(\varovee)$, downward closure was proved in [17] and [5]. For $\mathcal{EMDL}$, the claim follows from the fact that $\mathcal{EMDL} \leq \mathcal{ML}(\varovee)$. □

2.3 Bisimulation and definability in Kripke semantics

It is well known that the expressive power of basic modal logic $\mathcal{ML}$ with respect to Kripke semantics can be completely characterized in terms of k-bisimulation. Our aim is to give an analogous characterization for the expressive power of $\mathcal{ML}(\varovee)$ and $\mathcal{EMDL}$. For this purpose we need some basic concepts and results related to k-bisimulation.

The *modal depth* $\mathrm{md}(\varphi)$ of a formula of $\mathcal{ML}(\Phi)$ is defined in the obvious manner, i.e., $\mathrm{md}(p) = \mathrm{md}(\neg p) = 0$ for $p \in \Phi$, $\mathrm{md}(\varphi \wedge \psi) = \mathrm{md}(\varphi \vee \psi) = \max\{\mathrm{md}(\varphi), \mathrm{md}(\psi)\}$, and $\mathrm{md}(\Diamond\varphi) = \mathrm{md}(\Box\varphi) = \mathrm{md}(\varphi) + 1$.

A *pointed* Φ*-model* is a pair (K, w) such that K is a Kripke model over Φ, and w is a state in K. Let k be a natural number, and let (K, w) and (K', w') be pointed Φ-models. We say that (K, w) and (K', w') are k*-equivalent*, in symbols $K, w \equiv_k K', w'$, if for every $\varphi \in \mathcal{ML}(\Phi)$ with $\mathrm{md}(\varphi) \leq k$

$$K, w \models \varphi \quad \Leftrightarrow \quad K', w' \models \varphi.$$

Definition 2.8 Let $k \in \mathbb{N}$, and let (K, w) and (K', w') be pointed Φ-models. We write $K, w \rightleftarrows_k K', w'$ if (K, w) and (K', w') are k-bisimilar. The k-bisimilarity relation $\rightleftarrows_k$ can be defined recursively as follows:

- $K, w \rightleftarrows_0 K', w'$ if and only if the equivalence $K, w \models p \Leftrightarrow K', w' \models p$ holds for all $p \in \Phi$.
- $K, w \rightleftarrows_{k+1} K', w'$ if and only if $K, w \rightleftarrows_0 K', w'$, and
  - for every $v \in R[w]$ there is $v' \in R'[w']$ such that $K, v \rightleftarrows_k K', v'$, and
  - for every $v' \in R'[w']$ there is $v \in R[w]$ such that $K, v \rightleftarrows_k K', v'$.

  (Here $R[w]$ is a shorthand notation for $R[\{w\}]$. Thus, $v \in R[w] \Leftrightarrow wRv$.)

A class $\mathcal{K}$ of pointed Φ-models is *closed under* k*-bisimulation* if it satisfies the following condition:

- $(K, w) \in \mathcal{K}$ and $K, w \rightleftarrows_k K', w'$ implies that $(K', w') \in \mathcal{K}$.

We will also make use of the fact that for every pointed Φ-model (K, w) and every $k \in \mathbb{N}$ there is a formula that characterizes (K, w) completely up to k-equivalence. These *Hintikka formulas* (or *characteristic formulas*) are defined as follows (see e.g. [7]):

Definition 2.9 Assume that Φ is a finite set of proposition symbols. Let $k \in \mathbb{N}$ and let (K, w) be a pointed Φ-model. The k*-th Hintikka formula* $\chi^k_{K,w}$ of (K, w) is defined recursively as follows:

- $\chi^0_{K,w} := \bigwedge\{p \mid p \in \Phi, w \in V(p)\} \wedge \bigwedge\{\neg p \mid p \in \Phi, w \notin V(p)\}$.
- $\chi^{k+1}_{K,w} := \chi^k_{K,w} \wedge \bigwedge_{v \in R[w]} \Diamond \chi^k_{K,v} \wedge \Box \bigvee_{v \in R[w]} \chi^k_{K,v}$.

It is easy to see that $\mathrm{md}(\chi^k_{K,w}) = k$, and $K, w \models \chi^k_{K,w}$ for every pointed Φ-model (K, w). Moreover, the Hintikka formula $\chi^k_{K,w}$ captures the essence of k-bisimulation:

Proposition 2.10 *Let Φ be a finite set of proposition symbols, $k \in \mathbb{N}$, and (K, w) and (K', w') pointed Φ-models. Then*

$$K, w \equiv_k K', w' \quad \Leftrightarrow \quad K, w \rightleftarrows_k K', w' \quad \Leftrightarrow \quad K', w' \models \chi^k_{K,w}.$$

The characterization for the expressive power of $\mathcal{ML}$ with respect to Kripke-semantics can now be stated as follows:

Proposition 2.11 (van Benthem, Gabbay) *Assume that Φ is a finite set of proposition symbols. A class $\mathcal{K}$ of pointed Φ-models is definable in $\mathcal{ML}$ if and only if there is $k \in \mathbb{N}$ such that $\mathcal{K}$ is closed under k-bisimulation.*

3 $\mathcal{ML}(\oslash)$ and team bisimulation

In this section we prove a characterization for the expressive power of $\mathcal{ML}(\oslash)$. This characterization is based on a natural adaptation of the notion of k-bisimulation to logics with team semantics.

3.1 Bisimulation in team semantics

We start by defining k-bisimulation in the context of team semantics; the definition is directly based on the k-bisimulation relation $\rightleftarrows_k$ for Kripke semantics.

Definition 3.1 Let $(K,T),(K',T') \in \mathcal{KT}(\Phi)$ and $k \in \mathbb{N}$. We say that K,T and K',T' are *team k-bisimilar* and write $K,T\ [\rightleftarrows_k]\ K',T'$ if

(i) for every $w \in T$ there exists some $w' \in T'$ such that $K,w \rightleftarrows_k K,w'$, and

(ii) for every $w' \in T'$ there exists some $w \in T$ such that $K,w \rightleftarrows_k K,w'$.

It is well known that $K,w \rightleftarrows_k K',w'$ implies $K,w \rightleftarrows_n K',w'$ for all $n \leq k$. Using this it is easy to prove that the same holds also for team k-bisimilarity:

Lemma 3.2 *Let $(K,T),(K',T') \in \mathcal{KT}(\Phi)$ and $k \in \mathbb{N}$. If $K,T\ [\rightleftarrows_k]\ K',T'$, then $K,T\ [\rightleftarrows_n]\ K',T'$ for all $n \leq k$.*

We say that a class $\mathcal{K} \subseteq \mathcal{KT}(\Phi)$ is *closed under team k-bisimulation* if it satisfies the condition:

- $(K,T) \in \mathcal{K}$ and $K,T\ [\rightleftarrows_k]\ K',T'$ implies that $(K',T) \in \mathcal{K}$.

The next lemma shows that team k-bisimulation satisfies the natural counterparts of the back-and-forth properties that we used in defining $\rightleftarrows_k$, as well as a couple of other useful properties related to team semantics.

Lemma 3.3 *Let $k \in \mathbb{N}$, and assume that $(K,T),(K',T') \in \mathcal{KT}(\Phi)$ are such that $K,T\ [\rightleftarrows_{k+1}]\ K',T'$. Then*

(i) *for every S s.t. $T[R]S$ there is S' s.t. $T'[R']S'$ and $K,S\ [\rightleftarrows_k]\ K',S'$;*

(ii) *for every S' s.t. $T'[R']S'$ there is S s.t. $T[R]S$ and $K,S\ [\rightleftarrows_k]\ K',S'$;*

(iii) *$K,S\ [\rightleftarrows_k]\ K',S'$ for $S = R[T]$ and $S' = R'[T']$;*

(iv) *for all $T_1,T_2 \subseteq T$ s.t. $T = T_1 \cup T_2$ there are $T_1',T_2' \subseteq T'$ s.t. $T' = T_1' \cup T_2'$, and $K,T_i\ [\rightleftarrows_{k+1}]\ K',T_i'$ for $i \in \{1,2\}$.*

Proof (i) Assume that $T[R]S$. We define

$$S' := \{v' \in R'[T'] \mid \exists v \in S : K,v \rightleftarrows_k K',v'\}.$$

We will first show that $K,S\ [\rightleftarrows_k]\ K',S'$. By the definition of S', we have $\forall v' \in S' \exists v \in S : K,v \rightleftarrows_k K',v'$. On the other hand, if $v \in S$, then there is $w \in T$ such that wRv. Furthermore, since $K,T\ [\rightleftarrows_{k+1}]\ K',T'$, there is $w' \in T'$ such that $K,w \rightleftarrows_{k+1} K',w'$, whence by the definition of $\rightleftarrows_{k+1}$, there is $v' \in W'$ such that $w'R'v'$ and $K,v \rightleftarrows_k K',v'$. By the definition of S', v' is in S'. Thus we see that $\forall v \in S \exists v' \in S' : K,v \rightleftarrows_k K',v'$.

To see that $T'[R']S'$ holds, note first that $S' \subseteq R'[T']$ by its definition. Assume then that $w' \in T'$. Since $K,T\ [\rightleftarrows_{k+1}]\ K',T'$, there is $w \in T$ such that $K, w \rightleftarrows_{k+1} K', w'$. Furthermore, since $T[R]S$, there is $v \in S$ such that wRv, and consequently there is $v' \in R'[w']$ such that $K, v \rightleftarrows_k K', v'$. By the definition of S' we have now $v' \in S'$. Thus we conclude that $w' \in R'^{-1}[S']$.

(ii) The claim is proved in the same way as (i).

(iii) If $v \in R[T]$, then there is $w \in T$ such that wRv. By the assumption $K, T\ [\rightleftarrows_{k+1}]\ K', T'$, there is $w' \in T'$ such that $K, w \rightleftarrows_{k+1} K', w'$. Hence, there is v' such that $w'R'v'$ and $K, v \rightleftarrows_k K', v'$. As $w'R'v'$, we have $v' \in R'[T']$. Thus, we conclude that $\forall v \in R[T] \exists v' \in R'[T'] : K, v \rightleftarrows_k K', v'$. Using a symmetrical argument, we see that $\forall v' \in R'[T'] \exists v \in R[T] : K, v \rightleftarrows_k K', v'$.

(iv) Let $T_1, T_2 \subseteq T$ be such that $T = T_1 \cup T_2$. Define now

$$T_i' := \{w' \in T' \mid \exists w \in T_i : K, w \rightleftarrows_{k+1} K', w'\},$$

for $i \in \{1, 2\}$. Then by the definition of T_i', $\forall w' \in T_i' \exists w \in T_i : K, w \rightleftarrows_{k+1} K', w'$. On the other hand, if $w \in T_i$, then $w \in T$, whence there is $w' \in T'$ such that $K, w \rightleftarrows_{k+1} K', w'$. By the definition of T_i', then w' is in T_i'. Thus we conclude that $\forall w \in T_i \exists w' \in T_i' : K, w \rightleftarrows_{k+1} K', w'$, as desired. □

3.2 Characterizing the expressive power of $\mathcal{ML}(\varovee)$

Our goal is to prove that definability in $\mathcal{ML}(\varovee)$ can be characterized by downward closure and closure under team k-bisimulation. We already know that all $\mathcal{ML}(\varovee)$-definable classes are downward closed (see Proposition 2.7). The next step is to prove that $\mathcal{ML}(\varovee)$-definable classes are closed under team k-bisimulation for some k.

Theorem 3.4 *Let Φ be a set of proposition symbols, and let $\mathcal{K} \subseteq \mathcal{KT}(\Phi)$. If $\mathcal{K}$ is definable in $\mathcal{ML}(\varovee)$, then there is a $k \in \mathbb{N}$ such that $\mathcal{K}$ is closed under k-bisimulation.*

Proof Assume that $\varphi \in \mathcal{ML}(\varovee)$. We prove by induction on φ that the class $\|\varphi\|$ is closed under k-bisimulation, where $k = \mathrm{md}(\varphi)$.

- Let $\varphi = p \in \Phi$, and assume that $K, T \models \varphi$ and $K, T\ [\rightleftarrows_k]\ K', T'$ for $k = 0$. Then $K, w \models p$ for all $w \in T$, and for each $w' \in T'$ there is $w \in T$ such that $K, w \rightleftarrows_0 K', w'$. Thus, for all $w' \in T'$, $K', w' \models p$, whence $K', T' \models \varphi$.
- The case $\varphi = \neg p$ is similar to the previous one.
- Let $\varphi = \psi \vee \theta$, and assume that $K, T \models \varphi$ and $K, T\ [\rightleftarrows_k]\ K', T'$, where $k = \mathrm{md}(\varphi) = \max\{\mathrm{md}(\psi), \mathrm{md}(\theta)\}$. Then there are $T_1, T_2 \subseteq T$ such that $T = T_1 \cup T_2$, $K, T_1 \models \psi$ and $K, T_2 \models \theta$.

  By Lemma 3.3(iv), there are subteams $T_1', T_2' \subseteq T'$ such that $T' = T_1' \cup T_2'$ and $K, T_i\ [\rightleftarrows_k]\ K', T_i'$ for $i \in \{1, 2\}$, whence $K, T_1\ [\rightleftarrows_m]\ K', T_1'$ and $K, T_2\ [\rightleftarrows_n]\ K', T_2'$, where $m = \mathrm{md}(\psi)$ and $n = \mathrm{md}(\theta)$. By induction hypothesis, $K', T_1' \models \psi$ and $K', T_2' \models \theta$. Thus, we conclude that $K', T' \models \varphi$.
- The cases $\varphi = \psi \wedge \theta$ and $\varphi = \psi \varovee \theta$ are straightforward.
- Let $\varphi = \Diamond\psi$, and assume that $K, T \models \varphi$ and $K, T\ [\rightleftarrows_k]\ K', T'$, where $k = \mathrm{md}(\varphi) = \mathrm{md}(\psi) + 1$. Then there is a team S on K such that $T[R]S$ and $K, S \models \psi$. By Lemma 3.3(i), there is a team S' such that $T'[R']S'$ and

$K, S\ [\rightleftarrows_{k-1}]\ K', S'$. By induction hypothesis, $K', S' \models \psi$, and consequently $K', T' \models \varphi$.

- Let $\varphi = \Box\psi$, and assume that $K, T \models \varphi$ and $K, T\ [\rightleftarrows_k]\ K', T'$, where $k = \mathrm{md}(\varphi) = \mathrm{md}(\psi) + 1$. Then $K, R[T] \models \psi$, and by Lemma 3.3(iii), $K, R[T]\ [\rightleftarrows_{k-1}]\ K', R'[T']$. Thus, by induction hypothesis, $K', R'[T'] \models \psi$, and consequently $K', T' \models \varphi$.

□

Next we prove that downward closure and closure under team k-bisimulation are together a sufficient condition for $\mathcal{ML}(\mathbin{\rotatebox[origin=c]{90}{$\otimes$}})$-definability.

Theorem 3.5 *Let Φ be a finite set of proposition symbols and let $\mathcal{K} \subseteq \mathcal{KT}(\Phi)$. Assume that $\mathcal{K}$ is downward closed and closed under k-bisimulation for some $k \in \mathbb{N}$. Then $\mathcal{K}$ is definable in $\mathcal{ML}(\mathbin{\rotatebox[origin=c]{90}{$\otimes$}})$.*

Proof Let φ be the formula

$$\bigotimes_{(K,T)\in\mathcal{K}} \bigvee_{w\in T} \chi^k_{K,w},$$

where $\chi^k_{M,w}$ is the k-th Hintikka-formula of the pair (K, w). Note that since Φ is finite, there are only finitely many different Hintikka-formulas $\chi^k_{K,w}$. Thus, the disjunction $\bigvee_{w\in T}$ and the intuitionistic disjunction $\bigotimes_{(K,T)\in\mathcal{K}}$ in φ are essentially finite, whence $\varphi \in \mathcal{ML}(\mathbin{\rotatebox[origin=c]{90}{$\otimes$}})$. We will now prove that φ defines $\mathcal{K}$.

Assume first that $(K_0, T_0) \in \mathcal{K}$. By Proposition 2.5, $K_0, \{v\} \models \chi^k_{K_0,v}$ for each $v \in T_0$. Thus, $K_0, T_0 \models \bigvee_{w\in T_0} \chi^k_{K_0,w}$, and consequently, $K_0, T_0 \models \varphi$.

Assume for the other direction that $K_0, T_0 \models \varphi$. Then there is a pair $(K, T) \in \mathcal{K}$ such that $K_0, T_0 \models \bigvee_{w\in T} \chi^k_{K,w}$. Thus, there are subsets T_w, $w \in T$, of T_0 such that $T_0 = \bigcup_{w\in T} T_w$, and $K_0, T_w \models \chi^k_{K,w}$. By Proposition 2.5, $K_0, v \models \chi^k_{K,w}$ for every $v \in T_w$. Let $T' = \{w \in T \mid T_w \neq \emptyset\}$. Since $\mathcal{K}$ is downward closed, we have $(K, T') \in \mathcal{K}$. Observe now that for every $v \in T_0$ there is $w \in T'$ such that $K_0, v \models \chi^k_{K,w}$, and for every $w \in T'$ there is $v \in T_0$ such that $K_0, v \models \chi^k_{K,w}$. By Proposition 2.10 this means that $K, T'\ [\rightleftarrows_k]\ K_0, T_0$. Since $\mathcal{K}$ is closed under k-bisimulation, we conclude that $(K_0, T_0) \in \mathcal{K}$. □

Putting Proposition 2.7, Theorem 3.4 and Theorem 3.5 together, we finally get the promised characterization for the expressive power of $\mathcal{ML}(\mathbin{\rotatebox[origin=c]{90}{$\otimes$}})$.

Corollary 3.6 *A class $\mathcal{K} \subseteq \mathcal{KT}(\Phi)$ is definable in $\mathcal{ML}(\mathbin{\rotatebox[origin=c]{90}{$\otimes$}})$ if and only if $\mathcal{K}$ is downward closed and there exists $k \in \mathbb{N}$ such that $\mathcal{K}$ is closed under k-bisimulation.*

Note that from the proof of Theorem 3.5 we obtain the following normal form for $\mathcal{ML}(\mathbin{\rotatebox[origin=c]{90}{$\otimes$}})$-formulas: every formula $\varphi \in \mathcal{ML}(\mathbin{\rotatebox[origin=c]{90}{$\otimes$}})$ is equivalent with a formula of the form $\bigotimes \Psi$, where Ψ is a finite set of $\mathcal{ML}$-formulas. This normal form was proved in [12], but the idea goes back to [15]. Note further that each formula in Ψ can be assumed to be a disjunction of Hintikka formulas $\chi^k_{K,w}$, where k is the modal depth of φ.

4 $\mathcal{EMDL}$ is equivalent to $\mathcal{ML}(\varovee)$

By Proposition 2.6, we know that $\mathcal{ML}(\varovee)$ is at least as expressive as $\mathcal{EMDL}$. In this section we show that the converse is also true, thus solving the problem that was left open in [3].

Theorem 4.1 $\mathcal{ML}(\varovee) \leq \mathcal{EMDL}$.

The proof we give for Theorem 4.1 is an adaptation of the proof in [18] of the corresponding result for propositional logic with intuitionistic disjunction and propositional dependence atoms. The main idea (Lemma 4.3) is originally due to Taneli Huuskonen.

Before proving Theorem 4.1, we introduce some auxiliary concepts, and prove a couple of lemmas concerning them.

Let Ψ be a finite set of $\mathcal{ML}(\Phi)$-formulas, and let K be a Kripke model over Φ and w a state in K. The Ψ-*type* of w in K is defined as

$$\mathrm{tp}_\Psi(K, w) := \{\psi \in \Psi \mid K, w \models \psi\}.$$

Furthermore, the Ψ-*type* of a team T of K is just the set of Ψ-types of its elements:

$$\mathrm{Tp}_\Psi(K, T) := \{\mathrm{tp}_\Psi(K, w) \mid w \in T\}.$$

Each Ψ-type $\Gamma \subseteq \Psi$ can be defined by a formula: Let

$$\theta_\Gamma := \bigwedge_{\psi \in \Gamma} \psi \wedge \bigwedge_{\psi \in \Psi \setminus \Gamma} \psi^\neg$$

where $\psi^\neg$ denotes the formula obtained from $\neg\psi$ by pushing the negations in front of proposition symbols. Then it is easy to see that $\mathrm{tp}_\Psi(K, w) = \Gamma$ if and only if $K, w \models \theta_\Gamma$.

Lemma 4.2 *Assume that* $(K, T), (K', T') \in \mathcal{KT}(\Phi)$, *and let* Ψ *be a finite set of* $\mathcal{ML}(\Phi)$-*formulas.*

(i) *For each* $\psi \in \Psi$, $K, T \models \psi$ *if and only if* $\psi \in \bigcap \mathrm{Tp}_\Psi(K, T)$.

(ii) *If* $K, T \models \bigovee \Psi$ *and* $\mathrm{Tp}_\Psi(K', T') \subseteq \mathrm{Tp}_\Psi(K, T)$, *then* $K', T' \models \bigovee \Psi$.

Proof (i) If $K, T \models \psi$, then by Proposition 2.5, $K, w \models \psi$ for every $w \in T$, which means that $\psi \in \mathrm{tp}_\Psi(K, w)$ for every $w \in T$. On the other hand, if $\psi \in \bigcap \mathrm{Tp}_\Psi(K, T)$, then $K, w \models \psi$ for every $w \in T$. By Proposition 2.5, it follows that $K, T \models \psi$.

(ii) Assume that $K, T \models \bigovee \Psi$ and $\mathrm{Tp}_\Psi(K', T') \subseteq \mathrm{Tp}_\Psi(K, T)$. Thus, $K, T \models \psi$ for some $\psi \in \Psi$, and by claim (i), $\psi \in \bigcap \mathrm{Tp}_\Psi(K, T)$. Since $\mathrm{Tp}_\Psi(K', T') \subseteq \mathrm{Tp}_\Psi(K, T)$, it follows that $\psi \in \bigcap \mathrm{Tp}_\Psi(K', T')$. Thus, $K', T' \models \psi$, and consequently $K', T' \models \bigovee \Psi$. □

Consider next the formula $\gamma := \bigwedge_{\psi \in \Psi} {=}(\psi)$. It says that the truth value of each ψ in Ψ is constant, whence $K, T \models \gamma$ if and only if $|\mathrm{Tp}_\Psi(K, T)| \leq 1$. Define now recursively

$$\gamma^0 := p \wedge \neg p, \qquad \gamma^{k+1} := (\gamma^k \vee \gamma)$$

It is straightforward to show by induction that for all $k \in \mathbb{N}$, $K, T \models \gamma^k$ if and only if $|\mathrm{Tp}_\Psi(K,T)| \leq k$.

Lemma 4.3 *Let Ψ be a finite set of $\mathcal{ML}(\Phi)$-formulas. If $(K,T) \in \mathcal{KT}(\Phi)$, $T \neq \emptyset$, then there is a formula $\xi_{K,T} \in \mathcal{EMDL}(\Phi)$ such that for every $(K',T') \in \mathcal{KT}(\Phi)$*

$$K', T' \models \xi_{K,T} \quad \Leftrightarrow \quad \mathrm{Tp}_\Psi(K,T) \not\subseteq \mathrm{Tp}_\Psi(K',T').$$

Proof Let $|\mathrm{Tp}_\Psi(K,T)| = k+1$. We define

$$\xi_{K,T} := \Big(\bigvee_{\Gamma \in X} \theta_\Gamma\Big) \vee \gamma^k,$$

where $X = \mathcal{P}(\Psi) \setminus \mathrm{Tp}_\Psi(K,T)$. Now given a pair $(K',T') \in \mathcal{KT}(\Phi)$ we have

$$\begin{aligned} K', T' \models \xi_{K,T} &\Leftrightarrow \text{there are } T_1, T_2 \text{ such that } T_1 \cup T_2 = T' \text{ and} \\ &\qquad \mathrm{Tp}_\Psi(K',T_1) \subseteq X \text{ and } |\mathrm{Tp}_\Psi(K',T_2)| \leq k \\ &\Leftrightarrow |\mathrm{Tp}_\Psi(K,T) \cap \mathrm{Tp}_\Psi(K',T')| \leq k \\ &\Leftrightarrow \mathrm{Tp}_\Psi(K,T) \not\subseteq \mathrm{Tp}_\Psi(K',T'). \end{aligned}$$

□

Proof of Theorem 4.1. Let φ be an $\mathcal{ML}(\oslash)(\Phi)$-formula. By the normal form derived in the proof of Theorem 3.5, we may assume that φ is of the form $\bigoslash \Psi$, where Ψ is a finite set of $\mathcal{ML}(\Phi)$-formulas.

Let η be the formula

$$\bigwedge_{(K,T) \in \overline{\|\varphi\|}} \xi_{K,T},$$

where $\overline{\|\varphi\|} = \mathcal{KT}(\Phi) \setminus \|\varphi\|$ and $\xi_{K,T}$ is as in Lemma 4.3. Since Ψ is finite, there are finitely many different formulas of the form $\xi_{K,T}$. Thus, the conjunction in η is essentially finite, and hence η is in $\mathcal{EMDL}$.

To prove that $\|\eta\| = \|\varphi\|$, let $(K_0,T_0) \in \mathcal{KT}(\Phi)$. Assume first that $(K_0,T_0) \in \|\varphi\|$, and consider any pair $(K,T) \in \overline{\|\varphi\|}$. It follows from Lemma 4.2 that $\mathrm{Tp}_\Psi(K,T) \not\subseteq \mathrm{Tp}_\Psi(K_0,T_0)$, whence by Lemma 4.3, $K_0, T_0 \models \xi_{K,T}$. Thus we see that $(K_0,T_0) \in \|\eta\|$.

Assume then that $(K_0,T_0) \notin \|\varphi\|$. Since $\mathrm{Tp}_\Psi(K_0,T_0) \subseteq \mathrm{Tp}_\Psi(K_0,T_0)$, it follows from Lemma 4.3 that $K_0, T_0 \not\models \xi_{K_0,T_0}$. Thus we conclude that $(K_0,T_0) \notin \|\eta\|$. □

Combining Proposition 2.6 and Theorem 4.1, we see that the expressive power of $\mathcal{EMDL}$ and $\mathcal{ML}(\oslash)$ coincide. This means that the characterization for the expressive power of $\mathcal{ML}(\oslash)$ given in Corollary 3.6 is true for $\mathcal{EMDL}$, too.

Corollary 4.4 $\mathcal{EMDL} \equiv \mathcal{ML}(\oslash)$.

Corollary 4.5 *A class $\mathcal{K} \subseteq \mathcal{KT}(\Phi)$ is definable in $\mathcal{EMDL}$ if and only if $\mathcal{K}$ is downward closed and there is a $k \in \mathbb{N}$ such that $\mathcal{K}$ is closed under k-bisimulation.*

5 Dimensions for modal formulas

In this section we introduce two semantical invariants for formulas of $\mathcal{EMDL}$ and $\mathcal{ML}(\varovee)$. We will will first show that the truth of a formula φ in a team T of a Kripke model K can be determined by considering only subteams $T' \subseteq T$ of a fixed size n; we define the *lower dimension* of φ to be the least n such that this holds. Thus, lower dimension is a natural measure that can be used for classifying formulas with respect to their semantical complexity. We also believe that lower dimension can be useful in analyzing the computational complexity of the model checking problem of modal formulas.

The other semantical invariant we introduce, the *upper dimension* of a formula φ, is defined as the largest number of maximal teams T that satisfy φ in any single Kripke model K. We will show that the lower dimension of φ is always less than or equal to the upper dimension. Moreover, we will show that the upper dimension admits well-behaved estimates that are defined compositionally. These estimates are very useful in establishing upper bounds for lower dimension as well, since finding good estimates for the lower dimension directly is not straightforward.

As we proved in the previous section, the expressive power of $\mathcal{EMDL}$ and $\mathcal{ML}(\varovee)$ coincide. However, there can be a considerable difference in the sizes of equivalent formulas under any translation. It was already pointed out in [3] that there is an intrinsic difference in the complexity of $\mathcal{EMDL}$ and $\mathcal{ML}(\varovee)$: the satisfiability problem for the former is NEXP-complete ([3]), while for the latter it is PSPACE-complete ([15]). This strongly hints to the possibility that there is no polynomially bounded translation from $\mathcal{EMDL}$ to $\mathcal{ML}(\varovee)$. Using the upper dimension, we will prove that this is indeed the case: any translation from $\mathcal{EMDL}$ to $\mathcal{ML}(\varovee)$ introduces an exponential blow-up for the size of formulas.

5.1 Lower and upper dimension

Let φ be a formula in $\mathcal{ML}(\varovee)(\Phi)$, and let $n \in \mathbb{N}$. Adapting a notion that was introduced by Jarmo Kontinen in [11] for first-order dependence logic, we say that φ is *n-coherent* if the condition

$$K, T \models \varphi \Leftrightarrow K, T' \models \varphi \text{ for all } T' \subseteq T \text{ such that } |T'| \leq n$$

holds for all $(K, T) \in \mathcal{KT}(\Phi)$.

It follows from Corollary 3.6 that for every $\mathcal{ML}(\varovee)(\Phi)$-formula φ there is a natural number n such that φ is n-coherent. This can be seen as follows: Let $k \in \mathbb{N}$ be such that $\|\varphi\|$ is closed under team k-bisimulation, and let n be the number of $\rightleftarrows_k$-equivalence classes of pointed Φ-models (K, w). If $K, T \models \varphi$, then by downward closure, $K, T' \models \varphi$ for every subteam $T' \subseteq T$. On the other hand, if $K, T \not\models \varphi$, then $K, T' \not\models \varphi$ for any subteam T' of T such that for every $w \in T$ there is $w' \in T'$ with $K, w \rightleftarrows_k K, w'$. Clearly there is such a subteam T' with $|T'| \leq n$.

Intuitively, the lower dimension of a formula $\varphi \in \mathcal{ML}(\varovee)(\Phi)$ can be defined as the least n such that φ is n-coherent. However, due to technical reasons, we formulate the definition of lower dimension in a bit different, but equivalent way. Given a Kripke model K over Φ, let $N(\varphi, K)$ denote the family of minimal teams T of K such that $T \notin \|\varphi\|^K$.

Definition 5.1 Let $\varphi \in \mathcal{ML}(\oslash)(\Phi)$. The *lower dimension* $\dim(\varphi)$ of φ is the least $n \in \mathbb{N}$ such that for every Kripke model K over Φ and every $T \in N(\varphi, K)$ we have $|T| \leq n$.

We will next define the upper dimension for $\mathcal{ML}(\oslash)$-formulas. Let K be a Kripke model over Φ and let φ an $\mathcal{ML}(\oslash)(\Phi)$-formula. As $\|\varphi\|^K$ is downward closed, it is natural to study the family $M(\varphi, K)$ consisting of maximal elements of $\|\varphi\|^K$. We will see below that $\|\varphi\|^K$ is *generated* by $M(\varphi, K)$ in the sense that every team $T \in \|\varphi\|^K$ is contained in some team $S \in M(\varphi, K)$.

Definition 5.2 Let $\varphi \in \mathcal{ML}(\oslash)(\Phi)$. The *upper dimension* $\mathrm{Dim}(\varphi)$ of φ is the least $m \in \mathbb{N}$ such that for every Kripke model K over Φ we have $|M(\varphi, K)| \leq m$.

Note that it is not a priori clear that the upper dimension is *well-defined*: if there is no uniform bound $m \in \mathbb{N}$ for the size of $M(\varphi, K)$ over all Kripke models K, then $\mathrm{Dim}(\varphi)$ does not exist. In particular, the definition of $\mathrm{Dim}(\varphi)$ requires that $\|\varphi\|^K$ is always *finitely generated by* $M(\varphi, K)$, i.e., that $M(\varphi, K)$ is finite and generates $\|\varphi\|^K$ for all K.

Lemma 5.3 $\mathrm{Dim}(\varphi)$ *is well-defined for all* $\varphi \in \mathcal{ML}(\oslash)(\Phi)$. *Moreover, we have the following estimates for* $\varphi, \psi \in \mathcal{ML}(\oslash)(\Phi)$:

(i) $\mathrm{Dim}(p) = \mathrm{Dim}(\neg p) = 1$.

(ii) $\mathrm{Dim}(\varphi \wedge \psi) \leq \mathrm{Dim}(\varphi)\,\mathrm{Dim}(\psi)$.

(iii) $\mathrm{Dim}(\varphi \vee \psi) \leq \mathrm{Dim}(\varphi)\,\mathrm{Dim}(\psi)$.

(iv) $\mathrm{Dim}(\varphi \oslash \psi) \leq \mathrm{Dim}(\varphi) + \mathrm{Dim}(\psi)$.

(v) $\mathrm{Dim}(\Diamond\varphi) \leq \mathrm{Dim}(\varphi)$.

(vi) $\mathrm{Dim}(\Box\varphi) \leq \mathrm{Dim}(\varphi)$.

Proof We prove the first claim and the dimension estimates simultaneously by induction on φ. Let $K = (W, R, V)$ be an arbitrary Kripke model over Φ. We omit the cases for (i), (iii) and (vi), since (i) is trivial, and (iii) and (vi) are analogous to (ii) and (v), respectively.

(ii) We first notice that $\|\varphi \wedge \psi\|^K = \|\varphi\|^K \cap \|\psi\|^K$. By induction hypothesis, $\|\varphi\|^K$ and $\|\psi\|^K$ are finitely generated by $M(\varphi, K)$ and $M(\psi, K)$, respectively. Moreover, $|M(\varphi, K)| \leq \mathrm{Dim}(\varphi)$ and $|M(\psi, K)| \leq \mathrm{Dim}(\psi)$. It is immediate that $M(\varphi \wedge \psi, K) \subseteq \{T \cap U \mid T \in M(\varphi, K), U \in M(\psi, K)\}$.

Clearly, by the induction hypothesis the right-hand side of the inclusion above also generates the family $\|\varphi \wedge \psi\|^K$. The inclusion now implies $|M(\varphi \wedge \psi, K)| \leq |M(\varphi, K) \times M(\psi, K)| \leq \mathrm{Dim}(\varphi)\,\mathrm{Dim}(\psi)$. Hence, $\mathrm{Dim}(\varphi \wedge \psi) \leq \mathrm{Dim}(\varphi)\,\mathrm{Dim}(\psi)$.

(iv) For the intuitionistic disjunction, it holds that

$$M(\varphi \oslash \psi, K) \subseteq M(\varphi, K) \cup M(\psi, K)$$

and the right-hand side of the inclusion generates the family $\|\varphi \oslash \psi\|^K$. The dimension estimate follows immediately.

(v) For the diamond, we have that $M(\Diamond\psi, K) \subseteq \{R^{-1}[T] \mid T \in M(\varphi, K)\}$, and that $\{R^{-1}[T] \mid T \in M(\varphi, K)\}$ generates $\|\Diamond\psi\|^K$. Thus we get that $|M(\Diamond\psi, K)| \leq |M(\varphi, K)|$, which implies that $\mathrm{Dim}(\Diamond\varphi) \leq \mathrm{Dim}(\varphi)$.

□

Remark 5.4 In [1], Ciardelli gave estimates, that he calls *Groenendijk's inequalities*, for the size of *inquisitive meanings* of formulas. These estimates are essentially equivalent to (i), (ii) and (iv) above. In addition, he gave a similar estimate for the case of (intuitionistic) implication.

The estimates given in Lemma 5.3 are sharp in the sense that we cannot improve the upper bounds. For conjunction (and implicitly also for the intuitionistic disjunction), the following example demonstrates this sharpness.

Example 5.5 Let m and n be positive integers. We show that there are $\varphi, \psi \in \mathcal{ML}(\otimes)$ such that $\mathrm{Dim}(\varphi) = m$, $\mathrm{Dim}(\psi) = n$ and $\mathrm{Dim}(\varphi \wedge \psi) = mn$. Let $p_0, \ldots, p_{m-1}, q_0, \ldots, q_{n-1}$ be distinct propositional symbols. Put

$$\varphi_i := p_i \wedge \bigwedge_{k<m, k\neq i} \neg p_k \quad \text{and} \quad \psi_j := q_j \wedge \bigwedge_{l<n, l\neq i} \neg q_l,$$

for $i < m$ and $j < n$. Note that the formulas φ_i, $i < m$, are satisfiable, but mutually contradictory in the classical sense, and similarly for ψ_j's. If $K = (W, R, V)$ is a Kripke model over $\{p_0, \ldots, p_{m-1}, q_0, \ldots, q_{n-1}\}$, then

$$\|\varphi_i\|^K = \mathcal{P}(T_i) \quad \text{and} \quad \|\varphi_j\|^K = \mathcal{P}(U_j)$$

for appropriate teams T_i and U_j. Clearly we can pick K such that the intersections $T_i \cap U_j$ are all non-empty, for $i < m$ and $j < n$. Define

$$\varphi := \bigotimes_{i<m} \varphi_i \quad \text{and} \quad \psi := \bigotimes_{j<n} \psi_j.$$

The previous lemma gives the estimates $\mathrm{Dim}(\varphi) \leq m$ and $\mathrm{Dim}(\psi) \leq n$ for the upper dimensions. However, in the Kripke model we have chosen,

$$\|\varphi\|^K = \bigcup_{i<m} \mathcal{P}(T_i) \quad \text{and} \quad \|\psi\|^K = \bigcup_{j<n} \mathcal{P}(U_j),$$

so $M(\varphi, K) = \{T_0, \ldots T_{m-1}\}$ and $M(\psi, K) = \{U_0, \ldots, U_{n-1}\}$, which implies $\mathrm{Dim}(\varphi) = m$ and $\mathrm{Dim}(\psi) = n$. Consider now the sentence $\varphi \wedge \psi$. We have

$$\|\varphi \wedge \psi\|^K = \bigcap_{i<m, j<n} \mathcal{P}(T_i \cap U_j),$$

so $M(\varphi \wedge \psi, K) = \{T_i \cap U_j \mid i < m, j < n\}$. Consequently, $\mathrm{Dim}(\varphi \wedge \psi) = mn$.

We will now prove that the upper dimension $\mathrm{Dim}(\varphi)$ is always a uniform upper bound for $|N(\varphi, K)|$, whence $\dim(\varphi)$ is less than or equal to $\mathrm{Dim}(\varphi)$.

Lemma 5.6 *Assume that* $\varphi \in \mathcal{ML}(\otimes)(\Phi)$. *Then* $\dim(\varphi) \leq \mathrm{Dim}(\varphi)$.

Proof Let K be a Kripke model, and let $U \in N(\varphi, K)$. We need to prove that $|U| \leq \mathrm{Dim}(\varphi)$ (if there are no such sets U, there is nothing to prove). For each $T \in M(\varphi, K)$, pick a state $w_T \in U \backslash T$. Then the set $U_0 = \{w_T \mid T \in M(\varphi, K)\}$ is a subset of U, but not included in any $T \in M(\varphi, K)$. Hence, $U_0 \in N(\varphi, K)$ and by the minimality of U, we get $U = U_0$ and $|U| = |U_0| \leq |M(\varphi, K)| \leq \mathrm{Dim}(\varphi)$. Hence, $\dim(\varphi) \leq \mathrm{Dim}(\varphi)$. □

The next example shows that the gap between upper and lower dimension may be arbitrarily large.

Example 5.7 For $j < n$, let the formulas ψ_j, as well as the Kripke model K and sets U_j, be as in Example 5.5, Assume that $n \geq 4$. To simplify notation, write $\psi_n = \psi_0$ and $U_n = U_0$. Consider the sentence

$$\theta := \bigotimes_{j<n} (\psi_j \vee \psi_{j+1}).$$

Lemma 5.3 gives the estimate $\mathrm{Dim}(\theta) \leq n$. In the Kripke model K, it is easy to see that $M(\theta, K) = \{U_j \cup U_{j+1} \mid j < n\}$. Hence, $\mathrm{Dim}(\theta) = n$. However, if a team T is such that $K, T \not\models \theta$, then there is either a single point $w \in T$ such that $K, \{w\} \not\models \theta$, or there are $w \in U_j$, $w' \in U_k$ with $j \not\equiv k \pmod{n}$. In the latter case, $K, \{w, w'\} \not\models \theta$. The same reasoning applies to other Kripke models than K, so $\dim(\theta) = 2$.

5.2 The dimension of dependence atoms

As $\mathcal{EMDL} \equiv \mathcal{ML}(\otimes)$ and the definition of the upper and lower dimensions is purely semantical, $\mathrm{Dim}(\varphi)$ and $\dim(\varphi)$ are defined for every $\mathcal{EMDL}$-formula φ. Moreover, the estimates given in Lemma 5.3 are valid also for $\mathcal{EMDL}$-formulas. For the modal dependence atoms, we have the following estimate for the upper dimension:

Lemma 5.8 *For the dependence atoms of* $\mathcal{EMDL}(\Phi)$, *we have that* $\mathrm{Dim}(=(\psi_1, \ldots, \psi_n, \theta)) \leq 2^{2^n}$. *Moreover, equality holds if* ψ_i, $1 \leq i \leq n$, *and* θ *are distinct proposition symbols.*

Proof Denote the set $\{\psi_1, \ldots, \psi_n\}$ by Ψ and the dependence atom $=(\psi_1, \ldots, \psi_n, \theta)$ by φ. let $K = (W, R, V)$ be a Kripke model over Φ, and let $X = \{\mathrm{tp}_\Psi(K, w) \mid w \in W\}$, where $\mathrm{tp}_\Psi(K, w)$ is the Ψ-type of w in K (see Section 4). If $T \in M(\varphi, K)$, then there is a function $f_T : X \to \{\bot, \top\}$ such that for all $w \in W$

$$M, w \models \theta \quad \Leftrightarrow \quad f_T(\mathrm{tp}_\Psi(K, w)) = \top.$$

If T and U are different elements of $M(\varphi, K)$, then $T \cup U \notin \|\varphi\|^K$, whence there are states $w \in T$ and $u \in U$ such that $\mathrm{tp}_\Psi(K, w) = \mathrm{tp}_\Psi(K, u)$, but $K, w \models \theta \Leftrightarrow K, u \not\models \theta$. This means that $f_T \neq f_U$. Thus, we see that $M(\varphi, K)$ has at most $2^{|X|}$ elements. Since $X \subseteq \mathcal{P}(\Psi)$ and $|\Psi| = n$, we arrive at the upper bound 2^{2^n} for $|M(\varphi, K)|$.

For the second claim, note that if $\psi_i \in \Phi$, $1 \leq i \leq n$, and $\theta \in \Phi$ are distinct, then there is a Kripke model such that every $\Gamma \subseteq \Psi$ is the Ψ-type of some w in K, and for every $f : X \to \{\bot, \top\}$ there is a team $T \in M(\varphi, K)$ such that $f = f_T$. Then $|X| = 2^n$, and hence $|M(\varphi, K)| = 2^{|X|} = 2^{2^n}$. □

Thus, the upper dimension of dependence atoms can be doubly exponential with respect to the number of formulas occurring in it. On the other hand, any $\mathcal{ML}(\varovee)$-formula can reach only single exponential upper dimension with respect to its size. We prove this by considering the number $\mathrm{occ}_{\varovee}(\varphi)$ of occurrences of $\varovee$-symbols in the formula φ.

Proposition 5.9 *Let* $\varphi \in \mathcal{ML}(\varovee)$. *Then* $\mathrm{Dim}(\varphi) \leq 2^{\mathrm{occ}_{\varovee}(\varphi)}$.

Proof The proof is a straightforward application of Lemma 5.3 and induction. For the literals, we have

$$\mathrm{Dim}(p) = \mathrm{Dim}(\neg p) = 1 = 2^0 = 2^{\mathrm{occ}_{\varovee}(p)} = 2^{\mathrm{occ}_{\varovee}(\neg p)}.$$

Suppose $\mathrm{Dim}(\varphi) \leq 2^{\mathrm{occ}_{\varovee}(\varphi)}$ and $\mathrm{Dim}(\psi) \leq 2^{\mathrm{occ}_{\varovee}(\psi)}$. Then

$$\begin{aligned}
\mathrm{Dim}(\varphi \wedge \psi) &\leq \mathrm{Dim}(\varphi) \cdot \mathrm{Dim}(\psi) \\
&\leq 2^{\mathrm{occ}_{\varovee}(\varphi)} \cdot 2^{\mathrm{occ}_{\varovee}(\psi)} = 2^{\mathrm{occ}_{\varovee}(\varphi)+\mathrm{occ}_{\varovee}(\psi)} = 2^{\mathrm{occ}_{\varovee}(\varphi\wedge\psi)}, \\
\mathrm{Dim}(\varphi \vee \psi) &\leq \mathrm{Dim}(\varphi) \cdot \mathrm{Dim}(\psi) \leq 2^{\mathrm{occ}_{\varovee}(\varphi)} \cdot 2^{\mathrm{occ}_{\varovee}(\psi)} = 2^{\mathrm{occ}_{\varovee}(\varphi\vee\psi)} \text{ and} \\
\mathrm{Dim}(\varphi \varovee \psi) &\leq \mathrm{Dim}(\varphi) + \mathrm{Dim}(\psi) \leq 2^{\mathrm{occ}_{\varovee}(\varphi)} + 2^{\mathrm{occ}_{\varovee}(\psi)} \\
&\leq 2^{\mathrm{occ}_{\varovee}(\varphi)} \cdot 2^{\mathrm{occ}_{\varovee}(\psi)} + 1 \leq 2^{\mathrm{occ}_{\varovee}(\varphi)} \cdot 2^{\mathrm{occ}_{\varovee}(\psi)} \cdot 2 \\
&= 2^{\mathrm{occ}_{\varovee}(\varphi)+\mathrm{occ}_{\varovee}(\psi)+1} = 2^{\mathrm{occ}_{\varovee}(\varphi \varovee \psi)}.
\end{aligned}$$

The case of the modal operators is trivial. □

Theorem 5.10 *Assume that* $\varphi \in \mathcal{ML}(\varovee)$ *is a formula such that* $\|\varphi\| = \|{=}(p_1, \ldots, p_n, q)\|$. *Then* φ *contains more than* 2^n *symbols.*

Proof By Lemma 5.8, $\mathrm{Dim}(\varphi) = \mathrm{Dim}({=}(p_1, \ldots, p_n, q)) = 2^{2^n}$. Thus, by Proposition 5.9, $2^{2^n} \leq 2^{\mathrm{occ}_{\varovee}(\varphi)}$ implying $2^n \leq \mathrm{occ}_{\varovee}(\varphi)$. This means that φ contains at least 2^n intuitionistic disjunction symbols. □

Thus, any translation from $\mathcal{EMDL}$ to $\mathcal{ML}(\varovee)$ necessarily leads to an exponential blow-up in the size of formulas.

6 Summary

We studied the expressive power of various modal logics with team semantics: modal logic with intuitionistic disjunction $\mathcal{ML}(\varovee)$, modal dependence logic $\mathcal{MDL}$, and extended modal dependence logic $\mathcal{EMDL}$. We introduced the notion of team bisimulation and showed that a class $\mathcal{K}$ of Kripke structures with teams is definable by a sentence of $\mathcal{ML}(\varovee)$ if and only if $\mathcal{K}$ is downward closed and closed under team k-bisimulation. In addition, we established that the expressive power of $\mathcal{ML}(\varovee)$ and $\mathcal{EMDL}$ coincide and thus answered an open problem from [3]. Furthermore, we introduced novel semantical invariants for formulas of $\mathcal{EMDL}$ and $\mathcal{ML}(\oslash)$, i.e., the notions of upper and lower dimension. By using these invariants, we obtained that the translations from $\mathcal{MDL}$ and $\mathcal{EMDL}$ into $\mathcal{ML}(\oslash)$ are always worst-case exponential.

The characterization of the expressive power of $\mathcal{EMDL}$ and $\mathcal{ML}(\oslash)$ gives rise to the question whether similar characterizations can be found for other modal logics with team semantics. In particular, is there such a characterization for the extension of $\mathcal{ML}$ with inclusion atoms or independence atoms? For the definitions of these atoms, see the Ph.D. thesis [18] of Fan Yang.

References

[1] Ciardelli, I., "Inquisitive Semantics and Intermediate Logics," Master's thesis, University of Amsterdam (2009).

[2] Ciardelli, I. and F. Roelofsen, *Inquisitive logic*, J. Philosophical Logic **40** (2011), pp. 55–94.

[3] Ebbing, J., L. Hella, A. Meier, J.-S. Müller, J. Virtema and H. Vollmer, *Extended modal dependence logic*, in: *WoLLIC*, 2013, pp. 126–137.

[4] Ebbing, J. and P. Lohmann, *Complexity of model checking for modal dependence logic*, in: M. Bieliková, G. Friedrich, G. Gottlob, S. Katzenbeisser and G. Turán, editors, *SOFSEM*, Lecture Notes in Computer Science **7147** (2012), pp. 226–237.

[5] Ebbing, J., P. Lohmann and F. Yang, *Model checking for modal intuitionistic dependence logic*, in: G. Bezhanishvili, S. Löbner, V. Marra and F. Richter, editors, *Logic, Language, and Computation*, Lecture Notes in Computer Science **7758**, Springer, 2013 pp. 231–256.

[6] Galliani, P., *The dynamification of modal dependence logic*, Journal of Logic, Language and Information **22** (2013), pp. 269–295.

[7] Goranko, V. and M. Otto, *Model theory of modal logic*, in: P. Blackburn, J. Van Benthem and F. Wolter, editors, *Handbook of Modal Logic*, Studies in Logic and Practical Reasoning **3**, Elsevier, 2007 pp. 249–329.

[8] Groenendijk, J., *Inquisitive semantics: Two possibilities for disjunction*, in: P. Bosch, D. Gabelaia and J. Lang, editors, *TbiLLC*, Lecture Notes in Computer Science **5422** (2007), pp. 80–94.

[9] Hintikka, J. and G. Sandu, *Informational independence as a semantical phenomenon*, in: *Logic, methodology and philosophy of science, VIII (Moscow, 1987)*, Stud. Logic Found. Math. **126**, North-Holland, Amsterdam, 1989 pp. 571–589.

[10] Hodges, W., *Compositional semantics for a language of imperfect information*, Logic Journal of the IGPL **5** (1997), pp. 539–563.

[11] Kontinen, J., "Coherence and Complexity in Fragments of Dependence Logic," Ph.D. thesis, University of Amsterdam (2010).

[12] Lohmann, P. and H. Vollmer, *Complexity results for modal dependence logic*, Studia Logica **101** (2013), pp. 343–366.

[13] Müller, J.-S. and H. Vollmer, *Model checking for modal dependence logic: An approach through post's lattice*, in: L. Libkin, U. Kohlenbach and R. Queiroz, editors, *Logic, Language, Information, and Computation*, Lecture Notes in Computer Science **8071**, 2013 pp. 238–250.

[14] Sano, K., *First-order inquisitive pair logic*, in: M. Banerjee and A. Seth, editors, *Logic and Its Applications*, Lecture Notes in Computer Science **6521**, 2011 pp. 147–161.

[15] Sevenster, M., *Model-theoretic and computational properties of modal dependence logic*, J. Log. Comput. **19** (2009), pp. 1157–1173.

[16] Väänänen, J., "Dependence Logic - A New Approach to Independence Friendly Logic," London Mathematical Society student texts **70**, Cambridge University Press, 2007.

[17] Väänänen, J., *Modal dependence logic*, in: K. R. Apt and R. van Rooij, editors, *New Perspectives on Games and Interaction*, Texts in Logic and Games **4**, 2008 pp. 237–254.

[18] Yang, F., "On Extensions and Variants of Dependence Logic," Ph.D. thesis, University of Helsinki (2014).

Partiality and Adjointness in Modal Logic

Wesley H. Holliday

Department of Philosophy &
Group in Logic and the Methodology of Science
University of California, Berkeley

Abstract

Following a proposal of Humberstone, this paper studies a semantics for modal logic based on partial "possibilities" rather than total "worlds." There are a number of reasons, philosophical and mathematical, to find this alternative semantics attractive. Here we focus on the construction of possibility models with a finitary flavor. Our main completeness result shows that for a number of standard modal logics, we can build a canonical possibility model, wherein every logically consistent formula is satisfied, by simply taking each *individual finite formula* (modulo equivalence) to be a possibility, rather than each infinite maximally consistent set of formulas as in the usual canonical world models. Constructing these locally finite canonical models involves solving a problem in general modal logic of independent interest, related to the study of *adjoint* pairs of modal operators: for a given modal logic $\mathbf{L}$, can we find for every formula φ a formula $\mathsf{f}_a^{\mathbf{L}}(\varphi)$ such that for every formula ψ, $\varphi \to \Box_a\psi$ is provable in $\mathbf{L}$ if and only if $\mathsf{f}_a^{\mathbf{L}}(\varphi) \to \psi$ is provable in $\mathbf{L}$? We answer this question for a number of standard modal logics, using model-theoretic arguments with world semantics. This second main result allows us to build for each logic a canonical possibility model out of the lattice of formulas related by provable implication in the logic.

Keywords: possibility semantics, adjointness, completeness, canonical models.

1 Introduction

Humberstone [17] has proposed a semantics for modal logics based on partial "possibilities" rather than total "worlds." One difference between possibility models and world models is that each possibility provides a partial assignment of truth values to atomic sentences, which may leave the truth values of some atomic sentences indeterminate. Unlike standard three-valued semantics, however, Humberstone's semantics still leads to a classical logic because the connectives $\neg$, $\vee$, and $\to$ quantify over *refinements* of the current possibility that resolve its indeterminacies in various ways. Another difference between possibility models and world models, raised toward the end of Humberstone's paper, is that in possibility models a modal operator $\Box$ does not need to quantify over multiple accessible points—a single possibility will do, because a single possibility can leave matters indeterminate in just the way that a set of total worlds can. This idea is especially natural for doxastic and epistemic logic: an agent believes φ at possibility X if and only if φ is true at the single possibility Y that represents *the world as the agent believes it to be* in X.

There are a number of reasons, philosophical and mathematical, to find an alternative semantics based on possibilities attractive. Here we focus on the construction of possibility models with a finitary flavor. Our main completeness result shows that for a number of standard modal logics, we can build a canonical possibility model, wherein every logically consistent formula is satisfied, by simply taking each *individual finite formula* (modulo equivalence) to be a possibility, rather than each infinite maximally consistent set of formulas as in the usual canonical world models.[1] Constructing these locally finite canonical models involves first solving a problem in general modal logic of independent interest, related to the study of *adjoint* pairs[2] of modal operators: for a given modal logic $\mathbf{L}$, can we find for every formula φ a formula $\mathsf{f}_a^{\mathbf{L}}(\varphi)$ such that for every formula ψ, $\varphi \to \Box_a\psi$ is provable in $\mathbf{L}$ if and only if $\mathsf{f}_a^{\mathbf{L}}(\varphi) \to \psi$ is provable in $\mathbf{L}$? We answer this question in §3 for a number of standard modal logics, using model-theoretic arguments with world semantics.[3] This second main result allows us in §4 to build for each logic a canonical possibility model out of the lattice of formulas related by provable implication in the logic.

Given a normal modal logic $\mathbf{L}$, it is a familiar step to consider the lattice $\langle L, \leq\rangle$ where L is the set of equivalence classes of formulas under provable equivalence in $\mathbf{L}$, i.e., $[\varphi] = [\psi]$ iff $\vdash_{\mathbf{L}} \varphi \leftrightarrow \psi$, and $\leq$ is the relation of provable implication in $\mathbf{L}$ lifted to the equivalence classes, i.e., $[\varphi] \leq [\psi]$ iff $\vdash_{\mathbf{L}} \varphi \to \psi$.[4] (Below we will flip and change the relation symbol from '$\leq$' to '$\geqslant$' to match Humberstone.) What we will show is that for a number of modal logics $\mathbf{L}$, we can add to such a lattice functions $f_a\colon L \to L$ such that for all formulas φ and ψ, $[\varphi] \leq [\Box_a\psi]$ iff $f_a([\varphi]) \leq [\psi]$, and that the resulting structure serves as a canonical model for $\mathbf{L}$ according to the functional possibility semantics of §2.

2 Functional Possibility Semantics

We begin with a standard propositional polymodal language. Given a countable set $\mathsf{At} = \{p, q, r, \dots\}$ of atomic sentences and a finite set $I = \{a, b, c, \dots\}$ of

---

[1] Humberstone [17, p. 326] states a similar result without proof, but see the end of §2 for a problem. For world models, the idea of proving (weak) completeness by constructing models whose points are individual formulas has been carried out in [10,22]. The formulas used there as *worlds* are modal analogues of "state descriptions," characterizing a pointed world model up to n-bisimulation [6, §2.3] for a finite n and a finite set of atomic sentences. By contrast, in our §4, *any* formula (or rather equivalence class thereof) will count as a *possibility*.

[2] If for all formulas φ and ψ, $\vdash_{\mathbf{L}} \varphi \to \Box_1\psi$ iff $\vdash_{\mathbf{L}} \Diamond_2\varphi \to \psi$, then $\Box_1$ and $\Diamond_2$ form an adjoint pair of modal operators (also called a *residuated* pair as in [8, §12.2]). An example is the future box operator G and past diamond operator P of temporal logic. Exploiting such adjointness (or residuation) is the basis of *modal display calculi* (see, e.g., [24]).

[3] After writing a draft of this paper, I learned from Nick Bezhanishvili that Ghilardi [13, Theorem 6.3] proved a similar result for the modal logic $\mathbf{K}$ in an algebraic setting, showing that the finitely generated free algebra of $\mathbf{K}$ is a so-called *tense algebra*, which corresponds to $\mathbf{K}$ having *internal adjointness* as in Definition 3.1 below. Also see [5, Theorem 6.7].

[4] This is the lattice structure of the *Lindenbaum algebra* for $\mathbf{L}$ (see [8]).

modal operator indices, the language $\mathcal{L}$ is defined by

$$\varphi ::= p \mid \neg\varphi \mid (\varphi \wedge \varphi) \mid \Box_a\varphi,$$

where $p \in \mathsf{At}$ and $a \in I$. We define $(\varphi \vee \psi) := \neg(\neg\varphi \wedge \neg\psi)$, $(\varphi \to \psi) := \neg(\varphi \wedge \neg\psi)$, $\Diamond_a\varphi := \neg\Box_a\neg\varphi$, and $\bot := (p \wedge \neg p)$ for some $p \in \mathsf{At}$.

To fix intuitions, it helps to have a specific interpretation of the modal operators in mind. We will adopt a doxastic or epistemic interpretation, according to which $\Box_a$ is the belief or knowledge operator for agent a. This interpretation will also help in thinking about the semantics, but it should be stressed that the approach to follow can be applied to modal logic in general.

By *relational world models*, I mean standard relational structures $\mathfrak{M} = \langle \mathrm{W}, \{\mathrm{R}_a\}_{a\in I}, \mathrm{V}\rangle$ used to interpret $\mathcal{L}$ in the usual way [6]. By *relational possibility models*, I mean Humberstone's [17, §3] models, which we do not have room to review here. Modifying his models, we obtain the following (see [16]).

Definition 2.1 A *functional possibility model* for $\mathcal{L}$ is a tuple $\mathcal{M}$ of the form $\langle W, \geqslant, \{f_a\}_{a\in I}, V\rangle$ where:

1. W is a nonempty set with a distinguished element $\bot_{\mathcal{M}}$;

   Notation: we will use upper-case italic letters for elements of W and upper-case **bold** italic letters for elements of $W - \{\bot_{\mathcal{M}}\}$;

2. $f_a\colon W \to W$;
3. V is a partial function from $\mathsf{At} \times W$ to $\{0, 1\}$;[5]
4. $\geqslant$ is a weak partial order on W satisfying the following conditions:[6]
   (a) *persistence* – if $V(p, \boldsymbol{X})\downarrow$ and $\boldsymbol{X'} \geqslant \boldsymbol{X}$, then $V(p, \boldsymbol{X'}) = V(p, \boldsymbol{X})$;
   (b) *refinability* – if $V(p, \boldsymbol{X})\uparrow$, then $\exists \boldsymbol{Y}, \boldsymbol{Z} \geqslant \boldsymbol{X}$: $V(p, \boldsymbol{Y}) = 0$, $V(p, \boldsymbol{Z}) = 1$;
   (c) *f-persistence (monotonicity)* – if $\boldsymbol{X'} \geqslant \boldsymbol{X}$, then $f_a(\boldsymbol{X'}) \geqslant f_a(\boldsymbol{X})$;
   (d) *f-refinability* – if $\boldsymbol{Y} \geqslant f_a(\boldsymbol{X})$, then $\exists \boldsymbol{X'} \geqslant \boldsymbol{X}$ such that $\forall \boldsymbol{X''} \geqslant \boldsymbol{X'}$: $\boldsymbol{Y}$ and $f_a(\boldsymbol{X''})$ are *compatible*,

   where possibilities $\boldsymbol{Y}$ and $\boldsymbol{Z}$ are compatible iff $\exists \boldsymbol{U}$: $\boldsymbol{U} \geqslant \boldsymbol{Y}$ and $\boldsymbol{U} \geqslant \boldsymbol{Z}$.[7]

These models are defined in the same way as Humberstone's, except where f_a and $\bot_{\mathcal{M}}$ appear. W is the set of *possibilities*, and $\bot_{\mathcal{M}}$ is the totally incoherent "possibility."[8] (Often I will write '$\bot$' instead of '$\bot_{\mathcal{M}}$'.) Unlike worlds, possibilities can be indeterminate in certain respects, so V is a partial function. If $V(p, X)$ is undefined, then possibility X does not determine the truth

[5] As usual, every (total) function is a partial function. To indicate that $V(p, X)$ is defined, I write '$V(p, X)\downarrow$', and to indicate that $V(p, X)$ is undefined, I write '$V(p, X)\uparrow$'.

[6] Another natural condition, though not needed: *unrefinability* – if $Y \geqslant \bot_{\mathcal{M}}$, then $Y = \bot_{\mathcal{M}}$.

[7] So *f-refinability* says: if $\boldsymbol{Y} \geqslant f_a(\boldsymbol{X})$, then $\exists \boldsymbol{X'} \geqslant \boldsymbol{X}\ \forall \boldsymbol{X''} \geqslant \boldsymbol{X'}\ \exists \boldsymbol{Y'} \geqslant \boldsymbol{Y}$: $\boldsymbol{Y'} \geqslant f_a(\boldsymbol{X''})$.

[8] We include $\bot_{\mathcal{M}}$ in order to give semantics for logics that do not extend **KD**. Alternatively, we could drop $\bot_{\mathcal{M}}$ and allow the functions f_a to be partial, so instead of having $f_a(X) = \bot_{\mathcal{M}}$, we would have $f_a(X)\uparrow$. (Then we would modify Definition 2.2 to say that $\mathcal{M}, \boldsymbol{X} \Vdash \Box_a\varphi$ iff $f_a(\boldsymbol{X})\uparrow$ or $\mathcal{M}, f_a(\boldsymbol{X}) \Vdash \varphi$.) However, the approach with $\bot_{\mathcal{M}}$ seems to be more convenient.

or falsity of p. For each agent a, the doxastic/epistemic *function* f_a in functional possibility models replaces the doxastic/epistemic accessibility *relation* R_a from relational world models. At any possibility X, $f_a(X)$ represents *the world as agent a believes/knows it to be.* Inspired by Humberstone [17, p. 334], we call $f_a(X)$ agent a's *belief-possibility* at X. As officially stated in Definition 2.2 below, agent a believes/knows φ at X iff φ is true at $f_a(X)$.

All that remains to explain about the models is the *refinement* relation $\geqslant$. Intuitively, $Y \geqslant X$ means that Y is a refinement of X, in the sense that Y makes determinate whatever X makes determinate, and maybe more. (If $Y \geqslant X$ but $X \ngeqslant Y$, then Y is a *proper* refinement of X, written '$Y > X$'.) This explains Humberstone's *persistence* condition, familiar from Kripke semantics for intuitionistic logic [21]. The second condition, *refinability*, says that if a possibility $\boldsymbol{X}$ leaves the truth value of p indeterminate, then some coherent refinement of $\boldsymbol{X}$ decides p negatively and some coherent refinement of $\boldsymbol{X}$ decides p affirmatively. Intuitively, if there is no possible refinement $\boldsymbol{Y}$ of $\boldsymbol{X}$ with $V(p,\boldsymbol{Y}) = 1$ (resp. $V(p,\boldsymbol{Y}) = 0$), then $\boldsymbol{X}$ already determines that p is false (resp. true), so we should already have $V(p,\boldsymbol{X}) = 0$ (resp. $V(p,\boldsymbol{X}) = 1$).

Next are the conditions relating $\geqslant$ to f_a, which simply extend persistence and refinability from atomic to modal facts. First, just as *persistence* ensures that as we go from a possibility $\boldsymbol{X}$ to one of its refinements $\boldsymbol{X'}$, $\boldsymbol{X'}$ determines all of the atomic facts that $\boldsymbol{X}$ did, *f-persistence* ensures that $\boldsymbol{X'}$ determines all of the modal facts that $\boldsymbol{X}$ did, which is just to say that $f_a(\boldsymbol{X'})$ is a refinement of $f_a(\boldsymbol{X})$ for all $a \in I$. Second, just as *refinability* ensures that when $\boldsymbol{X}$ leaves an atomic formula p indeterminate, there are refinements of $\boldsymbol{X}$ that decide p each way, *f-refinability* ensures that when $\boldsymbol{X}$ leaves a modal formula $\Box_a\varphi$ indeterminate, there are refinements of $\boldsymbol{X}$ that decide $\Box_a\varphi$ each way. In fact, just the truth clause for $\neg$ in Definition 2.2 below ensures that if $\mathcal{M},\boldsymbol{X} \nVdash \neg\Box_a\varphi$, then there is a refinement of $\boldsymbol{X}$ that makes $\Box_a\varphi$ true. What *f-refinability* adds is that if $\mathcal{M},\boldsymbol{X} \nVdash \Box_a\varphi$, then there is a refinement of $\boldsymbol{X}$ that makes $\neg\Box_a\varphi$ true. Although it may not be initially obvious that this is the content of *f-refinability*, the proof of Lemma 2.3.2 together with Fig. 1 should make it clear.[9]

We now define truth for formulas of $\mathcal{L}$ in functional possibility models, following Humberstone's clauses for p, $\neg$, and $\wedge$, but changing the clause for $\Box_a$ to use f_a. The idea of using such a function instead of an accessibility relation to give the semantic clause for a modal operator appears in Fine's [9, p. 359] study of relevance logic (also see [18, p. 418], [19, p. 899], and cf. [4]).

Definition 2.2 Given a functional possibility model $\mathcal{M} = \langle W, \geqslant, \{f_a\}_{a\in I}, V\rangle$ with $X \in W$ and $\varphi \in \mathcal{L}$, define $\mathcal{M}, X \Vdash \varphi$ ("φ is true at X in $\mathcal{M}$") as follows:

1. $\mathcal{M}, \bot \Vdash \varphi$ for all φ;
2. $\mathcal{M}, \boldsymbol{X} \Vdash p$ iff $V(p,\boldsymbol{X}) = 1$;

---

[9] It is noteworthy that the *f-refinability* assumption is considerably weaker than the functional analogue of Humberstone's [17, 324] relational refinability assumption (**R**), explained at the end of this section. We discuss different strengths of modal refinability in [16].

3. $\mathcal{M}, \boldsymbol{X} \Vdash \neg\varphi$ iff $\forall \boldsymbol{Y} \geqslant \boldsymbol{X}$: $\mathcal{M}, \boldsymbol{Y} \nVdash \varphi$;
4. $\mathcal{M}, \boldsymbol{X} \Vdash (\varphi \wedge \psi)$ iff $\mathcal{M}, \boldsymbol{X} \Vdash \varphi$ and $\mathcal{M}, \boldsymbol{X} \Vdash \psi$;
5. $\mathcal{M}, \boldsymbol{X} \Vdash \Box_a\varphi$ iff $\mathcal{M}, f_a(\boldsymbol{X}) \Vdash \varphi$.

Given $(\varphi \vee \psi) := \neg(\neg\varphi \wedge \neg\psi)$, $(\varphi \to \psi) := \neg(\varphi \wedge \neg\psi)$, and $\Diamond_a\varphi := \neg\Box_a\neg\varphi$, one finds that the truth clauses for $\vee$, $\to$, and $\Diamond_a$ are equivalent to:

1. $\mathcal{M}, \boldsymbol{X} \Vdash (\varphi \vee \psi)$ iff $\forall \boldsymbol{Y} \geqslant \boldsymbol{X}\ \exists \boldsymbol{Z} \geqslant \boldsymbol{Y}$: $\mathcal{M}, \boldsymbol{Z} \Vdash \varphi$ or $\mathcal{M}, \boldsymbol{Z} \Vdash \psi$;
2. $\mathcal{M}, \boldsymbol{X} \Vdash (\varphi \to \psi)$ iff $\forall \boldsymbol{Y} \geqslant \boldsymbol{X}$ with $\mathcal{M}, \boldsymbol{Y} \Vdash \varphi$, $\exists \boldsymbol{Z} \geqslant \boldsymbol{Y}$: $\mathcal{M}, \boldsymbol{Z} \Vdash \psi$;
3. $\mathcal{M}, \boldsymbol{X} \Vdash \Diamond_a\varphi$ iff $\forall \boldsymbol{X'} \geqslant \boldsymbol{X}\ \exists \boldsymbol{Y} \geqslant f_a(\boldsymbol{X'})$: $\mathcal{M}, \boldsymbol{Y} \Vdash \varphi$.

The truth clause for $\vee$ crucially allows a possibility to determine that a disjunction is true without determining which disjunct is true; the clause for $\to$ can be further simplified, as in Lemma 2.3.3 below; and although the clause for $\Diamond_a$ appears unfamiliar, it is quite intuitive—a possibility $\boldsymbol{X}$ determines that φ is compatible with agent a's beliefs iff for any refinement $\boldsymbol{X'}$ of $\boldsymbol{X}$, a's belief-possibility at $\boldsymbol{X'}$ can be refined to a possibility where φ is true.

To get a feel for the semantics, it helps to consider simple models for concrete epistemic examples (see [16]), but we do not have room to do so here. We proceed to general properties of the semantics such as the following from [17].

Lemma 2.3 For any model $\mathcal{M}$, possibilities $\boldsymbol{X}$, $\boldsymbol{Y}$, and formulas φ, ψ:

1. Persistence: if $\mathcal{M}, \boldsymbol{X} \Vdash \varphi$ and $\boldsymbol{Y} \geqslant \boldsymbol{X}$, then $\mathcal{M}, \boldsymbol{Y} \Vdash \varphi$;
2. Refinability: if $\mathcal{M}, \boldsymbol{X} \nVdash \varphi$, then $\exists \boldsymbol{Z} \geqslant \boldsymbol{X}$: $\mathcal{M}, \boldsymbol{Z} \Vdash \neg\varphi$;
3. Implication: $\mathcal{M}, \boldsymbol{X} \Vdash \varphi \to \psi$ iff $\forall \boldsymbol{Z} \geqslant \boldsymbol{X}$: if $\mathcal{M}, \boldsymbol{Z} \Vdash \varphi$, then $\mathcal{M}, \boldsymbol{Z} \Vdash \psi$.

Proof. We treat only the $\Box_a$ case of an inductive proof of part 2 to illustrate *f-refinability* with Fig. 1. If $\mathcal{M}, \boldsymbol{X} \nVdash \Box_a\varphi$, then $\mathcal{M}, f_a(\boldsymbol{X}) \nVdash \varphi$, so by the inductive hypothesis there is some $\boldsymbol{Y} \geqslant f_a(\boldsymbol{X})$ such that $\mathcal{M}, \boldsymbol{Y} \Vdash \neg\varphi$. Now *f-refinability* implies that there is some $\boldsymbol{X'} \geqslant \boldsymbol{X}$ such that for all $\boldsymbol{X''} \geqslant \boldsymbol{X'}$, $\boldsymbol{Y}$ is compatible with $f_a(\boldsymbol{X''})$, which means there is a $\boldsymbol{Y'} \geqslant \boldsymbol{Y}$ with $\boldsymbol{Y'} \geqslant f_a(\boldsymbol{X''})$, which with $\mathcal{M}, \boldsymbol{Y} \Vdash \neg\varphi$ and part 1 implies $\mathcal{M}, f_a(\boldsymbol{X''}) \nVdash \varphi$ and hence $\mathcal{M}, \boldsymbol{X''} \nVdash \Box_a\varphi$. Since this holds for all $\boldsymbol{X''} \geqslant \boldsymbol{X'}$, we have $\mathcal{M}, \boldsymbol{X'} \Vdash \neg\Box_a\varphi$. $\Box$

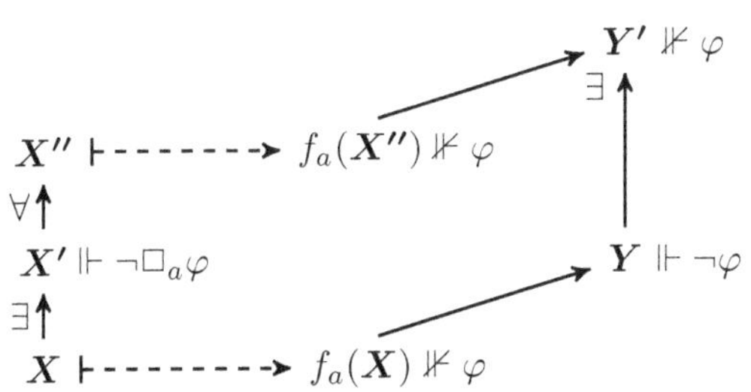

Fig. 1. *f-refinability* as used in the proof of Lemma 2.3.2, assuming $\mathcal{M}, \boldsymbol{X} \nVdash \Box_a\varphi$. Solid arrows are for the refinement relation $\geqslant$ and dashed are for the function f_a.

The definition of consequence over possibility models is as for world models.

Definition 2.4 Given a class S of possibility models, $\Sigma \subseteq \mathcal{L}$, and $\varphi \in \mathcal{L}$: $\Sigma \Vdash_{\mathsf{S}} \varphi$ ("φ is a consequence of Σ over S") iff for all $\mathcal{M} \in \mathsf{S}$ and X in $\mathcal{M}$, if $\mathcal{M}, X \Vdash \sigma$ for all $\sigma \in \Sigma$, then $\mathcal{M}, X \Vdash \varphi$; $\Vdash_{\mathsf{S}} \varphi$ ("φ is valid over S") iff $\emptyset \Vdash_{\mathsf{S}} \varphi$; and φ is *satisfiable* in S iff $\nVdash_{\mathsf{S}} \neg\varphi$ (iff there are $\mathcal{M} \in \mathsf{S}$ and $\boldsymbol{X}$ with $\mathcal{M}, \boldsymbol{X} \Vdash \varphi$).

One of Humberstone's insights was that by giving negation the intuitionistic-style clause in Definition 2.2.3, while at the same time defining $\vee$ and $\rightarrow$ in terms of $\neg$ and $\wedge$, contrary to intuitionistic logic, we obtain a classical logic from models based on possibilities. Van Benthem [1] also observed that by starting with Kripke models for intuitionistic logic and imposing a *cofinality* condition on the ordering (i.e., if $\forall \boldsymbol{Y} \geqslant \boldsymbol{X}\ \exists \boldsymbol{Z} \geqslant \boldsymbol{Y}$ such that $V(p, \boldsymbol{Z}) = 1$, then $V(p, \boldsymbol{X}) = 1$), one obtains a possibility semantics for (non-modal) classical logic by retaining the intuitionistic semantic clauses for $\neg$, $\rightarrow$, and $\wedge$, and defining $(\varphi \vee \psi)$ as $\neg(\neg\varphi \wedge \neg\psi)$. (In fact, van Benthem showed this for first-order logic.) For more on comparisons between classical and intuitionistic logic using possibility models, see [2], [3, Chs. 7-8], [7], [23], and [12, Ch. 8].

Lemma 2.5 If φ is a substitution instance of a classical propositional tautology, then φ is valid over functional possibility models.

Proof. Suppose φ is an instance of a propositional formula δ, where δ contains only the atomic sentences $q_1, \dots, q_n$. Let $\mathcal{L}_{\mathsf{PL}}(q_1, \dots, q_n)$ be the propositional language generated from $q_1, \dots, q_n$. Since φ is an instance of δ, there is some $s\colon \{q_1, \dots, q_n\} \rightarrow \mathcal{L}$ such that $\varphi = \hat{s}(\delta)$, where $\hat{s}\colon \mathcal{L}_{\mathsf{PL}}(q_1, \dots, q_n) \rightarrow \mathcal{L}$ is the usual extension of s such that $\hat{s}(q_i) = s(q_i)$, $\hat{s}(\neg\alpha) = \neg\hat{s}(\alpha)$, and $\hat{s}((\alpha \wedge \beta)) = (\hat{s}(\alpha) \wedge \hat{s}(\beta))$. Now suppose that φ is not valid, so there is some possibility model $\mathcal{M}$ and $\boldsymbol{X}$ in $\mathcal{M}$ such that $\mathcal{M}, \boldsymbol{X} \nVdash \varphi$, which by Lemma 2.3.2 implies there is a $\boldsymbol{X'} \geqslant \boldsymbol{X}$ such that $\mathcal{M}, \boldsymbol{X'} \Vdash \neg\varphi$. Also by Lemma 2.3.2, for any $\boldsymbol{Y} \in W$ and $\psi \in \mathcal{L}$, we can choose a $\boldsymbol{Y}^{\psi} \geqslant \boldsymbol{Y}$ with $\mathcal{M}, \boldsymbol{Y}^{\psi} \Vdash \psi$ or $\mathcal{M}, \boldsymbol{Y}^{\psi} \Vdash \neg\psi$. Enumerating the formulas of $\mathcal{L}$ as $\psi_1, \psi_2, \dots$, define a sequence $\boldsymbol{X}_0, \boldsymbol{X}_1, \boldsymbol{X}_2, \dots$ such that $\boldsymbol{X}_0 = \boldsymbol{X'}$ and $\boldsymbol{X}_{n+1} = \boldsymbol{X}_n^{\psi_{n+1}}$. Thus, $\boldsymbol{X}_0 \leqslant \boldsymbol{X}_1 \leqslant \boldsymbol{X}_2 \dots$ is a "generic" chain that decides every formula eventually. Define a propositional valuation $v\colon \{q_1, \dots, q_n\} \rightarrow \{0, 1\}$ such that $v(q_i) = 1$ if for some $k \in \mathbb{N}$, $\mathcal{M}, \boldsymbol{X}_k \Vdash s(q_i)$, and $v(q_i) = 0$ otherwise. Where $\overline{v}\colon \mathcal{L}_{\mathsf{PL}}(q_1, \dots, q_n) \rightarrow \{0, 1\}$ is the usual classical extension of v, one can prove that for all $\alpha \in \mathcal{L}_{\mathsf{PL}}(q_1, \dots, q_n)$,

$$\overline{v}(\alpha) = 1 \text{ iff } \exists k \in \mathbb{N}\colon \mathcal{M}, \boldsymbol{X}_k \Vdash \hat{s}(\alpha), \tag{1}$$

by induction on α. From above, $\mathcal{M}, \boldsymbol{X}_0 \Vdash \neg\hat{s}(\delta)$, i.e., $\mathcal{M}, \boldsymbol{X}_0 \Vdash \hat{s}(\neg\delta)$, and $\neg\delta \in \mathcal{L}_{\mathsf{PL}}(q_1, \dots, q_n)$, so (1) implies $\overline{v}(\neg\delta) = 1$. Thus, δ is not a tautology. $\square$

Not only is classical propositional logic sound over functional possibility models, but also standard normal modal logics are sound and complete over functional possibility models with constraints on f_a and $\geqslant$ corresponding to the logic's additional axioms. Throughout we adopt the standard nomenclature for normal modal logics, borrowing the names of monomodal logics for their polymodal (fusion) versions. Thus, each $\Box_a$ operator has the same axioms.

The following result raises obvious questions about general correspondence theory for possibility semantics, but we do not have room to discuss them here.

Theorem 2.6 (Soundness and Completeness) For any subset of the axioms $\{\mathrm{D}, \mathrm{T}, 4, \mathrm{B}, 5\}$,[10] the extension of the minimal normal modal logic **K** with that set of axioms is sound and strongly complete for the class of functional possibility models satisfying the corresponding constraint for each axiom:

1. D axiom: for all $\boldsymbol{X}$, $f_a(\boldsymbol{X}) \neq \bot$;
2. T axiom: for all $\boldsymbol{X}$, $\boldsymbol{X} \geqslant f_a(\boldsymbol{X})$;
3. 4 axiom: for all $\boldsymbol{X}$, $f_a(f_a(\boldsymbol{X})) \geqslant f_a(\boldsymbol{X})$;
4. B axiom: for all $\boldsymbol{X}, \boldsymbol{Y}$, if $\boldsymbol{Y} \geqslant f_a(\boldsymbol{X})$ then $\exists \boldsymbol{X'} \geqslant \boldsymbol{X}$: $\boldsymbol{X'} \geqslant f_a(\boldsymbol{Y})$;
5. 5 axiom: for all $\boldsymbol{X}, \boldsymbol{Y}$, if $\boldsymbol{Y} \geqslant f_a(\boldsymbol{X})$, then $\exists \boldsymbol{X'} \geqslant \boldsymbol{X}$: $f_a(\boldsymbol{X'}) \geqslant f_a(\boldsymbol{Y})$.

The proof of soundness is straightforward. First, by Lemma 2.5, all tautologies are valid. Second, by Lemma 2.3.3, if φ and $\varphi \to \psi$ are valid, then ψ is valid, so modus ponens is sound; and obviously if φ is valid, then $\Box_a\varphi$ is valid, so the necessitation rule is sound. Next, we check that the K axiom is valid:

> Suppose for reductio that $\mathcal{M}, \boldsymbol{X} \nVdash \Box_a(\varphi \to \psi) \to (\Box_a\varphi \to \Box_a\psi)$, so by Lemma 2.3.3, there is some $\boldsymbol{Y} \geqslant \boldsymbol{X}$ such that $\mathcal{M}, \boldsymbol{Y} \Vdash \Box_a(\varphi \to \psi)$ but $\mathcal{M}, \boldsymbol{Y} \nVdash \Box_a\varphi \to \Box_a\psi$, so by Lemma 2.3.3 again there is some $\boldsymbol{Z} \geqslant \boldsymbol{Y}$ with $\mathcal{M}, \boldsymbol{Z} \Vdash \Box_a\varphi$ but $\mathcal{M}, \boldsymbol{Z} \nVdash \Box_a\psi$, so $\mathcal{M}, f_a(\boldsymbol{Z}) \Vdash \varphi$ but $\mathcal{M}, f_a(\boldsymbol{Z}) \nVdash \psi$. By Lemma 2.3.1, $\mathcal{M}, \boldsymbol{Y} \Vdash \Box_a(\varphi \to \psi)$ and $\boldsymbol{Z} \geqslant \boldsymbol{Y}$ together imply $\mathcal{M}, \boldsymbol{Z} \Vdash \Box_a(\varphi \to \psi)$, so $\mathcal{M}, f_a(\boldsymbol{Z}) \Vdash \varphi \to \psi$. But by Lemma 2.3.3 and the reflexivity of $\geqslant$, we cannot have all of $\mathcal{M}, f_a(\boldsymbol{Z}) \Vdash \varphi \to \psi$, $\mathcal{M}, f_a(\boldsymbol{Z}) \Vdash \varphi$, and $\mathcal{M}, f_a(\boldsymbol{Z}) \nVdash \psi$. Thus, $\mathcal{M}, \boldsymbol{X} \Vdash \Box_a(\varphi \to \psi) \to (\Box_a\varphi \to \Box_a\psi)$.

Using Lemma 2.3.3, it is also easy to check the validity of D, T, 4, B, and 5 over the classes of models with the corresponding constraints.

Completeness can be proved by taking advantage of completeness with respect to relational world models and then showing how to transform any relational world model obeying constraints on R_a corresponding to the axioms into a functional possibility model satisfying the same formulas and obeying constraints of f_a and $\geqslant$ corresponding to the axioms (see [16]). Or completeness can be proved directly with a canonical model construction where the domain is the set of all (equivalence classes of) *sets of formulas* of $\mathcal{L}$ (see [16]).

Here we will prove weak completeness for a selection of the logics covered by Theorem 2.6 using a canonical model construction where the domain is simply the set of all (equivalence classes of) *formulas* of $\mathcal{L}$. In this way, we will prove weak completeness for classes of models obeying the following constraint.

Definition 2.7 A functional possibility model $\mathcal{M}$ is *locally finite* iff for all $\boldsymbol{X} \in W$, the set $\{p \in \mathsf{At} \mid V(p, \boldsymbol{X})\downarrow\}$ is finite.

---

[10] As usual, D is $\Box_a\varphi \to \neg\Box_a\neg\varphi$, T is $\Box_a\varphi \to \varphi$, 4 is $\Box_a\varphi \to \Box_a\Box_a\varphi$, B is $\neg\varphi \to \Box_a\neg\Box_a\varphi$ $(\psi \to \Box_a\Diamond_a\psi)$, and 5 is $\neg\Box_a\varphi \to \Box_a\neg\Box_a\varphi$ $(\Diamond_a\psi \to \Box_a\Diamond_a\psi)$.

If At is infinite, then every locally finite model contains infinitely many finite possibilities by *refinability*. Hence the term '*locally* finite', which leads to a distinction. All of the logics considered here have the "finite model property" with respect to possibility semantics (see [16]): any consistent formula is satisfied in a model where W is a finite set. But the elements of such a W are *infinite* possibilities, i.e., each deciding infinitely many atomic sentences. With Definition 2.7, we move from a finite set of infinite possibilities to an infinite set of finite possibilities, a move that has certain philosophical attractions (see [17,16]) and mathematical interest. For the latter, if we wish to build a model in which every consistent formula is satisfied, this inevitably requires an infinite W for the logics with no bound on modal depth. Yet, in a finitary spirit, we may at least aspire to construct such a model to be locally finite, as in §4.

In §3-4, we will work up to Theorem 2.8 below. In [16], we also prove the completeness of **K45** and **KD45** with respect to locally finite models satisfying the appropriate constraints, but space does not permit the proof here.

Theorem 2.8 (Completeness for Locally Finite Models)
Let **L** be one of the logics **K**, **KD**, **T**, **K4**, **KD4**, **S4**, or **S5**, and let $\mathsf{S}_{\mathbf{L}}^{LF}$ be the class of *locally finite* functional possibility models satisfying the constraints on f_a and $\geqslant$ corresponding to the axioms of **L**, as listed in Theorem 2.6. Then **L** is weakly complete with respect to $\mathsf{S}_{\mathbf{L}}^{LF}$: for all $\varphi \in \mathcal{L}$, if $\Vdash_{\mathsf{S}_{\mathbf{L}}^{LF}} \varphi$, then $\vdash_{\mathbf{L}} \varphi$.

Humberstone [17, p. 326] also states that one can prove the completeness of some modal logics with respect to classes of his *relational* possibility models, using a canonical model construction in which each possibility is the set of syntactic consequences of a consistent finite set of formulas, but he does not write out a proof. Relational possibility models have relations R_a, instead of functions f_a, so that $\mathcal{M}, \boldsymbol{X} \Vdash \Box_a\varphi$ iff $\mathcal{M}, \boldsymbol{Y} \Vdash \varphi$ for all $\boldsymbol{Y}$ with $\boldsymbol{X} R_a \boldsymbol{Y}$. Here it is relevant to consider Humberstone's [17, p. 324-5] refinability condition (**R**). According to (**R**), if $\boldsymbol{X} R_a \boldsymbol{Y}$, then $\exists \boldsymbol{X}' \geqslant \boldsymbol{X}\ \forall \boldsymbol{X}'' \geqslant \boldsymbol{X}'$: $\boldsymbol{X}'' R_a \boldsymbol{Y}$. This is very strong. Given $\boldsymbol{X}'' R_a \boldsymbol{Y}$, it must be that for every formula φ that is not true at $\boldsymbol{Y}$, $\Box_a\varphi$ is not true at $\boldsymbol{X}''$. Then since this holds for all $\boldsymbol{X}'' \geqslant \boldsymbol{X}'$, $\neg\Box_a\varphi$ must be true at $\boldsymbol{X}'$. Thus, if $\boldsymbol{Y}$ makes only finitely many atomic sentences p true, then $\boldsymbol{X}'$ must make infinitely many formulas $\neg\Box_a p$ true. But then $\boldsymbol{X}'$ cannot be the set of consequences of a consistent finite set of formulas, because no consistent finite set entails infinitely many formulas of the form $\neg\Box_a p$.

3 Internal Adjointness

Our goal is to construct a canonical model for **L** in which each possibility is the equivalence class of a single formula φ, such that $\mathcal{M}, [\varphi] \Vdash \psi$ iff $\vdash_{\mathbf{L}} \varphi \to \psi$. To do so, we need to define functions f_a such that $\mathcal{M}, [\varphi] \Vdash \Box_a\psi$ iff $\mathcal{M}, f_a([\varphi]) \Vdash \psi$, which means we need functions f_a such that $\vdash_{\mathbf{L}} \varphi \to \Box_a\psi$ iff $\vdash_{\mathbf{L}} \mathsf{f}_a(\varphi) \to \psi$. Intuitively, for a finite "possibility" φ, we want a finite "belief-possibility" $\mathsf{f}_a(\varphi)$ such that whatever is *believed* according to φ is *true* according to $\mathsf{f}_a(\varphi)$. It is an independently natural question whether such functions f_a exist for **L**.

Definition 3.1 A modal logic **L** has *internal adjointness* iff for all $\varphi \in \mathcal{L}$ and

$a \in I$, there is a $\mathsf{f}_a^{\mathbf{L}}(\varphi) \in \mathcal{L}$ such that for all $\psi \in \mathcal{L}$:

$$\vdash_{\mathbf{L}} \varphi \to \Box_a \psi \text{ iff } \vdash_{\mathbf{L}} \mathsf{f}_a^{\mathbf{L}}(\varphi) \to \psi.$$

Not every modal logic has internal adjointness. For example:

Proposition 3.2 **K5**, **K45**, **KD5**, and **KD45** lack internal adjointness.

Proof. Let $\mathbf{L}$ be any of the logics listed. Suppose there is a formula $\mathsf{f}_a^{\mathbf{L}}(\top)$ such that for all formulas ψ, $\vdash_{\mathbf{L}} \top \to \Box_a \psi$ iff $\vdash_{\mathbf{L}} \mathsf{f}_a^{\mathbf{L}}(\top) \to \psi$. Then since $\nvdash_{\mathbf{L}} \top \to \Box_a \bot$, we have $\nvdash_{\mathbf{L}} \mathsf{f}_a^{\mathbf{L}}(\top) \to \bot$, which we will show below to imply $\nvdash_{\mathbf{L}} \mathsf{f}_a^{\mathbf{L}}(\top) \to (\Box_a p \to p)$ for an atomic p that does not occur in $\mathsf{f}_a^{\mathbf{L}}(\top)$. But $\vdash_{\mathbf{L}} \Box_a(\Box_a p \to p)$ and hence $\vdash_{\mathbf{L}} \top \to \Box_a(\Box_a p \to p)$, so taking $\psi := \Box_a p \to p$ refutes the supposition. Thus, $\mathbf{L}$ lacks internal adjointness.

Given $\nvdash_{\mathbf{L}} \mathsf{f}_a^{\mathbf{L}}(\top) \to \bot$, it follows by the completeness of $\mathbf{L}$ with respect to the class $\mathsf{C}_{\mathbf{L}}$ of Euclidean/transitive/serial relational world models that there is a world model $\mathfrak{M} \in \mathsf{C}_{\mathbf{L}}$ such that $\mathfrak{M}, x \vDash \mathsf{f}_a^{\mathbf{L}}(\top)$. Define a new model $\mathfrak{M}'$ to be like $\mathfrak{M}$ except that (a) there is a new world x' that can "see" via each R_b for $b \in I$ all and only the worlds that x can see via R_b and (b) p is true everywhere in $\mathfrak{M}'$ except at x', which otherwise agrees with x on atomic sentences. Since p does not occur in $\mathsf{f}_a^{\mathbf{L}}(\top)$, and $\mathfrak{M}, x$ and $\mathfrak{M}', x'$ are bisimilar with respect to the language without p, from $\mathfrak{M}, x \vDash \mathsf{f}_a^{\mathbf{L}}(\top)$ we have $\mathfrak{M}', x' \vDash \mathsf{f}_a^{\mathbf{L}}(\top)$, and by construction we have $\mathfrak{M}', x' \vDash \Box_a p \wedge \neg p$. Also by construction, $\mathfrak{M}' \in \mathsf{C}_{\mathbf{L}}$ (for if $\mathrm{R}_b^{\mathfrak{M}}$ is Euclidean/transitive/serial, then so is $\mathrm{R}_b^{\mathfrak{M}'}$), so by the soundness of $\mathbf{L}$ with respect to $\mathsf{C}_{\mathbf{L}}$, we have $\nvdash_{\mathbf{L}} \mathsf{f}_a^{\mathbf{L}}(\top) \to (\Box_a p \to p)$, as claimed above. □

The problem is that with the logics in Proposition 3.2, the shift-reflexivity axiom $\Box_a(\Box_a p \to p)$ is derivable for all p, but a consistent formula entails $\Box_a p \to p$ only if it contains p, and no formula contains infinitely many p. If we wish to overcome Proposition 3.2, we must extend our language and logics.[11]

Yet Theorem 3.9 will show that for a number of logics $\mathbf{L}$, we already have the ability to find an appropriate $\mathsf{f}_a^{\mathbf{L}}(\varphi)$ in our original language $\mathcal{L}$.[12] In order to define $\mathsf{f}_a^{\mathbf{L}}(\varphi)$, we first need the following standard definition and result.

[11] For example, consider an expanded language that includes all the formulas of $\mathcal{L}$ plus a new formula $\mathbf{loop}_a$ for each $a \in I$, such that in relational world models, $\mathfrak{M}, w \vDash \mathbf{loop}_a$ iff $w\mathrm{R}_a w$ (as in [11, §3]), and in functional possibility models, $\mathcal{M}, \boldsymbol{X} \Vdash \mathbf{loop}_a$ iff $\boldsymbol{X} \geqslant f_a(\boldsymbol{X})$. Intuitively, $\mathbf{loop}_a$ says that agent a's beliefs are compatible with the facts. The key axiom schema for $\mathbf{loop}_a$ is $\mathbf{loop}_a \to (\Box_a \varphi \to \varphi)$, and the shift-reflexivity axiom $\Box_a(\Box_a \varphi \to \varphi)$ can be captured by $\Box_a \mathbf{loop}_a$, which says that the agent believes that her beliefs are compatible with the facts. As shown in [16], with logics **K45loop** and **KD45loop**, the problem of Proposition 3.2 does not arise. Moreover, a detour through these logics for the expanded language allows one to prove the completeness with respect to locally finite models of **K45** and **KD45** for $\mathcal{L}$, despite Proposition 3.2 [16]. Another approach to overcoming Proposition 3.2, which has greater generality but less doxastic/epistemic motivation than the approach with $\mathbf{loop}_a$, is to add a backward-looking operator $\Diamond_a^{\leftarrow}$ to our language with the truth clause: $\mathcal{M}, \boldsymbol{X} \Vdash \Diamond_a^{\leftarrow} \varphi$ iff for some $\boldsymbol{Y} \in W$, $\boldsymbol{X} \geqslant f_a(\boldsymbol{Y})$ and $\mathcal{M}, \boldsymbol{Y} \Vdash \varphi$. In world models, the clause is: $\mathfrak{M}, w \vDash \Diamond_a^{\leftarrow} \varphi$ iff for some $v \in \mathrm{W}$, $v\mathrm{R}_a w$ and $\mathfrak{M}, v \vDash \varphi$. So $\Diamond_a^{\leftarrow}$ is the existential modality for the converse relation. Then it is easy to see that $\varphi \to \Box_a \psi$ is valid iff $\Diamond_a^{\leftarrow} \varphi \to \psi$ is.

[12] Compare our definition of *internal adjointness* to that of *indigenous inverses* in [20, §6.2].

Definition 3.3 A $\varphi \in \mathcal{L}$ is in *modal disjunctive normal form* (DNF) iff it a disjunction of conjunctions, each conjunct of which is either (a) a propositional formula α (whose form will not matter here), (b) of the form $\Box_a\beta$ for some $a \in I$ and β in DNF, or (c) of the form $\Diamond_a\gamma$ for some $a \in I$ and γ in DNF.

Lemma 3.4 For any normal modal logic **L** and $\varphi \in \mathcal{L}$, there is a $\varphi' \in \mathcal{L}$ in DNF such that $\vdash_{\mathbf{L}} \varphi \leftrightarrow \varphi'$.

Another useful definition and result that will help us prove Theorem 3.9 for logics with the T axiom involves the idea of a *T-unpacked* formula from [15].

Definition 3.5 If $\varphi \in \mathcal{L}$ is in DNF, then a disjunct δ_φ of φ is T-unpacked iff for all $a \in I$ and formulas β, if $\Box_a\beta$ is a conjunct of δ_φ, then there is a disjunct δ_β of β such that every conjunct of δ_β is a conjunct of δ_φ.

The formula φ itself is T-unpacked iff every disjunct of φ is T-unpacked.

For example, one can check that $\varphi := \Box_a p \vee \big(p \wedge \Box_a(q \vee \Box_b r) \wedge \Diamond_a s\big)$ is not T-unpacked. By contrast, the following formula, which is equivalent to φ in the logic **T**, is T-unpacked, as highlighted by the boldface type:

$$\begin{aligned}\varphi^* := {} & \big(\Box_a p \wedge \boldsymbol{p}\big) \vee \big(p \wedge \Box_a(q \vee \Box_b r) \wedge \boldsymbol{q} \wedge \Diamond_a s\big) \vee \\ & \big(p \wedge \Box_a(q \vee \Box_b r) \wedge \Box_b \boldsymbol{r} \wedge \boldsymbol{r} \wedge \Diamond_a s\big).\end{aligned}$$

This kind of transformation between φ and φ^* can be carried out in general.

Lemma 3.6 For every extension **L** of **K** containing the T axiom and $\varphi \in \mathcal{L}$, there is a T-unpacked DNF $\varphi^* \in \mathcal{L}$ such that $\vdash_{\mathbf{L}} \varphi \leftrightarrow \varphi^*$.

Proof. Transform φ into DNF and then apply to each disjunct the following provable equivalences in **L**:

$$\begin{aligned}(\psi \wedge \cdots \wedge \Box_a(\bigvee_{\delta\in\Delta}\delta) \wedge \cdots \wedge \chi) & \Leftrightarrow (\psi \wedge \cdots \wedge \Box_a(\bigvee_{\delta\in\Delta}\delta) \wedge (\bigvee_{\delta\in\Delta}\delta) \wedge \cdots \wedge \chi) \\ & \Leftrightarrow \bigvee_{\delta'\in\Delta}(\psi \wedge \cdots \wedge \Box_a(\bigvee_{\delta\in\Delta}\delta) \wedge \delta' \wedge \cdots \wedge \chi),\end{aligned}$$

where the first step uses the T axiom and the second uses propositional logic. Repeated transformations of this kind produce a T-unpacked DNF formula. □

Henceforth, for every logic **L** and $\varphi \in \mathcal{L}$, we fix an **L**-equivalent $NF_{\mathbf{L}}(\varphi)$ in DNF which is T-unpacked if **L** contains the T axiom and each disjunct of which is **L**-consistent if φ is, since we may always drop inconsistent disjuncts.

We can now define the belief-possibility $\mathfrak{f}_a^{\mathbf{L}}(\varphi)$ of agent a according to φ and logic **L**. Since the definition of $\mathfrak{f}_a^{\mathbf{L}}(\varphi)$ depends on the specific logic **L**, for the sake of space I will restrict attention to the standard doxastic and epistemic logics not excluded by Proposition 3.2: **K**, **KD**, **T**, **K4**, **KD4**, **S4**, and **S5** (and any other extension of **KB**). For each **L** that does not extend **KB**, our $\mathfrak{f}_a^{\mathbf{L}}$ is a *non-connectival operation* on formulas in the terminology of [19, p. 49]

Definition 3.7 Consider an **L**-consistent formula $NF_{\mathbf{L}}(\varphi) := \delta_1 \vee \cdots \vee \delta_n$. For $a \in I$ and $\mathbf{L} \in \{\mathbf{K}, \mathbf{KD}, \mathbf{T}\}$, define

$$\mathfrak{f}_a^{\mathbf{L}}(\delta_i) := \bigwedge\{\beta \mid \Box_a\beta \text{ a conjunct of } \delta_i\}.$$

For $\mathbf{L} \in \{\mathbf{K4}, \mathbf{KD4}, \mathbf{S4}\}$, define

$$\mathsf{f}_a^{\mathbf{L}}(\delta_i) := \bigwedge\{\beta, \Box_a\beta \mid \Box_a\beta \text{ a conjunct of } \delta_i\}.$$

For all of the above $\mathbf{L}$,[13] define

$$\mathsf{f}_a^{\mathbf{L}}(\delta_1 \vee \dots \vee \delta_n) := \mathsf{f}_a^{\mathbf{L}}(\delta_1) \vee \dots \vee \mathsf{f}_a^{\mathbf{L}}(\delta_n).$$

For any $\mathbf{L}$-consistent formula φ not in normal form, let $\mathsf{f}_a^{\mathbf{L}}(\varphi) = \mathsf{f}_a^{\mathbf{L}}(NF_{\mathbf{L}}(\varphi))$.[14]

For any $\mathbf{L}$-inconsistent formula φ, let $\mathsf{f}_a^{\mathbf{L}}(\varphi) = \bot$.

Finally, for any extension of $\mathbf{KB}$ and any φ, simply let $\mathsf{f}_a^{\mathbf{L}}(\varphi) = \Diamond_a\varphi$.

To see the need for the assumption in Definition 3.7 that $NF_{\mathbf{L}}(\varphi)$ is T-unpacked if $\mathbf{L}$ contains the T axiom, suppose $\mathbf{L}$ is $\mathbf{T}$ and δ is $\neg q \wedge \Box_a(\Box_a p \vee \Box_a q)$, which is not T-unpacked. Then $\mathsf{f}_a^{\mathbf{L}}(\delta)$ would be $\Box_a p \vee \Box_a q$, and we would have $\vdash_{\mathbf{T}} \delta \to \Box_a p$ but $\nvdash_{\mathbf{T}} \mathsf{f}_a^{\mathbf{L}}(\delta) \to p$, contrary to our desired Theorem 3.9. If we T-unpack $\neg q \wedge \Box_a(\Box_a p \vee \Box_a q)$, we first obtain

$$\big(\neg q \wedge \Box_a(\Box_a p \vee \Box_a q) \wedge \Box_a p \wedge p\big) \vee \big(\neg q \wedge \Box_a(\Box_a p \vee \Box_a q) \wedge \Box_a q \wedge q\big),$$

the right disjunct of which is inconsistent, so we drop it to obtain the T-unpacked $\delta' := \neg q \wedge \Box_a(\Box_a p \vee \Box_a q) \wedge \Box_a p \wedge p$. Now $\mathsf{f}_a^{\mathbf{L}}(\delta')$ is $(\Box_a p \vee \Box_a q) \wedge p$, so we have $\vdash_{\mathbf{T}} \delta' \to \Box_a p$ and $\vdash_{\mathbf{T}} \mathsf{f}_a^{\mathbf{L}}(\delta') \to p$, as desired.

Next, note that each "possibility" φ determines that a believes $\mathsf{f}_a^{\mathbf{L}}(\varphi)$.

Lemma 3.8 For every $\mathbf{L}$ in Definition 3.7, $\varphi \in \mathcal{L}$, and $a \in I$: $\vdash_{\mathbf{L}} \varphi \to \Box_a \mathsf{f}_a^{\mathbf{L}}(\varphi)$.

Proof. For extensions of $\mathbf{KB}$, $\varphi \to \Box_a \mathsf{f}_a^{\mathbf{L}}(\varphi)$ is $\varphi \to \Box_a\Diamond_a\varphi$, which is the B axiom. For the other logics, it suffices to show $\vdash_{\mathbf{L}} \varphi \to \Box_a \mathsf{f}_a^{\mathbf{L}}(\varphi)$ where φ is $NF_{\mathbf{L}}(\psi)$ for some ψ. So φ is of the form $\delta_1 \vee \dots \vee \delta_n$. For each disjunct δ_i of φ,

$$\vdash_{\mathbf{L}} \delta_i \to \bigwedge_{\psi \text{ a conjunct of } \mathsf{f}_a^{\mathbf{L}}(\delta_i)} \Box_a\psi,$$

which for any normal modal logic implies

$$\vdash_{\mathbf{L}} \delta_i \to \Box_a \bigwedge_{\psi \text{ a conjunct of } \mathsf{f}_a^{\mathbf{L}}(\delta_i)} \psi, \text{ i.e., } \vdash_{\mathbf{L}} \delta_i \to \Box_a \mathsf{f}_a^{\mathbf{L}}(\delta_i),$$

which for any normal modal logic implies

$$\vdash_{\mathbf{L}} \delta_i \to \Box_a(\mathsf{f}_a^{\mathbf{L}}(\delta_1) \vee \dots \vee \mathsf{f}_a^{\mathbf{L}}(\delta_n)), \text{ i.e., } \vdash_{\mathbf{L}} \delta_i \to \Box_a \mathsf{f}_a^{\mathbf{L}}(\varphi).$$

Since the above holds for all disjuncts δ_i of φ, we have $\vdash_{\mathbf{L}} \varphi \to \Box_a \mathsf{f}_a^{\mathbf{L}}(\varphi)$. □

---

[13] For the logics **K45loop** and **KD45loop** mentioned in footnote 11, we would define

$$\mathsf{f}_a^{\mathbf{L}}(\delta_i) := \mathbf{loop}_a \wedge \bigwedge\{\beta, \Box_a\beta \mid \Box_a\beta \text{ a conjunct of } \delta_i\} \cup \{\Diamond_a\gamma \mid \Diamond_a\gamma \text{ a conjunct of } \delta_i\}.$$

[14] Note that by Lemma 3.8 and Theorem 3.9, if $\vdash_{\mathbf{L}} \varphi \leftrightarrow \psi$, then $\vdash_{\mathbf{L}} \mathsf{f}_a^{\mathbf{L}}(\varphi) \leftrightarrow \mathsf{f}_a^{\mathbf{L}}(\psi)$.

We are now ready to prove our first main result, Theorem 3.9, which shows that our selected logics have internal adjointness. The proof involves the gluing together of relational world models in the style of the completeness proofs in [15]. These constructions are interesting, as are the ways that the definition of $\mathsf{f}_a^{\mathbf{L}}(\varphi)$ is used, but for the sake of space we give the proof in the Appendix.

Theorem 3.9 (Internal Adjointness) For any $\mathbf{L}$ among $\mathbf{K}$, $\mathbf{KD}$, $\mathbf{T}$, $\mathbf{K4}$, $\mathbf{KD4}$, $\mathbf{S4}$, and $\mathbf{S5}$ (or any other extension of $\mathbf{KB}$), $\varphi, \psi \in \mathcal{L}$, and $a \in I$:

$$\vdash_{\mathbf{L}} \varphi \to \Box_a \psi \text{ iff } \vdash_{\mathbf{L}} \mathsf{f}_a^{\mathbf{L}}(\varphi) \to \psi.$$

Note that the right to left direction is straightforward: if $\vdash_{\mathbf{L}} \mathsf{f}_a^{\mathbf{L}}(\varphi) \to \psi$, then $\vdash_{\mathbf{L}} \Box_a \mathsf{f}_a^{\mathbf{L}}(\varphi) \to \Box_a \psi$ since $\mathbf{L}$ is normal, so $\vdash_{\mathbf{L}} \varphi \to \Box_a \psi$ by Lemma 3.8.

The following Lemma will be used to prove Lemma 4.4 in §4.

Lemma 3.10 Let $\mathbf{L}$ be one of the logics in Definition 3.7 and $\varphi, \psi \in \mathcal{L}$.

1. If $\mathbf{L}$ contains the D axiom and φ is $\mathbf{L}$-consistent, then $\mathsf{f}_a^{\mathbf{L}}(\varphi)$ is $\mathbf{L}$-consistent;
2. If $\mathbf{L}$ contains the T axiom, then $\vdash_{\mathbf{L}} \varphi \to \mathsf{f}_a^{\mathbf{L}}(\varphi)$;
3. If $\mathbf{L}$ contains the 4 axiom, then $\vdash_{\mathbf{L}} \mathsf{f}_a^{\mathbf{L}}(\varphi) \to \Box_a \mathsf{f}_a^{\mathbf{L}}(\varphi)$, which by Theorem 3.9 is equivalent to $\vdash_{\mathbf{L}} \mathsf{f}_a^{\mathbf{L}}(\mathsf{f}_a^{\mathbf{L}}(\varphi)) \to \mathsf{f}_a^{\mathbf{L}}(\varphi)$;
4. If $\mathbf{L}$ contains the B axiom, φ and ψ are $\mathbf{L}$-consistent, and $\vdash_{\mathbf{L}} \varphi \to \mathsf{f}_a^{\mathbf{L}}(\psi)$, then $\psi \wedge \mathsf{f}_a^{\mathbf{L}}(\varphi)$ is $\mathbf{L}$-consistent;
5. If $\mathbf{L}$ contains the 5 axiom, then $\vdash_{\mathbf{L}} \Diamond_a \varphi \to \Box_a \mathsf{f}_a^{\mathbf{L}}(\varphi)$, which by Theorem 3.9 is equivalent to $\vdash_{\mathbf{L}} \mathsf{f}_a^{\mathbf{L}}(\Diamond_a \varphi) \to \mathsf{f}_a^{\mathbf{L}}(\varphi)$.

Proof. For part 1, for any normal modal logic $\mathbf{L}$, if $\vdash_{\mathbf{L}} \mathsf{f}_a^{\mathbf{L}}(\varphi) \to \bot$, then $\vdash_{\mathbf{L}} \Box_a \mathsf{f}_a^{\mathbf{L}}(\varphi) \to \Box_a \bot$, which with Lemma 3.8 implies $\vdash_{\mathbf{L}} \varphi \to \Box_a \bot$, which for $\mathbf{L}$ with the D axiom implies $\vdash_{\mathbf{L}} \varphi \to \bot$. For part 2, given $\vdash_{\mathbf{L}} \varphi \to \Box_a \mathsf{f}_a^{\mathbf{L}}(\varphi)$ by Lemma 3.8, it follows for any $\mathbf{L}$ with the T axiom that $\vdash_{\mathbf{L}} \varphi \to \mathsf{f}_a^{\mathbf{L}}(\varphi)$.

For part 3, if $\mathbf{L}$ is $\mathbf{S5}$, then the claim is immediate given $\mathsf{f}_a^{\mathbf{L}}(\varphi) = \Diamond_a \varphi$ from Definition 3.7. Let us consider the other logics in Definition 3.7 with the 4 axiom. We can assume without loss of generality that φ is a formula in DNF of the form $\delta_1 \vee \cdots \vee \delta_n$, and $\mathsf{f}_a^{\mathbf{L}}(\varphi) = \mathsf{f}_a^{\mathbf{L}}(\delta_1) \vee \cdots \vee \mathsf{f}_a^{\mathbf{L}}(\delta_n)$ by Definition 3.7. Observe that for each of the $\mathbf{L}$ in Definition 3.7 with the 4 axiom and each δ_i,

$$\vdash_{\mathbf{L}} \mathsf{f}_a^{\mathbf{L}}(\delta_i) \to \bigwedge_{\psi \text{ a conjunct of } \mathsf{f}_a^{\mathbf{L}}(\delta_i)} \Box_a \psi.$$

Now the proof that $\vdash_{\mathbf{L}} \mathsf{f}_a^{\mathbf{L}}(\varphi) \to \Box_a \mathsf{f}_a^{\mathbf{L}}(\varphi)$ follows the pattern for Lemma 3.8.

Given Definition 3.7, part 4 is equivalent to the claim that for any $\mathbf{L}$-consistent φ and ψ, if $\vdash_{\mathbf{L}} \varphi \to \Diamond_a \psi$, then $\psi \wedge \Diamond_a \varphi$ is $\mathbf{L}$-consistent. If $\psi \wedge \Diamond_a \varphi$ is $\mathbf{L}$-inconsistent, then $\vdash_{\mathbf{L}} \Diamond_a \varphi \to \neg \psi$, which implies $\vdash_{\mathbf{L}} \Box_a \Diamond_a \varphi \to \Box_a \neg \psi$ for a normal $\mathbf{L}$. Then since $\mathbf{L}$ has the B axiom, $\vdash_{\mathbf{L}} \varphi \to \Box_a \Diamond_a \varphi$, so we have $\vdash_{\mathbf{L}} \varphi \to \Box_a \neg \psi$, which with $\vdash_{\mathbf{L}} \varphi \to \Diamond_a \psi$ contradicts the $\mathbf{L}$-consistency of φ.

For part 5, by Lemma 3.8, $\vdash_{\mathbf{L}} \varphi \to \Box_a \mathsf{f}_a^{\mathbf{L}}(\varphi)$, which for a normal $\mathbf{L}$ implies $\vdash_{\mathbf{L}} \Diamond_a \varphi \to \Diamond_a \Box_a \mathsf{f}_a^{\mathbf{L}}(\varphi)$, so $\vdash_{\mathbf{L}} \Diamond_a \varphi \to \Box_a \mathsf{f}_a^{\mathbf{L}}(\varphi)$ for $\mathbf{L}$ with the 5 axiom. $\Box$

4 Canonical Models of Finite Possibilities

We can now construct locally finite canonical possibility models for the logics $\mathbf{L}$ in Theorem 2.8. For $\varphi \in \mathcal{L}$, let $[\varphi]_{\mathbf{L}} = \{\psi \in \mathcal{L} \mid \vdash_{\mathbf{L}} \varphi \leftrightarrow \psi\}$. Fix an enumeration $\varphi_1, \varphi_2, \ldots$ of the formulas of $\mathcal{L}$, and for every $\varphi \in \mathcal{L}$, let $\varphi_{\mathbf{L}}$ be the member of $[\varphi]_{\mathbf{L}}$ that occurs first in the enumeration. We do this so that our possibilities can simply be formulas, rather than their equivalence classes.[15] This simplifies the presentation, but nothing important turns on it.

Definition 4.1 For each logic $\mathbf{L}$ in Theorem 2.8, define the canonical functional finite-possibility model $\mathbb{M}^{\mathbf{L}} = \langle W^{\mathbf{L}}, \geqslant^{\mathbf{L}}, \{f_a^{\mathbf{L}}\}_{a \in I}, V^{\mathbf{L}} \rangle$ as follows:

1. $W^{\mathbf{L}} = \{\sigma_{\mathbf{L}} \mid \sigma \in \mathcal{L}\}$; $\bot_{\mathbb{M}^{\mathbf{L}}} = \bot_{\mathbf{L}}$;
2. $\sigma' \geqslant^{\mathbf{L}} \sigma$ iff $\vdash_{\mathbf{L}} \sigma' \to \sigma$;
3. $f_a^{\mathbf{L}}(\sigma) = \mathsf{f}_a^{\mathbf{L}}(\sigma)_{\mathbf{L}}$;
4. $V^{\mathbf{L}}(p, \sigma) = 1$ iff $\vdash_{\mathbf{L}} \sigma \to p$; $V^{\mathbf{L}}(p, \sigma) = 0$ iff $\vdash_{\mathbf{L}} \sigma \to \neg p$.

Following our earlier convention, we will use boldface letters for the consistent formulas in $W^{\mathbf{L}} - \{\bot_{\mathbf{L}}\}$. In some of the text in the rest of this section, to reduce clutter we will leave the sub/superscript for $\mathbf{L}$ implicit.

Our first job is to check that $\mathbb{M}^{\mathbf{L}}$ is indeed a functional possibility model.

Lemma 4.2 (Canonical Model is a Model) For each logic $\mathbf{L}$ in Theorem 2.8, $\mathbb{M}^{\mathbf{L}}$ is a functional possibility model according to Definition 2.1, and $\mathbb{M}^{\mathbf{L}}$ is locally finite according to Definition 2.7.

Proof. The conditions of *persistence*, *refinability*, and *f-persistence* are all easy to check for $\mathbb{M}^{\mathbf{L}}$. It is also clear that for any $\boldsymbol{\sigma} \in W$, $\{p \in \mathsf{At} \mid \vdash \boldsymbol{\sigma} \to \pm p\}$ is finite, so $\mathbb{M}^{\mathbf{L}}$ is *locally finite*. Let us verify that *f-refinability* holds:

> For all consistent $\boldsymbol{\sigma}, \boldsymbol{\gamma} \in W$, if $\boldsymbol{\gamma} \geqslant f_a(\boldsymbol{\sigma})$, then there is a $\boldsymbol{\sigma}' \geqslant \boldsymbol{\sigma}$ such that for all $\boldsymbol{\sigma}'' \geqslant \boldsymbol{\sigma}'$ there is a $\boldsymbol{\gamma}' \geqslant \boldsymbol{\gamma}$ such that $\boldsymbol{\gamma}' \geqslant f_a(\boldsymbol{\sigma}'')$.

Given $\boldsymbol{\gamma} \geqslant f_a(\boldsymbol{\sigma})$, we have $\vdash \boldsymbol{\gamma} \to \mathsf{f}_a(\boldsymbol{\sigma})$. Now since $\boldsymbol{\gamma}$ is consistent, $\nvdash \boldsymbol{\gamma} \to \neg\boldsymbol{\gamma}$, which with $\vdash \boldsymbol{\gamma} \to \mathsf{f}_a(\boldsymbol{\sigma})$ implies $\nvdash \mathsf{f}_a(\boldsymbol{\sigma}) \to \neg\boldsymbol{\gamma}$, which with Theorem 3.9 implies $\nvdash \boldsymbol{\sigma} \to \Box_a \neg\boldsymbol{\gamma}$. Thus, $\boldsymbol{\sigma}' = \boldsymbol{\sigma} \wedge \Diamond_a \boldsymbol{\gamma}$ is consistent. Now for any consistent $\boldsymbol{\sigma}'' \geqslant \boldsymbol{\sigma}'$, i.e., $\vdash \boldsymbol{\sigma}'' \to \boldsymbol{\sigma}'$, we claim that $\boldsymbol{\gamma}' = \boldsymbol{\gamma} \wedge \mathsf{f}_a(\boldsymbol{\sigma}'')$ is consistent. If not, then $\vdash \mathsf{f}_a(\boldsymbol{\sigma}'') \to \neg\boldsymbol{\gamma}$, which for any normal modal logic implies $\vdash \Box_a \mathsf{f}_a(\boldsymbol{\sigma}'') \to \Box_a \neg\boldsymbol{\gamma}$, which with Lemma 3.8 implies $\vdash \boldsymbol{\sigma}'' \to \Box_a \neg\boldsymbol{\gamma}$. But given $\boldsymbol{\sigma}' = \boldsymbol{\sigma} \wedge \Diamond_a \boldsymbol{\gamma}$ and $\vdash \boldsymbol{\sigma}'' \to \boldsymbol{\sigma}'$, we have $\vdash \boldsymbol{\sigma}'' \to \Diamond_a \boldsymbol{\gamma}$, which with $\vdash \boldsymbol{\sigma}'' \to \Box_a \neg\boldsymbol{\gamma}$ contradicts the consistency of $\boldsymbol{\sigma}''$. Thus, $\boldsymbol{\gamma}'$ is consistent. Then since $\vdash \boldsymbol{\gamma}' \to \mathsf{f}_a(\boldsymbol{\sigma}'')$, we have $\boldsymbol{\gamma}' \geqslant f_a(\boldsymbol{\sigma}'')$. Hence we have shown that there is a $\boldsymbol{\sigma}' \geqslant \boldsymbol{\sigma}$ such that for all $\boldsymbol{\sigma}'' \geqslant \boldsymbol{\sigma}$ there is a $\boldsymbol{\gamma}' \geqslant \boldsymbol{\gamma}$ such that $\boldsymbol{\gamma}' \geqslant f_a(\boldsymbol{\sigma}'')$. □

Our next job is to show that for any formulas φ and σ, φ being *true* at the possibility σ in $\mathbb{M}^{\mathbf{L}}$ is equivalent to $\sigma \to \varphi$ being *derivable* in $\mathbf{L}$.

---

[15] The reason for dealing with equivalence classes and representatives at all is so that the relation $\geqslant$ in the canonical model will be antisymmetric, as Humberstone [17, p. 318] requires. If we had instead allowed $\geqslant$ to be a preorder—which would not have changed any of our results—then we could take our domain to be the set of all consistent formulas plus $\bot$.

Lemma 4.3 (Truth) For any logic $\mathbf{L}$ in Theorem 2.8, $\sigma \in W^{\mathbf{L}}$, and $\varphi \in \mathcal{L}$: $\mathbb{M}^{\mathbf{L}}, \sigma \Vdash \varphi$ iff $\vdash_{\mathbf{L}} \sigma \to \varphi$.

Proof. The claim is immediate for $\sigma = \bot$, given Definition 2.2.1. For $\sigma \neq \bot$, we prove the claim by induction on φ. The atomic case is by definition of $V^{\mathbf{L}}$, and the $\wedge$ case is routine. For the $\neg$ case, if $\nvdash \boldsymbol{\sigma} \to \neg\varphi$, then $(\boldsymbol{\sigma} \wedge \varphi)$ is consistent. Then since $\vdash (\boldsymbol{\sigma} \wedge \varphi) \to \boldsymbol{\sigma}$, i.e., $(\boldsymbol{\sigma} \wedge \varphi) \geqslant \boldsymbol{\sigma}$, we have a $\boldsymbol{\sigma'} \geqslant \boldsymbol{\sigma}$ such that $\vdash \boldsymbol{\sigma'} \to \varphi$, which by the inductive hypothesis implies that $\mathbb{M}, \boldsymbol{\sigma'} \Vdash \varphi$, which implies $\mathbb{M}, \boldsymbol{\sigma} \nVdash \neg\varphi$. In the other direction, if $\vdash \boldsymbol{\sigma} \to \neg\varphi$, then for all $\boldsymbol{\sigma'} \geqslant \boldsymbol{\sigma}$, i.e., $\vdash \boldsymbol{\sigma'} \to \boldsymbol{\sigma}$, we have $\vdash \boldsymbol{\sigma'} \to \neg\varphi$, so $\nvdash \boldsymbol{\sigma'} \to \varphi$ by the consistency of $\boldsymbol{\sigma'}$, so $\mathbb{M}, \boldsymbol{\sigma'} \nVdash \varphi$ by the inductive hypothesis. Thus, $\mathbb{M}, \boldsymbol{\sigma} \Vdash \neg\varphi$.

For the $\Box_a$ case, given $f_a(\boldsymbol{\sigma}) = \mathsf{f}_a(\boldsymbol{\sigma})$, we have the following equivalences: $\vdash \boldsymbol{\sigma} \to \Box_a\varphi$ iff $\vdash \mathsf{f}_a(\boldsymbol{\sigma}) \to \varphi$ (by Theorem 3.9) iff $\mathbb{M}, f_a(\boldsymbol{\sigma}) \Vdash \varphi$ (by the inductive hypothesis) iff $\mathbb{M}, \boldsymbol{\sigma} \Vdash \Box_a\varphi$ (by the truth definition). □

If we only wished to prove the case of Theorem 2.8 for $\mathbf{K}$, then with Lemmas 4.2 and 4.3 we would be done. However, to prove Theorem 2.8 for the various extensions of $\mathbf{K}$, we need to make sure that $\mathbb{M}^{\mathbf{L}}$ satisfies the conditions on f_a and $\geqslant$ corresponding to the extra axioms of $\mathbf{L}$, given in Theorem 2.6.

Lemma 4.4 (Canonicity) The model $\mathbb{M}^{\mathbf{L}}$ is such that:

1. If $\mathbf{L}$ contains the D axiom, then for all $\boldsymbol{\sigma} \in W^{\mathbf{L}}$, $f_a(\boldsymbol{\sigma}) \neq \bot$;
2. If $\mathbf{L}$ contains the T axiom, then for all $\boldsymbol{\sigma} \in W^{\mathbf{L}}$, $\boldsymbol{\sigma} \geqslant f_a(\boldsymbol{\sigma})$;
3. If $\mathbf{L}$ contains the 4 axiom, then for all $\boldsymbol{\sigma} \in W^{\mathbf{L}}$, $f_a(f_a(\boldsymbol{\sigma})) \geqslant f_a(\boldsymbol{\sigma})$;
4. If $\mathbf{L}$ contains the B axiom, then for all $\boldsymbol{\sigma}, \gamma \in W^{\mathbf{L}}$, if $\gamma \geqslant f_a(\boldsymbol{\sigma})$, then $\exists \boldsymbol{\sigma'} \geqslant \boldsymbol{\sigma} \colon \boldsymbol{\sigma'} \geqslant f_a(\gamma)$.
5. If $\mathbf{L}$ contains the 5 axiom, then for all $\boldsymbol{\sigma}, \gamma \in W^{\mathbf{L}}$, if $\gamma \geqslant f_a(\boldsymbol{\sigma})$, then $\exists \boldsymbol{\sigma'} \geqslant \boldsymbol{\sigma} \colon f_a(\boldsymbol{\sigma'}) \geqslant f_a(\gamma)$.

Proof. Each part follows from the corresponding part of Lemma 3.10. For part 1, we need that if $\boldsymbol{\sigma} \in W^{\mathbf{L}}$ is $\mathbf{L}$-consistent, then so is $\mathsf{f}_a^{\mathbf{L}}(\boldsymbol{\sigma})$, which is given by Lemma 3.10.1. For part 2, we need that for all $\boldsymbol{\sigma} \in W^{\mathbf{L}}$, $\vdash_{\mathbf{L}} \boldsymbol{\sigma} \to \mathsf{f}_a^{\mathbf{L}}(\boldsymbol{\sigma})$, which is given by Lemma 3.10.2. For part 3, we need that for all $\boldsymbol{\sigma} \in W^{\mathbf{L}}$, $\vdash_{\mathbf{L}} \mathsf{f}_a^{\mathbf{L}}(\mathsf{f}_a^{\mathbf{L}}(\boldsymbol{\sigma})) \to \mathsf{f}_a^{\mathbf{L}}(\boldsymbol{\sigma})$, which is given by Lemma 3.10.3.

For part 4, we need that for all $\mathbf{L}$-consistent $\boldsymbol{\sigma}, \gamma \in W^{\mathbf{L}}$, if $\vdash_{\mathbf{L}} \gamma \to \mathsf{f}_a^{\mathbf{L}}(\boldsymbol{\sigma})$, then there is some $\mathbf{L}$-consistent $\boldsymbol{\sigma'}$ with (i) $\vdash_{\mathbf{L}} \boldsymbol{\sigma'} \to \boldsymbol{\sigma}$ and (ii) $\vdash_{\mathbf{L}} \boldsymbol{\sigma'} \to \mathsf{f}_a^{\mathbf{L}}(\gamma)$. Setting $\boldsymbol{\sigma'} := \boldsymbol{\sigma} \wedge \mathsf{f}_a^{\mathbf{L}}(\gamma) = \boldsymbol{\sigma} \wedge \Diamond_a\gamma$, then (i) and (ii) are immediate, and the $\mathbf{L}$-consistency of $\boldsymbol{\sigma'}$ is given by Lemma 3.10.4.

For part 5, we need that for all $\mathbf{L}$-consistent $\boldsymbol{\sigma}, \gamma \in W^{\mathbf{L}}$, if $\vdash_{\mathbf{L}} \gamma \to \mathsf{f}_a^{\mathbf{L}}(\boldsymbol{\sigma})$, then there is some $\mathbf{L}$-consistent $\boldsymbol{\sigma'}$ such that (iii) $\vdash_{\mathbf{L}} \boldsymbol{\sigma'} \to \boldsymbol{\sigma}$ and (iv) $\vdash_{\mathbf{L}} \mathsf{f}_a^{\mathbf{L}}(\boldsymbol{\sigma'}) \to \mathsf{f}_a^{\mathbf{L}}(\gamma)$. Setting $\boldsymbol{\sigma'} := \boldsymbol{\sigma} \wedge \Diamond_a\gamma$, then (iii) is immediate. By Lemma 3.10.5, we have $\vdash_{\mathbf{L}} \Diamond_a\gamma \to \Box_a\mathsf{f}_a^{\mathbf{L}}(\gamma)$ and hence $\vdash_{\mathbf{L}} \boldsymbol{\sigma'} \to \Box_a\mathsf{f}_a^{\mathbf{L}}(\gamma)$, which by Theorem 3.9 implies (iv). Finally, suppose for reductio that $\boldsymbol{\sigma'}$ is $\mathbf{L}$-inconsistent, so $\vdash_{\mathbf{L}} \boldsymbol{\sigma} \to \neg\Diamond_a\gamma$. Then $\vdash_{\mathbf{L}} \boldsymbol{\sigma} \to \Box_a\neg\gamma$, which by Theorem 3.9 implies $\vdash_{\mathbf{L}} \mathsf{f}_a^{\mathbf{L}}(\boldsymbol{\sigma}) \to \neg\gamma$, which with $\vdash_{\mathbf{L}} \gamma \to \mathsf{f}_a^{\mathbf{L}}(\boldsymbol{\sigma})$ implies that γ is $\mathbf{L}$-inconsistent, contradicting our initial assumption. Thus, $\boldsymbol{\sigma'}$ is $\mathbf{L}$-consistent. □

We have now shown that lattices $\langle L, \leq \rangle$ as in §1, equipped with functions f_a exhibiting **L**'s internal adjointness, can be viewed as canonical possibility models. This illustrates the closeness of possibility *semantics* to modal *syntax*.

Finally, we put all of the pieces together for our culminating result.[16]

Theorem 2.8 (Completeness for Locally Finite Models)
Let **L** be one of the logics **K**, **KD**, **T**, **K4**, **KD4**, **S4**, or **S5**, and let $\mathsf{S}_{\mathbf{L}}^{LF}$ be the class of *locally finite* functional possibility models satisfying the constraints on f_a and $\geqslant$ corresponding to the axioms of **L**, as listed in Theorem 2.6. Then **L** is weakly complete with respect to $\mathsf{S}_{\mathbf{L}}^{LF}$: for all $\varphi \in \mathcal{L}$, if $\Vdash_{\mathsf{S}_{\mathbf{L}}^{LF}} \varphi$, then $\vdash_{\mathbf{L}} \varphi$.

Proof. By Lemmas 4.2 and 4.4, $\mathbb{M}^{\mathbf{L}} \in \mathsf{S}_{\mathbf{L}}^{LF}$, and by the definition of $\mathbb{M}^{\mathbf{L}}$, $\neg\varphi_{\mathbf{L}} \in W^{\mathbf{L}}$. Assuming $\nvdash_{\mathbf{L}} \varphi$, we have $\neg\varphi_{\mathbf{L}} \neq \bot_{\mathbf{L}}$. Then by Lemma 4.3, $\mathbb{M}^{\mathbf{L}}, \neg\varphi_{\mathbf{L}} \nVdash \varphi$, which with $\mathbb{M}^{\mathbf{L}} \in \mathsf{S}_{\mathbf{L}}^{LF}$ implies $\nVdash_{\mathsf{S}_{\mathbf{L}}^{LF}} \varphi$ by Definition 2.4. □

5 Conclusion

A Humberstonian model theory for modal logic, based on partial possibilities instead of total worlds, involves not only a different intuitive picture of modal models, but also a different mathematical approach to their construction. The infinitary staples of completeness proofs for world semantics—maximally consistent sets, Lindenbaum's Lemma—are not needed for possibility semantics. This may be considered an advantage,[17] but the purpose of this paper was not to advocate for possibilities *over* worlds. Nor was it to advocate for functions *over* relations. Modal reasoning with relational world models is natural and powerful, as our own Appendix shows. The purpose of this paper was instead to suggest how modal reasoning with functional possibility models is also natural and powerful, and how this reasoning leads to the independently interesting issue of *internal adjointness* for modal logics. There are many other interesting issues around the corner, such as the study of transformations between possibility models and world models (see [16,14]). Hopefully, however, we have already seen enough to motivate further study of possibilities for modal logic.

Acknowledgement

For helpful comments, I thank Johan van Benthem, Nick Bezhanishvili, Russell Buehler, Davide Grossi, Matthew Harrison-Trainor, Lloyd Humberstone, Alex Kocurek, James Moody, Lawrence Valby, and three anonymous AiML referees.

[16] Of course, we do not have *strong* completeness over locally finite models. For an infinite set $\Sigma \subseteq \mathsf{At}$ of atomic sentences, there is no locally finite model containing a possibility that makes all of Σ true, so $\Sigma \Vdash_{\mathsf{S}_{\mathbf{L}}^{LF}} \bot$; yet for every finite subset $\Sigma_0 \subseteq \Sigma$, we have $\Sigma_0 \nVdash_{\mathsf{S}_{\mathbf{L}}^{LF}} \bot$ (so the consequence relation $\Vdash_{\mathsf{S}_{\mathbf{L}}^{LF}}$ is not compact) and hence $\Sigma_0 \nvdash_{\mathbf{L}} \bot$ by soundness, so $\Sigma \nvdash_{\mathbf{L}} \bot$.

[17] Van Benthem [3, p. 78] remarks: "There is something inelegant to an ordinary Henkin argument. One has a consistent set of sentences S, perhaps quite small, that one would like to see satisfied semantically. Now, some arbitrary *maximal* extension S^+ of S is to be taken to obtain a model (for S^+, and hence for S)—but the added part $S^+ - S$ plays no role subsequently. We started out with something partial, but the method forces us to be total." This "problem of the 'irrelevant extension'" [1, p. 1] is solved by possibility semantics.

Appendix

In this appendix, we prove Theorem 3.9 from §3. In the proof, which uses standard relational semantics, we invoke the completeness of the logics listed with respect to their corresponding classes of relational world models. Let $\mathsf{C}_{\mathbf{L}}$ be the class of relational world models determined by logic $\mathbf{L}$, so, e.g., $\mathsf{C}_{\mathbf{K}}$ is the class of all relational world models, $\mathsf{C}_{\mathbf{T}}$ is the class of reflexive relational world models, $\mathsf{C}_{\mathbf{K4}}$ is the class of transitive relational world models, etc.

Theorem 3.9 (Internal Adjointness) For any $\mathbf{L}$ among $\mathbf{K}$, $\mathbf{KD}$, $\mathbf{T}$, $\mathbf{K4}$, $\mathbf{KD4}$, $\mathbf{S4}$, and $\mathbf{S5}$ (or any other extension of $\mathbf{KB}$), $\varphi, \psi \in \mathcal{L}$, and $a \in I$:

$$\vdash_{\mathbf{L}} \varphi \to \Box_a \psi \text{ iff } \vdash_{\mathbf{L}} \mathsf{f}_a^{\mathbf{L}}(\varphi) \to \psi.$$

Proof. From right to left, if $\vdash \mathsf{f}_a^{\mathbf{L}}(\varphi) \to \psi$, then $\vdash \Box_a \mathsf{f}_a^{\mathbf{L}}(\varphi) \to \Box_a \psi$ since $\mathbf{L}$ is normal, so $\vdash \varphi \to \Box_a \psi$ by Lemma 3.8.

From left to right, the proof for logics that contain the B axiom, such as $\mathbf{S5}$, is simple: if $\nvdash_{\mathbf{L}} \mathsf{f}_a^{\mathbf{L}}(\varphi) \to \psi$, then by the completeness of $\mathbf{L}$ with respect to $\mathsf{C}_{\mathbf{L}}$, there is a relational world model $\mathfrak{M} \in \mathsf{C}_{\mathbf{L}}$ with world w such that $\mathfrak{M}, w \vDash \mathsf{f}_a^{\mathbf{L}}(\varphi) \wedge \neg\psi$, which by the definition of $\mathsf{f}_a^{\mathbf{L}}(\varphi)$ for logics containing B means $\mathfrak{M}, w \vDash \Diamond_a \varphi \wedge \neg\psi$. Hence there is some v with $w\mathrm{R}_a v$ such that $\mathfrak{M}, v \vDash \varphi$. By the symmetry of R_a, we also have $v\mathrm{R}_a w$, so $\mathfrak{M}, v \vDash \Diamond_a \neg\psi$. Finally, by the soundness of $\mathbf{L}$ with respect $\mathsf{C}_{\mathbf{L}}$, $\mathfrak{M}, v \vDash \varphi \wedge \Diamond_a \neg\psi$ implies $\nvdash_{\mathbf{L}} \varphi \to \Box_a \psi$.

From left to right for logics without the B axiom, if φ is $\mathbf{L}$-inconsistent, then $\vdash_{\mathbf{L}} \varphi \to \Box_a \psi$ for all ψ, and $\mathsf{f}_a^{\mathbf{L}}(\varphi) = \bot$ by Definition 3.7, so $\vdash_{\mathbf{L}} \mathsf{f}_a^{\mathbf{L}}(\varphi) \to \psi$ for all ψ. So suppose that φ is $\mathbf{L}$-consistent. Given our definition of $\mathsf{f}_a^{\mathbf{L}}$ in terms of $NF_{\mathbf{L}}(\varphi)$, we can assume φ is in DNF and each of its disjuncts is $\mathbf{L}$-consistent; moreover, we can assume that φ is T-unpacked if $\mathbf{L}$ contains the T axiom.

Now if $\nvdash_{\mathbf{L}} \mathsf{f}_a^{\mathbf{L}}(\varphi) \to \psi$, then there is a $\mathbf{L}$-consistent disjunct δ of φ such that $\nvdash_{\mathbf{L}} \mathsf{f}_a^{\mathbf{L}}(\delta) \to \psi$. Since δ is $\mathbf{L}$-consistent, by the completeness of $\mathbf{L}$ with respect to $\mathsf{C}_{\mathbf{L}}$ there is a relational world model $\mathfrak{A} = \langle \mathrm{W}^{\mathfrak{A}}, \{\mathrm{R}_a^{\mathfrak{A}}\}_{a \in I}, \mathrm{V}^{\mathfrak{A}} \rangle \in \mathsf{C}_{\mathbf{L}}$ with $x \in \mathrm{W}^{\mathfrak{A}}$ such that $\mathfrak{A}, x \vDash \delta$. Now define a model $\mathfrak{A}' = \langle \mathrm{W}^{\mathfrak{A}'}, \{\mathrm{R}_a^{\mathfrak{A}'}\}_{a \in I}, \mathrm{V}^{\mathfrak{A}'} \rangle$, shown in Fig. 2 below, that is just like $\mathfrak{A}$ except with one new world x' that can "see" all and only the worlds that x can see (so x' cannot see itself):

- $\mathrm{W}^{\mathfrak{A}'} = \mathrm{W}^{\mathfrak{A}} \cup \{x'\}$ for $x' \notin \mathrm{W}^{\mathfrak{A}}$;
- for all $b \in I$, $w\mathrm{R}_b^{\mathfrak{A}'} v$ iff either $w\mathrm{R}_b^{\mathfrak{A}} v$ or $[w = x'$ and $x\mathrm{R}_b^{\mathfrak{A}} v]$;
- $\mathrm{V}^{\mathfrak{A}'}(p, w) = 1$ iff either $\mathrm{V}^{\mathfrak{A}}(p, w) = 1$ or $[w = x'$ and $\mathrm{V}^{\mathfrak{A}}(p, x) = 1]$.

Define $E \subseteq \mathrm{W}^{\mathfrak{A}'} \times \mathrm{W}^{\mathfrak{A}}$ such that wEv iff $[w = x'$ and $v = x]$ or $[w \neq x'$ and $w = v]$. Then E is a *bisimulation* relating $\mathfrak{A}', x'$ and $\mathfrak{A}, x$, so by the invariance of modal truth under bisimulation [6, §2.2], $\mathfrak{A}, x \vDash \delta$ implies $\mathfrak{A}', x' \vDash \delta$.

Since $\nvdash_{\mathbf{L}} \mathsf{f}_a^{\mathbf{L}}(\delta) \to \psi$, by the completeness of $\mathbf{L}$ with respect to $\mathsf{C}_{\mathbf{L}}$ there is a relational world model $\mathfrak{B} = \langle \mathrm{W}^{\mathfrak{B}}, \{\mathrm{R}_a^{\mathfrak{B}}\}_{a \in I}, \mathrm{V}^{\mathfrak{B}} \rangle \in \mathsf{C}_{\mathbf{L}}$ with $y \in \mathrm{W}^{\mathfrak{B}}$ such that $\mathfrak{B}, y \vDash \mathsf{f}_a^{\mathbf{L}}(\delta) \wedge \neg\psi$. Without loss of generality, we can assume that the domains of $\mathfrak{A}'$ and $\mathfrak{B}$ are disjoint. Define a new model $\mathfrak{C} = \langle \mathrm{W}^{\mathfrak{C}}, \{\mathrm{R}_a^{\mathfrak{C}}\}_{a \in I}, \mathrm{V}^{\mathfrak{C}} \rangle$, shown in Fig. 2 below, by first taking the disjoint union of $\mathfrak{A}'$ and $\mathfrak{B}$ and then connecting x' from $\mathfrak{A}'$ to y from $\mathfrak{B}$ by an accessibility arrow for agent a:

- $W^{\mathfrak{C}} = W^{\mathfrak{A}'} \cup W^{\mathfrak{B}}$; $V^{\mathfrak{C}}(p, w) = 1$ iff $V^{\mathfrak{A}'}(p, w) = 1$ or $V^{\mathfrak{B}}(p, w) = 1$.
- for a in the lemma, $wR_a^{\mathfrak{C}}v$ iff either $wR_a^{\mathfrak{A}'}v$, $wR_a^{\mathfrak{B}}v$, or $[w = x'$ and $v = y]$;
- for $b \neq a$, $wR_b^{\mathfrak{C}}v$ iff either $wR_b^{\mathfrak{A}'}v$ or $wR_b^{\mathfrak{B}}v$.

The identity relation on $W^{\mathfrak{A}'} \setminus \{x'\}$ is a bisimulation between $\mathfrak{A}'$ and $\mathfrak{C}$, so

$$\forall w \in W^{\mathfrak{A}'} \setminus \{x'\}\ \forall \chi \in \mathcal{L}\colon \mathfrak{A}', w \vDash \chi \text{ iff } \mathfrak{C}, w \vDash \chi. \tag{2}$$

Similarly, the identity relation on $W^{\mathfrak{B}}$ is a bisimulation between $\mathfrak{B}$ and $\mathfrak{C}$, so $\forall w \in W^{\mathfrak{B}}\ \forall \chi \in \mathcal{L}\colon \mathfrak{B}, w \vDash \chi$ iff $\mathfrak{C}, w \vDash \chi$. Given $\mathfrak{B}, y \vDash \mathsf{f}_a^{\mathbf{L}}(\delta) \wedge \neg\psi$, it follows that $\mathfrak{C}, y \vDash \mathsf{f}_a^{\mathbf{L}}(\delta) \wedge \neg\psi$. Then since $x'R_a^{\mathfrak{C}}y$, we have $\mathfrak{C}, x' \vDash \Diamond_a\neg\psi$.

Now we claim that given $\mathfrak{A}', x' \vDash \delta$, also $\mathfrak{C}, x' \vDash \delta$. Recall that δ is a conjunction that has as conjuncts zero or more propositional formulas α, formulas of the form $\Box_b\beta$, and formulas of the form $\Diamond_b\gamma$ for various $b \in I$, including a. The propositional part of δ is still true at x' in $\mathfrak{C}$, since the valuation on x' has not changed from $\mathfrak{A}'$ to $\mathfrak{C}$. For the modal parts, we use the following facts:[18]

$$R_b^{\mathfrak{C}}[x'] = R_b^{\mathfrak{A}'}[x'] \text{ for all } b \in I \setminus \{a\}; \tag{3}$$

$$R_a^{\mathfrak{C}}[x'] = R_a^{\mathfrak{A}'}[x'] \cup \{y\}. \tag{4}$$

For any $j \in I$ and conjunct of δ of the form $\Diamond_j\gamma$, given $\mathfrak{A}', x' \vDash \Diamond_j\gamma$, there is a $v \in R_j^{\mathfrak{A}'}[x']$ such that $\mathfrak{A}', v \vDash \gamma$, which implies $\mathfrak{C}, v \vDash \gamma$ by (2), given $x' \notin R_j^{\mathfrak{A}'}[x']$. Then since $R_j^{\mathfrak{A}'}[x'] \subseteq R_j^{\mathfrak{C}}[x']$ by (3) and (4), $\mathfrak{C}, x' \vDash \Diamond_j\gamma$.

For any $b \in I \setminus \{a\}$ and conjunct of δ of the form $\Box_b\beta$, given $\mathfrak{A}', x' \vDash \Box_b\beta$, we have that for all $v \in R_b^{\mathfrak{A}'}[x']$, $\mathfrak{A}', v \vDash \beta$, which implies $\mathfrak{C}, v \vDash \beta$ by (2), given $x' \notin R_b^{\mathfrak{A}'}[x']$. Then since $R_b^{\mathfrak{C}}[x'] \subseteq R_b^{\mathfrak{A}'}[x']$ by (3), $\mathfrak{C}, x' \vDash \Box_b\beta$.

Finally, for any conjunct of δ of the form $\Box_a\beta$, given $\mathfrak{A}', x' \vDash \Box_a\beta$, we have that for all $v \in R_a^{\mathfrak{A}'}[x']$, $\mathfrak{A}', v \vDash \beta$, which implies $\mathfrak{C}, v \vDash \beta$ by (2), given $x' \notin R_a^{\mathfrak{A}'}[x']$. Now since $\Box_a\beta$ is a conjunct of δ, β is a conjunct of $\mathsf{f}_a^{\mathbf{L}}(\delta)$, so given $\mathfrak{C}, y \vDash \mathsf{f}_a^{\mathbf{L}}(\delta)$, we have $\mathfrak{C}, y \vDash \beta$. Combining this with the fact that for all $v \in R_a^{\mathfrak{A}'}[x']$, $\mathfrak{C}, v \vDash \beta$, it follows by (4) that $\mathfrak{C}, x' \vDash \Box_a\beta$. Thus, $\mathfrak{C}, x' \vDash \delta$.

Putting together the previous arguments, we have shown $\mathfrak{C}, x' \vDash \delta \wedge \Diamond_a\neg\psi$ and hence $\mathfrak{C}, x' \vDash \varphi \wedge \Diamond_a\neg\psi$. Now if $\mathbf{L}$ is $\mathbf{K}$ (resp. $\mathbf{KD}$), then given that $\mathfrak{C} \in \mathsf{C}_{\mathbf{K}}$ (resp. $\mathfrak{C} \in \mathsf{C}_{\mathbf{KD}}$) by construction, it follows by soundness that $\nvdash_{\mathbf{L}} \varphi \to \Box_a\psi$. This completes the proof of the theorem for $\mathbf{K}$ and $\mathbf{KD}$.

Now for $\mathbf{L} \in \{\mathbf{T}, \mathbf{K4}, \mathbf{KD4}, \mathbf{S4}\}$, define $\mathfrak{C}_{\mathbf{L}}$ to be exactly like $\mathfrak{C}$ except that for every $b \in I$, $R_b^{\mathfrak{C}_{\mathbf{L}}}$ is the *reflexive* and/or *transitive closure* of $R_b^{\mathfrak{C}}$, depending on whether T and/or 4 are axioms of $\mathbf{L}$.[19] Thus, $\mathfrak{C}_{\mathbf{L}} \in \mathsf{C}_{\mathbf{L}}$. For example, see $\mathfrak{C}_{\mathbf{K4}}$ at the bottom of Fig. 2. Now we must check that we still have $\mathfrak{C}_{\mathbf{L}}, x' \vDash \delta$ and $\mathfrak{C}_{\mathbf{L}}, y \vDash \mathsf{f}_a^{\mathbf{L}}(\delta) \wedge \neg\psi$. Since $W^{\mathfrak{C}} = W^{\mathfrak{C}_{\mathbf{L}}}$ and $\forall z \in W^{\mathfrak{C}} \setminus \{x'\}\colon x' \notin R_b^{\mathfrak{C}_{\mathbf{L}}}[z]$, the identity relation on $W^{\mathfrak{C}} \setminus \{x'\}$ is a bisimulation between $\mathfrak{C}$ and $\mathfrak{C}_{\mathbf{L}}$, so

$$\forall w \in W^{\mathfrak{C}} \setminus \{x'\}\ \forall \chi \in \mathcal{L}\colon \mathfrak{C}, w \vDash \chi \text{ iff } \mathfrak{C}_{\mathbf{L}}, w \vDash \chi. \tag{5}$$

---

[18] For a world model $\mathfrak{M}$, $w \in W^{\mathfrak{M}}$, and $i \in I$, let $R_i^{\mathfrak{M}}[w] = \{v \in W^{\mathfrak{M}} \mid wR_iv\}$.

[19] Note that for $b \neq a$, the transitive closure of $R_b^{\mathfrak{C}}$ is just $R_b^{\mathfrak{C}}$ itself. However, the reflexive closure of $R_b^{\mathfrak{C}}$ is not $R_b^{\mathfrak{C}}$ itself, because we do not have $x'R_b^{\mathfrak{C}}x'$.

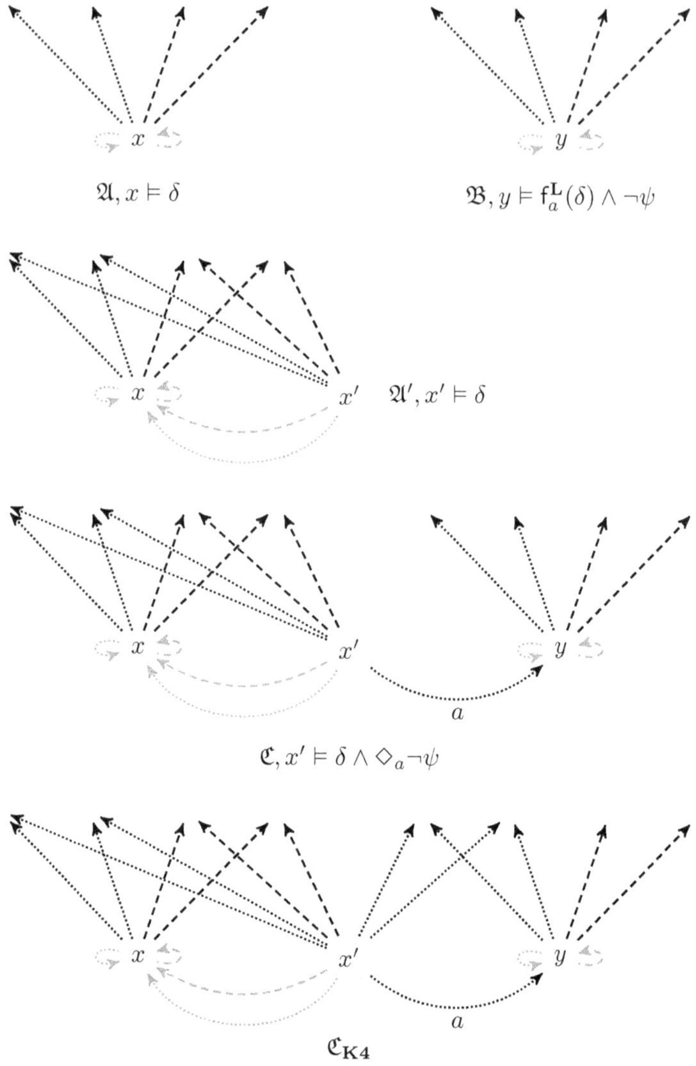

Fig. 2. models $\mathfrak{A}$ (upper left), $\mathfrak{B}$ (upper right), $\mathfrak{A}'$ (below $\mathfrak{A}$), $\mathfrak{C}$ (below $\mathfrak{A}'$), and $\mathfrak{C}_{\mathbf{K4}}$ (below $\mathfrak{C}$). Gray arrows might be included, depending on the initial models $\mathfrak{A}$ and $\mathfrak{B}$.

Thus, $\mathfrak{C}, y \models \mathsf{f}^{\mathbf{L}}_a(\delta) \wedge \neg\psi$ implies $\mathfrak{C}_{\mathbf{L}}, y \models \mathsf{f}^{\mathbf{L}}_a(\delta) \wedge \neg\psi$. It remains to show $\mathfrak{C}_{\mathbf{L}}, x' \models \delta$.

The propositional part of δ is still true at x' in $\mathfrak{C}_{\mathbf{L}}$, since the valuation on x' has not changed from $\mathfrak{C}$ to $\mathfrak{C}_{\mathbf{L}}$. From (5) and the fact that for every $j \in I$, $x' \notin \mathrm{R}^{\mathfrak{C}}_j[x']$ and $\mathrm{R}^{\mathfrak{C}}_j[x'] \subseteq \mathrm{R}^{\mathfrak{C}_{\mathbf{L}}}_j[x']$, it follows that the conjuncts of δ of the form $\Diamond_j \gamma$ are still true at x'. We need only check that every conjunct of δ of the form $\Box_j \beta$ is still true at x'. The argument depends on the choice of $\mathbf{L}$.

Let us begin with $\mathbf{T}$, so we can assume by Lemma 3.6 that φ is T-unpacked. Since for all $j \in I$, $\mathrm{R}_j^{\mathfrak{C}_{\mathbf{T}}}$ is the reflexive closure of $\mathrm{R}_j^{\mathfrak{C}}$, it follows that

$$\mathrm{R}_j^{\mathfrak{C}_{\mathbf{T}}}[x'] = \mathrm{R}_j^{\mathfrak{C}}[x'] \cup \{x'\}. \tag{6}$$

For any $j \in I$ and conjunct of δ of the form $\Box_j\beta$, given $\mathfrak{C}, x' \vDash \Box_j\beta$, we have that for all $v \in \mathrm{R}_j^{\mathfrak{C}}[x']$, $\mathfrak{C}, v \vDash \beta$. It follows given (5) and $x' \notin \mathrm{R}_j^{\mathfrak{C}}[x']$ that

$$\text{for any } j \in I, \text{ conjunct of } \delta \text{ of the form } \Box_j\beta, \text{ and } v \in \mathrm{R}_j^{\mathfrak{C}}[x']\colon \mathfrak{C}_{\mathbf{T}}, v \vDash \beta. \tag{7}$$

Thus, by (6), to show $\mathfrak{C}_{\mathbf{T}}, x' \vDash \Box_j\beta$ it only remains to show that $\mathfrak{C}_{\mathbf{T}}, x' \vDash \beta$. Since φ is T-unpacked, for each $\Box_j\beta$ conjunct of δ, there is some disjunct δ_β of β such that every conjunct of δ_β is a conjunct of δ. Given this fact, we can prove by induction on the modal depth $d(\beta)$ of β that $\mathfrak{C}_{\mathbf{T}}, x' \vDash \beta$.

If $d(\beta) = 0$, so β is propositional, then δ_β is a propositional conjunct of δ, so $\mathfrak{C}, x' \vDash \delta$ implies $\mathfrak{C}, x' \vDash \delta_\beta$, which implies $\mathfrak{C}_{\mathbf{T}}, x' \vDash \delta_\beta$, since δ_β is propositional, which implies $\mathfrak{C}_{\mathbf{T}}, x' \vDash \beta$, since δ_β is a disjunct of β.

If $d(\beta) = n + 1$, then by the inductive hypothesis, for every $\Box_j\chi$ conjunct of δ with $d(\chi) \leq n$, $\mathfrak{C}_{\mathbf{T}}, x' \vDash \chi$. Since φ is T-unpacked, there is a disjunct δ_β of β such that every conjunct of δ_β is a conjunct of δ. As shown above, every propositional conjunct of δ is true at $\mathfrak{C}_{\mathbf{T}}, x'$, and every conjunct of δ of the form $\Diamond_j\gamma$ is true at $\mathfrak{C}_{\mathbf{T}}, x'$, so every propositional conjunct of δ_β and every conjunct of δ_β of the form $\Diamond_j\gamma$ is true at $\mathfrak{C}_{\mathbf{T}}, x'$. Thus, to establish $\mathfrak{C}_{\mathbf{T}}, x' \vDash \delta_\beta$, it only remains to show that every conjunct of δ_β of the form $\Box_j\chi$ is true at $\mathfrak{C}_{\mathbf{T}}, x'$. Since $d(\beta) = n+1$ and $\Box_j\chi$ is a conjunct of δ_β, $d(\chi) \leq n$, so by the inductive hypothesis, $\mathfrak{C}_{\mathbf{T}}, x' \vDash \chi$; and since $\Box_j\chi$ is a conjunct of δ, we have from (7) that for all $v \in \mathrm{R}_j^{\mathfrak{C}}[x'] = \mathrm{R}_j^{\mathfrak{C}_{\mathbf{T}}}[x'] \setminus \{x'\}$, $\mathfrak{C}_{\mathbf{T}}, v \vDash \chi$. Putting these two facts together, it follows from (6) that $\mathfrak{C}_{\mathbf{T}}, x' \vDash \Box_j\chi$. This completes the proof of $\mathfrak{C}_{\mathbf{T}}, x' \vDash \delta_\beta$ and hence $\mathfrak{C}_{\mathbf{T}}, x' \vDash \beta$, which is all that was left to show $\mathfrak{C}_{\mathbf{T}}, x' \vDash \Box_j\beta$.

Let us now show for **K4**/**KD4** that every conjunct of δ of the form $\Box_j\beta$ is true at x'. Since for all $j \in I$, $\mathrm{R}_j^{\mathfrak{C}_{\mathbf{K4}}}$ is the transitive closure of $\mathrm{R}_j^{\mathfrak{C}}$, we have:

$$\mathrm{R}_b^{\mathfrak{C}_{\mathbf{K4}}}[x'] = \mathrm{R}_b^{\mathfrak{C}}[x'] \text{ for } b \in I \setminus \{a\}; \tag{8}$$

$$\mathrm{R}_a^{\mathfrak{C}_{\mathbf{K4}}}[x'] = \mathrm{R}_a^{\mathfrak{C}}[x'] \cup \mathrm{R}_a^{\mathfrak{C}}[y]. \tag{9}$$

For any $b \in I \setminus \{a\}$ and conjunct of δ of the form $\Box_b\beta$, given $\mathfrak{C}, x' \vDash \Box_b\beta$, we have that for all $v \in \mathrm{R}_b^{\mathfrak{C}}[x']$, $\mathfrak{C}, v \vDash \beta$, which implies $\mathfrak{C}_{\mathbf{K4}}, v \vDash \beta$ by (5), given $x' \notin \mathrm{R}_b^{\mathfrak{C}}[x']$. Then by (8), $\mathfrak{C}_{\mathbf{K4}}, x' \vDash \Box_b\beta$.

For any conjunct of δ of the form $\Box_a\beta$, given $\mathfrak{C}, x' \vDash \Box_a\beta$, we have that for all $v \in \mathrm{R}_a^{\mathfrak{C}}[x']$, $\mathfrak{C}, v \vDash \beta$, which implies $\mathfrak{C}_{\mathbf{K4}}, v \vDash \beta$ by (5), given $x' \notin \mathrm{R}_a^{\mathfrak{C}}[x']$. Now since $\Box_a\beta$ is a conjunct of δ, $\Box_a\beta$ is also a conjunct of $\mathfrak{f}_a^{\mathbf{K4}}(\delta)$, so given $\mathfrak{C}, y \vDash \mathfrak{f}_a^{\mathbf{K4}}(\delta)$, we have $\mathfrak{C}, y \vDash \Box_a\beta$. Thus, for all $u \in \mathrm{R}_a^{\mathfrak{C}}[y]$, $\mathfrak{C}, u \vDash \beta$, which implies $\mathfrak{C}_{\mathbf{K4}}, u \vDash \beta$ by (5), given $x' \notin \mathrm{R}_a^{\mathfrak{C}}[y]$. Combining this with the fact that for all $v \in \mathrm{R}_a^{\mathfrak{C}}[x']$, $\mathfrak{C}_{\mathbf{K4}}, v \vDash \beta$, it follows by (9) that $\mathfrak{C}_{\mathbf{K4}}, x' \vDash \Box_a\beta$.

This completes the proof for **K4**, and the same applies to **KD4**.

For **S4**, a combination of the arguments above for **KT** and **K4** works.

We have now shown that for all $\mathbf{L} \in \{$**K**, **KD**, **T**, **K4**, **KD4**, **S4**$\}$, there is a model $\mathfrak{M} \in \mathsf{C}_{\mathbf{L}}$ (i.e., $\mathfrak{C}$ for **K**/**KD** or $\mathfrak{C}_{\mathbf{L}}$ for the others) such that $\mathfrak{M}, x' \vDash \delta$,

$\mathfrak{M}, y \vDash \neg\psi$, and $x' \mathrm{R}_a^{\mathfrak{M}} y$, which implies $\mathfrak{M}, x' \vDash \varphi \wedge \Diamond_a \neg\psi$. Then since $\mathbf{L}$ is sound with respect to $\mathsf{C}_\mathbf{L}$ and $\mathfrak{M} \in \mathsf{C}_\mathbf{L}$, it follows that $\nvdash_\mathbf{L} \varphi \to \Box_a \psi$. $\Box$

References

[1] van Benthem, J., *Possible Worlds Semantics for Classical Logic*, Technical Report ZW-8018, Department of Mathematics, Rijksuniversiteit, Groningen (1981).

[2] van Benthem, J., *Partiality and Nonmonotonicity in Classical Logic*, Logique et Analyse **29** (1986), pp. 225–247.

[3] van Benthem, J., "A Manual of Intensional Logic," CSLI Publications, Stanford, 1988, 2nd revised and expanded edition.

[4] van Benthem, J., *Beyond Accessibility: Functional Models for Modal Logic*, in: M. de Rijke, editor, *Diamonds and Defaults*, Kluwer Academic Publishers, Dordrecht, 1993 pp. 1–18.

[5] Bezhanishvili, N. and A. Kurz, *Free Modal Algebras: A Coalgebraic Perspective*, in: T. Mossakowski, U. Montanari and M. Haveraaen, editors, *Algebra and Coalgebra in Computer Science*, Lectures Notes in Computer Science **4624**, Springer, 2007 pp. 143–157.

[6] Blackburn, P., M. de Rijke and Y. Venema, "Modal Logic," Cambridge University Press, New York, 2001.

[7] Cresswell, M., *Possibility Semantics for Intuitionistic Logic*, Australasian Journal of Logic **2** (2004), pp. 11–29.

[8] Dunn, J. M. and G. M. Hardegree, "Algebraic Methods in Philosophical Logic," Oxford University Press, New York, 2001.

[9] Fine, K., *Models for Entailment*, Journal of Philosophical Logic **3** (1974), pp. 347–372.

[10] Fine, K., *Normal Forms in Modal Logic*, Notre Dame Journal of Formal Logic **16** (1975), pp. 229–237.

[11] Gargov, G., S. Passy and T. Tinchev, *Modal Environment for Boolean Speculations*, in: D. Skordev, editor, *Mathemtatical Logic and Its Applications*, Plenum Press, New York, 1987 pp. 253–263.

[12] Garson, J. W., "What Logics Mean: From Proof Theory to Model-Theoretic Semantics," Cambridge University Press, Cambridge, 2013.

[13] Ghilardi, S., *An algebraic theory of normal forms*, Annals of Pure and Applied Logic **71** (1995), pp. 189–245.

[14] Harrison-Trainor, M., *First-Order Possibility Models and their Worldizations* (2014), manuscript.

[15] Holliday, W. H., *Epistemic Closure and Epistemic Logic I: Relevant Alternatives and Subjunctivism*, Journal of Philosophical Logic (2014), DOI 10.1007/s10992-013-9306-2.

[16] Holliday, W. H., *From Worlds to Possibilities for Knowledge and Belief* (2014), manuscript.

[17] Humberstone, I. L., *From Worlds to Possibilities*, Journal of Philosophical Logic **10** (1981), pp. 313–339.

[18] Humberstone, I. L., *Heterogeneous Logic*, Erkenntnis **10** (1988), pp. 395–435.

[19] Humberstone, L., "The Connectives," MIT Press, Cambridge, Mass., 2011.

[20] Humberstone, L. and T. Williamson, *Inverses for Normal Modal Operators*, Studia Logica **59** (1997), pp. 33–64.

[21] Kripke, S. A., *Semantical Analysis of Intuitionistic Logic I*, in: J. Crossley and M. Dummett, editors, *Formal Systems and Recursive Functions*, North-Holland Publishing Company, Amsterdam, 1965 pp. 92–130.

[22] Moss, L. S., *Finite Models Constructed from Canonical Formulas*, Journal of Philosophical Logic **36** (2007), pp. 605–640.

[23] Rumfitt, I., *On A Neglected Path to Intuitionism*, Topoi **31** (2012), pp. 101–109.

[24] Wansing, H., "Displaying Modal Logic," Springer, Dordrecht, 1998.

Filtration Safe Operations on Frames

Stanislav Kikot [1]

Institute for Information Transmission Problems, Russian Academy of Sciences
Bolshoy Karetniy 19, Moscow GSP-4, 101 447, Russia

Ilya Shapirovsky [2]

Institute for Information Transmission Problems, Russian Academy of Sciences
Bolshoy Karetniy 19, Moscow GSP-4, 101 447, Russia

Evgeny Zolin [3]

Department of Mathematical Logic and Theory of Algorithms, Faculty of Mechanics and Mathematics, Moscow State University,
Leninskie Gory 1, Moscow GSP-1, 119 991, Russia

Abstract

Filtration is a standard tool for establishing the finite model property of modal logics. We consider logics and classes of frames that admit filtration, and identify some operations on them that preserve this property. In particular, the operation of adding the inverse or the transitive closure of a relation is shown to be safe in this sense. These results are then used to prove that every regular grammar logic with converse admits filtration. We present filtration constructions for right-linear and left-linear grammar logics. We also give a simple example of a grammar modal logic that is undecidable and hence does not admit filtration.

Keywords: Modal logic, tense logic, finite model property, filtration, transitive closure, universal modality, grammar modal logic, Horn closure, regular grammar, propositional dynamic logic.

Introduction

Filtration is a method of collapsing an infinite model into a finite one while preserving the truth values of a given finite set of formulas. In modal logic, it is widely used as a tool for establishing the finite model property (FMP) and decidability. It dates back to the pioneering works of Scott, Lemmon [15] and Segerberg [18]. This technique was used by Fischer and Ladner who designed

[1] staskikotx@gmail.com

[2] ilya.shapirovsky@gmail.com. The research is supported by RFBR grant 14-01-31442.

[3] ezolin@gmail.com. The research is supported by RFBR grant 14-01-00127.

a filtration for **PDL** [7], Shehtman who developed a filtration for products of modal logics (see e.g. [20]), and many others.

Gabbay [9] was perhaps the first who introduced the term "*a logic* **L** *admits filtrations*", which means that any model over an **L**-frame can be "filtrated" into a finite model again over an **L**-frame. He summarized that the logics **K**, **T**, **B**, **S4**, **S5**, **S4.1** admit filtration, and extended this list with the logics $\mathbf{K} + \Box p \to \Box^m p$, $m \geq 2$; the latter are the simplest examples of the so called *regular grammar logics* [5].

Our aim here is to investigate modal logics and classes of Kripke frames that admit filtration (or have the *AF property*), from the following viewpoint. We tackle the problem of identifying the cases when it is possible to transfer the AF property from a modal logic (or a class of frames) to its enrichment with new kinds of modalities (or accessibility relations).

Specifically, given a class $\mathcal{F}$ of frames of the form $F = (W, (R_e)_{e \in \Sigma})$, we consider the corresponding class $\mathcal{F}^{\circledast} = \{F^{\circledast} \mid F \in \mathcal{F}\}$ of their expansions $F^{\circledast} = (W, (R_e)_{e \in \Sigma}, S)$, where the relation $S \subseteq W \times W$ is obtained by an operation $\circledast$ on (the relations in) F. We investigate for which operations $\circledast$, the AF property for $\mathcal{F}$ implies that for $\mathcal{F}^{\circledast}$; we call such operations *filtration safe.*

This question has already been addressed before. In particular, Goranko and Passy [11] proved that adding the universal relation is filtration safe. Bezhanishvili and ten Cate [1] proved that taking the hybrid companion of a logic is filtration safe in [8], the AF property was applied to products of modal logics.

In general, the AF property is not only a tool for obtaining the FMP, but also a strong sufficient condition for the decidability of many derived modal logics, and so we believe it is worthy of studying per se. In particular, when we prove the AF property for some logic for which the decidability and FMP had already been established by other methods, this still increases our knowledge, because it allows us to make conclusions about the derived logics.

In this paper, we identify several frame expanding operations that preserve the AF property, namely: adding the union $R_a \cup R_b$, the composition $R_a \circ R_b$, the diagonal relation $\{(x, x) \mid x \in W\}$, and most interestingly – the inverse relation R_a^{-1} and the transitive closure relation R_a^{+}. Consequently, if a modal logic **L** admits filtration, then so does its tense counterpart $\mathbf{L}^t$, or extension with the transitive closure modality $\mathbf{L}^{\boxplus}$ (despite being "extremely dangerous" in general, cf. [2, p. 373]), provided that the logics obtained are Kripke complete. The result about $\mathbf{L}^t$ seems particularly interesting to us in context of Wolter's program of "temporalization" of modal logics [23,24,25,26,27].

We then apply these results to grammar modal logics [5]. We show that, for a given regular grammar Π, starting from the class of all frames, one can apply a sequence of operations of the above kind and arrive at the class of all Π-frames, i.e., frames that satisfy the set of inclusions corresponding to the rules in Π. This yields a simple proof of the result that every *regular grammar logic* (with converse) admits filtration (and hence has the FMP and is decidable). Note that the FMP for these logics was already known from [5,6].

Finally, we give two examples of logics that do not admit filtrations. To this end, we show that the global satisfiability problem is undecidable for the logics $\mathbf{K}.2 = \mathbf{K}+\Diamond\Box p \to \Box\Diamond p$ and $\mathbf{K}_2+[a]p \to [b][a][b]p$. The latter corresponds to the *irregular* grammar $a \to bab$. This enables us to build a simple undecidable context-free grammar logic, which complements Demri's paper [5].

Section 1 introduces the notion of a logic (or a class of frames) that admits filtration. In Section 2, we identify some filtration safe operations on frames and obtain the corresponding transfer results for logics (Section 2.4). In Section 3, we recall the notion of a grammar logic and give a short proof (using the results of Section 2) of the fact that every regular grammar logic (with converse) admits filtration. Section 4 presents explicit filtration constructions for logics that correspond to some subfamilies of regular grammars. Finally, Section 5 contains the above mentioned undecidability results. The paper concludes with the discussion of open questions and further directions of research.

1 Preliminaries

We assume the reader to be familiar with syntax and semantics of multi-modal logic [2,4], so we only briefly recall some notions and fix notation. Let Σ be a finite alphabet (of indices for modalities). The set $\mathsf{Fm}(\Sigma)$ of *modal formulas over* Σ is defined from propositional letters $\mathsf{Var} = \{p_0, p_1, \dots\}$ using Boolean connectives and the modalities $[e]$, for $e \in \Sigma$, according to the syntax:

$$\varphi \ ::= \ \bot \ \mid \ p_i \ \mid \ \varphi \to \psi \ \mid \ [e]\varphi.$$

We use standard abbreviations (e.g., $\top, \wedge$); in particular, $\langle e\rangle\varphi := \neg[e]\neg\varphi$. For a set of formulas Γ, by $\mathsf{Sub}(\Gamma)$ we denote the set of all subformulas of formulas from Γ. We say that Γ is Sub-*closed* if $\mathsf{Sub}(\Gamma) \subseteq \Gamma$.

A (Σ-)*frame* is a pair $F = (W, (R_e)_{e\in\Sigma})$, where $W \neq \varnothing$ and $R_e \subseteq W\times W$ for $e \in \Sigma$. A *model* based on F is a pair $M = (F, V)$, where $V(p) \subseteq W$, for all $p \in \mathsf{Var}$. The *truth relation* $M, x \models \varphi$ is defined in the usual way, e.g.

$$M, x \models [e]\varphi \quad \leftrightharpoons \quad \text{for all } y \in W, \ \text{ if } x\, R_e\, y \ \text{ then } \ M, y \models \varphi.$$

A formula φ is *valid* in F, notation $F \models \varphi$, if $M, x \models \varphi$ for all M based on F and all worlds x in F. For a class of frames $\mathcal{F}$, an $\mathcal{F}$*-model* is a model based on a frame from $\mathcal{F}$. A formula φ is *satisfiable* in $\mathcal{F}$ if it is true in some world of some $\mathcal{F}$-model; φ is *globally satisfiable* in $\mathcal{F}$ if it is true in some $\mathcal{F}$-model.

A (*modal*) *logic* (*over* Σ) is a set of formulas $\mathbf{L}$ that contains all classical tautologies, the axioms $[e](p \to q) \to ([e]p \to [e]q)$, for each $e \in \Sigma$, and is closed under the rules of modus ponens, substitution, and necessitation (from φ, infer $[e]\varphi$, for each $e \in \Sigma$). An $\mathbf{L}$-*frame* is a frame in which $\mathbf{L}$ is valid. *The logic of a class of frames* $\mathcal{F}$ is the set of all formulas that are valid in $\mathcal{F}$. A logic is *Kripke complete* if it is the logic of some class of frames. A logic $\mathbf{L}$ has the *finite model property* (FMP) if it is the logic of some class of finite frames; or equivalently (see [2, Th. 3.28]) if, for every formula $\varphi \notin \mathbf{L}$, there is a finite $\mathbf{L}$-frame F such that $F \not\models \varphi$. If, additionally, the size of F is at most exponential in size of φ, we say that $\mathbf{L}$ has the *exponential model property* (ExpMP).

1.1 Filtration

The notion of a filtration we introduce below slightly generalizes the standard one (cf. [2, Def. 2.36], [4, Sect. 5.3]) in the following aspect: given a finite set of formulas Γ, we define a filtration as a model obtained by factorizing a given model w.r.t. an equivalence relation that we allow to be *finer* than the one induced by Γ. This modification seems to first appear in [19]; see also [20].

Let $M = (W, (R_e)_{e\in\Sigma}, V)$ be a model and Γ a finite Sub-closed set of Σ-formulas. An equivalence relation $\sim$ on W is *of finite index* if the quotient set $W/{\sim}$ is finite. The equivalence relation *induced by* Γ is defined as follows:

$$x \sim_\Gamma y \quad \leftrightharpoons \quad \forall \varphi \in \Gamma \ \big(M, x \models \varphi \ \Leftrightarrow \ M, y \models \varphi\big).$$

Clearly, $\sim_\Gamma$ is of finite index. We say that an equivalence relation $\sim$ *respects* Γ if $\sim \, \subseteq \, \sim_\Gamma$; in other words, if for every $\sim$-class $\alpha \subseteq W$ and every formula $\varphi \in \Gamma$, φ is either true in all worlds of α or false in all worlds of α.

Definition 1.1 (Filtration) A *filtration* of a model M that *respects* a set of formulas Γ (or a Γ-*filtration of* M) is any model $\widehat{M} = (\widehat{W}, (\widehat{R}_e)_{e\in\Sigma}, \widehat{V})$ satisfying the following conditions:

- $\widehat{W} = W/{\sim}$, for some equivalence relation of finite index $\sim$ on W;
- the equivalence relation $\sim$ respects Γ;
- the valuation $\widehat{V}$ is defined on the variables $p \in \Gamma$ canonically: $\widehat{x} \models p \Leftrightarrow x \models p$, for all worlds $x \in W$, where $\widehat{x}$ denotes the $\sim$-class of x;
- $R_e^{\mathsf{min}} \subseteq \widehat{R}_e \subseteq \Gamma_e$, for each $e \in \Sigma$. Here R_e^{min} is the e-th *minimal filtered relation* on $\widehat{W}$, and Γ_e is the e-th *maximal filtered relation*[4] on $\widehat{W}$ induced by the set of formulas Γ; they are defined in the usual way:

$$\begin{aligned}
\widehat{x}\, R_e^{\mathsf{min}}\, \widehat{y} \ &\leftrightharpoons \ \exists x' \sim x \ \exists y' \sim y\colon \ x' R_e\, y',\\
\widehat{x} \ \Gamma_e \ \widehat{y} \ &\leftrightharpoons \ \text{for every formula } [e]\varphi \in \Gamma \ \big(M, x \models [e]\varphi \ \Rightarrow \ M, y \models \varphi\big).
\end{aligned}$$

Note that the relations R_e^{min} and Γ_e are well-defined on $\sim$-classes, and that $R_e^{\mathsf{min}} \subseteq \Gamma_e$ always holds. The condition $R_e^{\mathsf{min}} \subseteq \widehat{R}_e$ is equivalent to that $\forall x, y \in W \ (x\, R_e\, y \Rightarrow \widehat{x}\, \widehat{R}_e\, \widehat{y})$. A filtration is always a finite model. The following lemma states the main property of filtrations (cf. [2, Th. 2.39], [4, Th. 5.23]).

Lemma 1.2 (Filtration lemma) *Let Γ be a finite* Sub*-closed set of formulas. Suppose that $\widehat{M}$ is a Γ-filtration of a model M. Then, for all worlds $x \in W$ and all formulas $\varphi \in \Gamma$, the equivalence holds: $M, x \models \varphi \Leftrightarrow \widehat{M}, \widehat{x} \models \varphi$.*

Definition 1.3 (AF) We say that a class of frames $\mathcal{F}$ *admits filtration* if, for every finite Sub-closed set of formulas Γ and every $\mathcal{F}$-model M, there exists a $\mathcal{F}$-model that is a Γ-filtration of M. A logic $\mathbf{L}$ *admits filtration* if it is Kripke complete and the class of all $\mathbf{L}$-frames admits filtration.

---

[4] In literature, it is sometimes denoted by $(R_e)^{\mathsf{max}}$. In fact, however, it depends not (only) on the relation R_e, but on Γ (and M). Later we use maximal filtered relations induced by different sets of formulas, so we need a notation that allows us to distinguish between them.

Example 1.4 **([9,4]).** The logics **K**, **T**, **K4**, **S4**, **B**, **S5**, **S4.1**, **K**+$\Box p \to \Box^m p$, for $m \geq 0$, the multi-modal **K** (i.e., $\mathbf{K}_n$) admit filtration. The classes of point-generated **S4.2**-frames and point-generated **S4.3**-frames admit filtration.

Theorem 1.5 (AF implies FMP) *If a logic admits filtration then it has the FMP. If additionally it is finitely axiomatizable, then it is decidable.*

Proof. By the standard argument, cf. [4, Corollary 5.26]. □

A trivial (but important) remark is that any class of *finite* frames admits filtration, since any finite model is a filtration of itself. Hence, one cannot say that **L** admits filtration if it is the logic of *some* class of frames that admits filtration (in the sense of Definition 1.3), otherwise the notions of AF and FMP would coincide. Indeed, AF implies FMP; conversely, an FMP logic is the logic of the class of all finite **L**-frames, which trivially admits filtration.

Below, we need the following lemma on the relationship between some operations on relations (or on frames) and minimal filtered relations.

Lemma 1.6 *For any equivalence relation* $\sim$ *on* W *and any relations on* W,

(1) $(\mathtt{Id}(W))^{\min} = \mathtt{Id}(\widehat{W})$	(4) $(R^{-1})^{\min} = (R^{\min})^{-1}$
(2) $(W \times W)^{\min} = \widehat{W} \times \widehat{W}$	(5) $(R \circ S)^{\min} \subseteq R^{\min} \circ S^{\min}$
(3) $\left(\bigcup_{i \in I} R_i\right)^{\min} = \bigcup_{i \in I} R_i{}^{\min}$	(6) $(R^+)^{\min} \subseteq (R^{\min})^+$

Here $\mathtt{Id}(W) := \{(x,x) \mid x \in W\}$, *and* R^+ *is the transitive closure of* R.

Proof. (1), (2), and (4) are trivial. (3) follows from that $\exists$ distributes over $\vee$. It remains to prove (5), because (3) and (5) together imply (6).

(5) If $(\widehat{x}, \widehat{y}) \in (R \circ S)^{\min}$ then $\exists x' \sim x \; \exists y' \sim y$ with $x' \, (R \circ S) \, y'$, hence $x' \, R \, z \, S \, y'$, for some $z \in W$. Thus $\widehat{x} \, R^{\min} \, \widehat{z} \, S^{\min} \, \widehat{y}$, so $(\widehat{x}, \widehat{y}) \in (R^{\min} \circ S^{\min})$. □

2 Transferring the 'admits filtration' property

Here we point out several operations on frames that preserve the AF property, in the sense that if a class of frames admits filtration, then so does the class of all frames transformed by this operation.

2.1 Simple operations on frames

Let $\Sigma' \subset \Sigma$. Given a frame $F = (W, (R_e)_{e \in \Sigma})$, denote by $F|_{\Sigma'} = (W, (R_e)_{e \in \Sigma'})$ its Σ'-*reduct*. For a class of frames $\mathcal{F}$, denote $\mathcal{F}|_{\Sigma'} = \{F|_{\Sigma'} \mid F \in \mathcal{F}\}$. The reader can easily prove the following lemma.

Lemma 2.1 *If* $\mathcal{F}$ *admits filtration, then so does* $\mathcal{F}|_{\Sigma'}$, *for any* $\Sigma' \subset \Sigma$.

For a class $\mathcal{F}$ of frames of the form [5] $F = (W, R)$, let $\mathcal{F}^u := \{F^u \mid F \in \mathcal{F}\}$, where $F^u = (W, R, W \times W)$ is the frame F enriched by the *universal relation*.

Lemma 2.2 ([11, Th. 5.9]) *If* $\mathcal{F}$ *admits filtration then so does* $\mathcal{F}^u$.

---

[5] Althouth we often consider the uni-modal case below, this is for simplicity only; one can always assume that frames have additional relations, as they do not influence the proof much.

Although the notion of a logic that admits filtration used in [11] differs from ours, the same proof works for our case: simply substitute $\top$ or $\bot$ for all formulas of the form $[*]\varphi$, where $[*]$ is the *universal modality*, depending on whether φ is true or false in the given model, and then filtrate the $\mathcal{F}$-model.

Lemma 2.3 *If a class $\mathcal{F}$ of Σ-frames admits filtration, then so do the following classes of frames, for any fixed $a, b \in \Sigma$:*

$$\begin{array}{lll} (i) & \mathcal{F}^{\cup} & = \{ (W,(R_e)_{e\in\Sigma}, R_a \cup R_b) \mid (W,(R_e)_{e\in\Sigma}) \in \mathcal{F} \}, \\ (ii) & \mathcal{F}^{\circ} & = \{ (W,(R_e)_{e\in\Sigma}, R_a \circ R_b) \mid (W,(R_e)_{e\in\Sigma}) \in \mathcal{F} \}, \\ (iii) & \mathcal{F}^{=} & = \{ (W,(R_e)_{e\in\Sigma}, \mathtt{Id}(W)) \mid (W,(R_e)_{e\in\Sigma}) \in \mathcal{F} \}. \end{array}$$

Proof. (i) To prove that $\mathcal{F}^{\cup}$ admits filtration, given a finite Sub-closed set of formulas $\Gamma \subset \mathsf{Fm}(\Sigma \cup \{c\})$, where $c \notin \Sigma$, and an $\mathcal{F}^{\cup}$-model $M^{\cup} = (W,(R_e)_{e\in\Sigma}, R_c, V)$, with $R_c = R_a \cup R_b$ and $M = (W,(R_e)_{e\in\Sigma}, V)$ an $\mathcal{F}$-model, let us show how to build an $\mathcal{F}^{\cup}$-model that is a Γ-filtration of $M^{\cup}$.

Let us introduce a translation $(\cdot)^*$ from $\mathsf{Fm}(\Sigma\cup\{c\})$ to $\mathsf{Fm}(\Sigma)$ that preserves variables, $\to$, $\bot$, $[e]$ for each $e \in \Sigma$, and satisfies: $([c]\varphi)^* = [a]\varphi^* \wedge [b]\varphi^*$. One can easily show that, for any $x \in W$ and any formula $\varphi \in \mathsf{Fm}(\Sigma \cup \{c\})$,

$$M^{\cup}, x \models \varphi \iff M, x \models \varphi^*. \tag{$*$}$$

Consider the set of formulas $\Phi = \mathsf{Sub}\,\Gamma^* = \mathsf{Sub}\{\varphi^* \mid \varphi \in \Gamma\} \subseteq \mathsf{Fm}(\Sigma)$. Since $\mathcal{F}$ admits filtration, there is an $\mathcal{F}$-model $\widehat{M} = (\widehat{W},(\widehat{R}_e)_{e\in\Sigma},\widehat{V})$ that is a Φ-filtration of M. Here $\widehat{W} = W/{\sim}$, where $\sim$ respects Φ, $R_e^{\mathsf{min}} \subseteq \widehat{R}_e \subseteq \Phi_e$ for each $e \in \Sigma$, and $\widehat{x} \models p \Leftrightarrow x \models p$, for every variable p from $\mathsf{Var}(\Phi) = \mathsf{Var}(\Gamma)$.

Now extend $\widehat{M}$ to $N = (\widehat{W},(\widehat{R}_e)_{e\in\Sigma},\widehat{R}_c,\widehat{V})$ by putting $\widehat{R}_c := \widehat{R}_a \cup \widehat{R}_b$. Clearly, N is an $\mathcal{F}^{\cup}$-model. Let us show that N is a Γ-filtration of $M^{\cup}$. From $(*)$ it easily follows that $\sim$ respects Γ. It remains to prove the inclusions:

$$R_e^{\mathsf{min}} \overset{(1)}{\subseteq} \widehat{R}_e \overset{(2)}{\subseteq} \Gamma_e \quad \text{for all } e \in \Sigma, \qquad R_c^{\mathsf{min}} \overset{(3)}{\subseteq} \widehat{R}_c \overset{(4)}{\subseteq} \Gamma_c.$$

Here we already have (1); the inclusion $(R_a \cup R_b)^{\mathsf{min}} \subseteq R_a^{\mathsf{min}} \cup R_b^{\mathsf{min}}$ from Lemma 1.6(3) implies (3); (2) follows from (a) and (4) follow from (b) below.

(a) $\Phi_e \subseteq \Gamma_e$ for each $e \in \Sigma$. (Therefore, $\widehat{R}_e \subseteq \Phi_e \subseteq \Gamma_e$.)
Assume $\widehat{x}\,\Phi_e\,\widehat{y}$. To show $\widehat{x}\,\Gamma_e\,\widehat{y}$, take any $[e]\varphi \in \Gamma$. Then $[e]\varphi^* \in \Phi$ and so

$$x \models [e]\varphi \iff x \models [e]\varphi^* \implies y \models \varphi^* \iff y \models \varphi.$$

(b) $\Phi_a \cup \Phi_b \subseteq \Gamma_c$. (Here we use that $\mathcal{F}^{\cup} \models [c]p \to [a]p \wedge [b]p$.)
Assume $\widehat{x}\,(\Phi_a \cup \Phi_b)\,\widehat{y}$. Without loss of generality, $\widehat{x}\,\Phi_a\,\widehat{y}$. To prove that $\widehat{x}\,\Gamma_c\,\widehat{y}$, take any $[c]\varphi \in \Gamma$. Then $([c]\varphi)^* \in \Phi$ and so $[a]\varphi^* \in \Phi$, hence:

$$x \models [c]\varphi \iff x \models [c]\varphi^* \implies x \models [a]\varphi^* \implies y \models \varphi^* \iff y \models \varphi.$$

(ii) Use $([c]\varphi)^* = [a][b]\varphi^*$. Now (b) is: $\Phi_a \circ \Phi_b \subseteq \Gamma_c$ by Lemma 1.6(5).
(iii) Use $([c]\varphi)^* = \varphi^*$. Now $\widehat{R}_c := \mathtt{Id}(\widehat{W}) = R_c^{\mathsf{min}} \subseteq \Gamma_c$ by Lemma 1.6(1). □

2.2 Inverse relation

Given a class $\mathcal{F}$ of frames of the form $F = (W, R)$, denote $\mathcal{F}^t = \{F^t \mid F \in \mathcal{F}\}$, where $F^t = (W, R, R^{-1})$ is called the *tense expansion* of the frame F.

Theorem 2.4 *If $\mathcal{F}$ admits filtration then so does $\mathcal{F}^t$.*

Proof. Given a finite Sub-closed set of formulas $\Gamma \subset \mathsf{Fm}(\Box, \boxminus)$ and an $\mathcal{F}^t$-model $M^t = (W, R, R^{-1}, V)$, where $\Box$ refers to R and $\boxminus$ to R^{-1}, with $M = (W, R, V)$ an $\mathcal{F}$-model, we build an $\mathcal{F}^t$-model that is a Γ-filtration of M^t.

Let us introduce fresh variables $\{q_\varphi \mid \varphi \in \Gamma\}$ and extend the valuation V to them by putting:[6] $x \models q_\varphi \leftrightharpoons x \models \varphi$. Thus, $\varphi \leftrightarrow q_\varphi$ and hence $\Box\varphi \leftrightarrow \Box q_\varphi$ and $\Box\neg\varphi \leftrightarrow \Box\neg q_\varphi$ are true in M^t, for all $\varphi \in \Gamma$. Now consider the set of formulas:

$$\Phi \ := \ \mathsf{Sub}\{\ \Box q_\varphi, \Box\neg q_\varphi \ \mid \ \varphi \in \Gamma\ \} \ \subset \ \mathsf{Fm}(\Box).$$

Since the class $\mathcal{F}$ admits filtration, there exists an $\mathcal{F}$-model $\widehat{M} = (\widehat{W}, \widehat{R}, \widehat{V})$ that is a Φ-filtration of M. Here $\widehat{W} = W/{\sim}$, where $\sim$ respects Φ, $R^{\mathsf{min}} \subseteq \widehat{R} \subseteq \Phi_\Box$, and $\widehat{x} \models q \Leftrightarrow x \models q$, for all variables q from Φ. Let us extend $\widehat{V}$ to the variables p from Γ by putting $\widehat{x} \models p \leftrightharpoons \widehat{x} \models q_p$.

We claim that $\widehat{M}^t := (\widehat{W}, \widehat{R}, (\widehat{R})^{-1}, \widehat{V})$ is an $\mathcal{F}^t$-model (this is obvious) and a Γ-filtration of M^t, i.e., that $\sim$ respects Γ and the inclusions $R^{\mathsf{min}} \subseteq \widehat{R} \subseteq \Gamma_\Box$ and $(R^{-1})^{\mathsf{min}} \subseteq \widehat{R}^{-1} \subseteq \Gamma_\boxminus$ hold.

(a) *The equivalence relation $\sim$ respects the set of formulas Γ.*
Assume that $x \sim y$. Then, since $\sim$ respects Φ, we have the equivalences:

$$x \models \varphi \iff x \models q_\varphi \iff y \models q_\varphi \iff y \models \varphi.$$

(b) $(R^{-1})^{\mathsf{min}} = (R^{\mathsf{min}})^{-1}$. (By Lemma 1.6(4).)

(c) $\Phi_\Box \subseteq \Gamma_\Box$.
Assume $\widehat{x}\,\Phi_\Box\,\widehat{y}$. To show $\widehat{x}\,\Gamma_\Box\,\widehat{y}$, take any $\Box\varphi \in \Gamma$. Then $\Box q_\varphi \in \Phi$ and so

$$x \models \Box\varphi \iff x \models \Box q_\varphi \implies y \models q_\varphi \iff y \models \varphi.$$

(d) $(\Phi_\Box)^{-1} \subseteq \Gamma_\boxminus$. (The proof of this item contains the main trick.)
Assume $\widehat{x}\,\Phi_\Box\,\widehat{y}$. To show $\widehat{y}\,\Gamma_\boxminus\,\widehat{x}$, take any $\varphi := \boxminus\alpha \in \Gamma$. Then $\Box\neg q_\varphi \in \Phi$. We prove the required implication: $y \models \boxminus\alpha \Rightarrow x \models \alpha$, by contraposition:

$$x \not\models \alpha \overset{(1)}{\implies} x \models \Box\neg\boxminus\alpha \overset{(2)}{\iff} x \models \Box\neg q_\varphi \overset{(3)}{\implies} y \models \neg q_\varphi \overset{(4)}{\iff} y \not\models \boxminus\alpha$$

Here (1) follows from that $F \models \neg p \to \Box\neg\boxminus p$; (2) and (4) hold since $\varphi \leftrightarrow q_\varphi$ and hence $\boxminus\alpha \leftrightarrow q_\varphi$ are true in M^t; (3) holds since $\Box\neg q_\varphi \in \Phi$ and $\widehat{x}\,\Phi_\Box\,\widehat{y}$.

Thus, $R^{\mathsf{min}} \subseteq \widehat{R} \subseteq \Phi_\Box \subseteq \Gamma_\Box$ and $(R^{-1})^{\mathsf{min}} = (R^{\mathsf{min}})^{-1} \subseteq \widehat{R}^{-1} \subseteq (\Phi_\Box)^{-1} \subseteq \Gamma_\boxminus$. □

As a corollary, we obtain the following. Given a frame $F = (W, (R_e)_{e\in\Sigma})$, let $F^\ominus := (W, (R_e^{-1})_{e\in\Sigma})$. For a class of frames $\mathcal{F}$, put $\mathcal{F}^\ominus = \{F^\ominus \mid F \in \mathcal{F}\}$.

Theorem 2.5 (Inverting) *If $\mathcal{F}$ admits filtration, then so does $\mathcal{F}^\ominus$.*

Proof. First, add inverse relations, using Theorem 2.4. Secongly, drop the original relations, i.e., take the $\overline{\Sigma}$-reduct, using Lemma 2.1. □

---

[6] Throughout the proof, we write $x \models \varphi$ instead of $M, x \models \varphi$ or $M^t, x \models \varphi$. This is unambiguous, since the truth values of $\Box$-formulas in M and M^t coincide, while for formulas involving $\boxminus$, the shortcut $x \models \varphi$ simply means $M^t, x \models \varphi$.

2.3 Transitive closure

Given a class $\mathcal{F}$ of frames of the form $F = (W, R)$, denote $\mathcal{F}^{\oplus} = \{F^{\oplus} \mid F \in \mathcal{F}\}$, where $F^{\oplus} = (W, R, R^{+})$ and $R^{+} = \bigcup_{n\geq 1} R^{n}$ is the *transitive closure* of R.

Theorem 2.6 *If $\mathcal{F}$ admits filtration then so does $\mathcal{F}^{\oplus}$.*

Proof. Given a finite Sub-closed set of formulas $\Gamma \subset \mathsf{Fm}(\Box, \boxplus)$ and an $\mathcal{F}^{\oplus}$-model $M^{\oplus} = (W, R, R^{+}, V)$, where $\Box$ refers to R and $\boxplus$ to R^{+}, with $M = (W, R, V)$ an $\mathcal{F}$-model, we build an $\mathcal{F}^{\oplus}$-model that is a Γ-filtration of $M^{\oplus}$.

Let us introduce fresh variables $\{q_{\varphi} \mid \varphi \in \Gamma\}$ and extend the valuation V to them by putting: $x \models q_{\varphi} \leftrightharpoons x \models \varphi$. Thus, $\varphi \leftrightarrow q_{\varphi}$ and hence $\Box\varphi \leftrightarrow \Box q_{\varphi}$ are true in $M^{\oplus}$. Now consider the following set of formulas:

$$\Phi := \{q_{\varphi}, \Box q_{\varphi} \mid \varphi \in \Gamma\} \subset \mathsf{Fm}(\Box).$$

Since the class $\mathcal{F}$ admits filtration, there exists an $\mathcal{F}$-model $\widehat{M} = (\widehat{W}, \widehat{R}, \widehat{V})$ that is a Φ-filtration of M. Here $\widehat{W} = W/{\sim}$, where $\sim$ respects Φ, $R^{\mathsf{min}} \subseteq \widehat{R} \subseteq \Phi_{\Box}$, and $\widehat{x} \models q \Leftrightarrow x \models q$, for all variables q from Φ. Let us extend $\widehat{V}$ to the variables p from Γ by putting: $\widehat{x} \models p \leftrightharpoons \widehat{x} \models q_{p}$.

We claim that the model $\widehat{M}^{\oplus} := (\widehat{W}, \widehat{R}, (\widehat{R})^{+}, \widehat{V})$ is an $\mathcal{F}^{\oplus}$-model (this is obvious) and a Γ-filtration of $M^{\oplus}$, i.e., that $\sim$ respects Γ and the inclusions $R^{\mathsf{min}} \subseteq \widehat{R} \subseteq \Gamma_{\Box}$ and $(R^{+})^{\mathsf{min}} \subseteq \widehat{R}^{+} \subseteq \Gamma_{\boxplus}$ hold.

(a) *The equivalence relation $\sim$ respects the set Γ.* (As in Theorem 2.4.)
(b) $(R^{+})^{\mathsf{min}} \subseteq (R^{\mathsf{min}})^{+}$. (By Lemma 1.6(6).)
(c) $\Phi_{\Box} \subseteq \Gamma_{\Box}$. (As in Theorem 2.4.)
(d) $\Phi_{\Box} \subseteq \Gamma_{\boxplus}$. (Here we will use that $F^{\oplus} \models \boxplus p \to \Box p$.)
Assume $\widehat{x}\,\Phi_{\Box}\,\widehat{y}$. To prove $\widehat{x}\,\Gamma_{\boxplus}\,\widehat{y}$, take any $\boxplus\varphi \in \Gamma$, then $\Box q_{\varphi} \in \Phi$ and so:
$x \models \boxplus\varphi \;\Rightarrow\; x \models \Box\varphi \;\Leftrightarrow\; x \models \Box q_{\varphi} \;\Rightarrow\; y \models q_{\varphi} \;\Leftrightarrow\; y \models \varphi$. Thus $\widehat{x}\,\Gamma_{\boxplus}\,\widehat{y}$.
(e) $\Phi_{\Box} \circ \Gamma_{\boxplus} \subseteq \Gamma_{\boxplus}$. (Here we will use that $F^{\oplus} \models \boxplus p \to \Box\boxplus p$.)
Assume $\widehat{x}\,\Phi_{\Box}\,\widehat{y}\,\Gamma_{\boxplus}\,\widehat{z}$. To show $\widehat{x}\,\Gamma_{\boxplus}\,\widehat{z}$, take any $\boxplus\varphi \in \Gamma$. Then $\Box q_{\boxplus\varphi} \in \Phi$, so:
$x \models \boxplus\varphi \Rightarrow x \models \Box\boxplus\varphi \Leftrightarrow x \models \Box q_{\boxplus\varphi} \Rightarrow y \models q_{\boxplus\varphi} \Leftrightarrow y \models \boxplus\varphi \Rightarrow z \models \varphi$.
(f) $(\Phi_{\Box})^{+} \subseteq \Gamma_{\boxplus}$.
It suffices to prove $(\Phi_{\Box})^{n} \subseteq \Gamma_{\boxplus}$, by induction on $n \geq 1$. Induction base is (d); induction step: $(\Phi_{\Box})^{n+1} = \Phi_{\Box} \circ (\Phi_{\Box})^{n} \subseteq \Phi_{\Box} \circ \Gamma_{\boxplus} \subseteq \Gamma_{\boxplus}$, by (e).

Thus, $R^{\mathsf{min}} \subseteq \widehat{R} \subseteq \Phi_{\Box} \subseteq \Gamma_{\Box}$ and $(R^{+})^{\mathsf{min}} \subseteq (R^{\mathsf{min}})^{+} \subseteq \widehat{R}^{+} \subseteq (\Phi_{\Box})^{+} \subseteq \Gamma_{\boxplus}$. □

An analogue of Theorem 2.6 holds for the *reflexive-transitive closure* $R^{*} = \mathtt{Id}(W) \cup R^{+}$, where additionally one needs to use Lemmas 2.3 and 2.1.

2.4 Operations on logics

Let $\mathbf{L}$ be a logic $\mathbf{L}$ over Σ. Denote by $\mathbf{L}^{u}$ its extension with the universal modality $[*]$ and the axioms, for all $e \in \Sigma$ (the last three are **S5**-axioms for $[*]$):

$$[*]p \to [e]p, \qquad [*]p \to p, \qquad [*]p \to [*][*]p, \qquad \neg[*]p \to [*]\neg[*]p.$$

The *tense counterpart* $\mathbf{L}^{t}$ of $\mathbf{L}$ is the logic over $\Sigma \cup \overline{\Sigma}$ that extends $\mathbf{L}$ with the following *tense axioms*, for all $e \in \Sigma$:

$$p \to [e]\langle\overline{e}\rangle p, \qquad p \to [\overline{e}]\langle e\rangle p.$$

Given a logic $\mathbf{L}$ over $\Sigma = \{\Box\}$, denote by $\mathbf{L}^{\boxplus}$ the extension of $\mathbf{L}$ with the *transitive closure* modality $\boxplus$ and the following axioms:

$$\boxplus p \to \Box p, \qquad \boxplus p \to \Box\boxplus p, \qquad \boxplus(p \to \Box p) \to (\Box p \to \boxplus p).$$

It is known that extending a logic even with a seemingly "harmless" universal modality is not safe: we can lose the FMP [22], decidability [21], and even Kripke completeness [14, Corollary 9.6.5]. A number of negative results are also known for the tense extension and the transitive closure extension, see e.g. [23,24], [2, Theorem 6.34]. However, if the extended logic is Kripke complete, then we can obtain the desired transfer results for logics.

Theorem 2.7 (Filtration safe operations on logics) *Suppose that a logic* $\mathbf{L}$ *admits filtration.*

(i) *If the logic* $\mathbf{L}^u$ *is Kripke complete then it admits filtration [11].*
(ii) *If the logic* $\mathbf{L}^t$ *is Kripke complete then it admits filtration.*
(iii) *If the logic* $\mathbf{L}^{\boxplus}$ *is Kripke complete then it admits filtration.*

Proof. Let $\mathcal{F}$ be the class of all $\mathbf{L}$-frames. As shown in [11], if $\mathbf{L}^u$ is complete then it is the logic of $\mathcal{F}^u$. Furthermore, it is known (see e.g. [14]) that $F \models \mathbf{L}$ iff $F^t \models \mathbf{L}^t$. Similarly, $F \models \mathbf{L}$ iff $F^{\oplus} \models \mathbf{L}^{\boxplus}$. Now, (i)–(iii) follow from the Kripke completeness and Lemma 2.2, Theorems 2.4 and 2.6. □

Corollary 2.8 *If a logic* $\mathbf{L}$ *admits filtration and is finitely axiomatizable, and* $\mathbf{L}^u$ *is Kripke complete, then the global satisfiability problem for* $\mathbf{L}$ *is decidable.*

Proof. Let $\mathcal{F}$ be the class of all $\mathbf{L}$-frames. Then, for any formula φ, we have: φ is globally satisfiable in $\mathbf{L}$ iff the formula $[*]\varphi$ is satisfiable in $\mathbf{L}^u$. □

The next lemma gives a sufficient condition for the completeness of $\mathbf{L}^u$ and $\mathbf{L}^t$.

Lemma 2.9 *If a logic* $\mathbf{L}$ *is canonical then* $\mathbf{L}^t$ *and* $\mathbf{L}^u$ *are Kripke complete.*

Proof. The axioms for $[*]$ and the tense axioms are canonical formulas. □

Wolter [23,24,25,26,27] obtained a lot of general transfer results for $\mathbf{L}^t$. However, they seem not to cover our Theorem 2.7(ii), so we believe the latter is new. At the same time, we are not aware of any transfer results for $\mathbf{L}^{\boxplus}$, nor general sufficient conditions for it to be Kripke complete.

3 Grammar logics

Here we show that the above results easily imply that every regular grammar logic (with converse) admits filtration and hence has the FMP. The result on FMP is not new, as there is a translation from regular grammar logics (with converse) into **PDL** with finite automata as modalities [5] (into the guarded fragment of the first-order logic with two variables $\mathbf{GF}^2$ [6], respectively), therefore, the FMP result for regular grammar logics (with converse) follows from the FMP for **PDL** and $\mathbf{GF}^2$ obtained in [7] and [17], respectively.

3.1 Grammars

By a *grammar*[7] over an alphabet Σ we mean a finite set Π of (*production*) *rules* of the form $u \to v$, where $u, v \in \Sigma^*$, $u \neq \varepsilon$ (here ε is the empty word). Below, we only deal with *context-free* grammars, whose rules have the form $e \to v$, where $e \in \Sigma$, although some definitions are applicable to arbitrary grammars. We say that a rule $u \to v$ transforms, for any $x, y \in \Sigma^*$, the word xuy into xvy, and denote this relation on words by $xuy \overset{\Pi}{\longmapsto} xvy$. Let[8] $\overset{\Pi}{\Longmapsto}$ be the reflexive-transitive closure of the relation $\overset{\Pi}{\longmapsto}$. The set of all words producible from a given word u is denoted by $\Pi(u) = \{v \in \Sigma^* \mid u \overset{\Pi}{\Longmapsto} v\}$.

A grammar Π is called *regular* if, for every $e \in \Sigma$, the language $\Pi(e)$ is regular. Recall that *regular languages* are obtained from the empty language[9] $\varnothing$, the singleton languages $\{\varepsilon\}$ and $\{e\}$, for all $e \in \Sigma$, using the operations of union $L_1 \cup L_2$, composition $L_1 \circ L_2$ and Kleene star L^* on languages (cf. [13]).

We also consider grammars over the alphabet $\Sigma \cup \overline{\Sigma}$, where the symbol $\overline{e}$ in $\overline{\Sigma}$ is called the *inverse* of the corresponding symbol $e \in \Sigma$. In a frame, we have $R_{\overline{e}} = (R_e)^{-1}$. The *inverse* of a word $u = e_1 \dots e_n$ is defined as $\overline{u} := \overline{e}_n \dots \overline{e}_1$.

3.2 Grammar modal logics

In the modal language over Σ, we have the modality $[e]$ for each $e \in \Sigma$. For any word $u = e_1 \dots e_n$ over Σ, we denote the modal operator $[u] := [e_1] \dots [e_n]$. To a rule $u \to v$, we associate the formula $[u]p \to [v]p$. For a grammar Π, its *grammar (modal) logic* $\mathbf{K}\Pi$ is the extension of the minimal normal multi-modal Σ-logic $\mathbf{K}_\Sigma$, or its tense extension $\mathbf{K}^t_\Sigma$ (see Section 2.4) in case we have a grammar over $\Sigma \cup \overline{\Sigma}$, with the axioms $[u]p \to [v]p$, for all rules $(u \to v) \in \Pi$.

In a Σ-frame $F = (W, (R_e)_{e \in \Sigma})$, any word $u = e_1 \dots e_n$ in Σ^* gives rise to the relation $R_u := R_{e_1} \circ \dots \circ R_{e_n}$. It is easily seen that $F \models [u]p \to [v]p$ iff $R_u \supseteq R_v$ (notice the converse inclusion!); in this case we write $F \models (u \to v)$. We say that F is a Π-*frame* and write $F \models \Pi$ if $F \models (u \to v)$, for all rules $(u \to v) \in \Pi$. A Π-*model* is a model based on a Π-frame. Since $R_{\overline{u}} = (R_u)^{-1}$, we have $F \models u \to v$ iff $F \models \overline{u} \to \overline{v}$, for any frame F over $\Sigma \cup \overline{\Sigma}$.

For any grammar Π, even over $\Sigma \cup \overline{\Sigma}$, the modal logic $\mathbf{K}\Pi$ is Kripke complete w.r.t. the class of Π-frames, since axioms of the form $[u]p \to [v]p$ are Sahlqvist formulas, see [2, Th. 4.42]. One of the main problems in this field is to determine for which grammars Π the logic $\mathbf{K}\Pi$ is decidable, or even has the FMP. Below, we show that, for every *regular* grammar Π, the logic $\mathbf{K}\Pi$ admits filtration and hence has the FMP and is decidable.

3.3 Regular grammar logics admit filtration

Let Π be a (context-free) grammar over Σ and $F = (W, (R_e)_{e \in \Sigma})$ a Σ-frame. The Π-*closure* of F is the frame $F^\Pi := (W, (R^\Pi_e)_{e \in \Sigma})$, where $R^\Pi_e = \bigcup_{v \in \Pi(e)} R_v$. Since $e \overset{\Pi}{\Longmapsto} e$, we have $R_e \subseteq R^\Pi_e$.

---

[7] Our definition does not include a start symbol, in which aspect it is closer to a *semi-Thue system*; however, we will need a distinction between terminal and non-terminal symbols.

[8] In texts on formal language theory, this relation is often denoted by $u \Rightarrow^*_\Pi v$.

[9] In fact, we do not need $\varnothing$ below, since we always have $\Pi(e) \neq \varnothing$ due to that $e \overset{\Pi}{\Longmapsto} e$.

Lemma 3.1 (a) *The* Π*-closure of any frame* F *is a* Π*-frame:* $F^\Pi \models \Pi$.
(b) *If* F *is a* Π*-frame then* $F^\Pi = F$.

Proof. (a) For each $(e \to u) \in \Pi$, let us show that $R_e^\Pi \supseteq R_u^\Pi$. Let $u = e_1 \dots e_n$. If $x\, R_u^\Pi\, y$, i.e., $x\,(R_{e_1}^\Pi \circ \dots \circ R_{e_n}^\Pi)\, y$, then $x\,(R_{v_1} \circ \dots \circ R_{v_n})\, y$, for some $v_i \in \Pi(e_i)$, hence $x\,(R_{v_1 \dots v_n})\, y$. Since $e \overset{\Pi}{\longmapsto} e_1 \dots e_n$ and $e_i \overset{\Pi}{\Longmapsto} v_i$, we have $e \overset{\Pi}{\Longmapsto} v_1 \dots v_n$. Thus, $x\, R_v\, y$ for the word $v := v_1 \dots v_n \in \Pi(e)$. Therefore, $x\, R_e^\Pi\, y$.

(b) Assume $F \models \Pi$. Let us show that $F^\Pi = F$, i.e., $R_e^\Pi = R_e$. Here '$\supseteq$' is trivial. To prove '$\subseteq$', note that $R_u \subseteq R_a$, for all rules $(a \to u) \in \Pi$. Then $R_v \subseteq R_a$, for all $v \in \Pi(a)$, by induction on derivation in Π. Thus, $R_e^\Pi \subseteq R_e$. □

By the above lemma, taking the Π-closure of all frames yields exactly the class of *all* Π-frames. Next, we show that in case Π is a regular grammar, the Π-closure can be obtained by finitely many operations $\cup, \circ, *$.

Let $\Sigma = \{e_1, \dots, e_n\}$ and let $E = E(e_1, \dots, e_n)$ be a regular expression over Σ, i.e., it is built up from ε and e_i using $\cup, \circ, *$. Denote by $\mathbb{L}(E)$ the (regular) language represented by E, see [13]. In a frame $F = (W, R_{e_1}, \dots, R_{e_n})$, we can "substitute" $\mathtt{Id}(W)$ for ε and R_{e_i} for e_i into the expression E, thus obtaining a relation $E(R_{e_1}, \dots, R_{e_n})$ built up from $\mathtt{Id}(W)$ and R_{e_i} using $\cup, \circ, *$.

Lemma 3.2 $E(R_{e_1}, \dots, R_{e_n}) = \bigcup_{v \in \mathbb{L}(E)} R_v$, *for any regular expression* E.

Proof. By an easy induction on the complexity of the expression E. □

Now, if Π is regular, then each language $\Pi(e)$ is represented by some regular expression E_e, i.e., $\Pi(e) = \mathbb{L}(E_e)$. Then $R_e^\Pi = E_e(R_{e_1}, \dots, R_{e_n})$ by Lemma 3.2. So, $\mathcal{F}^\Pi$ is obtained from $\mathcal{F}$ by applying the frame operations $\cup, \circ, *$ (see Sections 2.1 and 2.3) finitely many times. Thus we obtain the following result.

Theorem 3.3 *For every regular grammar* Π *over* $\Sigma \cup \overline{\Sigma}$, *the class of* Π*-frames admits filtration. Consequently, every regular grammar logic (with converse) admits filtration, has the FMP, and is decidable; moreover, it has the ExpMP.*

Proof. Starting from the class of all Σ-frames, first add the inverse relations $R_{\overline{e}}$, for each $e \in \Sigma$, by applying the tense expansion operation from Section 2.2. Then apply the operations $\cup, \circ, *$ on frames from Sections 2.1 and 2.3 to obtain the regular expressions E_e that represent the languages $\Pi(e)$, for each $e \in \Sigma$. Finally, in the resulting frames, drop the initial relations R_e and all the relations built at intermediate steps, using Lemma 2.1, and arrange the final relations R_e^Π in the same order as the initial relations R_e. Thus we obtain the class of all Π-frames. Since the class of all Σ-frames admits filtration (see Example 1.4) and the above operations are filtration safe, it follows that the class of all Π-frames admits filtration as well.

Moreover, at each step, the set of formulas grows linearly: $|\Phi| \le 3|\Gamma|$ (see Lemma 2.3, Theorems 2.4 and 2.6). Hence, starting with a filtration for the class of all $(\Sigma \cup \overline{\Sigma})$-frames through the set $\mathsf{Sub}(\varphi)$ of size $O(|\varphi|)$, after a fixed number of operations, we arrive at a filtration through a set of formulas of size $O(|\varphi|)$. Thus we obtain a countermodel of the size exponential in $|\varphi|$. □

4 Filtration for some regular grammar logics

Here we present filtration constructions for some familiar classes of regular grammar logics (with converse). First, for *right-linear* grammar logics, we give two different constructions of filtration. Then we adjust them to cover the converse modalities for terminal symbols. Next, we adapt the construction to *left-linear* grammar logics. Finally, we consider grammar logics with *left-* and *right-recursive* rules.

4.1 On maximal filtered relations

In Definition 1.1, we introduced the minimal R_e^{min} and the (Γ-)maximal Γ_e filtered relations on $\widehat{W}$, for every symbol $e \in \Sigma$. Consequently, for any word $u = e_1 \dots e_n \in \Sigma^*$, the relations $R_u^{\mathsf{min}} = R_{e_1}^{\mathsf{min}} \circ \dots \circ R_{e_n}^{\mathsf{min}}$ and $\Gamma_u = \Gamma_{e_1} \circ \dots \circ \Gamma_{e_n}$ are well-defined. Let us also introduce the maximal filtered relation $\Gamma_{[u]}$ on $\widehat{W}$ induced by the set Γ and the "compound" modality $[u] = [e_1] \dots [e_n]$:

$$\widehat{x}\, \Gamma_{[u]}\, \widehat{y} \quad \leftrightharpoons \quad \text{for every formula } [u]\varphi \in \Gamma \ \big(x \models [u]\varphi \ \Rightarrow \ y \models \varphi\big).$$

Lemma 4.1 $\Gamma_{[u]} \circ \Gamma_{[v]} \subseteq \Gamma_{[uv]}$, *for all words* $u, v \in \Sigma^*$.

Proof. Assume $\widehat{x}\, \Gamma_{[u]}\, \widehat{y}\, \Gamma_{[v]}\, \widehat{z}$. To prove $\widehat{x}\, \Gamma_{[uv]}\, \widehat{z}$, take any $[uv]\varphi \in \Gamma$. Since Γ is Sub-closed, we have $[v]\varphi \in \Gamma$ and $\varphi \in \Gamma$. Therefore, $x \models [u][v]\varphi$ implies $y \models [v]\varphi$, which in turn implies $z \models \varphi$, as required. □

If we deal with the modal language over $\Sigma \cup \overline{\Sigma}$, the following relations are useful: $\Gamma_u^\sharp := (\Gamma_{\overline{u}})^{-1}$ and $\Gamma_{[u]}^\sharp := (\Gamma_{[\overline{u}]})^{-1}$. Explicitly, for $e \in \Sigma$, we define:

$$\widehat{x}\, \Gamma_e^\sharp\, \widehat{y} \quad \leftrightharpoons \quad \text{for every formula } [\overline{e}]\varphi \in \Gamma \ \big(y \models [\overline{e}]\varphi \ \Rightarrow \ x \models \varphi\big).$$

Lemma 4.2 (Minimax) *For any words* $u, v \in (\Sigma \cup \overline{\Sigma})^*$, *we have:*

(a) $R_u^{\mathsf{min}} \subseteq \Gamma_u \subseteq \Gamma_{[u]}$,
(b) $R_u^{\mathsf{min}} \subseteq \Gamma_u^\sharp \subseteq \Gamma_{[u]}^\sharp$.

Proof. (a) The first inclusion, $R_u^{\mathsf{min}} \subseteq \Gamma_u$, follows from the trivial inclusion $R_{e_i}^{\mathsf{min}} \subseteq \Gamma_{e_i}$ for all i, by monotonicity of the composition. As for the second inclusion, we have $\Gamma_u = \Gamma_{e_1} \circ \dots \circ \Gamma_{e_n} = \Gamma_{[e_1]} \circ \dots \circ \Gamma_{[e_n]} \subseteq \Gamma_{[u]}$, where we used the trivial equality $\Gamma_{e_i} = \Gamma_{[e_i]}$ and Lemma 4.1.

(b) This follows from **(a)**, once we observe that $(R_u^{\mathsf{min}})^{-1} = R_{\overline{u}}^{\mathsf{min}}$. □

4.2 Filtration for right-linear grammar logics

Let Π be a *right-linear grammar* over[10] $\Sigma = T \cup N$, which means that Π consists of rules of the form $a \to uc$, where $a, c \in N$ and $u \in T^*$. Given a model $M = (F, V)$ based on a Π-frame $F = (W, (R_e)_{e \in \Sigma})$ and a finite Sub-closed set of formulas $\Gamma \subseteq \mathsf{Fm}(\Sigma)$, we will build a Π-model $\widehat{M}$ that is a Γ-filtration of M.

We introduce the following operators on (finite Sub-closed) sets of formulas:

[10] We partition the set of symbols Σ into two disjoint sets of the so-called *terminal* and *non-terminal* symbols. In any rule $a \to u$ of any grammar, we always assume that $a \in N$.

$$N(\Psi) := \mathsf{Sub}\{\,[c]\varphi \mid a,c \in N,\ [a]\varphi \in \Psi\,\},$$
$$\Pi(\Psi) := \mathsf{Sub}\{\,[v]\varphi \mid [a]\varphi \in \Psi,\ (a \to v) \in \Pi\,\}.$$

Now put $\Delta := \Gamma \cup N(\Gamma)$ and $\Phi := \Delta \cup \Pi(\Delta)$. We filter the model M *through* Φ, i.e., we consider the equivalence relation $\sim_\Phi$ (see Section 1.1), which obviously has the finite index and respects the set Γ, and put $\widehat{W} := W/{\sim_\Phi}$.

Lemma 4.3 (Max-frame) $\Delta_a \supseteq \Phi_u \circ \Delta_c$, *for every rule* $(a \to uc) \in \Pi$.

Proof. Assume $\widehat{x}\,\Phi_u\,\widehat{y}\,\Delta_c\,\widehat{z}$. To show that $\widehat{x}\,\Delta_a\,\widehat{z}$, take any $[a]\varphi \in \Delta$. Then:

$$x \models [a]\varphi \overset{(1)}{\Longrightarrow} x \models [u][c]\varphi \overset{(2)}{\Longrightarrow} y \models [c]\varphi \overset{(3)}{\Longrightarrow} z \models \varphi.$$

Here **(1)** is due to that $F \models (a \to uc)$, so that $M, x \models [a]\varphi \to [u][c]\varphi$; **(2)** holds since $[u][c]\varphi \in \Pi(\Delta) \subseteq \Phi$ and $(\widehat{x}, \widehat{y}) \in \Phi_u \subseteq \Phi_{[u]}$ by the Minimax Lemma; finally, **(3)** holds since $\widehat{y}\,\Delta_c\,\widehat{z}$ and $[c]\varphi \in N(\Delta) \subseteq \Delta$. □

Next, we define the valuation $\widehat{V}$ on the variables p from $\mathsf{Var}(\Phi) = \mathsf{Var}(\Gamma)$ canonically: $\widehat{x} \models p$ iff $x \models p$. In order to obtain the frame $\widehat{F} = (\widehat{W}, (\widehat{R}_e)_{e\in\Sigma})$ and the model $\widehat{M} = (\widehat{F}, \widehat{V})$, it remains to define the relations $\widehat{R}_e$ so that $\widehat{F} \models \Pi$ and $\widehat{R}_e \subseteq \Gamma_e$, for each $e \in \Sigma$. We give two different constructions for this.

4.2.1 Right-linear grammars: Mini-maximal frame

Let $\widehat{R}_e$ be the minimal (for $e \in T$) or the Δ-maximal (for $e \in N$) relation:

$$\widehat{R}_e := \begin{cases} R_e^{\mathsf{min}}, & \text{if } e \in T; \\ \Delta_e, & \text{if } e \in N. \end{cases}$$

Lemma 4.4 $\widehat{F} \models \Pi$.

Proof. Take any rule $(a \to uc) \in \Pi$, where $a, c \in N$ and $u \in T^*$. Then the required inclusion $\widehat{R}_a \supseteq \widehat{R}_u \circ \widehat{R}_c$ follows from Lemma 4.3, since $\widehat{R}_a = \Delta_a$, $\widehat{R}_c = \Delta_c$, and $\widehat{R}_u = R_u^{\mathsf{min}} \subseteq \Phi_u$, for any $u \in T^*$, by the Minimax Lemma. □

Finally, $\widehat{R}_e \subseteq \Gamma_e$, for each $e \in \Sigma$. Indeed, for $e \in T$ this is obvious, and for $e \in N$, we have that $\widehat{R}_e = \Delta_e \subseteq \Gamma_e$, because $\Gamma \subseteq \Delta$.

Remark 4.5 The above proof remains valid if, for $e \in T$, we put $\widehat{R}_e := \Phi_e$ (or even any relation between R_e^{min} and Φ_e). However, this does not generalize to the logic with converse terminals, while the relations R_e^{min} still work for them.

4.2.2 Right-linear grammars: Π-closure

This time, we put $\widehat{R}_e := R_e^{\mathsf{min}}$ for $e \in T$, while for $a \in N$, we define $\widehat{R}_a$ using the Π-*closure*. That is, we define, simultaneously for all $a \in N$, a tower of relations $R_a^{(0)} \subseteq R_a^{(1)} \subseteq R_a^{(2)} \subseteq \ldots$ by induction:

$$R_a^{(0)} := R_a^{\mathsf{min}}, \quad R_a^{(n+1)} := R_a^{(n)} \cup \bigcup_{(a\to uc)\in\Pi} \big(\widehat{R}_u \circ R_c^{(n)}\big), \quad \widehat{R}_a := \textstyle\bigcup_n R_a^{(n)}. \quad (*)$$

Note that in $(*)$ we have $u \in T^*$, so that $\widehat{R}_u$ is already defined via $\widehat{R}_e$, $e \in T$.

Lemma 4.6 $\widehat{F} \models \Pi$.

Proof. Since $\widehat{W}$ is finite, we can find the stage n at which the sequence in $(*)$ stabilizes for all $a \in N$. Now, for any rule $(a \to uc) \in \Pi$, we have $\widehat{R}_c = R_c^{(n)}$ and $\widehat{R}_a = R_a^{(n+1)}$. Then $(*)$ implies the required inclusion $\widehat{R}_a \supseteq \widehat{R}_u \circ \widehat{R}_c$. □

Lemma 4.7 $\widehat{R}_a \subseteq \Delta_a$, *for every non-terminal* $a \in N$.

Proof. It suffices to prove, by induction on n, that, for all $a \in N$, $R_a^{(n)} \subseteq \Delta_a$. Induction base is trivial: $R_a^{\mathsf{min}} \subseteq \Delta_a$. Induction step: in the expression $(*)$ for $R_a^{(n+1)}$, all terms are contained in Δ_a. Indeed, $R_a^{(n)} \subseteq \Delta_a$ by I.H., and for every rule $(a \to uc) \in \Pi$, by the Max-frame Lemma,

$$\widehat{R}_u \circ R_c^{(n)} \;\subseteq\; \Phi_u \circ \Delta_c \;\subseteq\; \Delta_a,$$

where $\widehat{R}_u \subseteq \Phi_u$, by the Minimax Lemma, and $R_c^{(n)} \subseteq \Delta_c$, by I.H. for $c \in N$.□

Thus, $\widehat{R}_e \subseteq \Gamma_e$ for $e \in T$; and for $e \in N$, we have $\widehat{R}_e \subseteq \Delta_e \subseteq \Gamma_e$, as $\Gamma \subseteq \Delta$.

4.3 Right-linear grammar logics with converse terminals

Let Π be a right-linear grammar with converse terminals, i.e., its rules have the form $a \to uc$, where $a, c \in N$ and $u \in (T \cup \overline{T})^*$. We adjust the "mini-maximal frame" construction. (We could adjust the "Π-closure" construction as well.)

In addition to the operators $N(\Psi)$ and $\Pi(\Psi)$ defined above, we introduce

$$S(\Psi) \;:=\; \mathsf{Sub}\{\,[a]\neg[\overline{a}]\varphi \;\mid\; [\overline{a}]\varphi \in \Psi\,\}.$$

Now, given a set Γ, we put $\Lambda = \Gamma \cup S(\Gamma)$, $\Delta = \Lambda \cup N(\Lambda)$, and $\Phi = \Delta \cup \Pi(\Delta)$.

Then the proof proceeds as in Section 4.2.1, so we have: (a) $R_e^{\mathsf{min}} \subseteq \widehat{R}_e$, and (b) $\widehat{R}_e \subseteq \Gamma_e$, for all $e \in \Sigma$. However, in presence converse modalities, we also need to prove: (c) $R_{\overline{e}}^{\mathsf{min}} \subseteq \widehat{R}_{\overline{e}}$, and (d) $\widehat{R}_{\overline{e}} \subseteq \Gamma_{\overline{e}}$, for all $e \in \Sigma$. Here (c) follows trivially from (a); while (d) for $e \in T$ is trivial, since $\widehat{R}_e = R_e^{\mathsf{min}}$. The next lemma proves (d) for $e \in N$; its proof resembles the one for Theorem 2.4.

Lemma 4.8 $\Delta_a \;\subseteq\; \Gamma^{\sharp}_{\overline{a}}$, *for every non-terminal* $a \in N$.

Proof. Assume $\widehat{x}\,\Delta_a\,\widehat{y}$. To prove that $\widehat{x}\,\Gamma^{\sharp}_{\overline{a}}\,\widehat{y}$, we take any formula $[\overline{a}]\varphi \in \Gamma$ and show that $y \models [\overline{a}]\varphi$ implies $x \models \varphi$. The proof is by contraposition:

$$x \not\models \varphi \;\overset{(1)}{\Longrightarrow}\; x \models [a]\neg[\overline{a}]\varphi \;\overset{(2)}{\Longrightarrow}\; y \not\models \neg[\overline{a}]\varphi.$$

Here **(1)** is due to that $F \models \neg p \to [a]\neg[\overline{a}]p$, while **(2)** follows from that $\widehat{x}\,\Delta_a\,\widehat{y}$ and $[a]\neg[\overline{a}]\varphi \in S(\Gamma) \subseteq \Delta$. □

4.4 Left-linear grammars (with converse terminals)

For left-linear grammars, the claim follows from the one for the corresponding right-linear grammars, due to Theorem 2.5. Indeed, inverting all relations in frames corresponds to replacing each rule $a \to w$ in a grammar with the rule $a \to w^{\mathsf{R}}$, where w^{R} is the word w written in reverse order, which, in turn, transforms a right-linear grammar into a left-linear one and vice versa. However, let us sketch an explicit filtration construction for left-linear grammars, since its ideas will be used later in Section 4.5.

Assume that Π is a left-linear grammar with converse terminals over $\Sigma \cup \overline{\Sigma}$, where $\Sigma = T \cup N$, so it consists of rules of the form $a \to cu$, where $a, c \in N$ and $u \in (T \cup \overline{T})^*$. To simplify notation, below we interpret $(\Sigma \cup \overline{\Sigma})$-formulas in Σ-frames and Σ-models in an obvious way.

Given a Π-model $M = (W, (R_e)_{e\in\Sigma}, V)$ and a finite Sub-closed set of formulas Γ over $\Sigma \cup \overline{\Sigma}$, we will build a Π-model that is a Γ-filtration of M. For this, we introduce the operators that are in a sense dual to the operators N, Π, S defined in Sections 4.2 and 4.3:

$$\begin{array}{rcl} \overline{N}(\Psi) & := & \mathsf{Sub}\{\, [\overline{c}]\varphi \mid a, c \in N,\ [\overline{a}]\varphi \in \Psi \,\}, \\ \overline{\Pi}(\Psi) & := & \mathsf{Sub}\{\, [\overline{v}]\varphi \mid [\overline{a}]\varphi \in \Psi,\ (a \to v) \in \Pi \,\}, \\ \overline{S}(\Psi) & := & \mathsf{Sub}\{\, [\overline{a}]\neg[a]\varphi \mid [a]\varphi \in \Psi \,\}. \end{array}$$

Next we put $\Lambda = \Gamma \cup \overline{S}(\Gamma)$, $\Delta = \Lambda \cup \overline{N}(\Lambda)$, and $\Phi = \Delta \cup \overline{\Pi}(\Delta)$.

Now we set $\widehat{R}_e$ to be R_e^{min} if $e \in T$, and $\Delta_e^\sharp$ if $e \in N$. Then we prove an analogue of the Max-frame lemma: $\Delta_a^\sharp \supseteq \Delta_c^\sharp \circ \Phi_u^\sharp$, for each rule $(a \to cu) \in \Pi$. This allows us to prove that $\widehat{F} \models \Pi$, $\widehat{R}_e \subseteq \Gamma_e^\sharp$ and $\widehat{R}_e \subseteq \Gamma_e$, for all $e \in \Sigma$.

4.5 Bi-recursive grammar logics (with converse terminals)

We call a grammar *bi-recursive* if it consists of rules of two kinds: *right-recursive* $a \to ua$ and *left-recursive* $a \to av$, for $a \in N$ and $u, v \in T^*$. So these grammars combine right and left rules, but in every rule, the non-terminal in the body (ua or av) is always the same as in the head.

Let $\Pi = \Pi^r \cup \Pi^\ell$, where Π^r (resp., Π^ℓ) consists of rules of the form $a \to ua$ (resp., $a \to av$) with $a \in N$ and $u, v \in (T \cup \overline{T})^*$. The filtration for Π proceeds in two stages: first, we build the mini-maximal frame for Π^ℓ (as in Section 4.4) and then take its Π^r-closure (as in Section 4.2.2). In order for this construction to work, the sets of formulas Λ, Δ, Φ must be chosen properly.

Given a Π-model $M = (W, (R_e)_{e\in\Sigma}, V)$ and a finite Sub-closed set of $(\Sigma \cup \overline{\Sigma})$-formulas Γ, we will build a Π-model that is a Γ-filtration of M. Put

$$\Lambda \;=\; \Gamma \cup S(\Gamma), \qquad \Delta \;=\; \Lambda \cup \overline{S}(\Lambda), \qquad \Phi \;=\; \Delta \cup \Pi^r(\Delta) \cup \overline{\Pi^\ell}(\Delta).$$

Again, $\widehat{W} := W/{\sim_\Phi}$ and the valuation $\widehat{V}$ is canonical. It remains to build $\widehat{R}_e$.

Stage 1. Take $\mathbb{F} = (\widehat{W}, (\mathbb{R}_e)_{e\in\Sigma})$, where $\mathbb{R}_e$ is R_e^{min} if $e \in T$ and $\Delta_e^\sharp$ if $e \in N$. As in Section 4.4, we show that $(\mathbb{F}, \widehat{V})$ is a Π^ℓ-model and a Λ-filtration of M.

Stage 2. Take $\widehat{F} = (\widehat{W}, (\widehat{R}_e)_{e\in\Sigma})$ to be the Π^r-closure of $\mathbb{F}$.
As in Section 4.2.2, we show that $(\widehat{F}, \widehat{V})$ is a Γ-filtration of M. By Lemma 3.1(a), $\widehat{F}$ is a Π^r-frame. Finally, $\widehat{F}$ is a Π^ℓ-frame. In fact,[11] the Π^r-closure of a Π^ℓ-frame is again a Π^ℓ-frame. This follows from a simple fact: if a relation R satisfies $R \supseteq R \circ S$, then the relation $R' = Q \circ R$ satisfies $R' \supseteq R' \circ S$ as well.

Remark 4.9 To any grammar considered above, we can add rules of the form $a \to u$, where $a \in N$ and $u \in (T \cup \overline{T})^*$. The Π-closure construction also works for Π consisting of rules $a \to (ax)^k a$, where $a \in N$, $x \in T^*$, and $k \geq 0$.

[11] This does not generalize to arbitrary right-linear and left-linear grammars Π^r and Π^ℓ.

5 Logics that do not admit filtration

Here we give examples of logics that do not admit filtration. Perhaps, the logic **GL** is the simplest example: if a **GL**-model $M = (W, \prec, V)$ has an infinite descending chain $\ldots \prec x_2 \prec x_1 \prec x_0$, then any filtration of M has a reflexive point and hence is not a **GL**-model. Despite of this, **GL** has the FMP.

We will prove that the logics $\mathbf{K} + \Diamond\Box p \to \Box\Diamond p$ and $\mathbf{K}_2 + [a]p \to [b][a][b]p$ do not admit filtration by showing the undecidability of the global satisfiability problem. The former logic is known to be decidable and to have FMP, which is proved by embedding it into $\mathbf{K}\times\mathbf{K}$ and then appealing to the decidability and FMP for $\mathbf{K}\times\mathbf{K}$ shown in [10]. It is open whether the latter logic is decidable. Arguing similary, one can prove the same negative results for $[b][a]p \to [a][b]p$ and even for $\langle a\rangle[b]p \to [a]p$ ([16, p. 19]).

A *domino system* is a triple $\mathcal{D} = (D, H, V)$, where $D \neq \varnothing$ is a set of *tile types* and $H, V \subseteq D\times D$ are horizontal and vertical matching relations. We say that $\mathcal{D}$ *tiles* $\mathbb{N}\times\mathbb{N}$ if there exists a function $t\colon \mathbb{N}\times\mathbb{N} \to D$ such that, for all $i, j \in \mathbb{N}$, we have $(t(i,j), t(i+1,j)) \in H$ and $(t(i,j), t(i,j+1)) \in V$. The following *domino tiling* problem is known [3] to be undecidable: determine whether a given domino system $\mathcal{D}$ tiles $\mathbb{N}\times\mathbb{N}$; similarly for $\mathbb{N}\times\mathbb{Z}$ and $\mathbb{Z}\times\mathbb{Z}$.

Given a domino system $\mathcal{D}$, we define $\lambda^{\mathcal{D}} = \lambda_0 \wedge \lambda^H \wedge \lambda^V$, where

$$\begin{aligned}
\lambda_0 &= (\bigvee_{d\in D} q_d) \wedge \bigwedge_{d\neq d'} \neg(q_d \wedge q_{d'}),\\
\lambda^H &= \bigwedge_{d\in D} (q_d \to [h](\bigvee_{d'\in H(d)} q_{d'})),\\
\lambda^V &= \bigwedge_{d\in D} (q_d \to [v](\bigvee_{d'\in V(d)} q_{d'})).
\end{aligned}$$

Theorem 5.1 *The global satisfiability problem for the logic* $\mathbf{L} := \mathbf{K}_2 + [h]p \to [v][h][v]p$ *is undecidable, and hence* $\mathbf{L}$ *does not admit filtration.*

Consider a frame $G = (W, R_h, R_v)$, where $W = \mathbb{N}\times\mathbb{Z}$ and

$$\begin{aligned}
R_h &= \{((i,j),(i+1,j)) \mid i\in\mathbb{N},\ j\in\mathbb{Z}\},\\
R_v &= \{((i,j),(i,j-(-1)^i) \mid i\in\mathbb{N}, j\in\mathbb{Z}\}.
\end{aligned}$$

Let G' be the restriction of G to the sector $S = \{(i,j) \mid 0 \le j < (2i+3)/4\}$, see Figure 1. Denote the formula $\xi = \langle h\rangle\top \wedge \langle v\rangle\top$.

Lemma 5.2 *For any* $\mathbf{L}$*-frame* F*, if* $F, \theta \models \xi$ *for some valuation* θ *on* F*, then there is a homomorphism* $f\colon G' \to F$.

Proof. 1) We set $f(0,0) = c$, for an arbitrary point c in F.

2) Suppose that f is already defined on $\{(i,j) \mid j < \frac{2i+3}{4}; i \le i_0\}$ for some even $i_0 = 2k_0$. By ξ, there is a point a_0 in F with $F \models f(i_0,0)R_h a_0$. We set $f(i_0+1,0) = a_0$. Now suppose that $f(i_0+1,j)$ is defined for some $j \le k_0$. By ξ, there exists a point a_{j+1} in F with $F \models f(i_0,j)R_v a_{j+1}$. Set $f(i_0+1,j+1) = a_{j+1}$. If $j < k_0$, then by $F \models R_v \circ R_h \circ R_v \subseteq R_h$, we have $F \models f(i_0,j+1)R_h a_{j+1}$, and so f is still a homomorphism.

Now suppose that f is defined on (i_0+1, k_0+1). By ξ, there exists b_{k_0+1} in F such that $F \models f(i_0+1,k_0+1)R_h b_{k_0+1}$. We set $f(i_0+2,k_0+1) = b_{k_0+1}$.

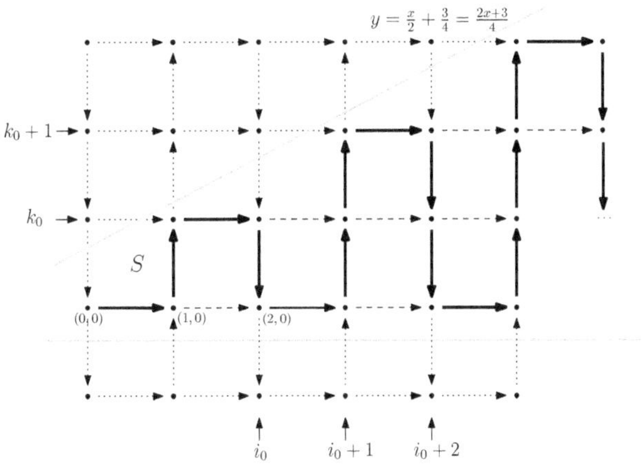

Fig. 1. The snake frame.

Suppose that $f(i_0 + 2, j + 1)$ is defined, for some $j \geq 0$. Then, by ξ, there is a point b_j in F such that $F \models f(i_0 + 2, j + 1)R_v b_j$. We set $f(i_0 + 2, j) = b_j$. Since $F \models R_v \circ R_h \circ R_v \subseteq R_h$, we have $F \models f(i_0 + 1, j)R_h b_j$ and so f is still a homomorphism. After $j = 0$ we have f defined on $\{(i, j) \mid j < \frac{2i+3}{4}; i \leq i_0 + 2\}$.

Iterating 2) yields the required homomorphism $f\colon G' \to F$. □

Proof. (of Theorem 5.1). Take a domino system $\mathcal{D}$. Without loss of generality, we can assume that

(tc) $\mathcal{D}$ tiles $\mathbb{N} \times \mathbb{Z}$ iff $\mathcal{D}$ tiles the sector S.

Indeed, recall that the tiling problem is undecidable because it models some Turing machine T. But after performing each instruction, the head of T can move to the right by at most one cell. This means that, in order to model T, it suffices to tile the sector $S' = \{(i, j) \mid i, j \in \mathbb{N},\ j \leq i\}$. If, additionally, without loss of generality, we assume that T idles after each right-move instruction, then we can model T by tiling the sector S. So, given a Turing machine T, we can generate a domino system $\mathcal{D}$ that satisfies (tc) — we only have to care how to tile $(\mathbb{N}\times\mathbb{Z}) \setminus S$ with special blank tiles without affecting the work of T.

Assuming (tc), we can show that $\xi \wedge \lambda^{\mathcal{D}}$ is globally **L**-satisfiable iff $\mathcal{D}$ tiles $\mathbb{N} \times \mathbb{Z}$. ($\Rightarrow$) If $\mathcal{D}$ tiles $\mathbb{N} \times \mathbb{Z}$, then we can use this tiling to define a model for $\xi \wedge \lambda^D$ based on G. ($\Leftarrow$) Suppose that $M \models \xi \wedge \lambda^{\mathcal{D}}$. Using Lemma 5.2, we can construct a homomorphism $f\colon G' \to M$. Now, from M we can read off a $\mathcal{D}$-tiling of S, and by (*tc*) conclude that there is a $\mathcal{D}$-tiling of $\mathbb{N}\times\mathbb{Z}$. □

Corollary 5.3 *The grammar logic corresponding to the context-free grammar* $\{h \to vhv,\ u \to uh,\ u \to uv\}$ *is undecidable.*

Proof. For every modal formula φ containing only $[h]$ and $[v]$, we have: φ is globally satisfiable in the logic $\mathbf{K}\{h \to vhv\}$ iff the formula $\langle u\rangle\top \wedge [u]\varphi$ is (locally) satisfiable in the logic $\mathbf{K}\{h \to vhv,\ u \to uh,\ u \to uv\}$. □

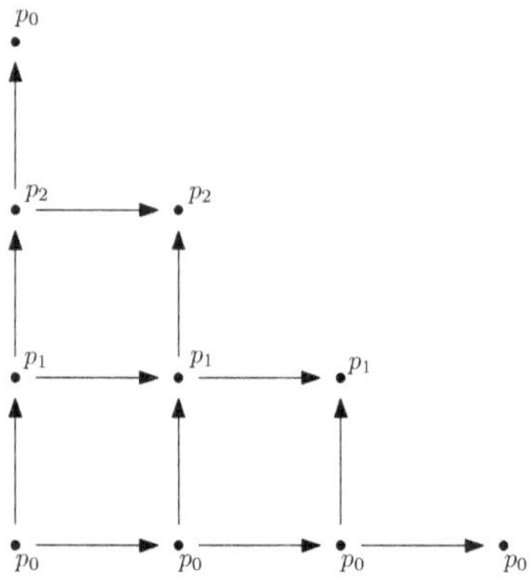

Fig. 2. A fragment of a "grid" model.

This gives a simple example of an undecidable context-free grammar logic without converse. Compare this with the undecidability of **PDL** enriched with the language $\{a^n ba^n \mid n \in \mathbb{N}\}$ shown in [12]. However, their formulas essentially use the **PDL**-constructs, and it is not clear if the latter can be eliminated so that the undecidability proof worked for context-free grammar logics.

Theorem 5.4 *The global satisfiability for* $\mathbf{K} + \Diamond\Box p \to \Box\Diamond p$ *is undecidable.*

Proof. Let $M = (W^G, R^G, \theta^G)$ be an infinite Kripke model based on a frame G (see Figure 2), where $W^G = \mathbb{N} \times \mathbb{N}$, $\theta^G(p_i) = \{(m,n) \mid n \equiv i \pmod 3\}$, and

$$R^G = \{((m,n),(m+1,n)) \mid m,n \in \mathbb{N}\} \cup \{((m,n),(m,n+1)) \mid m,n \in \mathbb{N}\}.$$

Let ξ be the conjunction of the following formulas (which capture some properties of G) for $0 \le i \ne j \le 2$ (subscripts of p's are understood modulo 3):

$$\begin{array}{ll} (X1) & p_i \wedge p_j \to \bot \\ (X2) & p_i \to \Box(p_i \vee p_{i+1}) \\ (X3) & p_i \to (\Diamond p_i \wedge \Diamond p_{i+1}) \\ (X4) & p_0 \vee p_1 \vee p_2 \end{array}$$

Lemma 5.5 *Assume that* $F \models \Diamond\Box p \to \Box\Diamond p$. *If* $F, \theta^F \models \xi$ *for some* θ^F, *then there is a homomorphism* $f\colon G \to F$ *in the following sense:*

$$\begin{array}{ll} (\mathsf{Homo}) & \textit{if } G \models xRy \textit{ then } F \models f(x)Rf(y) \textit{ and} \\ & x \in \theta^G(p_i) \;\; \textit{iff} \;\; f(x) \in \theta^F(p_i). \end{array}$$

Proof. Denote $N = (F, \theta^F)$. *Step 0.* From $(X3)$ and $(X4)$ it follows that there exists a point x in F such that $N, x \models p_0$. We set $f(0,0) = x$.

Step n. Suppose that, for some n, f is defined on $\{(i,j) \mid i,j \ge 0;\ i+j \le n\}$ and satisfies (Homo). We extend f to $\{(i,j) \mid i,j \ge 0;\ i+j = n+1\}$ as follows. By (X3), there is a point x in F such that $F \models f(n,0)Rx$ and $N, x \models p_0$. We set $f(n+1,0) = x$. Similarly, there is y in F such that $F \models f(0,n)Ry$ and $N, y \models p_{n+1}$. We set $f(0,n+1) = y$. Since $F \models \Diamond\Box p \to \Box\Diamond p$, the fact that, by I.H., for all i,j such that $i+j = n-1$, $F \models f(i,j)Rf(i,j+1)$

and $F \models f(i,j)Rf(i+1,j)$, implies that there exist points z_{ij} in F such that $F \models f(i,j+1)Rz_{ij} \wedge f(i+1,j)Rz_{ij}$. Now we define $f(i+1,j+1) = z_{ij}$. Then (X2) and I.H. imply that $N, z_{ij} \models p_j$. We claim that f is now defined on $\{(i,j) \mid 1 \leq i+j \leq n+1\}$ and satisfies (Homo). □

For a domino system $\mathcal{D} = (D,H,V)$, we define $\lambda^{\mathcal{D}} = \lambda_0 \wedge \lambda^H \wedge \lambda^V$, where

$$\lambda_0 = (\bigvee_{d \in D} q_d) \wedge \bigwedge_{d \neq d'} \neg(q_d \wedge q_{d'}),$$

$$\lambda^H = \bigwedge_{0 \leq i \leq 2} \bigwedge_{d \in D} \big(p_i \wedge q_d \to \Box(p_i \to \bigvee_{d' \in H(d)} q_{d'})\big),$$

$$\lambda^V = \bigwedge_{0 \leq i \leq 2} \bigwedge_{d \in D} \big(p_i \wedge q_d \to \Box(p_{i+1} \to \bigvee_{d' \in V(d)} q_{d'})\big).$$

We claim that $\xi \wedge \lambda^{\mathcal{D}}$ is globally satisfiable on some frame F validating $\Diamond\Box p \to \Box\Diamond p$ iff $\mathcal{D}$ tiles $\mathbb{N} \times \mathbb{N}$. This gives us the desired reduction. □

6 Conclusion and further research

In this paper we investigate classes of frames and modal logics that admit filtration and operations on them that preserve this property (they are called filtration safe). On the one hand, this notion is useful for obtaining results on the FMP and the decidability of logics. On the other, it appears robust, for many interesting operations turn out to be filtration safe (see Section 2).

We used filtration safe operations to show that every regular grammar logic (with converse) admits filtration and hence is decidable (Theorem 3.3). Note that all known examples of undecidable grammar logics (see e.g. Corollary 5.3) correspond to irregular grammars. The following question arises naturally.

Open problem. *Are the following claims equivalent, for every grammar* Π:

(i) Π *is a regular grammar,*
(ii) *the logic* $\mathbf{K}\Pi$ *is decidable,*
(iii) *the logic* $\mathbf{K}\Pi$ *admits filtration,*
(iv) *the logic* $\mathbf{K}\Pi$ *has the finite model property,*
(v) *the logic* $\mathbf{K}\Pi$ *has the exponential model property?*

The approach taken in this paper could be developed in various directions. One can examine other interesting operations on relations for filtration safety. Note that all operations on classes of frames considered above are in fact induced by operations on frames. It is reasonable to consider more general operations. For example, the following operation is filtration safe: given a class $\mathcal{F}$ of frames of the form (W,R), build the class $\{(W,R,S) \mid (W,R) \in \mathcal{F},\ R \subseteq S\}$. One could also seek for general conditions on operations that are sufficient for filtration safety; Lemma 1.6 gives an idea what such conditions might look like.

A 'relativized' notion also makes sense: one can say that a logic $\mathbf{L}$ *admits filtration in a class of frames* $\mathcal{F}$ if $\mathbf{L}$ is the logic of $\mathcal{F}$ and $\mathcal{F}$ admits filtration. As we mentioned in Example 1.4, the logics **S4.2** and **S4.3** admit filtration in the classes of their point-generated frames. Of course, in order to use the relativized notion for obtaining decidability results, we need to assume that the subclass of finite frames of the class $\mathcal{F}$ is recognizable.

References

[1] Bezhanishvili, N. and B. ten Cate, *Transfer results for hybrid logic. part I: The case without satisfaction operators*, J. of Logic and Computation **16** (2006), pp. 177–197.

[2] Blackburn, P., M. de Rijke and Y. Venema, "Modal Logic," Cambridge Tracts in Theoretical Computer Science **53**, Cambridge University Press, 2002.

[3] Börger, E., E. Grädel and Y. Gurevich, "The Classical Decision Problem," Perspectives in Mathematical Logic, Springer, 1997.

[4] Chagrov, A. and M. Zakharyaschev, "Modal Logic," Oxford Logic Guides **35**, Oxford University Press, 1997.

[5] Demri, S., *The complexity of regularity in grammar logics and related modal logics*, Journal of Logic and Computation **11** (2001), pp. 933–960.

[6] Demri, S. and H. De Nivelle, *Deciding regular grammar logics with converse through first-order logic*, Journal of Logic, Language and Information **14** (2005), pp. 289–329.

[7] Fischer, M. J. and R. E. Ladner, *Propositional dynamic logic of regular programs*, Journal of computer and system sciences **18** (1979), pp. 194–211.

[8] Gabbay, D., I. Shapirovsky and V. Shehtman, *Products of modal logics and tensor products of modal algebras*, Journal of Applied Logic (2014), to appear.

[9] Gabbay, D. M., *A general filtration method for modal logics*, Journal of Philosophical Logic **1** (1972), pp. 29–34.

[10] Gabbay, D. M. and V. B. Shehtman, *Products of modal logics, part 1*, Logic Journal of IGPL **6** (1998), pp. 73–146.

[11] Goranko, V. and S. Passy, *Using universal modality: Gains and questions*, Journal of Logic and Computation **2** (1992), pp. 5–30.

[12] Harel, D., A. Pnueli and J. Stavi, *Propositional dynamic logic of nonregular programs*, Journal of Computer and System Sciences **26** (1983), pp. 222–243.

[13] Hopcroft, J. and J. Ullman, "Introduction to Automata Theory, Languages, and Computation," Addison-Wesley, 1979.

[14] Kracht, M., "Tools and Techniques in Modal Logic," Elsevier, 1999.

[15] Lemmon, E. J. and D. Scott, "Intensional Logics," Stanford, 1966.

[16] Michaliszyn, J., "Decidability of modal logics with particular emphasis on the interval temporal logics," Ph.D. thesis, University of Wroclaw (2012).

[17] Mortimer, M., *On languages with two variables*, Math. Logic Quarterly **21** (1975), pp. 135–140.

[18] Segerberg, K., *Decidability of four modal logics*, Theoria **34** (1968), pp. 21–25.

[19] Shehtman, V., *On some two-dimensional modal logics*, in: *8th Congress on Logic Methodology and Philosophy of Science, vol. 1*, Abstracts **1** (1987), pp. 326–330.

[20] Shehtman, V. B., *Filtration via bisimulation*, Advances in Modal Logic **5** (2004), pp. 289–308.

[21] Spaan, E., "Complexity of Modal Logics," Ph.D. thesis, Department of Mathematics and Computer Science, University of Amsterdam (1993).

[22] Wolter, F., *Solution to a problem of Goranko and Passy*, Journal of Logic and Computation **4** (1994), pp. 21–22.

[23] Wolter, F., *The finite model property in tense logic*, Journal of Symbolic Logic **60** (1995), pp. 757–774.

[24] Wolter, F., *A counterexample in tense logic*, Notre Dame Journal of Formal Logic **37** (1996), pp. 167–173.

[25] Wolter, F., *Properties of tense logics*, Math. Logic Quarterly **42** (1996), pp. 481–500.

[26] Wolter, F., *Tense logic without tense operators*, Math. Logic Quarterly **42** (1996), pp. 145–171.

[27] Wolter, F., *Completeness and decidability of tense logics closely related to logics above K4*, Journal of Symbolic Logic **62** (1997), pp. 131–158.

Modal Independence Logic

Juha Kontinen* Julian-Steffen Müller† Henning Schnoor‡
Heribert Vollmer†

*University of Helsinki, Department of Mathematics and Statistics, P.O. Box 68, 00014 Helsinki, Finland.

†Leibniz Universität Hannover, Institut für Theoretische Informatik Appelstr. 4, 30167 Hannover

‡Institut für Informatik, Christian-Albrechts-Universität zu Kiel 24098 Kiel

---

Abstract

This paper introduces modal independence logic MIL, a modal logic that can explicitly talk about independence among propositional variables. Formulas of MIL are not evaluated in worlds but in sets of worlds, so called *teams*. In this vein, MIL can be seen as a variant of Väänänen's modal dependence logic MDL. We show that MIL embeds MDL and is strictly more expressive. However, on singleton teams, MIL is shown to be not more expressive than usual modal logic, but MIL is exponentially more succinct. Making use of a new form of bisimulation, we extend these expressivity results to modal logics extended by various generalized dependence atoms. We demonstrate the expressive power of MIL by giving a specification of the anonymity requirement of the *dining cryptographers* protocol in MIL. We also study complexity issues of MIL and show that, though it is more expressive, its satisfiability and model checking problem have the same complexity as for MDL.

Keywords: dependence logic, team semantics, independence, expressivity over finite models, computational complexity.

---

1 Introduction

The concept of independence is ubiquitous in many scientific disciplines such as experimental physics, social choice theory, computer science, and cryptography. Dependence logic D, introduced by Jouko Väänänen in [14], is a new logical framework in which various notions of dependence and independence can be formalized and studied. It extends first-order logic by so called dependence atoms

$$=(x_1, \ldots, x_{n-1}, x_n),$$

expressing that the value of the variable x_n depends (only) on the values of $x_1, \ldots, x_{n-1}$, in other words, that x_n is functionally dependent of $x_1, \ldots, x_{n-1}$. Of course, such a dependency does not make sense when talking about single assignments; therefore dependence logic formulas are evaluated for so called *teams*, i. e., sets of assignments. A team can for example be a relational database table, a collection of plays of a game, or a set of agents with features. It is this *team semantics*, together with dependence atoms, that gives dependence logic its expressive power: it is known that D is as expressive as Σ_1^1, that is, the properties of finite structures that can be expressed in dependence logic are exactly the NP-properties.

In a slightly later paper Väänänen [15] introduced dependence atoms into (propositional) modal logic. Here, teams are sets of worlds, and a dependence atom $=(p_1, \ldots, p_{n-1}, p_n)$ holds in a team T if there is a Boolean function that determines the value of p_n from those of $p_1, \ldots, p_{n-1}$ in all worlds in T. The so obtained modal dependence logic MDL was studied from the point of view of expressivity and complexity in [13].

In this article we introduce a novel modal variant of dependence logic called *modal independence logic*, MIL, extending the formulas of modal logic ML by so-called *independence* atoms

$$(p_1, \ldots, p_\ell) \perp_{(r_1, \ldots, r_m)} (q_1, \ldots, q_n),$$

the meaning of which is that the propositional sequences $\vec{p}$ and $\vec{q}$ are independent of each other for any fixed value of $\vec{r}$. Modal independence logic thus has its roots in modal dependence logic MDL [15] and first-order independence logic [8]. In modal independence logic, dependencies between propositions can be expressed, and thus, analogously to the first-order case, MDL can be embedded as a sublogic into MIL, and it is easy to see that MIL is strictly more expressive than MDL.

The aim of this paper is to initiate a study of the expressiveness and the computational complexity of modal independence logic. For this end, we first study the computational complexity of the satisfiability and the model checking problem for MIL. We show that, though MIL is more expressive than MDL, the complexity of these decision problems stays the same, i. e., the satisfiability problem is complete for nondeterministic exponential time (NEXP-complete, [13]) and the model checking problem is NP-complete [6]. In order to settle the complexity of satisfiability for MIL, we give a translation of MIL-formulas to existential second-order logic formulas the first-order part of which is in the Gödel–Kalmár–Schütte prefix class. Our result then follows from the classical result that the satisfiability problem for this prefix class is NEXP-complete [3]. We will also show that the same upper bound on satisfiability can be obtained for a whole range of variants of MIL via the notion of a generalized (modal) dependence atom (a notion introduced in the first-order framework in [11]).

The expressive power of MDL was first studied by Sevenster [13], where he showed that MDL is equivalent to ML on singleton teams. In this paper we prove a general result showing that MIL, and in fact any variant of it whose generalized dependence atoms are FO-definable, is bound to be equivalent to ML over singleton teams. Interestingly, it was recently shown in [5] that a so-called extended modal dependence logic EMDL is strictly more expressive than

MDL even on singletons.

To demonstrate the potential applications of MIL, we consider the *dining cryptographers* protocol [4], a classic example for anonymous broadcast which is used as a benchmark protocol in model checking of security protocols [1]. We show how the anonymity requirement of the protocol can be formalized in modal independence logic, where—unlike in the usual approaches using epistemic logic—we do not need to use the Kripke model's accessibility relation to encode knowledge, but to express the "possible future" relation of branching-time models. In addition to demonstrating MIL's expressivity, we also derive a succinctness result from our modeling of the dining cryptographers: While MIL and ML are equally expressive on singletons, MIL is exponentially more succinct.

2 Modal Independence Logic

Definition 2.1 The syntax of *modal logic* ML is inductively defined by the following grammar in extended Backus–Naur form:

$$\phi ::= p \mid \overline{p} \mid \phi \wedge \phi \mid \phi \vee \phi \mid \Diamond\phi \mid \Box\phi.$$

The syntax of *modal dependence logic* MDL is defined by

$$\phi ::= p \mid \overline{p} \mid {=}(\vec{q}, p) \mid \phi \wedge \phi \mid \phi \vee \phi \mid \Diamond\phi \mid \Box\phi,$$

where p is a propositional variable and $\vec{q}$ a sequence of propositional variables.

The syntax of *modal independence logic* MIL is defined by

$$\phi ::= p \mid \overline{p} \mid \vec{p} \bot_{\vec{r}} \vec{q} \mid \phi \wedge \phi \mid \phi \vee \phi \mid \Diamond\phi \mid \Box\phi,$$

where p is a propositional variable and $\vec{p}, \vec{r}, \vec{q}$ are sequences of propositional variables. The sequence $\vec{r}$ may be empty.

It is worth noting that the negation of dependence logic, and that of MDL, is not the classical negation but a so-called "game theoretic" negation that still satisfies the usual De Morgan laws. As now customary with team semantics, we think of negation as a defined operation and restrict attention to formulas in which negation only appears in front of proposition symbols (formulas $\neg{=}(\vec{q}, p)$ are logically equivalent to falsum and hence can also be dispensed without loss of generality [15]). A more detailed account of the role of negation in dependence logic can be found in [10].

A Kripke structure is a tuple $\mathcal{M} = (W, R, \pi)$, where W is a non-empty set of worlds, R is a binary relation over W and $\pi\colon W \to \mathcal{P}(\mathrm{V})$ is a labeling function for a set V of propositional variables. A *team* is a (possibly empty) set $T \subseteq W$. As usual, in a Kripke structure $\mathcal{M}$ the set of all successors of $T \subseteq W$ is defined as $R(T) = \{s \in W \mid \exists s' \in T : (s', s) \in R\}$. Furthermore we define $R\langle T\rangle = \{T' \subseteq R(T) \mid \forall s \in T\ \exists s' \in T' : (s, s') \in R\}$, the set of legal successor teams.

Definition 2.2 Let $\vec{p} = (p_1, \ldots, p_n)$ be a sequence of variables and w, w' be worlds of a Kripke model $\mathcal{M} = (W, R, \pi)$. Then w and w' are equivalent under π over $\vec{p}$, denoted by $w \equiv_{\pi,\vec{p}} w'$, if the following holds:

$$\pi(w) \cap \{p_1, \ldots, p_n\} = \pi(w') \cap \{p_1, \ldots, p_n\}.$$

Definition 2.3 (Semantics of ML, MDL, and MIL) Let $\mathcal{M} = (W, R, \pi)$ be a Kripke structure, T be a team over $\mathcal{M}$ and ϕ be a formula. The semantic evaluation (denoted as $\mathcal{M}, T \models \phi$) is defined inductively as follows.

$$\begin{array}{lcl}
\mathcal{M}, T \models p & \Leftrightarrow & \forall w \in T\colon p \in \pi(w) \\
\mathcal{M}, T \models \overline{p} & \Leftrightarrow & \forall w \in T\colon p \notin \pi(w) \\
\mathcal{M}, T \models \phi_1 \wedge \phi_2 & \Leftrightarrow & \mathcal{M}, T \models \phi_1 \text{ and } \mathcal{M}, T \models \phi_2 \\
\mathcal{M}, T \models \phi_1 \vee \phi_2 & \Leftrightarrow & \exists T_1, T_2\colon T_1 \cup T_2 = T, \mathcal{M}, T_1 \models \phi_1 \text{ and } \mathcal{M}, T_2 \models \phi_2 \\
\mathcal{M}, T \models \Diamond\phi & \Leftrightarrow & \exists T' \in R\langle T\rangle\colon \mathcal{M}, T' \models \phi \\
\mathcal{M}, T \models \Box\phi & \Leftrightarrow & \mathcal{M}, R(T) \models \phi \\
\mathcal{M}, T \models {=}(\vec{q}, p) & \Leftrightarrow & \forall w, w' \in T\colon w \equiv_{\pi,\vec{q}} w' \text{ implies } w \equiv_{\pi,p} w' \\
\mathcal{M}, T \models \vec{p}_1 \bot_{\vec{q}}\, \vec{p}_2 & \Leftrightarrow & \forall w, w' \in T\colon w \equiv_{\pi,\vec{q}} w' \text{ implies } \exists w'' \in T\colon \\
& & w'' \equiv_{\pi,\vec{p}_1} w \text{ and } w'' \equiv_{\pi,\vec{p}_2} w' \text{ and } w'' \equiv_{\pi,\vec{q}} w
\end{array}$$

Note that for modal logic formulas ϕ we have $\mathcal{M}, \{w\} \models \phi$ iff $\mathcal{M}, w \models \phi$ (where in the latter case, $\models$ is defined as in any textbook for usual modal logic). In fact it is easy to see that without dependence or independence atom, our logic has the so called *flatness property*, stating that team semantics and usual semantics essentially do not make a difference:

Lemma 2.4 *For every* ML*-formula ϕ and all models M and teams T, $M, T \models \phi$ iff $M, w \models \phi$ for all $w \in T$.*

Another simple observation is that the empty team satisfies all formulas.

Lemma 2.5 *For every* MIL*-formula ϕ and all models M, $M, \emptyset \models \phi$.*

Team semantics and independence atoms together will lead to a richer expressive power, as we will prove in Section 6. However, we will also show that over teams T consisting of one world only, ML and MIL have the same expressive power.

Definition 2.6 Formulas φ and φ' are *equivalent on singletons*, if for every model M and every $w \in M$, we have $M, \{w\} \models \varphi$ if and only if $M, \{w\} \models \varphi'$.

Note that on singleton teams, the independence atom trivially always evaluates to true. However, using the modal operators $\Box$ and $\Diamond$, a formula that is evaluated on singleton teams as a starting point clearly is able to talk about nontrivial teams as well. An example for this is the formula constructed in Section 5: The formula is evaluated on a singleton team—the starting point of the protocol—but specifies independence properties for much larger teams, namely subsets of all possible protocol outcomes.

3 Complexity Results

In this section we will study the computational complexity of the model checking and the satisfiability problem for MIL. In [8], it was observed that in first-order team semantics, $=(\vec{p}, q)$ is equivalent to $q \perp_{\vec{p}} q$. This observation clearly carries over to MIL, and hence in particular shows that MIL is a generalization of MDL.

Lemma 3.1 *Let $\mathcal{M}$ be a model and T a team over $\mathcal{M}$, let $\vec{p}$ and q be variables. Then $\mathcal{M}, T \models =(\vec{p}, q)$ if and only if $\mathcal{M}, T \models q \perp_{\vec{p}} q$.*

We now define the two decision problems whose complexity we wish to study, namely the model checking and the satisfiability problem for modal independence logic.

Problem:	MIL-SAT
Input:	MIL formula ϕ
Question:	Does there exists a Kripke model $\mathcal{M}$ and a team $T \neq \emptyset$ with $\mathcal{M}, T \models \phi$?

Problem:	MIL-MC
Input:	Kripke model $\mathcal{M}$, team T and MIL formula ϕ
Question:	$\mathcal{M}, T \models \phi$?

The corresponding problems for modal dependence logic are denoted by MDL-SAT and MDL-MC.

It is easy to see that model checking for MIL is not more difficult than model checking for MDL, namely NP-complete.

Theorem 3.2 MIL-MC *is* NP*-complete.*

Proof. The lower bound follows immediately from Lemma 3.1 and NP-completeness of MDL-MC [6]. The upper bound follows from a simple extension of the well-known model checking algorithm for modal logic, see Algorithm 1.□

Next we will consider the complexity of the satisfiability problem MIL-SAT for modal independence logic. From Lemma 3.1 and the hardness of MDL-SAT for nondeterministic exponential time [13] we immediately obtain the following lower bound:

Lemma 3.3 MIL-SAT *is* NEXP*-hard.*

In order to show containment in NEXP, we need to recall the following classical result. Recall that the so-called Gödel–Kalmár–Schütte prefix class $[\exists^*\forall^2\exists^*, all]$ contains sentences of FO, in a relational vocabulary without equality, which are in prenex normal form and have a quantifier prefix of the form $\exists^*\forall^2\exists^*$.

Proposition 3.4 ([3]) *Satisfiability of formulas in prefix class $[\exists^*\forall^2\exists^*, all]$ can be decided in* $\mathrm{NTIME}(2^{O(n/\log n)})$.

Next we will show that MIL-SAT $\in$ NEXP with the help of Proposition 3.4. We will first define a variant of the standard translation of ML into FO that

Algorithm 1 NP algorithm for MIL-MC

```
function MILMC(𝓜, T, φ)
    if φ = □ψ then
        return MILMC(M, ψ, R(T))
    if φ = ◇ψ then
        existentially guess T' ∈ R⟨T⟩
        return MILMC(M, ψ, T')
    else if φ = ψ1 ∧ ψ2 then
        return MILMC(M, ψ1, T) and MILMC(M, ψ2, T)
    else if φ = ψ1 ∨ ψ2 then
        existentially guess T1 ∪ T2 = T
        return MILMC(M, ψ1, T1) and MILMC(M, ψ2, T2)
    else if φ = p then
        for s ∈ T do
            if p ∉ π(s) then
                return false
        return true
    else if φ = p̄ then
        for s ∈ T do
            if p ∈ π(s) then
                return false
        return true
    else if φ = p⃗ ⊥_r⃗ q⃗ then
        for s ∈ T do
            for s' ∈ T do
                if π(s') ∩ r⃗ = π(s') ∩ r⃗ then
                    found ← false
                    for s'' ∈ T do
                        agreeP ← π(s'') ∩ p⃗ = π(s) ∩ p⃗
                        agreeQ ← π(s'') ∩ q⃗ = π(s') ∩ q⃗
                        agreeR ← π(s'') ∩ r⃗ = π(s) ∩ r⃗
                        if agreeP and agreeQ and agreeR then
                            found ← true
                    if not found then
                        return false
        return true
```

maps MIL-formulas to formulas of monadic existential second-order logic. For a Kripke structure (W, R, π), and a team $T \subseteq W$, we denote by $(W, \{A_i\}_i, R, T)$ the first-order structure of vocabulary $\{R, T\} \cup \{A_i\}_{i \in \mathrm{N}}$ encoding (W, R, π) in the obvious way.

Lemma 3.5 *For any formula $\phi \in$ MIL there is a sentence ϕ^* of monadic existential second-order logic of the form*

$$\exists Y_1 \ldots \exists Y_m \forall x \forall y \exists z_1 \ldots \exists z_k \theta, \tag{1}$$

where θ is quantifier-free, and such that for all (W, R, π) and T it holds

$$(W, R, \pi), T \models \phi \Leftrightarrow (W, \{A_i\}_{i\in\mathrm{N}}, R, T) \models \phi^*.$$

Proof. We first define an auxiliary translation $\phi \mapsto \phi'$ for which correctness is obvious and then indicate how to go from ϕ' to ϕ^*.

(i) Suppose ϕ is p_i. Then ϕ' is defined as

$$\phi' := \forall x(T(x) \to A_{p_i}(x)).$$

(ii) Suppose ϕ is $\overline{p}_i$. Then ϕ' is defined as

$$\phi' := \forall x(T(x) \to \neg A_{p_i}(x)).$$

(iii) Suppose ϕ is $\psi_1 \vee \psi_2$. Then ϕ' is defined as

$$\phi' := \exists Y_1 \exists Y_2(\forall x(T(x) \leftrightarrow (Y_1(x) \vee Y_2(x))) \wedge \psi_1'(T/Y_1) \wedge \psi_2'(T/Y_2)).$$

(iv) Suppose ϕ is $\psi_1 \wedge \psi_2$. Then ϕ' is defined as

$$\phi' := \psi_1' \wedge \psi_2'.$$

(v) Suppose ϕ is $\Diamond\psi$. Then ϕ' is defined as

$$\phi' := \exists Y(\forall x(T(x) \to \exists z(Y(z) \wedge E(x,z)) \wedge (Y(x) \to \exists u(T(u) \wedge E(u,x)))) \wedge \psi'(T/Y)).$$

(vi) Suppose ϕ is $\Box\psi$. Then ϕ' is defined as

$$\phi' := \exists Y(\forall x \forall y(((T(x) \wedge E(x,y)) \to Y(y)) \wedge (Y(x) \to \exists z(T(z) \wedge E(z,x)))) \wedge \psi'(T/Y)).$$

(vii) Suppose ϕ is $\vec{p}_1 \perp_{\vec{p}_2} \vec{p}_3$. Then ϕ' is defined as

$$\phi' := \forall x \forall y((T(x) \wedge T(y) \wedge EQ_{\vec{p}_2}(x,y)) \to \exists z(T(z) \wedge EQ_{\vec{p}_2}(x,z) \wedge EQ_{\vec{p}_1}(x,z) \wedge EQ_{\vec{p}_3}(y,z))),$$

where $EQ_{\vec{p}_i}(v,w)$ is a shorthand for the formula

$$\bigwedge_{p\in\vec{p}_i} A_p(v) \leftrightarrow A_p(w).$$

It remains to define the translation $\phi \mapsto \phi^*$. This translation is defined by modifying the above clauses by essentially moving all quantifiers to the left of the formula, and by possibly renaming some of the bound variables. We will indicate these modifications by considering the case of disjunction. The other cases are analogous. Assume that ψ_1^* and ψ_2^* are defined already:

$$\psi_i^* = \exists \bar{Y}_i \forall x \forall y \exists \vec{z}_i \theta_i,$$

where θ_i is quantifier free, and $\psi_i^* \equiv \psi_i'$. By renaming of bound variables, we may assume that $\bar{Y}_2 = Y_3 \dots Y_k$, and $\bar{Y}_1 = Y_{k+1} \dots Y_m$, and that $\vec{z}_1$ and $\vec{z}_2$ do not have any common variables either. Then $(\psi_1 \vee \psi_2)^*$ is defined by replacing ψ_i' by ψ_i^* in the definition of $(\psi_1 \vee \psi_2)'$ (see clause iii), and by extending the scopes of the quantifiers:

$$(\psi_1 \vee \psi_2)^* := \exists Y_1 \dots \exists Y_m \forall x \forall y \exists \vec{z}_2 \exists \vec{z}_1 ((T(x) \leftrightarrow (Y_1(x) \vee Y_2(x)) \wedge \theta_1(T/Y_1) \wedge \theta_2(T/Y_2)).$$

□

Theorem 3.6 MIL-SAT *is in* NEXP.

Proof. Let $\phi \in$ MIL. Then ϕ is satisfiable by a Kripke model $\mathcal{M}$ and a team $T \neq \emptyset$ if and only if $\phi^* \wedge \exists w T(w)$ is satisfiable. This follows from the previous lemma and the fact that there is a 1-1 correspondence with Kripke structures (W, R, π), and teams T for ϕ and $\{R, T\} \cup \{A_i\}_{1 \leq i \leq n}$-structures $(W, \{A_i\}_{1 \leq i \leq n}, R, T)$, where n is large enough such that all p_i appearing in ϕ satisfy $i \leq n$.

Recall now that ϕ^* has the form

$$\exists Y_1 \dots \exists Y_m \forall x \forall y \exists z_1 \dots \exists z_k \theta,$$

hence $\phi^* \wedge \exists w T(w)$ is logically equivalent to

$$\exists Y_1 \dots \exists Y_m \forall x \forall y \exists z_1 \dots \exists z_k \exists w (\theta \wedge T(w)),$$

which is satisfiable if and only if the first-order sentence

$$\forall x \forall y \exists z_1 \dots \exists z_k \exists w (\theta \wedge T(w)) \tag{2}$$

of vocabulary $\{Y_1, \dots, Y_m\} \cup \{R, T\} \cup \{A_i\}_{1 \leq i \leq n}$ is satisfiable. The sentence (2) is contained in prefix class $[\exists^* \forall^2 \exists^*, all]$, hence the satisfiability of it, and also of $\phi^* \wedge \exists w T(w)$, can be decided in time NTIME$(2^{O(|\phi^*|)})$. The claim now follows from the fact the mapping $\phi \mapsto \phi^*$ can be computed in time polynomial in $|\phi|$. □

Corollary 3.7 MIL-SAT *is* NEXP-*complete.*

It is interesting to note that Theorem 3.6 and Lemma 3.1 directly imply the result of Sevenster [13] that MDL-SAT is contained in NEXP. On the other hand, it seems that the original argument of Sevenster does not immediately generalize to MIL.

Corollary 3.8 MDL-SAT *is* NEXP-*complete.*

4 Generalized Dependency Notions

MIL can be seen as an extension of modal logic with team semantics by the independence atom—let us denote such an extension by ML($\perp$). Similarly, we can extend modal logic with other atoms, so-called *generalized dependence atoms*, which we define now.

Definition 4.1 Let $\mathcal{M} = (W, R, \pi)$ be a Kripke model and $T = (w_1, \ldots, w_m)$ be a team over $\mathcal{M}$. Then for any propositional variable p, $T(p)$ is defined as the tuple $(s_1, \ldots, s_m)$, where s_i for $1 \leq i \leq m$ is defined as:

$$s_i = \begin{cases} 1 & w_i \in \pi(p) \\ 0, & \text{otherwise.} \end{cases}$$

For a set of propositions $\vec{q} = (q_1, \ldots, q_k)$, we define $T(\vec{q})$ analogously as $(T(q_1), \ldots, T(q_k))$.

Similar to Kuusisto's [11] definition of generalized first-order dependence atoms we give a definition of generalized modal dependence atoms. In the following, a set of matrices D is invariant under permutations of rows, if for every matrix $M \in D$, if M' is obtained from M by permuting M's rows, then M' is an element of D as well.

Definition 4.2 Let D be a set of Boolean n-column matrices that is invariant under permutation of rows. The semantics of the *generalized dependence atom defined by* D is given as follows:

Let $\mathcal{M}$ be a Kripke model, T be a team over $\mathcal{M}$ and $p_1, \ldots, p_n$ atomic propositions. Then

$$\mathcal{M}, T \models D(p_1, \ldots, p_n) \iff \langle T(p_1), \ldots, T(p_n) \rangle \in D.$$

The *width* of D is defined to be n.

Note that for simplicity we do not distinguish in notation between the logical atom D and the set D of Boolean matrices.

The Boolean matrix $\langle T(p_1), \ldots, T(p_n) \rangle$ contains one column for each of the variables $p_1, \ldots, p_n$; each row of the matrix corresponds to one world from T. The entry for variable p_i and world $w \in T$ is 1 if and only if the variable p_i is satisfied in the world w. We require that D is invariant under permutation of rows in order to ensure that whether $\mathcal{M}, T \models D(p_1, \ldots, p_n)$ holds does not depend on the ordering of the worlds in T that is used in computing the tuple $T(p)$.

In the following we will mainly be interested in generalized dependence atom *definable* by first-order formulae. For this purpose let D be an atom of width n as above, and ϕ be a first-order sentence over signature $\langle A_1, \ldots, A_n \rangle$. Then ϕ *defines* D if for all Kripke models $\mathcal{M} = (W, R, \pi)$ and teams T over $\mathcal{M}$,

$$\mathcal{M}, T \models D(p_1, \ldots, p_n) \iff \mathcal{A} \models \phi,$$

where $\mathcal{A}$ is the first-order structure with universe T and relations $A_i^{\mathcal{A}}$ for $1 \leq i \leq n$, where for all $w \in T$, $w \in A_i^{\mathcal{A}} \Leftrightarrow p_i \in \pi(w)$.

We say that a generalized dependence atom D is FO-*definable* if there exists a FO-formula ϕ defining D as above. Strictly speaking, the dependence atoms considered in the literature are *families* of dependence atoms for different width, e.g., the simple dependence $=\!(p_1, \ldots, p_n)$ is defined for arbitrary values of n.

Let us say that such a family is (P-uniformly) FO-definable if there exists a family of defining first-order formulae ϕ_n such that ϕ_n defines the atom of width n and the mapping $1^n \mapsto \langle\phi_n\rangle$ is computable in polynomial time; that is, an encoding of formula ϕ_n is computable in time polynomial in n. Note that in particular this implies that $|\phi_n| = p(n)$ for some polynomial p.

As examples let us show how to define some well-studied generalized dependence atoms as follows.

$$
\begin{aligned}
=\!(\vec{p}, q) &\Leftrightarrow \forall w \forall w' \; ((\textstyle\bigwedge_{1\leq i\leq n} A_{p_i}(w) \leftrightarrow A_{p_i}(w')) \rightarrow (A_q(w) \leftrightarrow A_q(w'))) \\
\vec{p} \subseteq \vec{q} &\Leftrightarrow \forall w \exists w' \; (\textstyle\bigwedge_{1\leq i\leq n} A_{p_i}(w) \leftrightarrow A_{q_i}(w')) \\
\vec{p} \mid \vec{q} &\Leftrightarrow \forall w \forall w' \; (\textstyle\bigvee_{1\leq i\leq n} A_{p_i}(w) \leftrightarrow \neg A_{q_i}(w'))
\end{aligned}
$$

The latter two so-called *inclusion* and *exclusion* atoms were introduced by Galliani in [7]. In particular all above atoms are FO-definable. The independence atom $\vec{p}_1 \bot_{\vec{q}} \vec{p}_2$ is also FO-definable in the obvious way. [1]

We use ML(D) to denote the extension of ML by a generalized dependence atom D. We will next show that our complexity upper bounds from Section 3 can be generalized to cover ML(D) for certain FO-definable dependence atoms D. For model checking, we simply use the fact that first-oder formulas can be verified in polynomial time and obtain the following corollary:

Corollary 4.3 *Let D be a P-uniformly* FO*-definable generalized dependence atom. Then* ML(D)-MC *is in* NP.

For the satisfiability problem, we generalize the proof of Theorem 3.5 in case (vii) to dependence atoms which are definable by a $[\exists^*\forall^2\exists^*]$ formula.

Corollary 4.4 *Let D be a generalized dependence atom that is P-uniformly* FO*-definable by a (family of) first-order formula(e) in the prefix class $[\exists^*\forall^2\exists^*]$. Then* ML($D$)-SAT *is in* NEXP.

5 Example: The Dining Cryptographers

The dining cryptographers [4], a standard example for anonymous broadcast, is the following problem: A group of cryptographers $\{c_0, \ldots, c_{n-1}\}$ with $n \geq 3$ sit in a restaurant, where c_i sits between c_{i-1} and c_{i+1}. (Indices of the cryptographers are always modulo n, and i always ranges over $0, \ldots, n-1$). After dinner, it turns out that someone already paid. There are only two possibilities: Either one of the cryptographers secretly paid, or the NSA did. Naturally, they want to know which of these is the case, but without revealing the paying cryptographer if one of them paid. They use the following protocol (in the protocol, $\oplus$ defines the exclusive-or of two bits, where $b_1 \oplus b_2 = b_1 + b_2 \bmod 2$):

[1] Since the FO-formula ϕ may only depend on the width, we restrict ourselves to occurrences of $\vec{p}_1 \bot_{\vec{q}} \vec{p}_2$ where $|\vec{p}_1| = |\vec{p}_2| = |\vec{q}|$, if these sets are nonempty, which we can always assume without loss of generality by repeating variable occurrences, the case that one of these sets is empty can then be encoded into widths that are not multiples of 3 in a straightforward manner.

- For each i, let p_i be 1 iff c_i paid. Each c_i knows the value of p_i, but not of p_j for $j \neq i$. There is at most one i with $p_i = 1$. The protocol computes the value $p_0 \oplus p_1 \oplus \cdots \oplus p_{n-1}$, which is the same as $p_0 \vee p_1 \vee \cdots \vee p_{n-1}$.
- Each adjacent pair $\{c_i, c_{i+1}\}$ computes a random bit $bit_{\{i,i+1\}}$.
- Each c_i publicly announces the value $announce_i = p_i \oplus bit_{\{i,i-1\}} \oplus bit_{\{i,i+1\}}$.
- Then, $p_0 \oplus p_1 \oplus \cdots \oplus p_{n-1} = announce_0 \oplus announce_1 \oplus \cdots \oplus announce_{n-1}$.

The protocol clearly computes the correct answer, the interesting aspect is the *anonymity requirement*: No cryptographer c_i should learn anything about the values p_j for $j \neq i$ except for what follows from the values p_i or the result (if c_i or the NSA paid then c_j did not). The protocol models anonymous broadcast, since the message "1" is, if sent, received by all cryptographers, but the sender remains anonymous. We formalize this using modal independence logic. We start by capturing the protocol in the following Kripke model:

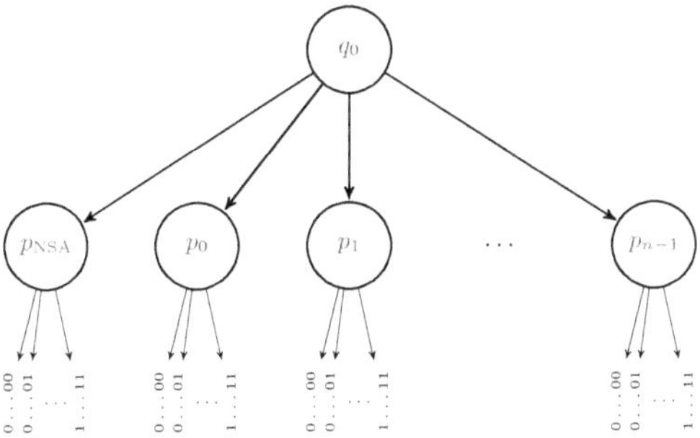

The protocol starts in q_0, the model then branches into states $p_{\mathrm{NSA}}, p_0, \ldots, p_{n-1}$, depending on whether the NSA or some c_i paid. Each of these states has 2^n successor states, for the 2^n possible random bit values, these states are *final*. The relation R is as indicated. We use the following variables:

- p_{NSA} and p_i are true if the NSA, resp. cryptographer c_i paid, i.e., in the states denoted with the same name as the variable and in their successors.
- each of the n variables $bit_{\{i,i+1\}}$ is true in the states where the bit shared between c_i and c_{i+1} is 1.
- each $announce_i$ is true in all final states which satisfy $p_i \oplus bit_{\{i,i-1\}} \oplus bit_{\{i,i+1\}}$ (this encodes that the cryptographers follow the protocol).

For each c_i, we define the set $\vec{k}_i$ of the variables whose values c_i knows after the protocol run as $\vec{k}_i := \{p_i, bit_{\{i,i-1\}}, bit_{\{i,i+1\}}\} \cup \{announce_j \mid j \neq i\}$. Clearly, c_i also knows the value $announce_i$, but since this can be computed from p_i, $bit_{\{i,i-1\}}$ and $bit_{\{i,i+1\}}$, we omit it from $\vec{k}_i$.

The formula expressing the anonymity requirement consists of several parts, one *global* part and then, for each combination of cryptographers, a *local* part. We start with the global part, which merely expresses that none of the individual bits that some cryptographer knows determines the value of any p_i on its own, with the exception that if $p_i = 1$, then of course cryptographer c_i knows that $p_j = 0$ for all $j \neq i$. The global part φ_g is as follows:

$$\varphi_g = \bigwedge_{\substack{v \in \{bit_{\{i,i+1\}}, announce_i\}, \\ k \in \{0,\ldots,n-1\}}} \Diamond\Diamond(v \wedge p_k) \wedge \Diamond\Diamond(v \wedge \overline{p_k}) \wedge \Diamond\Diamond(\overline{v} \wedge p_k) \wedge \Diamond\Diamond(\overline{v} \wedge \overline{p_k})$$
$$\wedge \bigwedge\nolimits_{i \neq j} \Diamond\Diamond(p_i \wedge \overline{p_j}) \wedge \Diamond\Diamond(\overline{p_i} \wedge p_j) \wedge \Diamond\Diamond(\overline{p_i} \wedge \overline{p_j})$$

The first line of the formula requires that, for every variable v of the $bit_{\{i,i+1\}}$ or $announce_i$-variables, and every cryptographer c_k, every combination of truth values of v and p_k appears. This encodes that the value of a *single* variable v does not give away any information about the value of any p_k. The second line is a similar requirement for the value p_i: If c_i paid, then she knows that c_j did not pay, for $i \neq j$. However, the combination "$p_i \wedge p_j$" for $i \neq j$ should be the only one not appearing. Hence the formula requires that all other combinations appear in some final state. The global part φ_g hence ensures that each individual bit that c_i knows does not tell him whether c_j paid, unless of course $i = j$ or $p_i = 1$.

The more interesting part is to encode that even the *combination* of the above bits does not lead to additional knowledge; this is where the independence atom is crucial. We introduce some notation to enumerate the variables in $\vec{k}_i$:

- for each i, let $\vec{k}_i = \{v_1^i, \ldots, v_{n+2}^i\}$,
- for $j \leq k$, let $V_{j\to k}^i = \{v_j^i, \ldots, v_k^i\}$,
- let $V_j^i = V_{j\to j}^i$.

We now use modal independence logic to express that if each single variable from $\vec{k}_i$ does not tell c_i anything about the value of p_k, then their combination does not, either. This is achieved with the following formula:

$$\varphi^{i,k} = \Box\Box\left((V_1^i \bot_{p_k} V_2^i) \wedge (V_{1\to 2}^i \bot_{p_k} V_3^i) \wedge \cdots \wedge (V_{1\to n+1}^i \bot_{p_k} V_{n+2}^i)\right).$$

This formula requires that for each j, each pair of variable assignments I_1 to $V_{1\to j-1}^i$ and I_2 to V_j^i that is "locally compatible" with some truth value $P(p_k)$—in other words, neither of these assignments by itself implies that the actual value of p_k is not $P(p_k)$—is also compatible with that value for the combination of I_1 and I_2, i.e., there is some state satisfying $I_1 \cup I_2 \cup P$ (where the notion of a state satisfying a propositional assignment is defined as expected and the union of these assignments is well-defined since their domains are disjoint). As a consequence, the formula requires that for each $I\colon \vec{k}_i \to \{0,1\}$ and each $P\colon \{p_k\} \to \{0,1\}$, if for each $v \in \vec{k}_i$, there is a world w such that $w \models I|_{\{v\}}$ and $w \models P$, then there is a world w such that $w \models I \cup P$.

The following proposition formally states that our above-developed formulas indeed express the anonymity property of the protocol as intended. From a single cryptographer c_i's point of view, it says that every observation I which can arise when c_i follows the protocol, as long as some cryptographer different from c_i paid for the dinner, then for every k different from i, both possibilities—c_k paid for the dinner, or c_k did not pay—cannot be ruled out by the observation I. We say that an assignment $I\colon \vec{k}_i \cup \{announce_i\} \to \{0,1\}$ is *consistent* if i follows the protocol, i.e., if $I(announce_i) = I(p_i) \oplus I(bit_{\{i,i-1\}}) \oplus I(bit_{\{i,i+1\}})$. Note that in the models we are interested in, only consistent assignments appear in final states.

Proposition 5.1 *If a Kripke model $M = (W, R, \pi)$ satisfies the formula $\varphi_g \wedge \bigwedge_{i,k\in\{0,\dots,n-1\}, i\neq k} \varphi^{i,k}$ at the world q_0, then the team $T = R(R(\{q_0\}))$ satisfies the following condition: For each $i \neq k \in \{0,\dots,n-1\}$ and each consistent $I\colon \vec{k}_i \cup \{announce_i\} \to \{0,1\}$ with $I(p_i) = 0$ and $\oplus_{j=0}^{n-1} I(announce_j) = 1$, there are worlds $w_1^I, w_2^I \in T$ with $w_1^I \models I, p_k$ and $w_2^I \models I, \overline{p_k}$.*

We omit the easy proof; the proposition immediately follows from the semantics of the independence atom.

Our discussion only treats the anonymity property of the protocol. For a complete treatment, one also has to address other aspects as e.g., correctness, we omit this discussion here.

Note that in comparison to express the anonymity requirement using epistemic logic (see, e.g., [1,12]), we do not use the relation of the Kripke model to represent knowledge, but to express branching time. In particular, our approach only uses a single modality.

6 Expressiveness

We now compare the expressiveness of MIL and classical modal logic, which we abbreviate with ML. We show that MIL is strictly more expressive than ML on teams (simply because MIL is not downwards closed, i.e., from $\mathcal{M}, T \models \varphi$ and $T' \subseteq T$, it does not follow that $\mathcal{M}, T' \models \varphi$), but that their expressiveness coincides on singleton teams. However, on singletons, MIL is exponentially more succinct than ML. We then study the expressiveness of MIL with a generalized dependence atom as introduced in Section 4 instead of the independence atom.

6.1 Expressiveness of MIL and ML

Clearly, since MIL is not downward-closed, we obtain the following:

Proposition 6.1 *There is an* MIL*-formula φ_{MIL} such that there is no* ML*-formula φ_{ML} with the property that $M, T \models \varphi_{\mathsf{MIL}}$ if and only if $M, T \models \varphi_{\mathsf{ML}}$ for all models M and all teams T.*

Proof. This is true for every formula φ_{MIL} that is not downwards closed: In this case we have teams $T' \subsetneq T$ of the same model M with $M, T \models \varphi_{\mathsf{MIL}}$ and $M, T' \not\models \varphi_{\mathsf{MIL}}$. However, for any modal formula φ_{ML}, clearly if $M, w \models \varphi_{\mathsf{ML}}$ for all $w \in T$, then the same is true for all $w \in T'$ as $T' \subseteq T$. An easy example for a formula that is not downwards closed is $x \bot_{\emptyset} y$. This formula is satisfied on a team T in which every combination of truth values of x and y is realized in some world, but not on its subset T' containing only worlds w and w' with assignments $x \wedge y$ and $\overline{x} \wedge \overline{y}$, respectively. □

The proposition remains true for classical modal logic extended with a global modality, or for MDL, since these logics remain downward-closed. In [13], it was shown that MDL is as expressive as classical modal logic on singletons. Therefore, a natural question to ask is whether on singletons, MIL is still more expressive than ML. We show that this is not the case, but we will also see that MIL is exponentially more succinct than ML, even on singletons. For our proof, we use bisimulations, which are a well-established tool to compare expressiveness of different concepts. We recall the classical definition of bisimulation for modal logic:

Definition 6.2 Let $M = (W, R, \pi)$ and $M' = (W', R', \pi')$ be Kripke models. A relation $Z \subseteq W \times W'$ is a *modal bisimulation* if for every $(w, w') \in Z$, the following holds:

- $\pi(w) = \pi'(w')$, i.e., w and w' satisfy the same propositional variables,
- if u is an R-successor of w, then there is an R'-successor u' of w' such that $(u, u') \in Z$ (forward condition),
- if u' is an R'-successor of w', then there is an R-successor u of w such that $(u, u') \in Z$ (backward condition).

It is well-known and easy to see that modal logic is invariant under bisimulation, i.e., if Z is a bisimulation and $(w, w') \in Z$, then w and w' satisfy the same modal formulas. We now "lift" this property to modal independence logic by considering a bisimulation Z as above on the team level:

Definition 6.3 Let $M = (W, R, \pi)$ and $M' = (W', R', \pi')$ be models, let $T \subseteq W$ and $T' \subseteq W'$ be teams. Let $Z \subseteq W \times W'$ be a modal bisimulation. Then T and T' are Z-bisimilar if the following is true:

- for each $w \in T$, there is a $w' \in T'$ such that $(w, w') \in Z$,
- for each $w' \in T'$, there is a $w \in T$ such that $(w, w') \in Z$.

We now show that on the team level, bisimulation for modal independence logic plays the same role as it does on the world level for modal logic: Simply stated, bisimilar teams satisfy the same formulas. Due to Lemma 3.1, the result also applies to modal dependence logic. This lemma may be of independent interest (for example, it implies a "family-of-trees"-like model property), we use it to compare the expressiveness of MIL and ML.

Lemma 6.4 *Let $M = (W, R, \pi)$ and $M' = (W', R', \pi')$ be Kripke models, let $T \subseteq W$ and $T' \subseteq W'$ be teams that are Z-bisimilar for a modal bisimulation Z. Then for any* MIL*-formula φ, we have that $M, T \models \varphi$ if and only if $M', T' \models \varphi$.*

Proof. We show the lemma by induction on φ. Clearly it suffices to show that if $M, T \models_{\mathsf{MIL}} \varphi$, then $M', T' \models_{\mathsf{MIL}} \varphi$. Hence assume $M, T \models_{\mathsf{MIL}} \varphi$.

- Let $\varphi = x$ for some propositional variable x, and let $w' \in T'$. Since T and T' are Z-bisimilar, there is a world $w \in T$ with $(w, w') \in Z$. Since $M, T \models_{\mathsf{MIL}} x$, the variable x is true at w in M. Since Z is a modal bisimulation, it follows that x is true at w' in M', and hence every world $w' \in T$ satisfies x. Therefore, it follows that $M', T' \models \varphi$.

- If $\varphi = \neg x$, the proof is the same as above.
- Let $\varphi = \varphi_1 \wedge \varphi_2$. This case trivially follows inductively.
- Let $\varphi = \varphi_1 \vee \varphi_2$. Since $M, T \models \varphi$, it follows that $T = T_1 \cup T_2$ for teams T_1 and T_2 with $M, T_1 \models \varphi_1$ and $M, T_2 \models \varphi_2$. We define teams T_1' and T_2' as follows:
  · $T_1' = \{w' \in T' \mid (w, w') \in Z \text{ for some } w \in T_1\}$,
  · $T_2' = \{w' \in T' \mid (w, w') \in Z \text{ for some } w \in T_2\}$.
  We prove the following:
  (i) $T' = T_1' \cup T_2'$
  (ii) T_1 and T_1' are Z-bisimilar,
  (iii) T_2 and T_2' are Z-bisimilar.
  By induction, it then follows that $M', T_1' \models \varphi_1$ and $M', T_2' \models \varphi_2$, which, since $T' = T_1' \cup T_2'$ implies that $M', T' \models \varphi$. We prove these points:
  (i) By construction, $T_1' \cup T_2' \subseteq T'$. Hence let $w' \in T'$. Since T and T' are Z-bisimilar, there is some $w \in T$ such that $(w, w') \in Z$. Since $T = T_1 \cup T_2$, we can, without loss of generality, assume that $w \in T_1$. By definition of T_1', it follows that $w' \in T_1'$.
  (ii) First let $w \in T_1 \subseteq T$. Since T and T' are Z-bisimilar, there is some $w' \in T'$ such that $(w, w') \in Z$. Due to the definition of T_1', it follows that $w' \in T_1'$. For the converse, assume that $w' \in T_1'$. By definition, there is some $w \in T_1$ such that $(w, w') \in Z$.
  (iii) This follows with the same proof as for T_1 and T_1'.
  Hence $M', T' \models \varphi$ as required.
- Let $\varphi = \Diamond\psi$. Since $M, T \models \varphi$, there exists a team $U \subseteq R(T)$ such that for each $w \in T$, the set $u(w) := R(\{w\}) \cap T$ is not empty, and $M, U \models \psi$. We define a corresponding team U' of M' as follows: Start with $U' = \emptyset$ and then for each $(w, w') \in (T \times T') \cap Z$, do the following:
  · For each R-successor v of w that is an element of U, since Z is a modal bisimulation and $(w, w') \in Z$, there is at least one R'-successor v' of w' with $(v, v') \in Z$. Add all such v' to the set U'.
  By construction, U' only contains worlds that are R'-successors of worlds in T'. Hence to show that $M', T' \models \Diamond\psi$, it remains to show that
  (i) for each $w' \in T'$, the team U' contains a world v' that is an R'-successor of w',
  (ii) the teams U and U' are Z-bisimilar.
  The claim then follows by induction, since $M, U \models \psi$. We now show the above two points:
  (i) Let $w' \in T'$. Since T and T' are Z-bisimilar, there is some $w \in T$ with $(w, w') \in Z$. Due to the choice of U, there is some $v \in U$ which is an R-successor of w. By construction of the set U', a world v' that is an R'-successor of w' has been added to U'.
  (ii) By construction, for every R-successor v of some w in T, at least one v'

has been added to U' with $(v, v') \in Z$. For the converse, by construction of U', for every v' added to U' there is a $v \in U$ with $(v, v') \in Z$.
Hence $M', T' \models \Diamond\psi$ as required.

- Now assume that $\varphi = \Box\psi$, and let U be the set of all R-successors of worlds in T, let U' be the set of all R'-successors of worlds in T'. By induction, it suffices to show that U and U' are Z-bisimilar. Hence let $v \in U$ be the R-successor of some $w \in T$. Since T and T' are Z-bisimilar, there is some $w' \in T'$ such that $(w, w') \in Z$. Since v is an R-successor of w, and since Z is a modal bisimulation, there is some v' which is an R'-successor of w' such that $(v, v') \in Z$. Since v' is an R'-successor of w', it follows that $v' \in U'$. The converse direction follows analogously.
- Let $\varphi = \vec{p_1} \perp_{\vec{q}} \vec{p_2}$. To show that $M', T' \models \varphi$, let $u, u' \in T'$ with the same truth values of the variables in q. Since T and T' are bisimilar, there are worlds $w, w' \in T$ such that $(w, u) \in Z$ and $(w', u') \in Z$. Since Z is a modal bisimulation, the Z-related worlds have the same propositional truth assignment. In particular, w and w' agree on the values for the variables in q. Since $M, T \models \varphi$, there is some world $w'' \in T$ such that
  · $w'' \equiv_{\vec{q}} w' \equiv_{\vec{q}} w$, and since Z is a modal bisimulation it follows that w'' and both u and u' have the same $\vec{q}$-assignment,
  · $w'' \equiv_{\vec{p_1}} w$, and hence w'' and u have the same $\vec{p_1}$-assignment,
  · $w'' \equiv_{\vec{p_2}} w'$, and hence w'' and u' have the same $\vec{p_2}$-assignment.
  Since T and T' are Z-bisimilar, there is a world $u'' \in T'$ such that $(w'', u'') \in Z$. Since Z is a modal bisimulation, w'' and u'' satisfy the same propositional variables, and hence for u'' we have that
  · $u'' \equiv_{\vec{q}} u$, $u'' \equiv_{\vec{q}} u'$
  · $u'' \equiv_{\vec{p_1}} u$
  · $u'' \equiv_{\vec{p_2}} u'$.
  Therefore, $M', T' \models \varphi$ as required.

□

With Lemma 6.4 and an application of van Benthem's Theorem [2], it follows directly that MIL and ML are in fact equivalent in expressiveness *on singletons*:

Theorem 6.5 *For each* MIL*-formula* φ_{MIL}*, there is an* ML*-formula* φ_{ML} *such that* φ_{ML} *and* φ_{MIL} *are equivalent on singletons.*

Proof. Due to Lemma 6.4, we know that MIL is invariant under bisimulation of teams. Since for singleton teams, bisimulation on teams and bisimulation on worlds coincide, it follows that MIL on singletons is invariant under modal bisimulation. Clearly, when evaluating an MIL-formula φ_{MIL} on a singleton team $\{w\}$, all worlds in the model that have a distance from w which exceeds the modal depth (i.e., maximal nesting degree of modal operators) of φ, are irrelevant for the question whether $M, \{w\} \models \varphi$. Therefore, φ_{MIL}, evaluated on singletons, captures a property of Kripke models that is invariant under modal bisimulation and only depends on the worlds that can be reached in at most $md(\varphi_{\mathsf{MIL}})$ steps. Due to [2], such a property can be encoded by a standard modal logic formula φ_{ML}. (The formula φ_{ML} can be obtained, for example, as the disjunction of formulas $\varphi_{M,w}$ which, for each model M and world w with $M, \{w\} \models_{\mathsf{MIL}} \varphi_{\mathsf{MIL}}$, encodes the finite tree unfolding of M, w up to depth $md(\varphi_{\mathsf{MIL}})$, up to bisimulation, and only taking into account the variables appearing in φ_{MIL}. This unfolding is finite and hence first-order definable, therefore we can apply the result from [2].) □

Since the application of van Benthem's Theorem yields a potentially very large formula, the above result does not give a "efficient" translation from MIL to ML. It turns out that one cannot do much better: MIL is exponentially more succinct than ML.

Theorem 6.6 *There is a family of* MIL*-formulas $(\varphi_i)_{i\in\mathbb{N}}$ such that the length of φ_i is quadratic in i, and for any family of* ML*-formulas $(\psi_i)_{i\in\mathbb{N}}$ such that for all i, φ_i and ψ_i are equivalent on singletons, the length of ψ_i grows exponentially in i.*

Proof. Let φ_i be the formula describing the security property of the dining cryptographers protocol in Section 5, with every sequence $\Diamond\Diamond$ replaced with $\Diamond$. As argued in that section, if φ_i is satisfied at $M, \{w\}$, then the number of propositional assignments appearing in the set of worlds that can be reached from w in one step (two steps for the original formula) is exponential in i. Now let $(\psi_i)_{i\in\mathbb{N}}$ a family of ML-formulas such that φ_i and ψ_i are locally equivalent for each i. Since ψ_i is locally equivalent to φ_i, we can without loss of generality assume that $md(\psi_i) = 1$ for all i (if ψ_i contains deeper nestings of modal operators, the formula can be simplified since the truth value of ψ_i cannot depend on worlds reachable in 2 or more steps). Therefore, modal operators do not appear nested in ψ_i. Without loss of generality, we can assume that only $\Diamond$ appears in ψ. It is clear that if $M, w \models \psi_i$, then there is a submodel of M which contains w, and in which the number of successors of w is bounded by the number of $\Diamond$-operators appearing in ψ_i. Therefore, ψ_i must have an exponential number of $\Diamond$-operators, which proves the theorem. □

As far as we know, the analogue of Theorem 6.6 for MDL has not yet been showed. Sevenster [13] has shown the analogue of Theorem 6.5 for MDL, but he did not show that going from MDL to ML (over singletons) inevitably leads to an exponential blow-up in the formula size. On the other hand, a closely related result showing that any formula of ML($\varovee$), where $\varovee$ is the classical disjunction) that is logically equivalent to

$$=(p_1, \dots, p_n, q),$$

has to have length exponential in n was obtained in [9].

6.2 Expressiveness of ML with Generalized Dependence Atoms

Theorem 6.5 applies not only to MIL, but (with the same proof) to all extensions of MIL with a dependence operator that, evaluated on a team T, depends only on the set of propositional assignments that occur in some world of the team. This is because the set of assignments is clearly invariant under bisimulation. As examples, operators like -those from exclusion logic or inclusion logic can be added to MIL without increasing its expressiveness. Hence, in light of Section 4, a natural question to ask is the following: For which generalized dependence atoms D is the logic $\mathsf{ML}(D)$ as expressive as ML on singletons, and which atoms do in fact add expressiveness?

-It is easy to see that there are generalized dependence atoms D that, even on singletons, add expressiveness beyond classical modal logic (and hence, beyond MIL). This is because with no restriction on the dependence operator D, one can express properties that depend on the *number* of worlds in a team, which clearly cannot be done in ML. As an example, consider the following:

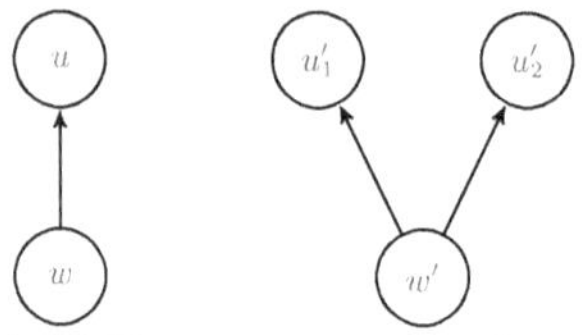

Model M Model M'

Example 6.7 Let D be the relation $\{(0)\}$, and consider the models M and M' above, where the single variable x is false in every world of both models. It is easy to see that $M, \{w\} \models \Box D(x)$, while $M', \{w'\} \not\models \Box D(x)$: For the model M, the set of successors of w is the team $T = \{u\}$, since x is false in u, it follows that $T(x) = (0)$, hence $T \models D(x)$. On the other hand, the set of successors of w' in M' is $T' = \{u_1, u_2\}$, hence $T'(x) = (0, 0)$ and $T' \not\models D(x)$.

Clearly, no ML-formula can distinguish M, w and M', w', since the relation $Z = \{(w, w'), (u, u_1), (u, u_2)\}$ is a bisimulation.

However, as mentioned earlier, the proof of Theorem 6.5 can be generalized to handle the generalized dependence atoms discussed earlier. In general, we obtain the following result: For a D which is definable using first-order formulas without equality, the expressiveness of $\mathsf{ML}(D)$ coincides with that on ML for singletons. For the proof, we show that $\mathsf{ML}(D)$ remains invariant under bisimulation, and then apply the proof of Theorem 6.5 again. On the other hand, clearly if D is not FO-definable, then D cannot be expressed in modal logic, as modal logic can be translated to first-order logic with the standard translation, and if D cannot be expressed without equality, then D cannot be invariant under bisimulation. Hence we obtain the following theorem:

Theorem 6.8 *Let D be a generalized dependence atom. Then the following statements are equivalent:*

(i) *D can be expressed in first-order logic without equality,*

(ii) *$\mathsf{ML}(D)$ and ML are equally expressive over singletons, i.e., for each $\mathsf{ML}(D)$-formula φ_{MIL}, there is an ML-formula φ_{ML} such that φ_{MIL} and φ_{ML} are equivalent on singletons.*

Proof. We first assume that D is FO-definable. As mentioned above, it suffices to adapt the proof of Lemma 6.4 to $\mathsf{ML}(D)$. Clearly, in the induction, we only need to cover the case that $\varphi = D(p_1, \ldots, p_n)$ for propositional variables $p_1, \ldots, p_n$. Hence assume that $M, T \models D(p_1, \ldots, p_n)$, we show that $M', T' \models D(p_1, \ldots, p_n)$, where T and T' are Z-bisimilar for a modal bisimulation Z.

Let ϕ be the first-order formula defining D. We prove the claim inductively over the formula, where we only cover the key cases explicitly. Let the free variables of ϕ be $\omega_1, \ldots, \omega_t$. We show that if $w_1, \ldots, w_t \in T$ and $w'_1, \ldots, w'_n \in T'$ such that $(w_i, w'_i) \in Z$ for all i, then $\phi(w_1, \ldots, w_n)$ evaluates to true if and only if $\phi(w'_1, \ldots, w'_n)$ does. If ϕ is quantifier-free, the claim is clear: Since Z is a modal bisimulation, w_i and w'_i satisfy the same propositional variables, and due to the choice of ϕ, the truth value of ϕ only depends on the propositional assignments of the worlds instantiating the variables $\omega_1, \ldots, \omega_n$. The second relevant case is when $\phi = \exists\omega\psi(\omega, \omega_1, \ldots, \omega_t)$. If $M, T \models \phi$, then there is a world $w \in T$ such that $\psi(w, w_1, \ldots, w_t)$ is true. Since T and T' are Z-bisimilar, it follows that there is a world $w' \in T'$ such that $(w, w') \in Z$. Due to induction, it follows that if $\psi(w, w_1, \ldots, w_t)$ is true, then so is $\psi(w', w'_1, \ldots, w'_t)$. This completes the proof that $\mathsf{ML}(D)$ and ML are equally expressive over singletons.

For the converse, assume that $\mathsf{ML}(D)$ is as expressive as ML over singletons. In particular, then for every sequence $x_1, \ldots, x_n$ of variables, there is a modal formula φ such that for every model M and every world $w \in M$, we have that $M, w \models \varphi$ if and only if $M, \{w\} \models \Box D(x_1, \ldots, x_n)$. By the standard translation from modal logic to first-order logic, this implies that $D(x_1, \ldots, x_n)$ can be expressed as a FO-formula ϕ_D.

Since whether $\mathcal{M}, T \models D(x_1, \ldots, x_n)$ is invariant under bisimulation (this can be seen by adding a single world w_0, connected to all $w' \in T$ and evaluating the formula $\Box D(x_1, \ldots, x_n)$ at the singleton team $\{w_0\}$—since $\mathsf{ML}(D)$ on singletons is equivalent to ML, which is invariant under bisimulation, it follows that $\mathsf{ML}(D)$ on singletons is invariant under bisimulation as well), it follows that whether $\mathcal{M}, T \models D(x_1, \ldots, x_n)$ is in particular invariant under adding/removing identical copies of worlds in T. Therefore, ϕ_D can be rewritten into a formula without equality. $\Box$

7 Conclusion and Open questions

In this paper we introduced modal independence logic MIL and settled the computational complexity of its satisfiability and model checking problem. Furthermore we compared the expressivity of MIL with that of classical modal logic. It turned out that most of our results can be generalized to modal logic extended with various so called generalized dependence atoms.

We end this paper by the following interesting open questions:

(i) Are there classes of frames definable in MIL that cannot be defined with ML?

(ii) Is it possible to formulate and prove a version of van Benthem's Theorem for our generalization of bisimilarity from the world level to the team level (see Definition 6.3)?

Acknowledgement

The authors thank the anonymous referees for very useful comments. The first author was supported by the Academy of Finland grants 264917 and 275241.

References

[1] Al-Bataineh, O. I. and R. van der Meyden, *Abstraction for epistemic model checking of dining cryptographers-based protocols*, in: K. R. Apt, editor, *Theoretical Aspects of Rationality and Knowledge (TARK)* (2011), pp. 247–256.

[2] Benthem, J. V., "Modal Logic and Classical Logic," Bibliopolis, 1985.

[3] Börger, E., E. Grädel and Y. Gurevich, "The Classical Decision Problem," Springer Verlag, Berlin Heidelberg, 2001.

[4] Chaum, D., *The dining cryptographers problem: Unconditional sender and recipient untraceability*, J. Cryptology **1** (1988), pp. 65–75.

[5] Ebbing, J., L. Hella, A. Meier, J.-S. Müller, J. Virtema and H. Vollmer, *Extended modal dependence logic*, in: *Workshop on Logic, Language, Information, and Computation (WoLLIC)*, number 8071 in Lecture Notes in Computer Science (2013), pp. 126–137.

[6] Ebbing, J. and P. Lohmann, *Complexity of model checking for modal dependence logic*, in: *Theory and Practice of Computer Science (SOFSEM)*, number 7147 in Lecture Notes in Computer Science (2012), pp. 226–237.

[7] Galliani, P., *Inclusion and exclusion dependencies in team semantics – on some logics of imperfect information*, Annals of Pure and Applied Logic **163** (2012), pp. 68–84.

[8] Grädel, E. and J. Väänänen, *Dependence and independence*, Studia Logica **101** (2013), pp. 399–410.

[9] Hella, L., K. Luosto, K. Sano and J. Virtema, *The expressive power of modal dependence logic*, in: *Advances in Modal Logic 2014*, 2014, to appear.

[10] Kontinen, J. and J. A. Väänänen, *A remark on negation in dependence logic*, Notre Dame Journal of Formal Logic **52** (2011), pp. 55–65.

[11] Kuusisto, A., *A double team semantics for generalized quantifiers*, CoRR, http://arxiv.org/abs/1310.3032 (2013).

[12] Schnoor, H., *Deciding epistemic and strategic properties of cryptographic protocols*, in: *European Symposium on Research in Computer Security (ESORICS)*, Lecture Notes in Computer Science **7459** (2012), pp. 91–108.

[13] Sevenster, M., *Model-theoretic and computational properties of modal dependence logic*, J. Log. Comput. **19** (2009), pp. 1157–1173.

[14] Väänänen, J., "Dependence Logic," Cambridge University Press, 2007.

[15] Väänänen, J., *Modal dependence logic*, New Perspectives on Games and Interaction **5** (2009), pp. 237–254.

Neighbourhood Frame Product KxK

Andrey Kudinov [1]

Institute for Information Transmission Problems, RAS, Moscow, Russia,
National Research University Higher School of Economics, Moscow, Russia
Moscow Institute of Physics and Technology

Abstract

We consider modal logics of products of neighborhood frames and find the modal logic of all products of normal neighborhood frames.

Keywords: Modal logic, neighborhood frame, product of modal logics, product neighborhood frame.

1 Introduction

In this paper we continue the research of [5] and study modal logics of products of neighborhood frames.

Neighborhood frames, as a generalization of Kripke semantics for modal logic, were introduced independently by Dana Scott [9] and Richard Montague [7]. Neighborhood semantics is more general than Kripke semantics, and in case of normal reflexive and transitive logics, coincides with topological semantics. In this paper we consider the product of neighborhood frames introduced by Sano in [8]. It is a generalization of the product of topological spaces [2] presented in [1].

The product of neighborhood frames is defined in the vein of the product of Kripke frames (see [11] and [12]). But there are some differences. Axioms of commutativity and Church-Rosser property are valid in any product of Kripke frames. Whereas in [1] it was shown that the logic of the products of all topological spaces is the fusion of logics $\mathsf{S4} * \mathsf{S4}$. Even more, $\mathsf{S4} * \mathsf{S4}$ is complete w.r.t. the product $\mathbb{Q} \times_t \mathbb{Q}$ ($\times_t$ stands for product of topological spaces, defined in [1]).

In [5] this result was extended. It was proven that for any pair L and L' of logics from set $\{\mathsf{S4}, \mathsf{D4}, \mathsf{D}, \mathsf{T}\}$ modal logic of products of L-neighborhood frames and L'-neighborhood frames is the fusion of L and L'. But it was unclear how to proceed in case of logics that do not contain axiom $\Diamond\top$ (correspond to seriality). In this paper we show that any product of neighborhood frames in

[1] This work was supported by RFBR grant N 14-01-31442-mol-a.

[2] "Product of topological spaces" is a well-known notion in Topology but it is different from what we use here (for details see [1]).

fact satisfy axiom $\psi \to \Box_2\psi$, where ψ is a variable-free and $\Box_2$-free formula (and similar for $\Box_1$). We prove that $\mathsf{K} * \mathsf{K}$ plus all such axioms will be the logic of all products of neighborhood frames.

Neighborhood frames are often considered in the context of non-normal modal logics, since, unlike Kripke semantics, it is complete w.r.t. many non-normal logics. As for normal modal logic, neighborhood frames rarely gives anything new in comparison to Kripke frames. This paper, however, shows that normal neighborhood frames, that correspond to normal modal logics, give different results than Kripke frames in case of products.

The results of this paper (and others: [1], [5], [8]) show that "neighborhood" product, in general, gives weaker logic than "Kripke" product of modal logics. From this we can conclude that neighborhood semantics is a finer tool for products of modal logics even for normal modal logics. It also shows that notion of the product of modal logics depend on the semantics.

We should also mention the possibility of adding the third modality that, in topological context, correspond to classical product topology. That was done in [1] for $\mathsf{S4}$ and topological semantics with the interior operator and in [6] for $\mathsf{D4}$ and topological semantics with the derivational operator. It may be possible to consider similar construction for other logics.

2 Language and logics

In this paper we study propositional modal logics. A formula is defined recursively as follows:

$$\phi ::= p \mid \bot \mid \phi \to \phi \mid \Box_i\phi,$$

where $p \in \mathrm{PROP}$ is a propositional letter and $\Box_i$ is a modal operator. Other connectives are introduced as abbreviations: classical connectives are expressed through $\bot$ and $\to$, dual modal operators $\Diamond_i$ are expressed as follows: $\Diamond_i = \neg\Box_i\neg$.

Definition 2.1 A *normal modal logic* (or *a logic,* for short) is a set of modal formulae closed under Substitution $\left(\frac{A(p_i)}{A(B)}\right)$, Modus Ponens $\left(\frac{A,\, A\to B}{B}\right)$ and Generalization rules $\left(\frac{A}{\Box_i A}\right)$; containing all classic tautologies and the following axioms:

$$\Box_i(p \to q) \to (\Box_i p \to \Box_i q).$$

$\mathsf{K_n}$ denotes *the minimal normal modal logic with n modalities* and $\mathsf{K} = \mathsf{K_1}$.

Let L be a logic and let Γ be a set of formulae, then $\mathsf{L} + \Gamma$ denotes the minimal logic containing L and Γ. If $\Gamma = \{A\}$, then we write $\mathsf{L} + A$ rather than $\mathsf{L} + \{A\}$.

Definition 2.2 Formula ϕ is called *closed* if it does not contain any variables.

Definition 2.3 Let $\mathsf{L_1}$ and $\mathsf{L_2}$ be two modal logics with one modality $\Box$, then *fusion* of these logics is

$$\mathsf{L_1} * \mathsf{L_2} = \mathsf{K_2} + \mathsf{L_1'} + \mathsf{L_2'};$$

where L_i' is the set of all formulae from L_i where all $\Box$ are replaced by $\Box_i$.

3 Products of neighborhood frames

The notions of Kripke frames and Kripke models are well known (see [2]).

Definition 3.1 Let $R \subseteq W \times W$ be a relation on $W \neq \varnothing$, then for $k \geq 1$ and $w \in W$ we define

$$\begin{aligned} R^0 &= Id_W; \\ R^{n+1} &= R^n \circ R; \\ R^* &= \bigcup_{k=0}^{\infty} R^k; \\ R(w) &= \{u \mid wRu\}. \end{aligned}$$

For a Kripke frame $F = (W, R)$ we define the submodel generated by $w \in W$ as the frame $F^w = (W', R|_{W'})$, where $W' = R^*(w)$ and $R|_{W'} = R \cap W' \times W'$.

Definition 3.2 Let $F_i = (W_i, R_i)$ $(i = 1, 2)$ be two Kripke frames. We define their *product* (see [4]) as a bimodal frame $F_1 \times F_2 = (W_1 \times W_2, R_1^h, R_2^v)$, where

$$\begin{aligned} (x,y)R_1^h(z,t) &\iff xR_1z \;\&\; y = t, \\ (x,y)R_2^v(z,t) &\iff yR_2t \;\&\; x = z. \end{aligned}$$

Furthermore, we consider neighborhood frames (see [10] and [3]).

Definition 3.3 Let X be a nonempty set, then $F \subseteq 2^X$ is a *filter* on X if

(i) $X \in F$;

(ii) if $U, V \in F$, then $U \cap V \in F$;

(iii) if $U \in F$ and $U \subseteq V$, then $V \in F$.

Note, it is usually demanded that $\varnothing \notin F$ (F is a proper filter), but in this paper we will not demand this.

Definition 3.4 A *(normal) neighborhood frame* (or an n-frame) is a pair $\mathfrak{X} = (X, \tau)$, where X is a nonempty set and $\tau : X \to 2^{2^X}$ such that $\tau(x)$ is a filter on X for any x. We call function τ the *neighborhood function* of $\mathfrak{X}$ and sets from $\tau(x)$ we call *neighborhoods of x. The neighborhood model* (n-model) is a pair $(\mathfrak{X}, \theta)$, where $\mathfrak{X} = (X, \tau)$ is an n-frame and $\theta : PROP \to 2^X$ is a *valuation.* In a similar way, we define *neighborhood 2-frame* (n-2-frame) as (X, τ_1, τ_2) such that $\tau_i(x)$ is a filter on X for any x, and a *n-2-model.*

Definition 3.5 *The valuation of a formula* φ at a point of an n-model $M = (\mathfrak{X}, \theta)$ is defined by induction. For Boolean connectives the definition is usual, so we omit it. For modalities the definition is as follows:

$$M, x \models \Box_i\psi \iff \exists\theta \in \tau_i(x) \forall y \in \theta(M, y \models \psi).$$

Formula is valid in an n-model M if it is valid at all points of M (notation $M \models \varphi$). Formula is valid in an n-frame $\mathfrak{X}$ if it is valid in all models based on $\mathfrak{X}$ (notation $\mathfrak{X} \models \varphi$). We write $\mathfrak{X} \models L$ if for any $\varphi \in L$, $\mathfrak{X} \models \varphi$. Logic of a class of n-frames $\mathcal{C}$ as $Log(\mathcal{C}) = \{\varphi \,|\, \mathfrak{X} \models \varphi \text{ for all } \mathfrak{X} \in \mathcal{C}\}$. For logic L we also define $nV(L) = \{\mathfrak{X} \,|\, \mathfrak{X} \text{ is an n-frame and } \mathfrak{X} \models L\}$. Note, that if there is no $\mathfrak{X}$ such that $\mathfrak{X} \models L$, then $nV(L) = \varnothing$.

Definition 3.6 Let $F = (W, R)$ be a Kripke frame. We define n-frame $\mathcal{N}(F) = (W, \tau)$ in the following way

$$\tau(w) = \{U \,|\, R(w) \subseteq U \subseteq W\}.$$

Lemma 3.7 *Let $F = (W, R)$ be a Kripke frame. Then*

$$Log(\mathcal{N}(F)) = Log(F).$$

The proof is straightforward (see [3]).

Definition 3.8 Let $\mathfrak{X} = (X, \tau_1, \ldots)$ and $\mathcal{Y} = (Y, \sigma_1, \ldots)$ be n-frames. Then function $f : X \to Y$ is a *p-morphism* if

(i) f is surjective;

(ii) for any $x \in X$ and $U \in \tau_i(x)$ $f(U) \in \sigma_i(f(x))$;

(iii) for any $x \in X$ and $V \in \sigma_i(f(x))$ there exists $U \in \tau_i(x)$ such that $f(U) \subseteq V$.

In notation $f : \mathfrak{X} \twoheadrightarrow \mathcal{Y}$.

Remark 3.9 According to Lemma 3.7, a Kripke frame is a particular case of a neighborhood frame. There is a notion of p-morphism for Kripke frames. It is easy to check that for any two Kripke frames F and G function f is a p-morphism from F to G iff f is a p-morphism from $\mathcal{N}(F)$ to $\mathcal{N}(G)$. So, p-morphism for n-frames is a natural generalization of p-morphism for Kripke frames.

Lemma 3.10 *Let $\mathfrak{X} = (X, \tau_1, \ldots)$, $\mathcal{Y} = (Y, \sigma_1, \ldots)$ be n-frames and $f : \mathfrak{X} \twoheadrightarrow \mathcal{Y}$. Let θ' be a valuation on $\mathcal{Y}$. We define $\theta(p) = f^{-1}(\theta'(p))$. Then*

$$\mathfrak{X}, \theta, x \models \varphi \iff \mathcal{Y}, \theta', f(x) \models \varphi.$$

The proof is by standard induction on the length of formula φ.

Corollary 3.11 *If $f : \mathfrak{X} \twoheadrightarrow \mathcal{Y}$, then $Log(\mathfrak{X}) \subseteq Log(\mathcal{Y})$.*

Definition 3.12 Let $\mathfrak{X}_1 = (X_1, \tau_1)$ and $\mathfrak{X}_2 = (X_2, \tau_2)$ be two n-frames. Then *the product* of these n-frames is an n-2-frame defined as follows:

$$\begin{aligned} &\mathfrak{X}_1 \times \mathfrak{X}_2 = (X_1 \times X_2, \tau_1', \tau_2'), \\ &\tau_1'(x_1, x_2) = \{U \subseteq X_1 \times X_2 \,|\, \exists V (V \in \tau_1(x_1) \ \&\ V \times \{x_2\} \subseteq U)\}, \\ &\tau_2'(x_1, x_2) = \{U \subseteq X_1 \times X_2 \,|\, \exists V (V \in \tau_2(x_2) \ \&\ \{x_1\} \times V \subseteq U)\}. \end{aligned}$$

Note that for normal n-frames $\mathfrak{X}_1$ and $\mathfrak{X}_2$ their product $\mathfrak{X}_1 \times \mathfrak{X}_2$ is also normal.

Definition 3.13 For two unimodal logics L_1 and L_2, so that $nV(L_1) \neq \varnothing$ and $nV(L_2) \neq \varnothing$, we define *n-product* of them as follows:

$$\mathsf{L}_1 \times_n \mathsf{L}_2 = Log(\{\mathfrak{X}_1 \times \mathfrak{X}_2 \,|\, \mathfrak{X}_1 \in nV(L_1) \ \&\ \mathfrak{X}_2 \in nV(L_2)\}).$$

If we forget about one of its neighborhood functions, say τ_2', then $\mathfrak{X}_1 \times \mathfrak{X}_2$ will be a disjoint union of L_1 n-frames. Hence,

Proposition 3.14 ([8]) *For two unimodal normal logics* L_1 *and* L_2

$$\mathsf{L}_1 * \mathsf{L}_2 \subseteq \mathsf{L}_1 \times_n \mathsf{L}_2.$$

From [5] we know that n-product of any two logics from set $\{\mathsf{S4}, \mathsf{D4}, \mathsf{D}, \mathsf{T}\}$ equals to the fusion of corresponding logics. But this is not the case for K.

Proposition 3.15 $\mathsf{K} \times_n \mathsf{K} \neq \mathsf{K} * \mathsf{K}$.

Proof. Let $\mathfrak{X}_1 = (X_1, \tau_1)$ and $\mathfrak{X}_2 = (X_2, \tau_2)$ be two n-frames and $\mathfrak{X}_1 \times \mathfrak{X}_2 = (X_1 \times X_2, \tau_1', \tau_2')$. Consider formula $\Box_1\bot \to \Box_2\Box_1\bot$. Since this formula has no variables the truth of this formula does not depend on the valuation. So

$$\begin{aligned} \mathfrak{X}_1 \times \mathfrak{X}_2, (x,y) \models \Box_1\bot \iff \varnothing \in \tau_1'(x,y) \iff \\ \varnothing \in \tau_1(x) \iff \forall y' \in X_2\ (\varnothing \in \tau_1'(x,y')) \iff \\ \forall y' \in X_2\ (\mathfrak{X}_1 \times \mathfrak{X}_2, (x,y') \models \Box_1\bot) \implies \mathfrak{X}_1 \times \mathfrak{X}_2, (x,y) \models \Box_2\Box_1\bot. \end{aligned}$$

Hence, $\mathfrak{X}_1 \times \mathfrak{X}_2 \models \Box_1\bot \to \Box_2\Box_1\bot$. □

Moreover,

Lemma 3.16 *For any two n-frames* $\mathfrak{X}_1$ *and* $\mathfrak{X}_2$ *1) if* ϕ *is a closed formula without* $\Box_2$*, then for any two n-frames* $\mathfrak{X}_1$ *and* $\mathfrak{X}_2$

$$\mathfrak{X}_1 \times \mathfrak{X}_2 \models \phi \to \Box_2\phi,$$

2) if ϕ *is a closed formula without* $\Box_1$*, then*

$$\mathfrak{X}_1 \times \mathfrak{X}_2 \models \phi \to \Box_1\phi.$$

Proof. We prove only 1) because 2) can be proved analogously. Since ϕ does not contain neither $\Box_2$, nor variables, its value does not depend on the second coordinate. Let $F = \mathfrak{X}_1 \times \mathfrak{X}_2$. So if $F, (x,y) \models \phi$, then $\forall y'(F, (x,y') \models \phi)$, hence, $F, (x,y) \models \Box_2\phi$. □

We put

$$\Delta = \{\phi \to \Box_2\phi \,|\, \phi \text{ is closed and } \Box_2\text{-free}\} \cup \{\psi \to \Box_1\psi \,|\, \psi \text{ is closed and } \Box_1\text{-free}\}.$$

Definition 3.17 For two unimodal logics L_1 and L_2, we define

$$\langle L_1, L_2 \rangle = L_1 * L_2 + \Delta.$$

From Lemma 3.16 and Proposition 3.14 follows:

Proposition 3.18 *For any two normal modal logics* L_1 *and* L_2 $\langle L_1, L_2\rangle \subseteq L_1 \times_n L_2$.

Corollary 3.19 $\langle \mathsf{K}, \mathsf{K}\rangle \subseteq \mathsf{K} \times_n \mathsf{K}$.

The rest of the paper is dedicated to proving the converse inclusion.

4 Weak product of Kripke frames

In order to prove completeness of $\langle \mathsf{K}, \mathsf{K}\rangle$ w.r.t. n-frames, we first establish completeness w.r.t. special kind of Kripke frames. For this purpose we use "weak product" of Kripke frames which basically is the result of unraveling (c.f. [2]) of the usual product of two Kripke frames.

Definition 4.1 Let $G = (W, R_1, R_2) = G^{w_0}$ be a 2-modal Kripke frame with root w_0. A *path* in G is a tuple $\delta = w_0 R_{i_1} w_1 \dots R_{i_k} w_k$, so that for any $j > 0$ $w_{j-1} R_{i_j} w_j$. Path δ is called an (n, m)-path if set $\{j \mid i_j = 1\}$ has no more than n elements and set $\{j \mid i_j = 2\}$ has no more than m elements.

Definition 4.2 Let F_1 and F_2 be two Kripke frames with roots x_0 and y_0 respectively. A *path* in the product $F_1 \times F_2$ is a sequence of the following type

$$(x_0, y_0) S_1 (x_1, y_1) S_2 \dots S_n (x_n, y_n),$$

where $S_i \in \{R_1^h, R_2^v\}$ and for any $i \leq n$ $(x_{i-1}, y_{i-1}) S_i (x_i, y_i)$ holds.

Let $\mathcal{P}(F_1 \times F_2)$ be the set of all paths in $F_1 \times F_2$.

We define relations on $\mathcal{P}(F_1 \times F_2)$ in the following way: for any two paths α and β

$$\alpha R_1' \beta \iff \beta = \alpha R_1^h (a, b)$$
$$\alpha R_2' \beta \iff \beta = \alpha R_2^v (a, b)$$

We will call the following Kripke frame *weak product* of F_1 and F_2

$$\langle F_1, F_2\rangle = (\mathcal{P}(F_1 \times F_2), R_1', R_2').$$

Lemma 4.3 *For any two Kripke frames* F_1 *and* F_2 $\langle F_1, F_2\rangle \models \Delta$.

Proof. Let $\phi \to \Box_2 \phi \in \Delta$, i.e. ϕ is closed, $\Box_2$-free and $\alpha \models \phi$. Since ϕ is variable-free and $\Box_2$-free, its truth in $\langle F_1, F_2\rangle$ depends only on the structure of frame $G_\alpha^1 = (W_1 \times W_2, R_1')^\alpha = (R_1'^*(\alpha), R_1'|_{R_1'^*(\alpha)})$.

Due to the construction of $\langle F_1, F_2\rangle$ for any β, so that $\alpha R_2' \beta$, G_β^1 is isomorphic to G_α^1, so $\beta \models \phi$ and, hence, $\alpha \models \Box_2 \phi$.

Similarly, we prove that $\langle F_1, F_2\rangle \models \psi \to \Box_1 \psi$ for any closed $\Box_1$-free formula ψ. □

The aim of this section is to prove the following theorem:

Theorem 4.4 *Logic* $\langle \mathsf{K}, \mathsf{K}\rangle$ *is complete with respect to the class of all weak products of Kripke frames.*

In order to prove this theorem, we introduce some notions and constructions.

From here on in this section we rewrite all formulae using only $\Diamond$ instead of $\Box$.

Definition 4.5 For a modal formula ψ, we define its modal depth $d(\psi)$ as follows:

$$\begin{aligned} d(\bot) &= d(p) = 0; \\ d(\psi_1 \to \psi_2) &= \max(d(\psi_1), d(\psi_2)); \\ d(\Diamond\psi) &= d(\psi) + 1. \end{aligned}$$

Let Σ be a consistent set of closed 1-modal formulae maximal up to depth n. We define frame $\mathcal{F}(\Sigma, n)$ of depth n by induction:

Base:

$$\mathcal{F}(\Sigma, 0) = (\{\bullet\}, \varnothing) \text{ — an irreflexive one-point frame.}$$

Step: assume that $\mathcal{F}(\Omega, n)$ is defined for any maximal up to depth n set of closed formulae Ω; and Σ is maximal up to depth $n+1$ set of closed formulae.

Let $\mathcal{O}_n$ be the set of all maximal consistent sets of closed formulae of depth not greater than n. Note that there are only finitely many nonequivalent closed formulae of depth not greater than n. Let $\Omega \in \mathcal{O}_n$, since Ω is finite, then we can define

$$\zeta_\Omega = \Diamond(\bigwedge \Omega).$$

Note that $d(\zeta_\Omega) \leq n+1$ and due to maximality of Σ either $\zeta_\Omega \in \Sigma$ or $\neg\zeta_\Omega \in \Sigma$.

Let $F^0_\Omega = \mathcal{F}(\Omega, n)$ and F^i_Ω be a copy of F^0_Ω for each $i \in \mathbb{N}$.

Definition 4.6 Let $G_0 = (W_0, R_0), G_1 = (W_1, R_1), \ldots$ be a finite or infinite set of Kripke frames, so that all W_i are disjoint. Then by $H = (\bullet) + \bigsqcup G_i$ we define Kripke frame $H = (V, S)$ in the following way:

$$\begin{aligned} V &= \{r\} \cup \bigcup W_i, \\ S &= \Big(\{r\} \times \bigcup W_i\Big) \cup \bigcup R_i. \end{aligned}$$

Now we can define $\mathcal{F}(\Sigma, n+1)$

$$\mathcal{F}(\Sigma, n+1) = (\bullet) + \bigsqcup_{\zeta_\Omega \in \Sigma} \bigsqcup_{i \in \mathbb{N}} F^i_\Omega.$$

Definition 4.7 A Kripke frame $F = (W, R)$ is called *tree* with root r, if for any point $w \in W$, $w \neq r$, there is only one immediate predecessor and r has no predecessors.

Using standard unraveling method (see [2]) one can prove

Lemma 4.8 *For any countable Kripke frame F there exists a countable tree G such that $G \twoheadrightarrow F$.*

Let G be a tree of depth not greater than n. For $w \in G$ we define

$$\Sigma_n(w) = \{\psi \mid \psi \text{ is closed, } d(\psi) \leq n \text{ and } G, w \models \psi\}.$$

Note, that here we can write $G, w \models \psi$ (without valuation) because ψ is closed and does not depend on the valuation.

Lemma 4.9 *Let G be a countable tree, then for any $w \in G$ there exists $f : \mathcal{F}(\Sigma_n(w), n) \twoheadrightarrow G^w\lceil n$. Where $G^w\lceil n$ is the subframe of G^w, so that all points of depth greater than n are eliminated.*

Proof.

We construct f by induction.

Let I be the set of all successors of w. We split I into classes $I = \bigcup I_j$, so that for any j and any $u, u' \in I_j$ $\Sigma_{n-1}(u) = \Sigma_{n-1}(u')$. For each I_j. For each j we fix a surjective map $h_j : \mathbb{N} \to I_j$.

Remember, that

$$\mathcal{F}(\Sigma_n(w), n) = (\bullet) + \bigsqcup_{\zeta_\Omega \in \Sigma_n(w)} \bigsqcup_{i \in \mathbb{N}} F^i_\Omega, \text{ where } (\bullet) = (\{r\}, \varnothing). \tag{1}$$

We put

$$f(r) = w. \tag{2}$$

By induction for each j and $i \in \mathbb{N}$ there exists

$$g_{j,i} : F^i_{\Sigma_{n-1}(h_j(i))} \twoheadrightarrow G^{h_j(i)}\lceil(n-1).$$

For each Ω, so that $\zeta_\Omega \in \Sigma_n(w)$, there exist $u \in I_j$ such that $\Omega = \Sigma_{n-1}(u)$, and there exists i such that $h_j(i) = u$. So for any $x \in F^i_\Omega$ we put

$$f(x) = g_{j,i}(x). \tag{3}$$

Thus, (2) and (3) define f completely. Now we need to show that f is indeed a p-morphism. Let (V, S) be the frame from (1).

(i) Surjectiveness of f is obvious.

(ii) Let xSy and $x \neq r$, then $f(x)Rf(y)$ because for corresponding i and j $g_{j,i}$ is a p-morphism.

Now assume that $x = rSy$, then $y \in I_j$ for some j and $f(y)$ is a successor of w.

(iii) For $x \neq r$ it follows from the fact that all $g_{j,i}$ are p-morphisms.

Assume that $x = r$ and $f(r) = wRu$, then $u \in I_j$ for some j, and there exist an i such that $h_j(i) = u$. Hence, for the root r' of frame $F^i_{\Sigma_{n-1}(h_j(i))}$ $f(r') = u$ and rSr'.

□

Lemma 4.10 *If $\phi \notin \mathsf{K}$, then there is a set of closed formulae Σ such that $\mathcal{F}(\Sigma, d(\phi)) \not\models \phi$.*

Proof. It is well-known that logic K has countable (even finite) model property (see [2]). By Lemma 4.8 there is a countable tree $G = (W, R)$ with root w_0, so that $G, \theta, w_0 \models \neg\phi$ for some valuation θ.

Since the truth of ϕ depends only on points of depth not greater than $n = d(\phi)$ then $G\lceil n, \theta|_{G\lceil n}, w_0 \models \neg\phi$. Let $\Sigma = \Sigma_n(w_0)$. Σ is obviously a maximal consistent set of closed formulae up to depth n. By Lemma 4.9 $\mathcal{F}(\Sigma, n) \twoheadrightarrow G\lceil n$. Hence, $\mathcal{F}(\Sigma, n) \not\models \phi$. □

Corollary 4.11 *Logic* K *is complete with respect to the following class of frames:* $\{\mathcal{F}(\Sigma, n) \mid \Sigma \in \mathcal{O}_n,\ n \in \mathbb{N}\}$.

Let us go back to proving Theorem 4.4. We define $\Diamond_1$-depth $d_{\Diamond_1}$ and $\Diamond_2$-depth $d_{\Diamond_2}$ for any 2-modal formula:

$$\begin{array}{rr} d_{\Diamond_i}(\bot) = d_{\Diamond_i}(p) = 0; & d_{\Diamond_i}(\psi_1 \to \psi_2) = \max(d_{\Diamond_i}(\psi_1), d_{\Diamond_i}(\psi_2)); \\ d_{\Diamond_1}(\Diamond_1\psi) = d_{\Diamond_1}(\psi) + 1; & d_{\Diamond_2}(\Diamond_1\psi) = d_{\Diamond_2}(\psi); \\ d_{\Diamond_1}(\Diamond_2\psi) = d_{\Diamond_1}(\psi); & d_{\Diamond_2}(\Diamond_2\psi) = d_{\Diamond_2}(\psi) + 1; \end{array}$$

for any $i \in \{1, 2\}$.

Since the standard translation of a closed formula produces a first-order condition on frames, $\langle \mathsf{K}, \mathsf{K} \rangle$ is Δ-elementary. Therefore, by [4, Prop. 5.4], $\langle \mathsf{K}, \mathsf{K} \rangle$ is complete with respect to its countable rooted Kripke frames.

Assume that $\phi \notin \langle \mathsf{K}, \mathsf{K} \rangle$, then for a countable rooted Kripke frame F with root r and valuation θ, $F, \theta, r \models \neg\phi$. By Lemma 4.8 there is a 2-modal tree G such that $G \not\models \phi$.

For a 2-modal tree $G = (W, R_1, R_2)$ with root w_0 and $w \in W$, we define

$$\Sigma_n^1(w) = \{\psi \mid \psi \text{ is closed and } \Box_2\text{-free, } d(\psi) \le n \text{ and } w \models \psi\};$$
$$\Sigma_n^2(w) = \{\psi \mid \psi \text{ is closed and } \Box_1\text{-free, } d(\psi) \le n \text{ and } w \models \psi\}.$$

Let $F_1 = \mathcal{F}(\Sigma_n^1(w), n) = (V_1, S_1)$, $F_2 = \mathcal{F}(\Sigma_m^2(w), m) = (V_2, S_2)$, where $n = d_{\Diamond_1}(\phi)$ and $m = d_{\Diamond_2}(\phi)$. Then

Definition 4.12 Tree $G = (W, R_1, R_2)$ is called an *(n, m)-tree with root w_0* if any point in W can be accessed from w_0 with an (n, m)-path.

Lemma 4.13 *Let G be an (n, m)-tree with root w_0, then there exist two unimodal frames F_1, F_2 and a p-morphism $f : \langle F_1, F_2 \rangle \twoheadrightarrow G$.*

Proof. We will use induction on $n + m$. Let r_1 be the root of F_1 and r_2 be the root of F_2. We define

$$f(r_1, r_2) = w_0.$$

Let $H_w^1 = (R_1^*(w), R_1|_{R_1^*(w)})$ and $H_w^2 = (R_2^*(w), R_2|_{R_2^*(w)})$. By Lemma 4.9 there are $g_1 : F_1 \twoheadrightarrow H_{w_0}^1$ and $g_2 : F_2 \twoheadrightarrow H_{w_0}^2$.

Consider a path α in $F_1 \times F_2$ (an element of $\langle F_1, F_2 \rangle$). There are two possibilities:

1) $\alpha = (r_1, r_2)S_1'(u, r_2) \ldots = (r_1, r_2)S_1'\gamma$ and $r_1 S_1 u$, $g_1(u) = x$.

By induction, there is a p-morphism

$$h_u = \langle \mathcal{F}(\Sigma^1_{n-1}(x), n-1), \mathcal{F}(\Sigma^2_m(x), m)\rangle \twoheadrightarrow G^x.$$

Note that $\Sigma^1_{n-1}(x) = \Sigma_{n-1}(u)$. Let us show that $\Sigma^2_m(x) = \Sigma^2_m(w_0)$.

Indeed, if $\psi \in \Sigma^2_m(w_0)$, then $w_0 \models \psi$, but by Lemma 4.3 $w_0 \models \Box_1\psi$. Since $w_0 R_1 x$, then $x \models \psi$. So $\Sigma^2_m(w_0) \subseteq \Sigma^2_m(x)$ and, due to maximality, they are actually equal.

Therefore, cone of $\langle F_1, F_2\rangle$ with root in (u, r_2) is isomorphic to $\langle \mathcal{F}(\Sigma^1_{n-1}(x), n-1), \mathcal{F}(\Sigma^2_m(x), m)\rangle$. Let t be this isomorphism.

We put

$$f(\alpha) = h_u(t(\alpha)).$$

2) $\alpha = (r_1, r_2)S'_2(r_1, v)\ldots = (r_1, r_2)S'_2\gamma$ and $r_2 S_2 v$, $g_2(v) = y$.

By induction, there is a p-morphism

$$h'_v = \langle \mathcal{F}(\Sigma^1_n(y), n), \mathcal{F}(\Sigma^2_{m-1}(y), m-1)\rangle \twoheadrightarrow G^y.$$

Similar to the previous case, there is an isomorphism

$$t' : \langle \mathcal{F}(\Sigma^1_n(y), n), \mathcal{F}(\Sigma^2_{m-1}(y), m-1)\rangle \to \langle F_1, F_2\rangle^{(r_1, v)}.$$

So, we put

$$f(\alpha) = h'_v(t'(\alpha)).$$

Let us check that f is p-morphism also by induction:

Surjectiveness. Take any $y \in G_{(n,m)}$. If $y = w_0$, then its preimage is (r_1, r_2). Assume that $y \neq w_0$, then there is an (n, m)-path $\delta = w_0 R_k x \ldots y$ in $G_{(n,m)}$. Without loss of generality, we assume that $k = 1$ (case $k = 2$ is similar). By the construction, there exists u such that $f(u, r_2) = x$. Path $\eta = x \ldots y$ is an $(n-1, m)$-path, so that $\delta = w_0 R_k \eta$. h_u is surjective, hence there is a h_u-preimage of η and corresponding f-preimage of δ.

Monotonisity. Assume that δ and η are related in $\langle F_1, F_2\rangle$ via the 1st relation, i.e. $\eta = \delta S'_1(v_1, v_2)$. If $\delta \neq (r_1, r_2)$, then monotonisity follows from monotonisity of h_u.

If $\delta = (r_1, r_2)$, then $\eta = (r_1, r_2)S'_1(v_1, r_2)$. By construction, $f(\delta) = w_0$, $f(\eta) = g_1(v_1)$ and $w_0 R_1 g_1(v_1)$.

For S'_2 the argument is the same.

Lifting. Assume that $f(\delta)R_1 y$. Since G is a tree, then there is only one predecessor of y, that is $f(\delta)$. If $f(\delta) \neq w_0$, then $f(\delta) = h_u(t(\delta))$ for some u. Since h_u is a p-morphism, then there exists γ such that $h_u(\gamma) = y$ and $t^{-1}(\gamma) = \delta S'_1(v_1, v_2)$.

If $f(\delta) = w_0$, then $g_1(u) = y$ for some u. So $\eta = (r_1, r_2)S'_1(u, r_2)$ satisfies the lifting condition.

□

To finish the proof of Theorem 4.4, note that $G_{(m,n)} \not\models \phi$ and by Lemma 4.13 there are F_1 and F_2 such that $\langle F_1, F_2\rangle \not\models \phi$.

5 Completeness theorem

In this section we explain how, given two Kripke frames F_1 and F_2, to construct n-frames $\mathfrak{X}_1$ and $\mathfrak{X}_2$, so that $\mathfrak{X}_1 \times \mathfrak{X}_2 \twoheadrightarrow \mathcal{N}(\langle F_1, F_2\rangle)$. This is only possible if points in $\mathfrak{X}_1$ and $\mathfrak{X}_2$ do not have minimal neighborhoods or, in other words, each point should have arbitrary small neighborhoods. Because, otherwise, n-frames will be equivalent to Kripke frames, and we know that any product of Kripke frames satisfies commutativity axioms and Church-Rosser axiom. In order to construct such an n-frame, we introduce pseudo-infinite paths with stops.

Definition 5.1 For a frame $F = (W, R)$ with root a_0 we define *a path with stops* as a tuple $a_0 a_1 \dots a_n$, so that $a_i \in W$ or $a_i = 0$ and after eliminating zeros each point is related to the next one by relation R. We also consider infinite paths with stops that end with infinitely many zeros. We call these sequences *pseudo-infinite paths (with stops)*. Let W_ω be the set of all pseudo-infinite paths in W.

Define $f_F : W_\omega \to W$ in the following way: for $\alpha = a_0 a_1 \dots a_n 0^\omega$, where 0^ω is an infinite sequence of zeros and $a_n \neq 0$, we put

$$f_F(\alpha) = a_n.$$

We also define

$$\begin{aligned} st(\alpha) &= \min\{N \mid \forall k \geq N (a_k = 0)\}; \\ \alpha|_k &= a_1 \dots a_k; \\ U_i^k(\alpha) &= \{\beta \in W_\omega \mid \alpha|_m = \beta|_m \;\&\; f_F(\alpha) R_i f_F(\beta), \text{ where } m = \max(k, st(\alpha))\}. \end{aligned}$$

Lemma 5.2 *$U_i^k(\alpha) \subseteq U_i^m(\alpha)$ whenever $k \geq m$ for any $i \in \{1, 2\}$.*

Proof. Let $\beta \in U_i^k(\alpha)$. Since $\alpha|_k = \beta|_k$ and $k \geq m$, then $\alpha|_m = \beta|_m$. Hence, $\beta \in U_i^m(\alpha)$. □

Definition 5.3 Due to Lemma 5.2, sets $U_n(\alpha)$ form a filter base. So we can define

$$\begin{aligned} \tau(\alpha) &- \text{the filter with base } \{U_n(\alpha) \mid n \in \mathbb{N}\}; \\ \mathcal{N}_\omega(F) = (W_\omega, \tau) &— \text{is } \textit{a dense n-frame based on } F. \end{aligned}$$

Frame $\mathcal{N}_\omega(F)$ is dense in a sense that the intersection of all neighborhoods of a point is empty. So, there are no minimal neighborhoods unlike $\mathcal{N}(F)$.

Lemma 5.4 *Let $F = (W, R)$ be a Kripke frame with root a_0, then*

$$f_F : \mathcal{N}_\omega(F) \twoheadrightarrow \mathcal{N}(F).$$

Proof. From now on in this proof we will omit the subindex in f_F. Since for any $b \in W$ there is a path $a_0 a_1 \dots b$ and, hence for pseudo-infinite path $\alpha = a_0 a_1 \dots b 0^\omega \in X$, $f(\alpha) = b$ and f is surjective.

Assume, that $\alpha \in W_\omega$ and $U \in \tau(\alpha)$. We need to prove that $R(f(\alpha)) \subseteq f(U)$. There exists m such that $U_m(\alpha) \subseteq U$ and since $f(U_m(\alpha)) = R(f(\alpha))$, then

$$R(f(\alpha)) = f(U_m(\alpha)) \subseteq f(U).$$

Assume that $\alpha \in W_\omega$ and V is a neighborhood of $f(\alpha)$, i.e. $R(f(\alpha)) \subseteq V$. We need to prove that there exists $U \in \tau(\alpha)$ such that $f(U) \subseteq V$. As U we take $U_m(\alpha)$ for some $m \geq st(\alpha)$, then

$$f(U_m(\alpha)) = R(f(\alpha)) \subseteq V.$$

□

Corollary 5.5 *For any frame F $Log(\mathcal{N}_\omega(F)) \subseteq Log(F)$.*

Proof. It follows from Lemmas 3.7, 5.4 and Corollary 3.11

$$Log(\mathcal{N}_\omega(F)) \subseteq Log(\mathcal{N}(F)) = Log(F).$$

□

Let $F_1 = (W_1, R_1) = F_1^{r_1}$ and $F_2 = (W_2, R_2) = F_2^{r_2}$ be two Kripke frames with roots. We assume that $W_1 \cap W_2 = \varnothing$. Consider the product of n-frames $\mathfrak{X}_1 = (X_1, \tau_1) = \mathcal{N}_\omega(F_1)$ and $\mathfrak{X}_2 = (X_2, \tau_2) = \mathcal{N}_\omega(F_2)$

$$\mathfrak{X} = (X_1 \times X_2, \tau_1', \tau_2') = \mathcal{N}_\omega(F_1) \times_n \mathcal{N}_\omega(F_2).$$

We define function $g : \mathfrak{X}_1 \times \mathfrak{X}_2 \to \langle F_1, F_2 \rangle$ by induction, as follows.

Let $(\alpha, \beta) \in \mathfrak{X}_1 \times \mathfrak{X}_2$, so that $\alpha = x_1 x_2 \ldots$ and $\beta = y_1 y_2 \ldots$, $x_i \in W_1 \cup \{0\}$, $y_j \in W_2 \cup \{0\}$. We define $s(\alpha, \beta)$ to be the finite sequence that we get after eliminating all zeros from the infinite sequence $x_1 y_1 x_2 y_2 \ldots$. Now note, that we can uniquely map finite sequence $s(\alpha, \beta)$ to a path in $F_1 \times F_2$, because in $F_1 \times F_2$ we can only go up or right. Going up corresponds to adding a point from F_2, whereas going right corresponds to adding a point from F_1.

To be more precise, let $s(\alpha, \beta) = \mathbf{c} = w_1 w_2 \ldots w_n$. We define $h(\mathbf{c})$ by induction. If $\mathbf{c} = \varepsilon$ (empty string), then $h(\mathbf{c}) = (r_1, r_2)$. Assume that we already define $h(\mathbf{c}) = (x, y)$ and $\mathbf{b} = \mathbf{c}u$, then

$$h(\mathbf{b}) = \begin{cases} h(\mathbf{c}) R_1'(u, y) & \text{if } u \in W_1 \\ h(\mathbf{c}) R_2'(x, u) & \text{if } u \in W_2. \end{cases}$$

This definition is correct since, in the first case $x R_1 u$ and, in the second case, $y R_2 u$.

So we put $g(\alpha, \beta) = h(s(\alpha, \beta))$.

Lemma 5.6 *Function g defined above is a p-morphism: $g : \mathfrak{X} \twoheadrightarrow \mathcal{N}(\langle F_1, F_2 \rangle)$.*

Proof. Let $\mathbf{z} = (r_1, r_2) S_1 (z_1, t_1) S_2 \ldots S_n (z_n, t_n) \in \langle F_1, F_2 \rangle$. Define for $i \leq n$

$$x_i = \begin{cases} z_i, & \text{if } S_i = R_1'; \\ 0, & \text{if } S_i = R_2'; \end{cases} \qquad y_i = \begin{cases} 0, & \text{if } S_i = R_1'; \\ t_i, & \text{if } S_i = R_2'. \end{cases}$$

Let $\alpha = x_1 x_2 \dots x_n 0^\omega$ and $\beta = y_1 y_2 \dots y_n 0^\omega$, then $g(\alpha, \beta) = \mathbf{z}$. Hence g is surjective.

The next two conditions we check only for τ_1, since for τ_2 it is similar. Assume that $(\alpha, \beta) \in X_1 \times X_2$ and $U \in \tau_1(\alpha, \beta)$. We need to prove that $R_1'(g(\alpha, \beta)) \subseteq g(U)$. There exist $m > \max\{st(\alpha), st(\beta)\}$ such that $U_1^m(\alpha) \times \{\beta\} \subseteq U$ and, since $g(U_1^m(\alpha) \times \{\beta\}) = R_1'(g(\alpha, \beta))$, then

$$R_1'(g(\alpha, \beta)) = g(U_1^m(\alpha) \times \{\beta\}) \subseteq g(U);$$

where $U_1^m(\alpha)$ is the corresponding neighborhood from $\mathfrak{X}_1$.

Assume that $(\alpha, \beta) \in X_1 \times X_2$ and $R_1'(g(\alpha, \beta)) \subseteq V$. We need to prove that there exists $U \in \tau_1'(\alpha, \beta)$ such that $g(U) \subseteq V$. As U we take $U_m'(\alpha) \times \{\beta\}$ for some $m > \max\{st(\alpha), st(\beta)\}$, then

$$g(U_m'(\alpha) \times \{\beta\}) = R_1'(g(\alpha, \beta)) \subseteq V.$$

□

Corollary 5.7 *Let $F_1 = (W_1, R_1)$ and $F_2 = (W_2, R_2)$, then $Log(\mathcal{N}_\omega(F_1) \times \mathcal{N}_\omega(F_2)) \subseteq Log(\langle F_1, F_2 \rangle)$.*

It immediately follows from Lemma 5.6 and Corollary 3.11.

Theorem 5.8 *Logic $\langle \mathsf{K}, \mathsf{K} \rangle$ is complete with respect to products of normal neighborhood frames, i.e.*

$$\langle \mathsf{K}, \mathsf{K} \rangle = \mathsf{K} \times_n \mathsf{K}. \tag{4}$$

Proof. The inclusion from left to rignt of (4) was proved in Corollary 3.19.

The converse inclusion follows from Theorem 4.4 and Corollary 5.7. Indeed

$$\begin{aligned} \mathsf{K} \times_n \mathsf{K} = & \bigcap_{\mathfrak{X}_1, \mathfrak{X}_2 \in nV(\mathsf{K})} Log(\mathfrak{X}_1 \times \mathfrak{X}_2) \subseteq \\ \subseteq & \bigcap_{F_1, F_2 - \text{Kripke frames}} Log(\mathcal{N}_\omega(F_1) \times \mathcal{N}_\omega(F_2)) \subseteq \\ \subseteq & \bigcap_{F_1, F_2 - \text{Kripke frames}} Log(\langle F_1, F_2 \rangle) \subseteq \langle \mathsf{K}, \mathsf{K} \rangle. \end{aligned}$$

□

6 Conclusion

Even though the logic $\langle \mathsf{K}, \mathsf{K} \rangle$ has infinite axiomatization, it is decidable. We will not go into details, but the argument is similar to the ones in [4]. To refute a formula ϕ we only need to consider frames of bounded depth, and standard argument shows that we can also assume bounded branching.

Even more, it seems that logic $\langle \mathsf{K}, \mathsf{K} \rangle$ has fmp in the class of weak products of Kripke frames. If the bounds turn out to be polinomial it, probably, will give us PSPACE completeness for $\langle \mathsf{K}, \mathsf{K} \rangle$.

The obvious next step is to apply these methods to other logics and try to prove completeness results for products of neighborhood frames for other logics. For example, we conjecture that $\langle \mathsf{K4}, \mathsf{K4} \rangle$ is the d-logic of all products of topological spaces.

References

[1] Benthem, J., G. Bezhanishvili, B. Cate and D. Sarenac, *Multimodal logics of products of topologies*, Studia Logica **84** (2006), pp. 369–392.

[2] Blackburn, P., M. de Rijke and Y. Venema, "Modal Logic," Cambridge University Press, 2002.

[3] Chellas, B., "Modal Logic: An Introduction." Cambridge University Press, Cambridge, 1980.

[4] Gabbay, D. and V. Shehtman, *Products of modal logics. part i*, Journal of the IGPL **6** (1998), pp. 73–146.

[5] Kudinov, A., *Modal logic of some products of neighborhood frames*, in: T. Bolander, T. Braüner, S. Ghilardi and L. S. Moss, editors, *Advances in Modal Logic* (2012), pp. 386–394.

[6] Kudinov, A., *D-logic of product of rational numbers*, in: *Procedings of Information Technology and Systems*, Moscow, 2013, pp. 95–99.
URL `http://itas2013.iitp.ru/pdf/1569763701.pdf`

[7] Montague, R., *Universal grammar*, Theoria **36** (1970), pp. 373–398.

[8] Sano, K., *Axiomatizing hybrid products of monotone neighborhood frames*, Electr. Notes Theor. Comput. Sci. **273** (2011), pp. 51–67.

[9] Scott, D., *Advice on modal logic*, in: *Philosophical Problems in Logic: Some Recent Developments*, D. Reidel, 1970 pp. 143–173.

[10] Segerberg, K., "An essay in classical modal logic," Filosofiska föreningen och Filosofiska institutionen vid Uppsala universitet (Uppsala), 1971.

[11] Segerberg, K., *Two-dimensional modal logic*, Journal of Philosophical Logic **2** (1973), pp. 77–96.

[12] Shehtman, V., *Two-dimensional modal logic*, Mathematical Notices of USSR Academy of Science **23** (1978), pp. 417–424, (Translated from Russian).

Label-free Modular Systems for Classical and Intuitionistic Modal Logics

Sonia Marin

ENS, Paris, France

Lutz Straßburger

Inria, Palaiseau, France

Abstract

In this paper we show for each of the modal axioms d, t, b, 4, and 5 an equivalent set of inference rules in a nested sequent system, such that, when added to the basic system for the modal logic K, the resulting system admits cut elimination. Then we show the same result also for intuitionistic modal logic. We achieve this by combining structural and logical rules.

Keywords: Modal logic, cut elimination, nested sequents, Hilbert axioms.

1 Introduction

It very often happens that a new logic is introduced in terms of axioms in a Hilbert system. It is then a tedious task for proof theorists to find a cut-free deductive system in a sequent-like calculus. This is usually done by "trial and error" since there is no general method. It should be a goal of structural proof theory to automate this process, and to find general criteria for determining when a set of Hilbert axioms can be transformed into an equivalent set of inference rules such that cut elimination is preserved.

Recently, this goal has been achieved for substructural logics: In [4] it has been unveiled which classes of axioms can be transformed into equivalent structural rules in the sequent calculus, respectively hypersequent calculus, such that the resulting system admits cut elimination. In [5] a similar result has been obtained in the display calculus.

It is a natural question to ask whether this can also be done for modal logics. The work in [9] shows how certain classes of axioms in modal-tense logics can be transformed into logical rules in the display calculus and in nested sequents. Unfortunately, the established correspondence between axioms and logical rules works well only in the presence of the tense modalities. For modal logics without tense modalities, nested sequents have been used to give cut-free deductive systems for all logics in the classical modal S5-cube [2], as well as

d: $\Box A \supset \Diamond A$
t: $(A \supset \Diamond A) \wedge (\Box A \supset A)$
b: $(A \supset \Box\Diamond A) \wedge (\Diamond\Box A \supset A)$
4: $(\Diamond\Diamond A \supset \Diamond A) \wedge (\Box A \supset \Box\Box A)$
5: $(\Diamond A \supset \Box\Diamond A) \wedge (\Diamond\Box A \supset \Box A)$

Fig. 1. Modal axioms d, t, b, 4, and 5

for all logics in the intuitionistic modal S5-cube [18]. This concerns the modal axioms d, t, b, 4, and 5, shown in Figure 1. In classical logic only one of the two conjuncts in each axiom shown in that Figure is needed because the other follows from De Morgan duality. However, in the intuitionistic setting both conjuncts are needed. With these five axioms one can, *a priori*, obtain 32 logics but some coincide, such that there are only 15, which can be arranged in a cube as shown in Figure 2. This cube has the same shape in the classical as well as in the intuitionistic setting.

However, the two papers [2] and [18] have one drawback: Although they provide cut-free systems for all logics in the cube, they do not provide cut-free systems for all possible combinations of axioms. For example, the logic S5 can be obtained by adding b and 4, or by adding t and 5, to the modal logic K, but a complete cut-free system could only be obtained by adding rules for b, 4, and 5, or for t, 4, and 5 (in both the classical and the intuitionistic case).

This might be sufficient for someone interested in a cut-free system for a particular logic, but it is not sufficient for our goal—we do not want different rules for axioms t and 5, depending on whether we have only one or both of them in the system.

The works in [9,2,18] all use logical rules for the $\Diamond$-modality. An alternative route is taken in [3] where the authors use structural rules, which is closer in spirit to the work in substructural logics [4], mentioned above. However, the work in [3] does not cover all possible axiom combinations either (although it claims to do so).

In the present paper we achieve full modularity, for classical and intuitionistic modal logic, by using the logical rules of [2] and [18] together with the structural rules of [3]. Interestingly, the structural rules are the same in the classical and the intuitionistic setting.

This paper is organized as follows. In the next section we recall the nested sequent system for classical modal logic presented in [2]. Then, in Section 3, we show the structural rules of [3] and discuss the mistake in that paper. In Section 4, we then show our modularity result for classical modal logics. Section 5 recalls how nested sequents can be used for intuitionistic modal logics, as done in [18]. Finally, in Section 6, we show our modularity result for intuitionistic modal logics.

2 Nested Sequents for Classical Modal Logics

For simplicity, we consider here only formulas in negation normal form, generated by the grammar:

$$A, B, \ldots ::= p \mid \bar{p} \mid A \wedge B \mid A \vee B \mid \Box A \mid \Diamond A$$

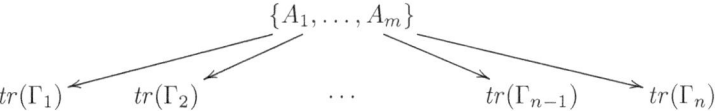

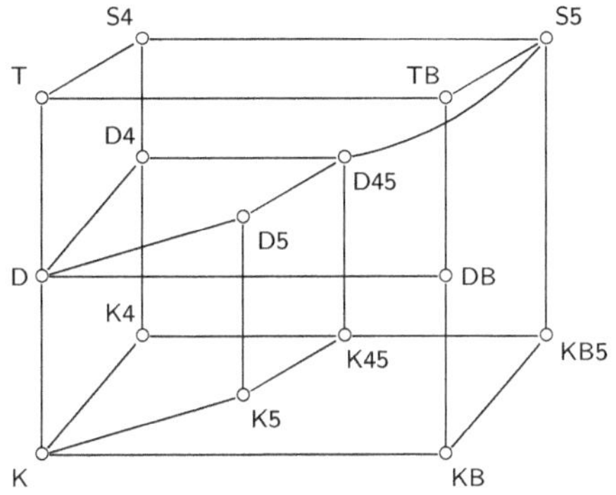

Fig. 2. The modal S5-cube

where p stands for a propositional variable and $\bar{p}$ its dual. Then the negation $\bar{A}$ of a formula A is defined in the usual way using the De Morgan duality, and implication $A \supset B$ is an abbreviation for $\bar{A} \vee B$.

Recall that a Hilbert system for the modal logic K can be obtained by taking some complete set of axioms for classical propositional logic extended with the k-axiom:

$$\mathsf{k}\colon\ \Box(A \supset B) \supset (\Box A \supset \Box B) \tag{1}$$

and the rules of *modus ponens* and *necessitation*, shown below:

$$\mathsf{mp}\ \frac{A \quad A \supset B}{B} \qquad\qquad \mathsf{nec}\ \frac{A}{\Box A} \tag{2}$$

For $\mathsf{X} \subseteq \{\mathsf{d}, \mathsf{t}, \mathsf{b}, \mathsf{4}, \mathsf{5}\}$ we write $\mathsf{K} + \mathsf{X}$ to denote the logic obtained from K by adding the axioms in X.

Let us now turn to the deductive system defined by Brünnler in [2] using nested sequents. Nested sequents have independently also been conceived by Kashima [10] and Poggiolesi [14]. Fitting [7] observed that nested sequents have the same data structure as prefixed tableaux.

A *nested sequent* (or simply a *sequent*) is a finite multiset of formulas and *boxed sequents*; that is, expressions like $[\Gamma]$ where Γ is also a sequent. Therefore a sequent is of the form:

$$\Gamma ::= A_1, \ldots, A_m, [\Gamma_1], \ldots, [\Gamma_n]$$

The *corresponding formula* of a sequent Γ, denoted by $fm(\Gamma)$, is defined as:

$$fm(\Gamma) = A_1 \vee \ldots \vee A_m \vee \Box fm(\Gamma_1) \vee \ldots \vee \Box fm(\Gamma_n)$$

Nested sequents can also be conceived as trees. For example, to the sequent $\Gamma = A_1, \ldots, A_m, [\Gamma_1], \ldots, [\Gamma_n]$ corresponds the tree $tr(\Gamma)$ defined as:

$$\begin{array}{ccccc} & & \{A_1, \ldots, A_m\} & & \\ \swarrow & \swarrow & & \searrow & \searrow \\ tr(\Gamma_1) & tr(\Gamma_2) & \cdots & tr(\Gamma_{n-1}) & tr(\Gamma_n) \end{array}$$

$$\mathsf{id}\ \frac{}{\Gamma\{a,\bar{a}\}} \qquad \vee\ \frac{\Gamma\{A,B\}}{\Gamma\{A\vee B\}} \qquad \wedge\ \frac{\Gamma\{A\}\quad \Gamma\{B\}}{\Gamma\{A\wedge B\}}$$

$$\mathsf{c}\ \frac{\Gamma\{A,A\}}{\Gamma\{A\}} \qquad \Box\ \frac{\Gamma\{[A]\}}{\Gamma\{\Box A\}} \qquad \Diamond\ \frac{\Gamma\{[A,\Delta]\}}{\Gamma\{\Diamond A,[\Delta]\}}$$

Fig. 3. System NK

$$\mathsf{d}^\Diamond\ \frac{\Gamma\{[A]\}}{\Gamma\{\Diamond A\}} \qquad \mathsf{t}^\Diamond\ \frac{\Gamma\{A\}}{\Gamma\{\Diamond A\}} \qquad \mathsf{b}^\Diamond\ \frac{\Gamma\{[\Delta],A\}}{\Gamma\{[\Delta,\Diamond A]\}}$$

$$\mathsf{4}^\Diamond\ \frac{\Gamma\{[\Diamond A,\Delta]\}}{\Gamma\{\Diamond A,[\Delta]\}} \qquad \mathsf{5}^\Diamond\ \frac{\Gamma\{\emptyset\}\{\Diamond A\}}{\Gamma\{\Diamond A\}\{\emptyset\}}\ \mathit{depth}(\Gamma\{\ \}\{\emptyset\})\geq 1$$

Fig. 4. Modal $\Diamond$-rules for axioms $\mathsf{d},\mathsf{t},\mathsf{b},\mathsf{4},\mathsf{5}$

Sometimes we will use for a sequent the vocabulary that would apply to the corresponding formula or to the corresponding tree without mentioning it.

To be able to apply inference rules deeply inside a sequent, we need the notion of context.

Definition 2.1 A *context* is a sequent with one or several holes; we distinguish *unary* context if there is exactly one hole, and *binary* context if there are exactly two. A *hole* $\{\ \}$ takes the place of a formula in the sequent but does not occur inside a formula. Finally, we write $\Gamma\{\Delta\}$ when we replace the hole in $\Gamma\{\ \}$ by Δ.

Definition 2.2 The *depth* of a unary context is defined inductively as:

$$\begin{aligned} \mathit{depth}(\{\ \}) &= 0 \\ \mathit{depth}(\Delta,\Gamma\{\ \}) &= \mathit{depth}(\Gamma\{\ \}) \\ \mathit{depth}([\Gamma\{\ \}]) &= \mathit{depth}(\Gamma\{\ \})+1 \end{aligned}$$

Example 2.3 Let $\Gamma\{\ \}\{\ \} = A,[B,\{\ \},[\{\ \}],C]$. For any sequents Δ_1 and Δ_2, we get: $\Gamma\{\Delta_1\}\{\Delta_2\} = A,[B,\Delta_1,[\Delta_2],C]$. In particular, $\Gamma\{\emptyset\}\{\Delta_2\} = A,[B,[\Delta_2],C]$ and $\Gamma\{\Delta_1\}\{\emptyset\} = A,[B,\Delta_1,[\emptyset],C]$. Moreover, we can compute $\mathit{depth}(\Gamma\{\ \}\{\Delta\}) = 1$ and $\mathit{depth}(\Gamma\{\Delta\}\{\ \}) = 2$.

The inference rules shown in Figure 3 form the *system* NK. Then, Figure 4 shows the $\Diamond$-rules for the axioms d, t, b, 4, and 5. For $\mathsf{X}\subseteq\{\mathsf{d},\mathsf{t},\mathsf{b},\mathsf{4},\mathsf{5}\}$ we write $\mathsf{X}^\Diamond$ for the corresponding subset of $\{\mathsf{d}^\Diamond,\mathsf{t}^\Diamond,\mathsf{b}^\Diamond,\mathsf{4}^\Diamond,\mathsf{5}^\Diamond\}$.

In the course of this paper we also need the *weakening*- and *cut-rule*, shown below:

$$\mathsf{w}\ \frac{\Gamma\{\emptyset\}}{\Gamma\{\Delta\}} \qquad \mathsf{cut}\ \frac{\Gamma\{\bar{A}\}\quad \Gamma\{A\}}{\Gamma\{\emptyset\}} \tag{3}$$

Lemma 2.4 *Let* $\mathsf{X}\subseteq\{\mathsf{d},\mathsf{t},\mathsf{b},\mathsf{4},\mathsf{5}\}$. *Then the* w*-rule is height-preserving admissible for* $\mathsf{NK}\cup\mathsf{X}^\Diamond$. [2]

Remark 2.5 In Brünnlers original formulation [2] of system $\mathsf{NK}\cup\mathsf{X}^\Diamond$, contraction was not given as explicit rule, but was absorbed in the $\Diamond$-rule and the

$$\mathsf{d}^{[]}\ \frac{\Gamma\{[\emptyset]\}}{\Gamma\{\emptyset\}} \qquad \mathsf{t}^{[]}\ \frac{\Gamma\{[\Delta]\}}{\Gamma\{\Delta\}} \qquad \mathsf{b}^{[]}\ \frac{\Gamma\{[\Sigma,[\Delta]]\}}{\Gamma\{[\Sigma],\Delta\}}$$

$$\mathsf{4}^{[]}\ \frac{\Gamma\{[\Delta],[\Sigma]\}}{\Gamma\{[[\Delta],\Sigma]\}} \qquad \mathsf{5}^{[]}\ \frac{\Gamma\{[\Delta]\}\{\emptyset\}}{\Gamma\{\emptyset\}\{[\Delta]\}}\ \mathit{depth}(\Gamma\{\ \}\{[\Delta]\}) \geq 1$$

$$\mathsf{m}^{[]}\ \frac{\Gamma\{[\Delta],[\Sigma]\}}{\Gamma\{[\Delta,\Sigma]\}}$$

Fig. 5. **Left:** Structural modal rules for axioms $\mathsf{d}, \mathsf{t}, \mathsf{b}, \mathsf{4}, \mathsf{5}$ – **Right:** Structural medial

rules in $\mathsf{X}^\diamond$. It is easy to see that both formulations are equivalent. In this paper we have an explicit contraction in the system because in the presence of the structural rules (introduced in the next section), contraction is no longer admissible.

As already observed in [2], not all combinations of modal rules lead to complete cut-free systems. For example, the 5-axiom $\Diamond A \supset \Box\Diamond A$ is valid in any $\{\mathsf{b}, \mathsf{4}\}$-frame, but it is not possible to prove it in $\mathsf{NK} \cup \{\mathsf{b}^\diamond, \mathsf{4}^\diamond\}$ without cut. Therefore, to get a cut-elimination proof, Brünnler [2] introduced the notion of 45-closure.

Definition 2.6 The *45-closure* of X is defined as:

$$\hat{\mathsf{X}} = \begin{cases} \mathsf{X} \cup \{\mathsf{4}\} & \text{if } \{\mathsf{b},\mathsf{5}\} \subseteq \mathsf{X} \text{ or if } \{\mathsf{t},\mathsf{5}\} \subseteq \mathsf{X} \\ \mathsf{X} \cup \{\mathsf{5}\} & \text{if } \{\mathsf{b},\mathsf{4}\} \subseteq \mathsf{X} \\ \mathsf{X} & \text{otherwise} \end{cases}$$

We say that X is *45-closed*, if $\mathsf{X} = \hat{\mathsf{X}}$.

Proposition 2.7 *Let* $\mathsf{X} \subseteq \{\mathsf{d}, \mathsf{t}, \mathsf{b}, \mathsf{4}, \mathsf{5}\}$. *We have that* X *is 45-closed, if and only if the following two conditions hold:*

- *whenever* $\mathsf{4}$ *is derivable in* $\mathsf{K} + \mathsf{X}$, *then* $\mathsf{4} \in \mathsf{X}$, *and*
- *whenever* $\mathsf{5}$ *is derivable in* $\mathsf{K} + \mathsf{X}$, *then* $\mathsf{5} \in \mathsf{X}$.

Now we can state Brünnler's [2] main results:

Theorem 2.8 *Let* $\mathsf{X} \subseteq \{\mathsf{d}, \mathsf{t}, \mathsf{b}, \mathsf{4}, \mathsf{5}\}$. *If a sequent* Γ *is derivable in* $\mathsf{NK} \cup \mathsf{X}^\diamond \cup \{\mathsf{cut}\}$ *then it is also derivable in* $\mathsf{NK} \cup \hat{\mathsf{X}}^\diamond$. [2]

Corollary 2.9 *Let* $\mathsf{X} \subseteq \{\mathsf{d}, \mathsf{t}, \mathsf{b}, \mathsf{4}, \mathsf{5}\}$ *be 45-closed. Then a formula* A *is a theorem of* $\mathsf{K} + \mathsf{X}$ *if and only if it is derivable in* $\mathsf{NK} \cup \mathsf{X}^\diamond$. [2]

The goal of this paper is to find a way to drop the 45-closed condition.

3 Structural Rules

The first attempt to drop the 45-closed condition was made in [3] where the authors suggest to use the structural rules shown in Figure 5. For $\mathsf{X} \subseteq \{\mathsf{d}, \mathsf{t}, \mathsf{b}, \mathsf{4}, \mathsf{5}\}$, we write $\mathsf{X}^{[]} \subseteq \{\mathsf{d}^{[]}, \mathsf{t}^{[]}, \mathsf{b}^{[]}, \mathsf{4}^{[]}, \mathsf{5}^{[]}\}$ for the corresponding set of rules from the left of that figure.

The work in [3] claims to prove cut elimination for $\mathsf{NK} \cup \{\mathsf{m}^{[]}\} \cup \mathsf{X}^{[]}$ (the $\mathsf{m}^{[]}$-rule is shown on the right of Figure 5), in order to obtain the following:

Claim 3.1 *Let* $\mathsf{X} \subseteq \{\mathsf{d}, \mathsf{t}, \mathsf{b}, \mathsf{4}, \mathsf{5}\}$. *A formula is a theorem of* $\mathsf{K}+\mathsf{X}$ *if and only if it is derivable in* $\mathsf{NK} \cup \{\mathsf{m}^{[]}\} \cup \mathsf{X}^{[]}$.

However, there is a mistake in the proof in [3], and the claim is not correct. For example, the formula $\Diamond\Box q \vee \Box(\Diamond\bar{p} \vee \Diamond\Diamond p)$ is a theorem of $\mathsf{K4}$ ($= \mathsf{K}+\mathsf{4}$), and also provable in $\mathsf{NK} \cup \{\mathsf{4}^{\Diamond}\}$, but it is not provable in $\mathsf{NK} \cup \{\mathsf{m}^{[]}, \mathsf{4}^{[]}\}$. The reason is that no rule in $\mathsf{NK} \cup \{\mathsf{m}^{[]}, \mathsf{4}^{[]}\}$ can increase the modal depth of the sequent (i.e., the maximal nesting of brackets and modalities) when read bottom-up.

The mistake in the cut elimination proof of [3] is rather subtle: In the cut reduction lemma (Lemma 10), the cut is permuted up, together with a stack of structural rule instances above the two premisses of the cut. If an instance of the $\Diamond$-rule is met, this $\Diamond$-rule instance is permuted down under the structural rules, using Lemma 7 and Lemma 8 of that paper, resulting in a derivation of structural rules above a derivation of logical rules (as shown in Figure 4 above), such that all rule instances in that derivation work on the same $\Diamond$-formula as the original $\Diamond$-rule instance. This stack of logical rules is then "reflected" at the cut (using Lemma 9), resulting in a stack of structural rules above the other premise of the cut.

The problem is that this only works if that $\Diamond$-formula is the cut-formula. Otherwise, the logical $\Diamond$-rules are not reflected at the cut but move under the cut as in a commutative case. This concerns the $\mathsf{4}^{\Diamond}$-rule and the $\mathsf{5}^{\Diamond}$-rule. Thus, the cut elimination proof of [3] breaks down if the $\mathsf{4}$- or $\mathsf{5}$-axiom is present.

For the convenience of the reader, we give an example in Appendix B.

4 Modularity for Classical Modal Logics

In this section, we show how the mistake of [3] can be corrected. We show that we can drop the 45-closure condition that appears in Theorem 2.8 if we use both the logical rules from [2] and the structural rules from [3].

Theorem 4.1 *Let* $\mathsf{X} \subseteq \{\mathsf{d}, \mathsf{t}, \mathsf{b}, \mathsf{4}, \mathsf{5}\}$. *A formula A is a theorem of* $\mathsf{K}+\mathsf{X}$ *if and only if it is derivable in* $\mathsf{NK} \cup \mathsf{X}^{\Diamond} \cup \mathsf{X}^{[]}$.

To be able to prove this theorem, we need to state first some lemmas. In particular, we need to show that weakening is still admissible.

Lemma 4.2 *For any* $\mathsf{X} \subseteq \{\mathsf{d}, \mathsf{t}, \mathsf{b}, \mathsf{4}, \mathsf{5}\}$ *the rule* w *is (contraction-preserving) admissible for* $\mathsf{NK} \cup \mathsf{X}^{\Diamond} \cup \mathsf{X}^{[]}$.

Proof. This is a straightforward induction on the height of the derivation. $\Box$

Lemma 4.3 *If* $\{\mathsf{t}, \mathsf{5}\} \subseteq \mathsf{X} \subseteq \{\mathsf{d}, \mathsf{t}, \mathsf{b}, \mathsf{4}, \mathsf{5}\}$ *then the* $\mathsf{4}^{\Diamond}$*-rule is admissible for* $\mathsf{NK} \cup \mathsf{X}^{\Diamond} \cup \mathsf{X}^{[]}$.

Proof. Any occurrence of the $\mathsf{4}^{\Diamond}$-rule can be replaced by the following derivation:

$$
\mathsf{t}^{[]}\,\dfrac{\mathsf{5}^{\Diamond}\,\dfrac{\mathsf{w}\,\dfrac{\Gamma\{[\Diamond A, \Delta]\}}{\Gamma\{[\emptyset], [\Diamond A, \Delta]\}}}{\Gamma\{[\Diamond A], [\Delta]\}}}{\Gamma\{\Diamond A, [\Delta]\}}
$$

Then we apply Lemma 4.2. □

Lemma 4.4 *If* $\{\mathsf{b},\mathsf{5}\} \subseteq \mathsf{X} \subseteq \{\mathsf{d},\mathsf{t},\mathsf{b},\mathsf{4},\mathsf{5}\}$ *then the* $\mathsf{4}^\diamond$*-rule is admissible for* $\mathsf{NK} \cup \mathsf{X}^\diamond \cup \mathsf{X}^{[]}$.

Proof. Any occurrence of the $\mathsf{4}^\diamond$-rule can be replaced by the following derivation:

$$\mathsf{b}^{[]}\frac{\mathsf{5}^\diamond\dfrac{\mathsf{w}\dfrac{\Gamma\{[\diamond A, \Delta]\}}{\Gamma\{[[\emptyset], \diamond A, \Delta]\}}}{\Gamma\{[[\diamond A], \Delta]\}}}{\Gamma\{\diamond A, [\Delta]\}}$$

Then we apply Lemma 4.2. □

To prove the admissibility of the $\mathsf{5}^\diamond$-rule, we decompose it into three rules that only use unary contexts, and are thus are easier to handle:

$$\mathsf{5}^\diamond_1\,\frac{\Gamma\{[\Delta], \diamond A\}}{\Gamma\{[\Delta, \diamond A]\}} \qquad \mathsf{5}^\diamond_2\,\frac{\Gamma\{[\Delta], [\diamond A, \Sigma]\}}{\Gamma\{[\Delta, \diamond A], [\Sigma]\}} \qquad \mathsf{5}^\diamond_3\,\frac{\Gamma\{[\Delta, [\diamond A, \Sigma]]\}}{\Gamma\{[\Delta, \diamond A, [\Sigma]]\}} \tag{4}$$

Clearly, each of $\mathsf{5}^\diamond_1$, $\mathsf{5}^\diamond_2$, and $\mathsf{5}^\diamond_3$ is a special case of $\mathsf{5}^\diamond$. Conversely, we have:

Lemma 4.5 *The* $\mathsf{5}^\diamond$*-rule is derivable from* $\{\mathsf{5}^\diamond_1, \mathsf{5}^\diamond_2, \mathsf{5}^\diamond_3\}$.

Proof. As in [2], but here the situation is a bit simpler since we do not have to deal with contraction. □

Lemma 4.6 *If* $\{\mathsf{4}\} \subseteq \mathsf{X} \subseteq \{\mathsf{d},\mathsf{t},\mathsf{b},\mathsf{4},\mathsf{5}\}$ *then the* $\mathsf{5}^\diamond_3$*-rule is admissible for* $\mathsf{NK} \cup \mathsf{X}^\diamond \cup \mathsf{X}^{[]}$.

Proof. Any occurrence of the $\mathsf{5}^\diamond_3$-rule is an instance of the $\mathsf{4}^\diamond$-rule. □

Lemma 4.7 *If* $\{\mathsf{b},\mathsf{4}\} \subseteq \mathsf{X} \subseteq \{\mathsf{d},\mathsf{t},\mathsf{b},\mathsf{4},\mathsf{5}\}$ *then the* $\mathsf{5}^\diamond_2$*-rule is (contraction-preserving) admissible for* $\mathsf{NK} \cup \mathsf{X}^\diamond \cup \mathsf{X}^{[]}$.

Proof. Any occurrence of the $\mathsf{5}^\diamond_2$-rule can be replaced by

$$\mathsf{b}^{[]}\frac{\mathsf{4}^\diamond\dfrac{\mathsf{4}^\diamond\dfrac{\mathsf{4}^{[]}\dfrac{\mathsf{4}^{[]}\dfrac{\mathsf{w}\dfrac{\Gamma\{[\Delta], [\diamond A, \Sigma]\}}{\Gamma\{[\Delta], [\emptyset], [\diamond A, \Sigma]\}}}{\Gamma\{[\Delta], [[\diamond A, \Sigma]]\}}}{\Gamma\{[\Delta, [[\diamond A, \Sigma]]]\}}}{\Gamma\{[\Delta, [\diamond A, [\Sigma]]]\}}}{\Gamma\{[\Delta, \diamond A, [[\Sigma]]]\}}}{\Gamma\{[\Delta, \diamond A], [\Sigma]\}}$$

As before, we conclude by applying Lemma 4.2. □

Lemma 4.8 *If* $\{\mathsf{b},\mathsf{4}\} \subseteq \mathsf{X} \subseteq \{\mathsf{d},\mathsf{t},\mathsf{b},\mathsf{4},\mathsf{5}\}$ *then the* $\mathsf{5}^\diamond_1$*-rule is admissible for* $\mathsf{NK} \cup \mathsf{X}^\diamond \cup \mathsf{X}^{[]}$.

Proof. If $\mathsf{5} \in \mathsf{X}$ then there is nothing to prove, so assume $\mathsf{5} \notin \mathsf{X}$. There is no simple derivation that can replace $\mathsf{5}^\diamond_1$. We consider the topmost instance of $\mathsf{5}^\diamond_1$, and let π be the derivation above it. We proceed by induction on the pair

$\langle c_\pi, h_\pi\rangle$ (under lexicographic ordering), where c_π is the number of c-instances in π, and h_π is the height of π. We have to carry out a case analysis on the bottommost rule instance r of π. If r only affects the context of our $\mathsf{5}^\Diamond_1$, we speak of a *trivial* case, because we can immediately apply the induction hypothesis:

$$\mathsf{5}^\Diamond_1\,\dfrac{\mathsf{r}\,\dfrac{\Gamma'\{\Diamond A,[\Delta']\}}{\Gamma\{\Diamond A,[\Delta]\}}}{\Gamma\{[\Diamond A,\Delta]\}} \quad\rightsquigarrow\quad \mathsf{r}\,\dfrac{\mathsf{5}^\Diamond_1\,\dfrac{\Gamma'\{\Diamond A,[\Delta']\}}{\Gamma'\{[\Diamond A,\Delta']\}}}{\Gamma\{[\Diamond A,\Delta]\}}$$

- $\mathsf{r}\in\{\mathsf{id},\wedge,\vee,\Box,\mathsf{d}^{[\,]}\}$: There are only trivial cases.
- $\mathsf{r}=\mathsf{c}$: There is one nontrivial case:

$$\mathsf{5}^\Diamond_1\,\dfrac{\mathsf{c}\,\dfrac{\Gamma\{\Diamond A,\Diamond A,[\Delta]\}}{\Gamma\{\Diamond A,[\Delta]\}}}{\Gamma\{[\Diamond A,\Delta]\}} \quad\rightsquigarrow\quad \mathsf{c}\,\dfrac{\mathsf{5}^\Diamond_1\,\dfrac{\mathsf{5}^\Diamond_1\,\dfrac{\Gamma\{\Diamond A,\Diamond A,[\Delta]\}}{\Gamma\{\Diamond A,[\Diamond A,\Delta]\}}}{\Gamma\{[\Diamond A,\Diamond A,\Delta]\}}}{\Gamma\{[\Diamond A,\Delta]\}}$$

  We can proceed by applying the induction hypothesis twice. This is possible because the number of c-instances above both $\mathsf{5}^\Diamond_1$ has decreased. (Note that none of our cases increases the number of contractions in the proof.)
- $\mathsf{r}=\Diamond$: There are two nontrivial cases:

$$\mathsf{5}^\Diamond_1\,\dfrac{\Diamond\,\dfrac{\Gamma\{[A,\Delta]\}}{\Gamma\{\Diamond A,[\Delta]\}}}{\Gamma\{[\Diamond A,\Delta]\}} \quad\rightsquigarrow\quad \mathsf{b}^{[\,]}\,\dfrac{\mathsf{4}^\Diamond\,\dfrac{\mathsf{b}^\Diamond\,\dfrac{\mathsf{w}\,\dfrac{\Gamma\{[A,\Delta]\}}{\Gamma\{[A,[\emptyset],\Delta]\}}}{\Gamma\{[[\Diamond A],\Delta]\}}}{\Gamma\{[\Diamond A,[\emptyset],\Delta]\}}}{\Gamma\{[\Diamond A,\Delta]\}}$$

$$\mathsf{5}^\Diamond_1\,\dfrac{\Diamond\,\dfrac{\Gamma\{[\Delta],[A,\Sigma]\}}{\Gamma\{[\Delta],\Diamond A,[\Sigma]\}}}{\Gamma\{[\Delta,\Diamond A],[\Sigma]\}} \quad\rightsquigarrow\quad \mathsf{b}^{[\,]}\,\dfrac{\mathsf{4}^{[\,]}\,\dfrac{\mathsf{b}^\Diamond\,\dfrac{\mathsf{4}^{[\,]}\,\dfrac{\mathsf{w}\,\dfrac{\Gamma\{[\Delta],[A,\Sigma]\}}{\Gamma\{[\Delta],[[\emptyset],A,\Sigma]\}}}{\Gamma\{[[\emptyset],[\Delta],A,\Sigma]\}}}{\Gamma\{[[\emptyset],[\Delta,\Diamond A],\Sigma]\}}}{\Gamma\{[[[\Delta,\Diamond A]],\Sigma]\}}}{\Gamma\{[\Delta,\Diamond A],[\Sigma]\}}$$

  In both cases, we can apply Lemma 4.2.
- $\mathsf{r}=\mathsf{d}^\Diamond$: There is one nontrivial case.

$$\mathsf{5}^\Diamond_1\,\dfrac{\mathsf{d}^\Diamond\,\dfrac{\Gamma\{[\Delta],[A]\}}{\Gamma\{[\Delta],\Diamond A\}}}{\Gamma\{[\Delta,\Diamond A]\}} \quad\rightsquigarrow\quad \mathsf{d}^\Diamond\,\dfrac{\mathsf{4}^{[\,]}\,\dfrac{\Gamma\{[\Delta],[A]\}}{\Gamma\{[\Delta,[A]]\}}}{\Gamma\{[\Delta,\Diamond A]\}}$$

- $\mathsf{r}=\mathsf{t}^\Diamond$: There is one nontrivial case.

$$\mathsf{5}^\Diamond_1\,\dfrac{\mathsf{t}^\Diamond\,\dfrac{\Gamma\{[\Delta],A\}}{\Gamma\{[\Delta],\Diamond A\}}}{\Gamma\{[\Delta,\Diamond A]\}} \quad\rightsquigarrow\quad \mathsf{b}^\Diamond\,\dfrac{\Gamma\{[\Delta],A\}}{\Gamma\{[\Delta,\Diamond A]\}}$$

- $\mathsf{r} = \mathsf{b}^{\diamond}$: There is one nontrivial case.

$$\mathsf{5}_1^{\diamond}\,\dfrac{\mathsf{b}^{\diamond}\,\dfrac{\Gamma\{[\Sigma,[\Delta]],A\}}{\Gamma\{[\Sigma,[\Delta],\diamond A]\}}}{\Gamma\{[\Sigma,[\Delta,\diamond A]]\}} \quad\leadsto\quad \mathsf{4}^{[]}\,\dfrac{\mathsf{b}^{\diamond}\,\dfrac{\mathsf{b}^{[]}\,\dfrac{\mathsf{4}^{[]}\,\dfrac{\mathsf{w}\,\dfrac{\Gamma\{[\Sigma,[\Delta]],A\}}{\Gamma\{[\Sigma,[\emptyset],[\Delta]],A\}}}{\Gamma\{[\Sigma,[[\Delta]]],A\}}}{\Gamma\{[\Sigma],[\Delta],A\}}}{\Gamma\{[\Sigma],[\Delta,\diamond A]\}}}{\Gamma\{[\Sigma,[\Delta,\diamond A]]\}}$$

And we apply Lemma 4.2.

- $\mathsf{r} = \mathsf{4}^{\diamond}$: There are two nontrivial cases.

$$\mathsf{5}_1^{\diamond}\,\dfrac{\mathsf{4}^{\diamond}\,\dfrac{\Gamma\{[\diamond A,\Delta]\}}{\Gamma\{\diamond A,[\Delta]\}}}{\Gamma\{[\diamond A,\Delta]\}} \quad\leadsto\quad \Gamma\{[\diamond A,\Delta]\}$$

$$\mathsf{5}_1^{\diamond}\,\dfrac{\mathsf{4}^{\diamond}\,\dfrac{\Gamma\{[\Delta],[\diamond A,\Sigma]\}}{\Gamma\{\diamond A,[\Delta],[\Sigma]\}}}{\Gamma\{[\diamond A,\Delta],[\Sigma]\}} \quad\leadsto\quad \mathsf{5}_2^{\diamond}\,\dfrac{\Gamma\{[\Delta],[\diamond A,\Sigma]\}}{\Gamma\{[\diamond A,\Delta],[\Sigma]\}}$$

In the second case, we need Lemma 4.7.

- $\mathsf{r} = \mathsf{t}^{[]}$: There is one nontrivial case.

$$\mathsf{5}_1^{\diamond}\,\dfrac{\mathsf{t}^{[]}\,\dfrac{\Gamma\{[\Delta],[\diamond A,\Sigma]\}}{\Gamma\{[\Delta],\diamond A,\Sigma\}}}{\Gamma\{[\Delta,\diamond A],\Sigma\}} \quad\leadsto\quad \mathsf{t}^{[]}\,\dfrac{\mathsf{5}_2^{\diamond}\,\dfrac{\Gamma\{[\Delta],[\diamond A,\Sigma]\}}{\Gamma\{[\Delta,\diamond A],[\Sigma]\}}}{\Gamma\{[\Delta,\diamond A],\Sigma\}}$$

Again, we can apply Lemma 4.7.

- $\mathsf{r} = \mathsf{b}^{[]}$: There are three nontrivial cases.

$$\mathsf{5}_1^{\diamond}\,\dfrac{\mathsf{b}^{[]}\,\dfrac{\Gamma\{[\Delta,[\diamond A,\Sigma]]\}}{\Gamma\{[\Delta],\diamond A,\Sigma\}}}{\Gamma\{[\Delta,\diamond A],\Sigma\}} \quad\leadsto\quad \mathsf{b}^{[]}\,\dfrac{\mathsf{4}^{\diamond}\,\dfrac{\Gamma\{[\Delta,[\diamond A,\Sigma]]\}}{\Gamma\{[\Delta,\diamond A,[\Sigma]]\}}}{\Gamma\{[\Delta,\diamond A],\Sigma\}}$$

$$\mathsf{5}_1^{\diamond}\,\dfrac{\mathsf{b}^{[]}\,\dfrac{\Gamma\{[\Delta],[\Sigma,[\diamond A,\Theta]]\}}{\Gamma\{[\Delta],\diamond A,[\Sigma],\Theta\}}}{\Gamma\{[\Delta,\diamond A],[\Sigma],\Theta\}} \quad\leadsto\quad \mathsf{5}_2^{\diamond}\,\dfrac{\mathsf{b}^{[]}\,\dfrac{\mathsf{4}^{\diamond}\,\dfrac{\Gamma\{[\Delta],[\Sigma,[\diamond A,\Theta]]\}}{\Gamma\{[\Delta],[\Sigma,\diamond A,[\Theta]]\}}}{\Gamma\{[\Delta],[\Sigma,\diamond A],\Theta\}}}{\Gamma\{[\Delta,\diamond A],[\Sigma],\Theta\}}$$

$$\mathsf{5}_1^{\diamond}\,\dfrac{\mathsf{b}^{[]}\,\dfrac{\Gamma\{[\Sigma,[\Theta,[\Delta]]],\diamond A\}}{\Gamma\{[\Sigma],\Theta,[\Delta],\diamond A\}}}{\Gamma\{[\Sigma],\Theta,[\Delta,\diamond A]\}} \quad\leadsto\quad \mathsf{5}_2^{\diamond}\,\dfrac{\mathsf{b}^{[]}\,\dfrac{\mathsf{5}_1^{\diamond}\,\dfrac{\Gamma\{[\Sigma,[\Theta,[\Delta]]],\diamond A\}}{\Gamma\{[\Sigma,[\Theta,[\Delta]],\diamond A]\}}}{\Gamma\{[\Sigma,\diamond A],\Theta,[\Delta]\}}}{\Gamma\{[\Sigma],\Theta,[\Delta,\diamond A]\}}$$

In the second and third case, we need Lemma 4.7. In the last case, we also apply the induction hypothesis.

- $\mathsf{r} = \mathsf{4}^{[]}$: There is one nontrivial case.

$$\mathsf{5}_1^{\diamond}\,\dfrac{\mathsf{4}^{[]}\,\dfrac{\Gamma\{[\Delta],[\diamond A,\Sigma]\}}{\Gamma\{[[\Delta],\diamond A,\Sigma]\}}}{\Gamma\{[[\Delta,\diamond A],\Sigma]\}} \quad \leadsto \quad \mathsf{4}^{[]}\,\dfrac{\mathsf{5}_2^{\diamond}\,\dfrac{\Gamma\{[\Delta],[\diamond A,\Sigma]\}}{\Gamma\{[\Delta,\diamond A],[\Sigma]\}}}{\Gamma\{[[\Delta,\diamond A],\Sigma]\}}$$

Then we can apply Lemma 4.7. □

We can now put Lemmas 4.3–4.8 together to prove our first main result:

Proof of Theorem 4.1. All rules in $\mathsf{NK} \cup \mathsf{X}^{\diamond} \cup \mathsf{X}^{[]}$ are sound wrt. $\mathsf{K}+\mathsf{X}$. This has already been shown in [2,3] and can easily be verified. Thus, any formula that is derivable in $\mathsf{NK} \cup \mathsf{X}^{\diamond} \cup \mathsf{X}^{[]}$ is also a theorem of $\mathsf{K}+\mathsf{X}$. Conversely, if A is a theorem of $\mathsf{K}+\mathsf{X}$, then by Corollary 2.9 we have a proof of A in $\mathsf{NK} \cup \hat{\mathsf{X}}^{\diamond}$. If $\hat{\mathsf{X}} = \mathsf{X}$, then a proof in $\mathsf{NK} \cup \hat{\mathsf{X}}^{\diamond}$ is trivially a proof in $\mathsf{NK} \cup \mathsf{X}^{\diamond} \cup \mathsf{X}^{[]}$, and we are done. Otherwise, we must have one of the following three cases:

- If $\{\mathsf{t},\mathsf{5}\} \subseteq \mathsf{X}$ then $\hat{\mathsf{X}} = \mathsf{X} \cup \{\mathsf{4}\}$. Then, by Lemma 4.3, we can construct a proof of A in $\mathsf{NK} \cup \mathsf{X}^{\diamond} \cup \mathsf{X}^{[]}$.
- If $\{\mathsf{b},\mathsf{5}\} \subseteq \mathsf{X}$ then $\hat{\mathsf{X}} = \mathsf{X} \cup \{\mathsf{4}\}$. We can use Lemma 4.4 similarly to get a proof of A in $\mathsf{NK} \cup \mathsf{X}^{\diamond} \cup \mathsf{X}^{[]}$.
- If $\{\mathsf{b},\mathsf{4}\} \subseteq \mathsf{X}$ then $\hat{\mathsf{X}} = \mathsf{X} \cup \{\mathsf{5}\}$. We can replace the $\mathsf{5}^{\diamond}$-rule with $\mathsf{5}_1^{\diamond}, \mathsf{5}_2^{\diamond}, \mathsf{5}_3^{\diamond}$ using Lemma 4.5. Then we get a proof of Γ in $\mathsf{NK} \cup \mathsf{X}^{\diamond} \cup \mathsf{X}^{[]}$ using Lemma 4.8, Lemma 4.7 and Lemma 4.6. □

5 Nested Sequents for Intuitionistic Modal Logics

Let us now turn to intuitionistic modal logics. The set of formulas is generated by

$$A, B, \ldots ::= p \mid \bot \mid A \wedge B \mid A \vee B \mid A \supset B \mid \Box A \mid \diamond A$$

where p stands for a propositional variable. The constant $\top$ can be recovered via $\bot \supset \bot$. Since $\Box$ and $\diamond$ are no longer De Morgan duals, it is not enough to just add the k-axiom (1) to intuitionistic propositional logic. In fact, there have been many different proposals of what should be added, e.g., [6,15,16,13,17,1,12]. Here, we consider the variant proposed in [16,13] and studied in detail by Simpson [17]. We add the following five axioms to intuitionistic propositional logic:

$$\begin{array}{ll} \mathsf{k}_1\colon\ \Box(A \supset B) \supset (\Box A \supset \Box B) & \mathsf{k}_3\colon\ \diamond(A \vee B) \supset (\diamond A \vee \diamond B) \\ \mathsf{k}_2\colon\ \Box(A \supset B) \supset (\diamond A \supset \diamond B) & \mathsf{k}_4\colon\ (\diamond A \supset \Box B) \supset \Box(A \supset B) \\ & \mathsf{k}_5\colon\ \diamond\bot \supset \bot \end{array} \qquad (5)$$

In a classical setting the axioms k_2–k_5 would follow from k_1 and the De Morgan laws. The theorems of the intuitionistic version of K, denoted by IK, are obtained from the axioms using the rules modus ponens and necessitation (2). As in the classical case, we write $\mathsf{IK}+\mathsf{X}$ for the logic obtained by adding a set of axioms $\mathsf{X} \subseteq \{\mathsf{d},\mathsf{t},\mathsf{b},\mathsf{4},\mathsf{5}\}$, shown in Figure 1.

Let us now recall how nested sequents can be used to give deductive systems for all logics in the intituionistic modal S5-cube, as done in [18]. A similar data structure is used in [8]. The sequents are essentially the same as in the

$$\bot^{\bullet}\,\frac{}{\Gamma\{\bot^{\bullet}\}} \qquad \mathsf{c}\,\frac{\Gamma\{A^{\bullet},A^{\bullet}\}}{\Gamma\{A^{\bullet}\}} \qquad \mathsf{id}\,\frac{}{\Gamma\{a^{\bullet},a^{\circ}\}}$$

$$\wedge^{\bullet}\,\frac{\Gamma\{A^{\bullet},B^{\bullet}\}}{\Gamma\{A\wedge B^{\bullet}\}} \qquad \wedge^{\circ}\,\frac{\Gamma\{A^{\circ}\} \quad \Gamma\{B^{\circ}\}}{\Gamma\{A\wedge B^{\circ}\}}$$

$$\vee^{\bullet}\,\frac{\Gamma\{A^{\bullet}\} \quad \Gamma\{A^{\bullet}\}}{\Gamma\{A\vee B^{\bullet}\}} \qquad \vee^{\circ}\,\frac{\Gamma\{A^{\circ}\}}{\Gamma\{A\vee B^{\circ}\}} \quad \vee^{\circ}\,\frac{\Gamma\{B^{\circ}\}}{\Gamma\{A\vee B^{\circ}\}}$$

$$\supset^{\bullet}\,\frac{\Gamma^{\downarrow}\{A^{\circ}\} \quad \Gamma\{B^{\bullet}\}}{\Gamma\{A\supset B^{\bullet}\}} \qquad \supset^{\circ}\,\frac{\Gamma\{A^{\bullet},B^{\circ}\}}{\Gamma\{A\supset B^{\circ}\}}$$

$$\Box^{\bullet}\,\frac{\Gamma\{[A^{\bullet},\Delta]\}}{\Gamma\{\Box A^{\bullet},[\Delta]\}} \qquad \Box^{\circ}\,\frac{\Gamma\{[A^{\circ}]\}}{\Gamma\{\Box A^{\circ}\}}$$

$$\Diamond^{\bullet}\,\frac{\Gamma\{[A^{\bullet}]\}}{\Gamma\{\Diamond A^{\bullet}\}} \qquad \Diamond^{\circ}\,\frac{\Gamma\{[A^{\circ},\Delta]\}}{\Gamma\{\Diamond A^{\circ},[\Delta]\}}$$

Fig. 6. System NIK

classical case, with the difference that formulas carry a polarity—there are two polarities, *input polarity* (marked with a black dot $\bullet$) and *output polarity* (marked with a white dot $\circ$)—such that exactly one formula in the whole sequent has the output polarity. More formally, a (*full*) *nested sequent* Γ for intuitionistic modal logic has two distinct parts: an *LHS-sequent* Λ in which all formulas have input polarity and an *RHS-sequent* Π which is either an output formula or a boxed sequent: given by:

$$\Gamma ::= \Lambda, \Pi \qquad \Lambda ::= A_1^{\bullet}, ..., A_m^{\bullet}, [\Lambda_1], ..., [\Lambda_n] \qquad \Pi ::= A^{\circ} \mid [\Gamma]$$

The *corresponding formula* of a sequent Γ is now defined as:

$$\begin{aligned} fm(\Lambda,\Pi) &= fm(\Lambda) \supset fm(\Pi) \\ fm(A_1^{\bullet}, ..., A_m^{\bullet}, [\Lambda_1], ..., [\Lambda_n]) &= A_1 \wedge ... \wedge A_m \wedge \Diamond fm(\Lambda_1) \wedge ... \wedge \Diamond fm(\Lambda_n) \\ fm(A^{\circ}) &= A \\ fm([\Gamma]) &= \Box fm(\Gamma) \end{aligned}$$

The notion of context is here again crucial. Since there are two different polarities, we also need two types of contexts: an *input context* (resp. an *output context*) is a sequent with one or several holes that should be filled with an input formula or an LHS-sequent (resp. an output formula, an RHS-sequent or a full sequent) to give a full sequent. The *depth* of a context is defined similarly to the classical case by induction.

As only one output formula is allowed in a sequent, we need, in some inference rules, to remove the output.

Definition 5.1 For an input context $\Gamma\{\ \}$ we obtain its *output pruning* $\Gamma^{\downarrow}\{\ \}$ by removing the unique output formula from it. For an output context $\Gamma\{\ \}$ we have $\Gamma^{\downarrow}\{\ \} = \Gamma\{\ \}$.

$$\mathsf{d}^\circ\ \frac{\Gamma\{[A^\circ]\}}{\Gamma\{\Diamond A^\circ\}} \quad \mathsf{t}^\circ\ \frac{\Gamma\{A^\circ\}}{\Gamma\{\Diamond A^\circ\}} \quad \mathsf{b}^\circ\ \frac{\Gamma\{[\Delta], A^\circ\}}{\Gamma\{[\Delta, \Diamond A^\circ]\}} \quad \mathsf{4}^\circ\ \frac{\Gamma\{[\Diamond A^\circ, \Delta]\}}{\Gamma\{\Diamond A^\circ, [\Delta]\}} \quad \mathsf{5}^\circ\ \frac{\Gamma\{\emptyset\}\{\Diamond A^\circ\}}{\Gamma\{\Diamond A^\circ\}\{\emptyset\}}$$

$$\mathsf{d}^\bullet\ \frac{\Gamma\{[A^\bullet]\}}{\Gamma\{\Box A^\bullet\}} \quad \mathsf{t}^\bullet\ \frac{\Gamma\{A^\bullet\}}{\Gamma\{\Box A^\bullet\}} \quad \mathsf{b}^\bullet\ \frac{\Gamma\{[\Delta], A^\bullet\}}{\Gamma\{[\Delta, \Box A^\bullet]\}} \quad \mathsf{4}^\bullet\ \frac{\Gamma\{[\Box A^\bullet, \Delta]\}}{\Gamma\{\Box A^\bullet, [\Delta]\}} \quad \mathsf{5}^\bullet\ \frac{\Gamma\{\emptyset\}\{\Box A^\bullet\}}{\Gamma\{\Box A^\bullet\}\{\emptyset\}}$$

Fig. 7. Intuitionistic $\Diamond^\circ$- and $\Box^\bullet$-rules; $\mathsf{5}^\circ$ and $\mathsf{5}^\bullet$ have proviso $depth(\Gamma\{\ \}\{\emptyset\}) \geq 1$.

Example 5.2 Let $\Gamma_1\{\ \} = A^\bullet, [B^\bullet, \{\ \}]$ and $\Gamma_2\{\ \} = A^\bullet, [B^\circ, \{\ \}]$. Then $\Gamma_1^\downarrow\{\ \} = A^\bullet, [B^\bullet, \{\ \}]$ and $\Gamma_2^\downarrow\{\ \} = A^\bullet, [\{\ \}]$.

The inference rules for intuitionistic modal logic are essentially the same as for classical modal logic. But since we are in an intuitionistic framework, each connective needs to be introduced by two rules, one for the input polarity and one for the output polarity, which doubles the number of rules. The system shown in Figure 6 is called *system* NIK.

Then, Figure 7 shows the rules for the axioms $\mathsf{d}, \mathsf{t}, \mathsf{b}, \mathsf{4}, \mathsf{5}$. Again, because we are intuitionistic now, the number of rules is doubled. For $\mathsf{X} \subseteq \{\mathsf{d}, \mathsf{t}, \mathsf{b}, \mathsf{4}, \mathsf{5}\}$, we write X° and $\mathsf{X}^\bullet$ for the corrseponding subset of $\{\mathsf{d}^\circ, \mathsf{t}^\circ, \mathsf{b}^\circ, \mathsf{4}^\circ, \mathsf{5}^\circ\}$ and $\{\mathsf{d}^\bullet, \mathsf{t}^\bullet, \mathsf{b}^\bullet, \mathsf{4}^\bullet, \mathsf{5}^\bullet\}$, respectively.

As in the classical case, we have the rules for weakening and cut:

$$\mathsf{w}\ \frac{\Gamma\{\emptyset\}}{\Gamma\{\Lambda\}} \qquad \mathsf{cut}\ \frac{\Gamma^\downarrow\{A^\circ\} \quad \Gamma\{A^\bullet\}}{\Gamma\{\emptyset\}}$$

Note that in the w-rule, the Λ has to be an LHS-sequent, i.e., must not contain the output formula. In the cut-rule we use the output pruning as for $\supset^\bullet$.

Remark 5.3 As in the classical case, the original formulation of NIK in [18] had no explicit contraction, but contraction was absorbed in into the rules of $\supset^\bullet$, $\Box^\bullet$, and the rules in $\mathsf{X}^\bullet$, instead. As before, we need explicit contraction here because of the structural rules. However, as in the classical case, both formulations are equivalent.

Remark 5.4 It is easy to see that we can use the two polarities $\circ$ and $\bullet$ to present a classical system in which negation $\neg$ is a primitive, as follows:

- allowing an arbitrary number of output-formulas in a sequent, and allow "contraction on the right", i.e., also for output formulas,
- add the two negation rules to NIK:

$$\neg^\bullet\ \frac{\Gamma\{A^\circ\}}{\Gamma\{\neg A^\bullet\}} \qquad \neg^\circ\ \frac{\Gamma\{A^\bullet\}}{\Gamma\{\neg A^\circ\}} \tag{6}$$

- and drop the output pruning from the left premiss in the $\supset^\bullet$- and cut-rules.

From this classical system, one could obtain an alternative intuitionistic system by allowing *at most one* output formula in the sequent and keeping the negation rules (6). However, we think that the systems presented here are simpler.

The notion of *45-closure* is also justified in the intuitionistic case:

Proposition 5.5 *Let* $\mathsf{X} \subseteq \{\mathsf{d}, \mathsf{t}, \mathsf{b}, 4, 5\}$. *We have that* X *is 45-closed iff*

- *whenever* 4 *is derivable in* $\mathsf{IK} + \mathsf{X}$, *then* $4 \in \mathsf{X}$, *and*
- *whenever* 5 *is derivable in* $\mathsf{IK} + \mathsf{X}$, *then* $5 \in \mathsf{X}$.

The following has been shown in [18]:

Theorem 5.6 *Let* $\mathsf{X} \subseteq \{\mathsf{d}, \mathsf{t}, \mathsf{b}, 4, 5\}$. *If a sequent* Γ *is derivable in* $\mathsf{NIK} \cup \mathsf{X}^{\bullet} \cup \mathsf{X}^{\circ} \cup \{\mathsf{cut}\}$ *then it is also derivable in* $\begin{cases} \mathsf{NIK} \cup \hat{\mathsf{X}}^{\bullet} \cup \hat{\mathsf{X}}^{\circ} & \text{if } \mathsf{d} \notin \mathsf{X} \\ \mathsf{NIK} \cup \hat{\mathsf{X}}^{\bullet} \cup \hat{\mathsf{X}}^{\circ} \cup \{\mathsf{d}^{[]}\} & \text{if } \mathsf{d} \in \mathsf{X} \end{cases}$

Remark 5.7 In the statement of Theorem 5.6, we distinguish whether d is or is not present in X rather than make use of Lemma 6.3 (ii) of [18] because it actually remains unclear how to permute the rule $\mathsf{d}^{[]}$ over the rules 4° and 5°, respectively, since the contraction-rule is not available for output formulas. Furthermore, with this formulation of Theorem 5.6, we do not need to extend the notion of 45-closure to t45-closure, as done in [18].

Corollary 5.8 *Let* $\mathsf{X} \subseteq \{\mathsf{d}, \mathsf{t}, \mathsf{b}, 4, 5\}$, *and let* $\mathsf{Z} = \mathsf{NIK} \cup \hat{\mathsf{X}}^{\bullet} \cup \hat{\mathsf{X}}^{\circ}$ *if* $\mathsf{d} \notin \mathsf{X}$, *and let* $\mathsf{Z} = \mathsf{NIK} \cup \hat{\mathsf{X}}^{\bullet} \cup \hat{\mathsf{X}}^{\circ} \cup \{\mathsf{d}^{[]}\}$ *if* $\mathsf{d} \in \mathsf{X}$. *Then a formula* A *is a theorem of* $\mathsf{IK} + \mathsf{X}$ *iff it is derivable in* Z.

6 Modularity for Intuitionistic Modal Logics

In this section, we prove a similar result as Theorem 4.1 for the intuitionistic setting. After our preparatory work of making the intuitionistic system look almost the same as the classical system, this work now becomes almost trivial. The key observation is that the structural rules in $\mathsf{X}^{[]}$ are also sound in the intuitionistic case, independently of the position of the output formula [18].

Theorem 6.1 *Let* $\mathsf{X} \subseteq \{\mathsf{d}, \mathsf{t}, \mathsf{b}, 4, 5\}$. *A formula* A *is a theorem of* $\mathsf{IK} + \mathsf{X}$ *if and only if it is derivable in* $\mathsf{NIK} \cup \mathsf{X}^{\bullet} \cup \mathsf{X}^{\circ} \cup \mathsf{X}^{[]}$.

Lemma 6.2 *For any* $\mathsf{X} \subseteq \{\mathsf{d}, \mathsf{t}, \mathsf{b}, 4, 5\}$ *the* w*-rule is height-preserving and contraction-preserving admissible for* $\mathsf{NK} \cup \mathsf{X}^{\circ} \cup \mathsf{X}^{\bullet} \cup \mathsf{X}^{[]}$.

Proof. This is a straightforward induction on the height of the derivation. □

Lemma 6.3 *If* $\{\mathsf{t}, 5\} \subseteq \mathsf{X} \subseteq \{\mathsf{d}, \mathsf{t}, \mathsf{b}, 4, 5\}$ *then the rules* 4° *and* $4^{\bullet}$ *are admissible for* $\mathsf{NIK} \cup \mathsf{X}^{\circ} \cup \mathsf{X}^{\bullet} \cup \mathsf{X}^{[]}$.

Proof. This is similar to Lemma 4.3. Any occurrence of the 4°-rule (respectively the $4^{\bullet}$-rule) can be replaced by the derivation on the left (respectively on the right) below:

$$\mathsf{t}^{[]}\dfrac{5^{\circ}\dfrac{\mathsf{w}\dfrac{\Gamma\{[\Diamond A^{\circ}, \Delta]\}}{\Gamma\{[\emptyset], [\Diamond A^{\circ}, \Delta]\}}}{\Gamma\{[\Diamond A^{\circ}], [\Delta]\}}}{\Gamma\{\Diamond A^{\circ}, [\Delta]\}} \qquad \mathsf{t}^{[]}\dfrac{5^{\bullet}\dfrac{\mathsf{w}\dfrac{\Gamma\{[\Box A^{\bullet}, \Delta]\}}{\Gamma\{[\emptyset], [\Box A^{\bullet}, \Delta]\}}}{\Gamma\{[\Box A^{\bullet}], [\Delta]\}}}{\Gamma\{\Box A^{\bullet}, [\Delta]\}}$$

We then apply Lemma 6.2. □

$$5_1^\circ \frac{\Gamma\{[\Delta], \Diamond A^\circ\}}{\Gamma\{[\Delta, \Diamond A^\circ]\}} \qquad 5_2^\circ \frac{\Gamma\{[\Delta], [\Diamond A^\circ, \Sigma]\}}{\Gamma\{[\Delta, \Diamond A^\circ], [\Sigma]\}} \qquad 5_3^\circ \frac{\Gamma\{[\Delta, [\Diamond A^\circ, \Sigma]]\}}{\Gamma\{[\Delta, \Diamond A^\circ, [\Sigma]]\}}$$

$$5_1^\bullet \frac{\Gamma\{[\Delta], \Box A^\bullet\}}{\Gamma\{[\Delta, \Box A^\bullet]\}} \qquad 5_2^\bullet \frac{\Gamma\{[\Delta], [\Box A^\bullet, \Sigma]\}}{\Gamma\{[\Delta, \Box A^\bullet], [\Sigma]\}} \qquad 5_3^\bullet \frac{\Gamma\{[\Delta, [\Box A^\bullet, \Sigma]]\}}{\Gamma\{[\Delta, \Box A^\bullet, [\Sigma]]\}}$$

Fig. 8. Variants of the rules $5^\bullet$ and 5°

Lemma 6.4 *If* $\{b, 5\} \subseteq X \subseteq \{d, t, b, 4, 5\}$ *then the rules* 4° *and* $4^\bullet$ *are admissible for* $NIK \cup X^\circ \cup X^\bullet \cup X^{[]}$.

Proof. For the 4°-rule, the proof is the same as for Lemma 4.4, and for $4^\bullet$-rule we use $5^\bullet$ instead of 5°. □

As in the classical case, to prove the admissibility of the rules 5° and $5^\bullet$, we need again to decompose them into variants asking for unary context, shown in Figure 8. The rules 5_1°, 5_2°, 5_3°, are special cases of 5°, and the rules $5_1^\bullet$, $5_2^\bullet$, $5_3^\bullet$, are special cases of $5^\bullet$.

Lemma 6.5 *The* 5°*-rule is derivable from* $\{5_1^\circ, 5_2^\circ, 5_3^\circ\}$, *and the* $5^\bullet$*-rule is derivable from* $\{5_1^\bullet, 5_2^\bullet, 5_3^\bullet\}$. [18]

Lemma 6.6 *If* $\{4\} \subseteq X \subseteq \{d, t, b, 4, 5\}$ *then the rules* 5_3° *and* $5_3^\bullet$ *are admissible for* $NIK \cup X^\circ \cup X^\bullet \cup X^{[]}$.

Proof. Any occurrence of the 5_3°-rule (resp. $5_3^\bullet$-rule) is an instance of the 4°-rule (resp. $4^\bullet$-rule). □

Lemma 6.7 *If* $\{b, 4\} \subseteq X \subseteq \{d, t, b, 4, 5\}$ *then the rules* 5_2° *and* $5_2^\bullet$ *are admissible for* $NIK \cup X^\circ \cup X^\bullet \cup X^{[]}$.

Proof. For the 5_2°-rule this is similar to Lemma 4.7. For the $5_2^\bullet$-rule, we use $4^\bullet$ instead of 4°. □

Lemma 6.8 *If* $\{b, 4\} \subseteq X \subseteq \{d, t, b, 4, 5\}$ *then the rules* 5_1° *and* $5_1^\bullet$ *are admissible for* $NIK \cup X^\circ \cup X^\bullet \cup X^{[]}$.

Proof. For the 5_1°-rule this is similar to Lemma 4.8. For the $5_1^\bullet$-rule, we use the corresponding $\Box^\bullet$-rules instead of the $\Diamond^\circ$-rules. □

Proof of Theorem 6.1. All rules in $NIK \cup X^\bullet \cup X^\circ \cup X^{[]}$ are sound wrt. $IK + X$ (see [18] and Appendix A). Hence, the first direction is trivial. Conversely, if A is a theorem of $IK+X$, then by Corollary 5.8, it is derivable in $NIK \cup \hat{X}^\bullet \cup \hat{X}^\circ \cup X^{[]}$. If $\hat{X} = X$, we are done. Otherwise, we have the same three cases as in the proof of Theorem 4.1, and we use Lemmas 6.3–6.8 instead of Lemmas 4.3–4.8. □

7 Future Work

We have used in this paper a combination of logical and structural rules, but for some axioms only the structural or/and only the logical rules would be sufficent, depending on the system, i.e., depending on which other axioms are present. This is a rather strange observation, and in strong contrast to what happens with substructural logics.

In order to better understand this phenomenon, we need to find a general pattern for translating axioms into structural and/or logical rules. In particular, it is an important question for future research, for which type of axioms such a translation is possible. Given the nature of nested sequents, we conjecture that this is possible for all Scott-Lemmon axioms [11], which are of the shape

$$\Diamond^h \Box^i A \supset \Box^j \Diamond^k A$$

where $h, i, j, k \geq 0$. However, for obtaining a general result, it might first be necessary to collect more evidence, as we provide it in this paper.

Another direction of future research is to investigate constructive modal logics [1], which reject axioms k_3, k_4, and k_5, shown in (5). The challenge here lies in the fact that some of the structural rules, for example $\mathsf{4}^{[]}$ and $\mathsf{5}^{[]}$, and some of the logical rules, for example $\mathsf{b}^\bullet$ and $\mathsf{5}^\bullet$, are not sound anymore.

References

[1] Bierman, G. M. and V. de Paiva, *On an intuitionistic modal logic*, Studia Logica **65** (2000), pp. 383–416.

[2] Brünnler, K., *Deep sequent systems for modal logic*, Archive for Mathematical Logic **48** (2009), pp. 551–577.

[3] Brünnler, K. and L. Straßburger, *Modular sequent systems for modal logic*, in: M. Giese and A. Waaler, editors, *Automated Reasoning with Analytic Tableaux and Related Methods, TABLEAUX'09*, LNCS **5607** (2009), pp. 152–166.

[4] Ciabattoni, A., N. Galatos and K. Terui, *From axioms to analytic rules in nonclassical logics*, in: *LICS*, 2008, pp. 229–240.

[5] Ciabattoni, A. and R. Ramanayake, *Structural extensions of display calculi: A general recipe*, in: L. Libkin, U. Kohlenbach and R. J. G. B. de Queiroz, editors, *Logic, Language, Information, and Computation – WoLLIC 2013*, LNCS **8071** (2013), pp. 81–95.

[6] Fitch, F., *Intuitionistic modal logic with quantifiers*, Portugaliae Mathematica **7** (1948), pp. 113–118.

[7] Fitting, M., *Prefixed tableaus and nested sequents*, APAL **163** (2012), pp. 291–313.

[8] Galmiche, D. and Y. Salhi, *Label-free natural deduction systems for intuitionistic and classical modal logics*, Journal of Applied Non-Classical Logics **20** (2010), pp. 373–421.

[9] Goré, R., L. Postniece and A. Tiu, *On the correspondence between display postulates and deep inference in nested sequent calculi for tense logics*, LMCS **7** (2011).

[10] Kashima, R., *Cut-free sequent calculi for some tense logics*, Studia Logica **53** (1994), pp. 119–136.

[11] Lemmon, E. J. and D. S. Scott, "An Introduction to Modal Logic," Blackwell, 1977.

[12] Pfenning, F. and R. Davies, *A judgmental reconstruction of modal logic*, Mathematical Structures in Computer Science **11** (2001), pp. 511–540.

[13] Plotkin, G. D. and C. P. Stirling, *A framework for intuitionistic modal logic*, in: J. Y. Halpern, editor, *Theoretical Aspects of Reasoning About Knowledge*, 1986.

[14] Poggiolesi, F., *The method of tree-hypersequents for modal propositional logic*, in: D. Makinson, J. Malinowski and H. Wansing, editors, *Towards Mathematical Philosophy*, Trends in Logic **28** (2009), pp. 31–51.

[15] Prawitz, D., "Natural Deduction, A Proof-Theoretical Study," Almq. and Wiksell, 1965.

[16] Servi, G. F., *Axiomatizations for some intuitionistic modal logics*, Rend. Sem. Mat. Univers. Politecn. Torino **42** (1984), pp. 179–194.

[17] Simpson, A., "The Proof Theory and Semantics of Intuitionistic Modal Logic," Ph.D. thesis, University of Edinburgh (1994).

[18] Straßburger, L., *Cut elimination in nested sequents for intuitionistic modal logics*, in: F. Pfenning, editor, *FoSSaCS'13*, LNCS **7794** (2013), pp. 209–224.

A Soundness of the Structural Rules

The soundness of the logical rules has been shown directly in [2] and [18]. The soundness of the structural rules follows only indirectly from these papers. For the convenience of the reader we give here a direct proof of the soundness of the structural rules in Figure 5 for the intuitionistic systems. Then their soundness in the classical systems follows immediately.

For simplicity, we show soundness with respect to the Hilbert system. Let us begin with two lemmas from [18], justifying the use of deep inference:

Lemma A.1 *Let* $\mathsf{X} \subseteq \{\mathsf{d}, \mathsf{t}, \mathsf{b}, \mathsf{4}, \mathsf{5}\}$, *let* Δ *and* Σ *be full sequents, and let* $\Gamma\{\ \}$ *be an output context. If* $fm(\Delta) \supset fm(\Sigma)$ *is a theorem of* $\mathsf{IK} + \mathsf{X}$, *then so is* $fm(\Gamma\{\Delta\}) \supset fm(\Gamma\{\Sigma\})$.

Lemma A.2 *Let* $\mathsf{X} \subseteq \{\mathsf{d}, \mathsf{t}, \mathsf{b}, \mathsf{4}, \mathsf{5}\}$, *let* Δ *and* Σ *be LHS-sequents, and let* $\Gamma\{\ \}$ *be an input context. If* $fm(\Sigma) \supset fm(\Delta)$ *is a theorem of* $\mathsf{IK} + \mathsf{X}$, *then so is* $fm(\Gamma\{\Delta\}) \supset fm(\Gamma\{\Sigma\})$.

Both lemmas are shown by an induction on the structure of $\Gamma\{\ \}$, using the following:

Lemma A.3 *Let* $\mathsf{X} \subseteq \{\mathsf{d}, \mathsf{t}, \mathsf{b}, \mathsf{4}, \mathsf{5}\}$. *For any formulas* A, B, *and* C *we have:*

(i) *If* $A \supset B$ *is a theorem of* $\mathsf{IK} + \mathsf{X}$, *then so is* $(C \supset A) \supset (C \supset B)$.

(ii) *If* $A \supset B$ *is a theorem of* $\mathsf{IK} + \mathsf{X}$, *then so is* $\Box A \supset \Box B$.

(iii) *If* $A \supset B$ *is a theorem of* $\mathsf{IK} + \mathsf{X}$, *then so is* $(C \wedge A) \supset (C \wedge B)$.

(iv) *If* $A \supset B$ *is a theorem of* $\mathsf{IK} + \mathsf{X}$, *then so is* $\Diamond A \supset \Diamond B$.

(v) *If* $A \supset B$ *is a theorem of* $\mathsf{IK} + \mathsf{X}$, *then so is* $(B \supset C) \supset (A \supset C)$.

Now, for showing soundness of a rule, we have to show that for every instance of the rule, if the premiss is a theorem of $\mathsf{IK}+\mathsf{X}$, then so is the conclusion. For this, we often use the following lemma:

Lemma A.4 *For all formulas* A, B, *the following are theorems of* IK:

(i) $\Diamond(A \wedge B) \supset \Diamond A \wedge \Diamond B$,

(ii) $\Diamond A \wedge \Box B \supset \Diamond(A \wedge B)$, *and*

(iii) $(\Box A \wedge \Box B) \supset \Box(A \wedge B)$.

The proofs of the Lemmas A.3 and A.4 are straightforward and left to the reader. We are now ready to see the main result of this appendix:

Proposition A.5 *Let* $\mathsf{X} \subseteq \{\mathsf{d}, \mathsf{t}, \mathsf{b}, \mathsf{4}, \mathsf{5}\}$ *and* $\mathsf{x} \in \mathsf{X}$. *The corresponding structural rule* $\mathsf{x}^{[]}$ *shown on the left of Figure 5 is sound with respect to* $\mathsf{IK} + \mathsf{X}$.

Proof. For each $\mathsf{x} \in \{\mathsf{d}, \mathsf{t}, \mathsf{b}, \mathsf{4}, \mathsf{5}\}$, let $\mathsf{x}^{[]}\,\dfrac{\Gamma_1}{\Gamma_2}$ denote the corresponding structural rule. We show that $fm(\Gamma_1) \supset fm(\Gamma_2)$ is a theorem of $\mathsf{IK} + \mathsf{x}$.

- $\mathsf{x} = \mathsf{d}$: We have that $\top$ is the unit for $\wedge$ (i.e., $\top = \bigwedge \emptyset$). Therefore, we have that $fm([\emptyset]) = \Diamond\top$ and $fm(\emptyset) = \top$ while $\top \supset \Diamond\top$ is a theorem of $\mathsf{IK}+\mathsf{d}$. Thus, by applying Lemma A.2, we get that $fm(\Gamma\{[\emptyset]\}) \supset fm(\Gamma\{\emptyset\})$ is a theorem of $\mathsf{IK}+\mathsf{d}$.

- $\mathsf{x}=\mathsf{t}$: We proceed by a case analysis on the position of the output formula in the sequent (see Figure 5).
  - If the output formula is in $\Gamma\{\ \}$, then $fm([\Delta]) = \Diamond fm(\Delta)$. Since $D \supset \Diamond D$ is a theorem of $\mathsf{IK}+\mathsf{t}$, so is $fm(\Gamma\{[\Delta]\}) \supset fm(\Gamma\{\Delta\})$ by Lemma A.2.
  - If the output formula is in Δ, then $fm([\Delta]) = \Box fm(\Delta)$. Since $\Box D \supset D$ is a theorem of $\mathsf{IK}+\mathsf{t}$, so is $fm(\Gamma\{[\Delta]\}) \supset fm(\Gamma\{\Delta\})$ by Lemma A.1.
- $\mathsf{x}=\mathsf{b}$: We proceed as in the previous case by a case analysis on the position of the output formula in the sequent (see Figure 5).
  - If the output formula is in $\Gamma\{\ \}$ we use the fact that $(\Diamond S \wedge D) \supset \Diamond(S \wedge \Diamond D)$ is a theorem of $\mathsf{IK}+\mathsf{b}$, together with Lemma A.2.
  - If the output formula is in Σ we use the fact that $\Box(\Diamond D \supset S) \supset (D \supset \Box S)$ is a theorem of $\mathsf{IK}+\mathsf{b}$, together with Lemma A.1.
  - If the output formula is in Δ we use the fact that $\Box(S \supset \Box D) \supset (\Diamond S \supset D)$ is a theorem of $\mathsf{IK}+\mathsf{b}$, together with Lemma A.1.

  The three formulas can be shown using the following three derivations, where each line stands for a valid implication in $\mathsf{IK}+\mathsf{b}$:

$$\cfrac{\cfrac{\Diamond S \wedge D}{\Diamond S \wedge \Box\Diamond D}\ \mathsf{b}+A.3.(iii)}{\Diamond(S \wedge \Diamond D)}\ A.4.(ii) \qquad \cfrac{\cfrac{\Box(\Diamond D \supset S)}{\Box\Diamond D \supset \Box S}\ \mathsf{k}_1}{D \supset \Box S}\ \mathsf{b}+A.3.(v) \qquad \cfrac{\cfrac{\Box(S \supset \Box D)}{\Diamond S \supset \Diamond\Box D}\ \mathsf{k}_2}{\Diamond S \supset D}\ \mathsf{b}+A.3.(i)$$

- $\mathsf{x}=4$: We proceed by a case analysis on the position of the output formula in the sequent (see Figure 5).
  - If the output formula is in $\Gamma\{\ \}$ we use the fact that $\Diamond(\Diamond S \wedge D) \supset (\Diamond S \wedge \Diamond D)$ is a theorem of $\mathsf{IK}+4$, together with Lemma A.2.
  - If the output formula is in Δ we use the fact that $(\Diamond S \supset \Box D) \supset \Box(S \supset \Box D)$ is a theorem of $\mathsf{IK}+4$, together with Lemma A.1.
  - If the output formula is in Σ we use the fact that $(\Diamond D \supset \Box S) \supset \Box(\Diamond D \supset S)$ is a theorem of $\mathsf{IK}+4$, together with Lemma A.1.

  As before, we can show the three formulas by simple derivations:

$$\cfrac{\cfrac{\Diamond(\Diamond S \wedge D)}{\Diamond\Diamond S \wedge \Diamond D}\ A.4.(i)}{\Diamond S \wedge \Diamond D}\ 4+A.3.(iii) \qquad \cfrac{\cfrac{\Diamond S \supset \Box D}{\Diamond S \supset \Box\Box D}\ 4+A.3.(i)}{\Box(S \supset \Box D)}\ \mathsf{k}_4 \qquad \cfrac{\cfrac{\Diamond D \supset \Box S}{\Diamond\Diamond D \supset \Box S}\ 4+A.3.(v)}{\Box(\Diamond D \supset S)}\ \mathsf{k}_4$$

- $\mathsf{x}=5$: For showing soundness of $5^{[]}$, we observe that it is derivable using the following three rules and show soundness for each of them individually:

$$5_1^{[]}\ \frac{\Gamma\{[\Theta,[\Delta]]\}}{\Gamma\{[\Theta],[\Delta]\}} \qquad 5_2^{[]}\ \frac{\Gamma\{[\Theta,[\Delta]],[\Sigma]\}}{\Gamma\{[\Theta],[[\Delta],\Sigma]\}} \qquad 5_3^{[]}\ \frac{\Gamma\{[\Theta,[\Delta],[\Sigma]]\}}{\Gamma\{[\Theta,[[\Delta],\Sigma]]\}}$$

For each of $5_1^{[]}$, $5_2^{[]}$, and $5_3^{[]}$, we proceed by a case analysis on the position of the output formula in the sequent. The cases for $5_1^{[]}$ are the following:

- If the output formula is in $\Gamma\{\ \}$ we use Lemma A.2, together with the fact that $(\Diamond T \wedge \Diamond D) \supset \Diamond(T \wedge \Diamond D)$ is a theorem of $\mathsf{IK}+\mathsf{5}$.
- If the output formula is in Θ we use Lemma A.1, together with the fact that $\Box(\Diamond D \supset T) \supset (\Diamond D \supset \Box T)$ is a theorem of $\mathsf{IK}+\mathsf{5}$.
- If the output formula is in Δ we use Lemma A.1, together with the fact that $\Box(T \supset \Box D) \supset (\Diamond T \supset \Diamond D)$ is a theorem of $\mathsf{IK}+\mathsf{5}$.

The following derivations show that the three formulas are theorems of $\mathsf{IK}+\mathsf{5}$:

$$\dfrac{\dfrac{\Diamond T \wedge \Diamond D}{\Diamond T \wedge \Box\Diamond D}\,\mathsf{5}+A.3.(\text{iii})}{\Diamond(T \wedge \Diamond D)}\,A.4.(\text{ii}) \qquad \dfrac{\dfrac{\Box(\Diamond D \supset T)}{\Box\Diamond D \supset \Box T}\,\mathsf{k}_1}{\Diamond D \supset \Box T}\,\mathsf{5}+A.3.(\text{v}) \qquad \dfrac{\dfrac{\Box(T \supset \Box D)}{\Diamond T \supset \Diamond\Box D}\,\mathsf{k}_2}{\Diamond T \supset \Box D}\,\mathsf{5}+A.3.(\text{i})$$

Let us now consider the cases for $\mathsf{5}^{[]}_2$:

- If the output formula is in $\Gamma\{\ \}$ we use Lemma A.2, together with the fact that $(\Diamond T \wedge \Diamond(\Diamond D \wedge S)) \supset (\Diamond(T \wedge \Diamond D) \wedge \Diamond S)$ is a theorem of $\mathsf{IK}+\mathsf{5}$.
- If the output formula is in Δ we use Lemma A.1 together with the fact that $(\Diamond S \supset \Box(T \supset \Box D)) \supset (\Diamond T \supset \Box(S \supset \Box D))$ is a theorem of $\mathsf{IK}+\mathsf{5}$.
- If the output formula is in Σ we use Lemma A.1 together with the fact that $(\Diamond(T \wedge \Diamond D) \supset \Box S) \supset (\Diamond T \supset \Box(\Diamond D \supset S))$ is a theorem of $\mathsf{IK}+\mathsf{5}$.
- If the output formula is in Θ we use Lemma A.1 together with the fact that $(\Diamond S \supset \Box(\Diamond D \supset T)) \supset (\Diamond(S \wedge \Diamond D) \supset \Box T)$ is a theorem of $\mathsf{IK}+\mathsf{5}$.

These formulas are shown by the following derivations:

$$\dfrac{\dfrac{\dfrac{\dfrac{\Diamond T \wedge \Diamond(\Diamond D \wedge S)}{\Diamond T \wedge \Diamond\Diamond D \wedge \Diamond S}\,A.4.(\text{i})+A.3.(\text{iii})}{\Diamond T \wedge \Diamond\Box\Diamond D \wedge \Diamond S}\,\mathsf{5}+A.3.(\text{iii},\text{iv})}{\Diamond T \wedge \Box\Diamond D \wedge \Diamond S}\,\mathsf{5}+A.3.(\text{iii})}{\Diamond(T \wedge \Diamond D) \wedge \Diamond S}\,A.4.(\text{ii})+A.3.(\text{iii})$$

$$\dfrac{\dfrac{\dfrac{\dfrac{\dfrac{\Diamond S \supset \Box(T \supset \Box D)}{\Diamond S \supset \Diamond T \supset \Diamond\Box D}\,\mathsf{k}_2+A.3.(\text{i})}{\Diamond S \supset \Diamond T \supset \Box\Diamond\Box D}\,\mathsf{5}+A.3.(\text{i})}{\Diamond S \supset \Diamond T \supset \Box\Box D}\,\mathsf{5}+A.3.(\text{i},\text{ii})}{\Diamond T \supset \Diamond S \supset \Box\Box D}}{\Diamond T \supset \Box(S \supset \Box D)}\,\mathsf{k}_4+A.3.(\text{i})$$

$$\dfrac{\dfrac{\dfrac{\dfrac{\dfrac{\Diamond(T \wedge \Diamond D) \supset \Box S}{(\Diamond T \wedge \Box\Diamond D) \supset \Box S}\,A.4.(\text{ii})+A.3.(\text{v})}{(\Diamond T \wedge \Diamond\Box\Diamond D) \supset \Box S}\,\mathsf{5}+A.3.(\text{iii},\text{v})}{(\Diamond T \wedge \Diamond\Diamond D) \supset \Box S}\,\mathsf{5}+A.3.(\text{iii},\text{iv},\text{v})}{\Diamond T \supset \Diamond\Diamond D \supset \Box B}}{\Diamond T \supset \Box(\Diamond D \supset S)}\,\mathsf{k}_4+A.3.(\text{i})$$

$$\dfrac{\dfrac{\dfrac{\dfrac{\dfrac{\Diamond S \supset \Box(\Diamond D \supset T)}{\Diamond S \supset \Box\Diamond D \supset \Box T}\,\mathsf{k}_1+A.3.(\text{i})}{(\Diamond S \wedge \Box\Diamond D) \supset \Box T}}{(\Diamond S \wedge \Diamond\Box\Diamond D) \supset \Box T}\,\mathsf{5}+A.3.(\text{iii},\text{v})}{(\Diamond S \wedge \Diamond\Diamond D) \supset \Box T}\,\mathsf{5}+A.3.(\text{iii},\text{iv},\text{v})}{\Diamond(S \wedge \Diamond D) \supset \Box T}\,A.4.(\text{i})+A.3.(\text{v})$$

Finally, let us consider the cases for $\mathsf{5}^{[]}_3$:

- If the output formula is in $\Gamma\{\ \}$, we use Lemma A.2, together with the fact that $\Diamond(T \wedge \Diamond(\Diamond D \wedge S)) \supset \Diamond(T \wedge \Diamond D \wedge \Diamond S)$ is a theorem of $\mathsf{IK}+\mathsf{5}$.
- If the output formula is in Θ, we use Lemma A.1, together with the fact that $\Box((\Diamond D \wedge \Diamond S) \supset T) \supset \Box(\Diamond(\Diamond D \wedge S) \supset T)$ is a theorem of $\mathsf{IK}+\mathsf{5}$.
- If the output formula is in Δ, we use Lemma A.1, together with the fact that $\Box((T \wedge \Diamond S) \supset \Box D) \supset \Box(T \supset \Box(S \supset \Box D))$ is a theorem of $\mathsf{IK}+\mathsf{5}$.
- If the output formula is in Σ, we use Lemma A.1, together with the fact that $\Box((T \wedge \Diamond D) \supset \Box S) \supset \Box(T \supset \Box(\Diamond D \supset S))$ is a theorem of $\mathsf{IK}+\mathsf{5}$.

Below are the derivations showing that these formulas are indeed theorems of $\mathsf{IK}+\mathsf{5}$:

$$
\begin{array}{ll}
\Diamond(T \wedge \Diamond(\Diamond D \wedge S)) & \\ \hline
\Diamond T \wedge \Diamond\Diamond(\Diamond D \wedge S) & A.4.(\mathrm{i}) \\ \hline
\Diamond T \wedge \Diamond(\Diamond\Diamond D \wedge \Diamond S) & A.4.(\mathrm{i}) + A.3.(\mathrm{iii}) \\ \hline
\Diamond T \wedge \Diamond(\Diamond\Box\Diamond D \wedge \Diamond S) & \mathsf{5} + A.3.(\mathrm{iii},\mathrm{iv}) \\ \hline
\Diamond T \wedge \Diamond(\Box\Diamond D \wedge \Diamond S) & \mathsf{5} + A.3.(\mathrm{iii},\mathrm{iv}) \\ \hline
\Diamond T \wedge \Diamond\Box\Diamond D \wedge \Diamond\Diamond S & A.4.(\mathrm{i}) + A.3.(\mathrm{iii}) \\ \hline
\Diamond T \wedge \Diamond\Box\Diamond D \wedge \Diamond\Box\Diamond S & \mathsf{5} + A.3.(\mathrm{iii}) \\ \hline
\Diamond T \wedge \Box\Diamond D \wedge \Box\Diamond S & \mathsf{5} + A.3.(\mathrm{iii}) \\ \hline
\Diamond(T \wedge \Diamond D) \wedge \Box\Diamond S & A.4.(\mathrm{ii}) + A.3.(\mathrm{iii}) \\ \hline
\Diamond(T \wedge \Diamond D \wedge \Diamond S) & A.4.(\mathrm{ii})
\end{array}
\qquad
\begin{array}{ll}
\Box((\Diamond D \wedge \Diamond S) \supset T) & \\ \hline
\Box(\Diamond D \wedge \Diamond S) \supset \Box T & \mathsf{k}_1 \\ \hline
(\Box\Diamond D \wedge \Box\Diamond S) \supset \Box T & A.4.(\mathrm{iii}) + A.3.(\mathrm{v}) \\ \hline
(\Diamond\Box\Diamond D \wedge \Diamond\Box\Diamond S) \supset \Box T & \mathsf{5} + A.3.(\mathrm{iii},\mathrm{v}) \\ \hline
(\Diamond\Box\Diamond D \wedge \Diamond\Diamond S) \supset \Box T & \mathsf{5} + A.3.(\mathrm{iii},\mathrm{v}) \\ \hline
(\Diamond\Diamond\Box\Diamond D \wedge \Diamond\Diamond S) \supset \Box T & \mathsf{5} + A.3.(\mathrm{iii},\mathrm{v}) \\ \hline
(\Diamond\Diamond\Diamond D \wedge \Diamond\Diamond S) \supset \Box T & \mathsf{5} + A.3.(\mathrm{iii},\mathrm{v}) \\ \hline
\Diamond(\Diamond\Diamond D \wedge \Diamond S) \supset \Box T & A.4.(\mathrm{i}) + A.3.(\mathrm{v}) \\ \hline
\Diamond\Diamond(\Diamond D \wedge S) \supset \Box T & A.4.(\mathrm{i}) + A.3.(\mathrm{v}) \\ \hline
\Box(\Diamond(\Diamond D \wedge S) \supset T) & \mathsf{k}_4
\end{array}
$$

$$
\begin{array}{ll}
\Box((T \wedge \Diamond S) \supset \Box D) & \\ \hline
\Diamond(T \wedge \Diamond S) \supset \Diamond\Box D & \mathsf{k}_2 \\ \hline
(\Diamond T \wedge \Box\Diamond S) \supset \Diamond\Box D & A.4.(\mathrm{ii}) + A.3.(\mathrm{v}) \\ \hline
(\Diamond T \wedge \Diamond\Box\Diamond S) \supset \Diamond\Box D & \mathsf{5} + A.3.(\mathrm{iii},\mathrm{v}) \\ \hline
(\Diamond T \wedge \Diamond\Diamond S) \supset \Diamond\Box D & \mathsf{5} + A.3.(\mathrm{iv},\mathrm{iii},\mathrm{v}) \\ \hline
(\Diamond T \wedge \Diamond\Diamond S) \supset \Box\Diamond\Box D & \mathsf{5} + A.3.(\mathrm{i}) \\ \hline
(\Diamond T \wedge \Diamond\Diamond S) \supset \Box\Box\Diamond\Box D & \mathsf{5} + A.3.(\mathrm{i},\mathrm{ii}) \\ \hline
(\Diamond T \wedge \Diamond\Diamond S) \supset \Box\Box\Box D & \mathsf{5} + A.3.(\mathrm{i},\mathrm{ii}) \\ \hline
\Diamond T \supset \Diamond\Diamond S \supset \Box\Box\Box D & \\ \hline
\Diamond T \supset \Box(\Diamond S \supset \Box\Box D) & \mathsf{k}_4 + A.3.(\mathrm{i}) \\ \hline
\Diamond T \supset \Box\Box(S \supset \Box D) & \mathsf{k}_4 + A.3.(\mathrm{i},\mathrm{ii}) \\ \hline
\Box(T \supset \Box(S \supset \Box D)) & \mathsf{k}_4
\end{array}
\qquad
\begin{array}{ll}
\Box((T \wedge \Diamond D) \supset \Box S) & \\ \hline
\Diamond(T \wedge \Diamond D) \supset \Diamond\Box S & \mathsf{k}_2 \\ \hline
(\Diamond T \wedge \Box\Diamond D) \supset \Diamond\Box S & A.4.(\mathrm{ii}) + A.3.(\mathrm{v}) \\ \hline
(\Diamond T \wedge \Box\Diamond D) \supset \Box\Diamond\Box S & \mathsf{5} + A.3.(\mathrm{i}) \\ \hline
(\Diamond T \wedge \Box\Diamond D) \supset \Box\Box S & \mathsf{5} + A.3.(\mathrm{ii},\mathrm{i}) \\ \hline
(\Diamond T \wedge \Diamond\Box\Diamond D) \supset \Box\Box S & \mathsf{5} + A.3.(\mathrm{iii},\mathrm{v}) \\ \hline
(\Diamond T \wedge \Diamond\Diamond\Box\Diamond D) \supset \Box\Box S & \mathsf{5} + A.3.(\mathrm{iv},\mathrm{iii},\mathrm{v}) \\ \hline
(\Diamond T \wedge \Diamond\Diamond\Diamond D) \supset \Box\Box S & \mathsf{5} + A.3.(\mathrm{iv},\mathrm{iii},\mathrm{v}) \\ \hline
\Diamond T \supset \Diamond\Diamond\Diamond D \supset \Box\Box S & \\ \hline
\Diamond T \supset \Box(\Diamond\Diamond D \supset \Box S) & \mathsf{k}_4 + A.3.(\mathrm{i}) \\ \hline
\Diamond T \supset \Box\Box(\Diamond D \supset S) & \mathsf{k}_4 + A.3.(\mathrm{i},\mathrm{ii}) \\ \hline
\Box(T \supset \Box(\Diamond D \supset S)) & \mathsf{k}_4
\end{array}
$$

□

B Addendum to Section 3

In this appendix we use a concrete example to explain the error in [3]. The example is due to an anonymous reviewer who first observed the problem. Let us consider the formula $\Diamond\Box q \vee \Box(\Diamond\bar{p} \vee \Diamond\Diamond p)$, which is a theorem of $\mathsf{K4}$ (it is derivable in $\mathsf{NK} \cup \{\mathsf{4}^{\Diamond}\}$, as the reader can easily verify).

Let us now argue why this formula is not derivable in $\mathsf{NK} \cup \{\mathsf{4}^{[]}, \mathsf{m}^{[]}\}$. For this, observe that the $\mathsf{m}^{[]}$-rule becomes admissible if we replace the c-rule by

$$
\hat{\mathsf{c}}\,\frac{\Gamma\{\Delta,\Delta\}}{\Gamma\{\Delta\}}
$$

which allows contraction on arbitrary sequents, and which is derivable for $\{\mathsf{c}, \mathsf{m}^{[]}\}$. Additionally, observe that the rules for $\wedge$, $\vee$, and $\Box$ are invertible and can therefore be applied eagerly. We can also apply the $\Diamond$-rule and $\mathsf{4}^{[]}$-rule eagerly, if we first apply the $\hat{\mathsf{c}}$-rule on the formula/subsequent that is moved by the $\Diamond/\mathsf{4}^{[]}$-rule. It can also be shown that there is no other need fo the $\hat{\mathsf{c}}$-rule (see [2] for a proof of admissibility of $\hat{\mathsf{c}}$ for such a system). This means we can do an exhaustive proof search without the need of backtracking. The following derivation shows our attempt to prove $\Diamond\Box q \vee \Box(\Diamond\bar{p} \vee \Diamond\Diamond p)$ in $\mathsf{NK} \setminus \{\mathsf{c}\} \cup \{\hat{\mathsf{c}}, \mathsf{4}^{[]}\}$:

$$
{\scriptstyle\vee,\Box,\vee}\dfrac{{\scriptstyle\hat{\mathsf{c}},\Diamond}\dfrac{{\scriptstyle\Box}\dfrac{{\scriptstyle\hat{\mathsf{c}},\Diamond,\Diamond}\dfrac{{\scriptstyle\hat{\mathsf{c}},4^{[]}}\dfrac{{\scriptstyle\hat{\mathsf{c}},\Diamond}\dfrac{{\scriptstyle\Box}\dfrac{{\scriptstyle\hat{\mathsf{c}},\Diamond}\dfrac{{\scriptstyle\hat{\mathsf{c}},4^{[]}}\dfrac{{\scriptstyle\hat{\mathsf{c}},\Diamond}\dfrac{{\scriptstyle\Box}\dfrac{{\scriptstyle\hat{\mathsf{c}},4^{[]}}\dfrac{\Diamond\Box q,[q],[[q],q,p],[[q,p],q,\bar p,\Diamond p],[[q,\bar p,\Diamond p],\Diamond\bar p,\Diamond\Diamond p]}{\Diamond\Box q,[[q],q,p],[[q,p],q,\bar p,\Diamond p],[[q,\bar p,\Diamond p],\Diamond\bar p,\Diamond\Diamond p]}}{\Diamond\Box q,[\Box q,q,p],[[q,p],q,\bar p,\Diamond p],[[q,\bar p,\Diamond p],\Diamond\bar p,\Diamond\Diamond p]}}{\Diamond\Box q,[q,p],[[q,p],q,\bar p,\Diamond p],[[q,\bar p,\Diamond p],\Diamond\bar p,\Diamond\Diamond p]}}{\Diamond\Box q,[[q,p],q,\bar p,\Diamond p],[[q,\bar p,\Diamond p],\Diamond\bar p,\Diamond\Diamond p]}}{\Diamond\Box q,[[q],q,\bar p,\Diamond p],[[q,\bar p,\Diamond p],\Diamond\bar p,\Diamond\Diamond p]}}{\Diamond\Box q,[\Box q,q,\bar p,\Diamond p],[[q,\bar p,\Diamond p],\Diamond\bar p,\Diamond\Diamond p]}}{\Diamond\Box q,[q,\bar p,\Diamond p],[[q,\bar p,\Diamond p],\Diamond\bar p,\Diamond\Diamond p]}}{\Diamond\Box q,[[q,\bar p,\Diamond p],\Diamond\bar p,\Diamond\Diamond p]}}{\Diamond\Box q,[[q],\Diamond\bar p,\Diamond\Diamond p]}}{\Diamond\Box q,[\Box q,\Diamond\bar p,\Diamond\Diamond p]}}{\Diamond\Box q,[\Diamond\bar p,\Diamond\Diamond p]}}{\Diamond\Box q\vee\Box(\Diamond\bar p\vee\Diamond\Diamond p)}
$$

Now we are essentially stuck. If we proceed, we can only obtain copies of existing sequent nodes, and none of them contains both p and $\bar{p}$. Thus, we will never be able to conclude with the id-rule.

However, if we allow the cut-rule, we can prove our formula as follows:

$$
\begin{array}{c}
{\scriptstyle\mathsf{cut}}\dfrac{{\scriptstyle\Box}\dfrac{{\scriptstyle\Diamond}\dfrac{{\scriptstyle 4^{[]}}\dfrac{{\scriptstyle\Diamond}\dfrac{{\scriptstyle\mathsf{id}}\dfrac{}{\Diamond\Box q,[[q],[q,p,\bar p]]}}{\Diamond\Box q,[[q],[q,p],\Diamond\bar p]}}{\Diamond\Box q,[[[q,p],q],\Diamond\bar p]}}{\Diamond\Box q,[[[q],q,\Diamond p],\Diamond\bar p]}}{\Diamond\Box q,[[\Box q,q,\Diamond p],\Diamond\bar p]} \qquad {\scriptstyle 4^{[]}}\dfrac{{\scriptstyle\Diamond}\dfrac{{\scriptstyle\Box}\dfrac{{\scriptstyle\Diamond}\dfrac{{\scriptstyle\mathsf{id}}\dfrac{}{[[q,\bar q],q,\Diamond p],[\Diamond\bar p]}}{[[q],\Diamond\bar q,q,\Diamond p],[\Diamond\bar p]}}{[\Box q,\Diamond\bar q,q,\Diamond p],[\Diamond\bar p]}}{\Diamond\Box q,[\Diamond\bar q,q,\Diamond p],[\Diamond\bar p]}}{\Diamond\Box q,[[\Diamond\bar q,q,\Diamond p],\Diamond\bar p]}}{\Diamond\Box q,[[q,\Diamond p],\Diamond\bar p]} \\
\Big\Vert\,{\scriptstyle\mathsf{NK}} \\
\Diamond\Box q\vee\Box(\Diamond\bar p\vee\Diamond\Diamond p)
\end{array}
\tag{B.1}
$$

Then, [3, Lemma 8] suggests to transform the branch on the right in (B.1) as shown on the left below:

$$
\begin{array}{c}
{\scriptstyle\mathsf{cut}}\dfrac{{\scriptstyle\Box}\dfrac{{\scriptstyle\Diamond}\dfrac{{\scriptstyle 4^{[]}}\dfrac{{\scriptstyle\Diamond}\dfrac{{\scriptstyle\mathsf{id}}\dfrac{}{\Diamond\Box q,[[q],[q,p,\bar p]]}}{\Diamond\Box q,[[q],[q,p],\Diamond\bar p]}}{\Diamond\Box q,[[[q,p],q],\Diamond\bar p]}}{\Diamond\Box q,[[[q],q,\Diamond p],\Diamond\bar p]}}{\Diamond\Box q,[[\Box q,q,\Diamond p],\Diamond\bar p]} \qquad {\scriptstyle 4^{\Diamond}}\dfrac{{\scriptstyle\Diamond}\dfrac{{\scriptstyle 4^{[]}}\dfrac{{\scriptstyle\Box}\dfrac{{\scriptstyle\Diamond}\dfrac{{\scriptstyle\mathsf{id}}\dfrac{}{[[q,\bar q],q,\Diamond p],[\Diamond\bar p]}}{[[q],\Diamond\bar q,q,\Diamond p],[\Diamond\bar p]}}{[\Box q,\Diamond\bar q,q,\Diamond p],[\Diamond\bar p]}}{[[\Box q,\Diamond\bar q,q,\Diamond p],\Diamond\bar p]}}{[\Diamond\Box q,[\Diamond\bar q,q,\Diamond p],\Diamond\bar p]}}{\Diamond\Box q,[[\Diamond\bar q,q,\Diamond p],\Diamond\bar p]}}{\Diamond\Box q,[[q,\Diamond p],\Diamond\bar p]} \\
\Big\Vert\,{\scriptstyle\mathsf{NK}} \\
\Diamond\Box q\vee\Box(\Diamond\bar p\vee\Diamond\Diamond p)
\end{array}
\quad\rightsquigarrow\quad
\begin{array}{c}
{\scriptstyle 4^{\Diamond}}\dfrac{{\scriptstyle\Diamond}\dfrac{{\scriptstyle\Box}\dfrac{{\scriptstyle\Diamond}\dfrac{{\scriptstyle 4^{[]}}\dfrac{{\scriptstyle\Diamond}\dfrac{{\scriptstyle\mathsf{id}}\dfrac{}{[[q],[q,p,\bar p]]}}{[[q],[q,p],\Diamond\bar p]}}{[[[q,p],q],\Diamond\bar p]}}{[[[q],q,\Diamond p],\Diamond\bar p]}}{[[\Box q,q,\Diamond p],\Diamond\bar p]}}{[\Diamond\Box q,[q,\Diamond p],\Diamond\bar p]}}{\Diamond\Box q,[[q,\Diamond p],\Diamond\bar p]} \\
\Big\Vert\,{\scriptstyle\mathsf{NK}} \\
\Diamond\Box q\vee\Box(\Diamond\bar p\vee\Diamond\Diamond p)
\end{array}
$$

However the $4^{\Diamond}$-rule does not apply to the cut-formula so it must be moved under the cut-rule as in a commutative case, and thus, Lemma 9 of [3] cannot be applied. This means that the proof of the reduction lemma (Lemma 10) in [3] is incorrect, and the instance of $4^{\Diamond}$ remains in the derivation after cut elimination, as shown on the right above.

A Hennessy-Milner Property for Many-Valued Modal Logics

Michel Marti[1]

Institute of Computer Science and Applied Mathematics, University of Bern
Neubrückstrasse 10, CH-3012 Bern, Switzerland
mmarti@iam.unibe.ch

George Metcalfe[2]

Mathematical Institute, University of Bern
Sidlerstrasse 5, CH-3012 Bern, Switzerland
george.metcalfe@math.unibe.ch

Abstract

A Hennessy-Milner property, relating modal equivalence and bisimulations, is defined for many-valued modal logics that combine a local semantics based on a complete MTL-chain (a linearly ordered commutative integral residuated lattice) with crisp Kripke frames. A necessary and sufficient algebraic condition is then provided for the class of image-finite models of these logics to admit the Hennessy-Milner property. Complete characterizations are obtained in the case of many-valued modal logics based on BL-chains (divisible MTL-chains) that are finite or have universe [0,1], including crisp Łukasiewicz, Gödel, and product modal logics.

Keywords: Modal Logics, Many-Valued Logics, Hennessy-Milner Property.

1 Introduction

Many-valued modal logics combine the Kripke semantics of modal logics with a local many-valued semantics to model epistemic, spatio-temporal, and other modalities in the presence of vagueness or uncertainty (including, e.g., fuzzy belief [19,14], fuzzy similarity measures [15], many-valued tense logics [10], and spatial reasoning with vague predicates [24]). As in the classical setting, fuzzy description logics may also be interpreted as many-valued multi-modal logics (see [25,18]). General approaches to many-valued modal logics are described in [12,13,23,5]. Here, for convenience, we follow [5] in assuming that the underlying many-valued algebras of the logics are complete MTL-chains: that is,

[1] Supported by Swiss National Science Foundation grants 200020_134740 and 200020_153169.
[2] Supported by Swiss National Science Foundation grant 20002_129507.

complete linearly ordered integral commutative residuated lattices. This framework spans, in particular, the families of Gödel and Łukasiewicz modal logics studied in [22,8,7] and [20], respectively. However, we restrict our attention in this paper to crisp many-valued modal logics where accessibility is a binary relation, leaving for future work, the case where accessibility is interpreted as a binary map from states to values of the algebra.

Theoretical studies of many-valued modal logics have concentrated to date mostly on issues of axiomatization, decidability, and complexity. Other topics from the rich theory of modal logics, such as first-order correspondence theory, canonical models, etc. have not as yet received much attention. In particular, the general question of the expressivity of many-valued modal logics has largely been ignored (although, see [2] for a coalgebraic approach to this topic). For classical modal logic, Van Benthem's theorem tells us that the modal logic K may be viewed as the bisimulation-invariant fragment of first-order logic, and it may be asked if similar results hold in the many-valued setting. A less demanding but still interesting question is whether analogues of the Hennessy-Milner property (modal equivalence coincides with bisimilarity) hold for image-finite models of many-valued modal logics. Modal equivalence between two states means in this context that each formula takes the same value in both states; the definition of a bisimulation matches the classical notion except that variables must take the same value in bisimilar states. Informally, our goal is to determine whether the language is expressive enough to distinguish image-finite models of many-valued modal logics.

More concretely, we define a (crisp) many-valued modal logic $\mathsf{K}(\mathbf{A})^{\mathsf{C}}$ based on one complete MTL-chain $\mathbf{A}$. The first main result of this paper is a necessary and sufficient algebraic condition on $\mathbf{A}$ for the class of image-finite models of $\mathsf{K}(\mathbf{A})^{\mathsf{C}}$ to admit the Hennessy-Milner property. We then obtain a complete classification when $\mathbf{A}$ is a divisible MTL-chain, called a BL-chain, that is finite or has universe $[0,1]$. In both cases, the property holds exactly when $\mathbf{A}$ is an MV-chain (a BL-chain where the negation is involutive) or the ordinal sum of two MV-chains. This means in particular that the class of image-finite models for (crisp) Łukasiewicz modal logics and the (crisp) three-valued Gödel modal logic admit the Hennessy-Milner property, but not (crisp) product modal logic or the (crisp) Gödel modal logics with more than three truth values.

Let us note finally that the approach to bisimulations and Hennessy-Milner properties for Heyting algebra based modal logics described by Eleftheriou et al. in [11] differs from the approach reported in this paper in several significant respects. Not only are (not necessarily linearly ordered) Heyting algebras considered, rather than the broad family of linearly ordered algebras investigated in this paper, but the Kripke frames are many-valued rather than crisp. This more general setting requires substantially different notions of bisimulation. Moreover, in order to obtain suitable Hennessy-Milner properties, it is assumed that the language contains constants for every element of the algebra, an assumption that would trivialize the approach taken here.

2 Many-Valued Modal Logics

For convenience, we follow [5] and restrict our attention to many-valued modal logics defined over *commutative integral residuated lattices*, algebraic structures

$$\mathbf{A} = \langle A, \wedge, \vee, \cdot, \to, \bot, \top \rangle$$

satisfying

(i) $\langle A, \wedge, \vee, \bot, \top \rangle$ is a bounded lattice where $a \leq b$ if and only if $a \wedge b = a$.

(ii) $\langle A, \cdot, \top \rangle$ is a commutative monoid.

(iii) $a \cdot b \leq c$ if and only if $a \leq b \to c$ for all $a, b, c \in A$.

We will be particularly interested in the case where $\mathbf{A}$ is both a *chain* (i.e., $\leq$ is a linear order on A) and *complete* (i.e., $\bigwedge B$ and $\bigvee B$ exist in A for all $B \subseteq A$). Such an algebra satisfies the prelinearity identity $\top \approx (x \to y) \vee (y \to x)$ and is called a complete *MTL-chain* (where MTL stands for monoidal t-norm logic).

Example 2.1 If the universe of $\mathbf{A}$ is the real unit interval $[0, 1]$, then the monoidal operation $\cdot$ is a t-norm with unit 1 and residual $\to$. Most notably, such algebras provide standard semantics for Łukasiewicz logic, Gödel logic, and product logic when $\cdot$ is the Łukasiewicz t-norm $\max(0, x+y-1)$, the minimum t-norm $\min(x, y)$, or the product t-norm xy (multiplication), respectively (see [17] for further details). Many-valued modal logics based on these and other algebras based on continuous t-norms will be considered in some detail in Section 5.

Our many-valued modal logics will be defined based on a language consisting of binary connectives $\to$, $\wedge$, $\vee$, constants $\bot$, $\top$, and unary (modal) connectives $\Box$ and $\Diamond$. The set of *formulas* $\mathrm{Fm}_{\Box\Diamond}$ of this language, with arbitrary members denoted $\varphi, \psi, \chi, \ldots$ is defined inductively over a fixed countably infinite set Var of (propositional) variables, denoted $p, q, \ldots$. We also denote the set of (purely) *propositional formulas* by Fm. *Subformulas* are defined in the usual way, and we let $\neg\varphi = \varphi \to \bot$, $\varphi \leftrightarrow \psi = (\varphi \to \psi) \wedge (\psi \to \varphi)$, $\varphi^0 = \top$, and $\varphi^{n+1} = \varphi \cdot \varphi^n$ for $n \in \mathbb{N}$. We fix the *length* $\ell(\varphi)$ of $\varphi \in \mathrm{Fm}_{\Box\Diamond}$ to be the number of symbols occurring in φ.

The many-valued modal logic $\mathsf{K}(\mathbf{A})^{\mathsf{C}}$ is defined over a fixed complete MTL-chain $\mathbf{A}$ as follows. A *(crisp) frame* is a pair $\langle W, R \rangle$ where W is a non-empty set of *states* and $R \subseteq W \times W$ is a binary *accessibility relation* on W.

A $\mathsf{K}(\mathbf{A})^{\mathsf{C}}$*-model* is a triple $\mathfrak{M} = \langle W, R, V \rangle$ where $\langle W, R \rangle$ is a frame and $V\colon \mathrm{Var} \times \mathrm{W} \to \mathrm{A}$ is a mapping, called a *valuation*. The valuation V is extended to $V\colon \mathrm{Fm}_{\Box\Diamond} \times \mathrm{W} \to \mathrm{A}$ by

$$\begin{aligned}
V(\bot, w) &= \bot & V(\top, w) &= \top \\
V(\varphi \wedge \psi, w) &= V(\varphi, w) \wedge V(\psi, w) & V(\varphi \vee \psi, w) &= V(\varphi, w) \vee V(\psi, w) \\
V(\varphi \cdot \psi, w) &= V(\varphi, w) \cdot V(\psi, w) & V(\varphi \to \psi, w) &= V(\varphi, w) \to V(\psi, w) \\
V(\Box\varphi, w) &= \bigwedge\{V(\varphi, v) : Rwv\} & V(\Diamond\varphi, w) &= \bigvee\{V(\varphi, v) : Rwv\}.
\end{aligned}$$

A formula $\varphi \in \mathrm{Fm}_{\Box\Diamond}$ is *valid* in a $\mathsf{K}(\mathbf{A})^{\mathsf{C}}$-model $\mathfrak{M} = \langle W, R, V\rangle$ if $V(\varphi, w) = \top$ for all $w \in W$. If φ is valid in all $\mathsf{K}(\mathbf{A})^{\mathsf{C}}$-models, then φ is said to be $\mathsf{K}(\mathbf{A})^{\mathsf{C}}$-*valid*. (Note, however, that $\mathsf{K}(\mathbf{A})^{\mathsf{C}}$-validity will not play any role in the remainder of this paper; we consider the values taken by formulas at states without giving $\top$ any special importance.)

Let us fix some useful notation. Given a frame $\langle W, R\rangle$, we define $R[w] = \{v \in W : Rwv\}$. We call a $\mathsf{K}(\mathbf{A})^{\mathsf{C}}$-model $\mathfrak{M} = \langle W, R, V\rangle$ a *tree-model* with *root* w and *height* n if $\langle W, R\rangle$ is a tree with root w and height n. We also write $\boldsymbol{a}$ for $a_1, \ldots, a_n \in A^n$ and given a propositional formula $\varphi(p_1, \ldots, p_n)$, we write $\varphi[\boldsymbol{a}]$ for the value of φ (understood as a term function) at $\boldsymbol{a}$ in $\mathbf{A}$. Given $\psi(p_1, \ldots, p_n) \in \mathrm{Fm}$ and $\varphi_1, \ldots, \varphi_n \in \mathrm{Fm}_{\Box\Diamond}$, the formula $\psi[\varphi_1/p_1, \ldots, \varphi_n/p_n]$ is obtained by replacing all occurrences of p_i in ψ with φ_i for $i \in \{1, \ldots, n\}$. The following useful lemma is then proved by a straightforward induction on formula length.

Lemma 2.2 *Let* $\psi(p_1, \ldots, p_n) \in \mathrm{Fm}$ *and* $\varphi_1, \ldots, \varphi_n \in \mathrm{Fm}_{\Box\Diamond}$. *Then for any* $\mathsf{K}(\mathbf{A})^{\mathsf{C}}$-*model* $\mathfrak{M} = \langle W, R, V\rangle$ *and* $w \in W$:

$$V(\psi[\varphi_1/p_1, \ldots, \varphi_n/p_n], w) = \psi[V(\varphi_1, w), \ldots, V(\varphi_n, w)].$$

As suggested by the superscript ^{C}, we can also define, more generally, $\mathsf{K}(\mathbf{A})$-models based on "$\mathbf{A}$-frames" where R is an "A-valued" accessibility relation $R\colon W \times W \to A$ and the above clauses for $\Box$ and $\Diamond$ are revised accordingly (see [5] for details). However, this requires also a significant revision of the definition of a bisimulation and such cases are therefore left for future work. (See also the concluding remarks.)

3 Modal Equivalence and Bisimulations

Let us fix again a complete MTL-chain $\mathbf{A}$ and consider two $\mathsf{K}(\mathbf{A})^{\mathsf{C}}$-models $\mathfrak{M} = \langle W, R, V\rangle$ and $\mathfrak{M}' = \langle W', R', V'\rangle$. We will say that $w \in W$ and $w' \in W'$ are *modally equivalent*, written $w \leftrightsquigarrow w'$, if $V(\varphi, w) = V'(\varphi, w')$ for all $\varphi \in \mathrm{Fm}_{\Box\Diamond}$.

A non-empty binary relation $Z \subseteq W \times W'$ will be called a *bisimulation* between $\mathfrak{M}$ and $\mathfrak{M}'$ if the following conditions are satisfied:

(i) If wZw', then $V(p, w) = V'(p, w')$ for all $p \in \mathrm{Var}$.

(ii) If wZw' and Rwv, then there exists $v' \in W'$ such that vZv' and $R'w'v'$ (the forth condition).

(iii) If wZw' and $R'w'v'$, then there exists $v \in W$ such that vZv' and Rwv (the back condition).

We say that $w \in W$ and $w' \in W'$ are *bisimilar*, written $w \equiv w'$, if there exists a bisimulation Z between $\mathfrak{M}$ and $\mathfrak{M}'$ such that wZw'.

Observe that the notions of modal equivalence and bisimulation defined here follow very closely the standard classical notions, the only distinguishing detail being that agreement of propositional variables in bisimilar states and formulas in modally equivalent states means that they take exactly the same values in these states. Note, moreover, that the proof that bisimilarity implies modal equivalence, is very similar to the classical proof (see, e.g., [4]).

Lemma 3.1 *Let $\mathfrak{M} = \langle W, R, V\rangle$ and $\mathfrak{M}' = \langle W', R', V'\rangle$ be $\mathsf{K}(\mathbf{A})^{\mathsf{C}}$-models. If $w \in W$ and $w' \in W'$ are bisimilar, then they are modally equivalent.*

Proof. We prove that for all $\varphi \in \mathrm{Fm}_{\Box\Diamond}$, $w \in W$, and $w' \in W'$, it holds that $w \equiv w'$ implies $V(\varphi, w) = V'(\varphi, w')$, proceeding by induction on $\ell(\varphi)$. For the case $\varphi \in \mathrm{Var}$, the claim follows directly from the definition of a bisimulation. The case where φ is a constant is immediate, and the cases of the propositional connectives follow immediately using the induction hypothesis.

Now let $\varphi = \Diamond\psi$, the case $\varphi = \Box\psi$ being very similar. Using $w \equiv w'$, it follows by the forth condition that for each $v \in R[w]$, there exists $v' \in R'[w']$ such that $v \equiv v'$ and, by the induction hypothesis, $V(\psi, v) = V'(\psi, v')$. So $V(\Diamond\varphi, w) \leq V'(\Diamond\varphi, w')$. But also by the back condition, for each $v' \in R'[w']$, there exists $v \in R[w]$ such that $v \equiv v'$ and, by the induction hypothesis, $V(\psi, v) = V'(\psi, v')$. So $V(\Diamond\psi, w) \geq V'(\Diamond\psi, w')$. Hence $V(\Diamond\psi, w) = V'(\Diamond\psi, w')$ as required. □

Of course modal equivalence does not in general imply the existence of a bisimulation even in the classical case. Rather we may consider certain classes of models for which this implication holds. Let us say that a class $\mathcal{K}$ of $\mathsf{K}(\mathbf{A})^{\mathsf{C}}$-models has the *Hennessy-Milner property* if for any models $\mathfrak{M} = \langle W, R, V\rangle$ and $\mathfrak{M}' = \langle W', R', V'\rangle$ in $\mathcal{K}$, whenever the states $w \in W$ and $w' \in W'$ are modally equivalent, they are bisimilar.

We call a $\mathsf{K}(\mathbf{A})^{\mathsf{C}}$-model *image-finite* if $R[w]$ is finite for each $w \in W$. A central aim of this paper will be to investigate when exactly (i.e., for which $\mathbf{A}$) the class of image-finite $\mathsf{K}(\mathbf{A})^{\mathsf{C}}$-models has the Hennessy-Milner property.

Example 3.2 Consider the family of (crisp) Gödel modal logics $\mathsf{K}(\mathbf{A})^{\mathsf{C}}$ where A is a complete subset of $[0,1]$ containing 0 and 1 and

$$\mathbf{A} = \langle A, \min, \max, \min, \rightarrow_{\mathsf{G}}, 0, 1\rangle$$

with $x \rightarrow_{\mathsf{G}} y = y$ if $x > y$ and 1 otherwise. Suppose that $|A| > 3$ where $0 < a < b < c$. Consider the $\mathsf{K}(\mathbf{A})^{\mathsf{C}}$-models $\mathfrak{M} = \langle W, R, V\rangle$ and $\mathfrak{M}' = \langle W', R', V'\rangle$ displayed in Fig. 1 with $W = \{w_0, w_1, w_2, w_3\}$ and $W' = \{v_0, v_1, v_2\}$, R and R' as indicated by the arrows, and V and V' with the given values of p and all other values 1. Then it is easily shown (e.g., by considering the non-equivalent one-variable formulas) that w_0 and v_0 are modally equivalent. However, they are clearly not bisimilar, as there is no state in W' corresponding to w_2. So the class of image-finite $\mathsf{K}(\mathbf{A})^{\mathsf{C}}$-models does not have the Hennessy-Milner property.

The standard proof that classical modal logic K has the Hennessy-Milner property for image-finite Kripke models proceeds (roughly) as follows (see [4]). Suppose for a contradiction that there are image-finite Kripke models $\mathfrak{M} = \langle W, R, V\rangle$ and $\mathfrak{M}' = \langle W', R', V'\rangle$ such that modal equivalence is not a bisimulation. Assume (without loss of generality) that the forth condition does not hold. Then there are $w, v \in W$ and $w' \in W'$ such that $w \leftrightsquigarrow w'$ and Rwv, but for each $v_i' \in R'[w'] = \{v_1', \ldots, v_n'\}$, there is a formula φ_i satisfying $V(\varphi_i, v) = 0$

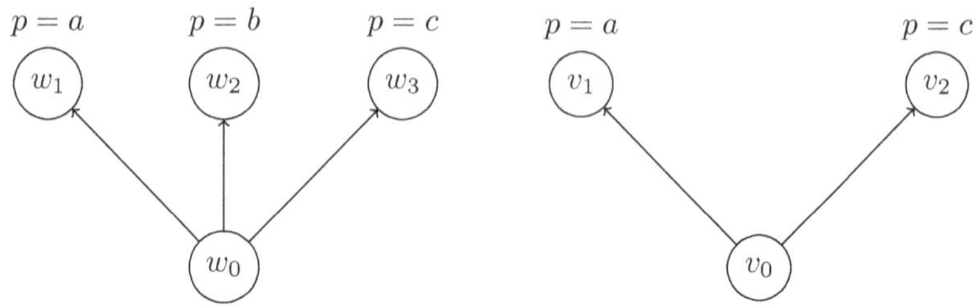

Fig. 1. Failure of the Hennessy-Milner property

but $V'(\varphi_i, v_i') = 1$. But then defining $\varphi = \Box(\varphi_1 \vee \ldots \vee \varphi_n)$, we have $V(\varphi, w) = 0$ and $V'(\varphi, w') = 1$, contradicting $w \leftrightsquigarrow w'$.

The above proof can be carried through for any logic $\mathsf{K}(\mathbf{A})^{\mathsf{C}}$ extended with additional constants for each $a \in A$. However, a more general condition suffices.

Lemma 3.3 *Suppose that for any distinct $a, b \in A$, there is a one-variable propositional formula $\psi_{a,b}(p) \in \mathrm{Fm}$ such that $\psi_{a,b}[a] = \top$ and $\psi_{a,b}[b] \neq \top$. Then the class of image-finite $\mathsf{K}(\mathbf{A})^{\mathsf{C}}$-models has the Hennessy-Milner property.*

Proof. We revisit the proof for the classical case. Suppose again for a contradiction that there are image-finite $\mathsf{K}(\mathbf{A})^{\mathsf{C}}$-models $\mathfrak{M} = \langle W, R, V\rangle$ and $\mathfrak{M}' = \langle W', R', V'\rangle$ such that modal equivalence is not a bisimulation. Assuming (without loss of generality) that the forth condition does not hold, there are $w, v \in W$ and $w' \in W'$ such that $w \leftrightsquigarrow w'$ and Rwv, but for each $v_i' \in R'[w'] = \{v_1', \ldots, v_n'\}$, there exists $\varphi_i \in \mathrm{Fm}_{\Box\Diamond}$ satisfying $V(\varphi_i, v) \neq V'(\varphi_i, v_i')$. Let $a_i = V'(\varphi_i, v_i')$ and $b_i = V(\varphi_i, v)$ for $i \in \{1, \ldots, n\}$. Then, by assumption, there is a one-variable propositional formula $\psi_i(p) \in \mathrm{Fm}$ for each $i \in \{1, \ldots, n\}$ such that $\psi_i[a_i] = \top$ and $\psi_i[b_i] \neq \top$. We define

$$\varphi = \Box(\psi_1[\varphi_1/p] \vee \ldots \vee \psi_n[\varphi_n/p]).$$

Then, using Lemma 2.2 and the linearity of $\mathbf{A}$,

$$\begin{aligned} V(\varphi, w) &\leq V(\psi_1[\varphi_1/p], v) \vee \ldots \vee V(\psi_n[\varphi_n/p], v) \\ &= \psi_1[b_1] \vee \ldots \vee \psi_n[b_n] \\ &< \top, \end{aligned}$$

but also

$$\begin{aligned} V'(\varphi, w') &= \textstyle\bigwedge_{i=1}^{n} V'(\psi_1[\varphi_1/p] \vee \ldots \vee \psi_n[\varphi_n/p], v_i') \\ &\geq \textstyle\bigwedge_{i=1}^{n} V'(\psi_i[\varphi_i/p], v_i') \\ &= \textstyle\bigwedge_{i=1}^{n} \psi_i[a_i] \\ &= \top. \end{aligned}$$

This contradicts $w \leftrightsquigarrow w'$. $\Box$

Example 3.4 Consider the three-valued Gödel modal logic $\mathsf{K}(\mathbf{G_3})^{\mathsf{C}}$ where

$$\mathbf{G_3} = \langle \{0, \tfrac{1}{2}, 1\}, \min, \max, \min, \to_{\mathsf{G}}, 0, 1\rangle$$

with $x \to_{\mathsf{G}} y = y$ if $x > y$ and 1 otherwise. Then we use the following formulas to distinguish values in $\{0, \frac{1}{2}, 1\}$:

$$\psi_{1,0} = \psi_{1,\frac{1}{2}} = (p \leftrightarrow \top), \qquad \psi_{0,\frac{1}{2}} = \psi_{0,1} = (p \leftrightarrow \bot),$$
$$\psi_{\frac{1}{2},0} = \neg(p \leftrightarrow \bot), \qquad \psi_{\frac{1}{2},1} = \neg(p \leftrightarrow \top).$$

So, by Lemma 3.3, the class of image-finite $\mathsf{K}(\mathbf{G_3})^{\mathsf{C}}$-models has the Hennessy-Milner property.

Finding formulas $\psi_{a,b}$, as described in Lemma 3.3, that distinguish distinct elements $a, b \in A$ is sufficient to establish that $\mathsf{K}(\mathbf{A})^{\mathsf{C}}$ has the Hennessy-Milner property for the class of image-finite $\mathsf{K}(\mathbf{A})^{\mathsf{C}}$-models, but does not appear to be necessary. To obtain a complete characterization, we introduce a more complicated but still purely algebraic condition for $\mathbf{A}$.

Let $\boldsymbol{a} \in A^n$ and $\boldsymbol{C} = (\boldsymbol{c}_1, \ldots, \boldsymbol{c}_n) \in A^{n \times n}$. We call a propositional formula $\psi(p_1, \ldots, p_n) \in \mathrm{Fm}$ an *$\boldsymbol{a}/\boldsymbol{C}$-distinguishing formula* if

$$\text{either} \quad \psi[\boldsymbol{a}] < \psi[\boldsymbol{c}_i] \ \text{ for } i \in \{1, \ldots, n\} \qquad \text{or} \quad \psi[\boldsymbol{a}] > \psi[\boldsymbol{c}_i] \ \text{ for } i = \{1, \ldots, n\}.$$

We say that $\mathbf{A}$ has the *distinguishing formula property* if for all $n \in \mathbb{N}$, $\boldsymbol{a} \in A^n$, and $\boldsymbol{C} = (\boldsymbol{c}_1, \ldots, \boldsymbol{c}_n) \in A^{n \times n}$ such that $\boldsymbol{a} \neq \boldsymbol{c}_i$ for $i \in \{1, \ldots, n\}$, there is an $\boldsymbol{a}/\boldsymbol{C}$-distinguishing formula.

In the next section, we establish the following characterization.

Theorem 3.5 *The following are equivalent for any complete MTL-chain* $\mathbf{A}$*:*

(1) *The class of image-finite* $\mathsf{K}(\mathbf{A})^{\mathsf{C}}$*-models has the Hennessy-Milner property.*

(2) $\mathbf{A}$ *has the distinguishing formula property.*

We note that this theorem also holds (with an almost identical proof) for the box and diamond fragments of $\mathsf{K}(\mathbf{A})^{\mathsf{C}}$, restricting the distinguishing formula property to the first (either) and second (or) condition, respectively.

4 Proof of Theorem 3.5

We first prove a useful lemma.

Lemma 4.1 *Let* $\mathfrak{M} = \langle W, R, V\rangle$ *and* $\mathfrak{M}' = \langle W', R', V'\rangle$ *be* $\mathsf{K}(\mathbf{A})^{\mathsf{C}}$*-tree-models of height one with roots* w *and* w'*, respectively. Suppose that* $V(p, w) = V'(p, w')$ *for all* $p \in \mathrm{Var}$ *and* $V(\Box\varphi, w) = V'(\Box\varphi, w')$ *and* $V(\Diamond\varphi, w) = V'(\Diamond\varphi, w')$ *for all* $\varphi \in \mathrm{Fm}$*. Then* w *and* w' *are modally equivalent.*

Proof. We prove the claim by induction on $\ell(\varphi)$. The base case is immediate and for the inductive step, the cases for the propositional connectives follow easily using the induction hypothesis. Suppose now that $\varphi = \Box\psi$, the case $\varphi = \Diamond\psi$ being very similar. Let ψ^* be the propositional formula obtained from ψ by replacing (iteratively) all subformulas of the form $\Box\psi'$ by $\top$ and all subformulas of the form $\Diamond\psi'$ by $\bot$. Then, using the fact that $\mathfrak{M}$ and $\mathfrak{M}'$ are tree-models of height one, it follows by an easy induction that $V(\psi, v) = V(\psi^*, v)$

for all $v \in W$ such that Rwv, and $V'(\psi, v') = V'(\psi^*, v')$ for all $v' \in W'$ such that $R'w'v'$. But then $V(\Box\psi, w) = V(\Box\psi^*, w) = V'(\Box\psi^*, w') = V'(\Box\psi, w')$ as required. □

Now we establish the implication (2) ⇒ (1) of Theorem 3.5. Assume that **A** has the distinguishing formula property and suppose for a contradiction that there are two image-finite $\mathsf{K}(\mathbf{A})^{\mathsf{C}}$-models $\mathfrak{M} = \langle W, R, V \rangle$ and $\mathfrak{M}' = \langle W', R', V' \rangle$ such that modal equivalence is not a bisimulation. If $w \leftrightsquigarrow w'$ for $w \in W$ and $w' \in W'$, then by definition $V(p, w) = V'(p, w')$ for all $p \in \mathrm{Var}$, so the back condition or forth condition must be violated.

Let us suppose that the forth condition fails, the back condition being very similar. Then there exist $w, v \in W$ and $w' \in W'$ such that

(i) $w \leftrightsquigarrow w'$ and Rwv.

(ii) No $v' \in W'$ satisfies both $R'w'v'$ and $v \leftrightsquigarrow v'$.

If $R'[w'] = \emptyset$, then consider $\Diamond\top$. We have $V(\Diamond\top, w) = \top$, but $V'(\Diamond\top, w') = \bigvee\emptyset = \bot$, which contradicts $w \leftrightsquigarrow w'$. Suppose then that $R'[w']$ is non-empty and (because of image-finiteness) finite, say $R'[w'] = \{v'_1, \ldots, v'_n\}$. So there are formulas φ_i such that $V(\varphi_i, v) \neq V'(\varphi_i, v'_i)$ for $i \in \{1, \ldots, n\}$.

We define $\boldsymbol{a} = (a_1, \ldots, a_n)$ and $\boldsymbol{C} = (\boldsymbol{c}_1, \ldots, \boldsymbol{c}_n)$ with $\boldsymbol{c}_i = c_{i,1}, \ldots, c_{i,n}$ as follows:

$$a_i = V(\varphi_i, v) \quad \text{and} \quad c_{i,j} = V'(\varphi_j, v'_i) \quad \text{for } 1 \leq i, j \leq n.$$

Note that $\boldsymbol{a} \neq \boldsymbol{c_i}$ for $i \in \{1, \ldots, n\}$ (because $a_i \neq c_{i,i}$).

By the distinguishing formula property, there exists an $\boldsymbol{a}/\boldsymbol{C}$-distinguishing propositional formula $\psi(p_1, \ldots, p_n) \in \mathrm{Fm}$. Suppose that

$$\psi[\boldsymbol{a}] < \psi[\boldsymbol{c}_i] \quad \text{for each } i \in \{1, \ldots, n\},$$

the case where $\psi[\boldsymbol{a}] > \psi[\boldsymbol{c}_i]$ for $i \in \{1, \ldots, n\}$ being very similar. Now define

$$\varphi = \Box\psi[\varphi_1/p_1, \ldots, \varphi_n/p_n].$$

Then using Lemma 2.2 and the linearity of **A**,

$$\begin{aligned} V(\varphi, w) &\leq V(\psi[\varphi_1/p_1, \ldots, \varphi_n/p_n], v) \\ &= \psi[V(\varphi_1, v), \ldots, V(\varphi_n, v)] \\ &= \psi[\boldsymbol{a}] \\ &< \textstyle\bigwedge_{i=1}^n \psi[\boldsymbol{c}_i] \\ &= \textstyle\bigwedge_{i=1}^n \psi[V'(\varphi_1, v'_i), \ldots, V'(\varphi_n, v'_i)] \\ &= \textstyle\bigwedge_{i=1}^n V'(\psi[\varphi_1/p_1, \ldots, \varphi_n/p_n], v'_i) \\ &= V'(\varphi, w'). \end{aligned}$$

This contradicts $w \leftrightsquigarrow w'$.

We turn our attention now to the implication (1) $\Rightarrow$ (2). Let us assume that the class of image-finite $\mathsf{K}(\mathbf{A})^{\mathsf{C}}$-models has the Hennessy-Milner property. Given $n \in \mathbb{N}$, $\boldsymbol{a} \in A^n$, and $\boldsymbol{C} = (\boldsymbol{c}_1, \ldots, \boldsymbol{c}_n) \in A^{n\times n}$ such that $\boldsymbol{a} \neq \boldsymbol{c}_i$ for $i \in \{1, \ldots, n\}$, we seek an $\boldsymbol{a}/\boldsymbol{C}$-distinguishing formula $\psi(p_1, \ldots, p_n)$.

Consider the two image-finite $\mathsf{K}(\mathbf{A})^{\mathsf{C}}$-models $\mathfrak{M} = \langle W, R, V\rangle$ and $\mathfrak{M}' = \langle W', R', V'\rangle$ satisfying

(i) $W = \{w, v_1, \ldots, v_n, v\}$ and $W' = \{w', v'_1, \ldots, v'_n\}$

(ii) $R = \{(w, v_i) : 1 \leq i \leq n\} \cup \{(w, v)\}$ and $R' = \{(w', v'_i) : 1 \leq i \leq n\}$

(iii) $V(p_j, v_i) = V'(p_j, v'_i) = c_{i,j}$ and $V(p_j, v) = a_j$ for $1 \leq i, j \leq n$ (all other variables and all variables at w take value $\top$).

We observe first that w and w' are not bisimilar: there is no state in W' that is accessible from w and agrees with v on all propositional variables because $\boldsymbol{a} \neq \boldsymbol{c}_i$ for $i \in \{1, \ldots, n\}$.

Hence by the Hennessy-Milner property for image-finite $\mathsf{K}(\mathbf{A})^{\mathsf{C}}$-models, w and w' are not modally equivalent. By Lemma 4.1, it follows that $V(\varphi, w) \neq V'(\varphi, w')$ for some formula $\varphi = \Box\psi$ or $\varphi = \Diamond\psi$ where $\psi \in \mathrm{Fm}$. Moreover, we may assume that ψ contains only the variables $p_1, \ldots, p_n$ as all other variables take the value $\top$.

Suppose that $\varphi = \Box\psi$ where $\psi(p_1, \ldots, p_n) \in \mathrm{Fm}$, the case $\varphi = \Diamond\psi$ being very similar. Clearly $V(\Box\psi, w) \leq V'(\Box\psi, w')$, so for each $i \in \{1, \ldots, n\}$,

$$\psi[\boldsymbol{a}] = V(\psi, v) = V(\Box\psi, w) < V'(\Box\psi, w') \leq V'(\psi, v'_i) = \psi[\boldsymbol{c}_i].$$

That is, ψ is the required $\boldsymbol{a}/\boldsymbol{C}$-distinguishing formula.

5 Divisible Chain Based Modal Logics

A commutative integral residuated lattice $\mathbf{A}$ is called *divisible* if for all $a, b \in A$ and $a \leq b$, there exists $c \in A$ such that $b \cdot c = a$. Equivalently, $\mathbf{A}$ is divisible if and only if $a \cdot (a \to b) = b \cdot (b \to a)$ for all $a, b \in A$. Divisible MTL-chains are also known (up to term equivalence) in the mathematical fuzzy logic literature as *BL-chains* (see [17,1,6]). In the case where $A = [0,1]$, the monoidal operation $\cdot$ is a *continuous t-norm* and $\mathbf{A}$ is called a *standard BL-chain*. For convenience, in this section, we exploit the fact that $a \wedge b = a \cdot (a \to b)$ and $a \vee b = ((a \to b) \to b) \wedge ((b \to a) \to a)$ for all $a, b \in A$, and restrict to the (traditional) language with operation symbols $\cdot, \to, \bot, \top$.

Our goal in this section is to obtain a complete characterization of the finite and standard BL-chains $\mathbf{A}$ for which the class of image-finite $\mathsf{K}(\mathbf{A})^{\mathsf{C}}$-models has the Hennessy-Milner property. First, we consider the special case where $\mathbf{A}$ is an *MV-chain*, defined here (up to term equivalence) as a BL-chain satisfying the involution property $\neg\neg a = a$ for all $a \in A$. Consider in particular the MV-chains

$$\mathbf{Ł_{n+1}} = \langle \{0, \tfrac{1}{n}, \ldots, \tfrac{n-1}{n}, 1\}, \cdot_{Ł}, \to_{Ł}, 0, 1\rangle \quad (n \in \mathbb{Z}^+)$$

$$\mathbf{Ł_\infty} = \langle [0,1], \cdot_{Ł}, \to_{Ł}, 0, 1\rangle$$

where $x \cdot_{\text{Ł}} y = \min(1, x + y - 1)$ and $x \to_{\text{Ł}} y = \max(0, 1 - x + y)$. Also, for convenience, let $\mathbf{Ł_1}$ be the trivial MV-chain with one element. Then every finite MV-chain $\mathbf{A}$ is isomorphic to $\mathbf{Ł}_{|A|}$ and every standard MV-chain is isomorphic to $\mathbf{Ł}_\infty$ (see [9] for proofs and a wealth of further information on MV-algebras).

We show here that for each $\alpha \in \mathbb{Z}^+ \cup \{\infty\}$, the class of image-finite $\mathsf{K}(\mathbf{Ł}_\alpha)^{\mathsf{C}}$-models has the Hennessy-Milner property, and hence that the same holds for any finite or standard MV-chain.

Example 5.1 Consider the algebra $\mathbf{Ł_3}$ for three-valued Łukasiewicz logic. We can distinguish values using the following formulas:

$$\psi_{1,0} = (p \leftrightarrow \top), \qquad \psi_{1,\frac{1}{2}} = (p \cdot p), \qquad \psi_{\frac{1}{2},0} = (\neg p \to p)$$

$$\psi_{0,1} = (p \leftrightarrow \bot), \qquad \psi_{\frac{1}{2},1} = (p \to \neg p), \qquad \psi_{0,\frac{1}{2}} = (\neg p \cdot \neg p).$$

Hence, by Lemma 3.3, the class of image-finite $\mathsf{K}(\mathbf{Ł_3})^{\mathsf{C}}$-models has the Hennessy-Milner property.

More generally, we may distinguish between rational values in $[0, 1]$ using unary *McNaughton functions*: that is, continuous functions $f \colon [0, 1] \to [0, 1]$ with the property that there exist linear functions $g_1, \dots, g_k$ with integer coefficients such that for any $x \in [0, 1]$, there is an $i \in \{1, \dots, k\}$ satisfying $f(x) = g_i(x)$.

Theorem 5.2 (McNaughton [21]) *The free one-generated MV-algebra is isomorphic to the algebra of unary McNaughton functions equipped with pointwise defined operations.*

Lemma 5.3 *For each $\alpha \in \mathbb{Z}^+ \cup \{\infty\}$, the class of image-finite $\mathsf{K}(\mathbf{Ł}_\alpha)^{\mathsf{C}}$-models has the Hennessy-Milner property.*

Proof. Consider distinct $a, b \in [0, 1]$. Then we can define a McNaughton function f such that $f(a) = 1$ and $f(b) = 0$. Suppose that $a < b$, the case $a > b$ being very similar. Then there exist $c, d \in \mathbb{Q}$ such that $a < c < d < b$ and we can define f to be 1 on the interval $[0, c]$, 0 on the interval $[d, 1]$, and linear on (c, d). Using Theorem 5.2, there exists a propositional formula $\psi_{a,b}(p)$ such that in the algebra $\mathbf{Ł}_\infty$, we have $\psi[x] = f(x)$ for all $x \in [0, 1]$. Because $\mathbf{Ł}_\alpha$ is a subalgebra of $\mathbf{Ł}_\infty$ for each $\alpha \in \mathbb{Z}^+ \cup \{\infty\}$, it follows that $\psi_{a,b}[a] = 1$ and $\psi_{a,b}[b] \neq 1$ whenever a, b are distinct elements of the algebra. Hence, by Lemma 3.3, $\mathbf{Ł}_\alpha$ has the Hennessy-Milner property. □

We now turn our attention to BL-chains, recalling a useful characterization of these algebras in terms of linearly ordered hoops (referring to [1,6] for further details). A *hoop* is an algebraic structure $\mathbf{H} = \langle H, \cdot, \to, \top \rangle$ such that $\langle H, \cdot, \top \rangle$ is a commutative monoid and for all $a, b, c \in H$:

(i) $a \to a = \top$.

(ii) $a \cdot (a \to b) = b \cdot (b \to a)$.

(iii) $a \to (b \to c) = (a \cdot b) \to c$.

Defining $a \leq b$ if and only if $a \to b = \top$ provides a semilattice order with meet operation $a \wedge b = a \cdot (a \to b)$ such that $\cdot$ and $\to$ are a residuated pair; i.e., $a \leq b \to c$ if and only if $a \cdot b \leq c$. If the order is linear, then $\mathbf{H}$ is called a *linearly ordered hoop* (*o-hoop* for short). Again, an o-hoop is *standard* if $H = [0, 1]$.

Now consider a linearly ordered set I with bottom element i_0 and suppose that $\mathbf{A}_i = \langle A_i, \cdot_i, \to_i, \top \rangle$ is a non-trivial o-hoop for each $i \in I$. Suppose, moreover, that $A_i \cap A_j = \{\top\}$ for distinct $i, j \in I$ and that $\mathbf{A}_{i_0}$ has a bottom element $\bot$. Then the *(bounded) ordinal sum* of $(\mathbf{A}_i)_{i \in I}$ is defined as

$$\bigoplus_{i \in I} \mathbf{A}_i = \langle \bigcup_{i \in I} A_i, \cdot, \to, \bot, \top \rangle$$

with operations

$$x \cdot y = \begin{cases} x \cdot_i y & \text{if } x, y \in A_i \\ x & \text{if } x \in A_i \setminus \{\top\},\, y \in A_j, \text{ and } i < j \\ y & \text{if } y \in A_i \setminus \{\top\},\, x \in A_j, \text{ and } i < j \end{cases}$$

$$x \to y = \begin{cases} \top & \text{if } x \in A_i \setminus \{\top\},\, y \in A_j, \text{ and } i < j \\ x \to_i y & \text{if } x, y \in A_i \\ y & \text{if } y \in A_i,\, x \in A_j, \text{ and } i < j. \end{cases}$$

We also write $\mathbf{A}_1 \oplus \ldots \oplus \mathbf{A}_n$ when $I = \{1, \ldots, n\}$ has the usual order.

Every ordinal sum of o-hoops is a BL-chain. Moreover, each "irreducible BL-chain" $\mathbf{A}$ (those that cannot be expressed as proper ordinal sums of o-hoops) are of exactly two types: either $\mathbf{A}$ is the *hoop reduct* $\langle A, \cdot, \to, \top \rangle$ of an MV-chain $\langle A, \cdot, \to, \bot, \top \rangle$, or $\mathbf{A}$ satisfies $a \to (a \cdot b) = b$ for all $a, b \in A$ and is called a *cancellative o-hoop*. Note that there are no finite cancellative o-hoops and that every standard cancellative o-hoop is isomorphic to the o-hoop

$$\mathbf{C} = \langle (0, 1], \cdot_{\mathsf{C}}, \to_{\mathsf{C}}, 1 \rangle$$

where $x \cdot_{\mathsf{C}} y = xy$ (multiplication) and $x \to_{\mathsf{C}} y$ is $\frac{y}{x}$ for $x > y$ and 1 otherwise.

Theorem 5.4 (Aglianò and Montagna [1]) *Every non-trivial BL-chain is the unique ordinal sum of a family of o-hoops each of which is either the hoop reduct of an MV-chain or a cancellative o-hoop.*

The next two lemmas identify ordinal sums $\mathbf{A}$ of o-hoops such that the class of image-finite $\mathsf{K}(\mathbf{A})^{\mathsf{C}}$-models does or does not have the Hennessy-Milner property. For convenience, we let $\mathbf{A}^{\mathbf{h}}$ denote the hoop reduct of a BL-chain $\mathbf{A}$.

Lemma 5.5 *Suppose that* $\mathbf{A}$ *is the ordinal sum of a family of (non-trivial) o-hoops* $(\mathbf{A}_i)_{i \in I}$. *If* $|I| \geq 3$ *or* $\mathbf{A}_i$ *is cancellative for some* $i \in I$, *then the class of image-finite* $\mathsf{K}(\mathbf{A})^{\mathsf{C}}$*-models does not have the Hennessy-Milner property.*

Proof. Suppose first that $\mathbf{A}_i$ is cancellative for some $i \in I$. Choose any element $c \in \mathbf{A}_i$ such that $c \neq \top$ and let $b = c \cdot c$ and $a = c \cdot c \cdot c$, noting that, by cancellativity, $a < b < c < \top$. Consider the $\mathsf{K}(\mathbf{A}_i)^{\mathsf{C}}$-models $\mathfrak{M} = \langle W, R, V\rangle$ and $\mathfrak{M}' = \langle W', R', V'\rangle$ displayed in Fig. 1 with $W = \{w_0, w_1, w_2, w_3\}$ and $W' = \{v_0, v_1, v_2\}$, R and R' as indicated by the arrows, and V and V' with the given values of p and all other values $\top$. Clearly, w_0 and v_0 are not bisimilar. To see that they are modally equivalent, consider any propositional formula $\psi(p) \in \mathrm{Fm}$. An easy induction on $\ell(\psi)$ establishes that ψ restricted to A_i is equivalent to $\bot$ or p^k for some $k \in \mathbb{N}$. But then $V(\Box\psi, w_0) = V'(\Box\psi, v_0)$ and $V(\Diamond\psi, w_0) = V'(\Diamond\psi, v_0)$ for all $\psi \in \mathrm{Fm}$. By Lemma 4.1, w_0 and v_0 are modally equivalent. So the class of image-finite $\mathsf{K}(\mathbf{A})^{\mathsf{C}}$-models does not have the Hennessy-Milner property.

Now consider the case where $|I| \geq 3$ and no $\mathbf{A}_i$ is cancellative for $i \in I$. Then, by Theorem 5.4, each $\mathbf{A}_i$ is an MV-chain and has a bottom element distinct from the top element of $\mathbf{A}$ for $i \in I$. But these bottom elements and also the top element are idempotents (i.e., $a \cdot a = a$) and hence $\mathbf{A}$ contains as a subalgebra, a Gödel algebra with more than three elements. The failure of the Hennessy-Milner property then follows exactly as described in Example 3.2. □

Lemma 5.5 establishes the failure of the Hennessy-Milner property for the class of image-finite models of the product modal logic $\mathsf{K}(\mathbf{P})^{\mathsf{C}}$ for $\mathbf{P} = \langle [0,1], \cdot_{\mathsf{C}}, \to_{\mathsf{C}}, 0, 1\rangle$ where $x \cdot_{\mathsf{C}} y = xy$ (multiplication) and $x \to_{\mathsf{C}} y$ is $\frac{y}{x}$ for $x > y$ and 1 otherwise. Just observe that $\mathbf{P}$ is isomorphic to the ordinal sum of $\mathbf{Ł}^{\mathbf{h}}_{\mathbf{2}}$ and $\mathbf{C}$.

Lemma 5.6 *Let $\mathbf{A} = \mathbf{Ł}^{\mathbf{h}}_{\alpha} \oplus \mathbf{Ł}^{\mathbf{h}}_{\beta}$ with $\alpha, \beta \in \mathbb{Z}^+ \cup \{\infty\}$. Then the class of image-finite $\mathsf{K}(\mathbf{A})^{\mathsf{C}}$-models has the Hennessy-Milner property.*

Proof. We use Lemma 3.3. Consider distinct elements $a, b \in A$. There are three cases. Suppose first that $a, b \in \mathrm{Ł}_{\alpha}$. We consider $\mathbf{Ł}_{\alpha}$ as an MV-chain with $\bot$ added to the language. As in the proof of Lemma 5.3, we obtain a distinguishing formula $\psi_{a,b}(p)$. Now suppose that $a, b \in \mathrm{Ł}_{\beta}$. We consider $\mathbf{Ł}_{\beta}$ as an MV-chain with $\bot'$ added to the language. Again we obtain a distinguishing formula $\psi_{a,b}(p)$ as in the proof of Lemma 5.3; however, in this case we must also replace $\bot'$ in $\psi_{a,b}(p)$ by p^k where $k \in \mathbb{Z}$ is sufficiently large that $p^k[a] = p^k[b] = \bot$. For the final case, consider $a \in \mathrm{Ł}_{\beta}$ and $b \in \mathrm{Ł}_{\alpha}$ (the converse is very similar). We fix $\psi_{a,b}(p) = \neg\neg p$ and observe that $\psi_{a,b}[a] = 1$ and $\psi_{a,b}[b] = b$. That is, $\psi_{a,b}$ is the required distinguishing formula. □

Combining these two lemmas with Theorem 5.4, we obtain the following characterization theorems for many-valued modal logics based on finite and standard BL-chains.

Theorem 5.7 *The following are equivalent for any finite BL-chain $\mathbf{A}$:*

(1) *The class of image-finite $\mathsf{K}(\mathbf{A})^{\mathsf{C}}$-models has the Hennessy-Milner property.*

(2) *$\mathbf{A}$ is isomorphic to $\mathbf{Ł}^{\mathbf{h}}_{\mathbf{n+1}}$ or $\mathbf{Ł}^{\mathbf{h}}_{\mathbf{n+1}} \oplus \mathbf{Ł}^{\mathbf{h}}_{\mathbf{m+1}}$ for some $m, n \in \mathbb{N}$.*

Theorem 5.8 *The following are equivalent for any standard BL-chain $\mathbf{A}$:*

(1) *The class of image-finite $\mathsf{K}(\mathbf{A})^{\mathsf{C}}$-models has the Hennessy-Milner property.*

(2) *$\mathbf{A}$ is isomorphic to $\mathbf{Ł}^{\mathbf{h}}$ or $\mathbf{Ł}^{\mathbf{h}} \oplus \mathbf{Ł}^{\mathbf{h}}$.*

Note that Lemma 5.5 and Theorem 5.4 actually tell us a little more: namely, that MV-chains and ordinal sums of two hoop reducts of MV-chains are the only candidates for BL-chains whose classes of image-finite models has the Hennessy-Milner property. We know by Lemmas 5.3 and 5.6 that this is the case if the MV-chain or hoop reducts of MV-chains are finite or standard, but not what happens for other (hoop reducts of) MV-chains.

6 Concluding Remarks

We have provided here a purely algebraic necessary and sufficient condition for the class of image-finite models of a many-valued modal logic based on an MTL-chain to have the Hennessy-Milner property. This result can be extended in a number of directions. Note first that from an algebraic perspective, there is no particular reason (other than convenience and readability) to limit our attention to residuated lattices. A similar characterization may be obtained for many-valued modal logics based on complete chains with extra operations, although for this general case, valid equations rather than formulas should be considered. In fact, alternative quantifiers (e.g., expressing the average truth value at accessible worlds) may also be considered by adding corresponding conditions to the characterization. A more challenging problem, as our proofs rely at certain crucial steps on linearity, is to extend the approach beyond chains to arbitrary complete lattices with additional operations. We may also seek to establish Hennessy-Milner properties for broader classes of models: in particular, for models admitting some version of the modal saturation property used in the classical setting.

The many-valued modal logics investigated in this paper are all based on crisp Kripke frames, but useful many-valued modal logics are also considered (e.g., in connection with fuzzy description logics) that are based on many-valued Kripke frames where the accessibility relation is replaced by a binary map from worlds to elements of the algebra. Extending our approach to this family of logics clearly requires alternative and appropriate definitions of bisimulation. Notably, the paper [11] considers two different notions of a bisimulation in the context of many-valued modal logics based on Heyting algebras extended with additional constants for elements of the algebra. Similarly, we expect that to obtain Hennessy-Milner properties for the models of many-valued modal logics with many-valued accessibility relations, we will also require constants or additional modal operators in the language.

Finally, let us remark that bisimulations and Hennessy-Milner properties have been investigated extensively in the setting of coalgebra and coalgebraic modal logics (see, e.g, [16]). We might therefore hope or expect that recent advances towards defining many-valued coalgebraic logics [3,2] will allow methods and theorems developed in the coalgebraic setting to be used also in studying many-valued modal logics.

References

[1] P. Aglianò and F. Montagna. Varieties of BL-algebras I: General properties. *Journal of Pure and Applied Algebra*, 181:105–129, 2003.

[2] M. Bílková and M. Dostal. Many-valued relation lifting and Moss' coalgebraic logic. In *Proceedings of CALCO 2013*, volume 8089 of *LNCS*, pages 66–79. Springer, 2013.

[3] M. Bílková, A. Kurz, D. Petrisan, and J. Velebil. Relation lifting, with an application to the many-valued cover modality. *Logical Methods in Computer Science*, 9(4), 2013.

[4] P. Blackburn, M. de Rijke, and Y. Venema. *Modal logic*. Cambridge University Press, Cambridge, 2001.

[5] F. Bou, F. Esteva, L. Godo, and R. Rodríguez. On the minimum many-valued logic over a finite residuated lattice. *Journal of Logic and Computation*, 21(5):739–790, 2011.

[6] M. Busaniche and F. Montagna. Hájek's BL and BL-algebras. In P. Cintula, P. Hájek, and C. Noguera, editors, *Handbook of Mathematical Fuzzy Logic Volume 1*, pages 355–447. College Publications, 2011.

[7] X. Caicedo, G. Metcalfe, R. Rodríguez, and Jonas Rogger. A finite model property for Gödel modal logics. In *Proceedings of WoLLIC 2013*, volume 8701 of *LNCS*, pages 226–237. Springer, 2013.

[8] X. Caicedo and R. Rodríguez. Bi-modal Gödel logic over [0,1]-valued Kripke frames. To appear in *Journal of Logic and Computation*, 2012.

[9] R. Cignoli, I. M. L. D'Ottaviano, and D. Mundici. *Algebraic Foundations of Many-Valued Reasoning*, volume 7 of *Trends in Logic*. Kluwer, Dordrecht, 1999.

[10] D. Diaconescu and G. Georgescu. Tense operators on MV-algebras and Łukasiewicz-Moisil algebras. *Fundamenta Informaticae*, 81(4):379–408, 2007.

[11] P. E. Eleftheriou, C. D. Koutras, and C. Nomikos. Notions of bisimulation for Heyting-valued modal languages. *Journal of Logic and Computation*, 22(2):213–235, 2012.

[12] M. C. Fitting. Many-valued modal logics. *Fundamenta Informaticae*, 15(3–4):235–254, 1991.

[13] M. C. Fitting. Many-valued modal logics II. *Fundamenta Informaticae*, 17:55–73, 1992.

[14] L. Godo, P. Hájek, and F. Esteva. A fuzzy modal logic for belief functions. *Fundamenta Informaticae*, 57(2–4):127–146, 2003.

[15] L. Godo and R. Rodríguez. A fuzzy modal logic for similarity reasoning. In *Fuzzy Logic and Soft Computing*, pages 33–48. Kluwer, 1999.

[16] R. Goldblatt. Final coalgebras and the Hennessy-Milner property. *Annals of Pure and Applied Logic*, 138(1–3):77–93, 2006.

[17] P. Hájek. *Metamathematics of Fuzzy Logic*. Kluwer, Dordrecht, 1998.

[18] P. Hájek. Making fuzzy description logic more general. *Fuzzy Sets and Systems*, 154(1):1–15, 2005.

[19] P. Hájek, D. Harmancová, F. Esteva, P. Garcia, and L. Godo. On modal logics for qualitative possibility in a fuzzy setting. In *Proceedings of UAI 1994*, pages 278–285, 1994.

[20] G. Hansoul and B. Teheux. Extending Łukasiewicz logics with a modality: Algebraic approach to relational semantics. *Studia Logica*, 101(3):505–545, 2013.

[21] R. McNaughton. A theorem about infinite-valued sentential logic. *Journal of Symbolic Logic*, 16(1):1–13, 1951.

[22] G. Metcalfe and N. Olivetti. Towards a proof theory of Gödel modal logics. *Logical Methods in Computer Science*, 7(2):1–27, 2011.

[23] G. Priest. Many-valued modal logics: a simple approach. *Review of Symbolic Logic*, 1:190–203, 2008.

[24] S. Schockaert, M. De Cock, and E. Kerre. Spatial reasoning in a fuzzy region connection calculus. *Artificial Intelligence*, 173(2):258–298, 2009.

[25] U. Straccia. Reasoning within fuzzy description logics. *Journal of Artificial Intelligence Research*, 14:137–166, 2001.

Recent Advances in Proof Systems for Modal Logic

Sara Negri

Department of Philosophy
P.O.Box 24 (Unioninkatu 40 A),
00014 University of Helsinki, Finland

In recent years, a number of challenges has been faced in the proof theory of modal and related logics, especially in relation to their applications in the widening field of philosophical logic. However, as discussed in [6], uniform methods for generating analytic calculi for these logics have been developed in full generality only for normal modal logics with geometric frame conditions. On the one hand, frame conditions beyond geometric implications are common in multimodal logics such as systems used in formal epistemology [4] and in intermediate logics [1]. On the other hand, a relaxation of the framework of normal modal logics imposed by Kripke semantics is essential for avoiding the presuppositions of a logically omniscient and perfect reasoner and for capturing a non-monotonic notion of conditionals. The talk will focus on the extension of the methodology of labelled sequent calculi in the aforementioned two directions.

First, the method of extension of sequent calculi with rules will be generalized to allow for *systems of rules* [8] to cover the class of generalised geometric implications, a class that includes the frame properties that correspond to formulas in the Sahlqvist fragment. Further, it will be shown how to turn arbitrary first-order frame conditions into a finite collection of geometric rules, through an appropriate conservative semidefinitional extension of the language. In particular, this "geometrization of first order logic" yields complete sequent calculi for all modal logics defined by first-order frame conditions [1].

Second, a revision of the standard relational semantics in the direction of a general topological semantics will be shown amenable of a formal proof-theoretical treatment, in particular through an analysis of Lewis conditional logic as a labelled sequent calculus based on similarity relations and indexed modalities [10]. At the same time, the various stages of the methodology will be illustrated: a process of abduction to obtain the specification of the semantic framework; the use of the guidelines of proof-theoretic semantics to obtain calculi with good structural properties [9]; a direct Tait-Schütte-Takeuti style completeness proof [5]; and finally, a syntactic counterpart of semantic filtration to extract finite countermodels and thereby to turn non-constructive completeness proofs into effective decision procedures [2,7].

References

[1] Dyckhoff, R. and Negri, S. Proof analysis in intermediate logics. *Archive for Mathematical Logic*, vol. 51, pp. 71–92, 2012.
[1] Dyckhoff, R. and Negri, S. Geometrization of first-order logic, ms., 2014.
[2] Garg, D., Genovese, V. and Negri, S. Countermodels from Sequent Calculi in Multi-Modal Logics. *Proceedings of LICS 2012*, IEEE Computer Society, pp. 315-324, 2012.
[3] Lewis, D. *Counterfactuals*. Blackwell, 1973.
[4] Maffezioli, P., Naibo, A. and Negri, S. The Church-Fitch knowability paradox in the light of structural proof theory. *Synthese*, vol. 190, pp. 2677-2716, 2013.
[5] Negri, S. Kripke completeness revisited. In G. Primiero and S. Rahman (eds), *Acts of Knowledge - History, Philosophy and Logic*, College Publications, 2009.
[6] Negri, S. Proof theory for modal logic. *Philosophy Compass*, vol. 6, pp. 523-538, 2011.
[7] Negri, S. Proofs and countermodels in non-classical logics. *Logica Universalis*, vol. 8, pp. 25-60, 2014
[8] Negri, S. Proof analysis beyond geometric theories: from rule systems to systems of rules. *Journal of Logic and Computation*, in press.
[9] Negri, S. and von Plato, J. Meaning in use. In H. Wansing (ed), *Dag Prawitz on Proofs and Meaning*, Trends in Logic, Springer, in press.
[10] Negri, S. and Sbardolini, G. Proof analysis for Lewis counterfactuals, ms., 2014.

A Duality for Distributive Unimodal Logic

Adam Přenosil [1]

Institute of Computer Science, Academy of Sciences of the Czech Republic
Pod vodárenskou věží 271/2, 182 00 Praha

Abstract

We introduce distributive unimodal logic as a modal logic of binary relations over posets which naturally generalizes the classical modal logic of binary relations over sets. The relational semantics of this logic is similar to the relational semantics of intuitionistic modal logic and positive modal logic, but it generalizes both of these by placing no restrictions on the accessibility relation. We introduce a corresponding quasivariety of distributive lattices with modal operators and prove a completeness theorem which embeds each such algebra in the complex algebra of its canonical modal frame. We then extend this embedding to a duality theorem which unifies and generalizes the duality theorems for intuitionistic modal logic obtained by A. Palmigiano and for positive modal logic obtained by S. Celani and A. Jansana. As a corollary to this duality theorem, we obtain a Hennessy-Milner theorem for bi-intuitionistic unimodal logic, which is the expansion of distributive unimodal logic by bi-intuitionistic connectives.

Keywords: Modal logic, distributive modal logic, intuitionistic modal logic, positive modal logic, bi-intuitionistic modal logic, duality theory.

1 Introduction

Suppose that we are given a semantically defined expansion of classical propositional logic, such as classical modal propositional logic, and we want to add some intuitionistic flavour to it. How could we go about doing that?

Let us adopt the simple perspective that a logic is given by a category of *frames* and a contravariant functor from the category of frames to some category of algebras which assigns a *complex algebra* to each frame. The consequence relation of the logic is identified with the quasiequational logic of this class of complex algebras. A completeness theorem then consists in axiomatizing the quasivariety generated by this class.

In a set-based semantics, the frames are sets possibly with additional structure and the complex algebras are expansions of the Boolean algebra of all subsets of a frame. In poset-based semantics, the frames are posets possibly with additional structure and the complex algebras are expansions of the dis-

[1] This research was supported by grant GAP202/10/1826 of the Czech Science Foundation.

tributive lattice of all *up*sets of a frame. Our task is thus to define a natural poset-based companion to a given set-based logic.

From an algebraic point of view, the natural thing to do is to drop the Boolean negation from the signature and consider the (quasi)variety generated by such reducts. This is the strategy followed by Dunn [5]. By contrast, our strategy will be to start on the semantic side.

We wish to extend an operation on the powerset of a set-based frame, say the binary operation $\circ$, to an operation on the set of all upsets of a poset-based frame with the same underlying set. There are two natural ways of doing this. One option is to define the poset-based operation as the *upper interior* of the set-based operation, that is, $u \in a \circ_+ b$ if and only if $v \in a \circ b$ for all $v \geq u$. The other option is to define the poset-based operation as the *upper closure* of $a \circ b$, that is, $u \in a \circ_- b$ if and only if $v \in a \circ b$ for some $v \leq u$.

The first of these alternatives is used in the semantics of intuitionistic logic: the intuitionistic implication $\rightarrow$ is defined as the upper interior of classical implication. The second alternative is used in bi-intuitionistic logic, introduced by Rauszer [11,12]. This logic expands intuitionistic logic by a *co-implication* connective, denoted as $a \succ\!\!- b$ here, such that $u \in a \succ\!\!- b$ if and only if there is some $v \leq u$ such that $v \notin a$ and $v \in b$. It is easy to see that for the lattice connectives $\wedge$ and $\vee$, the two extensions coincide.

The goal of the present paper is to investigate the poset-based companion of classical unimodal logic given by the connectives $\wedge$, $\vee$, $\top$, $\bot$, $\Box_+$, $\Diamond_-$ and its expansion by $\rightarrow$ and $\succ\!\!-$. We call this logic *distributive unimodal logic (DUML)* and we call the expansion *bi-intuitionistic unimodal logic (BiIUML)*. Our main result is a completeness theorem for DUML with respect to a suitable quasi-variety of modal algebras and its extension to a duality between these algebras and suitably topologized modal frames. In combination with known results, this also yields a completeness and duality theorems for BiIUML.

These results generalize and unify known completeness and duality results for intuitionistic modal logic and positive modal logic. We therefore briefly introduce these to provide some context for the present work. We also describe the relationship between the distributive unimodal logic presented here and the distributive modal logic of Gehrke et al. [8].

Intuitionistic modal logic (IML), introduced and axiomatized by Fischer Servi [6,7], expands intuitionistic propositional logic by a pair of modalities $\Box$ and $\Diamond$ which generalize the box and diamond modalities of classical modal logic. The frames for this logic are Kripke frames for intuitionistic propositional logic (that is, posets) equipped with a binary accessibility relation R required to satisfy the conditions $\geq \circ R \subseteq R \circ \geq$ and $R \circ \leq \subseteq \leq \circ R$. The box and diamond modalities are then defined asymmetrically: $u \in \Box A$ if and only if $u(\leq \circ R)v$ implies $v \in A$, whereas $u \in \Diamond A$ if and only if uRv for some $v \in A$. Observe that the box operator is persistent by definition, while the condition $\geq \circ R \subseteq R \circ \geq$ is needed to ensure the persistence of the diamond operator.

Unlike in classical modal logic, the box and diamond modalities of IML are not mutually interdefinable. Their interaction is captured by the axioms

$\Box(\varphi \to \psi) \to (\Diamond\varphi \to \Diamond\psi)$ and $(\Diamond\varphi \to \Box\psi) \to \Box(\varphi \to \psi)$. The completeness theorem for IML was later extended to a duality by Palmigiano [9]. For a (slightly dated) overview of research on IML, see [14] or [13].

Positive modal logic (PML), introduced and axiomatized by Dunn [5], is the negation-free fragment of classical modal logic. A poset-based [2] relational semantics for PML analogical to the semantics of IML was provided by Celani and Jansana [3], who formulated a Priestley-style duality for PML [4]. This semantics requires that the accessibility relation R satisfy the conditions $\geq \circ R \subseteq R \circ \geq$ and $\leq \circ R \subseteq R \circ \leq$, which ensure that both of the modal operators are local with respect to partial order.

Both IML and PML thus place non-trivial requirements on the accessibility relation, and furthermore that IML does not retain the symmetry between $\Box$ and $\Diamond$ present in classical modal logic. The present paper then answers the natural question: what is the modal logic of *semantically dual* box and diamond operators defined by *arbitrary* binary relations over posets?

Finally, a modal logic which does not fit the above semantic template of posets equipped with a single relation is distributive modal logic (DML) introduced by Gehrke et al. [8], which in addition to $\Box$ and $\Diamond$ contains primitive modal operators corresponding to the classical modalities $\Box\neg$ and $\Diamond\neg$. (We chose the name distributive *uni*modal logic precisely to differentiate the present approach from DML.) This logic has a poset-based relational semantics, which differs from the semantics of IML and PML in that each of the modal operators has *its own* accessibility relation, denoted $R_\Box$ and $R_\Diamond$ in the case of $\Box$ and $\Diamond$.

As regards, the relationship between DML and DUML, one can adopt either of the following positions. Either one can consider DML to be a fragment of a generalization of classical *multimodal* logic and consider DUML to be a generalization of classical *unimodal* logic, ideally with the proviso that the *full* language of DUML should include connectives corresponding to $\Box\neg$ and $\Diamond\neg$ which are not considered here. Alternatively, one can view DUML as a special case of DML for frames where $R_\Box$ and $R_\Diamond$ are uniquely determined by their intersection, that is, $\leq \circ (R_\Box \cap R_\Diamond) \circ \leq \subseteq R_\Box$ and $\geq \circ (R_\Box \cap R_\Diamond) \circ \geq \subseteq R_\Diamond$.

The outline of the paper is as follows. In Section 2, we introduce some basic notation and briefly overview known facts concerning the representation of distributive lattice and (bi-)Heyting algebras. In Sections 3 and 4, we introduce DUML via its relational semantics, define the corresponding quasivariety of modal algebras and prove the soundness of the algebraic semantics. In Section 5 defines the canonical frame of a modal algebra and obtain a completeness theorem which embeds each modal algebra in the complex algebra of its canonical frame. In Section 6, we use the completeness theorem to show that the class of all frames for DUML is not definable by modal quasiequations relative to the class of all frames for DML. In Section 7, we consider the conditions that IML and PML place on R and prove that in our setting they correspond to canon-

[2] Strictly speaking, their semantics is formulated in terms of pre-ordered sets rather than posets, but the distinction is irrelevant for our purposes.

ical equations. The known completeness and duality theorems for these logics are therefore covered by the completeness and duality theorems for DUML. Finally, in Section 8, we formulate a duality for DUML based on the bitopological duality for distributive lattices by Bezhanishvili et al. [1] and derive a Hennessy-Milner theorem for BiIUML as a corollary.

2 Preliminaries

By relations we will mean *binary* relations, and by distributive lattices and their homomorphisms we will mean *bounded* distributive lattices and bound-preserving homomorphisms. The category of posets and monotone functions will be denoted Pos and the category of distributive lattices and their homomorphisms will be denoted DLat.

A *Heyting algebra* is a distributive lattice expanded by a binary operation $\to$ such that $a \wedge b \leq c$ if and only if $b \leq a \to c$. Dually, a *co-Heyting algebra* is a distributive lattice expanded by binary operation $\succ\!-$ (called co-implication) such that $a \leq b \vee c$ if and only if $c \succ\!- a \leq b$. A *bi-Heyting algebra* is then both a Heyting algebra and a co-Heyting algebra. It is well known that all of these classes of algebras are varieties.

Let $(W, \leq)$ be a poset and R be a relation on W. The *opposite* of $(W, \leq)$ is the poset $(W, \geq)$. We denote the diagonal relation on W by Δ_W. Given $U \subseteq W$, $R_{|U}$ denotes the restriction of R to U, $R^{-1}[U] = \{w \in W | wRu$ for some $u \in U\}$ and $R^{-1}[u] = R^{-1}[\{u\}]$ for $u \in W$. By a *pre-order on* $(W, \leq)$ we mean a pre-order on the W which extends $\leq$.

The upward (downward) closure of $U \subseteq W$ will be denoted $U^{\uparrow}$ ($U^{\downarrow}$). We say that U is *convex* if $U^{\uparrow} \cap U^{\downarrow} \subseteq U$. We call $U^{\uparrow} \cap U^{\downarrow}$ the *convex closure* of U.

Given any relation $R \subseteq U \times V$, we use the abbreviations $R^{\uparrow} = \leq \circ R \circ \leq$ and $R^{\downarrow} = \geq \circ R \circ \geq$. This notation will be used often in the following. We call a relation *convex* if $R = R^{\uparrow} \cap R^{\downarrow}$.

Let $(U, \leq)$ and $(V, \sqsubseteq)$ be posets. A relation $R \subseteq U \times V$ is *monotone (antitone)* if it is an upset (downset) of $U^{op} \times V$. A *monotone relation pair* $(R_{\uparrow}, R_{\downarrow})$ between U and V is a pair consisting of a monotone relation $R_{\uparrow} \subseteq U \times V$ and an antitone relation $R_{\downarrow} \subseteq U \times V$.

The *kernel* of a monotone function $f : (U, \leq) \to (V, \sqsubseteq)$ is the pre-order $\leq_f$ on $(U, \leq)$ such that $u \leq_f v$ if and only if $f(u) \sqsubseteq f(v)$. A *monotone function pair* from U and V is a pair of monotone functions from U and V with a common codomain. The *kernel pair* of a monotone function pair $f : U \to W$, $g : V \to W$ is the pair consisting of a monotone relation $\sigma_{\uparrow} : U \times V$ such that $u\sigma_{\uparrow}v$ if and only if $f(u) \leq g(v)$ holds in W and an antitone relation $\sigma_{\downarrow} : U \times V$ such that $u\sigma_{\downarrow}v$ if and only if $f(u) \geq g(v)$ holds in W. The kernels of monotone functions from a poset W are precisely the pre-orders on W and the kernel pairs of monotone function pairs from U and V are precisely the monotone relation pairs between U and V.

We now formulate known representation theorems for distributive lattices and (bi-)Heyting algebras in the notation which we will later use to formulate the completeness theorem for distributive unimodal logic. Given a poset W, let

W^+ be the distributive lattice of all upsets of W (called the *complex algebra* of W), and given a monotone function $f : U \to V$, let f^+ be the homomorphism of distributive lattices $f^{-1} : V^+ \to U^+$. This assignment yields a functor $(-)^+ : \mathsf{Pos} \to \mathsf{DLat}$.

In the opposite direction, given a distributive lattice $\mathbf{A}$, let $\mathbf{A}_\bullet$ be the poset of all prime filters on $\mathbf{A}$ ordered by inclusion, and given a homomorphism of distributive lattices $h : \mathbf{A} \to \mathbf{B}$, let $h_\bullet$ be the monotone function $h^{-1} : \mathbf{B}_\bullet \to \mathbf{A}_\bullet$. This assignment again yields a functor $(-)_\bullet : \mathsf{DLat} \to \mathsf{Pos}$.

Given a distributive lattice $\mathbf{A}$, define the function $\eta_{\mathbf{A}} : \mathbf{A} \to (\mathbf{A}_\bullet)^+$ such that $\mathcal{U} \in \eta_{\mathbf{A}}(a)$ if and only if $a \in \mathcal{U}$. Then $\eta_{\mathbf{A}}$ is an embedding of distributive lattices. Each distributive lattice is thus a sublattice of the complex algebra of some poset, hence the semantically defined quasiequational logic of complex algebras of posets distributive lattices coincides with the algebraically defined quasiequational logic of distributive lattices.

These constructions extend to Heyting (bi-Heyting) algebras and posets with suitably defined morphisms. We say that a *Heyting morphism* of posets $f : (U, \leq) \to (V, \sqsubseteq)$ is a monotone function such that $f(u) \sqsubseteq v'$ implies that $u \leq v$ and $f(v) = v'$ for some $v \in U$, and a *bi-Heyting morphism* of posets is a function $f : (U, \leq) \to (V, \sqsubseteq)$ such that both $f : (U, \leq) \to (V, \sqsubseteq)$ and $f : (U, \geq) \to (V, \sqsupseteq)$ are Heyting morphisms of posets. The functors $(-)^+$ and $(-)_\bullet$ then extend to contravariant functors between the category of posets with Heyting (bi-Heyting) morphisms and the category of Heyting (bi-Heyting) algebras and $\eta_{\mathbf{A}}$ is an embedding of Heyting (bi-Heyting) algebras if $\mathbf{A}$ is a Heyting (bi-Heyting) algebra.

3 Modal frames

Let us start by introducing the relational semantics for distributive unimodal logic which was outlined in the introduction. We will define frames for this logic, their complex algebras and their p-morphisms. We then describe subframes, simulation pairs and bisimulations.

Definition 3.1 A *(modal) frame* $\mathcal{F} = (W, \leq, R)$ is a poset $(W, \leq)$ equipped with a convex relation R.

Recall that R is convex if $R^\uparrow \cap R^\downarrow \subseteq R$. It is easily seen from the definition of p-morphisms of modal frames given below, which only cares about $R^\uparrow$ and $R^\downarrow$, that a poset equipped with a non-convex relation R is in fact p-isomorphic to the same poset equipped with the convex closure of R. No loss of generality is thus involved in assuming that R is convex. This assumption will later allow us to obtain the right correspondence results for modal axioms.

The distributive lattice $(W, \leq)^+$ of upsets of a modal frame $\mathcal{F} = (W, \leq, R)$ can now be equipped with a box operator

$$\Box_R(A) = \{u \in X \mid \text{ if } uR^\uparrow v, \text{ then } v \in A\}$$

and with a diamond operator

$$\Diamond_R(A) = \{u \in X \mid uR^\downarrow v \text{ for some } v \in A\}.$$

More compactly, $\Diamond_R(A) = (R^{\downarrow})^{-1}[A]$ and $W \setminus \Box_R(A) = (R^{\uparrow})^{-1}[W \setminus A]$. The *complex algebra* of $\mathcal{F}$ is the expansion $\mathcal{F}^+$ of the distributive lattice $(W, \leq)^+$ by these two operators . The *Heyting (bi-Heyting) complex algebra* $\mathcal{F}^+_{\rightarrow}$ $(\mathcal{F}^+_{\rightarrow,\succ\!\!-})$ of $\mathcal{F}$ is the unique expansion of $\mathcal{F}^+$ by Heyting (bi-Heyting) connectives.

A *(Heyting, bi-Heyting) modal quasiequation* is a quasiequation in the language of distributive lattices expanded by $\Box$ and $\Diamond$ (and $\rightarrow$, and $\rightarrow$ and $\succ\!\!-$). A modal frame $\mathcal{F}$ is a *model* of a set of (Heyting, bi-Heyting) modal quasiequations Σ if $\mathcal{F}^+ \vDash \Sigma$. The class of all models of Σ will be denoted $\mathrm{Mod}(\Sigma)$.

For any class of modal frames K, let $\mathsf{K}^+ = \{\mathcal{F}^+ \mid \mathcal{F} \in \mathsf{K}\}$ and likewise for $\mathsf{K}^+_{\rightarrow}$ and $\mathsf{K}^+_{\rightarrow,\succ\!\!-}$. The *distributive unimodal logic* of K is then the quasiequational logic of K^+, the *intuitionistic unimodal logic* of K is the quasiequational logic of $\mathsf{K}^+_{\rightarrow}$, and the *bi-intuitionistic unimodal logic* of K is the quasiequational logic of $\mathsf{K}^+_{\rightarrow,\succ\!\!-}$. If K is not specified, we take it to be the class of *all* modal frames. These quasiequational logics can be translated into the form of sequent calculi via the correspondence between the sequent $\Gamma \vdash \Delta$ and the inequality $\bigwedge \Gamma \leq \bigvee \Delta$.

Definition 3.2 Given modal frames $\mathcal{F} = (U, \leq, R)$ and $\mathcal{G} = (V, \sqsubseteq, S)$, a *p-morphism* $f : \mathcal{F} \rightarrow \mathcal{G}$ is a monotone function $f : (U, \leq) \rightarrow (V, \sqsubseteq)$ such that

- $uR^{\uparrow}v$ implies $f(u)S^{\uparrow}f(v)$,
- $uR^{\downarrow}v$ implies $f(u)S^{\downarrow}f(v)$,
- $f(u)S^{\uparrow}v'$ implies $uR^{\uparrow}v$ for some $v \in U$ such that $f(v) \leq v'$,
- $f(u)S^{\downarrow}v'$ implies $uR^{\downarrow}v$ for some $v \in U$ such that $f(v) \geq v'$.

A *Heyting (bi-Heyting) p-morphism* of modal frames is a p-morphism of modal frames which is a Heyting (bi-Heyting) morphism of posets.

Proposition 3.3 *If $f : \mathcal{F} \rightarrow \mathcal{G}$ is a p-morphism of modal frames, then $f^+ = f^{-1} : \mathcal{G}^+ \rightarrow \mathcal{F}^+$ is a homomorphism of their complex algebras.*

Proof. The proof is standard. $\Box$

The category of modal frames and their p-morphisms will be denoted ModPos. We now describe the images and kernels of p-morphisms.

Definition 3.4 A *subframe* of a frame $\mathcal{F} = (W, \leq, R)$ is a frame $(U, \leq_{|U}, R_{|U})$ for some $U \subseteq W$ such that

- if $u \in U$ and $uR^{\uparrow}v \in W$, then $w \leq v$ for some $w \in U$,
- if $u \in U$ and $uR^{\downarrow}v \in W$, then $w \geq v$ for some $w \in U$.

A *Heyting subframe* of $\mathcal{F}$ is a frame $(U, \leq_{|U}, R_{|U})$ for some $U \subseteq W$ such that U is closed under $\leq$ and R and

- if $u \in U$ and $uR^{\downarrow}v \in W$, then $w \geq v$ for some $w \in U$.

A *bi-Heyting subframe* of $\mathcal{F}$ is a frame $(U, \leq_{|U}, R_{|U})$ for some $U \subseteq W$ such that U is closed under $\leq$, $\geq$ and R.

Proposition 3.5 *The (Heyting, bi-Heyting) subframes of a modal frame $\mathcal{F}$ are precisely the images of (Heyting, bi-Heyting) p-morphisms into $\mathcal{F}$.*

Proof. It suffices to inspect the definition of p-morphisms. $\square$

Definition 3.6 Let $\mathcal{F} = (U, \leq, R)$ and $\mathcal{G} = (V, \sqsubseteq, S)$ be modal frames. A *simulation pair* between modal frames $\mathcal{F} = (U, \leq, R)$ and $\mathcal{G} = (V, \sqsubseteq, S)$ is a monotone relation pair $(\sigma_\uparrow, \sigma_\downarrow)$ between $(U, \leq)$ and $(V, \sqsubseteq)$ such that

$$\sigma_\uparrow \circ S^\uparrow \subseteq R^\uparrow \circ \sigma_\uparrow,$$
$$\sigma_\downarrow \circ S^\downarrow \subseteq R^\downarrow \circ \sigma_\downarrow.$$

A *Heyting simulation pair* is a simulation pair such that

$$\sigma_\uparrow \circ \sqsubseteq \subseteq \leq \circ \sigma_\uparrow.$$

A *bi-Heyting simulation pair* is a Heyting simulation pair such that

$$\sigma_\downarrow \circ \sqsupseteq \subseteq \geq \circ \sigma_\downarrow.$$

A *bisimulation* is a convex relation σ such that $(\sigma^\uparrow, \sigma^\downarrow) = (\leq \circ \sigma \circ \leq, \geq \circ \sigma \circ \geq)$ is a bi-Heyting simulation pair.

Observe that bi-Heyting simulation pairs are precisely pairs of the form $(\sigma^\uparrow, \sigma^\downarrow)$ for some convex σ.

Proposition 3.7 *The (Heyting, bi-Heyting) simulation pairs between modal frames $\mathcal{F}$ and $\mathcal{G}$ are precisely the kernel pairs of (Heyting, bi-Heyting) p-morphism pairs from $\mathcal{F}$ and $\mathcal{G}$.*

Proof. The p-morphism pairs from $\mathcal{F}$ and $\mathcal{G}$ are in bijective correspondence with the p-morphisms from the naturally defined disjoint union of $\mathcal{F}$ and $\mathcal{G}$. The definition of (Heyting, bi-Heyting) simulation pairs is then just a repackaging of the definition of (Heyting, bi-Heyting) p-morphisms. $\square$

4 Modal algebras

Having described the relational semantics of distributive unimodal logic, we introduce the corresponding class of modal expansions of distributive lattices and show that it includes the complex algebras of modal frames.

Definition 4.1 A *box operator* on a distributive lattice $\mathbf{A}$ is a unary function $\Box : \mathbf{A} \to \mathbf{A}$ such that $\Box(a \wedge b) = \Box a \wedge \Box b$ and $\Box \top = \top$. A *diamond operator* on a distributive lattice $\mathbf{A}$ is a unary function $\Diamond : \mathbf{A} \to \mathbf{A}$ such that $\Diamond(a \vee b) = \Diamond a \vee \Diamond b$ and $\Diamond \bot = \bot$.

A *modal algebra* is an algebra $\mathbf{A} = (A, \wedge, \vee, \top, \bot, \Box, \Diamond)$ such that $\mathbf{A}_{\mathsf{DLat}} = (A, \wedge, \vee, \top, \bot)$ is a distributive lattice, $\Box$ is a box operator on $\mathbf{A}_{\mathsf{DLat}}$, $\Diamond$ is a diamond operator on $\mathbf{A}_{\mathsf{DLat}}$, and $\mathbf{A}$ satisfies the *positive modal law*

$$\Diamond b \leq \Box a \vee c \Rightarrow \Diamond b \leq \Diamond(a \wedge b) \vee c$$

and the *negative modal law*

$$\Diamond a \wedge c \leq \Box b \Rightarrow \Box(a \vee b) \wedge c \leq \Box b.$$

The *opposite* of $\mathbf{A}$ is the modal algebra $\mathbf{A}^{op} = (A, \vee, \wedge, \bot, \top, \Diamond, \Box)$.

A *Heyting (bi-Heyting) modal algebra* is an algebra which is both a modal algebra and a Heyting (bi-Heyting) algebra.

Modal algebras form a quasivariety. On bi-Heyting modal algebras, the positive and negative modal laws are equivalent to the equations

$$\Diamond(a \wedge b) \prec \Diamond b \leq \Box a \prec \Diamond b,$$
$$\Diamond a \to \Box b \leq \Box(a \vee b) \to \Box b,$$

hence bi-Heyting modal algebras in fact form a variety. The category of modal algebras and their homomorphisms will be denoted ModDLat.

Note that moving from $\mathbf{A}$ to $\mathbf{A}^{op}$ transforms the positive modal law into the negative one and vice versa. We will often implicitly appeal to this symmetry between boxes and diamonds to cut our proofs down to half.

Proposition 4.2 (Soundness) *Every complex algebra $\mathcal{F}^+$ is a modal algebra.*

Proof. It suffices to verify that the positive modal law holds in $\mathcal{F}^+$. Suppose that $\Diamond_R a \subseteq \Box_R b \cup C$ and $u \in \Diamond_R a$, $u \notin c$ for some $a, b, c \in \mathcal{F}^+$. Then there are some v, w such that $u \geq vRw$ and $w \in a$. But then $v \in \Diamond_R a$ and $v \notin c$, therefore $v \in \Box_R b$, $w \in a \cap b$, and $u \in \Diamond_R(a \cap b)$. $\Box$

5 Canonical frames

We now show that each modal algebra can be embedded in the complex algebra of its suitably defined canonical frame. This means that distributive unimodal logic is precisely the quasiequational logic of modal algebras. Since by [10] quasiequational consequence can be captured by a simple calculus, this result deserves to be called a completeness theorem.

To define the *canonical frame* $\mathbf{A}_\bullet$ of a modal algebra $\mathbf{A}$, we equip the poset of prime filters on $\mathbf{A}$ with the accessibility relation $R_\mathbf{A}$ such that

$$\mathcal{U} R_\mathbf{A} \mathcal{V} \text{ if and only if } \Box a \in \mathcal{U} \text{ implies } a \in \mathcal{V} \text{ and } a \in \mathcal{V} \text{ implies } \Diamond a \in \mathcal{U}.$$

We use the notation $R_\mathbf{A}^{\supseteq} = \supseteq \circ R_\mathbf{A} \circ \supseteq$ and $R_\mathbf{A}^{\subseteq} = \subseteq \circ R_\mathbf{A} \circ \subseteq$.

The algebra $(\mathbf{A}_\bullet)^+$ is called the *canonical extension* of the algebra $\mathbf{A}$. We say that a quasiequation σ is *canonical* if $\mathbf{A} \vDash \sigma$ implies $(\mathbf{A}_\bullet)^+ \vDash \sigma$. A set of quasiequations Σ is canonical if each $\sigma \in \Sigma$ is canonical. The canonicity of Σ then implies that Σ axiomatizes the logic of $\mathrm{Mod}(\Sigma)$.

We will need some basic constructions to prove a crucial lemma about prime filters on modal algebras. Given filters $\mathcal{U}$ and $\mathcal{V}$ on $\mathbf{A}$, we define the filter $\mathcal{U} \wedge \mathcal{V} = \{u \wedge v \in \mathbf{A} \mid u \in \mathcal{U} \text{ and } v \in \mathcal{V}\}$, the filter $\Diamond \mathcal{V} = \{a \in \mathbf{A} \mid \Diamond v \leq a \text{ for some } v \in \mathcal{V}\}$, the filter $\Box^{-1}\mathcal{V} = \{a \in \mathbf{A} \mid \Box a \in \mathcal{V}\}$, and the filter $\mathcal{U}_\downarrow(\mathcal{V}) = \{a \in \mathbf{A} \mid v \leq a \vee u_- \text{ for some } v \in \mathcal{V}, u_- \notin \mathcal{U}\}$. A $\mathcal{U}_\downarrow$*-filter* is then a filter $\mathcal{V}$ such that $\mathcal{U}_\downarrow(\mathcal{V}) = \mathcal{V}$. Observe that a directed union of $\mathcal{U}_\downarrow$-filters is again a $\mathcal{U}_\downarrow$-filter.

Lemma 5.1 *Let $\mathcal{U}$ be a prime filter and $\mathcal{V}$ be a filter on $\mathbf{A}$. If $\Diamond \mathcal{V} \subseteq \mathcal{U}$, then $\mathcal{U} R_\mathbf{A}^{\supseteq} \mathcal{V}'$ for some prime $\mathcal{V}' \supseteq \mathcal{V}$.*

Proof. We need to find suitable prime filters $\mathcal{U}'$ and $\mathcal{V}'$ such that $\mathcal{U} \supseteq \mathcal{U}' R_{\mathbf{A}} \mathcal{V}' \supseteq \mathcal{V}$. For $\mathcal{V}'$, we use Zorn's lemma to take a maximal filter extending $\mathcal{V}$ such that $\Diamond\mathcal{V}' \subseteq \mathcal{U}$. Such a filter is prime: if $a, b \notin \mathcal{V}'$, then there are some $v_1', v_2' \in \mathcal{V}'$ such that $\Diamond(a \wedge v_1'), \Diamond(b \wedge v_2') \notin \mathcal{U}$, hence $\Diamond(a \wedge v'), \Diamond(b \wedge v') \notin \mathcal{U}$ for $v' = v_1' \wedge v_2'$, $\Diamond(a \wedge v') \vee \Diamond(b \wedge v') = \Diamond((a \vee b) \wedge (v_1' \wedge v_2')) \notin \mathcal{U}$ and $a \vee b \notin \mathcal{V}'$.

The inclusion $\Diamond(\mathcal{V}' \wedge \Box^{-1}\mathcal{U}_{\downarrow}(\Diamond\mathcal{V}')) \subseteq \mathcal{U}$ now holds: if $v_1' \in \mathcal{V}'$ and $\Diamond v_2' \leq \Box a \vee u^-$ for some $v_2' \in \mathcal{V}'$, $u^- \notin \mathcal{U}$, then $\Diamond(v_1' \wedge v_2') \leq \Diamond(a \wedge v_1' \wedge v_2') \vee u^-$, hence $\Diamond(a \wedge v_1' \wedge v_2') \in \mathcal{U}$. Since $\mathcal{V}'$ was chosen to be a maximal filter such that $\Diamond\mathcal{V}' \subseteq \mathcal{U}$, it follows that in fact $\Box^{-1}\mathcal{U}_{\downarrow}(\Diamond\mathcal{V}') \subseteq \mathcal{V}'$.

It now suffices to use Zorn's lemma to extend $\mathcal{U}_{\downarrow}(\Diamond\mathcal{V}')$ to a maximal $\mathcal{U}_{\downarrow}$-filter $\mathcal{U}'$ such that $\Box^{-1}\mathcal{U}' \subseteq \mathcal{V}'$. Such a filter is prime: if $a, b \notin \mathcal{U}'$, then $a \wedge u' \leq \Box v_1^-$ and $b \wedge u' \leq \Box v_2^-$ for some $v_1^-, v_2^- \notin \mathcal{V}'$, $u' \in \mathcal{U}'$, hence $(a \vee b) \wedge u' \leq \Box v_1^- \vee \Box v_2^- \leq \Box(v_1^- \vee v_2^-)$. Since $\mathcal{V}'$ is prime, $v_1^- \vee v_2^- \notin \mathcal{V}'$, thus $a \vee b \notin \mathcal{U}'$. $\Box$

Corollary 5.2 *$\mathcal{U} R_{\mathbf{A}}^{\supseteq} \mathcal{V}$ if and only if $a \in \mathcal{V}$ implies $\Diamond a \in \mathcal{U}$. $\mathcal{U} R_{\mathbf{A}}^{\subseteq} \mathcal{V}$ if and only if $\Box a \in \mathcal{U}$ implies $a \in \mathcal{V}$.*

Embedding a modal algebra into the complex algebra of its canonical frame is now straightforward. Recall that the function $\eta_{\mathbf{A}} : \mathbf{A} \to (\mathbf{A}_\bullet)^+$ such that $\mathcal{U} \in \eta_{\mathbf{A}}(a)$ if and only if $a \in \mathcal{U}$ for each prime filter $\mathcal{U}$ on $\mathbf{A}$ is an embedding of distributive lattices.

Theorem 5.3 (Completeness) *For every modal algebra $\mathbf{A}$, $\mathbf{A}_\bullet$ is a modal frame and $\eta_{\mathbf{A}} : \mathbf{A} \to (\mathbf{A}_\bullet)^+$ is an embedding of modal algebras.*

Proof. The convexity of $R_{\mathbf{A}}$ follows from Corollary 5.2. It thus suffices to prove that $\eta_{\mathbf{A}}$ preserves diamonds. It is clear that $\Diamond_{R_{\mathbf{A}}} \eta_{\mathbf{A}}(a) \subseteq \eta_{\mathbf{A}}(\Diamond a)$ by the definition of $R_{\mathbf{A}}$. The opposite inclusion is precisely Lemma 5.1. $\Box$

Lemma 5.1 also shows that the assignment $(-)_\bullet$ in fact extends to a functor $(-)_\bullet : \mathsf{ModDLat} \to \mathsf{ModPos}$. Recall that for any homomorphism of distributive lattices $h : \mathbf{A} \to \mathbf{B}$, we define the monotone function $h_\bullet : \mathbf{B}_\bullet \to \mathbf{A}_\bullet$ as h^{-1}.

Proposition 5.4 *If $h : \mathbf{A} \to \mathbf{B}$ is a homomorphism of modal algebras, then $h_\bullet : \mathbf{B}_\bullet \to \mathbf{A}_\bullet$ is a p-morphism of modal frames.*

Proof. We only verify that $\mathcal{U} R_{\mathbf{B}}^{\supseteq} \mathcal{V}$ implies $h_\bullet(\mathcal{U}) R_{\mathbf{A}}^{\supseteq} h_\bullet(\mathcal{V})$ and that $h_\bullet(\mathcal{U}) R_{\mathbf{A}}^{\supseteq} \mathcal{V}'$ implies $\mathcal{U} R_{\mathbf{B}}^{\supseteq} \mathcal{V}$ for some prime filter $\mathcal{V}$ on $\mathbf{B}$ such that $h_\bullet(\mathcal{V}) \supseteq \mathcal{V}'$.

Suppose that $\mathcal{U}_{\mathbf{B}} R_{\mathbf{B}} \mathcal{V}_{\mathbf{B}}$. If $a \in h^{-1}[\mathcal{V}_{\mathbf{B}}]$, then $h(a) \in \mathcal{V}_{\mathbf{B}}$, $\Diamond h(a) = h(\Diamond a) \in \mathcal{U}_{\mathbf{B}}$, $\Diamond a \in h^{-1}[\mathcal{U}_{\mathbf{B}}]$ and dually for $\Box$. Therefore $h^{-1}[\mathcal{U}_{\mathbf{B}}] R_{\mathbf{A}} h^{-1}[\mathcal{V}_{\mathbf{B}}]$.

Now suppose that $h^{-1}[\mathcal{U}_{\mathbf{B}}] R_{\mathbf{A}}^{\supseteq} \mathcal{V}_{\mathbf{A}}$. Then $\Diamond\mathcal{V}_{\mathbf{A}} \subseteq h^{-1}[\mathcal{U}_{\mathbf{B}}]$, hence $\Diamond h[\mathcal{V}_{\mathbf{A}}] \subseteq h[\Diamond\mathcal{V}_{\mathbf{A}}] \subseteq \mathcal{U}_{\mathbf{B}}$. By Lemma 5.1, there is a prime filter $\mathcal{W}_{\mathbf{B}} \supseteq h[\mathcal{V}_{\mathbf{A}}]$ such that $\mathcal{U}_{\mathbf{B}} R_{\mathbf{B}}^{\supseteq} \mathcal{W}_{\mathbf{B}}$. Clearly $h^{-1}[\mathcal{W}_{\mathbf{B}}] \supseteq \mathcal{V}_{\mathbf{A}}$ for any such $\mathcal{W}_{\mathbf{B}}$. $\Box$

Theorem 5.3 and Proposition 5.4 extend to (bi-)Heyting modal algebras and (bi-)Heyting p-morphisms: if $\mathbf{A}$ and $\mathbf{B}$ are (bi-)Heyting algebras and $h : \mathbf{A} \to \mathbf{B}$ is a homomorphism of (bi-)Heyting algebras, then $\eta_{\mathbf{A}}$ is in fact an embedding of (bi-)Heyting algebras and $h_\bullet$ is a (bi-)Heyting morphism of posets.

6 Relationship with distributive modal logic

As a corollary to the completeness theorem, we can show that the class of all frames for DML such that $R_\Diamond$ and $R_\Box$ are generated by the same relation is not definable by (bi-Heyting) modal quasiequations. Recall that a frame for DML was introduced by Gehrke et al. [8] as a poset $(W, \leq)$ equipped with a pair of accessibility relations $R_\Diamond$ and $R_\Box$ such that $\geq \circ R_\Diamond \circ \geq \subseteq R_\Diamond$ and $\leq \circ R_\Box \circ \leq$.[3] The box (diamond) operator of DML is defined in terms of $R_\Box$ ($R_\Diamond$) precisely as in classical modal logic. We say that $R_\Diamond$ and $R_\Box$ are generated by the same underlying relation if and only if $(R_\Diamond \cap R_\Box)^\uparrow = R_\Diamond$ and $(R_\Box \cap R_\Diamond)^\downarrow = R_\Box$.

Proposition 6.1 *The positive modal law $\Diamond b \leq \Box a \vee c \Rightarrow \Diamond b \leq \Diamond(a \wedge b) \vee c$ holds in a frame for DML if and only if for each $uR_\Diamond v$ there are $u' \leq u$ and $v' \geq v$ such that $uR_\Diamond v'$, $u'R_\Diamond v$ and $u'R_\Box v'$.*

Proof. We only show the harder (left-to-right) direction. Suppose that $uR_\Diamond v$ but there are no $u' \leq u$ and $v' \geq v$ such that $u'R_\Diamond v$, $uR_\Diamond v'$ and $u'R_\Box v'$. Then let $w \in b$ if and only if $w \geq v$, let $w \in a$ if and only if there is some $u' \leq u$ such that $u'R_\Diamond v$ and $u'R_\Box w$, and let $w \notin c$ if and only if $w \leq u$. It follows that $\Diamond b \leq \Box a \vee c$, but $u \in \Diamond b$, $u \notin \Diamond(a \wedge b)$ and $u \notin c$. □

Proposition 6.2 *There is no set of (bi-Heyting) modal quasiequations Σ such that Σ holds in a frame for DML if and only if $R_\Diamond$ and $R_\Box$ are generated by the same underlying relation.*

Proof. By Theorem 5.3, any such Σ is a quasiequational consequence of the axioms of modal algebras. It therefore suffices to build a frame for DML where $R_\Diamond$ and $R_\Box$ are not generated by the same underlying relation but which satisfies the relational conditions of Proposition 6.1 and its dual. This can be done in a brute-force way.

Let $\mathcal{F} = \mathcal{F}_0$ be any frame for DML which at least contains some pair of points connected by some accessibility relation. Let $\mathcal{F}_{i+1}$ be the frame obtained from $\mathcal{F}_i$ by adding for each $uR_\Diamond v$ a pair of points u', v' which satisfy exactly the condition of Proposition 6.1 and dually for each $uR_\Box v$. Finally, let $\mathcal{F}_\omega = \bigcup_{i \in \omega} \mathcal{F}_i$. The frame $\mathcal{F}_\omega$ satisfies the condition of Proposition 6.1 by construction and it is easy to see that $u(R_\Diamond \cap R_\Box)v$ in $\mathcal{F}_\omega$ if and only if $u(R_\Diamond \cap R_\Box)v$ already in $\mathcal{F}$. The relations $R_\Diamond$ and $R_\Box$ on $\mathcal{F}_\omega$ are therefore *never* generated by the same underlying relation. □

7 Modal locality conditions

We have obtained a completeness via canonicity theorem for the logic of modal frames with arbitrary (without loss of generality convex) relations. Let us now investigate what happens when we impose some conditions relating $\leq$ and R.

[3] They in fact use the opposite order convention, that is, $\leq \circ R_\Diamond \circ \leq \subseteq R_\Diamond$.

In particular, we consider the following conditions:

$$\geq \circ R \subseteq R \circ \geq$$
$$R \circ \leq \subseteq \leq \circ R$$
$$\leq \circ R \subseteq R \circ \leq$$
$$R \circ \geq \subseteq \geq \circ R$$

The first two of these are exactly the conditions that IML imposes on R, while PML requires the first and third conditions. The following observation made already by Gehrke et al. [8] justifies calling them *locality conditions.*

Proposition 7.1 *A frame satisfies the condition* $\geq \circ R \subseteq R \circ \geq$ *if and only if* $\Diamond_R(A) = \{u \in W \mid uRv \text{ for some } v \in A\}$. *A frame satisfies the condition* $\leq \circ R \subseteq R \circ \leq$ *if and only if* $\Box_R(A) = \{u \in W \mid uRv \text{ implies } v \in A\}$.

The other two conditions can be seen as locality conditions for the backward-looking box and diamond operators which we are not considered in this paper.

In the present framework, these classes of modal frames can be axiomatized by canonical modal equations. The completeness and duality theorems proved here therefore generalize known completeness and duality theorems established for IML by Fischer-Servi [6] and Palmigiano [9] and for PML by Dunn [5] and Celani and Jansana [3].

In the propositions below, we use the notation $\mathcal{F} \vDash \geq \circ R \subseteq R \circ \geq$ to mean that the inclusion holds in $\mathcal{F}$.

Proposition 7.2 $\mathcal{F} \vDash \geq \circ R \subseteq R \circ \geq$ *if and only if* $\mathcal{F}^+ \vDash \Box a \wedge \Diamond b \leq \Diamond(a \wedge b)$.

Proof. If $u \geq vRw$ but uRx implies $x \not\geq w$, let $y \in A$ if and only if $uR^{\uparrow}y$ and let $y \in B$ if and only if $y \geq w$. Then $u \in \Box A$ and $u \in \Diamond B$. If $u \in \Diamond(A \cap B)$, then there is some $z \geq w$ such that $u(\geq \circ R)z$ and $u(\leq \circ R \circ \leq)z$, hence uRz by the convexity of R, contradicting the assumption that uRz implies $z \not\geq w$.

Vice versa, if $u \in \Box A$ and $u \in \Diamond B$, then $u(\geq \circ R)v$ for some $v \in B$, hence $uRw \geq v$ for some w. But then $w \in A$ and $w \in B$, hence $u \in \Diamond(A \cap B)$. □

Proposition 7.3 $\mathcal{F} \vDash R\circ\leq \subseteq \leq\circ R$ *if and only if* $\mathcal{F}^+ \vDash \Diamond a \to \Box b \leq \Box(a \to b)$.

Proof. If $uRv \leq w$ but xRw implies $x \not\geq u$, let $y \in A$ if and only if $y \geq w$ and $y \in B$ if and only if $y \not\leq w$. Clearly $u \notin \Box(A \to B)$. But if $z \in \Diamond A$ and $z \notin \Box B$ for some $z \geq u$, then $z(\geq \circ R \circ \geq)w$ and $z(\leq \circ R \circ \leq)w$, hence zRw by the convexity of R, contradicting the assumption that zRw implies $z \not\geq u$.

Vice versa, if $u \notin \Box(A \to B)$, then $u(R \circ \leq)v$ for some $v \in A$, $v \notin B$, hence $u \leq wRv$ for some w. But then $w \in \Diamond A$, $w \notin \Box B$, hence $u \notin \Diamond A \to \Box B$. □

Dualizing these two propositions yields the following.

Proposition 7.4 $\mathcal{F} \vDash \leq \circ R \subseteq R \circ \leq$ *if and only if* $\mathcal{F}^+ \vDash \Box(a \vee b) \leq \Box a \vee \Diamond b$.

Proposition 7.5 $\mathcal{F} \vDash R\circ\geq \subseteq \geq\circ R$ *if and only if* $\mathcal{F}^+ \vDash \Diamond(a \succ\!\!- b) \leq \Box a \succ\!\!- \Diamond b$.

To prove that the above equations are canonical, it now suffices to show that if $\mathbf{A}$ satisfies the equation, $\mathbf{A}_\bullet$ satisfies the corresponding relational condition.

Proposition 7.6 *If* $\mathbf{A} \models \Box a \wedge \Diamond b \leq \Diamond(a \wedge b)$, *then* $\mathbf{A}_\bullet \models \geq \circ R \subseteq R \circ \geq$.

Proof. If $\mathcal{U}(\geq \circ R)\mathcal{V}$, then $\Diamond\mathcal{V} \subseteq \mathcal{U}$. Extend $\mathcal{V}$ to a maximal filter $\mathcal{W}$ such that $\Diamond\mathcal{W} \subseteq \mathcal{U}$. The filter $\mathcal{W}$ is prime and if $a \notin \mathcal{W}$, then $\Diamond(a \wedge w) \notin \mathcal{U}$ for some $w \in \mathcal{W}$. But then $\Diamond w \in \mathcal{U}$, hence $\Box a \notin \mathcal{U}$ because $\Box a \wedge \Diamond w \leq \Diamond(a \wedge w)$. □

Proposition 7.7 *If* $\mathbf{A} \models \Diamond a \to \Box b \leq \Box(a \to b)$, *then* $\mathbf{A}_\bullet \models R \circ \leq \subseteq \leq \circ R$.

Proof. If $\mathcal{U}(R \circ \leq)\mathcal{V}$, then $\Box^{-1}\mathcal{U} \subseteq \mathcal{V}$. Extend $\mathcal{U}$ to a maximal filter $\mathcal{W}$ such that $\Box^{-1}\mathcal{W} \subseteq \mathcal{V}$. The filter $\mathcal{W}$ is prime and if $\Diamond a \notin \mathcal{W}$, then $\Diamond a \wedge w \leq \Box b$ for some $w \in \mathcal{W}$, $b \notin \mathcal{V}$, hence $w \leq \Diamond a \to \Box b \leq \Box(a \to b)$. But then $a \to b \in \mathcal{V}$, hence $a \notin \mathcal{V}$. □

Again, dually we obtain the following two propositions.

Proposition 7.8 *If* $\mathbf{A} \models \Box(a \vee b) \leq \Box a \vee \Diamond b$, *then* $\mathbf{A}_\bullet \models \leq \circ R \subseteq R \circ \leq$.

Proposition 7.9 *If* $\mathbf{A} \models \Diamond(a \succ b) \leq \Box a \succ \Diamond b$, *then* $\mathbf{A}_\bullet \models R \circ \geq \subseteq \geq \circ R$.

Observe that the equation $\Box a \wedge \Diamond b \leq \Diamond(a \wedge b)$ implies the positive modal law and the equation $\Diamond a \to \Box b \leq \Box(a \to b)$ implies the negative modal law. We conjecture that the variety of modal algebras relatively axiomatized by $\Box a \wedge \Diamond b \leq \Diamond(a \wedge b)$ and $\Box(a \vee b) \leq \Box a \wedge \Diamond b$ (that is, the variety of positive modal algebras introduced by Dunn [5]) is in fact the largest variety of modal algebras, and the variety relatively axiomatized by $\Box a \wedge \Diamond b \leq \Diamond(a \wedge b)$ and $\Diamond a \to \Box b \leq \Box(a \vee b) \to \Box b$ is the largest variety of Heyting modal algebras.

8 Duality for modal algebras

We now extend the completeness theorem for modal algebras to a duality based on the Priestley duality for distributive lattices. In order to derive a Hennessy-Milner theorem as a corollary, we in fact formulate a dual *adjunction* between modal algebras and "compactly branching" modal spaces, which restricts to a dual equivalence if we require the spaces to be compact.

The *bitopological* framework of Bezhanishvili et al. [1] will turn out to be suitable for this purpose. Bezhanishvili et al. formulate a dual equivalence between the category of distributive lattices (Heyting, bi-Heyting algebras) and suitable categories of spaces equipped with a pair of topologies. We slightly diverge from their framework in two ways. Firstly, we take the partial order $\leq$ to be part of the signature of such spaces, even though it is uniquely determined by the topologies. We therefore call them Priestley spaces rather than Stone spaces. Secondly, we generalize this dual equivalence to a dual adjunction. This involves no substantial novelty, only checking that the proof of the dual adjunction goes through without the assumption of compactness.

Definition 8.1 A *bitopological poset* $\mathcal{X} = (W, \leq, \tau_\pm)$ is a poset $(W, \leq)$ such that the *upspace* (W, τ_+) is a topological space, the *downspace* (W, τ_-) is a topological space, each $U \in \tau_+$ is an upset of $(W, \leq)$, and each $U \in \tau_-$ is a downset of $(W, \leq)$. The *opposite* of $\mathcal{X}$ is the bitopological poset $\mathcal{X}^{op} = (W, \geq, \tau_\mp)$. A *bicontinuous* function $f : (U, \leq, \tau_\pm) \to (V, \sqsubseteq, \upsilon_\pm)$ both a continuous function $f : (U, \tau_+) \to (V, \upsilon_+)$ and a continuous function $f : (U, \tau_-) \to (V, \upsilon_-)$.

Let $\mathcal{X} = (W, \leq, \tau_\pm)$ be a bitopological poset. The *join topology* of $\mathcal{X}$ is the topology $\tau = \tau_+ \vee \tau_-$. A subset of $\mathcal{X}$ is *upopen*, *upclosed* or *upcompact* if it is open, closed or compact in the *uptopology* τ_+, and it is *downopen*, *downclosed* or *downcompact* if it is open, closed or compact in the *downtopology* τ_-. A set is *upclopen* (*downclopen*) if it is upopen and downclosed (downopen and upclosed). An *upbasis* (*downbasis*) is a basis for the upspace (downspace).

A convex subset U of $\mathcal{X}$ is *closed* if $U^\uparrow$ is downclosed and $U^\downarrow$ is upclosed. The bitopological poset $\mathcal{X}$ is *compact* if it is compact in the join topology, or equivalently if and only if each cover of $\mathcal{X}$ by elements from $\tau_+ \cup \tau_-$ has a finite subcover. It is *Hausdorff* if the diagonal relation Δ_W is a closed subset of $\mathcal{X}^{op} \times \mathcal{X}$. It is a *pre-Priestley space* if it is Hausdorff, has an upbasis of upclopens, and has a downbasis of downclopens. It is a *Priestley space* if it is a compact pre-Priestley space. The category of pre-Priestley (Priestley) spaces and bicontinuous monotone functions will be denoted $\mathsf{PrePries}$ (Pries).

We now set up a dual adjunction between $\mathsf{PrePries}$ and DLat. Given a pre-Priestley space $\mathcal{X}$, let $\mathcal{X}^*$ be the distributive lattice of all upclopen subsets of $\mathcal{X}$, and given a bicontinuous monotone function $f : \mathcal{X} \to \mathcal{Y}$, let f^+ be the homomorphism of distributive lattices $f^{-1} : V^+ \to U^+$.

Let $\eta^+_{\mathbf{A}}(a) = \eta_{\mathbf{A}}(a)$ be the set of all prime filters $\mathcal{U}$ on $\mathbf{A}$ such that $a \in \mathcal{U}$ and let $\eta^-_{\mathbf{A}}(a)$ be the set of all prime filters $\mathcal{U}$ on $\mathbf{A}$ such that $a \notin \mathcal{U}$. Given a distributive lattice $\mathbf{A}$, let $\mathbf{A}_*$ be the poset of prime filters $\mathbf{A}_\bullet$ equipped with the uptopology generated by $\eta^+_{\mathbf{A}}(a)$ for $a \in \mathbf{A}$ and with the downtopology generated by $\eta^-_{\mathbf{A}}(a)$ for $a \in \mathbf{A}$. Given a homomorphism of modal algebras $h : \mathbf{A} \to \mathbf{B}$, let h_* be the function $h^{-1} : \mathbf{B}_* \to \mathbf{A}_*$.

It remains to define the co-unit of the dual adjunction. Given any pre-Priestley space $\mathcal{X}$, we define the function $\varepsilon_{\mathcal{X}} : \mathcal{X} \to (\mathcal{X}^*)_*$ such that $U \in \varepsilon_{\mathcal{X}}(u)$ if and only if $u \in U$ for each upclopen subset U of $\mathcal{X}$.

Theorem 8.2 (Bitopological Priestley dual adjunction) $(-)_* \dashv (-)^* : \mathsf{PrePries}^{op} \to \mathsf{DLat}$ *is an adjunction with unit* η *and co-unit* ε *which restricts to an equivalence between* Pries^{op} *and* DLat.

Proof. We know from [1] that restricting to Priestley spaces yields a dual equivalence. What we need to prove is that the assignment $(-)^*$ defines a functor $(-)^* : \mathsf{PrePries}^{op} \to \mathsf{DLat}$, that $\varepsilon_{\mathcal{X}} : \mathcal{X} \to (\mathcal{X}^*)_*$ is a bicontinuous monotone function for any pre-Priestley space $\mathcal{X}$, and that the triangle equality $(\varepsilon_{\mathcal{X}})^* \circ \eta_{\mathcal{X}^*} = 1_{\mathcal{X}^*}$ holds. The first claim is straightforward to prove.

To prove the second claim, we know that $(\mathcal{X}^*)_*$ has an upbasis of upclopen sets and that each upclopen subset V of $(\mathcal{X}^*)_*$ consists of all prime filters of upclopens on $\mathcal{X}$ which contain some upclopen subset U of $\mathcal{X}$. It then follows that $u \in U$ if and only if $\varepsilon_{\mathcal{X}}(u) \in V$.

To prove the third claim, $\eta_{\mathcal{X}^*}$ sends an upclopen subset U of $\mathcal{X}$ to the set V of all prime filters of upclopens on $\mathcal{X}$ which contain U. But $\varepsilon_{\mathcal{X}}$ sends a point $u \in \mathcal{X}$ to a filter which contains U if and only if $u \in U$, hence the pre-image of V under $\varepsilon_{\mathcal{X}}$ is precisely U. □

We now extend this dual adjunction to modal algebras. A convex relation

R on a bitopological poset $\mathcal{X}$ is said to be *continuous* if $(R^{\downarrow})^{-1}[U]$ is upopen for U upopen and $(R^{\uparrow})^{-1}[U]$ is downopen for U downopen. It is *compact* if $R^{\downarrow}[u]$ is downcompact and $R^{\uparrow}[u]$ is upcompact for each $u \in \mathcal{X}$.

Definition 8.3 A *pre-modal space* $\mathcal{X} = (W, \leq, R, \tau_{\pm})$ is both a modal frame $(W, \leq, R)$ and a pre-Priestley space $(W, \leq, \tau_{\pm})$ such that R is closed, $(R^{\downarrow})^{-1}[U]$ is upclopen for U upclopen, and $(R^{\uparrow})^{-1}[U]$ is downclopen for U downclopen. A *modally compact space* is a pre-modal space such that R is compact. A *modal space* is a compact pre-modal space.

Modal frames can be viewed as modal spaces with a discrete bitopology, that is, the uptopology of all upsets and the downtopology of all downsets. Modally compact frames are then frames such that for each point u, $R^{\downarrow}[u]$ is the lower closure of a finite set and $R^{\uparrow}[u]$ is the upper closure of a finite set. It is easily seen that each modal space is a modally compact space. (Since the relation R is closed, $R^{\downarrow}[u]$ is downclosed, hence also compact, for each u.)

We define the complex algebra $\mathcal{X}^*$ of a pre-modal space $\mathcal{X}$ as the expansion of the complex algebra of the underlying pre-Priestley space by the operations $\Box_R$ and $\Diamond_R$. The set of all upclopens on a pre-modal space is closed under these operations. The functions $\eta_{\mathbf{A}}$ and $\varepsilon_{\mathcal{X}}$ are defined for modal algebras and pre-modal spaces as for distributive lattices and pre-Priestley spaces.

Proposition 8.4 *If $\mathcal{X}$ is a modally compact space, then the function $\varepsilon_{\mathcal{X}} : \mathcal{X} \to (\mathcal{X}^*)_*$ is a bicontinuous p-embedding of modally compact spaces.*

Proof. Let $\mathcal{X} = (W, \leq, \tau_{\pm}, R)$. We know that $\varepsilon_{\mathcal{X}}$ is a bicontinuous monotone embedding of pre-Priestley spaces. Corollary 5.2 implies that if $uR^{\downarrow}v$, then $\varepsilon_{\mathcal{X}}(u) R^{\supseteq}_{\mathcal{X}^*} \varepsilon_{\mathcal{X}}(v)$. Vice versa, suppose that $\varepsilon_{\mathcal{X}}(u) R^{\supseteq}_{\mathcal{X}^*} \mathcal{V}$. The inclusion $\Diamond \mathcal{V} \subseteq \varepsilon_{\mathcal{X}}(u)$ holds by Corollary 5.2, hence each upclopen set in $\mathcal{V}$ intersects with $R^{\downarrow}[u]$. By the downcompactness of $R^{\downarrow}[u]$, so does their intersection. There is therefore some $v \in R^{\downarrow}[u]$ such that $\varepsilon_{\mathcal{X}}(v) \supseteq \mathcal{V}$. □

We denote the category of modally compact spaces and their bicontinuous p-morphisms by $\mathsf{ModKSpace}$ and the full subcategory of modal spaces by $\mathsf{ModSpace}$. Using Proposition 8.4, we obtain the following duality theorem.

Theorem 8.5 $(-)_* \dashv (-)^* : \mathsf{ModKSpace}^{op} \to \mathsf{ModDLat}$ *is a dual adjunction with unit η and co-unit ε which restricts to a dual equivalence between* $\mathsf{ModSpace}$ *and* $\mathsf{ModDLat}$.

Proof. Given Theorem 8.2, it suffices to show that η is preserves $\Box$ and $\Diamond$ and that ε is a morphism in $\mathsf{ModKSpace}$. Theorem 5.3 proves the former claim and Proposition 8.4 proves the latter claim. □

We now derive a Hennessy-Milner theorem as a corollary. See [2] for a proof of the Hennessy-Milner theorem for classical modal logic.

Definition 8.6 A *modal model* over a set of atomic propositions *Prop* consists of a pre-modal space $\mathcal{X}$ and a valuation function $val_{\mathcal{X}} : Prop \to \mathcal{X}^*$. A *modally compact model* is a modal model such that $\mathcal{X}$ is a modally compact space.

A *bisimulation* between modal models $(\mathcal{X}, val_{\mathcal{X}})$ and $(\mathcal{Y}, val_{\mathcal{Y}})$ is a bisimulation σ between $\mathcal{X}$ and $\mathcal{Y}$ such that $u\sigma v$ implies that $u \in val_{\mathcal{X}}(p)$ if and only if $v \in val_{\mathcal{X}}(p)$ for all $p \in Prop$.

The valuation function extends to a unique homomorphism $h_{\mathcal{X}} : \mathbf{F}(Prop) \to \mathcal{X}^*$, where $\mathbf{F}(Prop)$ is the free bi-Heyting modal algebra generated by *Prop*. Given a pair of modal models $(\mathcal{X}, val_{\mathcal{X}})$ and $(\mathcal{Y}, val_{\mathcal{Y}})$ and a pair of points $u \in \mathcal{X}$, $v \in \mathcal{Y}$, define $u\sigma_{\mathcal{X},\mathcal{Y}}v$ to hold in case $u \in h_{\mathcal{X}}(a)$ if and only if $v \in h_{\mathcal{X}}(a)$ for all $a \in \mathbf{F}(Prop)$. In other words, $u\sigma_{\mathcal{X},\mathcal{Y}}$ if and only if $h_{\mathcal{X}}^{-1}[\varepsilon_{\mathcal{X}}(u)] = h_{\mathcal{Y}}^{-1}[\varepsilon_{\mathcal{Y}}(v)]$.

Theorem 8.7 (Hennessy-Milner theorem) *Let $(\mathcal{X}, val_{\mathcal{X}})$ and $(\mathcal{Y}, val_{\mathcal{Y}})$ be modally compact models. Then $\sigma_{\mathcal{X},\mathcal{Y}}$ is the largest bisimulation between these models.*

Proof. By Proposition 8.4, $(h_{\mathcal{X}})_* \circ \varepsilon_{\mathcal{X}}(u)$ and $(h_{\mathcal{Y}})_* \circ \varepsilon_{\mathcal{Y}}(v)$ are p-morphisms. By Proposition 3.7, σ is thus a bisimulation, and clearly the largest one. □

9 Conclusion

We have introduced distributive unimodal logic as a semantically motivated generalization of classical set-based modal logic to a poset-based setting. We defined a suitable quasivariety of modal algebras and proved a completeness theorem embedding each modal algebra in the complex algebra its canonical frame. We then discussed the relationship between distributive unimodal logic and the distributive modal logic of Gehrke et al. [8] and showed that the existing completeness and duality theorems for intuitionistic modal logic and positive modal logic are subsumed by the completeness and duality theorems for distributive unimodal logic. Finally, the completeness theorem was extended to a duality between the category of modal algebras and a cat/egory of suitably topologized modal frames and a Hennessy-Milner theorem for bi-intuitionistic unimodal logic was proved as a corollary.

The completeness and duality theorems proved here can in fact be extended to the full language of distributive modal logic (which includes modal operators corresponding to the classical modalities $\Box\neg$ and $\Diamond\neg$). Extending them to modalities of higher arity, however, seems to be substantially more difficult.

Apart from other standard areas of investigation which were left untouched (such the finite model property, decidability and correspondence theory), we can also pose a question which does not arise in any of the other modal logics considered here, namely: given some choice of connectives, is there a largest variety of modal algebras in this language? In other words, is there a most general equational condition in a given language which ensures the validity of the quasiequations defining the class of modal algebras? It seems natural to conjecture that the positive modal algebras introduced by Dunn in fact form the largest variety of modal algebras.

References

[1] Bezhanishvili, G., N. Bezhanishvili, D. Gabelaia and A. Kurz, *Bitopological duality for distributive lattices and Heyting algebras*, Mathematical Structures in Computer Science **20** (2010), pp. 359–393.

[2] Blackburn, P., M. de Rijke and Y. Venema, "Modal Logic," Cambridge University Press, New York, NY, USA, 2001.

[3] Celani, S. and R. Jansana, *A new semantics for positive modal logic*, Notre Dame Journal of Formal Logic **38** (1997), pp. 1–18.

[4] Celani, S. and R. Jansana, *Priestley duality, a Sahlqvist theorem and a Goldblatt-Thomason theorem for positive modal logic*, Logic Journal of IGPL **7** (1999), pp. 683–715.

[5] Dunn, J. M., *Positive modal logic*, Studia Logica **55** (1995), pp. 301–317.

[6] Fischer Servi, G., *On modal logics with an intuitionistic base*, Studia Logica **36** (1977), pp. 141–149.

[7] Fischer Servi, G., *Semantics for a class of intuitionistic modal calculi*, in: M. Chiara, editor, *Italian Studies in the Philosophy of Science*, Boston Studies in the Philosophy of Science **47**, Springer Netherlands, 1981 pp. 59–72.

[8] Gehrke, M., H. Nagahashi and Y. Venema, *A Sahlqvist theorem for distributive modal logic*, Annals of Pure and Applied Logic **131** (2005), pp. 65–102.

[9] Palmigiano, A., *Dualities for some intuitionistic modal logics* (2004).

[10] Quackenbush, R. W., *Completeness theorems for universal and implicational logics of algebras via congruences*, Proceedings of the American Mathematical Society **103** (1988), pp. 1015–1021.

[11] Rauszer, C., *A formalization of the propositional calculus of H-B logic*, Studia Logica **33** (1974), pp. 23–34.

[12] Rauszer, C., "An algebraic and Kripke-style approach to a certain extension of intuitionistic logic," Number 168 in Dissertationes Mathematicae, Institute of Mathematics, Polish Academy of Sciences, 1980.

[13] Simpson, A. K., *The proof-theory and semantics of intuitionistic modal logic*, Technical report, University of Edinburgh (1994).

[14] Wolter, F. and M. Zakharyashchev, *Intuitionistic modal logic*, in: A. Cantini, E. Casari and P. Minari, editors, *Logic and Foundations of Mathematics*, Kluwer Academic Publishers, 1999 pp. 227–238.

A Tableau for Temporal Logic over the Reals

Mark Reynolds [1]

School of Computer Science and Software Engineering
University of Western Australia
Crawley, Perth Western Australia 6009

Abstract

We provide a simple, sound, complete and terminating tableau decision procedure for the temporal logic of until and since over the real numbers model of time. This logic is an important basis for reasoning about concurrency, metric constraints and planning. Despite its usefulness and long history, there are no existing implementable reasoning techniques for it.
The tableau uses a mosaic-based technique to translate the satisfiability problem into a question about the way that intervals of a real-flowed model relate to each other. It builds on top of recently developed reasoning tools for general linear time by applying some interesting but computationally simple checks.

Keywords: Temporal Logic, Reals, Tableau.

1 Introduction

Although discrete time temporal logics are the most common, there has been a separate thread of steady development of continuous time alternatives since the earliest beginnings. Being able to reason about events and processes unfolding continuously has an enormous range of applications from concurrency and refinement in reactive systems, as a basis for the metric temporal logics used for model checking automated systems, to artificial planning, natural language semantics and philosophical arguments.

In this paper we investigate the most natural and useful such temporal logic: RTL, the propositional temporal logic over real-numbers time using the Until and Since connectives introduced in [6]. RTL is as expressive as first-order logics over linear structures [6]. It is decidable [2,10,8] (in PSPACE) and has complete axioms systems [5,9].

Currently there is no satisfiability or validity checking procedure for RTL that looks remotely amenable to implementation. In this paper we build on the results and techniques of [10] and present what seems to be an intuitive tableau style decision procedure for RTL which will not be hard to implement (albeit only to work with sufficiently small formulas).

[1] mark.reynolds@uwa.edu.au partially supported by ARC

The proof of correctness here uses the *mosaics* which were used to prove PSPACE decidability of RTL in [10]. Mosaics are small pieces of a model. We can decide whether a finite set of mosaics is sufficient to be used to build a real-numbers model of a given formula by considering something like a game tree which can also be viewed as a tableau. Such an idea was suggested for general dense time reasoning in [11] and those ideas have led to recent tableaux [12] and more streamlined implementations [1].

Narrowing our focus to the reals, we have to look carefully at shapes of sub-graphs within the tree to enforce the peculiar properties of the reals: such as density, Dedekind completeness and separability.

The contribution here is presenting a sound and complete mosaic-based tableau system to decide satisfiability in RTL. We aim mainly to show clearly how mosaics can be the building blocks of a tableau with this logic here: the system is not at all streamlined and is intended to provide the foundation of more intelligent tableau construction techniques in future work.

Below we define the logic RTL in section 2, explain mosaics in section 3, show how to make a sufficient set of mosaics in section 4, lay out the basic mosaic tableau in section 5, adjust it for the case of the reals flow in Section 6, soundness in Section 7 and prove completeness in Section 8. Section 9 has a quick overview of complexity and implementation issues.

2 The logic

Fix a countable set $\mathbf{L}$ of atoms. Here, *frames* $(T,<)$, or flows of time, will be irreflexive linear orders. *Structures* $\mathbf{T} = (T,<,h)$ will have a frame $(T,<)$ and a valuation h for the atoms, i.e. for each atom $p \in \mathbf{L}$, $h(p) \subseteq T$. Of particular importance will be *real* structures $\mathbf{T} = (\mathbb{R},<,h)$, which have the real numbers flow (with their usual irreflexive linear ordering).

The language $L(U,S)$ is generated by the 2-place connectives U and S along with classical $\neg$ and $\wedge$. That is, we define the set of formulas recursively to contain the atoms and for formulas α and β we include $\neg\alpha$, $\alpha \wedge \beta$, $U(\alpha,\beta)$ and $S(\alpha,\beta)$.

Each formula is evaluated at a point in a structure $\mathbf{T} = (T,<,h)$. We write $\mathbf{T}, x \models \alpha$ when α is true at the point $x \in T$. This is defined recursively as follows. Suppose that we have defined the truth of formulas α and β at all points of $\mathbf{T}$. See Figure 1 for the semantics.

Common temporal abbreviations are: $F\alpha = U(\alpha,\top)$, "α will be true (sometime in the future)"; $G\alpha = \neg F(\neg\alpha)$, "$\alpha$ will always hold (in the future)"; and their mirror images P and H. Particularly for dense time applications we also have: $C^+\alpha = U(\top,\alpha)$, "$\alpha$ will be constantly true for a while after now"; and $K^+\alpha = \neg C^+\neg\alpha$, "$\alpha$ will be true arbitrarily soon". They have mirror images C^- and K^-.

A formula ϕ is $\mathbb{R}$-*satisfiable* if it has a real model: i.e. there is a real structure $\mathbf{S} = (\mathbb{R},<,h)$ and $x \in \mathbb{R}$ such that $\mathbf{S}, x \models \phi$. A formula is $\mathbb{R}$-*valid* iff it is true at all points of all real structures. Of course, a formula is $\mathbb{R}$-valid iff its negation is not $\mathbb{R}$-satisfiable. We will refer to the logic of L(U,S) over real

For all points x:

$\mathbf{T}, x \models p$	iff	$x \in h(p)$, for p atomic;
$\mathbf{T}, x \models \neg\alpha$	iff	$\mathbf{T}, x \not\models \alpha$;
$\mathbf{T}, x \models \alpha \wedge \beta$	iff	both $\mathbf{T}, x \models \alpha$ and $\mathbf{T}, x \models \beta$;
$\mathbf{T}, x \models U(\alpha, \beta)$	iff	there is $y > x$ in T such that $\mathbf{T}, y \models \alpha$ and for all $z \in T$ such that $x < z < y$, we have $\mathbf{T}, z \models \beta$; and
$\mathbf{T}, x \models S(\alpha, \beta)$	iff	there is $y < x$ in T such that $\mathbf{T}, y \models \alpha$ and for all $z \in T$ such that $y < z < x$, we have $\mathbf{T}, z \models \beta$.

Fig. 1. Sematics

structures as RTL.

Let RTL-SAT be the problem of deciding whether a given formula of $L(U, S)$ is $\mathbb{R}$-satisfiable or not. [10] proves:

Theorem 2.1 *RTL-SAT is PSPACE-complete.*

3 Mosaics for U and S

Each mosaic is a syntactic object intended to represent a small piece, or interval, of a model, i.e. sets of formulas for a pair of points indicating which formulas are true there and in between in the whole model. There will be *coherence* conditions on the mosaic which are necessary for it to be part of a larger model. Full details, definitions and proofs can be found in [10].

Our mosaics will only be concerned with a finite set of formulas:

Definition 3.1 For each formula ϕ, define the *closure* of ϕ to be $\text{Cl}\phi = \{\psi, \neg\psi \mid \psi \leq \phi\}$ where $\chi \leq \psi$ means that χ is a subformula of ψ.

We can think of $\text{Cl}\phi$ as being closed under negation: we treat $\neg\neg\alpha$ as if it was α.

Often we will intend that a set of formulas will be exactly the set of formulas which hold at a particular point in a model. Such a set should at least be consistent in terms of classical propositional logic:

Definition 3.2 Suppose $\phi \in L(U, S)$ and $S \subseteq \text{Cl}\phi$. Say S is *propositionally consistent* (PC) iff there is no substitution instance of a tautology of classical propositional logic of the form $\neg(\alpha_1 \wedge ... \wedge \alpha_n)$ with each $\alpha_i \in S$. Say S is *maximally propositionally consistent* (MPC) iff S is maximal in being a subset of $\text{Cl}\phi$ which is PC.

We will define a mosaic to be a triple (A, B, C) of sets of formulas. The intuition is that this corresponds to two points from a structure: A is the set of formulas (from $\text{Cl}\phi$) true at the earlier point, C is the set true at the later point and B is the set of formulas which hold at all points strictly in between. Look ahead to definition 3.12 to see how mosaics can be found in a real structure.

Definition 3.3 Suppose ϕ is from $L(U, S)$. A *ϕ-mosaic* is a triple (A, B, C) of subsets of $\text{Cl}\phi$ such that:

C0. A and C are maximally propositionally consistent,

and the following four *coherency* conditions hold:

C1. if $\neg U(\alpha, \beta) \in A$ and $\beta \in B$ then we have both:
C1.1. $\neg\alpha \in C$ and either $\neg\beta \in C$ or $\neg U(\alpha, \beta) \in C$; and
C1.2. $\neg\alpha \in B$ and $\neg U(\alpha, \beta) \in B$.

C2. if $U(\alpha, \beta) \in A$ and $\neg\alpha \in B$ then we have both:
C2.1 either $\alpha \in C$ or both $\beta \in C$ and $U(\alpha, \beta) \in C$; and
C2.2. $\beta \in B$ and $U(\alpha, \beta) \in B$.

C3-4 mirror images of C1-C2.

Definition 3.4 If $m = (A, B, C)$ is a mosaic then $\text{start}(m) = A$ is its *start*, $\text{cover}(m) = B$ is its *cover* and $\text{end}(m) = C$ is its *end*.

If we start to build a model using mosaics as building blocks then we may realise that the inclusion of one mosaic necessitates the inclusion of others: defects need curing.

Definition 3.5 A *defect* in a mosaic (A, B, C) is either types 1, 2 or 3:

1. a formula $U(\alpha, \beta) \in A$ with either
   1.1 $\beta \notin B$,
   1.2 ($\alpha \notin C$ and $\beta \notin C$), or
   1.3 ($\alpha \notin C$ and $U(\alpha, \beta) \notin C$);
2. mirror image for S;
3. a formula $\beta \in \text{Cl}\phi$ with $\neg\beta \notin B$.

We will need to string mosaics together to build linear orders. This can only be done under certain conditions. We introduce the idea of composition of mosaics and present some results which are straightforward to prove.

Definition 3.6 We say that ϕ-mosaics (A', B', C') and (A'', B'', C'') *compose* iff $C' = A''$. In that case, their *composition* is $(A', B' \cap C' \cap B'', C'')$.

Lemma 3.7 *If mosaics m and m' compose then their composition is a mosaic.*

Lemma 3.8 *Composition of mosaics is associative.*

Thus we can talk of sequences of mosaics composing and then find their composition. We define the composition of a sequence of length one to be just the mosaic itself and we leave the composition of an empty sequence undefined. Write $\sigma = \langle m_1, m_2, ..., m_n \rangle$ for a sequence and $\sigma^\wedge\tau$ for the concatenation of two sequences.

Definition 3.9 A *decomposition* for a mosaic m is any finite sequence $\langle m_1, ..., m_n \rangle$ of mosaics which composes to m.

It will be useful to introduce an idea of fullness of decompositions. This is intended to be a decomposition which provides witnesses to the cure of every defect in the decomposed mosaic.

Definition 3.10 The decomposition above is *full* iff the following three conditions all hold:

1. for all $U(\alpha, \beta) \in A$ we have
   1.1. $\beta \in B$ and either ($\beta \in C$ and $U(\alpha, \beta) \in C$) or $\alpha \in C$,
   1.2. or there is some i with $1 \leq i < n$ such that $\alpha \in C_i$, $\beta \in B_j$ (all $j \leq i$) and $\beta \in C_j$ (all $j < i$);
2. the mirror image of 1.; and
3. for each $\beta \in \mathrm{Cl}\phi$ such that $\neg\beta \notin B$ there is some i such that $1 \leq i < n$ and $\beta \in C_i$.

If 1.2 above holds in the case that $U(\alpha, \beta) \in A$ is a type 1 defect in (A, B, C) then we say that *a cure for the defect is witnessed* (in the decomposition) by the end of (A_i, B_i, C_i) (or equivalently by the start of $(A_{i+1}, B_{i+1}, C_{i+1})$). Similarly for the mirror image for $S(\alpha, \beta) \in C$. If $\beta \in C_i$ is a type 3 defect in (A, B, C) then we also say that *a cure for this defect is witnessed* (in the decomposition) by the end of (A_i, B_i, C_i). If a cure for any defect is witnessed then we say that the defect is cured.

Lemma 3.11 *If $\langle m_1, ..., m_n \rangle$ is a full decomposition of m, then every defect in m is cured in the decomposition.*

For the reals we do not allow full decompositions of length one, although they are allowed in general linear time contexts for mosaics with no defects.

In the rest of this section we define a notion of satisfiability for mosaics and relate the satisfiability of formulas (which is our ultimate interest) to that of mosaics.

Because mosaics represent linear orders with end points, it is inconvenient for us to continue to work directly with $\mathbb{R}$ and because we want to make use of some simple tricks with convergence of sequences in the metric at several places in the proof, we will move to work in the unit interval $[0, 1]$ instead.

If $x < y$ from $\mathbb{R}$ then let $]x, y[$ denote the open interval $\{z \in \mathbb{R} \mid x < z < y\}$ and $[x, y]$ denote the closed interval $\{z \in \mathbb{R} \mid x \leq z \leq y\}$. Similarly for half open intervals.

One can get a mosaic (you can check it is a mosaic) from any two points in a structure.

Definition 3.12 If $\mathbf{T} = (T, <, h)$ is a structure and ϕ a formula then for each $x < y$ from T we define $\mathrm{mos}^{\phi}_{\mathbf{T}}(x, y) = (A, B, C)$ where:

$A = \{\alpha \in \mathrm{Cl}\phi \mid \mathbf{T}, x \models \alpha\}$,
$B = \{\beta \in \mathrm{Cl}\phi \mid$ for all $z \in T$, if $x < z < y$ then $\mathbf{T}, z \models \beta\}$, and
$C = \{\gamma \in \mathrm{Cl}\phi \mid \mathbf{T}, y \models \gamma\}$.

We will now relate the satisfiability of a formula ϕ to that of certain mosaics. Obviously, a formula will be satisfiable over the reals iff it is satisfiable over the $]0, 1[$ flow. Furthermore, this happens iff a relativized version of the formula is satisfiable somewhere in the interior of a model over $[0, 1]$. To define this relativization we need to use a new atom to indicate points in the interior. Hence the next few definitions.

Definition 3.13 Given ϕ and an atom q which does not appear in ϕ, we define a map $* = *_q^\phi$ on formulas in $\text{Cl}\phi$ recursively: $*p = p \wedge q$, $*\neg\alpha = \neg(*\alpha) \wedge q$, $*(\alpha \wedge \beta) = *(\alpha) \wedge *(\beta) \wedge q$, $*U(\alpha, \beta) = U(*\alpha, *\beta) \wedge q$, and similarly S.

With the relativization machinery we can then define a relativized mosaic to be one which could correspond to the whole of a $[0, 1]$ structure in which q is true of exactly the interior $]0, 1[$ and the interior is a model of ϕ.

Definition 3.14 We say that a $*_q^\phi(\phi)$-mosaic (A, B, C) is (ϕ, q)-*relativized* iff

1. $\neg q$ is in A and C, no $S(\alpha, \beta)$ is in A, no $U(\alpha, \beta)$ in C; and
2. $q \in B$ and $\neg *_q^\phi(\phi) \notin B$.

Here we confirm that ϕ is satisfiable over the reals exactly when we can find such a relativized mosaic.

Lemma 3.15 (Lemma 29 from [10]) *Suppose that ϕ is a formula of $L(U, S)$ and q is an atom not appearing in ϕ. Then ϕ is $\mathbb{R}$-satisfiable iff there is a (ϕ, q)-relativized $*_q^\phi(\phi)$-mosaic satisfied on the whole of* $[0, 1]$.

Our satisfiability procedure in [10] was to guess a relativized mosaic (A, B, C) and then check that (A, B, C) is satisfied on the whole of $[0, 1]$. Thus we now turn to the question of deciding whether a relativized mosaic is satisfiable.

4 Real Mosaic Systems

In this section we define a concept of a collection or system of mosaics in which each member is decomposable in terms of simpler members. We will later show that being in such a system is (roughly) equivalent to satisfiability. First two of the simpler tactics for decomposition.

4.1 Tactics Lead and Trail

The mirror image tactics *lead* and *trail* allow mosaics which can be fully decomposed in terms of themselves along with some other mosaics. In a game setting this is a legitimate way for the game to be won: the player who has to keep providing full decompositions can keep supplying a full decomposition $\langle m \rangle^\wedge \sigma$ for m if the other player keeps choosing m to be decomposed. The tactic trail corresponds to an operation in [7] for building a new linear order from a simpler one by laying ω copies of it one after the other towards the future. The tactic lead corresponds to laying the copies towards the past.

Definition 4.1 Suppose $\phi \in L(U, S)$, m is a ϕ-mosaic and σ is a non-empty sequence of ϕ-mosaics. Then, we say that m is fully decomposed by the *tactic* lead(σ) iff $\langle m \rangle^\wedge \sigma$ is a full decomposition of m. We say that m is fully decomposed by the *tactic* trail(σ) iff $\sigma^\wedge \langle m \rangle$ is a full decomposition of m.

4.2 Shuffles

The term shuffle has been used in the literature (see, for example, [7], [2] or [14]) to refer to certain methods of constructing a linear structure (often a monadic one) from a thorough mixture of copies of members of a finite set of other linear structures.

Suppose that $(T_1, ..., T_n)$ are linear structures. A shuffle of the T_i is any linear order made from intervals which are each copies of one of the T_i such that between any two of the intervals lies a copy of each of the T_i. To be precise,

Definition 4.2 Suppose that $(T_1, <_1), ..., (T_n, <_n)$ are linear structures. $(K, <_K)$ is a *shuffle* of $T = \{(T_i, <_i) \mid i = 1, ..., n\}$ iff there is a linear order $(B, <_B)$ and a map $\pi : B \to \{1, ..., n\}$ such that

- $K = \bigcup_{b \in B} \{(b, t) \mid t \in T_{\pi(b)}\}$ and
- for all $b, b' \in B$, for all $t \in T_{\pi(b)}$, for all $t' \in T_{\pi(b')}$, $(b, t) <_K (b', t')$ iff either $b <_B b'$ or $b = b'$ and $t <_{\pi(b)} t'$, and
- if $b <_B b'$ then for all $i \in \{1, ..., n\}$ there is $b'' \in B$ such that $b <_B b'' <_B b'$ and $\pi(b'') = i$.

The intention here is similar except we need to deal with mosaics corresponding to linear structures instead of structures themselves. We consider (a mosaic corresponding to) a shuffle S of linear structures $U_0, U_1, \ldots, U_s$, $V_1, V_2, \ldots, V_r$ where each U_i is a singleton structure and each V_i is a non-singleton structure consisting of the concatenation of a finite sequence of (one or more) mosaics representing other structures. Thus, we actually only consider an MPC set P_i instead of U_i and a non-empty composing sequence λ_i of mosaics instead of V_i. In this case it is possible to construct a certain set of mosaics such that one, o, corresponds to S and each one in the set has a full decomposition in terms of others in the set and/or the mosaics which decompose each λ_i.

In [10], in this vein, there is a rather complex definition of when a mosaic o is *fully decomposed by the tactic shuffle* $(\langle P_0, \ldots, P_s \rangle, \langle \lambda_1, \ldots, \lambda_r \rangle)$. See Definition 31 of that paper. We will not repeat it here to save space and also to save the reader effort.

Instead we present a slightly shorter alternative characterisation that also appeared (and was proved equivalent) in that paper.

The forward $K(m)$ property is supposed to hold of an MPC set if that set could be the end of the last mosaic in some λ_i where mosaic m is fully decomposed by the tactic shuffle $(\langle P_0, \ldots, P_s \rangle, \langle \lambda_1, \ldots, \lambda_r \rangle)$. This is the set of formulas from $\mathrm{Cl}(\phi)$ true at the end point of one of the structures V_i referred to above.

Definition 4.3 (Definition 32 from [10]) Suppose $\phi \in L(U, S)$ and m is a ϕ-mosaic. We say that a set $Q \subseteq \mathrm{Cl}\phi$ satisfies the forward $K(m)$ property iff Q is MPC and for any $U(\alpha, \beta) \in \mathrm{Cl}\phi$ we have $U(\alpha, \beta) \in Q$ iff both $\beta \in \mathrm{cover}(m)$ and (at least) one of the following holds:

K1 $\neg\alpha \notin \mathrm{cover}(m)$;

K2 $\alpha \in \mathrm{end}(m)$; or

K3 $\beta \in \mathrm{end}(m)$ and $U(\alpha, \beta) \in \mathrm{end}(m)$.

The mirror image is the backwards $K(m)$ property.

Lemma 4.4 *(Lemma 33 from [10]) Suppose $\phi \in L(U,S)$, $m = (A,B,C)$ is a ϕ-mosaic, and each $P_i \subseteq \mathrm{Cl}\phi$ $(0 \leq i \leq s)$ and each λ_i $(1 \leq i \leq r)$ is a composing non-empty sequence of ϕ-mosaics.*

Then, m is fully decomposed by the tactic shuffle $(\langle P_0, \ldots, P_s\rangle, \langle \lambda_1, \ldots, \lambda_r\rangle)$ iff the following seven conditions hold:

- *S0* *B is a subset of each P_i and of the start, end and cover of each mosaic in each λ_i;*
- *S1* *each P_i satisfies both the forward and backwards $K(m)$ property;*
- *S2* *the start of the first mosaic in each λ_i satisfies the backwards $K(m)$ property;*
- *S3* *the end of the last mosaic in each λ_i satisfies the forwards $K(m)$ property;*
- *S4* *A satisfies the forward $K(m)$ property;*
- *S5* *C satisfies the backwards $K(m)$ property;*
- *S6* *if $\beta \in \mathrm{Cl}\phi$ but $\neg\beta \notin B$ then either β is contained in some P_i or β is contained in the start or end of some mosaic in some λ_i.*

Note that as $s \geq 0$ there is at least one P_i involved in the shuffle. This corresponds to a one point structure. In a general linear order setting we could define a shuffle with no P_is (provided that then $r > 0$) but over the reals it turns out to be crucial to require at least one P_i. This is because, as it is not too hard to see, a shuffle of only non-singleton closed intervals of the reals can not be both Dedekind complete and separable (i.e. having a countable dense suborder).

4.3 The levels that make an RMS

Now we define the hierarchy of membership of the system of mosaics which we need. Mosaics at one level of membership will be constructed from ones at lower levels of membership by concatenation or some combination of the tactics we have introduced above. As we build up, we only want to allow a limited use of leads and trails before a shuffle takes us to the next highest level. As we will only allow nesting of trails and/or leads of depth 2 within shuffles we define some intermediate levels between levels n and $n+1$. So, as we will see now, the levels, in increasing order are actually $0, 0^+, 1^-, 1, 1^+, 2^-, 2, 2^+, \ldots$.

Definition 4.5 For $\phi \in L(U,S)$, suppose S is a set of ϕ-mosaics and $n \geq 0$.

A ϕ-mosaic $m \in S$ is a *level n^+ member of S* iff m is the composition of a sequence of mosaics, each of them being either a level n member of S or fully decomposed by the tactics lead(σ) or trail(σ) with each mosaic in σ being a level n member of S.

A ϕ-mosaic $m \in S$ is a *level $(n+1)^-$ member of S* iff m is the composition of a sequence of mosaics, each of them being either a level n^+ member of S or fully decomposed by the tactics lead(σ) or trail(σ) with each mosaic in σ being a level n^+ member of S.

A ϕ-mosaic $m \in S$ is a *level n member of S* iff m is the composition of a sequence of mosaics with each of them being either a level n^- member of S or a mosaic which is fully decomposed by the tactic shuffle($\langle P_0, ..., P_s\rangle, \langle \sigma_1, ..., \sigma_r\rangle$) with each mosaic in each σ_i being a level n^- member of S.

Note that it is generally possible for mosaics to be level 0 members of some S provided that they are compositions of mosaics which can be fully decomposed by shuffles in which there are no sequences (i.e. $r = 0$). Thus these mosaics will have an interior which is a dense mixture of points where $P_0, ..., P_s$ hold. These are the only mosaics which can be level 0 members of any S.

Also note that if m is a level n member of S then m is the composition of $\langle m \rangle$ so m is clearly a level n^+ member of S. Similarly, level n^+ implies level $(n+1)^-$ and level n^- implies level n.

Finally, note that the set S in the definition above may not be closed under composition. It is even possible that a mosaic is a member of S at a certain level by virtue of being a composition of other mosaics each of which, although being fully decomposed by tactics involving only members of S, is not itself a members of S. Later we will see that for our purposes we mostly work with sets S which are closed under composition.

Definition 4.6 For $\phi \in L(U, S)$, a *real mosaic system* (RMS) of ϕ-mosaics is a set S of ϕ-mosaics such that, for every $m \in S$, there exists some n such that m is a level n member of S. For any n, we say that S is a real mosaic system of depth n iff every $m \in S$ is a level n member of S.

Theorem 4.7 (Theorem 75 in [10]) *Suppose ϕ is a formula of $L(U, S)$ and q is an atom not appearing in ϕ. Suppose $\psi = *_q^\phi(\phi)$ has length N.*

Then the following are equivalent:

1. *ϕ is $\mathbb{R}$-satisfiable;*
2. *there is a (ϕ, q)-relativized ψ-mosaic which appears in some RMS.*

5 Tableaux

In this section we see how the mosaics and RMS machinery can be the basis of a tableau-style decision procedure. We will start with a formula ϕ and determine whether ϕ is satisfiable in RTL or not.

The tableaux we construct will be roughly tree-shaped, albeit the traditional upside down tree with a root at the top: predecessors and ancestors above, successors and descendants below. They can be thought of as structures for organising and representing iterative full decompositions in the RMS.

We imagine trees growing downwards from the root. A node may have children immediately below it, every node except the root has a unique parent. Each node itself and its parent and the parent's parent and the parent's parent's parent etc. form the set of ancestors of the node. We will also impose an earlier-later relation between siblings (children of the same parent) on some trees and represent it by left-to-right ordering in diagrams.

Here are the basic definitions.

Definition 5.1
1. A *tree* here is just a set (of *nodes*), with a *successor* relation determining (as its transitive closure) a derived, reflexive, anti-symmetric, transitive, *ancestor* relation such that the set of ancestors of any node is finite and well-ordered (by the ancestor relation) and there is a unique *root* with no ancestors (apart from itself).

2. If node x has a successor y then we say that x is the *parent* of y (it is unique) and y is a *child* of x. Any other child of x is called a *sibling* of y. A node with no children will be called a *leaf* node.

3. The *depth* of a node with n ancestors is n.

4. An *ordered tree* is a tree with finite numbers of children for each node and a left-right relation which totally orders siblings. The left-right relation does not relate non-siblings.

5. A ϕ-mosaic labelled tree is a map from nodes of a tree to ϕ-mosaics.

The idea, as we will see, is that the labels of the children of a node form a full decomposition for the label of the node.

Definition 5.2 A (ϕ-) *tableau* (for ϕ-mosaic m) is a ϕ-mosaic labelled ordered tree with root labelled by m; and each node having the labels on the children nodes taken in order forming a full decomposition of the label on the node.

Definition 5.3 Define a leaf node to be a *clone* iff it has the same label as one of its other ancestors. Define a *complete* node of a tableau to be either a non-leaf, or a clone leaf node. Define a *successful* tableau as one in which all nodes are complete (otherwise the tableau is incomplete).

As an example see the successful $U(p,q)$-tableau in Figure 2. The three sets of formulas appearing are: $A = \{p, q, U(p,q)\}$, $B = \{\neg p, q, U(p,q)\}$ and $C = \{p, \neg q, U(p,q)\}$.

Definition 5.4 Suppose that ϕ is a formula of $L(U,S)$ and q is an atom not appearing in ϕ. Say $\psi = *_q^\phi(\phi)$. A ψ-tableau is a *tableau* for ϕ iff the root is labelled by a (ϕ, q)-relativized $*_q^\phi(\phi)$-mosaic.

6 The Reals

The mosaic tableaux of the last section were quite simple and quite general but they are not adequate for the special properties of the reals. Thus, in this section we define a $\mathbb{R}$-tableau to be a type of mosaic tableau. However, we impose some subtle restrictions on the labelling as we travel around the tree. They are essentially simple graph-theoretic properties of the labels on the decomposition tree.

First, we specify that in an $\mathbb{R}$-tableau we do not allow tableau nodes with a single child. Mosaics which have singleton sequences of themselves as full decompositions, are possible in general linear time, they are called *units*, but not allowed in the reals.

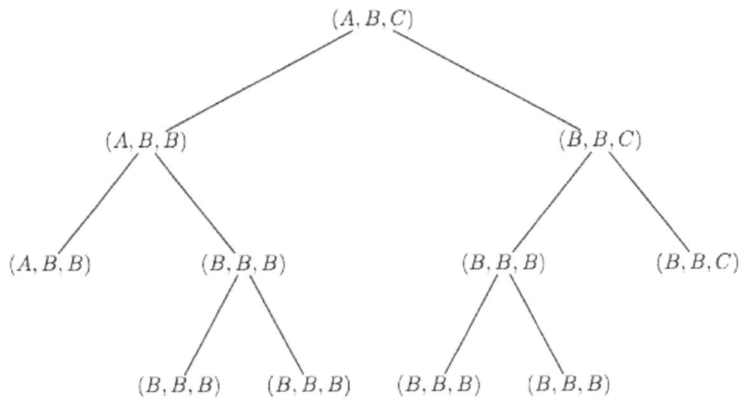

Fig. 2. A successful $U(p,q)$-tableau

6.1 Approval of the labels

Next we need some machinery to enable the other properties to be defined properly. Assume $\psi \in L(U,S)$ and suppose that T is a successful tableau of ψ-mosaics.

In order to determine whether T is a successful $\mathbb{R}$-tableau we will define an iterative process of *approving* individual mosaics in the tree. We approve the mosaic labels themselves regardless of how many times a particular label appears in the tableau. Once it is approved, it is approved everywhere that it appears.

The simplest criterion for approval is that a parent label can be approved whenever all the labels of its child nodes are approved. There are a couple of other ways to gain approval that we will outline below.

If, after some iterations, the root label in the tableau is approved then the tableau is a successful $\mathbb{R}$-tableau.

If at some stage there are no applicable rules to approve any more nodes, and the root mosaic remains unapproved then the tableau has failed to be a $\mathbb{R}$-tableau. We can terminate the check.

6.2 Trails and Leads

The following pattern in the tableau corresponds to a lead tactic and allows the mosaic m to be approved. Suppose $m = m_0$ is decomposed as $\langle m_1 \rangle^\wedge \sigma_0$ in T, i.e. m is the label of a parent node and $\langle m_1 \rangle^\wedge \sigma_0$ are the labels of the children in order. Suppose further that for all $i = 1, 2, ...$, m_i is decomposed as

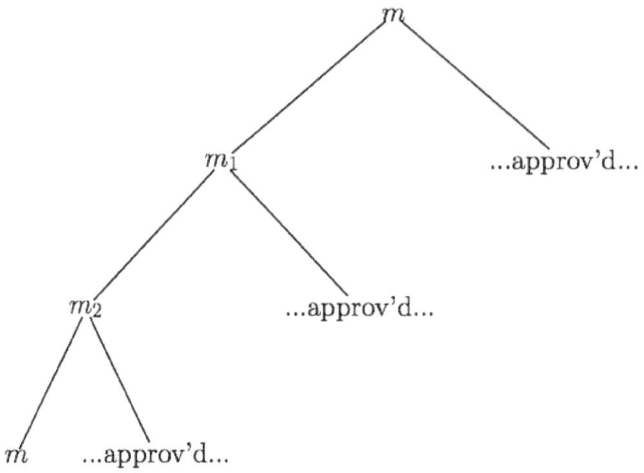

Fig. 3. Approval as a lead

$\langle m_{i+1}\rangle^\wedge\sigma_i$ in T. Suppose that all the mosaics appearing in each σ_i are already approved.

Finally suppose that some $m_i = m$.

Then we can approve m. We say that we approve m as a lead. See Figure 3 for an example of a sub-tree which leads to the approval of a mosaic as a lead.

Similarly we can approve mosaics as trails if we have a looping sequence of decompositions all ending in the m_i.

6.3 Shuffles

The pattern to allow approval of a mosaic as a shuffle is a little bit more complicated to describe and identify. It can involve a set of more than one (as yet) unapproved mosaic labels.

Because the covers of mosaics in decompositions are supersets of the cover of the parent, if there is a sequence $u = m_0, m_1, ..., m_n = v$ of mosaics in respective decompositions such that each m_i is fully decomposed (somewhere in T) as $\sigma_i{}^\wedge\langle m_{i+1}\rangle^\wedge\pi_i$ then the cover of v is a superset of the cover of u.

The conditions for approving a mosaic as a shuffle are SH1-SH6 as set out below. Consider the mosaic m appearing as a label in a tableau.

(SH1) m is an unapproved mosaic.

(SH2) Every unapproved (label of a) descendent of (a node labelled by) m, including m itself, has some descendent which has at least two separate child nodes labelled by unapproved mosaics.

(SH3) All descendants of m which are unapproved have the same cover as m.

(SH4) is the requirement that every unapproved descendent u of m (including m itself) has a "crisp start". That is, there is a sequence $u = m_0, m_1, ...$ of

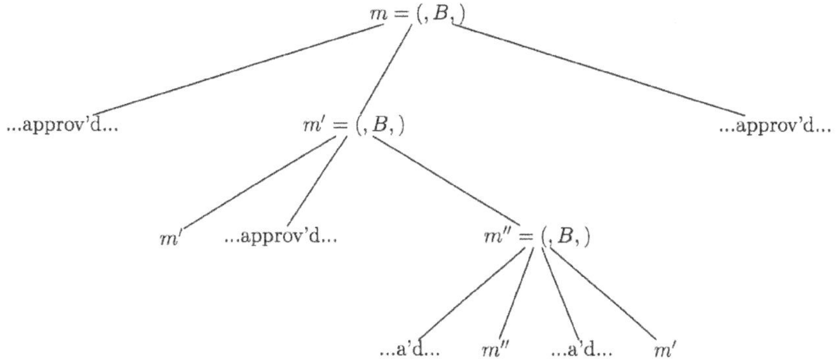

Fig. 4. Almost a shuffle

unapproved mosaics in respective decompositions as follows. Each m_i is fully decomposed as $\sigma_i{}^\wedge\langle m_{i+1}\rangle^\wedge\pi_i$ where each mosaic in σ_i is approved already (and we do not care what the π_i are). We require that $m_i = m_j$ for some $i < j$ and further, we require that for each $k = i, i+1, ..., j-1$, σ_k is actually empty.

We will see below that this condition allows us to identify a start of a possible shuffle involving u.

Similarly, (SH5), we require the unapproved descendants of m to have crisp ends using the mirror image construction.

The last check (SH6) before we approve m as a shuffle is to find an unapproved descendent u of m such that u has two adjacent children with unapproved labels v and w that further satisfy the following pattern.

We have a sequence $v = m_0, m_1, ...$ of unapproved mosaics in respective decompositions as follows. Each m_i is fully decomposed as $\sigma_i{}^\wedge\langle m_{i+1}\rangle$ where σ_i is any (even perhaps empty) sequence of any mosaics, approved or not. However, note that m_{i+1} is always the last mosaic in the decompositions for each m_i. We also have $m_i = m_j$ for some $i < j$.

The mirror image condition is required of w.

In this case it is easy to see that the end of v will be the same as the start of w. SH6 corresponds to making sure that there is a point structure taking part in the shuffle, a condition which we have seen ensures Dedekind completeness.

If SH1-6 hold then we can be sure that the shuffle is acceptable and we can approve m as a shuffle.

In Figure 4 there is a sub-tree which almost allows m to be approved as a shuffle except that condition SH6 is not established.

In Figure 5, however, m can be approved as a shuffle: condition SH6 is established with m' and m'' witnesses.

6.4 ℝ-tableau by approval

This concludes our account of the approval process that defines a successful ℝ-tableau.

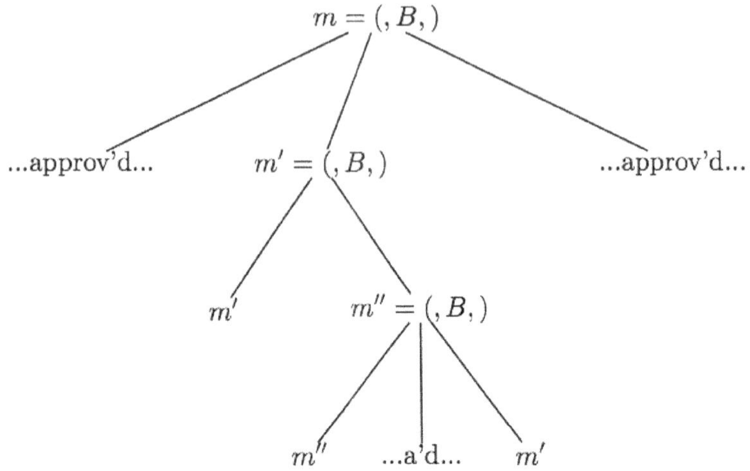

Fig. 5. Approved as a shuffle

Definition 6.1 A successful tableau is a successful $\mathbb{R}$-tableau iff all mosaic labels can be approved according to the iterative process above.

As an example, we find that the tableau in Figure 2 is a successful $\mathbb{R}$-tableau. The mosaic (B, B, B) can be approved as a shuffle, then (A, B, B) and (B, B, C) separately as a lead and trail, then (A, B, C) because its two children are approved.

The main work here is mostly in the appendices (see below). Then we can put the soundness and completeness lemmas together and get our desired overall theorem.

Theorem 6.2 *$L(U, S)$ formula ϕ is $\mathbb{R}$-satisfiable iff ϕ has a successful $\mathbb{R}$-tableau.*

7 Soundness

In [10], we define a concept of realization intended to capture the idea of a mosaic being satisfiable (over the reals) as far as internal information is concerned: i.e. we ignore formulas of the form $U(\alpha, \beta)$ in the end or $S(\alpha, \beta)$ in the start. To do so we generalise the idea of the semantic valuation function— the map which maps a time point to the set of formulas true then— to a more general class of functions (realization maps) which only have some of their properties. This is [10], definition 39:

Definition 7.1 Suppose that $x < y$ from $[0, 1]$. We say that ϕ-mosaic m is realized by the map μ on the closed interval $[x, y]$ iff the following conditions all hold:

R1. for each $z \in [x, y]$, $\mu(z)$ is a maximally propositionally consistent subset of $\text{Cl}\phi$;

R2. Suppose $z \in [x, y[$. Then $U(\alpha, \beta) \in \mu(z)$ iff either
R2.1, there is u such that $z < u \leq y$ and $\alpha \in \mu(u)$ and
for all v, if $z < v < u$ then $\beta \in \mu(v)$ or
R2.2, $\beta \in \mu(y)$, $U(\alpha, \beta) \in \mu(y)$ and
for all v, if $z < v < y$, then $\beta \in \mu(v)$;

R3. the mirror image of R2 for $S(\alpha, \beta)$;

R4. $\mu(x)$ is the start of m;

R5. $\mu(y)$ is the end of m; and

R6. for each $\beta \in \text{Cl}\phi$, β is in the cover of m iff for all u, if $x < u < y$, $\beta \in \mu(u)$.

A mosaic m is said to be realized on $[x, y]$ iff there exists a map μ such that m is realized on $[x, y]$ by μ. Say mosaic m is realised in $[0, 1]$ iff for all $x < y$ from $[0, 1]$, there is μ such that m is realised by μ on $[x, y]$. Say that m is realised iff it is realised on $[0, 1]$.

Consider the mosaic corresponding to an interval in a structure in the sense of definition 3.12. It should be clear that this mosaic is realized by the semantic valuation function for formulas at points within the interval, i.e. the semantic valuation function is a type of realization map.

Some lemmas from [10]:

Lemma 7.2 (Lemma 41 from [10]) *Suppose ψ is a $L(U, S)$ formula, and ψ-mosaic m is the composition of m' and m'' with each of m' and m'' being realised.*

Then m is realised.

Lemma 7.3 (Lemma 42 from [10]) *Suppose ψ is a $L(U, S)$ formula, ψ-mosaic m is fully decomposed by the tactic lead σ (and similarly trail) and each mosaic in σ is realised. Then m is realised.*

Lemma 7.4 (Lemma 44 from [10]) *Suppose ψ is a $L(U, S)$ formula, ψ-mosaic m is fully decomposed by the tactic shuffle $(\langle P_0, ..., P_s \rangle, \langle \lambda_1, ..., \lambda_r \rangle)$ and each mosaic in each λ_i is realised. Then m is realised.*

7.1 Approval Implies Realised

In the next few lemmas we show that approval in a $\mathbb{R}$-tableau implies being realised.

Lemma 7.5 *Suppose ψ is a $L(U, S)$ formula and ψ-mosaic m is approved in a $\mathbb{R}$-tableau because its children are approved. Then m is realised.*

Proof. By Lemma 7.2. □

Lemma 7.6 *Suppose ψ is a $L(U, S)$ formula and ψ-mosaic m is approved in a $\mathbb{R}$-tableau as a lead. Then m is realised. Similarly trail.*

Proof. By Lemma 7.3. □

7.2 Shuffle

The final possibility is that m is approved as a shuffle. Thus we have SH1-6 as follows.

(SH1) m is an unapproved mosaic.

(SH2) Every unapproved (label of a) descendent of (a node labelled by) m, including m itself, has some descendent which has at least two separate child nodes labelled by unapproved mosaics.

(SH3) All descendants of m which are unapproved have the same cover as m.

(SH4) is the "crisp start" requirement that we outlined above. Thus, there is a sequence $u = m_0, m_1, ...$ of unapproved mosaics in respective decompositions as follows. Each m_i is fully decomposed as $\sigma_i{}^\wedge\langle m_{i+1}\rangle^\wedge\pi_i$ where each mosaic in σ_i is approved already. We have $m_i = m_j$ for some $i < j$ such that for each $k = i, i+1, ..., j-1$, σ_k is empty.

Now some useful terminology when dealing with SH4. We have let σ_i be the (possibly empty) sequence of approved mosaics in the decomposition of m_i before m_{i+1} appears. We put $pre(u) = \sigma_0{}^\wedge\sigma_1{}^\wedge...{}^\wedge\sigma_{i-1}$ which ay be empty.

Similarly, (SH5), we require the unapproved descendants of m to have crisp ends using the mirror image construction.

Call the corresponding sequence $post(u)$.

The last check (SH6) before we approved m as a shuffle was to find an unapproved descendent u of m such that u has two adjacent children with unapproved labels v and w that further satisfy the following pattern.

We have a sequence $v = m_0, m_1, ...$ of unapproved mosaics in respective decompositions as follows. Each m_i is fully decomposed as $\sigma_i{}^\wedge\langle m_{i+1}\rangle$ and $m_i = m_j$ for some $i < j$. Note that in that case $post(v)$ is empty. The mirror image condition applied to w and we have $pre(w)$ empty as well with the end of v being the same as the start of w.

All the above (SH1-6) were checked before we approved m.

Let K be the set of unapproved mosaics v below m.

We claim that this set defines a shuffle as follows. We define a new set Σ of mosaics and *point-structures*, i.e. MCS subsets from $\mathrm{Cl}(\phi)$.

Suppose $w \in K$ and choose a full decomposition $F(w) = \langle v_1, ..., v_k\rangle$ of w from the tree with $v_i \in K$, $v_j \in K$ and $v_k \notin K$ for all $i < k < j$, for some $i \neq j$. Say that σ is the possibly empty sequence of mosaics $v_{i+1}, v_{i+2}, ..., v_{j-1}$. For all such w, i, j, we include in Σ a mosaic or point-structure corresponding to the composition of $post(v_i)^\wedge\sigma^\wedge pre(v_j)$ if that is non-empty, or a point structure being the start of v_j otherwise.

Note that by the shuffle restriction SH6 on $\mathbb{R}$-tableaux, there will be at least one such point structure in Σ.

If we look at a decompositions of m that are deep enough below m then we can find one of the form $pre(m)^\wedge\pi^\wedge post(m)$. Just keep decomposing mosaics at the start and end.

Let s be the composition of π. By SH3 and SH2, s will have cover the same as m. In fact we will have the following: $start(s)$ is the end of $pre(m)$ (or the start of m if $pre(m)$ is empty); $cover(s)$ is the cover of m; and $end(s)$ is the start of $post(m)$ (or the ned of m if $post(m)$ is empty).

We can also show that s is fully decomposed by the tactic shuffle $(\langle \Sigma_0 \rangle, \langle \Sigma' \rangle)$ where Σ_0 is the sequence of point-structures in Σ in any order and Σ' is the rest of Σ in any order.

To do so we use Lemma 4.4. Say $m = (A, B, C)$, $\Sigma_0 = \langle P_0, ..., P_s \rangle$ and $\Sigma' = \langle \lambda_1, ..., \lambda_r \rangle$.

S0) B is a subset of each P_i and of the start, end and cover of each mosaic in each λ_i. S0 holds as each point and each mosaic appears as a descendent of m which has cover B.

S1) each P_i satisfies both the forward and backwards $K(m)$ property. S1 holds as each element of Σ_0 is the start of a B cover mosaic.

Consider why $P_{i'}$ was put in Σ. There was $w \in K$ with a full decomposition $F(w) = \langle v_1, ..., v_k \rangle$ of w from the tree with $v_i \in K$, $v_j \in K$ and $v_k \notin K$ for all $i < k < j$, for some $i \neq j$. The sequence $v_{i+1}, v_{i+2}, ..., v_{j-1}$ is empty so that $j = i + 1$. We include in Σ the point-structure $P_{i'}$ when $post(v_i)$, σ and $pre(v_j)$ are all empty. In that case the end of v_i and the start of v_j are the same, and that is $P_{i'}$.

For $post(v_i)$ to be empty, there is a sequence $v_i = m_0, m_1, ...$ of unapproved mosaics in respective decompositions as follows. Each m_k is fully decomposed as $\pi_k{}^\wedge\langle m_{k+1} \rangle$. Suppose $m_l = m_j$ for some $l < j$.

Thus m_l is fully decomposed as $\pi_l{}^\wedge\langle m_{l+1} \rangle$ and v_i is the composition of $\pi_0{}^\wedge\langle m_1 \rangle$ which is he composition of $\pi_0{}^\wedge\pi_1{}^\wedge\langle m_2 \rangle$, etc which is the composition of $\pi_0{}^\wedge\pi_1{}^\wedge...{}^\wedge\pi_{k-1}\langle m_k \rangle$.

Thus the end of v_i and the end of m_k are the same.

By noting that the cover of m_{k+1}, the last mosaic in the full decomposition $\pi_l{}^\wedge\langle m_{l+1} \rangle$ for m_k, has the same cover as m, we can deduce that the end of m_k satisfies the backward $K(m)$ condition as required.

S2) the start of the first mosaic in each λ_i satisfies the backwards $K(m)$ property. S2 holds as the first mosaic starts with the end of B mosaic.

Ditto S3. S3) the end of the last mosaic in each λ_i satisfies the forwards $K(m)$ property.

S4) A satisfies the forward $K(m)$ property. S4 holds as A starts the mosaic which starts the shuffle.

Similarly S5. S5) C satisfies the backwards $K(m)$ property.

S6) if $\beta \in \text{Cl}\phi$ but $\neg\beta \notin B$ then either β is contained in some P_i or β is contained in the start or end of some mosaic in some λ_i. S6 holds as each B mosaic gets fully decomposed and therefore there is a witness to each such β in one of the sequences that we put together to get Σ.

Recall that m is just the composition of $pre(m){}^\wedge s^\wedge post(m)$ and so is realised as well.

In this subsection we have proved the following.

Lemma 7.7 *Suppose ψ is a $L(U, S)$ formula and ψ-mosaic m is approved in a $\mathbb{R}$-tableau as a shuffle. Then m is realised.*

7.3 Putting it all together

Now the main new lemma.

Lemma 7.8 *Suppose ψ is a $L(U,S)$ formula and ψ-mosaic m is approved in a $\mathbb{R}$-tableau. Then m is realised.*

Proof. We show by induction on the order of approving mosaic labels in a successful $\mathbb{R}$-tableau that all such mosaics are realised. Suppose that all mosaics so far approved are realised. Now suppose that ψ-mosaic m appears in a successful tableau T and gets approved.

There are four ways that m can get approved and we consider them case by case.

The simplest way that m is approved is when it labels a node and all the children nodes are approved. In this case we know that m is the composition of the child mosaic labels and all of those are realisable. Then m is approved by Lemma 7.5.

Another possibility is that m is approved as a lead. Use lemma 7.6 and we are done.

Similarly trail and shuffle (Lemma 7.7). □

Lemma 7.9 *Suppose ϕ is a $L(U,S)$ formula, not containing the atom q and $\psi = *_q^\phi(\phi)$. Say that there is a successful tableau for the ψ-mosaic m and it is (ϕ,q)-relativized.*

Then m is satisfied in a structure on the whole of $[0,1]$.

Proof. By Lemma 7.8, as m appears in a successful tableau then there is μ such that m is realised by μ on $[0,1]$.

As m is (ϕ,q)-relativized, m is satisfied in a structure on the whole of $[0,1]$.□

Lemma 7.10 *If $L(U,S)$ formula ϕ has a successful $\mathbb{R}$-tableau then ϕ is $\mathbb{R}$-satisfiable.*

Proof. Suppose $L(U,S)$ formula ϕ has a successful tableau.

Then there is an atom q not appearing in ϕ and $\psi = *_q^\phi(\phi)$ and ψ-mosaic m that is (ϕ,q)-relativized and has a successful tableau. It is the root of the tableau.

By Lemma 7.9, m is satisfied in a structure on the whole of $[0,1]$.

By Lemma 3.15, m, ϕ has a $\mathbb{R}$-flowed model. □

8 Completeness

Showing that satisfiable formulas have successful tableaux is not too hard when we can use the levels of an RMS and the way that we can use leads, trails and shuffles to get to the next level. In [10] it was quite clear that these operations correspond to simple repetitive patterns in a decomposition tree. They translate directly to good behaviour in tableaux.

For example, if m is fully decomposed by tactic lead applied to the sequence σ of mosaics at lower levels, then m has a tableau starting with a root with children m and then the mosaics in σ in order. There will be no central sticks because of the way leads and trails are defined. An induction takes care of the lower level σ mosaics and we are done.

Lemma 8.1 *Suppose ψ is a formula of $L(U,S)$, ψ-mosaic m is fully decomposed by the tactic lead σ (or trail) and each mosaic in σ has a successful $\mathbb{R}$-tableau in which m does not appear.*
Then m has a successful $\mathbb{R}$-tableau.

Equally, a shuffle tells us about a set of mutual decompositions which end up leaving a tableau with only lower level mosaics. See Definition 31, page 16/17 of [10]. The $\mathbb{R}$-tableau conditions can be checked directly on these decompositions.

Lemma 8.2 *Suppose ψ is a formula of $L(U,S)$, ψ-mosaic m is fully decomposed by the tactic shuffle $(\langle P_0, ..., P_s\rangle, \langle \lambda_1, ..., \lambda_r\rangle)$ and each mosaic in each λ_i has a successful $\mathbb{R}$-tableau in which m does not appear.*
Then m has a successful $\mathbb{R}$-tableau.

Put these two lemmas together in an induction and we get:

Lemma 8.3 *Suppose ψ is a formula of $L(U,S)$ and ψ-mosaic m appears in an RMS. Then m is the root of a successful $\mathbb{R}$-tableau.*

Then use the relativisation results to translate from mosaics to formulas:

Lemma 8.4 *If $L(U,S)$ formula ϕ is $\mathbb{R}$-satisfiable then ϕ has a successful $\mathbb{R}$-tableau.*

9 Termination, Complexity and Implementation Issues

It is easy to see that because we can, without loss of generality, stop at clone nodes, and limit branching factors, only a finite number of different tableaux need be considered for a formula. However, that is the end of the good news. There is an exponential bound on the number of different mosaics for a formula (in terms of its length). This also bounds the length of branches in a tableau. With a linear bound on the branching factor (—the defects need to be cured and any mosaics in between can be composed—) we thus have a double exponential bound on the size of any tableau in terms of number of nodes. There is thus a triple exponential bound on the number of tableaux which would govern the complexity of any exhaustive search through the tableaux.

However, by guessing a tableau of double exponential size we have a decision procedure that runs in 2-NEXPTIME.

Lemma 9.1 *In terms of the length of the input formula ϕ, there is a finite triple exponential bound on the number of tableaux for ϕ. A decision procedure runs in 2-NEXPTIME.*

The complexity of reasoning using such tableaux is thus 2-NEXPTIME.

In future work (joint with others) we will report on the possibilities for implementation of this technique. Early Java implementations [12] of a mosaic tableau for the logic US/LIN of $L(U,S)$ over general linear time show that any direct implementation of this tableau technique is quickly overwhelmed by the multi-exponential blow-up in data structures. The number of mosaics for a formula is a particular problem if they all need to be generated and checked. Clearly, more intelligent techniques are needed to make practical use of this basic framework. Our latest work [1] uses a notion of partial mosaics for the US/LIN case shows that there is great potential for speed-ups in practice.

Note that an implementation of the tableau reasoner for RTL would need two parts. First there is the tableau of mosaic decompositions which has a similar task to that of the US/LIN tableau in [12,1]. The second part is a much less computationally complex check through the successful tableau for the graph restrictions corresponding to approval.

References

[1] J. Bian, T. French and M. Reynolds. An Efficient Tableau for Linear Time Temporal Logic. In S Cranefield and A Nayak, editors, *26th Australasian Joint Conference on Artificial Intelligence, AI 2013, 1-6 December 2013, Dunedin, New Zealand*, Volume 8272. Pages 289–300, 2013.

[2] J. P. Burgess and Y. Gurevich. The decision problem for linear temporal logic. *Notre Dame J. Formal Logic*, 26(2):115–128, 1985.

[3] E. Emerson and E. C. Clarke. Using branching time temporal logic to synthesise synchronisation skeletons. *Sci. of Computer Programming*, 2, 1982.

[4] E. Emerson and J. Halpern. Decision procedures and expressiveness in the temporal logic of branching time. *J. Comp and Sys. Sci*, 30(1):1–24, 1985.

[5] D. M. Gabbay and I. M. Hodkinson. An axiomatisation of the temporal logic with until and since over the real numbers. *J. Logic and Computation*, 1(2):229 – 260, 1990.

[6] H. Kamp. *Tense logic and the theory of linear order.* PhD thesis, University of California, Los Angeles, 1968.

[7] H. Läuchli and J. Leonard. On the elementary theory of linear order. *Fundamenta Mathematicae*, 59:109–116, 1966.

[8] A. Rabinovich. Temporal logics over linear time domains are in PSPACE. *Inf. Comput*, 201:40–67, 2012.

[9] M. Reynolds. An axiomatization for Until and Since over the reals without the IRR rule. *Studia Logica*, 51:165–193, May 1992.

[10] M. Reynolds. The complexity of the temporal logic over the reals. *Annals of Pure and Applied Logic*, 161(8):1063–1096, 2010. Online at doi:10.1016/j.apal.2010.01.002.

[11] M. Reynolds. Dense time reasoning via mosaics. In *TIME '09: Proceedings of the 2009 16th International Symposium on Temporal Representation and Reasoning*, pages 3–10, Washington, DC, USA, 2009. IEEE Computer Society.

[12] M. Reynolds. A Tableau for General Linear Time. *Journal of Logic and Computation*, 23(5):1057–1080, 2013.

[13] M. Reynolds. A tableau for RTL (long version). Report online 2013 at `http://www.csse.uwa.edu.au/~mark/research/Online/rtltab.html`.

[14] J. G. Rosenstein. *Linear Orderings.* Academic Press, New York, 1982.

Bilattice Public Announcement Logic

Umberto Rivieccio [1]

Delft University of Technology
Jaffalaan 5, 2628 BX Delft
The Netherlands

Abstract

Building on recent work on bilattice modal logic and extensions of public announcement logic to a non-classical setting, we introduce a dynamic epistemic logic having the logic of modal bilattices as propositional support. Bilattice logic is both inconsistency-tolerant and paracomplete, thus suited for applications in contexts with multiple sources of information, where one may have to deal with lacking as well as potentially contradictory evidence. We introduce an algebra-based semantics for bilattice public announcement logic as well as a relational semantics based on many-valued Kripke models. We show via duality that the two semantics are equivalent and axiomatize the resulting logic by means of a Hilbert-style calculus. Our results and methodology extend recent work on non-classical dynamic epistemic logics such as intuitionistic public announcement logic.

Keywords: Bilattices, public announcement, epistemic updates, dynamic logic, modal logic, inconsistency-tolerant logic, many-valued logic.

1 Introduction

Dynamic logics are language expansions of classical (modal) logic designed to reason about changes induced by actions of different kinds, e.g. updates on the memory state of a computer, displacements of a moving robot, belief-revisions changing the common ground among different cognitive agents, knowledge update. Semantically, an action is represented as a transformation of a model describing a given state of affairs into a new one that represents the state of affairs after the action has been performed.

The logic of public announcements [15], [2], [7], [3] is a simple and well-known dynamic logic that models the epistemic change on the cognitive state of a group of agents resulting from a given proposition becoming publicly known. To each proposition α one associates a *dynamic* modal operator $\langle\alpha\rangle$ whose

[1] This research has been supported by Vidi grant 016.138.314 of the Netherlands Organization for Scientific Research (NWO). The author would like to thank Alessandra Palmigiano and Guiseppe Greco for several helpful discussions on earlier versions of the paper.

semantic interpretation is given by the transformation of models corresponding to its action-parameter α.

The present paper builds on the logic of public announcements (PAL) developed in [14],[13], [2] on the one hand, and on the bilattice-valued modal logic [12] on the other. [14], [13] introduce a semantically justified definition of dynamic epistemic logic on a base that is weaker than classical logic: the main methodological novelty of these papers is the dual characterization of epistemic updates via Stone-type dualities.

It is well known that epistemic updates induced by public announcements are formalized in relational models by means of the relativization construction, which creates a submodel of the original model. In [14] the corresponding submodel injection map is dually represented as a quotient construction between the complex algebras of the original model and of the updated one. This construction allows one to study epistemic updates within mathematical environments having a support that is weaker than classical logic.

Here we develop a similar study in a context that is yet more general. As propositional base we take the bilattice logic introduced by Arieli and Avron [1], which is both an inconsistency-tolerant and a paracomplete logic. Epistemic (i.e. *static*) modalities are modeled using the framework of the bilattice modal logic introduced in [12].

The algebraic framework of bilattices [10] and their associated logic builds on seminal ideas of Belnap [4], [5] motivated by the issue of dealing with incomplete and potentially inconsistent information. This setting has been further developed in [1] and generalized to weaker logics in, e.g., [11], [6]. In particular, [12] expands the language of bilattice logic with modal operators that are interpreted in many-valued analogues of Kripke frames.

In the present paper we generalize the quotient construction of [14] to the algebraic semantics of bilattice modal logic, which allows us to define a natural interpretation of the language of PAL on modal bilattices. In this way we establish which interaction axioms among dynamic modalities are sound with respect to our intended semantics. The resulting calculus defines a bilattice-based version of public announcement logic (called *bilattice public announcement logic*, BPAL), which we prove to be complete with respect to our algebra-based semantics analogously to classical PAL. We also introduce an equivalent relational semantics for BPAL based on many-valued Kripke frames, which is obtained from the algebraic semantics via a Stone-type duality. Preliminary results on BPAL are contained in [16], to which we will sometimes refer in order to shorten our proofs.

The main aim of our work is to pave the way to a semantically-grounded analysis of epistemic updates in the presence of incomplete and/or inconsistent information. It is also a contribution to the research line initiated in [14], [13], which aims at introducing methods of algebraic logic, duality and proof theory in the study of mathematical foundations of dynamic logic (see also [8], [9]).

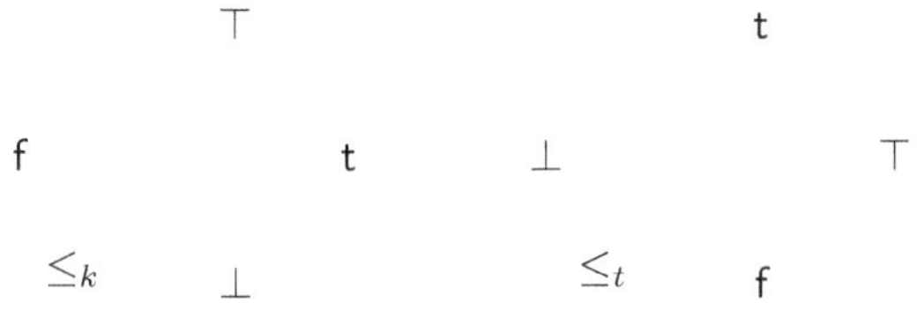

Fig. 1. The four-element Belnap bilattice $\mathcal{FOUR}$ in its two orders

2 Bilattice modal logic

In this section we introduce the setting of bilattice modal logic and recall facts and definitions that will be needed to develop a bilattice public announcement logic. We refer the reader to [12] for proofs and further details. The non-modal basis of bilattice modal logic is the logic introduced by Arieli and Avron [1], which can be defined through Belnap's (bi)lattice FOUR (Figure 1). We view FOUR as an algebra having operations $\langle \wedge, \vee, \otimes, \oplus, \supset, \neg, \mathsf{f}, \mathsf{t}, \bot, \top \rangle$ of type $\langle 2, 2, 2, 2, 2, 1, 0, 0, 0, 0 \rangle$. Both $\langle \mathsf{FOUR}, \wedge, \vee, \mathsf{f}, \mathsf{t} \rangle$ and $\langle \mathsf{FOUR}, \otimes, \oplus, \bot, \top \rangle$ are bounded distributive lattices, as shown in Figure 1, whose lattice orders are denoted, respectively, by $\leq_t$ (*truth order*) and $\leq_k$ (*knowledge order*). We have, moreover, a binary *weak implication* operation $\supset$ defined by $x \supset y := y$ if $x \in \{\mathsf{t}, \top\}$ and $x \supset y := \mathsf{t}$ otherwise. *Negation* is a unary operation $\neg$ having $\bot$ and $\top$ as fixed points and such that $\neg \mathsf{f} = \mathsf{t}$ and $\neg \mathsf{t} = \mathsf{f}$.

We have included the operations $\otimes$ and $\oplus$ in the primitive signature as they are essential ingredients of bilattices as they were originally introduced, and of the motivation behind them. In the present context, however, they can be retrieved as terms in the language $\langle \wedge, \vee, \supset, \neg, \mathsf{f}, \mathsf{t}, \bot, \top \rangle$. We will thus consider them as abbreviations of the terms shown below, together with the following defined operations:

$$
\begin{aligned}
x \otimes y &:= (x \wedge \bot) \vee (y \wedge \bot) \vee (x \wedge y) \\
x \oplus y &:= (x \wedge \top) \vee (y \wedge \top) \vee (x \wedge y) \\
\sim x &:= x \supset \mathsf{f} \\
x \to y &:= (x \supset y) \wedge (\neg y \supset \neg x) \\
x * y &:= \neg (y \to \neg x) \\
x \equiv y &:= (x \supset y) \wedge (y \supset x) \\
x \leftrightarrow y &:= (x \to y) \wedge (y \to x).
\end{aligned}
$$

The operation $\sim$ provides an alternative negation, while $\to$ is an alternative implication called *strong implication*, which is adjoint to the operation $*$, called strong conjunction or *fusion*. The operations $\langle *, \to \rangle$ form a *residuated pair* (in the residuated lattice sense [11]) w.r.t. the truth order of FOUR, and so $\to$ can be seen as a truth-implication. It might be possible to consider, dually, a knowledge-implication, but we will not pursue this here; as mentioned in [12], this option seems to be technically less viable in a modal logic setting.

The bilattice logic of [1] can be introduced as the propositional logic defined by the matrix $\langle \mathsf{FOUR}, \{\mathsf{t}, \top\}\rangle$ as follows. Starting from a countable set of propositional variables Var, one constructs the formula algebra $\mathbf{Fm} = \langle Fm, \wedge, \vee, \supset, \neg, \mathsf{f}, \mathsf{t}, \bot, \top\rangle$ in the usual way. Given formulas $\Gamma, \{\varphi\} \subseteq Fm$, one sets $\Gamma \models_{\mathsf{FOUR}} \varphi$ iff, for all homomorphisms $v\colon \mathbf{Fm} \to \mathsf{FOUR}$, if $v(\gamma) \in \{\mathsf{t}, \top\}$ for all $\gamma \in \Gamma$, then also $v(\varphi) \in \{\mathsf{t}, \top\}$. This logic can be axiomatized through the Hilbert-style calculus introduced in [12, Section III B]. It is sufficient to take all axioms of classical logic in the language $\langle \wedge, \vee, \supset, \mathsf{f}, \mathsf{t}\rangle$ plus the following:

$\top \wedge \neg\top$	$\neg(p \supset q) \equiv (p \wedge \neg q)$	$\neg(p \wedge q) \equiv (\neg p \vee \neg q)$
$(\bot \vee \neg\bot) \supset \mathsf{f}$		$\neg(p \vee q) \equiv (\neg p \wedge \neg q)$

The only rule is *modus ponens* (mp): $p, p \supset q \vdash q$. Notice that the above axioms involving the bilattice negation $\neg$ are not derivable from those of classical logic, because $\neg$ is not defined from the falsum constant and implication in the usual way.

This logic can be semantically expanded with modal operators by considering *four-valued Kripke models*. These are structures $\langle W, R, v\rangle$ such that both R and v are four-valued. That is, one defines $R\colon W \times W \to \mathsf{FOUR}$ and $v\colon \mathbf{Fm} \times W \to \mathsf{FOUR}$. We then call $\langle W, R\rangle$ a *four-valued Kripke frame*. Valuations are required to be homomorphisms in their first argument, so they preserve all non-modal connectives (including the four constants) of the logic of FOUR. The modal operator $\Box$ is defined as follows: for every $w \in W$ and every $\varphi \in Fm$,

$$v(\Box\varphi, w) := \bigwedge\{R(w, w') \to v(\varphi, w') : w' \in W\}$$

where $\bigwedge$ denotes the infinitary version of $\wedge$ in FOUR and $\to$ is the strong implication introduced above. If we replace FOUR by the two-element Boolean algebra and $\wedge, \to$ with classical conjunction and implication, this can be readily seen to be a generalization of the classical case. Notice that *all* worlds $w' \in W$ are taken into account to evaluate $v(\Box\varphi, w)$.

The dual operator $\Diamond$ is defined as

$$v(\Diamond\varphi, w) := \bigvee\{R(w, w') * v(\varphi, w') : w' \in W\}$$

where $\bigvee$ denotes the infinitary version of $\vee$ in FOUR and $*$ is the *fusion* operation introduced above. It is straightforward to check that $v(\Box\varphi, w) = v(\neg\Diamond\neg\varphi, w)$ for all $w \in W$ and all valuations v. Thus, as happens in the classical case (and unlike the intuitionistic one), the two modal operators are inter-definable. In the present paper we will take $\Diamond$ as primitive.

A modal consequence relation can now be defined in the usual way. We say that a point $w \in W$ of a four-valued model $M = \langle W, R, v\rangle$ satisfies a formula $\varphi \in Fm$ iff $v(\varphi, w) \in \{\mathsf{t}, \top\}$, and we write $M, w \models \varphi$. For a set of formulas $\Gamma \subseteq Fm$, we write $M, w \models \Gamma$ to mean that $M, w \models \gamma$ for each $\gamma \in \Gamma$. The (local) consequence $\Gamma \models \varphi$ holds if, for every model $M = \langle W, R, v\rangle$ and every $w \in W$, it is the case that $M, w \models \Gamma$ implies $M, w \models \varphi$.

Notice that this consequence relation inherits from the non-modal fragment the deduction-detachment theorem in the following form: $\Gamma \models \varphi$ if and only if $\emptyset \models \bigwedge \Gamma \supset \varphi$, where $\bigwedge \Gamma := \bigwedge \{\gamma \in \Gamma\}$. This, which will remain true about its dynamic expansion BPAL, implies that we can without loss of generality restrict our attention to valid formulas.

The above-defined consequence is axiomatized in [12]. The set of axioms is the least set $\Sigma \subseteq Fm$ containing all substitution instances of the schemata axiomatizing non-modal bilattice logic plus the following ones:

$(\Box\ \mathsf{t})$	$\Box \mathsf{t} \leftrightarrow \mathsf{t}$
$(\Box\ \wedge)$	$\Box(p \wedge q) \leftrightarrow (\Box p \wedge \Box q)$
$(\Box\ \bot)$	$\Box(\bot \to p) \leftrightarrow (\bot \to \Box p)$

Moreover, Σ must satisfy: (val-mp) if φ and $\varphi \supset \psi$ are in Σ, then so is ψ; (val-mono) if $\varphi \to \psi$ is in Σ, then so is $\Box\varphi \to \Box\psi$. The only inference rule is (mp). We notice that (val-mono) replaces the more common *necessitation* rule (if $\varphi \in \Sigma$, then $\Box\varphi \in \Sigma$) because the latter would not be sound in our setting [12, Section III.A].

This calculus is complete not only with respect to the semantics of four-valued Kripke models, but also with respect to an algebra-based semantics given by the class of *modal bilattices*. We briefly recall these results in the remaining part of this section, as we will build on them later on. We begin with completeness with respect to Kripke models [12, Theorem 19].

Theorem 2.1 (Relational completeness) *For all* $\Gamma, \{\varphi\} \subseteq Fm$, $\Gamma \vdash \varphi$ *iff* $M, w \models \Gamma$ *implies* $M, w \models \varphi$ *for every four-valued Kripke model* $M = \langle W, R, v\rangle$ *and every* $w \in W$.

In order to state the algebraic completeness theorem we need to introduce a class of algebras providing an alternative semantics for our calculus. A *modal bilattice* is an algebra $\mathbf{B} = \langle B, \wedge, \vee, \supset, \neg, \Diamond, \mathsf{f}, \mathsf{t}, \bot, \top\rangle$ such that the $\Diamond$-free reduct of $\mathbf{B}$ is an *implicative bilattice* [2], that is, the algebra $\langle B, \wedge, \vee, \supset, \neg, \mathsf{f}, \mathsf{t}, \bot, \top\rangle$ belongs to the variety generated by FOUR, and moreover the following identities are satisfied:

(i)	$\Diamond \mathsf{f} = \mathsf{f}$
(ii)	$\Diamond(p \vee q) = (\Diamond p \vee \Diamond q)$
(iii)	$\Box(x \supset \bot) = \Diamond x \supset \bot$

Thus, in particular, $\langle B, \wedge, \vee, \mathsf{f}, \mathsf{t}\rangle$ is a bounded distributive lattice. It is easy to show that identities (i)-(iii) correspond, respectively, to axioms (i)-(iii) of our calculus, and that the presentation of modal bilattices given here is equivalent to that of [12].

Given a modal bilattice $\mathbf{B}$ and a subset $F \subseteq B$, we say that F is a *bifilter* if F is a lattice filter of $\langle B, \wedge, \vee, \mathsf{f}, \mathsf{t}\rangle$ such that $\top \in F$. Given a pair $\langle \mathbf{B}, F\rangle$

[2] An abstract equational of implicative bilattices can be found in [6].

and formulas $\Gamma, \{\varphi\} \subseteq Fm$, we write $\Gamma \models_{\langle B,F \rangle} \varphi$ to mean that, for every modal bilattice homomorphism $v \colon \mathbf{Fm} \to \mathbf{B}$, if $v(\gamma) \in F$ for all $\gamma \in \Gamma$, then also $v(\varphi) \in F$. A valid formula φ is one such that $v(\varphi) \geq_t \top^{\mathbf{B}}$ for every $\mathbf{B}$ and v. We can then state the algebraic completeness result [12, Theorem 10] as follows.

Theorem 2.2 (Algebraic completeness) *For all $\Gamma, \{\varphi\} \subseteq Fm$, $\Gamma \vdash \varphi$ iff $\Gamma \models_{\langle B,F \rangle} \varphi$ for any modal bilattice $\mathbf{B}$ and any bifilter $F \subseteq B$.*

Just as in the case of classical modal logic, the relational and the algebraic semantics of bilattice modal logic are interrelated via a Stone-type duality [12, Theorem 18]. In the case of bilattices, another essential ingredient is the so-called *twist-structure* representation. Let $\mathbf{A} = \langle A, \wedge, \vee, \sim, \Diamond_+, \Diamond_-, 0, 1 \rangle$ be a *bimodal Boolean algebra* [12, Definition 11], i.e. a structure such that $\langle A, \wedge, \vee, \sim, 0, 1 \rangle$ is a Boolean algebra and $\Diamond_+$ and $\Diamond_-$ are unary operators that preserve finite joins (no relation between the two is required). The dual operators $\Box_+$ and $\Box_-$ are defined in the usual way as $\Box_+ x := \sim \Diamond_+ \sim x$ and $\Box_- x := \sim \Diamond_- \sim x$. The *twist-structure over* $\mathbf{A}$ is the algebra $\mathbf{A}^{\bowtie} = \langle A \times A, \wedge, \vee, \supset, \neg, \Diamond, \mathsf{f}, \mathsf{t}, \bot, \top \rangle$ with operations given, for all $\langle a_1, a_2 \rangle, \langle b_1, b_2 \rangle \in A \times A$, by:

$$
\begin{aligned}
\langle a_1, a_2 \rangle \wedge \langle b_1, b_2 \rangle &:= \langle a_1 \wedge b_1,\ a_2 \vee b_2 \rangle \\
\langle a_1, a_2 \rangle \vee \langle b_1, b_2 \rangle &:= \langle a_1 \vee b_1,\ a_2 \wedge b_2 \rangle \\
\langle a_1, a_2 \rangle \supset \langle b_1, b_2 \rangle &:= \langle \sim a_1 \vee b_1, a_1 \wedge b_2 \rangle \\
\neg \langle a_1, a_2 \rangle &:= \langle a_2,\ a_1 \rangle \\
\Diamond \langle a_1, a_2 \rangle &:= \langle \Diamond_+ a_1,\ \Box_+ a_2 \wedge \sim \Diamond_- a_1 \rangle \\
\mathsf{f} &:= \langle 0, 1 \rangle \\
\mathsf{t} &:= \langle 1, 0 \rangle \\
\bot &:= \langle 0, 0 \rangle \\
\top &:= \langle 1, 1 \rangle
\end{aligned}
$$

It is straightforward to check that any twist-structure is a modal bilattice. More interestingly, *any modal bilattice is isomorphic to a twist-structure* [12, Theorem 12]. This means that instead of working directly with modal bilattices, one can (and we will) without loss of generality focus only on twist-structures.

The twist-structure construction allows us to relate four-valued Kripke frames and modal bilattices via Jónsson-Tarski duality for classical modal logic. Given a modal bilattice $\mathbf{B}$ viewed as a twist-structure $\mathbf{A}^{\bowtie}$, we can consider the structure $\langle \mathbf{A}_\bullet, R_+, R_- \rangle$, where $\langle \mathbf{A}_\bullet, R_+ \rangle$ and $\langle \mathbf{A}_\bullet, R_- \rangle$ are the classical Kripke frames associated to the modal Boolean algebras $\langle A, \wedge, \vee, \sim, \Diamond_+, 0, 1 \rangle$ and $\langle A, \wedge, \vee, \sim, \Diamond_-, 0, 1 \rangle$ according to Jónsson-Tarski duality. The relations R_+ and R_- can obviously be combined into one four-valued relation R_4 by letting, for instance, $R(w, w') = \mathsf{t}$ iff $\langle w, w' \rangle \in R_+ \cap R_-$, $R(w, w') = \top$ iff $\langle w, w' \rangle \in R_+ \backslash R_-$, $R(w, w') = \bot$ iff $\langle w, w' \rangle \in R_- \backslash R_+$ and $R(w, w') = \mathsf{f}$

iff $\langle w, w' \rangle \notin R_+ \cup R_-$. In this way[3] we obtain a four-valued Kripke frame $\langle \mathbf{A}_\bullet, R \rangle$. Conversely, every four-valued Kripke frame $F = \langle W, R \rangle$ can be viewed as a pair of Kripke frames $\langle W, R_+ \rangle$, $\langle W, R_- \rangle$ by defining $\langle w, w' \rangle \in R_+$ iff $R(w, w') \in \{\mathsf{t}, \top\}$ and $\langle w, w' \rangle \in R_-$ iff $R(w, w') \in \{\mathsf{t}, \bot\}$. Thus we obtain classical Kripke frames $F_+ = \langle W, R_+ \rangle$ and $F_- = \langle W, R_- \rangle$, to which one associates modal Boolean algebras $(F_+)^\bullet$ and $(F_-)^\bullet$ according to Jónsson-Tarski duality. Since $(F_+)^\bullet$ and $(F_-)^\bullet$ share the same carrier set, we actually have a bimodal Boolean algebra $F^\bullet$, from which a modal bilattice $(F^\bullet)^{\bowtie}$ can be obtained via the twist-structure construction. It is shown in [12] that the correspondence between four-valued Kripke frames and modal bilattices extends to Kripke models and algebraic models, which implies that the relational and the algebraic semantics for bilattice modal logic are indeed equivalent.

3 Pseudo-quotients on modal bilattices

When considering epistemic updates in the context of bilattice logic, we have to take into account that validity of a formula in our logic only depends on its "positive part". In fact, any two formulas φ, ψ are logically equivalent if and only if, for every valuation $v\colon Fm \to \mathsf{FOUR}$, it holds that $\pi_1(v(\varphi)) = \pi_1(v(\psi))$, where π_1 denotes first component projection defined by the twist-structure representation of FOUR as $\{0,1\} \times \{0,1\}$. For instance, t and $\top$ (viewed as propositional constants) are both valid formulas (hence, logically equivalent) because $\pi_1(\mathsf{t}) = \pi_1(\top) = 1$. Thus, in particular, the public announcements of t or $\top$ should both be vacuous. This unusual feature, which depends only on the non-modal support of the logic, can be traced back to Belnap's proposal that derivations should preserve (only) positive evidence (see [4], [5] for a discussion of the intuitions justifying this choice). An alternative characterization of logical equivalence is the following: any two formulas φ, ψ are logically equivalent if and only if $v(\sim\sim\varphi) = v(\sim\sim\psi)$ for any valuation v. This remark motivates the definition of pseudo-quotients that we are going to introduce below, but before we proceed let us make one more observation that may help avoiding misunderstandings.

The fact that logical equivalence (and hence validity) of a formula only depends on its positive part does not mean that the negative part does not play any role in bilattice logic. For instance, announcing the negation of t (that is, f) does not have the same effect as announcing the negation of $\top$ (which is $\top$ itself): the latter announcement remains vacuous and thus does not produce any change in the original model, while the former, as in the classical case, makes the model collapse. This is due to an essential feature of bilattice negation, namely the fact that φ and ψ being logically equivalent does not entail that $\neg\varphi$ is equivalent to $\neg\psi$. As mentioned in [12, Section VIII], using the twist-structure representation it may indeed be possible to embed bilattice

[3] Although it is obviously possible to combine the information conveyed by R_+ and R_- in many alternative ways, the work in [12] indicates that the one suggested above is, at least from a technical point of view, the most suitable one.

modal logic in classical (bi)modal logic, but this is not as straightforward as it may seem, for in order to account for the negative part too one would need to translate formulas of bilattice logic into *pairs* of formulas of classical logic.

Given a modal bilattice $\mathbf{B}$ and an element $a \in B$, we define a relation $\equiv_a$ by the following prescription: for all $b, c \in B$,

$$b \equiv_a c \qquad \text{iff} \qquad b \wedge \sim\sim a = c \wedge \sim\sim a.$$

This definition is adapted from (and can indeed be seen as a special case of) that of [14] [4]. The only difference is that, as noted above, here we need to consider only the positive part of $a \in B$, hence the term $\sim\sim a$. We are now going to see that the above-defined relation is indeed a congruence of the non-modal reduct of any modal bilattice.

Lemma 3.1 ([16], Lemma 2.1) *Let* $\mathbf{A}^{\bowtie}$ *be a twist-structure over a Boolean algebra* $\mathbf{A}$. *Then, for all* $\langle a_1, a_2\rangle, \langle b_1, b_2\rangle, \langle c_1, c_2\rangle \in A \times A$,

$$\langle b_1, b_2\rangle \equiv_{\langle a_1, a_2\rangle} \langle c_1, c_2\rangle \qquad \textit{iff} \qquad b_1 \equiv_{a_1} c_1 \quad \textit{and} \quad b_2 \equiv_{a_1} c_2$$

where $\equiv_{a_1}$ *is defined as in [14, Section 3.2], i.e.,* $x \equiv_{a_1} y$ *iff* $x \wedge a_1 = y \wedge a_1$.

Notice that in the preceding lemma, as mentioned above, the negative part a_2 of the pair $\langle a_1, a_2\rangle$ does not play any role. The following result is a straightforward consequence.

Fact 3.2 ([16], Fact 2.2) *For any modal bilattice* $\mathbf{B}$ *and any* $a \in B$, *the relation* $\equiv_a$ *is a congruence of the non-modal reduct of* $\mathbf{B}$.

As happened in [14], our relation $\equiv_a$ is in general not compatible with the modal operator(s). The next step is thus to find a suitable definition for modal operators on the pseudo-quotient. We begin with the following observation (cf. [14, Fact 6]).

Fact 3.3 *Let* $\mathbf{B}$ *be a modal bilattice and* $a \in B$. *Then*

(i) $[b \wedge \sim\sim a] = [b]$ *for every* $b \in B$. *Hence, for every* $b \in B$, *there exists a unique* $c \in B$ *such that* $c \in [b]_a$ *and* $c \leq_t \sim\sim a$.

(ii) $[b] \leq_t [c]$ *iff* $b \wedge \sim\sim a \leq_t c \wedge \sim\sim a$ *for all* $b, c \in B$.

Proof. Essentially the same as [14, Fact 6], replacing a by $\sim\sim a$. □

Item (i) of Fact 3.3 implies that for each equivalence class modulo $\equiv_a$ we can choose a canonical representative, namely the unique element in the given class that is below $\sim\sim a$ in the truth order. Hence we can define an (injective) map $i' = i'_a \colon \mathbf{B}^a \to \mathbf{B}$ given, for every $[b] \in B^a$, by $i'[b] := b \wedge \sim\sim a$. Notice, moreover, that the composition $\pi \cdot i'$ is the identity on $\mathbf{B}^a$.

---

[4] Notice however that, while the equation $\neg\neg x = x$ is valid in any bilattice, in general it is not the case that $\sim\sim x = x$. This explains why our definition does not coincide with that of [14], and is also the reason why, even if two formulas φ and ψ are logically equivalent (hence $v(\sim\sim\varphi) = v(\sim\sim\psi)$ for any valuation v), it may well happen that $v(\varphi) \neq v(\psi)$.

At this point we are ready to introduce modal operator(s) on the pseudo-quotient. Given $a, b \in \mathbf{B}$, we let

$$\Diamond^a[b] := [\Diamond(b \wedge \sim\sim a)] = [\Diamond(b \wedge \sim\sim a) \wedge \sim\sim a]$$

The dual operator is defined as $\Box^a[b] := \neg\Diamond^a\neg[b]$. Using Fact 3.2 and the identities of modal bilattices, it is easy to check that, in keeping with [14, Section 3.3.2], we have

$$\Box^a[b] = [\Box(a \supset b)] = [a \supset \Box(a \supset b)].$$

This could thus be taken as an alternative but equivalent definition.

The following result shows that our definition indeed suits our purpose (cf. [14, Fact 10]).

Fact 3.4 ([16], Fact 2.4) *For every modal bilattice* $\mathbf{B}$ *and all* $a, b, c \in B$*:*

(i) $\Diamond^a[\mathsf{f}] = [\mathsf{f}]$

(ii) $\Diamond^a([b] \vee [c]) = \Diamond^a[b] \vee \Diamond^a[c]$

(iii) $\Box^a([b] \supset [\bot]) = \Diamond^a[b] \supset [\bot]$

(iv) Hence, $(\mathbf{B}^a, \Diamond^a)$ *is a modal bilattice.*

The following lemma relates the pseudo-quotient construction and the twist-structure representation of modal bilattices. This will be used in Section 5.2.

Lemma 3.5 *Let* $\mathbf{A}^{\bowtie}$ *be a modal twist-structure over a bimodal Boolean algebra* $\mathbf{A}$*, and let* $\langle a_1, a_2\rangle \in A \times A$*. Then* $(\mathbf{A}^{\bowtie})^{\langle a_1, a_2\rangle} \cong (\mathbf{A}^{a_1})^{\bowtie}$.

4 Axiomatization fo BPAL

Our calculus for bilattice public announcement logic is defined over the language $\langle \wedge, \vee, \supset, \neg, \Diamond, \langle\alpha\rangle, \mathsf{f}, \mathsf{t}, \bot, \top\rangle$, where $\alpha \in Fm$. Derived connectives $\langle \sim, \Box, \otimes, \oplus, \rightarrow, *, \leftrightarrow\rangle$ are introduced as before. Moreover, we let $[\alpha]\varphi := \neg\langle\alpha\rangle\neg\varphi$. BPAL is axiomatically defined by the axioms and rules of the above-mentioned (local) calculus for bilattice modal logic [12] augmented with the following axioms:

Interaction with logical constants	$\langle\alpha\rangle\mathsf{f} \leftrightarrow \mathsf{f}$	$\langle\alpha\rangle\mathsf{t} \leftrightarrow \sim\sim\alpha$
	$\langle\alpha\rangle\top \leftrightarrow (\alpha \wedge \top)$	$\langle\alpha\rangle\bot \leftrightarrow \neg(\alpha \supset \bot)$
Interaction with $\wedge$	$\langle\alpha\rangle(\varphi \wedge \psi) \leftrightarrow (\langle\alpha\rangle\varphi \wedge \langle\alpha\rangle\psi)$	
Interaction with $\vee$	$\langle\alpha\rangle(\varphi \vee \psi) \leftrightarrow (\langle\alpha\rangle\varphi \vee \langle\alpha\rangle\psi)$	
Interaction with $\supset$	$\langle\alpha\rangle(\varphi \supset \psi) \leftrightarrow (\sim\sim\alpha \wedge (\langle\alpha\rangle\varphi \supset \langle\alpha\rangle\psi))$	
Interaction with $\neg$	$\langle\alpha\rangle\neg\varphi \leftrightarrow (\sim\sim\alpha \wedge \neg\langle\alpha\rangle\varphi)$	
Interaction with $\Diamond$	$\langle\alpha\rangle\Diamond\varphi \leftrightarrow (\sim\sim\alpha \wedge \Diamond\langle\alpha\rangle\varphi)$	
Preservation of facts	$\langle\alpha\rangle p \leftrightarrow (\sim\sim\alpha \wedge p)$	

where φ, ψ, α are arbitrary formulas, while p is a propositional variable. We observe that, using the rules and axioms of the non-modal basis of BPAL, it is easy to establish that the following formulas are derivable:

Interaction with logical constants	$[\alpha]\mathsf{f} \leftrightarrow {\sim}\alpha$	$[\alpha]\mathsf{t} \leftrightarrow \mathsf{t}$
	$[\alpha]\top \leftrightarrow \alpha \supset \top$	$[\alpha]\bot \leftrightarrow (\alpha \supset \bot)$
Interaction with $\wedge$	$[\alpha](\varphi \wedge \psi) \leftrightarrow ([\alpha]\varphi \wedge [\alpha]\psi)$	
Interaction with $\vee$	$[\alpha](\varphi \vee \psi) \leftrightarrow (\alpha \supset ([\alpha]\varphi \vee [\alpha]\varphi))$	
Interaction with $\supset$	$[\alpha](\varphi \supset \psi) \leftrightarrow (\langle\alpha\rangle\varphi \supset \langle\alpha\rangle\psi)$	
Interaction with $\neg$	$[\alpha]\neg\varphi \leftrightarrow \neg\langle\alpha\rangle\varphi$	
Interaction with $\Diamond$	$[\alpha]\Diamond\varphi \leftrightarrow (\alpha \supset \Diamond\langle\alpha\rangle\varphi)$	
Interaction with $\Box$	$[\alpha]\Box\varphi \leftrightarrow (\alpha \supset \Box[\alpha]\varphi)$	
Preservation of facts	$[\alpha]p \leftrightarrow (\alpha \supset p)$	

5 Algebraic and relational models of BPAL

In this section we introduce two kinds of semantics that will be proven to be (equivalent and) complete with respect to the calculus introduced in Section 4. The first kind is the algebraic semantics for BPAL, which we define as indicated by the algebraic analysis of pseudo-quotients on modal bilattices developed in Section 3. The second kind is the relational semantics based on Kripke models. We are then going to use duality to see that the two semantics are indeed equivalent.

5.1 Algebraic semantics

We define an *algebraic model* as a tuple $M = (\mathbf{B}, v)$ where $\mathbf{B}$ is a modal bilattice and $v \colon Var \to B$. The *extension* map $[\![\cdot]\!]_M \colon Fm \to \mathbf{B}$ is defined as follows:

$$\begin{aligned}
[\![p]\!]_M &:= v(p) \\
[\![c]\!]_M &:= c^{\mathbf{B}} && \text{for } c \in \{\mathsf{f}, \mathsf{t}, \bot, \top\} \\
[\![\circ\varphi]\!]_M &:= \circ^{\mathbf{B}}[\![\varphi]\!]_M && \text{for } \circ \in \{\neg, \Diamond\} \\
[\![\varphi \bullet \psi]\!]_M &:= [\![\varphi]\!]_M \bullet^{\mathbf{B}} [\![\psi]\!]_M && \text{for } \bullet \in \{\wedge, \vee, \supset\} \\
[\![\langle\alpha\rangle\varphi]\!]_M &:= {\sim}{\sim}[\![\alpha]\!]_M \wedge^{\mathbf{B}} i'([\![\varphi]\!]_{M^\alpha}) \\
[\![[\alpha]\varphi]\!]_M &:= [\![\alpha]\!]_M \supset^{\mathbf{B}} i'([\![\varphi]\!]_{M^\alpha})
\end{aligned}$$

where $M^\alpha = (\mathbf{B}^\alpha, v^\alpha)$ is given by $\mathbf{B}^\alpha = \mathbf{B}^{[\![\alpha]\!]_M}$ and $v^\alpha = \pi \circ v \colon Var \to \mathbf{B}^\alpha$. That is, $[\![p]\!]_{M^\alpha} = v^\alpha(p) = \pi(v(p)) = \pi([\![p]\!]_M)$ for every $p \in Var$.

We define $\Gamma \vDash_{BPAL} \varphi$ iff, for every algebraic model $M = (\mathbf{B}, F, v)$, it holds that $[\![\gamma]\!] \in F$ for all $\gamma \in \Gamma$ implies $[\![\varphi]\!] \in F$. We will see in the next section that the calculus introduced in Section 4 is sound and complete with respect to the semantics provided by the above-defined algebraic models. We are now going to use duality theory and algebraic semantics to introduce a relational semantics for BPAL.

5.2 Relational semantics and duality

Consider a four-valued Kripke frame $\mathcal{F}$. For simplicity we view the four-valued accessibility relation R as split into two standard relations, so we let $\mathcal{F} =$

$\langle W, R_+, R_- \rangle$ and let $s \subseteq W$. We define $\mathcal{F}^s = \langle W^s, R_+^s, R_-^s \rangle$, *the subframe of* $\mathcal{F}$ *relativized to* s, as follows (cf. [14, Definition 19]): $W^s = s$, $R_+^s = R_+ \cap (s \times s)$ and $R_-^s = R_- \cap (s \times s)$. Given a four-valued Kripke model $M = \langle W, R_+, R_-, v \rangle$ and $\alpha \in Fm$, we define $v_+(\alpha) := \{w \in W : v(\alpha, w) \in \{\mathsf{t}, \top\}\}$. Analogously, we can define $v_-(\alpha) := \{w \in W : v(\alpha, w) \in \{\mathsf{f}, \top\}\}$ but notice that $v_-(\alpha) = v_+(\neg\alpha)$. The submodel $M^\alpha = \langle W^\alpha, R_+^\alpha, R_-^\alpha, v^\alpha \rangle$ is then defined as follows. $W^\alpha := v_+(\alpha)$, $R_+^\alpha := R^+ \cap (W^\alpha \times W^\alpha)$, $R_-^\alpha := R^- \cap (W^\alpha \times W^\alpha)$, and for all $p \in Var$ and $w \in W^\alpha$,

$$v^\alpha(p, w) = \begin{cases} \mathsf{t} \text{ iff } w \in v(p) \text{ and } w \notin v(\neg p) \\ \top \text{ iff } w \in v(p) \text{ and } w \in v(\neg p) \\ \bot \text{ iff } w \notin v(p) \text{ and } w \notin v(\neg p) \\ \mathsf{f} \text{ iff } w \notin v(p) \text{ and } w \in v(\neg p) \end{cases}$$

Extending v^α to arbitrary formulas in the usual way, we can introduce a notion of satisfaction for BPAL formulas of type $\langle \alpha \rangle \varphi$ as follows:

$$M, w \vDash \langle \alpha \rangle \varphi \quad \text{iff} \quad M, w \vDash \alpha \text{ and } M^\alpha, w \vDash \varphi.$$

Noticing that $M, w \vDash \alpha$ iff $M, w \vDash \sim\sim\alpha$, one easily sees that the above definition is in keeping with the algebraic one given in the preceding section.

In order to prove equivalence between the algebraic and the relational semantics for BPAL, we consider *complex algebras* of four-valued Kripke frames as defined in [12]. For any four-valued Kripke frame $\mathcal{F} = \langle W, R_+, R_- \rangle$, following Jónsson-Tarski duality for classical modal logic, we can construct the complex algebras of the two frames $\langle W, R_+ \rangle$ and $\langle W, R_- \rangle$, which are the structures $\langle P(W), \cap, \cup, \sim, \Diamond_+ \rangle$ and $\langle P(W), \cap, \cup, \sim, \Diamond_- \rangle$, where $\sim$ is the Boolean complement operation and $\Diamond_+ U := R_+^{-1}[U]$, $\Diamond_- U := R_-^{-1}[U]$ for all $U \subseteq W$. These are not only modal Boolean algebras, they are also *perfect* (see below). The structure $\langle P(W), \cap, \cup, \sim, \Diamond_+, \Diamond_- \rangle$ is thus a bimodal Boolean algebra. We can then apply the twist-structure construction introduced in Section 2 to obtain a modal bilattice. We define the *complex algebra* of $\mathcal{F} = \langle W, R_+, R_- \rangle$ as the twist-structure $\mathcal{F}^\bullet = \langle P(W), \cap, \cup, \sim, \Diamond_+, \Diamond_- \rangle^{\bowtie}$.

Given a four-valued Kripke model $M = \langle \mathcal{F}, v \rangle$, we can define a valuation $v^\bullet \colon Var \to \mathcal{F}^\bullet$, for every $p \in Var$, as $v^\bullet(p) := \langle v_+(p), v_+(\neg p) \rangle$, where $v_+(p) := \{w \in W : v(p, w) \in \{\mathsf{t}, \top\}\}$ as before. We then extend $v^\bullet$ homomorphically to any formula φ in the language of bilattice modal logic and we set $\langle \mathcal{F}^\bullet, v^\bullet \rangle \vDash \varphi$ iff $v^\bullet \geq_t \top^{\mathcal{F}^\bullet}$. The following result follows from the duality developed in [12].

Proposition 5.1 *For every four-valued Kripke model* $M = \langle \mathcal{F}, v \rangle$ *and every formula* φ *of bilattice modal logic,* $M \vDash \varphi$ *iff* $\langle \mathcal{F}^\bullet, v^\bullet \rangle \vDash \varphi$.

The proof of the following proposition is analogous to the classical case (cf. [14], Proposition 5]).

Proposition 5.2 *Let* $M = \langle \mathcal{F}, v \rangle$ *be a four-valued Kripke model,* α *a BPAL formula and* $\mathbf{B} = \mathcal{F}^\bullet$ *the complex algebra of* $\mathcal{F}$. *Let* $M^\alpha = \langle W^\alpha, R_+^\alpha, R_-^\alpha, v^\alpha \rangle$ *be defined as above and denote* $a := \langle v_+(\alpha), v_+(\neg\alpha) \rangle \in B$. *Then the complex* *algebra* $\mathbf{B}^\alpha$ *of* $\langle W^\alpha, R_+^\alpha, R_-^\alpha\rangle$ *can be identified up to modal bilattice isomorphism with* $(\mathbf{B}/Ker(\pi), \Diamond^\alpha)$*, where* $\pi : \mathbf{B} \to \mathbf{B}^\alpha$ *is defined by* $\pi(b) = b \wedge \sim\sim a$*, and* $\Diamond^\alpha[b]_{Ker(\pi)} = [\Diamond^{\mathbf{B}}(b \wedge \sim\sim a)]_{Ker(\pi)}$ *for all* $b \in B$*. The isomorphism* $\mu\colon (\mathbf{B}/Ker(\pi), \Diamond^\alpha) \to \mathbf{B}^\alpha$ *is defined by* $\mu[\langle X_1, X_2\rangle] := \langle X_1 \cap v_+(\alpha), X_2 \cap v_+(\alpha)\rangle$ *for all* $X_1, X_2 \subseteq W$*.*

Recall that, given a modal bilattice $\mathbf{B}$ with associated pseudo-quotient $\mathbf{B}^a$, we define the map $i'\colon \mathbf{B}^a \to \mathbf{B}$ by $i'[b] := b \wedge \sim\sim a$ for all $[b] \in B^a$. For any four-valued Kripke frame $\mathcal{F} = \langle W, R_+, R_-\rangle$ and $s \subseteq W$, we also have an injective map from $\mathcal{F}^s$ to $\mathcal{F}$ given by the inclusion $j\colon W^s \to W$. For a pair $\langle Y_1, Y_2\rangle \in W^s \times W^s$, we let $i\langle Y_1, Y_2\rangle := \langle j(Y_1), j(Y_2)\rangle$. It is easy to check that the map $\nu\colon \mathbf{B}^\alpha \to (\mathbf{B}/Ker(\pi), \Diamond^\alpha)$ defined by $\nu\langle Y_1, Y_2\rangle := [i(\langle Y_1, Y_2\rangle)]_{Ker(\pi)}$ is the inverse of the map μ of Proposition 5.2. Using this, the following proposition can be proved similarly to [14], Proposition 7].

Proposition 5.3 *If* $\mathbf{B} = \mathcal{F}^\bullet$ *for some four-valued Kripke frame* $\mathcal{F}$ *and* $a = \langle a_1, a_2\rangle \in \mathbf{B}$*, then* $i'(c) = i(\mu(c))$ *for every* $c \in \mathbf{B}^a$*, where* $\mu : \mathbf{B}^a \to (\mathcal{F}^{a_1})^\bullet$ *is the modal bilattice isomorphism identifying the two algebras. It follows that* $i(c) = i'(\nu(c))$ *for every* $c \in (\mathcal{F}^{a_1})^\bullet$*, where* $\nu\colon (\mathcal{F}^{a_1})^\bullet \to \mathbf{B}^a$ *is the inverse of* μ*.*

In light of the above results, we are going to take a closer look at the modal bilattices that arise as complex algebras of Kripke frames. As we will see, these are the *perfect modal bilattices.*

In general, a lattice $\langle L, \wedge, \vee, 0, 1\rangle$ endowed with a modal operator $\Diamond$ is *perfect* when it is: (i) complete, (ii) completely distributive (infinitary $\bigwedge$ distributes over infinitary $\bigvee$), (iii) completely $\wedge$-generated by its completely $\wedge$-prime members, and (iv) when $\Diamond$ preserves infinitary $\bigvee$. Property (iii) means the following. An element $x \in L$ is *completely* $\wedge$*-prime* if $y \neq 1$ and, for every $S \subseteq L$ such that $\bigwedge S \leq x$, there is $s \in S$ such that $s \leq x$. Dually, $y \in L$ is *completely* $\vee$*-prime* if $y \neq 0$ and, whenever $y \leq \bigvee S$ for some $S \subseteq L$, there is $s \in S$ such that $y \leq s$. We say that L is completely $\wedge$-generated (resp., completely $\vee$-generated) by $S \subseteq L$ if for every $x \in L$ there is $S' \subseteq S$ such that $x = \bigwedge S'$ (resp., $x = \bigvee S'$). In the context of distributive lattices, the two properties are equivalent.

It is well-known that a Boolean algebra $\mathbf{A}$ is perfect if and only if $\mathbf{A}$ is complete as a lattice and *atomic.* The latter means that $\mathbf{A}$ is completely $\vee$-generated by the set of its *atoms* $\mathsf{At}(\mathbf{A})$, defined as follows:

$$\mathsf{At}(\mathbf{A}) := \{x \in A : x \neq 0 \text{ and, for all } y \in A,\ y < x \text{ implies } y = 0\}.$$

We define a *perfect bimodal Boolean algebra* as a bimodal Boolean algebra $(\mathbf{A}, \Diamond_+, \Diamond_-)$ such that $(\mathbf{A}, \Diamond_+)$ and $(\mathbf{A}, \Diamond_-)$ are both perfect modal Boolean algebras, i.e., $\mathbf{A}$ is a complete atomic Boolean algebra and, moreover, both $\Diamond_+$ and $\Diamond_-$ preserve arbitrary joins. It follows from duality for classical modal logic that $\langle P(W), \cap, \cup, \sim, \Diamond_+, \Diamond_-\rangle$ is a perfect bimodal Boolean algebra. We are going to see that twist-structures over perfect bimodal Boolean algebras are exactly the algebraic objects that correspond via duality to four-valued Kripke frames. We say that a modal bilattice $\mathbf{B} = \langle B, \wedge, \vee, \supset, \neg, \Diamond, \mathsf{f}, \mathsf{t}, \bot, \top\rangle$

is *perfect* when (i) $\langle B, \wedge, \vee, \mathsf{f}, \mathsf{t}\rangle$ is a perfect lattice and (ii) $\Diamond$ preserves $\bigvee$. The following result, which is easily proved, shows an alternative condition that we could have taken as our definition of perfect modal bilattices.

Fact 5.4 *A modal bilattice* $\mathbf{B}$ *is perfect if and only if* $\mathbf{B} = \mathbf{A}^{\bowtie}$ *with* $\mathbf{A}$ *a perfect bimodal Boolean algebra.*

It follows from the above remarks that to each four-valued Kripke frame $\mathcal{F}$ corresponds a perfect modal bilattice $\mathcal{F}^{\bullet}$. We are now going to see that, conversely, to each perfect modal bilattice $\mathbf{B}$ we can associate a four-valued Kripke frame $\mathbf{B}_{\bullet}$.

We can assume without loss of generality $\mathbf{B} = \mathbf{A}^{\bowtie}$, where $\mathbf{A} = \langle A, \wedge, \vee, \sim, \Diamond_+, \Diamond_-, 0, 1\rangle$ is a bimodal Boolean algebra. We take $\mathsf{At}(\mathbf{A})$ as the set of points of our Kripke frame, on which we define relations R_+ and R_- given, for all $x, y \in \mathsf{At}(\mathbf{A})$, by:

$$xR_+y \quad \text{iff} \quad x \leq \Diamond_+ y, \qquad\qquad xR_-y \quad \text{iff} \quad x \leq \Diamond_- y.$$

Thus, we define the *prime structure of* $\mathbf{B}$ as the four-valued Kripke frame $\mathbf{B}_{\bullet} = \langle \mathsf{At}(\mathbf{A}), R_+, R_-\rangle$. The following results summarizes the duality between perfect modal bilattices and four-valued Kripke frames (cf. [14, Proposition 18]).

Proposition 5.5 *For every four-valued Kripke frame* $\mathcal{F}$ *and every perfect modal bilattice* $\mathbf{B}$*, we have* $\mathcal{F} \cong (\mathcal{F}^{\bullet})_{\bullet}$ *and* $\mathbf{B} \cong (\mathbf{B}_{\bullet})^{\bullet}$.

The correspondence of objects established by the preceding proposition extends to morphisms and can thus be formulated as a categorical duality. We will not pursue this here, but we are going to see how the correspondence sketched above allows us to translate epistemic updates from the algebraic into the relational setting.

Given a perfect modal bilattice $\mathbf{B} = \mathbf{A}^{\bowtie}$ and $a = \langle x, x'\rangle \in B$, we let $\bar{a} := \{y \in \mathsf{At}(\mathbf{A}) : y \leq x\}$. Thus, the subframe $(\mathbf{B}_{\bullet})^{\bar{a}}$ of the prime structure $\mathbf{B}_{\bullet}$ is $\langle \bar{a}, R_+ \cap (\bar{a} \times \bar{a}), R_- \cap (\bar{a} \times \bar{a})\rangle$. We then have the following.

Proposition 5.6 *For every perfect modal bilattice* $\mathbf{B}$ *and every* $a \in B$*, we have* $(\mathbf{B}^a)_{\bullet} \cong (\mathbf{B}_{\bullet})^{\bar{a}}$.

Rephrasing a remark in [14, Section 4.3], we can say that the identification between the two relational structures above shows that the mechanism of epistemic updates for public announcements is essentially unchanged when moving from the classical to an intuitionistic and even to a bilattice setting.

Joining the above results, it is easy to see that the definition of satisfaction for formulas of type $\langle\alpha\rangle\varphi$,

$$M, w \vDash \langle\alpha\rangle\varphi \quad \text{iff} \quad M, w \vDash \alpha \text{ and } M^{\alpha}, w \vDash \varphi$$

can be rewritten as follows: $w \in v_+(\langle\alpha\rangle\varphi)$ iff $\exists w' \in W^{\alpha}$ such that $j(w') = w \in v_+(\alpha)$ and $w' \in v_+^{\alpha}(\varphi)$. Since the map $j : W^{\alpha} \hookrightarrow W$ is injective, we have $w' \in v_+^{\alpha}(\varphi)$ iff $w = j(w') \in j(v_+^{\alpha}(\varphi))$.

Hence we have $w \in v_+(\langle\alpha\rangle\varphi)$ iff $w \in v_+(\alpha) \cap j(v_+^{\alpha}(\varphi))$, i.e. $v_+(\langle\alpha\rangle\varphi) = v_+(\alpha) \cap j(v_+^{\alpha}(\varphi))$.

Since $v_+(\alpha) = v_+({\sim}{\sim}\alpha)$ for any $\alpha \in Fm$ and any valuation v and, as observed earlier, satisfaction of a formula in bilattice modal logic only depends, for each valuation v, on its "positive part" $v_+(\alpha)$, we have that the result of Proposition 5.1 indeed extends to any BPAL formula.

Proposition 5.7 *For every four-valued Kripke model $M = \langle \mathcal{F}, v\rangle$ and every formula φ of BPAL, $M \vDash \varphi$ iff $\langle \mathcal{F}^{\bullet}, v^{\bullet}\rangle \vDash \varphi$.*

6 Soundness and completeness

The following lemmas are needed to establish that the calculus of Section 4 is sound with respect to the above-introduced algebraic semantics (cf. [14, Lemmas 29-34]).

Lemma 6.1 ([16], Lemma 4.1) *Let $M = (\mathbf{B}, v)$ be an algebraic model and φ a formula such that $[\![\varphi]\!]_{M^{\alpha}} = \pi([\![\varphi]\!]_M)$ for any $\alpha \in Fm$. Then $[\![\langle\alpha\rangle\varphi]\!]_M = {\sim}{\sim}[\![\alpha]\!]_M \wedge [\![\varphi]\!]_M$ and $[\![[\alpha]\varphi]\!]_M = [\![\alpha]\!]_M \supset [\![\varphi]\!]_M$.*

Fact 6.2 ([16], Lemma 4.2) *Let $\mathbf{B}$ be modal bilattice, $a \in B$, and let $i' =: \mathbf{B}^a \to \mathbf{B}$ be given, for every $[b] \in B^a$, by $i'[b] := b \wedge {\sim}{\sim}a$. Then, for every $[b], [c] \in B^a$,*

(i) $i'([b] \wedge [c]) = i'[b] \wedge i'[c]$

(ii) $i'([b] \vee [c]) = i'[b] \vee i'[c]$

(iii) $i'([b] \supset [c]) = {\sim}{\sim}a \wedge (i'[b] \supset i'[c])$

(iv) $i'(\neg[b]) = {\sim}{\sim}a \wedge \neg i'[b]$

(v) $i'(\Diamond^a[b]) = {\sim}{\sim}a \wedge \Diamond(i'[b]) = {\sim}{\sim}a \wedge \Diamond({\sim}{\sim}a \wedge i'[b])$

(vi) $i'(\Box^a[b]) = {\sim}{\sim}a \wedge \Box(a \supset i'[b])$.

Lemma 6.3 ([16], Lemma 4.3) *For any algebraic model $M = (\mathbf{B}, v)$ with underlying modal bilattice $\mathbf{B} = \langle B, \wedge, \vee, \supset, \neg, \Diamond, \mathsf{f}, \mathsf{t}, \bot, \top\rangle$ and for and all formulas $\alpha, \varphi, \psi \in Fm$,*

(i) $[\![\langle\alpha\rangle(\varphi \wedge \psi)]\!]_M = [\![\langle\alpha\rangle\varphi]\!]_M \wedge [\![\langle\alpha\rangle\psi]\!]_M$

(ii) $[\![\langle\alpha\rangle(\varphi \vee \psi)]\!]_M = [\![\langle\alpha\rangle\varphi]\!]_M \vee [\![\langle\alpha\rangle\psi]\!]_M$

(iii) $[\![\langle\alpha\rangle(\varphi \supset \psi)]\!]_M = {\sim}{\sim}[\![\alpha]\!]_M \wedge ([\![\langle\alpha\rangle\varphi]\!]_M \supset [\![\langle\alpha\rangle\psi]\!]_M)$

(iv) $[\![\langle\alpha\rangle\neg\varphi]\!]_M = {\sim}{\sim}[\![\alpha]\!]_M \wedge \neg[\![\langle\alpha\rangle\varphi]\!]_M$

(v) $[\![[\alpha]\varphi]\!]_M = [\![\neg\langle\alpha\rangle\neg\varphi]\!]_M$

(vi) $[\![\langle\alpha\rangle\Diamond\varphi]\!]_M = {\sim}{\sim}[\![\alpha]\!]_M \wedge \Diamond[\![\langle\alpha\rangle\varphi]\!]_M$

(vii) $[\![\langle\alpha\rangle\Box\varphi]\!]_M = {\sim}{\sim}[\![\alpha]\!]_M \wedge \Box[\![[\alpha]\varphi]\!]_M$.

Item (v) of the preceding lemma shows that the choice of considering the formula $[\alpha]\varphi$ as an abbreviation for $\neg\langle\alpha\rangle\neg\varphi$ is indeed sound. The following result easily follows from Lemma 6.3.

Fact 6.4 *For any algebraic model* $M = (\mathbf{B}, v)$ *with underlying modal bilattice* $\mathbf{B} = \langle B, \wedge, \vee, \supset, \neg, \Diamond, \mathsf{f}, \mathsf{t}, \bot, \top \rangle$ *and for all formulas* $\alpha, \varphi, \psi \in Fm$,

(i) $[\![[\alpha](\varphi \wedge \psi)]\!]_M = [\![[\alpha]\varphi]\!]_M \wedge [\![[\alpha]\psi]\!]_M$

(ii) $[\![[\alpha](\varphi \vee \psi)]\!]_M = [\![\alpha]\!]_M \supset ([\![\langle\alpha\rangle\varphi]\!]_M \vee [\![\langle\alpha\rangle\psi]\!]_M)$

(iii) $[\![[\alpha](\varphi \supset \psi)]\!]_M = [\![\langle\alpha\rangle\varphi]\!]_M \supset [\![\langle\alpha\rangle\psi]\!]_M$

(iv) $[\![[\alpha]\neg\varphi]\!]_M = \neg [\![\langle\alpha\rangle\varphi]\!]_M$

(v) $[\![[\alpha]\Diamond\varphi]\!]_M = [\![\alpha]\!]_M \supset \Diamond [\![\langle\alpha\rangle\varphi]\!]_M$

(vi) $[\![[\alpha]\Box\varphi]\!]_M = [\![\alpha]\!]_M \supset \Box [\![[\alpha]\varphi]\!]_M$.

We are now ready to state the announced completeness result.

Theorem 6.5 *The calculus for BPAL is sound and complete with respect to algebraic and relational models.*

As a potential direction for future work, we would like here to mention the possibility of extending BPAL in order to define a bilattice version of the logic of epistemic actions and knowledge of [2], along the line, e.g., of [13] which extends this logic to an intuitionistic setting. Another interesting development would be to formalize in BPAL a concrete example of multi-agent reasoning, such as the muddy children puzzle (see [14, Section 5]): this may prove useful in order to better appreciate the potentiality and limits of our new formalism.

References

[1] Arieli, O. and A. Avron, *Reasoning with logical bilattices*, Journal of Logic, Language and Information **5** (1996), pp. 25–63.

[2] Baltag, A., Moss, L. and A. Solecki, *The logic of public announcements, common knowledge, and private suspicions*, CWI technical report SEN-R9922, 1999.

[3] van Benthem, J., *Logical Dynamics of Information and Interaction*, Cambridge University Press, 2011.

[4] Belnap, N. D., *How a computer should think*, in: G. Ryle, editor, *Contemporary Aspects of Philosophy*, Oriel Press, Boston, 1976 pp. 30–56.

[5] Belnap, Jr., N. D., *A useful four-valued logic*, in: J. M. Dunn and G. Epstein, editors, *Modern uses of multiple-valued logic (Fifth Internat. Sympos., Indiana Univ., Bloomington, Ind., 1975)*, Reidel, Dordrecht, 1977 pp. 5–37. Episteme, Vol. 2.

[6] Bou, F. and U. Rivieccio, *Bilattices with implications*, Studia Logica **101** (2013), pp. 651–675.

[7] van Ditmarsch, H., van der Hoek, W. and B. Kooi, *Dynamic Epistemic Logic*, Springer, 2007.

[8] Frittella, S., Greco, G., Kurz, A., Palmigiano, A. and V. Sikimic, *A Proof-Theoretic Semantic Analysis of Dynamic Epistemic Logic*, Journal of Logic and Computation, Special issue on Sub-structural logic and information dynamics, forthcoming.

[9] Frittella, S., Greco, G., Kurz, A., Palmigiano, A. and V. Sikimic, *Multi-type Display Calculus for Propositional Dynamic Logic*, Journal of Logic and Computation, Special issue on Sub-structural logic and information dynamics, forthcoming.

[10] Ginsberg, M. L., *Multivalued logics: A uniform approach to inference in artificial intelligence*, Computational Intelligence **4** (1988), pp. 265–316.

[11] Jansana, R. and U. Rivieccio, *Residuated bilattices*, Soft Computing **16** (2012), pp. 493–504.

[12] Jung, A. and U. Rivieccio, *Kripke semantics for modal bilattice logic*, Proceedings of the 28th Annual ACM/IEEE Symposium on Logic in Computer Science, IEEE Computer Society Press, 2013, pp. 438–447.

[13] Kurz, A. and A. Palmigiano, *Epistemic Updates on Algebras*, Logical Methods in Computer Science **9** (2013), DOI: 10.2168/LMCS-9(4:17)2013.

[14] Ma, M., Palmigiano, A. and M. Sadrzadeh, *Algebraic semantics and model completeness for Intuitionistic Public Announcement Logic*, Annals of Pure and Applied Logic, **165** (2014), pp. 963–995.

[15] Plaza, J., *Logics of Public Communications*, Proceedings 4th International Symposium on Methodologies for Intelligent Systems, 1989, pp. 201–216.

[16] Rivieccio, U., *Algebraic semantics for bilattice public announcement logic*, Proceedings of the 13th Trends in Logic international conference, Łodz (Poland), 2-5 July 2014, forthcoming.

Appendix

Proof of Lemma 3.2. It is sufficient to check that the statement holds in a twist-structure $\mathbf{B} = \mathbf{A}^{\bowtie}$. Assume $\langle b_1, b_2 \rangle \equiv_{\langle a_1, a_2 \rangle} \langle c_1, c_2 \rangle$ and $\langle d_1, d_2 \rangle \equiv_{\langle a_1, a_2 \rangle} \langle e_1, e_2 \rangle$. By Lemma 3.1, this is equivalent to $b_1 \equiv_{a_1} c_1$, $b_2 \equiv_{a_1} c_2$, $d_1 \equiv_{a_1} e_1$, $d_2 \equiv_{a_1} e_2$. Since $\equiv_{a_1}$ is a congruence of the Boolean algebra $\mathbf{A}$, we have, for instance, $\sim b_1 \vee d_1 \equiv_{a_1} \sim c_1 \vee e_1$ and $b_1 \wedge d_2 \equiv_{a_1} c_1 \wedge e_2$. By Lemma 3.1 again, this means that $\langle b_1, b_2 \rangle \supset \langle d_1, d_2 \rangle \equiv_{\langle a_1, a_2 \rangle} \langle c_1, c_2 \rangle \supset \langle e_1, e_2 \rangle$. Compatibility with all the other bilattice operations can be shown in a similar way.

Proof of Lemma 3.5. We claim that the map $h \colon (\mathbf{A}^{\bowtie})^{\langle a_1, a_2 \rangle} \to (\mathbf{A}^{a_1})^{\bowtie}$ defined, for all $\langle b_1, b_2 \rangle \in \mathbf{A}^{\bowtie}$, by $h([\langle b_1, b_2 \rangle]_{\langle a_1, a_2 \rangle}) := \langle [b_1]_{a_1}, [b_2]_{a_1} \rangle$ is a modal bilattice isomorphism. Surjectivity of h is immediate. To prove that h is one-to-one, assume $[\langle b_1, b_2 \rangle]_{\langle a_1, a_2 \rangle} \neq [\langle c_1, c_2 \rangle]_{\langle a_1, a_2 \rangle}$. This means that $\langle b_1, b_2 \rangle \wedge \sim\sim\langle a_1, a_2 \rangle = \langle b_1 \wedge a_1, b_2 \vee \sim a_1 \rangle \neq \langle c_1 \wedge a_1, c_2 \vee \sim a_1 \rangle = \langle c_1, c_2 \rangle \wedge \sim\sim\langle a_1, a_2 \rangle$. Notice that $b_2 \vee \sim a_1 = c_2 \vee \sim a_1$ iff $\sim(b_2 \vee \sim a_1) = \sim b_2 \wedge \sim\sim a_1 = \sim c_2 \wedge \sim\sim a_1 = \sim(c_2 \vee \sim a_1)$ iff $[\sim b_2]_{a_1} = [\sim c_2]_{a_1}$ iff $[b_2]_{a_1} = [c_2]_{a_1}$. Thus we have either $[b_1]_{a_1} \neq [c_1]_{a_1}$ or $[b_2]_{a_1} \neq [c_2]_{a_1}$. Hence $\langle [b_1]_{a_1}, [b_2]_{a_1} \rangle \neq \langle [c_1]_{a_1}, [c_2]_{a_1} \rangle$, as required. Thus h is a bijection. Moreover, using Fact 3.2 and [14, Fact 7.4], one can check that, for instance,

$$
\begin{aligned}
& h([\langle b_1, b_2 \rangle]_{\langle a_1, a_2 \rangle} \supset [\langle c_1, c_2 \rangle]_{\langle a_1, a_2 \rangle}) = \\
& = h([\langle b_1, b_2 \rangle \supset \langle c_1, c_2 \rangle]_{\langle a_1, a_2 \rangle}) \\
& = h([\langle \sim b_1 \vee c_1, b_1 \wedge c_2 \rangle]_{\langle a_1, a_2 \rangle}) \\
& = \langle [\sim b_1 \vee c_1]_{a_1}, [b_1 \wedge c_2]_{a_1} \rangle \\
& = \langle \sim [b_1]_{a_1} \vee [c_1]_{a_1}, [b_1]_{a_1} \wedge [c_2]_{a_1} \rangle \\
& = \langle [b_1]_{a_1}, [b_2]_{a_1} \rangle \supset \langle [c_1]_{a_1}, [c_2]_{a_1} \rangle \\
& = h([\langle b_1, b_2 \rangle]_{\langle a_1, a_2 \rangle}) \supset h([\langle c_1, c_2 \rangle]_{\langle a_1, a_2 \rangle}).
\end{aligned}
$$

The cases of the other non-modal connectives are similar, so we omit the proof. We can use Fact 3.2 and [14, Fact 7.4] to check that the modal operator is also preserved:

$$
\begin{aligned}
&h(\Diamond^{\langle a_1,a_2\rangle}[\langle b_1, b_2\rangle]_{\langle a_1,a_2\rangle}) = \\
&= h([\Diamond(\langle b_1, b_2\rangle \wedge \sim\sim\langle a_1, a_2\rangle)]_{\langle a_1,a_2\rangle}) \\
&= h([\Diamond\langle b_1 \wedge a_1, b_2 \vee \sim a_1\rangle]_{\langle a_1,a_2\rangle}) \\
&= h([\langle \Diamond_+(b_1 \wedge a_1), \Box_+(b_2 \vee \sim a_1) \wedge \Box_- \sim(b_1 \wedge a_1)\rangle]_{\langle a_1,a_2\rangle}) \\
&= \langle [\Diamond_+(b_1 \wedge a_1)]_{a_1}, [\Box_+(b_2 \vee \sim a_1) \wedge \Box_- \sim(b_1 \wedge a_1)]_{a_1}\rangle \\
&= \langle [\Diamond_+(b_1 \wedge a_1)]_{a_1}, [\Box_+(b_2 \vee \sim a_1)]_{a_1} \wedge [\Box_- \sim(b_1 \wedge a_1)]_{a_1}\rangle \\
&= \langle [\Diamond_+(b_1 \wedge a_1)]_{a_1}, [\Box_+(\sim a_1 \vee b_2)]_{a_1} \wedge [\Box_-(\sim a_1 \vee \sim b_1)]_{a_1}\rangle \\
&= \langle \Diamond_+[b_1]_{a_1}, \Box_+[b_2]_{a_1} \wedge \Box_-[\sim b_1]_{a_1}\rangle \\
&= \langle \Diamond_+[b_1]_{a_1}, \Box_+[b_2]_{a_1} \wedge \Box_- \sim[b_1]_{a_1}\rangle \\
&= \Diamond\langle [b_1]_{a_1}, [b_2]_{a_1}\rangle \\
&= \Diamond(h([\langle b_1, b_2\rangle]_{\langle a_1,a_2\rangle})).
\end{aligned}
$$

Proof of Proposition 5.5. $\mathcal{F} \cong (\mathcal{F}^\bullet)_\bullet$ is the first claim. We have $(\mathcal{F}^\bullet)_\bullet = \langle \mathsf{At}(P(W)), R'_+, R'_-\rangle$. Clearly the map $\epsilon\colon \mathcal{F} \to (\mathcal{F}^\bullet)_\bullet$ given by $\epsilon(x) := \{x\}$ for all $x \in W$ is a bijection. Moreover, for all $x, y \in W$, we have

$$
\begin{aligned}
\epsilon(x)R'_+\epsilon(y) \quad &\text{iff} \quad \{x\}R'_+\{y\} && \text{iff} \quad \{x\} \subseteq \Diamond_+\{y\} \\
&\text{iff} \quad \{x\} \subseteq R_+^{-1}[\{y\}] && \text{iff} \quad x \in R_+^{-1}[\{y\}] \\
&\text{iff} \quad xR_+y.
\end{aligned}
$$

Similarly one proves that $\epsilon(x)R'_-\epsilon(y)$ if and only if xR_-y. We now prove $\mathbf{B} \cong (\mathbf{B}_\bullet)^\bullet$ for $\mathbf{B} = \mathbf{A}^{\bowtie}$. Consider the map $\eta\colon \mathbf{B} \to (\mathbf{B}_\bullet)^\bullet$ defined, for all $\langle a_1, a_2\rangle \in A \times A = B$, by $\eta\langle a_1, a_2\rangle := \langle \iota(a_1), \iota(a_2)\rangle$, where $\iota\colon \mathbf{A} \to P(\mathsf{At}(\mathbf{A}))$ is given by $\iota(a) := \{b \in \mathsf{At}(\mathbf{A}) : b \leq a\}$. It easily follows from [14, Proposition 18] that ι is a Boolean algebra isomorphism and, moreover, $\iota(\Diamond_+a) = R_+^{-1}[\iota(a)]$ and $\iota(\Diamond_-a) = R_-^{-1}[\iota(a)]$ for all $a \in A$. That is, ι is an isomorphism of bimodal Boolean algebras. It is then straightforward to check that η is a modal bilattice isomorphism. For instance, we have

$$
\begin{aligned}
\eta(\Diamond\langle a_1, a_2\rangle) &= \eta\langle \Diamond_+a_1, \Box_+a_2 \wedge \sim\Diamond_-a_1\rangle \\
&= \langle \iota(\Diamond_+a_1), \iota(\Box_+a_2 \wedge \sim\Diamond_-a_1)\rangle \\
&= \langle \iota(\Diamond_+a_1), \iota(\Box_+a_2) \wedge \sim\iota(\Diamond_-a_1)\rangle \\
&= \langle \Diamond_+(\iota(a_1)), \Box_+(\iota(a_2)) \wedge \sim\Diamond_-(\iota(a_1))\rangle \\
&= \Diamond\langle \iota(a_1), \iota(a_2)\rangle \\
&= \Diamond(\eta\langle a_1, a_2\rangle).
\end{aligned}
$$

Proof of Proposition 5.6. Let $\mathbf{B} = \mathbf{A}^{\bowtie}$ and $a = \langle x, x'\rangle \in A \times A$. In this case, by Lemma 3.5, we have $\mathbf{B}^a \cong (\mathbf{A}^x)^{\bowtie}$. We are thus going to define an isomorphism $\kappa\colon ((\mathbf{A}^x)^{\bowtie})_\bullet \to (((\mathbf{A}^{\bowtie})_\bullet)^{\bar{a}}$ given by $\kappa([y]_x) := y \wedge x$ for all $y \in A$ with $[y]_x \in \mathsf{At}(\mathbf{A}^x)$. By [14, Fact 20] we have $\kappa([y]_x) \in \mathsf{At}(\mathbf{A})$, and it is easy to see that κ is a bijection. It remains to check κ preserves R_+ and R_-. For all $y, z \in A$ such that $[y]_x, [z]_x \in \mathsf{At}(\mathbf{A}^x)$,

$$
\begin{array}{lll}
[y]_x R_+ [z]_x & \text{iff} & [y]_x \leq \Diamond_+^x [z]_x \\
& \text{iff} & [y]_x \leq [\Diamond_+(z \wedge x)]_x \\
& \text{iff} & y \wedge x \leq \Diamond_+(z \wedge x) \wedge x \qquad \text{[14, Fact 6.2]} \\
& \text{iff} & y \wedge x \leq \Diamond_+(z \wedge x) \\
& \text{iff} & (y \wedge x) R_+ (z \wedge x) \\
& \text{iff} & \kappa([y]_x) R_+ \kappa([z]_x).
\end{array}
$$

Proof of Lemma 6.1. Concerning the first statement:

$$
\begin{aligned}
[\![\langle\alpha\rangle\varphi]\!]_M &= {\sim}{\sim}[\![\alpha]\!]_M \wedge i'([\![\varphi]\!]_{M^\alpha}) = {\sim}{\sim}[\![\alpha]\!]_M \wedge i'(\pi([\![\varphi]\!]_M)) = \\
&= {\sim}{\sim}[\![\alpha]\!]_M \wedge ([\![\varphi]\!]_M \wedge [\![{\sim}{\sim}\alpha]\!]_M) = {\sim}{\sim}[\![\alpha]\!]_M \wedge [\![\varphi]\!]_M.
\end{aligned}
$$

Concerning the second:

$$
\begin{aligned}
[\![[\alpha]\varphi]\!]_M &= [\![\alpha]\!]_M \supset i'([\![\varphi]\!]_{M^\alpha}) \\
&= [\![\alpha]\!]_M \supset i'(\pi([\![\varphi]\!]_M)) \\
&= [\![\alpha]\!]_M \supset ([\![\varphi]\!]_M \wedge [\![{\sim}{\sim}\alpha]\!]_M) \\
&= ([\![\alpha]\!]_M \supset [\![\varphi]\!]_M) \wedge ([\![\alpha]\!]_M \supset [\![{\sim}{\sim}\alpha]\!]_M) && (1) \\
&= ([\![\alpha]\!]_M \supset [\![\varphi]\!]_M) \wedge ([\![\alpha]\!]_M \supset {\sim}{\sim}[\![\alpha]\!]_M) \\
&= ([\![\alpha]\!]_M \supset [\![\varphi]\!]_M) \wedge \mathsf{t} && \mathsf{t} = x \supset {\sim}{\sim}x \\
&= [\![\alpha]\!]_M \supset [\![\varphi]\!]_M. && x \leq_t \mathsf{t}
\end{aligned}
$$

Here (1) holds because the equation $x \supset (y \wedge z) = (x \supset y) \wedge (x \supset z)$ is satisfied by every modal bilattice.

Proof of Fact 6.2.

(i)

$$
\begin{aligned}
i'([b] \wedge [c]) &= i'([b \wedge c]) && \text{Fact 3.2} \\
&= (b \wedge c) \wedge {\sim}{\sim}a \\
&= (b \wedge {\sim}{\sim}a) \wedge (c \wedge {\sim}{\sim}a) \\
&= i'[b] \wedge i'[c].
\end{aligned}
$$

(ii)

$$
\begin{aligned}
i'([b] \vee [c]) &= i'([b \vee c]) && \text{Fact 3.2} \\
&= (b \vee c) \wedge {\sim}{\sim}a \\
&= (b \wedge {\sim}{\sim}a) \vee (c \wedge {\sim}{\sim}a) && \text{distributivity} \\
&= i'[b] \vee i'[c].
\end{aligned}
$$

(iii) We are going to use Fact 3.2 together with the following identities:

$$
\begin{aligned}
{\sim}{\sim}x \wedge (y \supset z) &= {\sim}{\sim}x \wedge ((y \wedge {\sim}{\sim}x) \supset z) \\
\mathsf{t} &= (x \wedge y) \supset {\sim}{\sim}y \\
(x \supset y) \wedge (x \supset z) &= x \supset (y \wedge z)
\end{aligned}
$$

which are valid in any modal bilattice. We have:

$$\begin{aligned} i'([b] \supset [c]) &= i'[b \supset c] \\ &= {\sim}{\sim} a \wedge (b \supset c) = {\sim}{\sim} a \wedge ((b \wedge {\sim}{\sim} a) \supset c) = \\ &= {\sim}{\sim} a \wedge (((b \wedge {\sim}{\sim} a) \supset c) \wedge \mathsf{t}) \\ &= {\sim}{\sim} a \wedge (((b \wedge {\sim}{\sim} a) \supset c) \wedge ((b \wedge {\sim}{\sim} a) \supset {\sim}{\sim} a)) \\ &= {\sim}{\sim} a \wedge ((b \wedge {\sim}{\sim} a) \supset (c \wedge {\sim}{\sim} a)) \\ &= {\sim}{\sim} a \wedge (i'[b] \supset i'[c]). \end{aligned}$$

(iv)

$$\begin{aligned} i'(\neg[b]) &= i'([\neg b]) = {\sim}{\sim} a \wedge \neg b = && \text{Fact 3.2} \\ &= ({\sim}{\sim} a \wedge \neg b) \vee \mathsf{f} && \mathsf{f} \leq_t x \\ &= ({\sim}{\sim} a \wedge \neg b) \vee ({\sim}{\sim} a \wedge \neg {\sim}{\sim} a) && \mathsf{f} = {\sim}{\sim} x \wedge \neg {\sim}{\sim} x \\ &= {\sim}{\sim} a \wedge (\neg b \vee \neg {\sim}{\sim} a) && \text{distributivity} \\ &= {\sim}{\sim} a \wedge \neg (b \wedge {\sim}{\sim} a) && \text{De Morgan law} \\ &= {\sim}{\sim} a \wedge \neg i'[b]. \end{aligned}$$

(v) Straightforward, because we have on the one hand $i'(\Diamond^a[b]) = i'[\Diamond(b \wedge {\sim}{\sim} a)] = {\sim}{\sim} a \wedge \Diamond(b \wedge {\sim}{\sim} a) = {\sim}{\sim} a \wedge \Diamond(i'[b])$, and on the other $i'(\Diamond^a[b]) = i'[\Diamond(b \wedge {\sim}{\sim} a)] = i'[\Diamond(b \wedge {\sim}{\sim} a) \wedge {\sim}{\sim} a] = {\sim}{\sim} a \wedge {\sim}{\sim} a \wedge \Diamond(b \wedge {\sim}{\sim} a) = {\sim}{\sim} a \wedge \Diamond(i'[b])$.

Proof of Theorem 6.5. Soundness of the preservation of facts and logical constants axioms follow from Lemma 6.1. For the remaining axioms we only need to invoke Lemma 6.3. The proof of completeness is similar to those for classical and intuitionistic PAL [14, Theorem 22] and follows from the reducibility of BPAL to the bilattice modal logic of [12] via reduction axioms. By the deduction-detachment theorem, we can without loss of generality limit ourselves to single formulas. Let φ be a valid BPAL formula. Consider some innermost occurrence of a dynamic modality in φ. Hence, the subformula ψ having that occurrence labeling the root of its generation tree has the form $\langle \alpha \rangle \psi'$ for some formula ψ' in the static language. The distribution axioms make it possible to equivalently transform ψ by pushing the dynamic modality down the generation tree, through the static connectives, until it attaches to a proposition letter or to a constant symbol. Here the dynamic modality disappears by applying the appropriate 'preservation of facts' or 'interaction with constant' axiom. The process is repeated for all dynamic modalities of φ, so as to obtain a formula φ' which is provably equivalent to φ. Since φ is valid by assumption, and since the process preserves provable equivalence, by soundness we can conclude that φ' is valid. By Theorem 2.2, we can conclude that φ' is provable in bilattice modal logic and thus in BPAL. This, together with the provable equivalence of φ and φ', concludes the proof.

Axiomatic and Tableau-Based Reasoning for Kt(H,R)

Renate A. Schmidt [1]

School of Computer Science
University of Manchester, Manchester, UK

John G. Stell

School of Computing
University of Leeds, Leeds, UK

David Rydeheard

School of Computer Science
University of Manchester, Manchester, UK

Abstract

We introduce a tense logic, called $Kt(H, R)$, arising from logics for spatial reasoning. $Kt(H, R)$ is a multi-modal logic with two modalities and their converses defined with respect to a pre-order and a relation stable over this pre-order. We show $Kt(H, R)$ is decidable, it has the effective finite model property and reasoning in $Kt(H, R)$ is PSPACE-complete. Two complete Hilbert-style axiomatisations are given. The main focus of the paper is tableau-based reasoning. Our aim is to gain insight into the numerous possibilities of defining tableau calculi and their properties. We present several labelled tableau calculi for $Kt(H, R)$ in which the theory rules range from accommodating correspondence properties closely, to accommodating Hilbert axioms closely. The calculi provide the basis for decision procedures that have been implemented and tested on modal and intuitionistic problems.

1 Introduction

In this paper we consider a variety of different deduction approaches in the spectrum between the purely axiomatic approach and the explicitly semantic approach. Our investigation is focussed on a tense logic, called $Kt(H, R)$. $Kt(H, R)$ has forward and backward looking modal operators defined by two accessibility relations H and R. The frame conditions are reflexivity and transitivity of H, and *stability* of R with respect to H. The stability condition is

[1] Much of the work was conducted while visiting the Max-Planck-Institut für Informatik, Saarbrücken. Partial support from UK EPSRC research grant EP/H043748/1 is gratefully acknowledged.

defined as $H;R;H \subseteq R$, where ; denotes relational composition. This means in a Kripke frame, for any two states u and v, whenever there is an H-transition from u, followed by an R-transition and an H-transition, to v, then there is also an R-transition from u to v.

The logic $Kt(H, R)$ originates with recent work on a bi-intuitionistic tense logic, called BISKT, which is studied with the motivation to develop a theory of relations on graphs and applications to spatial reasoning [23]. Given an undirected graph G, we can consider Kripke frames where the set of states is the set of all edges and all nodes in G. On these states we make an H-transition from u to v when either $u = v$ or when u is an edge which is incident with the node v. The significance of relations R, which are stable with respect to H is that they correspond exactly to the union-preserving functions on the lattice of subgraphs of G. This justifies viewing these stable relations as 'relations on G'. One motivation for investigating relations on graphs comes from mathematical morphology as used in image processing [4]. In its basic form this uses relations on sets of pixels to generate operations that approximate images. These approximations are designed to emphasise significant features and to reduce other features (such as noise). Mathematical morphology on graphs is currently being developed in image processing [6], although without explicitly using relations on graphs. In a Kripke frame for BISKT, constructed from a graph, formulae are interpreted as subgraphs and the box and diamond modal operators arising from R are operations on subgraphs providing forms of the erosion and dilation operations in mathematical morphology. The precise relationship between these modal operators and the morphological operators described in [6] is still under investigation. Using the standard embedding of intuitionistic logics into modal logic, the logic BISKT can be embedded into $Kt(H, R)$ and properties such as decidability, the finite model property and complexity of $Kt(H, R)$ carry over to BISKT. Moreover, deduction methods for $Kt(H, R)$ and implementations can be used for BISKT.

$Kt(H, R)$ is of independent interest because the modal axiom(s) corresponding to the stability condition can be used to ascribe levels of awareness to agents in a multi-agent setting. The standard model for formalising knowledge and actions performed by agents, or events happening in an agent environment, uses the $S5$-modality as knowledge operator and K modalities as action operators. In $Kt(H, R)$, the $[H]$-modality and the $[R]$-modality can be seen as modelling knowledge and action operators. [2] $[H]\phi$ is read to mean 'the agent knows ϕ' and $[R]\phi$ is read to mean 'always after executing action R, ϕ holds'. The Axiom $S = [R]\phi \to [H][R][H]\phi$ corresponding to the stability condition $H ; R ; H \subseteq R$, can then be viewed as saying 'the agent knows that, after performing an action R, it knows the effects of the action'. Thus, it states the agent has (strong) awareness of performing action R and its effects.

[2] The formalisation is slightly more general, because the negative introspection axiom is not assumed for the $[H]$-modality but this is not critical because it can be easily added to the logic. Also, allowing multiple knowledge operators and multiple action operators does not pose any technical difficulties.

The logic $Kt(H,R)$ has an alternative axiomatisation in which the stability axiom S is equivalent to the two axioms $A = [R]\phi \rightarrow [H][R]\phi$ and $P = [R]\phi \rightarrow [R][H]\phi$. From an agent perspective, Axiom A says 'the agent knows, when action R is performed, then ϕ necessarily holds'; in other words, the agent is aware of action R. Axiom P says 'after performing action R the agent knows ϕ holds', i.e., it knows the post-condition has been realised. In some sense, Axioms A and P can be viewed as weak forms of no learning and perfect recall. No learning is typically formalised as $[R][H]\phi \rightarrow [H][R]\phi$, and perfect recall as $[H][R]\phi \rightarrow [R][H]\phi$ [26].

A contribution of this paper is a series of *labelled semantic tableau calculi*, also referred as *explicit tableau systems* [11], for the logic $Kt(H,R)$. Labelled semantic tableau systems are widely studied, cf. [13,8,5,7,22], and are related to labelled sequent and natural deduction systems, cf. [14,17,27]. Labelled semantic tableau systems are proof confluent, which means committing to an inference step never requires backtracking over the proof search for an unsatisfiable formula. Proof-confluent calculi provide more flexibility in designing and experimenting with search strategies, and they are easier to implement while preserving soundness and completeness. For the purposes of our theoretical and practical analyses and comparisons in this paper this is useful.

Labelled semantic tableau calculi of the pure semantic kind explicitly and directly construct Kripke models during the inference process. They use *structural rules* which are direct reflections of the background theory given by a set of characterising frame conditions. For example, for Axiom $4 = [H]\phi \rightarrow [H][H]\phi$ the structural rule is $H(s,t),\ H(t,u)\ /\ H(s,u)$ and ensures H will be a transitive relation. For logics with semantic characterisations, labelled tableau calculi using structural rules may be developed by systematic methods. A general method is described in [22,24].

Alternatively, the background theory can be accommodated as *propagation rules* [5]. The propagation rule for Axiom 4 is $s : [H]\phi,\ H(s,t)\ /\ t : [H]\phi$. Propagation rules accommodate the background theory not by representations of the correspondence properties, but by representations of inferences with the Hilbert axioms [18]. Propagation rules can be seen to attempt to speed up the inference process by not returning complete concrete models but only skeleton models and performing just enough inferences to determine both satisfiability and unsatisfiability.

In this paper we also explore the extreme case of basing the tableau rules of the background theory on direct representations of Hilbert axioms, e.g., using the rule $s : [H]\phi\ /\ s : [H][H]\phi$ for Axiom 4. This is an example of what we call an *axiomatic rule*. Calculi with such rules are seldom seen in the literature (but [14] is an exception), and some authors have suggested completeness and termination cannot be guaranteed with such rules. We show however complete and terminating tableau calculi based on such rules can be obtained.

After formally defining $Kt(H,R)$ in Section 2, we give two Hilbert-style axiomatisations in Section 3, which will form the basis for deriving various semantic labelled tableau calculi. Section 4 recalls standard notions of labelled tableau reasoning and presents a tableau calculus with structural rules, derived from the semantics of $Kt(H, R)$. With one of the Hilbert axiomatisations as a basis, a tableau calculus $Tab^{prop}_{A,P}$ using propagation rules is presented in Section 5. The underlying proof idea of the completeness of the calculus is the same as for the completeness of the axiomatic translation principle in [18]. A reduction of satisfiability problems in $Kt(H, R)$ to the guarded fragment, defined as a partial evaluation of the calculus $Tab^{prop}_{A,P}$, is presented in Section 6. This enables us to give decidability and complexity results for $Kt(H, R)$ and implies the effective finite model property. The various possibilities of mixing structural and propagation rules yield more sound, complete and terminating tableau calculi in Section 7. We also present sound, complete and terminating tableau calculi using axiomatic rules, including $s : [H]\phi \;/\; s : [H][H]\phi$. Implementations of the presented tableau calculi and experimental results are discussed in Section 8. The proofs may be found in the long version [19] of this paper.

2 The modal logic $Kt(H, R)$

$Kt(H, R)$ is an extension of a normal bi-modal logic with two pairs of tense operators. The connectives of $Kt(H, R)$ are those of propositional logic, we take as primitives the operators $\bot$, $\wedge$, $\neg$, as well as the four box operators $[H]$, $[R]$, $[\breve{H}]$ and $[\breve{R}]$. These are standard box operators interpreted over two relations H and R and their converses $\breve{H}$ and $\breve{R}$. Other Boolean operators including $\top$, $\vee$, $\rightarrow$ and the respective diamond operators can be defined as expected: $\top = \neg\bot$, $\phi \vee \psi = \neg(\neg\phi \wedge \neg\psi)$, $\phi \rightarrow \psi = \neg(\phi \wedge \neg\psi)$ and $\Diamond\phi = \neg\Box\neg\phi$ for each $\Diamond \in \{\langle H\rangle, \langle R\rangle, \langle\breve{H}\rangle, \langle\breve{R}\rangle\}$ and the corresponding $\Box \in \{[H], [R], [\breve{H}], [\breve{R}]\}$.

The semantics of $Kt(H, R)$ is defined over Kripke models of the form $\mathcal{M} = (W, H, R, \mathcal{V})$, where W is any non-empty set (the set of worlds), H and R are binary relations over W, and $\mathcal{V}$ is a valuation mapping defining where propositional variables hold. The semantics of formulae in $Kt(H, R)$ is inductively defined as follows.

$$
\begin{array}{ll}
\mathcal{M}, w \Vdash p & \text{iff } w \in \mathcal{V}(p) \\
\mathcal{M}, w \nVdash \bot & \\
\mathcal{M}, w \Vdash \neg\phi & \text{iff } \mathcal{M}, w \nVdash \phi \\
\mathcal{M}, w \Vdash \phi \wedge \psi & \text{iff } \mathcal{M}, w \Vdash \phi \text{ and } \mathcal{M}, w \Vdash \psi \\
\mathcal{M}, w \Vdash [H]\phi & \text{iff } \mathcal{M}, v \Vdash \phi \text{ for all } H\text{-successors } v \text{ of } w \\
\mathcal{M}, w \Vdash [R]\phi & \text{iff } \mathcal{M}, v \Vdash \phi \text{ for all } R\text{-successors } v \text{ of } w \\
\mathcal{M}, w \Vdash [\breve{H}]\phi & \text{iff } \mathcal{M}, v \Vdash \phi \text{ for all } H\text{-predecessors } v \text{ of } w \\
\mathcal{M}, w \Vdash [\breve{R}]\phi & \text{iff } \mathcal{M}, v \Vdash \phi \text{ for all } R\text{-predecessors } v \text{ of } w
\end{array}
$$

We further impose that

(i) H is reflexive and transitive, and

(ii) R is *stable with respect to* H, i.e., $H\,;R;H \subseteq R$, where ; denotes relational composition.

K	axiomatisation of propositional logic, modus ponens, axioms K and necessitation for all four modalities, and substitutivity		
$\breve{H}$	$\neg[H]\neg[\breve{H}]\phi \to \phi$	$\breve{\breve{H}}$	$\neg[\breve{H}]\neg[H]\phi \to \phi$
$\breve{R}$	$\neg[R]\neg[\breve{R}]\phi \to \phi$	$\breve{\breve{R}}$	$\neg[\breve{R}]\neg[R]\phi \to \phi$
T	$[H]\phi \to \phi$	$\breve{T}$	$[\breve{H}]\phi \to \phi$
4	$[H]\phi \to [H][H]\phi$	$\breve{4}$	$[\breve{H}]\phi \to [\breve{H}][\breve{H}]\phi$
S	$[R]\phi \to [H][R][H]\phi$	$\breve{S}$	$[\breve{R}]\phi \to [\breve{H}][\breve{R}][\breve{H}]\phi$

Table 1
Axiomatisation $\mathcal{H}_S$ of $Kt(H, R)$.

Ax $\mathcal{A}$	Frame conditions	
T	H is reflexive	$\forall x\, H(x, x)$
4	H is transitive	$\forall xyz\, (H(x, y) \wedge H(y, z) \to H(x, z))$
S	R is stable wrt. H	$\forall xyzu\, (H(x, y) \wedge R(y, z) \wedge H(z, u) \to R(x, u))$
A	R is ante-stable wrt. H	$\forall xyz\, (H(x, y) \wedge R(y, z) \to R(x, z))$
P	R is post-stable wrt. H	$\forall xyz\, (R(x, y) \wedge H(y, z) \to R(x, z))$

Table 2
Axioms and frame conditions.

As usual, $\mathcal{F} = (W, H, R)$ is referred to as the Kripke frame of $\mathcal{M}$. Any Kripke frame (W, H, R) for which (i) and (ii) hold is called a $Kt(H, R)$*-frame* and any model $(W, H, R, \mathcal{V})$ for which (i) and (ii) hold is called a $Kt(H, R)$*-model*. We refer to Kripke models (frames) defined over relations and their converses as *tense Kripke models (frames)*.

It follows from results in Section 6 below that:

Theorem 2.1 *(i) $Kt(H, R)$ is decidable and has the effective finite model property.*

(ii) Satisfiability in $Kt(H, R)$ is PSPACE-complete.

3 Axiomatisation and alternative characterisation

Table 1 presents an axiomatisation $\mathcal{H}_S$ of $Kt(H, R)$. $\mathcal{H}_S$ is given as an extension of basic multi-modal logic with four modal operators. The axioms $\breve{H}$, $\breve{\breve{H}}$, $\breve{R}$ and $\breve{\breve{R}}$ define the pairs of tense operators. Axioms T and 4 define $[H]$ as an *S4*-modality. Similarly, $[\breve{H}]$ is defined as an *S4*-modality. S and $\breve{S}$ are the *stability axioms*. This means $Kt(H, R)$ is an extension of the basic tense logic Kt with two pairs of modalities: an *S4*-modality and a modality stable with respect to the *S4*-modality.

$Kt(H, R)$ is first-order definable, because the extra axioms are expressible by first-order conditions on tense Kripke frames: in particular, $\breve{H}$, $\breve{\breve{H}}$, $\breve{R}$ and $\breve{\breve{R}}$ by tautologies, and the remaining axioms have first-order correspondence properties, as given in Table 2. T and 4 means that H is a reflexive, transitive relation and the frame condition for S is (H, R)-stability, i.e., $H ; R ; H \subseteq R$.

K	axiomatisation of propositional logic, modus ponens, axioms K and necessitation for all four modalities, substitutivity		
$\breve{H}$	$\neg[H]\neg[\breve{H}]\phi \to \phi$	$\breve{\breve{H}}$	$\neg[\breve{H}]\neg[H]\phi \to \phi$
$\breve{R}$	$\neg[R]\neg[\breve{R}]\phi \to \phi$	$\breve{\breve{R}}$	$\neg[\breve{R}]\neg[R]\phi \to \phi$
T	$[H]\phi \to \phi$	$\breve{T}$	$[\breve{H}]\phi \to \phi$
4	$[H]\phi \to [H][H]\phi$	$\breve{4}$	$[\breve{H}]\phi \to [\breve{H}][\breve{H}]\phi$
A	$[R]\phi \to [H][R]\phi$	$\breve{A}$	$[\breve{R}]\phi \to [\breve{H}][\breve{R}]\phi$
P	$[R]\phi \to [R][H]\phi$	$\breve{P}$	$[\breve{R}]\phi \to [\breve{R}][\breve{H}]\phi$

Table 3
Axiomatisation $\mathcal{H}_{A,P}$ of $Kt(H, R)$.

(The frame conditions for the converse versions should be clear.)

Generalisations of Sahlqvist's correspondence and completeness results [25] give us:

Theorem 3.1 *The axiomatisation $\mathcal{H}_S$ of $Kt(H, R)$ is sound and complete with respect to the class of $Kt(H, R)$-frames.*

The following properties provide the basis for an alternative characterisation of $Kt(H, R)$-frames and models.

Lemma 3.2 *Let (W, H, R) be any relational structure where H is reflexive.*

(i) The following are equivalent:
  a. R is stable with respect to H, i.e., $H\,;R\,;H \subseteq R$.
  b. R is ante- and post-stable with respect to H, i.e., $H\,;R \subseteq R$ and $R\,;H \subseteq R$.

(ii) If R is H-stable, then $[R]$ is monotone with respect to H, i.e., for any $w, v \in W$, if $\mathcal{M}, w \Vdash [R]\phi$ and $H(w, v)$, then $\mathcal{M}, v \Vdash [R]\phi$.

Lemma 3.3 *In any $Kt(H, R)$-frame $\breve{R}$ has the same properties with respect to $\breve{H}$, as R has with respect to H. For example:*

(i) $\breve{H}$ is reflexive and transitive.

(ii) $\breve{R}$ is stable with respect to $\breve{H}$, i.e., $\breve{H}\,;\breve{R}\,;\breve{H} \subseteq \breve{R}$.

(iii) $\breve{R}$ is ante- and post-stable with respect to $\breve{H}$, i.e., $\breve{H}\,;\breve{R} \subseteq \breve{R}$. and $\breve{R}\,;\breve{H} \subseteq \breve{R}$.

(iv) $[\breve{R}]$ is monotone with respect to $\breve{H}$, i.e., for any $w, v \in W$, if $\mathcal{M}, w \Vdash [\breve{R}]\phi$ and $H(v, w)$, then $\mathcal{M}, v \Vdash [\breve{R}]\phi$.

These results imply $Kt(H, R)$-frames, in which (H, R)-stability holds, and tense (W, H, R)-frames, in which H is a pre-order and (H, R)-ante and post-stability hold, are equivalent.

Since ante- and post-stability of (H, R) are correspondence properties of the Axioms $A = [R]\phi \to [H][R]\phi$ and $P = [R]\phi \to [R][H]\phi$, an alternative axiomatisation of $Kt(H, R)$ is $\mathcal{H}_{A,P}$ as given in Table 3. In $\mathcal{H}_{A,P}$ the stability axioms S and $\breve{S}$ have been replaced by the axioms A, $\breve{A}$, P and $\breve{P}$.

Theorem 3.4 *(i) The axiomatisation $\mathcal{H}_{A,P}$ is sound and complete with respect to the class of tense (H,R)-frames, where H is a pre-order, and ante- and post-stability of (H,R) hold (cf. Table 2).*

(ii) $\mathcal{H}_{A,P}$ is equivalent to $\mathcal{H}_S$.

The axiomatisation $\mathcal{H}_{A,P}$ is the basis for the tableau calculus presented in Section 5 and the axiomatic translation presented in Section 6.

4 A semantic tableau calculus for $Kt(H,R)$

Tableau formulae in our calculi have one of the forms $\bot$, $s:\phi$, $H(s,t)$, $R(s,t)$, $s \approx t$ or $s \not\approx t$. s and t denote *labels* which are terms of a freely generated term algebra over a finite set of constants (denoted by $a, b, \ldots$) and four unary function symbols $f_{\neg\Box\phi}$, one for each modality $\Box \in \{[H],[R],[\breve{H}],[\breve{R}]\}$. $\approx$ is the equality symbol.

The semantics of tableau formulae is an appropriately defined extension of the semantics of modal formulae. The extension $(\mathcal{M},\iota)$ of a $Kt(H,R)$-model $\mathcal{M}$ with an assignment ι mapping labels to worlds in W is called an *extended $Kt(H,R)$-model.* Satisfiability of tableau formulae in $(\mathcal{M},\iota)$ is defined by:

$\mathcal{M},\iota \not\Vdash \bot$		$\mathcal{M},\iota \Vdash s:\phi$	iff $\mathcal{M},\iota(s) \Vdash \phi$
$\mathcal{M},\iota \Vdash H(s,t)$	iff $(\iota(s),\iota(t)) \in H$	$\mathcal{M},\iota \Vdash R(s,t)$	iff $(\iota(s),\iota(t)) \in R$
$\mathcal{M},\iota \Vdash s \approx t$	iff $\iota(s) = \iota(t)$	$\mathcal{M},\iota \Vdash s \not\approx t$	iff $\iota(s) \neq \iota(t)$

Let Tab_S^{str} be the tableau calculus consisting of the *basic rules* and the *theory rules* given respectively in Figures 1 and 2. The basic rules are the standard decomposition rules for labelled modal formulae; as usual, there is one pair of rules for each primitive logical operator, plus the closure rule (cl). In the rules for negated box formulae we see how the function symbols are used to create new successors represented by Skolem terms. (Instead, new constants could be created, but an advantage of using Skolem terms is that no inference steps need to be recomputed when blocking occurs.) The theory rules are the reflexivity rule for H, the transitivity rule for H and the stability rule for (H,R). Since they are direct reflections of the frame conditions, following [5], they are referred to as *structural rules.*

A general form of blocking is provided by the *unrestricted blocking mechanism* [20,21], which is based on the use of the (ub) rule and an appropriate form of equality reasoning, for example, the equality rules in Figure 3. Adding the unrestricted blocking mechanism to a sound and complete labelled tableau calculus forces termination, when the logic has the (effective) finite model property. We denote the calculus extended with the unrestricted blocking mechanism by $Tab_S^{str}(ub)$.

The tableau inference process constructs derivation trees. Starting with a set of tableau formulae, the rules are applied in a top-down manner. This leads to the formulae being decomposed into smaller formulae. The application of

$$\frac{s:\phi,\ s:\neg\phi}{\bot}(\mathsf{cl}) \qquad \frac{s:\bot}{\bot}(\bot) \qquad \frac{s:\neg\neg\phi}{s:\phi}(\neg\neg)$$

$$\frac{s:\phi\wedge\psi}{s:\phi,\ s:\psi}(\wedge) \qquad \frac{s:(\neg\phi\wedge\psi)}{s:\neg\phi \mid s:\neg\psi}(\neg\wedge)$$

$$\frac{s:[H]\phi,\ H(s,t)}{t:\phi}([H]) \qquad \frac{s:\neg[H]\phi}{H(s,f_{\neg[H]\phi}(s),\ f_{\neg[H]\phi}(s):\neg\phi}(\neg[H])$$

$$\frac{s:[\breve{H}]\phi,\ H(t,s)}{t:\phi}([\breve{H}]) \qquad \frac{s:\neg[\breve{H}]\phi}{H(f_{\neg[\breve{H}]\phi}(s),s),\ f_{\neg[\breve{H}]\phi}(s):\neg\phi}(\neg[\breve{H}])$$

$$\frac{s:[R]\phi,\ R(s,t)}{t:\phi}([R]) \qquad \frac{s:\neg[R]\phi}{R(s,f_{\neg[R]\phi}(s)),\ f_{\neg[R]\phi}(s):\neg\phi}(\neg[R])$$

$$\frac{s:[\breve{R}]\phi,\ R(t,s)}{t:\phi}([\breve{R}]) \qquad \frac{s:\neg[\breve{R}]\phi}{R(f_{\neg[\breve{R}]\phi}(s),s),\ f_{\neg[\breve{R}]\phi}(s):\phi}(\neg[\breve{R}])$$

Fig. 1. The basic tableau rules.

$$\frac{}{H(s,s)}(T_c) \qquad \frac{H(s,t),\ H(t,u)}{H(s,u)}(4_c) \qquad \frac{H(s,t),\ R(t,u),\ H(u,v)}{R(s,v)}(S_c)$$

Fig. 2. Structural theory rules of Tab_S^{str}.

Unrestricted blocking rule:

$$\frac{}{s\approx t \mid s\not\approx t}(\mathsf{ub})$$

Paramodulation equality rules:

$$\frac{s\not\approx s}{\bot} \qquad \frac{s\approx t}{t\approx s} \qquad \frac{s\approx t,\ G[s]_\lambda}{G[\lambda/t]}$$

Fig. 3. Unrestricted blocking and equality rules. G denotes any tableau formula. $G[s]_\lambda$ means s occurs as a subterm at position λ in G, and $G[\lambda/t]$ denotes the formula obtained by replacing s at position λ with t.

the $(\neg\wedge)$-rule splits the current tableau branch into two branches. As soon as $\bot$ is derived in a branch, the branch is regarded as *closed* and the expansion of this branch stops. The inference process continues with the extension of a not yet closed or not yet fully expanded branch. A branch is *open* when it is not closed. When, in an open branch, no more rules are applicable, the derivation stops because a model can be read off from the branch.

A tableau calculus is *sound* when for a satisfiable set of tableau formulae any fully expanded tableau derivation has an open branch. (A tableau derivation is fully expanded if all branches are either closed, or open and fully expanded.) A tableau calculus is *(refutationally) complete* if for any unsatisfiable set of tableau formulae there is a closed tableau derivation. Though we do not emphasise it or show it explicitly, the tableau calculi we present are in fact *constructively complete*, by which we mean for every fully expanded open branch a model of the input set exists (that can either be read off from the branch, or, for the tableau calculi using propagation rules, constructed from it). A tableau calculus is *terminating*, if any fully expanded tableau derivation has a finite open branch if the input set is satisfiable.

Theorem 4.1 *(i) The tableau calculus* Tab_S^{str} *is sound and complete.*

$$\frac{s : [H]\phi}{s : \phi}(T) \qquad \frac{s : [\breve{H}]\phi}{s : \phi}(\breve{T})$$

$$\frac{s : [H]\phi,\ H(s,t)}{t : [H]\phi}(4) \qquad \frac{s : [\breve{H}]\phi,\ H(t,s)}{t : [\breve{H}]\phi)}(\breve{4})$$

$$\frac{s : [R]\phi,\ H(s,t)}{t : [R]\phi}(A) \qquad \frac{s : [\breve{R}]\phi,\ H(t,s)}{t : [\breve{R}]\phi}(\breve{A})$$

$$\frac{s : [R]\phi,\ R(s,t)}{t : [H]\phi}(P) \qquad \frac{s : [\breve{R}]\phi,\ R(t,s)}{t : [\breve{H}]\phi}(\breve{P})$$

Fig. 4. Propagation theory rules of $Tab^{prop}_{A,P}$.

(ii) So is the extension $Tab^{str}_S(ub)$ *with unrestricted blocking. Moreover:*

(iii) $Tab^{str}_S(ub)$ *is terminating and provides a decision procedure for* $Kt(H, R)$.

The calculus Tab^{str}_S provides the baseline for the completeness proofs of the tableau calculi defined in the next two sections.

5 Using propagation rules

Applying the ideas of the axiomatic translation principle [18] to the axiomatisation $\mathcal{H}_{A,P}$, based on the ante- and post-stability axioms, produces the calculus $Tab^{prop}_{A,P}$ consisting of the basic rules in Figure 1 and the theory rules in Figure 4. The basic rules are the same as for the calculus in the previous section. They form the core also for the calculi defined in the next section. Only the theory rules are varied. In $Tab^{prop}_{A,P}$ the theory rules are propagation rules. Box formulae defined over H are propagated by the rules (4) and ($\breve{4}$) to H-successors and predecessors, while box formulae defined over R are propagated by the rules (A) and ($\breve{A}$) to H-successors and predecessors, and the rules (P) and ($\breve{P}$) propagate them over R-links but turn them into box formulae defined over H.

Proving soundness of the calculus is routine. The creative and more difficult part is proving completeness. Our proof uses a simulation argument in which we show every refutation in Tab^{str}_S can be mapped to a refutation in $Tab^{prop}_{A,P}$. For lack of space the proof appears only in the long version [19], but we note the proof gives useful insight into what the essential inference steps are, and has inspired the definition of the calculi in the next section.

Theorem 5.1 *The tableau calculus* $Tab^{prop}_{A,P}$ *is sound and complete.*

We refer to the left-most premises of any rule as the *main premises.* With two exceptions the modal formulae in the conclusions of all rules of $Tab^{prop}_{A,P}$ are subformulae of the main premise, or are negations of subformulae of the main premise. The exceptions are the rules (P) and ($\breve{P}$), which produce new $[H]\psi$ and $[\breve{H}]\psi$ formulae, but where ψ occurs in the input formula immediately below $[R]$ and $[\breve{R}]$ operators. This means indefinite formula growth does not occur. This observation is exploited in the proof of Theorem 6.1.

Because the unrestricted blocking rule is sound we can add the unrestricted blocking mechanism to the calculus while preserving soundness and complete-

$\mathcal{A}$	Axiom	Schema formulae $\mathrm{Ax}^{\mathcal{A}}(p)$	$\mathfrak{X}_{\mathcal{A}}$
T	$[H]p \rightarrow p$	$\forall x(\neg Q_{[H]p}(x) \vee Q_p(x))$	$\mathfrak{X}^{\epsilon}_{[H],\varphi} \cup \mathfrak{X}^{\epsilon}_{[R],\varphi}$
$\breve{T}$	$[\breve{H}]p \rightarrow p$	$\forall x(\neg Q_{[\breve{H}]p}(x) \vee Q_p(x))$	$\mathfrak{X}^{\epsilon}_{[H],\varphi} \cup \mathfrak{X}^{\epsilon}_{[R],\varphi}$
4	$[H]p \rightarrow [H][H]p$	$\forall x \forall y\, (\neg Q_{[H]p}(x) \vee \neg H(x,y) \vee Q_{[H]p}(y))$	$\mathfrak{X}^{\epsilon}_{[H],\varphi} \cup \mathfrak{X}^{\epsilon}_{[R],\varphi}$
$\breve{4}$	$[\breve{H}]p \rightarrow [\breve{H}][\breve{H}]p$	$\forall x \forall y\, (\neg Q_{[\breve{H}]p}(x) \vee \neg H(y,x) \vee Q_{[\breve{H}]p}(y))$	$\mathfrak{X}^{\epsilon}_{[\breve{H}],\varphi} \cup \mathfrak{X}^{\epsilon}_{[\breve{R}],\varphi}$
A	$[R]p \rightarrow [H][R]p$	$\forall x \forall y\, (\neg Q_{[R]p}(x) \vee \neg H(x,y) \vee Q_{[R]p}(y))$	$\mathfrak{X}^{\epsilon}_{[R],\varphi}$
$\breve{A}$	$[\breve{R}]p \rightarrow [\breve{H}][\breve{R}]p$	$\forall x \forall y\, (\neg Q_{[\breve{R}]p}(x) \vee \neg H(y,x) \vee Q_{[\breve{R}]p}(y))$	$\mathfrak{X}^{\epsilon}_{[\breve{R}],\varphi}$
P	$[R]p \rightarrow [R][H]p$	$\forall x \forall y\, (\neg Q_{[R]p}(x) \vee \neg R(x,y) \vee Q_{[H]p}(y))$	$\mathfrak{X}^{\epsilon}_{[R],\varphi}$
$\breve{P}$	$[\breve{R}]p \rightarrow [\breve{R}][\breve{H}]p$	$\forall x \forall y\, (\neg Q_{[\breve{R}]p}(x) \vee \neg R(y,x) \vee Q_{[\breve{H}]p}(y))$	$\mathfrak{X}^{\epsilon}_{[\breve{R}],\varphi}$

Table 4
Schema formulae.

ness. Termination is a consequence of the effective finite model property of $Kt(H,R)$ (shown below) and results in [21] (cf. also [22]).

Theorem 5.2 *(i) The extension* $\mathrm{Tab}^{prop}_{A,P}(ub)$ *with unrestricted blocking is sound and complete.*

(ii) $\mathrm{Tab}^{prop}_{A,P}(ub)$ *is terminating and provides a decision procedure for* $Kt(H,R)$.

6 Axiomatic translation

In this section we show the tableau calculus $Tab^{prop}_{A,P}$ of the previous section can serve as a basis for translating problems in $Kt(H,R)$ to the guarded fragment from which decidability and the finite model property of $Kt(H,R)$ then follow.

Let φ be any $Kt(H,R)$-formula. We assume φ is in a normal form using only the primitive operators of the logic. We define a mapping $\Pi^{\Delta}_{\mathfrak{X}}$ from $Kt(H,R)$-formulae to first-order formulae, called the *axiomatic translation* of $Kt(H,R)$. The definition follows the axiomatic translation principle in [18] and is in accordance with the tableau rules of $Tab^{prop}_{A,P}$ modulo one small variation. The variation is that the rule of double negation is worked into the definition.

The definition of $\Pi^{\Delta}_{\mathfrak{X}}$ is based on the axiomatisation $\mathcal{H}_{A,P}$, so we let Δ, which is the set of extra axioms, be the set of the axioms T, $\breve{T}$, 4, $\breve{4}$, A, $\breve{A}$, P and $\breve{P}$. $\mathfrak{X}$ is the set of instantiation sets for each extra axiom. Formally, $\mathfrak{X} = \{\mathfrak{X}_{\mathcal{A}}\}_{\mathcal{A}\in\Delta}$, where $\mathfrak{X}_{\mathcal{A}}$ is defined in the right-most column of Table 4. By definition, $\mathfrak{X}_{[\alpha],\varphi} = \{\psi \mid [\alpha]\psi \in \mathrm{Sf}(\varphi)\}$, where $\alpha \in \{H,R\}$, and $\mathrm{Sf}(\varphi)$ denotes the set of all subformulae of φ. This means that $\mathfrak{X}_{[\alpha],\varphi}$ is the set of subformulae occurring immediately below $[\alpha]$ in φ.

Now, let $\Pi^{\Delta}_{\mathfrak{X}}(\varphi)$ be the conjunction of (1)–(3).

$$\exists x\, Q_{\varphi}(x) \wedge \bigwedge\{\mathrm{Def}(\psi) \mid \psi \in \mathrm{Sf}(\varphi)\} \tag{1}$$

$$\bigwedge\{\mathrm{Ax}^{\mathcal{A}}(\psi) \mid \mathcal{A} \in \Delta,\ \psi \in \mathfrak{X}_{\mathcal{A}}\} \tag{2}$$

$$\bigwedge\{\mathrm{Def}(\psi) \mid \psi \in \mathrm{Sf}(X)\} \tag{3}$$

$\pi(\bot, x) = \bot$	$\pi(p, x) = \top$	$\pi(\neg p, x) = \neg Q_p(x)$
$\pi(\psi \wedge \phi, x) = Q_\psi(x) \wedge Q_\phi(x)$		$\pi(\neg(\psi \wedge \phi), x) = Q_{\sim\psi}(x) \vee Q_{\sim\phi}(x)$
$\pi([H]\psi, x) = \forall y\,(H(x,y) \rightarrow Q_\psi(y))$		$\pi(\neg[H]\psi, x) = \exists y\,(H(x,y) \wedge Q_{\sim\psi}(y))$
$\pi([\breve{H}]\psi, x) = \forall y\,(H(y,x) \rightarrow Q_\psi(y))$		$\pi(\neg[\breve{H}]\psi, x) = \exists y\,(H(y,x) \wedge Q_{\sim\psi}(y))$
$\pi([R]\psi, x) = \forall y\,(R(x,y) \rightarrow Q_\psi(y))$		$\pi(\neg[R]\psi, x) = \exists y\,(R(x,y) \wedge Q_{\sim\psi}(y))$
$\pi([\breve{R}]\psi, x) = \forall y\,(R(y,x) \rightarrow Q_\psi(y))$		$\pi(\neg[\breve{R}]\psi, x) = \exists y\,(R(y,x) \wedge Q_{\sim\psi}(y))$

Table 5
Definition of the basic translation mapping π.

$\mathrm{Def}(\psi)$ is defined by:

$$\mathrm{Def}(\psi) = \forall x\,(Q_\psi(x) \rightarrow \pi(\psi, x)) \wedge \forall x\,(Q_\psi(x) \rightarrow \neg Q_{\sim\psi}(x)) \wedge \forall x\,(Q_{\sim\psi}(x) \rightarrow \pi(\sim\psi, x)).$$

π is the basic translation mapping inductively defined in Table 5. Each unary predicate symbol Q_ψ represents the translation of modal formula ψ indicated in the index. Their purpose is to make the translation more effective through structure sharing (it is clear that further optimisations are possible). $\sim$ denotes complementation, i.e., $\sim\psi = \phi$ if $\psi = \neg\phi$, and $\sim\psi = \neg\psi$, otherwise. $\mathrm{Ax}^{\mathcal{A}}(\psi)$ in (2) is the conjunction of instances of all schema formulae $F\{p/\psi\}$ associated with each axiom $\mathcal{A}$. The schema formulae for $Kt(H,R)$ and the instantiation sets $\mathfrak{X}_{\mathcal{A}}$ for each axiom $\mathcal{A}$ are given in Figure 4. X in (3) is the set $\{[H]\psi \mid \psi \in \mathfrak{X}^{\epsilon}_{[R],\varphi}\} \cup \{[\breve{H}]\psi \mid \psi \in \mathfrak{X}^{\epsilon}_{[\breve{R}],\varphi}\}$. This concludes the definition of $\Pi^{\Delta}_{\mathfrak{X}}(\varphi)$.

Intuitively, $\Pi^{\Delta}_{\mathfrak{X}}(\varphi)$ is an encoding of the calculus $Tab^{prop}_{A,P}$ for a given formula φ. (1) and (3) are partial evaluations of applications of the basic rules, and (2) is the partial evaluation of applications of the theory propagation rules with respect to the instantiation sets for each axiom.

Theorem 6.1 *Let φ be any $Kt(H,R)$-formula. Then:*

(i) φ is satisfiable in $Kt(H,R)$ iff $\Pi^{\Delta}_{\mathfrak{X}}(\varphi)$ is first-order satisfiable.

(ii) $\Pi^{\Delta}_{\mathfrak{X}}(\varphi)$ can be computed in linear time and the size of $\Pi^{\Delta}_{\mathfrak{X}}(\varphi)$ is linear in the size of φ.

(iii) $\Pi^{\Delta}_{\mathfrak{X}}(\varphi)$ is equivalent to a guarded formula.

Thus, $\Pi^{\Delta}_{\mathfrak{X}}$ defines an effective translation of any $Kt(H,R)$-formula into the guarded fragment [1,12]. It defines, in fact, a mapping to the subfragment $GF1^-$ of the guarded fragment, which has been shown to be PSPACE-complete if the arity of predicates is finitely bounded [16]. Therefore, carrying over properties of the guarded fragment and $GF1^-$ give us decidability, the effective finite model property and complexity results for $Kt(H,R)$, as summarised in Theorem 2.1.

The guarded fragment can be decided by ordered resolution [10]; therefore, one further consequence is:

Theorem 6.2 *Both ordered resolution and ordered resolution with selection of binary literals as defined in [10] (see also [18]) decide the axiomatic translation*

$$\frac{H(s,t),\ R(t,u)}{R(s,u)}(A_c) \qquad\qquad \frac{R(s,t),\ H(t,u)}{R(s,u)}(P_c)$$

Fig. 5. Structural theory rules for ante- and post-stability.

of satisfiability problems in $Kt(H, R)$.

Because the definition incorporates the needed number of modal formula instantiations of the propagation rules, the axiomatic translation can be viewed and reformulated as an encoding in basic tense logic Kt with global satisfiability (and two tense operators) of the $Kt(H, R)$-satisfiability of a formula φ. This encoding can be viewed as a global reduction function in the sense of [15] for $Kt(H, R)$, however a crucial variation is the signature extension with propositional symbols corresponding to the Q_ψ symbols. This makes further manipulation more efficient [18].

The calculus Tab_S^{str} of Section 4 based on structural rules also provides a basis for a translation to first-order logic, namely, the standard (relational) translation of $Kt(H, R)$ with structural transformation. However, it is not a mapping to the guarded fragment or any other known solvable fragment of first-order logic.

7 Other terminating tableau calculi

As is already apparent from Sections 3–5 there are several quite different deduction approaches for the logic $Kt(H, R)$. Further possibilities involve tableau systems based on a mixture of structural and propagation rules. Replacing the propagation rules for the axioms T and $\breve{T}$ in $Tab_{A,P}^{prop}$ by the reflexivity rule (T_c) preserves soundness and completeness. The proof is a small adaptation of the proof of Theorem 5.1.

Theorem 7.1 *The calculus* $Tab_{A,P}^{mix}$ *consisting of the basic tableau rules of Figure 1, the reflexivity rule* (T_c) *for* H *and the propagation rules* (4), $(\breve{4})$, (A), $(\breve{A})$, (P) *and* $(\breve{P})$ *for* 4, A *and* P *is sound and complete.*

Basing the rules on the frame conditions of the semantics of the alternative axiomatisation $\mathcal{H}_{A,P}$ is another (obvious) possibility:

Theorem 7.2 *The calculus* $Tab_{A,P}^{str}$ *consisting of the basic tableau rules, the reflexivity and transitivity rules as well as the structural rules* (A_c) *and* (P_c) *for ante- and post-stability (see Figure 5) is sound and complete.*

Propagation rules can be viewed as partial expansions of the corresponding axioms, and the results of the previous two sections show these partial expansions are sufficient for completeness. It also means the way the axioms are used can be accordingly restricted. This is the idea underlying the next result.

Theorem 7.3 *The calculi* $Tab_{A,P}^{ax}$ *and* Tab_S^{ax} *consisting of the basic tableau rules, the rules* (T) *and* $(\breve{T})$, *and the rules* (4^*), $(\breve{4}^*)$, (A^*), $(\breve{A}^*)$, (P^*) *and* $(\breve{P}^*)$, *respectively* (4^*), $(\breve{4}^*)$, (S^*) *and* $(\breve{S}^*)$ *(as in Figure 6), are sound and complete.*

$$\frac{s:[H]\phi}{s:[H]^*[H]\phi}(4^*) \qquad \frac{s:[\breve{H}]\phi}{s:[\breve{H}]^*[\breve{H}]\phi)}(\breve{4}^*)$$

$$\frac{s:[R]\phi}{s:[H]^*[R]\phi}(A^*) \qquad \frac{s:[\breve{R}]\phi}{s:[\breve{H}]^*[\breve{R}]\phi}(\breve{A}^*)$$

$$\frac{s:[R]\phi}{s:[R]^*[H]\phi}(P^*) \qquad \frac{s:[\breve{R}]\phi}{s:[\breve{R}]^*[\breve{H}]\phi}(\breve{P}^*)$$

$$\frac{s:[R]\phi}{s:[H][R]^*[H]\phi}(S^*) \qquad \frac{s:[\breve{R}]\phi}{s:[\breve{H}][\breve{R}]^*[\breve{H}]\phi}(\breve{S}^*)$$

Fig. 6. Axiomatic theory rules. * binds with the box operator preceding it.

The meaning of the marker * is that box formulae annotated with it are not expanded with any theory rules, only with the standard expansion rules, namely the standard box rules and the closure rule. Though the starred rules cause formulae to grow in size, the formula growth is only temporary because of the restriction. The restriction defines a refinement, which is immediate from the remark before the theorem and is explicit in the completeness proof (cf. [19]).

$Tab^{ax}_{A,P}$ and Tab^{ax}_{S} can be flexibly varied by using the structural rules or propagation rules for subsets of the axioms. When using propagation rules for stability some care is needed. We can show:

(i) The calculus consisting of the basic tableau rules and the following rules is sound and complete, where $Y = \{[H][R][H]\psi \mid [R]\psi \in \mathrm{Sf}(\varphi)\} \cup \{[\breve{H}][\breve{R}][\breve{H}]\psi \mid [\breve{R}]\psi \in \mathrm{Sf}(\varphi)\}$.

$$\frac{}{s:\phi \mid s:\neg\phi}(\mathsf{cut})\ \phi \in Y \qquad \frac{}{H(s,s)}(T_c)$$

$$\frac{s:[H]\phi,\ H(s,t)}{t:[H]\phi}(4) \qquad \frac{s:[\breve{H}]\phi,\ H(t,s)}{t:[\breve{H}]\phi)}(\breve{4})$$

$$\frac{s:[R]\phi,\ H(s,t),\ R(t,u)}{u:[H]\phi}(S) \qquad \frac{s:[\breve{R}]\phi,\ H(t,s),\ R(u,t)}{u:[\breve{H}]\phi}(\breve{S})$$

(ii) The calculus as in (i) but with propagation rules for T is sound and complete.

In both cases omitting the cut rule leads to incompleteness. There is a connection between these calculi and the calculus Tab^{ax}_{S} of Theorem 7.3, from which it is clear that the $[R]$ and $[\breve{R}]$-formulae occurring in the cut formulae do not need to be expanded with the (S) or $(\breve{S})$-rules. For completeness, in fact, the rule $s:[R]\psi \ /\ s:[H][R]^*[H]\psi \mid s:\neg[H][R]^*[H]\psi$ and the converse version, with the same restrictions for the starred boxes, are sufficient. However, the right branch can always be almost immediately closed, so that no gain is apparent over the calculus Tab^{ax}_{S}. This was confirmed in experiments.

Each of the calculi in this section is terminating when endowed with the unrestricted blocking mechanism. Even the calculus $Tab^{ax}_{S}(ub)$ without the restrictions to starred box formulae can be shown to be sound, complete and terminating. An important assumption is that the rules are applied fairly, i.e., no non-redundant application of a rule is postponed indefinitely.

Finally we note that each of the presented tableau calculi provides the basis for a reduction to first-order logic and their soundness and completeness is a consequence of the soundness and completeness of the calculi, and the fact that derivations are defined over a bounded number of modal formulae. Then, in the case of the propagation and axiomatic rules corresponding, effective partial evaluations as in the axiomatic translation in Section 6 can be defined and proved sound and complete. The reductions for structural rules involve, as expected, the corresponding frame conditions.

8 Implementation and experiments

We implemented the tableau calculi by encoding them into first-order logic and using the SPASS-YARRALUMLA system. SPASS-YARRALUMLA is a bottom-up model generator based on the SPASS theorem prover (Version 3.8d) [28]. SPASS-YARRALUMLA emulates the behaviour of semantic labelled tableau provers [2,3]. The resolution refinement used is ordered resolution and selection of at least one negative literal in every clause. The inference loop of SPASS was slightly modified so that it always takes the least complex clause as the given clause, ground clauses with positive equality literals are eagerly split, and a branch with a positive equality literal is always explored first. Equality reasoning is realised by ordered forward and backward rewriting. SPASS-YARRALUMLA implements several blocking techniques. We used four forms: (i) sound ancestor blocking (i.e., blocking is applied to distinct terms s and t if one is a subterm of the other, flag `-bld`); (ii) unrestricted blocking as defined in Figure 3 (flag `-bld -ubl`); (iii) sound ancestor blocking on non-disjoint worlds (i.e., blocking is restricted to subterms on unary predicates, flag `-blu`); and (iv) sound anywhere blocking on non-disjoint worlds (flag `-blu -ubl`).

The encodings of the tableau calculi are implemented as an extension of the `ml2dfg` tool used for the empirical evaluation of the axiomatic translation principle in [18]. Because this earlier work was limited to the evaluation of extensions of basic modal logic K, we extended the implementation to handle multiple modalities and backward looking modalities, and we implemented the encodings of the structural, propagation and axiomatic tableau rules for Axioms S, A and P, and extended the implementation of the encodings for T and 4. Thirteen encodings were evaluated. These include encodings of $Tab^{str}_{A,P}$ and Tab^{str}_{S} based correspondence properties (named `KtAcPcTc4c` and `KtScTc4c` in the results tables), the encoding of the tableau calculus $Tab^{prop}_{A,P}$ using propagation rules which was implemented via the axiomatic translation as defined in Section 6 (`KtAPT4`), the encodings of $Tab^{ax}_{A,P}$ and Tab^{ax}_{S} (`KtA*P*T4*` and `KtS*T4*`) as well as mixes of encodings of correspondence properties, the axiomatic translation, and almost purely axiomatic encodings. All tested encodings are sound and complete.

Evaluations were performed on problems created for the investigation of the logic BISKT in [23], and modal logic problems consisting predominantly of problems used in the experiments of [18]. The BISKT problems include intuitionistic propositional logic and intuitionistic modal logic problems. The

Spass-yarralumla.												
Blocking	(i) -bld			(ii) -bld -ubl			(iii) -blu			(iv) -blu -ubl		
Encoding	S	U	S/U	S	U	S/U	S	U	S/U	S	U	S/U
KtA*P*T4*	9.1	8.7	17.8	15.1	76.7	91.8	18.8	16.9	35.6	15.0	12.2	27.2
KtA*PT4*	8.3	9.2	17.5	17.7	97.7	115.4	15.5	16.6	32.1	12.3	12.6	24.9
KtAP*T4*	8.0	8.5	16.5	16.7	75.0	91.7	15.8	15.9	31.8	11.3	12.4	23.8
KtA*P*T4	7.4	6.7	14.1	14.7	44.5	59.2	14.1	11.9	25.9	11.2	9.6	20.9
KtA*PT4	6.7	6.4	13.2	13.9	42.0	55.9	11.6	11.0	22.6	10.4	9.8	20.2
KtAP*T4	6.5	6.6	13.0	12.4	42.6	55.1	12.1	11.4	23.6	9.7	9.6	**19.3**
KtAPT4	**6.2**	**6.2**	**12.4**	11.7	**40.6**	**52.3**	9.8	**10.5**	**20.3**	9.9	**9.4**	**19.3**
KtAPTc4	6.3	7.2	13.6	**10.2**	58.5	68.7	10.1	11.1	21.2	10.2	11.8	21.9
KtAcPcTc4c	6.4	28.4	34.7	14.9	220.3	235.2	**9.5**	60.9	70.4	**9.5**	32.8	42.3
KtS*T4*	10.6	9.2	19.8	16.2	80.0	96.2	24.6	18.3	42.9	14.9	12.5	27.4
KtS*T4	7.9	6.7	14.6	13.1	45.1	58.2	16.1	11.6	27.7	10.2	9.7	19.9
KtS*Tc4	8.2	7.7	15.8	11.2	61.8	73.0	15.7	12.1	27.8	11.7	12.3	23.9
KtScTc4c	6.5	34.7	41.1	14.3	180.1	194.4	9.8	76.5	86.4	9.8	34.8	44.6

Table 6
Average running times in 10 ms. S = satisfiable, U = unsatisfiable, S/U = satisfiable or unsatisfiable.

average size of the SPASS files generated by ml2dfg varied between 5.4 KB and 5.5 KB for KtScTc4c and KtAcPcTc4c to 12.9 KB and 16.1 KB for KtA*P*T4* and KtS*T4*. This range is plausible because for structural rules the encoding is smallest and for the rules closest to axiom form the partial evaluation results in larger encodings. The input files were the same for the tests done with SPASS-YARRALUMLA and SPASS in auto mode. In total there were 240 satisfiable and 150 unsatisfiable problems.

The tests were run on a Linux PC with a 3.30GHz Intel Core i3-2120 CPU and 10 GB RAM. Each problem was run with a timeout of 600 seconds. The problems and detailed results are available at http://staff.cs.manchester.ac.uk/~schmidt/publications/kthr14/.

Table 6 summarises the results obtained for runs with SPASS-YARRALUMLA. The best results in each column are highlighted in bold dark grey. To account for variability in measurement, results within 10% of the best values are highlighted in light grey. Looking at the table for SPASS-YARRALUMLA, on the whole, the encoding KtAPT4 of propagation rules fared best for all forms of blocking tested. Similarly good results were obtained for the encodings KtA*P*T4, KtA*PT4, KtAP*T4, KtAPTc4, and to some extent KtS*T4. For satisfiable problems the encodings based on correspondence properties fared well, too, in two cases giving best results for KtAcPcTc4c. For unsatisfiable problems the performance was always significantly worse, especially for unrestricted blocking. In terms of blocking, for all encodings unrestricted blocking was most expensive on unsatisfiable problems. In contrast to other blocking techniques, unrestricted blocking generates models with domains of minimal size, which is a much harder problem than determining if models exist. Sound ancestor blocking produced best results for all encodings on both satisfiable and unsatisfiable problems.

Table 7 gives the results of SPASS (Version 3.8d) [28] in auto mode. In auto

Spass in auto mode.				
Encoding	S	U	S/U	M
KtA*P*T4*	75.4(2)	5.3	80.7(2)	12.9
KtA*PT4*	120.8(2)	5.4	126.3(2)	12.6
KtAP*T4*	69.9(2)	5.3	75.2(2)	11.7
KtA*P*T4	102.5(1)	4.8	107.3(1)	9.8
KtA*PT4	82.2	4.8	87.0	9.6
KtAP*T4	32.2(2)	**4.7**	36.9(2)	8.6
KtAPT4	**80.3**	4.9	**85.2**	8.4
KtAPTc4	33.2(2)	4.9	38.1(2)	7.8
KtAcPcTc4c	4.6(131)	685.1(11)	689.7(142)	5.5
KtS*T4*	754.2(12)	5.8	760.1(12)	16.1
KtS*T4	669.3(6)	4.9	674.3(6)	11.6
KtS*Tc4	353.3(12)	5.0	358.4(12)	10.8
KtScTc4c	4.6(131)	54.7	59.3(142)	5.4

Table 7
Average running times in 10 ms. S = satisfiable, U = unsatisfiable, S/U = satisfiable or unsatisfiable, M = average size in KB of input files.

mode SPASS used a form of ordered resolution with dynamic selection. The number of timeouts or unclean exits is indicated in brackets. For two encodings there were no timeouts: KA*PT4 and KAPT4, and their performances were very close. Since SPASS in auto mode is not a decision procedure for problems with chaining laws such as transitivity or the stability properties, the many timeouts for the encodings KAcPcTc4c and KScTc4c are no surprise. For all encodings, apart from these two, the performances were very close for unsatisfiable problems, with KAP*T4 performing best. It is interesting how much faster these performances were than the best performances for SPASS-YARRALUMLA, but not unexpected. For unsatisfiable problems, tableau-like approaches need to construct a complete derivation tree in which every formula is grounded, and this is generally larger than the non-ground clause set derived with ordered resolution. For satisfiable problems, tableau approaches have an advantage because there is no need to explore the entire search space. SPASS computes clause set completions, which are compact representations of all possible models, not just one model. With the axiomatic translation, back-translation is not a big obstacle [18], and the ability to compute entailments is useful.

The problems used in the evaluation can be divided into three groups. One group are problems of the logic BISKT from the investigation in [23]. Essentially these are intuitionistic propositional logic and intuitionistic modal logic problems that have been translated to $Kt(H, R)$. We also used the problems from the investigation of [18]. Because these are predominantly uni-modal problems we have used them as problems for the modality $[R]$ and the modality $[H]$ in separate runs.

Table 8 presents the experimental results differentiated by problem group. For the BISKT problems, results in very close proximity were shown for the encodings KtA*P*T4, KtA*PT4, KtAP*T4, KtAPT4, KtAPTc4 and KtS*T4. Most best times were observed for KtAPT4. Sound ancestor blocking and unrestricted blocking produced best and worst results respectively. Interestingly we see

BISKT problems. Number of problems: 43 S, 103 U, 146 S/U.												
Blocking	(i) -bld			(ii) -bld -ubl			(iii) -blu			(iv) -blu -ubl		
Encoding	S	U	S/U	S	U	S/U	S	U	S/U	S	U	S/U
KtA*P*T4*	4.2	10.1	14.3	4.1	84.4	88.6	5.8	19.9	25.7	4.2	13.2	17.4
KtA*PT4*	4.1	10.5	14.5	4.1	119.4	123.5	5.8	20.4	26.2	4.2	14.0	18.2
KtAP*T4*	4.0	9.3	13.3	4.2	84.0	88.2	5.7	19.2	24.9	4.2	13.0	17.2
KtA*P*T4	3.5	7.0	10.5	3.8	61.5	65.3	4.6	12.8	17.4	3.8	10.9	14.7
KtA*PT4	3.7	7.0	10.6	3.9	59.9	63.8	4.5	12.2	16.6	**3.7**	11.2	14.9
KtAP*T4	**3.4**	6.9	10.3	**3.7**	58.6	62.3	4.4	12.0	16.4	4.0	10.8	14.8
KtAPT4	3.5	**6.6**	**10.1**	3.8	**57.0**	**60.8**	4.4	**11.2**	**15.6**	**3.7**	**10.4**	**14.1**
KtAPTc4	3.6	8.0	11.6	3.8	85.6	89.4	4.4	12.1	16.5	3.8	14.1	17.9
KtAcPcTc4c	3.6	19.1	22.7	4.2	323.1	327.3	4.3	43.1	47.3	3.7	34.7	38.4
KtS*T4*	4.3	10.1	14.4	4.1	90.6	94.7	6.0	22.1	28.1	4.3	13.8	18.0
KtS*T4	3.7	7.2	10.9	4.0	62.4	66.4	4.9	13.0	17.8	4.0	11.1	15.1
KtS*Tc4	3.7	8.6	12.3	4.0	89.3	93.3	4.6	14.6	19.2	3.8	15.3	19.1
KtScTc4c	3.5	28.3	31.9	4.3	261.3	265.6	4.4	64.9	69.3	4.0	37.6	41.6

Modal [R] problems. Number of problems: 118 S, 4 U, 122 S/U.												
Blocking	(i) -bld			(ii) -bld -ubl			(iii) -blu			(iv) -blu -ubl		
Encoding	S	U	S/U	S	U	S/U	S	U	S/U	S	U	S/U
KtA*P*T4*	11.4	4.8	16.1	15.9	5.0	20.9	28.3	10.0	38.3	19.7	5.8	25.4
KtA*PT4*	10.1	6.0	16.1	20.2	4.8	25.0	21.0	9.8	30.7	13.9	5.5	19.4
KtA*P*T4	9.5	4.5	14.0	18.7	5.0	23.7	21.7	10.0	31.7	14.6	5.8	20.3
KtAP*T4*	9.2	5.2	14.5	18.1	5.0	23.1	22.4	10.0	32.4	11.5	5.8	17.3
KtA*PT4	8.4	4.5	12.9	17.5	4.8	22.3	16.3	9.2	25.6	12.4	5.8	18.1
KtAP*T4	7.8	4.5	12.3	14.9	4.8	19.7	17.1	9.2	26.3	10.7	5.5	16.2
KtAPT4	7.3	4.2	11.5	12.8	4.5	17.3	12.8	9.2	22.1	11.4	5.8	17.1
KtAPTc4	7.8	4.5	12.3	**11.1**	4.5	**15.6**	13.0	9.0	22.0	11.0	5.5	16.5
KtAcPcTc4c	**5.6**	4.5	**10.1**	15.8	4.8	20.5	**8.5**	9.0	**17.5**	**9.6**	6.2	**15.9**
KtS*T4*	14.9	5.0	19.9	18.1	5.5	23.6	39.7	10.5	50.2	19.7	7.2	27.0
KtS*T4	10.8	4.8	15.5	14.7	5.0	19.7	25.5	9.8	35.3	11.8	6.2	18.1
KtS*Tc4	11.1	4.5	15.6	14.2	5.0	19.2	24.7	10.8	35.4	14.5	6.2	20.8
KtScTc4c	5.7	4.5	10.2	14.1	4.8	18.9	8.6	9.5	18.1	10.9	5.8	16.7

Modal [H] problems. Number of problems: 79 S, 43 U, 122 S/U.												
Blocking	(i) -bld			(ii) -bld -ubl			(iii) -blu			(iv) -blu -ubl		
Encoding	S	U	S/U	S	U	S/U	S	U	S/U	S	U	S/U
KtA*P*T4*	9.5	8.7	18.3	22.1	81.0	103.1	14.5	12.8	27.4	16.8	13.2	30.0
KtA*PT4*	9.6	9.0	18.5	22.0	80.7	102.7	14.7	12.5	27.2	17.6	13.0	30.6
KtAP*T4*	9.8	8.9	18.6	23.5	81.6	105.0	14.6	12.7	27.4	17.0	12.9	29.9
KtA*P*T4	7.2	6.9	14.1	16.7	15.9	32.6	10.5	11.5	22.0	13.4	10.1	23.4
KtA*PT4	**7.0**	7.0	14.0	15.7	**15.8**	31.5	10.3	11.7	22.0	13.3	**9.5**	22.9
KtAP*T4	7.1	6.9	14.0	15.6	16.2	31.9	10.2	**11.4**	**21.6**	**13.2**	**9.5**	**22.7**
KtAPT4	**7.0**	**6.8**	**13.9**	15.6	**15.8**	31.3	10.2	11.5	21.8	**13.2**	9.7	22.8
KtAPTc4	**7.0**	7.1	14.1	13.4	16.7	**30.1**	**10.0**	11.7	21.7	15.3	10.0	25.3
KtAcPcTc4c	9.8	58.1	67.9	21.5	47.4	68.9	16.4	121.4	137.8	14.2	38.6	52.8
KtS*T4*	9.5	8.7	18.3	22.1	81.2	103.3	14.6	13.0	27.6	17.1	13.0	30.2
KtS*T4	7.2	**6.8**	**13.9**	16.3	17.1	33.5	10.2	11.9	22.1	**13.2**	9.7	22.9
KtS*Tc4	7.2	7.4	14.6	**13.2**	16.9	**30.1**	10.2	11.5	**21.6**	14.3	10.0	24.3
KtScTc4c	10.4	61.6	71.9	21.7	50.8	72.5	17.0	128.9	145.9	13.3	42.9	56.2

Table 8

Average running times in 10 ms for the different problem sets. S = satisfiable, U = unsatisfiable, S/U = satisfiable or unsatisfiable.

that for satisfiable problems the performances were very close for all blocking techniques.

For the modal problems with mainly one modality, the $[R]$ modality, the sample of unsatisfiable problems is very small so we focus just on the results for satisfiable problems. Here, the standard translation or structural rules gave best results, except for the case of `-bld -ubl` and `-blu -ubl`. That the structural rules showed better performance is explained by the fact that the problems contain no $[H]$ modalities, only $[R]$ modalities. This means while especially the starred rules are applicable, the rules (A_c), (P_c) and (S_c) are not. Investigation of the results for unrestricted blocking has revealed that for `KtAcPcTc4c` the results are typically better than for `KtAPTc4`, except for one particular large and difficult problem where `KtAcPcTc4c` was about two times slower than `KtAPTc4` (1378 ms as opposed to 711 ms), which has affected the average results.

The results for the modal $[H]$ problems are interesting because although good performances are again obtained for `KtA*P*T4`, `KtA*PT4`, `KtAP*T4`, `KtAPT4` and `KtS*T4`, good performances in the same range were also obtained for `KtAPTc4` and `KtS*Tc4`. Here the systems based on the starred rules for A, P and S fared very well. Since the problems contain no $[R]$ modalities, these rules are not applicable resulting in fewer inference steps.

Overall, the results confirm that different performances should be expected for different methods on problem classes with different characteristics.

The reasons for implementing the tableau calculi as described were twofold. First, to get insight into the relative performances of different approaches and the properties of different techniques, we wanted a *fair* comparison. Second, using the `ml2dfg` tool and SPASS-YARRALUMLA was an easy way to test different sets of tableau rules and different rule refinements. That models can be read off from the output, aided quick discovery of less effective rule sets and counter-examples for incomplete rule sets, which was extremely useful during the development process. SPASS allowed us to confirm answers with a completely different approach.

9 Conclusion

We have introduced a tense logic $Kt(H, R)$ with two modalities interacting in a non-trivial way. We defined a range of different tableau calculi emerging in a systematic way from axiomatisations and the semantics of the logic. Via effective encodings these calculi can be mapped in various ways to the guarded fragment. This means any decision procedure for the guarded fragment can be used as decision procedure for $Kt(H, R)$. The results of the experiments with implementations of the tableau calculi with SPASS-YARRALUMLA, and SPASS using ordered resolution, give useful insights into the practical properties and relative efficiency of the different deduction approaches.

A more comprehensive empirical investigation needs to be done, but already several observations can be made. First, there are many more ways of deciding modal logics than is usually assumed. Second, we have gained useful insight into how and to what extent different approaches fit together and map to each other. Third, the behaviour of procedures depends a lot on the inference rules (or the transformations in the encodings), rule refinements, termination techniques, and what kind of deduction approaches are used. Fourth, a detailed analysis of the results revealed different performances can be observed for different approaches on problems with different properties (e.g., problems that are predominantly satisfiable, or are predominantly unsatisfiable, or have one of modalities dominate).

Though the focus has been on $Kt(H,R)$, the techniques and ideas presented in this paper are of general nature and provide a useful methodology for developing practical decision procedures for modal logics. Some aspects are completely routine. In particular, the structural rules can be obtained from the Hilbert axioms using methods of automated correspondence theory (cf. [9]) and tableau synthesis [22,24], and soundness and completeness of the calculi Tab_S^{str}, $Tab_S^{str}(ub)$, $Tab_{A,P}^{str}$ and $Tab_{A,P}^{str}(ub)$ are easily obtained. The main aspect for which creativity is required and is specific to $Kt(H,R)$ is the development of the tableau calculi based on propagation or axiomatic rules and the axiomatic translation to the guarded fragment. Here the contribution of the paper has been to extend the ideas of the axiomatic translation principle from [18]. Key is finding effective refinements and showing completeness and termination, which is in general non-trivial and will not always be possible. This gave us also the effective finite model property and termination of the presented tableau systems via the results in [21] and unrestricted blocking.

All in all, because of the ubiquity of modal logics, we believe this kind of systematic research of decidability, proof theory, refinements and relative efficiency is widely applicable and useful, and should be extended to more logics, more types of tableau approaches, other deduction approaches, different provers and more problem sets.

References

[1] Andréka, H., I. Németi and J. van Benthem, *Modal languages and bounded fragments of predicate logic*, J. Philos. Logic **27** (1998), pp. 217–274.

[2] Baumgartner, P. and R. A. Schmidt, *Blocking and other enhancements for bottom-up model generation methods*, in: *Proc. IJCAR 2006*, LNAI **4130** (2006), pp. 125–139.

[3] Baumgartner, P. and R. A. Schmidt, *Blocking and other enhancements for bottom-up model generation methods* (2008), manuscript.

[4] Bloch, I., H. J. A. M. Heijmans and C. Ronse, *Mathematical morphology*, in: M. Aiello, I. Pratt-Hartmann and J. van Benthem, editors, *Handbook of Spatial Logics*, Springer, 2007 pp. 857–944.

[5] Castilho, M. A., L. Fariñas del Cerro, O. Gasquet and A. Herzig, *Modal tableaux with propagation rules and structural rules*, Fund. Inform. **3–4** (1997), pp. 281–297.

[6] Cousty, J., L. Najman, F. Dias and J. Serra, *Morphological filtering on graphs*, Computer Vision and Image Understanding **117** (2013), pp. 370–385.

[7] Fariñas del Cerro, L. and O. Gasquet, *A general framework for pattern-driven modal tableaux*, Logic J. IGPL **10** (2002), pp. 51–83.

[8] Fitting, M., *Tableau methods of proof for modal logics*, Notre Dame J. Formal Logic **13** (1972), pp. 237–247.

[9] Gabbay, D. M., R. A. Schmidt and A. Szałas, "Second-Order Quantifier Elimination: Foundations, Computational Aspects and Applications," College Publ., 2008.

[10] Ganzinger, H. and H. de Nivelle, *A superposition decision procedure for the guarded fragment with equality*, in: *Proc. LICS 1999* (1999), pp. 295–303.
[11] Goré, R., *Tableau methods for modal and temporal logics*, in: M. D'Agostino, D. Gabbay, R. Hähnle and J. Posegga, editors, *Handbook of Tableau Methods*, Kluwer, 1999 pp. 297–396.
[12] Grädel, E., *On the restraining power of guards*, J. Symbolic Logic **64** (1999), pp. 1719–1742.
[13] Hughes, G. E. and M. J. Cresswell, "An Introduction to Modal Logic," Routledge, London, 1968.
[14] Indrzejczak, A., "Natural Deduction, Hybrid Systems and Modal Logics," Trends in Logic **30**, Springer, 2010.
[15] Kracht, M., "Tools and Techniques in Modal Logic," Elsevier, 1999.
[16] Lutz, C., U. Sattler and S. Tobies, *A suggestion of an n-ary description logic*, in: *Proc. DL 1999* (1999), pp. 81–85.
[17] Negri, S., *On the duality of proofs and countermodels in labelled sequent calculi*, in: *Proc. TABLEAUX 2013*, LNCS **8123**, Springer, 2013 pp. 5–9.
[18] Schmidt, R. A. and U. Hustadt, *The axiomatic translation principle for modal logic*, ACM Trans. Comput. Log. **8** (2007), pp. 1–55.
[19] Schmidt, R. A., J. G. Stell and D. Rydeheard, *Axiomatic and tableau-based reasoning for Kt(H,R)* (2014), available from `http://www.cs.man.ac.uk/~schmidt/publications/SchmidtStellRydeheard14b.html`.
[20] Schmidt, R. A. and D. Tishkovsky, *Using tableau to decide expressive description logics with role negation*, in: *Proc. ISWC 2007 + ASWC 2007*, LNCS **4825** (2007), pp. 438–451.
[21] Schmidt, R. A. and D. Tishkovsky, *A general tableau method for deciding description logics, modal logics and related first-order fragments*, in: *Proc. IJCAR 2008*, LNCS **5195** (2008), pp. 194–209.
[22] Schmidt, R. A. and D. Tishkovsky, *Automated synthesis of tableau calculi*, Logical Methods in Comput. Sci. **7** (2011), pp. 1–32.
[23] Stell, J. G., R. A. Schmidt and D. Rydeheard, *Tableau development for a bi-intuitionistic tense logic*, in: *Proc. RAMiCS 14*, LNCS **8428** (2014), pp. 412–428.
[24] Tishkovsky, D. and R. A. Schmidt, *Refinement in the tableau synthesis framework*, arXiv e-Print 1305.3131v1 (2013).
[25] van Benthem, J., *Correspondence theory*, in: D. Gabbay and F. Guenther, editors, *Handbook of Philosophical Logic*, Reidel, 1984 pp. 167–247.
[26] van der Hoek, W., *Logical foundations of agent-based computing*, in: *Muti-Agent Systems and Applications*, LNAI **2086**, Springer, 2001 pp. 50–73.
[27] Viganò, L., "Labelled Non-Classical Logics," Kluwer, 2000.
[28] Weidenbach, C., R. A. Schmidt, T. Hillenbrand, R. Rusev and D. Topic, *System description: SPASS version 3.0*, in: *Proc. CADE-21*, LNAI **4603** (2007), pp. 514–520.

Canonical Filtrations and Local Tabularity

Valentin Shehtman [1]

Institute for Information Transmission Problems
B. Karetny 19
127994, Moscow, Russia

Abstract

The paper deals with a special type of filtration in modal logic called "canonical". This filtration has been known since the 1970s, but was used only occasionally. Applying it in a systematic way allows us to prove new results on finite model property (and in some cases — local tabularity) for different polymodal logics. In particular, we consider products of logics of finite depth with **S5** and **DL**, and also temporal logics of finite depth.

Keywords: modal logic, temporal logic, the finite model property, local tabularity, filtration,canonical model, modal product.

1 Introduction

The filtration method is a standard and powerful instrument in modal logic. Filtrations for Kripke models were first introduced and studied by John Lemmon [9] and (in a general form) by Krister Segerberg [11], [12]

A filtration of a Kripke model M through a set of formulas Ψ is given by a truth-preserving map $h : M \longrightarrow M'$ onto another Kripke model M'. This map is monotonic for all relations in M. So if we can prove that a certain modal logic L is complete w.r.t a class of frames $\mathcal{C}$ and for any model over a $\mathcal{C}$-frame there exists a finite filtration M', also over a $\mathcal{C}$-frame, then L has the finite model property.

This definitely holds if we can construct finite filtrations, for which the filtration map h is a p-morphism. For example, such an argument works in model-theoretic proofs of the well-known Bull's theorem, cf. [4]. In general h need not be a p-morphism, but in some cases p-morphic filtrations can be obtained in a regular way. The corresponding procedure was discovered also by Segerberg [13] [2]. Viz., consider a Kripke model $M = (W, R, \theta)$ and a modal logic L such that $M \vDash L$. Let Ψ be the set of all modal m-formulas (i.e., formulas in proposition letters $p_1, \ldots, p_m$), and let M' be the greatest filtration

[1] shehtman@netscape.net

[2] Segerberg used this filtration to show that every Kripke model is equivalent to a distinguished one. Applications to the fmp proofs were not realized at that time.

of M through Ψ. We call such a filtration *canonical*. In this case M' can be identified with a submodel of the weak canonical model $M_{L\lceil m}$, and one can easily show that h is p-morphic whenever M' is finite.

So to prove the fmp of a certain logic L, we can try to find a class of models characterizing L and prove that their canonical filtrations are finite. So far this method has been used only occasionally [7], [8], but in this paper we will show different situations when it is applicable. For the proofs of finiteness of canonical filtrations we shall use the following strategy. Let $\equiv$ be the equivalence relation modulo Ψ. We construct its *stratification*, i.e., we present $\equiv$ as the intersection of a decreasing sequence of equivalence relations $\equiv_0 \supseteq \equiv_1 \supseteq \ldots$, for which the quotient sets $W/\equiv_n$ are finite. Now if the sequence $(\equiv_n)$ stabilizes, this readily implies the finiteness of M'.

Actually for many logics this argument proves not only the fmp, but *local tabularity*, i.e., finiteness of all weak canonical models. Traditional proofs of local tabularity were just by examining points in the canonical model; cf. for example, the proof of Segerberg's theorem on local tabularity of transitive logics of finite depth [13]. However, canonical filtrations may simplify the job. To this end, we first unravel a canonical model into a tree, then go back by the canonical filtration and prove finiteness by stratification.

The plan of the paper is as follows. Section 2 contains basic material on modal logic. The main result of section 3 is the local tabularity of products of (finitely many) modal logics of finite depth with **S5**; hence we deduce the fmp for $\mathbf{S5} \times \mathbf{K}^r$. A similar argument is applied to $\mathbf{DL} \times \mathbf{K}^r$. In section 4 we consider temporal logics of finite depth and show their local tabularity.

2 Preliminaries

We begin with recalling some standard notions and facts.

In this paper we consider normal polymodal propositional logics understood as usual, as sets of polymodal formulas. *r-modal formulas* are built from a countable set $PL = \{p_1, p_2, \ldots\}$ of proposition letters, the classical connectives $\rightarrow, \bot$, and the modal connectives $\Box_1, \ldots, \Box_n$. A *k-formula* is a formula using only proposition letters from the set $PL\lceil k := \{p_1, p_2, \ldots p_k\}$. A formula without proposition letters is called *closed*.

$\mathcal{L}_r$ (respectively, $\mathcal{L}_r\lceil k$) denotes the set of all r-modal formulas (respectively, r-modal k-formulas).

An *r-modal logic* is a set of r-modal formulas containing the classical tautologies, the axioms $\Box_i(p_1 \rightarrow p_2) \rightarrow (\Box_i p_1 \rightarrow \Box_i p_2)$, closed under Substitution, Modus Ponens, and Necessitation. The *k-restriction* of a modal logic L is $L\lceil k := L \cap \mathcal{L}_n\lceil k$. These sets $L\lceil k$ are called *k-weak modal logics*.

$\mathbf{K}_r$ denotes the minimal r-modal logic; $\mathbf{K} = \mathbf{K}_1$.

An *r-temporal logic* is a $2r$-modal logic (with the modal connectives $\Box_1, \ldots, \Box_r, \Box_{-1}, \ldots, \Box_{-r}$) containing the axioms $\Diamond_i \Box_{-i} p \rightarrow p$, $\Diamond_{-i} \Box_i p \rightarrow p$. $\mathbf{K.t}_r$ denotes the minimal r-temporal logic; $\mathbf{K.t} = \mathbf{K.t}_1$.

For a modal formula A, $md(A)$ denotes its *modal depth* defined by induction:

$$md(\bot) = md(p_i) = 0,\ md(A \to B) = \max(md(A), md(B)),$$
$$md(\Box_j A) = md(A) + 1.$$

Recall that the *fusion* of two logics, r-modal L_1 and m-modal L_2 is $L_1 * L_2 := \mathbf{K}_{r+m} + L_1 + L_2^{+r}$, where L_2^{+r} is obtained from L_2 by replacing every occurrence of any $\Box_j$ with $\Box_{j+r}$.

An *(r-modal) Kripke frame* is a tuple $F = (W, R_1, \ldots, R_r)$, where $W \neq \varnothing$, $R_i \subseteq W \times W$.

We use the standard notation $R_i(x) := \{y \in W \mid xR_iy\}$.

A *Kripke model* over F is a pair $M = (F, \theta)$, where $\theta : PL \longrightarrow 2^W$ is a *valuation.*

$M, x \vDash A$ denotes that a formula A is true at a point x in a Kripke model M; the definition is standard.

A *submodel* of a Kripke model is its restriction to some subset of F. A submodel M' of M is called *reliable* if $M, x \vDash A \Leftrightarrow M', x \vDash A$ for any modal formula A and x in M'.

A *k-weak Kripke model* is $M = (F, \theta)$, where $\theta : PL\lceil k \longrightarrow 2^W$ is a *k-valuation*; in this case we can find truth values only for k-formulas.

A formula A is *valid in a frame* F (notation: $F \vDash A$) if it is true at every world of every Kripke model over F. A set of formulas Γ is *valid* in F (notation: $F \vDash \Gamma$) if every $A \in \Gamma$ is valid. In the latter case we also say that F is a Γ-*frame.* The logic *determined* by a class of frames $\mathcal{C}$ is the set of all formulas valid in all frames from $\mathcal{C}$; it is denoted by $\mathbf{L}(\mathcal{C})$.

In particular, $\mathbf{K}_r$ is determined by all r-modal frames; $\mathbf{K.t}_r$ by all n-*temporal frames* of the form $(W, R_1, \ldots, R_r, R_1^{-1}, \ldots, R_r^{-1})$. The well-known logic **S5** is determined by *clusters*, i.e., frames of the form $(W, W \times W)$.

Definition 2.1 A *p-morphism* from a frame $F = (W, R_1, \ldots, R_r)$ onto $F' = (W', R'_1, \ldots, R'_r)$ is a surjective map $f : W \longrightarrow W'$ satisfying the conditions

- $xR_iy \Longrightarrow f(x)R'_if(y)$ (monotonicity),
- $f(x)R'_iz \Longrightarrow \exists y\ (xR_iy\ \&\ f(y) = z)$ (the lift property).

A p-morphsim of a Kripke model $M = (F, \theta)$ onto $M' = (F', \theta')$ should also satisfy the condition

$$\theta(p) = f^{-1}(\theta'(p))$$

for any $i \leq r$, $p \in PL$ (or $PL\lceil k$ if the models are k-weak).

$f : F \twoheadrightarrow F'$ denotes that f is a p-morphism from F onto F'; the same notation is used for Kripke frames.

Lemma 2.2 *Let F, F' be r-modal Kripke frames, M, M' Kripke models over them, A an r-modal formula.*

(i) If $f : F \twoheadrightarrow F'$, then $\mathbf{L}(F) \subseteq \mathbf{L}(F')$.

(ii) For any x in F, $M, x \vDash A$ iff $M', f(x) \vDash A$.

(iii) If A is closed, then $F \vDash A$ iff $F' \vDash A$.

Definition 2.3 Let $F = (W, R_1, \dots, R_r)$ be a frame, $u, v \in W$, $m \geq 1$. A *path of length m* from u to v is a sequence $(u_0, j_0, u_1, \dots, j_{m-1}, u_m)$ such that $u = u_0$, $v = u_m$ and for all $i < m$, $u_i R_{j_i} u_{i+1}$. A singleton sequence (u) is the *path of length 0* (from u to u).

An r-temporal frame $(W, R_1, \dots, R_r, R_1^{-1}, \dots, R_r^{-1})$ will be denoted by $(W, R_1, \dots, R_r, R_{-1}, \dots, R_{-r})$ (where $R_{-j} := R_j^{-1}$). Then paths are sequences $(u_0, j_0, u_1, \dots, j_{m-1}, u_m)$, in which $j_0, \dots, j_m$ are integers.

Definition 2.4 A path $(u_0, j_0, u_1, \dots, j_{m-1}, u_m)$ in an r-temporal frame $(W, R_1, \dots, R_r, R_{-1}, \dots, R_{-r})$ is called *reduced* if it does not contain adjacent opposite arrows; speaking precisely, if there is no j such that $u_{j-1} = u_{j+1}$ and $i_j = -i_{j+1}$.

Definition 2.5 The *depth* of a point x in a frame F (denoted by $d(x)$) is the maximum of lengths of paths in F beginning from x (if this maximum exists), or ∞ otherwise.

The depth of x w.r.t. to the relation R_i (denoted by $d_i(x)$) is the depth of x in the frame (W, R_i).

Similarly in a temporal frame we define the *reduced depth* of x (denoted by $rd(x)$) as the maximum of lengths of reduced paths beginning from x.

The depth of a frame F (denoted by $d(F)$) is the maximal depth of its points (if it exists) and ∞ otherwise; similarly for the reduced depth in a temporal frame.

Definition 2.6 A *cone in a frame F with root u* (notation: $F{\uparrow}u$) is the restriction of F to the set of all points, to which there exists a path from u; similarly a *cone in a Kripke model* $M{\uparrow}u$ is defined.

Lemma 2.7 *(Generation Lemma)*

(i) $\mathbf{L}(F) = \bigcap_{u \in F} \mathbf{L}(F{\uparrow}u)$.

(ii) $M{\uparrow}u$ is a reliable submodel of M.

Definition 2.8 A *tree* with root u is a frame F such that $F = F{\uparrow}u$ and for every $v \in F$ there exists a unique path from u to v. The length of this path is called the *height* of v and denoted by $h(v)$. The height of F ($h(F)$) is the maximal $h(v)$ (if it exists), or ∞ otherwise.

Definition 2.9 For a $2r$-modal tree $G = (W, S_1, \dots, S_{2r})$, the frame $F = (W, R_1, \dots, R_r, R_1^{-1}, \dots, R_r^{-1})$, where $R_i = S_i \cup S_{r+i}^{-1}$, is called the *$r$-temporal tree (with the pattern G)*. The height function in F is then defined as the height function in G.

Speaking informally, a temporal tree is a modal tree, in which some of the arrows are inverted.

There is an equivalent definition: an n-temporal tree with root r is an n-temporal frame, in which for every point x there exists a unique reduced path from r to x.

Recall the standard unravelling construction (cf. [6]).

Definition 2.10 Let $F = (W, R_1, \ldots, R_r)$ be a cone with root u. The *unravelling* of F is the frame $F^\sharp = (W^\sharp, R_1^\sharp, \ldots, R_r^\sharp)$, in which $W^\sharp$ is the set of all paths from u to points in F, and $\alpha R_i^\sharp \beta$ iff $\beta = (\alpha, i, x)$ for some x.

Lemma 2.11 *$F^\sharp$ is a tree. The map π sending every path to its endpoint is a p-morphism $F^\sharp \twoheadrightarrow F$.*

A similar construction exists in the temporal case [14]:

Definition 2.12 Let $F = (W, R_1, \ldots, R_r, R_1^{-1}, \ldots, R_r^{-1})$ be a cone with root u. The *temporal unravelling* of F is the frame

$$F^{t\sharp} = (W^\sharp, R_1^{t\sharp}, \ldots, R_r^{t\sharp}, R_{-1}^{t\sharp}, \ldots, R_{-r}^{t\sharp}),$$

in which $W^{t\sharp}$ is the set of all reduced paths from u to points in F, and $\alpha R_i^{t\sharp} \beta$ iff ($\beta = (\alpha, i, x)$ or $\alpha = (\beta, -i, x)$) for some x.

Lemma 2.13 *$F^{t\sharp}$ is a temporal tree. The map π sending every path to its endpoint is a p-morphism $F^{t\sharp} \twoheadrightarrow F$.*

Definition 2.14 The *canonical frame* for an r-modal logic (maybe weak) L is $F_L = (W_L, R_{1,L}, \ldots, R_{r,L})$, where W_L is the set of all maximal L-consistent sets of formulas in the language of L; $xR_{i,L}y$ iff for any A, $\Box_i A \in x$ implies $A \in y$.

The *canonical model* for L is $M_L = (F_L, \theta_L)$, where $\theta_L(p_i) = \{x \mid p_i \in x\}$.

Theorem 2.15 *(Canonical model theorem) For any formula A in the language of L,*

(1) $M_L, x \vDash A$ iff $A \in x$;

(2) $M_L \vDash A$ iff $A \in L$.

Lemma 2.16 *(Rigidity lemma) In a canonical model, if there is an isomorphism of two cones $M_L{\uparrow}x$, $M_L{\uparrow}y$ sending x to y, then $x = y$.*

Proof. Since an isomorphism preserves the truth values of formulas, the same formulas (in the language of L) are true in x and y. Hence $x = y$ by 2.15(1). ∎

Definition 2.17 A modal logic L is called *canonical* if $F_L \vDash L$ (or equivalently, $L = \mathbf{L}(F_L)$) and *weakly canonical* if $F_{L\lceil k} \vDash L$ for any finite k.

Definition 2.18 An r-modal logic L is called *locally tabular* if for any finite k there exist finitely many r-modal k-formulas up to equivalence in L.

The local tabularity of L is obviously equivalent to the local finiteness of the variety of L-algebras (which means finiteness of all finitely generated L-algebras, cf. [10], Ch.6, Sec. 14).

Definition 2.19 An r-modal logic L is called *tabular* if $L = \mathbf{L}(F)$ for some finite r-modal frame F. L has the *finite model property (fmp)* if it is an intersection of tabular logics.

The following simple facts are well-known:

Lemma 2.20 *(1) A modal logic L is locally tabular iff every weak canonical model $M_{L\lceil k}$ is finite.*

(2) Every extension of a locally tabular modal logic in the same language is locally tabular.

(3) Every tabular logic is locally tabular.

(4) Every locally tabular logic has the fmp.

Definition 2.21 The *product* of Kripke frames $F = (W, R_1, \dots, R_r)$, $G = (V, S_1, \dots, S_m)$ is the frame

$$F \times G = (W \times V, R_{11}, \dots, R_{r1}, S_{12}, \dots, S_{m2})$$

such that

$$(x, y)R_{i1}(x', y') \Leftrightarrow xR_i x' \ \&\ y = y';$$

$$(x, y)S_{j2}(x', y') \Leftrightarrow x = x' \ \&\ yS_j y'.$$

Definition 2.22 The *product* of an r-modal logic L_1 and an m-modal logic L_2 is the $(r+m)$-modal logic

$$L_1 \times L_2 := \mathbf{L}(\{F_1 \times F_2 \mid F_1 \vDash L_1,\ F_2 \vDash L_2\}).$$

Definition 2.23 The *commutative join* of an r-modal logic L_1 and an m-modal logic L_2 is obtained from their fusion $L_1 * L_2$ by adding the axioms

$$\Diamond_i \Box_{r+j} p \to \Box_{r+j} \Diamond_i p,\ \Box_i \Box_{r+j} p \leftrightarrow \Box_{r+j} \Box_i p$$

for $1 \leq i \leq r$, $1 \leq j \leq m$.

Recall that the corresponding frame conditions are:

$$R_i^{-1} \circ R_{r+j} \subseteq R_{r+j} \circ R_i^{-1},\ R_{r+j} \circ R_i = R_i \circ R_{r+j}.$$

Definition 2.24 Logics L_1, L_2 are called *product-matching* if $L_1 \times L_2 = [L_1, L_2]$.

Recall a sufficient condition for the product matching property.

Definition 2.25 A modal formula is called *Horn* if the class of its frames is first-order definable by a universal Horn sentence.

A modal logic is *Horn axiomatizable* if it is axiomatized by by adding closed or Horn modal formulas.

Theorem 2.26 *Every two complete Horn axiomatizable modal logics are product-matching.*

For the proof cf. Theorem 7.12 from [6] (a slightly weaker claim) or Theorem 5.9 from [5] (for 1-modal logics).

Definition 2.27 Let $M = (W, R_1, \ldots, R_r, \theta)$ be an n-modal Kripke model, Ψ a set of r-modal formulas closed under subformulas. For $x \in W$ let $\Psi_x := \{A \in \Psi \mid M, x \vDash A\}$. Two worlds $x, y \in W$ are called *Ψ-equivalent in M* (notation: $(M, x) \equiv_\Psi (M, y)$, or just $x \equiv_\Psi y$) if $\Psi_x = \Psi_y$. The map $h : x \mapsto x/\equiv_\Psi$ sending every world to its Ψ-equivalence class is called the *filtration map* (through Ψ).

Definition 2.28 (cf. [6]) Under the assumptions of Definition 2.27, a Kripke model $M' = (W', R'_1, ..., R'_r, \theta')$ is called a *filtration of M through Ψ* if for any $x, y \in W$, for any formula A, $1 \leq i \leq r$:

(f1) $W' = W/\equiv_\Psi$;

(f2) $xR_iy \implies h(x)R'_ih(y)$;

(f3) $h(x)R'_ih(y)\ \&\ M, x \vDash \Box_iA\ \&\ \Box_iA \in \Psi \implies M, y \vDash A$;

(f4) if $q \in \Psi \cap PL$, then $M, x \vDash q \iff M', h(x) \vDash q$.

The *greatest filtration* of M through Ψ is defined by the conditions (f1), (f4), and

(f3$^+$) $h(x)R'_ih(y)$ iff for any A,
$M, x \vDash \Box_iA\ \&\ \Box_iA \in \Psi \implies M, y \vDash A$.

Lemma 2.29 *(Filtration Lemma). Let M' be a filtration of M through Ψ. Then for any $x \in W$, for any $A \in \Psi$*

$$M, x \vDash A \text{ iff } M', h(x) \vDash A.$$

Definition 2.30 Let M be an n-modal Kripke model, Ψ the set of all n-modal k-formulas. The greatest filtration of M through Ψ is called *canonical.*

For the canonical filtration we can obviously identify $h(x)$ with Ψ_x, i.e., the set of all k-formulas true at M, x. For any modal logic L true in M, the set Ψ_x is maximal L-consistent, i.e., $\Psi_x \in W_L$ (cf. Definition 2.14). So M' is isomorphic to a reliable submodel of the canonical model $M_{L\lceil k}$.

In fact every p-morphism onto a reliable submodel of a canonical model is a canonical filtration:

Lemma 2.31 *Suppose $h : M \twoheadrightarrow M'$ for a reliable submodel M' of a weak canonical model $M_{L\lceil k}$ for some modal logic L. Then M' is a canonical filtration of M through $\Psi = \mathcal{L}_n\lceil k$ and h is the filtration map.*

Proof. For any k-formula A

$$M, x \vDash A \text{ iff } M', h(x) \vDash A \text{ iff } M, h(x) \vDash A \text{ iff } A \in h(x)$$

by Lemma 2.2, the reliability of M', and Theorem 2.15. Thus $h(x) = \Psi_x$. ■

Proposition 2.32 *If the canonical filtration M' is finite, then the filtration map $h : M \longrightarrow M'$ is a p-morphism.*

Proof. By definition and the Filtration Lemma, $h(x) = h(y)$ iff for any $A \in \Psi$, $M', h(x) \vDash A \Leftrightarrow M', h(y) \vDash A$. So every two different points in M' are distinguished by a formula from Ψ. Since M' is finite, for any $u \in M'$ there exists $A_u \in \Psi$ which is true exactly at u.

Now suppose $h(x)R'_i h(y)$. Then $h(x) \vDash \Diamond_i A_{h(y)}$, and so $x \vDash \Diamond_i A_{h(y)}$. Thus there exists $z \in R_i(x)$ such that $z \vDash A_{h(y)}$. Hence $h(z) \vDash A_{h(y)}$ implying that $h(z) = h(y)$. ■

3 Modal logics of finite depth

In the r-modal language we introduce the total box and diamond as abbreviations:

$$\Box A := \Box_1 A \wedge \ldots \wedge \Box_r A, \quad \Diamond A := \Diamond_1 A \vee \ldots \vee \Diamond_r A.$$

Lemma 3.1 *Let $F = (W, R_1, \ldots, R_r)$ be a frame, $k \geq 1$. Then (in any model over F), for any $x \in W$*

$$x \vDash \Box^{k+1}\bot \text{ iff } d(x) \leq k.$$

The proof is by induction, cf. [6], Lemma 9.2.

Theorem 3.2 *Every logic $\mathbf{K}_r + \Box^k\bot$ is locally tabular.*

This theorem was proved in [6] by examining weak canonical models: a simple inductive argument shows that for any finite d there are finitely many points of depth d in every weak canonical model $M_{L\lceil m}$. Since $\Box^k\bot$ holds in M, all points are of depth less than k; therefore M is finite.

However, let us sketch another proof in the style of the present paper. Consider a cone M' in $M_{L\lceil m}$ and its unravelling M (which is a model over an r-modal tree of depth $(k-1)$ $(W, R_1, \ldots, R_r)$). The canonical map $h : M \longrightarrow M'$ is a p-morphism.

We define the equivalence relations $\equiv_n$ on W by induction:

- $x \equiv_0 y$ iff $x \vDash q \Leftrightarrow y \vDash q$ for any $q \in PL\lceil m$,
- $x \equiv_{n+1} y$ iff $x \equiv_0 y$ & $\forall i \; (R_i(x)/\equiv_n) = (R_i(y)/\equiv_n)$.

Also put

$$x \sim_n y := (x \vDash A \Leftrightarrow y \vDash A \text{ for any } m\text{-formula } A \text{ of depth } \leq n).$$

Lemma 3.3 *If $x \equiv_n y$, then $x \sim_n y$.*

Proof. The proof is straightforward by induction. ■

Lemma 3.4 *If $d(x), d(y) \leq n$ and $x \equiv_n y$, then $x \equiv_{n+1} y$.*

Proof. By induction on n.

The case $n = 0$ is trivial, since $R_i(x) = R_i(y) = \varnothing$.

Suppose the claim holds for $n > 0$, and consider points x, y of depth $\leq n+1$ such that $x \equiv_{n+1} y$. Then for any $z \in R_i(x)$ there exists $z' \in R_i(y)$ such that $z \equiv_n z'$. Then $d(z), d(z') \leq n$, so by IH $z \equiv_{n+1} z'$. It follows that $(R_i(x)/\equiv_{n+1}) \subseteq (R_i(y)/\equiv_{n+1})$. The converse follows by symmetry. ■

Thus by Lemma 3.4 $x \equiv_k y$ implies $x \equiv_n y$ for any $n \geq k$; hence $h(x) = h(y)$ whenever $x \equiv_k y$. But the number of $\equiv_k$-classes is finite; this follows easily by induction.

So every cone in the weak canonical model is finite of limited size. Since this model is distinguishable, it is rigid in the following sense: every two points with isomorphic cones and the same truth values of the proposition letters $p_1, \ldots, p_m$ must coincide. Therefore the whole model is finite.

Let us use a similar method to prove a stronger result.

Consider the logics $\mathbf{S5} \times (\mathbf{K}_r + \Box^s \bot)$. First note that $\mathbf{S5} \times (\mathbf{K}_r + \Box^s \bot) = [\mathbf{S5}, \mathbf{K}_r + \Box^s \bot]$ by Theorem 2.26. The axioms of this logic are Sahlqvist formulas, so it is canonical.

Proposition 3.5 *Every cone validating $\mathbf{K}_n \times \mathbf{K}_m = [\mathbf{K}_n, \mathbf{K}_m]$ is a p-morphic image of a product of an n-modal tree and an m-modal tree.*

Proof. The proof is by applying a transfinite version of the "rectification game". Such a game for the countable case is constructed in the proof Lemma 5.2 from [5]. For a transfinite game just add the requirement that the network at the limit stage is the union of all earlier networks. ■

Lemma 3.6 *Every cone validating $\mathbf{S5} \times (\mathbf{K}_r + \Box^s \bot)$ is a p-morphic image of a product of a cluster and an r-modal tree of depth $\leq s - 1$.*

Proof. The argument is the same as in the proof of theorem 7.2 from [6]. Let $F = (W, R_0, R_1, \ldots, R_r)$ be a given cone; then $F \vDash [\mathbf{K}, \mathbf{K}_r]$. So by Proposition 3.5, there is a p-morphism $f : F_1 \times F_2 \twoheadrightarrow F$, where F_1, F_2 are trees (respectively, 1-modal and r-modal). Let C be the cluster with the set of worlds of F_1. Since R_0 is an equivalence, it follows that $f : C \times F_2 \twoheadrightarrow F$. By lemma 2.2, the validity of the closed formula $\Box^s \bot$ is preserved in $F_1 \times F_2$. Hence by Lemma 3.1, F_2 is of depth $\leq s - 1$. ■

Theorem 3.7 *Every logic $\mathbf{S5} \times (\mathbf{K}_r + \Box^s \bot)$ is locally tabular.*

Proof. Let $L = \mathbf{S5} \times (\mathbf{K}_r + \Box^s \bot)$, and again let us show that all the cones in $M_{L\lceil m}$ are finite.

Consider a cone $M_1 = M_{L\lceil m} \uparrow u$; let $M_1 = (F_1, \theta_1)$. By Lemma 3.6, F_1 is a p-morphic image of a product $C \times F$ of a cluster C and an r-modal tree F of depth $\leq s - 1$. So M_1 is a p-morphic image of a model M over $C \times F$.

Let R_0, $R_1, \ldots, R_r$ be the relations in M (so R_0 is an equivalence).

We define the equivalence relations $\equiv_n$ on M by induction:

- $x \equiv_0 y$ iff $x \vDash q \Leftrightarrow y \vDash q$ for any $q \in PL\lceil m$,
- $x \equiv_{2n+1} y$ iff $x \equiv_{2n} y$ & $(R_0(x)/\equiv_{2n}) = (R_0(y)/\equiv_{2n})$.
- $x \equiv_{2n+2} y$ iff $x \equiv_{2n+1} y$ & $\forall i > 0$ $(R_i(x)/\equiv_{2n+1}) = (R_i(y)/\equiv_{2n+1})$.

Lemma 3.8 *The number of $\equiv_n$-classes in M is finite.*

Proof. By induction we show that the set $W_n := W/\equiv_n$ is finite (where W is the set of worlds in M).

Obviously W_0 is finite, of cardinality at most 2^m.

Suppose W_{2n} is finite. Note that every class $x/\equiv_{2n+1}$ is fully determined by the pair $(x/\equiv_{2n}, R_0(x)/\equiv_{2n})$. Thus

$$|W_{2n+1}| \leq |W_{2n}| \cdot 2^{|W_{2n}|}.$$

(where $|\ldots|$ denotes the cardinality).

Similarly, $(x/\equiv_{2n+2})$ is fully determined by the tuple $(x/\equiv_{2n+1}, R_1(x)/\equiv_{2n+1}, \ldots, R_r(x)/\equiv_{2n+1})$. Hence

$$|W_{2n+2}| \leq |W_{2n+1}| \cdot 2^{r|W_{2n+1}|}.$$

■

Lemma 3.9 *If $x \equiv_{2n} y$, then $x \sim_n y$.*

Proof. By induction. The base is trivial. For the step, suppose $x \equiv_{2n+2} y$.

If $x \vDash \Diamond_i A$, $i > 0$, $md(\Diamond_i A) \leq n+1$, then $z \vDash A$ for some $z \in R_i(x)$, and $d(A) \leq n$. Since $x \equiv_{2n+2} y$, there is $z' \in R_i(y)$ such that $z' \equiv_{2n+1} z$, and so $z' \equiv_{2n} z$. By the IH, $z' \vDash A$. It follows that $y \vDash \Diamond_i A$.

If $x \vDash \Diamond_0 A$, $md(\Diamond_0 A) \leq n+1$, then $z \vDash A$ for some $z \in R_0(x)$, and $d(A) \leq n$. $x \equiv_{2n+2} y$ implies $x \equiv_{2n+1} y$, so there is $z' \in R_0(y)$ such that $z' \equiv_{2n} z$. By the IH, $z' \vDash A$. It follows that $y \vDash \Diamond_0 A$. ■

By $d(x)$ we denote the depth of a point of a point $x \in M$ over the second coordinate. More precisely, if $x = (a, b)$, then $d(x)$ is $d(b)$ (in the tree F).

Lemma 3.10 *If $d(x), d(y) \leq n$ and $x \equiv_{2n+1} y$, then $x \equiv_k y$ for any $k > 2n+1$.*

Proof. By induction on n.

(a) Consider the case $n = 0$. Suppose $d(x) = d(y) = 0$, $x \equiv_1 y$ and show that $x \equiv_k y$ for any $k > 1$ by induction on k.

$x \equiv_{2j+1} y$ clearly implies $x \equiv_{2j+2} y$, since $R_i(x) = R_i(y) = \varnothing$ for $i > 0$.

On the other hand, if $x \equiv_{2j+2} y$, then $x \equiv_{2j+1} y$, so $R_0(x)/\equiv_{2j} = R_0(y)/\equiv_{2j}$. Let us show that $R_0(x)/\equiv_{2j+2} = R_0(y)/\equiv_{2j+2}$. In fact, since $x \equiv_{2j+1} y$, for any $z \in R_0(x)$ there is $z' \in R_0(y)$ such that $z \equiv_{2j} z'$. Since $R_0(z) = R_0(x)$ and $R_0(z') = R_0(y)$, we also have $z \equiv_{2j+1} z'$. But $d(z) = d(z') = 0$, so as we have noticed above, $z \equiv_{2j+2} z'$. It follows that $R_0(x)/\equiv_{2j+2} \subseteq R_0(y)/\equiv_{2j+2}$, and we obtain the converse by symmetry. Therefore $x \equiv_{2j+2} y$ implies $x \equiv_{2j+3} y$.

(b) Now consider the induction step for the main induction on n. Suppose $d(x), d(y) \leq n$, $x \equiv_{2n+1} y$ and show that $x \equiv_k y$ for any $k > 2n+1$ by induction on k.

Suppose $k = 2j+1 > 2n+1$ and the claim is proved for less k. $x \equiv_{2n+1} y$ implies $R_0(x)/\equiv_{2n} = R_0(y)/\equiv_{2n}$, so for any $z \in R_0(x)$ there is $z' \in R_0(y)$ such that $z \equiv_{2n} z'$. Then $z \equiv_{2n+1} z'$ (since $R_0(z) = R_0(x)$, $R_0(z') = R_0(y)$). Since $d(z), d(z') \leq n$, by the IH (applied to z, z') it follows that $z \equiv_{2j} z'$. Thus we have proved $R_0(x)/\equiv_{2j} \subseteq R_0(y)/\equiv_{2j}$, and the converse follows by symmetry. So we obtain $x \equiv_k y$.

Suppose $k = 2j+2 > 2n+1$ and the claim is proved for less k. $x \equiv_{2n+1} y$ implies $x \equiv_{2n} y$ and thus $R_i(x)/\equiv_{2n-1} = R_i(y)/\equiv_{2n-1}$ for any $i > 0$. Now for any $z \in R_i(x)$ there is $z' \in R_i(y)$ such that $z \equiv_{2n-1} z'$. But $d(z), d(z') \leq n-1$, so by the IH of the main induction, $z \equiv_{2n-1} z'$ implies $z \equiv_{2j+1} z'$. So we obtain $R_i(x)/\equiv_{2j+1} \subseteq R_i(y)/\equiv_{2j+1}$, and the converse follows by symmetry. Thus $x \equiv_k y$. ■

The argument in the proof of the theorem is now as in the case of $\mathbf{K}_r + \Box^s\bot$. Since $d(x), d(y) \leq s-1$, by Lemma 3.10 $x \equiv_{2s} y$ implies $x \equiv_k y$ for any $k \geq 2s$; hence by Lemma 3.9, $h(x) = h(y)$ whenever $x \equiv_{2s} y$. Now Lemma 3.8 implies that M_1 is finite of limited size. Therefore up to isomorphism, there are finitely many cones in the weak canonical model, so it is finite by rigidity. ■

Theorem 3.11 *Every logic* $\mathbf{S5} \times \mathbf{K}^m$ *has the fmp.*

Proof. First note that

$$\mathbf{S5} \times \mathbf{K}^m = \mathbf{L}(\{C \times F_1 \times \ldots \times F_m \mid C \text{ is a cluster}, F_1, \ldots, F_m \text{ are trees}\}.$$

This is proved similarly to proposition 4.10 from [6]. In fact, every cone in a product $G_0 \times G_1 \times \ldots G_m$ has the form $C \times H_1 \times \ldots \times H_m$, where C is a cluster, H_i are cones, so it is a p-morphic image of $C \times F_1 \times \ldots \times F_m$, where $F_i = H_i^\sharp$.

Every formula refutable in $C \times F_1 \times \ldots \times F_m$ is also refutable in a product $C \times F_1^- \times \ldots \times F_m^-$, where F_i^- is a tree of finite depth obtained by truncation of F_i; this is similar to Lemma 9.11 from [6]. The product $F_1^- \times \ldots \times F_m^-$ is also of finite depth: if $d(F_i^-) < s$, then $d(F_1^- \times \ldots \times F_m^-) < ms$. Therefore

$$\mathbf{S5} \times \mathbf{K}^m = \bigcap_s (\mathbf{S5} \times (\mathbf{K}^m + \Box^s\bot)).$$

Note that the logic $\mathbf{S5} \times (\mathbf{K}^m + \Box^s\bot)$ contains $\mathbf{S5} \times (\mathbf{K}_m + \Box^s\bot)$, so it is locally tabular by Theorem 3.2 and Lemma 2.20. Then it has the fmp (Lemma 2.20) and eventually $\mathbf{S5} \times \mathbf{K}^m$ has the fmp as an intersection of logics with the fmp. ■

The fmp for the logic $\mathbf{S5} \times \mathbf{K}$ was proved in [6] by another method giving a better upper bound for the size of countermodels. The above theorem for $m > 1$ seems new; however, all these logics are undecidable (this follows from a general result by R. Hirsch, I. Hodkinson, and A. Kurucz, cf. Theorem 8.28 from [5]).

Now consider the difference logic **DL**. Recall that

$$\mathbf{DL} = \mathbf{K} + \Diamond\Box p \to p + p \wedge \Box \to \Box\Box p.$$

DL-cones are of the form (W, R), where R contains the inequality relation $\neq_W := \{(x, y) \in W^2 \mid x \neq y\}$ (cf. [3]).

Lemma 3.12 *Every* **DL**-*cone* (W, R) *is a p-morphic image of some inequality frame* $(V, \neq_V)$.

Proof. By a well-known construction: to obtain V duplicate the reflexive points of W and make them irreflexive. ■

Theorem 3.13 *Every logic* $\mathbf{DL} \times (\mathbf{K}_r + \Box^s\bot)$ *has the fmp.*

Proof. Almost the same as in Theorem 3.7, but with another starting point. Suppose a formula A in letters $p_1, \ldots, p_m$ is not in $L = \mathbf{DL} \times (\mathbf{K}_r + \Box^s\bot)$. Then it is refuted in a weak model M_0 over a product of cones $G \times F$, where $G \vDash \mathbf{DL}$, F is of depth $< s$.

By Lemma 3.12 G is a p-morphic image of an inequality frame C; so A is refuted in $C \times F$, and thus in $C \times F^\sharp$; $F^\sharp$ is a tree of depth $< s$.

Next, we define the relations $\equiv_n$ in the corresponding model M exactly as in the proof of 3.7 and repeat the further proof (with a slight change in the proof of 3.10: at the induction step for $k = 2j+1$ instead of $R_0(z) = R_0(x)$, $R_0(z') = R_0(y)$ we have $R_0(z) \cup \{z\} = R_0(x) \cup \{x\}$, $R_0(z') \cup \{z\} = R_0(y) \cup \{y\}$). Therefore the canonical filtration M' of M is finite, and we can apply Proposition 2.32. Thus A is refuted in a finite L-frame. ∎

Note that now we cannot claim the local tabularity, because we obtain a p-morphism onto some submodel of $M_{L\lceil m}$, but not onto an arbitrary cone.

Theorem 3.14 *Every logic* $\mathbf{DL} \times \mathbf{K}^m$ *has the fmp.*

Proof. Similar to 3.11. By truncation, refutability of a formula in a product $C \times F_1 \times \ldots \times F_m$, where C is an inequality frame, F_i are trees, is reduced to refutability in $C \times F_1^- \times \ldots \times F_m^-$, where F_i^- are trees of finite depth. Hence

$$\mathbf{DL} \times \mathbf{K}^m = \bigcap_s (\mathbf{DL} \times (\mathbf{K}^m + \Box^s\bot)),$$

and we can apply Theorem 3.13. ∎

4 Temporal logics of finite depth

Now let us modify some results of the previous section for temporal logics.

Definition 4.1 Consider the following r-temporal formulas

$$Rd_{i_1 \ldots i_n} := \neg(P_0 \wedge \Diamond_{i_1}(P_1 \wedge \Diamond_{i_2}(P_2 \wedge \ldots \wedge \Diamond_{i_n} P_n) \ldots))),$$

where $i_j \in \{\pm 1, \ldots, \pm r\}$,

$$P_0 := p_0,$$

$$P_{j+1} := \begin{cases} p_{j+1} \wedge \neg p_{j-1} & \text{if } i_{j+1} = -i_j, \\ p_{j+1} & \text{otherwise,} \end{cases}$$

$$Rd_n := \bigwedge \{Rd_{i_1 \ldots i_n} \mid i_1, \ldots, i_n \in \{\pm 1, \ldots, \pm r\}\}.$$

Proposition 4.2 *For an* r*-temporal frame* F

$$F \vDash Rd_n \text{ iff } rd(F) < n.$$

Proof. (If.) Suppose in a model over F we have $u_0 \models \neg Rd_{i_1\dots i_n}$. Then there are $u_1, \dots, u_n$ such that for any $j \geq 0$, $u_j \models P_j$ and $u_j R_{i_j} u_{j+1}$. So $u_0, \dots, u_n$ is a path in F. Note that $i_{j+1} = -i_j$ implies $u_{j+1} \neq u_{j-1}$, since in this case $P_{j+1} = p_{j+1} \wedge \neg p_{j-1}$ and P_{j-1} implies p_{j-1}. Thus $(u_0, i_1 \dots, i_n, u_n)$ is a reduced path of length n.

(Only if.) Suppose there is a reduced path $(u_0, i_1 \dots, i_n, u_n)$. Consider a valuation θ in F such that $\theta(p_j) = \{u_j\}$. Then we obtain a Kripke model, in which $u_j \models P_j$ and $u_0 \models \neg Rd_{i_1\dots i_n}$. Thus $F \not\models Rd_n$. ■

Now consider the logics $\mathbf{K.t}_r + Rd_n$.

Proposition 4.3 $\mathbf{K.t}_r + Rd_n$ *is weakly canonical.*

Proof. Let $L = \mathbf{K.t}_r + Rd_n$. Consider a weak canonical frame $F_{L\lceil k} = (W, R_1, \dots, R_r, R_{-1}, \dots, R_{-r})$ and suppose it has a reduced path $(u_0, i_1 \dots, i_n, u_n)$. Let A_j be a formula true at u_j and false at all the u_m differing from u_j. Then in $M_{L\lceil k}$ for any j $u_j \models A_j$, and $u_{j+1} \models A_{j+1} \wedge \neg A_{j-1}$. It follows that $u_0 \models \neg Rd_{i_1\dots i_n}(A_0, \dots, A_n)$. At the same time $L\lceil k \vdash Rd_{i_1\dots i_n}(A_0, \dots, A_n)$ contradicting the Canonical model theorem 2.15. Therefore $d(F_{L\lceil k}) < n$, and thus L is weakly canonical by Proposition 4.2. ■

Theorem 4.4 *Every logic* $\mathbf{K.t}_r + Rd_n$ *is locally tabular.*

Proof. As we have just proved, the weak canonical frames are of reduced depth $< n$. So by Lemma 2.13, every cone M' in a weak canonical model can be unravelled into a model M over an r-temporal tree of height $< n$.

It remains to show that the canonical filtration of M (which coincides with M' by Lemma 2.31) is finite. We do this again by an appropriate stratification.

Let R_i, $i = \pm 1, \dots \pm r$, be the accessibility relations in M. For a point $x \in M$ (which is not a root) let x^- be its predecessor in the tree, $R_i^\bullet(x) := R_i(x) - \{x^-\}$. For $x, y \in M$ we put

$$x \approx y \text{ iff } \forall i \in \{\pm 1, \dots, \pm r\}(x^- R_i x \Leftrightarrow y^- R_i y).$$

For $x, y \in M$ we define $x \equiv_n y$ by induction: $x \equiv_0 y$ is the same as above; $x \equiv_{n+1} y$ iff

$$x \equiv_n y \ \&\ x \approx y \ \&\ x^- \equiv_n y^- \ \&\ \forall i\ (R_i^\bullet(x)/\equiv_n) = (R_i^\bullet(y)/\equiv_n).$$

Lemma 4.5 $x \equiv_n y$ *implies* $x \sim_n y$

Proof. By induction. For the induction step: suppose the claim holds for n and $x \equiv_{n+1} y$; consider a formula $\Diamond_i A$ of depth $(n+1)$. If $x \models \Diamond_i A$, then $z \models A$ for some $z \in R_i(x)$. Note that then there is $z' \in R_i(y)$ such that $z \equiv_n z'$. In fact, if $x R_i^\bullet z$, this follows from $(R_i^\bullet(x)/\equiv_n) = (R_i^\bullet(y)/\equiv_n)$. If $x R_i z = x^-$, then $x \approx y$ implies $y R_i y^-$; since $x^- \equiv_n y^-$, we can take $z' = y^-$.

Thus we have $z' \models A$, $y \models \Diamond_i A$.

The proof for $\Diamond_{-i} A$ is similar. ■

$d(x)$ denotes the depth and $h(x)$ the height of a point x in the tree M (more precisely, in its pattern).

Lemma 4.6 *If $h(x) = h(y)$, $\max(d(x), d(y)) \leq n$, $x \equiv_n y$, $x \approx y$ and $x^- \equiv_n y^-$, then $x \equiv_{n+1} y$.*

Proof. By induction on $k := \max(d(x), d(y))$. The case $k = 0$ is obvious.

Suppose the claim holds for k, and $\max(d(x), d(y)) = k + 1$. To show that $x \equiv_{n+1} y$, we have to check $(R_i^\bullet(x)/\equiv_n) = (R_i^\bullet(y)/\equiv_n)$.

First suppose $n > 0$. If $xR_i^\bullet z$, then $d(z) < d(x)$ and there exists $z' \in R_i^\bullet(y)$ such that $z \equiv_{n-1} z'$ (since $x \equiv_n y$). Then $\max(d(z), d(z')) \leq k$, so by IH, $z \equiv_n z'$ (note that $z \approx z'$, since $z^- = x$, $z'^- = y$, xR_iz, yR_iz'). Thus $(R_i^\bullet(x)/\equiv_n) \subseteq (R_i^\bullet(y)/\equiv_n)$; and the converse holds by symmetry.

Now suppose $n = 0$, i.e., $d(x) = d(y) = 0$. Then $R_i^\bullet(x) = R_i^\bullet(y) = \varnothing$, and we readily obtain $x \equiv_1 y$. ∎

Next we define x^{-k} as the $k-th$ predecessor of x (and $x^{-0} = x$). For x, y of the same height h put

$$x \approx^+ y \text{ iff } \forall m < h(x^{-m} \approx y^{-m} \;\&\; x^{-m} \equiv_r y^{-m})$$

Lemma 4.7 *If $x \approx^+ y$, then $x \equiv_\Psi y$.*

Proof. By induction on h. If $h = 0$, then we have the root $x = y$.

Suppose the claim holds for $h - 1$. By induction we prove that $x \equiv_n y$ for $n \geq r$. The base is given. Supposing $x \equiv_n y$ let us check $x \equiv_{n+1} y$.

In fact, since $x \approx^+ y$, we have $x \approx y$ and $x^- \equiv_r y^-$. Hence $x^- \equiv_n y^-$ by the IH of the main induction (on h). Therefore $x \equiv_{n+1} y$ by Lemma 4.6. ∎

Since the height of M is finite, Lemma 4.7 implies the finiteness of M'. ∎

5 Conclusion

The method developed in this paper can probably be modified for different kinds of logics: intuitionistic, intuitionistic modal and maybe others. There are more applications within modal logic as well; for example, theorems 3.7, 3.13 can be extended to the temporal case. We hope to publish further results in the sequel.

The study of locally tabular logics can be made within a general context of locally finite varieties of algebras. Cf. [1], where in particular, an algebraic proof of Segerberg's theorem is proposed. It is likely that the results of the present paper can also be proved using the technique from [1].

The paper [2] proves other interesting results on local tabularity using an algebraic technique; in particular, it shows that every proper extension of $\mathbf{S5}^2$ is locally tabular. Our model-theoretic method is probably applicable to this case as well, but this not so obvious.

I would like to thank the anonymous referees for useful comments and references.

References

[1] Bezhanishvili, G., *Locally finite varieties*, Algebra Universalis **46** (2001), pp. 531–548.

[2] Bezhanishvili, N., *Varieties of two-dimensional cylindric algebras. I. Diagonal-free case*, Algebra Universalis **48** (2002), pp. 11–42.
[3] de Rijke, M., *The modal logic of inequality*, Journal of Symbolic Logic **57** (1992), pp. 566–584.
[4] Fine, K., *The logics containing S4.3*, Zeitschrift für mathematische Logik und Grundlagen der Mathematik **17** (1971), pp. 371–376.
[5] Gabbay, D., A. Kurucz, F. Wolter and M. Zakharyaschev, "Many-dimensional modal logics : theory and applications," Elsevier, 2003.
[6] Gabbay, D. and V. Shehtman, *Products of modal logics. part 1*, Journal of the IGPL **6** (1998), pp. 73–146.
[7] Gabbay, D. and V. Shehtman, *Products of modal logics, part 2*, Logic Journal of the IGPL **8** (2000), pp. 165–210.
[8] Gabbay, D. and V. Shehtman, *Products of modal logics, part 3*, Studia Logica **72** (2002), pp. 157–183.
[9] Lemmon, E., *Algebraic semantics for modal logics, part 2*, Journal of Symbolic Logic **31** (1966), pp. 191–218.
[10] Malcev, A., "Algebraic systems," Springer, Berlin, 1973.
[11] Segerberg, K., *Decidability of S4.1*, Theoria **34** (1968), pp. 7–30.
[12] Segerberg, K., *Modal logics with linear alternative relations*, Theoria **36** (1970), pp. 301–322.
[13] Segerberg, K., "An essay in classical modal logic," Filosofiska Studier, Uppsala, 1971.
[14] van Benthem, J., "The logic of time," Synthese Library, v. 156. Kluwer Academic Publishers, 1971.

Paraconsistent Justification Logic: a Starting Point

Che-Ping Su [1]

School of Historical and Philosophical Studies
The University of Melbourne

Abstract

In his 2003 paper, *Ten philosophical problems in belief revision*, Sven Ove Hansson argues that sometimes belief revision might essentially involve inconsistent epistemic states, and that to better model belief revision requires well modeling inconsistent epistemic states. In this paper, we are going to develop a type of justification logic, which is intended to model the agent's justification structure, when she is in an inconsistent epistemic state. We call the logic to be developed *paraconsistent justification logic*. The hope is that this logic could help us better model belief revision.
More specifically, we will construct a three-valued justification logic system, which serves as the starting point of the research. Roughly speaking, the system reflects the idea that committing to a contradiction does not imply committing to everything. This idea is taken to be our basic assumption about inconsistent epistemic states. In addition, the main technical result of the paper is that quasi-realization theorem – which holds for standard two-valued justification logic systems – also holds for this three-valued system.

Keywords: justification logic, paraconsistent logic, inconsistent epistemic state, quasi-realization, belief revision.

1 Introduction

The classic justification logic system JT4 (the *Logic of Proofs*) [1] is the justification counterpart of the classic epistemic logic system S4. Similarly, *paraconsistent justification logic* is intended to be the justification counterpart of *paraconsistent epistemic logic*. To introduce the former, let us start with the latter.

By 'paraconsistent epistemic logic', I mean the family of epistemic logic systems that have this property: $(\Box\phi \wedge \Box\neg\phi) \rightarrow \Box\psi$ is not a valid scheme. System K and all extensions of K do not have this property. In [5,8,10,13], systems with the property are introduced. And, in the literature, some works interpret $\Box$ occurring in the property as the belief operator. Hence, the property then

[1] sucheping@gmail.com is my e-mail. And, I would especially like to thank Graham Priest, Sergei Artemov, Melvin Fitting, Tudor Protopopescu and three anonymous referees.

expresses: having inconsistent beliefs does not imply believing everything. In the context of this paper, I would like to interpret $\Box\phi$ as 'the agent (explicitly) commits to ϕ'. In general, paraconsistent epistemic logic is intended to better model inconsistent epistemic states.

Paraconsistent justification logic is to be developed in this paper.[2] It shall refer to the family of justification logic systems that satisfy the property: it is not the case that for all justification terms s_1, s_2, there exists a justification term t such that $(s_1 : \phi \wedge s_2 : \neg\phi) \rightarrow t : \psi$ is a valid scheme.[3] $s_1 : \phi$ will be intuitively interpreted as 'the agent (explicitly) commits to ϕ for evidence s_1'. And, $s_2 : \neg\phi$ and $t : \psi$ will be interpreted in a similar way. Then, roughly speaking, the property expresses: having pieces of evidence that force us to (explicitly) commit to a contradiction does not imply that for each statement ψ, we have corresponding evidence forcing us to commit to ψ. In general, paraconsistent justification logic is intended to model the agent's justification structure, when she is in an inconsistent epistemic state.

How to construct paraconsistent justification logic is suggested by Fitting models [6] for justification logic. In standard justification logics such as J, JT4 and JT45, formula $t : \phi$ intuitively means that 'the agent believes/knows ϕ for evidence t'. And, informally, under Fitting models, the truth condition of $t : \phi$ is analyzed as the conjunction of:

- t is admissible evidence for ϕ;
- the agent believes/knows ϕ.

In Fitting models, the first condition is formally handled by evidence functions, which are syntactic functions that map worlds and justification terms to sets of formulas. And, Fitting models formalize the second condition by appealing to some epistemic logic. If we use, for example, K to handle the second condition and put corresponding constraints on evidence functions, roughly speaking we get J. Similarly, if we use S4, then we get JT4. Hence, the above suggests that using a paraconsistent epistemic logic system to handle the second condition, we might get a corresponding paraconsistent justification logic. This is how in this paper we are going to construct a paraconsistent justification logic.

There are three main tasks in this paper. First of all, we will try to persuade readers that paraconsistent justification logic is worth developing. Secondly, based on a specific paraconsistent epistemic logic, called $\mathsf{PE_b}$, we are going to construct a paraconsistent justification logic system, called $\mathsf{PJ_b}$. $\mathsf{PJ_b}$ serves as a starting point of the project of developing paraconsistent justification logic. Finally, we will prove: as modal logics, for example, K, S4, S5 can be embedded into justification logics J, JT4, JT45, respectively, $\mathsf{PE_b}$ can be embedded into $\mathsf{PJ_b}$, too. More specifically, we will follow Fitting's non-constructive way [6] to

[2] Joseph Lurie also independently comes up with the idea of merging justification logic with paraconsistent logic. Readers could go to [11] to see more details about how Lurie develops the idea.

[3] Actually, the paraconsistent justification logic system $\mathsf{PJ_b}$, which we are going to construct, satisfies stronger properties, that is, Fact 4.4's (iii) and (iv).

show that quasi-realization theorem holds for $\langle \mathsf{PE_b}, \mathsf{PJ_b} \rangle$.

2 Why Paraconsistent Justification Logic

In this section, I hope to persuade readers that paraconsistent justification logic is worth developing.

It was advertised in Introduction that in the context of this paper, paraconsistent epistemic logic is intended to model the epistemic state of 'explicitly committing to a contradiction'. And, paraconsistent justification logic is intended to model the agent's justification structure, when she is in this kind of epistemic state. Hence, to well motivate paraconsistent justification logic, answers to the following six questions should be provided:

(i) what the epistemic state – explicitly committing to a contradiction – is;

(ii) whether such sort of epistemic states really exist;

(iii) what the basic properties of the epistemic state are;

(iv) why talk about explicit commitments, rather than beliefs;

(v) why care about such sort of epistemic states;

(vi) why care about the agent's justification structure, when she is in this kind of epistemic state.

To the first question, let us start with distinguishing *explicit beliefs*, *implicit commitments* and *explicit commitments*. If an agent explicitly believes some statement, then the agent accepts this statement. Examples of explicit beliefs are my belief that $2 + 2 = 4$ and my belief that I am a human being. In contrast, implicit commitments are the agent's explicit beliefs' consequences that the agent is not aware of. For example, let ϕ be a deep consequence of Peano arithmetic that has not been proved by any mathematician. Then, all people believing PA implicitly commit to ϕ. On the other hand, explicit commitments are the agent's explicit beliefs consequences that the agent is aware of, but the agent might/might not accept.

Explicitly believing a contradiction is not usual. Implicitly committing to a contradiction sometimes happens. For instance, before knowing Russel's paradox of his set theory, Frege implicitly committed to a contradiction. Explicitly committing to a contradiction is the sort of epistemic state that paraconsistent justification logic intends to model.

Secondly, I will argue by example that the epistemic state of explicitly committing to a contradiction does occur. In 1859, Urbain Le Verrier discovered that the actual movement of the planet Mercury is not like what the traditional theory of Newtonian gravity predicts. In other words, the Newtonian gravity plus Le Verrier's obervation leads to a contradiction. The successful revision of the theory of gravity had to wait till Einstein's theory of general relativity (1915). Between 1859 and 1915, physicists could not simply give up/'freeze' Newtonian gravity to keep consistency, because it was the best available theory of the time. Physicists of this period were aware of the contradiction. The contradiction is a consequence of their explicit beliefs in Newtonian gravity and

Le Verrier's obervation. Hence, physicists of this period explicitly committed to the contradiction. However, they did not accept the contradiction to be true.

To the third question, in this paper, we assume that two properties hold for the epistemic state of explicitly committing to a contradiction. One is that explicitly committing to a contradiction does not imply explicitly committing to everything. The other property assumed is: having pieces of evidence that force us to explicitly commit to a contradiction does not imply that for each ψ, we have corresponding evidence forcing us to commit to ψ.

Fourth, in the context of this paper, to simplify things we assume that the agent we model is ideal in the sense that her awareness is logical omniscient. Hence, there is no implicit commitment for the agent; she can only either explicitly believe a contradiction or explicitly commit to a contradiction. Since it is rare that an agent explicitly believes a contradiction, we really need to argue for the category of explicit commitments to contradictions.

The fifth question is why we need to care about modeling the epistemic state of explicitly committing to a contradiction. In the following, I would like to answer the question from the angle of belief revision.

In Sven Ove Hansson's paper [9], it is argued that sometimes belief revision might essentially involve inconsistent epistemic states. Here is a summery of Hansson's point. In the literature of belief revision, there is one approach called *belief base belief revision*, where the belief set is not required to be closed under a consequence relation. In belief base belief revision, there are two ways to define the revision operator:

- revision = expansion + contraction
- revision = contraction + expansion

At least formally, these two ways of defining the revision operator do not collapse into the same operator. When the agent must accept the new information but it is unclear to give up which piece of old information, the first way – revision = expansion + contraction – seems to fit real psychology more. In the first way, there is an intermediate inconsistent epistemic state that occurs after expansion.

To explain Hansson's point, let us recall the example of Le Verrier's observation about Mercury. The physicists of that period ought to accept the new information, that is, Le Verrier's observation. However, it was unclear which part of the Newtonian gravity (or background postulates) we should give up/change. Before the successful revision, physicists explicitly committed to a contradiction. We might take the history as: first expanding with the new information (Le Verrier's observation); secondly, reaching the inconsistent epistemic state of explicitly committing to a contradiction; third, changing the original theory. Therefore, from the angle of belief revision, it seems that this sort of inconsistent epistemic states – explicitly committing to a contradiction

– should be paid attention to[4].

To the final question, I would like to answer it also from the angle of belief revision. My answer begins with an intuition: the same belief/commitment content with different justification structures might lead to different belief revision results. Let us compare two example cases to illustrate the intuition. First, assume that John believes all european swans are white. He forms this belief based on observing most part of Northern Europe and randomly checking the rest of Europe. However, one day, he sees a black swan in Germany. The other case goes as follows. Here, Miles also believes all european swans are white. However, he forms the belief because he watches one program in the Discovery channel and the program says so. And, one day Miles also sees a black swan in Germany. In the first case, John probably can still keep his belief that all swans in Finland are white, because John has done careful survey in Northern Europe. But, it is less clear that in the second case, Miles can keep the same belief. In both cases, each agent starts with the same belief content. But the results of revision are different. It is probably because agents' justification structures are different.

Hence, putting the following two ideas together:

- the epistemic state of explicitly committing to a contradiction sometimes plays an essential role in the process of belief revision;
- the agent's justification structure is a factor to determine the revision result,

we might conclude that we should pay attention to the agent's justification structure, when she explicitly commits to a contradiction.

3 Paraconsistent Epistemic Logic $\mathsf{PE_b}$

In this section, we are going to construct a three-valued epistemic logic system such that $(\Box\phi \wedge \Box\neg\phi) \rightarrow \Box\psi$ is not a valid scheme. We call this system $\mathsf{PE_b}$: the subscript b stands for the third truth value b. Our strategy to construct such a system is to take as the propositional part of $\mathsf{PE_b}$ a non-classical propositional logic where from $\phi \wedge \neg\phi$ we can not derive everything. The specific non-classical propositional logic we pick here has been studied in the paraconsistent logic literature [2,3].

Here is the syntax of $\mathsf{PE_b}$. $\mathsf{PE_b}$ *formulas* are inductively defined as:

$$\phi \ ::= \ p \mid \neg\phi \mid (\phi \wedge \phi) \mid (\phi \rightarrow \phi) \mid \Box\phi$$

Intuitively, $\Box\phi$ means that 'the agent explicitly commits to ϕ'.

Here is the semantics of $\mathsf{PE_b}$. A *frame* is a pair $\langle W, R\rangle$, where W is a non-empty set of possible worlds and R is an accessibility relation of type $W \times W$. At the current stage, to simplify things, we do not put any constraint on R. A $\mathsf{PE_b}$ *model*, $\mathcal{M}$ is a three-tuple $\langle W, R, v\rangle$, where $\langle W, R\rangle$ is a frame, and v is

---

[4] Note that up to this point, we might have provided sufficient reasons to motivate developing *paraconsistent* dynamic epistemic logic for belief revision. [8] is along this line.

a function from worlds and propositional variables to $\{1, b, 0\}$. In addition, 1 and b are designated values. b intuitively means 'being both true and false'.

[12] provides one way to make sense of value b. Each world with some propositional variables assigned value b is an *impossible world.* "Impossible worlds are just more than one possible world taken together" [12]. An impossible world represents a way that the world can not be. Different impossible worlds represent different ways of 'clashes' (in commitment contents).

We extend v in the following way.

Definition 3.1 Let $\mathcal{M} = \langle W, R, v\rangle$ be a $\mathsf{PE_b}$ model. $v^{\mathcal{M}}$ extends v in the following way. Let u be a world in $\mathcal{M}$.

(i)

$$v_u^{\mathcal{M}}(p) = v_u(p)$$

(ii)

$$v_u^{\mathcal{M}}(\neg\phi) = \begin{cases} 1 & \text{if } v_u^{\mathcal{M}}(\phi) = 0; \\ b & \text{if } v_u^{\mathcal{M}}(\phi) = b; \\ 0 & \text{otherwise.} \end{cases}$$

(iii)

$$v_u^{\mathcal{M}}(\phi \wedge \psi) = \begin{cases} 1 & \text{if } v_u^{\mathcal{M}}(\phi) = 1 \text{ and } v_u^{\mathcal{M}}(\psi) = 1; \\ b & \text{if } v_u^{\mathcal{M}}(\phi) = b \text{ and } v_u^{\mathcal{M}}(\psi) \neq 0, \\ & \text{or } v_u^{\mathcal{M}}(\phi) \neq 0 \text{ and } v_u^{\mathcal{M}}(\psi) = b; \\ 0 & \text{otherwise.} \end{cases}$$

(iv)

$$v_u^{\mathcal{M}}(\phi \to \psi) = \begin{cases} v_u^{\mathcal{M}}(\psi) & \text{if } v_u^{\mathcal{M}}(\phi) \in \{1, b\}; \\ 1 & \text{otherwise.} \end{cases}$$

(v)

$$v_u^{\mathcal{M}}(\Box\phi) = \begin{cases} 1 & \text{if for all } u' \text{ with } uRu',\ v_{u'}^{\mathcal{M}}(\phi) \in \{1, b\}; \\ 0 & \text{otherwise.} \end{cases}$$

Define $\phi \vee \psi$ as the abbreviation of $\neg(\neg\phi \wedge \neg\psi)$ [5]; $\phi \leftrightarrow \psi$ as the abbreviation of $(\phi \to \psi) \wedge (\psi \to \phi)$. In addition, note that the conditional of $\mathsf{PE_b}$ can not be defined by $\neg$, $\wedge$ [4]. Furthermore, $\Box\phi$ never gets value b. The rationale is that although the agent's commitment content could be inconsistent, it is never the case that the agent *both* explicitly commits to some statement and does not explicitly commit to the same statement [6].

Let us look at an example model of the system.

---

[5] Assume this order on the truth values: $0 < b < 1$. Then, $\vee$ corresponds to the operation of taking the maximum value. This justifies the abbreviation.

[6] Note that not committing to ϕ is different from committing to not-ϕ.

Example 3.2 Let us consider a restricted language, which only has two propositional variables, p, q. Let 10 stand for a world where p is true (only) and q is false (only). Let $b0$ stand for a world where p is *both* true and false, but q is false (only) . (So, $b0$ is an impossible world.) Then, the following is an example $\mathsf{PE_b}$ model. Let us call it $\mathcal{N}$.

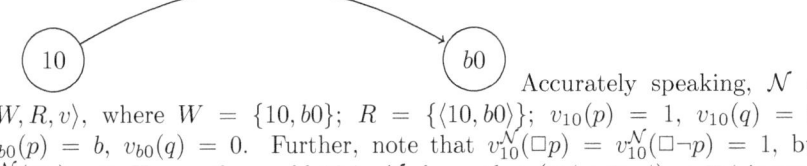

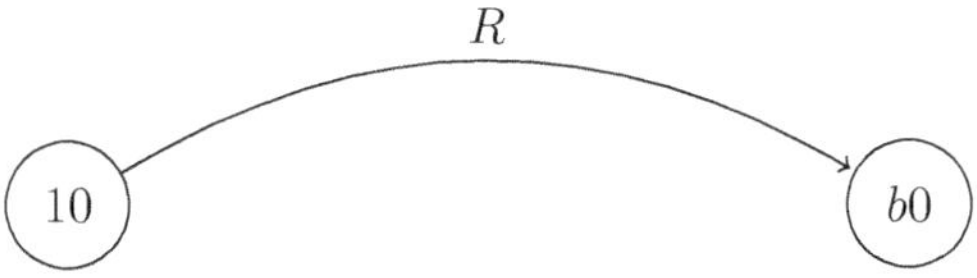

Accurately speaking, $\mathcal{N} = \langle W, R, v\rangle$, where $W = \{10, b0\}$; $R = \{\langle 10, b0\rangle\}$; $v_{10}(p) = 1$, $v_{10}(q) = 0$, $v_{b0}(p) = b$, $v_{b0}(q) = 0$. Further, note that $v^{\mathcal{N}}_{10}(\Box p) = v^{\mathcal{N}}_{10}(\Box\neg p) = 1$, but $v^{\mathcal{N}}_{10}(\Box q) = 0$. Hence, the world 10 in $\mathcal{N}$ shows that $(\Box\phi \wedge \Box\neg\phi) \rightarrow \Box\psi$ is not a valid scheme in $\mathsf{PE_b}$.

Here are some basic facts about $\mathsf{PE_b}$.

Fact 3.3 (i) *$(\Box p \wedge \Box\neg p)$ is satisfiable.*

(ii) *$(\Box\phi \wedge \Box\neg\phi) \rightarrow \Box\psi$ is not a valid scheme. That is, for some formulas ϕ and ψ, $(\Box\phi \wedge \Box\neg\phi) \rightarrow \Box\psi$ is invalid.*

(iii) *For all formulas ϕ, if ϕ contains no $\Box$, then there exist some formula ψ such that $(\Box\phi \wedge \Box\neg\phi) \rightarrow \Box\psi$ is invalid.*

(iv) *For all formulas ϕ and ψ, if both ϕ, ψ contain no $\Box$; ϕ and ψ share no propositional variable; and ψ is not valid, then $(\Box\phi \wedge \Box\neg\phi) \rightarrow \Box\psi$ is invalid.*

(v) *Modus Ponens preserves designated values.*

(vi) *$\Sigma \cup \{\phi\} \Vdash \psi$, iff $\Sigma \Vdash \phi \rightarrow \psi$.*[7]

(vii) *$\Box(\phi \rightarrow \psi) \rightarrow (\Box\phi \rightarrow \Box\psi)$ is a valid scheme.*

Proof. Please see the first half of Appendix A. □

The first four of Fact 3.3 form a group. It is about the paraconsistency of $\mathsf{PE_b}$. (i) says that an agent might explicitly commits to a contradiction. We explain (ii) and (iii) together. Maybe, the slogan for paraconsistent epistemic logic is that 'explicitly committing to a contradiction does not imply explicitly committing to everything'. However, the slogan is vague. There are two possible accurate readings of the slogan:

- for **some** contradiction $\phi \wedge \neg\phi$, explicitly committing to $\phi \wedge \neg\phi$ does not imply explicitly committing to everything;
- for **all** contradictions $\phi \wedge \neg\phi$, explicitly committing to $\phi \wedge \neg\phi$ does not imply explicitly committing to everything.

The first is weaker than the second. (ii) expresses the weaker reading of the

7 We define the semantical consequence relation $\Vdash$ for $\mathsf{PE_b}$ as follows: $\Gamma \Vdash \tau$, iff for all $\mathsf{PE_b}$ models $\mathcal{M}$ and all worlds u in $\mathcal{M}$, if $v^{\mathcal{M}}_u(\tau') \in \{1, b\}$ for all $\tau' \in \Gamma$, then $v^{\mathcal{M}}_u(\tau) \in \{1, b\}$.

slogan. What (iii) expresses is closer to the stronger reading. However, (iii) does not quantify over unrestrictedly all contradictions, but just all contradictions that do not involve $\Box$. Now, we explain (iv). First of all, if two formulas share no propositional variable, intuitively we could take these two formulas to be irrelevant to each other. Secondly, based on the previous point, (iv) intuitively means:

- for all contradictions $\phi \wedge \neg \phi$ (with some restriction), for all statements ψ (with some restriction), if $\phi \wedge \neg \phi$ and ψ are irrelevant to each other in the sense that $\phi \wedge \neg \phi$ and ψ share no propositional variable, then explicitly committing to $\phi \wedge \neg \phi$ does not imply explicitly committing to ψ.

In addition, note that (iv) implies (iii), but not vice versa; (iii) implies (ii), but not vice versa.

The last three items of Fact 3.3 form another group. This group tells us that $\mathsf{PE_b}$ has Modus Ponens, deduction theorem and the validity of K-axiom. The way we define the conditional gives us these three. If we define $\phi \to \psi$ as $\neg \phi \vee \psi$, the resulting system will not have these three. As readers will see in the next section, making our system have these three will be helpful for letting the corresponding justification logic behave nicely in the technical aspect.

4 Paraconsistent Justification Logic $\mathsf{PJ_b}$

Based on $\mathsf{PE_b}$, we are going to construct a justification logic system $\mathsf{PJ_b}$ by giving its Fitting models. $\mathsf{PJ_b}$ reflects the idea: explicitly committing to a contradiction for two conflicting pieces of evidence does not imply that for all statement ψ, we have corresponding evidence that forces us to commit to ψ. This idea is accurately formulated in (ii) of Fact 4.4. Actually, $\mathsf{PJ_b}$ satisfies stronger paraconsistent properties, that is, (iii) and (iv) of Fact 4.4.

4.1 Syntax of $\mathsf{PJ_b}$

$\mathsf{PJ_b}$ has the following symbols.

(i) propositional variables, $p, q, r, \ldots$

(ii) connectives, $\neg$, $\wedge$, $\to$

(iii) justification variables, $x, y, \ldots$

(iv) justification constants, $c, d \ldots$

(v) function symbol, $\cdot$, $+$

(vi) operator symbol of the type $\langle term \rangle : \langle formula \rangle$

Justification terms are inductively defined as:

$$t \quad ::= \quad x \mid c \mid (t \cdot t) \mid (t + t)$$

$\mathsf{PJ_b}$ *formulas* are inductively defined as:

$$\phi \quad ::= \quad p \mid \neg \phi \mid (\phi \wedge \phi) \mid (\phi \to \phi) \mid t : \phi$$

Intuitively, $t : \phi$ is interpreted as 'the agent explicitly commits to ϕ based on evidence t'.

4.2 Semantics of $\mathsf{PJ_b}$

We specify $\mathsf{PJ_b}$'s semantics by giving its Fitting models.

An *evidence function* on the frame $\langle W, R\rangle$ is a function $\mathcal{E}$ from worlds and justification terms to sets of formulas. In $\mathsf{PJ_b}$, we put the following constrains on the evidence function:

- **Application** $(\phi \to \psi) \in \mathcal{E}(u,s)$ and $\phi \in \mathcal{E}(u,t)$ implies $\psi \in \mathcal{E}(u, s\cdot t)$.
- **Sum** $\mathcal{E}(u,s) \cup \mathcal{E}(u,t) \subseteq \mathcal{E}(u, s+t)$.

A $\mathsf{PJ_b}$ *model*, $\mathcal{M}$ is a four-tuple $\langle W, R, \mathcal{E}, v\rangle$, where $\langle W, R\rangle$ is a frame, $\mathcal{E}$ is an *evidence function* on $\langle W, R\rangle$, and v is a function from worlds and propositional variables to $\{1, b, 0\}$. And, 1 and b are designated values.

We extend v in the following way.

Definition 4.1 Let $\mathcal{M} = \langle W, R, \mathcal{E}, v\rangle$ be a $\mathsf{PJ_b}$ model. $v^{\mathcal{M}}$ extends v in the following way. Let u be a world in $\mathcal{M}$.

(i) propositional variables, $\neg$, $\wedge$ and $\to$ are interpreted as in $\mathsf{PE_b}$

(ii)

$$v_u^{\mathcal{M}}(t:\phi) = \begin{cases} 1 & \text{if } \phi \in \mathcal{E}(u,t) \text{ and for all } u' \text{ with } uRu',\ v_{u'}^{\mathcal{M}}(\phi) \in \{1, b\}; \\ 0 & \text{otherwise.} \end{cases}$$

Definition 4.2 [constant specification] Let $X_0 = \{\phi \mid v_u^{\mathcal{M}}(\phi) = 1 \text{ or } b$ for every world u at every $\mathsf{PJ_b}$ models $\mathcal{M}\}$. Let $X_{n+1} = \{c : \phi \mid c$ is a justification constant and $\phi \in X_n\}$. Let $X = \bigcup_{i\in\mathbb{N}} X_i$. A *constant specification* $\mathcal{C}$ is a function from justification constants to subsets of X such that $\mathcal{C}$ satisfies the following two conditions:

- if $c_n : \cdots : c_1 : \phi \in \mathcal{C}(c_{n+1})$, then $c_{n-1} : \cdots : c_1 : \phi \in \mathcal{C}(c_n)$, where $n \geqslant 2$ and $\phi \in X_0$.
- if $c_1 : \phi \in \mathcal{C}(c_2)$, then $\phi \in \mathcal{C}(c_1)$, where $\phi \in X_0$.

A model $\mathcal{M} = \langle W, R, \mathcal{E}, v\rangle$ is said to *meet a constant specification* $\mathcal{C}$, iff for all $u \in W$, for all justification constants c, $\mathcal{C}(c) \subseteq \mathcal{E}(u,c)$.

Here is an example $\mathsf{PJ_b}$ model.

Example 4.3 Let $\mathcal{C}$ be a constant specification. Let $x, y, z_1, z_2, \ldots$ be an enumeration of all justification variables. We construct a $\mathsf{PJ_b}$ model $\mathcal{M}$ meeting $\mathcal{C}$, based on the $\mathsf{PE_b}$ model $\mathcal{N}$ defined in Example 3.2. Define $\mathcal{M} = \langle W, R, \mathcal{E}, v\rangle$, where $\langle W, R, v\rangle = \mathcal{N}$ and $\mathcal{E}$ is defined as follows:

- for all justification constants c, define $\mathcal{E}(u,c) = \mathcal{C}(c)$, where $u \in \{10, b0\}$;
- define $\mathcal{E}(10, x) = \{p\}$, $\mathcal{E}(10, y) = \{\neg p\}$ and $\mathcal{E}(10, z_i) = \emptyset$, where $i \in \mathbb{N}$;
- define $\mathcal{E}(b0, x) = \mathcal{E}(b0, y) = \mathcal{E}(10, z_i) = \emptyset$, where $i \in \mathbb{N}$;
- define $\mathcal{E}(u, s\cdot t) = \{\psi \mid \phi \to \psi \in \mathcal{E}(u,s) \text{ and } \phi \in \mathcal{E}(u,t)\}$, where $u \in \{10, b0\}$;
- define $\mathcal{E}(u, s+t) = \mathcal{E}(u,s) \cup \mathcal{E}(u,t)$, where $u \in \{10, b0\}$.

The fourth and fifth item guarantee that $\mathcal{E}$ satisfies **Application** and **Sum** condition on evidence function, respectively. The first item guarantees that $\mathcal{M}$ meets $\mathcal{C}$. Further, note that $v_{10}^{\mathcal{M}}(x : p) = v_{10}^{\mathcal{M}}(y : \neg p) = 1$, but $v_{10}^{\mathcal{M}}(t : q) = 0$ for all justification terms t.

4.3 Notations and Terminologies

Let $\mathcal{C}$ be a constant specification. A set Σ of formulas is said to be *$\mathcal{C}$-satisfiable*, if there are some model $\mathcal{M}$ that meets $\mathcal{C}$ and some world u in $\mathcal{M}$ such that $v_u^{\mathcal{M}}(\psi) \in \{1, b\}$ for all $\psi \in \Sigma$. A formula ϕ is *valid in a model* $\mathcal{M} = \langle W, R, \mathcal{E}, v \rangle$, if for all worlds u in $\mathcal{M}$, we have $v_u^{\mathcal{M}}(\phi) \in \{1, b\}$. A formula is *$\mathcal{C}$-valid*, if it is valid in every model that meets $\mathcal{C}$. Now, we define the semantical consequence relation. $\Sigma \Vdash_{\mathcal{C}} \phi$ iff for all models $\mathcal{M}$ that meet $\mathcal{C}$ and all worlds u in $\mathcal{M}$, if $v_u^{\mathcal{M}}(\psi) \in \{1, b\}$ for all $\psi \in \Sigma$, then $v_u^{\mathcal{M}}(\phi) \in \{1, b\}$.

Let Σ be a set of formulas. Define $v_u^{\mathcal{M}}[\Sigma] = \{v_u^{\mathcal{M}}(\phi) \mid \phi \in \Sigma\}$.

4.4 Basic Facts about $\mathsf{PJ_b}$

Here are some basic facts about $\mathsf{PJ_b}$. We divide them into two groups.

The following is the first group, which is about the paraconsistency of $\mathsf{PJ_b}$. To simplify the formulation, we use $\forall$, $\exists$ as the abbreviations of the meta-expressions, 'for all' and 'for some', respectively. s_1, s_2, t range over justification terms; x, y range over justification variables.

Fact 4.4 *Let $\mathcal{C}$ be a constant specification.*

(i) $(x : p \wedge y : \neg p)$ *is $\mathcal{C}$-satisfiable;*

(ii) $\exists s_1 \exists s_2 \exists \phi \exists \psi \forall t$, $\{s_1 : \phi, s_2 : \neg\phi\} \nVdash_{\mathcal{C}} t : \psi$;

(iii) $\forall s_1 \forall s_2 \forall \phi$, *if ϕ contains no justification term, then* $\exists \psi \forall t$, $\{s_1 : \phi, s_2 : \neg\phi\} \nVdash_{\mathcal{C}} t : \psi$;

(iv) $\forall s_1 \forall s_2 \forall \phi \forall \psi \forall t$, *if ϕ and ψ contain no justification term; ϕ and ψ share no propositional variable; and ψ is not $\mathcal{C}$-valid, then* $\{s_1 : \phi, s_2 : \neg\phi\} \nVdash_{\mathcal{C}} t : \psi$.

Proof. Please see the second half of Appendix A. □

The item (i) of Fact 4.4 says that an agent might explicitly commits to a contradiction based on two different pieces of evidence. The item (ii) intuitively means:

- for **some** two conflicting pieces s_1, s_2 of evidence, for **some** contradiction $\phi \wedge \neg\phi$, explicitly committing to $\phi \wedge \neg\phi$ based on s_1, s_2 does not imply that for all statement ψ, there is evidence t which forces us to explicitly commit to ψ.

The item (iii) expresses something stronger:

- for **every** two conflicting pieces s_1, s_2 of evidence, for **every** contradiction $\phi \wedge \neg\phi$ (with some restriction), explicitly committing to $\phi \wedge \neg\phi$ based on s_1, s_2 does not imply that for all statement ψ, there is evidence t which forces us to explicitly commit to ψ.

The item (iv) intuitively says:

- Given any two pieces s_1, s_2 of evidence, any contradiction $\phi \wedge \neg\phi$ (with some restriction) and any statement ψ (with some restriction), if $\phi\wedge\neg\phi$ is irrelevant to ψ (in the sense that $\phi \wedge \neg\phi$ and ψ share no propositional variable), then explicitly committing to $\phi \wedge \neg\phi$ based on s_1, s_2 does not imply that there is evidence t that forces us to explicitly commit to ψ.

And, note two things. First, (iv) implies (iii), but not vice versa; (iii) implies (ii), but not vice versa. Secondly, typically, in justification logic, we consider constant specifications that *entail internalization* in the sense that for all formulas ϕ, if ϕ is $\mathcal{C}$-valid, then $t : \phi$ is $\mathcal{C}$-valid for some justification term t. In standard two-valued justification logics, if the constant specification considered entails internalization, then all of (ii) – (iv) fail. However, in $\mathsf{PJ_b}$, (ii) – (iv) hold, even if the constant specification considered entails internalization.

Now, we move to the second group of facts about $\mathsf{PJ_b}$. They are good properties as a justification logic. Proofs for them are skipped.

Fact 4.5 *Let $\mathcal{C}$ be a constant specification.*

(i) *Modus Ponens preserves designated values;*

(ii) $\Sigma \cup \{\phi\} \Vdash_{\mathcal{C}} \psi$, *iff* $\Sigma \Vdash_{\mathcal{C}} \phi \to \psi$;

(iii) $s : (\phi \to \psi) \to (t : \phi \to (s \cdot t) : \psi)$ *is $\mathcal{C}$-valid;*

(iv) $s : \phi \to (s+t) : \phi$ *and* $t : \phi \to (s+t) : \phi$ *are $\mathcal{C}$-valid.*

The way we define the conditional gives us (i), (ii) and (iii). These three play important roles in proving quasi-realization theorem for $\langle \mathsf{PE_b}, \mathsf{PJ_b} \rangle$ (Theorem 6.3).

5 Axiomatic Soundness and Completeness of $\mathsf{PJ_b}$

In this section, we give an axiomatization of $\mathsf{PJ_b}$ and show its soundness and completeness. The machinery that we will use to prove completeness of $\mathsf{PJ_b}$ will help us prove quasi-realization theorem in the next section.[8]

5.1 An Axiomatic Proof System

Before giving the axiomatization, we list some setups. First, recall that $\phi \vee \psi$ is the abbreviation of $\neg(\neg\phi\wedge\neg\psi)$; $\phi \leftrightarrow \psi$ is the abbreviation of $(\phi \to \psi)\wedge(\psi \to \phi)$. Secondly, pick a specific propositional variable p. Define $\dot\neg\phi$ as the abbreviation of $\phi \to (\Box p \wedge \neg\Box p)$. Note that based on $\mathsf{PJ_b}$'s semantics, $\Box p \wedge \neg\Box p$ always gets value 0. Hence, $\dot\neg$ actually behaves like the classical negation.[9]

Let $\mathcal{C}$ be a constant specification. The $\mathsf{PJ_b}$ *axiomatic proof system w.r.t. $\mathcal{C}$* is defined by the following axiom schemes and inference rules.

[8] Accurately speaking, it is the truth lemma for $\mathsf{PJ_b}$ – the key lemma to completeness – that will help us prove quasi-realization theorem.

[9] Roughly speaking, we can take $\mathsf{PJ_b}$ as having two negations: $\neg$ and $\dot\neg$. The former does not lead to explosion; the latter does. That is, $\phi \to (\neg\phi \to \psi)$ is not a valid scheme; however, $\phi \to (\dot\neg\phi \to \psi)$ is. $\neg$ helps us model paraconsistency (Fact 4.4). And, $\dot\neg$ helps us give a complete axiomatization.

A1. $\phi \to (\psi \to \phi)$
A2. $(\tau \to (\phi \to \psi)) \to ((\tau \to \phi) \to (\tau \to \psi))$
A3. $(\phi \vee \psi) \to (\psi \vee \phi)$
A4. $(\phi \vee \psi) \to ((\phi \to \tau) \to (\tau \vee \psi))$
A5. $\phi \vee \neg\phi$
A6. $\phi \vee \dot\neg\phi$
A7. $\phi \to (\dot\neg\phi \to \psi)$
A8. $\phi \to (\psi \to (\phi \wedge \psi))$
A9. $(\phi \wedge \psi) \to \phi$
A10. $(\phi \wedge \psi) \to (\psi \wedge \phi)$
A11. $\neg\phi \to \neg(\phi \wedge \psi)$
A12. $(\dot\neg\neg\phi \wedge \dot\neg\neg\psi) \leftrightarrow \dot\neg\neg(\phi \wedge \psi)$
A13. $\neg(\phi \to \psi) \leftrightarrow (\phi \wedge \neg\psi)$
A14. $\neg\neg\phi \leftrightarrow \phi$
A15. $\dot\neg\neg\neg\phi \leftrightarrow \dot\neg\phi$
A16. $s : (\phi \to \psi) \to (t : \phi \to (s \cdot t) : \psi)$
A17. $s : \phi \to (s + t) : \phi$
A18. $s : \phi \to (t + s) : \phi$
A19. $t : \phi \to (\neg t : \phi \to \psi)$

R1.	*Modus Ponens*	$\phi, \phi \to \psi \Rightarrow \psi$
R2.	*$\mathcal{C}$ Axiom Necessitation*	$\Rightarrow c : \phi$ where $\phi \in \mathcal{C}(c)$ and either ϕ is an axiom **A1** – **A19** or ϕ is inferable using **R2**.

We use the notation $\vdash_{\mathcal{C}}$ to denote the proof-theoretic consequence relation with respect to a constant specification $\mathcal{C}$.

Following [6], when giving the axiomatization, actually we do not consider all constant specifications, but put further constraints. In this paper, we put the following constraint on constant specifications. A constant specification is said to be *strongly appropriate*, if every axiom has a justification constant; $c : \phi$ has a justification constant, whenever c is a justification constant for ϕ; and apart from these conditions, no other formulas have justification constants. This constraint helps, when we prove the axiomatic proof system is complete (Theorem 5.10) and satisfies internalization (Theorem 5.2).

We finish this subsections with two basic theorems. First, with the help of axiom schemes **A1** and **A2**, we can show the deduction theorem.

Theorem 5.1 *Let $\mathcal{C}$ be a constant specification. $\Sigma \cup \{\phi\} \vdash_{\mathcal{C}} \psi$, iff $\Sigma \vdash_{\mathcal{C}} \phi \to \psi$.*

Secondly, with the help of **A16**, **R2** and the strong appropriateness, we have the following theorem.

Theorem 5.2 (internalization) *Let $\mathcal{C}$ be a strongly appropriate constant specification. Then, if $\vdash_{\mathcal{C}} \phi$, then $\vdash_{\mathcal{C}} t : \phi$, for some justification term t.*

5.2 Soundness and Completeness

Proving the soundness of the axiomatic system is relatively simple, so we focus on completeness.

We prove completeness by the canonical model construction. One small difference from the classical case is that the notion of maximal consistent set is replaced by the notion of *$\mathcal{C}$-partition*. Before defining this notion, we need to introduce a notation: $\Gamma \vdash_{\mathcal{C}} \Gamma'$ holds, iff there are some formulas $\phi_1, \ldots, \phi_n$ in Γ' such that $\Gamma \vdash_{\mathcal{C}} \phi_1 \vee \cdots \vee \phi_n$. Now, we are ready to define the notion. Given a strongly appropriate constant specification $\mathcal{C}$ and two sets Γ, Γ' of formulas, $\langle \Gamma, \Gamma' \rangle$ is said to be a *$\mathcal{C}$-partition*, if the following three conditions hold:

(1) $\Gamma' \neq \emptyset$; (2) $\Gamma \nvdash_{\mathcal{C}} \Gamma'$;

(3) for all pairs $\langle \Omega, \Omega' \rangle$ with $\Gamma \subseteq \Omega$ and $\Gamma' \subseteq \Omega'$, if $\Gamma \subsetneq \Omega$ or $\Gamma' \subsetneq \Omega'$, then $\Omega \vdash_{\mathcal{C}} \Omega'$ holds.

By Zorn's lemma, we are able to prove the following lemma.

Lemma 5.3 *Let $\mathcal{C}$ be a strongly appropriate constant specification. Let $\langle \Sigma, \Sigma' \rangle$ be such that $\Sigma \not\vdash_{\mathcal{C}} \Sigma'$ and $\Sigma' \neq \emptyset$. Then, there exists a $\mathcal{C}$-partition $\langle \Gamma, \Gamma' \rangle$ such that $\Sigma \subseteq \Gamma$ and $\Sigma' \subseteq \Gamma'$.*

The axiom scheme **A4** helps us prove the first item of the following fact. In addition, **A5** is for the second item; **A6** and **A7** are for the third item. Furthermore, the fourth item can be shown by using the first item.

Fact 5.4 *Let $\mathcal{C}$ be a strongly appropriate constant specification. Let $\langle \Gamma, \Gamma' \rangle$ be a $\mathcal{C}$-partition. Then, the following hold:*

(i) *for all formulas ϕ, either $\phi \in \Gamma$ or $\phi \in \Gamma'$.*

(ii) *for all formulas ϕ, either $\phi \in \Gamma$ and $\neg\phi \in \Gamma'$, or $\phi \in \Gamma'$ and $\neg\phi \in \Gamma$, or $\phi \in \Gamma$ and $\neg\phi \in \Gamma$.*

(iii) *for all formulas ϕ, either $\phi \in \Gamma$ and $\dot{\neg}\phi \in \Gamma'$, or $\dot{\neg}\phi \in \Gamma$ and $\phi \in \Gamma'$;*

(iv) *if $\Gamma \vdash_{\mathcal{C}} \phi$, then $\phi \in \Gamma$.*

Fact 5.4 is used, when we prove each item of the following fact. In addition, **A8**, **A9**, **A10** are for handling the first item of the following fact; **A10**, **A11**, **A12** are for the second item. Furthermore, **R1** is used in dealing with the third item; **A13** is for the fourth item.

Fact 5.5 *Let $\mathcal{C}$ be a strongly appropriate constant specification. Let $\langle \Gamma, \Gamma' \rangle$ be a $\mathcal{C}$-partition. Then, for all formulas ϕ and ψ, the following hold:*

(i) *$\phi \wedge \psi \in \Gamma$, iff $\phi \in \Gamma$ and $\psi \in \Gamma$.*

(ii) *$\neg(\phi \wedge \psi) \in \Gamma$, iff $\neg\phi \in \Gamma$ or $\neg\psi \in \Gamma$.*

(iii) *$\phi \rightarrow \psi \in \Gamma$, iff $\phi \in \Gamma'$ or $\psi \in \Gamma$.*

(iv) *$\neg(\phi \rightarrow \psi) \in \Gamma$, iff $\phi \in \Gamma$ and $\neg\psi \in \Gamma$.*

Definition 5.6 [canonical model] Let $\mathcal{C}$ be a strongly appropriate constant specification. The *canonical model in* $\mathsf{PJ_b}$ *w.r.t.* $\mathcal{C}$ is a four-tuple $\langle W, R, \mathcal{E}, v \rangle$, where

(i) W is the set of all $\mathcal{C}$-partitions.

(ii) for all $\langle \Gamma_1, \Gamma_1' \rangle$, $\langle \Gamma_2, \Gamma_2' \rangle \in W$, $\langle \Gamma_1, \Gamma_1' \rangle R \langle \Gamma_2, \Gamma_2' \rangle$ iff $\Gamma_1^\sharp \subseteq \Gamma_2$, where $\Gamma_1^\sharp$ is defined to be the following set:

$$\{\phi \mid t : \phi \in \Gamma_1, \text{ for some } t\}$$

(iii) for all $\langle \Gamma, \Gamma' \rangle \in W$, for all justification terms t, for all formulas ϕ, $\phi \in \mathcal{E}(\langle \Gamma, \Gamma' \rangle, t)$ iff $t : \phi \in \Gamma$.

(iv) for all $\langle \Gamma, \Gamma' \rangle \in W$ and all propositional variables p,

$$v_{\langle\Gamma,\Gamma'\rangle}(p) = \begin{cases} 1 & \text{if } p \in \Gamma \text{ and } \neg p \in \Gamma'; \\ b & \text{if } p \in \Gamma \text{ and } \neg p \in \Gamma; \\ 0 & \text{otherwise.} \end{cases}$$

Axiom schemes **A16**, **A17**, **A18** help us prove (i) of the following lemma. In addition, the inference rule **R2** helps us prove (ii).

Lemma 5.7 *Let $\mathcal{C}$ be a strongly appropriate constant specification.*

(i) *The canonical model in $\mathsf{PJ_b}$ w.r.t. $\mathcal{C}$ is a model in $\mathsf{PJ_b}$.*

(ii) *The canonical model in $\mathsf{PJ_b}$ w.r.t. $\mathcal{C}$ meets $\mathcal{C}$.*

The way we define the accessibility relation for the canonical model yields the following fact.

Fact 5.8 *Let $\mathcal{C}$ be a strongly appropriate constant specification. Let $\mathcal{M} = \langle W, R, \mathcal{E}, v\rangle$ be the canonical model in $\mathsf{PJ_b}$ w.r.t. $\mathcal{C}$. Let $\langle\Gamma,\Gamma'\rangle \in W$. Then, for all justification terms t and for all formulas ϕ, if $t : \phi \in \Gamma$, then for all $\langle\Omega,\Omega'\rangle \in W$ with $\langle\Gamma,\Gamma'\rangle R \langle\Omega,\Omega'\rangle$, it holds that $\phi \in \Omega$.*

With the help of Fact 5.5, Fact 5.8 and **A14**, **A15** and **A19**, we can show the truth lemma.

Lemma 5.9 (truth lemma for $\mathsf{PJ_b}$) *Let $\mathcal{C}$ be a strongly appropriate constant specification and $\mathcal{M} = \langle W, R, \mathcal{E}, v\rangle$ be the canonical model in $\mathsf{PJ_b}$ w.r.t. $\mathcal{C}$. Then, for all formulas ϕ, for all $\langle\Gamma,\Gamma'\rangle \in W$, the following hold:*

(i) *$\phi \in \Gamma$ and $\neg\phi \in \Gamma'$, iff $v^{\mathcal{M}}_{\langle\Gamma,\Gamma'\rangle}(\phi) = 1$.*

(ii) *$\phi \in \Gamma$ and $\neg\phi \in \Gamma$, iff $v^{\mathcal{M}}_{\langle\Gamma,\Gamma'\rangle}(\phi) = b$.*

(iii) *$\phi \in \Gamma'$ and $\neg\phi \in \Gamma$, iff $v^{\mathcal{M}}_{\langle\Gamma,\Gamma'\rangle}(\phi) = 0$.*

With the help of Lemma 5.3 and Lemma 5.9, we can prove completeness.

Theorem 5.10 (completeness of $\mathsf{PJ_b}$) *Let $\mathcal{C}$ be a strongly appropriate constant specification. If $\Sigma \Vdash_{\mathcal{C}} \phi$, then $\Sigma \vdash_{\mathcal{C}} \phi$.*

6 Quasi-Realization Theorem for $\langle\mathsf{PE_b}, \mathsf{PJ_b}\rangle$

We follow Fitting's non-constructive way [6] to prove quasi-realization theorem for $\langle\mathsf{PE_b}, \mathsf{PJ_b}\rangle$. This theorem establishes an embedding from $\mathsf{PE_b}$ into $\mathsf{PJ_b}$. In proving quasi-realization theorem for $\langle\mathsf{PE_b}, \mathsf{PJ_b}\rangle$, the only part that is not totally obviously suggested by Fitting's proof in [6] is how to handle the negation.

Before going to proofs, here are some setups. Let ϕ be a formula in $\mathsf{PE_b}$. Assume that ϕ is fixed for the rest of the section. In the following, for the sake of simplicity, when we talk about some *subformula* of ϕ, we actually mean some *occurrence* of this subformula of ϕ. In addition, *positive* subformulas and *negative* subformulas are defined as usual. More specifically, ϕ itself is a positive subformula of ϕ. Given a subformula α of ϕ, if α is of the form $\Box\beta$ or $\beta \wedge \gamma$,then the polarity of β and the polarity of γ are the same as α. If α is of the form $\beta \to \gamma$, then the polarity of β is different from α and the polarity of

γ is the same as α. If α is of the form $\neg\beta$, then the polarity of β is different from α. Furthermore, A is any assignment of a justification variable to each subformula of ϕ of the form $\Box\alpha$ that is at the negative position. It is assumed that A is one-one. Relative to A, we define one mapping π_A as follows.

Definition 6.1 π_A assigns a set of $\mathsf{PJ_b}$ formulas to each subformula of ϕ:

(i) if p is an atomic subformula of ϕ, then $\pi_A(p) = \{p\}$;

(ii) if $\neg\alpha$ is a subformula of ϕ,
$\pi_A(\neg\alpha) = \{\neg\alpha' \mid \alpha' \in \pi_A(\alpha)\}$;

(iii) if $\alpha * \beta$ is a subformula of ϕ,
$\pi_A(\alpha * \beta) = \{\alpha' * \beta' \mid \alpha' \in \pi_A(\alpha) \text{ and } \beta' \in \pi_A(\beta)\}$, where $* \in \{\wedge, \rightarrow\}$;

(iv) if $\Box\alpha$ is a negative subformula of ϕ,
$\pi_A(\Box\alpha) = \{x : \alpha' \mid A(\Box\alpha) = x \text{ and } \alpha' \in \pi_A(\alpha)\}$;

(v) if $\Box\alpha$ is a positive subformula of ϕ,
$\pi_A(\Box\alpha)= \{t : (\alpha_1 \vee \cdots \vee \alpha_n) \mid \alpha_1, \ldots, \alpha_n \in \pi_A(\alpha)$ and t is any justification term$\}$.

Finally, for the rest of the section, we fix a constant specification $\mathcal{C}$. Assume $\mathcal{C}$ is strongly appropriate. Let $\mathcal{M} = \langle W, R, \mathcal{E}, v\rangle$ be the canonical model w.r.t $\mathcal{C}$ in $\mathsf{PJ_b}$. Define a $\mathsf{PE_b}$ model $\mathcal{N} = \langle W, R, v\rangle$. So, $\mathcal{N}$ is $\mathcal{M}$ dropping the evidence function $\mathcal{E}$. In addition, recall the following notation. Given a world u in $\mathcal{M}$ and a set Σ of formulas, $v_u^{\mathcal{M}}[\Sigma] = \{v_u^{\mathcal{M}}(\phi) \mid \phi \in \Sigma\}$.

Now, we are ready proving things. In [6], its Proposition 7.7 is the key to the quasi-realization theorem for $\langle\mathsf{S4}$, the *Logic of Proofs (without Plus)*$\rangle$. Here, the following lemma is the key to the quasi-realization theorem for $\langle\mathsf{PE_b}, \mathsf{PJ_b}\rangle$.

Lemma 6.2 *For all formulas ψ in $\mathsf{PE_b}$, for all $\langle\Gamma, \Gamma'\rangle \in W$, the following hold:*

(i) *if ψ is a positive subformula of ϕ and $v^{\mathcal{M}}_{\langle\Gamma,\Gamma'\rangle}[\pi_A(\psi)] \subseteq \{b, 0\}$, then $v^{\mathcal{N}}_{\langle\Gamma,\Gamma'\rangle}(\psi) \in \{b, 0\}$.*

(ii) *if ψ is a positive subformula of ϕ and $v^{\mathcal{M}}_{\langle\Gamma,\Gamma'\rangle}[\pi_A(\psi)] = \{0\}$, then $v^{\mathcal{N}}_{\langle\Gamma,\Gamma'\rangle}(\psi) = 0$.*

(iii) *if ψ is a negative subformula of ϕ and $v^{\mathcal{M}}_{\langle\Gamma,\Gamma'\rangle}[\pi_A(\psi)] \subseteq \{1, b\}$, then $v^{\mathcal{N}}_{\langle\Gamma,\Gamma'\rangle}(\psi) \in \{1, b\}$.*

(iv) *if ψ is a negative subformula of ϕ and $v^{\mathcal{M}}_{\langle\Gamma,\Gamma'\rangle}[\pi_A(\psi)] = \{1\}$, then $v^{\mathcal{N}}_{\langle\Gamma,\Gamma'\rangle}(\psi) = 1$.*

Proof. Please see Appendix B. □

The following three paragraphs might let readers see the rough shape of the proof for Lemma 6.2.

First of all, I would like to address one small difference between the statement of [6]'s Proposition 7.7 and the statement of Lemma 6.2 (of this paper). If looking at the statement of [6]'s Proposition 7.7, readers will find: under this classical two-valued context, positive subformulas correspond to and only correspond to the non-designated value 0, and negative subformulas correspond to and only correspond to the designated value 1. However, the correspondence between polarity and truth value is less perfect in the statement of Lemma 6.2.. More specifically, in the item (i) of Lemma 6.2, positive subformulas correspond to $\{b, 0\}$, so do not only correspond to the non-designated value. Besides, in the item (iv) of of Lemma 6.2, negative subformulas correspond to $\{1\}$, so do not correspond to all designated values.

The root that leads to the small difference just described is the non-classical negation used by both $\mathsf{PE_b}$ and $\mathsf{PJ_b}$. First, consider a simplified version of Lemma 6.2 that does not contain the items (i) and (iv). Actually, this simplified version of Lemma 6.2 is already sufficient for helping us prove the quasi-realization theorem for $\langle \mathsf{PE_b}, \mathsf{PJ_b} \rangle$. Second, the reason why we prove the more complicated statement, rather than the simplified version is that without the items (i) and (iv), the inductive proof will get stuck at the case of negation. Hence, the items (i) and (iv) of Lemma 6.2 are not redundant. In short, to handle the non-classical negation, we need Lemma 6.2's item (i) and item (iv), which breaks the perfect correspondence between polarity and truth value.

Now, we turn to the similarity between [6]'s proof for its Proposition 7.7 and the proof for Lemma 6.2. Both are proofs by induction on formulas. Except the case of negation, these two inductive proofs are very similar. The most complicated part of [6]'s proof for its Proposition 7.7 is the case of positive necessity, which relies on: the *Logic of Proofs* has Modus Ponens, deduction theorem and the validity of $s : (\phi \to \psi) \to (t : \phi \to (s \cdot t) : \psi)$. The way we interpret the conditional of $\mathsf{PJ_b}$ gives us these three. Therefore, we can also handle the case of positive necessity, when proving Lemma 6.2.

Applying Lemma 6.2 (and the truth lemma for $\mathsf{PJ_b}$), we can then prove the quasi-realization theorem.

Theorem 6.3 *Let $\mathcal{C}$ be a strongly appropriate constant specification. Let ϕ be a formula in $\mathsf{PE_b}$. If ϕ is valid in $\mathsf{PE_b}$, then there are $\phi_1, \ldots, \phi_n \in \pi_A(\phi)$ such that $\phi_1 \vee \cdots \vee \phi_n$ is $\mathcal{C}$-valid in $\mathsf{PJ_b}$.*

7 Conclusion

Based on $\mathsf{PE_b}$, we have constructed a paraconsistent justification logic system $\mathsf{PJ_b}$ by giving its Fitting models. $\mathsf{PJ_b}$ is paraconsistent in the sense that it is not the case that for all justification terms s_1, s_2, there exists a justification term t such that $(s_1 : \phi \wedge s_2 : \neg\phi) \to t : \psi$ is a valid scheme. Actually, $\mathsf{PJ_b}$ satisfies stronger paraconsistent properties (Fact 4.4's (iii), (iv)). Informally speaking, $\mathsf{PJ_b}$ reflects the idea: explicitly committing to a contradiction for two conflicting pieces of evidence does not imply that for all statement ψ, we have a corresponding evidence that forces us to commit to ψ. In addition, we motivate $\mathsf{PJ_b}$ from the angle of belief revision. We have argued that to better model belief revision, we might need to be able to model (1) inconsistent epistemic states and (2) the agent's justification structure. $\mathsf{PJ_b}$ is able to handle these two, so might help us better model belief revision. Furthermore, the main technical result of the paper is quasi-realization theorem for $\langle \mathsf{PE_b}, \mathsf{PJ_b} \rangle$, which establishes an embedding from $\mathsf{PE_b}$ into $\mathsf{PJ_b}$.

Possible future work: First of all, in standard justification logics, there is an algorithmic conversion from quasi-realization to realization [7], which is a simpler form of embedding from a modal logic into a justification logic. Hence, one natural next step is to check whether this conversion also works for $\langle \mathsf{PE_b}, \mathsf{PJ_b} \rangle$, so we can also have realization for $\langle \mathsf{PE_b}, \mathsf{PJ_b} \rangle$. Secondly, $\mathsf{PJ_b}$ performs nicely in the technical aspect because of the way we define the conditional. To avoid ad-hoc-ness, either a good story for motivating the conditional used by $\mathsf{PJ_b}$ should be further provided, or we should change to a conditional that is well-motivated. However, if we go for the second option, for the project to succeed, it is better that the resulting system still performs nicely in the technical aspect. Finally, we motivate paraconsistent justification logic from the angle of belief revision. Therefore, to finish the whole story, *dynamic* paraconsistent justification logic for belief revision should be developed.

Appendix

A Proofs for Fact 3.3 and Fact 4.4

Proof. [Fact 3.3] Proofs for (v)–(vii) are skipped. Example 3.2 provides a model for showing (i) and (ii). (iv) implies (iii), so for the rest, we focus on proving (iv).

Let ϕ, ψ be such that both ϕ, ψ contain no $\Box$; ϕ and ψ share no propositional variable; and ψ is not valid. Our goal is to construct a model showing $(\Box\phi \wedge \Box\neg\phi) \to \Box\psi$ is invalid.

Let $p_1, \ldots, p_n$ be all of the propositional variables occurring in ϕ; let $q_1, \ldots, p_m$ be those occurring in ψ. Let $(\mathcal{N}', u_3)$ be such that $\mathcal{N}' = \langle W', R', v' \rangle$ is a $\mathsf{PE_b}$ model, $u_3 \in W'$ and $v'^{\mathcal{N}'}_{u_3}(\psi) = 0$. Note that such a $(\mathcal{N}', u_3)$ exists, since ψ is assumed to be invalid. Define a $\mathsf{PE_b}$ model $\mathcal{N} = \langle W, R, v \rangle$, where $W = \{u_1, u_2\}$; $R = \{\langle u_1, u_2 \rangle\}$; $v^{\mathcal{N}}_{u_2}(p_i) = b$ for each p_i; $v^{\mathcal{N}}_{u_2}(q_i) = v'^{\mathcal{N}'}_{u_3}(q_i)$ for each q_i. How v assigns values to other propositional variables at u_2 and how v assigns values (to all propositional variables) at u_1 could be arbitrary.

Claim that $v^{\mathcal{N}}_{u_1}((\Box\phi \wedge \Box\neg\phi) \to \Box\psi) = 0$. Here is a proof for the claim. First, since ϕ contains no $\Box$, by the way $\mathsf{PE_b}$ interprets propositional connectives, it holds that $v^{\mathcal{N}}_{u_2}(\phi) = v^{\mathcal{N}}_{u_2}(\neg\phi) = b$. Secondly, because ψ contains no $\Box$, ψ's getting value 0 at u_3 is totally determined by how v' assign values to $q_1, \ldots, q_m$ at u_3. Since at u_2, v mimics how v' assigns values to $q_1, \ldots, q_m$ at u_3, it holds that $v^{\mathcal{N}}_{u_2}(\psi) = 0$. Third, by the previous two points, it follows that $v^{\mathcal{N}}_{u_1}(\Box\phi \wedge \Box\neg\phi) = 1$ and $v^{\mathcal{N}}_{u_1}(\Box\psi) = 0$. Therefore, the claim is proved. $\Box$

Proof. [Fact 4.4] Example 4.3 provides a model for showing (i) and (ii). (iv) implies (iii), so for the rest, we focus on proving (iv).

Let s_1, s_2, t be three justification terms. Assume that ϕ, ψ satisfy conditions listed in the antecedent of the conditional occurring in (iv). Our goal is to construct a model showing $\{s_1 : \phi, s_2 : \neg\phi\} \not\Vdash_{\mathcal{C}} t : \psi$.

Let $p_1, \ldots, p_n$ be all of the propositional variables occurring in ϕ; let

$q_1, \ldots, p_m$ be those occurring in ψ. Let $(\mathcal{N}', u_3)$ be such that $\mathcal{N}' = \langle W', R', \mathcal{E}', v' \rangle$ is a $\mathsf{PJ_b}$ model, $u_3 \in W'$ and $v'^{\mathcal{N}'}_{u_3}(\psi) = 0$. Note that such a $(\mathcal{N}', u_3)$ exists, since ψ is assumed to be not $\mathcal{C}$-valid. Define a $\mathsf{PJ_b}$ model $\mathcal{N} = \langle W, R, \mathcal{E}, v \rangle$, where $W = \{u_1, u_2\}$; $R = \{\langle u_1, u_2 \rangle\}$; $\mathcal{E}(u_1, j) = \mathcal{E}(u_2, j) = \{\tau \mid \tau \text{ is a } \mathsf{PJ_b} \text{ formula}\}$, for all justification terms j; $v^{\mathcal{N}}_{u_2}(p_i) = b$ for each p_i; $v^{\mathcal{N}}_{u_2}(q_i) = v'^{\mathcal{N}'}_{u_3}(q_i)$ for each q_i.

First, by the definition of $\mathcal{E}$ and a reasoning similar to the one we employ in proving (iv) of Fact 3.3, we can show that $v^{\mathcal{N}}_{u_1}(s_1 : \phi \wedge s_2 : \neg\phi) = 1$ and $v^{\mathcal{N}}_{u_1}(t : \psi) = 0$. Secondly, by the definition of $\mathcal{E}$, $\mathcal{M}$ meets any constant specification. By the previous two points, it follows that $\{s_1 : \phi, s_2 : \neg\phi\} \not\Vdash_{\mathcal{C}} t : \psi$. □

B Proof for Lemma 6.2

Proof. [Lemma 6.2] We prove the lemma by induction on formulas. The atomic case is trivial, so is skipped.

Let α and β be two formulas in $\mathsf{PE_b}$. Assume the following as the induction hypothesis (**IH**): for all $\langle \Gamma, \Gamma' \rangle \in W$,

- if α is a positive subformula of ϕ and $v^{\mathcal{M}}_{\langle \Gamma, \Gamma' \rangle}[\pi_A(\alpha)] \subseteq \{b, 0\}$, then $v^{\mathcal{N}}_{\langle \Gamma, \Gamma' \rangle}(\alpha) \in \{b, 0\}$.
- if α is a positive subformula of ϕ and $v^{\mathcal{M}}_{\langle \Gamma, \Gamma' \rangle}[\pi_A(\alpha)] = \{0\}$, then $v^{\mathcal{N}}_{\langle \Gamma, \Gamma' \rangle}(\alpha) = 0$.
- if α is a negative subformula of ϕ and $v^{\mathcal{M}}_{\langle \Gamma, \Gamma' \rangle}[\pi_A(\alpha)] \subseteq \{1, b\}$, then $v^{\mathcal{N}}_{\langle \Gamma, \Gamma' \rangle}(\alpha) \in \{1, b\}$.
- if α is a negative subformula of ϕ and $v^{\mathcal{M}}_{\langle \Gamma, \Gamma' \rangle}[\pi_A(\alpha)] = \{1\}$, then $v^{\mathcal{N}}_{\langle \Gamma, \Gamma' \rangle}(\alpha) = 1$.
- for ψ, we also assume similar four items of conditions as part of **IH**.
- case 1. $\neg\alpha$
  Let $\langle \Gamma, \Gamma' \rangle \in W$.
  - case 1(a). Positive Negation (I)
    The goal conditional we want to show is: if $\neg\alpha$ is a positive subformula of ϕ and $v^{\mathcal{M}}_{\langle \Gamma, \Gamma' \rangle}[\pi_A(\neg\alpha)] \subseteq \{b, 0\}$, then $v^{\mathcal{N}}_{\langle \Gamma, \Gamma' \rangle}(\neg\alpha) \in \{b, 0\}$. The third item of **IH** helps us prove this.

    First, the statement that $\neg\alpha$ is a positive subformula of ϕ implies the statement that α is a negative subformula of ϕ.

    Secondly, the statement that $v^{\mathcal{M}}_{\langle \Gamma, \Gamma' \rangle}[\pi_A(\neg\alpha)] \subseteq \{b, 0\}$ is equivalent to the statement that $v^{\mathcal{M}}_{\langle \Gamma, \Gamma' \rangle}[\pi_A(\alpha)] \subseteq \{1, b\}$.

    Assume that $\neg\alpha$ is a positive subformula of ϕ and $v^{\mathcal{M}}_{\langle \Gamma, \Gamma' \rangle}[\pi_A(\neg\alpha)] \subseteq \{b, 0\}$. By the first and the second point, the assumption implies that α is a negative subformula of ϕ and $v^{\mathcal{M}}_{\langle \Gamma, \Gamma' \rangle}[\pi_A(\alpha)] \subseteq \{1, b\}$. Then, by the third item of **IH**, we can derive that $v^{\mathcal{N}}_{\langle \Gamma, \Gamma' \rangle}(\alpha) \in \{1, b\}$. Therefore, $v^{\mathcal{N}}_{\langle \Gamma, \Gamma' \rangle}(\neg\alpha) \in \{b, 0\}$.
  - case 1(b). Positive Negation (II)

The goal conditional we want to show is: if $\neg\alpha$ is a positive subformula of ϕ and $v^{\mathcal{M}}_{\langle\Gamma,\Gamma'\rangle}[\pi_A(\neg\alpha)] = \{0\}$, then $v^{\mathcal{N}}_{\langle\Gamma,\Gamma'\rangle}(\neg\alpha) = 0$. By the fourth item of **IH** and reasoning similar to the previous case, this can be shown.

· case 1(c). Negative Negation (I)
The goal conditional we want to show is: if $\neg\alpha$ is a negative subformula of ϕ and $v^{\mathcal{M}}_{\langle\Gamma,\Gamma'\rangle}[\pi_A(\neg\alpha)] \subseteq \{1,b\}$, then $v^{\mathcal{N}}_{\langle\Gamma,\Gamma'\rangle}(\neg\alpha) \in \{1,b\}$. By the first item of **IH** and reasoning similar to case 1-(a), this can be shown.

· case 1(d). Negative Negation (II)
The goal conditional we want to show is: if $\neg\alpha$ is a negative subformula of ϕ and $v^{\mathcal{M}}_{\langle\Gamma,\Gamma'\rangle}[\pi_A(\neg\alpha)] = \{1\}$, then $v^{\mathcal{N}}_{\langle\Gamma,\Gamma'\rangle}(\neg\alpha) = 1$. By the second item of **IH** and reasoning similar to case 1-(a), this can be shown.

- case 2. $\Box\alpha$
Let $\langle\Gamma,\Gamma'\rangle \in W$.

· case 2(a). Positive Necessity (I)
Assume that $\Box\alpha$ is a positive subformula of ϕ and $v^{\mathcal{M}}_{\langle\Gamma,\Gamma'\rangle}[\pi_A(\Box\alpha)] \subseteq \{b,0\}$. Note that here α is a positive subformula of ϕ. Our goal is to show that $v^{\mathcal{N}}_{\langle\Gamma,\Gamma'\rangle}(\Box\alpha) \in \{b,0\}$. This case is similar the the next case, that is, the case 2-(b). Hence, we skip this case.

· case 2(b). Positive Necessity (II)
Assume that $\Box\alpha$ is a positive subformula of ϕ and $v^{\mathcal{M}}_{\langle\Gamma,\Gamma'\rangle}[\pi_A(\Box\alpha)] = \{0\}$. Our goal is to show that $v^{\mathcal{N}}_{\langle\Gamma,\Gamma'\rangle}(\Box\alpha) = 0$. Note that here α is a positive subformula of ϕ.

Claim that $\Gamma^{\sharp} \nvdash_{\mathcal{C}} \pi_A(\alpha)$ holds. Here is the proof for the claim. Suppose not. Then, there exist some finite subset $\Sigma \subseteq \Gamma^{\sharp}$ and some formulas $V_1,\ldots,V_m \in \pi_A(\alpha)$ such that $\Sigma \vdash_{\mathcal{C}} V_1 \vee \cdots \vee V_m$. Either $\Sigma \neq \emptyset$ or $\Sigma = \emptyset$.

In this paragraph, we consider the situation that $\Sigma \neq \emptyset$. Say $\Sigma = \{U_1,\ldots,U_n\}$. Then, it holds that $\{U_1,\ldots,U_n\} \vdash_{\mathcal{C}} V_1 \vee \cdots \vee V_m$. First, by the deduction theorem (Theorem 5.1), we have that $\vdash_{\mathcal{C}} U_1 \to (U_2 \to \cdots (U_n \to (V_1 \vee \cdots \vee V_m)\cdots)$. Let Λ stand for $U_1 \to (U_2 \to \cdots (U_n \to (V_1 \vee \cdots \vee V_m)\cdots)$. So, we have that $\vdash_{\mathcal{C}} \Lambda$. By internalization theorem (Theorem 5.2) and the assumption that $\mathcal{C}$ is strongly appropriate, it holds that $\vdash_{\mathcal{C}} t : \Lambda$, for some justification term t. Secondly, since $\{U_1,\ldots,U_n\} \subseteq \Gamma^{\sharp}$, we have that $\{s_1 : U_1,\ldots,s_n : U_n\} \subseteq \Gamma$, for some $s_1,\ldots,s_n$. Third, because our axiomatization of $\mathsf{PJ_b}$ has Modus Ponens (**R1**) and takes $j : (\tau_1 \to \tau_2) \to (j' : \tau_1 \to (j \cdot j') : \tau_2)$ as an axiom scheme (**A16**), it holds that $\{t : \Lambda, s_1 : U_1,\ldots,s_n : U_n\} \vdash_{\mathcal{C}} ((\ldots(t \cdot s_1)\cdot s_2\ldots)\cdot s_n) : (V_1 \vee \cdots \vee V_m)$. Fourth, by the previous three points, we can conclude that $\Gamma \vdash_{\mathcal{C}} ((\ldots(t \cdot s_1)\cdot s_2\ldots)\cdot s_n) : (V_1 \vee \cdots \vee V_m)$. Fifth, since $V_1,\ldots,V_m \in \pi_A(\alpha)$, it holds that $\{j : (V_1 \vee \cdots \vee V_m) \mid j \text{ is a term}\}$ is a subset of $\pi_A(\Box\alpha)$. Therefore, $(\ldots(t \cdot s_1)\cdot s_2\ldots)\cdot s_n) : (V_1 \vee \cdots \vee V_m)$ is in $\pi_A(\Box\alpha)$. By the assumption that $v^{\mathcal{M}}_{\langle\Gamma,\Gamma'\rangle}[\pi_A(\Box\alpha)] = \{0\}$ and the truth lemma, we have that $((\ldots(t \cdot s_1)\cdot s_2\ldots)\cdot s_n) : (V_1 \vee \cdots \vee V_m)$ is in Γ'. Sixth, by the fourth and fifth points, we can conclude that $\Gamma \vdash_{\mathcal{C}} \Gamma'$. This contradicts that Γ, Γ' is a $\mathcal{C}$-partition.

The situation that $\Sigma = \emptyset$ can be handled in a similar way, so is skipped. By the claim we just prove and Lemma 5.3, there exists a $\mathcal{C}$-partition $\langle \Delta, \Delta' \rangle$ such that $\Gamma^\sharp \subseteq \Delta$ and $\pi_A(\alpha) \subseteq \Delta'$. Then, we can show that $v^{\mathcal{N}}_{\langle \Gamma,\Gamma' \rangle}(\Box\alpha) = 0$. Here is the proof. First, since $\Gamma^\sharp \subseteq \Delta$, we have that $\langle \Gamma, \Gamma' \rangle R \langle \Delta, \Delta' \rangle$. And, because $\mathcal{N}$ and $\mathcal{M}$ share the same R, in $\mathcal{N}$ it also holds that $\langle \Gamma, \Gamma' \rangle R \langle \Delta, \Delta' \rangle$. Second, since $\pi_A(\alpha) \subseteq \Delta'$, by the truth lemma it holds that $v^{\mathcal{M}}_{\langle \Delta,\Delta' \rangle}[\pi_A(\alpha)] = \{0\}$. Note that α is positive subformula of $\Box\alpha$. By **IH**, it follows that $v^{\mathcal{N}}_{\langle \Delta,\Delta' \rangle}(\alpha) = 0$. Third, by the previous two points, we can conclude that $v^{\mathcal{N}}_{\langle \Gamma,\Gamma' \rangle}(\Box\alpha) = 0$.

- case 2(c). Negative Necessity (I)
  Assume that $\Box\alpha$ is a negative subformula of ϕ and $v^{\mathcal{M}}_{\langle \Gamma,\Gamma' \rangle}[\pi_A(\Box\alpha)] \subseteq \{1, b\}$. Our goal is to show that $v^{\mathcal{N}}_{\langle \Gamma,\Gamma' \rangle}(\Box\alpha) \in \{1, b\}$. Note that here α is a negative subformula of ϕ. This case is similar to the corresponding case in [6].
- case 2(d). Negative Necessity (II)
  Assume that $\Box\alpha$ is a negative subformula of ϕ and $v^{\mathcal{M}}_{\langle \Gamma,\Gamma' \rangle}[\pi_A(\Box\alpha)] = \{1\}$. Note that here α is a negative subformula of ϕ. Our goal is to show that $v^{\mathcal{N}}_{\langle \Gamma,\Gamma' \rangle}(\Box\alpha) = 1$. This case is similar to the previous case, that is, the case 2-(c). Therefore, we skip it.

The case of $\to$ is similar to [6]; the case of $\wedge$ is similar to the case of $\to$. □

References

[1] Artemov, S., *Explicit provability and constructive semantics*, Bulletin of Symbolic logic **7** (2001), pp. 1–36.

[2] Avron, A., *On an implication connective of RM*, Notre Dame Journal of Formal Logic **27** (1986), pp. 201–209.

[3] Batens, D., *Paraconsistent extensional propositional logics*, Logique et Analyse Louvain **23** (1980), pp. 195–234.

[4] Beall, J., T. Forster and J. Seligman, *A note on freedom from detachment in the logic of paradox*, Notre Dame Journal of Formal Logic **54** (2013), pp. 15–20.

[5] Fagin, R. and J. Y. Halpern, *Belief, awareness, and limited reasoning*, Artificial intelligence **34** (1987), pp. 39–76.

[6] Fitting, M., *The logic of proofs, semantically*, Annals of Pure and Applied Logic **132** (2005), pp. 1–25.

[7] Fitting, M., *Realization implemented*, Technical Report TR-2013005, CUNY Ph.D Program in Computer Science (May 2013).

[8] Girard, P. and K. Tanaka, *Paraconsistent dynamics* (February 2014), manuscript, 15 pages.

[9] Hansson, S. O., *Ten philosophical problems in belief revision*, Journal of Logic and Computation **13** (2003), pp. 37–49.

[10] Jaspars, J. O., *Logical omniscience and inconsistent belief*, in: *Diamonds and Defaults*, Springer, 1993 pp. 129–146.

[11] Lurie, J., *Non-applicative justification logic* (June 2014), manuscript, 12 pages.

[12] Restall, G., *Ways things can't be*, Notre Dame Journal of Formal Logic **38** (1997), pp. 583–596.

[13] Van der Hoek, W. and J.-J. C. Meyer, *Possible logics for belief*, Logique et analyse **127** (1989), pp. 177–194.

On Polarity Frames: Applications to Substructural and Lattice-based Logics

Tomoyuki Suzuki [1]

Institute of Computer Science, Academy of Sciences of the Czech Republic
Pod Vodarenskou vezi 271/2,
182 07, Prague 8, Czech Republic

Abstract

In this paper, on one hand, we address topology on polarities via general polarity frames by analogy of the relationship between topology on sets and general Kripke frames. Based on the topology on polarities, we provide the topological characterisation of descriptive polarity frames. On the other hand, we introduce disjoint unions and amalgamations of polarity frames with additional relations and constants. As applications of these constructions, we establish the Goldblatt-Thomason's theorem for (distributive) substructural logic and the amalgamation property for some lattice-based algebras.

Keywords: Relational semantics, substructural and lattice-based logics, lattice expansions, topological characterisation, Goldblatt-Thomason's theorem and amalgamation property.

1 Introduction

The study of *polarities*, i.e. triples of two sets and a binary relation, is already found in the first edition of [2]. We can find the same structures in the context of formal concept analysis e.g. [7,18]. Interestingly, in non-classical logics, polarities have attracted attentions from the generalisation of the canonicity problem on the setting of modal logics to lattice-based logics such as substructural logic and distributive modal logics. Whilst the method in [21] was extended mainly by [12,11,13], it was not clear whether we had to restrict the Sahlqvist argument for non-Boolean based logics. However, by means of the Ghilardi-Meloni canonicity methodology [14], we found that the same argument works for lattice-based logics [40] and even for poset expansions with minor restriction [42].

---

[1] *Email:tomoyuki.suzuki@cs.cas.cz*
The author would like to thank Marta Bílková, Rostislav Horčík, Adam Přenosil, Jiří Velebil, and anonymous referees for their valuable comments. The author is supported by grant GAP202/10/1826 of the Czech Science Foundation and RV067985807.

Bi-approximation semantics for substructural logic was introduced in [39] to explicate the Ghilardi-Meloni canonicity methodology [14] from the relational semantic viewpoints. On one hand, as the semantics enjoys the Sahlqvist-type canonicity and elementary results [44], many lattice-based logics are sound and complete with respect to elementary classes. On the other hand, it evaluates not just formulae but also sequents (logical consequences) based on a polarity. Note that the idea of semantics on polarities are also found in e.g. [10,9], although their main targets are duality and algebraic proof theory.

In contrast, the motivation of our research for bi-approximation semantics is to introduce logical properties or well established proof methods on Kripke semantics to substructural and lattice-based logics. The main question of this research line may boil down to the following:

By replacing sets by polarities, can we obtain universal relational semantics?

More precisely, we are interested in whether the relationships between modal logics and Kripke semantics can be universally extended to the ones between lattice-based logics and bi-approximation semantics. Towards the goal, the current author has introduced morphisms [43] for bi-approximation semantics and addressed the interpretations between Kripke semantics for distributive lattice-based logics and distributive polarity frames [38,41]. In the current paper, we will address topology on polarities, definable classes of polarity frames and amalgamations of polarities.

Topology. The Stone representation and topological duality of Boolean algebras and sets was already established in [36,37]. Later, by [32,33], the relationship was generalised to distributive lattices and ordered Stone spaces, see also [19,8]. A further generalisation can be found, e.g. in [47].

However, unlike what happens in the setting over distributive lattices, the notion of topology does not seem clear for lattices, because topology forms algebraically a distributive lattice with respect to $\cap$ and $\cup$, *a priori*. Hence, it is natural to ask what is the appropriate notion of topology over lattice-based algebras. For the question, we give a possible answer by pulling back the connection between general Kripke frames and topological spaces in [34] via general polarity frames.

Goldblatt-Thomason's theorem. The Goldblatt-Thomason's theorem was originally shown in [46,17,15] to account for classes of first-order models (with a single binary relation) which are definable by modal formulae, see also [3,16]. The theorem was also applied to other languages: hybrid language [45], graded modal language [35] or coalgebraic logic [23,1].

Since our motivation is to justify that polarities are sufficiently qualified as a replacement of sets, in the present paper, we introduce the notion of disjoint unions of polarities and, via the Birkhoff's variety theorem [4], we establish the Goldblatt-Thomason's theorem for substructural and lattice-based languages on polarity models.

Amalgamation property. In non-classical logics, the amalgamation property is known as a foundational property to explain logical properties in an algebraic term e.g. [31,28,24]. In substructural logic, the correspondence between logical properties and algebraic properties is presented in [22]. Along the same research direction, we can also find in the correspondence over lattice-based algebras and logics in [26].

Whilst these correspondences between algebras and logics have been provided, it must be interesting if a systematic proof method can universally account for logics and algebras which have the amalgamation property. In fact, [25] found a sufficient condition for this question. However, considering non-distributive lattice-based algebras, the condition seems too strong. In this paper, we establish a basic scheme to explicate the amalgamation property for lattice-based algebras by introducing amalgamations of polarity frames via the dual representation. Note that our results include existing results such as the amalgamation property of lattices [20], distributive lattices [30], see also [18]. However, our main concern for this topic is a "schematic" account. [2]

Outline. In Section 2, we briefly recall necessary definitions, properties and theorems on bi-approximation semantics in the current author's previous publications. In Section 3, we provide topology on polarities and general polarity frames. Based on the setting, we establish the topological characterisation of descriptive polarity frames. In Section 4, we introduce disjoint unions of polarity frames and show their fundamental properties. Accordingly, by means of the Birkhoff's variety theorem, we achieve the Goldblatt-Thomason's theorem on polarity models. In Section 5, we explain our methodology to prove the amalgamation property via the dual representation and amalgamations of polarity frames. In Section 6, as concluding remarks, we list forthcoming work.

2 Preliminaries for polarity frames

Polarity frames. Details are found for example in [2,7]. A *polarity frame* is a triple $\mathbb{F} = \langle X, Y, B\rangle$ with non-empty sets X and Y, and a binary relation $B \subseteq X \times Y$. Note that, X and Y are not necessarily disjoint. However, if the equality is in the frame language, we must guarantee that, if $x = y$ then xBy for $x \in X$ and $y \in Y$. Given a polarity $\mathbb{P}$, a natural pre-order $\leq_B$ on $X \cup Y$ is introduced as follows: for all $x, x_1, x_2 \in X$ and $y, y_1, y_2 \in Y$,

(i) $x_1 \leq_B x_2 \iff \forall y' \in Y. [x_2 B y' \Longrightarrow x_1 B y']$,

(ii) $y_1 \leq_B y_2 \iff \forall x' \in X. [x' B y_1 \Longrightarrow x' B y_2]$,

(iii) $x \leq_B y \iff xBy$,

(iv) $y \leq_B x \iff \forall x' \in X, y' \in Y. [xBy' \text{ and } x'By \Longrightarrow x'By']$.

It is straightforward to check that the equality, if exists, is subsumed by $\leq_B$-equivalence. We may omit the subscript $_B$ when it is clear from the context. Note that, throughout this paper, we treat non-trivial polarity frames (i.e. $B \neq X \times Y$) only.

[2] In the literature, we can also find the amalgamation property for distributive lattice-based algebras such as *tetravalent modal algebras* [5] or varieties of lattice-ordered commutative groups and many-valued algebras [27].

Galois stable lattice. Given a polarity frame $\mathbb{F} = \langle X, Y, B\rangle$, let $\wp(X)$ and $\wp(Y)^\partial$ be the powerset poset of X ordered by the set-theoretic inclusion $\subseteq$ and the powerset poset of Y ordered by the set-theoretic *reverse* inclusion $\supseteq$. We introduce two functions $\lambda\colon \wp(X) \to \wp(Y)^\partial$ and $\upsilon\colon \wp(Y)^\partial \to \wp(X)$ as follows: for all $\mathfrak{X} \in \wp(X)$ and $\mathfrak{Y} \in \wp(Y)^\partial$, $\lambda(\mathfrak{X}) := \{y \in Y \mid \forall x \in \mathfrak{X}.\ xBy\}$ and $\upsilon(\mathfrak{Y}) := \{x \in X \mid \forall y \in \mathfrak{Y}.\ xBy\}$. It is known that λ and υ form a Galois connection, hence the images $\lambda[\wp(X)]$ and $\upsilon[\wp(Y)^\partial]$ are isomorphic. A pair of $\mathfrak{X} \in \wp(X)$ and $\mathfrak{Y} \in \wp(Y)^\partial$ is called a *Galois stable pair*, a.k.a *Dedekind-cut*, if they satisfy $\lambda(\mathfrak{X}) = \mathfrak{Y}$ and $\upsilon(\mathfrak{Y}) = \mathfrak{X}$.

Definition 2.1 [Galois stable lattice] *The Galois stable lattice* $\mathsf{G}_\mathbb{F}$ *on a polarity frame* $\mathbb{F}$ is the subposet of all Galois stable pairs of the product poset $\wp(X) \times \wp(Y)^\partial$, where the lattice operations are defined as follows: for all $(\mathfrak{X}_1, \mathfrak{Y}_1)$ and $(\mathfrak{X}_2, \mathfrak{Y}_2)$,

(i) $(\mathfrak{X}_1, \mathfrak{Y}_1) \vee (\mathfrak{X}_2, \mathfrak{Y}_2) := (\upsilon(\mathfrak{Y}_1 \cap \mathfrak{Y}_2), \mathfrak{Y}_1 \cap \mathfrak{Y}_2)$,

(ii) $(\mathfrak{X}_1, \mathfrak{Y}_1) \wedge (\mathfrak{X}_2, \mathfrak{Y}_2) := (\mathfrak{X}_1 \cap \mathfrak{X}_2, \lambda(\mathfrak{X}_1 \cap \mathfrak{X}_2))$.

Note that both $\mathsf{G}_\mathbb{F}$ and $\wp(X) \times \wp(Y)^\partial$ are complete lattices, but their lattice operations do not coincide in general. Also, note that the Galois stable lattice of $\mathbb{F}$ is isomorphic to the dual algebra of $\mathbb{F}$, denoted by $\mathbb{F}^+$ (so, throughout this paper, we think of $\mathsf{G}_\mathbb{F}$ as the definition of $\mathbb{F}^+$).

Distributivity. Details are in [38]. Given a polarity frame $\mathbb{F} = \langle X, Y, B\rangle$, an element $x \in X$ (resp. $y \in Y$) is a *splitter*, if there exists y_x (called a splitting counterpart of x) such that xBy_x does not hold and, for each $y \in Y$, if xBy does not hold, $y \leq y_x$ holds. (resp. there exists x_y such that x_yBy does not hold and, for each $x \in X$, if xBy does not hold, $x_y \leq x$ holds). It is known that every splitting counterpart is also a splitter, and splitting counterparts are unique up to $\leq$-equivalence. We call a pair of a splitter and its splitting counterpart *splitting pair*.

Definition 2.2 [Distributive polarity frame] A polarity frame $\mathbb{F} = \langle X, Y, B\rangle$ is *distributive*, if it satisfies

Splitting: for all $x \in X$ and $y \in Y$, if xBy does not hold, there exists a splitting pair (x_s, y_s) such that $x_s \leq x$ and $y \leq y_s$.

Note that the splitting condition is a first-order sentence.

Theorem 2.3 *For a distributive polarity frame* $\mathbb{F}$*, the dual algebra* $\mathbb{F}^+$ *satisfies the distributive law.*

Unary modality. For diamond $\Diamond$ and box $\Box$ (adjoints pair), polarity frames are extended with a binary relation $S \subseteq X \times Y$. On a polarity frame $\mathbb{F} = \langle X, Y, B\rangle$, the binary relation S is extended to a binary relation $S^\Diamond$ on X and a binary relation $S^\Box$ on Y as follows: $S^\Diamond(x, x') \iff \forall y \in Y. [S(x, y) \Longrightarrow x'By]$ and $S^\Box(y', y) \iff \forall x \in X. [S(x, y) \Longrightarrow xBy']$. We assume that the binary relation S on $\mathbb{F}$ satisfies the following conditions:

S-transitivity: $\forall x, x' \in X, y, y' \in Y. [x' \leq x, y \leq y' \;\&\; S(x, y) \Longrightarrow S(x', y')]$,

◇-adjoint: $\forall x \in X, \forall y \in Y. [\forall x' \in X. [S^{\Diamond}(x, x') \Longrightarrow x'By] \Longrightarrow S(x, y)]$,

□-adjoint: $\forall x \in X, y \in Y. [\forall y' \in Y. [S^{\Box}(y', y) \Longrightarrow xBy'] \Longrightarrow S(x, y)]$.

Note that the adjoint conditions [3] are intuitively to obtain the following adjointness condition: $S^{\Diamond}(x, x')By \iff xBS^{\Diamond}(y', y)$, and S is the generator of the adjointness.

Theorem 2.4 *The dual algebra is a lattice with adjoint unary modality of* $\Diamond$ *and* $\Box$.

Example 2.5 A quadruple $\langle X, Y, B, S\rangle$ forms a polarity frame for lattice-based modal logic. Lattice-based modal formulae are given by $\phi ::= p \mid \phi \vee \phi \mid \phi \wedge \phi \mid \Diamond\phi \mid \Box\phi$.

De Morgan negation. For the de Morgan negation $\neg$, polarity frames are extended with two binary relations $C \subseteq X \times X$ and $D \subseteq Y \times Y$. On a polarity frame $\mathbb{F} = \langle X, Y, B\rangle$, these binary relations C and D are extended to binary relations $\tilde{C}, \tilde{D} \subseteq X \times Y$ as follows: $\tilde{C}(x, y) \iff \forall x' \in X. [C(x, x') \Longrightarrow x'By]$ and $\tilde{D}(x, y) \iff \forall y' \in Y. [D(y', y) \Longrightarrow xBy']$. We assume that these binary relations satisfy the following conditions:

C-symmetry: $\forall x, x' \in X. [C(x, x') \Longrightarrow C(x', x)]$,

D-symmetry: $\forall y, y' \in Y. [D(y', y) \Longrightarrow D(y, y')]$,

C-transitivity: $\forall x_1, x_1', x_2, x_2' \in X. [x_1' \leq x_1, x_2' \leq x_2 \,\&\, C(x_1, x_2) \Rightarrow C(x_1', x_2')]$,

D-transitivity: $\forall y_1, y_1', y_2, y_2' \in Y. [y_1 \leq y_1', y_2 \leq y_2' \,\&\, D(y_1, y_2) \Rightarrow D(y_1', y_2')]$,

C-interdefinability: $\forall x, x' \in X. \left[\exists y \in Y. \left[\tilde{D}(x, y) \,\&\, x'By\right] \Longrightarrow C(x, x')\right]$,

D-interdefinability: $\forall y, y' \in Y. \left[\exists x \in X. \left[\tilde{C}(x, y) \,\&\, xBy'\right] \Longrightarrow D(y', y)\right]$,

C-duality: $\forall x \in X, \exists y \in Y. \left[\tilde{C}(x, y) \,\&\, \tilde{D}(x, y)\right]$,

D-duality: $\forall y \in Y, \exists x \in X. \left[\tilde{D}(x, y) \,\&\, \tilde{C}(x, y)\right]$.

Note that these conditions are dependent on each other. Also, we mention that the binary relations C and D are interdefinable. However, to keep nice symmetry, we choose the above conditions.

Theorem 2.6 *The dual algebra is a lattice with the de Morgan negation* $\neg$.

Example 2.7 A quintuple $\langle X, Y, B, C, D\rangle$ forms a polarity frame for (unbounded) ortholattice logic. Ortholattice formulae are given by $\phi ::= p \mid \phi \vee \phi \mid \phi \wedge \phi \mid \neg\phi$.

Residuality. Details are in [39]. For fusion $\circ$ and the residuals $\rightarrow$ and $\leftarrow$, polarity frames are extended by a ternary relation $R \subseteq X \times X \times Y$. In addition, for constants **t** and **f**, polarity frames are extended with two subsets of X and two subsets of Y, i.e. $O_X(\neq \emptyset)$, O_Y, N_X and N_Y. We

[3] Adjoint conditions were called tightness in the current author's papers.

assume that (O_X, O_Y) and (N_X, N_Y) form Galois stable pairs. On a polarity frame $\mathbb{F} = \langle X, Y, B\rangle$, the ternary relation R is extended to three ternary relations $R^\circ \subseteq X \times X \times X$, $R^\rightarrow \subseteq X \times Y \times Y$ and $R^\leftarrow \subseteq Y \times X \times Y$ as follows: $R^\circ(x_1, x_2, x) \iff \forall y \in Y.\, [R(x_1, x_2, y) \Longrightarrow xBy]$, $R^\rightarrow(x_1, y_2, y) \iff \forall x_2 \in X.\, [R(x_1, x_2, y) \Longrightarrow x_2 B y_2]$ and $R^\leftarrow(y_1, x_2, y) \iff \forall x_1 \in X.\, [R(x_1, x_2, y) \Longrightarrow x_1 B y_1]$. We assume that the ternary relation R satisfies the following conditions:

R-order: $\forall x, x' \in X.\, [x' \leq x \iff \exists o \in O_X.\, [R^\circ(x, o, x') \text{ or } R^\circ(o, x, x')]]$,

R-identity: $\forall x \in X.\, [\exists o \in O_X.\, [R^\circ(x, o, x)] \,\&\, \exists o' \in O_X.\, [R^\circ(o', x, x)]]$,

R-transitivity: $\forall x_1, x_1', x_2, x_2' \in X, y, y' \in Y.$

$$[x_1' \leq x_1, x_2' \leq x_2, y \leq y' \,\&\, R(x_1, x_2, y) \Longrightarrow R(x_1', x_2', y')]\,,$$

R-associativity: for all $x_1, x_2, x_3, x \in X$,

$$\begin{aligned} &\exists x' \in X.\, [R^\circ(x_1, x_2, x') \,\&\, R^\circ(x', x_3, x)] \\ &\qquad \iff \exists x'' \in X.\, [R^\circ(x_1, x'', x) \,\&\, R^\circ(x_2, x_3, x'')]\,, \end{aligned}$$

$\circ$-adjoint: $\forall x_1, x_2 \in X, y \in Y.\, [\forall x{\in}X.\, [R^\circ(x_1, x_2, x) \Rightarrow xBy] \Rightarrow R(x_1, x_2, y)]$,

$\rightarrow$-adjoint: $\forall x_1, x_2{\in}X, y{\in}Y.\, [\forall y_2{\in}Y.\, [R^\rightarrow(x_1, y_2, y) \Rightarrow x_2 B y_2] \Rightarrow R(x_1, x_2, y)]$,

$\leftarrow$-adjoint: $\forall x_1, x_2{\in}X, y{\in}Y.\, [\forall y_1{\in}Y.\, [R^\leftarrow(y_1, x_2, y) \Rightarrow x_1 B y_1] \Rightarrow R(x_1, x_2, y)]$.

Note that these conditions are not independent. The adjoint conditions intuitively tell

$$R^\circ(x_1, x_2, x)By \iff x_2 B R^\rightarrow(x_1, y_2, y) \iff x_1 B R^\leftarrow(y_1, x_2, y)$$

and R is the generator of this residuality.

Example 2.8 An octuple $\langle X, Y, B, R, O_X, O_Y, N_X, N_Y\rangle$ forms a polarity frame for substructural logic. Substructural formulae are given by $\phi ::= p \mid \mathbf{t} \mid \mathbf{f} \mid \phi \vee \phi \mid \phi \wedge \phi \mid \phi \circ \phi \mid \phi \rightarrow \phi \mid \phi \leftarrow \phi$.

Dedekind-cut preserving morphism. Details are in [43]. Given two polarity frames $\mathbb{F} = \langle X_1, Y_1, B_1\rangle$ and $\mathbb{G} = \langle X_2, Y_2, B_2\rangle$, a pair of functions $\sigma\colon X_1 \rightarrow X_2$ and $\tau\colon Y_1 \rightarrow Y_2$ forms a *Dedekind-cut preserving morphism* (*d-morphism* for short), denoted by $\langle\sigma|\tau\rangle$, if it satisfies

(i) $\forall x \in X_1, y \in Y_1.\, [\sigma(x) B_2 \tau(y) \Longrightarrow x B_1 y]$,

(ii) $\forall x \in X_1, y' \in Y_2.\, [\forall y \in Y_1.\, [y' \leq_2 \tau(y) \Longrightarrow x B_1 y] \Longrightarrow \sigma(x) B_2 y']$,

(iii) $\forall x' \in X_2, y \in Y_1.\, [\forall x \in X_1.\, [\sigma(x) \leq_2 x' \Longrightarrow x B_1 y] \Longrightarrow x' B_2 \tau(y)]$.

A d-morphism $\langle\sigma|\tau\rangle\colon \mathbb{F} \rightarrow \mathbb{G}$ is called

***B*-embedding:** if $\forall x \in X_1, y \in Y_1.\, [x B_1 y \Longrightarrow \sigma(x) B_2 \tau(y)]$,

***B*-separating:** if, for all $x' \in X_2$ and $y' \in Y_2$,

$$\forall x \in X_1, y \in Y_1.\, [\sigma(x) \leq_2 x' \text{ and } y' \leq_2 \tau(y) \Longrightarrow x B_1 y] \Longrightarrow x' B_2 y'.$$

B-reflecting: if B-embedding and B-separating.

Notice that B-separating and B-reflecting hold the same invariance of validity of logical formulae (sequents) as surjective and isomorphic, hence we may depict the arrows for B-embedding and B-separating with $\rightarrowtail$ and $\twoheadrightarrow$.

For additional relations S, C, D and R, we require the following conditions:

(iv) $\forall x \in X_1, y \in Y_1.\, [S_2(\sigma(x), \tau(y)) \Longrightarrow S_1(x, y)]$,

(v) $\forall x \in X_1, y' \in Y_2.\, [\forall y \in Y_1.\, [y' \leq_2 \tau(y) \Longrightarrow S_1(x, y)] \Longrightarrow S_2(\sigma(x), y')]$,

(vi) $\forall x' \in X_2, y \in Y_1.\, [\forall x \in X_1.\, [\sigma(x) \leq_2 y' \Longrightarrow S_1(x, y)] \Longrightarrow S_2(x', \tau(y))]$,

(vii) $\forall x_1, x_2 \in X_1.\, [C_2(\sigma(x_1), \sigma(x_2)) \Longrightarrow C_1(x_1, x_2)]$,

(viii) $\forall x_1 \in X_1, x_2' \in X_2.\, [\forall x_2 {\in} X_1.\, [\sigma(x_2) \leq_2 x_2' \Rightarrow C(x_1, x_2)] \Rightarrow C_2(\sigma(x_1), x_2')]$,

(ix) $\forall y_1, y_2 \in Y_1.\, [D_2(\tau(y_1), \tau(y_2)) \Longrightarrow D_1(y_1, y_2)]$,

(x) $\forall y_1' \in Y_2, y_2 \in Y_2.\, [\forall y_1 \in Y_1.\, [y_1' \leq_2 \tau(y_1) \Rightarrow D_1(y_1, y_2)] \Rightarrow D_2(y_1', \tau(y_2))]$,

(xi) $\forall x_1, x_2 \in X_1, y \in Y_1.\, [R_2(\sigma(x_1), \sigma(x_2), \tau(y)) \Longrightarrow R_1(x_1, x_2, y)]$,

(xii) $\forall x_1', x_2' \in X_2, y \in Y_1$,

$$\forall x_1, x_2 {\in} X_1.\, [\sigma(x_1) {\leq_2} x_1' \,\&\, \sigma(x_2) {\leq_2} x_2' \Rightarrow R_1(x_1, x_2, y)] \Rightarrow R_2(x_1', x_2', \tau(y)),$$

(xiii) $\forall x_1' \in X_2, x_2 \in X_1, y' \in Y_2$,

$$\forall x_1 {\in} X_1, y {\in} Y_1.\, [\sigma(x_1) {\leq_2} x_1' \,\&\, y' {\leq_2} \tau(y) {\Rightarrow} R_1(x_1, x_2, y)] \Rightarrow R_2(x_1', \sigma(x_2), y'),$$

(xiv) $\forall x_1 \in X_1, x_2' \in X_2, y' \in Y_2$,

$$\forall x_2 {\in} X_1, y {\in} Y_1.\, [\sigma(x_2) {\leq_2} x_2' \,\&\, y' {\leq_2} \tau(y) {\Rightarrow} R_1(x_1, x_2, y)] \Rightarrow R_2(\sigma(x_1), x_2', y').$$

Note that the conditions for C and D can be reduced by means of interdefinability. If we include constants, we also need to preserve their Dedekind-cuts.

Theorem 2.9 *Every d-morphism preserves Dedekind-cuts.*

Dual representation. Details are in [39]. Given a lattice $\mathbb{L} = \langle L, \vee, \wedge \rangle$, we let $\mathcal{F}$ and $\mathcal{I}$ be the set of filters and the set of ideals. Also, between $\mathcal{F}$ and $\mathcal{I}$, we define a binary relation $\sqsubseteq$ as follows: $F \sqsubseteq I \iff F \cap I \neq \emptyset$. Then, $\mathbb{L}_+ = \langle \mathcal{F}, \mathcal{I}, \sqsubseteq \rangle$ forms a polarity frame. We call $\mathbb{L}_+$ the dual frame.

For additional logical operations $\Diamond$, $\Box$, $\neg$, $\circ$, $\rightarrow$ and $\leftarrow$, we define their relations S, C, D and R as follows: for all $F, G \in \mathcal{F}$ and $I, J \in \mathcal{I}$, we let

(i) $S(F, I) \iff \Diamond F \sqsubseteq I \iff F \sqsubseteq \Box I$,

(ii) $C(F, G) \iff F \sqsubseteq \neg G$,

(iii) $D(I, J) \iff \neg I \sqsubseteq J$,

(iv) $R(F, G, I) \iff F \circ G \sqsubseteq I \iff G \sqsubseteq F \rightarrow I \iff F \sqsubseteq I \leftarrow G$,

where $\Diamond F := \{a \in L \mid \exists f \in F.\ \Diamond f \leq a\}$, $\Box I := \{a \in L \mid \exists i \in I.\ a \leq \Box i\}$, $\neg F := \{a \in L \mid \exists f \in F.\ a \leq \neg f\}$, $\neg I := \{a \in L \mid \exists i \in I.\ \neg i \leq a\}$, $F \circ G := \{a \in$

$L \mid \exists f \in F, g \in G.\ f \circ g \leq a\}$, $F \to I := \{a \in L \mid \exists f \in F, i \in I.\ a \leq f \to i\}$ and $I \leftarrow G := \{a \in L \mid \exists g \in G, i \in I.\ a \leq i \leftarrow g\}$.

Remark 2.10 Note that, even if we do not have full adjoint pairs in our language, on the dual frame, we can introduce them: see [40, Lemma 5.8].

Theorem 2.11 *The dual frame satisfies the appropriate relational conditions for each logical connectives.*

For a strict homomorphism $h\colon \mathbb{L} \to \mathbb{M}$, we define two functions $h_+\colon \mathcal{F}_2 \to \mathcal{F}_1$ and $h_-\colon \mathcal{I}_2 \to \mathcal{I}_1$ with $h_+(F) := h^{-1}[F]$ and $h_-(I) := h^{-1}[I]$.

Theorem 2.12 *For a strict homomorphism $h\colon \mathbb{L} \to \mathbb{M}$, the functions h_+ and h_- form a d-morphism, i.e. $\langle h_+ | h_- \rangle\colon \mathbb{M}_+ \to \mathbb{L}_+$. Furthermore, we have*

(i) *if h is injective, $\langle h_+ | h_- \rangle$ is B-separating,*

(ii) *if h is surjective, $\langle h_+ | h_- \rangle$ is B-embedding.*

For a polarity frame $\mathbb{F}$, additional structures S, C, D and R (and constants) yield appropriate lattice operations $\Diamond$, $\Box$, $\neg$, $\circ$, $\to$ and $\leftarrow$ (and constants) on the dual algebra $\mathbb{F}^+$ as follows: recall that $\mathbb{F}^+$ is isomorphic to the Galois stable lattice $\mathsf{G}_\mathbb{F}$. For all $(\mathfrak{X}, \mathfrak{Y})$, $(\mathfrak{X}_1, \mathfrak{Y}_1)$ and $(\mathfrak{X}_2, \mathfrak{Y}_2)$, we let

(i) $\Diamond(\mathfrak{X}, \mathfrak{Y}) := (\upsilon(\mathfrak{X}^\Diamond), \mathfrak{X}^\Diamond)$ where $\mathfrak{X}^\Diamond := \{y \in Y \mid \forall x \in \mathfrak{X}.\ S(x, y)\}$,

(ii) $\Box(\mathfrak{X}, \mathfrak{Y}) := (\mathfrak{Y}^\Box, \lambda(\mathfrak{Y}^\Box))$ where $\mathfrak{Y}^\Box := \{x \in X \mid \forall y \in \mathfrak{Y}.\ S(x, y)\}$,

(iii) $\neg(\mathfrak{X}, \mathfrak{Y}) := (\mathfrak{X}^\neg, \mathfrak{Y}^\neg)$
where $\mathfrak{X}^\neg := \{x{\in}X \mid \forall x'{\in}X.\ C(x, x')\}$, $\mathfrak{Y}^\neg := \{y{\in}Y \mid \forall y'{\in}Y.\ D(y, y')\}$,

(iv) $(\mathfrak{X}_1, \mathfrak{Y}_1) \circ (\mathfrak{X}_2, \mathfrak{Y}_2) := (\upsilon(\mathfrak{X}_1 \circ \mathfrak{X}_2), \mathfrak{X}_1 \circ \mathfrak{X}_2)$
where $\mathfrak{X}_1 \circ \mathfrak{X}_2 := \{y \in Y \mid \forall x_1 \in \mathfrak{X}_1, x_2 \in \mathfrak{X}_2.\ R(x_1, x_2, y)\}$,

(v) $(\mathfrak{X}_1, \mathfrak{Y}_1) \to (\mathfrak{X}_2, \mathfrak{Y}_2) := (\mathfrak{X}_1 \to \mathfrak{Y}_2, \lambda(\mathfrak{X}_1 \to \mathfrak{Y}_2))$
where $\mathfrak{X}_1 \to \mathfrak{Y}_2 := \{x_2 \in X \mid \forall x_1 \in \mathfrak{X}_1, y \in \mathfrak{Y}_2.\ R(x_1, x_2, y)\}$,

(vi) $(\mathfrak{X}_2, \mathfrak{Y}_2) \leftarrow (\mathfrak{X}_1, \mathfrak{Y}_1) := (\mathfrak{Y}_2 \leftarrow \mathfrak{X}_1, \lambda(\mathfrak{Y}_2 \leftarrow \mathfrak{X}_1))$
where $\mathfrak{Y}_2 \leftarrow \mathfrak{X}_1 := \{x_1 \in X \mid \forall x_2 \in \mathfrak{X}_1, y \in \mathfrak{Y}_2.\ R(x_1, x_2, y)\}$.

Theorem 2.13 *For a d-morphism $\langle \sigma | \tau \rangle\colon \mathbb{F} \to \mathbb{G}$, the function $\langle \sigma^+ | \tau^- \rangle\colon \mathbb{G}^+ \to \mathbb{F}^+$, defined by $\langle \sigma^+ | \tau^- \rangle(\mathfrak{X}_2, \mathfrak{Y}_2) := (\sigma^{-1}[\mathfrak{X}_2], \tau^{-1}[\mathfrak{Y}_2])$ for $(\mathfrak{X}_2, \mathfrak{Y}_2) \in \mathfrak{G}^+$, is well-defined and homomorphic. Moreover, we have*

(i) *if $\langle \sigma | \tau \rangle$ is B-embedding, $\langle \sigma^+ | \tau^- \rangle$ is surjective,*

(ii) *if $\langle \sigma | \tau \rangle$ is B-separating, $\langle \sigma^+ | \tau^- \rangle$ is injective.*

3 Topology and general polarity frames

In this section, we will extend the notion of topology from sets to polarities, address the topological representation between lattices and general polarity frames, and establish the topological characterisation of descriptive polarity frames: see e.g. [6] for the arguments on Kripke frames.

Definition 3.1 [B-topology] Let $\mathbb{P} = \langle X, Y, B \rangle$ be a polarity. A *B-topology* $\mathcal{O}$ of $\mathbb{P}$ is a collection of Galois stable pairs of $\mathbb{P}$, i.e. $\mathcal{O} \subseteq \mathsf{G}_\mathbb{P}$, satisfying

(i) for all $(\mathfrak{X}_1, \mathfrak{Y}_1), \ldots, (\mathfrak{X}_n, \mathfrak{Y}_n) \in \mathcal{O}$. $(\mathfrak{X}_1 \cap \cdots \cap \mathfrak{X}_n, \lambda(\mathfrak{X}_1 \cap \cdots \cap \mathfrak{X}_n)) \in \mathcal{O}$,

(ii) for all $(\mathfrak{X}_1, \mathfrak{Y}_1), \ldots, (\mathfrak{X}_n, \mathfrak{Y}_n) \in \mathcal{O}$. $(\upsilon(\mathfrak{Y}_1 \cap \cdots \cap \mathfrak{Y}_n), \mathfrak{Y}_1 \cap \cdots \cap \mathfrak{Y}_n) \in \mathcal{O}$,

(iii) $\bigcup\{\mathfrak{X} \mid (\mathfrak{X}, \mathfrak{Y}) \in \mathcal{O}\} = X$,

(iv) $\bigcup\{\mathfrak{Y} \mid (\mathfrak{X}, \mathfrak{Y}) \in \mathcal{O}\} = Y$.

We may call the pair $\langle \mathbb{P}, \mathcal{O} \rangle$ a *B-topological space* on $\mathbb{P}$, or simply *B-topology*.

One may feel that Definition 3.1 is far from the topology on sets: for example, there is no condition for arbitrary unions nor for X and $\emptyset$. This is because Galois stable pairs are not closed under unions and we do not know whether X and $\emptyset$ can be natural requirements for topological representations (in particular unbounded cases). Instead, a B-topology employs a possible generalisation of the bound condition, called *covering conditions*: items (iii) and (iv). From the topological viewpoints, the covering conditions tell us that (iii) for each point in X, there exists at least one open set to which the point belongs, and (iv) for each point in Y, there exists at least one open set to which the point belongs. It is obvious that, if $(X, \lambda(X))$ and $(\upsilon(Y), Y)$ are in $\mathcal{O}$, it satisfies the covering conditions. Note that the non-trivial difference disappears over distributive polarity frames (with one-side infinitary extension), hence it could be a natural generalisation of topology on sets.

We introduce the following notions: A B-topology $\mathcal{O}$ on $\mathbb{P}$ is

differentiated: if $\forall x \in X, y \in Y. [xBy \iff \exists(\mathfrak{X}, \mathfrak{Y}) \in \mathcal{O}. [x \in \mathfrak{X} \,\&\, y \in \mathfrak{Y}]]$,

compact: for all subfamilies $\mathcal{X}, \mathcal{Y} \subseteq \mathcal{O}$, if $\pi_1[\mathcal{X}](:= \{\mathfrak{X} \mid (\mathfrak{X}, \mathfrak{Y}) \in \mathcal{X}\})$ and $\pi_2[\mathcal{Y}](:= \{\mathfrak{Y} \mid (\mathfrak{X}, \mathfrak{Y}) \in \mathcal{Y}\})$ have the finite intersection property and $\mathcal{X} \cap \mathcal{Y} = \emptyset$, there are $x \in X$ and $y \in Y$ such that xBy does not hold, $x \in \bigcap \pi_1[\mathcal{X}]$ and $y \in \bigcap \pi_2[\mathcal{Y}]$.

Given two B-topological spaces $\langle \mathbb{P}, \mathcal{O}_P \rangle$ and $\langle \mathbb{Q}, \mathcal{O}_Q \rangle$, a *continuous map* is defined as an extension of a d-morphism $\langle \sigma | \tau \rangle : \mathbb{P} \to \mathbb{Q}$ with

(i) $\forall(\mathfrak{X}', \mathfrak{Y}') \in \mathcal{O}_Q$. $(\sigma^{-1}[\mathfrak{X}'], \tau^{-1}[\mathfrak{Y}']) \in \mathcal{O}_P$,

(ii) $\forall(\mathfrak{X}, \mathfrak{Y}) \in \mathcal{O}_P, \exists(\mathfrak{X}', \mathfrak{Y}'), (\mathfrak{X}'', \mathfrak{Y}'') \in \mathcal{O}_Q. [\sigma[\mathfrak{X}] \subseteq \mathfrak{X}'$ and $\mathfrak{Y}'' \supseteq \tau[\mathfrak{Y}]]$.

Note that item (i) is exactly the same as the continuity of topology on sets and item (ii) is the strictness condition of lattice homomorphisms. Since B-topology does not always include $(X, \lambda(X))$ or $(\upsilon(Y), Y)$, it is necessary to introduce the condition. We also mention the algebraic view of B-topology.

Definition 3.2 [Covering sublattice] Let $\mathbb{L} = \langle L, \vee, \wedge, \bot, \top \rangle$ be a complete lattice. A sublattice $\mathbb{L}' = \langle L', \vee, \wedge \rangle$ of $\mathbb{L}$ is *covering*, if it satisfies $\top = \bigvee_{a \in L'} a$ and $\bot = \bigwedge_{b \in L'} b$.

Therefore, a B-topology on a polarity $\mathbb{P}$ is a covering sublattice of $\mathsf{G}_{\mathbb{P}}$.

Definition 3.3 [General polarity frame & continuous d-morphism] A quadruple $\mathbb{F} = \langle X, Y, B, P \rangle$ is a *general polarity frame* if $\langle X, Y, B \rangle$ is a polarity frame and P is a covering sublattice of $\mathsf{G}_{\mathbb{F}}$. Given two general polarity frames $\mathbb{F} = \langle X_1, Y_1, B_1, P \rangle$ and $\mathbb{G} = \langle X_2, Y_2, B_2, Q \rangle$, a continuous map from $\mathbb{F}$ to $\mathbb{G}$ is a *continuous d-morphism*: the terminology (B-embedding, B-separating and B-reflecting) is inherited in the obvious way. Note that, for the B-embedding, the following condition is also necessary as in the case of general Kripke frames:

$$\forall(\mathfrak{X},\mathfrak{Y}) \in P, \exists(\mathfrak{X}',\mathfrak{Y}') \in Q. \left[\sigma^{-1}[\mathfrak{X}'] = \mathfrak{X} \,\&\, \tau^{-1}[\mathfrak{Y}'] = \mathfrak{Y}\right].$$

The dual representation is also defined in a usual way. Given a lattice $\mathbb{L} = \langle L, \vee, \wedge\rangle$, the dual general frame $\mathbb{L}_*$ is a pair of the dual frame $\mathbb{L}_+$ with $\hat{L} := \{(\lfloor a\rfloor, \lceil a\rceil) \mid a \in L\}$, where $\lfloor a\rfloor := \{F \in \mathcal{F} \mid a \in F\}$ and $\lceil a\rceil := \{I \in \mathcal{I} \mid a \in I\}$: i.e. $\mathbb{L}_* = \langle \mathcal{F}, \mathcal{I}, \sqsubseteq, \hat{L}\rangle$. For a strict homomorphism $h\colon \mathbb{L} \to \mathbb{M}$, the dual morphism $\langle h_\times | h_\div\rangle\colon \mathbb{M}_* \to \mathbb{L}_*$ is obtained as the dual morphism of the underlying frame$\langle h_+ | h_-\rangle$.

Theorem 3.4 *For lattices $\mathbb{L}$ and $\mathbb{M}$, and a strict homomorphism $h\colon \mathbb{L} \to \mathbb{M}$, the dual general frames $\mathbb{L}_*$ and $\mathbb{M}_*$ are general polarity frames and the dual morphism $\langle h_\times | h_\div\rangle$ is a continuous d-morphism from $\mathbb{M}_*$ to $\mathbb{L}_*$.*

For a general polarity frame $\mathbb{F} = \langle X, Y, B, P\rangle$, the dual algebra $\mathbb{F}^*$ is the covering sublattice P of $\mathsf{G}_{\mathbb{F}}$ itself. For a continuous d-morphism $\langle \sigma | \tau\rangle\colon \mathbb{F} \to \mathbb{G}$, the dual homomorphism $\langle \sigma^\times | \tau^\div\rangle\colon \mathbb{G}^* \to \mathbb{F}^*$ is the same as $\langle \sigma^+ | \tau^-\rangle$.

Theorem 3.5 *For general polarity frames $\mathbb{F}$ and $\mathbb{G}$, and a continuous d-morphism $\langle \sigma | \tau\rangle\colon \mathbb{F} \to \mathbb{G}$, the dual algebras $\mathbb{F}^*$ and $\mathbb{G}^*$ are lattices and the dual homomorphism $\langle \sigma^\times | \tau^\div\rangle$ is a strict homomorphism, i.e. $\langle \sigma^\times | \tau^\div\rangle\colon \mathbb{G}^* \to \mathbb{F}^*$.*

Theorem 3.6 *For a strict homomorphism $h\colon \mathbb{L} \to \mathbb{M}$ and a continuous d-morphism $\langle \sigma | \tau\rangle\colon \mathbb{F} \to \mathbb{G}$, we have*

(i) *if h is injective then $\langle h_\times | h_\div\rangle$ is B-separating,*

(ii) *if h is surjective then $\langle h_\times | h_\div\rangle$ is B-embedding,*

(iii) *if $\langle \sigma | \tau\rangle$ is B-embedding then $\langle \sigma^\times | \tau^\div\rangle$ is surjective,*

(iv) *if $\langle \sigma | \tau\rangle$ is B-separating then $\langle \sigma^\times | \tau^\div\rangle$ is injective.*

Topological characterisation of descriptive general polarity frames is addressed as follows: for a general polarity frame $\mathbb{F} = \langle X, Y, B, P\rangle$, we introduce two functions $\mathfrak{s}\colon X \to \wp(X)$ and $\mathfrak{t}\colon Y \to \wp(Y)$ with $\mathfrak{s}(x) := \{(\mathfrak{X},\mathfrak{Y}) \in P \mid x \in \mathfrak{X}\}$ and $\mathfrak{t}(y) := \{(\mathfrak{X},\mathfrak{Y}) \in P \mid y \in \mathfrak{Y}\}$.

Proposition 3.7 *$\mathfrak{s}(x)$ and $\mathfrak{t}(y)$ are a filter and an ideal over P, hence $\mathfrak{s}\colon X \to \mathcal{F}(P)$ and $\mathfrak{t}\colon Y \to \mathcal{I}(P)$ are well-defined.*

Definition 3.8 [Descriptive polarity frame] A general polarity frame $\mathbb{F} = \langle X, Y, B, P\rangle$ is *descriptive*, if the maps $\mathfrak{s}$ and $\mathfrak{t}$ form a B-reflecting continuous d-morphism from $\mathbb{F}$ to $(\mathbb{F}^*)_*$, i.e. $\langle \mathfrak{s} | \mathfrak{t}\rangle\colon \mathbb{F} \to (\mathbb{F}^*)_*$.

To obtain the topological characterisation of descriptive polarity frames, [4] we show two lemmata.

[4] The result on the similar setting can be found in [29].

Lemma 3.9 *For each lattice* $\mathbb{L} = \langle L, \vee, \wedge \rangle$, *the bi-dual lattice* $(\mathbb{L}_*)^* = \langle \hat{L}, \vee, \wedge \rangle$ *is isomorphic to* $\mathbb{L}$ *via the canonical map* $\mathbb{L} \to (\mathbb{L}_*)^*$ *with* $(\lfloor a \rfloor, \lceil a \rceil) \in \hat{L}$ *for each* $a \in L$.

Lemma 3.10 *A differentiated B-topological space on a polarity* $\langle \mathbb{P}, \mathcal{O} \rangle$ *satisfies*

(i) *for all* $x_1, x_2 \in X$, $x_1 \leq x_2 \iff \forall (\mathfrak{X}, \mathfrak{Y}) \in \mathcal{O}. [x_2 \in \mathfrak{X} \Longrightarrow x_1 \in \mathfrak{X}]$,

(ii) *for all* $y_1, y_2 \in Y$, $y_1 \leq y_2 \iff \forall (\mathfrak{X}, \mathfrak{Y}) \in \mathcal{O}. [y_1 \in \mathfrak{Y} \Longrightarrow y_2 \in \mathfrak{Y}]$.

Theorem 3.11 *A general polarity frame* $\mathbb{F} = \langle X, Y, B, P \rangle$ *is descriptive if and only if it is differentiated and compact as a B-topological space.*

4 Goldblatt-Thomason's theorem

In this section, we introduce disjoint unions of polarity frames. Also, based on the setting, we establish the Goldblatt-Thomason's theorem for substructural logic via the Birkhoff's variety theorem. Hence, throughout this section, we consider polarity frames for substructural logic. However, the results can be naturally applied for the other relational structures and variants of distributive lattice-based logics.

The disjoint union. To discuss disjoint unions of polarity frames, it seems natural to introduce the following notions: for a polarity frame $\mathbb{F} = \langle X, Y, B \rangle$, an element $x \in X$ is a *bottom element*, denoted by $\bot$, if $\forall y \in Y. [xBy]$, and an element $y \in Y$ is a *top element*, denoted by $\top$, if $\forall x \in X. [xBy]$. In general we do not assume the existence of bottom elements and top elements, but, if they exist, they are unique up to $\leq_B$-equivalence.

Let $\mathbb{F}_i$ for $i \in I$ be polarity frames for substructural logic. *The disjoint union* $\biguplus_{i \in I} \mathbb{F}_i$ consists of the set-theoretical disjoint unions $\biguplus_{i \in I} X_i$, $\biguplus_{i \in I} Y_i$, $\biguplus_{i \in I} O_{X\,i}$, $\biguplus_{i \in I} O_{Y\,i}$, $\biguplus_{i \in I} N_{X\,i}$, $\biguplus_{i \in I} N_{Y\,i}$, and the following relations $B_\uplus$ and $R_\uplus$: for all $x, x_1, x_2 \in \biguplus_{i \in I} X_i$ and $y \in \biguplus_{i \in I} Y_i$, we let

$$xB_\uplus y \iff \begin{cases} xB_i y & x \in X_i, y \in Y_i \\ \text{always holds} & x \in X_i, y \in Y_j, i \neq j \end{cases}$$

$$R_\uplus(x_1, x_2, y) \iff \begin{cases} R_i(x_1, x_2, y) & x_1, x_2 \in X_i, y \in Y_i \\ \text{always holds} & x_1 \in X_i, x_2 \in X_j, y \in Y_k, \neq\{i, j, k\} \end{cases}$$

where (and hereafter) $\neq\{i, j, k\}$ means that at least one index is different from the others.

On the disjoint union $\biguplus_{i \in I} \mathbb{F}_i$, the extended relations satisfy the following.

Proposition 4.1 *For* $x, x_1, x_2 \in \biguplus_{i \in I} X_i$ *and* $y, y_1, y_2 \in \biguplus_{i \in I} Y_i$, *we have*

(i) $x_1 \leq_\uplus x_2 \iff \begin{cases} x_1 \leq_i x_2 & x_1, x_2 \in X_i \\ x_1 = \bot_i & x_1 \in X_i, x_2 \in X_j, i \neq j \end{cases}$

(ii) $y_1 \leq_\uplus y_2 \iff \begin{cases} y_1 \leq_i y_2 & y_1, y_2 \in Y_i \\ y_2 = \top_j & y_1 \in Y_i, y_2 \in Y_j, i \neq j \end{cases}$

(iii) $R^{\circ}_{\uplus}(x_1, x_2, x) \iff \begin{cases} R^{\circ}_i(x_1, x_2, x) & x, x_1, x_2 \in X_i \\ x = \bot_k & x_1 \in X_i, x_2 \in X_j, x \in X_k, \neq\{i, j, k\} \end{cases}$

(iv) $R^{\rightarrow}_{\uplus}(x_1, y_2, y) \iff \begin{cases} R^{\rightarrow}_i(x_1, y_2, y) & x_1 \in X_i, y_2, y \in Y_i \\ y_2 = \top_j & x_1 \in X_i, y_2 \in Y_i, y \in Y_k, \neq\{i, j, k\} \end{cases}$

(v) $R^{\leftarrow}_{\uplus}(y_1, x_2, y) \iff \begin{cases} R^{\leftarrow}_i(y_1, x_2, y) & x_2 \in X_i, y_1, y \in Y_i \\ y_1 = \top_i & y_1 \in Y_i, x_2 \in X_j, y \in Y_k, \neq\{i, j, k\} \end{cases}$

where $=$ is a shorthand for $\leq_B$-equivalence.

Theorem 4.2 *For polarity frames for substructural logic $\mathbb{F}_i$ (for $i \in I$), the disjoint union $\biguplus_{i \in I} \mathbb{F}_i$ is also a polarity frame for substructural logic. Hence, the class of polarity frames for substructural logic is closed under disjoint unions.*

We can also have the natural canonical embeddings as well.

Theorem 4.3 *Let $\mathbb{F}_i$ (for $i \in I$) be polarity frames for substructural logic. For each index $i \in I$, the canonical functions $\iota^X_i : X_i \to \biguplus_{i \in I} X_i$ (i.e. $\iota^X_i(x) = x$) and $\iota^Y_i : Y_i \to \biguplus_{i \in I} Y_i$ (i.e. $\iota^Y_i(y) = y$) form a B-embedding d-morphism from $\mathbb{F}_i$ to the disjoint union $\biguplus_{i \in I} \mathbb{F}_i$, i.e. $\langle \iota^X_i | \iota^Y_i \rangle : \mathbb{F}_i \to \biguplus_{i \in I} \mathbb{F}_i$ for each $i \in I$.*

Goldblatt-Thomason's theorem. To state the theorem, we introduce the following: for polarity frames (with additional structures) $\mathbb{F}$ and $\mathbb{G}$, $\mathbb{F}$ is a *subframe* of $\mathbb{G}$, if there exists a B-embedding d-morphism from $\mathbb{F}$ to $\mathbb{G}$, i.e. $\mathbb{F} \rightarrowtail \mathbb{G}$; $\mathbb{G}$ is a *separating image* of $\mathbb{F}$, if there exists a B-separating d-morphism from $\mathbb{F}$ to $\mathbb{G}$, i.e. $\mathbb{F} \twoheadrightarrow \mathbb{G}$, and $(\mathbb{F}^+)_+$ is *the filter-ideal extension* of $\mathbb{F}$.

Lemma 4.4 *Let $\mathbb{F}_i$ for $i \in I$ be polarity frames for substructural logic.*

$$\left(\biguplus_{i \in I} \mathbb{F}_i \right)^+ \cong \prod_{i \in I} (\mathbb{F}_i^+)$$

Theorem 4.5 *The first-order definable class of polarity frames for substructural logic is definable by substructural formulae, if and only if it is closed under subframes, separating images and disjoint unions, and reflects filter-ideal extensions.*

The above statements hold for distributive polarity frames, the results also apply for distributive substructural and lattice-based logics as well.

5 Amalgamation property

In this section, we will discuss the amalgamation property based on the dual representation for canonical lattice-based logics. As we shall see below, the argument goes from the basic structure, i.e. lattices. But, surprisingly, the amalgamation property of the other variants of lattice-based algebras are also schematically proved on the base result as well.

Definition 5.1 [Amalgamation property] A class $\mathcal{C}$ of lattice-based algebras has *the amalgamation property*, if for all $\mathbb{A}, \mathbb{B}, \mathbb{C} \in \mathcal{C}$ with injections $f : \mathbb{A} \to \mathbb{B}$

and $g\colon \mathbb{A} \to \mathbb{C}$, there are an algebra $\mathbb{D} \in \mathcal{C}$ and two injections $i\colon \mathbb{B} \to \mathbb{D}$ and $j\colon \mathbb{C} \to \mathbb{D}$ such that $i \circ f = j \circ g$.

The recipe is as follows:

(i) Given algebras $\mathbb{A}, \mathbb{B}, \mathbb{C}$ and two injections $f\colon \mathbb{A} \to \mathbb{B}$ and $g\colon \mathbb{A} \to \mathbb{C}$, dualise the algebras and injections, i.e. $\langle f_+|f_-\rangle\colon \mathbb{B}_+ \to \mathbb{A}_+$ and $\langle g_+|g_-\rangle\colon \mathbb{C}_+ \to \mathbb{A}_+$ (they are B-separating: see Theorem 2.13). Note that if given algebras are unbounded or injections are not strict, we embed them into bounded algebras and strict injections in the standard way.

(ii) Construct, by amalgamating polarity frames $\mathbb{B}_+$ and $\mathbb{C}_+$, a polarity frame $\mathbb{F}_D$ endowed with two B-separating d-morphisms $\langle \sigma_B|\tau_B\rangle$ and $\langle \sigma_C|\tau_C\rangle$ to $\mathbb{B}_+$ and $\mathbb{C}_+$.

(iii) Check the commutativity $\langle \sigma_B|\tau_B\rangle \circ \langle f_+|f_-\rangle = \langle \sigma_C|\tau_C\rangle \circ \langle g_+|g_-\rangle$.

(iv) Dualise the commutative diagram to the dual algebras.

(v) Connect the original algebras to the bi-dual algebras with the canonical embeddings, i.e. $\mathfrak{c}_A\colon \mathbb{A} \to (\mathbb{A}_+)^+$. Note that the canonical embedding is injective, and concatenations of injections are injective, hence $\langle \sigma_B^+|\tau_B^-\rangle \circ \mathfrak{c}_B \circ f = \langle \sigma_C^+|\tau_C^-\rangle \circ \mathfrak{c}_C \circ g$.

Theorem 5.2 *Lattices admit the amalgamation property. Also, lattices extended with the distributivity, adjoint unary modality ($\Diamond \dashv \Box$), de Morgan negation $\neg$ admit the amalgamation property.*

Instead of the proof of this theorem, which requires a lot of space, we show the construction of the amalgamation $\mathbb{F}_D$ and B-separating d-morphisms.

Let $\mathbb{A}_+$, $\mathbb{B}_+$, $\mathbb{C}_+$ be the dual polarity frames and $\langle f_+|f_-\rangle\colon \mathbb{B}_+ \to \mathbb{A}_+$, $\langle g_+|g_-\rangle\colon \mathbb{C}_+ \to \mathbb{A}_+$ the dual morphisms of injections $f\colon \mathbb{A} \to \mathbb{B}$ and $g\colon \mathbb{A} \to \mathbb{C}$. The amalgamation $\mathbb{F}_D$ is constructed as the disjoint union of $\mathbb{B}_+$ and $\mathbb{C}_+$ with additional requirements for $\mathbb{A}_+$. That is, $\mathbb{F}_D = \langle \mathcal{F}_B \uplus \mathcal{F}_C, \mathcal{I}_B \uplus \mathcal{I}_C, \sqsubseteq_D\rangle$ with

$$F \sqsubseteq_D I \iff \begin{cases} F \sqsubseteq_B I & \text{if } F \in \mathcal{F}_B, I \in \mathcal{I}_B \\ F \sqsubseteq_C I & \text{if } F \in \mathcal{F}_C, I \in \mathcal{I}_C \\ f_+(F) \sqsubseteq_A g_-(I) & \text{if } F \in \mathcal{F}_B, I \in \mathcal{I}_C \\ g_+(F) \sqsubseteq_A f_-(I) & \text{if } F \in \mathcal{F}_C, I \in \mathcal{I}_B \end{cases}$$

For adjoint unary modality $\Diamond \dashv \Box$, the relation S_D on $\mathbb{F}_D$ is defined as follows:

$$S_D(F, I) \iff \begin{cases} S_B(F, I) & \text{if } F \in \mathcal{F}_B, I \in \mathcal{I}_B \\ S_C(F, I) & \text{if } F \in \mathcal{F}_C, I \in \mathcal{I}_C \\ S_A(f_+(F), g_-(I)) & \text{if } F \in \mathcal{F}_B, I \in \mathcal{I}_C \\ S_A(g_+(F), f_-(I)) & \text{if } F \in \mathcal{F}_C, I \in \mathcal{I}_B \end{cases}$$

For the de Morgan negation $\neg$, the two relations C_D and D_D on $\mathbb{F}_D$ are defined as follows: $C_D(F, G)$ and $D_D(I, J)$ are

$$\begin{cases} C_B(F, G) & F, G \in \mathcal{F}_B \\ C_C(F, G) & F, G \in \mathcal{F}_C \\ C_A(f_+(F), g_+(G)) & F \in \mathcal{F}_B, G \in \mathcal{F}_C \\ C_A(g_+(F), f_+(G)) & F \in \mathcal{F}_C, G \in \mathcal{F}_B \end{cases} \quad \begin{cases} D_B(I, J) & I, J \in \mathcal{I}_B \\ D_C(I, J) & I, J \in \mathcal{I}_C \\ D_A(f_-(I), g_-(J)) & I \in \mathcal{I}_B, J \in \mathcal{I}_C \\ D_A(g_-(I), f_-(J)) & I \in \mathcal{I}_C, J \in \mathcal{I}_B \end{cases}$$

The d-morphisms $\langle \sigma_B | \tau_B \rangle \colon \mathbb{F}_D \to \mathbb{B}_+$ and $\langle \sigma_C | \tau_C \rangle \colon \mathbb{F}_D \to \mathbb{C}_+$ are defined as follows:

$$\sigma_B(F) := \begin{cases} F & \text{if } F \in \mathcal{F}_B \\ \uparrow f[g_+(F)] & \text{if } F \in \mathcal{F}_C \end{cases} \qquad \tau_B(I) := \begin{cases} I & \text{if } I \in \mathcal{I}_B \\ \downarrow f[g_-(I)] & \text{if } I \in \mathcal{I}_C \end{cases}$$

where $\uparrow f[g_+(F)]$ and $\downarrow f[g_-(I)]$ are the generated filter of the image of f of $g_+(F)$ and the generated ideal of the image of f of $g_-(I)$. The definition of $\langle \sigma_C | \tau_C \rangle$ is analogous.

The (distributive) polarity frame $\mathbb{F}_D$ (endowed with the above relations) and the d-morphisms $\langle \sigma_B | \tau_B \rangle$ and $\langle \sigma_C | \tau_C \rangle$ satisfy our requirements. Therefore Theorem 5.2 holds.

6 Conclusion

In this paper, we have introduced the notion of topology on polarities, named B-topology, and general polarity frames, the disjoint union of polarity frames and amalgamation of polarity frames. Based on these notions and constructions, we have provided the topological characterisation of descriptive polarity frames as in the case of modal logics, established the Goldblatt-Thomason's theorem on polarity frames, and shown the amalgamation property for some lattice-based algebras.

As concluding remarks, we shortly list the current author's forthcoming work (with collaborators). For the topological representation, we will study the persistence properties for substructural and lattice-based formulae. For the Goldblatt-Thomason's theorem, we will also provide the model-theoretic proof and model theory on polarities. For the amalgamation property, we are investigating sufficient conditions for the amalgamation property for lattice-base algebras. Note that, unfortunately, our approach for the amalgamation property seems containing strong requirements to amalgamate polarity frames with (more than) ternary relations, hence for substructural logics, we can universally discuss the amalgamation property only above intuitionistic logic.

Appendix

A Proof of Theorem 3.11

Proof. ($\Rightarrow$). Assume that $\langle \mathfrak{s} | \mathfrak{t} \rangle$ forms a B-reflecting d-morphism. For all $x \in X$ and $y \in Y$ satisfying that xBy does not hold and $(\mathfrak{X}, \mathfrak{Y}) \in P$ does, if

$x \in \mathfrak{X}$ then $y \notin \mathfrak{Y}$, otherwise, as $\mathfrak{X} = \upsilon(\mathfrak{Y})$, we obtain xBy which contradicts to the assumption that xBy does not hold. Conversely, suppose xBy for $x \in X$ and $y \in Y$. Since $\langle \mathfrak{s}|\mathfrak{t}\rangle$ is d-embedding, we have $\mathfrak{s}(x) \sqsubseteq \mathfrak{t}(y)$. So there exists $(\mathfrak{X}, \mathfrak{Y}) \in P$ such that $(\mathfrak{X}, \mathfrak{Y}) \in \mathfrak{s}(x) \cap \mathfrak{t}(y)$. By definition, $x \in \mathfrak{X}$ and $y \in \mathfrak{Y}$, hence $\mathbb{F}$ is differentiated. For all subfamilies $\mathcal{X}, \mathcal{Y} \subseteq P$ satisfying $\pi_1[\mathcal{X}]$ and $\pi_2[\mathcal{Y}]$ have finite intersection property, and $\mathcal{X} \cap \mathcal{Y} = \emptyset$. Let $F_{\mathcal{X}}$ and $I_{\mathcal{Y}}$ be the generated filter by $\mathcal{X}$ and the generated ideal by $\mathcal{Y}$ over P, i.e. $F_{\mathcal{X}} := \uparrow\mathcal{X}$ and $I_{\mathcal{Y}} := \downarrow\mathcal{Y}$. Suppose $F_{\mathcal{X}} \sqsubseteq I_{\mathcal{Y}}$. It contradicts to $\mathcal{X} \cap \mathcal{Y} = \emptyset$, so $F_{\mathcal{X}} \not\sqsubseteq I_{\mathcal{I}}$. As $\langle \mathfrak{s}|\mathfrak{t}\rangle$ is d-separating, there exist $x \in X$ and $y \in Y$ such that $\mathfrak{s}(x) \sqsubseteq F_{\mathcal{X}}$ and $I_{\mathcal{Y}} \sqsubseteq \mathfrak{t}(y)$ hold, but xBy does not. For arbitrary $(\mathfrak{X}_f, \mathfrak{Y}_f) \in F_{\mathcal{X}}$ and $(\mathfrak{X}_i, \mathfrak{Y}_i) \in I_{\mathcal{Y}}$, since $F_{\mathcal{X}} \subseteq \mathfrak{s}(x)$ and $I_{\mathcal{Y}} \subseteq \mathfrak{t}(y)$, we have $(\mathfrak{X}_f, \mathfrak{Y}_f) \in \mathfrak{s}(x)$ and $(\mathfrak{X}_i, \mathfrak{Y}_i) \in \mathfrak{t}(y)$. By definition, $x \in \mathfrak{X}_f$ and $y \in \mathfrak{Y}_i$. So $x \in \bigcap \pi_1[\mathcal{X}]$ and $y \in \bigcap \pi_2[\mathcal{Y}]$. Therefore $\mathbb{F}$ is compact.

($\Leftarrow$). Because of Proposition 3.7 and Lemma 3.9, it suffices to show that $\langle \mathfrak{s}|\mathfrak{t}\rangle$ is a d-morphism, i.e. items (i) - (iii), B-embedding and B-separating for differentiated and compact B-topological spaces. **Item (i)**: for all $x \in X$ and $y \in Y$, suppose that $\mathfrak{s}(x) \sqsubseteq \mathfrak{t}(y)$. Then there exists $(\mathfrak{X}, \mathfrak{Y}) \in P$ such that $x \in \mathfrak{X}$ and $y \in \mathfrak{Y}$. Since $(\mathfrak{X}, \mathfrak{Y})$ is a Galois stable pair, we have $X = \upsilon(\mathfrak{Y})$ and $\lambda(\mathfrak{X}) = \mathfrak{Y}$, which concludes xBy. **Item (ii)**: we prove the contraposition. Suppose $\mathfrak{s}(x) \not\sqsubseteq I$ for arbitrary $x \in X$ and $I \in \mathcal{I}_P$. Since $\mathfrak{s}(x)$ is a filter over P and I is an ideal over P, they have intersection property. Because $\mathbb{F}$ is compact, there exist $x_c \in X$ and $y_c \in Y$ such that $x_c \in \bigcap \pi_1[\mathfrak{s}(x)]$ and $y_c \in \bigcap \pi_2[I]$ hold, but $x_c B y_c$ does not. By definition, for each $\mathfrak{X} \in \pi_1[P]$, if $\mathfrak{X} \in \pi_1[\mathfrak{s}(x)]$ then $\mathfrak{X} \in \pi_1[\mathfrak{s}(x_c)]$. Moreover, for each $\mathfrak{Y} \in \pi_2[P]$, if $\mathfrak{Y} \in \pi_2[I]$ then $\mathfrak{Y} \in \pi_2[\mathfrak{t}(y_c)]$. Hence $\mathfrak{s}(x_c) \sqsubseteq \mathfrak{s}(x)$ and $I \sqsubseteq \mathfrak{t}(y_c)$. As $\mathbb{F}$ is differentiated, by Lemma 3.10, we have $x_c \leq x$. Further, since $x_c B y_c$ does not hold, neither does xBy_c. **Item (iii)**: this is analogous to item (ii). B-**embedding**: for arbitrary $x \in X$ and $y \in Y$, if xBy, since it is differentiated, there exists $(\mathfrak{X}, \mathfrak{Y}) \in P$ such that $x \in \mathfrak{X}$ and $y \in \mathfrak{Y}$. Because of the definitions $\mathfrak{s}(x)$ and $\mathfrak{t}(y)$, we have $(\mathfrak{X}, \mathfrak{Y}) \in \mathfrak{s}(x) \cap \mathfrak{t}(y)$, hence $\mathfrak{s}(x) \sqsubseteq \mathfrak{t}(y)$. B-**separating**: we prove the contraposition. for arbitrary $F \in \mathcal{F}_P$ and $I \in \mathcal{I}_P$, assume $F \not\sqsubseteq I$. Since F and I are subfamilies of P, they have finite intersection property. In addition, by our assumption, $F \cap I = \emptyset$. Because of the compactness, there exist $x \in X$ and $y \in Y$ such that $x \in \bigcap \pi_1[F]$ and $y \in \bigcap \pi_2[I]$ hold, but xBy does not. As $x \in \bigcap \pi_1[F]$, each $\mathfrak{X} \in \pi_1[F]$ is in $\mathfrak{s}(x)$, that is, $\mathfrak{s}(x) \sqsubseteq F$. Also, as $y \in \bigcap \pi_2[I]$, each $\mathfrak{Y} \in \pi_2[I]$ is in $\mathfrak{t}(y)$, that is, $I \sqsubseteq \mathfrak{t}(y)$. □

B Hints for Section 4

Theorem 4.2 follows from Propositions 4.1, B.1 and B.2.

Proposition B.1 *For all $x, x_1, x_2 \in X$ and $y, y_1, y_2 \in Y$, we have*

(i) *if y is a top element then $R(x_1, x_2, y)$,*

(ii) *if x_2 is a bottom element then $R(x_1, x_2, y)$,*

(iii) *if x_1 is a bottom element then $R(x_1, x_2, y)$,*

(iv) *if* x *is a bottom element then* $R^{\circ}(x_1, x_2, x)$,

(v) *if* y_2 *is a top element then* $R^{\rightarrow}(x_1, y_2, y)$,

(vi) *if* y_1 *is a top element then* $R^{\leftarrow}(y_1, x_2, y)$.

Proposition B.2 *For all* $x, x_1, x_2 \in X$ *and* $y, y_1, y_2 \in Y$, *we have*

(i) *if* $x \neq \bot$ *and* $R^{\circ}(x_1, x_2, x)$ *then* $x_1 \neq \bot$ *and* $x_2 \neq \bot$,

(ii) *if* $y_2 \neq \top$ *and* $R^{\rightarrow}(x_1, y_2, y)$ *then* $x_1 \neq \bot$ *and* $y \neq \top$,

(iii) *if* $y_1 \neq \top$ *and* $R^{\leftarrow}(y_1, x_2, y)$ *then* $x_2 \neq \bot$ *and* $y \neq \top$.

For Lemma 4.4, it suffices to show that the following function η is a well defined isomorphism. We define $\eta \colon \left(\biguplus_{i\in I} \mathbb{F}_i\right)^+ \to \prod_{i\in I}(\mathbb{F}_i^+)$ as follows: for each Galois stable pair $(\mathfrak{X}, \mathfrak{Y}) \in \mathsf{G}_{\uplus\mathbb{F}_i}$,

$$\eta(\mathfrak{X}, \mathfrak{Y}) := \left((\mathfrak{X} \cap X_1, \mathfrak{Y} \cap Y_1), \ldots, (\mathfrak{X} \cap X_i, \mathfrak{Y} \cap Y_i), \ldots\right).$$

A proof of Theorem 4.5 is sketched as follows: the ($\Rightarrow$)-direction follows from the invariance of validity of sequents via B-embedding d-morphisms and B-separating d-morphisms. The ($\Leftarrow$)-direction is as follows: let $\mathcal{P}$ be a class of polarity frames satisfying the condition. For any polarity frame $\mathbb{F}$ validating the substructural formulae of $\mathcal{P}$, the dual algebra $\mathbb{F}^+$ is a model of the equational theory of the class of dual algebras of $\mathcal{P}$.[5] Due to the Birkhoff's variety theorem, $\mathbb{F}^+$ is in the variety, which means $\mathbb{F}^+$ is a homomorphic image of a subalgebra of a product of dual algebras of polarity frames in $\mathcal{P}$. Because of Lemma 4.4, products of dual algebras of polarity frames are isomorphic to dual algebras of disjoint unions of the same polarity frames. By dualising the HSP conditions, the filter-ideal extension $(\mathbb{F}^+)_+$ is a subframe of a separating image of the filter-ideal extension of the disjoint union of polarity frames in $\mathcal{P}$. Hence, $\mathbb{F}$ is in $\mathcal{P}$.

Note that, for variants of distributive lattice-based logics, it is easy to check the construction of disjoint unions straightforwardly admits the splitting condition. For the connection between distributive polarity frames and Kripke frames can be found in [38].

C Hints for Section 5

Theorem C.1 *Let* $\mathbb{A}$, $\mathbb{B}$ *and* $\mathbb{C}$ *be lattices, and* $f \colon \mathbb{A} \to \mathbb{B}$ *and* $g \colon \mathbb{A} \to \mathbb{C}$ *be injective. There exists a polarity frame* $\mathbb{F}_D$ *with two* B*-separating d-morphisms* $\langle \sigma_B | \tau_B \rangle \colon \mathbb{F}_D \to \mathbb{B}_+$ *and* $\langle \sigma_C | \tau_C \rangle \colon \mathbb{F}_D \to \mathbb{C}_+$ *satisfying* $\langle f_+ | f_- \rangle \circ \langle \sigma_B | \tau_B \rangle = \langle g_+ | g_- \rangle \circ \langle \sigma_C | \tau_C \rangle$.

Lemma C.2 *Let* $\mathbb{A}$, $\mathbb{B}$ *and* $\mathbb{C}$ *be lattices, and* $f \colon \mathbb{A} \to \mathbb{B}$ *and* $g \colon \mathbb{A} \to \mathbb{C}$ *injective homomorphisms. On the dual frames, we have*

(i) *for arbitrary* $F \in \mathcal{F}_B$ *and* $I \in \mathcal{I}_C$,

$$F \sqsubseteq_B {\downarrow} f[g_-(I)] \iff f_+(F) \sqsubseteq_A g_-(I) \iff {\uparrow} g[f_+(F)] \sqsubseteq_C I,$$

[5] Each sequent corresponds to an inequality, but on lattices, it is naturally translated to an equality.

(ii) *for arbitrary* $F \in \mathcal{F}_C$ *and* $I \in \mathcal{I}_B$,

$$\uparrow f[g_+(F)] \sqsubseteq_B I \iff g_+(F) \sqsubseteq_A f_-(I) \iff F \sqsubseteq_C \downarrow g[f_-(I)],$$

(iii) *for each* $F \in \mathcal{F}_A$, $F = f_+(\uparrow f[F])$ *and* $F = g_+(\uparrow g[F])$,

(iv) *for each* $I \in \mathcal{I}_A$, $I = f_-(\downarrow f[I])$ *and* $I = g_-(\downarrow g[I])$,

(v) *for each* $F \in \mathcal{F}_B$, $F \sqsubseteq_B \uparrow f[f_+(F)]$,

(vi) *for each* $F \in \mathcal{F}_C$, $F \sqsubseteq_C \uparrow g[g_+(F)]$,

(vii) *for each* $I \in \mathcal{I}_B$, $\downarrow f[f_-(I)] \sqsubseteq_B I$,

(viii) *for each* $I \in \mathcal{I}_C$, $\downarrow g[g_-(I)] \sqsubseteq_C I$.

Proposition C.3 *For all filters* $F, G \in X_D$ *and ideals* $I, J \in Y_D$, *we have*

$$F \sqsubseteq_D G \iff \begin{cases} F \sqsubseteq_B G & F, G \in \mathcal{F}_B \\ F \sqsubseteq_C G & F, G \in \mathcal{F}_C \\ F \sqsubseteq_B \uparrow f[g_+(G)] \ \&\ \uparrow g[f_+(F)] \sqsubseteq_C G & F \in \mathcal{F}_B, G \in \mathcal{F}_C \\ \uparrow f[g_+(F)] \sqsubseteq_B G \ \&\ F \sqsubseteq_C \uparrow g[f_+(G)] & F \in \mathcal{F}_C, G \in \mathcal{F}_B \end{cases}$$

$$I \sqsubseteq_D J \iff \begin{cases} I \sqsubseteq_B J & I, J \in \mathcal{I}_B \\ I \sqsubseteq_C J & I, J \in \mathcal{I}_C \\ I \sqsubseteq_B \downarrow f[g_-(J)] \ \&\ \downarrow g[f_-(I)] \sqsubseteq_C J & I \in \mathcal{I}_B, J \in \mathcal{I}_C \\ \downarrow f[g_-(I)] \sqsubseteq_B J \ \&\ I \sqsubseteq_C \downarrow g[f_-(J)] & I \in \mathcal{I}_C, J \in \mathcal{I}_B \end{cases}$$

Proposition C.4 *The dualised commutative diagram also commutes.*

Distributive lattices. We extend the previous result to distributive lattice along completely the same construction and the method.

Theorem C.5 *For distributive lattices* $\mathbb{A}$, $\mathbb{B}$ *and* $\mathbb{C}$ *and injective homomorphisms* $f\colon \mathbb{A} \to \mathbb{B}$ *and* $g\colon \mathbb{A} \to \mathbb{C}$, *there exists a distributive polarity frame* $\mathbb{F}_D$ *equipped with two d-separating morphisms* $\langle \sigma_B | \tau_B \rangle\colon \mathbb{F}_D \to \mathbb{B}_+$ *and* $\langle \sigma_C | \tau_C \rangle\colon \mathbb{F}_D \to \mathbb{C}_+$ *such that* $\langle f_+ | f_- \rangle \circ \langle \sigma_B | \tau_B \rangle = \langle g_+ | g_- \rangle \circ \langle \sigma_C | \tau_C \rangle$.

Theorem C.6 *Distributive lattices admit the amalgamation property.*

Modal operators. Now we consider adjoint unary modal operators $\Diamond \dashv \Box$.

Theorem C.7 *For lattices with adjoint unary modality* ($\Diamond \dashv \Box$) $\mathbb{A}$, $\mathbb{B}$ *and* $\mathbb{C}$, *and two injective homomorphisms* $f\colon \mathbb{A} \to \mathbb{B}$ *and* $g\colon \mathbb{A} \to \mathbb{C}$, *there exists a polarity frame with adjoint unary modality* $\mathbb{F}_D$ *endowed with two d-separating morphisms* $\langle \sigma_B | \tau_B \rangle\colon \mathbb{F}_D \to \mathbb{B}_+$ *and* $\langle \sigma_C | \tau_C \rangle\colon \mathbb{F}_D \to \mathbb{C}_+$ *such that* $\langle f_+ | f_- \rangle \circ \langle \sigma_B | \tau_B \rangle = \langle g_+ | g_- \rangle \circ \langle \sigma_C | \tau_C \rangle$.

To prove the main statement, Lemma C.8 and Proposition C.9 are useful.

Lemma C.8 *Let* $\mathbb{A}$, $\mathbb{B}$ *and* $\mathbb{C}$ *be lattices with adjoint unary modality* $\Diamond \dashv \Box$, *and* $f\colon \mathbb{A} \to \mathbb{B}$ *and* $g\colon \mathbb{A} \to \mathbb{C}$ *injective homomorphisms. On the dual frames,*

(i) *for arbitrary* $F \in \mathcal{F}_B$ *and* $I \in \mathcal{I}_C$,

$$S_B(F, \downarrow f[g_-(I)]) \iff S_A(f_+(F), g_-(I)) \iff S_C(\uparrow g[f_+(F)], I),$$

(ii) *for arbitrary* $F \in \mathcal{F}_C$ *and* $I \in \mathcal{I}_B$,

$$S_B(\uparrow f[g_+(F)], I) \iff S_A(g_+(F), f_-(I)) \iff S_C(F, \downarrow g[f_-(I)]),$$

(iii) *for arbitrary* $F \in \mathcal{F}_B$ *and* $I \in \mathcal{I}_B$,

$$S_B(F, I) \Longleftarrow S_A(f_+(F), f_-(I)) \iff S_C(\uparrow g[f_+(F)], \downarrow g[f_-(I)]),$$

(iv) *for arbitrary* $F \in \mathcal{F}_C$ *and* $I \in \mathcal{I}_C$,

$$S_B(\uparrow f[g_+(F)], \downarrow f[g_-(I)]) \iff S_A(g_+(F), g_-(I)) \Longrightarrow S_C(F, I).$$

Proposition C.9 *For all* $F, G \in X_D$ *and* $I, J \in Y_D$, *we have*

$$S_D^{\Diamond}(F,G) \iff \begin{cases} S_B^{\Diamond}(F,G) & F, G \in \mathcal{F}_B \\ S_C^{\Diamond}(F,G) & F, G \in \mathcal{F}_C \\ S_B^{\Diamond}(F, \uparrow f[g_+(G)]) \ \&\ S_C^{\Diamond}(\uparrow g[f_+(F)], G) & F \in \mathcal{F}_B, G \in \mathcal{F}_C \\ S_B^{\Diamond}(\uparrow f[g_+(F)], G) \ \&\ S_C^{\Diamond}(F, \uparrow g[f_+(G)]) & F \in \mathcal{F}_C, G \in \mathcal{F}_B \end{cases}$$

$$S_D^{\Box}(J,I) \iff \begin{cases} S_B^{\Box}(J,I) & I, J \in \mathcal{I}_B \\ S_C^{\Box}(J,I) & I, J \in \mathcal{I}_C \\ S_B^{\Box}(\downarrow f[g_-(J)], I) \ \&\ S_C^{\Box}(J, \downarrow g[f_-(I)]) & I \in \mathcal{I}_B, J \in \mathcal{I}_C \\ S_B^{\Box}(J, \downarrow f[g_-(I)]) \ \&\ S_C^{\Box}(\downarrow g[f_-(J)], I) & I \in \mathcal{I}_C, J \in \mathcal{I}_B \end{cases}$$

Theorem C.10 *Lattices with adjoint unary modality* ($\Diamond \dashv \Box$) *admit the amalgamation property.*

Corollary C.11 *Distributive lattices with adjoint unary modality admit the amalgamation property.*

De Morgan negation. Now we consider the de Morgan negation $\neg$.

Theorem C.12 *For lattices with the de Morgan negation* ($\neg$) $\mathbb{A}$, $\mathbb{B}$ *and* $\mathbb{C}$, *and two injective homomorphisms* $f\colon \mathbb{A} \to \mathbb{B}$ *and* $g\colon \mathbb{A} \to \mathbb{C}$, *there exists a polarity frame with the de Morgan negation* $\mathbb{F}_D$ *endowed with two d-separating morphisms* $\langle \sigma_B | \tau_B \rangle\colon \mathbb{F}_D \to \mathbb{B}_+$ *and* $\langle \sigma_C | \tau_C \rangle\colon \mathbb{F}_D \to \mathbb{C}_+$ *such that* $\langle f_+ | f_- \rangle \circ \langle \sigma_B | \tau_B \rangle = \langle g_+ | g_- \rangle \circ \langle \sigma_C | \tau_C \rangle$.

Lemma C.13 *Let* $\mathbb{A}$, $\mathbb{B}$ *and* $\mathbb{C}$ *be lattices with the de Morgan negation, and* $f\colon \mathbb{A} \to \mathbb{B}$ *and* $g\colon \mathbb{A} \to \mathbb{C}$ *injective homomorphisms. On the dual frame,*

(i) *for arbitrary* $F \in \mathcal{F}_B$, $G \in \mathcal{F}_C$,

$$C_B(F, \uparrow f[g_+(G)]) \iff C_A(f_+(F), g_+(G)) \iff C_C(\uparrow g[f_+(F)], G),$$

(ii) *for arbitrary* $I \in \mathcal{I}_B$, $J \in \mathcal{I}_C$,

$$D_B(I, \downarrow f[g_-(I)]) \iff D_A(f_-(I), g_-(J)) \iff D_C(\downarrow g[f_-(I)], J),$$

(iii) *for arbitrary* $F, G \in \mathcal{F}_B$,

$$C_B(F, G) \Longleftarrow C_A(f_+(F), f_+(G)) \iff C_C(\uparrow g[f_+(F)], \uparrow g[f_+(G)]),$$

(iv) *for arbitrary* $F, G \in \mathcal{F}_C$,

$$C_B(\uparrow f[g_+(F)], \uparrow f[g_+(G)]) \iff C_A(g_+(F), g_+(G)) \Longrightarrow C_C(F, G),$$

(v) *for arbitrary* $I, J \in \mathcal{I}_B$,

$$D_B(I, J) \Longleftarrow D_A(f_-(I), f_-(J)) \iff D_C(\downarrow g[f_-(I)], \downarrow g[f_-(J)]),$$

(vi) *for arbitrary* $I, J \in \mathcal{I}_C$,

$$D_B(\downarrow g[f_-(I)], \downarrow g[f_-(J)]) \iff D_A(f_-(I), f_-(J)) \Longrightarrow D_C(I, J).$$

Proposition C.14 *For all* $F \in X_D$, $I \in Y_D$, *we have*

$$\tilde{C}_D(F, I) \iff \begin{cases} \tilde{C}_B(F, I) & F \in \mathcal{F}_B, I \in \mathcal{I}_B \\ \tilde{C}_C(F, I) & F \in \mathcal{F}_C, I \in \mathcal{F}_C \\ \tilde{C}_B(F, \downarrow f[g_-(I)]) \ \& \ \tilde{C}_C(\uparrow g[f_+(F)], I) & F \in \mathcal{F}_B, I \in \mathcal{I}_C \\ \tilde{C}_B(\uparrow f[g_+(F)], I) \ \& \ \tilde{C}_C(F, \downarrow g[f_-(I)]) & F \in \mathcal{F}_C, I \in \mathcal{I}_B \end{cases}$$

$$\tilde{D}_D(F, I) \iff \begin{cases} \tilde{D}_B(F, I) & F \in \mathcal{F}_B, I \in \mathcal{I}_B \\ \tilde{D}_C(F, I) & F \in \mathcal{F}_C, I \in \mathcal{I}_C \\ \tilde{D}_B(F, \downarrow f[g_-(I)]) \ \& \ \tilde{D}_C(\uparrow g[f_+], I) & F \in \mathcal{F}_B, I \in \mathcal{I}_C \\ \tilde{D}_B(\uparrow f[g_+(F)], I) \ \& \ \tilde{D}_C(F, \downarrow g[f_-(I)]) & F \in \mathcal{F}_C, I \in \mathcal{I}_B \end{cases}$$

Theorem C.15 *Lattices with the de Morgan negation admit the amalgamation property.*

Corollary C.16 *Distributive lattices with the de Morgan negation, lattices with the de Morgan negation and adjoint unary modality and distributive lattices with the de Morgan negation and adjoint unary modality admit the amalgamation property.*

References

[1] Bílková, M., R. Horčík and J. Velebil, *Distributive substructural logics as coalgebraic logics over posets.*, in: *AiML*, 2012.

[2] Birkhoff, G., "Lattice Theory," AMS Colloquium Publications, 1973.

[3] Blackburn, P., M. de Rijke and Y. Venema, "Modal logic," Cambridge UP, 2002.

[4] Burris, S. and H. Sankappanavar, "A course in universal algebra," Springer-Verlag, 1981.

[5] Celani, S. A., *Classical modal de morgan algebras.*, Studia Logica **98** (2011).

[6] Chagrov, A. and M. Zakharyaschev, "Modal logic," Oxford Science Publications, 1997.

[7] Davey, B. and H. Priestley, "Introduction to Lattices and Order," Cambridge UP, 2002.

[8] Esakia, L., *Topological Kripke models*, Soviet Math. Dokl. **15** (1974).

[9] Galatos, N. and P. Jipsen, *Residuated frames with applications to decidability*, Trans. AMS **365** (2013).

[10] Gehrke, M., *Generalized Kripke frames*, Studia Logica **84** (2006).

[11] Gehrke, M. and J. Harding, *Bounded lattice expansions*, J. Alg. **239** (2001).

[12] Gehrke, M. and B. Jónsson, *Bounded distributive lattice expansions*, Math. Scand. **94** (2004).

[13] Gehrke, M., H. Nagahashi and Y. Venema, *A Sahlqvist theorem for distributive modal logic*, Ann. Pure Appl. Log. **131** (2005).
[14] Ghilardi, S. and G. Meloni, *Constructive canonicity in non-classical logics*, Ann. Pure Appl. Log. **86** (1997).
[15] Goldblatt, R., "Mathematics of modality," CSLI Publications, 1993.
[16] Goldblatt, R., *Fine's theorem on first-order complete modal logics*, in: *Metaphysics, Meaning, and Modality. Themes From Kit Fine*, Oxford UP, to appear .
[17] Goldblatt, R. and S. Thomason, *Axiomatic classes in propositional modal logic*, in: *Algebra and Logic*, Springer Berlin Heidelberg, 1975 .
[18] Grätzer, G., "General Lattice Theory," Birkhäuser Verlag, 1996.
[19] Johnstone, P. T., "Stone spaces," Cambridge UP, 1982.
[20] Jónsson, B., *Universal relational systems*, Math. Scand. **4** (1956).
[21] Jónsson, B., *On the canonicity of Sahlqvist identities*, Studia Logica **53** (1994).
[22] Kihara, H. and H. Ono, *Interpolation properties, Beth definability properties and amalgamation properties for substructural logics*, J. Log. Comp. **20** (2010).
[23] Kurz, A. and J. Rosicky, *The Goldblatt-Thomason theorem for coalgebras*, in: *Proc. CALCO'07* (2007).
[24] Maksimova, L., *Definability and interpolation in non-classical logics*, Studia Logica **82** (2006).
[25] Maksimova, L., *Amalgamation, interpolation, and implicit definability in varieties of algebras*, Proc. Steklov Institute Math. **278** (2012).
[26] Maksimova, L. and E. Orlowska, *The Beth property and interpolation in lattice-based algebras and logics*, Algebra and Logic **47** (2008).
[27] Metcalfe, G., F. Montagna and C. Tsinakis, *Amalgamation and interpolation in ordered algebras*, J. Alg. **402** (2014).
[28] Ono, H., *Interpolation and the Robinson property for logics not closed under the boolean operations*, Alg. Univ. **23** (1986).
[29] Paoli, F., "Substructural Logics: A Primer," Kluwer Academic Publishers, 2002.
[30] Pierce, R., "Introduction to the Theory of Abstract Algebras," Holt, Rinehart and Winston, 1968.
[31] Pigozzi, D., *Amalgamation, congruence-extension, and interpolation properties in algebras*, Alg. Univ. **1** (1971).
[32] Priestley, H. A., *Representation of distributive lattices by means of ordered stone spaces*, Bull. London Math. Soc. **2** (1970).
[33] Priestley, H. A., *Ordered topological spaces and the representation of distributive lattices*, Proc. London Math. Soc. **24** (1972).
[34] Sambin, G. and V. Vaccaro, *Topology and duality in modal logic*, Ann. Pure Appl. Log. **37** (1988).
[35] Sano, K. and M. Ma, *Goldblatt-Thomason-style theorems for graded modal language.*, in: *AiML*, 2010.
[36] Stone, M. H., *The theory of representations for Boolean algebras*, Trans. AMS **40** (1936).
[37] Stone, M. H., *Topological representations of distributive lattices and Brouwerian logics*, Casopis Pest. Mat. Fys. **67** (1937).
[38] Suzuki, T., *The distributivity on bi-approximation semantics*, NDJFL Accepted.
[39] Suzuki, T., *Bi-approximation semantics for substructural logic at work.*, in: *AiML*, 2010.
[40] Suzuki, T., *Canonicity results of substructural and lattice-based logics*, Rev. Symb. Log. **4** (2011).
[41] Suzuki, T., *First-order definability for distributive modal substructural logics: an indirect method* (2011), draft: available on *http://www2.cs.cas.cz/~suzuki/papers.html.*
[42] Suzuki, T., *On canonicity of poset expansions*, Alg. Univ. **66** (2011).
[43] Suzuki, T., *Morphisms on bi-approximation semantics.*, in: *AiML*, 2012.
[44] Suzuki, T., *A Sahlqvist theorem for substructural logic*, Rev. Symb. Log. **6** (2013).
[45] ten Cate, B., "Model Theory for Extended Modal Languages," Ph.D. thesis, University of Amsterdam (2005).
[46] Thomason, S., *Reduction of second-order logic to modal logic*, Math. Log. Quar. **21** (1975).
[47] Urquhart, A., *A topological representation theory for lattices*, Alg. Univ. **8** (1978).

Reasoning About Obligations in *Obligationes*: A Formal Approach

Sara L. Uckelman [1]

Exzellenzcluster "Asia and Europe"
Ruprecht-Karls Universität Heidelberg

Abstract

Despite the appearance of 'obligation' in their name, medieval obligational disputations between an Opponent and a Respondent seem to many to be unrelated to deontic logic. However, given that some of the example disputations found in medieval texts involve Respondent reasoning about his obligations within the context of the disputation, it is clear that some sort of deontic reasoning is involved. In this paper, we explain how the reasoning differs from that in ordinary basic deontic logic, and define dynamic epistemic semantics within which the medieval obligations can be expressed and the examples evaluated. Obligations in this framework are history-based and closely connected to action, thus allowing for comparisons with, e.g., the knowledge-based obligations of Pacuit, Parikh, and Cogan, and stit-theory. The contributions of this paper are twofold: The introduction of a new type of obligation into the deontic logic family, and an explanation of the precise deontic concepts involved in *obligationes*.

Keywords: deontic logic, dynamic epistemic logic, obligation, *obligationes*, stit.

1 Introduction

Deontic logicians who are interested in the history of their field may upon first introduction to the medieval genre of *disputationes de obligationibus* think they have found their ancestor: For what else could treatises on "disputations concerning obligations" be about other than reasoning about obligation and permission, i.e., deontic logic? Closer inspection of these disputations, however, may lead the deontic logician to a place of puzzlement, for the example disputations which can be found in the treatises often have little or nothing to do with obligation, permission, commitment, or any of the related notions which make up the core of deontic logic. On an initial survey, it is unclear what the obligation involved in these disputations is or how they are related to reasoning about deontic principles in general—if at all. In fact, many contemporary scholars of *obligationes* explicitly disavow any connection between obligational

[1] sara.uckelman@asia-europe.uni-heidelberg.de

disputations and deontic logic [18,19,25].[2] Those who do take seriously the notion of 'obligation' in terms of discursive commitments primarily focus on how these obligations function at the meta-level of the disputation and disputational norms, [8,9,10,13,14,23]. To date, little work has been done on the formal nature of the obligations involved in *obligationes*.

There are two participants in an obligational disputation, Opponent and Respondent.[3] Opponent begins by putting forth ('positing') a proposition which Respondent either admits or does not admit. (If he does not admit it, then no disputation begins.) Propositions in the disputation are divided into two types, relevant (*pertinens*) and irrelevant (*impertinens*). A proposition is relevant if either it or its negation is a logical consequence of the set of propositions which have already been admitted (in the initial round of the disputation) or conceded (in a later round of the disputation) along with the negations of those denied, and it is irrelevant otherwise. The relevant propositions are typically further divided into those which are 'relevant and following' (*pertinens sequens*) and those which are 'relevant and contradictory' (*pertinens repugnans*). The typical rule is that any relevant following proposition must be conceded, any relevant contradictory proposition must be denied; and any irrelevant proposition must be conceded if it is known to be true, denied if known to be false, and doubted if neither (where doubting is taken as a neutral action).

These disputations, so simple to describe, have nevertheless been a matter of contention amongst contemporary scholars—and not just because it is unclear what (if anything) the relation is between them and deontic logic. Another puzzling feature of these disputations is that they often appear to be empty of content [18]; the examples that occur in the medieval treatises do not involve any substantive doctrinal issues. Instead, the propositions which appear in many examples have a feeling of genericity to them: When Opponent puts forward that Socrates is white or that Plato is black, or that Socrates and Plato have the same color, it is clear that what is at stake is not the substantive question of what color two ancient Greek philosophers were. Likewise, when Respondent denies that a human being is a donkey or concedes that God exists, he is not making a point about biology or theology. Instead, these propositions are functioning as arbitrary contingent, impossible, and necessary propositions, whose content is less important than their modal status.

But not all of the disputations are like this. In this paper we examine some examples where the propositions put forward are not about Socrates, Plato,

[2] [12] does not explicitly reject a connection, but does so implicitly by not mentioning *obligationes* at all.

[3] Because we do not have the space to go into the complexities of the historical situation, we vastly oversimplify here. What we say here is true of the type of *obligatio* called *positio* 'positing', and of the rules for *positio* given by Walter Burley in his treatise on *obligationes*, c.1302. Readers interested in the historical information are directed to [21,22] for further information. In this paper, we present only those portions of the medieval theory which are necessary to motivate the proposed formalization.

donkey-humans, or God. Instead, the propositions Opponent posits have to do with Respondent's own responses, both the responses that he does make and the responses he ought to make. This brings deontic reasoning directly into the disputational framework, forcing Respondent to reason about his obligations explicitly within the object language of the disputation.

2 *Obligationes* and deontic logic

In the brief description above of the rules that Respondent must follow when responding to Opponent's propositions, it is clear from the outset that there are three different places where deontic concepts such as "must" or "ought to" come in—one for each of the different types of actions Respondent can make. What are these obligations rooted in? They arise when Respondent admits the *positum* and thus binds himself to play by the rules. That is, they are generated by his actions. This is true for the remainder of the disputation, in that it is his actions which give rise to further obligations which he must abide by. In this, the basic type of obligation that is involved in *obligationes* differs relevantly from those in ordinary deontic logic. In the dominant approach to deontic logic [2, p. 23], the operator 'it is obligatory that' attaches to propositions, and no agent is specified, because it is expressed in the passive voice. (Such approaches violate the Restricted Complement Thesis, which will appear below in §4.) When $\mathrm{O}\,\varphi$ is asserted in a deontic logic context, there is no indication of whose responsibility it is to see to it that φ is the case, or what actions can or should be taken in order to meet the obligation.[4] In contrast, in an *obligatio*, the obligation exists between an agent and an action.

Given this, there is a sense in which those who disavow any connection between *obligationes* and deontic logic are right; it is certainly true no direct comparison can be done. However, it is still possible to ask how these two types of obligation, albeit different, compare to each other, and the answer sheds interesting light on the question of what the obligation in an *obligatio* is. Boh briefly considered this question in [5]. He offers the following formalization of "the three most general constitutive rules of (the main type of) an obligational disputation called *positio*" [5, p. 112]:[5]

- $\forall\varphi(\mathrm{K}_a\,\varphi \to \mathrm{O}_a\,\mathrm{C}\,\varphi)$
- $\forall\varphi(\mathrm{K}_a\,\neg\varphi \to \mathrm{O}_a\,\mathrm{N}\,\varphi)$
- $\forall\varphi((\neg\,\mathrm{K}_a\,\varphi \wedge \neg\,\mathrm{K}_a\,\neg\varphi) \to \mathrm{O}_a\,\mathrm{D}\,\varphi)$

According to Boh, "Rule (1) is read: for any proposition which is or might be put forward in a disputation in which the person a takes part, if a knows that φ, it is obligatory that he grants it...'O', of course, represents the deontic operator of obligation" [5, p. 112]. The other rules are to be understood analogously. Unfortunately, inspection of the medieval texts will quickly show that these

---

4 Our use of "see to it that" here is not accidental. We discuss the relationship between our approach and stit-theory in §6.

5 We have slightly adapted his notation to be consistent with the notation used in the current paper (see Def. 4.1), as well as corrected errors in parentheses in the second two formulas.

are untenable as correct formulations of the principle duties of Respondent in *positio*. Virtually every author writing on this type of obligation points out explicitly that it is only worthwhile to pursue a *positio* when the *positum* is (known to be) false. Thus, Respondent upon admitting such a *positum* will immediately violate (2), by admitting and conceding a proposition which is known to be false.[6]

Let us take a step back and ask a more general question: In what way are the obligations in *obligationes* like the obligations in deontic logic? For a given deontic logical language $\mathcal{L}_d$, the minimal deontic logic is axiomatized by the following schemata [24, p. 274]:

(i) All axiom schemata and rules of classical logic over $\mathcal{L}_d$.

(ii) $\mathrm{O}\,\varphi \wedge \mathrm{O}\,\psi \rightarrow \mathrm{O}(\varphi \wedge \psi)$

(iii) $\neg\,\mathrm{O}\,\neg\top$

and the rule of inference:

$$\frac{\varphi \Rightarrow \psi}{\mathrm{O}\,\varphi \Rightarrow \mathrm{O}\,\psi}$$

If we interpret the O above as "ought to concede", then the second schema says that if Respondent ought to concede two conjuncts individually in their own right, then he also ought to concede their conjunction. This is true in the obligational setting only in certain cases. There are three possibilities: φ and ψ are both relevant and following, φ and ψ are both irrelevant and known to be true, and one of φ or ψ is relevant and following and the other is irrelevant and known to be true. In the first two cases, if Respondent ought to concede φ and he ought to concede ψ, then he also ought to concede $\varphi \wedge \psi$. However, the third case fails. Suppose that Respondent ought to concede φ because it is relevant and following, even though in reality it is known to be false. Suppose that Respondent ought to concede ψ because it is irrelevant and known to be true. Then consider $\varphi \wedge \psi$. The conjunction is neither relevant and following nor relevant and contradictory, since in either case, ψ would then also be relevant. Thus, $\varphi \wedge \psi$ is irrelevant, and since φ is known to be false, the conjunction is known to be false as well. As a result, not only is Respondent not obligated to concede the conjunction, he is obligated to deny it. A similar story shows that the rule of inference also fails. Suppose that $\varphi \rightarrow \psi$ and Respondent ought to concede φ. If he ought to concede φ because it is relevant and following, then it is true he also ought to concede ψ, because whatever φ follows from, ψ follows from as well, if $\varphi \rightarrow \psi$. However, suppose φ is irrelevant, but known to be true. If ψ is relevant in its own right, then it is true that Respondent ought to concede ψ, independently of his obligations towards φ. But if ψ is irrelevant, it is possible that $\varphi \rightarrow \psi$ but Respondent does not know that ψ is true. In such a case, his obligation to ψ is to doubt it, not to concede it. (In such a

[6] Burley gives an example which violates Boh's reconstruction: "It should be said that 'My hand is not closed' must be denied even though it is true. And that is because it is incompatible, since its opposite was previously granted" (earlier in the disputation) [7, 3.30].

way, logical omniscience does not arise in *obligationes*, as noted by Uckelman [21, p. 20].)

However, a modified version of this schema does hold: If Respondent *has in fact* conceded φ, and likewise he has also conceded ψ, then it is true that he ought to concede their conjunction: His previous actions have changed his obligations with respect to $\varphi \wedge \psi$. This turns out to be the crucial insight into understanding how the obligations in *obligationes* differ from the ones in ordinary deontic logic. To state the (perhaps) obvious: Actions change commitments. A person's commitments in a given situation are influenced both by what he does and what he knows. These obligations are, to borrow a term from game-theory, *history-dependent*. (We discuss this further in §6.)

3 An example

Before we turn to the technical details, let us consider a concrete example, due to Walter Burley. After he lists the rules for *positio*, Burley considers various objections to these rules in the forms of sophisms, or logically problematic sentences, where it appears that Respondent has no consistent way of responding in a disputation beginning with such a sophism. The third objection to the first rule[7] runs as follows:

Objection 3.1 *Let this be posited: 'You are in Rome or that you are in Rome must be granted'. Next, let 'That you are in Rome must be granted' be proposed. This is false and irrelevant; therefore it must be denied. Next, let 'That you are in Rome follows from the positum and the opposite of a proposition already correctly denied' be proposed. This is necessary, because this conditional is necessary: 'If either you are in Rome, or that you are in Rome must be granted, but that you are in Rome is not to be granted, then you are in Rome.' Once this has been granted—'That you are in Rome follows from the positum and the opposite of a proposition already correctly denied'—let this be proposed: 'That you are in Rome must be granted'. If you grant this, you have granted and denied the same thing; therefore [you have responded badly]. If you deny it, the disputation is over; you have denied what follows according to a rule. Because if the rule is good, this follows: 'That you are in Rome follows from the positum and the opposite of a proposition already correctly denied; therefore that you are in Rome must be granted' [7, 3.21].*

This is presented as an objection to the rule because—if it is a correct description of how to proceed in such a case—it conflicts with the general principle that a Respondent who follows the rules correctly will never be forced to concede a contradiction. We briefly sketch in an informal fashion why one might think the objection correctly describes how Respondent should proceed. The

---

[7] The first rule is: "Everything that follows from the positum must be granted. Everything that follows from the positum either together with an already granted proposition (or propositions), or together with the opposite of a proposition (or the opposites of propositions) already correctly denied and known to be such, must be granted" [7, 3.15]. Different translators use 'granted' and 'conceded' to translate Latin *concedendum*.

positum takes the form of a disjunction, $p \vee q$, with the assumption being that both disjuncts are false.[8] Next, one disjunct, q, is put forward. Neither p nor q follow from $p \vee q$ alone in the absence of other information (and we have been provided with no such additional information), and q is, by assumption, false. Thus, q should be denied. If q is denied, then p follows from this denial along with the earlier concession of $p \vee q$, by simple application of disjunctive syllogism. But $p \vee q$ is the *positum*, and if it is granted that p follows from this along with the opposite of something correctly denied, namely q, this is tantamount to saying that p ought to be granted—and this fact itself is q, which was previously denied.

Three points should be noted. First, we reiterate what we said above about the apparent contentless nature of the disputations. Whether Respondent is in Rome is a matter of little import.[9] This example should be understood not as an example of a specific dialogue, but of a template into which any contingent falsehood whatever can be substituted in for "You are in Rome".

Second, we equated the "opposite of something correctly denied" with that proposition's contradictory negation. This is not a problematic equation, but nevertheless it is one that should be noted explicitly.

Third, when we equated the *positum* with a simple disjunction of the form $p \vee q$, this overlooked the internal structure of the two disjuncts, in particular the fact that p occurs in the second disjunct as well. The only way that we can take this internal structure into account is if we can express the notion of obligation embedded in 'must be granted' and 'must be denied' within the object language itself, not only in the metalanguage. Extending the ordinary object language of *obligationes* so that it can handle cases like this one will shed light on the type of obligation involved, and also how one can use the *obligationes* framework in order to carry out (a certain limited type of) deontic reasoning. Because we must first have the sophisticated framework in place before we can understand how the example works, we defer discussion of Burley's solution to the objection to §5.

4 An abstract model for *obligationes*

In this section we define an abstract model, built on the framework defined in [21] for Burley-style *obligationes*, within which specific examples, such as the one above, can be analysed. The language is standard multi-agent dynamic epistemic language, with two agents O(pponent) and R(espondent) and three actions:

Definition 4.1 Let φ_n be a proposition put forward by O. R's available ac-

[8] Though it should be noted that the assumption is that Respondent is in fact not in Rome; Burley wrote the treatise in Oxford, and thus he and his potential opponents and respondents were located there as well.

[9] See footnote 8.

tions are C (for *concedendo*), N (for *negando*), and D (for *dubitando*):

$$\begin{aligned} \mathrm{C}\,\varphi_n &:= [\varphi_n]\top \\ \mathrm{N}\,\varphi_n &:= [\neg\varphi_n]\top \\ \mathrm{D}\,\varphi_n &:= [\top]\top \end{aligned}$$

Well-formed formulas of this language are defined in the usual way, and these formulas are interpreted on epistemic Kripke models, again as is standard. We give only the semantics for the three actions. For an epistemic Kripke model $\mathfrak{M}$, let

$$\mathfrak{M}\restriction\varphi = \langle W^{\mathfrak{M},\varphi}, \{\sim_a^{\mathfrak{M},\varphi} : a \in \mathcal{A}\}, V^{\mathfrak{M},\varphi}\rangle,$$

where $W^{\mathfrak{M},\varphi} := \{w \in W : \mathfrak{M}, w \models \varphi\}$, and the relations and valuation functions are just restrictions of the originals. For an ordered set of propositions Γ_n, let $\mathfrak{M}\restriction\Gamma_n = \mathfrak{M}\restriction\gamma_0\restriction\ldots\restriction\gamma_n$, that is, $\mathfrak{M}\restriction\Gamma_n$ is the result of the sequential restriction of $\mathfrak{M}$ by the elements of Γ_n. Then:

Definition 4.2 $\mathfrak{M}, w \models [\varphi]\psi$ iff $\forall v \in \mathfrak{M}\restriction\varphi, v \models \psi$.

When Respondent announces "I concede φ", the model is reduced to only those worlds where φ is true, and when he announces "I deny φ", the model is reduced to only those worlds where φ is false. The action of doubt is strongly agnostic: If Respondent is in doubt about a particular proposition, and he announces this $\mathfrak{M}\restriction\top = \mathfrak{M}$, i.e., nothing changes. Given these semantics, the actions defined above can be understood as tests for consistency. If Respondent's actions are correct, then after his announcement there will be at least one world left in the model. This captures the fact, noted by Burley himself, that if Respondent responds according to the rules (which we define formally below), then he will never be led into inconsistency, that is, into an empty model.

From an abstract perspective, an obligational disputation can be seen as a pair of sequences, one of propositions which are put forward by Opponent and the other is the actions of Respondent[10], along with a rule which indicates the specific type of disputation the *obligatio* is.

Definition 4.3 An *obligatio* is a quadruple $\mathfrak{O} = \langle\Theta, R, \Gamma^R, \Gamma\rangle$ where

- Θ is a sequence of propositions, such that $\theta_0 \in \Theta$ is the initial proposition and $\theta_n \in \Theta$ is the proposition put forward by O at round n.
- $R : \Theta \times \mathbb{N} \to \mathrm{Act}$ is a function determining R's correct response to each element of Θ. We write $R(\theta_n)$ for $R(\theta, n)$ to simplify notation.
- Γ^R is a sequence of elements from Act, formed by the *correct* response of R to each element in Θ, as given by R. That is:

$$\begin{aligned} \Gamma_0^R &= \langle R(\theta_0)\rangle \\ \Gamma_n^R &= \langle \gamma_0, \ldots, \gamma_{n-1}, R(\theta_n)\rangle \end{aligned}$$

Whether Γ^R is unique depends on R.

[10] We could also view it as a single sequence, formed of these two sequences interleaved.

- Γ is a sequence of elements from Act, formed by R's actual responses to each element of Θ.

Note that there are no constraints on Γ: In principle, Respondent is free to respond in any fashion that he likes (so long as it is one of the actions 'concede', 'deny', or 'doubt'). In practice, however, if Respondent's responses are not directed to the immediately preceding proposition of Opponent, then it makes more sense to think of him not actually participating in an *obligatio* disputation, rather than participating but failing wholly to play by the rules. We do not add this as a formal constraint, but simply note that we do not consider cases where Respondent acts in such a mulish, recalcitrant fashion.

We formalize Burley's rules for *positio*, presented informally in §1, as follows:

Definition 4.4 For a model $\mathfrak{M}$ and formula $\theta_0 \in \Theta$:

$$R^{\text{Bur}}(\theta_0) = \begin{cases} \text{C}\,\theta_0 & \text{iff } \mathfrak{M}, w \vDash \langle\theta_0\rangle\top \\ \text{N}\,\theta_0 & \text{otherwise} \end{cases}$$

For $\theta_n \in \Theta, n > 0$:

$$\begin{array}{ll} \text{If } \mathfrak{M}\lceil\Gamma_{n-1} \vDash \theta_n: & R^{\text{Bur}}(\theta_n) = \text{C}\,\theta_n \\ \text{If } \mathfrak{M}\lceil\Gamma_{n-1} \vDash \neg\theta_n: & R^{\text{Bur}}(\theta_n) = \text{N}\,\theta_n \\ \text{Otherwise, let } w^* \text{ be the actual world, and:} & \\ \quad \text{If } \mathfrak{M}, w^* \vDash \text{K}_\text{R}\,\theta_n: & R^{\text{Bur}}(\theta_n) = \text{C}\,\theta_n \\ \quad \text{If } \mathfrak{M}, w^* \vDash \text{K}_\text{R}\,\neg\theta_n: & R^{\text{Bur}}(\theta_n) = \text{N}\,\theta_n \\ \quad \text{If } \mathfrak{M}, w^* \vDash \neg(\text{K}_\text{R}\,\theta \vee \text{K}_\text{R}\,\neg\theta_n): & R^{\text{Bur}}(\theta_n) = \text{D}\,\theta_n \end{array}$$

If Respondent denies θ_0, then no *obligatio* begins. We represent this by having him deny the *positum*, which has the effect of 'canceling' the model and not allowing any further progress. Admitting the *positum*, which triggers the start of the disputation, is equivalent in effect, if not exactly in action, to conceding it. For an *obligatio* according to these rules, Γ^R will always be uniquely defined.

These rules provide the grounding for the obligations in the disputation. Respondent is obliged to follow the rules, or risk the charge of having responded badly, and hence 'losing' the disputation. Thus, Respondent's obligations in a *positio* according to the rule defined in Def. 4.4 are:

Definition 4.5 In an *obligatio* $\mathfrak{O}$ at round n

- R ought to concede φ iff either $\mathfrak{M}\lceil\Gamma_{n-1} \vDash \varphi$; or $\mathfrak{M}\lceil\Gamma_{n-1} \nvDash \varphi$, $\mathfrak{M}\lceil\Gamma_{n-1} \nvDash \neg\varphi$, and $\mathfrak{M}, w^* \vDash \text{K}_\text{R}\,\varphi$.
- R ought to deny φ iff either $\mathfrak{M}\lceil\Gamma_{n-1} \vDash \neg\varphi$; or $\mathfrak{M}\lceil\Gamma_{n-1} \nvDash \varphi$, $\mathfrak{M}\lceil\Gamma_{n-1} \nvDash \neg\varphi$, and $\mathfrak{M}, w^* \vDash \text{K}_\text{R}\,\neg\varphi$.
- R ought to doubt φ iff $\mathfrak{M}\lceil\Gamma_{n-1} \nvDash \varphi$, $\mathfrak{M}\lceil\Gamma_{n-1} \nvDash \neg\varphi$, and $\mathfrak{M}, w^* \vDash \neg(\text{K}_\text{R}\,\varphi \vee \text{K}_\text{R}\,\neg\varphi)$.

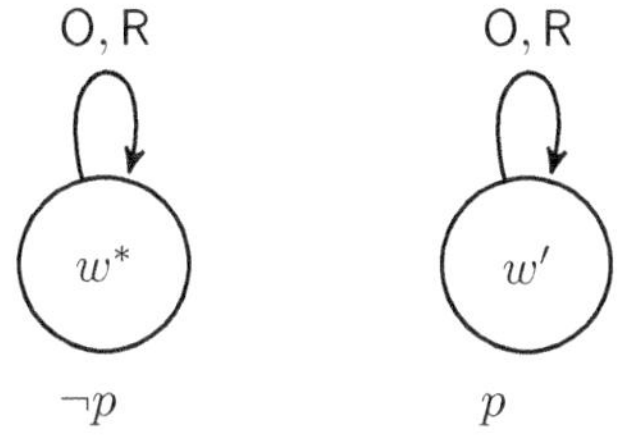

Fig. 1.

These obligations provide us with the truth conditions for an object-language obligation operator O_a 'a ought to' which can be applied to actions. That is, our O operator satisfies the Restricted Complement Thesis [3, p. 787], which requires that "the deontic constructions such as obligation, prohibition, and permission must take agentives as their complements". While only actions can be obligatory (and not propositions), for any proposition, it is in principle possible that Respondent be obliged to conceded, deny, or doubt that proposition.

Definition 4.6 Fix an *obligatio* $\mathfrak{O}$ and model $\mathfrak{M}$.

$$\begin{array}{lll} \mathfrak{M}\restriction\Gamma_{n-1}, w \models \mathrm{O}_{\mathsf{R}}\,\mathrm{C}\,\varphi_n & \text{iff} & \text{either } \mathfrak{M}\restriction\Gamma_{n-1} \models \varphi \\ & & \text{or } \mathfrak{M}\restriction\Gamma_{n-1} \not\models \varphi, \mathfrak{M}\restriction\Gamma_{n-1} \not\models \neg\varphi \\ & & \text{and } \mathfrak{M}, w^* \models \mathrm{K}_{\mathsf{R}}\,\varphi \\ \mathfrak{M}\restriction\Gamma_{n-1}, w \models \mathrm{O}_{\mathsf{R}}\,\mathrm{N}\,\varphi_n & \text{iff} & \text{either } \mathfrak{M}\restriction\Gamma_{n-1} \models \neg\varphi, \\ & & \text{or } \mathfrak{M}\restriction\Gamma_{n-1} \not\models \varphi, \mathfrak{M}\restriction\Gamma_{n-1} \not\models \neg\varphi \\ & & \text{and } \mathfrak{M}, w^* \models \mathrm{K}_{\mathsf{R}}\,\neg\varphi \\ \mathfrak{M}\restriction\Gamma_{n-1}, w \models \mathrm{O}_{\mathsf{R}}\,D\varphi_n & \text{iff} & \mathfrak{M}\restriction\Gamma_{n-1} \not\models \varphi, \mathfrak{M}\restriction\Gamma_{n-1} \not\models \neg\varphi \\ & & \text{and } \mathfrak{M}, w^* \models \neg(\mathrm{K}_{\mathsf{R}}\,\varphi \vee \mathrm{K}_{\mathsf{R}}\,\neg\varphi) \end{array}$$

Note that the obligations are global: They do not depend on the world of evaluation.

We are now in a position to formalize and evaluate the *positio* given in §3. Let $p :=$'R is in Rome'. Then, the *positum* $\theta_0 = p \vee \mathrm{O}_{\mathsf{R}}\,\mathrm{C}\,p$, and the rest of Θ for the *positio* is as follows:

$$\begin{aligned} \theta_1 &:= \mathrm{O}_{\mathsf{R}}\,\mathrm{C}\,p \\ \theta_2 &:= (\theta_0 \wedge \neg\theta_1) \rightarrow p \\ \theta_3 &:= \mathrm{O}_{\mathsf{R}}\,\mathrm{C}\,p \end{aligned}$$

Consider the model in Figure 1, with two worlds, one where Respondent is in Rome and one where he is not; the latter is the actual world, and both participants know this.

Proposition 4.7 *The* positum $\theta_0 = p \vee \mathrm{O}_{\mathsf{R}}\,\mathrm{C}\,p$ *is false at the actual world because both disjuncts are false.*

Proof. (1) $\mathfrak{M}, w^* \models \neg p$. (2) $\mathfrak{M} \not\models p$, because of w^*, and $\mathfrak{M} \not\models \neg p$, because of w'. $\mathfrak{M}, w^* \models \neg p$, so $\mathfrak{M}, w^* \models \neg\,\mathrm{K}_{\mathsf{R}}\,p$. Hence, $\mathfrak{M} \not\models \mathrm{O}_{\mathsf{R}}\,\mathrm{C}\,p$. □

However, the *positum* is not inconsistent (since $w' \models p$), so Respondent is correct in admitting it, and the resulting model $\mathfrak{M}|\theta_0$ is displayed in Figure 2.

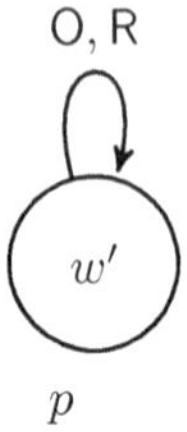

Fig. 2. $\mathfrak{M}|\theta_0$

Opponent then posits $\mathsf{O}_\mathsf{R}\,\mathsf{C}\,p$. Immediately we can see where the objection Burley considers has gone wrong. The objection says that $\mathsf{O}_\mathsf{R}\,\mathsf{C}\,p$ "is false and irrelevant; therefore it must be denied". However, once $p \vee \mathsf{O}_\mathsf{R}\,\mathsf{C}\,p$ has been admitted, $\mathsf{O}_\mathsf{R}\,\mathsf{C}\,p$ is no longer irrelevant. The only situations where Respondent ought to concede p are cases where either p is true in all remaining worlds of the model *or* where p is known at the actual world. Because we start from the assumption that p is false at the actual world, it cannot be known there, which means that after conceding θ_0, the only worlds that remain are worlds which were retained because they made p true. Hence, whatever the initial model was like, after conceding $p \vee \mathsf{O}_\mathsf{R}\,\mathsf{C}\,p$, $\mathsf{O}_\mathsf{R}\,\mathsf{C}\,p$ becomes relevant. Thus, the objection fails at round 1.

5 Burley's solution

It is of course interesting that we are able to use our formalization to identify a problem with the objection. But the only way this will go beyond merely *interesting* is if the problem that we have identified is the same problem that Burley identifies; and if not, does it tell us something about the problem he identifies? This is one test of the adequacy of the framework. It turns out that Burley solves the objection differently than we did in the previous section, in the following way:

Solution 5.1 (To Objection 3.1) *This must be denied: 'That you are in Rome follows from the positum and the opposite of a proposition correctly denied'; it is not necessary either. Even if it is necessary that from the posited disjunction together with the opposite of one disjunct it follows that you are in Rome, it is nonetheless not necessary that the disjunction be posited [7, 3.22].*

Few modern commentators have discussed either the objection or this response to it, and those who have find it puzzling. Stump says that this solution to the problem "looks bizarre", and that:

> On the face of it, then, Burley is saying that $[\theta_2]$ is to be denied because it is not necessary, and his reason for claiming that $[\theta_2]$ is not necessary is that one of the premisses it is derived from, namely, $[\theta_0]$ is not necessary [20, p. 324].

But Stump is correct in noting that "there is no obligational rule to the effect that we must deny any propositions which are not necessary" [20, p. 324], and this is clearly not what Burley is arguing here. One way to understand what he is saying is to look at another objection and response that he presents, since the second response makes much the same point:

Objection 5.2 *Let this be posited: 'Nothing is posited to you.' Next, let 'Everything that follows from the positum must be granted' be proposed. This must be granted because it is a rule. Next, let 'Something follows from the positum' be proposed. This follows and therefore must be granted. Next, let 'Something is posited' be proposed. If you grant this, you grant the opposite of the positum; therefore, [you have responded] badly. If you deny it, you deny something that follows, because this follows: 'Something follows from the positum; therefore something is posited' [7, 3.17].*

Solution 5.3 (To Objection 5.2) *One says that this must be denied: 'Everything that follows from the positum must be granted'; it is not necessary either. But this is necessary: If something follows from the positum, it must be granted [7, 3.19].*

In both solutions, Burley identifies a proposition in the disputation which Respondent had conceded but which in fact he ought to have denied, and justifies this fact by an appeal to a lack of necessity. Stump offers a different interpretation of the solution to Objection 3.1 on which "an unintelligible solution [turns] into an intelligible red herring" [20, p. 325]. When Burley says "'That you are in Rome' follows from the *positum* and the opposite of a proposition correctly denied" must be denied, Stump argues that Burley takes an "extreme" [20, p. 327] stance with respect to the phrase 'the *positum*', treating it as ambiguous because it may refer to the *positum* in the disputation at hand, or it may refer to any other potential *positum*. Thus, when Burley says "it is not necessary that the disjunct be posited either", he is pointing out that if another proposition had been the *positum*—which is perfectly reasonable—then none of the other propositions put forward would follow. This interpretation is bolstered by Burley's solution to yet another objection (which we do not discuss), in which he makes a distinction between "You have denied something that follows from the *positum*, therefore [you have responded] badly" and "You have denied something that follows from what was posited to you, therefore [you have responded] badly" [7, 3.20]. The only way that these can differ is if 'the *positum*' can be interpreted more generically than 'what was in fact posited to you'.[11]

There are two ways to react to this interpretation. On the one hand, calling this a red herring may not be incorrect, since on this interpretation the solution doesn't appear to address the issue we identified in the previous section, namely, that given the relationship between admittance and concession, once

[11] It is worth remembering here that Latin does not have definite or indefinite articles, and that occurrences of 'the *positum*' are better read as simply '*positum*' without any article.

the *positum* of Objection 3.1 has been admitted, the second disjunct becomes relevant.[12] On the other hand, calling it a mere red herring is dismissive, implying that what Burley is demonstrating here is in some sense beside the point. But this is not the case. In fact, what Burley is pointing out in each of the three solutions is a genuine feature of obligational disputations—namely, that until the *positum* is admitted, no obligation exists. The rules themselves that Respondent can bind himself to follow are not themselves necessary. They are not encoded in the models, and while Respondent ought to follow the rules, he is not forced to, because sometimes he can make mistakes. What Burley is pointing out is the fact that Respondent is not under obligation until he admits the *positum*; only after that does it become necessary that he grant whatever it is that does in fact follow from what was posited to him initially. It is precisely that the obligations can be violated—that they are weaker than necessity—which our approach illustrates so clearly. Thus, while the primary problem with the objections that our framework identifies is not the one Burley identifies, our model nevertheless sheds light on the solutions he does give, and provides an explanation which has hitherto been elusive.

6 Comparison with related work

In §2 we briefly explained how the axioms and inference rule of the minimal deontic logic fail to obtain in the context of obligational reasoning. In this section, we compare what we have introduced above to two other well-known approaches which combine action and obligation, knowledge-based obligations and stit-theory.

6.1 Knowledge-based obligations

In [15], Pacuit, Parikh, and Cogan (PPC) introduce a semantics for knowledge-based obligations which model the interaction between knowledge and obligation in history-based models such as those of [16,17]. These obligations bear a strong resemblance to the obligations in *obligationes*, and in fact much of what we developed above was inspired by the PPC approach. Nevertheless, there are some important differences, which we highlight.

PPC models are composed of sets of events indexed to each agent, and a set of all global possible histories of these event-sets. Obligations in these models are expressed by the introduction of a value function for histories which assigns a real number to each global infinite history such that higher-valued histories are considered 'better' than lower-valued ones [15, p. 321]. An agent is then obliged to perform a certain action at a certain time if any maximal extension of the agent's local history is one where that action is considered 'good' [15, Def. 4.2].

The first difference is conceptual. The goal of [15] was to explain how the creation of new knowledge can engender obligations for an agent that he did not previously have—that is what makes the obligations of PPC *knowledge-*

[12] Stump ultimately recognizes this fact [20, p. 326].

based. In our case, we also want to represent a type of dependency but it is based on action rather than knowledge. We could say that the obligations in *obligationes* are knowledge-based, but the knowledge is rooted in the actions of Respondent. Thus, the dependency on knowledge is only derivative; the dependency on action is primary.

The second is structural, relating to the shape of the models. History-based models are Kripke frames built from a set of events which are organized into sequences, linear histories which themselves make up a branching structure. The models distinguish between local (agent-indexed) histories and global (agent-indifferent) histories, and there is a global clock which is used to keep the histories synchronous. In our framework, there is no history parameter which is internal to the model. There is no temporal reasoning or actions which happen within a given model; all of the action happens at the level of transitioning from one model to another, via the public concession, denial, and doubting of propositions by Respondent. These transitions are tracked at the level of the *obligatio*, with the parameter Γ, rather than at the level of the model that the *obligatio* is evaluated against. If we consider the sequence of models that a particular Γ gives rise to as forming the history of the disputation in which Respondent's obligations are grounded, then it is clear that we do not have to distinguish between local histories and global histories, because both participants are aware of every move made in the disputation; there is no uncertainty as to whether, e.g., Opponent put forward a certain proposition, or whether Respondent conceded it. By keeping the 'histories' separate from the models, we are easily able to explain what grounds the high value of Γ_R, which by definition (Def. 4.3) is the history which follows the rules R.

Additionally, doing this obviates the need for a global clock: Each move in the disputation functions as a 'clock tick', in one sense. The fact that the 'clock' is external to the model reflects one of the structural properties of *obligationes* which we haven't otherwise discussed in this paper, namely, the often-stated rule that "all responses must be directed to the same instant" [7, 3.84]. That is, the determination of the correct action for irrelevant propositions should be done with respect to a single, fixed moment—in our set-up, the actual world in the initial model—so as to prevent Respondent from being forced to concede "Opponent is sitting" put forward while Opponent is in fact sitting, but then having to deny it when Opponent is no longer sitting.

6.2 stit-theory

Since its introduction in [4], the 'seeing to it that' approach to agency has proven a rich source of tools for reasoning about agency and action. It is no surprise that many people have used different flavors of stit-operators to define various explicit obligation operators. We survey only a few of these approaches here and do not make any claim to completeness.

Similar to the PPC approach, in branching stit frames obligations are generally expressed by the addition of an *Ought* function which maps each moment m to a subset of $H(m)$, the set of histories containing m. More fine-grained notions of obligation can be expressed by placing valuations on the histories, allowing histories to be compared to each other as being better or worse. Deontic operators defined in these ways lie between historical necessity and historical possibility [11, p. 616], that is, if it is necessary for Respondent to concede φ, then he ought to concede φ; and if he ought to concede φ, then it is possible for him to concede φ. In another approach ([6], adapted from [1]), "an agent is obliged to do something if and only if by not doing it, it performs a violation" [6, p. 55]. This informal description of obligation is consistent with the conception of obligation in *obligationes*, whereby Respondent is required to do that which will keep him from responding badly. Formally, the obligation operator is defined as follows:

$$O[a\mathsf{xstit}]\varphi := \Box(\neg[a\mathsf{xstit}]\varphi \to [a\mathsf{xstit}]V)$$

where V is a violation.

Belnap sees stit-theory as "a powerful alternative to two different programs". The first program we already saw in §2, and relates a deontic operator to a proposition. The second relates deontic operators to actions. Belnap criticizes this approach because it "does not offer us a logical point of view from which it is easy and natural to see that obligation, etc., can in fact make at least *subordinate* reference to declaratives" [2, p. 23]. He says that stit :

> make[s] it easy to see that obligation must take an imperative, and also easy to see the important truth that any declarative whatsoever can give rise to an imperative [2, p. 23].

The same can be said of the actions and obligations in *obligationes*: Since Opponent can put forward any sentence whatsoever, not only can any declarative give rise to an imperative, any (well-formed within the language) imperative can as well! Not only can Opponent put forward φ, giving rise to $\mathrm{O}_{\mathsf{R}}\, X\varphi$ for $X \in \{\mathrm{C}, \mathrm{N}, \mathrm{D}\}$, but he can also put forward $\mathrm{O}_{\mathsf{R}}\, X\varphi$, giving rise to $\mathrm{O}_{\mathsf{R}}\, X(\mathrm{O}_{\mathsf{R}}\, X\varphi)$.

Nevertheless, conceptually there is a large difference between obligations in stit-theory and obligations in *obligationes*, and that is that stit-theory, despite being about imperatives rather than declaratives, is still oriented towards the resulting proposition, not towards the action which causes the result. To put it simply, $\mathsf{Rstit}\varphi$ encodes an arbitrary action with a definite outcome—the sentence says nothing about what action R must take in order to see to it that φ is true, only that φ's being true must result, if the stit sentence is true. In an *obligatio*, on the other hand, $\mathrm{C}\,\varphi$ encodes an arbitrary outcome with a definite action—this sentence says nothing about what the result is, only what action Respondent must take. Given his preceding actions, at each round there is a definite action which Respondent is obligated to perform, regardless of the consequences performing that action might have.

7 Conclusion

Historically, many people have disavowed any connection between the medieval genre of *obligationes* and deontic logic in general. What we have shown here is that on the contrary, the obligational setting provides an interesting framework for reasoning about dialectical obligations in an agent-directed rather than a proposition-directed way. The approach has features in common with both knowledge-based obligations of PPC and with various types of obligations developed in conjunction with stit, but differs from both in important ways by focusing on the specific actions which give rise to the obligations. Thus, even if it were stripped of its medieval trappings, the present framework provides an interesting starting point for further exciting investigations.

References

[1] Bartha, P., *Conditional obligation, deontic paradoxes, and the logic of agency*, Annals of Mathematics and Artificial Intelligence **9** (1993), pp. 1–23.

[2] Belnap, N., *Declaratives are not enough*, Philosophical Studies **59** (1990), pp. 1–30.

[3] Belnap, N., *Backwards and forwards in the modal logic of agency*, Philosophy and Phenomenological Research **51** (1991), pp. 777–807.

[4] Belnap, N. and M. Perloff, *Seeing to it that: A canonical form for agentives*, Theoria **54** (1988), pp. 175–199.

[5] Boh, I., "Epistemic Logic in the Later Middle Ages," Routledge, 1993.

[6] Broersen, J., *A complete stit logic for knowledge and action, and some of its applications*, in: M. Baldoni, T. C. Son, M. B. van Riemsdijk and M. Winikoff, editors, *DALT 2008*, 2009, pp. 47–59.

[7] Burley, W., *Obligations (selections)*, in: N. Kretzmann and E. Stump, editors, *Logic and the Philosophy of Language*, Cambridge University Press, 1988 pp. 369–412.

[8] Dutilh Novaes, C., *Medieval* Obligationes *as a regimentation of 'the game of giving and asking for reasons'*, in: M. Palis, editor, *LOGICA Yearbook 2008*, (London: College Publications), 2009 .

[9] Dutilh Novaes, C., *Medieval* Obligationes *as a theory of discursive commitment management*, Vivarium **49** (2011), pp. 240–257.

[10] Dutilh Novaes, C., *Lessons in philosophy of logic from medieval* Obligationes, in: G. Restall and G. Russell, editors, *New Waves in Philosophical Logic*, Palgrave Macmillan, 2012 pp. 142–168.

[11] Horty, J. F. and N. Belnap, *The deliberative stit: A study of action, omission, ability, and obligation*, Journal of Philosophical Logic **24** (1995), pp. 583–644.

[12] Knuuttila, S., *The emergence of deontic logic in the fourteenth century*, in: R. Hilpinen, editor, *New Studies in Deontic Logic*, D. Reidel Publishing Company, 1981 pp. 225–248.

[13] Knuuttila, S. and O. Hallamaa, *Roger Roseth and medieval deontic logic*, Logique et Analyse **149** (1995), pp. 75–87.

[14] Knuuttila, S. and M. Yrjönsuuri, *Norms and action in obligational disputations*, in: J. Biard and O. Pluta, editors, *Die Philosophie im 14. und 15. Jahrhundert. In memoriam Konstanty Michalski (1879–1947)*, B.R. Grüner, 1988 pp. 191–202.

[15] Pacuit, E., R. Parikh and E. Cogan, *The logic of knowledge based obligations*, Synthese **149** (2006), pp. 311–341.

[16] Parikh, R. and R. Ramanujam, *Distributed processes and the logic of knowledge*, in: *Logic of Programs*, LNCS **193**, Springer, 1985 pp. 256–268.

[17] Parikh, R. and R. Ramanujam, *A knowledge based semantics of messages*, Journal of Logic, Language, and Information **12** (2003), pp. 453–467.

[18] Spade, P. V., *Why don't medieval logicians ever tell us what they're doing? Or, what is this, a conspiracy?* (2000), http://pvspade.com/Logic/docs/Conspiracy.pdf.

[19] Spade, P. V. and M. Yrjönsuuri, *Medieval theories of* obligationes, in: E. N. Zalta, editor, *Stanford Encyclopedia of Philosophy*, 2013, fall edition `http://plato.stanford.edu/archives/fall2013/entries/obligationes/`.

[20] Stump, E., *Obligations: From the beginning to the early fourteenth century*, in: N. Kretzman, A. Kenny and J. Pinborg, editors, *Cambridge History of Later Medieval Philosophy*, Cambridge University Press, 1982 pp. 315–334.

[21] Uckelman, S. L., *A dynamic epistemic logic approach to modeling* Obligationes, in: D. Grossi, S. Minica, B. Rodenhäuser and S. Smets, editors, *LIRa Yearbook*, Institute for Logic, Language & Computation, 2011 pp. 147–172.

[22] Uckelman, S. L., *Interactive logic in the Middle Ages*, Logic and Logical Philosophy **21** (2012), pp. 439–471.

[23] Uckelman, S. L., *Medieval* Disputationes de obligationibus *as formal dialogue systems*, Argumentation **27** (2013), pp. 143–166.

[24] Woleński, J., *Deontic logic and possible world semantics: A historical sketch*, Studia Logica **49** (1990), pp. 273–282.

[25] Yrjönsuuri, M., *Duties, rules and interpretations in obligational disputations*, in: M. Yrjönsuuri, editor, *Medieval Formal Logic*, Kluwer Academic Publishers, 2001 pp. 3–34.

Conditionally Knowing What

Yanjing Wang Jie Fan

Department of Philosophy
Peking University
{y.wang,fanjie}@pku.edu.cn

Abstract

Classic epistemic logic focuses on propositional knowledge expressed by "knowing that" operators. However, there are various types of knowledge used in natural language, in terms of "knowing how", "knowing whether", "knowing what", and so on. In [10], Plaza proposed an intuitive know-what operator which was generalized in [16] by introducing a condition. The latter know-what operator can express natural conditional knowledge such as "I know what your password is, if it is 4-digits", which is not simply a material implication. Essentially this know-what operator packages a first-order quantifier and an S5-modality together in a non-trivial way, thus making it hard to axiomatize. In [16] an axiomatization is given for the single-agent epistemic logic with both know-that and know-what operators, while leaving axiomatizing the multi-agent case open due to various technical difficulties. In this paper, we solve this open problem. The completeness proof is highly non-trivial, compared to the single-agent case, which requires different techniques inspired by first-order intensional logic.

Keywords: knowing what, first-order intensional logic, epistemic logic, conditional knowledge.

1 Introduction

Epistemic Logic (**EL**), since its birth, has been mainly focusing on reasoning about propositional knowledge, the knowledge expressed by "i knows that ϕ". However, besides "knowing that", there are apparently various types of knowledge used in everyday life, expressed by "knowing what" (i knows what d is), "knowing how" (i knows how to do ϕ), "knowing whether"(i knows whether ϕ) and so on. A natural question which keeps philosophers busy is to ask whether these types of knowledge can be reduced to propositional knowledge. For example, there is a long-lasting debate in philosophy on whether "knowing how" can be reduced to "knowing that" ever since the seminal work of Ryle [12]. Compared to this heated discussion, "knowing what" received relatively little attention despite efforts trying to unify "knowing wh-" (what, where, which, who, why) and "knowing how" in terms of "knowing that" (e.g., [13]).

On the other hand, in computer science and AI, "knowing what" plays an important role, as argued by McCarthy [9]. In particular, in a security setting, we need to express that "He knows that she knows [what] her own private key [is], but he does not know what exactly the key is." The literal translation of this sentence in terms of the usual know-that modal operator K_i does not work since $K_1 K_2 p \wedge \neg K_1 p$ is not consistent in the standard epistemic logic with T axiom. This fact has lead a number of authors to propose suitable new knowledge operators (e.g., [11,6]). In [10], one of the defining works of dynamic epistemic logic, Plaza introduced a very natural modal operator Kv_i to express knowing what in a dynamic epistemic setting. $Kv_i d$ expresses exactly that "i knows what d is". As for the semantics, $Kv_i d$ is true on a pointed epistemic model with world-dependent assignments for d iff d has the same value on all the epistemically accessible worlds for i. In this setting, it is perfectly possible that $K_1 Kv_2 d \wedge \neg Kv_1 d$ since two kinds of knowledge are treated differently.

In [16], we generalize the Kv_i operator to a *conditional* one and obtain a complete axiomatization of the single-agent public announcement logic with both know-that and know-what operators. [1] The resulting new formula $Kv_i(\phi, d)$ expresses that "agent i knows what d is, given ϕ." For example, as it happens a lot in this internet age, I may forget my own login password for some website, but I know if the password is 4-digit then it must be 1234, since I have never used another 4-digit password (though I have several 6-digit passwords). In such a case people often say "I know my password if it is 4-digit". This can be expressed as $Kv_i(p, d)$ where p denotes the proposition that the password is 4-digit. Note that this *conditional knowledge* is not an implication $p \to Kv_i d$ nor $K_i(p \to Kv_i d)$. The difference is that, according to the semantics, $Kv_i(\phi, d)$ essentially expresses what i *would know* if he were *informed* that ϕ. This distinction will become clear when we define the operator formally. In this light, there is clearly a connection to Public Announcement Logic [10]: to know what d is given ϕ is similar to knowing what d is after the announcement ϕ. [2]

This kind of conditional knowledge has a philosophical connection to the phenomenon of *elusive knowledge* studied by Lewis [8]: "Maybe we do know a lot in daily life; but maybe when we look hard at our knowledge, it goes away." One explanation is that what we claim to know is mainly *conditional knowledge* where the conditions are often implicit, e.g., "I know I have hands" can be viewed as an abbreviation of "I know I have hands, given that I am not a brain in the vat". This holds for all kinds of common sense "knowledge" that we have in every day life, which invites systematic logical study.

Coming back to the technical storyline, note that the original Kv_i operator is a special case of the conditional one since $Kv_i d$ is simply $Kv_i(\top, d)$. Then it is natural to ask whether the epistemic logic extended with conditional Kv_i (call it $\mathbf{ELKv}^r$) is more expressive than the epistemic logic with the standard Kv_i operator (call it **ELKv**). In [16] we show that $\mathbf{ELKv}^r$ is indeed strictly

[1] We called the conditional version of Kv_i the *relativized* Kv_i operator in [16], due to the similarity between it and the relativized common knowledge operator introduced in [15]

[2] Though there is still a difference: we do not require ϕ to be truthful in $Kv_i(\phi, d)$: ϕ can be just *hypothetical.*

more expressive than **ELKv**. More interestingly, we show that adding the public announcement operators to $\mathbf{ELKv}^r$ does not increase the expressive power, i.e., $\mathbf{ELKv}^r$ is closed under announcement updates. As in the standard epistemic logic, this is a good property for a logic as a foundation of epistemic reasoning (cf. e.g.,[14]).

To really layout the foundation for reasoning about both knowing that and knowing what, we need to axiomatize $\mathbf{ELKv}^r$. In [16], an interesting system is given to axiomatize the single-agent $\mathbf{ELKv}^r$. The completeness proof relies on a canonical model construction which consists of two copies of each maximal consistent set. However, such a method does not generalize to the multi-agent case thus leaving the axiomatization of the multi-agent $\mathbf{ELKv}^r$ open.

In this paper, we solve this open problem by showing that the multi-agent version of the system proposed in [16] is indeed complete for the multi-agent $\mathbf{ELKv}^r$. The techniques used here are quite different from the single-agent completeness proof and are inspired by the following observation: $\mathbf{ELKv}^r$ can be viewed as a fragment of first-order intensional logic (**FOIL**) proposed and studied in [4,5]. **FOIL** features two kinds of variables: the object (rigid) variables and the intension (non-rigid) variables where the latter variables range over the functions from the set of possible worlds to the set of objects. A technique called *predicate abstraction* is applied to abstract predicates from formulas. Now consider the following fragment ($\mathbf{FOIL}^-$) of **FOIL** where the non-rigid variables are not quantified, the only predicates are unary ones over rigid variables, and the only (implicit) predicate abstraction is applied to equalities between a rigid variable x and a non-rigid variable d: [3]

$$\phi ::= \top \mid Px \mid d = x \mid \neg\phi \mid \phi \wedge \phi \mid K_i\phi \mid \forall x\phi$$

$\mathbf{ELKv}^r$ can then be viewed as a small fragment of $\mathbf{FOIL}^-$ by recursively translating $Kv_i(\phi, d)$ into $\exists x K_i(\phi' \to d = x)$ where ϕ' is the $\mathbf{FOIL}^-$-translation of ϕ. Actually, this first-order formulation is also in accordance with the treatment of "knowing-wh" in terms of "knowing that" in [7,13]. In this way we can see clearly that Kv_i packages a first-order quantifier and a modality together.

This observation motivates our construction of the canonical model for multi-agent $\mathbf{ELKv}^r$. However, the method for axiomatizing **FOIL** as in [5] cannot be applied directly here, due to two reasons: first, our language is much weaker and we cannot express the desired first-order axioms in $\mathbf{ELKv}^r$; second, to our knowledge, it is unknown, how to axiomatize **FOIL** on S5 frames due to the diffcuities introduced by symmetry property as explained in [5]. In this work, we found a way to provide *just enough* extra information in the states of the canonical model to encode the "omitted" information expressible by potential $\mathbf{FOIL}^-$ formulas, while keeping it controlled by purely $\mathbf{ELKv}^r$ axioms. We believe that this method can be applied to other similar fragments of first-order intensional logic (over S5 frames).

[3] In Fitting's syntax of **FOIL**, $d = x$ should be formalized as $\langle \lambda y.y = x\rangle(d)$ (cf. [5]). Here, d may also be viewed as a constant since it is never quantified in the language.

In the rest of this paper, we first review the syntax and semantics of multi-agent $\mathbf{ELKv}^r$ and the proof system $\mathbb{ELKV}^r$ in Section 2. In Section 3, we prove our main result that $\mathbb{ELKV}^r$ completely axiomatizes multi-agent $\mathbf{ELKv}^r$. We conclude with future work in Section 4.

2 Preliminaries

Given a countably infinite set of proposition letters $\mathbf{P}$, a countably infinite set of agent names $\mathbf{I}$, and a countably infinite set of (non-rigid) constant symbols $\mathbf{D}$, the language of $\mathbf{ELKv}^r$ is defined as follows:

$$\phi ::= \top \mid p \mid \neg\phi \mid (\phi \wedge \phi) \mid K_i\phi \mid Kv_i(\phi, d)$$

where $p \in \mathbf{P}, i \in \mathbf{I}, d \in \mathbf{D}$.

$K_i\phi$ says that the agent i knows that ϕ. $Kv_i(\phi, d)$ says that the agent i knows what d is, given ϕ. More precisely, $Kv_i(\phi, d)$ says that the agent i would know what d is if he were informed that ϕ. The original (unconditional) $Kv_i d$ formulas proposed in [10] can be viewed as $Kv_i(\top, d)$. As usual, we define $\bot, (\phi \vee \psi), (\phi \to \psi), (\phi \leftrightarrow \psi), \hat{K}_i\phi$ as the abbreviations of, respectively, $\neg\top, \neg(\neg\phi \wedge \neg\psi), (\neg\phi \vee \psi), ((\phi \to \psi) \wedge (\psi \to \phi)), \neg K_i\neg\phi$. We omit parentheses from formulas unless confusion results.

$\mathbf{ELKv}^r$ is interpreted on epistemic models with assignments for the elements in $\mathbf{D}$: $\mathcal{M} = \langle S, O, \{\sim_i \mid i \in \mathbf{I}\}, V, V_{\mathbf{D}}\rangle$ where S is a non-empty set of possible worlds, O is a non-empty set of objects, $\sim_i$ is an equivalence relation over S, and V is a valuation function assigning a set of worlds $V(p) \subseteq S$ to each $p \in \mathbf{P}$, and $V_{\mathbf{D}} : \mathbf{D} \times S \to O$ is a function assigning each $d \in \mathbf{D}$ at each world an object. In terms of first-order intensional logic [4], $\mathcal{M}$ is an S5 intensional model with a constant domain O and assignments for non-rigid variables in $\mathbf{D}$. Note that for each $d \in \mathbf{D}$, $V_{\mathbf{D}}(d, \cdot)$ is a function from S to O which can be viewed as an *intension* as in [4]. The semantics is defined as follows:

$\mathcal{M}, s \vDash \top$		always holds
$\mathcal{M}, s \vDash p$	$\Leftrightarrow$	$s \in V(p)$
$\mathcal{M}, s \vDash \neg\phi$	$\Leftrightarrow$	$\mathcal{M}, s \nvDash \phi$
$\mathcal{M}, s \vDash \phi \wedge \psi$	$\Leftrightarrow$	$\mathcal{M}, s \vDash \phi$ and $\mathcal{M}, s \vDash \psi$
$\mathcal{M}, s \vDash K_i\psi$	$\Leftrightarrow$	for all t such that $s \sim_i t : \mathcal{M}, t \vDash \psi$
$\mathcal{M}, s \vDash Kv_i(\phi, d)$	$\Leftrightarrow$	for any $t_1, t_2 \in S$ such that $s \sim_i t_1$ and $s \sim_i t_2$: if $\mathcal{M}, t_1 \vDash \phi$ and $\mathcal{M}, t_2 \vDash \phi$, then $V_{\mathbf{D}}(d, t_1) = V_{\mathbf{D}}(d, t_2)$

Intuitively, $Kv_i(\phi, d)$ is true at s iff all the i-accessible ϕ-worlds agree on the value of d. In other words, i knows what d is given ϕ iff he is sure about d's value on ϕ-worlds. Based on this semantics, we can see clearly that $Kv_i(\phi, d)$ is indeed different from $\phi \to Kv_i d$ and $K_i(\phi \to Kv_i d)$. The condition ϕ restricts the accessible worlds to be considered, and we then check whether d has the same value on these "relative alternatives".

In [16], we give a complete axiomatization of the single agent $\mathbf{ELKv}^r$ and the following is the multi-agent version of that system which is an extension of the multi-modality $\mathbb{S}5$.

System $\mathbb{ELKV}^r$

Axiom Schemas		Rules	
TAUT	all the instances of tautologies		
DISTK	$K_i(\phi \to \psi) \to (K_i\phi \to K_i\psi)$	MP	$\dfrac{\phi, \phi \to \psi}{\psi}$
T	$K_i\phi \to \phi$		
4	$K_i\phi \to K_iK_i\phi$	NECK	$\dfrac{\phi}{K_i\phi}$
5	$\neg K_i\phi \to K_i\neg K_i\phi$		
DISTKv^r	$K_i(\phi \to \psi) \to (Kv_i(\psi, d) \to Kv_i(\phi, d))$	RE	$\dfrac{\psi \leftrightarrow \chi}{\phi \leftrightarrow \phi[\psi/\chi]}$
Kv^{r}4	$Kv_i(\phi, d) \to K_iKv_i(\phi, d)$		
Kv$^r\bot$	$Kv_i(\bot, d)$		
Kv$^r\vee$	$\hat{K}_i(\phi \wedge \psi) \wedge Kv_i(\phi, d) \wedge Kv_i(\psi, d) \to Kv_i(\phi \vee \psi, d)$		

where RE is the rule of replacement of equivalents, which plays an important role in the later proofs. In the rest of the paper, we use $\vdash$ to denote the derivation relation within $\mathbb{ELKV}^r$.

Note that Kv_i operators do not behave like modalities in a normal modal logic and the (obvious adaptations of) necessitation rule and the K axiom are not valid for Kv_i. Instead, we have the distribution axiom schema DISTKv^r(note the swap of ψ and ϕ in the consequent). Kv^{r}4 is the counter part of the positive introspection axiom 4, and Kv$^r\bot$ stipulates the effect of the absurd precondition. The most important axiom is Kv$^r\vee$ which handles the composition of the conditions: if all the possible ϕ-worlds agree on what d is and all the possible ψ-worlds also agree on d, then the overlap between ϕ possibilities and ψ possibilities implies that all the $\phi \vee \psi$ possibilities also agree on what d is. We can show that the above system is sound (cf. [16, Theorem 11]).

To facilitate the later proofs, we need the following propositions.

Proposition 2.1 (i) $\vdash \neg K_i\phi \leftrightarrow K_i\neg K_i\phi$

(ii) *The rule (RM):* $\dfrac{\phi \to \psi}{\hat{K}_i\phi \to \hat{K}_i\psi}$ *is admissible in* $\mathbb{ELKV}^r$.

(iii) $\vdash \neg Kv_i(\phi, d) \leftrightarrow K_i\neg Kv_i(\phi, d)$

Proof (i) and (ii) are standard exercises in S5. The $\leftarrow$ direction in (iii) is trivial due to T. We show the other way around:

$K_iKv_i(\phi, d) \leftrightarrow Kv_i(\phi, d)$	T, Kv^{r}4
$\neg K_iKv_i(\phi, d) \to K_i\neg K_iKv_i(\phi, d)$	5
$\neg Kv_i(\phi, d) \to K_i\neg Kv_i(\phi, d)$	RE

□

Note that the $\to$ half of (iii) can be viewed as the Kv_i counterpart of 5 thus we denote it by Kv^{r}5.

Another useful observation is that Kv$^r\vee$ can be generalized to arbitrary finite disjunctions as the following proposition shows.

Proposition 2.2 *For any non-empty finite set U of $\boldsymbol{ELKv^r}$ formulas:*

$$\vdash \hat{K}_i(\bigwedge U) \wedge \bigwedge_{\phi \in U} Kv_i(\phi, d) \to Kv_i(\bigvee U, d).$$

Proof If $|U| = 1$ then the statement holds trivially. Suppose $|U| \geq 2$, we prove the statement by an inductive proof on $|U|$. The case of $|U| = 2$ is immediate due to $\mathtt{Kv}^r\vee$. Suppose the claim holds when $|U| = k$. Now consider the case $|U| = k + 1$, and let $U = U' \cup \{\psi\}$ such that $|U'| = k$:

(i)	$\hat{K}_i(\bigwedge U') \wedge \bigwedge_{\phi \in U'} Kv_i(\phi, d) \to Kv_i(\bigvee U', d)$	IH
(ii)	$Kv_i(\bigvee U', d) \wedge Kv_i(\psi, d) \wedge \hat{K}_i(\bigvee U' \wedge \psi) \to Kv_i(\bigvee U, d)$	$\mathtt{Kv}^r\vee$
(iii)	$\hat{K}_i(\bigwedge U) \wedge \bigwedge_{\phi \in U} Kv_i(\phi, d)$	TAUT
	$\to \hat{K}_i(\bigwedge U') \wedge \bigwedge_{\phi \in U'} Kv_i(\phi, d) \wedge Kv_i(\psi, d) \wedge \hat{K}_i(\bigvee U' \wedge \psi)$	DISTK, NECK
(iv)	$\hat{K}_i(\bigwedge U) \wedge \bigwedge_{\phi \in U} Kv_i(\phi, d) \to Kv_i(\bigvee U, d)$	$(i)(ii)(iii)$

□

The following proposition essentially says if d has the same value over accessible ϕ- and ψ-worlds respectively, and there are a ϕ-world and a ψ-world sharing the same d value, then d has the same value over accessible $\phi \vee \psi$-worlds. This proposition allows us to relax the antecedent of $\mathtt{Kv}^r\vee$ a little bit to make it more useful.

Proposition 2.3 *For any $\phi, \psi, \chi \in \boldsymbol{ELKv^r}$, any $d \in \boldsymbol{D}$:*
$\vdash \hat{K}_i(\phi \wedge \chi) \wedge \hat{K}_i(\psi \wedge \chi) \wedge Kv_i(\phi, d) \wedge Kv_i(\psi, d) \wedge Kv_i(\chi, d) \to Kv_i(\phi \vee \psi, d)$

Proof

(i)	$\hat{K}_i(\phi \wedge \chi) \wedge Kv_i(\phi, d) \wedge Kv_i(\chi, d) \to Kv_i(\phi \vee \chi, d)$	$\mathtt{Kv}^r\vee$
(ii)	$\hat{K}_i(\psi \wedge \chi) \wedge Kv_i(\psi, d) \wedge Kv_i(\chi, d) \to Kv_i(\psi \vee \chi, d)$	$\mathtt{Kv}^r\vee$
(iii)	$\phi \wedge \chi \to (\phi \vee \chi) \wedge (\psi \vee \chi)$	TAUT
(iv)	$\hat{K}_i(\phi \wedge \chi) \to \hat{K}_i((\phi \vee \chi) \wedge (\psi \vee \chi))$	RM, (iii)
(v)	$\hat{K}_i((\phi \vee \chi) \wedge (\psi \vee \chi)) \wedge Kv_i(\phi \vee \chi, d) \wedge Kv_i(\psi \vee \chi, d)$	
	$\to Kv_i(\phi \vee \psi \vee \chi, d)$	$\mathtt{Kv}^r\vee$
(vi)	$\hat{K}_i(\phi \wedge \chi) \wedge \hat{K}_i(\psi \wedge \chi) \wedge Kv_i(\phi, d) \wedge Kv_i(\psi, d) \wedge Kv_i(\chi, d)$	
	$\to Kv_i(\phi \vee \psi \vee \chi, d)$	$(i)(ii)(iv)(v)$
(vii)	$\phi \vee \psi \to \phi \vee \psi \vee \chi$	TAUT
$(viii)$	$K_i(\phi \vee \psi \to \phi \vee \psi \vee \chi)$	NECK, (vii)
(ix)	$K_i(\phi \vee \psi \to \phi \vee \psi \vee \chi) \to Kv_i(\phi \vee \psi \vee \chi, d)$	
	$\to Kv_i(\phi \vee \psi, d)$	$\mathtt{DISTKv}^r$
(x)	$Kv_i(\phi \vee \psi \vee \chi, d) \to Kv_i(\phi \vee \psi, d)$	MP, $(viii)(ix)$
(xi)	$\hat{K}_i(\phi \wedge \chi) \wedge \hat{K}_i(\psi \wedge \chi) \wedge Kv_i(\phi, d) \wedge Kv_i(\psi, d) \wedge Kv_i(\chi, d)$	
	$\to Kv_i(\phi \vee \psi, d)$	$(vi)(x)$

□

3 Completeness of multi-agent $\mathbb{ELKV}^r$

To prove the completeness, we need to build a canonical model such that each maximal consistent set of $\mathbb{ELKV}^r$ is satisfied in it. The general difficulty is as in the single-agent case: just using the maximal consistent sets as the states in the canonical model is not enough, more information should be provided in the states of the canonical model. As we mentioned, the $Kv_i(\phi, d)$ formulas can be viewed as $\exists x K_i(\phi \to d = x)$ where x is a rigid variable and d is a non-rigid one. To build a canonical model for such a first-order intensional logic, we need to include atomic formulas such as $d = x$ and modal formulas such as $K_i(\phi \to d = x)$ and control their interactions by axioms. However, those formulas are not expressible in $\mathbf{ELKv}^r$ since we simply cannot say what d exactly is. Therefore, in the canonical model, we need to equip the $\mathbb{ELKV}^r$-maximal consistent sets with information which can function as those atomic formulas. Moreover, we need to specify how such extra information is related to the $\mathbb{ELKV}^r$-maximal consistent sets. Note that since $\mathbb{ELKV}^r$ is very limited we cannot enforce the extra information behave exactly like the intended first-order intensional formulas. The real difficulty is to find the requirements which are "just enough" to make sure the truth lemma holds and this is the most fundamental idea behind our definition of the canonical model. It will also become more clear why the single-agent case is much simpler (cf. Remark 3.6).

3.1 Canonical model

In the sequel we define our canonical model with the set of natural numbers $\mathbb{N}$ as the constant domain of objects. [4]

Definition 3.1 *Let* $\boldsymbol{MCS}$ *be the set of maximal consistent sets w.r.t.* $\mathbb{ELKV}^r$, *and let* $\mathbb{N}$ *be the set of natural numbers. The canonical model* $\mathcal{M}^c$ *of* $\mathbb{ELKV}^r$ *is a tuple* $\langle S^c, \mathbb{N}, \{\sim_i^c \mid i \in \boldsymbol{I}\}, V^c, V_{\boldsymbol{D}}^c\rangle$ *where:*

- S^c *consists of all the triples* $\langle \Gamma, f, g\rangle \in \boldsymbol{MCS} \times \mathbb{N}^{\boldsymbol{D}} \times (\mathbb{N} \cup \{\star\})^{\boldsymbol{I} \times \boldsymbol{ELKv}^r \times \boldsymbol{D}}$ *that satisfy the following three conditions for any* $i \in \boldsymbol{I}$, *any* $\psi, \phi \in \boldsymbol{ELKv}^r$, *and any* $d \in \boldsymbol{D}$:
  (i) $g(i, \psi, d) = \star$ *iff* $Kv_i(\psi, d) \land \hat{K}_i\psi \notin \Gamma$,
  (ii) *If* $g(i, \phi, d) \neq \star$ *and* $g(i, \psi, d) \neq \star$ *then:* $g(i, \phi, d) = g(i, \psi, d)$ *iff there exists a* χ *such that* $Kv_i(\chi, d)$ *and* $\hat{K}_i(\phi \land \chi)$ *and* $\hat{K}_i(\psi \land \chi)$ *are in* Γ.
  (iii) $\psi \land Kv_i(\psi, d) \in \Gamma$ *implies* $f(d) = g(i, \psi, d)$.
  For any $s \in S^c$, *we write* $\phi \in s$ *if* ϕ *is in the maximal consistent set of* s *and write* $\phi \in s \cap t$ *if* $\phi \in s$ *and* $\phi \in t$. f_s *and* g_s *are used as the corresponding functions in* s, *and* $g_s(i)$ *is the function from* $\boldsymbol{ELKv}^r \times \boldsymbol{D}$ *to* $\mathbb{N} \cup \{\star\}$ *induced by* g_s *fixing a particular* $i \in \boldsymbol{I}$.
- $s \sim_i^c t$ *iff* $\{\phi \mid K_i\phi \in s\} \subseteq t$ *and* $g_s(i) = g_t(i)$
- $V_{\boldsymbol{D}}^c(d, s) = f_s(d)$

Remark 3.2 Intuitively, f is roughly functioning as the collection of $d = x$ formulas, and g is roughly functioning as the collection of $K_i(\psi \to d = x)$ formulas. Now for the intuitive ideas behind the three conditions:

---

[4] Note that this countable set is indeed big enough, since the (countable) language of $\mathbf{ELKv}^r$ can be translated into first-order intensional logic, which can be again translated into 3-sorted first-order logic, which still enjoys Löwenheim-Skolem property (cf. [1]).

- (i): We use $\star$ to mark that the value of $g(i, \psi, d)$ is *irrelevant*. If $Kv_i(\psi, d) \notin \Gamma$ then of course the value of $g(i, \psi, d)$ is irrelevant. If $Kv_i(\psi, d) \in \Gamma$ but $K_i \neg\psi \in \Gamma$, then the condition ψ is never possible for i thus the value of $g(i, \psi, d)$ is also irrelevant. Condition (i) is mainly for the technical convenience.
- (ii): This condition handles how the g values are inter-related. Intuitively, it roughly says that $x = y$ iff ($K_i(\psi \to d = x)$ and $K_i(\phi \to d = y)$, and there are some accessible ψ-world and ϕ-world which share the same value of d).
- (iii): Intuitively, this condition says that if $K_i(\psi \to d = x)$ and ψ is indeed true then $d = x$.

The definition of $\sim_i^c$ is in spirit the same as in the canonical model for the standard epistemic logic. The extra condition $g_s(i) = g_t(i)$ says the i-indistinguishable worlds should satisfy the same $K_i(\psi \to d = x)$ formulas.

Condition (i), (ii), (iii) and the definition of $\sim_i^c$ specify the minimal requirements of the extra information attached to maximal consistent sets. We need to control them without using **FOIL**$^-$ formulas.

To show the above model is indeed an epistemic model, we need the following proposition:

Proposition 3.3 *For any $i \in \boldsymbol{I}$, $\sim_i^c$ is an equivalence relation.*

Proof As a standard exercise in modal logic, by using `T`, `4`, and `5`, we can prove the following claim:

$$\text{(for all } \phi : K_i\phi \in s \text{ implies } \phi \in t\text{) iff (for all } \phi : K_i\phi \in s \text{ iff } K_i\phi \in t\text{)} \qquad (*).$$

Thus $s \sim_i^c t$ iff $\{\phi \mid K_i\phi \in s\} = \{\phi \mid K_i\phi \in t\}$ and $g_s(i) = g_t(i)$. Then it is easy to see $\sim_i^c$ is an equivalence relation.

□

Based on the above claim (∗), using `T` and `Kv`^{r}`4`, the following is immediate:

Proposition 3.4 *For any two maximal consistent sets Δ and Γ, if $\{\phi \mid K_i\phi \in \Delta\} \subseteq \Gamma$, then the following hold for all ϕ:*

- *$K_i\phi \in \Delta$ iff $K_i\phi \in \Gamma$*
- *$\hat{K}_i\phi \in \Delta$ iff $\hat{K}_i\phi \in \Gamma$*
- *$Kv_i(\phi, d) \in \Delta$ iff $Kv_i(\phi, d) \in \Gamma$.*

3.2 Completeness

In order to establish the truth lemma, the most difficult things are the existence lemmas for K_i and Kv_i operators. Since the states in the canonical models are *not* merely maximal consistent sets, more efforts are required.

We first propose a general method to construct proper successors. This can be viewed as some kind of Lindenbaum's Lemma, though highly non-trivial in this case, if we view g and f as collections of "hidden formulas".

Proposition 3.5 *Given a state $s \in S^c$, an agent $i \in \boldsymbol{I}$, and a maximal consistent set Γ such that $\{\phi \mid K_i\phi \in s\} \subseteq \Gamma$, and any natural number x, we have a deterministic method to construct $t = \langle \Gamma, f, g \rangle$ based on s, Γ, and x such that $t \in S^c$ and $s \sim_i^c t$.*

Proof The construction and the proof are quite involved: we first build f and build g by using finite approximations and show that the constructions are well-defined; then we show that the $\langle \Gamma, f, g \rangle$ satisfy the three conditions of states in S^c and that $s \sim_i^c t$. In the following, we fix a natural number x.

Let $T^0 = \{\langle j, \phi, d \rangle \mid j = i \text{ or } \phi \wedge Kv_j(\phi, d) \in \Gamma\}$. Let $g^0 : T^0 \to \mathbb{N} \cup \{\star\}$ be defined as follows:

$$g^0(j, \phi, d) = \begin{cases} g_s(i, \phi, d) & \text{if } j = i \\ g_s(i, \psi, d) & \text{if } j \neq i \text{ and } \psi \wedge Kv_i(\psi, d) \in \Gamma \text{ for some } \psi \\ x & \text{if otherwise} \end{cases}$$

We need to show that the second case is well-defined: the choice of ψ does not affect the value of $g_s(i, \psi, d)$, namely $\psi \wedge Kv_i(\psi, d) \in \Gamma$ and $\psi' \wedge Kv_i(\psi', d) \in \Gamma$ implies $g_s(i, \psi, d) = g_s(i, \psi', d)$. Suppose that $\psi \wedge Kv_i(\psi, d) \in \Gamma$ and $\psi' \wedge Kv_i(\psi', d) \in \Gamma$, then the following formulas are also in Γ: $\hat{K}_i\psi'$, $\hat{K}_i\psi$, $\hat{K}_i(\psi' \wedge \psi)$ due to the contrapositive of Axiom $\mathtt{T}$. By Proposition 3.4, the following formulas are all in s: $\hat{K}_i(\psi' \wedge \psi)$, $\hat{K}_i\psi'$, $\hat{K}_i\psi$, $Kv_i(\psi', d)$, and $Kv_i(\psi, d)$. Now it is clear that $g_s(i, \psi, d) \neq \star$ and $g_s(i, \psi', d) \neq \star$ due to condition (i) of s. Let $\chi = \psi'$ now we have $Kv_i(\chi, d)$ and $\hat{K}_i(\psi' \wedge \chi)$ and $\hat{K}_i(\psi \wedge \chi)$ are all in s. By condition (ii) of s, $g_s(i, \psi', d) = g_s(i, \psi, d)$.

Now we define f as follows:

$$f(d) = \begin{cases} g^0(j, \phi, d) & \text{if } Kv_j(\phi, d) \wedge \phi \in \Gamma \text{ for some } \phi \text{ and } j \\ x & \text{if otherwise} \end{cases}$$

We need to show that the first case is well-defined: the choices of ϕ and j do not affect the value of $g^0(j, \phi, d)$, namely $\phi \wedge Kv_j(\phi, d) \in \Gamma$ and $\psi \wedge Kv_k(\psi, d) \in \Gamma$ implies $g^0(j, \phi, d) = g^0(k, \psi, d)$. To see this, consider four cases:

- $j = i$ and $k = i$. Then due to the above proof and the first clause of the definition of g^0, we have $g^0(j, \phi, d) = g_s(i, \phi, d) = g_s(i, \psi, d) = g^0(k, \psi, d)$.
- $j \neq i$ and $k \neq i$. If there exists χ such that $\chi \wedge Kv_i(\chi, d) \in \Gamma$, then by the second clause of g^0, we have $g^0(j, \phi, d) = g_s(i, \chi, d) = g^0(k, \psi, d)$; otherwise, by the third clause of g^0, we have $g^0(j, \phi, d) = x = g^0(k, \psi, d)$.
- $j = i$ and $k \neq i$. From the first clause of g^0, it follows that $g^0(j, \phi, d) = g_s(i, \phi, d)$. Due to the fact that $\phi \wedge Kv_i(\phi, d) \in \Gamma$ and the second clause of g^0, we have $g^0(k, \psi, d) = g_s(i, \phi, d)$, thus $g^0(j, \phi, d) = g^0(k, \psi, d)$.
- $j \neq i$ and $k = i$. Similar to the third case.

Now let Δ be the set of the remaining non-i-triples:

$$\Delta = \{\langle j, \phi, d \rangle \mid j \neq i, \phi \wedge Kv_j(\phi, d) \notin \Gamma, j \in \mathbf{I}, \phi \in \mathbf{ELKv}^r, d \in \mathbf{D}\}.$$

Due to the fact that $\mathbf{D}$, $\mathbf{I}$ and $\mathbf{ELKv}^r$ are countable, we can enumerate Δ as $\delta_1, \delta_2, \ldots$ and approximate g step by step by extending the domain of g^k with δ_{k+1}. Let Δ^k be $\{\delta_l \mid 1 \leq l \leq k\}$, and in particular $\Delta^0 = \emptyset$. Let Λ^k be $\{g^k(\delta) \mid \delta \in \Delta^k\}$ and let max be the function which assigns to each non-empty finite set of natural numbers its maximum. Let $Dom(g^k)$ be the domain of g^k. For $k \geq 0$, our construction of $g^{k+1} : T^0 \cup \Delta^{k+1} \to \mathbb{N} \cup \{\star\}$ is as follows: let $g^{k+1}(j, \phi, d) = g^k(j, \phi, d)$ if $\langle j, \phi, d\rangle \in Dom(g^k) = T^0 \cup \Delta^k$, and for the only new $\langle j, \phi, d\rangle \notin Dom(g^k)$ (thus $j \neq i$) we have:

$$g^{k+1}(j,\phi,d) = \begin{cases} \star & \text{if } \hat{K}_j\phi \wedge Kv_j(\phi,d) \notin \Gamma \\ g^k(j,\psi,d) & \text{if } \hat{K}_j\phi \wedge Kv_j(\phi,d) \in \Gamma \text{ and there are} \\ & \chi,\psi \text{ such that } \langle j,\psi,d\rangle \in Dom(g^k) \\ & \text{and the following formulas are all} \\ & \text{in } \Gamma: \ \hat{K}_j(\chi\wedge\phi),\ \hat{K}_j(\chi\wedge\psi), \\ & Kv_j(\psi,d),\ Kv_j(\chi,d) \\ max(\Lambda^k \cup \{f(d)\})+1 & \text{if otherwise} \end{cases}$$

Note that, we still need to show that the second case in the above definition of g^{k+1} is well-defined. More precisely, we need to show for any $k \geq 0$ any $j \neq i$, the following (1) implies (2):

(1) there exist ψ, ψ', χ, and χ' such that $\langle j, \psi, d\rangle$ and $\langle j, \psi', d\rangle$ are in $Dom(g^k)$ and the following formulas are all in Γ: $\hat{K}_j(\chi \wedge \psi), \hat{K}_j(\chi \wedge \phi), \hat{K}_j(\chi' \wedge \psi'), \hat{K}_j(\chi' \wedge \phi), Kv_j(\psi, d), Kv_j(\psi', d), Kv_j(\chi, d), Kv_j(\chi', d)$.

(2) $g^k(j, \psi, d) = g^k(j, \psi', d)$.

Induction on k:

- $k = 0$: $\langle j, \psi, d\rangle$ and $\langle j, \psi', d\rangle$ are both in $Dom(g^0) = T^0$ then according to the definition of g^0 and the fact that $j \neq i$, $g^0(j, \psi, d) = g^0(j, \psi', d)$.
- Induction Hypothesis: (1) implies (2) holds for all $k \leq n$.
- $k = n + 1$: w.l.o.g, we assume that at least one of (j, ψ, d) and (j, ψ', d) is not in $Dom(g^0)$, for otherwise the case is like the above one. Then we can assume that there exists an $m \leq n$ such that $\langle j, \psi, d\rangle \in Dom(g^m)$, $\langle j, \psi', d\rangle \notin Dom(g^m)$, and $\langle j, \psi', d\rangle$ is added into $Dom(g^{m+1})$ by our construction. Assuming (1), let θ be $\chi \vee \chi'$, we can show $\hat{K}_j(\theta \wedge \psi)$ and $\hat{K}_j(\theta \wedge \psi') \in \Gamma$ since $\hat{K}_j(\chi \wedge \psi)$ and $\hat{K}_j(\chi' \wedge \psi')$ are in Γ. Moreover, since $\hat{K}_j(\chi \wedge \phi), \hat{K}_j(\chi' \wedge \phi), Kv_j(\chi, d), Kv_j(\chi', d), Kv_j(\phi, d) \in \Gamma$, we have $Kv_j(\theta, d) \in \Gamma$ by Proposition 2.3. Now we have $\hat{K}_j(\theta \wedge \psi)$, $\hat{K}_j(\theta \wedge \psi')$, and $Kv_j(\theta, d)$ are all in Γ. According to our construction of g^{m+1}, $g^{m+1}(j, \psi', d) = g^m(j, \psi, d)$ and the induction hypothesis guarantees the uniqueness of $g^m(j, \psi, d)$ since $m \leq n$. Therefore $g^k(j, \psi', d) = g^{m+1}(j, \psi', d) = g^m(j, \psi, d) = g^k(j, \psi, d)$.

Viewing each g^k as a set of pairs $\langle\langle j, \phi, d\rangle, g^k(j, \phi, d)\rangle$, we let g be $\bigcup_{k<\omega} g^k$.

Now we need to verify conditions (i), (ii) and (iii). Condition (iii) is trivial by the definition of f. We verify condition (i) and (ii) below.

For condition (i): we first show that for the fixed i and any $\phi \in \mathbf{ELKv}^r$, any $d \in \mathbf{D}$: $g^0(i, \phi, d) \neq \star$ iff $\hat{K}_i\phi \wedge Kv_i(\phi, d) \in \Gamma$. From right to left: suppose that $g^0(i, \phi, d) = \star$ then we have $g_s(i, \phi, d) = \star$ thus $\hat{K}_j\phi \wedge Kv_i(\phi, d) \notin s$, i.e., $K_i\neg\phi \in s$ or $\neg Kv_i(\phi, d) \in s$. By Proposition 3.4, we have $K_i\neg\phi \in \Gamma$ or $\neg Kv_i(\phi, d) \in \Gamma$, i.e., $\hat{K}_i\phi \wedge Kv_i(\phi, d) \notin \Gamma$. From left to right: suppose that $g^0(i, \phi, d) \neq \star$ then $\hat{K}_i\phi \wedge Kv_i(\phi, d) \in s$ thus by Proposition 3.4 again, $\hat{K}_i\phi \wedge Kv_i(\phi, d) \in \Gamma$.

Now consider $\langle j, \phi, d\rangle \in Dom(g^0)$ where $j \neq i$. By definition, $\phi \wedge Kv_j(\phi, d) \in \Gamma$, thus $\hat{K}_j\phi \wedge Kv_j(\phi, d) \in \Gamma$. By the construction of g^0 it is clear that $g^0(j, \phi, d) \neq \star$, since $x \neq \star$ and the fact that $\hat{K}_i\psi \wedge Kv_i(\psi, d) \in \Gamma$ implies $g^0(i, \psi, d) \neq \star$ which we have just proved. This concludes the proof for the base case: for any $\langle j, \phi, d\rangle \in Dom(g^0)$: $g^0(j, \phi, d) \neq \star$ iff $\hat{K}_j\phi \wedge Kv_j(\phi, d) \in \Gamma$. The inductive case is obvious by the three cases of our construction of g^{k+1}.

Condition (ii) is more complicated to verify and it requires an inductive proof. We first claim the following:

Claim ($\circ$): For each $k \geq 0$, and any $\langle j, \psi, d\rangle$ and $\langle j, \phi, d\rangle$ in $Dom(g^k)$ such that $g^k(j, \psi, d) \neq \star$ and $g^k(j, \phi, d) \neq \star$, the following two are equivalent:

(1) $g^k(j, \phi, d) = g^k(j, \psi, d)$

(2) there exists a χ such that $Kv_j(\chi, d)$, $\hat{K}_j(\phi \wedge \chi)$ and $\hat{K}_j(\psi \wedge \chi)$ are in Γ.

If claim ($\circ$) holds then Condition (ii) holds too, since any $\langle j, \psi, d\rangle$ and $\langle j, \phi, d\rangle$ must both exist in $Dom(g^k)$ for some k. Now we prove the claim ($\circ$).

- If $k = 0$ then both $\langle j, \psi, d\rangle$ and $\langle j, \phi, d\rangle$ are in $Dom(g^0)$. There are two subcases:
  - If $j = i$ then we have $g(j, \phi, d) = g(j, \psi, d)$ iff $g^0(i, \phi, d) = g^0(i, \psi, d)$ iff $g_s(i, \phi, d) = g_s(i, \psi, d)$ iff there exists a χ such that $Kv_i(\chi, d)$ and $\hat{K}_i(\phi \wedge \chi)$ and $\hat{K}_i(\psi \wedge \chi)$ are all in s (by condition (ii) of s). According to Proposition 3.4, the last statement is equivalent to that there exists a χ such that $\{Kv_i(\chi, d), \hat{K}_i(\phi \wedge \chi), \hat{K}_i(\psi \wedge \chi)\} \subseteq \Gamma$.
  - If $j \neq i$ then clearly $g(j, \psi, d) = g^0(j, \psi, d) = g^0(j, \phi, d) = g(j, \phi, d)$ by the definition of g^0. Now since $g(j, \psi, d) \neq \star$, $Kv_j(\psi, d) \in \Gamma$ due to condition (i) of Γ which we have just verified. Since $\langle j, \psi, d\rangle$ and $\langle j, \phi, d\rangle$ are both in $Dom(g^0) = T^0$, we have $\phi, \psi \in \Gamma$, thus $\hat{K}_j(\phi \wedge \psi) \in \Gamma$ by axiom T. Finally we have $\{Kv_j(\chi, d), \hat{K}_j(\phi \wedge \chi), \hat{K}_j(\psi \wedge \chi)\} \subseteq \Gamma$ given $\chi = \psi$.
- Induction Hypothesis: the claim ($\circ$) holds for $k \leq m$.
- Suppose $k = m+1$ and at least one of $\langle j, \phi, d\rangle$ and $\langle j, \psi, d\rangle$ is not in $Dom(g^m)$, for otherwise it can be handled by IH. Then clearly $j \neq i$ for otherwise both triples are in $Dom(g^0)$ thus in $Dom(g^m)$. W.l.o.g, we can assume that $\langle j, \psi, d\rangle \in Dom(g^m)$ and $\langle j, \phi, d\rangle \notin Dom(g^m)$ but $\langle j, \phi, d\rangle \in Dom(g^{m+1})$, i.e., $\langle j, \phi, d\rangle$ is added at step $m + 1$. By assumption $g^{m+1}(j, \phi, d) \neq \star$ and $g^{m+1}(j, \psi, d) \neq \star$, then by condition (i) $\hat{K}_j\phi \wedge Kv_j(\phi, d) \in \Gamma$ and $\hat{K}_j\psi \wedge Kv_j(\psi, d) \in \Gamma$. Now if there exists a χ such that $\hat{K}_j(\chi \wedge \psi) \wedge \hat{K}_j(\chi \wedge \phi) \wedge$

$Kv_j(\chi, d) \in \Gamma$ then by the second clause of the definition of g^{m+1} we have $g^{m+1}(j, \phi, d) = g^m(j, \psi, d) = g^{m+1}(j, \psi, d)$. This proves that (2) implies (1).

For the other direction, suppose that $g^{m+1}(j, \phi, d) = g^{m+1}(j, \psi, d) = g^m(j, \psi, d) \neq \star$, due to the definition of g^{m+1}, $g^{m+1}(j, \phi, d)$ must be constructed according to the second clause, for the third clause can make sure $g^{m+1}(j, \phi, d) \neq g^m(j, \psi, d)$. To see this, note that if $\langle j, \psi, d\rangle \in Dom(g^0)$ then $g^m(j, \psi, d) = g^0(j, \psi, d) = f(d)$ by the definition of g^0 (note that $j \neq i$). The third clause guarantees that $g^{m+1}(j, \phi, d) > f(d) = g^m(j, \psi, d)$. If $\langle j, \psi, d\rangle \notin Dom(g^0)$ then the third clause guarantees that $g^{m+1}(j, \phi, d) > max(\Lambda^m) \geq g^m(j, \psi, d)$.

Now, if $g^{m+1}(j, \phi, d)$ is constructed by the second clause based on $g^m(j, \psi, d)$ then (2) is immediate. Suppose otherwise that $g^{m+1}(j, \phi, d)$ is constructed based on $g^m(j, \theta, d)$ for some $\theta \neq \psi$, such that $g^m(j, \theta, d) = g^m(j, \psi, d) \neq \star$, then there exists a ξ such that $\hat{K}_j(\xi \wedge \theta) \wedge \hat{K}_j(\xi \wedge \phi) \wedge Kv_j(\xi, d) \in \Gamma$. Since $g^m(j, \theta, d) \neq \star$, by condition (i) we also have $Kv_j(\theta, d) \in \Gamma$. Since $g^m(j, \theta, d) = g^m(j, \psi, d) \neq \star$, by IH, there exists ξ' such that $\hat{K}_j(\xi' \wedge \psi) \wedge \hat{K}_j(\xi' \wedge \theta) \wedge Kv_j(\xi', d) \in \Gamma$. Now we have $\hat{K}_j(\xi \wedge \theta), \hat{K}_j(\xi' \wedge \theta), Kv_j(\xi, d), Kv_j(\xi', d)$ and $Kv_j(\theta, d)$ all in Γ. By Proposition 2.3, $Kv_j(\xi \vee \xi', d) \in \Gamma$. Let $\chi = \xi \vee \xi'$. Since $\hat{K}_j(\xi \wedge \phi)$ and $\hat{K}_j(\xi' \wedge \psi)$ are in Γ, we have $\hat{K}_j(\chi \wedge \phi), \hat{K}_j(\chi \wedge \psi)$ and $Kv_j(\chi, d)$ are all in Γ, and this completes the proof of claim ($\circ$). Thus $\langle \Gamma, f, g\rangle$ satisfies the condition (ii).

In sum, $\langle \Gamma, f, g\rangle \in S^c$, and $s \sim_i^c \langle \Gamma, f, g\rangle$ due to the facts that $g(i) = g_s(i)$ (by the construction of g^0) and the assumption that $\{\phi \mid K_i\phi \in s\} \subseteq \Gamma$. □

Remark 3.6 To build an i-successor of s, we need to construct a proper g such that it takes care of the information not only about i but also about $j \neq i$. Note that if $\mathbf{I} = \{i\}$, then $g(i) = g_s(i)$ implies $g = g_s$. In this case we do not need the above construction, thus the single-agent case is much simper.

Important notation In the sequel, we refer to the above construction of f as $F(s, i, \Gamma, x)$ where x is a natural number as a parameter.

Now we are ready to prove two important existence lemmas:

Lemma 3.7 *For any $s \in S^c$, any $i \in \boldsymbol{I}$: $K_i\psi \notin s$ implies there is a world t such that $s \sim_i^c t$ and $\neg\psi \in t$.*

Proof It is a standard exercise in modal logic to show that $X = \{\neg\psi\} \cup \{\phi \mid K_i\phi \in s\}$ is consistent. Then by Lindenbaum Lemma for $\mathbf{ELKv}^r$, there exists an MCS Γ including X. Now from Proposition 3.5 we can equip Γ with some proper f and g, such that $\langle \Gamma, f, g\rangle \in S^c$ and $s \sim_i^c \langle \Gamma, f, g\rangle$. □

Lemma 3.8 *For any $s \in S^c$, any $i \in \boldsymbol{I}$: $\neg Kv_i(\phi, d) \in s$ implies there are two states w, v in S^c such that $s \sim_i^c w$, $s \sim_i^c v$, $\phi \in w \cap v$, and $f_w(d) \neq f_v(d)$.*

The proof of the above lemma is again quite involved, we break it into Proposition 3.9 and Proposition 3.10 below.

Proposition 3.9 *Given any $s \in S^c$ and any $i \in \boldsymbol{I}$, suppose there exist two (possibly identical) maximal consistent sets Γ_1 and Γ_2 such that:*

(a) $\{\psi \mid K_i\psi \in s\} \subseteq \Gamma_1 \cap \Gamma_2$

(b) *for any $Kv_i(\theta, d) \in s$, $\theta \notin \Gamma_1 \cap \Gamma_2$.*

then Γ_1 and Γ_2 can be extended into two states w, v in S^c such that $s \sim_i^c w$, $s \sim_i^c v$ and $f_w(d) \neq f_v(d)$.

Proof By condition (a) and Proposition 3.5, Γ_1 and Γ_2 can be extended into two i-accessible states by using $F(s, i, \Gamma_1, x)$ and $F(s, i, \Gamma_2, y)$. We argue that condition (b) and condition (ii) of s allow us to construct two states in S^c that differ in the value of d. Consider the following cases:

- Suppose that there is no $Kv_i(\chi, d) \wedge \chi \in \Gamma_1$ for any χ. Note that in this case if $f_w = F(s, i, \Gamma_1, x)$ then $f_w(d) = x$. Now let $f_v = F(s, i, \Gamma_2, 0)$ and let $f_w = F(s, i, \Gamma_1, f_v(d) + 1)$. Clearly $f_w(d) = f_v(d) + 1 \neq f_v(d)$. The symmetric case when there is no $Kv_i(\chi, d) \wedge \chi \in \Gamma_2$ for any χ is similar.
- Suppose there exists $Kv_i(\chi, d) \wedge \chi \in \Gamma_1$ for some χ and there exists $Kv_i(\chi', d) \wedge \chi' \in \Gamma_2$ for some χ'. Now let $f_w = F(s, i, \Gamma_1, 0)$ and $f_v = F(s, i, \Gamma_2, 0)$ we have $f_w(d) = g_s(i, \chi, d)$ and $f_v(d) = g_s(i, \chi', d)$. We need to show $g_s(i, \chi, d) \neq g_s(i, \chi', d)$. Towards contradiction suppose $g_s(i, \chi, d) = g_s(i, \chi', d)$ then by condition (ii) of s, there exists θ such that $Kv_i(\theta, d)$ and $\hat{K}_i(\theta \wedge \chi)$ and $\hat{K}_i(\theta \wedge \chi')$ are in s. Note that due to Proposition 3.4, $Kv_i(\chi, d)$ and $Kv_i(\chi', d)$ are in s. Now by Proposition 2.3, $Kv_i(\chi \vee \chi', d) \in s$. However, since $\chi \in \Gamma_1$ and $\chi' \in \Gamma_2$, $\chi \vee \chi' \in \Gamma_1 \cap \Gamma_2$, which contradicts to the assumption (b).

□

In [16], we proved the following proposition in the single agent case. The proof for the multi-agent version is almost the same.

Proposition 3.10 *Given any $s \in S^c$ and any $i \in \boldsymbol{I}$, suppose $\neg Kv_i(\phi, d) \in s$ then there are two (possibly identical) maximal consistent sets Γ_1 and Γ_2 such that:*

(a') $\{\phi\} \cup \{\psi \mid K_i\psi \in s\} \subseteq \Gamma_1 \cap \Gamma_2$

(b) *for any $Kv_i(\theta, d) \in s$, $\theta \notin \Gamma_1 \cap \Gamma_2$.*

Proof Let $Z = \{\psi \mid K_i\psi \in s\} \cup \{\phi\}$ and let $X = \{\neg\theta \mid Kv_i(\theta, d) \in s\}$. Note that due to $\mathtt{Kv}^r\bot$, X is non-empty.[5] We want to build two consistent sets B and C such that $Z \subseteq B \cap C$ and $X \subseteq B \cup C$. Then by a Lindenbaum-like argument over countable language, we can extend B and C into the desired Γ_1 and Γ_2: (a') is guaranteed by $Z \subseteq B \cap C$, and (b) is guaranteed by $X \subseteq B \cup C$

[5] $\mathtt{Kv}^r\bot$ is indispensable in the proof system $\mathbb{ELKV}^r$. We can show that it is not provable in $\mathbb{ELKV}^r - \mathtt{Kv}^r\bot$ by designing an alternative semantics which coincides with the standard semantics for Kv_i-free formulas but falsifies all the $Kv_i(\phi, d)$ formulas for any i, ϕ and d. It is not hard to see that $\mathbb{ELKV}^r - \mathtt{Kv}^r\bot$ is sound w.r.t. this new semantics but $Kv_i(\bot, d)$ is not valid, thus $\mathtt{Kv}^r\bot$ is not provable in $\mathbb{ELKV}^r$.

which says that for any $Kv_i(\theta, d) \in s$, $\neg\theta \in B$ or $\neg\theta \in C$ thus $\theta \notin \Gamma_1$ or $\theta \notin \Gamma_2$. In the following we build B and C.

The idea is straightforward: simply adding the formulas in X one by one into two copies of Z while keeping the consistency. Formally, we enumerate formulas in X as $\neg\theta_0, \neg\theta_1, \ldots$ and let $B_0 = Z \cup \{\neg\theta_0\}$ and let $C_0 = Z$ as the starting points. Then we build B_{n+1} and C_{n+1} based on the already defined B_n and C_n by adding $\neg\theta_{n+1}$ into one of them:

(i) if $\neg\theta_{n+1}$ is consistent with B_n then $B_{n+1} = B_n \cup \{\neg\theta_{n+1}\}$ and $C_{n+1} = C_n$;

(ii) if $\neg\theta_{n+1}$ is not consistent with B_n then $B_{n+1} = B_n$ and $C_{n+1} = C_n \cup \{\neg\theta_{n+1}\}$.

Let $B = \bigcup_{n<\omega} B_n$, $C = \bigcup_{n<\omega} C_n$ and we need to show that B and C are consistent. Note that B (C) is consistent iff B_n (C_n) is consistent for each n, since if B (C) is not consistent then there must be an n such that B_n (C_n) is not consistent, due to the finitary nature of logical consistency. In the following we show B_n and C_n are consistent by induction on n.

- $n = 0$: Suppose towards contradiction that B_0 is not consistent, then there exist $\psi_1, \ldots, \psi_m \in \{\psi \mid K_i\psi \in s\}$ such that $\vdash \psi_1 \wedge \cdots \wedge \psi_m \wedge \phi \to \theta_0$, i.e., $\vdash \psi_1 \wedge \cdots \wedge \psi_m \to (\phi \to \theta_0)$. Therefore $\vdash K_i\psi_1 \wedge \cdots \wedge K_i\psi_m \to K_i(\phi \to \theta_0)$ by `DISTK`, `NECK` and `RE`. Since $K_i\psi_1, \ldots, K_i\psi_m \in s$, $K_i(\phi \to \theta_0) \in s$. Now by `DISTKv`^{r} and the fact that $Kv_i(\theta_0, d) \in s$ (since X is non-empty), it follows that $Kv_i(\phi, d) \in s$, contradiction. Since $C_0 \subseteq B_0$, C_0 is also consistent.

- $n = k + 1$: by the induction hypothesis B_k and C_k are consistent. According to our construction of B_{k+1} we just need to show that if $\neg\theta_{k+1}$ is not consistent with B_k then it is consistent with C_k. Suppose not, then both $B_k \cup \{\neg\theta_{k+1}\}$ and $C_k \cup \{\neg\theta_{k+1}\}$ are inconsistent. In the sequel, to derive a contradiction, we adopt the proof in [16, Lemma 19] for the multi-agent setting .

  Let $\overline{U} = B_k \backslash Z$, $\overline{V} = C_k \backslash Z$, $U = \{\theta \mid \neg\theta \in \overline{U}\}$, and $V = \{\theta \mid \neg\theta \in \overline{V}\}$. Note that $U, V, \overline{U}, \overline{V}$ are all finite and each formula in $\overline{V}$ is not consistent with B_k due to the construction of B_k.

  We claim: there exist $\psi_1, \ldots, \psi_l, \psi'_1, \ldots, \psi'_m, \psi''_1, \ldots, \psi''_r \in \{\psi \mid K_i\psi \in s\}$ such that

$$
\begin{array}{ll}
(i) & \vdash \psi_1 \wedge \cdots \wedge \psi_l \wedge \phi \wedge \bigwedge \overline{U} \to \theta_{k+1}, \\
(ii) & \vdash \psi'_1 \wedge \cdots \wedge \psi'_m \wedge \phi \wedge \bigwedge \overline{V} \to \theta_{k+1}, \\
(iii) & \vdash \psi''_1 \wedge \cdots \wedge \psi''_r \wedge \phi \wedge \bigwedge \overline{U} \to \bigwedge V.
\end{array}
$$

  (i) and (ii) are immediate from the inconsistency of $B_k \cup \{\neg\theta_{k+1}\}$ and $C_k \cup \{\neg\theta_{k+1}\}$. For (iii), first recall that for any $\theta \in V$, $\{\neg\theta\} \cup B_k$ is inconsistent due to the construction of B_k. Therefore for each $\theta \in V$ there exist $\chi_1, \ldots, \chi_h \in \{\psi \mid K_i\psi \in s\}$ such that:

$$\vdash (\chi_1 \wedge \cdots \wedge \chi_h \wedge \phi \wedge \bigwedge \overline{U}) \to \theta$$

  Since V is a finite set, we can collect all such χ for each $\theta \in V$ to obtain (iii).

From $(i) - (iii)$, NECK, DISTK, RE and the fact that

$$K_i\psi_1, \dots, K_i\psi_l, K_i\psi'_1, \dots, K_i\psi'_m, K_i\psi''_1, \dots, K_i\psi''_n \in s,$$

we can show the following:

$$\begin{array}{ll} (iv) & K_i((\phi \wedge \bigwedge \overline{U}) \to \theta_{k+1}) \in s, \\ (v) & K_i((\phi \wedge \bigwedge \overline{V}) \to \theta_{k+1}) \in s, \\ (vi) & K_i((\phi \wedge \bigwedge \overline{U}) \to \bigwedge V) \in s. \end{array}$$

In the following, we will show that $\hat{K}_i(\theta_{k+1} \wedge \bigwedge V) \in s$. First we claim $\hat{K}_i(\phi \wedge \bigwedge \overline{U}) \in s$. Suppose not, then $K_i\neg(\phi \wedge \bigwedge \overline{U}) \in s$, thus $\neg(\phi \wedge \bigwedge \overline{U}) \in B_k$. Due to the construction of B_k we know ϕ and $\overline{U}$ are in B_k, thus B_k is inconsistent, contradicting the assumption. Therefore $\hat{K}_i(\phi \wedge \bigwedge \overline{U}) \in s$ thus by $(iv), (vi)$ we have $\hat{K}_i(\theta_{k+1} \wedge \bigwedge V) \in s$.

By our assumption, for any $\theta \in V \cup \{\theta_{k+1}\}$ we have $Kv_i(\theta, d) \in s$. Now based on this fact and $\hat{K}_i(\theta_{k+1} \wedge \bigwedge V) \in s$, we can use Proposition 2.2, and obtain the following:

$$(vii) \;\; Kv_i(\theta_{k+1} \vee \bigvee V, d) \in s.$$

Now using $\vdash \neg \bigwedge \overline{V} \leftrightarrow \bigvee V$, let us change the from of (v) to the following:

$$(v') \;\; K_i(\phi \to (\bigvee V \vee \theta_{k+1})) \in s,$$

Based on $(v'), (vii)$ and DISTKv^r, we have $Kv_i(\phi, d) \in s$, contradiction. Therefore, B_{k+1} and C_{k+1} are consistent and this concludes the the inductive proof.

In sum, B and C are consistent thus can be extended into Γ_1 and Γ_2 satisfying (a') and (b). □

Clearly, (a') in Proposition 3.10 implies (a) in Proposition 3.9, then Lemma 3.8 is immediate.

Now we are ready to prove the truth lemma:

Lemma 3.11 (Truth Lemma) *For any $\phi \in$ **ELKv**^{r} and $s \in S^c$, $\phi \in s$ iff $\mathcal{M}^c, s \vDash \phi$.*

Proof We only show the non-trivial cases of $K_i\psi$ and $Kv_i(\psi, d)$.

- $\phi = K_i\psi$: If $K_i\psi \in s$, then for any t such that $s \sim_i^c t$ we have $\psi \in t$ by the definition of $\sim_i^c$. Now by induction hypothesis (IH), $\mathcal{M}^c, s \vDash K_i\psi$. Now suppose $K_i\psi \notin s$, then by Lemma 3.7 and the IH, we have $\mathcal{M}^c, s \vDash \neg K_i\psi$.
- $\phi = Kv_i(\psi, d)$: Suppose that $Kv_i(\psi, d) \in s$, $s \sim_i^c t$, $s \sim_i^c t'$, $\psi \in t$ and $\psi \in t'$. It is easy to see that $Kv_i(\psi, d) \in t \cap t'$ and $g_t(i) = g_{t'}(i) = g_s(i)$. Since $\psi \in t$ and $\psi \in t'$, according to condition (iii) and the fact that $g_t(i) = g_{t'}(i)$:

$$V_{\mathbf{D}}(d, t) = f_t(d) = g_t(i, \psi, d) = g_s(i, \psi, d) = g_{t'}(i, \psi, d) = f_{t'}(d) = V_{\mathbf{D}}(d, t').$$

Now suppose $Kv_i(\psi, d) \notin s$ then by Lemma 3.8 and IH, $\mathcal{M}^c, s \vDash \neg Kv_i(\psi, d)$.

□

From the above truth lemma, the completeness theorem almost follows. The only missing piece is to show for each maximal consistent set there is indeed at least one corresponding state in $\mathcal{M}^c$.

Lemma 3.12 *For every maximal consistent set* Γ, *there exist* f *and* g *such that* $\langle \Gamma, f, g \rangle \in S^c$.

Proof The construction is very similar to the one in the proof of Proposition 3.5, though simpler. The only essential difference is the definition of g^0, thus the proofs related to g^0 need to be adapted.

Let x be a natural number, and let $T = \{\langle j, \phi, d\rangle \mid Kv_j(\phi, d) \wedge \phi \in \Gamma, j \in \mathbf{I}, \phi \in \mathbf{ELKv}^r, d \in \mathbf{D}\}$.[6] Let $g^0 : T \to \mathbb{N} \cup \{\star\}$ be the constant function such that $g^0(j, \phi, d) = x$ for all the triples in T. We define f as the constant function such that $f(d) = x$ for all $d \in \mathbf{D}$. Now redefine Δ as the set of the remaining triples:

$$\Delta = \{\langle j, \phi, d\rangle \mid \phi \wedge Kv_j(\phi, d) \notin \Gamma, j \in \mathbf{I}, \phi \in \mathbf{ELKv}^r, d \in \mathbf{D}\}.$$

As before, we can enumerate Δ and build g^{k+1} by adding the new $\langle j, \phi, d\rangle \notin Dom(g^k)$ into the domain (where Λ^k is defined as before w.r.t. the new Δ):

$$g^{k+1}(j, \phi, d) = \begin{cases} \star & \text{if } \hat{K}_j\phi \wedge Kv_j(\phi, d) \notin \Gamma \\ g^k(j, \psi, d) & \text{if } \hat{K}_j\phi \wedge Kv_j(\phi, d) \in \Gamma \text{ and there are} \\ & \chi, \psi \text{ such that } \langle j, \psi, d\rangle \in Dom(g^k) \\ & \text{and the following formulas are all} \\ & \text{in } \Gamma: \ \hat{K}_j(\chi \wedge \phi),\ \hat{K}_j(\chi \wedge \psi), \\ & Kv_j(\psi, d),\ Kv_j(\chi, d) \\ max(\Lambda^k \cup \{f(d)\}) + 1 & \text{if otherwise} \end{cases}$$

Similar to the corresponding proof of Proposition 3.5, we can show that the second clause in the above definition of g^{k+1} is well-defined ($k = 0$ case is now obvious due to the definition of g^0).

Now let $g = \bigcup_{k \in \mathbb{N}} g^k$, we need to verify conditions (i), (ii) and (iii). Condition (iii) is immediate from the definition of f.

For condition (i), if $\langle j, \phi, d\rangle \in Dom(g^0) = T$, then $g^0(j, \phi, d) = x \neq \star$, and we can see that $Kv_j(\phi, d) \wedge \hat{K}_j\phi \in \Gamma$ since $Kv_j(\phi, d) \wedge \phi \in \Gamma$. Thus we have for any $\langle j, \phi, d\rangle \in Dom(g^0)$, $g^0(j, \phi, d) \neq \star$ iff $Kv_j(\phi, d) \wedge \hat{K}_j\phi \in \Gamma$. The inductive case is obvious by the three clauses of g^{k+1}.

For condition (ii), suppose that $g(j, \phi, d) \neq \star$ and $g(j, \psi, d) \neq \star$, we need to prove claim ($\circ$) inductively as before. For that, we only need to revise the proof for the base case as follows:

Suppose $k = 0$ and thus both $\langle j, \phi, d\rangle$ and $\langle j, \psi, d\rangle$ are in $Dom(g^0)$. By definition of g^0, it is easy to see that $g(j, \phi, d) = g^0(j, \phi, d) = x = g^0(j, \psi, d) = g(j, \psi, d)$. Moreover, we have $\phi \wedge Kv_j(\phi, d) \in \Gamma$ and $\psi \wedge Kv_j(\psi, d) \in \Gamma$. Then

[6] Note that T may be empty. In that case we start from the empty function g^0.

setting $\chi = \phi$ gives us $Kv_j(\chi, d) \wedge \hat{K}_j(\phi \wedge \chi) \wedge \hat{K}_j(\psi \wedge \chi) \in \Gamma$. Then for any $\langle j, \phi, d \rangle$ and $\langle j, \psi, d \rangle$ in $Dom(g^0)$, $g^0(j, \phi, d) = g^0(j, \psi, d)$ iff there exists a χ such that $\{Kv_j(\chi, d), \hat{K}_j(\phi \wedge \chi), \hat{K}_j(\psi \wedge \chi)\} \subseteq \Gamma$.

In sum, $\langle \Gamma, f, g \rangle \in S^c$. □

Based on Lemma 3.12 and Lemma 3.11 we can show the completeness.

Theorem 3.13 $\mathbb{ELKV}^r$ *is sound and strongly complete for multi-agent* $\boldsymbol{ELKv^r}$.

Proof The soundness part can be found in [16, Theorem 11]. For the completeness part, we show that each consistent set of $\mathbf{ELKv}^r$ formulas is satisfiable. Given a consistent set Δ of $\mathbf{ELKv}^r$ formulas, by the Lindenbaum Lemma for the countable language $\mathbf{ELKv}^r$, there exists a maximal consistent set Γ such that $\Delta \subseteq \Gamma$. Now Lemma 3.12 tells us that there exist f, g such that $\langle \Gamma, f, g \rangle \in S^c$. From Lemma 3.11, it follows that $\mathcal{M}^c, \langle \Gamma, f, g \rangle \vDash \Gamma$ thus $\mathcal{M}^c, \langle \Gamma, f, g \rangle \vDash \Delta$. □

In [16], we also discussed the logic of $\mathbf{ELKv}^r$ extended with public announcement operators (call it $\mathbf{PALKv}^r$):

$$\phi ::= \top \mid p \mid \neg\phi \mid \phi \wedge \phi \mid K_i\phi \mid \langle\phi\rangle\phi$$

As an immediate corollary of the above completeness theorem and Theorem 10 in [16], we can axiomatize multi-agent $\mathbf{PALKv}^r$ by adding the following reduction axioms to $\mathbb{ELKV}^r$ (call the resulting system $\mathbb{PALKV}^r$):

`!ATOM`	$\langle\psi\rangle p \leftrightarrow (\psi \wedge p)$
`!NEG`	$\langle\psi\rangle\neg\phi \leftrightarrow (\psi \wedge \neg\langle\psi\rangle\phi)$
`!CON`	$\langle\psi\rangle(\phi \wedge \chi) \leftrightarrow (\langle\psi\rangle\phi \wedge \langle\psi\rangle\chi)$
`!K`	$\langle\psi\rangle K_i\phi \leftrightarrow (\psi \wedge K_i(\psi \to \langle\psi\rangle\phi))$
$\texttt{!Kv}^r$	$\langle\phi\rangle Kv_i(\psi, d) \leftrightarrow (\phi \wedge Kv_i(\langle\phi\rangle\psi, d))$

Corollary 3.14 $\mathbb{PALKV}^r$ *is sound and complete for multi-agent* $\boldsymbol{PALKv^r}$.

4 Future work

In this paper, we showed that $\mathbb{ELKV}^r$ is sound and complete for multi-agent $\mathbf{ELKv}^r$ (over S5 frames). This is just a starting point of an unfolding story about interesting modal fragments of first-order intensional logic.

For future work, the decidability of $\mathbf{ELKv}^r$ deserves a careful investigation. The single-agent case is particularly promising, since we do have a neat canonical model construction which only uses two copies of each maximal consistent set, which may facilitate a finite filtration leading to to the small model property. On the other hand, there are also hints for the undecidablity, for example, in [3], it is shown that the quantifier-free fragment of S5-**FOIL** is undecidable, where arbitrary relation symbols and arbitrary predicate abstractions are allowed. Of course we may study the logic of $\mathbf{ELKv}^r$ on other weaker frame classes, where decidability is more plausible according to [3].

Another natural question to ask is how to axiomatize the logic where Kv_i are the *only* primitive operators (call it $\mathbf{PLKv}^r$). In $\mathbb{ELKV}^r$, most of the axioms involve interactions between "knowing that" and "knowing what". We are unsure if the system without these axioms can axiomatize $\mathbf{PLKv}^r$, though it is unlikely.

As motivated in the introduction, $\mathbf{ELKv}^r$ can be used in a security setting where the interaction between "knowing what" and "knowing that" is important. To really handle epistemic reasoning in such scenarios, we need to express statements like "I know that the message I just received is indeed the private message that I sent before for authentication", where equality is inevitable. Due to our completeness proof method, we suspect that adding the equality symbol (between $d \in \mathbf{D}$) freely in $\mathbf{ELKv}^r$ may in turn ease the axiomatization, since we have a better grip on the information we need in the canonical model.

Last but not least, on the philosophical side, the conditional versions of other types of knowledge should be studied, probably in the context of relevant alternative theory [2], since they may capture the common sense use of knowledge better.

Acknowledgement

This work is partially supported by European Research Council (ERC) grant EPS 313360. Yanjing Wang thanks National Social Science Foundation of China (SSFC) for the grant 11CZX054 and the support from SSFC major project 11&ZD088. Jie Fan acknowledges the support from China Scholarship Council (CSC). We are also grateful to the anonymous referees of AiML2014 for their insightful comments on an early version of this paper.

References

[1] Braüner, T. and S. Ghilardi, *First-order modal logic*, in: P. Blackburn, J. van Benthem and F. Wolter, editors, *Handbook of Modal Logic*, Elsevier, 2007 pp. 549–620.

[2] Dretske, F., *Epistemic operators*, The Joutnal of Philosophy **67** (1970), pp. 1007–1023.

[3] Fitting, M., *Modal logics between propositional and first-order*, Journal of Logic and Computation **12** (2002), pp. 1017–1026.

[4] Fitting, M., *First-order intensional logic*, Annals of Pure and Applied Logic **127** (2004), pp. 171–193.

[5] Fitting, M., *FOIL axiomatized*, Studia Logica **84** (2006), pp. 1–22.

[6] Halpern, J. Y. and R. Pucella, *Modeling adversaries in a logic for security protocol analysis*, in: *Formal Aspects of Security*, 2003, pp. 87–100.

[7] Hintikka, J., "Knowledge and Belief: An Introduction to the Logic of the Two Notions," Cornell University Press, Ithaca N.Y., 1962.

[8] Lewis, D., *Elusive knowledge*, Australasian Journal of Philosophy **74** (1996), pp. 549–567, 418-446.

[9] McCarthy, J., *First-Order theories of individual concepts and propositions*, Machine Intelligence **9.** (1979), pp. 129–147.

[10] Plaza, J. A., *Logics of public communications*, in: M. L. Emrich, M. S. Pfeifer, M. Hadzikadic and Z. W. Ras, editors, *Proceedings of the 4th International Symposium on Methodologies for Intelligent Systems*, 1989, pp. 201–216.

[11] Ramanujam, R. and S. P. Suresh, *Decidability of secrecy for context-explicit security protocols*, Journal of Computer Security **13** (2005), pp. 135–165.

[12] Ryle, G., "The Concept of Mind," Penguin, 1949.

[13] Stanley, J. and T. Williamson, *Knowing how*, The Journal of Philosophy **98** (2001), pp. 411–444.

[14] van Benthem, J., "Logical Dynamics of Information and Interaction," Cambridge University Press, 2011.

[15] van Benthem, J., J. van Eijck and B. Kooi, *Logics of communication and change*, Information and Computation **204** (2006), pp. 1620–1662.

[16] Wang, Y. and J. Fan, *Knowing that, knowing what, and public communication: Public announcement logic with Kv operators*, in: *Proceedings of IJCAI*, 2013, pp. 1139–1146.

www.ingramcontent.com/pod-product-compliance
Lightning Source LLC
Chambersburg PA
CBHW060747230426
43667CB00010B/1474
9781848901513